About Island Press

Island Press is the only nonprofit organization in the United States whose principal purpose is the publication of books on environmental issues and natural resource management. We provide solutions-oriented information to professionals, public officials, business and community leaders, and concerned citizens who are shaping responses to environmental problems.

In 2002, Island Press celebrates its eighteenth anniversary as the leading provider of timely and practical books that take a multidisciplinary approach to critical environmental concerns. Our growing list of titles reflects our commitment to bringing the best of an expanding body of literature to the environmental community throughout North America and the world.

Support for Island Press is provided by The Jenifer Altman Foundation, The Bullitt Foundation, The Mary Flagler Cary Charitable Trust, The Nathan Cummings Foundation, The Geraldine R. Dodge Foundation, The Charles Engelhard Foundation, The Ford Foundation, The Vera I. Heinz Endowment, The W. Alton Jones Foundation, The John D. and Catherine T. MacArthur Foundation, The Andrew W. Mellon Foundation, The Charles Stewart Mott Foundation, The Curtis and Edith Munson Foundation, The National Fish and Wildlife Foundation, The National Science Foundation, The New-Land Foundation, The David and Lucile Packard Foundation, The Pew Charitable Trusts, The Surdna Foundation, The Winslow Foundation, and individual donors.

About World Wildlife Fund

Known worldwide by its panda logo, World Wildlife Fund is dedicated to protecting the world's wildlife and the rich biological diversity that we all need to survive. The leading privately supported international conservation organization in the world, WWF has invested in more than 13,100 projects in 157 countries and has more than 1 million members in the United States.

WWF directs its conservation efforts toward three global goals: protecting endangered spaces, saving endangered species, and addressing global threats. From working to save the giant panda, tiger, and rhino to helping establish and manage parks and reserves worldwide, WWF has been a conservation leader for forty years. Visit the WWF Web site at www.worldwildlife.org or write to us.

Terrestrial Ecoregions
of the Indo-Pacific

Terrestrial Ecoregions of the Indo-Pacific

A CONSERVATION ASSESSMENT

Eric Wikramanayake
Eric Dinerstein
Colby J. Loucks
David M. Olson
John Morrison
John Lamoreux
Meghan McKnight
and Prashant Hedao

Foreword by Stuart L. Pimm

WORLD WILDLIFE FUND—
UNITED STATES

ISLAND PRESS
Washington • Covelo • London

Copyright © 2002 Island Press

All rights reserved under International and Pan-American Copyright Conventions. No part of this book may be reproduced in any form or by any means without permission in writing from the publisher: Island Press, 1718 Connecticut Avenue, N.W., Suite 300, Washington, DC 20009.

ISLAND PRESS is a trademark of The Center for Resource Economics.

Library of Congress Cataloging-in-Publication Data
Terrestrial ecoregions of the Indo-Pacific : a conservation assessment / by Eric Wikramanayake...[et al.]
        p. cm.
      ISBN 1-55963-923-7 (paper : alk. paper)
      1. Biological diversity conservation—Indo-Pacific Region.
      2. Biotic communities—Indo-Pacific Region.   3. Ecology—Indo-Pacific Region.   I. Wikramanayake, Eric D.
QH77.I55 T47 2002
333.95'16'091823—dc21   2001006930

British Cataloguing-in-Publication Data available.

Printed on recycled, acid-free paper ✺
Using soy-based inks
Manufactured in the United States of America
10 9 8 7 6 5 4 3 2 1

This book is dedicated to all those individuals and institutions, both well known and anonymous, who have worked to conserve the Indo-Pacific region's biodiversity.

In memory of Derek Holmes
Ornithologist, land-use planner, and committed conservationist
Died in Indonesia on October 2, 2000

Contents

LIST OF SPECIAL ESSAYS   xiii

LIST OF FIGURES   xv

LIST OF TABLES   xix

LIST OF BOXES   xxi

ACRONYMS   xxiii

FOREWORD   xxv

PREFACE   xxvii

ACKNOWLEDGMENTS   xxix

1. NATURE'S END OR A NEW BEGINNING?   1
    - Assessment Overview   3
    - Structure of the Book   7

2. ASSEMBLING THE ECOLOGICAL JIGSAW PUZZLE: THE ECOREGIONS OF THE INDO-PACIFIC   17
    - Delineation of Ecoregions and Geographic Scope of the Study   17
    - Geographic Scope   18
    - Themes   19
    - Delineating Bioregions   21
    - Delineating Biomes   24
    - Delineating Ecoregions   25
    - Representation of Ecoregions and Biomes in the Indo-Pacific Region   26

3. REPRESENTATION OR TRIAGE? APPROACHES TO SETTING CONSERVATION PRIORITIES IN THE INDO-PACIFIC REGION   39
    - Biological Distinctiveness Index (BDI)   39
    - Conservation Status Index (CSI)   41
    - Threat Analysis   44
    - Integrating Biological Distinctiveness and Conservation Status   45
    - Guidelines for Using the Integration Matrix   46

- Use of Proxies to Develop Indices 46
- Biodiversity Representation or Triage? 47

4. **WHERE IS BIODIVERSITY DISTRIBUTED?** 61
    - Broad Patterns in Biological Distinctiveness 62
    - A Closer Look at the Indo-Pacific's Natural Wealth 65
    - Ecoregions, Hotspots, and EBAs: Concordance and Overlap 85
    - Concordance of Biological Priorities: The Overriding Importance of Indonesian Ecoregions 89

5. **WHERE IS BIODIVERSITY UNDER GREATEST THREAT?** 99
    - Snapshot Conservation Status: How Much Habitat Remains in Indo-Pacific Ecoregions? 100
    - Threat-Modified Conservation Status: Projecting Future Integrity of Ecoregions 109
    - Overarching Threats 109
    - Degradation Threats: Human Population Density and Distribution 111
    - High-Intensity Threats: Logging Concessions and Habitat Loss 115
    - Final Conservation Status 118

6. **WHAT TO SAVE FIRST? SETTING PRIORITIES FOR BIODIVERSITY CONSERVATION** 139
    - The Priority-Setting Matrix 140
    - Regional and Bioregional Trends 140
    - The Priority Portfolio 140
    - Summary 158

7. **SOLUTIONS TO THE BIODIVERSITY CRISIS FACING THE INDO-PACIFIC REGION** 165
    - Big Conservation: The Next Frontier 165
    - Conservation Planning and Action at the Ecoregion Scale 165
    - Landscape-Scale Conservation 168
    - Innovative Mechanisms to Achieve Conservation Gains 173
    - Reevaluating Existing Mechanisms in a New Light 175
    - Bold Leadership: Who Will Be the Next Indira Gandhi of the Indo-Pacific? 177
    - Nature's End or a New Beginning? 177

APPENDIXES
- A. Details on Ecoregion Delineation 193
- B. Terrestrial Biome Descriptions 203
- C. Methods for Assessing the Biological Distinctiveness of Terrestrial Ecoregions 209
- D. Methods for Assessing the Conservation Status of Terrestrial Ecoregions 213

APPENDIXES (*continued*)

    E. Summary of Ecoregion Scores for Species Richness, Endemism, and Biological Distinctiveness Index   219

    F. Summary of Ecoregion Scores for the Conservation Status Index and Integration Matrix   225

    G. Habitat Types by Ecoregion   231

    H. Near-Endemic and Strict Endemic Mammals by Ecoregion   243

    I. Near-Endemic and Strict Endemic Birds by Ecoregion   255

    J. Ecoregion Descriptions   281

GLOSSARY   601

LITERATURE CITED AND CONSULTED   607

AUTHORS   631

CONTRIBUTORS   633

INDEX   637

List of Special Essays

Essay 1. The Predicted Extinction of Lowland Forests in Indonesia　7
Derek Holmes

Essay 2. For Whom the Bell Tolls, or What Is the Future of the Tropical Timber Trade in the Face of a Probable Glut of Plantation Timber?　14
Alf J. Leslie

Essay 3. Island Life along Wallace's Line: Biogeography and Patterns of Endemism in the Philippines and Indonesia　28
Lawrence R. Heaney

Essay 4. Aquatic Systems: Neglected Biodiversity　30
Maurice Kottelat

Essay 5. Limestone Biodiversity: Treasure Houses of Rare Species　36
Tony Whitten

Essay 6. Amphibians of the Indo-Pacific Region: Conservation of Neglected Biodiversity　47
Indraneil Das

Essay 7. Reptile Diversity in the Indo-Pacific Region　49
Walter Erdelen

Essay 8. Mangroves of the Indo–West Pacific　55
David Olson

Essay 9. Tigers: Top Carnivores and Controlling Processes in Asian Forests　56
John Seidensticker

Essay 10. "New Species" and Other Endemics: Implications for Conserving Asia's Biodiversity　90
Eric Wikramanayake

Essay 11. Plant Richness and Endemism in the Indo-Pacific: Dipterocarpaceae　92
Gary A. Krupnick and June Rubis

Essay 12. From Many Trees to One: The Widespread Conversion of Tropical Forests to Oil Palm Plantations　122
Jason Clay

Essay 13. Into the Frying Pan and onto the Apothecary Shelves: The Fate of Tropical Forest Wildlife in Asia?　124
Elizabeth L. Bennett and John G. Robinson

Essay 14. Taking the Bounty off Wildlife in Asia 126
Judy Mills

Essay 15. Protected Areas: Keystones to the Conservation of Asia's Natural Wealth 127
Colby Loucks and Eric Dinerstein

Essay 16. Composition of the Alpine Himalayan Protected Areas Network and Its Contribution to Biodiversity Conservation 131
Tom Allnutt, Eric Wikramanayake, Eric Dinerstein, Colby Loucks, Rodney Jackson, Dan Hunter, and Chris Carpenter

Essay 17. Pastures of Plenty? Livestock Grazing and Conservation of Himalayan Rangelands 135
Daniel J. Miller

Essay 18. Saving the Last Remnants of Tropical Dry Forest of New Caledonia 162
Arnaud Greth, Arnaud Collin, and David Olson

Essay 19. The Terai Arc: Managing Tigers and Other Wildlife as Metapopulations 178
Anup Joshi, Eric Dinerstein, and David Smith

Essay 20. Forests and the Kyoto Protocol: Implications for Asia's Forestry Agenda 182
Joyotee Smith

Essay 21. Reconciling Long-Term Biodiversity Conservation and Local Communities: A Case for Conservation Performance Payments 185
Paul J. Ferraro

Essay 22. Reconciling the Long-Term Conservation Needs of Large Mammals and Local Communities: Integrating Landscape Conservation and Development 188
Kathy MacKinnon

List of Figures

1.1	Satellite Image of the Cambodian Plains along the Mekong River	1
1.2	Satellite Image of Riau Province on the Island of Sumatra	2
1.3	The 140 Terrestrial Ecoregions of the Indo-Pacific	4
1.4	Distribution of Fires in Sumatra and Borneo between 1997 and 2000	9
1.5	Changes in Forest Cover in Sumatra	10
1.6	Changes in Forest Cover in Kalimantan	11
1.7	Changes in Forest Cover in Sulawesi	11
2.1a	Terrestrial Ecoregions of the Indian Subcontinent Bioregion	18
2.1b	Terrestrial Ecoregions of the Indochina and Sunda Shelf and Philippines Bioregions	19
2.1c	Terrestrial Ecoregions of the Wallacea and New Guinea and Melanesia Bioregions	20
2.2	Biomes and Bioregions of the Terrestrial Ecoregions of the Indo-Pacific	22
2.3	Distribution of Ecoregions by Biome and Bioregion	23
2.4	Late Pleistocene Land Area in the Sundaic Region	29
2.5	Nonflying Endemic and Native Mammals of the Philippines	30
2.6	Distribution of Major Limestone Areas in Sumatra and Borneo	36
3.1	Integration Matrix of Biological Distinctiveness Index (BDI) and Conservation Status Index (CSI)	45
4.1	Biological Distinctiveness Status for Terrestrial Ecoregions of the Indo-Pacific	63
4.2	Comparison of Globally Outstanding Ecoregions with Overlapping Zones of Holarctic and Tropical Floras	64
4.3	Distribution of Ecoregions by Biological Distinctiveness Categories	66
4.4	Ecoregions That Are Globally Outstanding for Endemism, Richness, or Both	67
4.5	Distribution of Ecoregions by Richness Index and Endemism Index Categories	68

4.6	Correlation between Mammal and Bird Species Richness for Ecoregions of the Indo-Pacific	68
4.7	Mammal Richness for Terrestrial Ecoregions of the Indo-Pacific	69
4.8	Bird Richness for Terrestrial Ecoregions of the Indo-Pacific	70
4.9	Distribution of Biological Distinctiveness for Species Richness and Endemism among Island and Continental Ecoregions	71
4.10	Mammal and Bird Species Richness by Size of Ecoregion	72
4.11	Correlation between Mammal and Bird Endemism for Ecoregions of Indo-Pacific	73
4.12	Mammal Endemism for Terrestrial Ecoregions of the Indo-Pacific	74
4.13	Bird Endemism for Terrestrial Ecoregions of the Indo-Pacific	75
4.14	Plant Richness for Terrestrial Ecoregions of the Indo-Pacific	76
4.15	Plant Endemism for Terrestrial Ecoregions of the Indo-Pacific	77
4.16	Number of Strict Endemic Mammals of the Indo-Pacific	79
4.17	The Number of Strict Endemic Mammals in Island and Continental Ecoregions	79
4.18	Strict Endemic Mammals Expressed as a Percentage of the Total Mammal Fauna	81
4.19	Strict Endemic Birds Expressed as a Percentage of the Total Avifauna	82
4.20	Globally Outstanding Ecoregions for Biologically Important Phenomena	83
4.21	WWF's Global 200 Ecoregions	86
4.22	Hotspots Found in the Region of Analysis	87
4.23	Dipterocarpaceae Richness for Terrestrial Ecoregions of the Indo-Pacific	94
4.24	Dipterocarpaceae Endemism for Terrestrial Ecoregions of the Indo-Pacific	95
4.25	Percentage Dipterocarpaceae Endemism for Terrestrial Ecoregions of the Indo-Pacific	96
4.26	Number of IUCN *Red List* Dipterocarpaceae Species in the Terrestrial Ecoregions of the Indo-Pacific	97
5.1	Percentage of Ecoregions and Ecoregion Areas in the Five Conservation Status Categories	101
5.2	The Current Conservation Status of Ecoregions without the Inclusion of Future Threats	102
5.3	A Comparison of Remaining Habitat in Lowland and Montane Ecoregions of the Indo-Pacific	103
5.4	Distribution of Remaining Habitat by Elevation across the Indo-Pacific Region	105
5.5	Forest Loss and Fire Hotspots in Sumatra	106
5.6	Percentage of Intact Habitats Protected by Ecoregion	107
5.7	Percentage of Remaining Habitat and Area Protected in Indo-Pacific Ecoregions	108
5.8	Population Density in the Indo-Pacific Region	112

5.9	Mean Human Population Densities in Ecoregions of the Indo-Pacific Bioregions	113
5.10	Distribution of Mean Human Population Densities in the Different Biomes of the Indo-Pacific Bioregions	113
5.11	Percentage of Ecoregions of Each Bioregion in the Population Threat Categories	114
5.12	Relationship between Population Density and Habitat Loss for Forest Ecoregions of the Indo-Pacific Bioregions	115
5.13	The Consequences of Logging Concessions in the Forests of Cambodia, Sumatra, Borneo, and New Guinea	117
5.14	Ecoregions Facing High Threat Because of Population or Logging Concessions	119
5.15	The Final Conservation Status of Terrestrial Ecoregions of the Indo-Pacific	121
5.16	Creation of Protected Areas in Tropical Forest Regions	128
5.17	Creation of New Protected Areas in the Indo-Pacific Region	129
5.18	Average Size of Protected Areas Created in the Indo-Pacific from 1930 to 1997	129
5.19	Rock and Permanent Ice in Protected Areas of the Alpine Ecoregions of the Himalayas and Tibet	133
5.20	Some Intact Rangelands of the Himalayas	136
6.1	Integration Matrix for Conservation Action	141
6.2	Relative Proportions of Priority Ecoregions and All Ecoregions by Biome	160
7.1	Region of Analysis for the Forests of the Lower Mekong Ecoregion Conservation Workshop	167
7.2	Biological Priority Areas Identified in the Forests of the Lower Mekong Ecoregion Conservation Workshop	169
7.3	Original Range of Tiger, Asian Elephant, and Rhinoceros in the Region of Analysis	170
7.4	The Terai Arc	179
C.1	Flow Diagram Showing the Steps Taken to Create the Biological Distinctiveness Index (BDI)	211

List of Tables

1.1	Summary Table of Forest Conversions, 1985–1997	8
1.2	Priority Conservation Areas of Biodiversity Action Plan in Indonesia	13
3.1	Flagship Reptile Species for Bioregions in the Indo-Pacific Region	54
4.1	Number of Ecoregions That Are Globally Outstanding for Species Richness, Endemism, or Important Ecological and Evolutionary Phenomena by Bioregion and Biome	66
4.2	Richness and Endemism Status of Ecoregions That Were Elevated to Globally Outstanding Status for Extraordinary Phenomena and Rare Habitat Types	84
4.3	Breakdown of the BDI by Bioregion and Biome (Corrected for Ecoregions Elevated for Ecological Phenomena and Rare Habitat Types)	85
4.4	Biological Distinctiveness Status of Ecoregions That Overlap with the Biodiversity Hotspots	88
4.5	Number of Ecoregions that Overlap with Endemic Bird Areas in the Indian Subcontinent and Wallacea Bioregions of the Indo-Pacific and Their Biological Distinctiveness Status	89
5.1	Habitat Loss in Biomes of the Indo-Pacific	103
5.2	Extent of Lowland, Submontane, and Montane Forests Remaining in the Indo-Pacific	103
5.3	Results of the Logging Concession Threat Analysis	118
5.4	Natural World Heritage Sites That Overlap Tropical Moist or Dry Forest Ecoregions of the Indo-Pacific	130
5.5	Data Sources and Four Models Used to Establish the Vegetation Limit in the Montane Himalayas	132
5.6	Breakdown of Rock and Permanent Ice in Protected Areas by Ecoregion	132
5.7	Protected Areas with More Than 50 Percent Inhospitable Habitat	134
5.8	Comparison of Rock and Permanent Ice in Protected Areas with Rock and Permanent Ice in Ecoregion	134
6.1	The Distribution of Ecoregions by Biome and Class	142

6.2	Portfolio of Ecoregions in the Indo-Pacific for Third-Tier Priority Conservation Actions	159
6.3	Number of Priority Ecoregions in Each Bioregion and Biome	161
7.1	Original and Current Range Overlap of Tiger, Asian Elephant, and Three Asian Rhinoceros Species with Ecoregions	171
7.2	Degree of Protection for Class I Priority Ecoregions	176
7.3	Estimated Tiger Population in the Terai Protected Areas	180
7.4	Development-Based Conservation Interventions versus Conservation Performance Payments	186
C.1	References, Sources, and Expert Opinion Used to Derive the Biological Distinctiveness Index (BDI)	209
C.2	Species Richness Categories for Mammals, Birds, and Plants	210
C.3	Endemism Categories for Mammals, Birds, and Plants	210
D.1	Index for Calculating Habitat Loss	214
D.2	Decision Rules for Assigning Points for Size and Number of Habitat Blocks	215
D.3	Fragmentation Index	215
D.4	Index for Evaluating the Degree of Protection	216
D.5	Index for Assessing Threats from Logging Concessions	217

List of Boxes

2.1   Hierarchy of spatial units in the conservation assessment of the Indo-Pacific   23

3.1   Criteria for the biological distinctiveness and conservation status indices   39

C.1   Matrix to show how taxonomic scores for the richness (RI) and endemism (EI) Indices add up to provide REI status   211

Acronyms

ABC	Asian Bureau for Conservation
ACEP	Atlantic–Caribbean–East Pacific
BDI	Biological Distinctiveness Index
BO	Bioregionally Outstanding
CBD	Convention on Biodiversity
CDM	Clean Development Mechanism of the Kyoto Protocol
CI	Conservation International
CITES	Convention on International Trade in Endangered Species
CPD	Centre of Plant Diversity
CSI	Conservation Status Index
DEM	Digital Elevation Model
EBA	Endemic Bird Area
EIA	Environmental impact assessment
FAO	Food and Agriculture Organization of the United Nations
FFI	Fauna and Flora International
FONAFIFO	National Forestry Financial Fund of Costa Rica
GEF	Global Environment Facility
GIS	Geographic Information System
GO	Globally Outstanding
ICDP	Integrated Conservation and Development Project
ICIMOD	International Centre for Integrated Mountain Development
IUCN	International Union for the Conservation of Nature and Natural Resources (now called World Conservation Union)
IWP	Indo–West Pacific
LI	Locally Important
MHT	Major Habitat Type
MoFEC	Ministry of Forestry and Estate Crops (MoFEC)
NBCA	National Biodiversity Conservation Areas
NFI	National Forest Inventory
NGO	Nongovernment Organization
NTFP	Nontimber Forest Product
PNG	Papua New Guinea
RIL	Reduced-Impact Logging
RO	Regionally Outstanding
TCU	Tiger Conservation Unit
TNC	The Nature Conservancy
UNESCO	United Nations Educational, Scientific and Cultural Organization
USGS	U.S. Geological Survey
WAHLI	Wahana Lingkungan Hidup Indonesia
WCMC	World Conservation Monitoring Centre
WCS	Wildlife Conservation Society
WHS	World Heritage Site
WWF	World Wildlife Fund

Foreword

We hand down the tale of Charles Darwin and Alfred Russell Wallace, professor to student, just as we recount folk tales, parent to child. Please indulge me as I tell my variant of it.

Charles Darwin left England and arrived in Brazil a month later. Writing home, he peppered his letters with "striking," "luxuriant," "no person could imagine anything so beautiful," and "exquisite glorious pleasure." He went on like this for pages. When one is escaping an English winter, the tropics are a sensory shock.

HMS *Beagle* stopped in port as it sailed south. It allowed Darwin a stay in a cottage on Botofago Bay in Rio de Janeiro. He loved the cottage, but his letters, brimming with enthusiasm only weeks earlier, had changed. In letters home from remote places we all try to conceal the bad news from our families. Darwin did not do a good job. He knew that an infected knee that kept him in a hammock for eight days, a boating accident, a miserable two-week trip inland in torrential rain, and fevers that were killing other crew members could also kill him quickly, violently, far from home.

Perhaps understandably, he missed an extraordinary feature of the species he collected in the Atlantic coastal forests of Brazil. Many had tiny geographic ranges, some as small as a few mountaintops. He also had no reason to expect such species; the species he knew from England had geographic ranges that spanned all of Europe, and some spanned much of Asia or Africa too.

Three and a half years later, HMS *Beagle*'s mission of mapping South America was complete. The *Beagle* stopped for just a few weeks in the Galápagos. It, too, has a dense concentration of small-ranged species. You know the rest of the story. He did not miss them here.

For years, his diaries were filled with careful and detailed descriptions of the species he found on his voyage. Yet in the Galápagos he still mixed specimens from different islands. The islands are close together, and he had no idea that species could differ across such short distances. Island residents told him again and again that he should not mix his specimens. Finally, he recognized his mistake and marveled at what he discovered. Each island has a unique species of mockingbird, similar but separate, even though the islands are close. The same applies to giant tortoises and small black finches. They are "aboriginal productions," he wrote; *this is where species are born*, he realized.

He made it home the next year and compiled notebooks to express his thoughts. Then he waited. And then he waited some more. Twenty years later Alfred Russell Wallace had the same idea. He got it from the same experiences: looking at species with small, seemingly idiosyncratic geographic ranges on the islands of southeast Asia that are a major part of this book. Wallace wrote to Darwin with his idea, and Darwin could procrastinate no longer. They wrote short papers back-to-back in a scientific journal. Not much happened. The next year Darwin wrote *The Origin of Species*. Its initial print run sold out in a day and changed our world.

This book includes the islands that gave Wallace his ideas on evolution, ideas that forced Darwin to action and so to prominence. The facts and ideas this book contains should change our world. Indeed, it documents world changes, ones that are large and irreversible. Were it written for the readers of the tabloid press, it might well be titled *The Death of Species*.

The mostly tropical areas this book describes are home to an extraordinary array of species. Many of them have the small geographic ranges—particular islands or mountaintops—that gave Wallace and Darwin their ideas. Small-ranged species are Nature's eggs, her most vulnerable species, for it is easier to endanger a species with a small range than one with a large one. Nature has put her eggs in a few baskets, defying the conventional wisdom on spreading risk. These baskets are the concentrations of vulnerable species that we call hotspots. Examples include the Galápagos, the coastal rain forests of Brazil, and most of the regions this book describes, such as New Caledonia, the montane forests of New Guinea, and the lowland forests of Sri Lanka. Few other places in the world have so many endemic species clustered into so small an area. Few other places stand to witness the death of so many species if current trends of habitat destruction, hunting, and other threats continue.

The "eggs in a few baskets" analogy does offer hope. Defend those baskets vigorously and we can protect many species in small amounts of space. In such grand, simplifying assertions, the details are always the rub. Where exactly are the places we must protect? And against what do we protect them?

This book, along with its companions in this series, takes an ecoregional approach, dividing large regions into small, distinct units, each with its characteristic species, ecosystems, natural history, and threats. Therefore, it has no peers. It is *the* sourcebook for anyone who needs to know where and how to act to save the variety of life on Earth. I can dip into it to find out which regions have humid forest and which have grasslands, what species are in the forests, and how long those forests will last at current rates of deforestation. I can find out immediately which ecoregions are in the most urgent need of attention. In short, it is a handbook on maintaining species—*On the Continued Origin of Species*, we might say.

Stuart L. Pimm
Professor of Conservation Biology
Center for Environmental Research and Conservation

Preface

My love affair with Asian biodiversity began when I was a researcher studying the endangered swamp deer in Kanha National Park, India in the 1970s, a time when conservationists were filled with both optimism and concern for the fate of nature in Asia. In November 2000, I participated in an effort that rekindled my optimism for saving Asia's unique wildlife heritage: the translocation of a greater one-horned rhinoceros from Royal Chitwan National Park to Royal Bardia National Park in Nepal, one of the great conservation success stories documented in this book. In between, I have followed closely the destruction of irreplaceable habitats and populations across the region. It was clear to me that we were winning battles here and there to save wild places but losing the war. It was also clear that if we were to save Asian nature, we had to be more strategic in determining where to act first.

The ecoregion framework used in this book provides just such a strategy for priority setting. The authors take a hard look at the state of the Indo-Pacific's biodiversity and habitats, moving beyond endangered or charismatic species to quantify for the first time the number of mammal and bird species, including endemics, in each ecoregion. This in itself is a monumental contribution to our knowledge of biological patterns and processes. The analysis in this volume builds on this base and reviews the current status of each ecoregion's habitat, the essential matrix for all species. Conservation status analyses quantify the amount of habitat remaining, how it is distributed, and how much is protected, which are critical questions that must be answered if we are to act in the most expeditious and efficient manner to save the Indo-Pacific's biological legacy. This book concludes with a set of ecoregions that deserve immediate attention because they hold vast treasures of biodiversity that are being destroyed at alarming rates. It also highlights ecoregions that are still in pristine condition, where we can be proactive and achieve significant conservation success. Finally, the analysis looks to the future and presents methods such as ecoregion conservation to begin to work at a finer scale within ecoregions.

This analysis of the terrestrial ecoregions of the Indo-Pacific is the third installment of a series using ecoregions to identify biological and conservation priority areas. This study has built on and improved the lessons learned from the previous two volumes (Dinerstein et al. 1995; Ricketts et al. 1999) and is a guide to an upcoming assessment on Africa's biological and conservation status (Burgess et al. in press). Together these assessments form the backbone of WWF's global terrestrial, freshwater, and marine strategies for conserving the Global 200 ecoregions (Olson and Dinerstein 1998).

Dr. Claude Martin
Director General
WWF International

Acknowledgments

This book was made possible by the contributions of numerous people and institutions. Without the tireless work of many biogeographers, ecologists, taxonomists, conservation biologists, and cartographers over the past few decades, this analysis would not have been possible. Many have been cited throughout the text, and others provided information through personal communications and discussions. They have dedicated their lives to helping conserve the natural biodiversity of the Indo-Pacific region. In turn, we dedicate this book to these scientists.

We also want to acknowledge the numerous reviewers of the manuscript who provided important information and much-appreciated direction early on in this process. Discussions and reviews by Max van Balgooy, Robert Johns, Kathy MacKinnon, Tim Whitmore, and Tony Whitten vastly improved the analyses. The ecoregion descriptions, which make up a large part of the book, were greatly improved by the contributions of Nantiya Aggimarangsee, Bruce Beehler, Frank Bonaccorso, Ramesh Boonratana, Thomas Brooks, Chris Carpenter, Geoff Davison, Ajay Desai, Savithri Goonetilleke, Larry Heaney, Robert Johns, A. J. T. Johnsingh, Andy Maxwell, John Maxwell, Rohan Pethiyagoda, Gopal Rawat, Tom Roberts, Philip Round, Phil Rundel, Hema Somanathan, Sompoad Srikosamatara, Seng Teak, WWF-India and Meeta Vyas, Tony Whitten, Carey Yeager, and Pralad Yonzon. Many of these contributors are listed as coauthors for the ecoregion descriptions.

The team from the Conservation Science Program of WWF-US put in a tremendous effort in reviewing, editing, and providing technical assistance throughout the analysis, no mean task given the vast amount of data that had to be collected, collated, analyzed, and updated over the past three years. For this we especially want to thank Robin Abell for her timely editing of several versions of this manuscript; Meseret Taye for administrative support and for developing the exhaustive Literature Cited and Consulted list; and Wesley Wettengel, Tom Allnutt, Holly Strand, Michelle Thieme, Ken Kassem, Jolie Patterson, Matt Anderson, and Jen D'Amico for technical reviews of the ecoregion descriptions and administrative support. We would also like to acknowledge Diane Wood and Jim Leape, who provided encouragement to produce this book.

Walter Erdelen would like to acknowledge those who reviewed and provided comments on earlier drafts of his essay. Indraneil Das would like to thank the institutions that supported his research and all the colleagues who provided him information through museum and field work.

Finally, we would like to thank those who have provided support for this conservation effort: Environmental Systems Research Institute, Hewlett-Packard Company, and National Geographic Society.

Terrestrial Ecoregions
of the Indo-Pacific

CHAPTER 1

Nature's End or a New Beginning?

As we begin this new millennium, a glance at a satellite image in the Indo-Pacific region reveals a bleak panorama (figure 1.1). Swaths of brown areas bereft of vegetation dominate the picture. Scattered throughout are patches of green, representing natural habitats that harbor the rich biodiversity of the region. But many green patches are geometric shapes, of squares and rectangles that represent monoculture plantations. Substantial areas of natural land cover—oases of irreplaceable biodiversity—are rare. The few in evidence occur mostly along the higher elevations of rugged and inaccessible mountain ranges or in the few unexploited places such as New Guinea. Even the wisps of cone-shaped clouds dotting the view are an illusion. They are not clouds but smoke plumes from fires set to clear large tracts of forests for agriculture and plantations.

If it were possible to compare this image with one taken a century ago, we would see a very different picture. To be sure, the fires would still be visible—a legacy of the shifting cultivation practices that have prevailed for hundreds of years—but the wide swaths would be green and the scattered patches and specks would be brown. When human populations in the region were a fraction of what they are today, shifting cultivation played a prominent role in maintaining biodiversity. Now, permanent agriculture has marginalized shifting cultivation,

Figure 1.1. The Cambodian Plains northeast of Phnom Penh, Cambodia. The Mekong River descends onto the plains from southwest Laos, laden with silt. Deforested areas, especially in the lowlands, appear on both sides of the river. Remaining forested regions are often degraded forests or plantations. Image (STS61A-32-0026) courtesy of NASA Space Shuttle Earth Observations Photography (http://earth.jsc.nasa.gov).

but where it still occurs, the clearings are larger and the rotation cycles are shorter, preventing the forests from regenerating. Together, the agriculture, plantations, settlements, and other land uses that support the burgeoning human population have turned the tapestry that represents natural biodiversity in the region from green to brown.

If these current trends continue, even the remaining patches of green may disappear. From Myanmar to Indonesia, many uncut areas are slated for large-scale logging operations and conversion to vast plantations of oil palm, tea, rubber, and other crops. Huge mining operations will devastate the intact forests of New Guinea, where the few large verdant patches still exist. The demand for land from one of Earth's densest human populations will continue to erode away the edges of green until natural biodiversity is restricted to insular protected areas. Any animal that ventures beyond the bounds of these protective refuges will be vulnerable. Indeed, many of Asia's large vertebrate species, such as elephants, tigers, and rhinoceros, are now confined mostly to these small reserves. Unless urgent steps are taken to stop wholesale conversion of land, these wide-ranging species may not survive the twenty-first century.

A recent satellite image of Riau province on the island of Sumatra (figure 1.2) illustrates the key forces behind this transformation of landscapes. Visible are the telltale smoke plumes from fires that destroy large tracts of forests and create a poisonous haze in the atmosphere that has far-reaching effects on human populations. Also visible are the logging roads, flanked by extensive clearings as they run through intact forests. Following the loggers and logging trucks along these roads will be the hunters and settlers, who will have easy access to core areas of forests, leaving little opportunity for natural forest regeneration. And many of the concessions will invariably leave behind large clear-cuts that will never be reforested. No wonder conservation biologists, such as Dr. Michael Soulé, view roads in tropical forests as "daggers in the heart of nature."

There is an urgent need for action to save the Indo-Pacific's natural habitats. Some parts of this region remain forested and appear impenetrable, but this is a misperception: they are highly imperiled. Derek Holmes underscores this point in

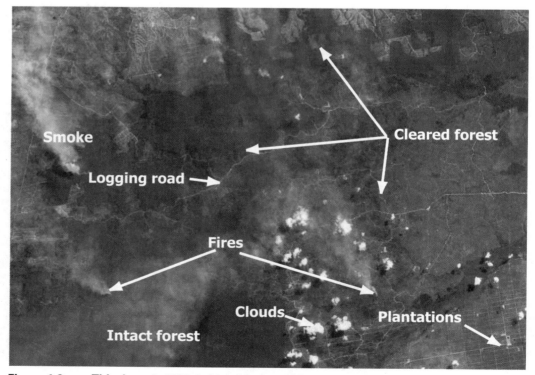

Figure 1.2. This June 1, 1999, infrared image of Riau, Sumatra, shows active burning of logging debris in logged forests. Intact forests appear as dark red in the bottom left of the image. Throughout Indonesia forests are being burned to convert the land to agricultural plantations such as oil palm (bottom right of image). Image courtesy of Centre for Remote Imaging, Sensing, and Processing (CRISP; www.crisp.wus.edu.sg).

essay 1, in which he concludes that if the current trends continue there will be no standing natural forest in the Indonesian part of Borneo (Kalimantan) by 2010.

Yet an obituary to nature in the Indo-Pacific region may be premature. A combination of new visionary approaches to conservation can halt and perhaps even reverse the trend of habitat destruction. Timber market trends could contribute by discouraging irresponsible and large-scale logging practices and consequently slow the destructive trajectory. In essay 2, Alf Leslie presents an exciting scenario of what could occur within the next five years if a predicted overproduction of timber from plantations leads to a market glut. The successes in community-based reforestation programs in southern Nepal (Dinerstein et al. 1999), where there has been a history of large-scale habitat degradation, provide a renewed sense of hope for the restoration of degraded lands elsewhere. And more importantly, the people who live adjacent to protected areas have embraced the wildlife that ventures beyond the reserve boundaries, indicating the possibility for landscapes where conservation can occur outside of strict protected areas (Dinerstein et al. 1999).

Which is the future for the Indo-Pacific? Is it nature's end, as intimated by Holmes's alarming statistics on forest loss during the next five years? Or is it Leslie's more optimistic and iconoclastic view that the tropical timber trade will collapse within two decades? Both predictions may be accurate. There may be a timber glut on the horizon, but it probably will not discourage local entrepreneurs from clearing intact lowland forests and replacing marvelously complex rain forests with oil palm plantations or tomorrow's next cash crop. If Leslie's predictions held true, then our strategy would be to protect the most biologically diverse forests until plantation wood floods the markets within the next decade. If pressure on natural forests is not reduced, we will need to redouble our efforts to secure the most irreplaceable habitats in a network of effective protected areas before a tidal wave of extinctions washes over the Indo-Pacific lowlands during the coming years.

But where do we start? Which green specks on the map of the Indo-Pacific are in most urgent need of protection and other conservation actions? Which ones can and should be reconnected? The answers to these questions are critical because the resources available to protect biological diversity across the region are limited. In short, we must prioritize where, when, and how we direct our conservation efforts.

Assessment Overview

This conservation assessment of the Indo-Pacific region begins the process of addressing these difficult questions. Our goal is to map regional patterns of biodiversity first and then to use this information to prioritize where and how conservation efforts should be directed in a regional strategy. Our approach gives weight not only to the most species-rich areas but also to ecosystems that are less rich but are distinctive and representative of this highly diverse and complex region.

Most of the terrestrial biodiversity in the Indo-Pacific, if measured in terms of species richness alone, is contained within the tropical rain forests that once covered a large part of this region. But the region is endowed with an incredible array of ecosystems beyond these rich tropical rain forests. These range from the high alpine meadows in the Himalayas and New Guinea to the alluvial grasslands along the Himalayan foothills, the vast dry forests of continental Asia, the deserts surrounding the Indus River, and the mangroves along the coastlines. All these ecosystems harbor different species and ecological processes, representing a broad range of biodiversity features that are nevertheless characteristic of the region. Priority setting on the basis of species richness alone will exclude many of the ecosystems, natural communities, species, and endemic features found across the Indo-Pacific region. Because our goal is to save all the pieces, and not merely one facet of the region's biodiversity, we have to look beyond tallies of species richness and strive for representation.

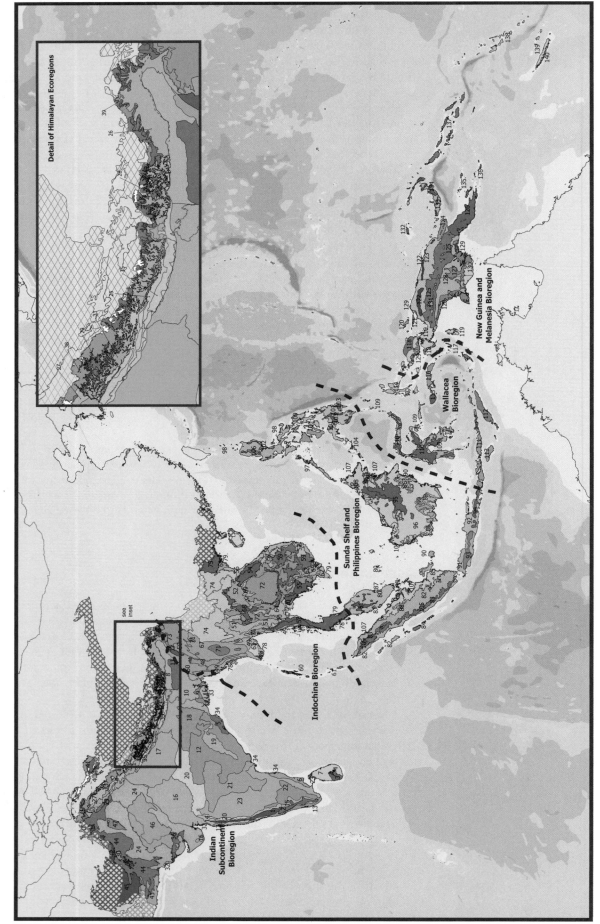

Figure 1.3. The 140 terrestrial ecoregions of the Indo-Pacific, nested within 5 bioregions (demarcated by black lines).

INDIAN SUBCONTINENT

1. TROPICAL AND SUBTROPICAL MOIST BROADLEAF FORESTS
1. North Western Ghats montane rain forests
2. South Western Ghats montane rain forests
3. Sri Lanka lowland rain forests
4. Sri Lanka montane rain forests
5. Orissa semi-evergreen rain forests
8. Brahmaputra Valley semi-evergreen rain forests
9. Sundarbans freshwater swamp forests
10. Lower Gangetic plains moist deciduous forests
11. Meghalaya subtropical forests
12. Eastern highlands moist deciduous forests
13. Malabar coast moist deciduous forests
14. North Western Ghats moist deciduous forests
15. South Western Ghats moist deciduous forests
17. Upper Gangetic plains moist deciduous forests
25. Himalayan subtropical broadleaf forests

2. TROPICAL AND SUBTROPICAL DRY BROADLEAF FORESTS
6. East Deccan dry-evergreen forests
7. Sri Lanka dry-zone dry evergreen forests
16. Khathiarbar-Gir dry deciduous forests
18. Chhota-Nagpur dry deciduous forests
19. Northern dry deciduous forests
20. Narmada valley dry deciduous forests
21. Central Deccan plateau dry deciduous forests
22. South Deccan plateau dry deciduous forests

3. TROPICAL AND SUBTROPICAL CONIFEROUS FORESTS
31. Himalayan subtropical pine forests

4. TEMPERATE BROADLEAF AND MIXED FORESTS
26. Eastern Himalaya broadleaf forests
27. Western Himalayan broadleaf forests

5. TEMPERATE CONIFEROUS FORESTS
28. Eastern Himalaya subalpine conifer forests
29. Western Himalaya subalpine conifer forests
30. East Afghan montane coniferous forests

7. TROPICAL AND SUBTROPICAL GRASSLANDS, SAVANNAS, AND SHRUBLANDS
35. Terai-Duar savanna and grasslands

9. FLOODED GRASSLANDS AND SAVANNAS
36. Rann of Kutch seasonal salt marsh

10. MONTANE GRASSLANDS AND SHRUBLANDS
37. Northwestern Himalayan alpine shrub and meadows
38. Western Himalayan alpine shrub and meadows
39. Eastern Himalayan alpine shrub and meadows
40. Karakoram-West Tibetan Plateau alpine steppe
41. Central Tibetan Plateau alpine steppe
42. Sulaiman Range alpine meadows

13. DESERTS AND XERIC SHRUBLANDS
23. Deccan thorn scrub forests
24. Northwestern thorn scrub forests
43. South Iran Nubo-Sindian desert and semi-desert
44. Baluchistan xeric woodlands
45. Rajasthan-North Pakistan sandy desert
46. Thar desert
47. Indus Valley desert

14. MANGROVES
32. Indus River delta-Arabian Sea mangroves
33. Sundarbans mangroves
34. Godavari-Krishna mangroves

INDOCHINA

1. TROPICAL AND SUBTROPICAL MOIST BROADLEAF FORESTS
48. Chin Hills-Arakan Yoma montane rain forests
49. Myanmar coastal rain forests
50. Mizoram-Manipur-Kachin rain forests
51. Kayah-Karen montane rain forests
52. Luang Prabang montane rain forests
53. Tenasserim-South Thailand semievergreen rain forests
54. Northern Annamites rain forests
55. Cardamom Mountains rain forests
56. Northern Vietnam lowland rain forests
57. Southern Annamites montane rain forests
60. Andaman Islands rain forests
61. Nicobar Islands rain forests
62. Tonle Sap-Mekong peat swamp forests
63. Irrawaddy freshwater swamp forests
64. Chao Phraya freshwater swamp forests
65. Tonle Sap freshwater swamp forests
66. Red River freshwater swamp forests
67. Irrawaddy moist deciduous forests
68. Northern Thailand-Laos moist deciduous forests
69. Northern Khorat plateau moist deciduous forests
70. Chao Phraya lowland moist deciduous forests
73. Northern Triangle subtropical forests
74. Northern Indochina subtropical forests
75. South China-Vietnam subtropical evergreen forests

2. TROPICAL AND SUBTROPICAL DRY BROADLEAF FORESTS
58. Southern Vietnam lowland dry forests
59. Southeastern Indochina dry evergreen forests
71. Irrawaddy dry forests
72. Central Indochina dry forests

3. TROPICAL AND SUBTROPICAL CONIFEROUS FORESTS
77. Northeast India-Myanmar pine forests

4. TEMPERATE BROADLEAF AND MIXED FORESTS
76. Northern Triangle temperate forests

14. MANGROVES
78. Myanmar coastal mangroves
79. Indochina mangroves

SUNDA SHELF AND PHILIPPINES

1. TROPICAL AND SUBTROPICAL MOIST BROADLEAF FORESTS
80. Peninsular Malaysian rain forests
81. Peninsular Malaysian montane rain forests
82. Sumatran lowland rain forests
83. Sumatran montane rain forests
84. Mentawai Islands rain forests
85. Sumatran peat swamp forests
86. Borneo peat swamp forests
87. Peninsular Malaysian peat swamp forests
88. Sumatran freshwater swamp forests
89. Southern Borneo freshwater swamp forests
90. Sundaland heath forests
91. Western Java rain forests
92. Eastern Java-Bali rain forests
93. Western Java montane rain forests
94. Eastern Java-Bali montane rain forests
95. Borneo montane rain forests
96. Borneo lowland rain forests
97. Palawan rain forests
98. Luzon rain forests
99. Mindoro rain forests
100. Greater Negros-Panay rain forests
101. Mindanao-Eastern Visayas rain forests
102. Luzon montane rain forests
103. Mindanao montane rain forests
104. Sulu Archipelago rain forests

3. TROPICAL AND SUBTROPICAL CONIFEROUS FORESTS
105. Sumatran tropical pine forests
106. Luzon tropical pine forests

10. MONTANE GRASSLANDS AND SHRUBLANDS
108. Kinabalu montane alpine meadows

14. MANGROVES
107. Sunda shelf mangroves

WALLACEA

1. TROPICAL AND SUBTROPICAL MOIST BROADLEAF FORESTS
109. Sulawesi lowland rain forests
110. Sulawesi montane rain forests
114. Halmahera rain forests
115. Buru rain forests
116. Seram rain forests
117. Banda Sea Islands moist deciduous forests

2. TROPICAL AND SUBTROPICAL DRY BROADLEAF FORESTS
111. Lesser Sundas deciduous forests
112. Timor and Wetar deciduous forests
113. Sumba deciduous forests

NEW GUINEA AND MELANESIA

1. TROPICAL AND SUBTROPICAL MOIST BROADLEAF FORESTS
118. Vogelkop montane rain forests
119. Vogelkop-Aru lowland rain forests
121. Yapen rain forests
122. Northern New Guinea montane rain forests
123. Northern New Guinea lowland rain and freshwater swamp forests
124. Huon Peninsula montane rain forests
125. Central Range montane rain forests
126. Southeastern Papuan rain forests
127. Southern New Guinea freshwater swamp forests
128. Southern New Guinea lowland rain forests
132. Admiralty Islands lowland rain forests
133. New Britain-New Ireland lowland rain forests
134. New Britain-New Ireland montane rain forests
135. Trobriand Islands rain forests
136. Louisiade Archipelago rain forests
137. Solomon Islands rain forests
138. Vanuatu rain forests
139. New Caledonia rain forests

2. TROPICAL AND SUBTROPICAL DRY BROADLEAF FORESTS
140. New Caledonia dry forests

7. TROPICAL AND SUBTROPICAL GRASSLANDS, SAVANNAS, AND SHRUBLANDS
130. Trans Fly savanna and grasslands

10. MONTANE GRASSLANDS AND SHRUBLANDS
131. Central Range sub-alpine grasslands

14. MANGROVES
120. Biak-Numfoor rain forests
129. New Guinea mangroves

- Water
- Rock, ice, and snow
- Portions of ecoregions not included in this analysis
- Ecoregion's full extent. This analysis only pertains to portions of those ecoregions with solid colors.
- Bioregion boundary
- International boundary
- Disputed boundaries, lines of control, or alignment unconfirmed

Note: The boundaries and names shown and the designations used on these maps do not imply official endorsement or acceptance by WWF.

To achieve our goal of representation, we use ecoregions as the fundamental conservation unit in this assessment. A shorthand definition of an ecoregion is "an ecosystem of regional extent" (Dinerstein et al. 1995). Ecoregions define distinct ecosystems that share broadly similar environmental conditions and natural communities, and therefore they are appropriate conservation units for regional-scale assessments that include representation as a primary goal. Because the boundaries of ecoregions reflect biogeographic patterns, they certainly make more sense for priority-setting efforts than do political units such as countries, provinces, or districts. Ecoregions are large enough to include the areas over which ecological processes operate—forces such as migrations of species, seed dispersal, predator-prey interactions, plant succession, floods, and forest fires—and shape the patterns of biodiversity (Orians 1993; Terborgh 1999).

This regional-scale analysis paints a big picture of regional priorities. But in doing so, it helps to put local conservation efforts into a broader perspective and shows where small-scale efforts can contribute most to larger conservation goals. Investing limited conservation resources in local projects without this regional perspective can lead to ad hoc or redundant efforts that contribute little to a sensible, holistic regional strategy. When the threats to biodiversity are dire, financial resources are limited, and time is short, this is an unaffordable luxury.

The Indo-Pacific region, as we define it, spans five bioregions and includes 140 ecoregions (figure 1.3). Two of the five bioregions are continental, and three represent island systems. Of the latter, two cover archipelagos and the third comprises mostly the large island of New Guinea and several satellite islands. Together, these five bioregions represent a diverse jigsaw puzzle of ecoregions that have varying degrees of species richness, endemism, and ecological and evolutionary phenomena. For instance, the continental ecoregions harbor communities of large vertebrate ungulates that include the Asian elephant, three rhinoceros species, and the most diverse assemblage of montane ungulates on earth. These ecoregions also support predator-prey systems where tigers, snow leopards, wolves, and wild dogs hunt the large herbivores. These species and processes need vast spaces. In contrast, most of the archipelago ecoregions lack large vertebrate communities and high species richness, but they harbor highly endemic and globally irreplaceable biotas. A conservation strategy for the region must include the full tapestry of biodiversity features.

Setting conservation priorities at regional scales entails measuring the biodiversity values of an ecoregion and assessing the threats to them. We analyze these parameters using two indices: the Biological Distinctiveness Index (BDI), which is based on species richness, endemism, and measures of ecological processes and evolutionary phenomena; and the Conservation Status Index (CSI), based on habitat-related variables and the degree of protection. We integrate these two indices to prioritize ecoregions. The highest-priority ecoregions—representative of all the habitat types found in Asia (e.g., rain forests, dry forests, alpine meadows, mangroves, flooded grasslands)—are the building blocks of a global conservation strategy, the Global 200 ecoregions (Olson and Dinerstein 1998; Olson et al. 2000).

We hope that this analysis will catalyze conservation efforts in the region and help to guide them in a meaningful direction to save the green patches and specks within the most distinct ecoregions because these are the biodiversity source pools that will help to restore the verdant landscape tapestry. Many of the biodiversity-rich nations of the Indo-Pacific are embarking on the fast track to development, and efforts to harmonize development with nature protection are falling by the wayside. Our attention can easily be diverted toward any of the numerous small crisis areas that in the long term will fail to serve conservation well. Instead, conservation must move beyond reactive efforts and into a realm of visionary, proactive approaches at ecoregional and landscape scales. We have set our sights high: to create a new regional landscape for the Indo-Pacific that will look far more encouraging from space than today's landscape appears. This new landscape will feature large, linked protected areas surrounded by a matrix of conservation-friendly land-use practices where critical habitat has been restored and maintained in part by transboundary cooperation. If this vision is realized within a few decades, satellite images of the Indo-Pacific will be dominated by swaths of green rather than brown.

Structure of the Book

This study involved the collection, synthesis, and analysis of an enormous amount of information. In chapter 2 we provide the background and basis for ecoregion delineation. In chapter 3 we define the objectives and the approach used in the analysis and provide more detailed explanations in appendix A. In chapters 4 and 5, respectively, we describe the biological distinctiveness and conservation status of ecoregions. In chapter 6 we integrate these two indices to set the conservation agenda and provide recommendations for where we should first concentrate our efforts in the Indo-Pacific. Chapter 7 offers an introduction to the next frontier for protecting biodiversity: Big Conservation. Big Conservation, as we define it, involves carrying out conservation at the ecoregional or landscape scale, which we argue is the best hope for saving nature in Asia.

In every chapter are short essays, contributed by regional experts. These essays address special topics focusing on finer-scale conservation issues or on ecological processes that are typically overlooked in a regional-scale analysis.

The appendices include more detailed results and information. To make the analyses as transparent and repeatable as possible, we provide details of the methodology, including the index values, in appendices A–F. We also provide descriptions of all of the ecoregions, with reference to the biodiversity and conservation status of each. Throughout the text, each ecoregion is indicated by its name followed by the map code in brackets (e.g., "New Caledonia Dry Forests [140]").

Essay 1

The Predicted Extinction of Lowland Forests in Indonesia

Derek Holmes

Conservation of the species-rich lowland forests of Indonesia has reached the crisis stage. Recent mapping of forest cover of Indonesia by the Ministry of Forestry and Estate Crops (MoFEC) reveals that the rate of deforestation over the past decade has increased from about 1.0 million ha to at least 1.7 million ha each year. Neither the presence, at least on paper, of a permanent forest estate nor national and international concern and donor assistance has had much effect in preventing this rapid increase in forest loss.

The Data Source

These startling results are the product of mapping conducted from satellite imagery by Badan Planologi (MoFEC). The main objective was to obtain a rapid overview to detect changes in forest cover. In the absence of field checking, the data must be regarded as provisional. Wherever available, the imagery dates from 1996 to 1998, but in some areas it was necessary to use 1994 or 1995 images. An average date of 1997 is assumed for the new maps, and some of the mapping predates the widespread forest fires of 1997–98 and the extensive illegal logging after the political crisis of 1998. The extent of forest on the new maps has been compared with the forest cover mapped by the Regional Physical Planning Project for Transmigration (RePPProT) program of the 1980s. Broadly interpreted, the earlier data are from around 1985, which assumes a period of twelve years over which the changes have occurred.

The methods (interpretation from digital Landsat satellite imagery) and the scale of mapping (1:500,000) are intended to provide information on forest cover only. Forest cover is defined as natural forest that can be recognized as such on satellite imagery. The presence of forest cover makes no assumptions about the quality of that forest. Timber plantations are excluded from the definition of natural forest.

The exercise has been conducted for most of the Outer Islands of Indonesia: Sumatra, Kalimantan, and Sulawesi (excluding Nusa Tenggara). The analysis is complete for Irian Jaya but is preliminary for Maluku.

The new forest cover maps can be inspected at the MoFEC Web site at http://mofrinet.cbn.net.id/e_informasi/e_nfi/GIS/vegetasi.htm, which also presents the maps derived from an earlier mapping program (National Forest Inventory). Area figures are presented for both datasets.

Rates of Deforestation

The gross figures of forest cover in the summary table (table 1.1) indicate that more than 20 million ha of forest have been lost over a twelve-year period, including 6.7 million ha in Sumatra and 8.5 million ha in Kalimantan. This amounts to an average annual rate of 1.7 million ha nationwide, or 4,658 ha per day (194 ha per hour). Of this total, the combined rate for the three islands of Sumatra, Kalimantan, and Sulawesi is 1.45 million ha per year. These are gross figures, and the average yearly rate is higher in years with widespread forest fires. However, this long-term average rate masks what has probably been a steep increase over the years, with probably more than 2 million ha being lost annually over the last three years.

TABLE 1.1. Summary Table of Forest Conversions, 1985–1997.

	1985		1997		Deforestation		
	Forest	% Total Land Area	Forest	% Total Land Area	Decrease 1985–1997	% Loss	Ha/Year
Sumatra	23,324,000	49%	16,632,000	35%	6,692,000	29%	558,000
Kalimantan	39,986,000	75%	31,512,000	60%	8,474,000	21%	706,000
Sulawesi	11,269,000	61%	9,000,000	49%	2,269,000	20%	189,000
Maluku	6,348,000	81%	[>5,544,000]*	?	>800,000	13%	>67,000
Irian Jaya	34,958,000	84%	33,160,000	81%	1,798,000	5%	150,000
Total	115,885,000	68.50%	c. 95,848,000	57%	20,033,000	17%	1,670,000

*Data for Maluku are preliminary.

The figures for forest cover quoted here differ slightly from those published on the MoFEC Web site because adjustments have been made to allow for areas that are presumed to have some forest cover but lack recent data (because of cloud cover or lack of imagery). The precise figure over the twelve years is less important in such a dynamic situation than the confirmation of a huge increase in the rate of deforestation.

The presence of hardly 57 million ha of forest in the three islands of Sumatra, Kalimantan, and Sulawesi in 1997 should not be cause for complacency because only 15 percent of this forest lies on the lowland nonswampy plains (about 8.5 million ha). This is the forest that is usually the richest source of timber and supports the highest biodiversity. Most of the remainder lies either in the hills and mountains (66 percent), which are generally too steep to log according to the ministry's own criteria, or in the alluvial swamps (15 percent).

Assuming constant rates of clearing, a further 3 million ha of forest is predicted to have been lost since 1997, and probably this has been mainly in the nonswampy lowlands. The rates are not constant. There is evidence that the rate has increased from 800,000 ha/year nationwide in the 1980s to around 1.2 million ha/year from 1990 through 1996. This implies that a steep increase occurred since 1996, probably to more than 2 million ha/year. This frightening rate of loss has been exacerbated by the enormous destruction of the 1997–98 forest fires. These fires were not a single event but rather the culmination of several years of major El Niño–induced fires. Clearly, forest fires will be a serious threat after future El Niño events, especially in view of the rampant logging that is reported to have occurred everywhere, including inside the boundaries of legally established protected areas (figure 1.4). The next drought probably will see the destruction through fire of many remaining areas of priceless heritage.

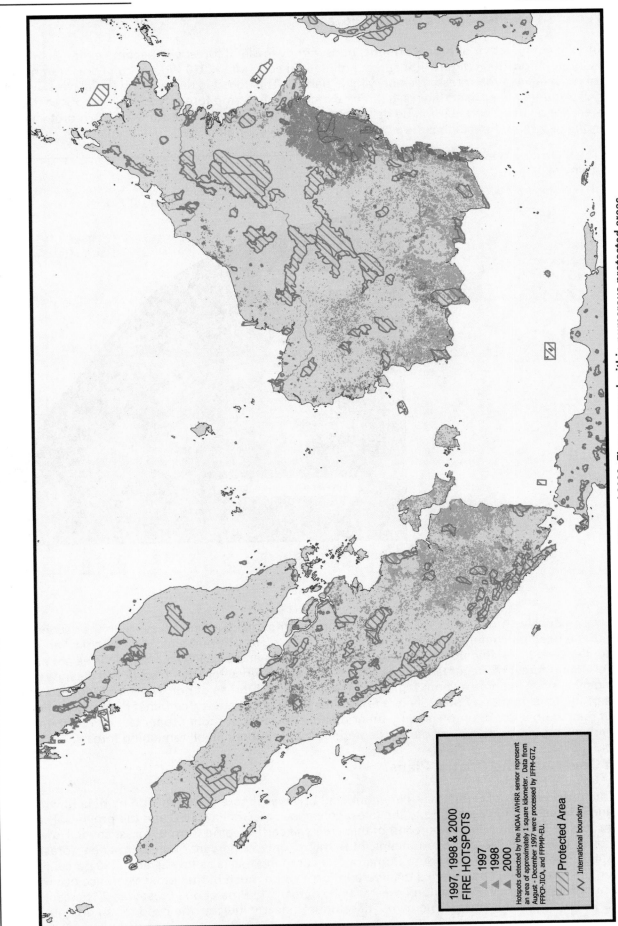

Figure 1.4. Distribution of fires in Sumatra and Borneo between 1997 and 2000. Fires occurred within numerous protected areas.

The Predicted Extinction of Lowland Forests

Assuming the continuation of present trends, nonswampy lowland forest will become extinct in Sumatra by or before 2005 (figure 1.5) and in Kalimantan soon after 2010 (figure 1.6). Forest remnants will be of no value as timber resources or wildlife habitat. This forest is nearly gone in Sulawesi, which is a largely mountainous island (figure 1.7). The extinction of swamp forests could follow about five years later. Not all this clearance will be deliberate; forest fires will have the same effect in protected areas and peat swamp forests that are now exposed to heavy logging.

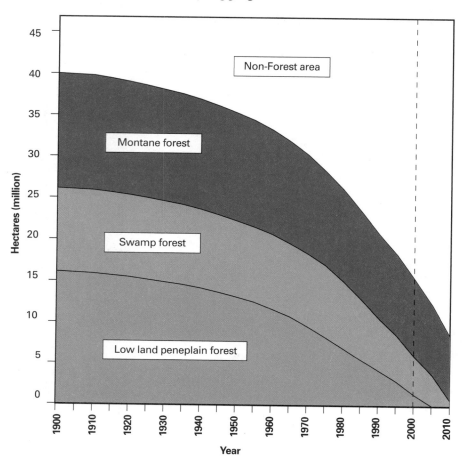

Figure 1.5. Changes in forest cover in Sumatra.

The only extensive forests that will remain in Sumatra, Kalimantan, and Sulawesi in the second decade of the new millennium will be the low stature forests of the mountains. Unless radical and far-reaching steps are taken to enforce existing laws, regulations, and policies and new policies are established for sound forest management, some of the most species-rich forests on earth will disappear. Forests may survive in some swamp regions in the high-rainfall zones of northwestern Indonesia that are less prone to drought and possibly in a few lowland protected areas that benefit from exceptional levels of management or are protected by other means. However, without proper law enforcement and a new management paradigm, steady degradation will continue in all remaining forests.

Forest Cover and the Spatial Plans

Within the three islands of Sumatra, Kalimantan, and Sulawesi, there are 69 million ha of land with "permanent forest" status according to the latest consensus between MoFEC and the provincial spatial plans. Some 57 million ha (83 percent) of this area still carries some form of forest cover. However, two-thirds of this forest lies in the mountains. In Sumatra, only 74 percent of the permanent forest estate still supports forest, and in south Sumatra and Lampung this is as low as about 30 percent. Based on the forest status boundaries that were in force when much of the forest clearance occurred, total area of forest cover in Production Forest was 66 percent, in Protection Forest was 77 percent, and in Conservation Forest was 82 percent. These figures clearly indicate the need for an urgent

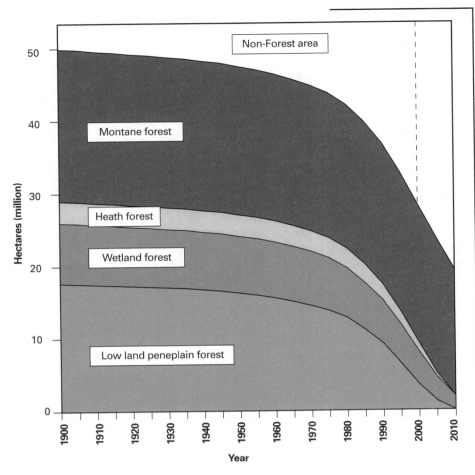

Figure 1.6. Changes in forest cover in Kalimantan.

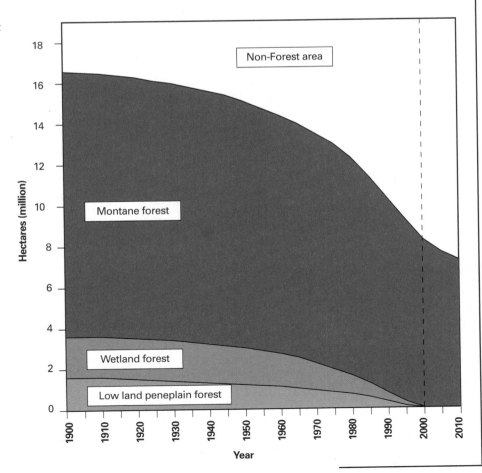

Figure 1.7. Changes in forest cover in Sulawesi.

re-evaluation of forest function in the context of spatial plans and the management of the natural resources of the Outer Islands.

Recommended Conservation Initiatives

All parties concerned with natural resource management and further revisions of provincial spatial plans must use up-to-date forest-cover maps. Unfortunately, the recent reconnaissance maps produced by MoFEC representing the 1996–1998 situation are already out of date in some areas, and a continuous process of monitoring and updating should become a routine activity. Forest management planning must include all stakeholders, including the traditional forest communities whose rights have hitherto been overlooked by planners and policymakers. The planning process should also include members of the scientific community who are concerned with biodiversity conservation, watershed management, and carbon sequestration. There must be no further approvals for estate-based forest conversion other than through a transparent process of participation of all stakeholders. Indeed, in Sumatra and Sulawesi especially, no further forest conversions should be authorized. This might also hold true for Maluku in view of the high rates of species endemism on the islands of this region.

The future of lowland forests in Sumatra has become so critical that a permanent moratorium on the conversion of any additional forest is needed. Spatial plans must reflect the reality of the present forest distribution, with all remaining hill forests, peat swamp forests, and mangrove forests being allocated permanent protection status (either as Protection Forest or as Conservation Areas). In Kalimantan, comprehensive review is now mandated within the forest function classification, with absolute limits being set on the extent and distribution of further conversions. In consideration of the topography of Sulawesi and the loss of its lowland forests, no further permits for conversion might legitimately be granted.

The remaining forests of these three regions have now become far too fragile and too precious to be managed by a single government department that has exploitation as its primary focus. The people of Indonesia need to be apprised of the current situation at the earliest opportunity, and they need to become stakeholders in the future management of the forests that remain. Recent experience has shown that the mere allocation of protection status on paper has little meaning in practice at the provincial or field level.

Forest loss has been so rapid that a review is now required of Indonesia's Biodiversity Action Plan. Some of the sites proposed have lost their conservation value, and the intrinsic value of others has increased as the forests around them have been progressively cleared. This may be a final opportunity to evaluate remaining options.

The provincial descriptions that follow identify a number of sites where such initiatives are considered to have immediate urgency. These are not the national parks that are already the subject of management interventions or are otherwise well known. Rather, they are the lesser-known areas that are believed to still have major importance for biodiversity conservation, especially in the highly critical lowlands.

Priority Areas of the Biodiversity Action Plan

The priority areas identified in the Biodiversity Action Plan for Sumatra, Kalimantan, and Sulawesi are listed here (table 1.2), with comment on the present forest cover situation. Again, this is not a statement of the quality of that forest.

TABLE 1.2. Priority Conservation Areas of Biodiversity Action Plan in Indonesia.

Site	Province	Comment
SUMATRA		
Gunung Leuser	Aceh	Forest cover mainly intact except along the margins
Perluasan Leuser	Aceh	Now very high priority
Singkil Barat	Aceh	Very high priority
Dolok Sembilin	North	Partially destroyed
Kambang Lubuk Niur	West	Now part of Kerinci-Seblat National Park
Taitai Batti	West	The Siberut National Park now exists
Kerinci Seblat	Four provinces	Forest cover mainly intact except along the margins of
Kerumutan	Riau	Still intact but under threat from adjacent drainage schemes
Seberida	Riau	Now Bukit Tiga Puluh National Park but under severe threat
Bukit Besar	Jambi	Now part of Bukit Tiga Puluh National Park
Berbak	Jambi	Under threat from fire damage during El Niño years
Sembilang	South	Under immediate threat, urgent action needed
Bentayan	South	Completely destroyed
Barisan Selatan	Lampung	Forest cover mainly intact except along margins
Way Kambas	Lampung	Forest cover still further reduced, high fire risk
KALIMANTAN		
Gn Palung	West	Forest cover mainly intact except some margins but high risk
Bentuang-Karimun	West	Forest cover mainly intact
Sentarum	West	Existing forested areas largely intact but at risk
Bukit Baka	West	Forest cover mainly intact except some margins
Bukit Raya	Central	As Bukit Baka
Bukit Raya Ext.	West Central	As Bukit Baka
Tanjung Puting	Central	Under immediate severe threat from conversion and logging
Tanjung. Puting Ext.	Central	As Tanjung Puting
Kayan-Mentarang	East	Mainly intact
Ulu Kayan Mutlak	East	Now part of Kayan-Mentarang National Park
Ulu Sembakung	East	Proposal accepted by Minister but far from secured; forest intact
Muara Sebuku	East	As Ulu Sembakung; mangroves especially may be at risk
Kutai	East	Value of remaining forest after fires/conversion must be assessed
Sangkulirang	East	High-priority area, intervention urgent
SULAWESI		
Tangkoko-Dua Saudara	North	Extensive deforestation around all the margins
Dumoga-Bone	North	Mainly intact except some narrow footslope margins
Marisa complex	North	No data (new imagery not obtained)
Lore Lindu	Central	Mainly intact except some narrow margins
Morowali	Central	Little change, but encroachment may increase on west side
Gn Latimojong	South	Data incomplete; probably heavy deforestation on lower slopes
Rawa Aopa	Southeast	Core region intact but with ongoing extensive deforestation in surrounds

ESSAY 2

For Whom the Bell Tolls, or What Is the Future of the Tropical Timber Trade in the Face of a Probable Glut of Plantation Timber?

Alf J. Leslie

The demand for timber and forest products over the current and next rotations, a period that takes us well into the twenty-first century, is a fundamental element in devising forest policy. It is especially important for the continued existence of tropical forests. Consequently, sustainable forest management has become critical to their future. Equally important, however, is the role of the global timber plantation resource, which has simultaneously undergone tremendous expansion.

These two trends have mutually reinforcing effects on the future of tropical forests and the tropical timber trade. The first, the sustainable forest management effect, inevitably will push up the cost of timber harvesting and management. The second, the plantation effect, will almost certainly put limits on the extent to which timber prices can be increased to cover higher production costs and, depending on its scale and timing, might even push the price ceiling lower.

In a broad sense, the plantation effect has been analyzed in a number of recent studies by or for the Food and Agriculture Organization (FAO) of the United Nations, the European Forest Institute, and the Intergovernmental Forum on Forests. However, few have examined specific implications for tropical forests or for the tropical timber trade. One such study is the recently conducted review prepared for the Malaysian state of Sarawak under an International Tropical Timber Organization project (Model Forest Management Area, Phase II). Although it is focused on Sarawak, some of its findings have a pan-tropical significance. As a preview to wood supply and demand in the new century, these results are presented here, modified slightly to fit the wider tropical context.

A Tidal Wave of Timber

The tropical timber market, after recovery from the east and southeast Asian economic crisis of 1997–1999, will never be the same. This is because the main factor that will change the market will not arise from economic restructuring but from a huge increase in the potential supply of plantation wood. This supply has recently begun to enter the Asia-Pacific region and world demand-supply balance.

Two features of this addition on the supply side are particularly relevant. First, the wood will be largely in the form of general-purpose, utility timber (commodity timbers), and pulpwood. Second, the age class distribution of the plantation resource is such that these wood products will arrive more as a tidal wave than as a steadily rising flow.

This tidal wave will add $35-40 \times 10^6$ m^3 of industrial roundwood to the supply within the Pacific Rim beginning around 2005. Representing no more than 2–2.5 percent of the present world consumption, it is unlikely to have much of an impact at the world level. But in the Pacific Rim markets, where the first tidal wave will hit (originating as it does in the maturation of plantations established in the planting boom of the 1970–1985 period in New Zealand, Australia, and Chile), the addition amounts to 10–15 percent of the regional demand. An impact of that scale cannot be disregarded or absorbed easily.

The Second Wave

Even more dramatic, a second tidal wave will follow ten or so years later as the $100-150 \times 10^6$ ha of industrial plantations already established worldwide are harvested. This resource has the potential to meet at least 70 percent of the world's present consumption of industrial wood. And this potential is being supplemented by the establishment of $5-8 \times 10^6$ ha of new plantations, with at least 100×10^6 m^3 of annual timber production each year.

Thus, in both the Pacific Rim and, before long, the world, the outlook is for oversupplied markets for commodity timbers and pulpwood. This would not occur if supply from natural forests falls or is reduced at a rate of 15–20 percent annually from the year 2000 or so onward or if demand for industrial roundwood increases after the year 2000 at a corresponding rate.

A reduction in timber extraction from natural forests as a result of these forests being allocated for nontimber use at an annual rate of 15–20 percent seems unlikely, although it could occur if competition from plantation wood virtually eliminates natural forest operations. Nor does an annual increase of 15–20 percent in the demand for industrial wood seem any more likely, given that the average annual rate of increase in the past has been less than 2 percent. Even with a combination of both trends, the necessary rates of change—10 percent per annum decline in output from natural forests and 5 percent per annum increase in demand—seem highly improbable.

Tropical Timber: Uncompetitive as a Commodity?

The conclusion is almost inescapable. The postrecovery market for commodity timber and pulpwood-based produce will be fiercely competitive from around 2005 onward in the Asia-Pacific region and from around 2010–2015 for the world as a whole. Much of the wood from the tropical forest resource and almost all of it from the existing and prospective plantations is in these categories. So from around 2010 to 2015, the bulk of tropical timber exports will face oversupplied markets.

Can tropical timber take on that competition and beat it? If it cannot, what else can be done?

The chances of tropical timber being successful in such a market environment cannot realistically be rated as high. Even countries with the domestic market base to support a predatory, marginal pricing campaign as a marketing ploy will be hard-pressed to succeed.

Seeking the High-Value Markets

Can anything else be done? The oversupply will be in the commodity pulpwood grades but not necessarily in the markets for specialty or decorative timbers. Nevertheless, they will be affected to some degree: the mass market for specialty and decorative products can be met through technological developments that eliminate the technical differences between softwoods and hardwoods and reproduce the decorative features through overlays. But beyond the mass market, only the genuine article will do.

Questions then arise:
- Are there such markets for genuine decorative timbers and products?
- If so, where are they, what exactly are the markets for, and how big are they?
- What are their key characteristics?
- Which tropical timbers can match or be matched to those characteristics?

At present, we have little more than generalities to work from. Teak, mahogany, rosewood, cherry, and walnut make up such markets. Prices much higher than those for commodity timbers are reported or rumored. Other supply characteristics are not known, at least not in the public domain. Therefore, research to find, measure, and assess these markets is a prerequisite, followed by market development to capture and hold them. All this will be very expensive because the key element in market research—uncovering trade secrets—cannot be accomplished by conventional market research methods.

Time to Start?

There could hardly be a worse time to start such market research, with cash flows to fund research and development depressed. However, the time is never right. When cash flows are good, supplying the buoyant markets takes up all the time and hides the need and urgency for it.

So the tropical timber trade is trapped. It will be hard to fund the market research and development programs that might enable it to avoid being overwhelmed by this plantation effect. A revival of the preslump market is needed to stimulate the necessary cash flows, but this postrecovery market is unlikely to be like the preslump market for long. To complicate matters, certification of sustainable forest management will be needed for access to many of the export markets affluent enough to want and afford the genuine article at prices that will carry the cost of sustainable forest management.

What is the solution for the tropical timber trade? The best route may be to disengage from its largely commodity timber strategy and develop itself as an exporter of high-value, high-quality, decorative timbers from sustainably managed natural tropical forests.

Somehow, tropical forestry must fairly quickly take a route in which
- A multispecies Brandis type of sustainable forest management with near-zero impact (helicopter) logging is the standard practice.
- The market research and development for identifying, characterizing, and implementing the high-value strategy is started immediately.

- The production, distribution, marketing, and quality control systems to capture the high-value markets become the industry norm without delay.

Failing that, the future for tropical forestry is highly likely to be one in which timber production is in rapid decline and tropical forestry is a low-output, low-return, weak, and fragile competitor in a cut-throat commodity timber export market. The main value of tropical forests increasingly will lie in their conservation, environmental, and ecotourism values; because sustainable forest management will be essential for this, and the tropical timber industry is in no shape to bear the cost, the forestry industry will all but disappear.

Postscript

The outlook summarized here is only one of many possible views of the way in which regional and global wood supply-demand balances are evolving. Lower estimates of the effective area of the established plantation resource, lower estimates of the rate of additional planting, and lower mean annual increment estimates, combined with higher forecasts of future demand levels, would reduce the impact of the plantation effect, even to the point of eliminating it.

The original version of this essay appeared in A. J. Leslie. 1999. For whom the bell tolls. *Tropical Forest Update* 9: 4. International Tropical Timber Organization, Yokohama.

CHAPTER 2

Assembling the Ecological Jigsaw Puzzle: The Ecoregions of the Indo-Pacific

Delineation of Ecoregions and Geographic Scope of the Study

Each year, herds of wild elephants cross from northern India to reserves in the lowlands of neighboring Nepal during their seasonal migrations. Although the pachyderms cross an international boundary, they remain in the same ecoregion: the Terai-Duar Savannas and Grasslands [35]. The recently discovered saola (*Pseudoryx nghetinhensis*), a forest ungulate in the ox family, moves across the border between Vietnam and Laos but never leaves the densely forested limestone mountains that define the Northern Annamites Rain Forests [54] ecoregion. In the high Himalayas, rich rhododendron forests form a belt from Nepal through Bhutan and northeast India into the Golden Triangle of Myanmar, harboring species whose distributions do not heed the several international boundaries that the forests intersect. Birds of paradise and palm cockatoos fly across the virtual vertical line that separates New Guinea into Irian Jaya and Papua New Guinea but never leave the ecoregion known as Southern New Guinea Lowland Rain Forests [128].

Biogeographers have long known that the distributions of plants, animals, habitats, and ecosystems rarely adhere to political boundaries, in the Indo-Pacific or in any region of the world. For the last one and half centuries, biogeographers have attempted to divide the earth into units that reflect these distribution patterns (Spellerberg and Sawyer 1999). Conservation planners in national and state agencies are now becoming increasingly aware of the incongruence between political boundaries and biogeographic distributions, as shown by the increasing attention being paid to transboundary conservation (Roca et al. 1996; Fall 1999; Dillon and Wikramanayake 1997).

Some of the more recent biogeographic classifications (Dasmann 1973; Udvardy 1975; Schultz 1995; Bailey 1998) are valuable contributions to understanding global patterns of biodiversity. However, their scales are too coarse to guide conservation planning and priority-setting in a region as biogeographically complex as the Indo-Pacific. Other classifications that are based on distributions of single taxa such as BirdLife International's Endemic Bird Areas (Stattersfield et al. 1998) have too narrow a focus to achieve representation of all distinct natural communities within conservation landscapes and a network of protected areas (after Noss 1991). However, these classifications are useful in places where the urgency for conservation is great but broad-based information about biodiversity is lacking. The classification most consistent with our approach is the delineation of biounits by MacKinnon and MacKinnon (1986) and MacKinnon (1997). But even the biounits combine vastly different habitat types, making representation of biodiversity a challenge.

In keeping with an approach taken for other biogeographic realms across the world (Dinerstein et al. 1995; Ricketts et al. 1999; Burgess et al. in press; Olson et al. in press), we developed a classification scheme for biodiversity conservation in the Indo-Pacific region using ecoregions as the fundamental conservation unit. An ecoregion is defined as "a relatively large area of land or water that contains a geographically distinct set of natural communities that share a majority of their species, ecological dynamics, and environmental conditions and function together as a conservation unit at global and regional scales" (Ricketts et al. 1999: 7). Ecoregions make suitable conservation planning units at regional and global scales because they

- represent the range of habitat types and ecological processes rather than the distribution of a single taxonomic group
- correspond to the major ecological and evolutionary processes that create and maintain biodiversity
- better address the conservation needs of populations, especially for species that need large habitat areas
- enable conservationists to determine the best places to invest scarce resources first to protect a representative sample of the region's biodiversity
- represent the dynamic arena within which restoration efforts should be undertaken

Geographic Scope

This assessment encompasses south and southeast Asia, through the Indonesian and Philippine archipelagos to New Caledonia and Vanuatu in the South Pacific, hereafter the Indo-Pacific region (figure 2.1a–c). We treat China and west Asia in separate analyses; thus, the northern boundary along the Himalayas and Indochina reflects a nonbiologically defined boundary, as does the western bound-

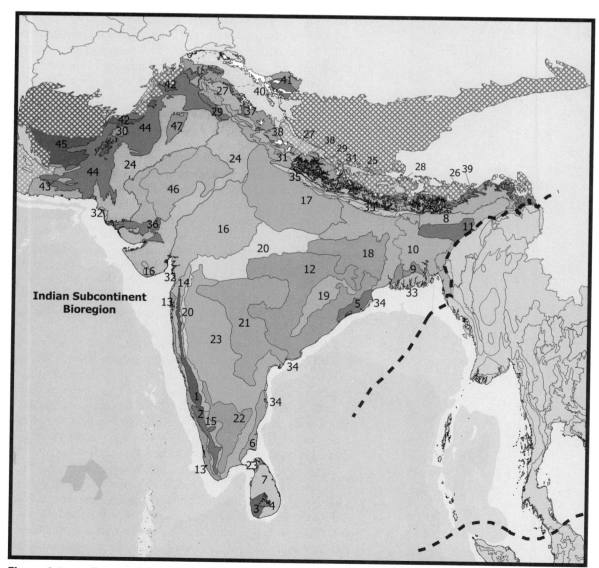

Figure 2.1a. Terrestrial ecoregions of the Indian Subcontinent bioregion.

ary between Pakistan and Afghanistan and Iran. However, the ecoregions that straddle these boundaries are defined on biological bases. At the eastern extreme of the region of analysis, the Torres Straits between New Guinea and northern Australia represents a floral transition zone (Whitmore 1984), and we treat Australian ecoregions elsewhere (Olson et al. in press).

Themes

Four overarching themes form the basis of our conservation assessment using ecoregions:

> Our objective is practical. We present a set of ecological units designed to improve conservation planning and prioritizing at regional or even global scales. It is not our intention to merely introduce yet another permutation of biogeographic units. Instead, we try to build on biogeographic units already recognized by regional and national biodiversity experts.

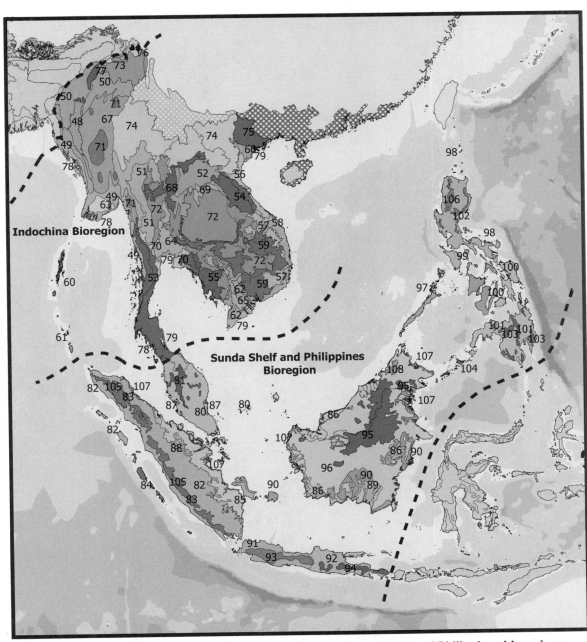

Figure 2.1b. Terrestrial ecoregions of the Indochina and Sunda Shelf and Philippines bioregions.

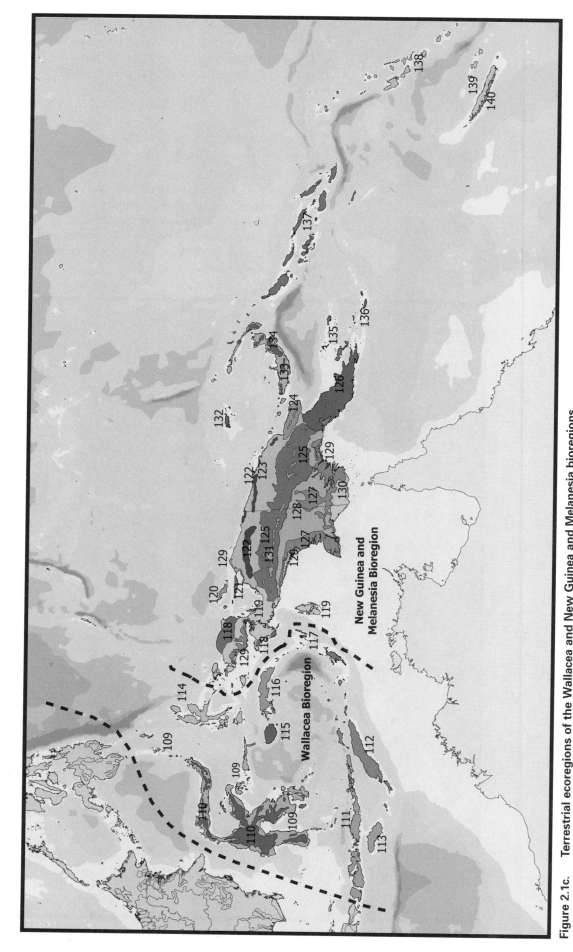

Figure 2.1c. Terrestrial ecoregions of the Wallacea and New Guinea and Melanesia bioregions.

Our approach is hierarchical. We first divide the Indo-Pacific into five bioregions. Within these bioregions we identify ecoregions and place them within biomes (figures 2.2 and 2.3; box 2.1). The approach is similar to Dasmann's (1974) hierarchy of biotic regions and provinces, which includes both composition and process. However, our ecoregions represent conservation units of much finer scale than Dasmann's biotic provinces (appendix A). For instance, in the Indian subcontinent, Dasmann recognizes three biotic provinces, whereas we identify seventeen ecoregions.

Our rationale is representation. The hierarchical framework allows ecoregions to be evaluated by biome within each bioregion. Therefore, the relatively species-poor tropical dry forests are not compared with the species-rich tropical moist forests, nor are tropical moist forests in different bioregions—where the species assemblages are different— compared and evaluated together. In this manner, ecoregions from all biomes can be represented in a regional portfolio for conservation investment despite differences in species richness, endemism, and other biological characteristics.

Our methods are both transparent and repeatable. Details of the methods, analyses, data, and sources of metadata are provided either in the following chapters, or as appendices at the end of the book. Using these methods, databases, and criteria, the analysis can be repeated.

Delineating Bioregions

Within the Indo-Pacific region we define broad bioregions, which are "clusters of ecoregions that share a similar biogeographic history and share many genera and families of plants and animals" (Ricketts et al. 1999). Bioregions ensure that ecoregions sharing the same dominant habitat type or biome but are biogeographically distinct are not compared. For example, the tropical dry forests in the Indian subcontinent bioregion are structurally similar to the tropical dry forests of New Caledonia (New Guinea and Melanesia bioregion), but these biomes have few species in common. Thus, assuming that tropical dry forests of the Indo-Pacific are well represented in a regional portfolio that includes only forests in the Indian subcontinent would not achieve representation. Instead, complete representation should include ecoregions from all biomes in each bioregion.

We demarcate five bioregions across the Indo-Pacific (figure 2.2):

- The Indian Subcontinent bioregion extends from west of the Indus River in Pakistan to the Indian and Nepalese Himalayas and foothills, and east to the Burmese transition zone. Although the higher Himalayan habitats (alpine scrub and meadow) have closer affinities with the Palearctic Realm, we considered them within the Indian Subcontinent bioregion because they fall within a broad transition zone between the Oriental and Palearctic Realms. The western border also represents a transition zone between the Palearctic and Oriental Realms. The boundary separating the Indian subcontinent bioregion from the Indochina bioregion is based on the western limit of the seasonal wet climates of the Indo-Burmese region (Ashton 1995). This biogeographical boundary lies to the west of the India-Myanmar national border and follows the Brahmaputra valley to the Himalayas west of the Golden Triangle in northern Myanmar. MacKinnon (1997) also uses this boundary to demarcate the Indian subcontinent subregion from the Indo-Chinese subregion.
- The Indochina bioregion is defined by MacKinnon and MacKinnon (1986) to include Myanmar and Thailand rather than being restricted to French Indochina (i.e., Laos, Cambodia, and Vietnam), and we follow this definition. It extends from the Indo-Burmese border to the Kangar-Pattani line that demarcates the northern limits of the Malesian flora (Van Steenis 1950; Whitmore 1984).

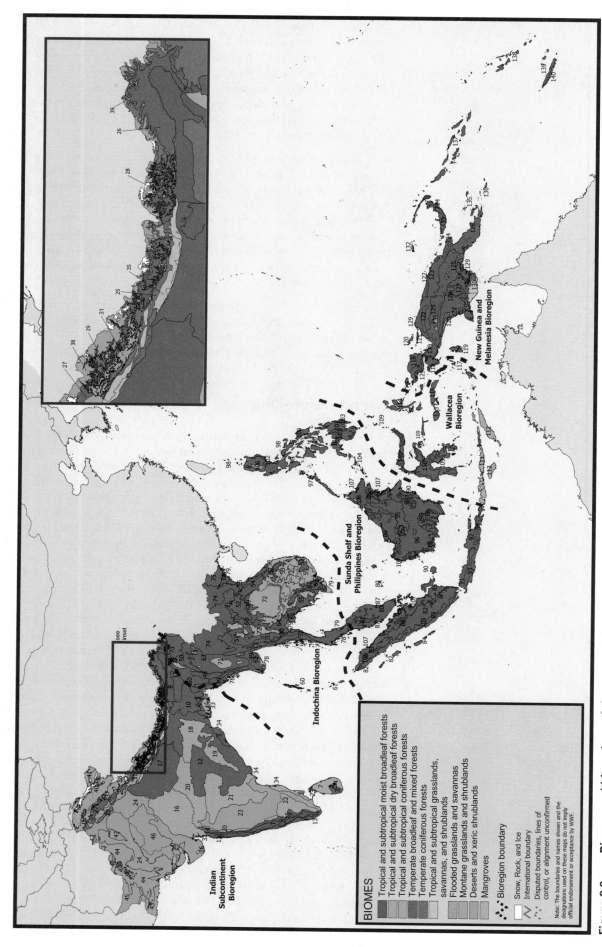

Figure 2.2. Biomes and bioregions of the terrestrial ecoregions of the Indo-Pacific. There are five bioregions in the Indo-Pacific, demarcated by the black dashed lines. Most of the ecoregions fall into tropical and subtropical moist broadleaf forest and tropical and subtropical dry forest biomes, which account for 72 percent of all ecoregions.

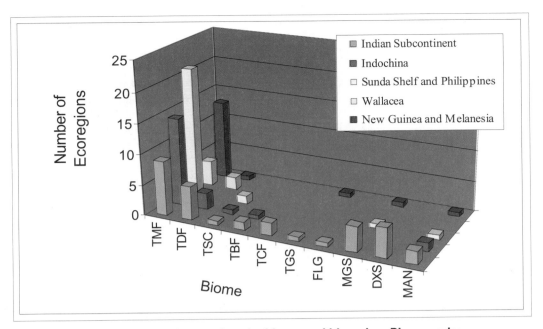

Figure 2.3. Distribution of ecoregions by biome and bioregion. Biome code: TMF = Tropical and subtropical moist broadleaf forests; TDF = Tropical and subtropical dry broadleaf forests; TSC = Tropical and subtropical coniferous forests; TBF = Temperate broadleaf and mixed forests; TCF = Temperate coniferous forests; TGS = Tropical and subtropical grasslands, savannas, and shrublands; FLG = Flooded grasslands and savannas; MGS = Montane grasslands and shrublands; DXS = Deserts and xenic shrublands; MAN = Mangroves.

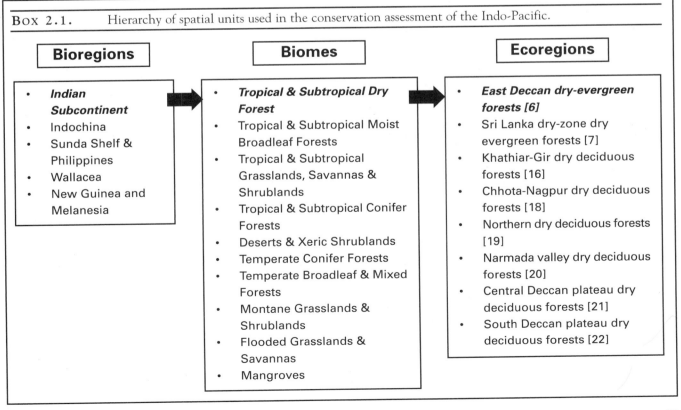

BOX 2.1. Hierarchy of spatial units used in the conservation assessment of the Indo-Pacific.

Bioregions
- ***Indian Subcontinent***
- Indochina
- Sunda Shelf & Philippines
- Wallacea
- New Guinea and Melanesia

Biomes
- ***Tropical & Subtropical Dry Forest***
- Tropical & Subtropical Moist Broadleaf Forests
- Tropical & Subtropical Grasslands, Savannas & Shrublands
- Tropical & Subtropical Conifer Forests
- Deserts & Xeric Shrublands
- Temperate Conifer Forests
- Temperate Broadleaf & Mixed Forests
- Montane Grasslands & Shrublands
- Flooded Grasslands & Savannas
- Mangroves

Ecoregions
- ***East Deccan dry-evergreen forests [6]***
- Sri Lanka dry-zone dry evergreen forests [7]
- Khathiar-Gir dry deciduous forests [16]
- Chhota-Nagpur dry deciduous forests [18]
- Northern dry deciduous forests [19]
- Narmada valley dry deciduous forests [20]
- Central Deccan plateau dry deciduous forests [21]
- South Deccan plateau dry deciduous forests [22]

- The Sunda Shelf and the Philippines bioregion extends from the Kangar-Pattani line in the Malaysian peninsula to Wallace's Line in the east. It includes peninsular Malaysia, Sumatra, Java, Bali, Borneo, and all the Philippine islands. Our delineation of this bioregion differs from the Sundaic subregion defined by MacKinnon and MacKinnon (1986), who use Weber's line to exclude the Philippines from this bioregion.
- The Wallacea bioregion includes Sulawesi, Nusa Tenggara (from Lombok eastward), and Maluku. The eastern demarcation of this bioregion is Lydekker's line, which runs along the Sahul shelf (see MacKinnon and MacKinnon 1986).
- The New Guinea and Melanesia bioregion consists of New Guinea and the satellite islands (including Aru) and extends into the southern Pacific to include Vanuatu and New Caledonia.

Delineating Biomes

We define biomes based on vegetation structure, ecological dynamics, and environmental conditions. Ecoregions within respective biomes in the same bioregion thus are more similar in terms of species community composition and dynamics than are ecoregions in different biomes. Therefore, comparing biodiversity values among ecoregions within their respective biomes is more relevant than comparing values between different biomes or within the same biome but in different bioregions. For example, tropical moist forests generally have many more species than do tropical dry forests, and species assemblages of the two biomes show little overlap. Because the goal is to conserve representative biodiversity across the region, we must conserve natural communities in both biomes.

Stratifying ecoregions within biomes also allows us to tailor the conservation status assessment and recommendations according to each biome's features. Biomes differ according to the minimum area requirements needed to conserve area-sensitive species, habitat specialists, and beta-diversity; the range of ecological processes and their thresholds of resiliency; and levels of protection and configuration of protected areas required to capture the ecoregion's biodiversity. For example, conifer forests that are fire-adapted, and even fire-maintained, will respond and recover from fire differently than tropical lowland moist forests. Conifer forests also have lower beta-diversity than tropical moist forests. (Beta-diversity is a measure of species diversity between habitats, reflecting changes in species assemblages along environmental gradients.) For these and other reasons, it is unwise to apply a single measure of habitat intactness or a single set of guidelines for reserve design to ecoregions in both biomes.

We distinguish ten biomes in the Indo-Pacific region (figure 2.2) that represent variations in ecological processes, patterns of biodiversity, and responses to disturbance (see appendix B for details on biomes). The biomes correspond to a global framework applied to other regional analyses (Dinerstein et al. 1995; Olson and Dinerstein 1998; Ricketts et al. 1999; Burgess et al. in press).

The biomes for the Indo-Pacific are
- tropical and subtropical moist broadleaf forests (including moist deciduous forests)
- tropical and subtropical dry broadleaf forests
- tropical and subtropical coniferous forests
- temperate broadleaf and mixed forests
- temperate coniferous forests
- tropical and subtropical grasslands, savannas, and shrublands
- flooded grasslands and savannas
- montane grasslands and shrublands
- deserts and xeric shrublands
- mangroves

To enhance transparency of our classification scheme, we present details of how we aggregate forest formations into the two most complex biomes: tropical and subtropical moist broadleaf forests and tropical and subtropical dry broadleaf forests (sensu Blasco et al. 1996; Whitmore 1984).

Tropical and subtropical moist broadleaf forests consist of
- perhumid (rainfall >1,500 mm, 0–3 dry months): tropical evergreen rain forests, tropical montane rain forests, subtropical montane rain forests
- humid (rainfall >2,000 mm, 5–6 dry months; or rainfall 1,000–2,000, 3–4 dry months): tropical lowland semi-evergreen rain forests
- sub-humid (rainfall >2,000 mm, 7–8 dry months): moist deciduous forests

Tropical and subtropical dry broadleaf forests consist of
- sub-humid (rainfall 1,000–2,000 mm, 5–7 dry months): dry deciduous dipterocarp woodlands
- dry (rainfall 1,000–1,500 mm, 5–7 dry months; or 500–1,000 mm, 6–9 dry months): dry deciduous forests
- arid (rainfall <500 mm, 9–11 dry months): thorny thickets

Delineating Ecoregions

A number of vegetation maps are available for the Indo-Pacific region. These are MacKinnon's (1997) reconstruction of the original forest-cover data from Pakistan to New Guinea (data provided in digital format by World Conservation Monitoring Centre [WCMC] and the Asian Bureau for Conservation [ABC]), forest-cover maps from Bouchet et al. (1995) for New Caledonia, the IUCN atlas of tropical forests (IUCN 1991), and the atlas of mangroves (Spalding et al. 1997). Using these maps as a foundation, we delineated ecoregion boundaries based on climate (wet forests vs. dry forests), elevation (lowland vs. montane), and analyses and reviews of published biogeographic zones (Udvardy 1974; MacKinnon and MacKinnon 1986; Rodgers and Panwar 1988; MacKinnon 1997; Stattersfield et al. 1998). We used Blasco et al. (1996) and Whitmore (1984) to help classify forest formations for the Indian subcontinent and Indochina and for Malesia, respectively. Several experts with extensive biogeographic knowledge of the Indo-Pacific region helped us to modify and refine ecoregion boundaries (appendix A).

MacKinnon's (1997) biounits are the latest biogeographic classification for this region and build on previous efforts. We stayed within the general framework of MacKinnon's biounits with our ecoregions but depart in three major ways.

First, ecoregion boundaries are based more closely on maps of original vegetation, whereas biounits are delineated largely according to the distribution of vertebrates. Vascular plants and invertebrates constitute a disproportionate percentage of species in natural communities, and the distributions of invertebrates typically are closely linked to host plants or plant communities. Thus, potential vegetation maps serve as a proxy for capturing the distributions of the most abundant taxon groups. Additionally, many vertebrate species show close affinity and ecological specialization to distinct vegetation types.

Second, the ecoregion approach avoids mixing species and natural communities that are characteristic of specific habitat types. MacKinnon's (1997) biounits often contain a mix of distinct biomes and habitat types, such as mangroves, tropical dry deciduous forests, thorn scrub, tropical semi-evergreen forests, and tropical moist forests within a single biounit. These habitat types differ not only in characteristic species but also in ecological dynamics such as the natural densities of large herbivores, vectors for seed dispersal, types of natural disturbance regimes, response to disturbance, and degree of resiliency. Mangroves are vastly different ecological systems from tropical dry thorn scrub. Conservation planning is improved for each by recognizing them as separate biomes. Furthermore, unless these biomes are differentiated, conducting a representation analysis is virtually impossible.

Third, we separate lowland from montane forests because of differences in assemblages of species, patterns of endemism, dynamics, and levels of threat. Montane habitats of the Indo-Pacific typically support different floras and faunas than do lowland areas. Throughout the Indo-Pacific region a large number of vertebrates show ranges restricted to a single mountaintop or small isolated mountain ranges (Corbet and Hill 1992; Heaney 1986; Phillips 1980; Schaller and Vrba

1996; Whitten and Whitten 1996; Flannery 1995). We use a digital elevation model to identify altitudinal bounds separating montane and lowland habitats.

There are examples in our approach where historical factors are given higher priority than present climatic features. As Lawrence Heaney describes (essay 3), in the Philippines ecoregions, the presence or absence of land bridges between island groups during the last Ice Age has had a more pronounced effect on the distributions of species than present rainfall gradients. For Philippine ecoregions in particular, we used bathymetry and the locations of land bridges to further refine our ecoregion delineation.

In summary, our decision rules for delineating ecoregions, presented in order, are as follows:
- Use patterns of biogeography to create the first separation.
- Use a digital elevation model to distinguish lowland from montane ecoregions where ecoregions are larger than very small islands.
- Use vegetation patterns as a proxy for climatic gradients where appropriate.

Using these rules, we identified 140 ecoregions in the Indo-Pacific region (figure 2.1a–c; details in appendix A). This analysis treats only terrestrial ecoregions. In general there is little concordance between terrestrial and freshwater ecoregions because the latter are based on river basins (Kottelat 1999). A comprehensive treatment of freshwater ecoregions for the Indo-Pacific is beyond the scope of this assessment. However, in essay 4 Maurice Kottelat provides an overview of aquatic biodiversity targets and acute threats to them and dispels the myth that wetlands conservation is equivalent to freshwater biodiversity conservation.

Representation of Ecoregions and Biomes in the Indo-Pacific Region

Of the 140 ecoregions, 79 are in continental bioregions (i.e., Indian Subcontinent and Indochina) and 61 are in archipelagic bioregions (Sunda Shelf and Philippines, Wallacea, and New Guinea and Melanesia). Five of the ecoregions in the continental bioregions are also islands (three ecoregions in Sri Lanka and the Andaman and Nicobar groups).

On average, the ecoregions in continental bioregions (average area = 79,779 km^2) are about 60 percent larger than archipelago ecoregions (average area = 50,134 km^2). The greater size of the continental ecoregions is driven by the large tropical dry forest, desert, thorn scrub, and subtropical broadleaf forest ecoregions.

The Indian subcontinent and Indochina bioregions contain vast dry forest ecoregions in the Deccan Plateau, the Irrawaddy River plains, and central Indochina, and these ecoregions are flanked by montane moist forests that lie along the north-south mountain ranges (figure 2.1a and b). Most of the archipelago ecoregions are tropical moist forest, with lowland forest ecoregions surrounding montane moist forest ecoregions (figure 2.1c).

All ten biomes of the Indo-Pacific are represented in the Indian Subcontinent bioregion, whereas the Indochina bioregion has five biomes, Sunda Shelf and Philippines has four, Wallacea two, and New Guinea and Melanesia five (figure 2.3).

> *Tropical and subtropical moist broadleaf forests.* In the Indian Subcontinent bioregion, the tropical moist forest ecoregions are distributed along the Western Ghats Mountains, on the Eastern Deccan Plateau, along the Gangetic Plains, and in southwestern Sri Lanka. These ecoregions include both rain forests (moist broadleaf forests) along the Western Ghats and southwestern Sri Lanka and moist deciduous forests (Eastern Deccan Plateau and Gangetic Plains). The subtropical broadleaf forest ecoregions are in a narrow band along the lower elevations of the Himalayan range.
>
> In the Indochina bioregion, the tropical moist forest ecoregions are distributed along mountain ranges. These include the Mizoram-

Manipur-Kachin range along the Indo-Myanmar border; Kayah-Karen and Tennasserim range along the Myanmar-Thailand border; the Annamite Mountains in Vietnam, Laos, and Cambodia; and the Cardamom Mountains in Cambodia and Thailand and the immediate foothills. The subtropical moist forest ecoregions form a broad swath along the mountains of northern Myanmar, Thailand, and Laos. In the three archipelago bioregions, the tropical moist forests account for fifty of the sixty-one ecoregions.

Tropical and subtropical dry broadleaf forests. This biome is widespread across the Indian Subcontinent. Most of the Deccan Plateau and northwestern Sri Lanka consist of dry forests. In the Indochina bioregion, the dry forests make up a large area of central Thailand, the lowlands of Laos, and large parts of Cambodia. Small patches of dry forests also exist in Myanmar. Small dry forest ecoregions are present in the Wallacea and the New Guinea and Melanesia bioregions but not in the Sunda Shelf and Philippines bioregion.

Tropical and subtropical coniferous forest. Ecoregions in this biome are found in the Indian Subcontinent and Sunda Shelf and Philippines bioregions. There are also smaller patches of tropical pine forests in the Indochina and Sunda Shelf and Philippines bioregions that were not delineated as ecoregions because of their smaller extents. Instead, these patches were subsumed into the larger broadleaf forest ecoregions within which they are located.

Temperate broadleaf and mixed forests. This biome is represented by ecoregions across the Himalayas and in northern Myanmar. It is found only in the Indian Subcontinent and Indochina bioregions.

Temperate coniferous forests. The three ecoregions in this biome are in the Indian Subcontinent bioregion, above the temperate broadleaf forest ecoregions of the Himalayan range.

Tropical and subtropical grasslands, savannas, and shrublands. This biome is represented by just two ecoregions: the Terai-Duar Savanna and Grasslands [35] in the Indian Subcontinent bioregion and the Trans Fly Savanna and Grasslands [130] in the New Guinea and Melanesia bioregion.

Flooded grasslands and savannas. This biome is represented by one ecoregion—Rann of Kutch Seasonal Salt Marsh [36]—in the Indus River delta of the Indian Subcontinent bioregion.

Montane grasslands and shrublands. This biome is most extensive in the Indian subcontinent bioregion, where it is represented by six ecoregions across the northern Himalayan range, and also in the high mountains of the Sunda Shelf and Philippines bioregion and the New Guinea and Melanesia bioregion.

Deserts and xeric shrublands. This biome is found only in the Indian Subcontinent bioregion. The thorn scrub ecoregions are thought to represent secondary habitat following large-scale disturbances of the dry forests (Puri et al. 1989).

Mangroves. This biome is found in four of the five bioregions. The largest extent of mangroves is represented by the Sundarbans [33].

Ecoregions are delineated on the basis of the regional distribution of biomes. Thus, many ecoregions include smaller, sub-regional patches of habitats from other biomes; for example, the Southern Annamites Montane Rain Forests [57], classified within the tropical moist broadleaf forest biome, contains small patches of tropical conifer forests of the Da Lat plateau. In other instances, ecoregions may also contain different habitat types that belong to the same biome; for example, forests on limestone substrates typically have unique biodiversity features but can be included within tropical moist forest ecoregions. Finer-scale analyses for sub-regional conservation planning must disaggregate these forest and vegetation types to ensure representation of their unique biodiversity and to address the specific threats they face. Tony Whitten provides an example of the threats facing the unique biodiversity of limestone habitats (essay 5) and the need to consider habitat types within ecoregions for conservation management at sub-ecoregional scales.

ESSAY 3

Island Life along Wallace's Line: Biogeography and Patterns of Endemism in the Philippines and Indonesia

Lawrence R. Heaney

When Alfred Russell Wallace published his description of the biogeographic line that now bears his name, he not only described a pattern of diversity and endemism but also made a prediction that deserves mention as one of the first testable biogeographic hypotheses. Wallace knew that the Malay Peninsula, Sumatra, Java, Bali, and Borneo share faunas that are generally similar and that Lombok, Sulawesi, and islands to the east have very different faunas. For example, marsupials live only to the east of the line, tree shrews and apes only to the west. From recent mapping carried out by the British Admiralty, Wallace also knew that the islands to the west (on what we now call the Sunda Shelf) were separated by shallow waters (not more than about 80 meters) but that those to the east were surrounded by water usually more than 200 meters deep. Wallace hypothesized that at some time in the past the land was higher or the water lower by 100 meters or more. This exposed the islands of the shallow Sunda Shelf as a continuous part of the Asian continent. However, the islands to the east remained isolated, with perhaps occasional connection to New Guinea, which lies still further east (Wallace 1860, 1880).

Wallace was startlingly correct, and his thoughtful observations are essential to anyone striving to understand the complex patterns of biodiversity in the Indo-Pacific. We now know that there were twenty-one or more cycles of continental glacial development during the Pleistocene, the period of more than 2 million years popularly known as the Ice Ages. During this period, the transfer of liquid water from the oceans to solid ice on land resulted in sea level declines that often exceeded 100 meters. The most recent glaciation, which peaked about 18,000 years ago, dropped sea level to 120 meters below the present level, resulting in exactly what Wallace envisioned: a massive peninsula extending south from Indochina to Borneo and Java, surrounded by deep water (figure 2.4; Heaney 1991a). To the east lay a smaller number of larger islands than we see today, but these islands remained isolated from one another by deep straits (Heaney 1985, 1986, 1991b). The repeated pattern of periodic connection between some—but not all—of the region's islands has shaped patterns of distribution and endemism among plants and animals. On Borneo, one of the largest islands in the world (743,000 km^2), the native, nonflying mammals include about 30 species that are endemic out of 130 (ca. 23 percent); Sumatra has fewer endemic species (about 10 out of 112; 9 percent) and Java fewer still (ca. 8 out of 62; 13 percent; Heaney 1986). Most of the endemic species are found in montane or mossy forest above 1,500 meters (Md. Nor, in press); in contrast, the majority of lowland mammals range widely from northwest Sumatra to eastern Borneo. At 125,000 km^2, Java is the smallest island on the Sunda Shelf with any endemic mammals.

Just a short distance away across Wallace's Line, things are dramatically different: on Sulawesi, with about seventy species of native nonflying mammals, all but two species are endemic (Musser 1987), and the many small islands east of Java and Sulawesi are rich in endemic species (Flannery 1990, 1995). The pattern remains the same in the Philippines, where the island faunas are well known: on the Ice Age islands of Greater Mindanao and Greater Luzon, respectively, 79 and 71 percent of their nonflying native mammals are endemic (figure 2.5; Heaney et al. 1999; Heaney and Regalado 1998), and even the medium-sized islands such as Mindoro (9,800 km^2) have 45 to 50 percent endemics. Sibuyan, an island of 463 km^2, has four endemic nonflying mammals (plus one bat; Goodman and Ingle 1993; Heaney et al. 1999), and Camiguin, with an area of 265 km^2, has two endemic species

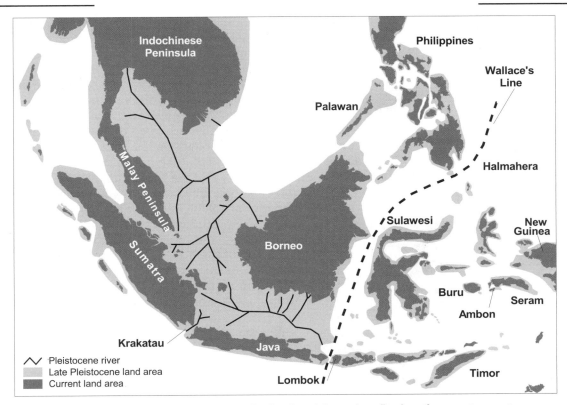

Figure 2.4. Late Pleistocene land area in the Sundaic region. During the most recent glaciation, the sea levels dropped approximately 120 meters below the present level. A massive peninsula extending from Indochina incorporated present-day Sumatra, Borneo, and Java.

(Heaney and Tabaranza 1997). We now recognize that every island that remained isolated during the Pleistocene periods of low sea is a unique center of biological diversity (Heaney and Regalado 1998). Species richness remains highly correlated with island area, so small islands have fewer species than large ones, but each Ice Age island is a distinctive center of biodiversity.

On the other hand, islands that were connected during the period of low sea level have highly similar faunas. All the islands that made up Greater Luzon Island share the same mammal fauna; there are no differences in the lowland fauna. Only the high elevations on Luzon have localized endemic species, where mossy forest habitats are isolated from one another (Rickart et al. 1998). The same is true, with only a few exceptions, throughout the rest of the Philippines and Indonesia.

These same patterns of endemism hold strongly among reptiles and amphibians in the Philippines (Alcala and Brown 1998), where endemism percentages can run even higher than among the nonflying mammals. Bats and birds show similar patterns to herpetofauna but with some interesting differences. The *places* where endemic species occur are similar, but the *percentage of endemism* is much less: usually 10–20 percent rather than 40–80 percent (Heaney 1991b; Stattersfield et al. 1998). The reasons seem clear: animals that are able to fly can maintain gene flow across permanent sea channels whereas nonflying animals cannot (Peterson and Heaney 1993). Thus, birds and bats are diverse and widespread, but greater degrees of isolation are needed to form distinct species than among the nonflying mammals, reptiles, or amphibians. Invertebrates such as butterflies in this region show patterns similar to those of birds and bats, confirming the observations based on vertebrates (Holloway 1987, 1998).

Much remains to be learned of details and exceptions, especially concerning the biogeographic processes by which the patterns are produced (Whitmore 1987; Hall and Holloway 1998). However, there is now a clear and consistent guide to the centers of endemism in the vast archipelagoes that lie south of Indochina and east to New Guinea. Islands that have been connected to one another during periods of low sea level share very similar faunas, whereas each island (or set of islands) that remained isolated is a unique center of endemism. On an island or set of islands that has never been surveyed, chances are that endemic species will be found if a survey were conducted. This was the case for Sibuyan, where mammals were almost undocumented and where field studies were predicted to turn up new endemics. Not surprisingly, recent field investigations discovered five new mammal species (Goodman and Ingle 1993; Heaney et al. 1999).

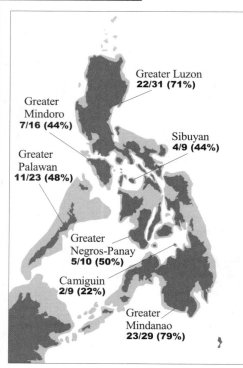

Figure 2.5. Non-flying endemic and native mammals of the Philippines. As a result of isolation, many Philippine islands have a high percentage of endemic mammal species.

There is one notable exception to this pattern, however: natural habitat must still be present for endemic species to be present. At the same time that Sibuyan and Camiguin were investigated, Siquijor, an island of similar size (235 km^2) off the southeast tip of Greater Negros-Panay, was also surveyed. No endemic mammals were found, nor was any of the old-growth rain forest that once covered the island (Lepiten 1997). If any endemic mammals once lived there—or reptiles or amphibians—loss of habitat has resulted in their extinction. Endemic mammals on other islands in the Philippines, the Lesser Sunda Islands, and the Moluccas may have suffered the same fate (Flannery 1990, 1995; Heaney and Regalado 1998; Kitchener et al. 1990). In the Philippines, old-growth forest has declined from originally covering about 97 percent of the country to 6–8 percent today, resulting in the country being listed as one of the hottest hotspots (Mittermeier et al. 1999). Whereas some islands in Indonesia (such as Sulawesi) still have fairly good montane forest cover, other islands such as in the Moluccas and the Lesser Sundas have been severely deforested.

This combination of factors makes conservation of biodiversity in the islands of Australasia especially challenging. First, a few large parks will miss much of the biodiversity because much of that biodiversity is scattered among many small centers of endemism, either in isolated mountain ranges or on scattered oceanic islands. Second, many of the potential centers of endemism have never been inventoried and can thus easily be overlooked. And third, for some of the smaller centers of endemism, it may be too late: when the last of the mature rain forest disappears, so does the endemic biota. Quick action is needed at all levels.

Essay 4

Aquatic Systems: Neglected Biodiversity

Maurice Kottelat

The diversity of inland fishes, especially in southeast Asia, is remarkable, with about 1,000 species known from western Indonesia (Kottelat et al. 1993; Kottelat and Whitten 1996) and more than 900 known from mainland southeast Asia (Kottelat 1989). However, our knowledge of fish biology over most of the region is still at an exploratory stage and is often restricted to species lists (Kottelat and Whitten 1996). Large areas are still unsurveyed, so total diversity is underestimated. For example, field exploration in March 1996 from the freshwater habitats of Laos recorded only 219 species. Subsequent surveys over the next two years boosted this number to 480 species (Kottelat 1998, 2000). The limits to our knowledge of aquatic biodiversity and the urgent threats to freshwater systems require us to pay attention to conservation needs. Recent studies suggest that rates of extinction of freshwater fauna are much higher than those of terrestrial species (Ricciardi and Rasmussen 1999). In this essay I comment on conservation considerations for aquatic habitats and identify specific problems—misconceptions and oversights—in the conservation of aquatic systems.

Aquatic Habitats of Conservation Concern

The habitat types within aquatic systems that should receive special conservation attention include the lowland main rivers and their floodplains; lakes, swamps, and marshes; middle reaches of main rivers;

rapids; small streams in lowlands and foothills; swamp forests; headwaters; caves; and hot springs. These habitats are interdependent, and impacts in one place may have influences throughout the watershed. Also, the transition between habitats is rarely abrupt in nature, and it is typical to observe that on one side of a river the habitat is rapids and on the other side it could be main river or middle reach habitat.

The distinction between middle reaches and headwaters is somewhat arbitrary, but the latter generally are overhung by forest canopy, are usually on hill slopes, and the streams are first to third order. (Strahler's stream ordering system classifies first-order streams as the uppermost channels in a drainage network (headwater channels) down to their first confluence; a second-order stream is formed below the confluence of two first-order streams, and third-order streams are created when two second-order streams form [Federal Interagency Stream Restoration Working Group 1998].) The lowland and middle reaches of main rivers are distinguished by the presence or absence of a floodplain, respectively. The lower reaches of rivers and floodplains are characterized by very low gradients, low oxygen concentrations, higher temperatures, high turbidity and nutrient loads, muddy bottoms, and cyclical floods—salient features of large river ecology. During floods, fish invade the floodplains where food is abundant for adults and fry, and adults spawn there (Taki 1978). Floods may also allow fish movements between basins; thus, assemblages do not show much interbasin variation in the lowlands.

The middle reaches of main rivers tend to have more heterogeneous habitats: deep water stretches, shallows, rapids, waterfalls, and gorges. In many rivers surveyed to date, the middle reaches have species endemic to that single basin. Thus any impact (e.g., severe pollution, the construction of a single dam, or large-scale deforestation) may lead to the immediate extinction of some species.

Swamp forests and the small streams in lowlands and foothills often are neglected as distinct habitats—especially by fishery biologists—but they usually host distinctive communities. In the lowlands and foothills, small streams are in forested areas with an overhead canopy. They are shallow in the dry season and usually run in a succession of pools and riffles on a gravel to earth substrate, with woody debris (logs, branches, leaf litter). The water usually is clear and cool. There is little light penetration and thus limited underwater plant growth, assuming that the canopy is still intact. The majority of the aquatic fauna relies on exogenous material, either vegetal debris or invertebrates, as food sources. In the wet season, the streams flood and the fish spawn in the flooded riparian forest where food is abundant for adults and fry. At this time of year, large fishes from the large rivers also enter these small streams for feeding, and these streams might also be the spawning grounds of a few migratory species.

These communities are threatened by deforestation, which increases water turbidity and temperature, suppresses food and shelter resources, and modifies the spawning grounds; by water diversion for agriculture; by pollution; and by overfishing (especially with poisons). This type of habitat is completely lost in many areas. For example, it is now completely missing in Java, which is probably one of the main reasons why half the species reported from the island until the mid–twentieth century have not been encountered in recent surveys (Kottelat 1995).

Headwaters are characterized by having high gradients, often higher altitudes, usually high dissolved oxygen and low temperature, and rocky substrates. This habitat shares many characteristics with rapids and also hosts many species adapted to swift waters. The headwaters are threatened by deforestation, which increases the temperature and sediment load; dam construction, which eradicates high gradient sectors in the reservoir and alters the flow pattern downriver; and channelization, which reduces the heterogeneity of the river bed and thus the number of available ecological niches. There is no study of the spawning season of the different species occupying these habitats, but field observations suggest that that there is no simple rule, even within a genus or a family. Data on fishes in headwaters are still very poor because large areas are accessible only by long, slow overland journeys or by air.

Rapids are a habitat type characterized by high gradient, usually high dissolved oxygen and low temperature (these two characters are seldom pronounced in lowland rapids), and rocky substrates. High friction occurs over and around the stones and boulders. There are few macrophytes. This habitat is present in the main river and middle reaches. In the headwaters, it may be present but usually is not as distinctive from the adjacent habitats, often is difficult to distinguish from large riffles, and is inhabited by about the same fauna. I discuss here large rapids or sets of rapids that are permanent physiographic features. Some smaller rapids may be perceived only seasonally: they may be completely covered at high waters, or some rocks along the shore in the dry season may become rapids during the high waters. Most observations on rapid communities have been obtained during the dry season, and we know almost nothing about these communities when the water level is high or about species' possible movements between rapids or between temporary and permanent rapids. Observations at the dry season indicate that large, complex rapids have a much more diverse and specialized fauna than small or temporary rapids and that the most specialized species are absent from isolated or temporary rapids. The complex habitat heterogeneity of rapids offers ecological niches to a variety of species, from extremely specialized rheophilic (current-loving) species found only in rapids to species from the adjacent river sector. Many fish and invertebrates living in these streams have adaptations such as suckers, flattened depressed bodies, ventral mouths, and laterally expanded fins to resist being swept away by the current. Often the water is clear and the light penetration promotes good algal growth on the rocks, an important food resource for fishes and grazing invertebrates. The latter in turn constitute food for fishes. The chaotic nature of these habitats results in a mosaic of microhabitats, and in rapids of large river basins, sampling a few hundred meters of a stream may yield more species than a comparable distance in any other habitat.

Like those in headwaters, the aquatic communities in rapids are threatened by dam construction. Rapids are also points along the river where fishing activities often concentrate because it is easier to catch migratory species as they try to find their way upstream. Areas immediately adjacent to rapids also are reported by local communities as spawning grounds of some species, migratory or not.

Lakes and swamps are bidimensional aquatic habitats, as opposed to the unidimensional or linear rivers, and they usually have extensive macrophyte growth. Marshes and swamps have high productivity because of intense light and resulting plant development. They can support diverse invertebrate communities. Most swamps are connected to the river system and in the rainy season are part of the flooded area, which is also an important spawning and nursery ground for riverine fishes. Most fish in these habitats have developed ways to breathe atmospheric air because dissolved oxygen is insufficient. These species are hardy fishes and can survive in adverse conditions; they are the last fish present (or last ones to die) in urban sewers. This is why most marshes and swamps—the classic wetlands—often are of little to no interest from the point of view of their strictly aquatic biodiversity. Noteworthy exceptions are peat swamp forests and swamp forests, which are inhabited by a large number of stenotopic species with high endemicity.

Organisms inhabiting subterranean habitats such as caves and hot springs often are overlooked in conservation plans. Few caves and aquifers in Asia have been investigated. Several caves have been found to host unique faunas, aquatic and terrestrial. Most cave aquatic organisms are restricted to a single cave or a single cave system (see essay 5 by Tony Whitten).

Most cave environments must be considered threatened. Threats include surface water pollution; surface waste disposal; water tapping for human, agricultural, or industrial use; and tourism. A reservoir was built in Funing County, Yunnan, China, and cave systems where geologists reported blind fishes are now submerged. Permanently flooding a cave suppresses the water flow in the cave, with the likely creation of an anoxic zone and silt deposition. Cave species usually are adapted for life in fast water, not standing water.

In contrast, human-made freshwater habitats are seldom interesting in a biodiversity context. They usually host a depauperate fauna with a number of introduced exotics, have a uniform underwater landscape, and are subject to negative human impacts. The fauna of most reservoirs consists mainly of stocked species, both native and exotic, but in most cases one has the feeling that no serious effort has been made to investigate the possibility of breeding and stocking with native species. However, reservoirs occasionally may benefit some of the native aquatic fauna in arid areas. Dams and effective forest conservation practice in the Chalakkudy River basin, Kerala, India, have contributed significantly to providing a continuous flow of water in the river, even during periods of drought.

Threats

The major threats affecting aquatic biodiversity—flow alteration, pollution, introductions, and increased sediment load—usually are manifested on large scales. Their characteristics and impacts are well known and have been studied in many places in the world. However, assessing threats is constrained by the lack of reliable data describing the situation before an impact (planned or accidental) has occurred. Dams are a good example. Despite the large number of dams constructed and planned in the region, the poor quality of the majority of environmental impact assessments (especially of their aquatic biodiversity sections) precludes serious monitoring of the impacts of the project. And, worse, it does not allow us to use any lessons learned for planning and managing similar projects.

Flow Alteration and Water Diversion

The negative impacts of flow regulation measures and water diversions include habitat destruction and modification, alteration of floods, creation of sterile habitat, and barriers to migration. By connecting different river basins, canals may allow the entry of exotic species and pathogens. I am not aware of a single case where flow regulation measures in themselves have been beneficial to aquatic communities (in combination with other factors, they may in some specific instances reduce the significance of these impacts). Reservoir construction may result in the destruction of rapids and associated aquatic communities in the reservoir itself as well as down river. Ironically, nature conservation areas often are established around reservoirs; they may make sense for land animals, but they are not a mitigation measure for the very part of the biodiversity that is most affected by the reservoir: the aquatic fauna.

At a much smaller scale, the development of so-called micro-hydroelectricity generators, as in hill streams of northern Laos and Vietnam, is a potentially serious problem. Major stretches of small or medium-sized streams have been transformed (at least temporarily in the dry season) into standing water where silt accumulates on the bottom, water temperature increases, and native fishes are quickly overfished (pers. obs. in Vietnam and northern Laos).

Pollution

The major source of aquatic pollution is domestic and industrial organic wastes. Paper pulp mills and food-processing plants are the main industries causing organic pollution. Algal blooms often accompany organic pollution. A particular pollutant may not be lethal to fish, but it may stress them so that they can no longer resist diseases (Menasveta 1985; Roberts 1993). Beside death, injury, and increased sensitivity to pathogens, pollution may induce intersexuality (an alteration of the sexuality in which both male and female characters are found in a single individual at some time during its life; Bortone and Davis 1994).

Increased Sediment Load

In developing countries, increased sediment load is one of the most serious threats to freshwater biodiversity. Its economic impacts have been quantified (e.g., silting-up causing increased flooding and waterlogging, navigation problems), but the impact on biodiversity and fish productivity has been largely ignored (although it is well documented in developed countries). Increased sediments negatively affect fishes by damaging their gill epithelium, causing stress, injuries, and death. When sediments redeposit, freshly laid eggs become asphyxiated. Silt deposition transforms heterogeneous stony substrate into homogeneous sandy substrate. Water turbidity reduces light penetration and plant survival and thus primary productivity. The main causes of increased sediments are deforestation, agriculture, mining, and road construction. Road construction often is ignored or overlooked, but it is a serious concern in hilly areas.

Introductions of Exotic Species

Introductions are almost always damaging for aquatic biodiversity because introduced species predate or compete with native species, disturb the ecosystem (or the existing fishery practices), and are vectors of diseases or parasites (of aquatic organisms but also humans). Poor aquaculture standards result in numerous species being accidentally released and established using the fry of the target species. The effects of these introductions on native species usually are grossly ignored or overlooked by fisheries and development agencies and their foreign sponsors and advisors; too often they perceive the importance of the fish fauna in terms of landed kilograms rather than number of species. Recommendations to introduce exotic fish species in waterbodies of protected areas still proliferate, which in a biodiversity conservation context is nonsense.

Habitat Loss

Land reclamation and wetland draining have a profound impact on aquatic biodiversity. For example, by 1994 80 percent of western Malaysian peat swamp forests had already been drained, burned, and converted into rice, oil palm, or pineapple plantations, thereby threatening many rich, diverse, and geographically restricted aquatic communities (Ng 1994). The rapid transformation and urbanization of the islands of Batam and Bintang (Indonesia) imperil the futures of many endemic aquatic species (Tan and Tan 1994; Ng and Lim 1994).

Overfishing

Mechanisms leading to overfishing, especially economic factors, are well documented (see Fairlie et al. 1995 for a recent summary). Overfishing occurs in many freshwaters, but in Asia as elsewhere, it remains uninvestigated. An analysis of overfishing must include data to identify stocks and productivity. For most countries, both are missing and are urgently needed. Overfishing often is easily identified where streams are empty. Stream morphology is normal, riparian vegetation is intact, the water is clear and pollution-free, but fish, tadpoles, and almost all insects are absent. Typically, such streams are poisoned regularly, sometimes weekly. Under these conditions, life quickly disappears. In many Asian countries, the recent availability of electric fishing has wiped out entire aquatic communities in small-order streams (e.g., Vietnam, south China).

Specific Problems of Aquatic Biodiversity Conservation

On the Distinction between Wetland Biodiversity and Freshwater Biodiversity

Many assume that wetland conservation projects encompass freshwater biodiversity conservation. (The term wetland here refers to land that is periodically submerged or whose soil contains a great deal of moisture.) This is seldom the case. Wetland conservation projects are inadequate surrogates for specific freshwater biodiversity conservation because until recently wetland concerns have been focused primarily on birds. Wetlands of ornithological interest usually are swamps and marshes; these habitats usually have low oxygen concentrations and host hardy fishes and other aquatic organisms with wide distribution ranges. Most of these organisms are resistant to extirpation, and few are of great conservation value. Most can survive in other habitats, including swampy river shores, irrigation canals feeding rice fields, and reservoirs.

I shall briefly comment on a few examples where wetland biodiversity interests are either congruent or incongruent with freshwater biodiversity. The main lesson from this overview is that when prioritizing wetlands (or whatever they are called), one should investigate freshwater ecology comprehensively rather than focusing on birds, scenic attractions, or other such aspects that may be important but are poor indicators of aquatic biodiversity.

Examples of Congruence

Specialized aquatic habitats—being of limited ornithological value by themselves—seldom attract attention in wetland conservation planning. For example, few rapids or foothill streams are protected for their intrinsic biodiversity value. However, they may be included in other conservation projects (usually not wetland-related) involving large geographic areas (usually focusing on large game). Even in the few cases where wetlands have a real conservation value for freshwater biodiversity, fish and other aquatic organisms are overlooked. When they are mentioned, they often are flatly listed as "fish." When precise species are reported, the list is then restricted to a few larger species (usually widely distributed), and the smaller and endemic ones are ignored. The *Directory of Asian Wetlands* (Scott 1989) provides a broad range of such examples (e.g., peat swamps of Malaysia, pp. 825–826, 844–845, 871–880).

Peat swamps probably are the wetland type where congruence is highest. They host a variety of aquatic species with specialized ecological needs, including the need for acidic waters. Because these habitats tend to be fragmented naturally, ancestral populations have been isolated, giving rise to many endemic, highly restricted species. The following peat swamps in Malaysia each have a set of endemic fish, crab, or shrimp species: North Selangor swamp forest (wetland #9 in Directory of Asian Wetlands), Southeast Pahang swamp forest (#1), Sedili Kecil swamp forest (#5), and Third Division and Sibu swamp forest (#30 and #31). In Thailand, this applies to Pa Phru (#35), in Indonesia to Berbak Reserve (#8) and probably to Tanjung Puting National Park (#74), and other yet unsurveyed areas of Borneo. Lake Inle, Myanmar (#5), with its nine endemic fish species and three endemic genera (Annandale 1918, and pers. obs.) is also an example of congruence. Scott (1989: 644) records only the carp *Cyprinus intha*, the largest but probably least interesting endemic.

Examples of Noncongruence and Consequences

Many wetlands are considered of low conservation value but are actually important in terms of freshwater biodiversity. The most striking example probably is the Malili lakes of Sulawesi. More than half of the sixty Sulawesi freshwater fishes known to date (including fifteen of the seventeen known species of Telmatherinidae) are endemic to this group of five lakes (Kottelat 1990a, 1990b, 1991, unpubl.). These species constitute one of the few known cases of parallel species flocks (a species flock is a group of closely related species, occupying a sharply delimited geographic area). Despite this uniqueness the lakes have been gazetted as a Tourist Park (Lake Towuti) and as a proposed Tourist Park or Nature Reserve (Lake Matano), both with low conservation value (data from Whitten, Mustafa, and Henderson 1987: 100). The most serious threat to aquatic biodiversity in this situation probably is the introduction of exotics, permitted in Indonesian tourist parks. Species richness was taken into account for computing conservation values, but it counted only birds and tree species. When most of the park area is covered by a lake, paying attention to aquatic organisms seems more appropriate. They are more likely to be unique to a lake than is the avifauna.

Second, wetlands of little importance for freshwater biodiversity often are considered a conservation priority. Among wetlands listed in the *Directory of Asian Wetlands* (Scott 1989) good examples could be Lake Toba (Sumatra), Nam Ngum reservoir (Laos), Tasek Cini (Malaysia), or the many reservoirs of Sri Lanka. Lake Toba has a great scenic value but lacks any special aquatic fauna. The human-made habitats of Nam Ngum and the reservoirs of Sri Lanka merit little attention for aquatic biodiversity because they host only a depauperate version of the original communities and are foci for the introduction of exotics. Many tropical reservoirs (including Nam Ngum) first went through a several-year phase of anoxic water (because of decomposition of flooded vegetation) so that the original aquatic life has been nearly eliminated. Admittedly, some old reservoirs such as those of Sri Lanka may constitute alternative habitats for species that were originally abundant in swamps, which are now turned into cultivated land. But the fish fauna in Nam Ngum reservoir is an impoverished version of the fauna of the Mekong floodplain, lacking any species of conservation interest.

Spatial Scale for Conservation

Conservation planners must recognize that freshwater conservation must occur over larger spatial scales than terrestrial conservation to be effective. The natural unit to manage aquatic biodiversity is the basin. For example, an impact on one branch of a river system may have major effects on other branches of this system: a factor that would affect the fish migrations in the Tonle Sap in Cambodia would have significant impacts on the fishes of the lower and middle Mekong basin. Therefore, as a general rule, habitats cannot be taken as the unit of conservation for riverine aquatic biodiversity. Conservation of a particular rapid community entails managing the basin upriver of the rapids correctly and, case by case, making the spawning grounds downstream accessible to local migratory species. A complicating factor in managing fish populations is that not all fish migrate or move in the same direction at the same time or season or for the same reason (mainly spawning or feeding). Aquatic habitats are not isolated boxes inhabited by different communities. The transition between communities usually is gradual, and there are significant movements between habitats: random movements, daily feeding movements, or real migrations.

Designation of River Basins to Remain Free of Harmful Development Projects

National and local authorities must designate river basins where projects that affect basin topography, hydrology, and aquatic biodiversity are excluded. If some basins are sacrificed for development, others should be strictly protected, with long-term commitments to enforce this strict protection. It is much less damaging to have five hydropower projects on the same river and no development on four others than to locate one project on each of five rivers. To be meaningful, the rivers designated for conservation should be those of greatest biodiversity interest, and determining which basins are of greatest biodiversity value necessitates for some sort of survey of comparable quality.

ESSAY 5

Limestone Biodiversity: Treasure Houses of Rare Species

Tony Whitten

Limestone habitats are famous for supporting rare plants and animals. They pose particular problems for regional biodiversity assessments because large parts are poorly mapped. Typically only the outcropping areas are known and mapped accurately, and these are often small (figure 2.6). Thus, in large-scale assessments such as this, they are encompassed within larger, more obvious ecoregions. However, limestone deserves special attention from conservationists because of the extreme biological specializations of species found there, high levels of endemism, and the high levels of alpha and betadiversity of limestone-associated flora and fauna.

A few examples from the region highlight the prominence of limestone habitats.

- Gunung Subis is a fossil coral reef in the vast lowlands of Sarawak, Borneo. It now stands as an impressive forested limestone outcrop, isolated in a matrix of sandstone and shale bedrock. The closest limestone outcrops are about 35 km away. But Gunung Subis—a rugged and inaccessible plateau of just 12 km²—hosts a remarkably large fauna of land snails, comprising far more than 100 species with dozens of these and other animal and plant species found nowhere else.
- Gua Salukkan Kallang–Gua Tanette is an extensive cave system in the Maros limestones in southern Sulawesi, Indonesia. There, a surface river dips underground for miles. In its most remote parts and in the ponds and puddles isolated from the main stream by boulders, a fragile fauna persists: minute blind crustaceans, delicate pink fishes with minute eyes, and pygmy crabs with long appendages. These animals are totally dependent on their cave environment, unable to survive in the outside world, and occur nowhere on Earth except in this cave system. Even here they are rare, with small, discrete populations. In Thailand, there is a cave system with four small ponds containing six aquatic arthropod species found nowhere else.
- The slipper orchid *Paphiopedilum sanderianum* is spectacularly beautiful, with corkscrew petals up to 1 m long. It is known only from two small populations in the limestone hills of Mulu National Park, Sarawak. The dramatic slipper orchid is one of many equally attractive plants, such as gesneriads and balsams, that grow on limestone in often restricted ranges.

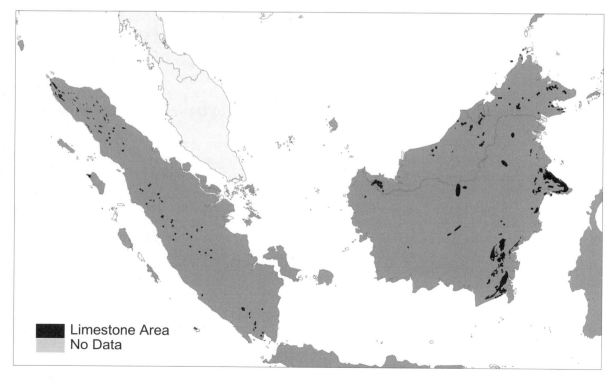

Figure 2.6. Distribution of major limestone areas in Sumatra and Borneo.

These examples illustrate some of the biological characteristics of limestone areas: high species richness but restricted distributions, often limited to a single mountain chain, a single limestone hill, or a single cave system.

The biodiversity of limestone has important direct and indirect economic benefits as well as cultural and aesthetic values. For example, the swiftlets and bats that dwell in the limestone caves contribute to the economic well-being of the surrounding farmlands. They consume thousands of kilograms of insects daily, thus assisting in pest control near agricultural land. Limestone reservoirs provide hundreds of millions of people with clean drinking water. Limestone landscapes and caves are spectacular, and many have become important tourist attractions. Such areas epitomize wild, untamed nature in Chinese art have served as the background for a variety of famous movies. Limestone areas also are valuable in other ways. Caves often are the sole source of archaeological information in the humid tropics because only in caves is the climate suitable for preserving debris from ancient households. Cave deposits are a treasure trove for geologists and paleontologists because they contain information about past climates as well as fossils.

Unfortunately, the taxonomic inventory of limestone areas in southeast Asia is patchy at best; hundreds and probably thousands of species await discovery and description. For example, during a recent survey in northern Vietnam, 270 snail species were collected. Eighty of these were new to science. Entomologists working in limestone areas in the region often find that about 90 percent of the species collected after a day's work are new. There are such pronounced differences between hills that thorough collecting is necessary. For example, in peninsular Malaysia no single hill supports more than 20 percent of the limestone flora. Also, the best-studied sites for snails on limestone in northern Vietnam are only 150 km apart and have a total species list of 270, with an overlap between the site lists of only 35 percent.

Throughout southeast Asia, limestone environments and limestone biodiversity are being irreversibly destroyed at an unprecedented rate. The main threats are limestone quarrying and destruction of the vegetative cover by fire. Quarrying may be a local activity that consumes entire limestone hills with the total obliteration of all plant and animal life within a few years, but small quarries can also be scattered over a large area, spreading the destruction. Limestone dust from the quarrying can spread beyond the boundaries of the quarries, choking vegetation. The blasting associated with quarrying destroys bat and swiftlet colonies. The cascading effects of exploitation and human colonization cause the collapse of sensitive cave systems from changes in the microclimate and underground hydrology.

The best-quality swiftlet nests—the major ingredient of bird's nest soup—sell at US$3,400/kg. At this price, greed promotes unsustainable harvest. Guano, or the dung deposits of cave bats and swiftlets, is so valuable that local governments license its exploitation, and guards and barbed wire fences are used to prevent theft. Unmanaged tourism, unsustainable hunting, and in some cases excessive collecting for scientific purposes also pose serious threats to limestone ecosystems.

Removal of the surface vegetation allows rainwater to disappear quickly through numerous crevices typical of limestone, leaving the surface dry even after heavy rain. This makes limestone vegetation in the wet tropics vulnerable to fire. Any fire started by farmers close to the perimeter of a limestone hill will sweep upward along its sides. Because the fire may burn both the vegetation and the organic fraction of the soil, subsequent rain washes away the soil, making recovery of the original vegetation difficult. The mixture of clay, ashes, and burned organic matter may also clog crevices in the rock, inhibiting the passage of water into cave systems underneath, thus affecting cave conditions and fauna. Whereas in dry tropical climates the vegetation cover adapted to regular fires may recover after such an event, in wet tropical climates the limestone forests and the biodiversity they harbor may be destroyed forever by a single fire.

Biodiversity loss in limestone systems from such disturbances is indisputable. For instance, 231 butterfly species were recorded along a transect through a primary forest on limestone in northern Vietnam. A similar transect through recently disturbed vegetation yielded only 65 species. Twenty-nine snail species were recorded from undisturbed limestone habitat in Kalimantan, Indonesia, against only eight in disturbed habitats.

The paucity of surveys and our inadequate taxonomic knowledge of limestone biodiversity inhibits assessment of the degree of species loss. However, extinctions have occurred. Snails and plants provide at least some estimate: seven limestone-restricted snail species in Sabah and Sarawak—all of a single family recently reviewed—are presumed extinct on the sites where they were collected for the first and only time a century ago. In peninsular Malaysia, seventeen species of limestone-restricted plants are regarded as extinct, and altogether 12 percent of its flora is endangered because limestone areas are not adequately protected. Nobody knows how many plant and animal species have already gone extinct unnoticed.

When devising measures to protect limestone biodiversity, it is important to consider that limestone environments are highly interactive ecological systems of water, air, soil, rock, life, and energy. The integrity of limestone environments depends on these interactions. No part of the environment can be preserved without the others; for instance, one cannot expect to preserve the health of a cave system, with all its fauna, without preserving the vegetation cover on the cave roof. Without vegetation the underground environment will change, resulting in loss of cave biodiversity. Any conservation action should recognize these interactions.

Regeneration of disturbed environments often is an extremely slow process. Thus, the best chances of *preserving* a representative part of limestone biodiversity lie in a careful selection of areas to be exploited. Biodiversity assessments—based on carefully compiled species lists of a range of taxa living on the site under investigation—help us attach a relative biodiversity value to sites considered for exploitation. The location and protection of areas with high biodiversity value should be part of any project that includes the exploitation of limestone resources. Next to this, other values—geological, geomorphological, archaeological, and economic—should be taken into consideration. Valuation should be done from the local and larger-scale perspectives. For instance, a cave may have cultural values to the local people but may also provide clean water to a larger population, and it may have aesthetic value to a still larger population.

The choice of limestone areas for exploitation should also be governed by a set of rules and through consultation with specialists. Two obvious rules: limestone deposits already affected by disturbance should be selected over undisturbed areas, and limestone hills with underground watercourses should be avoided. Other, less obvious rules should influence the nature conservation policies of various countries. An example is the rule that selection of isolated limestone hills (remote from other hills) for exploitation should be avoided because the most isolated hills tend to host the largest number of species endemic to one particular hill. Instead, an exploitation site should be located in the largest limestone areas and should never extend over the entire outcrop but always leave a large part of it untouched. In large limestone areas, species endemic to the area as a whole may well occur; however, species endemic to only a small portion of the area are less likely to occur. A well-contained and well-managed quarry in a part of a large limestone area is least likely to lead to the extinction of species.

A general awareness is growing that conservation of limestone areas in the tropics is much more important than has previously been thought. Although much is lost already, a large surface of undisturbed limestone environment remains. However, the destruction continues steadily. If measures to stem this tide are not implemented soon, a significant part of Earth's biodiversity will be destroyed.

CHAPTER 3

Representation or Triage? Approaches to Setting Conservation Priorities in the Indo-Pacific Region

There are a variety of approaches for setting conservation priorities at global and regional scales (Dinerstein and Wikramanayake 1993; Olson and Dinerstein 1998; Margules and Pressey 2000; Mittermeier and Mittermeier 1997; Myers et al. 2000). These efforts are guided by the same four principles:
- Biodiversity is unequally distributed around the globe.
- Some important areas for biodiversity are under more immediate threats than others.
- Financial and technical resources are currently too limited to save everything, so areas must be prioritized in a logical manner.
- Biodiversity and habitat integrity are becoming increasingly mappable.

A comprehensive database of species distributions from which to derive important areas for conservation of biodiversity would be invaluable. Unfortunately, we lack good information on the distribution and status for several of the larger vertebrates, let alone the smaller, cryptic groups such as amphibians, reptiles, and the invertebrates. We are thus forced to rely on proxies to represent overall biodiversity and set conservation priorities.

To address the first and second principles, we created two indices (box 3.1) using several familiar criteria (detailed descriptions are provided in appendices C and D).

Biological Distinctiveness Index (BDI)

Many biodiversity assessments typically focus on species richness. Our use of the term *biological distinctiveness* invokes a broader definition of biodiversity to include ecosystem diversity and ecological processes that sustain biodiversity. Specifically, biological distinctiveness is based on measures of species richness, endemism, higher taxonomic uniqueness, presence and maintenance of significant ecological and evolutionary processes, and the global rarity of habitat types.

The endemism and species richness parameters of the BDI are applied independently to each ecoregion (appendix C). Island communities tend to be less rich in species than mainland or continental communities, but because of their isolation, they tend to be higher in endemism (Brown and Gibson 1983; Cronk 1997; Lawlor 1986). Therefore, we consider it more appropriate to apply the species richness and endemism indices independently so that an ecoregion may score high on the BDI because of either high species richness or levels of endemism.

We used three richness and endemism parameters to create the BDI: plant species richness and endemism, mammal species richness and endemism, and bird

---

BOX 3.1. Criteria for the Biological Distinctiveness and Conservation Status Indices

Biological Distinctiveness Index	*Conservation Status Index*
1. Species richness	1. Habitat loss
2. Species endemism	2. Large habitat areas
3. Important ecological or evolutionary processes	3. Degree of fragmentation
	4. Degree of protection
4. Rare habitat type	5. Future threat

species richness and endemism. These taxonomic groups were chosen because of data availability for most of the ecoregions across the five bioregions and also because they represent a subset of the biota with diverse dispersal capabilities, a significant factor in distribution patterns of biodiversity (Lawlor 1986; Lomolino 1986). We constructed databases based on ecoregions for the three taxonomic groups using available literature, field guides, and expert assessments (appendix C). We also constructed databases of reptile and amphibian richness and endemism for all ecoregions but omitted them from the indices because survey efforts for many areas are inadequate or taxonomy is poorly resolved. However, we include detailed essays on the status of knowledge of the distributions and taxonomy of reptiles and amphibians in the Indo-Pacific (essays 6 and 7).

We believe that, in aggregate, the plant, mammal, and bird data can be used cautiously as an effective proxy for other taxa, including the more numerous and less well-known groups (e.g., arthropods, mollusks, annelids, bacteria). Consequently, these data can be taken as indicators of overall biodiversity patterns (Margules and Pressey 2000; Panzer and Schwartz 1998). We also emphasize that these databases are derived from information available from regional or country-scale maps that translate only crudely into habitat-based ecoregions. Furthermore, the experts' assessments for plants are only estimates because of the lack of any complete flora that can be easily translated into ecoregions. Therefore, in deriving the index for plants, we use broad categories that account for much of the uncertainty of the numbers. However, a more detailed enumeration of species in selected plant families by ecoregion is under way (see essay 11 by Krupnick and Rubis in chapter 4 for preliminary results).

We calculated mammal and bird richness for each ecoregion by tallying the number of species listed for the respective ecoregions. For endemism, we used a modified version of BirdLife International's rules for restricted range species (Stattersfield et al. 1998) as follows:

> If the total species range <50,000 km^2 and a species occurs in five or fewer ecoregions, the species is recorded as endemic to all ecoregions containing it; if the species occurs in more than five ecoregions (as many small, disjunct populations), the species is recorded as only present in all ecoregions containing it.

> If the total species range is >50,000 km^2 and a single ecoregion contains 75–100 percent of the species' range, the species is recorded as endemic to that ecoregion and recorded as present in all others containing it; if no single ecoregion contains more than 75 percent of the species' total range, the species is recorded as only present in all ecoregions containing it.

A mammal or bird restricted to a single ecoregion was considered a strict endemic, whereas species endemic to more than one ecoregion were considered near endemic (appendix C, appendix J).

Depending on the endemism or species richness scores, ecoregions were assigned to one of four categories of biological distinctiveness: globally outstanding, regionally outstanding, bioregionally outstanding, and locally important. We took the higher of the two categories—richness or endemism—as the biological distinctiveness value of the ecoregion.

We followed this method for ecoregions in all biomes except for mangroves. It proved difficult to assign species to mangroves based on field guides and other resources used to construct the databases. Moreover, the biological importance of mangroves is captured more in their ecological processes than their species diversity, as elaborated by David Olson (essay 8). For this reason we categorized all mangrove ecoregions as either globally or regionally outstanding.

Globally outstanding ecoregions are those that rank among the most biologically distinctive on Earth. Regionally outstanding ecoregions are distinctive at the level of biogeographic provinces (e.g., the Palearctic, Oriental, or Australasian), and bioregionally outstanding ecoregions are those that are distinctive at the level of bioregions (e.g., Sunda Shelf and the Philippines). The biodiversity of every ecoregion is at least locally important because every ecoregion not only harbors biodiversity but also provides essential ecosystem services, natural resources, and natural recreation opportunities for local human communities.

In a recent critique of global priority-setting efforts, Mace et al. (2000: 393) observed that "conservation practitioners and academic researchers alike are concerned that systematic prioritization focuses on patterns of species and community distribution, yet largely fails to address the conservation of key ecological and evolutionary processes which maintain those patterns. There is growing evidence that conserving the pattern will not by itself guarantee the conservation of these processes. Yet we currently lack robust measures for quantifying the extent to which different areas contribute to core processes, or for evaluating the overall performance of priority sets in terms of process maintenance. In consequence, although it is claimed that the hotspots approach is cost-effective, problems concerning selection of areas, scale of analysis, socioeconomic concerns and maintenance of key processes mean that areas of high conservation value may have been missed."

We agree with Mace et al. (2000). Thus, our definition of biodiversity includes both species and the supporting ecological and evolutionary processes that cannot be measured purely in terms of species richness and endemism (Smith et al. 1993). These include the presence of unique higher taxonomic groups, such as endemic genera and families; extraordinary adaptive species radiations; assemblages of intact vertebrate populations that fluctuate within natural ranges; large vertebrate migrations; globally rare habitat types; and examples of large, intact ecosystems (wilderness areas). If an ecoregion possesses any of these characteristics, it is elevated to globally outstanding status. We also address some aspects of the issue of process maintenance by incorporating landscape-scale features such as habitat block size and connectivity in our Conservation Status Index (discussed later in this chapter).

We pay particular attention to ecoregions that still maintain intact or near-intact large vertebrate assemblages, including the top predators (tigers and leopards). High diversity of large mammals is or was a conspicuous feature of the continental component of the Indo-Pacific and, as John Seidensticker discusses in essay 9, their role in maintaining biodiversity is just beginning to be appreciated.

Conservation Status Index (CSI)

The CSI is designed to estimate the present and future capability of an ecoregion to meet three fundamental goals of biodiversity conservation: to maintain viable species populations and communities, sustain ecological processes, and respond effectively to short- and long-term environmental change. Comprehensive ecological and biological information for all areas across the Indo-Pacific is largely unavailable, preventing a consistent and rigorous analysis. Therefore, we must rely on landscape-level features and assume that they are sound proxies for predicting the ability of an ecoregion to meet these fundamental conservation goals (see Noss 1992; Primack 1993; Meffe and Carroll 1994; Noss and Cooperrider 1994; Kareiva and Wennergren 1995; Soulé and Terborgh 1999).

Habitat integrity is becoming increasingly mappable, as indicated by the work of Derek Holmes and his colleagues (essay 1). We used maps and data of current land cover and land use to assess the following landscape-level criteria:
- amount of habitat lost and degraded
- the presence and number of large blocks of remaining habitat
- the degree of habitat fragmentation
- percentage of the ecoregion's intact habitat within protected areas

Using a combination of these features, we created an index to conduct an assessment of the conservation status at the time of this analysis, or a snapshot assessment. Details of the index and assessment are provided in appendix D.

The rationale for the choice of criteria is that habitat loss, fragmentation, and degradation are three important factors that contribute to the reduction and extinction of species populations and the disruption of ecosystem function (Groombridge 1992; Laurance 2000; Lynam 1997; Saunders et al. 1991; Sih et al. 2000; Turner 1996; Brooks et al. 1997, 1999). Habitat loss can result in elimination of geographically restricted species and area-sensitive species. Loss of habitat integrity results in elimination of habitat specialists.

The spatial pattern of habitat distribution and the size of habitat blocks are also important for maintaining native species, communities, and ecological processes across large landscapes. Large blocks of habitat generally contain larger and more stable species populations and can support viable populations of species that occur at naturally low population densities or have large home ranges (Noss and Cooperrider 1994; Soulé and Terborgh 1999). Larger habitat blocks are also more resilient to disturbances—both natural and anthropogenic—than are smaller habitat blocks, and they are less affected by edge effects (Soulé and Terborgh 1999; Laurance 2000; Woodroffe and Ginsberg 1998; Yahner 1988). For instance, large habitat blocks provide better core refuges from hunting than do smaller fragments (Peres and Terborgh 1995). Habitat blocks that are in close proximity to each other or are linked by corridors of natural habitat also enhance ecological processes by allowing species movements between blocks. The strategic location of intact habitats can provide critical watershed protection, stabilize upland areas, and increase the overall effectiveness of protected areas systems.

Fragmentation often results in many small blocks of habitat that are unable to maintain critical ecological functions or lack them altogether (Laurance and Bierregaard 1997). As populations become isolated, the ability of animals to forage over large areas for scattered resources and disperse from natal areas is constrained (Berger 1990; Wilcove et al. 1986). Fragmentation forces species that are intolerant of external disturbances into shrinking core areas that cannot support them. However, we should also recognize that small habitat blocks can be important for supporting source pools and providing refuges for geographically restricted species (e.g., rare plants and endemic species) and for maintaining metapopulations of widely distributed species (see essay 3, by Lawrence Heaney).

The degree of protection criterion evaluates whether the existing network of protected areas within the ecoregion includes an adequate number of sufficiently large blocks of habitat. Judging the quality of existing management within protected areas is beyond the scope of this assessment, although we recognize its importance. We strongly recommend that assessments at a finer scale (e.g., at ecoregion and subecoregion levels) include the quality of protected area management in analyses of conservation priorities.

We overlaid a digital land-cover map and accompanying database provided by MacKinnon (see MacKinnon 1997 for data sources) with ecoregion boundaries and calculated habitat loss, the number of remaining large habitat blocks for two size classes of ecoregions (using an arbitrary area of 20,000 km^2 as the threshold), and the degree of fragmentation. (Continental ecoregions tend to be much larger than island ecoregions. Island and montane ecoregions also tend to be higher in endemic species that have smaller range distributions. Habitat needs can be smaller for these range-restricted species than for the wide-ranging species in the larger ecoregions. Therefore, we used two size thresholds to reflect the spatial needs and distributions of species in continental and island ecoregions.)

We used an ArcInfo geographic information system (GIS) to conduct the intersection and area calculations. Even while this book was in preparation, habitat continued to be lost and fragmented, at rapid pace in places such as Indonesia. It is impossible to keep pace with these rates of habitat loss for the entire region, although we strove to update our information for Indochina, Sumatra, and Kalimantan. However, for the region, we feel that our analyses are sufficiently robust to show region-wide patterns. (Remaining habitat data for Sumatra and Kalimantan come from the World Bank, and remaining habitat data for Indochina were derived from gathered satellite imagery and expert opinion during a workshop on ecoregion-based conservation in Phnom Penh, Cambodia, March 21–24, 2000.)

We calculated the extent of area protected in each ecoregion using an updated database provided by the WCMC (MacKinnon 1997). Again, this database may not contain protected areas gazetted within the past five years, but it is the only region-wide protected area database (appendix D). The protected areas were overlaid on ecoregions in a GIS to determine the number, names, and extent of protected areas in each ecoregion. (In appendix G we provide information on the type and amount of remaining habitat in each ecoregion. In appendix J we list the protected areas we used in this analysis for each ecoregion. For both appendixes we used the WCMC 1997 database.)

We used the conservation status index to rank ecoregions according to five categories that reflect the current, or snapshot, conservation situation: critical, endangered, vulnerable, relatively stable, and relatively intact. An additional category—extinct—is used to classify ecoregions where no natural communities remain.

These categories follow Dinerstein et al. (1995) and were developed to reflect the status of species in the IUCN *Red Data Book* series (IUCN 1994; Mace et al. 1992). A qualitative description of each category in terms of landscape integrity and predicted ecological impacts within an ecoregion follows.

Extinct. No natural communities resembling the original ecosystems remain. Some of the original biota are still present but persist only in highly modified communities and landscapes. No opportunities for restoration of the original natural communities exist because of permanent alteration of physical conditions, loss of source pools of native species, substantive alteration of natural ecological processes, or an inability to eradicate or control aggressive alien species.

Critical. The remaining intact habitat is limited to isolated small fragments with low probabilities of persistence over the next five to ten years without immediate and intensive protection and restoration. Many species are already extirpated or extinct because of the loss of viable habitat. Remaining habitat fragments fail to meet the minimum area requirements for maintaining viable populations of many species and ecological processes. Land use in areas between remaining fragments often is incompatible with maintaining most native species and communities. Spread of alien species may be a serious ecological problem, particularly on islands. Top predators are or have almost been exterminated.

Endangered. The remaining intact habitat is limited to isolated fragments of varying size (a few larger blocks may be present) with medium to low probabilities of persistence over the next ten to fifteen years without immediate or continuing protection or restoration. Some species already are extirpated because of loss of viable habitat. Remaining habitat fragments are not expected to meet minimum area requirements for most species populations and large-scale ecological processes. Land use in areas between remaining fragments is largely incompatible with maintaining most native species and communities. Top predators are almost exterminated.

Vulnerable. The remaining intact habitat occurs in habitat blocks ranging from large to small; many intact clusters probably will persist over the next fifteen to twenty years, especially if given adequate protection and moderate restoration. In many areas, some sensitive or exploited species, particularly top predators, larger primates, and game species have been extirpated or are declining. Land use in areas between remaining fragments sometimes is compatible with maintaining most native species and communities.

Relatively stable. Natural communities have been altered in certain areas, causing local declines in exploited populations and disruption of ecosystem processes. These disrupted areas can be extensive but are still patchily distributed relative to the area of intact habitats. Ecological links between intact habitat blocks are largely functional. Guilds of species that are sensitive to human activities, such as top predators, the larger primates, and ground-dwelling birds, are present but at densities below the natural range of variation.

Relatively intact. Natural communities within an ecoregion are largely intact with species, populations, and ecosystem processes occurring within their natural ranges of variation. Guilds of species that are sensitive to human activities, such as top predators, the larger primates, and ground-dwelling birds, occur at densities within the natural range of variation. Species move and disperse naturally within the ecoregion. Ecological processes fluctuate naturally throughout largely contiguous natural habitats.

Threat Analysis

The pressures from the rapid pace of economic development in the region are causing the conservation status of some ecoregions to change rapidly. Even by the time this book reaches publication, newly deforested areas probably will be large enough to be detected by satellite imagery. Thus, besides understanding current threats we must project where future threats will be most intense.

After evaluating each ecoregion's snapshot conservation status, we modified that classification according to the degree of threat projected into the future, to produce the Final (Threat-Modified) Conservation Status. We identified two principal types of threat to remaining blocks of natural habitat. The first comprises chronic, low-intensity impacts from subsistence-level resource extraction and small-scale clearing for agriculture, plantations, and settlements. We used human population density as a proxy to assess these threats from chronic, low-intensity exploitation of forested areas. We assumed that areas with a high human population density would have a higher impact on the remaining forest fragments.

The second type is characterized by intense, catastrophic threats from large-scale logging concessions, large-scale plantations, hydropower schemes, and large developments. Where available, we used ongoing and planned logging concession maps to determine high-impact threats to large habitat blocks. We view large-scale logging to be most important in areas where large forest blocks make such operations economically viable and attractive. These trajectories are based on current threats and are projected over the next two to five decades. We acknowledge that these threats and their trajectories may change during this time, but there is much inertia associated with them. The analysis indicates where such intense threats will be cause for concern and where we must act to mitigate them. We welcome a reversal or decline in these trends.

Chronic and high-impact threats were evaluated separately for each ecoregion, and each was classified as high, medium, or low. An ecoregion rated high for chronic threat would have its final threat-modified conservation status elevated one level from its snapshot conservation status. For example, an endangered ecoregion with a high chronic threat would be elevated to critical status by this procedure. However, because of the catastrophic effects of high-impact threats, any ecoregion that ranked high for these threats was automatically elevated to critical status, regardless of its snapshot conservation status. The conservation status for ecoregions with moderate or low threat remains unchanged. Details of the assessment and the variables are provided elsewhere (appendix D).

Integrating Biological Distinctiveness and Conservation Status

The biological distinctiveness and conservation status indices represent an evaluation of the relative biological importance of ecoregions and a measure of current and projected anthropogenic impacts. Combined, the two indices provide a powerful tool for indicating appropriate conservation activities within ecoregions and for setting regional priorities when limited resources require careful and strategic planning.

We adopted a matrix developed by Dinerstein et al. (1995) and modified by Ricketts et al. (1999) to integrate the indices. In this matrix, the biological distinctiveness categories are arranged along the vertical axis, with the conservation status categories along the horizontal axis (figure 3.1). Ecoregions are assigned to one of the twenty cells in the matrix, based on their index categories. This integration is carried out independently for each biome to ensure representation of each biome in the final portfolio of highest-priority ecoregions.

	CRITICAL	ENDANGERED	VULNERABLE	RELATIVELY STABLE	RELATIVELY INTACT
GLOBALLY OUTSTANDING	I	I	I	III	III
REGIONALLY OUTSTANDING	II	II	II	III	III
BIOREGIONALLY OUTSTANDING	IV	IV	V	V	V
LOCALLY IMPORTANT	IV	IV	V	V	V

Figure 3.1. Integration matrix of biological distinctiveness index (BDI) and conservation status (CSI).

Each cell in the integration matrix represents one of five classes that reflect the nature and extent of the management activities likely to be needed for effective biodiversity conservation. The classes are as follows:

Class I. Globally outstanding ecoregions needing immediate protection of remaining habitat and extensive restoration. These ecoregions contain elements of biodiversity that are of extraordinary global value or rarity and are under extreme threat. Conservation actions in these ecoregions must be swift and immediate to protect the remaining source pools of native species and communities for restoration efforts that could be extensive and costly.

Class II. Regionally outstanding ecoregions needing immediate protection of remaining habitat and extensive restoration. These ecoregions have high regional biodiversity and are under serious threat. Conservation actions should be swift and may include extensive and costly habitat restoration.

Class III. These are globally and regionally outstanding ecoregions that present rare opportunities to conserve large blocks of intact habitat. (Block size is relative to the ecoregion size class, i.e., >20,000 km^2 or <20,000 km^2). Although immediate action is not warranted, they should receive attention before large-scale, intense threats are brought to bear and conservation options are lost.

Class IV. Bioregionally outstanding and locally important ecoregions needing protection of remaining habitat and extensive restoration. These ecoregions are under extreme threat, and proper stewardship or expansion of protected areas, conservation management across the broader landscape, and vigilant monitoring of ecological integrity are needed.

Class V. Bioregionally outstanding and locally important ecoregions needing protection of representative habitat blocks and proper management elsewhere for biodiversity conservation. Conservation actions include proper stewardship or expansion of protected areas, conservation management on public and private lands, and vigilant monitoring of ecological integrity.

Guidelines for Using the Integration Matrix

The integration matrix can help conservation planners and donors to identify high-priority activities and conservation needs for each ecoregion. But which ecoregions should receive highest priority? We suggest that prioritizing should take a two-pronged approach: choosing the Class I ecoregions allows us to focus on the ecoregions that support globally outstanding levels of biodiversity but are highly threatened, and the Class III ecoregions support globally and regionally outstanding biodiversity and contain relatively intact habitats that provide better opportunities for effective conservation efforts.

If there are no Class I ecoregions in a particular biome, ecoregions from the Class II cells are chosen to ensure representation of that biome in the high-priority conservation portfolio. The order of choice for representation is ecoregions that are vulnerable, endangered, and then critical for conservation status.

We urge that international donors ensure representation of biomes across bioregions in their overall portfolio. Regional agencies and donors should ensure representation across biomes within their area of jurisdiction and prioritize ecoregions in Classes I and III, and then II.

Use of Proxies to Develop Indices

The biological distinctiveness and conservation status indices rely on proxies to measure biological values and habitat status. Unlike some previous approaches that have relied on single taxonomic groups (e.g., Endemic Bird Areas [Stattersfield et al. 1998], Centers of Plant Diversity [WWF and IUCN 1995]), we have used several groups to make the index more robust and representative.

The use of variables that emphasize the presence of large habitat blocks reflects the need for larger landscapes that are more resilient to external perturbations and can also support large, space-dependent species. In the absence of comprehensive biodiversity inventories, some megavertebrates can be effective indicators of ecosystem integrity because they play a critical role in structuring ecological systems. However, these focal species must be chosen carefully and with knowledge of their ecological needs and the roles they play in ecosystems. In essay 9, John Seidensticker presents a cogent argument for the use of carefully chosen focal species as proxies in conservation priority-setting and planning. He discusses how a lack of tigers can indicate an ecosystem in decline, and in an ironic twist he points out how the region's second largest carnivore, the leopard, can also indicate degradation in ecosystem integrity.

Biodiversity Representation or Triage?

This analysis begins and ends with the premise that all ecoregions contain and represent unique elements of Earth's biodiversity and provide many services to nonhuman and human communities. Conservation of natural areas in all ecoregions ensures preservation of many species and distinct communities as well as the genetic and functional diversity of populations across species' ranges. Flood control, groundwater recharge, freshwater purification, and innumerable recreational opportunities are all examples of local ecosystem services that must be maintained within each ecoregion (Daily 1997). By ranking the ecoregions we merely represent an order of priority for addressing urgent conservation needs when faced with limited resources. It does not imply that some ecoregions are unimportant or unworthy of conservation attention. We discuss this issue in more detail in chapter 6.

ESSAY 6

Amphibians of the Indo-Pacific Region: Conservation of Neglected Biodiversity

Indraneil Das

The amphibians of the Oriental region are among the most diverse and fascinating anywhere. (The Oriental region is defined here as extending from the southern slopes of the Himalayas to the islands of the Sunda Shelf, equivalent to south Asia as defined by Inger 1999, except that it also includes Pakistan.) The strange gliding frog, *Pterorana khare*, has a patagium of skin that allows it to glide and parachute over waterbodies (Kiyasetuo and Khare 1986). The leaf-eating *Euphlyctis hexadactylus* eats mostly macrophytes (Das 1996b). Large (to 18 cm) river frogs, *Limnonectes blythi*, and their relatives make shallow moat nests of pebbles (Emerson 1992). Finally, there are the guardian frogs, *Rana palavanensis* and *R. finchi*, whose larvae ride piggyback to water (Inger et al. 1986).

This amphibian fauna is composed essentially of anurans (frogs and toads), with very few Gymnophiona (caecilians) and Caudata (salamanders and newts). Until recently, almost nothing was known of their microhabitat preferences, behaviors, diets, and reproduction. But it is now acknowledged that amphibians, often concealed in the leaf litter of a tropical forest, in the unexplored canopy, or underwater in streams, springs, pools, and seepages, play a major role in the web of cryptic ecological interactions.

Inger (1999) recorded approximately 650 nominal species from the region (this manuscript went to press in 1994; R. F. Inger, pers. comm., 2000), and since then major additions to the fauna have been recorded. For example, recent work in Vietnam has yielded at least six new frog species (Inger et al. 1999), and in Sri Lanka, sampling of the midcanopy tree frogs has increased the amphibian fauna from 53 to more than 250 species (Pethiyagoda and Manamendra-Arachchi 1998). The distribution patterns of these new species have important implications not only for systematics but also for conservation planning.

The high levels of amphibian diversity in tropical Asia are attributable primarily to the following factors: parts of south and southeast Asia were refugia during the Pleistocene glaciation; the region has a complex history of sea-level lowering that attached and detached islands to the Asian mainland, joining and severing populations in the process; the region has high geological and climatic diversity that supports diverse ecological conditions; and the region has had large unbroken tracts of primary forests, especially in the montane and submontane regions. Over the millennia these mountains and islands have acted as refugia and dispersal barriers.

Regional patterns of amphibian diversity are still poorly understood, given the uncertainties in systematic knowledge, which is primarily a result of poor sampling. However, some general trends can be discerned. On a coarse scale, the amphibian fauna is unequally distributed across the region. On a finer scale, we can only deduce patterns from what we know of the better-studied areas. One of these is the Western Ghats, a chain of mountains that runs along the west coast of the Indian Peninsula. The hill ranges are ecologically distinct from the intervening 100- to 150-m flat savanna. From a regional perspective, endemism is an important aspect of amphibian diversity of the Western Ghats range (Inger et al. 1987). However, Daniels (1992) showed that lowlands and mid-hills (0–1,000 m) in the southern extent of the range have higher species diversity and more endemic species than the northern and higher hills. This pattern illustrates some of the scale-related issues that must be considered in conservation plans for amphibians and probably other restricted-range species. Moreover, many species with wide ranges show patchy distributions, possibly occurring in appropriate habitats and microhabitats such as windward sides that receive more precipitation than ridges and leeward sides (Das 1997).

Amphibian Diversity in the Orient: How Many Frogs Are There?

An understanding of systematics and distributions is more than academic, with major implications for conservation. Daugherty et al. (1990) clearly showed how the failure to recognize a species could lead to its extinction. Several factors are responsible for our poor knowledge of amphibian diversity. Monographs prepared in the early twentieth century lack color illustrations and contain only terse descriptions that would equally fit several closely related species, potentially causing gross underestimation of biodiversity. Studies of systematics usually are published in scientific journals that are seldom distributed widely or read by the majority of the conservation community. As a result, new findings remain unknown to researchers in the field. For instance, the recent use of molecular techniques has resulted in dramatic generic reallocations (Vences et al. 2000), changes in synonymy (Matsui et al. 1996), and splitting of putative widespread species (Manthey and Steiof 1998; Matsui et al. 1999).

Mapping of morphological characters onto molecular phylogeny has shown that specific characters may have evolved in parallel in different lineages. Thus, many of the new species that will be discovered in the twenty-first century are predicted to be taxonomically cryptic species. Because of close similarities, revision of entire groups using modern laboratory (including gene sequencing) and field (ecological and behavioral) methods will add a host of new species to the list of amphibians.

Distribution of Frogs

Cryptic species often are localized, some restricted to small patches of forests a few dozen hectares in extent or to one or two adjacent hill streams, making their discovery difficult except through exhaustive inventories. Other species may show disjunct populations structured into well-defined phylogenetic assemblages or metapopulations, some with significant genetic variants and all warranting careful consideration for identification and conservation (Sites and Crandall 1997). Thus, preparing comprehensive inventories of cryptic amphibians that live in architecturally complex habitats such as rain forests is an extremely labor-intensive activity, and these species surely will be missed during rapid assessments. To make a complete inventory of even a small area in the tropics would take a lifetime of effort (Myers and Rand 1969). However, the use of modern collection or analysis methods such as canopy sampling (Vogt 1987), molecular techniques (Beebee 1990; Dessauer 1966; Narins et al. 1998), and the use of acoustic data (Blair 1958; Dubois 1975; Schneider et al. 1992) will undoubtedly reveal a much richer amphibian fauna.

The recognition of cryptic species increases the conservation burden, but it also emphasizes the importance and need to move away from taxon-based conservation to that emphasizing conservation of landscapes and ecosystems (Lovich and Gibbons 1997). Despite our poor knowledge of amphibian distribution and diversity, what we know of their biology, biogeography, and ecology, coupled with recent findings from extensive surveys, allows us to make some predictions, albeit tentative, of which ecoregions are likely to harbor the highest amphibian diversity. Inventories in tropical moist forest ecoregions in Indochina, Indonesian, and Philippine archipelagos, Himalayan foothills (especially the border areas between China, India, and Myanmar), and the Western Ghats of India are now grossly incomplete and probably will yield many new amphibians in comprehensive surveys.

There is also an urgent need to conserve these tropical moist habitats. Local and regional habitat loss and degradation (microhabitat changes such as drying of aquatic and moist habitat, loss of leaf litter, pesticides, and diseases) and large-scale environmental changes (such as acid deposition and ozone depletion) are causal factors associated with declining amphibian populations (Blaustein 1994; Blaustein and Wake 1990, 1995; Dodd 1997; Lawrance 1995; Vasudevan 1996). The loss of amphibian diversity that has probably accompanied the recent wave of deforestation (over the past 100 years) in the Indo-Pacific region will be immeasurable. There are few records of amphibian extinctions in the Indo-Pacific. However, this is probably because of lack of information and study, rather than a reflection of threat, because two recent studies based on extensive field work in Sri Lanka have documented several amphibian extinctions caused by habitat loss and degradation (Pethiyagoda, pers. comm., 1999).

Conclusions

Amphibians are an important component of many natural ecosystems in terms of both species numbers and biomass. Although the global herpetofauna constitutes roughly a third of all vertebrates (not including fish), amphibians and reptiles typically are excluded from consideration in habitat management and environment impact assessments. Few populations are managed, with management priorities typically set using nonbiological criteria such as economic value and the size and appeal of the species (Scott and Seigel 1991). Many amphibian species are habitat specialists, so their ecology and distributions can be used to monitor the quality and integrity of these habitats and microhabitats (Bury et al. 1980; Parent 1992). Where quantitative data on populations and habitats are available, they are potentially useful for constructing predictive models of the abundance of the target taxa (Clawson et al. 1984).

Identification of important biodiversity areas, whether centers of high diversity or endemism, are critical for reserve selection and design, helping focus scarce conservation money on areas of highest priority. Design of these conservation programs and priorities should be based on inputs from scientific databases. But information on amphibian distribution, ecology, and systematics is still grossly incomplete. For tropical Asia in general, the lack of data on distribution and abundance makes accurate estimates of levels of imperilment of the amphibian fauna impossible. These data gaps seriously constrain our ability to understand factors that potentially threaten species and populations. Therefore, active lobbying and conservation education are necessary for directing both public interest and institutional support toward the fauna. On the bright side, the publication of several new field guides and the increasing interest in the herpetofauna bodes well for conservation of the group as a whole. Global amphibian die-offs are one of the most intensively studied conservation projects at present, and although much remains to be learned, the increase in the number of workers in the last two decades and the resulting publications reveal a growing interest in tropical Asia's amphibian fauna.

Tropical Asia has fascinated generations of explorers with its natural wealth, charismatic megavertebrates, and ancient civilizations and cultures. It is still described as a new frontier for work on amphibian systematics (Dubois 1999). Thoughtless exploitation of the region's tropical forests increases the importance of efforts to understand the region's amphibian fauna. Therefore, intensive sampling of key habitats is urgently needed, especially in the tropical moist forest ecoregions, to integrate amphibian conservation into regional biodiversity conservation strategies.

ESSAY 7

Reptile Diversity in the Indo-Pacific Region

Walter Erdelen

Analyzing reptile diversity for the ecoregions of the Indo-Pacific region is a challenging task. Information on geographic distribution is too scanty for most taxa, and species lists normally are compiled on a national basis even though the underlying determinants of reptile diversities are biogeographic rather than political. Nonetheless, there are some important studies to draw from. Some thirty years ago, Brown and Alcala (1970) addressed biogeographic questions in their study on the Philippine herpetofauna, which was later followed by Brown's monograph on the skink genus *Emoia* (Brown 1991). A number of similar studies have been published since then. Later herpetological studies by Allen Allison in New Guinea and the Pacific and by Aaron Bauer in New Caledonia, for instance, have clearly demonstrated the usefulness of reptiles for addressing biogeographic questions. Modern systematic studies have shown that reptiles may represent key groups for understanding evolution on islands within the Indo-Pacific region (Adler et al. 1995; Das and Bauer 1998; How et al. 1996a, 1996b; How and Kitchener 1997; Lazell 1992). New genetic approaches have been used successfully to analyze reptilian evolution (Austin 1995; Bruna et al. 1996; Ota et al. 1996). Reptiles may turn out to be a key group in further analyses using phylogeographic approaches (overviews in Avise 1998, 2000; Bermingham and Moritz 1998).

Recent detailed studies in almost all parts of the globe, particularly in the tropics, have shown that we are far from knowing all extant species of reptiles. For instance, researchers are finding new reptiles in densely populated areas such as the island of Java or the wet zone of Sri Lanka, where human population densities exceed 800 inhabitants per km^2 (Erdelen 1996; Whitten et al. 1996), and even short-term expeditions are uncovering new species (e.g., Iskandar and Tjan 1996). New techniques in molecular systematics (Hillis et al. 1996) have revealed a number of cryptic reptile species, formerly assigned to a single taxon. For instance, in a study of the *cyt b* mtDNA gene, Bruna et al. (1996) have shown that morphologically very similar species of *Emoia* with extensive geographic ranges (*E. cyanura* and *E. impar*) are in fact clades of cryptic species. This may even be a more general phenomenon among reptile species with extensive geographic ranges. However, newly discovered species in the Asia-Pacific region are not limited to cryptic taxa detectable only with highly sophisticated molecular techniques. One of the best examples of species that have been overlooked are the monitor lizard species recently described from Indonesia (Böhme and Ziegler 1997; Harvey and Barker 1998; Philipp et al. 1999; Sprackland 1999; Ziegler et al. 1999).

For many of the known reptiles, we have a poor understanding of the species' geographic ranges and even their basic biology and ecology. This is of particular importance for restricted-range species, narrow endemics, or species with subspecifically differentiated populations. Such species should receive high priority in conservation measures. As the Indo-Pacific region includes the largest archipelago area on Earth, it also contains a large number of species restricted to small islands. As a result of human impacts on formerly extensive and continuous habitats or vegetation types, many species now are restricted to habitat islands.

Concern over human impacts, particularly in areas with high biodiversity, has led to approaches that prioritize areas on the basis of species richness, levels of endemism, numbers of rare or threatened species, and threat intensity. The significance of these approaches for reptile conservation is apparent; because of our paucity of knowledge of reptile distributions, ensuring that reptile species richness and endemism are represented in these priority areas is difficult. Any attempt to compile species lists or numbers for any area must be read with caution because the rapid description of new species suggests that any tallies will be low. We expect that future investigation will find new species for all extant reptile groups in the region (details in Das 1996; Iskandar 1996; Iskandar and Erdelen ms), not only for taxa with cryptic lifestyles like many of the fossorial reptile species (see Gans 1993).

Yet some generalities emerge. Species compilations tend to show a trend toward increased species numbers with area. This is partly attributable to the classic species-area relationship, but there are other factors as well. For example, there is high habitat heterogeneity, as in India, where the highest numbers of species are found in regions with the highest vegetational diversity. Also, isolation plays a role, as in the Philippines (for an overview, see the classic paper by Brown and Alcala [1970]) and Indonesia, where an archipelago setting has facilitated taxonomic differentiation (Mittermeier and Mittermeier 1997).

Because reptiles are arguably one of the most neglected vertebrate groups in conservation programs, knowledge of species diversity and geographic patterns would be conducive to their long-term successful conservation. A brief overview of current knowledge of reptile diversity and distribution across the region follows.

Reptile Diversity of South Asia

The best overviews of reptile diversities in the south Asian region are provided by Das (1996) in his study on the biogeography of the reptiles of south Asia and by de Silva (1998) in his proceedings volume on the biology and conservation of south Asian herpetofauna. The area commonly called south Asia covers more than 4 million km^2 and stretches over both the Palearctic and Oriental biogeographic realms. Accordingly, the terrestrial ecosystems in this enormous area range from deserts to tropical rain forests.

Generally, regions with higher and more equally distributed rainfall also contain higher numbers of reptile species. This probably is not related to precipitation itself but rather to the more complex vegetation communities found in areas that receive high amounts of rainfall. For instance, Das (1996), in his study on south Asian reptiles, found that the four moist tropical zones—the wet zone of Sri Lanka, the Western Ghats, northeastern India, and the Himalayas—contain the highest reptile diversities. Interestingly, the same study showed that snake species outnumber lizard species in forested areas, whereas the converse holds for areas with little tree cover. However, this phenomenon does not apply generally, as illustrated by species relations in New Guinea, where the largely forested island has about 2.7 times as many lizard species as snakes (Allison 1996). Proportions of lizards and snakes in a given herpetofaunal assemblage probably can be explained only partly in terms of extant abiotic and biotic factors. Historical factors obviously play a major role in shaping such patterns.

In total, Das (1996) listed 632 species of reptiles for south Asia, comprising 3 orders, 25 families, and 185 genera. Particularly speciose families in the region are batagurid turtles (16 spp.); the lizard families Gekkonidae (97 spp.), Agamidae (68 spp.), and Scincidae (87 spp.); and the snake families Typhlopidae (23 spp.), Uropeltidae (47 spp.), Colubridae (176 spp.), Elapidae (16 spp.), Hydrophiidae (21 spp.), and Viperidae (25 spp.). Das (1996) found a significant correlation between species richness and number of endemic species in the ten different physiographic zones he distinguished. Endemism at the species level is high for the region (64 percent). Even within physiographic zones such as in the Western Ghats and Sri Lanka, species occurrence may be highly localized (Erdelen 1993). Areas of endemism may be rather restricted, such as isolated mountain ranges in the Western Ghats or in Sri Lanka (see Erdelen 1993). One of the best examples is the agamid genus *Ceratophora*, endemic to Sri Lanka, of which most species have extremely restricted distributional ranges within the island (Pethiyagoda and Manamendra-Arachchi 1998b). As noted by Das (1996), several of the endemic reptile taxa have undergone distinctive radiations in the south Asian region. Examples are *Aspideretes* (Trionychidae); *Ristella, Lankascincus,* and *Nessia* (all Scincidae); *Uropeltis* (Uropeltidae); and *Aspidura* (Colubridae). Except for *Aspideretes*, all these taxa are largely or entirely limited to the Western Ghats and Sri Lanka.

At the other extreme in geographic ranges are widespread species. Most are commensals, such as the lizard species *Hemidactylus frenatus* (Gekkonidae), *Calotes versicolor* (Agamidae), *Mabuya macularia* (Scincidae), and *Varanus bengalensis* (Varanidae) and snakes such as *Ramphotyphlops braminus* (Typhlopidae), *Xenochrophis piscator* (Colubridae), and *Naja naja* (Elapidae).

At the country level in south Asia, endemism is highest in Sri Lanka, followed by India. In both countries endemics are limited to mountain massifs or areas with higher precipitation and complex vegetation such as tropical rain forests. Endemism in Bangladesh is low, possibly because the country has no effective barriers and therefore may have received immigrants from the west and the east. Bangladesh and northeastern India contain mainly Indochinese and Indo-Malayan elements in their reptile faunas (Das 1996). Similarly, Nepal and Bhutan are biogeographically parts of the Indian subcontinent and thus support herpetofaunas of Indian occurrence (Das 1996; Shrestha 1998).

Reptile Diversity of Southeast Asia

Southeast Asia is the region between Myanmar and Indonesia. In southeast Asian countries, endemism is low, although this may be an artifact of poor sampling effort in areas that might contain endemic taxa: montane regions or regions with high plant species diversity. The number of species endemic to mainland southeast Asian countries is certainly much higher than documented to date. Because of the lack of localized studies that refer to ecoregions or parts thereof, it has not been possible to relate species diversities of these countries to smaller geographic units. Most species lists are still derived from publications that cover former British India (e.g., Smith 1931, 1935, 1943).

Studies on the country level provide only gross information on the geographic distribution of individual reptile taxa. In terms of reptile fauna, possibly the best understood of southeast Asia are in ecoregions that fall within Thailand, largely as a result of Taylor's studies (1963, 1965, 1970). Without a doubt, the ecoregions in Cambodia, Laos, and Myanmar need more in-depth study of their reptile faunas. Hundley's (1964) checklist of Burmese reptiles, for instance, is based largely on previous work by Malcolm Smith and contains an outdated taxonomy. Evidence from Vietnam indicates that species numbers, and in particular numbers of endemic species, may be much higher than thought to date. Many new endemics from almost all families occurring in the region have only recently been described. For example, in Vietnam new species of the gekkonid genera *Gekko* (Günther 1994; Shcherbak and Nekrasova 1994) and *Goniurosaurus* (Orlov and Darevsky 1999), new specimens of virtually unknown agamid lizards such as *Japalura chapaensis* (Ota and Weidenhöfer 1992), new agamid species (e.g., *Leiolepis guentherpetersi*; Darevsky and Kupriyanova 1993), new colubrid snakes such as *Opisthotropis daovantieni* (Orlov et al. 1998), and new dibamid species such as *Dibamus greeri* (Darevsky 1992) have been described recently. Even new genera such as the scincid *Paralipinia* and *Vietnascincus* have been described by Darevsky and his co-workers (Darevsky and Orlov 1994, 1997). These new descriptions reflect both our poor understanding and the long neglect in carrying out herpetological studies in Vietnam, certainly a result of the political situation in the country since World War II. But most original forests in Vietnam have been lost, and we do not know how many reptile (and other) species might have already become extinct. Similarly, an explosion in knowledge would be expected from new studies on the reptile fauna of Laos. To date, for Laos and Cambodia, only single endemic species, the agamid lizard *Pseudocalotes poilani*, and the colubrid snake *Enhydris longicauda*, respectively, have been described from each country.

Reptile Diversity of the Indo-Pacific Region

Before 1800, only nine amphibian and reptile species had been described from the Indo-Pacific region. This figure increased to almost 300 by the end of the nineteenth century (Allison 1996). Seminal reptile studies of the early twentieth century include the work by van Denburgh (1917) on the herpetology of Guam and by Kinghorn (1928) on the herpetofauna of the Solomon Islands. Allison (1996) has presented one of the most modern studies in his compilation of the herpetofauna distributions of Melanesia, tropical Polynesia, and Micronesia.

Reptile diversity in the Pacific is linked to the relative capacities of different taxa to colonize islands (Gibbons 1985). For example, the combined abilities of geckos to survive on driftwood and to undergo parthenogenesis allowed them to drift and colonize islands with small founder populations. Skinks—species with continuous breeding, prolonged incubation times, the ability for sperm storage, and long gestation times—may be "preadapted" for dispersal between islands. For the iguanid genus *Brachylophus* there is even speculation that much of its rafting journey was undertaken in the egg stage (Gibbons 1981).

Although the Bismarck Archipelago and the Admiralty Archipelago have undergone complex geological histories (for details, see Allison 1996), the reptile faunas of these island groups are similar to that of neighboring New Guinea, sharing a large fraction of widespread species and having low levels of reptile endemicity. McCoy (1980, 2000) described the reptiles of the Solomon Islands, which have comparatively high reptile species richness and endemism. Allison (1996) assumed that these islands are older than the other two groups and that, through radiation and diversification, autochthonous species evolved on the Solomons. Even island hopping might have contributed to the evolution of the reptile fauna of the Solomons. The Solomon Islands nevertheless share many species with New Britain, New Ireland, and the Admiralty Archipelago.

Vanuatu generally is considered to have a depauperate reptile fauna with many widespread Pacific species, but as Allison (1996) notes, the herpetofauna of Vanuatu has not been studied recently. Earlier work by Bauer (1988) indicates that its fauna may have evolved through a complex array of geological-historical processes. Probably we will have to revise our simplified picture of Vanuatu's herpetofauna as soon as more recent and in-depth work becomes available.

To say that New Caledonia contains a unique herpetofauna is an understatement. Its indigenous herpetofauna consists only of geckos and skinks (Bauer and Vindum 1990), but these groups comprise two endemic gecko genera with nine species and more than seven endemic skink genera, represented by at least fourteen species (Allison 1996). There are no native amphibians on New Caledonia (the hylid *Litoria aurea* probably was introduced from Australia).

Bauer and Vindum (1990) distinguished four types of geographic distribution among New Caledonian reptiles: New Caledonian endemics, regional endemics occurring on New Caledonia and the Loyalty Islands, Melanesian-Polynesian elements with characteristics that make them good overwater dispersalists, and species that have been introduced by humans during historical times. The carphodactyline geckos of New Caledonia may represent a lineage that originated some 60–80 million years ago; the evolution of the endemic skink species may have resulted from both vicariance and dispersal events. In sum, the reptiles of New Caledonia are a heterogeneous group of species with different origins (see discussion in Bauer and Vindum 1990).

The biogeographic picture of New Caledonia becomes even more complicated if taxa such as meiolaniid turtles, mekosuchian crocodiles, and varanids, three groups now extinct on the island, are included in a comprehensive analysis. Bauer and Vindum (1990) suggest that these groups might even have occurred contemporaneously with humans on New Caledonia. In a more recent discussion of the origin and evolution of the terrestrial reptiles of New Caledonia, Bauer (1999) gives an impressive figure for reptile diversity of the New Caledonian region, including the Loyalty Islands and the Isle of Pines (for descriptions of their herpetofaunas, see Bauer and Sadlier 1994 and Sadlier and Bauer 1997), comprising a total of seventy-one taxa, of which 86 percent are endemic. New taxa, both on the species and genus levels, continue to be described from New Caledonia.

Reptile Distribution and the Ecoregion Concept: Issues of Scale

The ecoregional concept represents a hierarchy of levels of biogeographic organization that includes 140 ecoregions assigned to ten biomes stratified within five bioregions. For conservation efforts, the implication is that the same habitat types in different bioregions may be similar in terms of their structural characteristics but not in terms of their species compositions and biogeographic origins. Ecoregions are defined and delineated largely on the basis of vegetation maps. Thus, there are instances in which one island can have two or more ecoregions (e.g., Java), whereas in other cases an ecoregion can span multiple islands (e.g., New Britain and New Ireland). How are these divisions related to the distribution patterns of reptiles?

The linkage between ecoregions and reptile diversity is moderate. Many centers of reptile diversity are also areas where other taxonomic groups are exceptionally diverse. For instance, reptile species richness and plant species richness are highly correlated between south and southeast Asian countries ($r = 0.92$, $N = 12$) and among oceanic islands in the Indo-Pacific region, including the Nicobar and Andaman islands ($r = 0.77$, $N = 9$; data from WCMC 1992). These gross relationships indicate that conservation of plant species diversity might also contribute indirectly to preservation of reptiles. The usefulness of such an approach would need a careful analysis based on more accurate estimates of species diversities and patterns of geographic ranges of both reptiles and plants.

Flagship Reptiles

Although considered outdated by many conservationists, the use of flagship species to create awareness may be a useful approach for reptile conservation (table 3.1). Modern approaches to conservation have not increased awareness about reptiles. Rather, reptiles appear only as numbers in diversity lists, if even that much information is available. Without a thorough understanding of reptile faunas, focusing on individual flagship species may be the best short-term path to garnering attention for this group.

In conclusion, we are facing a seeming paradox: the more that studies of reptiles of the Indo-Pacific region become available, the clearer the picture of our ignorance becomes. This lack of knowledge is paralleled by an enormous rate of destruction of natural habitats, especially of forests, and an alarming increase in the exploitation of certain reptile groups. Most affected at present are the tortoises and freshwater turtles, particularly species occurring in south and southeast Asian countries. As a consequence, conservation efforts, in the past largely targeted at crocodilians and marine turtles, now must be reoriented toward other reptile groups that have been largely ignored. However, successful conservation must be based on sound data on the biology and ecology of the relevant species. Further research, including species inventories in poorly known regions and systematic-taxonomic and ecological studies, therefore is urgently needed. However, a quest for a complete inventory should not preclude conservation options that can be recommended using currently available information.

With increasing knowledge of reptile diversity, particularly in species-rich tropical regions, reptiles are estimated to make up nearly 25 percent of all vertebrate species. International conservation efforts are shifting to favor reptiles more than before. But detailed action or conservation plans for reptiles like the action plan for Australian reptiles (Cogger et al. 1993) and a forthcoming action plan for the south Asian herpetofauna (Das, in prep.) are almost nonexistent for most countries of the Indo-Pacific region. Such plans would be vital for overall biodiversity conservation in this region. Sustainable use of wildlife, protected area management, and other activities in the conservation sector should form part of a coherent conceptual framework that adequately considers biological and ecological diversity among the different taxonomic groups, including the largely ignored reptiles.

TABLE 3.1. Flagship Reptile Species for Bioregions in the Indo-Pacific Region. Marine turtles and the estuarine crocodile (*Crocodylus porosus*) may be considered flagship species for all bioregions listed here.

Family	Species	Common Name
Indian Subcontinent		
Agamidae	*Ceratophora* spp.	Horned lizards
Agamidae	*Lyriocephalus scutatus*	Hump-nosed lizard
Bataguridae	*Kachuga kachuga*	Red-crowned roofed turtle
Boidae	*Python molurus*	Indian python
Crocodylidae	*Crocodylus palustris*	Mugger
Elapidae	*Ophiophagus hannah*	King cobra
Gavialidae	*Gavialis gangeticus*	Gharial
Scincidae	*Barkudia insularis*	Legless skink
Testudinidae	*Indotestudo forstenii*	Travancore tortoise
Trionychidae	*Aspideretes nigricans*	Black softshell turtle
Trionychidae	*Chitra indica*	Narrow-headed softshell turtle
Indochina		
Agamidae	*Physignathus cocincinus*	Indochinese water dragon
Boidae	*Python molurus*	Burmese python
Boidae	*Python reticulatus*	Reticulated python
Crocodylidae	*Crocodylus siamensis*	Siamese crocodile
Elapidae	*Ophiophagus hannah*	King cobra
Platysternidae	*Platysternon megacephalum*	Big-headed turtle
Testudinidae	*Geochelone platynota*	Burmese starred tortoise
Trionychidae	*Chitra chitra*	Striped narrow-headed softshell turtle
Sunda Shelf and Philippines		
Agamidae	*Hydrosaurus pustulatus*	Philippine sailfin lizard
Bataguridae	*Batagur baska*	River terrapin
Bataguridae	*Callagur borneoensis*	Painted terrapin
Bataguridae	*Orlitia borneoensis*	Malaysian giant turtle
Boidae	*Python reticulatus*	Reticulated python
Crocodylidae	*Crocodylus mindorensis*	Philippine crocodile
Crocodylidae	*Tomistoma schlegelii*	False gharial
Elapidae	*Ophiophagus hannah*	King cobra
Gekkonidae	*Ptychozoon intermedium*	Intermediate flying gecko
Lanthanotidae	*Lanthanotus borneoensis*	Earless monitor
Varanidae	*Varanus olivaceus*	Gray's monitor
Wallacea		
Agamidae	*Hydrosaurus amboinensis*	Sailfin lizard
Boidae	*Python timorensis*	Timor python
Geoemydidae	*Heosemys yuwonoi*	Sulawesi forest turtle
Varanidae	*Varanus caerulivirens*	Turquoise monitor
Varanidae	*Varanus komodoensis*	Komodo dragon
Varanidae	*Varanus melinus*	Quince monitor
New Guinea and Melanesia		
Agamidae	*Chlamydosaurus kingii*	Frilled lizard
Agamidae	*Hypsilurus* spp.	Angle-headed dragons
Boidae	*Candoia* spp.	Pacific boas
Boidae	*Morelia boeleni*	Boelen's python
Boidae	*Morelia viridis*	Green tree python
Carettochelyidae	*Carettochelys insculpta*	Pignosed softshell turtle
Crocodylidae	*Crocodylus novaeguineae*	New Guinea crocodile
Elapidae	*Acanthophis* spp.	Death adders
Elapidae	*Oxyuranus scutellatus*	Taipan
Gekkonidae	*Rhacodactylus* spp.	Giant New Caledonian geckos
Iguanidae	*Brachylophus* spp.	Fiji iguanas
Scincidae	*Tribolonotus* spp.	Helmet skinks
Varanidae	*Varanus prasinus*	Emerald monitor
Varanidae	*Varanus salvadorii*	Crocodile monitor

ESSAY 8

Mangroves of the Indo–West Pacific

David Olson

The mangrove swamps and forests of the Indo-Malayan and Australasian realms are the world's most extensive. South and southeast Asia alone contain 42 percent of the total area of the world's mangroves (Spalding et al. 1997). The Sundarbans are the largest contiguous mangrove forest in the world. The vast floodplains of New Guinea also support extensive mangrove swamps unrivaled elsewhere in the world. Mangroves in this region also are distinctive from those of the Atlantic, Caribbean, and eastern Pacific in that they occur in a wider range of habitats upstream in rivers (Ricklefs and Latham 1993).

Mangrove swamps and forests occur in the intertidal ecotone found along coasts and extending inland along riverbanks (Tomlinson 1986). The most extensive mangrove forests occur in soft sediment habitats away from strong wave action, but mangroves can also become established along protected rocky shores. True mangrove plants are limited to tidal swamps and have morphological and physiological features that allow them to persist in inundated and saline environments (Tomlinson 1986). These plants are distinct at the generic or subfamily level from terrestrial relatives. Mangrove species represent a diverse group of plant taxa, including several families of trees (e.g., Avicenniaceae, Rhizophoraceae, Sonneratiaceae, Meliaceae, Euphorbiaceae, Bombacaceae, Combretaceae, Rubiaceae) and some palms (e.g., *Nypa fructicans, Calamus* sp.) and ferns (e.g., *Acrostichum aureum*) (Tomlinson 1986). Many mangrove species occur in distinct zones within mangrove forests depending on the salinity and aeration of the soil, water salinity, and degree of tidal inundation (Ricklefs and Latham 1993). For example, six distinct zones have been recorded in mangroves of eastern Australia (Macnae 1968). Mangrove forests in different zones can vary in height from 1- to 3-m *Avicennia marina* stands to 30-m forests of *Brugiera, Rhizophora*, and *Ceriops* species.

The diversity of mangroves in the Indo–West Pacific (IWP) region is much greater than in the Atlantic–Caribbean–East Pacific (ACEP) region; the former supports seventeen genera and forty to forty-two species of true mangroves and the latter has only four genera and seven species (Ricklefs and Latham 1993). A single site in the ACEP typically contains three or four true mangrove species, whereas thirty species have been recorded from one locality in the IWP region (Ricklefs and Latham 1993). The long stability of warm, wet climates in the Indo-Pacific region may have allowed more terrestrial species to become adapted to mangrove environments than did the occasionally arid conditions of the ACEP (Ricklefs and Latham 1993). Thus, the region may be as much a refugium as a center of origin for mangrove evolution. The reasons for this strong gradient in mangrove species richness are still debated (Tomlinson 1986; Ricklefs and Latham 1993; Hogarth 1999; Duke, Ball, and Ellison 1998), but other groups such as crustaceans and polychaetes also tend to be more species-rich in IWP mangroves (Ricklefs and Latham 1993). Richness gradients within the IWP region also occur. For example, thirty mangrove species are recorded for New Guinea, whereas only eleven are recorded from New Caledonia (Hogarth 1999). New Guinea may be a fusion zone between two formerly isolated and now distinct mangrove floras, with fourteen more species found along the island's southern coastline than in the north (Spalding et al. 1997).

If all the marine, freshwater, and terrestrial species that occur in mangroves are considered, these seemingly simple forests can be considered one of the most diverse ecosystems in the tropics. Epiphytes such as orchids, mistletoes, lichens, and ferns grow on mangrove trees. Some of these are known only from mangroves, such as the rare mistletoe *Amyema anisomeres*, known only from one mangrove area in southern Sulawesi (Whitten and Whitten 1996). Another mangrove plant, *Ixora timorensis*, has been recorded only from Nusa Tenggara (Whitten and Whitten 1996). Many invertebrate, reptile, bird, fish, and mammal species inhabit the mangrove forests of the region (Whitmore 1986). They often depend on mangroves for some part of their life cycle. Ants, crabs, and snails are particularly conspicuous residents of mangrove trees, and numerous species of fish, crustaceans, and other aquatic taxa use mangroves for shelter, feeding, nurseries, and breeding habitat. The proboscis monkey (*Nasalis larvatus*) of Borneo is largely restricted to mangrove habitats, feeding mainly on young leaves of *Rhizophora* and *Sonneratia*. A wide range of other mammals in the region also use mangroves, including the silvered langur (*Presbytis cristata*), lion-tailed macaque (*Macaca silenus*), Bengal tiger (*Panthera tigris*), Javan rhinoceros (*Rhinoceros sundaicus*), long-tongued fruit bat (*Macroglossus minimus*; pollinators of some *Sonneratia*), estuarine crocodile (*Crocodylus porosus*), fishing cat (*Felis viverrina*), and Gangetic and Irrawaddy dolphins (*Platanista gangetica* and *Orcaella brevirostris*) (Whitmore 1986; Hogarth 1999). The cave fruit bat, *Eoncyteris spelea*, relies on *Sonneratia* as a food source when inland species, such as the commercially harvested durian (*Durio zibethinus*), are not in flower (Hogarth 1999).

Mangroves are keystone habitats because they have an inordinately strong influence on species populations and ecosystems well beyond their limited area. Besides providing habitat and resources to a wide range of species, mangrove forests and swamps also protect inland habitats and shorelines from damage by damping storm waves and tidal action. Mangroves filter silt and pollutants from terrestrial runoff that would otherwise damage seagrass beds and coral reefs. Human communities benefit from all of these ecosystem services and also use a wealth of mangrove resources including timber, thatching materials, charcoal, fodder, fuelwood, and a host of foods including fish, mollusks, and crustaceans.

Mangrove habitats are threatened worldwide by a range of factors including clearing and channelization for shrimp ponds, aquaculture, and agriculture, the extraction of timber and fuelwood, pollution, and habitat loss caused by urban and industrial expansion. Those of the Indo-Pacific region are particularly threatened by the pressures of high population densities along the region's coasts and intensive commercial activities. Future development activities should not proceed at the expense of the remaining mangrove forests.

ESSAY 9

Tigers: Top Carnivores and Controlling Processes in Asian Forests

John Seidensticker

"Top carnivores—including tigers, eagles, and great white sharks—are at the apex of food webs are therefore predestined by their position to be big in size and sparse in numbers. They live on such a small portion of life's available energy as always to skirt the edge of extinction, and they are the first to suffer when the ecosystem around them starts to erode." (Wilson 1993: 36)

"The removal of a top carnivore from an ecosystem can have an impact on the relative abundance of herbivore species within a guild. In the absence of predation, usually one or two herbivore species come to dominate the community. The consequence is often a direct alteration of herbaceous vegetation near the base of the food web. Top carnivores have an important role to play in the structuring of communities and ultimately of ecosystems. Thus, the preservation of carnivores becomes an important consideration in the discipline of conservation biology." (Eisenberg 1989: 7)

These quotes from two eminent ecologists contrast the bottom-up and top-down control processes that regulate species abundance and structure the numbers and kinds of species that co-exist in an area (Power 1992). Most conservation biologists agree that top-down and bottom-up processes occur concurrently in many ecosystems. Investigating the role of top carnivores in structuring ecosystems has been an elusive goal because of the daunting task posed by the large spatial scales necessary to study these processes and the few places where we can do so (Seidensticker and McDougal 1993). These natural biotic processes sustain biodiversity, and conserving them is crucial. In this essay I examine the role of one of the Indo-Pacific region's most prominent top carnivores, the tiger (*Panthera tigris*). I discuss the tiger's food and spatial requirements, what the tiger needs to survive, and how the tiger and other top predators shape community structure.

Big Carnivores Need Big Prey to Survive

Among the Carnivora, there is an overall correlation between body mass and the mass of the most common prey item. In general, carnivores weighing 21.5 kg or less feed mostly on prey that is less than 45 percent of their own mass; carnivores above this size threshold feed mostly on prey that is more than 45 percent of their own mass. This dichotomy is the consequence of mass-related energetic needs, that is, the hunting times needed to obtain the prey (energy) to balance the net energy expenditure. As carnivore mass increases, the total energy expenditure increases, and so must hunting time to obtain prey to balance the energy budget (Carbone et al. 1999).

The Felidae can be divided into three clusters according to their mass. The cats in each cluster have diets and strategies particular to the size group to which they belong. Twenty-nine cat species are less than 20 kg in mass, and they eat small prey weighing less than 1 kg. The medium-sized cats such as leopards (40–60 kg) feed on larger prey, from 2 kg up to their own mass and larger. The two largest cats, tigers and lions (140–160 kg), are specialized predators of large hoofed mammals (ungulates). In Asian forests, tigers need wild swine (*Sus* spp.) and large deer (*Axis* spp., *Cervus* spp.), prey that are greater than or equal to one half the mass of tigers. Where wild cattle such as gaur (*Bos gaurus*) and banteng (*B. javanicus*) occur in the tiger's range, tigers select these megaherbivores (~1,000 kg) over smaller ungulates (Seidensticker and McDougal 1993; Karanth and Sunquist 1995).

Tigers cannot survive where they lack access to ungulate prey that is at least about half their own body mass because of mass-specific energy needs. For example, if only 20-kg muntjac deer (*Muntiacus muntjac*) are available, a tigress on a maintenance diet would need to kill one every two to three days. In contrast, a tigress feeding two large cubs would need to kill one muntjac every one to two days, or 183–365 muntjac a year (Sunquist et al. 1999). This scenario may approximate the situation in many tropical rain forests that support a low density of ungulate prey (Eisenberg and Seidensticker 1976) and thus support tiger densities of only 0.5 to 2 individuals/100 km^2 (Carbone et al. 2001). The importance of large ungulate prey is so pervasive that tiger distribution can be predicted accurately by mapping the preferred habitat of key prey species (Miquelle et al. 1999).

These calculations result in three conclusions. First, the top carnivores in Asian forests are at risk when their key prey items are depleted. Second, ungulate management is an integral component of tiger conservation. Third, depleted prey density caused by intensive legal and illegal harvests is a critical tiger conservation issue. When large ungulates decline through overhunting by humans, carrying capacity for breeding female tigers is reduced, cub survival rate decreases, and tiger population size declines rapidly (Karanth and Stith 1999). This is the norm today in much of the tiger's potential and actual range (Wikramanayake et al. 1999). This is one illustration of how ecosystems "erode," as Wilson described in the epigraph, and how this erosion of available energy has a bottom-up impact on ecosystem structure.

A principal result of the first scientific effort to understand tiger ecology (Schaller 1967: 326) was that "all evidence indicates that the tiger was the main factor limiting the growth of the populations of these species (chital *Axis axis*, gaur, sambar *Cervus unicolor*, barasingha *Cervus duvauceli*)." This is a top-down effect. Tigers are such splendid and powerful predators that our focus on this top-down process drives our notions and emotions about them and has obscured and delayed our recognizing the tiger's vulnerability to the depletion of key prey populations. Even the strong correlation between the tiger's large mass and its dependence on large ungulate prey has eluded consideration by many conservationists. The critical issue of key prey species depletion was absent on the tiger conservation agenda until 1997 (Karanth and Stith 1999; Miquelle et al. 1999). Then researchers recognized that within most of the remaining forest tracts through south and southeast Asia, the key prey populations necessary to sustain tiger populations were in decline or extirpated (Rabinowitz 1999; Karanth et al. 1999; Sunquist et al. 1999).

Top Carnivores Face Multiple Risks through Forest Fragmentation

The survival of tigers is further compromised by the source-sink conditions created at the hard and even soft edges of isolated reserves. Because of their high energy needs, the territories of wide-ranging, top carnivores are large, and the animals are therefore exposed to threats at reserve boundaries. Indeed, conflict with people on the borders of protected areas throughout the world is the major cause of mortality of wide-ranging, top carnivores, through hunting and poisoning, collisions with vehicles, and diseases from domestic animals. Unless the killing of carnivores that venture beyond park boundaries can be prevented through expanded no-kill zones, hunter education, and significantly enlarged protected areas (in most cases), these parks and reserves will not sustain intact wildlife assemblages over the long term (Woodroffe and Ginsberg 1998).

Unlike leopards, which can move at night across open and barren landscapes, tigers are relatively poor dispersers (Smith 1993; Sunquist et al. 1999). They dislike crossing open areas. Forest fragmentation creates barriers to dispersal that fragment tigers into many distinct population segments. The present estimate of about 160 distinct population segments (Wikramanayake et al. 1999) is a minimum estimate because the coarse forest-cover data layer used in these analyses underestimated the continual march of forest fragmentation.

Furthermore, the proliferation of roads through remaining forest tracts across Asia fragments habitats and places remaining tigers at risk. Roads provide poachers access, and tigers seem to prefer to travel along the edges of roads where they are flanked by forest. Because tigers are sight-oriented predators, open roadsides increase the area tigers can scan for prey. And open roadsides provide an environment where the tiger's key prey come to feed on the early successional vegetation that grows there (Seidensticker 1986; Kerley et al. in prep.). Where roads have been improved to enable higher vehicular speed, collisions are becoming more common (Bittu Sahgal, pers. comm., 1999). Improving road quality and decreasing transport times on national transportation networks is a high development priority through most of Asia and attracts major international donor backing. The barriers and sinks created by roads when they cut through reserves or through critical landscape-connecting corridors can in some circumstances be mitigated through features incorporated into roadway design, such as underpasses and fencing. Such features have been used in the transportation architecture in southern Florida to reduce mortality of the Florida panther (*Puma concolor coryi*) and reestablish landscape connectivity (Maehr 1997: 80). Creating incentives to close or limit access to logging roads into remote areas is a priority for top carnivore conservation efforts.

Top Carnivores Control Meso-Carnivores

The last track of a Javan tiger (*Panthera tigris sondaica*) was found in 1965 in the rain forest of Ujung Kulon National Park, a 600 km^2 peninsula on the west end of Java, Indonesia. Today, Ujung Kulon supports a leopard (*Panthera pardus*) population. When Hoogerwerf (1970) documented the lives of large mammals in Ujung Kulon in the 1930s, leopards were absent, but tigers occurred at low density, and they killed adult banteng. In the tropical moist evergreen forest of Khao Yai National Park, Thailand, tigers occur at low density, and there are no leopards (W. Brockelman, pers. comm., 1986). Tigers catch and kill leopards when they can. In case after case, a larger mammalian predator has been found to reduce or exclude populations of a smaller one (Palomares and Caro 1999). In the absence of large, dominant carnivores, smaller omnivores and carnivores can undergo population explosions, a phenomenon known as meso-predator release. This results in local extinctions of vulnerable prey populations (Soulé et al. 1988).

Resource limitations appear to drive interspecific competition (Weins 1993). If tigers eliminate leopards in habitats where appropriate prey are scarce, what happens where appropriate-sized prey is both diverse and abundant? In the tropical moist deciduous and dry forest mosaic in the Nagarhole National Park, India, the principal prey species of tigers, leopards, and dholes (*Cuon alpinus*) reached combined densities of sixty-six animals/km^2. This prey base comprised gaur, sambar (*Cervus unicolor*), wild swine (*Sus scrofa*), chital (*Axis axis*), muntjac (*Muntiacus muntjac*), and hanuman langur (*Presbytis entelleus*). With this abundant and diverse prey base, leopards, tigers, and dhole co-existed. The two solitary, stalking predators and the group-hunting coursing predator selectively killed different prey types in terms of species, size, and age-sex classes. This selective predation, which facilitated co-existence of the three predators, was mediated primarily by the adequate availability of prey in different size classes. This study by Karanth and Sunquist (1995, 2000) suggests that ecological factors such as adequate availability of appropriate-sized prey, dense cover, and high tree density are primary factors in structuring predator communities in these tropical forest types. In contrast, behavioral factors, such as differential habitat selection and interspecific social dominance, drive the structure of the large predator community in savanna habitats. In Nagarhole, however, both tigers and dholes were socially dominant over leopards, but there was no spatial exclusion of leopards as there is in rain forests with low abundance of key prey species.

In a survey of leopard food habits from areas in Asia and Africa, a pattern in the adaptability of leopard feeding emerged (Seidensticker 1991). Leopards switch prey easily. If one or more abundant ungulate species in the 20- to 50-kg size range are present in a habitat, the leopard focuses its hunting on those animals. Leopards mainly ignore primates in those habitats, even if primates are abundant, and leopards seldom pursue smaller prey animals. When ungulate prey in the 20- to 50-kg size class is present but not at high density, leopard do prey on primates, which may make up as much as 30 percent of their diet. In habitats where 20- to 50-kg ungulate prey is essentially absent, leopards switch and eat a high proportion of primates. Where prey of any kind occurs at very low density, leopards take just about any sized animal they can capture: small ungulates up to 50 kg, porcupines, foxes, jackals, and primates as encountered. In the absence of a larger predator such as the tiger, the large ungulates (>50 kg) escape predation by leopards, although their young do not.

From this review, it is apparent that the consequences of losing the top carnivore, followed by a meso-predator release, are mediated by the structure of the available prey assemblage. The structure of the available prey assemblage is mediated through vegetation structure, productivity of the habitat, and the extent to which humans have altered the habitat and affected the prey assemblage.

Top Carnivores Play a Pivotal Role in Maintaining Biodiversity in Asian Forests

Recent reviews (Soulé and Terborgh 1999) and research conducted in large natural systems (Clark et al. 1999) indicate the pivotal role of predation in structuring and preserving biodiversity in terrestrial communities. The top-down force that top carnivores exert on selected prey populations has been long appreciated. An understanding of the means by which top carnivores maintain the overall biodiversity in ecological systems is only recently emerging. This pivotal role has yet to be incorporated into the doctrine of conservation planning. Similarly, conservation planners have been slow to recognize and incorporate the ecological, behavioral, and security needs of top carnivores.

In our battle against extinction, we must come to understand how essential ecological processes such as predation work and how humans alter ecosystems in terms of the degree of change, the degree of sustained control, the spatial extent of change, and its abruptness (Angermeier 2000). The erosion of connectivity and habitat quality has been of concern to conservationists seeking to preserve biodiversity. Top carnivores are the first to disappear as a result of this degradation. We have not fully understood the impact of poachers' bullets and snares as they selectively removed top carnivores from relatively intact Asian forests. We now appreciate the importance of predation in structuring and maintaining biodiversity. Maintaining a reserve network that is large enough to sustain viable populations of the top carnivores has emerged as an overarching conservation objective (Soulé and Terborgh 1999). We must shift our conservation thinking from viewing top carnivores as an isolated part of ecosystem management to viewing their maintenance as an essential component.

Critically endangered populations of top carnivores are indicators of ecosystems in decline. As we set our conservation sights on recovery, top carnivores can be the stars in our ongoing efforts to restore and maintain biodiversity. But the star power of top carnivores, their flagship and umbrella role, is more than symbolic. Without top carnivores our efforts to stem the loss of biodiversity will ultimately fail.

CHAPTER 4

How Is Biodiversity Distributed?

The Indo-Pacific region is one of the most biologically rich on Earth. One reason for the concentration of such biological wealth in such a small terrestrial area is that the region is segmented by a series of dispersal barriers. Rugged mountain ranges that include the world's highest peaks, broad rivers, "islands" of wet forests surrounded by vast stretches of dry forests, and true islands separated by deep ocean trenches are important features that have isolated populations of plants and animals. As sea levels rose and fell over the millennia, land bridges were submerged and created, with the latter creating opportunities for major dispersal events. However, several islands, including the Philippines archipelago, New Caledonia, and some of the more recent islands of the Indonesian archipelago such as Nusa Tenggara and Maluku, remained isolated through these events.

Famous explorer and naturalist Alfred Russell Wallace first untangled the complex puzzle that characterizes the biogeography of the region. Wallace's Line runs along the deep water chasm at the edge of the Sunda Shelf (figure 2.4) and essentially demarcates the Indo-Malayan from the Australasian faunas. The tiger—the largest and most charismatic of the large vertebrate predators of the Indo-Malayan realm—roamed as far east as the small island of Bali, sitting at the edge of the Sunda Shelf (it became extinct from Bali recently). However, this powerful swimmer never reached Lombok, only a few kilometers away but across the deep strait that represents Wallace's Line. Similarly, apes and tree shrews stop to the west of this boundary, and marsupials and birds of paradise assume dominance in the east. This biogeographic line also serves as a boundary for plants. To the west of Wallace's Line, the island of Borneo supports nine genera and 287 species of trees in the Dipterocarpaceae family, which dominates the canopy tree community of southeast Asian rain forests. In contrast, just across the Makassar Strait and on the eastern side of Wallace's Line, the island of Sulawesi contains only two genera and four species of dipterocarps (Whitmore 1995).

Globally, Wallace's Line is perhaps the best-known biogeographic boundary. However, it is only one of several in the Indo-Pacific region. From the Palk Strait isolating Sri Lanka from the Indian subcontinent, to the Torres Straits between New Guinea and Australia, there are numerous such oceanic barriers that separate islands and continental landmasses. Over the eons, these barriers have constrained floral and faunal dispersal, thereby playing a critical role in determining the distribution of biodiversity in the region. For example, the botanical region known as Malesia extends from the Isthmus of Kra in the Malaysian Peninsula to the Torres Straits (Whitmore 1995). Within this region, about 40 percent of the plant genera and species are restricted to a single island or a cluster of a few islands (van Steenis 1987; van Welzen 1992). In south Asia, the small rain forest ecoregions of Sri Lanka have been separated from the continental landmass since the early Pleistocene. Many of the species have evolved in isolation, producing a high level of endemicity (Wikramanayake 1990) that includes an extraordinary radiation of more than 250 species of endemic frogs in the family Rhacophoridae (Pethiyagoda and Manamendra-Arachchi 1998). At the eastern extreme of the Indo-Pacific region is New Caledonia, a fragment of the ancient southern continent, Gondwanaland. The main island of New Caledonia has been isolated since it first split off more than 60 million years ago and now harbors an incredible level of endemism at higher taxonomic levels, including endemic plant families, bird genera, and a radiation of unique large geckos (Allison 1996; Bouchet et al. 1995; Mueller-Dombois 1998).

Oceanic straits are perhaps the most obvious biogeographic barriers. However, the steep, rugged mountain ranges of the Asian mainland have also limited dispersal, thus isolating species and assemblages and promoting endemism. The alpine and sub-alpine habitats of the high Himalayan mountain range harbor a number of endemic plants with limited ranges and speciation in the genera *Saxifraga* (Saxifragaceae), *Rhododendron* (Ericaceae), *Pedicularis* (Scrophulariaceae), *Meconopsis* (Papaveraceae), and *Primula* (Primulaceae) (Shrestha and Joshi 1996). The steep mountains also hold the greatest diversity of montane ungulates in the world, such as markhor (*Capra falconeri*), Himalayan tahr (*Hemitragus jemlahicus*), urial (*Ovis orientalis*), argali (*Ovis ammon*), bharal (*Pseudois nayaur*), takin (*Budorcus taxicolor*), serow (*Capricornis sumatraensis*), chiru (*Pantholops hodgsonii*), and other hoofed mammals that are preyed on by the near-mystical snow leopards (*Uncia uncia*), lynx (*Felis lynx*), and wolves (*Canis lupus*) (Schaller 1977). These species are specialized to live in high montane habitats and rarely venture into the lowlands, where an entirely different community of predators and their prey dwell. In the alluvial plains of the Ganges and Brahmaputra River valleys and further south into the Deccan Plateau of India, tigers, leopards, clouded leopards, and wild dogs hunt large ungulates such as wild water buffalo (*Bubalus arnee*), gaur (*Bos gaurus*), barasingha (*Cervus duvaucelii*), sambar (*Cervus unicolor*), nilgai (*Boselaphus tragocamelus*), blackbuck (*Antilope cervicapra*), chinkara (*Gazella bennettii*), chital (*Cervus axis*), hog deer (*Cervus porcinus*), muntjac (*Muntiacus muntjak*), and Nilgiri tahr (*Hemitragus hylocrius*). The larger islands of Indonesia and the Philippines harbor a number of endemic species that are limited to individual mountain tops (Heaney and Regalado 1998; Corbet and Hill 1992).

Some of these barriers have also helped to preserve the intactness of wildlife populations and habitats over time by limiting human access. For instance, over the past five years or so, biological explorations in the Annamite mountain range along the border between Laos and Vietnam have produced a wealth of new species in several taxonomic groups, including an impressive assemblage of new large mammals with restricted ranges. Essay 10, by Eric Wikramanayake, looks at these startling discoveries in the Annamites and discusses the historical patterns that may have contributed to the evolution of these specialized species within this rugged chain. In southern Vietnam, a population of Javan rhinoceros, one of Asia's largest mammals, was rediscovered only about seven years ago, adding to the only other population of about forty to fifty animals in the faraway Ujung Kulon peninsula of western Java, Indonesia.

In this chapter we begin quantifying the region's biodiversity at the ecoregion scale. We first present the results of the Biological Distinctiveness Index (BDI), which illustrates the biological contributions of each ecoregion in the Indo-Pacific and describe the broad distribution patterns of biodiversity in the Indo-Pacific. We also explore some interesting patterns of richness and endemism for mammals, birds, and plants, especially as they relate to the history of isolation, and discuss the implications for conservation planning and action. We illustrate how the most biologically valuable ecoregions that represent all biomes fit within a global strategy for conservation and compare our results with other approaches for setting conservation priorities.

Broad Patterns in Biological Distinctiveness

The map of biological distinctiveness of the Indo-Pacific ecoregions (figure 4.1) reveals several broad patterns. Among the ecoregions that are ranked as globally outstanding, several lie along the Himalayan chain, three occur in north-south swaths along the mountain ranges in the Indochina bioregion, and many are island ecoregions.

Interestingly, most globally outstanding ecoregions fall within broad ecotones between biogeographic realms. The range overlap of species from different realms and the high beta-diversity associated with the steep, dissected terrain of mountain

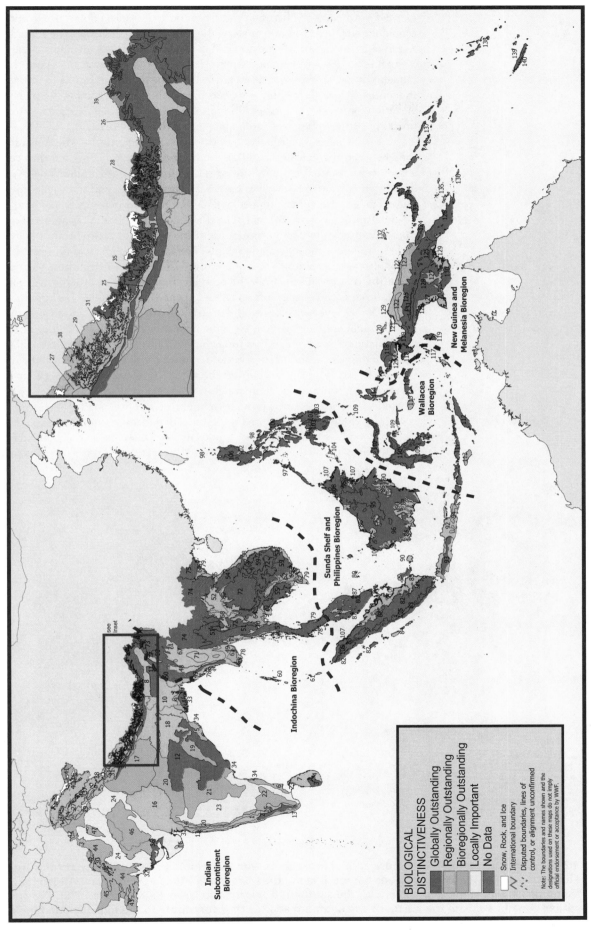

Figure 4.1. Biological distinctiveness status for terrestrial ecoregions of the Indo-Pacific. The Sunda Shelf and Philippines bioregion and Wallacea bioregion rank among the highest concentrations of globally outstanding ecoregions in the world.

ranges and archipelagos create species-rich zones. For instance, several mountain ranges form an ecotone between the Holarctic and Tropical floras: the Mizoram-Manipur and Chin Hills ranges along the India-Myanmar border, the Kayah-Karen range along the Myanmar-Thailand border, and the Annamites range along Vietnam's border with Laos and Cambodia. Here, the Holarctic flora consisting of Himalayan species dips southward and follows the mountain ranges to intrude into Indochina (figure 4.2). Thus, oaks, beech, and chestnuts dominate the higher-elevation forests of these mountain ranges, but the lower elevations have tropical floras consisting of figs, teak, and dipterocarps (Laubenfels 1975).

In the Himalayas, the globally outstanding ecoregions form the transition zone between the Palearctic and Oriental zoogeographic realms and thus contain a mix of species from both. Within this transition zone langur monkeys overlap broadly with red pandas, wolves co-exist with wild dogs, and snow leopards stalk prey in mountain habitats immediately above forests where tigers roam.

The third major area of faunal overlap occurs in the small island ecoregions in the Wallacean bioregion, situated within the ecotone of the Indo-Malayan and Australasian zoogeographic realms. The faunal assemblages of these islands are an interesting mix of Asian and Australasian elements. But here the biological distinctiveness results from the high levels of endemism where isolation of islands from each other and from the Sunda and Sahul shelves has resulted in extraordinary levels of speciation. For example, the island of Sulawesi has the highest number of ecoregional endemic mammals in the Indo-Pacific. The most charismatic of these probably is the babirusa (*Babyrousa babyrussa*), a wild pig with long upper canines that penetrate upward through the cheeks and curve over backward. The largest lizard in the world, the Komodo dragon (*Varanus komodoensis*), is also endemic to a small ecoregion in Wallacea. The essay by Lawrence Heaney (essay 3) provides a brief history of this archipelago's geologic isolation, which has contributed to patterns of endemism and richness.

Many of the smaller island ecoregions in other bioregions also qualified for globally outstanding status based on levels of endemism. These ecoregions support

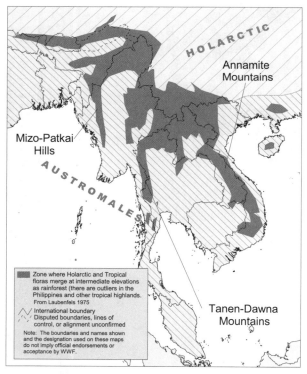 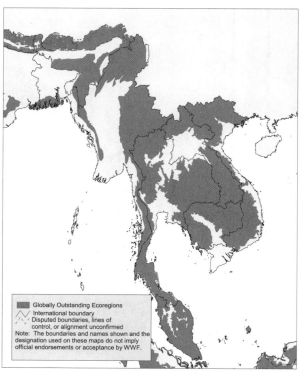

Figure 4.2. Comparison of globally outstanding ecoregions with overlapping zones of Holarctic and Tropical floras. The northern Holarctic flora extends far south along the Mizo-Patkai, Tanen-Dawna, and Annamite mountain ranges in Indochina to overlap with the southern tropical floras (from Launbenfels 1975). The species-rich ecotones contain a mix of the northern and southern species that contribute to the globally outstanding status of the ecoregions along these mountain ranges.

relatively depauperate biotas compared with those of some continental ecoregions, but a large proportion of the species they harbor are found nowhere else on Earth. An extreme example is New Caledonia, which is represented by two terrestrial ecoregions. This island is a Gondwana relict, unlike the other small Pacific islands that are of more recent volcanic origin, and its long history of isolation and ancient origin give it an incredibly high level of taxonomic endemism that includes five endemic plant families (Amborellaceae, Onocothecaceae, Paracryphiaceae, Phellinaceae, Strasburgiaceae), an endemic bird family (Rhynochetidae), and four endemic monotypic bird genera (Hannecart and Letocart 1980, 1983).

However, endemism is not confined to islands surrounded by deep-water trenches. Several small ecoregions along the continental mountain ranges that are surrounded by distinctly different habitat are "terrestrial islands." The terrain and the surrounding habitat have constrained dispersal so that the fauna and flora of these ecoregions have become isolated, leading to the evolution of new species with limited distributions and specialized ecological needs (Schaller and Vrba 1996; Giao et al. 1998). The South Western Ghats Montane Rain Forests [2] in the Indian subcontinent bioregion is a good example of such a terrestrial island. This tropical rain forest ecoregion lying along the western coast of the vast Deccan Peninsula is surrounded by semi-evergreen and dry forest. It harbors many species of endemic plants and animals, including two primates, the lion-tailed macaque (*Macaca silenus*), and the Nilgiri leaf monkey (*Trachypithecus johnii*) (Corbet and Hill 1992; WWF and IUCN 1995).

On average, island ecoregions are almost half as big as continental ecoregions and have depauperate floras and faunas. However, several of the larger islands, such as Sumatra, Borneo, and New Guinea, have ecoregions that are comparable in size to the continental ecoregions and are globally outstanding for species richness rather than endemism. This pattern is a result of the species-area effect, where larger land areas have higher species richness because of beta- and gamma-diversity (MacArthur and Wilson 1967). But as discussed by Heaney (essay 3), Pleistocene-era land bridges between these large islands, which lie within the Sunda (Sumatra, Java, Borneo) or Sahul (New Guinea) shelves, have also allowed species dispersal and exchanges that have contributed to species richness but have limited the evolution of higher-level endemic taxa.

In the continental bioregions, the large semi-deciduous or dry forest ecoregions support a relatively depauperate flora, but the fauna includes several large mammals and birds. In the Indian subcontinent bioregion, the globally outstanding ecoregions are in the northeastern Deccan plateau, whereas in the Indochina bioregion the large dry forest ecoregions covering most of central Thailand and the lowlands of Cambodia and Laos (Central Indochina Dry Forests [72] and Southeastern Indochina Dry Evergreen Forests [59]) have rich faunas. These ecoregions harbor many charismatic large vertebrates such as the tiger, Asian elephant, three rhinoceros species (all of Africa has only two species), several wild cattle species, and other ungulates.

A Closer Look at the Indo-Pacific's Natural Wealth

Overall, 60 (43 percent) of the 140 Asia-Pacific ecoregions are globally outstanding for biological distinctiveness (figure 4.1, figure 4.3, table 4.1).

Patterns of Richness and Endemism

Seven ecoregions are globally outstanding for both species richness and endemism (Sumatran Montane Rain Forests [83], Vogelkop-Aru Lowland Rain Forests [119], Eastern Himalayan Broadleaf Forests [26], Kinabalu Montane Alpine Meadows [108], Southeastern Papuan Rain Forests [126], Borneo Montane Rain Forests [95], Borneo Lowland Rain Forests [96]). Thirty-three (24 percent) ecoregions are ranked as globally outstanding for their species richness and thirty-seven (26 percent) for their endemism (figures 4.4 and 4.5). Fifty ecoregions are regionally outstanding for richness, compared with twenty-seven for endemism. However, at

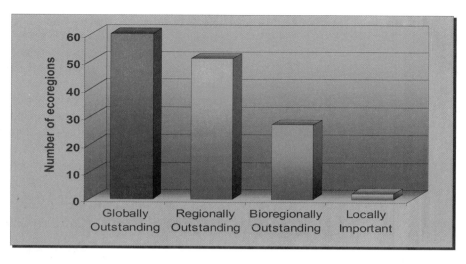

Figure 4.3. Distribution of ecoregions by biological distinctiveness categories.

TABLE 4.1. Number of Ecoregions That Are Globally Outstanding for Species Richness, Endemism, or Important Ecological and Evolutionary Phenomena by Bioregion and Biome.

	Richness	Endemism	Phenomena
Indian Subcontinent Bioregion			
Tropical and Subtropical Moist Broadleaf Forests	1	3	4 (1)
Temperate Broadleaf and Mixed Forests	1	1	0
Temperate Coniferous Forests	1	0	0
Tropical and Subtropical Grasslands, Savannas, and Shrublands	1	0	1
Montane Grasslands and Shrublands	1	0	0
Flooded Grasslands and Savannas	1	0	0
Mangroves	1	1	0
Indochina Bioregion			
Tropical and Subtropical Moist Broadleaf Forests	8	0	1 (1)
Tropical and Subtropical Dry Broadleaf Forests	2	0	0
Temperate Broadleaf and Mixed Forests	1	0	0
Mangroves	0	0	0
Sunda Shelf and Philippines Bioregion			
Tropical and Subtropical Moist Broadleaf Forests	6	10	0
Tropical and Subtropical Coniferous Forests	0	1	0
Montane Grasslands and Shrublands	1	1	1
Mangroves	1	1	0
Wallacea Bioregion			
Tropical and Subtropical Moist Broadleaf Forests	0	4	0
Tropical and Subtropical Dry Broadleaf Forests	0	2	0
New Guinea and Melanesia Bioregion			
Tropical and Subtropical Moist Broadleaf Forests	5	8	2
Tropical and Subtropical Dry Broadleaf Forests	0	1	1
Tropical and Subtropical Grasslands, Savannas, and Shrublands	0	1	0
Montane Grasslands and Shrublands	0	1	1
Mangroves	1	1	0

Only two ecoregions (in parentheses) were elevated to globally outstanding status because of outstanding phenomena alone; the rest were already globally outstanding for richness or endemism. Some ecoregions were also globally outstanding for both richness and endemism.

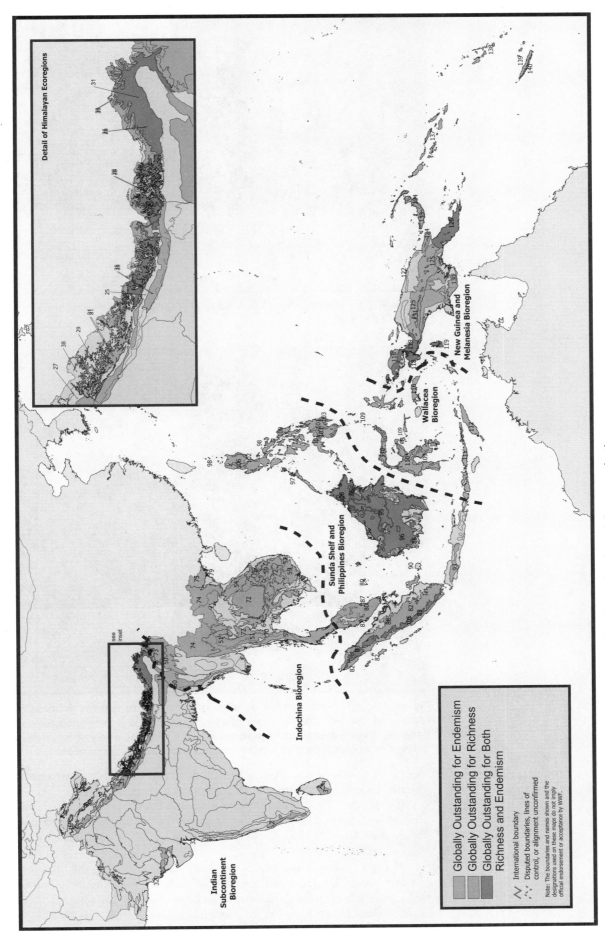

Figure 4.4. Ecoregions that are globally outstanding for endemism, richness, or both.

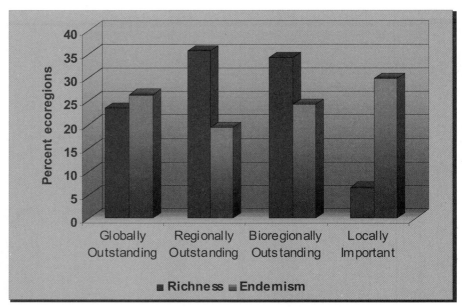

Figure 4.5. Distribution of ecoregions by richness index and endemism index categories.

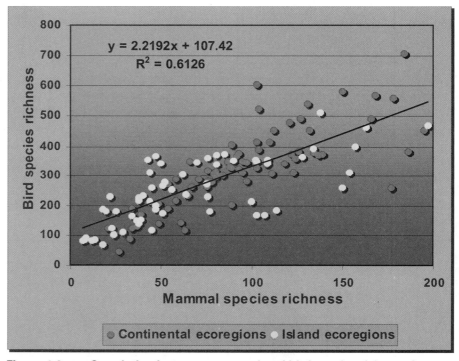

Figure 4.6. Correlation between mammal and bird species richness for ecoregions of the Indo-Pacific. The relationship indicates good concordance between the distributions of mammals and birds. The analysis excludes the mangrove ecoregions that were not scored for mammal richness.

the other end of the scale, almost five times as many ecoregions were classified as locally important by the endemism index as by the richness index.

There is good concordance between mammal and bird species richness (figure 4.6). Several ecoregions, especially in the Indochina and the Sunda Shelf and Philippines bioregions, exhibit high species richness values for both taxa (figures 4.7 and 4.8).

In Indochina, the montane moist forest ecoregions along the Mizoram-Manipur and Chin Hills, the Tenasserims, and the Annamite mountain ranges are especially rich in both mammals and birds. The vast dry forest ecoregion in central Indochina and the large subtropical ecoregion that extends along northern Myanmar, Thailand, Laos, and Vietnam is also high in species richness for

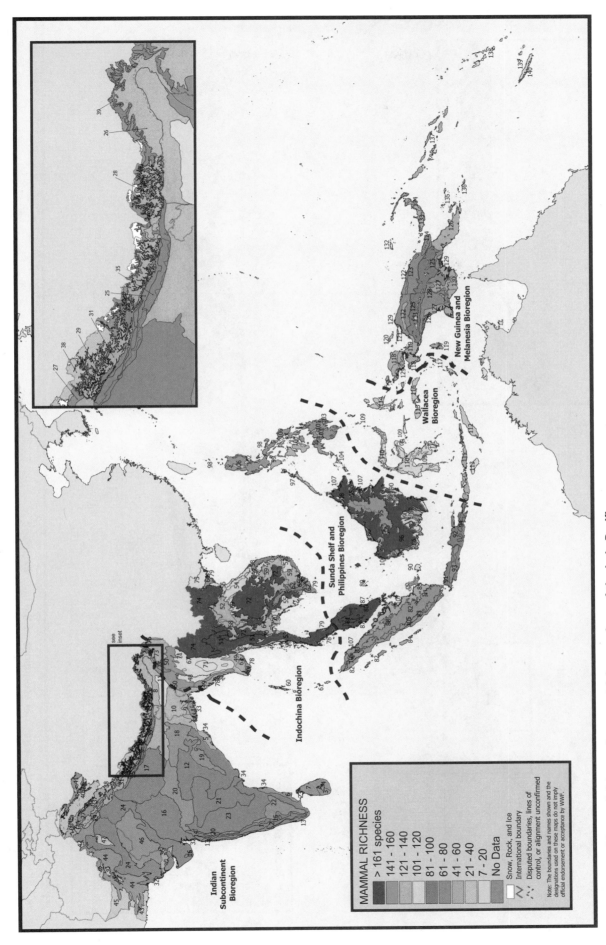

Figure 4.7. Mammal richness for terrestrial ecoregions of the Indo-Pacific.

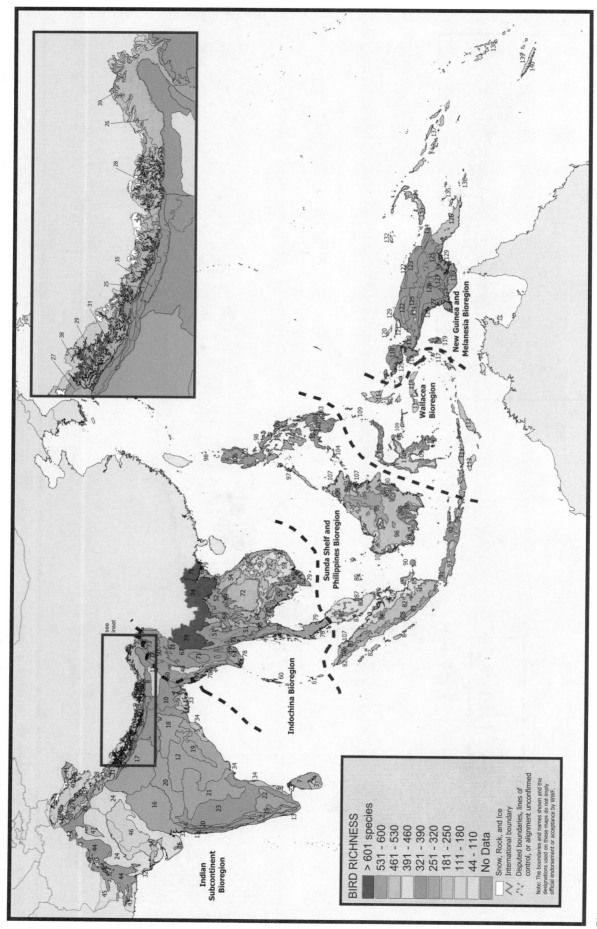

Figure 4.8. Bird richness for terrestrial ecoregions of the Indo-Pacific.

these two taxonomic groups. The broad trend in high species richness extends westward from the Naga Hills along the Indo-Myanmar border into the montane forests of Meghalaya and Assam, into the broadleaf forests of the Himalayan range where the Palearctic and Indo-Malayan faunas overlap. But the dry forest ecoregions of the vast Deccan Plateau are not exceptionally high in mammal or bird species richness.

The ecoregions of the large islands in the Sunda Shelf and the Philippines bioregion also stand out for their high species richness among both mammals and birds. These moist forests are arguably some of the most species-rich ecosystems in the Indo-Pacific region. Mammal richness is high in the ecoregions of the large island of Sulawesi, but bird richness is low. Further east, in the New Guinea and Melanesia bioregion, the ecoregions of the large island of New Guinea are less rich than comparable islands such as Borneo or Sumatra. However, New Guinea is also less explored scientifically, and its true biodiversity value is undoubtedly underestimated.

Continents versus Islands

Overall, thirty-six island ecoregions and twenty-one continental ecoregions are ranked as globally outstanding. Fifty-four percent of the sixty-three island ecoregions were globally outstanding by the endemism index, compared with only 4 percent of the seventy-seven continental ecoregions (figure 4.9; see appendix E for a complete list). And more continental ecoregions than island ecoregions were globally outstanding by the richness index. Several of the island ecoregions that were globally outstanding for richness are from the large islands: Sumatra, Borneo, and New Guinea. Almost half of the continental ecoregions ranked as locally important by the endemism index. The general pattern of high endemism and low species richness in the smaller island ecoregions and high species richness and low endemism in the continental ecoregions justifies our use of endemism and species richness independently to measure biological distinctiveness in the Indo-Pacific.

Many of the island ecoregions have lower species richness than the continental ecoregions. The relationships between species richness and area for both island and continental ecoregions indicate that island ecoregions are less species rich than similar-sized continental ecoregions (figure 4.10). Thus, the species-area effect (sensu MacArthur and Wilson 1967) alone cannot explain the low species richness in island ecoregions. Instead, the oceanic straits probably have acted as dispersal barriers, resulting in depauperate faunas in these islands.

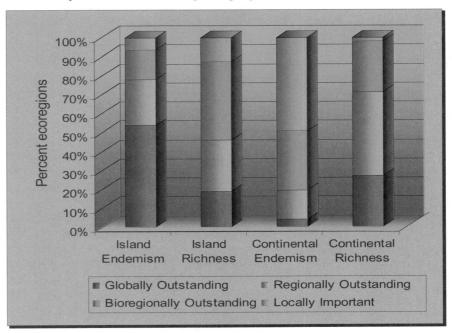

Figure 4.9. Distribution of biological distinctiveness for species richness and endemism among island and continental ecoregions.

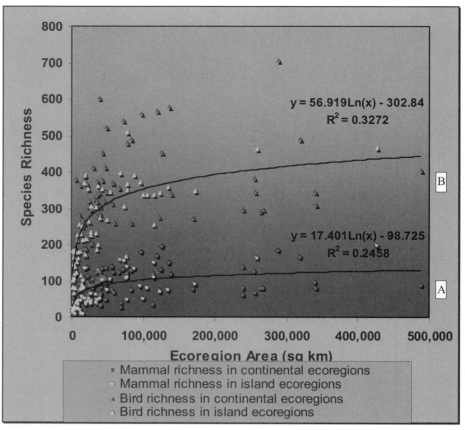

Figure 4.10. Mammal (A) and bird (B) species richness by size of ecoregion. The data shown as island and continental ecoregions indicate that ecoregion size accounts for little variance in species richness beyond approximately 50,000 km². Moreover, similar-sized continental ecoregions tend to be more species-rich than island ecoregions.

Furthermore, the species-area effect in both mammals and birds is evident until about 50,000 km², after which the addition of new species tapers off (figure 4.10). This relationship suggests that even the larger ecoregions contain distinct species assemblages and communities, confirming that our ecoregions represent distinct ecological units.

Concordance between mammal and bird endemism is also good (figure 4.11), although it is less strong for species richness. Endemism values for most of the continental ecoregions are tightly clustered in the low values, whereas the island ecoregions had much higher numbers of endemic species for both taxonomic groups. The maps of endemism for both mammals and birds show several broad trends (figures 4.12 and 4.13). For instance, the Philippines and Sulawesi stand out as high endemism areas for both mammals and birds, as does the central cordillera of New Guinea.

However, richness values for both the Philippines and Sulawesi are low compared with those of the ecoregions of the adjacent islands. In the Philippines archipelago, the islands remained isolated from the Sunda Shelf and from each other during the Pleistocene Ice Age (see essay 3). Thus, these islands did not benefit from species migrations and exchanges across the land bridges that connected the larger islands on the Sunda Shelf.

Lying off the eastern edge of the Sunda Shelf, Sulawesi was also isolated from the numerous Pleistocene-era land bridges. Thus, its fauna is markedly different from that of its close neighbors, Borneo and Irian Jaya (Whitten, Mustafa, and Henderson 1987). These species, which include a large number of rats and bats, have evolved in isolation, giving Sulawesi a high level of endemism. Surprisingly, a critical study of plant biogeography seems to point toward a connection between Java, the Lesser Sundas, and Sulawesi during the not-too-distant past (Whitten, Mustafa, and Henderson et al. 1987).

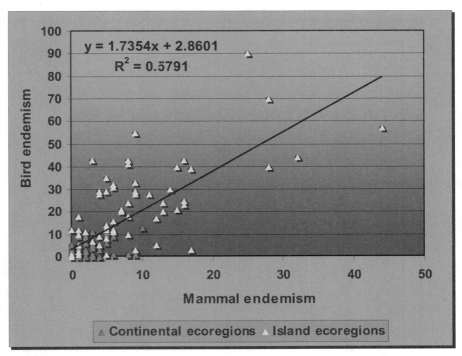

Figure 4.11. Correlation between mammal and bird endemism for ecoregions of the Indo-Pacific. The relationship indicates good concordance between the distributions of mammal and bird endemism. The analysis excludes the mangrove ecoregions that were not scored for mammal endemism.

Further east in the Wallacean bioregion, the islands in the Lesser Sundas and Maluku are younger and have low levels of species richness and of endemism at the species level but not at higher taxonomic levels (Monk et al. 1997). The modern mammal fauna of the islands of Nusa Tenggara and Maluku include a host of species that have been introduced by humans, to the extent that nine of the twenty-six families represented in the island probably are introduced (Monk et al. 1997).

The third area of high endemism is the largest island in the world, New Guinea. Here, the central cordillera that runs along the length of the island ranks highly for mammal and bird endemism. The more vagile birds show lower levels of endemism than do the mammals. But many of the species records for this rugged and little-explored mountain range come from limited collections. Experience from exploration in other mountain ranges, such as the Annamites, suggests that most of the species along this mountain range may still be undescribed and that the real levels of endemism probably are much higher.

Relative to the distribution data for mammals and birds, the data for plant richness and endemism for the region are estimates by regional experts. Therefore, rather than use actual numbers of species richness, we assigned richness and endemism values to broad categories that account for the large variance expected (appendix C).

The results show several broad but interesting patterns (figures 4.14 and 4.15). In general, the large, tropical lowland moist forest ecoregions in Borneo, Sumatra, the Malayan Peninsula, and New Guinea and the alpine meadows along the eastern Himalayan range are the most species-rich (figure 4.14). The montane moist forest ecoregions along the north-south mountain ranges in the Indian subcontinent and Indochina bioregions and the large moist deciduous forest ecoregions along the plains of the Ganges and Brahmaputra rivers are also high in plant richness.

Although there is good concordance between plant richness and mammal and bird richness in the archipelago bioregions, the relationship is less strong in the continental bioregions (compare figures 4.7, 4.8, and 4.14). The large dry forest and subtropical broadleaf forest ecoregions of Indochina are relatively high in mammal and bird richness but not in plant richness. In the Indian sub-

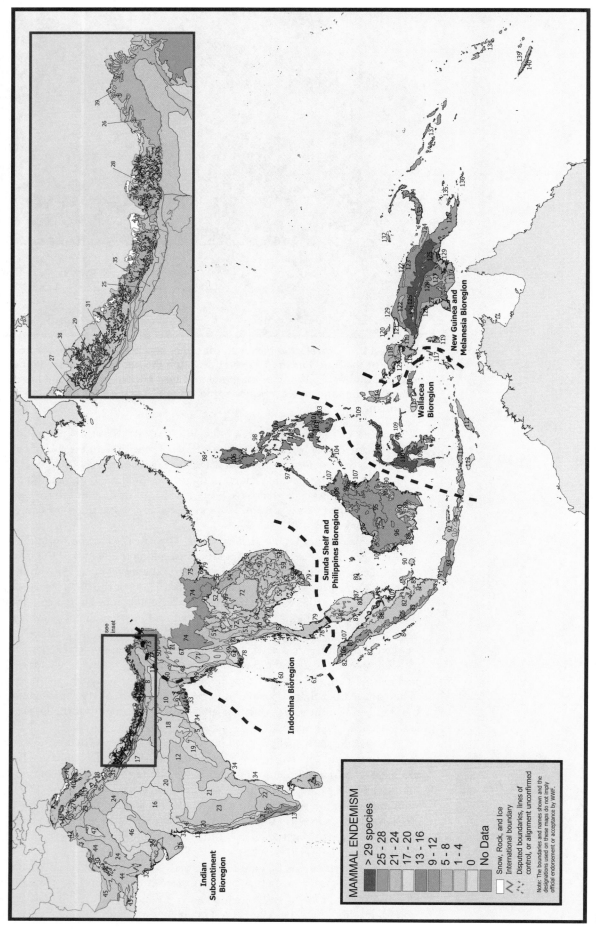

Figure 4.12. Mammal endemism for terrestrial ecoregions of the Indo-Pacific.

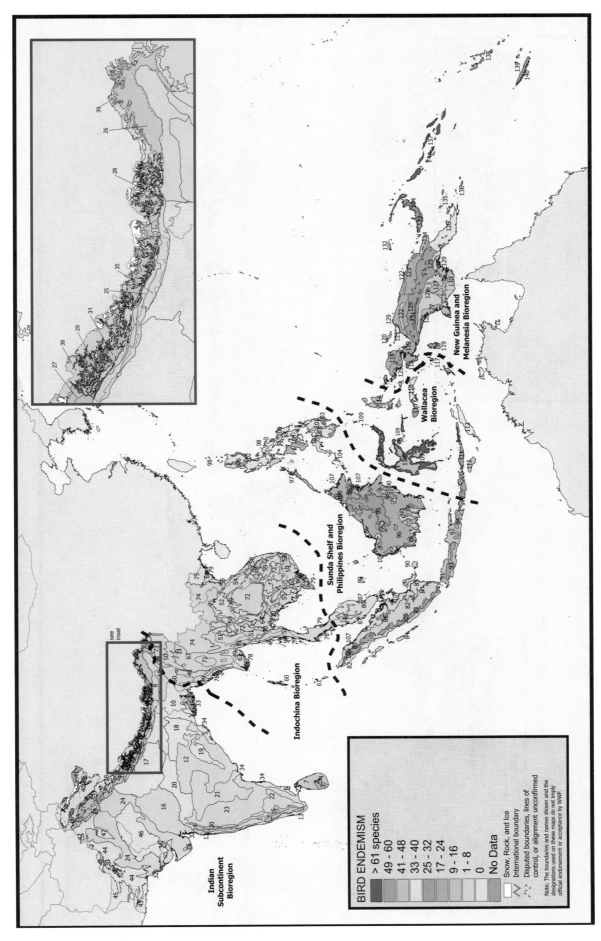

Figure 4.13. Bird endemism for terrestrial ecoregions of the Indo-Pacific.

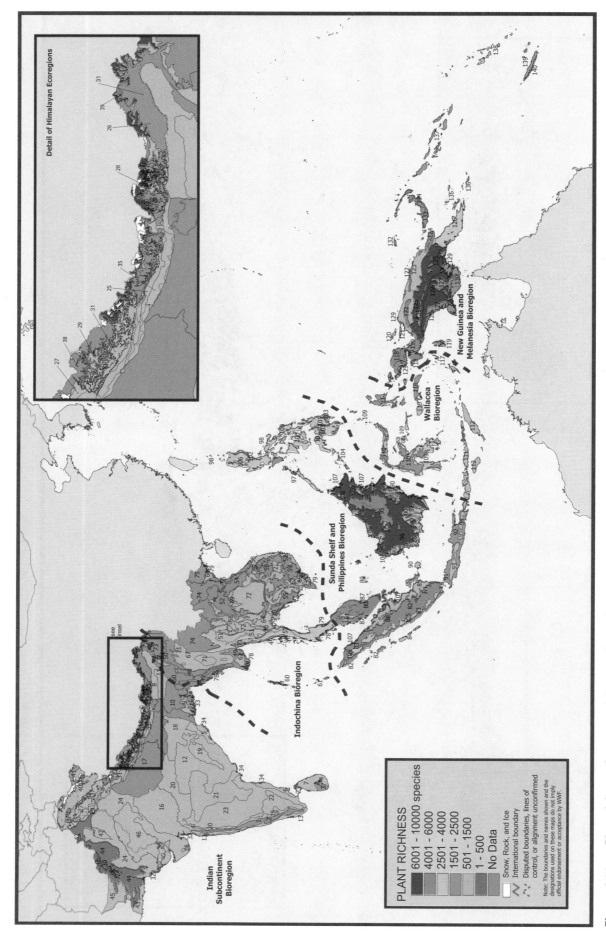

Figure 4.14. Plant richness for terrestrial ecoregions of the Indo-Pacific.

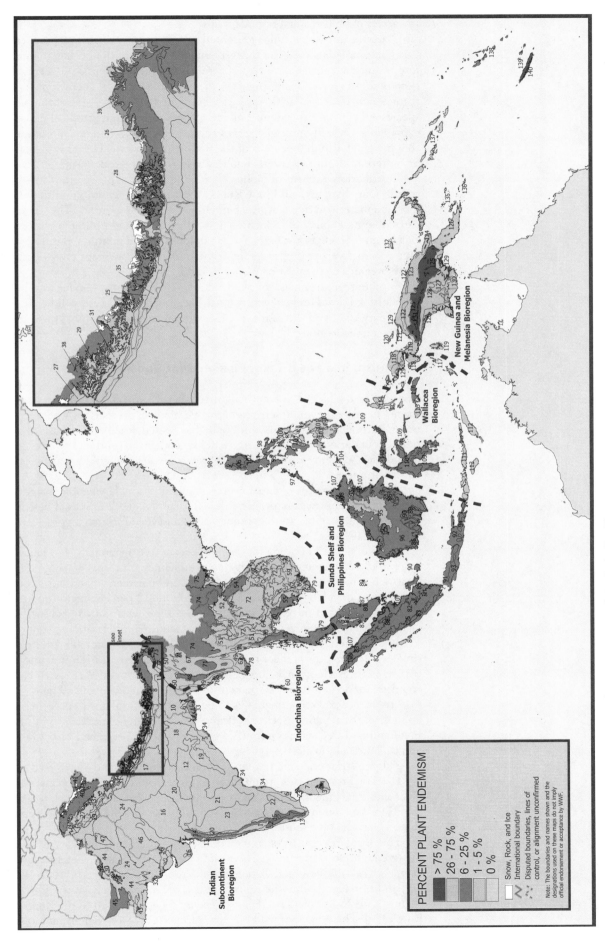

Figure 4.15. Plant endemism for terrestrial ecoregions of the Indo-Pacific.

continent bioregion, however, most of the dry forest ecoregions are depauperate in species of all three taxonomic groups.

Plant species endemism is presented as a percentage of richness (appendix C). The ecoregions with the highest endemism were in the Sri Lanka montane forests and the central cordillera of New Guinea, where over 75 percent of the flora is considered endemic (figure 4.15). There are several ecoregions scattered throughout the Indo-Pacific region where more than 25 percent of the plant species are considered endemic; all are lowland or montane tropical moist forest ecoregions. Overall, most of the moist forest ecoregions had endemism levels that are greater than 6 percent of the flora. Another broad trend was that the dry and semi-deciduous forests had less than 6 percent endemism.

Distribution patterns of higher taxonomic groups of plants show interesting trends. Gary Krupnick and June Rubis examine the distribution on the region's most characteristic tree families, Dipterocarpaceae, in essay 11. The dipterocarps dominate much of Indochina and Sunda Shelf and Philippines bioregions. But with few exceptions, (e.g., Sri Lanka lowland and montane rain forests), dipterocarp diversity drops rapidly in the Indian Subcontinent bioregion and in the bioregions to the west of Wallace's Line.

A more precise analysis of plant richness by ecoregion is under way, based on published floras and checklists for the region. This analysis will be completed by January 2002 and the results published soon thereafter online at www.worldwildlife.org/ecoregions.

Endemism in a Focal Taxon: Insights for Conservation

Conservation biologists are particularly concerned with protecting endemic species because they represent irreplaceable biodiversity features wherever they occur. Although endemism traditionally has been defined on the basis of national boundaries, in this analysis we define endemism on the basis of ecoregion boundaries, which renders endemic species more meaningful for setting conservation priorities.

The broad patterns of endemism in mammals provide valuable insights in determining conservation priorities. Ideally, similar analyses should be conducted for all taxa, but in the absence of data needed for such analyses, we can use mammals as proxies.

According to our criteria for near endemism (see appendix C for details), more than 460 mammal species qualify as endemics. Of these, 46 percent are rodents, 26 percent bats, 9 percent marsupials, and 6 percent insectivores. Primates (5 percent), ungulates (4 percent), and carnivores (2 percent) make up just about 10 percent of the endemic mammals, with other midsized taxa such as the lagomorphs, monotremes, and scandentia accounting for about 1 percent each.

The strict endemic mammals, defined as the species that are restricted to single ecoregions, also reflect the taxonomic breakdown of near-endemic species (figure 4.16). Rodents make up 52 percent of the strict endemics, with bats accounting for an additional 20 percent of the species. Marsupials (9 percent), insectivores (7 percent), ungulates (5 percent), and primates (3 percent) make up most of the remaining strict endemic species. Many of the marsupials are small and midsized species from the ecoregions of Wallacea and New Guinea and Melanesia bioregions.

The ten strict endemic primates include two tarsiers and a loris (see appendix H). The other species are midsized to large mammals: two gibbons, two macaques, two langurs, and the proboscis monkey (*Nasalis concolor*), which dwells in the mangroves of Borneo. The four strict endemic carnivores comprise three viverrids and a mustelid.

The ungulates consist of sixteen species, of which eight are cervids (including five muntjacs), five bovids, two wild pigs, and one equus. The two larger deer, *Axis calamianensis* and *Cervus alfredi*, are limited to Mindoro Rain Forests [99] and Mindanao Montane Rain Forests [103], respectively, in the Philippine islands. The Asiatic wild ass, *Equus hemionus*, is now limited to the Rann of Kutch [36]. The species of wild pig, *Sus salvanius* and *Sus cebifrons*, are found in the Terai-Duar Savanna and Grasslands [35] and Mindanao Montane Rain

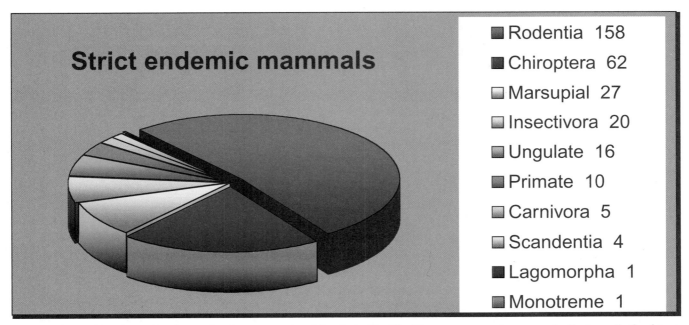

Figure 4.16. Number of strict endemic mammals of the Indo-Pacific. These represent mammals that are limited to a single ecoregion. The species are grouped into broad categories that represent higher taxonomic orders, and the number of species in each category is also shown.

Forests [103], respectively. Thus, overall, the vast majority of endemic mammals are small or midsized mammals.

Unlike large mammals, and especially the large carnivores that need large spaces, these smaller endemic mammals can be conserved even in small fragments of natural habitats. The ecoregions that contain the highest numbers of strict endemic species are lowland forest island ecoregions that have little remaining natural habitat (figure 4.17). Most of the montane island ecoregions that have high numbers of strict endemic mammals have also lost more than 50 percent of the habitat. It is thus important that the remaining habitat in these ecoregions,

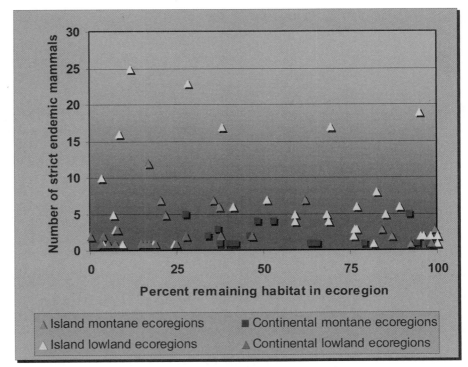

Figure 4.17. The number of strict endemic mammals in island and continental ecoregions. Many of the lowland forest island ecoregions, which have lost a considerable amount of natural habitat, have a high number of strict endemic species. It is thus critical that the remaining habitat in these ecoregions be protected.

notably the Sulu Archipelago Rain Forests [104], Mindoro Rain Forests [99], New Caledonia Dry Forests [140], Sumba Deciduous Forests [113], Timor and Wetar Deciduous Forests [112], Western Java Montane Rain Forests [93], Mindanao Montane Rain Forests [103], and Luzon Montane Rain Forests [102], be conserved because of the large number of strict endemic mammals they harbor.

Several island ecoregions, such as the Mentawai Islands Rain Forests [84], Southern New Guinea Lowland Rain Forests [128], Banda Sea Islands Moist Deciduous Forests [117], Vogelkop-Aru Lowland Rain Forests [119], and Sumatra Montane Rain Forests [83], harbor several endemic mammals and still have considerable intact habitat. However, many of these are threatened by logging and other forms of land clearing, as we shall see in chapter 5. Thus conservation actions in these ecoregions should focus on protecting suitable habitat areas to conserve endemic species.

The continental ecoregions have few strict endemic mammals. Continental montane ecoregions, such as the Northern Triangle Subtropical Forests [73], Northern Annamites Rain Forests [54], and Northern Indochina Subtropical Forests [74], have four to five endemic small mammal species that are limited to specialized habitats in these mountains. Protected areas and other conservation actions should target these centers of endemism within the ecoregions.

Expressing the strict endemic bird and mammal species as a percentage of the total richness for that taxa shows that many more of the fauna of island ecoregions are endemic than those in continental ecoregions (figures 4.18 and 4.19). This is even greater in the montane ecoregions of archipelagic bioregions (i.e., Sunda Shelf and Philippines, Wallacea, and New Guinea and Melanesia bioregions). Not surprisingly, island ecoregions, such as the Solomon Islands Rain Forests [137] and the Timor and Wetar Deciduous Forests [112], also display high percentages of strict endemics.

The Addition of Phenomena and Rare Habitats

Nine ecoregions qualified for globally outstanding status because of the presence of ecological or evolutionary phenomena, and two qualified for rare habitat type. However, seven of the former were already ranked as globally outstanding based on endemism or richness values, so only two (Eastern Highlands Moist Deciduous Forests [12] and Cardamom Mountains Rain Forests [55]) were actually elevated to globally outstanding status (table 4.2 and figure 4.20); these ecoregions harbor intact assemblages of large vertebrates. Two ecoregions that represent tropical alpine meadows were considered to represent rare habitat types (table 4.2 and figure 4.20).

Distribution of Globally Outstanding Ecoregions by Biome

Forty-two of the sixty ecoregions that were ranked as globally outstanding are in the tropical and subtropical moist broadleaf forest biome. Of these, twenty-eight occur in the archipelago bioregions (i.e, Sunda Shelf and the Philippines, Wallacea, and New Guinea and Melanesia). Two others (Sri Lanka Lowland Rain Forests [3] and Sri Lanka Montane Rain Forests [4]) in the Indian Subcontinent bioregions are also island ecoregions.

Many of the tropical moist forest ecoregions ranked as globally outstanding in the continental bioregions are along mountain ranges. These include the ecoregions along the Annamite range (Northern Annamites Rain Forests [54] and Southern Annamites Montane Rain Forests [57]), along the border of Vietnam with Laos and Cambodia; the Tenasserim–Naga Hills mountain ranges (Mizoram-Manipur-Kachin Rain Forests [50], Kayah-Karen Montane Rain Forests [51], and Tenasserim–South Thailand Semi-Evergreen Rain Forests [53]), which extend from the peninsula of southern Myanmar and Thailand up to the Myanmar-India border; the Meghalaya Hills in northeastern India (Meghalaya Subtropical Forests [11]); and the northern areas of Myanmar, Thailand, Laos, and Vietnam (Northern Triangle Subtropical Forests [73],

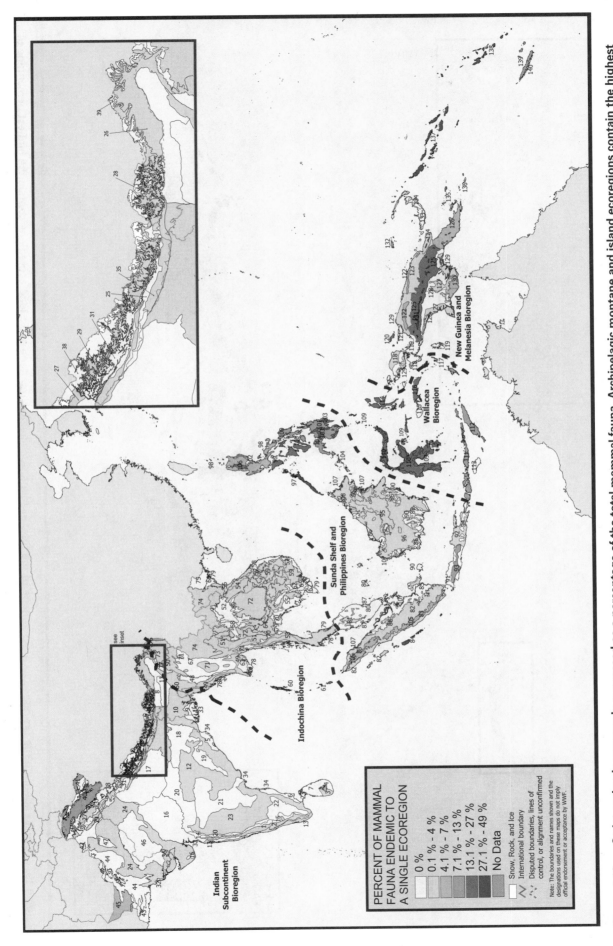

Figure 4.18. Strict endemic mammals expressed as a percentage of the total mammal fauna. Archipelagic montane and island ecoregions contain the highest percentage of their mammal fauna limited to one ecoregion.

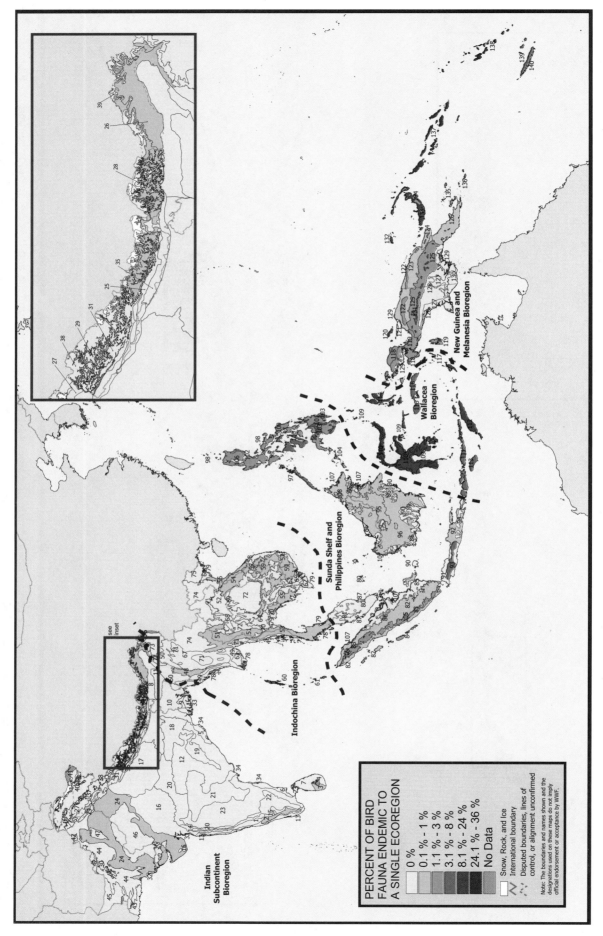

Figure 4.19. Strict endemic birds expressed as a percentage of the total avifauna. Archipelagic montane and island ecoregions contain the highest percentage of their avifauna limited to one ecoregion.

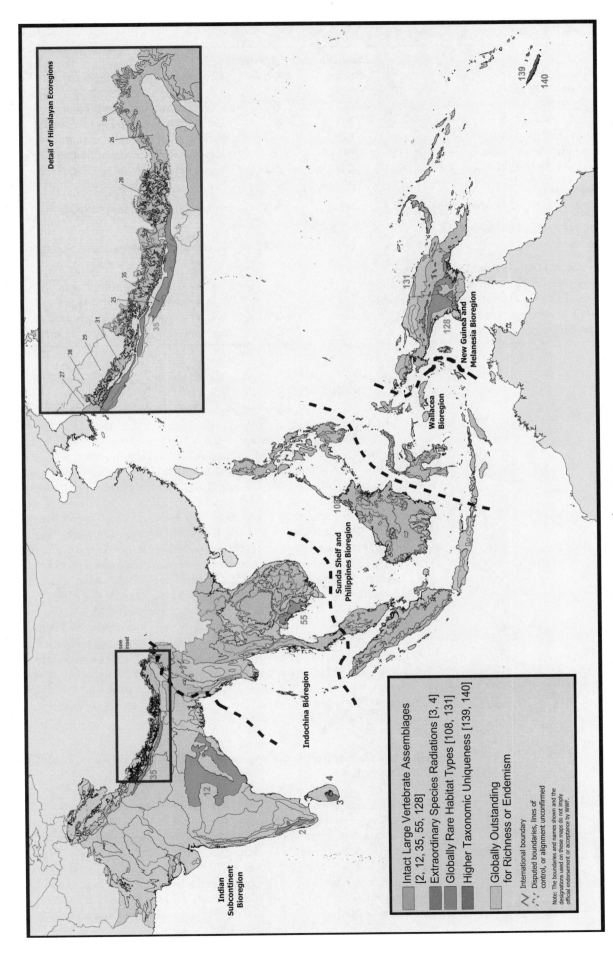

Figure 4.20. Globally outstanding ecoregions for biologically important phenomena.

TABLE 4.2. Richness and Endemism Status of Ecoregions That Were Elevated to Globally Outstanding Status for Extraordinary Phenomena and Rare Habitat Types.

	Richness Status	Endemism Status	Phenomenon
Indian Subcontinent Bioregion			
South Western Ghats Montane Rain Forests [2]	BO	GO	Intact vertebrate assemblages
Sri Lanka Lowland Rain Forests [3]	BO	GO	Extraordinary adaptive radiations
Sri Lanka Montane Rain Forests [4]	BO	GO	Extraordinary adaptive radiations
Eastern Highlands Moist Deciduous Forests [12]	BO	LI	Intact vertebrate assemblages
Terai-Duar Savanna and Grasslands [35]	GO	BO	Intact vertebrate assemblages
Indochina Bioregion			
Cardamom Mountains Rain Forests [55]	RO	BO	Intact vertebrate assemblages
Sunda Shelf and Philippines Bioregion			
Kinabalu Montane Alpine Meadows [108]	GO	GO	Rare habitat type (tropical alpine areas)
New Guinea and Melanesia Bioregion			
New Caledonia Rain Forests [139]	LI	GO	Higher taxonomic uniqueness
Southern New Guinea Lowland Rain Forests [128]	RO	GO	Intact vertebrate assemblages
New Caledonia Dry Forests [140]	BO	GO	Higher taxonomic uniqueness
Central Range Sub-Alpine Grasslands [131]	BO	GO	Rare habitat type (tropical alpine areas)

Northern Indochina Subtropical Forests [74], and South China–Vietnam Subtropical Evergreen Forests [75]).

Five (31 percent) of the ecoregions in the tropical and subtropical dry broadleaf forest biome are globally outstanding. Of these, two are in the continental bioregions (Southeastern Indochina Dry Evergreen Forests [59] and Central Indochina Dry Forests [72]) and three in the archipelago bioregions (Lesser Sundas Deciduous Forests [111], Timor and Wetar Deciduous Forests [112], and New Caledonia Dry Forests [140]). Globally outstanding ecoregions are also represented in the tropical and subtropical coniferous forest (Luzon Tropical Pine Forests [106]); temperate broadleaf and mixed forests (Eastern Himalayan Broadleaf Forests [26] and Northern Triangle Temperate Forests [76]); temperate coniferous forest (Eastern Himalaya Sub-Alpine Conifer Forests [28]); tropical and subtropical grasslands, savannas, and shrublands (Terai-Duar Savanna and Grasslands [35] and Trans Fly Savanna and Grasslands [130]); montane grasslands and shrublands (Eastern Himalayan Alpine Shrub and Meadows [39], Kinabalu Montane Alpine Meadows [108], and Central Range Sub-Alpine Grasslands [131]); flooded grasslands (Rann of Kutch Seasonal Salt Marsh [36]); and mangroves (Sundarbans Mangroves [33], Sunda Shelf Mangroves [107], and New Guinea Mangroves [129]) biomes (table 4.3). Only one biome, the deserts and xeric shrublands, lacked globally outstanding ecoregions. The mangrove biome ecoregions were not scored for biological distinctiveness but were considered either globally outstanding or regionally outstanding for their ecological processes.

Globally Outstanding Ecoregions: The Building Blocks of the Global 200 Ecoregions

Regional-scale assessments are important for the information they supply to regional decision-makers, but they also form the building blocks of a global conservation strategy. Several other regional-scale assessments across the globe use the same ecoregion delineation scheme and ranking criteria as this study (Dinerstein et al. 1995; Ricketts et al. 1999; Burgess et al. in press). These can be joined to create a map of the earth's globally outstanding ecoregions representing examples of all biomes (Olson et al. 2001). The Global 200 ecoregions analysis is one attempt to amalgamate the regional studies and identify the high-priority ecoregions for a global-scale conservation strategy (Olson and

TABLE 4.3. Breakdown of the BDI by Bioregion and Biome (Corrected for Ecoregions Elevated for Ecological Phenomena and Rare Habitat Types).

	Globally Outstanding	Regionally Outstanding	Bioregionally Outstanding	Locally Important
Indian Subcontinent Bioregion	11	18	16	2
Tropical and Subtropical Moist Broadleaf Forests	5	6	4	0
Tropical and Subtropical Dry Broadleaf Forests	0	3	5	0
Tropical and Subtropical Coniferous Forests	0	1	0	0
Temperate Broadleaf and Mixed Forests	1	1	0	0
Temperate Coniferous Forests	1	1	1	0
Tropical and Subtropical Grasslands, Savannas, and Shrublands	1	0	0	0
Flooded Grasslands and Savannas	1	0	0	0
Montane Grasslands and Shrublands	1	4	1	0
Deserts and Xeric Shrublands	0	0	5	2
Mangroves	1	2	0	0
Indochina Bioregion	14	13	5	0
Tropical and Subtropical Moist Broadleaf Forests	9	10	5	0
Tropical and Subtropical Dry Broadleaf Forests	2	2	0	0
Tropical and Subtropical Coniferous Forests	0	1	0	0
Temperate Broadleaf and Mixed Forests	1	0	0	0
Mangroves	2	0	0	0
Sunda Shelf and Philippines Bioregion	16	11	2	0
Tropical and Subtropical Moist Broadleaf Forests	13	10	2	0
Tropical and Subtropical Coniferous Forests	1	1	0	0
Montane Grasslands and Shrublands	1	0	0	0
Mangroves	1	0	0	0
Wallacea Bioregion	6	1	2	0
Tropical and Subtropical Moist Broadleaf Forests	4	1	1	0
Tropical and Subtropical Dry Broadleaf Forests	2	0	1	0
New Guinea and Melanesia Bioregion	15	6	2	0
Tropical and Subtropical Moist Broadleaf Forests	11	6	2	0
Tropical and Subtropical Dry Broadleaf Forests	1	0	0	0
Tropical and Subtropical Grasslands, Savannas, and Shrublands	1	0	0	0
Montane Grasslands and Shrublands	1	0	0	0
Mangroves	1	0	0	0

Dinerstein 1998; figure 4.21). The Global 200 is a priority portfolio of 238 terrestrial, freshwater, and marine ecoregions that represent the Earth's diverse ecosystems and species assemblages.

The Indo-Pacific region contains 37 (26 percent) of the world's 142 terrestrial Global 200 ecoregions. This incredible concentration of Global 200 terrestrial ecoregions in seven biomes is packed in only 6 percent of the earth's surface. Eleven of these Global 200 ecoregions are in the Indian subcontinent bioregion, seven in Indochina, eight each in the Sunda Shelf and Philippines and New Guinea and Melanesia bioregions, and three in Wallacea.

Ecoregions, Hotspots, and EBAs: Concordance and Overlap

Ecoregions provide a more fine-scale yet robust conservation unit for regional-scale conservation planning than many previous efforts. Among the more recent analyses of global biodiversity, Myers et al. (2000) identified twenty-five biodiversity hotspots that represent "areas featuring exceptional concentrations of endemic species and experiencing exceptional loss of habitat." But several of these hotspots are very broadly delineated and cover a number of different biomes under our classification scheme. In fact, some of Asia's hotspots are comparable to our bioregions in scale. For example, the huge Indo-Burma hotspot includes montane grasslands and shrublands, temperate coniferous forests, temperate broadleaf and

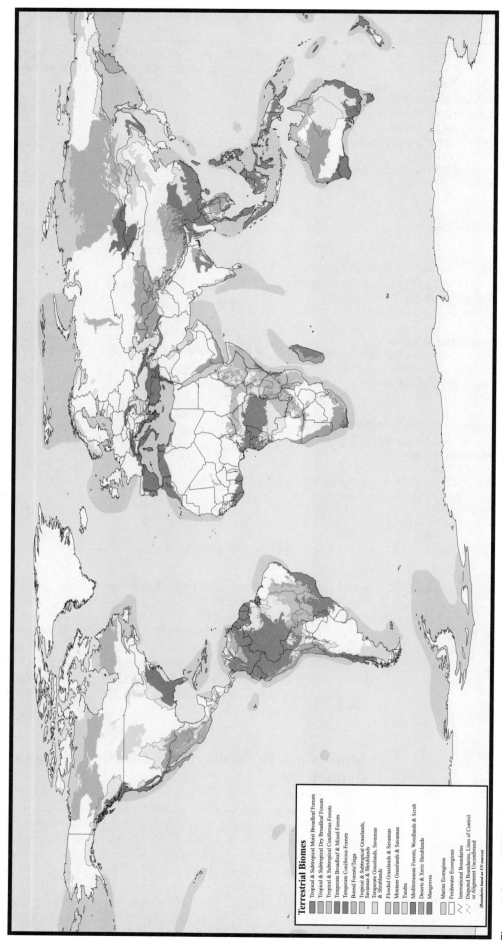

Figure 4.21. WWF's Global 200 Ecoregions. A portfolio of 238 terrestrial, freshwater, and marine priority ecoregions.

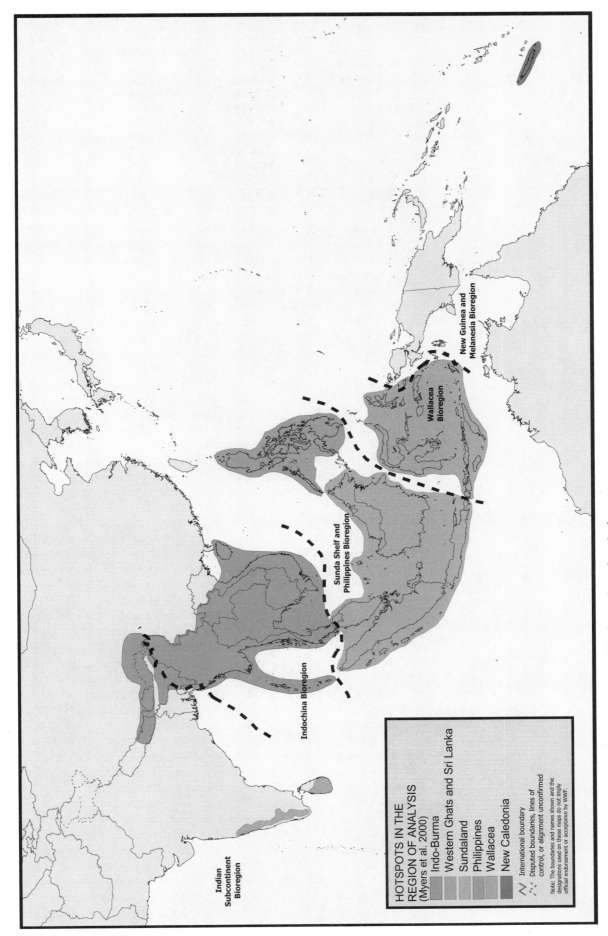

Figure 4.22. Hotspots (Myers et al. 2000) found in the region of analysis.

TABLE 4.4. Biological Distinctiveness of Ecoregions That Overlap with the Biodiversity Hotspots (after Myers et al. 2000) of the Indo-Pacific Region.

Hotspot	Number of Ecoregions in Hotspot	Ecoregions That Are Globally Outstanding	Ecoregions That Are Regionally Outstanding	Ecoregions That Are Bioregionally Outstanding
Sundaland (16)	20	9 (45%)	10 (50%)	1 (5%)
Wallacea (17)	9	6 (67%)	1 (11%)	2 (22%)
Philippines (18)	8	7 (88%)	1 (13%)	0
Indo-Burma (19)	37	17 (46%)	15 (41%)	5 (14%)
Western Ghats and Sri Lanka (21)	8	3 (38%)	5 (63%)	0
New Caledonia (23)	2	2 (100%)	0	0
Total	84	44 (52%)	32 (38%)	8 (10%)

mixed forests, tropical and subtropical moist broadleaf forests, tropical and subtropical grasslands, savannas, and shrublands, and tropical and subtropical dry broadleaf forests broken into thirty-seven ecoregions and extends across two different bioregions (figure 4.22; table 4.4).

Globally, there is 92 percent congruence between these twenty-five biodiversity hotspots and the Global 200 (Myers et al. 2000). However, among the nonoverlapping areas are four Global 200 ecoregions of the Indo-Pacific region: the Eastern Highlands Moist Deciduous Forests [12] and the Sunderbans Mangroves [33], which were ranked as globally outstanding for richness and for ecological processes; and the Chhota-Nagpur Dry Deciduous Forests [18] and Rann of Kutch Seasonal Salt Marsh [36], which were assessed as regionally outstanding for biological richness but were elevated to the Global 200 to ensure representation of the Indo-Pacific region's biodiversity.

Each of the six hotspots in the Indo-Pacific region overlaps with several ecoregions as delineated in this study (table 4.4). In all, the six hotspots overlap with eighty-four ecoregions. In five of the hotspots, overlapping ecoregions represent different levels of biological distinctiveness (table 4.4); only forty-four (52 percent) of the ecoregions represented within the hotspots are globally outstanding for biological distinctiveness. For example, the Sundaland hotspot overlaps with twenty ecoregions, of which only nine are globally outstanding. Ten other ecoregions are regionally outstanding, and one is bioregionally outstanding for biological distinctiveness. Because of the finer scale of resolution at which ecoregions are delineated, they enable us to identify the hot, warm, or lukewarm spots within the larger polygons of Myers et al. (2000). Thus, ecoregions allow better conservation planning at a finer scale.

Other approaches that identify high-priority conservation areas are based on single taxonomic groups (WWF and IUCN 1995; Stattersfield et al. 1998). BirdLife International's endemic bird areas (EBAs) use suites of restricted-range bird species as proxies to identify and rank biodiversity conservation areas (Stattersfield et al. 1998). There are fifty-two EBAs in the Indo-Pacific region, and fifty-one of the sixty globally outstanding ecoregions in the region overlap with these EBAs. There is fair one-to-one concordance between EBAs and ecoregions in the small islands of the archipelago bioregions, such as in Wallacea, where many of the small islands each represent both EBAs and ecoregions (table 4.5). But in the larger continental bioregions, such as the Indian Subcontinent, there is less concordance. In the latter, several of the EBAs are represented by a number of ecoregions that belong to different biomes and by ecoregions that are not all globally outstanding for biological distinctiveness. An EBA is based on the overlapping breeding ranges of two or more restricted range bird species (a land bird with a breeding range of less than 50,000 km^2 since historic times; Stattersfield et al. 1998). Both by combining several taxonomic groups to provide more breadth to the index that appraises biodiversity and by differentiating between biomes, we can discriminate better between biologically important areas than when using the single-species index.

Obviously the BDI used to rank ecoregions can be made even more robust by using a wider range of taxonomic groups that includes amphibians, reptiles, and

TABLE 4.5. Number of Ecoregions That Overlap with Endemic Bird Areas (EBAs; Stattersfield et al. 1998) in the Indian Subcontinent and Wallacea Bioregions of the Indo-Pacific and Their Biological Distinctiveness Status. The numbers in parentheses refer to the EBA's identification number (Stattersfield et al. 1998).

	Ecoregions That Overlap with EBA	Number of Ecoregions That Are			
		Globally Outstanding	Regionally Outstanding	Bioregionally Outstanding	Locally Important
INDIAN SUBCONTINENT BIOREGION	32	12	18	2	0
Western Ghats (123)	5	1	3	1	0
Sri Lanka (124)	3	2	1	0	0
Western Himalayas (128)	6	0	5	1	0
Central Himalayas (129)	4	2	2	0	0
Eastern Himalayas (130)	8	6	2	0	0
Assam Plains (131)	4	1	3	0	0
Irrawaddy Plains (132)	2	0	2	0	0
WALLACEA BIOREGION	11	8	2	1	0
Northern Nusa Tenggara (162)	1	1	0	0	0
Sumba (163)	1	0	0	1	0
Timor and Wetar (164)	1	1	0	0	0
Banda Sea Islands (165)	1	0	1	0	0
Sulawesi (166)	2	2	0	0	0
Sangihe and Talaud (167)	1	1	0	0	0
Banggai and Sula Islands (168)	1	1	0	0	0
Buru (169)	1	0	1	0	0
Seram (170)	1	1	0	0	0
Northern Maluku (171)	1	1	0	0	0

invertebrates. However, as Indraneil Das and Walter Erdelen point out in essays 6 and 7, the paucity of data for amphibians and reptiles—let alone invertebrates—and the cryptic nature of many species make even estimates of richness and endemism patterns difficult. Although new discoveries can change the distribution map for both groups, current information points to broad correlations between herptile and plant diversities. Both Erdelen and Das make tentative predictions of which ecoregions or biomes may yield the greatest herptile diversity if comprehensive inventories are undertaken. They also acknowledge that there is a great urgency to conserve the ecosystems and habitats that may harbor high diversity because of the wave of habitat loss throughout the region.

Concordance of Biological Priorities: The Overriding Importance of Indonesian Ecoregions

Indonesia is the most biologically diverse country in the region, ranking with Brazil as a global biological treasure chest. Using Indonesia as a backdrop, we compare our globally outstanding (GO) ecoregions against the other priority-setting exercises, notably the global hotspots of Myers et al. (2000), BirdLife International's EBAs (Stattersfield et al. 1998), and the Centres of Plant Diversity (WWF and IUCN 1995).

In our analysis, nineteen GO ecoregions cover over 90 percent of Indonesia's landmass, whereas the hotspots (of Myers et al. 2000) cover 79 percent of the landmass in only two units (figures 4.1 and figure 4.22). The latter excludes Irian Jaya, which is considered a high-priority wilderness area because of the large areas of intact natural habitat.

EBAs and Centres of Plant Diversity are more selective in their conservation targets. Centres of Plant Diversity are also smaller conservation units. The twenty-three EBAs cover only 47 percent of the landmass and miss high biodiversity habitats such as the lowland forests of Java, Sumatra, and Borneo. The eighteen small Centres of Plant Diversity include only 7 percent of the land area and miss areas with high faunal endemism, such as the islands of Nusa Tenggara and Maluku. Both sets of priorities capture particular facets of Indonesia's biodiversity but fall short of capturing the overall representative biodiversity. The globally outstanding ecoregions include all the areas captured by both EBAs and Centres of Plant Diversity and the important habitats that they miss. Thus, conservation units derived using broader taxonomic foci are better at capturing the tremendous biodiversity distributed throughout the more than 13,000 islands of Indonesia that span three bioregions.

A better understanding of the biological disturbance of Indo-Pacific region will come from intensive, strategically targeted surveys and the publications of floras and plant checklists. However, Derek Holmes (essay 1) calculates that in Indonesia alone we are losing natural forests at the rate of nearly 2,000 ha/day and 200 ha/hour. There is no time to lose to set priorities based on the data we have now. We believe that the ecoregions for which new species lists add greatly to the existing count are those that are already designated as globally outstanding based on our current level of knowledge.

Essay 10

"New Species" and Other Endemics: Implications for Conserving Asia's Biodiversity.

Eric Wikramanayake

About a decade ago, Dr. John MacKinnon and several Vietnamese colleagues were poring over satellite images of Vietnam to plan biodiversity surveys. Their attention was drawn to a large patch of forest along the northern Annamite Mountains that had a spectral signature distinct from the rest of the area, indicating it to be much wetter. Because this area was scientifically unexplored and seemed more intact, the Annamites became a priority for wildlife surveys.

Several months later, field biologists were rewarded beyond their greatest expectations. Surveys revealed several new species ranging from plants to fishes to butterflies, but most astounding were new two monotypic large mammal genera, the saola (*Pseudoryx nghetinhensis*) (Dung et al. 1993) and giant muntjac (*Megamuntiacus vuquangensis*) (Tuoc et al. 1994), discovered in quick succession. Subsequent surveys along the Annamites have turned up a wealth of other species including another new muntjac, the Truongson muntjac (*Muntiacus truongsonensis*) (Giao et al. 1998); a striped rabbit whose nearest relative is in faraway Sumatra (Surridge et al. 1999); and a bovid (*Pseudonovibos spiralis*) known locally as *linh duong* in Vietnam and *khting vor* in Cambodia and described on the basis of its spiral horns (Peter and Feiler 1994). According to local lore, the *khting vor* has an appetite for poisonous snakes. Although the animal has not been seen dead or alive by scientists, a seventeenth-century Chinese encyclopedia has an illustration of a goatlike animal with curly horns that could represent it (Schaller 1998).

Several other mammal species that were thought to have become extinct since their first discovery have also been rediscovered. For example, the Vietnamese warty pig (*Sus bucculentus*), first described more than a century ago (Heude 1892, in Schaller and Vrba 1996), remained unreported until a few years ago (Groves et al. 1997). Another small muntjac, *Muntiacus rooseveltorum*, last described after a hunting expedition led by President Theodore Roosevelt's son in 1929, has been reconfirmed in the Annamites.

Why is this spate of new mammals being discovered in the Annamites, especially considering that back in 1812, French naturalist Georges Cuvier asserted that "there is little hope of discovering new species of large quadrupeds" (in Schaller 1998).

One reason is that the region has been poorly explored scientifically because the rugged terrain has deterred all but the most intrepid biologists. Also, years of conflict in Indochina have relegated biological exploration to a low priority. Only now that political stability has recently returned to the region is national and international attention being directed to this "lost world."

A second and probably more interesting reason for these new discoveries is that we are only now beginning to discover and understand the geological and ecological processes that have created these biodiversity hotspots.

The collision between the Deccan Plate and Laurasia greatly affected the geology of continental southeast Asia, but the Annamites predate the Himalayan range by many millions of years (Hutchinson 1989). Besides geological upheavals, there have been numerous climatic fluctuations, from wetter and cooler conditions to drier and warmer periods. In fact, oxygen isotope analyses of fossil shells of foraminifers from the oceanic depths have shown that just during the past million years there have been at least twenty-two measurable climatic fluctuations in the Northern Hemisphere (Graham 1986).

These alternating wetter and drier conditions resulted in large-scale changes in vegetation. During the wet phases the moist forests were widespread and contiguous throughout the Annamite Range, including the lowlands, but during the dry phases they were replaced by dry forest. As the moist forest species retreated into the mountains during dry phases, they found refuge in the cooler, higher elevations, moister valleys, and the eastern slopes that received precipitation from the monsoon rains emanating from the eastern seaboard.

As the moist forests retreated, so did the fauna adapted to the moist forest habitats. Evidence indicates that the rate of Holocene-era tree migrations averaged about 10–45 km per century (Davis 1981). But animals are more vagile and can respond faster to changing climates; thus many of the specialized species migrated with the retreating forests rather than adapting to the dry conditions (Davis 1986).

Throughout the Holocene, beginning about 11,000 years ago, the rapid warming trend allowed dry forests to dominate in the lowlands and extend into the lower valleys and foothills of the western, leeward side of the Annamite Mountains. The eastern, windward side that bore the brunt of the monsoon winds continued to be clothed in moist forests until very recently, when human activity began to clear and fragment these species-rich forests.

During the Holocene, the moist forest-adapted animal species were confined to habitat fragments that served as refugia. In a terrain that was (and still is) rugged and highly dissected, the animals' dispersal was constrained and populations were reproductively isolated. The stage was set for speciation. Genetic drift through thousands of years of isolation led to the evolution of new species, each with a very limited distribution range.

Implications for Conservation

The new species discoveries in the Annamite Mountains are exciting events, but they raise some serious issues for conservation planning. Despite years of scientific exploration we still have made few inroads into identifying and inventorying Earth's biodiversity. If we are still discovering larger species in the twenty-first century, it implies that smaller species, which make up the vast majority of biodiversity, are barely known. Consider that it was less than a decade ago that a population of Javan rhinoceros (*Rhinoceros sondaicus*) was discovered in an isolated patch of forest not far from Ho Chi Minh City, in southern Vietnam (Santiapillai et al. 1993). That one of Asia's largest mammals could have survived so close to one of Vietnam's largest cities and yet escape detection for so long indicates how much biodiversity still remains to be discovered by scientists.

The finds in the Annamites have spurred scientists to survey other mountain ranges in Indochina and elsewhere in Asia, with similar success. Recent surveys in remote mountains in northern Myanmar have revealed another new species of muntjac—the leaf muntjac (*Muntiacus putaoensis*) (Amato et al. 1999)—and surveys in the Cardamom Mountains in southwestern Cambodia have turned up a plethora of new species of reptiles and amphibians (http://forests.org/forests/asia.html). Surveys along the mountains of New Guinea have yielded several mammals, ranging from bats to tree kangaroos (Flannery 1995) and even a poisonous bird, *Pitohui dicrous* (Dumbacher et al. 1992).

Even in places considered to have been historically well surveyed, recent surveys have resulted in astonishing finds. Many of the more than 250 species of endemic frogs recently discovered from the rain forests of Sri Lanka were not located through field surveys but through careful re-examination of museum specimens that were cataloged as serial syntypes (Pethiyagoda and Manamendra-Arachchi 1998).

The Indo-Pacific region is also characterized by hundreds of islands of various sizes, ranging from small atolls to the world's largest, New Guinea. The geological histories and periods of isolation of these islands also vary from recent volcanic origins (e.g., Krakatoa) to Gondwanaland relicts (e.g., New Caledonia). Depending on the degree of isolation and size, these islands harbor various levels of endemism. Because many have not been surveyed adequately, however, we are uncertain of the true levels of endemism in different taxonomic groups.

The implications of these findings for conservation, especially when prioritizing areas for long-term conservation, are immense. Among its mountain ranges and its islands, the Indo-Pacific region probably harbors the highest levels of endemism on Earth. And an important goal in conservation is to ensure that these endemic species, and especially the centers of endemism, are protected.

However, our knowledge of the distribution of biodiversity across the region is woefully incomplete. We cannot wait until we have a complete database from which to identify high-priority areas—habitat loss caused by intensive and extensive human activities is progressing too fast. Most of the mammals that are new to science have been long known by the local people who hunt them for food. George Schaller (1998) describes a field trip to Laos thus: "I feel like a nineteenth-century biologist—but with a difference. Discoveries, I soon learn, are made not in the depths of the forest but in village huts and markets. Wildlife is so intensively hunted with guns, deadfalls, and snares that signs or sightings of any mammals are uncommon."

If our knowledge of endemic species, and biodiversity in general, is only the tip of the proverbial iceberg, how can we represent and protect these communities and habitats?

We are faced with a situation in which urgent decisions to identify areas of high biodiversity and endemism must be made based on current knowledge, field experience, and tested principles of ecology. A combination of factors, such as the history of isolation, levels of precipitation, temperature regimes, habitat types, known levels of biodiversity, and geology, can be used to make informed decisions about where and how biodiversity might be distributed. MacKinnon's selection of the northern Annamites for surveys that paid rich dividends was made not by throwing darts at a map but by careful consideration of these variables. The ecoregions described in this book are delineated using the same variables. Thus, using ecoregions for conservation planning will enable conservation biologists and planners to ensure that known and predicted biodiversity is given priority in a regional conservation strategy.

ESSAY 11

Plant Richness and Endemism in the Indo-Pacific: Dipterocarpaceae

Gary A. Krupnick and June Rubis

Determining plant richness and endemism for the 866 terrestrial ecoregions is a challenging task (Olson et al. in press), considering that approximately 300,000 flowering plant species exist in the world today (Prance et al. 2000). Range maps typically are unavailable for the majority of plant species, so data must be extracted from published literature and herbaria collections. Records normally provide information about the collection sites and where the plant was observed or collected; therefore, we must estimate the entire range of each species. The laborious endeavor of categorizing plants by ecoregion is essential because the distributions of invertebrates, by far the most abundant terrestrial species, often are tied to the range occupied by their host plants. For botanists to make a major contribution to setting global and regional priorities for conservation, we must make the effort to map plant richness and endemism to complement similar efforts completed for vertebrates.

We used the ecoregions of the Indo-Pacific region as a test case in attempting to determine patterns of global plant richness and endemism. Specifically, we mapped out all plant species from a few families to each ecoregion in the Indo-Pacific using published floras. To illustrate our results, we focus attention on one family, Dipterocarpaceae. Dipterocarp species dominate the lowland moist evergreen forests of southeast Asia, and they have the dubious distinction of being the world's main source of hardwood timber. Many Dipterocarpaceae species also are used for their resin and fruit.

The Dipterocarpaceae are pantropical in distribution (Ashton 1982), with the center of distribution in the tropical rain forests of Malaysia. Three subfamilies have been described. The subfamily Dipterocarpoideae (Blume 1825) comprises around 500 species in fourteen genera. This subfamily is confined to the Seychelles and the Indo-Pacific, occurring within 109 of the 140 Indo-Pacific ecoregions. The two other subfamilies are found outside the Indo-Pacific. Subfamily Monotoideae (Gilg 1925), comprising three genera with approximately thirty species, is represented in Africa and Madagascar, and a monotypic genus has been recently discovered in Colombia, South America (Londoño et al. 1995). Finally, the monotypic subfamily Pakaraimoideae (Maguire and Ashton 1977) is described from the Guyana Highlands in South America.

We mapped the 525 taxa of the subfamily Dipterocarpoideae to 110 ecoregions of the Indo-Pacific using published flora (Ashton 1980; Smitinand et al. 1980; Ashton 1982; Rodgers and Panwar 1988; Puri et al. 1989; Smitinand et al. 1990; Kostermans 1992; Ramesh and Pascal 1999). Descriptions in the floras include the ecological characteristics and the distribution of the species. For some species, detailed accounts of collection sites were given, whereas for other species only a vague description of the habitat was given; in those cases ecoregions were estimated based on the description. A species is considered endemic to an ecoregion if it occurs in only one ecoregion (strict endemic). To obtain the most complete and accurate picture of plant richness and endemism, georeferenced herbarium specimens will be used in future efforts. With several million specimens in worldwide herbaria, these resources can be used in conjunction with published monographs and floras in determining global centers of plant richness and endemism. Specimen collections can provide detailed information about which sites have been studied, which sites warrant further study, and what changes have occurred at the sites over time.

Richness

Dipterocarp species are confined to a tropical climate where the mean annual rainfall exceeds 1,000 mm or where a dry season lasts less than six months. Most species grow below 1,000 m altitude. With the center of dipterocarp species richness in the rain forests of Malaysia, it is not surprising that the ecoregions with the highest number of species are the Borneo Lowland Rain Forests [96] (273 species), Peninsular Malaysian Rain Forests [80] (165 species), and Sumatran Lowland Rain Forests [82] (111 species) (figure 4.23). These three ecoregions have two to five times more species than the next highest ranked ecoregions. The next richest ecoregions in order are the Borneo Peat Swamp Forests [86] (56 species), Sri Lanka Lowland Rain Forests [3] (54 species), Tenasserim–South Thailand Semi-Evergreen Rain Forests [53] (50 species), and Sumatran Peat Swamp Forests [85] (50 species). Ecoregions in the Philippines and the dry forests of southeastern Indochina rank next. Ecoregion species richness declines markedly in southwestern India, eastern India and northern Myanmar, and east of Wallace's Line (Sulawesi, the Moluccas, and New Guinea).

The results contrast sharply with estimated patterns of total plant richness. The ecoregions of New Guinea are among the richest locations for total plant species but rank low in the number of dipterocarp species. The limited distribution of dipterocarps east of Wallace's Line explains the low number in New Guinea. New Guinea is influenced by Australasian or southern Gondwanaland flora. The montane rain forests of Borneo, the Malay Peninsula, Sumatra, and the Philippines also contain few dipterocarp species, despite containing biologically diverse vegetative communities. The majority of dipterocarp species occur below 1,000 m elevation, but the montane forests are dominated by Palearctic tree families. The one similarity between estimated total plant species richness and dipterocarp species richness is found in the Borneo lowland rain forest ecoregion. This site ranks as the richest ecoregion in the Indo-Pacific for both overall plant species and dipterocarp species. The rain forests of peninsular Malaysia and Sumatra are also high in terms of total plant species richness and dipterocarp species richness.

When comparing ecoregions within biomes, the tropical and subtropical moist forests of lowland Borneo, Sumatra, peninsular Malaysia, and Sri Lanka contain the greatest number of species for that biome. Among the tropical and subtropical dry forests biome, the Southeastern Indochina Dry Evergreen Forests [59], Central Indochina Dry Forests [72], and Southern Vietnam Lowland Dry Forests [58] are among the richest in dipterocarps but contain far fewer species than the moist forests. The Northern Indochina Subtropical Forests [74] and the South China–Vietnam Subtropical Evergreen Forests [75] contain the highest number of dipterocarps among the subtropical broadleaf forests. Only a handful of dipterocarp species grow in other biomes. Two species are described from Northeast India–Myanmar Pine Forests [77], one species from Northern Triangle Temperate Forests [76], and one species from Sumatran Tropical Pine Forests [105].

Endemism

Dipterocarpaceae show a high rate of endemism, probably because of their poor seed dispersal (Ashton 1982), and this pattern is evident in the ecoregion analysis. The Borneo Lowland Rain Forests [96] (126 species) have the highest number of endemic dipterocarp species. These forests have more than three times as many endemics as the next highest ecoregion, the Sri Lanka Lowland Rain Forests [3] (41 species) and six times as many endemics as the Peninsular Malaysian Rain Forests [80] (21

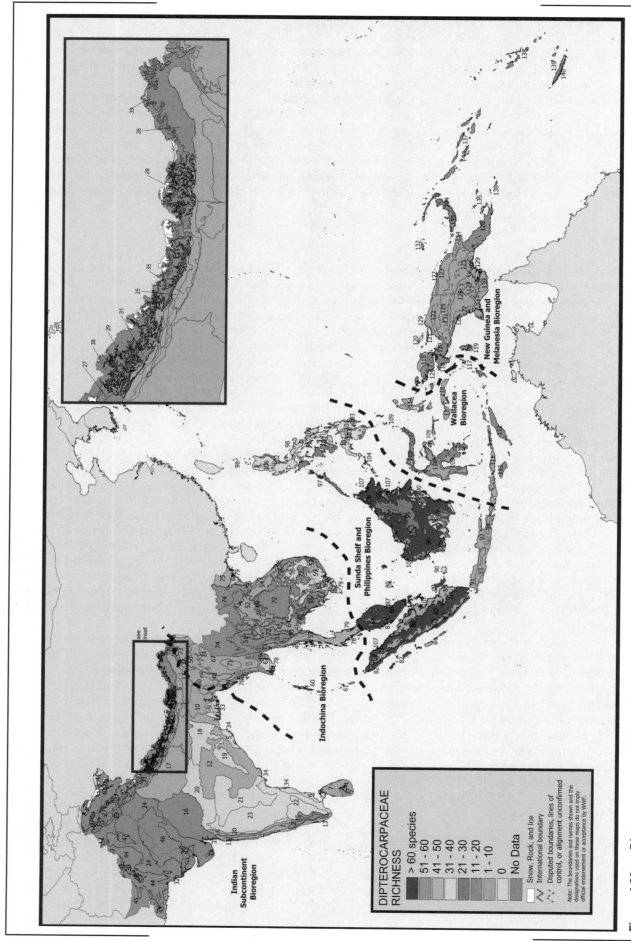

Figure 4.23. Dipterocarpaceae richness for terrestrial ecoregions of the Indo-Pacific.

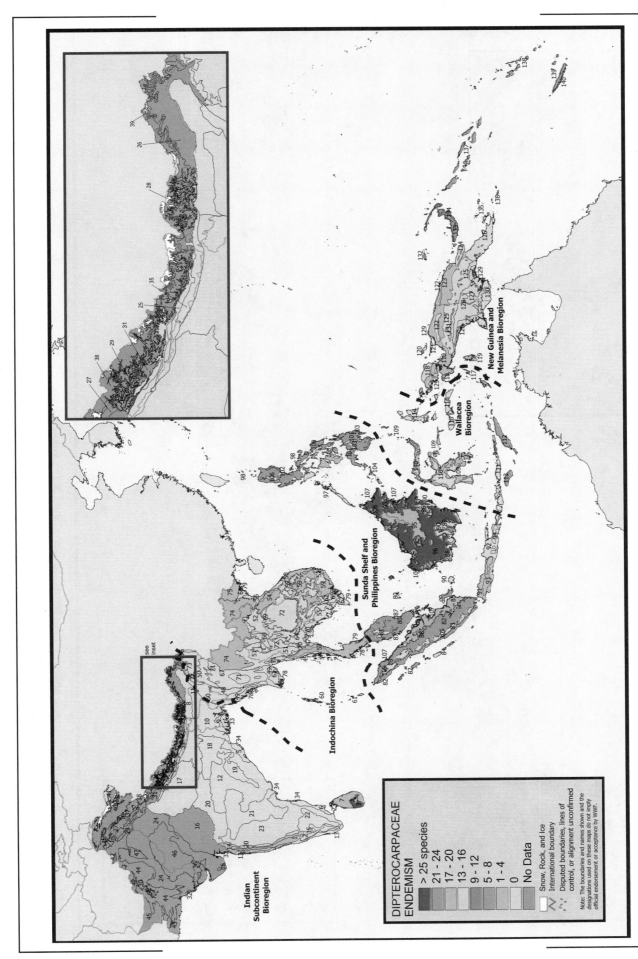

Figure 4.24. Dipterocarpaceae endemism for terrestrial ecoregions of the Indo-Pacific.

How Is Biodiversity Distributed?

95

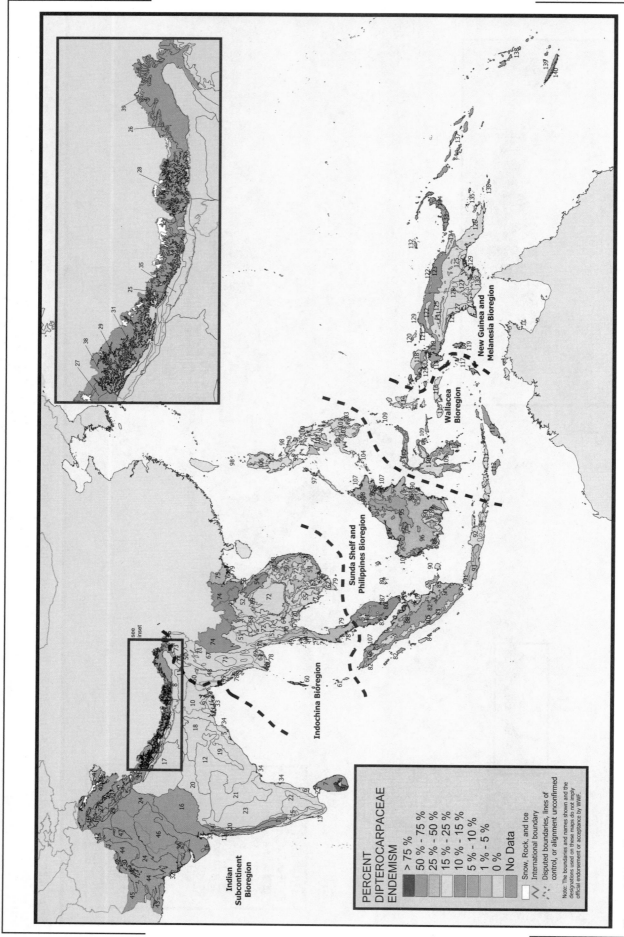

Figure 4.25. Percentage Dipterocarpaceae endemism for terrestrial ecoregions of the Indo-Pacific.

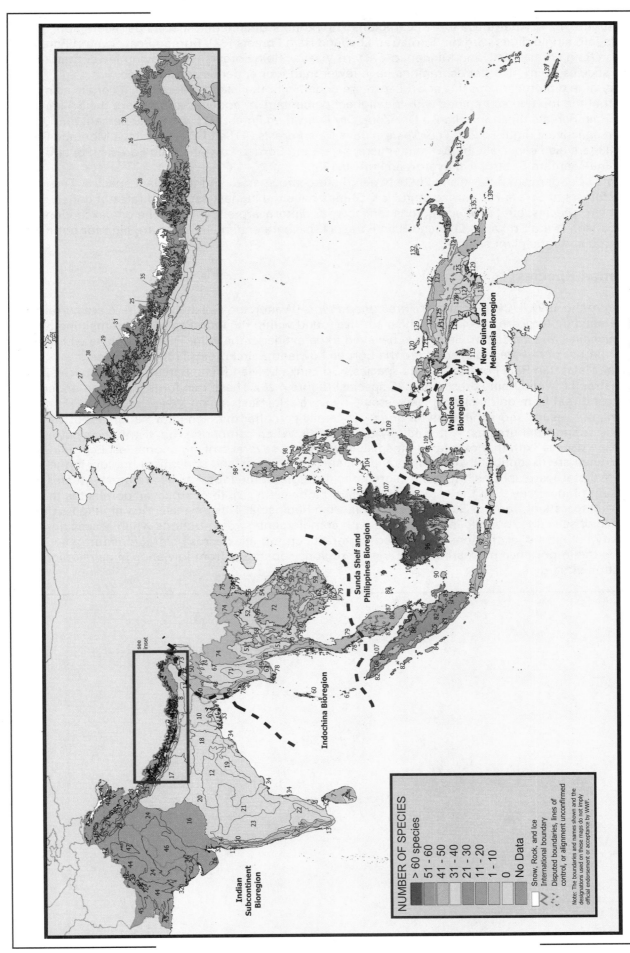

Figure 4.26. Number of *IUCN Red List* Dipterocarpaceae species in the terrestrial ecoregions of the Indo-Pacific.

How Is Biodiversity Distributed?

species) (figure 4.24). After these three ecoregions, the number of endemic species per ecoregion plummets into single digits with the Sumatran Lowland Rain Forests [82], Borneo Peat Swamp Forests [86], Luzon Rain Forests [98], and Mindanao–Eastern Visayas Rain Forests [101] having fewer than ten endemic species each. All other ecoregions have fewer than four endemic species each.

All fifty-nine dipterocarp species in Sri Lanka are endemic to the island. Therefore, it comes as no surprise that the top two ecoregions with the highest percentage of endemic species are the Sri Lanka Lowland Rain Forests [3] and Sri Lanka Dry-Zone Dry Evergreen Forests [7], with 75.9 percent and 66.7 percent, respectively (figure 4.25). The Western Java Rain Forests [91] and Western Java Montane Rain Forests [93] are also especially high in endemism, as are the Borneo Lowland Rain Forests [96] and South China–Vietnam Subtropical Evergreen Forests [75].

Patterns of endemism differ distinctly between dipterocarp species and total plant species. The montane forests of Sri Lanka, Borneo, and New Guinea have the highest estimated rates of endemism for total plant species, but these ecoregions rank low for dipterocarp endemism. The one exception in the comparison is the Sri Lanka Lowland Rain Forests [3], where endemism rates are high for both dipterocarps and total plant species.

Threatened Species

According to the 1997 IUCN *Red List of Threatened Plants* (Walter and Gillett 1998), there are currently 195 threatened dipterocarp species, of which 193 are found within the Indo-Pacific (the remaining two are in Seychelles and Tanzania). Seven are believed to be extinct in the wild. The ecoregions with the greatest number of threatened species are the Borneo Lowland Rain Forests [96] (66 species), Peninsular Malaysian Rain Forests [80] (42 species), Sri Lanka Lowland Rain Forests [3] (36 species), and Sumatran Lowland Rain Forests [82] (25 species) (figure 4.26). These rain forests are also under the greatest threat from deforestation (see essay 1 by Derek Holmes).

Overall, our results show that data on one plant family can offer detailed information about which ecoregions support the greatest within-family biodiversity. When comparing the distribution patterns of Dipterocarpaceae with that of total plant species, there are several striking anomalies. Ecoregions in New Guinea are exceptionally high in total plant species richness and endemism but low in the number of dipterocarp species. Montane ecoregions throughout the Indo-Pacific are much greater in overall plant biodiversity than they are in dipterocarp biodiversity. Dipterocarpaceae dominates the lowland rain forests of Borneo, Sumatra, and Malaysian Peninsula. This exercise thus highlights the importance of selecting multiple indicator taxa with complementary distributions when assessing biodiversity richness in ecoregions. Differences in forest communities across bioregions further emphasize the importance of treating montane ecoregions separately from lowlands in developing conservation strategies.

CHAPTER 5

Where Is Biodiversity under Greatest Threat?

Across the Indian subcontinent and most of Indochina, human populations continue to expand rapidly, exacting a heavy toll on the natural forests and the biodiversity they harbor. The few remaining frontier forests (intact forest blocks greater than 5,000 km^2 [Bryant et al. 1997]) in Cambodia and Borneo are falling victim to loggers and logging companies engaged in rampant felling. From peninsular Malaysia to Kalimantan, vast areas of some of the world's most species-rich rain forests are being replaced by monocultures of oil palm (essay 12). To the east of Kalimantan, the intact high-biodiversity forests of New Guinea also face transformation to depauperate wastelands as a result of logging and mining. In New Caledonia, within a few years the mining industry and cattle ranchers have already destroyed much of the extraordinary vegetation that has survived and evolved for more than 60 million years.

Conservationists are nearly unanimous in their agreement that land-use changes—as manifested by habitat loss, degradation, and fragmentation—are the most severe, overarching threats to terrestrial biodiversity. These threats stem from a variety of sources, but the primary sources are resource exploitation and land conversion to meet the demands of a rapidly growing human population, commercial logging and mining undertaken without regard for sustainable management practices, and poorly planned economic development projects. Where forests remain standing, wildlife exploitation for the market trade and medicinal use remain an insidious threat, creating hectares of empty forests devoid of vertebrate faunas (see essays 13 and 14).

These large-scale land-use changes are becoming increasingly easy to map in terms of their intensity and spatial extent, permitting assessment of the threats they pose not only to biodiversity but also to the welfare of the human societies that depend on intact ecosystems. For instance, logging and agricultural expansion is easily georeferenced, although other variables such as wildlife consumption are less so in a meaningful manner for conservation planning. Human demographics can also be mapped, but invariably the data are too coarse or archived in inappropriate units for use in conservation planning. Nevertheless, we can use demographic data to obtain a general picture of patterns and trends in threats to biodiversity. In this chapter, we analyze current and predicted land-use changes and the drivers behind them to forecast the trajectory of impacts to biodiversity in Indo-Pacific ecoregions.

Whereas habitat conversion can lead to outright elimination of most natural ecological processes that create and maintain biodiversity, habitat degradation and fragmentation disrupt the processes that create and maintain biodiversity. Measuring this erosion of ecological processes is difficult, but increasingly conservation biologists recognize that priority-setting efforts that neglect ecological processes are incomplete (Mace et al. 2000). In this analysis we use a combination of landscape features—the amount of remaining habitat in the ecoregion, the size of remaining habitat blocks, the degree of fragmentation of remaining forest blocks, and the amount of remaining habitat under formal protection—as a simple set of proxies to assess erosion and persistence of ecological processes. More details of these variables and the reasons for their selection are discussed in appendix D. The analysis of these variables forms the basis of the snapshot conservation status for all 140 ecoregions. After this overview of conservation status we examine several overarching threats to biodiversity and use two—the distribution of human populations and loss of forest cover to logging concessions—to show how future threats can affect the conservation status of ecoregions.

In the Indo-Pacific region, as elsewhere in the world, lowland forests face a disproportionate risk of being felled. Here, we analyze forest loss by three elevational strata—forests below 300 m, 300–1,000 m, and >1,000 m—to illustrate how many Indo-Pacific ecoregions with high topographic relief are being transformed into archipelagos of montane forests within oceans of cleared or converted land that was once covered by lowland forests. There is an urgent need to address the loss of both lowland forests and connectivity among different forest types.

Loss of forests and other habitats in areas of high endemism inevitably will result in species extinctions unless critical areas are protected (Brooks et al. 1997, 1999; see essay 6 by Das). We highlight the role of protected areas because many conservation biologists view them as the cornerstones of conservation. They are perhaps the only hope for the time being of maintaining source pools of species for future restoration efforts (Noss and Cooperrider 1994) and for the survival of species sensitive to disturbance and with restricted distributions.

Adding new protected areas and upgrading the management of the many "paper parks" in the region is part of an ambitious alliance between the World Bank and WWF to increase forest protection efforts worldwide. Many conservation organizations suggest a target of 10 percent land protection, and some national governments have acceded to this figure, but this one-size-fits-all target can have dangerous consequences (Soulé and Sanjayan 1998). For instance, in ecosystems where diversity is high or species and processes need large spatial scales, 10 percent may be inadequate. In their analysis of several species-area studies, Soulé and Sanjayan (1998) found that in tropical regions about 50 percent of natural habitat is necessary to protect "most elements of biodiversity," including wide-ranging animal species. To see how well the Indo-Pacific ecoregions meet the targets suggested by Soulé and Sanjayan, we provide a scorecard for the number of tropical moist and tropical dry forest ecoregions that attain the 50 percent threshold.

Despite the importance of protected areas, over the past decade there has been a decline in the establishment of protected areas in the tropics (Terborgh 1999). Some view this as an inevitable consequence of increasing demands for space and resources from the growing human enterprise. In essay 15 we look at efforts to create new protected areas over the past fifty years and identify trends and patterns. We then offer an ecoregion scorecard to show progress, or lack thereof, to attain the 10 percent minimum of protection of each ecoregion as advocated by some conservation biologists.

Finally, we apply the threat analysis to assess the expected changes to conservation status over the next several decades.

Snapshot Conservation Status: How Much Habitat Remains in Indo-Pacific Ecoregions?

Overall, 29 (21 percent) of the 140 ecoregions across the Indo-Pacific region are critical for conservation status. Another 24 (17 percent) are endangered, and 47 (34 percent) are vulnerable, leaving only 40 ecoregions (29 percent) as relatively stable or relatively intact (figures 5.1 and 5.2).

The summed areas of ecoregions in each conservation status category show good concordance with the numbers of ecoregions in the respective conservation status categories (figure 5.1). This relationship indicates that none of the conservation status categories are represented by one or a few very large ecoregions that account for a disproportionate area within any category.

The tropical and subtropical moist broadleaf forest biome ecoregions account for 73 percent of the critical ecoregions. Most of the critical or endangered ecoregions in this biome are located in the Gangetic and Brahmaputra River plains of northern India and Bangladesh, the Philippine archipelago, and the islands of Sumatra and Java (figure 5.2).

At the other end of the spectrum, many of the forty relatively intact or relatively stable moist forest ecoregions lie along the mountain ranges in the

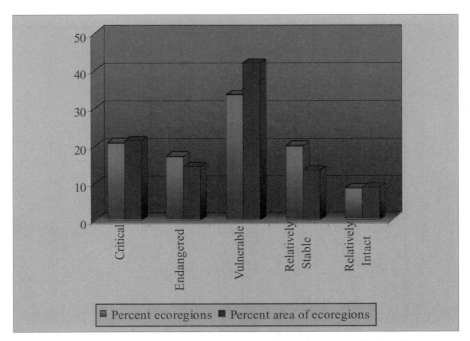

Figure 5.1. Percentage of ecoregions and ecoregion areas in the five conservation status categories for the snapshot conservation status.

Indochina bioregion and in the New Guinea and Melanesia bioregion. Fourteen tropical moist forest ecoregions in the New Guinea and Melanesia bioregion are relatively intact or relatively stable—more than in all other bioregions. A detailed breakdown of the conservation status of all ecoregions is found in appendix F.

The dry forest ecoregions are in the worst condition of any in the region. Fifty percent of these ecoregions are critical or endangered. Only one ecoregion (Southeastern Indochina Dry Evergreen Forests [59]) qualifies as relatively stable. The tropical dry forests are predominantly lowland, with soils well suited for agriculture, and contain valuable hardwood trees, making them particularly vulnerable to exploitation.

All the mangrove ecoregions are critical, endangered, or vulnerable. But the other forested biomes contain only a few ecoregions apiece, so broad trends are not evident. Among them no ecoregion is considered as critical, although eight are classified as endangered or vulnerable. All of the nonforested, montane grassland and shrubland ecoregions are relatively stable (figure 5.2). Conversely, of the seven deserts and xeric shrubland ecoregions, three are critical and another three are endangered or vulnerable.

General Patterns of Landscape Features: Habitat Loss, Size of Remaining Habitat Blocks, Fragmentation, and Protection

Habitat Loss

Habitat loss is prevalent across almost every biome (table 5.1). But encouragingly, 50 of the 140 ecoregions still retain more than 50 percent intact habitat and 36 ecoregions retain 26–50 percent of forest cover. Fifty-four ecoregions have less than 25 percent of forest cover remaining.

Among the forested biomes, the tropical and subtropical dry broadleaf forests have the highest percentage of habitat loss, followed by mangroves, reflecting the widespread destruction of these important habitats across the entire region.

The tropical and subtropical moist broadleaf forests biome still has approximately 40 percent of the original habitat intact, although most intact forests are montane; the lowland habitats have long been converted to other land uses. Overall, among the forested ecoregions, lowland forest loss is almost twice as extensive as in the montane ecoregions (table 5.2).

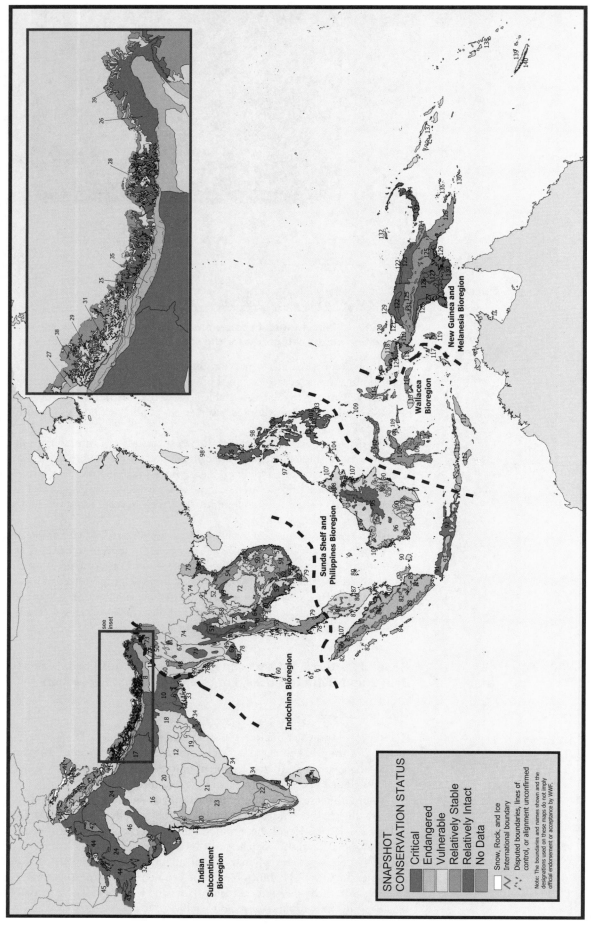

Figure 5.2. The current (1999) conservation status of ecoregions without the inclusion of future threats.

TABLE 5.1. Habitat Loss in Biomes of the Indo-Pacific.

Biome	Total Area (km²)	Percentage of Habitat Altered or Degraded
Tropical and subtropical moist broadleaf forests	4,997,500	57
Tropical and subtropical dry broadleaf forests	1,613,400	73
Tropical and subtropical coniferous forests	580,200	62
Temperate broadleaf and mixed forests	124,200	46
Temperate coniferous forests	73,600	59
Tropical and subtropical grasslands, savannas, and shrublands	61,100	51
Flooded grasslands and savannas	27,800	3
Montane grasslands and shrublands	7,623,400	61
Deserts and xeric shrublands	8,635,900	60
Mangroves	145,600	65

TABLE 5.2. Extent of Lowland, Submontane, and Montane Forests Remaining in the Indo-Pacific.

Forest Type	Total Area (km2)	Total Remaining Habitat (km2)	Percentage Habitat Lost
Lowland (<300 m)	5,230,703	1,384,501	74%
Submontane (300–1,000 m)	2,860,506	1,120,107	61%
Montane (>1,000 m)	1,308,048	750,344	43%

Most (91 percent) of the ecoregions that have 20 percent or less remaining forest cover are lowland ecoregions (figure 5.3). The montane ecoregions that have lost more than 20 percent forest cover include those of Java and Mindanao, which have dense human populations. The montane forest ecoregion in northern Indochina is less densely populated but is subject to heavy shifting cultivation. The lowland forest ecoregions that still retain more than 75 percent forest cover are in New Guinea and the smaller islands in the Wallacea and New Guinea and Melanesia bioregion (appendix G).

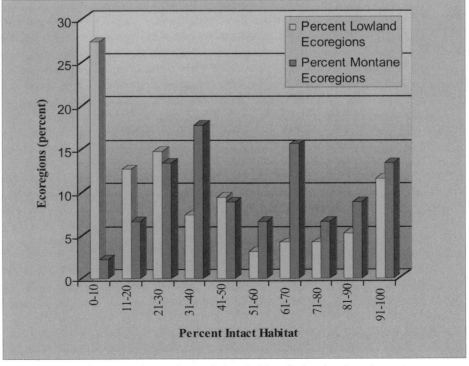

Figure 5.3. A comparison of remaining habitat in lowland and montane ecoregions of the Indo-Pacific.

Size of Remaining Habitat Blocks

The distribution of the remaining habitat blocks by elevation thus confirms a serious conservation concern (figure 5.4): large lowland forest blocks (under 300 m in elevation) are now rare. Focusing on one area suffering from severe forest loss over the past decade—Sumatra—makes this point even more clearly (figure 5.5). Here, a combination of forest fires, land clearing recently accelerated by the Asian economic crisis, and large-scale land clearing for forest concessions and oil palm plantations have resulted in a near extinction of lowland forest.

One of the goals of biodiversity conservation is the conservation of habitat blocks that are large enough to be resilient to large-scale disturbances and long-term changes and capable of supporting species and processes with large spatial requirements (Noss 1991). Among the island ecoregions, nine of fourteen with remaining habitat blocks greater than 1,000 km^2, a threshold chosen to distinguish between large and small habitat blocks among ecoregions less than 20,000 km^2, are in the New Guinea and Melanesia bioregion. Three additional ecoregions are in the Wallacea bioregion.

Twenty-five of the thirty small (<20,000 km^2) island ecoregions are in the tropical and subtropical moist broadleaf forests biome. Less than 50 percent of these ecoregions have large habitat blocks (>1,000 km^2). From among the small, continental ecoregions, only four (Northern Triangle Temperate Forests [76], Peninsular Malaysian Montane Rain Forests [81], Sundarbans Mangroves [33], and Indus Valley Desert [47]) have habitat blocks greater than 1,000 km^2.

Forty-two large ecoregions (>20,000 km^2) lack any habitat blocks greater than 5,000 km^2. Many of the other fifty-four ecoregions that do have large habitat blocks that exceed 5,000 km^2 are in large islands (on Borneo, Sumatra, and New Guinea) or are along the mountain ranges. Only the Sundarbans Mangroves [33] contain blocks of habitat greater than 1,000 km^2 among the mangroves, but thirty-five of the tropical and subtropical moist broadleaf forests, nine tropical dry forests, and five montane grasslands and shrublands ecoregions had at least one habitat block larger than 5,000 km^2.

Fragmentation

Fragmented habitat tends to be more susceptible to external threats because the reduced core area and increased edge effects result in loss of characteristic ecological processes (Gascon et al. 2000). Our analysis finds that more than one-third of the tropical and subtropical moist forest ecoregions are highly fragmented, and these are distributed primarily in the Indian Subcontinent, Indochina, and Sunda Shelf and Philippines bioregions. The ecoregions of the Philippine islands stand out as examples of severe fragmentation. Most of the relatively unfragmented moist forest ecoregions are in the Wallacea bioregion and the New Guinea and Melanesia bioregion.

The tropical dry forest ecoregions are also highly fragmented, the only exception being the Southeastern Indochina Dry Evergreen Forests [59], which snakes along the low hills of southern Indochina. In contrast, the montane grasslands and shrublands biome stands out as being the least fragmented and most intact habitat type found in the Indo-Pacific, although the ecoregions in this biome are widely degraded.

Habitat Protection

We analyzed habitat protection as the percentage of remaining intact habitat that has been protected in an ecoregion, rather than the entire extent of protected areas in the ecoregion. This metric provides a better indication of effective protection because it excludes the degraded and converted habitats that are sometimes included in protected areas. Nineteen (13 percent) ecoregions lack protected areas, and fifty-five (38 percent) ecoregions have less than 10 percent of their remaining habitat in protected areas (figure 5.6). Eleven ecoregions have more than 30 percent of the remaining habitat within protected areas.

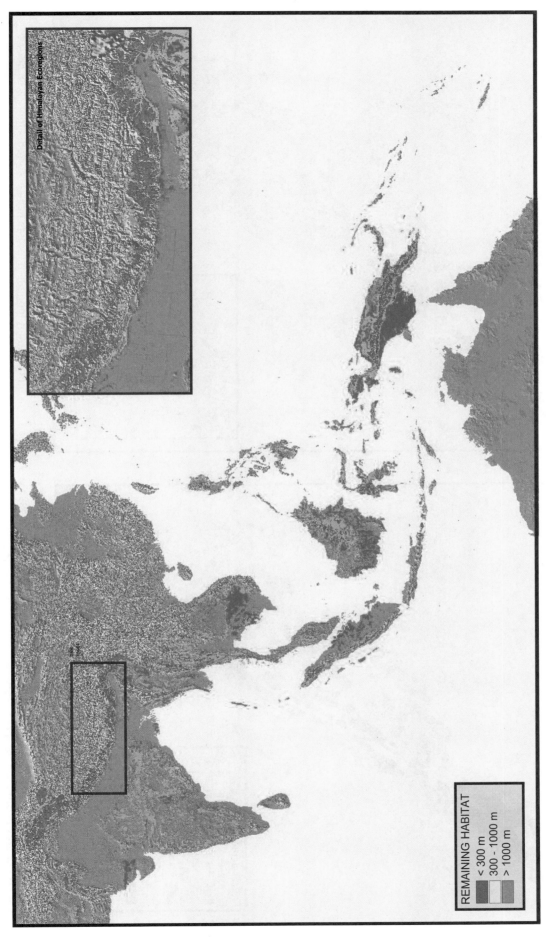

Figure 5.4. Distribution of remaining habitat by elevation across the Indo-Pacific region. Many of the large, lowland habitat (<300 m) landscapes are now in ecoregions that fall within Cambodia and the islands of Borneo and New Guinea. Several large patches of lowland forest also exist in eastern Sumatra, although these have been largely destroyed by the recent fires. The montane (>1,000 m) and submontane (300–1,000 m) are also intact and contiguous, except in the island of Java, where the montane forests have become isolated and limited to upper areas of the volcanic peaks that form the central cordillera of this densely populated island.

Where Is Biodiversity under Greatest Threat?

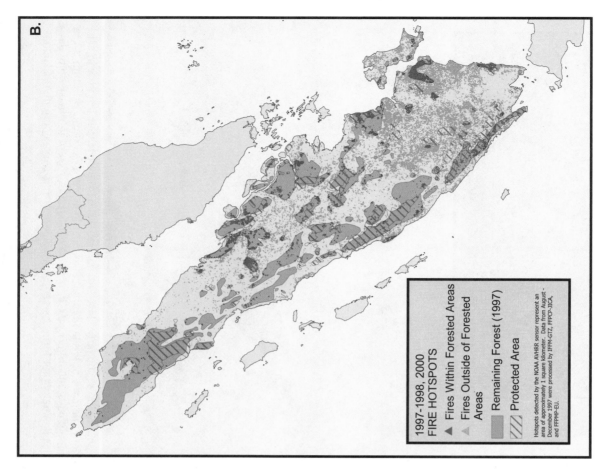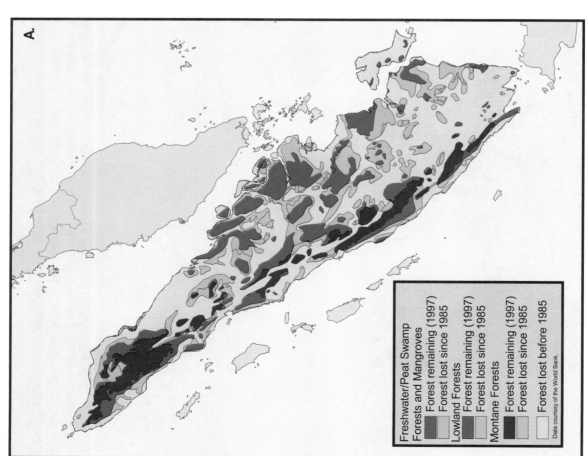

Figure 5.5. Forest loss and fire hotspots in Sumatra. Figure 5.5a documents forest loss from 1985 to 1997 for the island of Sumatra. Lowland forests (blue and green in the map) have been the most severely degraded. However, since 1997 widespread fires and rampant logging in forests have occurred. Figure 5.5b shows outbreaks of fires in Sumatra for three years.

Terrestrial Ecoregions of the Indo-Pacific

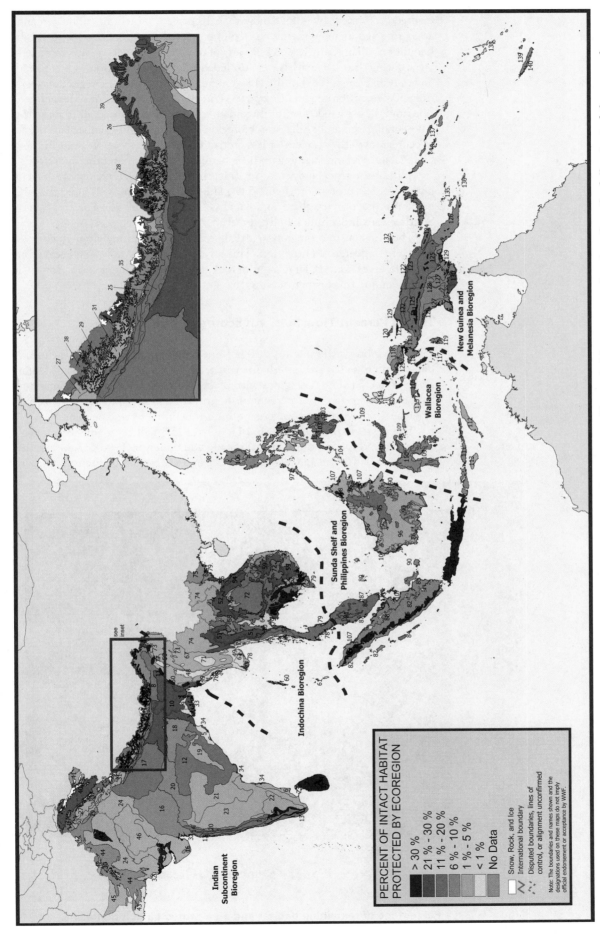

Figure 5.6. Percentage of intact habitat protected by ecoregion. Analyzing the amount of intact habitat protected, rather than the full extent of the ecoregion, provides an effective indicator of biodiversity protection because it excludes degraded and converted areas.

Where Is Biodiversity under Greatest Threat?

Among the ecoregions in the tropical and subtropical moist forests biome, only eighteen (of ninety-five) have more than 10 percent protection. (The percentage area protected includes the proportion of protected areas that overlap with intact habitat only. Thus, any reserves or parts of reserves that overlap with degraded areas during the GIS analysis were excluded from the calculation.) One ecoregion, Yapen Rain Forests [121] in the New Guinea and Melanesia bioregion, has more than 30 percent protection. Sixty-two ecoregions have 5 percent or less protection, and these include twenty-three that have no protection at all (figure 5.6). The Sunda Shelf and Philippines bioregion has the greatest number (twenty-four) of ecoregions with 5 percent or less protection. Overall, the ecoregions in the New Guinea and Melanesia bioregion have better protection than other bioregions.

Within the tropical dry forest biome, thirteen of fifteen ecoregions have 5 percent or less protection (figure 5.6). The two ecoregions with more than 10 percent protection are the Sri Lanka Dry-Zone Dry Evergreen Forests [7] and Southeastern Indochina Dry Evergreen Forests [59] ecoregions.

The Himalayan ecoregions, particularly those containing alpine meadows, have a high degree of protection. This level is deceiving, however, because many of the reserves contain large areas of rock, ice, and permanent snow. Tom Allnutt et al. examine this further in essay 16.

The 50 Percent Threshold: An Ecoregion Scorecard

The ecoregions with more than 50 percent intact habitat also have the most habitat in protected areas, with an average of 13 percent of the intact habitat already protected. The ecoregions with 25 percent or less forest cover remaining have an average of only 2 percent of the intact habitat in protected areas, whereas those with 26 to 50 percent forest remaining have 4 percent.

The ecoregions in cells 1A and 1B of figure 5.7 will be unable to achieve the 50 percent protection targets (sensu Soulé and Sanjayan 1998). However, the forty-eight ecoregions in cells 1C, 1D, 2C, and 2D have the potential to increase

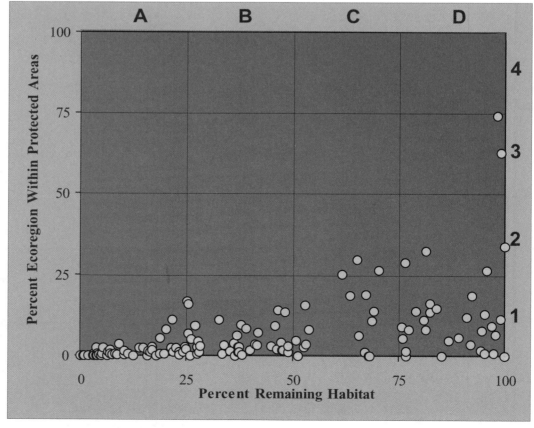

Figure 5.7. **Percentage of remaining habitat and area protected in Indo-Pacific ecoregions.**

the protected areas coverage to over 50 percent. Only two ecoregions have already achieved more than 50 percent protection, but these—Indus Valley Desert [47] and Rann of Kutch Seasonal Salt Marsh [36]—are two of the least biologically diverse ecoregions. The large percentage of protected areas in these less productive and generally inhospitable ecoregions reflects another conservation issue: the potential inflation of a conservation portfolio with low-biodiversity areas to meet percentage-based conservation targets.

Threat-Modified Conservation Status: Projecting Future Integrity of Ecoregions

The previous section presents a broad snapshot view of the current conservation status of Indo-Pacific ecoregions. But a multitude of forces, such as commercial timber extraction, human population growth, increasing natural resource demands, infrastructure development, and other land uses can increase the rates of habitat loss and commercial exploitation of wildlife that can degrade the ecological integrity of ecoregions. Conservation planners should consider these projected trends and the threats they pose to biodiversity. In this section we first provide an overview of some of the overarching threats and then estimate and project trends of a few selected major threats to show how they might modify the snapshot conservation status of Indo-Pacific ecoregions.

We stress that the analysis of projected threats is not meant to represent an exhaustive compilation of the threat variables that impinge on ecological integrity. The spatial scale of this analysis places a comprehensive treatment of threats in all ecoregions across the Indo-Pacific region beyond the scope of this book. Instead, we use two broad variables to evaluate how degradation and conversion threats might affect the future conservation status of ecoregions.

We use chronic, low-intensity anthropogenic impacts as degradation threats that contribute to biodiversity loss. These include natural resource exploitation from forests by people; small-scale clearing for agriculture, plantations, and villages; and wildlife exploitation. These threats erode intact habitat blocks and the wildlife within them, and over the long term they result in fragmentation and degradation of habitat. In this analysis, we use the distribution and density of human populations as a proxy to assess degradation threats within the remaining habitats of ecoregions.

Conversion threats occupy the other end of the continuum. These represent high-impact threats, such as those from large-scale logging concessions, plantations, hydropower schemes, and large infrastructure or development projects. Unlike degradation threats, the results from these threats will be manifested quickly and severely.

Overarching Threats

Large-Scale Commercial Logging and the Loss of Forest Frontiers

Large-scale commercial logging and its aftermath are a major threat to terrestrial biodiversity. Helicopter logging is rare in the region, so extensive road networks must be built and maintained to remove large quantities of logs. These road networks lead to extensive soil erosion and provide commercial hunters access to previously remote tropical forests. They also facilitate the establishment of human settlements and permanent conversion of the logged forests into agricultural lands and large-scale plantations (see essay 12, by Jason Clay), thus preventing natural forest regeneration.

The frontier forests of the Indo-Pacific region are especially vulnerable to large-scale logging. Frontier forests represent the world's large, ecologically intact and relatively undisturbed natural forest ecosystems that are big enough to maintain all of their biodiversity, including viable populations of wide-ranging species (Bryant et al. 1997). Of the more than 16 million km² of original forest that once covered Asia and Oceania, only a little over 1 million km² of frontier forests remain, mostly in Papua New Guinea, Indonesia, and to a lesser extent Cambodia and along the inaccessible mountain ranges of continental Asia. Logging and its secondary impacts pose by far the greatest danger to these frontier forests (Bryant et al. 1997). Later in this chapter we illustrate how threats to frontier forests can be identified by analyzing logging concession data.

Wildlife Exploitation

Wildlife exploitation is pervasive throughout the Indo-Pacific region. A diverse arsenal consisting of small snares baited with seeds to trap pheasants and jungle fowl, log-fall traps that crush small deer and mongooses, pitfall traps set for elephants, explosive-laden bait to blow up tigers, and home-made crossbows and automatic rifles in the hands of hunters spares no animal, big or small.

Consumption goes well beyond large vertebrates such as tigers, elephants, and rhinoceros, whose parts are in high demand in the thriving wildlife trade (Menon 1996; Mills and Jackson 1994). A glance at a restaurant menu in Vietnam or a walk in any Sunday morning market in Laos or Thailand will show that there is also a heavy toll on the less charismatic species such as rats, squirrels, turtles, snakes, and even small birds. The trade in songbirds in Asia is enormous (Nash 1993). The volume of wildlife exported from Vietnam into China is staggering; for instance, in June 1996 two tons of live animals, including 200 macaque monkeys, were headed for China when they were confiscated at the Noi Bai airport in Hanoi (WWF-Indochina 1998). Because interceptions are rare, this figure represents only the tip of the iceberg of the actual volume of animals that are smuggled. In essay 13, Elizabeth Bennett and John Robinson provide more examples of the extent of this consumptive trade throughout the region and discuss the consequences to biodiversity unless immediate conservation actions are undertaken.

The intensity of exploitation also varies widely across this diverse region. In south Asia, levels of commercial wildlife exploitation and trade are low, presumably because there has been a long history of wildlife protection there, and the distance to China, the main consumer of wildlife products, is greater. In contrast, most of Indochina, which is much closer both geographically and culturally to China and where the history of protection is more recent, suffers from the "empty forest syndrome" (Redford 1992), where even seemingly intact forests support little wildlife. Even within a given bioregion, there are high- and low-threat areas. In the Indochina bioregion, for instance, the ecoregions that fall within Vietnam are now almost empty forests, but there is relatively more wildlife in the ecoregions of neighboring Cambodia, where the hunting pressure is lower.

As Judy Mills points out in essay 14, ultimately the lure of big money is too great to resist for many people who live around wildlife. If conservation of wildlife, and by extension biodiversity in its broadest definition, is to be successful, these people must be presented with persuasive economic arguments and incentives to curb their hunting.

Quantifying the extent and consequences of wildlife exploitation is a difficult task and at these scales is nearly impossible. Nevertheless, some broad patterns emerge at the bioregion scale. The ecoregions in the Indian subcontinent are less threatened by wildlife exploitation than are the other bioregions. At the other extreme, the ecoregions in the Indochina bioregion are severely threatened. Along this continuum lies the Sunda Shelf and Philippines, Wallacea, and New Guinea and Melanesia bioregions, all closer to the Indochina bioregion in their degrees of exploitation.

Overgrazing of Alpine Meadows

Taken together, the Himalayan ecoregions are by far the wealthiest ecoregions on Earth for alpine plants (Mani 1978). People rarely visit many of these meadows that lie high above all but a few permanent settlements. However, the meadows are heavily grazed by large herds of domestic livestock (essay 17). More rational grazing management is needed to save montane biodiversity because current grazing levels, coupled with the commercial harvest of alpine plants for traditional medicines, have greatly increased the stress on these ecosystems in recent years.

Degradation Threats: Human Population Density and Distribution

In general, there is a strong relationship between habitat degradation threats and human densities (Cincotta and Engleman 2000). In this threat analysis we used the distribution of human population densities across ecoregions to assess where the potential threat hotspots are most likely to be manifested. We did this by mapping the density of human populations in all ecoregions and then using patterns of distribution immediately around large, intact habitat blocks to assess threat levels. We considered a high-density population to pose greater threats than a low-density population and an evenly distributed population to pose greater threats than a clumped population because the threats are more widely distributed in the former. These two parameters were combined to create a categorical index to measure degradation threat. Thus, ecoregions with a population that was of high or medium density and evenly distributed faced sufficient threat from population pressures to warrant being elevated one threat category above its snapshot conservation status. Ecoregions where population distributions are clumped are considered to face a lesser threat because the threats are concentrated in a few places. We present the details of the methodology and the index in appendix D.

This analysis makes several assumptions, notably that there is a good correlation between the intensity of threats and human population density, that population growth rates are similar across the region, and that anthropogenic threats from different land-use patterns are consistent across the landscape. We consider these assumptions to be reasonable at the bioregional and ecoregional scales of this analysis. However, we stress that these generalizations and assumptions do not hold at smaller scales. Analyses at smaller spatial scales must consider differences in population growth rates and in the relative impacts of human activities (e.g., forest clearing for swidden agriculture versus nontimber forest product harvest).

Patterns of Population Distributions

As a general trend, population densities within bioregions decrease from west to east (figures 5.8 and 5.9). Overall, all biomes in the Indian Subcontinent bioregion have higher human densities than the biomes in the other bioregions (figure 5.10). However, across the bioregions, the tropical and subtropical moist broadleaf forests, tropical and subtropical dry broadleaf forests, and mangrove biomes were the most densely populated.

The ecoregions of the Indian Subcontinent bioregion have the highest human population densities in the region, with the majority of the ecoregions being in categories 6 and 7 (these are ecoregions that have a high population density but have no large habitat blocks remaining; figure 5.11). The distribution of population density in the Indochina bioregion is variable; several areas (mostly in Laos, Cambodia, and northern Thailand and Burma) have low population densities, whereas other areas (central Thailand, Vietnam, and peninsular Malaysia) have high densities. Thus, although the mean human density for this bioregion is lower, the standard deviation is high (figure 5.9), reflecting the disparate human population densities between ecoregions. Several ecoregions in the Sunda Shelf and Philippines bioregion also have high population densities, especially those in western Indonesia. In both of these bioregions, most ecoregions were in category 2 or 3, although a number of ecoregions are in categories 5, 6, and 7 (figure 5.11). About 80 percent of the ecoregions of the Wallacea bioregion are category 3, and the New Guinea and Melanesia bioregion has the lowest population density (figure 5.11).

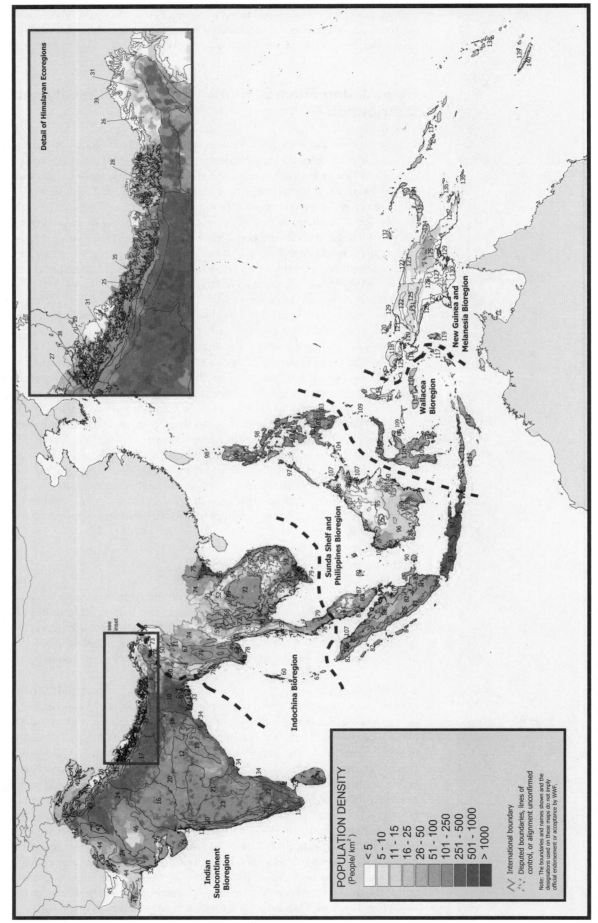

Figure 5.8. Population density in the Indo-Pacific region.

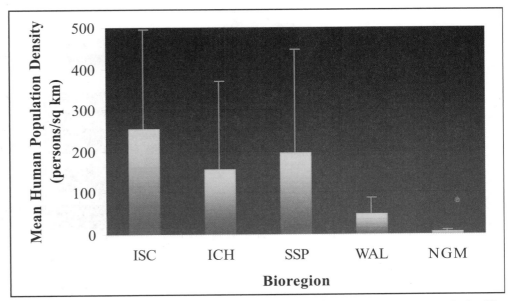

Figure 5.9. Mean human population density (+ *SD*) in ecoregions of the Indo-Pacific bioregions. Bioregions: ISC = Indian Subcontinent bioregion; ICH = Indochina bioregion; SSP = Sunda Shelf and Philippines bioregion; WAL = Wallacea bioregion; NGM = New Guinea and Melanesia bioregion.

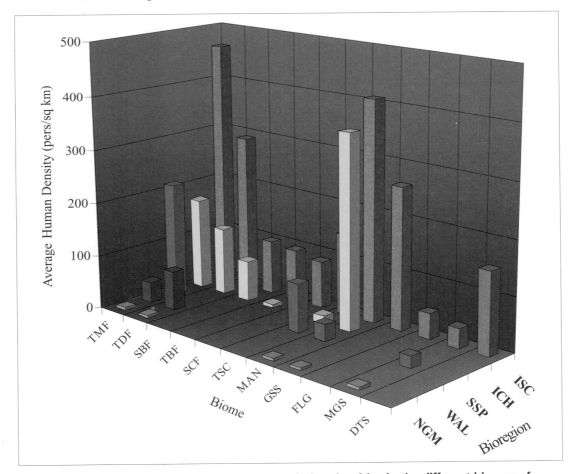

Figure 5.10. Distribution of mean human population densities in the different biomes of Indo-Pacific bioregions. Biomes: TMF = tropical and subtropical moist broadleaf forests; TDF = tropical and subtropical dry broadleaf forests; TSC = tropical and subtropical coniferous forests; TBF = temperate broadleaf and mixed forests; SCF = temperate coniferous forests; GSS = tropical and subtropical grasslands, savannas, and shrublands; FLG = flooded grasslands and savannas; MGS = montane grasslands and shrublands; DTS = deserts and xeric shrublands; MAN = mangroves. Bioregions: ISC = Indian Subcontinent; ICH = Indochina; SSP = Sunda Shelf and Philippines; WAL = Wallacea; NGM = New Guinea and Melanesia.

Where Is Biodiversity under Greatest Threat?

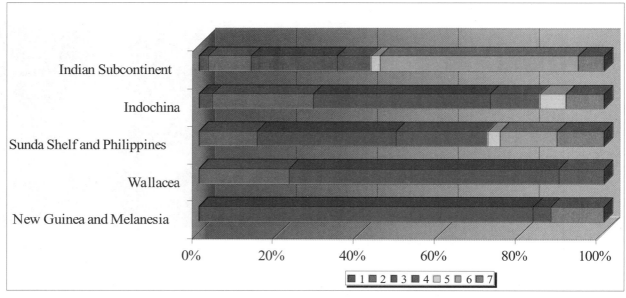

Figure 5.11. Percentage of ecoregions of each bioregion in the population threat categories. 1 = sparse population distribution; 2 = low density, clustered distribution; 3 = low density, even distribution; 4 = medium density, clustered distribution; 5 = medium density, even distribution; 6 = high population density; 7 = high population density, no large habitat blocks.

Population Density and Habitat Loss

The high correlation ($r^2 = 0.61$) between human population density (log people/km^2) and the percentage of habitat loss in forest ecoregions (the analysis excludes the nonforested alpine and sub-alpine, grassland, and desert ecoregions) of the Indo-Pacific bioregions (figure 5.12) support the assumption that ecoregions with a high human density experience greater levels of habitat loss. The ecoregions of the Indian Subcontinent and Sunda Shelf and Philippines bioregions that have higher human population densities have lost a greater percentage of forest cover. The three outliers from the New Guinea and Melanesia bioregion in the bottom right quadrant of figure 5.12 are the moist and dry forest ecoregions of New Caledonia and the moist forests of Solomon Islands. These ecoregions have lost more than 75 percent of forest cover despite the low human population density.

New Caledonia has several large nickel and cobalt mines that, combined with nutrient deficient soils and a long history of logging, burning, ranching, and agriculture, have effectively converted much of the island's forests despite the low human population density.

Population-Related Threats and Conservation Status

On the basis of human population–related threats, we elevated the conservation status of forty-five ecoregions across the five bioregions. One ecoregion, Indus Valley Desert [47], was changed from relatively stable to vulnerable, seventeen ecoregions were elevated from vulnerable to endangered, and eight went from endangered to critical status. The other nineteen ecoregions were already critical under the snapshot conservation status, and of these, nine ecoregions with high population densities lacked any large habitat blocks (category 7).

The distribution of the human population across biomes is unequal, and this is reflected in the percentages of ecoregions in each biome whose conservation status classifications changed because of population pressures. Thirty of the forty-five ecoregions that have high population-related threats are in the tropical and subtropical moist forests biome (although only fifteen actually changed status; the rest were critical already), and these ecoregions are distributed across all bioregions except Wallacea. They include lowland and montane rain forests and freshwater swamp forest ecoregions.

Eight of the sixteen tropical dry forest ecoregions have high population-related threats, and six changed snapshot conservation status. Except the New

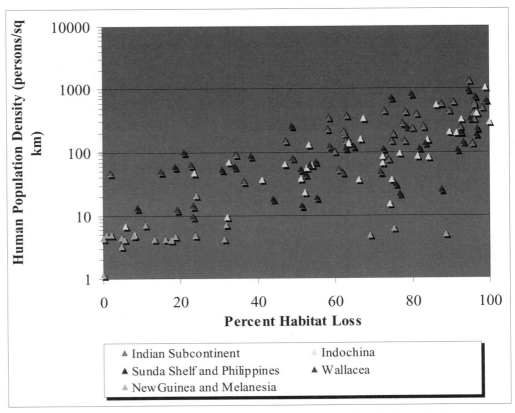

Figure 5.12. Relationship between population density and habitat loss for forest ecoregions of the Indo-Pacific bioregions. The relationship excludes grasslands, alpine, sub-alpine, and desert systems.

Caledonia Dry Forests [140], all ecoregions that underwent a change in conservation status are in the Indian Subcontinent bioregion. The remaining ecoregions in other biomes that underwent a change in snapshot conservation status are three mangrove ecoregions, three desert and xeric shrublands ecoregions, and the Terai-Duar Savanna and Grasslands [35] (appendix F).

None of the ecoregions in the subtropical and temperate forests or alpine habitats had a high enough human population density and distribution to warrant a change from the snapshot conservation status. The Himalayan alpine ecoregions are classified as relatively stable or intact for snapshot conservation status, but as Daniel Miller points out in essay 17, vast areas of these high-elevation meadows are heavily grazed by domestic livestock (especially yak, *Bos grunniens*), and the habitat is undergoing severe degradation. Thus, despite the low human population density, these habitats are highly threatened by anthropogenic impacts unrelated to population density. A similar analysis using domestic livestock densities probably would change the snapshot conservation status of these ecoregions.

High-Intensity Threats: Logging Concessions and Habitat Loss

Threats from logging concessions will be most serious in areas with large forest blocks, where large-scale operations and habitat conversion would be economically viable and attractive. We used logging concession data from Indonesia, parts of Malaysia, Papua New Guinea, and Cambodia to evaluate conversion threats. Concession data for many of the other ecoregions are difficult to obtain or are unavailable for a wider and more comprehensive analysis. But many of the ecoregions for which we do not have data have already been severely logged and are economically unattractive for future large-scale logging operations. The population threat analysis to evaluate degradation threats is more appropriate in these ecoregions.

The primary purpose of the analysis is to illustrate how high-impact threats can change the status of seemingly intact ecoregions, thereby highlighting the need for a proactive conservation strategy. But we attach two caveats to the analysis. First, legal logging concessions are not the only source of forest habitat loss. Illegal and uncontrolled logging practices, even within legal concessions (Global Witness 1999), are common in many countries with substantial timber resources. The extent and range of this type of logging ranges from small-scale isolated events to well-financed, large-scale operations. This type of logging is hard to quantify, and we are unable to account for it in our analysis. Second, logging concessions also change, are revoked, or are reallocated depending on the government and changes to the economies. We recognize that these dynamic factors can also change the trajectory of this analysis and consequently alter future conservation status.

In this analysis, the ecoregions under high threat levels from logging concessions were assigned to the critical conservation status regardless of the snapshot conservation status because of the intensity of the threats from commercial logging and its pervasive secondary effects. Details of the analysis and methodology are presented in appendix D.

Thirty-eight ecoregions from four bioregions were assessed for logging impacts. Of these, twenty-two were either relatively stable or relatively intact by the snapshot conservation status because they still have contiguous, large areas of intact habitat. After application of the logging threat analysis, the conservation status of twenty-two (of the thirty-eight) ecoregions was changed to critical (table 5.3). Logging concessions will fragment the large, intact blocks of habitat that now exist in Cambodia, Borneo, Sulawesi, and New Guinea (figure 5.13). In Cambodia, the Southeastern Indochina Dry Evergreen Forests [59] was assigned a critical status based on logging concessions. In Sumatra, every ecoregion, with the exception of the Sumatran Montane Rain Forests [83] and Sumatran Tropical Pine Forests [105], was changed to critical status. The ecoregions in Borneo and Sulawesi are similarly threatened by large-scale commercial logging (and oil palm plantations), and all but the Borneo Montane Rain Forests [95] changed to critical status. Because all the ecoregions in the Wallacea bioregion are under only moderate logging threats, none had their conservation status increased. In the New Guinea and Melanesia bioregion, most of the lowland forest ecoregions of New Guinea and surrounding islands were changed to critical status. Only the Southeastern Papuan Rain Forests [126], Southern New Guinea Freshwater Swamp Forests [127], Trobriand Islands Rain Forests [135], and the montane forest ecoregions of New Guinea's central spine remained unchanged.

We excluded some of the smaller ecoregions in the Pacific islands, such as the Solomon Islands and Vanuatu, from the analysis. Although they are also severely threatened by logging concessions, sufficient data were unavailable. Concession maps should be used to reanalyze the conservation status of these ecoregions to determine threat levels related to logging.

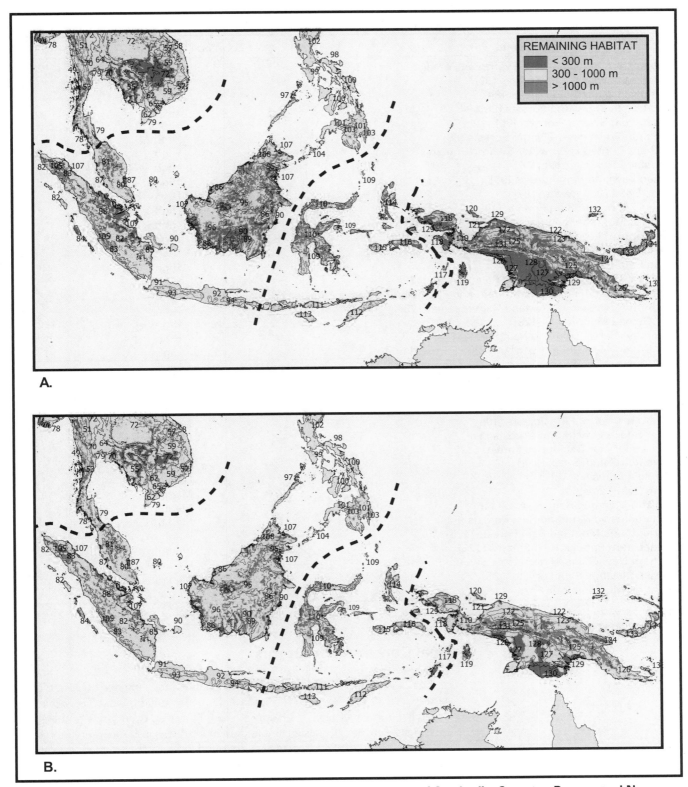

Figure 5.13. The consequences of logging concessions in the forests of Cambodia, Sumatra, Borneo, and New Guinea. The approximate extent of archipelagic forests in 1997 shows large swaths of lowland forest in Borneo and New Guinea (A). If the logging concessions are implemented, the large lowland forest patches will be fragmented and will resemble the landscapes of the ecoregions in the Indian Subcontinent bioregion, which were logged more than two centuries ago (B). Thus, the presence of relatively intact, large forest patches in several ecoregions is no cause for complacency. Large-scale logging presents imminent and severe threats to habitat and biodiversity of these ecoregions, threats that will be manifested within short time periods. Proactive conservation efforts should target these seemingly intact and safe ecoregions before conservation opportunities are lost.

TABLE 5.3. Results of the Logging Concession Threat Analysis.

Ecoregion Name	Logging Concession Threat
Southeastern Indochina Dry Evergreen Forests [59]	High
Sumatran Lowland Rain Forests [82]	High
Mentawai Islands Rain Forests [84]	High
Sumatran Peat Swamp Forests [85]	High
Borneo Peat Swamp Forests [86]	High
Sumatran Freshwater Swamp Forests [88]	High
Southern Borneo Freshwater Swamp Forests [89]	High
Sundaland Heath Forests [90]	High
Borneo Lowland Rain Forests [96]	High
Sunda Shelf Mangroves [107]	High
Sulawesi Lowland Rain Forests [109]	High
Sulawesi Montane Rain Forests [110]	High
Vogelkop-Aru Lowland Rain Forests [119]	High
Biak-Numfoor Rain Forests [120]	High
Yapen Rain Forests [121]	High
Northern New Guinea Montane Rain Forests [122]	High
Northern New Guinea Lowland Rain and Freshwater Swamp Forests [123]	High
Southern New Guinea Lowland Rain Forests [128]	High
New Guinea Mangroves [129]	High
Admiralty Islands Lowland Rain Forests [132]	High
New Britain–New Ireland Lowland Rain Forests [133]	High
New Britain–New Ireland Montane Rain Forests [134]	High
Cardamom Mountains Rain Forests [55]	Medium
Central Indochina Dry Forests [72]	Medium
Borneo Montane Rain Forests [95]	Medium
Sumatran Tropical Pine Forests [105]	Medium
Lesser Sundas Deciduous Forests [111]	Medium
Timor and Wetar Deciduous Forests [112]	Medium
Halmahera Rain Forests [114]	Medium
Buru Rain Forests [115]	Medium
Seram Rain Forests [116]	Medium
Vogelkop Montane Rain Forests [118]	Medium
Huon Peninsula Montane Rain Forests [124]	Medium
Central Range Montane Rain Forests [125]	Medium
Southeastern Papuan Rain Forests [126]	Medium
Southern New Guinea Freshwater Swamp Forests [127]	Medium
Trobriand Islands Rain Forests [135]	Medium
Sumatran Montane Rain Forests [83]	Low

The conservation status of ecoregions with a high logging concession threat was elevated to critical.

Final Conservation Status

In the final analysis, sixty-eight ecoregions qualified for a change in conservation status to reflect the projected threats from human population and large-scale logging-related threats (figure 5.14; appendix F). However, twenty-two of these ecoregions were already designated as critical for the snapshot conservation status. Of the other forty-four ecoregions that underwent a change in classification, eighteen were from the Indian Subcontinent bioregion. These are changed because of chronic, low-intensity threats. In the Indochina bioregion, two ecoregions underwent a change in snapshot conservation status, one for logging concession–related threats and one for population threats. However, a broader application of this threat analysis for other parts of Indochina (e.g., Laos, Thailand, Myanmar), using additional parameters such as wildlife exploitation, plantations, and hydropower schemes, probably would lead to a change in status for several other ecoregions as well. In the Sunda Shelf and Philippines bioregion, twelve ecoregions were elevated to a higher threat level from the snapshot conservation status. Of these, five are for population-related threats and seven are for threats from large-scale logging concessions. Most of the ecoregions in Borneo, Sumatra, and the endangered ecoregions in Java became critical be-

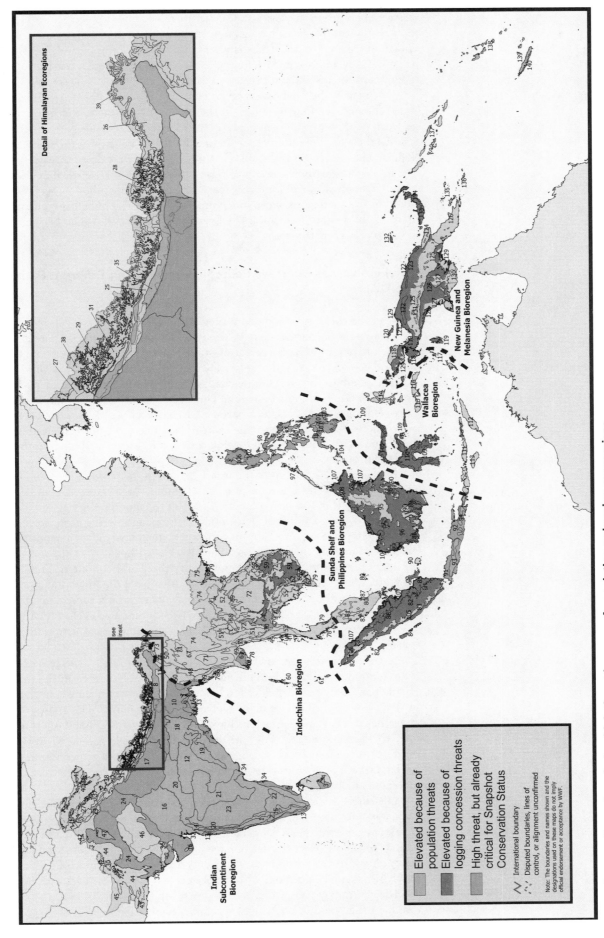

Figure 5.14. Ecoregions facing high threat because of population or logging concessions.

Where Is Biodiversity under Greatest Threat?

cause of projected threats. In Borneo and Sumatra in particular, the logging concessions place these ecoregions on the critical list. The two ecoregions of Sulawesi in the Wallacea bioregion warranted a change in conservation status based on logging threats. The snapshot conservation status of twelve ecoregions from the New Guinea and Melanesia bioregion were changed; two because of population threats and ten because of threats from logging concessions.

The map of the final conservation status for the Indo-Pacific region (figure 5.15) shows several broad trends. In the Indian subcontinent bioregion, the ecoregions along the alluvial plains of the Ganges and Brahmaputra rivers were critical for snapshot conservation status. But the ecoregions in the vast Deccan Plateau changed to critical status because of population-related threats, so that in the final status many of the ecoregions in this bioregion are critical or endangered. In contrast, there were very few changes to the snapshot conservation status in the ecoregions of the Indochina bioregion. The only major change was in the Southeastern Indochina Dry Evergreen Forests [59], which became critical because of logging threats.

Lowland versus Montane Ecoregions: Following Different Paths to the Future

None of the thirty-two montane ecoregions (figure 5.2) are critical for the snapshot conservation status, whereas twenty-seven lowland ecoregions are critical. At the other end of the conservation status scale, fourteen montane and fourteen lowland ecoregions are either relatively intact or stable. Ten of these lowland ecoregions are in the New Guinea and Melanesia bioregion, three in Wallacea, and one in the Indochina bioregion. However, when the threat indices are applied, eight lowland ecoregions become critical because of large-scale logging concessions. Among the montane ecoregions, three ecoregions changed to critical. Thus, as we would expect, logging concession threats are higher in the lowland ecoregions.

On average, 20 percent of the intact habitat of the montane ecoregions (which include terrestrial ecoregions in all biomes) are protected, compared with only 11 percent in lowland ecoregions. Two of the montane ecoregions along the border of Myanmar and India have no protection, and twelve ecoregions have less than 10 percent protection. In contrast, sixteen lowland ecoregions have no protection, and another forty-three have less than 10 percent of the intact habitat under protection. Twenty-five and twenty-three of the montane and lowland ecoregions, respectively, have more than 15 percent of the intact habitat within protected areas. But because there are ninety-eight lowland ecoregions (total area 7.2 million km^2) compared with forty-two montane ecoregions (2.1 million km^2), this represents a disproportionately lower amount of protection for the lowland habitats.

In general worldwide, lowland forests tend to have higher species richness than montane forests but lower degrees of endemism. However, in the Indo-Pacific region, there is a greater tendency toward higher endemism in the lowland ecoregions because many represent island habitats. The analysis indicates that the lowland ecoregions have suffered the greatest habitat loss and have the least protection. They are also the most densely populated and face the greatest threats from logging concessions. Thus, additional protection in the lowland island ecoregions is desperately needed to better conserve their irreplaceable biodiversity.

The threat analysis indicates that even ecoregions with large areas of remaining habitat and sparse human populations can be threatened by large-scale developments, which can be much more devastating and occur in a short time period. Therefore, the presence of large habitat blocks is no cause for complacency. A prioritized portfolio must consider a bifurcated approach in which highly threatened ecoregions with irreplaceable biodiversity (e.g., with high endemism and rare species and processes) are given high priority, as are ecoregions that have high biodiversity values and intact habitat landscapes.

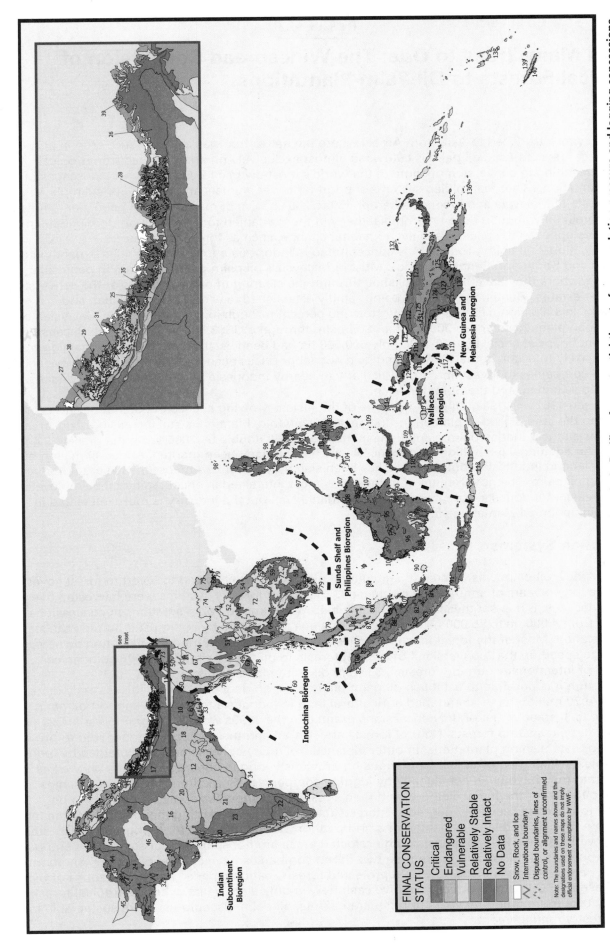

Figure 5.15. The final conservation status of the terrestrial ecoregions of the Indo-Pacific, after potential threats from population pressure and logging concessions are applied.

Where Is Biodiversity under Greatest Threat?

ESSAY 12

From Many Trees to One: The Widespread Conversion of Tropical Forests to Oil Palm Plantations

Jason Clay

Oil palm was introduced to Asia from Africa, where the native tree was a subsistence crop in agroforestry gardens. In Asia, oil palm is cultivated almost exclusively on areas cleared from tropical forests, resulting in conversion of some of the world's most diverse forests into vast landscapes of oil palm plantations. These simplified systems support far fewer animals and plants. For example, the rain forests of Malaysia and Indonesia harbor 150 to 200 species of mammals. Oil palm plantations may contain only about a dozen mammals, many of them adapted to life in human-dominated landscapes, as opposed to species sensitive to habitat change, such as forest primates.

Today, global oil palm production is concentrated in Indonesia and Malaysia and is expanding into Oceania and South and Central America. Most of Malaysia's oil palm plantations are in peninsular Malaysia, where they have been established through the clearing of natural forests or the conversion of rubber estates. Plantations are now being rapidly established in east Malaysia (Sabah and Sarawak). In 1998 alone, fifty foreign investors (80 percent through joint ventures with Malaysian companies) had plans to establish 900,000 ha of oil palm through a US$3 billion investment. Because of the recent financial crisis, by mid-1998 only 600,000 ha had been established, but the remaining plantings are likely to occur. In the 1990s, rising land and labor prices encouraged some fifty Malaysian palm oil companies to invest abroad, particularly in nearby Indonesia and in Cambodia, Thailand, Papua New Guinea, and the Solomon Islands.

Between 1967 and 1997, oil palm was one of the fastest-growing subsectors of Indonesia's economy, when the land placed in plantations increased twentyfold. Planting is still increasing. Conservative estimates suggest that oil palm plantations in Indonesia will double by 2005, from the present 2.4 million ha. An additional 5.5 million ha of concessions have already been granted. About 40 million ha of forested land in Indonesia's outer islands has been slated by the government for cash crop production, including oil palm. The government of Indonesia is most interested in encouraging oil palm development in eastern Indonesia (e.g., Kalimantan and Irian Jaya), but the industry is more interested in developing palm oil plantations in Sumatra.

Production Systems

In west Africa, oil palm was planted in agricultural plots that were allowed to revert to forest cover after one to two years of annual crop production. These agroforestry systems were harvested over several decades. But in southeast Asia, oil palm is grown on monocrop plantations in concessions ranging from 4,000 ha to 75,000 ha or more. Plantations are created by cutting and burning the standing vegetation. Most of the forest fires in Indonesia in 1997–98 were started by plantation owners who wanted to speed up the "conversion" of tropical forests to plant oil palm; of the 176 companies accused of intentionally burning forests, 133 were oil palm plantation companies.

Although it is possible to establish oil palm plantations on degraded areas, this is rarely done. There are 20 million ha of abandoned agricultural land in Indonesia that would support oil palm plantations. Instead, most of the concessions granted in the 1990s were in natural forest areas, particularly in peat swamp forests. Natural forests are cleared and their valuable timber sold to provide capital for establishing plantations. In other instances, oil palm concessions are obtained by logging companies, which then establish plantations on previously logged "degraded" lands (as well as in forests that may have been cut illegally by logging companies). Sometimes logging companies receive permits for the initial clearing, and then subsidiary companies apply for permits to be allowed to clear the "severely degraded" forests for oil palm, thus skirting the intent if not the letter of the law. The cleared land is planted in a grid pattern with little regard for topography (e.g., steep slopes, wetlands, or riparian areas). Even protected areas are not spared by unscrupulous and powerful companies. In Indonesia, more than two dozen companies have created oil palm plantations illegally on more than 100,000 ha within protected forests, national parks, and buffer zones of biosphere reserves. Oil palm produces positive cash flow in only eight years, whereas tropical forests typically take more than fifty years to regenerate. Thus, oil palm discourages more sustainable long-term forestry management.

Palm nuts spoil within forty-eight hours of harvest and must be processed quickly. Therefore, oil palm operations must be large to be economically viable. The minimum plantation size to supply a state-of-the-art processing plant is 50,000 ha. Because of this production constraint, small holdings that may be more environmentally benign are not economical except as subsidiary plantations that sell their palm nuts to large-scale plantations that have processing factories. To increase economic viability, small holders often clear additional forests around larger plantations.

The Malaysian government has successfully lobbied for rubber plantations to be classified as forest by the FAO, a move that could be extended to oil palm. In both Malaysia and Indonesia, oil palm is already integrated into development within what is called locally the permanent forest estate. However, planting oil palm in these forests would have an adverse affect on the area's biodiversity. People lured to work on the plantations will have a continued and profound negative affect on biodiversity through killing or harvesting of local flora and fauna.

Market Trends

The main traded products from oil palm plantations are crude palm oil, palm kernel oil, and stearin, a refined product used to make soap. Palm oil is substituted for other vegetable oils, depending on price and availability. It also substitutes for cocoa butter and coconut oil for most personal care and cosmetic products. Hydrogenated palm oil can be substituted for animal fats in many foods. Large quantities of heavier palm oil are used by the steel industry during the cold rolling of steel plate to lubricate and prevent surface corrosion.

Oil palm is very competitive in vegetable oil markets. The price of palm oil declined by 30 percent in 1999, probably because of increasing supplies. The increasing volume of oil on world markets and declining prices will be further exacerbated because Indonesia recently cut its export taxes on palm oil exports by 30 percent. Malaysia also has large palm oil surpluses and is trying to gain market access to India and China by delivering palm oil on credit. This will affect markets also and may establish a trend in which producers from countries with less working capital will find it hard to remain competitive. And recent plantings in Indonesia will begin to produce in a few years, adding to the glut.

But most vegetable oil analysts believe that the markets for vegetable oil and meal will continue to increase. In fact, palm oil consumption is predicted to increase by 10 million metric tons between 1996 and 2005 which would make it the most consumed oil in the world. From 1990 to 1996, global palm oil consumption increased by 5.44 percent per year, the fastest growth rate of any edible oil. Total consumption of palm oil is second only to soy oil.

Thus, despite the depressed prices, the oil palm industry is poised to expand. Planting targets are increasing. The Indonesian government has reduced the 60 percent export tax on oil palm, lowered interest rates, and changed regulations to encourage exports and the expansion of the industry. Cooperation between Indonesian and Malaysian producers will allow them to work together to push up the price of oil palm and to increase their share in the world vegetable oil market. It is clear that the expansions in plantations will pose severe threats to some of the most species-rich and diverse ecosystems in the world.

Conservation Strategies

Several strategies can reduce the overall impact of oil palm plantations. Given the rate of expansion of the industry and the market, the first would be to encourage the industry to identify and then adopt cost-effective strategies that reduce its impact.

The identification and mapping of priority areas for biodiversity, in conjunction with both existing and proposed plantations, will allow the implementation of appropriate conservation measures to ensure the establishment of adequate conservation areas, including core areas and habitat linkages to allow dispersal. Proper siting of plantations that will spare riparian habitat, avoid steep slopes, and protect wildlife habitat will address most of the environmental problems. This will include engaging the decision-makers, especially in the regional government, who control permits and concession agreements for expansion of oil palm plantations.

Viable oil palm operations depend on buyers, lenders, and investors. Conservationists should appeal to each of these players. An important strategy will be to develop industry-wide investment screens that could actually stimulate investments in more sustainable palm oil businesses. Expanding businesses need a great deal of capital, and at present all major Dutch banks are involved in oil palm investments in Indonesia. Approaching one of the more progressive banks about the problems associated with the unchecked expansion of oil palm plantations and the potential to create a viable investment screen could reduce the industry's impact.

Important work to be undertaken as part of an effective conservation strategy for oil palm plantations could include the following:
- enforcement of existing laws, regulations, and concession agreements
- creation and adoption of appropriate siting criteria
- identification of clearing practices that include zero-burning techniques
- identification of minimal-size, biologically significant and viable forest fragments and forest corridors within and between oil palm concessions
- investigation of the potential liability for flooding and fire caused by establishing oil palm plantations

A parallel strategy would be to work with large-scale vegetable oil purchasers to use similar best management practice–based screens as a condition for oil purchases. This would signal to producers that buyers are interested in purchasing palm oil only if it is produced with acceptable practices. Unilever and Ecover are companies that might be sympathetic to such a screen. In fact, Unilever is vertically integrated in palm oil and claims to control 1 percent of global palm oil production. Educating consumers about the companies they invest in or whose oil palm products they purchase could exert important pressure on an industry whose expansion may accelerate the decline of southeast Asia's biodiversity-rich lowland forests.

ESSAY 13

Into the Frying Pan and onto the Apothecary Shelves: The Fate of Tropical Forest Wildlife in Asia?

Elizabeth L. Bennett and John G. Robinson

No longer can we delude ourselves into thinking that we can ensure survival of all species of animals and plants simply by establishing representative protected areas throughout the world's ecoregions. Across the landscape, both inside and outside parks and reserves, people increasingly are harvesting wild species. Our voracious appetite for almost anything that is large enough to be eaten, potent enough to be turned into medicine, and lucrative enough to be sold is stripping wildlife from wild areas, leaving empty forest shells and an unnatural quiet.

Ever since they first inhabited Asia's forests some 40,000 years ago, humans have hunted wild animals. Wild meat has been the major source of animal protein for forest people throughout the region and remains so today in many areas. The hunted species must have been able to withstand such pressure in the past. But the situation has changed, and hunting of most species in most areas is no longer sustainable. What happened? Although it has played out with innumerable local variations, a general trend can be discerned. The forest estate itself shrank and became fragmented as forests were converted for human settlement and agriculture, including large-scale commercial ventures such as oil palm. Road and rail building, dam construction, and the like have opened up the forest to the pecuniary demands of the modern market state. With access into and out of the forest, forest-dwelling people have increasingly entered the cash economy, and immigrants have flooded in. Patterns of hunting have changed concomitantly. Local demand increased, traditional hunting practices and taboos were lost, efficient modern technology such as shotguns and wire snares was increasingly used, and, perhaps most of all, hunting was increasingly commercialized.

For forest-dwelling peoples, the distinction between subsistence and commercial hunting is rarely clear, with patterns ranging from an additional animal hunted for sale on rare occasions when the hunter goes to town, to frequent hunting to supply a regular trader, to full-scale professional hunting. Selling wildlife is an easy source of money for rural people, with the money used not only for essential and desired commodities but also to buy technologies to increase the efficiency of the hunt itself, thereby entering a spiral of increasing harvest rates. At the consumer end, the availability of new wildlife products creates and stimulates new markets, and the increased consumer spending power fosters the demand for wildlife meat and other products such as pets and traditional animal-based medicines.

The scale of the wildlife trade in Asia is difficult to assess accurately, but it is vast. In a single market in north Sulawesi, an estimated 3,848 wild pigs were sold every year from 1993 to 1995, of which a third to a half were the endangered and legally protected babirusa. Market sales also included macaques (50 to 200 a year), forest rats (50,000 to 75,000 per year), bats (up to 15,000 per year), and occasional sales of rare cuscus and tarsiers (Clayton and Milner-Gulland 2000). In the single Malaysian state of Sarawak, the wild meat trade in 1996 was conservatively estimated to be more than 1,000 tons per year and comprised mainly large ungulates: bearded pigs, sambar deer, and barking deer (Wildlife Conservation Society and Sarawak Forest Department 1996). In Indochina, the wildlife trade for food is more eclectic, with almost all vertebrate species being eaten. In southern Laos, for example, until recently there was probably little trade, with most villages isolated and largely outside mainstream market economies. By 1990, however, the Lao Theung people were deriving 30–60 percent of their income from selling wildlife and other forest products (Chazée 1990). In a single food market at That Luang in southern Laos, in 1992 annual sales involved 8,000 to 10,000 mammals, 6,000 to 7,000 birds, and 3,000 to 4,000 reptiles (Srikosamatara et al. 1992).

Wildlife is also sold for the trophy and pet trade: Bangkok weekend markets sold an estimated 42,840 native birds and 1,954 exotic birds from 276 different species in 1990 (Round 1990). The traditional medicine trade also consumes huge numbers of animals, but because much of this is illegal, data are difficult to obtain. However, some numbers are instructive. Between 1970 and 1993, key east Asian consuming countries imported at least 10 tons of tiger bones from other parts of Asia (Mills and Jackson 1994), probably representing some 500 to 1,000 tigers (Jackson and Kemf 1994). Although the large species get most of the publicity, small ones are also traded on a large scale; every year from 1984 to 1989, 1.2 million to 2.4 million geckoes (whose tails are believed to have medicinal properties) were exported from Bangkok's main port and airport (Srikosamatara et al. 1992). Again, this trade represents a spectrum from indigenous Bornean forest hunters harvesting additional porcupines and langurs to sell locally for their medicinal bezoar stones, to Laotian trappers hunting principally for food but selling any incidental catches of tigers, to commercial hunters targeting particular highly valuable species.

As a consequence of this vast trade, hunting is no longer sustainable in most forests in Asia. Out of ten species harvested in north Sulawesi, harvest levels of seven were unsustainable, and in eight sites across Asia, hunting had reduced large mammal densities by an average of 64 percent (Bennett and Robinson 2000). The usual eclectic nature of the hunt means that populations of a wide range of mammal, bird, and reptile species are affected. Sustained heavy hunting is extirpating vulnerable species, especially large-bodied species with low intrinsic rates of natural increase. In Sarawak, for example, hunting for a combination of subsistence and trade resulted in all diurnal primates being locally extirpated from three out of four heavily hunted sites surveyed and barking deer in two of them (Bennett et al. 2000), hunting of long-tailed porcupines for bezoar stones has caused severe local population declines, and the trade in straw-headed bulbuls for the Indonesian songbird market has decimated the species throughout southeast Asia. On the island of Sulawesi, babirusa and anoa are suffering range reductions as a result of unsustainable harvesting (Alvard 2000), and in Laos, populations of many large mammals have suffered dramatic declines because of hunting for trade (Rabinowitz 1998). As the preferred species in trade are extirpated from local areas, hunters shift to other, less preferred species. Some of these, such as primates and hornbills, have low productivity and little resilience to hunting.

The extirpation of hunted species has wider effects on the forests themselves. The species preferred by hunters generally are large-bodied, typically fruit eaters and herbivorous browsers. These species often play keystone roles in forest ecology as pollinators, seed dispersers, and seed predators and make up the majority of the vertebrate biomass. Their reduction or extirpation produces cascading effects throughout the biological community. In India, for example, human hunting can result in reductions of more than 90 percent of the ungulate prey eaten by tigers (Madhusudan and Karanth 2000). The consequent reduction in tiger densities and dietary shifts of the remaining tigers have further impacts on the biological community. Throughout Asia, the result is that forests are becoming increasingly devoid of wildlife, the classic "empty forest" syndrome (Redford 1992).

There are no easy ways to break this downward spiral of loss. Local forest people are doing much of the hunting and selling of wildlife, but they often have few other commodities that they can sell. They are also the first to suffer when the wildlife on which they depend for food and cash no longer exists. Management that both protects wildlife resources and modulates the boom-and-bust commercial markets must be put in place. The problem is multifaceted and diffuse, both geographically and socially, and solutions must be sensitive to this complexity. Management must include targeted trade bans, with proper legal, administrative, and logistical mechanisms for enforcing them and education programs at all levels of society. Commercial trade must be regulated, and valuation of wildlife products must incorporate ecological and social values. Proper land-use planning is essential, with a mosaic that includes strictly protected areas where hunting is prohibited both in law and in practice, areas that act as sources of wildlife for subsistence hunting in surrounding extractive areas. Productivity of tropical forests for wildlife is extremely low (Robinson and Bennett 2000), so even in the absence of significant commercial trade, hunting can no longer fully support the needs and aspirations of the rural people in much of Asia today. Therefore, further areas are needed where intensive agricultural or other forms of production can supply the food and cash needed to meet those needs and aspirations as part of a sustainable landscape (Robinson 1994).

Although the overall situation is appalling, progress is being made on local fronts. Sarawak has instituted a ban on all trade of wildlife and wildlife products taken from the wild and is enforcing it rigorously; rural communities and their leaders are supporting this ban strongly because hunting for subsistence is still allowed, and they see this as conserving their wildlife resources. India and Nepal strictly protect the large animals in many of their reserves from hunting. These examples offer rays of hope. As long as wildlife is seen as an open-access commodity to be harvested and traded, however, forests throughout Asia will become increasingly silent, their diversity gradually declining, until their former vibrant grandeur becomes a fading memory.

Essay 14

Taking the Bounty off Wildlife in Asia

Judy Mills

An insidious, pervasive threat to long-term conservation in Asian ecoregions is the rampant growth of the commercial wildlife trade. Though not as easily quantifiable as habitat loss or fragmentation or as easy to depict in a GIS, commercial wildlife trade has placed many wild animals and plants in jeopardy and by extension the ecosystems of which they are an integral part. The prices of wildlife products from tiger skins to ginseng roots to musk deer scent pods are worth small fortunes to poor rural hunters across the region. To complicate this economic dynamic, a conservation ethic has only recently germinated in the many countries and cultures involved. The crux of any solution will be taking the bounty off wildlife, somehow making it as valuable to protect as to poach.

The effects of international trade in wildlife and wildlife derivatives have been devastating in Asia, tipping the balance toward extinction for some populations of rhinoceros, tigers, bears, elephants, ginseng, and numerous other species. Although it should be noted that trade pressures are seldom stochastic in themselves, they clearly can be pivotal, deadly factors for already-fragmented populations suffering from habitat destruction and competition for space with humans.

Like most issues of wildlife conservation, the elimination of unsustainable trade rests with addressing complex human factors. Most formidable among these factors are poverty, unequal distribution of wealth, and a cultural allegiance to revered traditions of using wild animals and plants to enhance human health.

The tiger provides a representative example that applies to most wildlife "commodities" in Asia. The tiger lives exclusively in countries with low per-capita income and large rural populations. Thus, tigers live next to a populace in legitimate need of supplementary income but very few legitimate ways to earn it. An opportunistic villager in the forest on a hunting foray stands to make the equivalent of at least a year's wage if he poaches a tiger. He may even become a local hero because of his hunting prowess, for his newfound ability to provide for his family, and for making the forest a safer place for humans and livestock.

Networks of intermediaries stand close at hand and ready to buy the hunter's wares. Some are organized; some are not. All are somehow connected, often by black market channels, to consuming markets. Profits can escalate exponentially as tiger parts change hands again and again en route to end-use consumers, who have some of the highest per-capita incomes in the global economy.

Convincing consumers to stop the demand that drives poaching is equally complicated. Again using the tiger as an example, tiger bones have been valued for treating rheumatic disorders in traditional Asian medicine systems for centuries. When a poaching surge in the late 1980s and early 1990s was linked to the medicinal demand for tiger bone, international conservationists were quick to criticize traditional Chinese medicine for touting the use of tiger bone as medicine. This criticism peaked in 1993, when the United States government—prompted by conservationists—threatened China, South Korea, and Taiwan with trade sanctions unless they stopped all commerce in tiger bone. China, South Korea, and Taiwan did ban the trade, but the Chinese and Korean traditional medicine communities at first complied reluctantly. Many perceived the bans as fallout from Western cultural imperialism or a thinly veiled effort to destroy the traditional Asian medicine industry. Nonetheless, this international pressure prompted both sides to communicate and eventually cooperate. By 1999, the traditional medicine establishment in China had officially embraced its role and responsibility in ending all trade in tigers and their parts, a milestone for traditional Chinese medicine as a whole.

At the start of the twenty-first century, conservationists seem poised, at last, to sufficiently address the stubborn social, economic, and political reasons that tigers and other commercially valuable species of wild animals and plants are worth more dead than alive to the very people who could ensure their survival. This phenomenon is beginning to be addressed with success in places such as Nepal's Royal Chitwan National Park, where villagers living next to tigers, rhinoceros, and bears are receiving direct economic benefits for protecting and expanding wildlife habitats and refusing to poach (Dinerstein et al. 1999). An effective antipoaching network pays informants at least as much as a poacher would earn to hunt illegally. These components—economic incentives for rural people, effective antipoaching networks, and cooperation with the traditional medicine communities—are the keys to maintaining healthy wildlife populations rather than structurally intact forests empty of their native, large mammals (Redford 1992). Herein lies the hope for stopping unsustainable wildlife trade in Asia.

Essay 15

Protected Areas: Keystones to the Conservation of Asia's Natural Wealth

Colby Loucks and Eric Dinerstein

Few aspects of landscape-scale conservation are as controversial as the role of strict protected areas. Why this is so is puzzling. One only has to compare the impact of protected areas throughout the Indian subcontinent with nature reserves in Indochina or Indonesia. That large mammals are still holding on in many south Asian reserves, even in the face of intensive poaching pressure and degradation in some areas outside reserves, is truly a cause for optimism. From Sri Lanka to India to Nepal, reserves once set aside by maharajahs or by colonial rulers for their hunting exploits, which later became national parks, still maintain a visible if not vibrant megafauna. But beyond the Indian subcontinent, particularly in Indochina and Indonesia, the few megafauna reserves that exist are in name only; intense poaching has decimated the large mammal populations of this region. One is lucky to encounter tracks of large vertebrates in protected areas, let alone the animals themselves. Forests once alive with birds and mammals are eerily silent.

Why is this so? We believe that the answer lies in the tradition of strict protection of the existing nature reserves in the Indian subcontinent and the continued addition to that network. Both the addition of new protected areas and strict protection of existing ones are features absent in the management of Indochina and Indonesia's wildlands. As long as the countries in this region pay scant attention to strict protection, conservation of large mammals and other biodiversity features of commercial value will be impossible. Without immediate action to improve the number and management of protected areas in the Indo-Pacific region, species such as the Douc langur (*Pygathris namaeus*) or Javan gibbon (*Hylobates moloch*) may soon experience the same fate as west Africa's now-extinct Miss Waldron's red colobus monkey (*Procolobus badius waldroni*). John Oates and his fellow authors (2000) conclude that this primate species may have survived the intensive logging, road building, and poaching if there had been better management of game preserves in Ghana and the Ivory Coast. If protected areas can be properly policed and can provide a modicum of protection for a few years, the large megafauna down to the smallest reptile will recover rapidly. Thus, we feel that protected areas, if truly afforded protection, are indeed the cornerstone of biodiversity conservation.

The Creation of Protected Areas in the Indo-Pacific

If, as John Terborgh argues in his recent book, *Requiem for Nature,* protected areas are essential for nature conservation, what is the history of the creation of protected areas in the Indo-Pacific, and how does it compare with other tropical areas? Terborgh (1999) graphs the number of new protected areas and area size by five-year intervals for the entire tropical belt. We see a peak in both variables between 1980 and 1985, and then a steep decline after 1990 (figure 5.16).

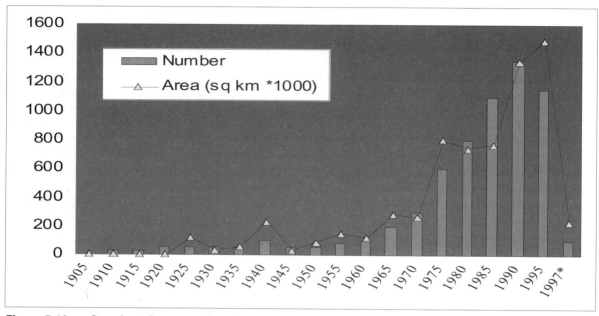

Figure 5.16. Creation of protected areas in tropical forest regions. Timescale is in five-year increments, with dates indicating the end of the five-year time period. The final time period is an exception: it only includes protected areas created during 1996–1997. Adapted from Terborgh (1997), based on data from UNEP-WCMC.

Our analysis of the creation of protected areas in the Indo-Pacific agrees with Terborgh's finding for the tropics as a whole: since 1990 there has been a precipitous drop in the designation of protected areas, the lowest amount gazetted since World War II (figure 5.17). (Our analysis considers only protected areas classified as IUCN categories I–IV, the most restrictive designations. We also rely on the WCMC 1997 as our source of data, which includes proposed reserves.) To gain further insights into this pattern, we focused on two countries: India, because it is the largest continental country in the region and has habitats that range from deserts and rain forests to temperate forests and alpine meadows; and Indonesia, because, with Brazil, it is considered one of the two most biologically diverse countries in the world. Both India and Indonesia have attempted to conserve their biodiversity by designating protected areas for more than fifty years. India has created more protected areas than Indonesia but has designated few since 1990 and none since 1995. India began creating numerous protected areas in the 1950s, but most of its protected areas were created in the late 1980s, driving the sharp increase in overall creation of protected areas in Asia. In contrast, Indonesia created few protected areas before 1970 and created most of theirs in the 1980s. As in India, the government of Indonesia has created only a few protected areas since 1990. The increase seen in Indonesia in 1995–2000 can be attributed to the designation of two large reserves: Gunung Lorentz (Irian Jaya) and Kayan Mentarang (Kalimantan).

The average size of protected areas for the Indo-Pacific as a whole remained small until the 1990s, when there was a substantial increase in average size (figure 5.18). This also coincides with a substantial decrease in the number of protected areas (a smaller sample size). Indonesia is one of the few countries analyzed that has consistently created larger than average protected areas since 1980, although since 1995 it has created only Gunung Lorentz and Kayan Mentarang. Up to 1990, India created a large number of protected areas that mirror the region's average protected area size. Since 1990, however, there has been a sharp drop in reserve size and number. However, India may be turning back to establishing protected areas to further protect its natural heritage. India recently proposed for designation more than 292 national parks and wildlife sanctuaries (IUCN status unknown) throughout the country. However, the average size of these reserves is only 117 km^2 (Rodgers et al. 2000).

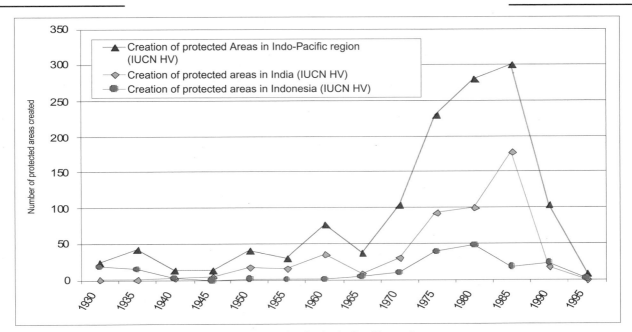

Figure 5.17. Creation of new protected areas in the Indo-Pacific region.

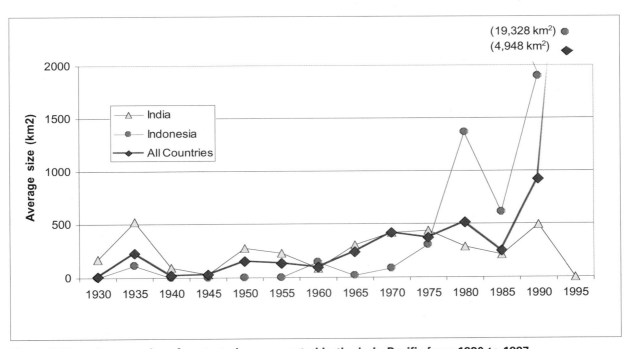

Figure 5.18. Average size of protected areas created in the Indo-Pacific from 1930 to 1997.

ICDPs and Protected Areas: Finding the Balance

Some biologists, including Terborgh, John Oates (1999), and Michael Soulé (pers. comm. to Eric Dinerstein, 2000), attribute the near abandonment of establishing new protected areas to the greater emphasis by development agencies and governments on financing expensive integrated conservation and development projects (ICDPs) or ecodevelopment projects. These projects seek to harmonize nature conservation with the use of wildlands by local people. This more development-oriented approach is more consistent with the mandates of aid agencies, and for a time many conservation organizations embraced this new approach.

In fairness, some nongovernment organizations (NGOs) accepted the new development-oriented paradigm as vital to taking the pressure off protected areas. A precious few ICDPs proved to be successful. Chitwan National Park in Nepal is one of the success stories. Through community develop-

TABLE 5.4. Natural World Heritage Sites That Overlap Tropical Moist or Dry Forest Ecoregions of the Indo-Pacific.

World Heritage Site	Date Established	Size (km²)	Ecoregions Represented by World Heritage Site
Tropical Moist Forest Biome			
Kaziranga National Park	1985	430	Brahmaputra Valley Semi-Evergreen Rain Forests [8]
Manas Wildlife Sanctuary	1985	391	Brahmaputra Valley Semi-Evergreen Rain Forests [8]
Sinharaja Forest Reserve	1988	112	Sri Lanka Lowland Rain Forests [3]
Thungyai-Huai Kha Khaeng Wildlife Sanctuaries	1991	2,480	Kayah-Karen Montane Rain Forests [51]
Puerto-Princesa Subterranean River National Park	1999	202	Palawan Rain Forests [97]
Ujung Kulon National Park	1991	1,120	Western Java Rain Forests [91]
Lorentz National Park	1999	21,190	Central Range Montane Rain Forests [125], Southern New Guinea Freshwater Swamp Forests [127], Southern New Guinea Lowland Rain Forests [128]
East Rennell	1998	370	Solomon Islands Rain Forests [137]
Tropical Dry Forest Biome			
Keoladeo (Bharatpur) National Park	1985	29	Khathiarbar-Gir Dry Deciduous Forests [16]
Komodo National Park	1991	1,735	Lesser Sundas Deciduous Forests [111]

ment projects, local communities have begun to reclaim pasturelands for forest. Rhinoceros and tigers now use the regeneration areas for breeding, and the local communities gain revenue from the lost pastures through ecotourism. The poaching pressure on Chitwan's megafauna has declined since this project was initiated, and the available habitat has increased. However, the success of this ICDP can be attributed only to the presence of tigers and rhinos to attract tourists. The fact that these animals still survived in the park resulted in part from the strict enforcement of Chitwan's boundaries from poachers or squatters.

On the other end of the spectrum, some propose that conservation resources should be directed to saving the region's protected areas that have also been identified as a World Heritage Sites (WHSs). Of these, ten lie in tropical moist or dry forest ecoregions (table 5.4). In the Indo-Pacific region six sites overlap with eight globally outstanding ecoregions. The remaining sites all lie within regionally outstanding ecoregions. However, these ten sites fail to capture the representation of tropical forests needed across the region to conserve their tremendous biodiversity.

Four sites lie within the Indian subcontinent bioregion. The globally outstanding forest ecoregions of the Western Ghats do not contain any sites, and the Brahmaputra Valley semi-evergreen forests contain two. The vast majority of India's tropical dry forests are represented by only one site and miss altogether the eastern Deccan region, famous for its tiger populations and several important tiger conservation units (Dinerstein et al. 1999).

The Indochina bioregion contains only one WHS in the Kayah-Karen Mountains of Myanmar and Thailand. Indochina's characteristic dry deciduous forests throughout Cambodia and southern Vietnam lack representation, as does one of the most unique mountain chains in Asia, the Annamites Mountains. Several recent large mammals, such as the Sao la, were discovered in these mountains, and this may be one of the last remaining places in Indochina to conserve tigers and other large megafauna.

The Sunda Shelf and Philippines bioregion also contains only one WHS, on Palawan island. The bioregion contains the most diverse assemblage of rain forest species in Asia, yet 99 percent are not represented in the WHS. The tremendous endemicity of the Philippines is missing. Sumatra, Borneo, and peninsular Malaysia's unique dipterocarp and montane forests lack WHS sites. The Mentawai Islands of Sumatra, with four endemic primates, lack representation.

The Wallacea bioregion contains two WHSs. These sites capture some of the last vitally important lowland habitat in Java and the lesser Sundas but fail to protect Sulawesi and the Maluku Islands, with their tremendous numbers of endemic birds and mammals.

The New Guinea and Melanesia bioregion contains the final two WHSs, including the newly designated, immense Lorentz National Park. This addition to the WHS list adds much to the protection of

New Guinea's montane, lowland, and freshwater swamp forests. The reserve's diverse habitats and elevational gradients support a multitude of fauna and flora. To properly represent the Indo-Pacific's tropical moist and dry forests in a suite of WHSs, numerous reserves must be added to the regional protected areas networks. Sayer et al. (2000) in their analysis of WHSs found that the most significant omissions in WHSs are the forests of the Western Ghats, Borneo, Sumatra, Sulawesi, and New Guinea. We would add New Caledonia to this list of globally outstanding forests without representation in the WHSs.

Conclusion

Unfortunately, individual parks alone cannot restore tigers and rhinoceros to Nepal's terai grasslands. In essay 19, by Joshi, Dinerstein, and Smith, the authors provide background about an ambitious project to link the current protected areas system of the entire Terai ecosystem with the intervening remaining habitat. The designation of some of these last remnants of forest as protected areas, and their strict enforcement, would greatly improve the chances for success for restoring tigers and rhinoceros. Recently the World Bank–WWF alliance (World Bank and WWF 1999) set as a goal the designation of 50 million ha of new protected areas and the improved management of another 50 million ha of existing protected areas by 2005. The conservation community should follow their lead and push for the increased designation, with strict enforcement, of protected areas throughout the region to prevent extinction of the region's unique flora and fauna.

ESSAY 16

Composition of the Alpine Himalayan Protected Areas Network and Its Contribution to Biodiversity Conservation

Tom Allnutt, Eric Wikramanayake, Eric Dinerstein, Colby Loucks, Rodney Jackson, Dan Hunter, and Chris Carpenter

Protected areas are key to the success of any regional biodiversity conservation strategy. However, in some cases, reasons other than biodiversity conservation, from scenic beauty to political expediency, have driven the establishment of protected areas. In this essay, we show how a large percentage of the high-elevation protected areas of the Himalayan mountains include rock and permanent ice rather than the habitats that are most important for conserving the rich and unique biodiversity of the largest and highest mountain range on Earth. Thus, inclusion and representation of the montane biodiversity will take additional protection of lower elevation alpine areas. We discuss initiatives already under way in several countries in the region that aim toward achieving these goals.

Methods and Data Sources

Our assessment assumes that biodiversity is not distributed uniformly throughout the montane grassland ecoregions. The complex topography, elevations, slopes, aspect, geology, and other biophysical parameters produce habitat matrices, including areas of bare rock and permanent ice that have limited value for biodiversity conservation. This vegetation limit is variable throughout the Himalaya and almost impossible to determine precisely. Because of the decreasing precipitation from east to west, the vegetation limit in the eastern Himalayas generally is higher than in the western Himalayas.

We used three sources to determine the areas of rock and permanent ice. The first was a 1-km-resolution digital elevation model (DEM) (USGS 1997). The second was MacKinnon's (1997) land-cover dataset for India, Bhutan, and Nepal, and the final dataset was the Digital Chart of the World (Environmental Systems Research Institute 1993). These datasets were combined in the following order to create the region of rock and permanent ice. First we created an elevation-based vegetation limit by selecting all areas above the vegetation boundary (table 5.5). Next, we extracted the areas defined as glaciers in MacKinnon's (1997) dataset and the areas defined as snowfields, glaciers, ice fields, and ice caps from the Digital Chart of the World drainage layer coverage. These three data layers were then combined to produce the rock and permanent ice coverage.

TABLE 5.5. Data Sources and the Four Models Used to Establish the Vegetation Limit in the Montane Himalayas.

Data Sources	Vegetation Limit [1]
[1] Digital Elevation Model, Limit Assigned by Ecoregion	
Montane Himalayan Ecoregion	
Central Tibetan Plateau Alpine Steppe	6,000 m
Eastern Himalayan Alpine Shrub and Meadows	5,500 m
Karakoram–West Tibetan Plateau Alpine Steppe	5,600 m
North Tibetan Plateau–Kunlun Mountains Alpine Desert	6,000 m
Northwestern Himalayan Alpine Shrub and Meadows	5,000 m
Pamir Alpine Desert and Tundra	6,000 m
Qilian Mountains Subalpine Meadow	5,500 m
Southeast Tibet Shrublands and Meadow	5,500 m
Tibetan Plateau Alpine Shrublands and Meadows	5,500 m
Western Himalayan Alpine Shrub and Meadows	5,500 m
Yarlung Zambo Arid Steppe	5,500 m

[2] MacKinnon (1997)
Glaciers (India, Nepal, and Bhutan only)

[3] Digital Chart of the World (DCW, Environmental Systems Research Institute 1993)
All areas defined as snowfields, glaciers, ice fields, and ice caps from the DCW drainage layer coverage (DNNET)

Vegetation limit established by combining vegetation limit [1] with glaciers [2] and [3]

To determine the extent of rock and permanent ice in the protected areas, we combined this coverage with a protected area dataset of the region. Protected area boundaries came from multiple sources: a global forest conservation CD-ROM (Iremonger et al. 1997), a snow leopard habitat study (Jackson 1998), and expert modifications (WWF and ICIMOD in prep.). We chose all protected areas with IUCN categories I–IV. We conducted a similar analysis to determine the extent of rock and permanent ice within each ecoregion.

Results

The analysis shows that approximately 17 percent of the protected system in the alpine region is covered by rock and permanent ice (table 5.6, figure 5.19). However, sixteen protected areas consist of more than 50 percent rock and permanent ice (table 5.7). In general, the protected areas of Tibet represent a higher proportion of alpine habitat, given that the alpine zone here extends much higher than in other ecoregions. The protected areas of the Eastern Himalayan Alpine Shrub and Meadows [39], the Karakoram–West Tibetan Plateau Alpine Steppe [40], the Northwestern Himalayan Alpine Shrub and Meadows [37], and the Western Himalayan Alpine Shrub and Meadows [38] contain the highest proportion of rock and ice.

These four ecoregions also contain the highest proportion of rock and ice overall (table 5.8). To assess whether rock and ice were disproportionately represented in protected areas, we calculated a

TABLE 5.6. Breakdown of Rock and Permanent Ice in Protected Areas by Ecoregion.

Ecoregion	Percentage Rock and Ice in Protected Areas System
Central Tibetan Plateau Alpine Steppe	1.1
Eastern Himalayan Alpine Shrub and Meadows	33.7
Karakoram–West Tibetan Plateau Alpine Steppe	36.7
North Tibetan Plateau–Kunlun Mountains Alpine Desert	2.0
Northwestern Himalayan Alpine Shrub and Meadows	50.7
Pamir Alpine Desert and Tundra	7.7
Qilian Mountains Subalpine Meadow	0
Southeast Tibet Shrublands and Meadow	3.6
Tibetan Plateau Alpine Shrublands and Meadows	0
Western Himalayan Alpine Shrub and Meadows	51.1
Yarlung Zambo Arid Steppe	2.4
Total	**17.2**

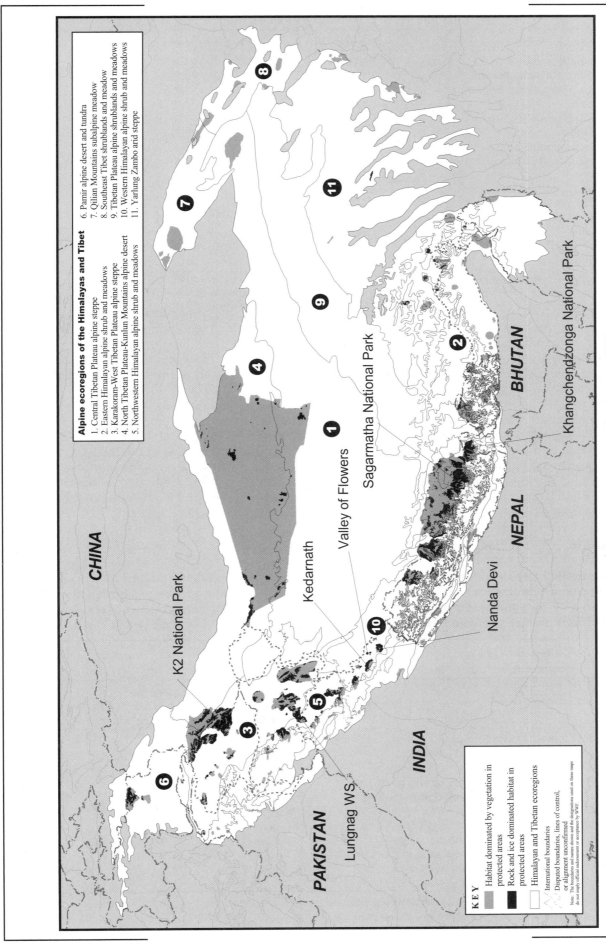

Figure 5.19. Rock and permanent ice in protected areas of the Alpine ecoregions of the Himalayas and Tibet.

Where Is Biodiversity under Greatest Threat?

TABLE 5.7. Protected Areas with More Than 50 Percent Inhospitable Habitat.

Protected Area; Country	Area (km^2)	Percentage Inhospitable
Tingri WS; India (Pakistan)*	47.2	84.8
K2 National Park; Pakistan (India)	2,333.8	76.8
Kilika Minteka GR; Pakistan (India)	500.7	75.2
Pin Valley; India	1,185.1	74.0
Sagarmatha National Park; Nepal	1,109.8	72.8
Nanda Devi; India	812.2	71.2
Valley of Flowers; India	64.9	68.2
Siachu Tuan Nalla WS; India (Pakistan)	875.2	66.4
Govind; India	495.0	64.8
Lungnag WS; India (Pakistan)	940.2	62.0
Khangchendzonga National Park; India	3,772.0	61.5
Khunjerab National Park; Pakistan (India)	3,332.4	60.1
Kedarnath; India	966.7	56.6
Pakura Nullah GR; Pakistan (India)	140.3	56.2
Manali WS; India	100.2	50.7

*Protected area is in first country but same land is claimed by country in parentheses.

simple index in which we divided the percentage of rock and ice in the protected areas of each ecoregion by the percentage of rock and ice in each ecoregion. A value of 1 would indicate that the rock and ice represented in the protected areas is proportionate to that in the ecoregion. A value less than 1 indicates that the protected areas have proportionally less rock and ice than occurs in the ecoregion as a whole, and a value greater than 1 indicates that rock and ice are overrepresented in the protected areas (table 5.8).

The Western Himalayan Alpine Shrub and Meadows [38] had almost twice as much rock and ice in protected areas than is found in the ecoregion. The two other alpine shrub and meadow ecoregions—Eastern Himalayan Alpine Shrub and Meadows and Northwestern Himalayan Alpine Shrub and Meadows—also had disproportionately high rock and ice in the protected areas (table 5.8).

Further north, the Pamir Alpine Desert and Tundra, the Southeast Tibet Shrublands and Meadow, and the Yarlung Zambo Arid Steppe have an overrepresentation of rock and ice habitat in their protected areas (table 5.8, figure 5.19). However, the overall percentages of rock and ice represented in these ecoregions and their protected areas are much less than in the alpine meadow ecoregions.

Conclusions

This coarse-scale analysis is intended to assess how well the protected areas in the Himalaya capture the most important habitats for biodiversity conservation. It shows that on average 11 percent of the protected area system consists of rock and ice, which is of low value for biodiversity conservation; the total area of rock and ice protected throughout the Himalaya is about 41,000 km^2, which is a substan-

TABLE 5.8. Comparison of Rock and Permanent Ice in Protected Areas with Rock and Permanent Ice in Ecoregion.

Ecoregion	I. Percentage Rock and Ice in Protected Areas	II. Percentage Rock and Ice in Ecoregion	Representation Index (I:II)*
Central Tibetan Plateau Alpine Steppe	1.1	1.3	0.8
Eastern Himalayan Alpine Shrub and Meadows	33.7	26.1	1.3
Karakoram–West Tibetan Plateau Alpine Steppe	36.7	29.2	1.3
North Tibetan Plateau–Kunlun Mountains Alpine Desert	2.0	3.7	0.5
Northwestern Himalayan Alpine Shrub and Meadows	50.7	29.8	1.7
Pamir Alpine Desert and Tundra	7.7	1.7	4.5
Qilian Mountains Subalpine Meadow	0	1.4	0
Southeast Tibet Shrublands and Meadow	3.6	1.4	2.6
Tibetan Plateau Alpine Shrublands and Meadows	0	2.2	0
Western Himalayan Alpine Shrub and Meadows	51.1	28.0	1.8
Yarlung Zambo Arid Steppe	2.4	1.0	2.4

*Values greater than 1 indicate areas in which the protected areas represent a greater proportion of rock and permanent ice than occurs in the ecoregion as a whole.

tial area that contributes little to biodiversity conservation; and several protected areas have overrepresentation of rock and ice habitat in comparison to areas not formally protected but within the same ecoregion.

Recent ecoregion-based conservation strategies have begun to reassess conservation efforts (WWF and ICIMOD in prep.). These conservation efforts include establishing large conservation landscapes, complete with altitudinal connectivity and large protected areas. However, core protected area systems cannot be anchored by large areas of habitat that are depauperate and unimportant for biodiversity conservation. Finer-scale analyses using the principles applied here will allow a more critical assessment of the protected areas of the high Himalaya, making them more effective in capturing and representing the unique flora and fauna that live at the top of the world.

ESSAY 17

Pastures of Plenty? Livestock Grazing and Conservation of Himalayan Rangelands

Daniel J. Miller

Himalayan grazing lands are rich in biodiversity. They include many endemic plants and a diverse mammalian fauna distinguished by forty-two species and subspecies of large Himalayan ungulates, plus magnificent large predators such as the near-mystical snow leopard (Fox 1997; Gupta 1994; Kachroo 1993; Mani 1978; Miller 1998b; Ram and Singh 1994; Rawat and Rodgers 1988; Rikhari et al. 1993; Schaller 1977, 1998; Wang 1988). This high diversity is explained in part by the location and isolation of the Himalayas; the peaks and deep valleys are situated at the confluence of five major biogeographic subregions and encompass a large expanse of altitudes and varied climates.

However, throughout the Himalayas, rangelands are heavily grazed by livestock; only the most inaccessible areas are spared. As economies in the region modernize and resource demands increase, the pressures on these sensitive ecosystems will mount (His Majesty's Government of Nepal/Asian Development Bank/ANZDEC 1992). Moreover, many of Asia's major rivers originate in Himalayan and Tibetan Plateau rangelands; thus, negative impacts in these headwater environments also have far-reaching effects on downstream areas. For these reasons, livestock grazing–related issues are of major concern in the Himalayan region.

Impacts of Livestock Grazing

Large livestock herds exert a wide range of effects on rangeland ecosystems (Ellison 1960; Detling 1988; Fleischner 1994; Laycock 1994; Risser 1988). Information from the Himalayas is limited, but work elsewhere indicates that livestock grazing can directly and indirectly affect plants, wildlife, and soils. It can also have secondary or ecosystem-level effects that can be immediate or take decades to be manifested (Belsky 1986; Briske and Richards 1995; Brown 1995; Coppedge and Shaw 1998; Dormaar et al. 1997; Lauenroth et al. 1994; Memmott et al. 1998; Miller et al. 1994; Noss and Cooperrider 1994). The degree of influence livestock has on vegetation is influenced by the intensity, frequency, seasonal timing, and duration of grazing, the spatial distribution of livestock, the level of selectivity of the livestock species, and site characteristics (Holechek et al. 1995; Lauenroth et al. 1994; Miller et al. 1994).

Researchers in the Himalayas have expressed concerns that heavy livestock concentrations have already led to overgrazing and trampling. These impacts can result in a reduction of palatable plant species, changes in plant species composition, soil erosion, and reduced wildlife and wildlife habitat (Bauer 1990, 1999; Bjonness 1990; Brower 1990; Fox et al. 1991, 1994; Fox 1997; Gyamtsho 1996; Heinen and Yonzen 1994; Mallon 1983; Miehe 1989; Miller 1997; Miller and Jackson 1994; Oli 1994; Rao and Casimir 1990; Rikhari et al. 1993; Saberwal 1996; Stevens 1996; Sundriyal and Joshi 1990; Tucker 1986; Yonzon and Hunter 1991). In Himalayan rangeland ecosystems, the change in plant species composition as a result of heavy grazing is a critical problem (Barbier et al. 1994); for instance, a shift in vegetation composition from palatable to unpalatable herbs and shrubs reduces the carrying capacity of rangeland ecosystems for domestic and wild grazers. Many nomads depend on rangelands for their livelihood, so a decrease in their capacity to support grazing has serious implications for their survival.

Intact Rangelands

Despite the pervasiveness of livestock in the Himalayas, there are still areas where rangelands are in good condition and where healthy populations of native wildlife can be found. Notable sites include (figure 5.20) the following:

- The Rumbak-Rumchang catchment in Ladakh, India, which supports four species of Caprinae: blue sheep (*Pseudois nayaur*), Ladakh urial (*Ovis orientalis vignei*), argali (*Ovis ammon hodgsoni*), and Asiatic ibex (*Capra ivex sibirica*). Here the ranges of these four species meet in a small area, and another three ungulate species—Tibetan antelope (*Pantholops hodgsoni*), Tibetan gazelle (*Procapra picticaudata*), and Tibetan wild ass (*Equus kiang*)—are also found in the region (Fox et al. 1991).
- The Dhorpatan region of Nepal, with good blue sheep populations and Himalayan tahr (*Hemitragus jemlahicus*), serow (*Nemorhaedus sumatraensis thar*), goral (*Nemorhaedus goral*), and musk deer (*Moschus chrysogaster*).
- Northeastern Mustang in Nepal, where argali and possibly brown bear (*Ursus arctos pruinosus*) were recently observed for the first time.
- The Kanchenjunga region in northeast Nepal, with good blue sheep habitat (Wegge 1991).
- The Soi and Lingshi regions of northwestern Bhutan, where some of the largest concentrations of blue sheep in the Himalayas are found.
- Some of the Chir Pine (*Pinus roxburghii*) savanna grasslands in the lower-elevation, dry, inner valleys of Bhutan. These *Themeda-Arundinella-Chrysopogon* grasslands in Bhutan are some of the last remaining examples of a subtropical savanna that were once common across the Himalaya and support ungulates such as sambar (*Cervus unicolor*) and barking deer (*Muntiacus muntjak*).
- The remaining patches of low-elevation Terai-Duar Savannas and Grasslands [35] in Nepal, such as in the Royal Chitwan and Royal Bardia National Parks, which support unique large mammal faunas.

Protecting and managing these relatively intact rangeland areas in the Himalayas should be a priority for conservation and development specialists alike.

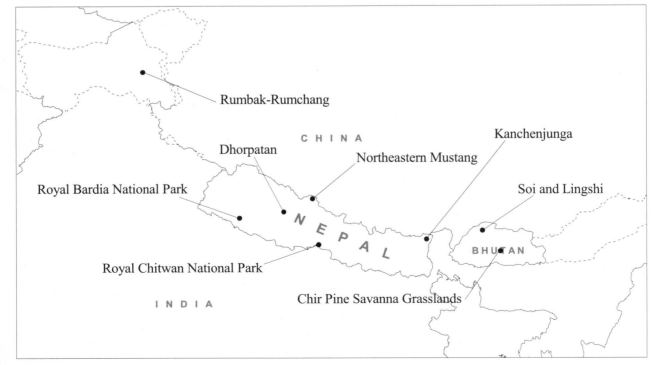

Figure 5.20. **Some intact rangelands of the Himalayas.**

Restoration of Degraded Himalayan Rangelands

Many of the degraded alpine meadows in the Himalayas are critically in need of restoration. Years of heavy livestock grazing have taken their toll. Yet these ecosystems are surprisingly resilient, and there are many noteworthy examples of rangeland plant communities that have recovered naturally from a disturbed condition after a year or two of respite from heavy livestock grazing. For example, heavily grazed *Andropogon-Stipa* rangelands in Mustang, Nepal recovered remarkably well when protected.

Restoration in many areas will take drastically improved livestock grazing management and, in many instances, restriction of livestock grazing for periods of time. Cutting and uprooting of shrubs in the alpine rangelands should also be halted. Grasses, protected from grazing within the shrubs, often provide a valuable seed bank for rangeland regeneration; if the shrubs are gone, the potential for regeneration is severely diminished. Restoration with native plant species is also possible in many Himalayan rangelands. In alpine areas in the Khumbu region of Nepal, native grasses such as *Elymus nutans* are used for developing hay meadows. This same grass is used throughout Tibetan Plateau areas of China in reseeding efforts.

Reconciling Domestic Livestock Grazing and Biodiversity Conservation in Himalayan Rangelands

Livestock grazing issues can be controversial, and they must be approached with scientific objectivity and rigor (Brown and McDonald 1995). This is especially true in the Himalayas, where grazing has been practiced for hundreds of years and is now considered a customary right. Yet the increasing intensity of grazing in recent years necessitates changes to make the practice sustainable over the long term.

The focus of rangeland research also must move away from the traditional studies of herbivore ecology, forage availability, and the effects of livestock grazing on wildlife toward research that seeks to understand rangelands as dynamic biological communities, ecosystems, and landscapes (Noss and Cooperrider 1994). Research should be more holistic to understand the ecological functions and processes of Himalayan rangeland ecosystems; as Noss and Cooperrider (1994) emphasize, ecological processes determine landscape patterns. Conservation and pastoral development plans that recognize the interdependency of pattern and process in rangeland landscapes and strive to maintain fundamental ecological and evolutionary processes will more successfully conserve biodiversity and allow sustained livestock grazing. Indigenous knowledge of rangeland environments held by the local nomads in the Himalayas will also provide many clues to incentives that can influence local people's behavior. It can also assist in the design of new incentive systems in places where traditional resource management systems have broken down or must be superseded (Barbier et al. 1994).

CHAPTER 6

What to Save First? Setting Priorities for Biodiversity Conservation

Setting priorities is an unpopular and difficult undertaking, for no matter what the approach, priority-setting amounts to triage. And triage is based on the principle that all areas will not receive attention first. At global, regional, or even ecoregional scales, some places will not make the list of top priorities even though these places may be highly important to those living and working in them. But the environment of uncertainty within which priorities are chosen contributes to a hesitation to engage in the process. In many parts of the world information about the extent and distribution of biodiversity is poor. This is also true of the Indo-Pacific region. In these instances we are forced to use proxies in lieu of precise information describing species, habitats, and threats, even though some biologists call for more thorough species inventories before there is a move to set priorities.

We believe that every ecoregion is worthy of biodiversity conservation effort, a principle we have articulated before and continue to reiterate. However, some ecoregions support outstanding levels of biological diversity, contain globally unique species and assemblages, or harbor the last viable representatives of formerly wide-ranging species or important ecological processes, and they obviously vie for priority attention. Whereas the natural habitats in some ecoregions have dwindled to scattered fragments, other ecoregions still possess large intact areas that present opportunities to create conservation landscapes complete with large, linked protected areas. By combining these biodiversity and landscape attributes in a representative framework of ecoregions and biomes, we believe that the resulting portfolio prioritizes the broadest array of the most distinctive examples of biodiversity and conservation opportunities across the region. In this chapter we introduce a decision-making matrix that addresses the question of where, given limited resources, we are likely to conserve the most distinct and representative biodiversity. The matrix also provides general guidelines as to the kinds of conservation actions that would best serve each ecoregion.

As for the uncertainty, we agree that exploration and species inventories are necessary to fill the information lacunae. However, we cannot afford to spend the next decade arguing about the accuracy, precision, types, and quantity of information that should be available before we begin to set priorities. Given the rapid loss of biodiversity throughout the region, there is an urgent need for concerted conservation action to save the most important reservoirs of biodiversity. The challenge is to act using the existing knowledge base to develop acceptable proxies that represent overall biodiversity and combine these with science-based conservation principles and good judgement based on experience. The essay by Arnaud Greth, Arnaud Collin, and David Olson on restoration of New Caledonia Dry Forests [140] documents one such effort (essay 18). New Caledonia is prototypical in this sense because it is on tropical islands where vertebrate species are likely to go extinct very quickly (Alcover, Sans, and Palmer 1998; Brooks et al. 1997, 1999).

In this analysis, the proxies we use for biodiversity, such as endemism, high levels of species richness, and large-scale and extraordinary ecological and evolutionary phenomena, are widely accepted within the conservation biology community. In our analysis of conservation status and future threat, the principles of conservation biology again shape the criteria, which include the presence and configuration of large habitat blocks, degree of habitat loss and fragmentation, and existing protection. We integrate the results of our conservation status and biodiversity analyses to produce a final prioritized portfolio of ecoregions.

The Priority-Setting Matrix

The integration matrix assigns each ecoregion to one of five priority classes (figure 3.1). The specific conservation actions needed in each class will differ among ecoregions because of differences in habitat type, levels of diversity, and ecosystem dynamics.

The Class I ecoregions are globally outstanding for biological distinctiveness but have highly threatened and relatively unprotected habitats (figure 6.1). This class also includes ecoregions that are now intact but are immediately threatened by large-scale logging concessions and other human population–related degradation threats.

The Class I ecoregions are obvious priorities. But we should not focus all attention on these "salvage operations" at the expense of overlooking the ecoregions with expansive landscapes of intact habitats. As we have shown in the analysis of conservation status, ecoregions with intact habitats can be susceptible to high-impact threats, resulting in extensive habitat loss and fragmentation in a short time. Thus, a more judicious approach would be to complement the salvage operations with equally heavy investments in Class III ecoregions, where opportunities remain to do conservation right. Class III eco-regions are those that are globally and regionally outstanding for biological distinctiveness and that still retain a large amount of intact habitat. They provide good opportunities to design conservation landscapes with protected-areas systems that adequately cover the areas with high biological values and ecological functions. As experience elsewhere has shown, restoration is a more expensive undertaking than foresighted conservation. Both Class I and Class III therefore are conservation imperatives.

Regional and Bioregional Trends

Across the five bioregions, 41 (29 percent) of the ecoregions are Class I, and 26 (19 percent) are Class III. Together, these two classes comprise about half of the ecoregions in the Indo-Pacific.

Within the tropical and subtropical moist broadleaf forests biome, which has eighty-nine ecoregions, thirty (34 percent) are Class I, and sixteen (18 percent) are Class III (table 6.1). This biome comprises about 64 percent of all ecoregions, and it makes up about 72 percent of Class I and 64 percent of Class III ecoregions. The second largest biome is the tropical and subtropical dry broadleaf forest biome, with sixteen ecoregions, of which five are Class I, but none are Class III.

The other biomes, with far fewer ecoregions apiece, were all represented within either Class I or Class III, with the exception of the deserts and xeric shrublands biome. Ascertaining trends with such small samples is difficult, but three of seven mangrove ecoregions are Class I, and six of eight montane grasslands and shrublands ecoregions are Class III. However, the montane grassland biome is not as amenable to this analysis because degradation threats are difficult to assess using remotely sensed data. Therefore, ground assessments of the Himalayan montane grasslands may warrant changes to the conservation status to reflect the degradation threats these ecoregions face. The integration classification for each ecoregion is provided in appendix F.

The Priority Portfolio

We created a priority portfolio using the Class I and Class III according to three levels of urgency: (1) ecoregions needing the most urgent attention, (2) high-priority ecoregions, and (3) priority ecoregions.

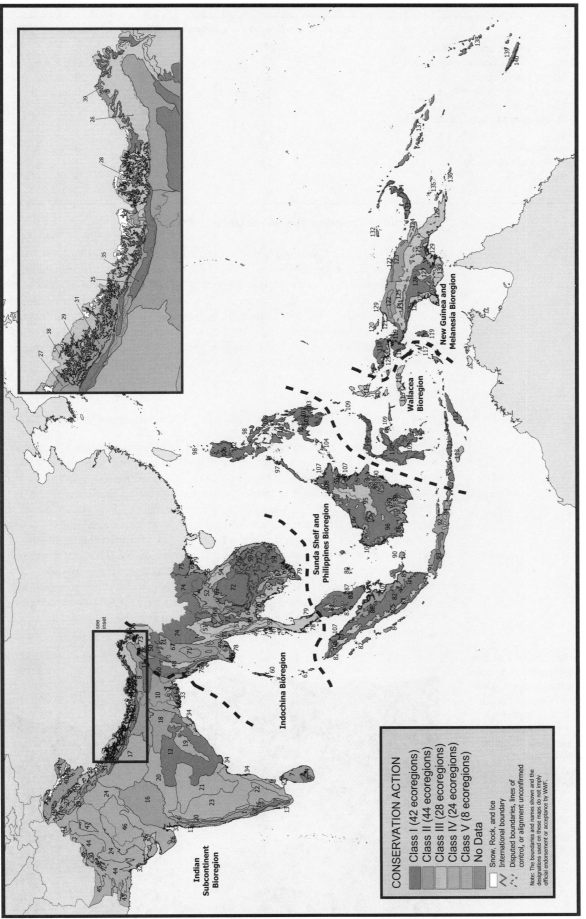

Figure 6.1. Results of integration matrix for conservation action. Class I ecoregions are globally outstanding ecoregions requiring immediate protection of remaining habitat and extensive restoration. Class II ecoregions are regionally outstanding ecoregions requiring immediate protection of remaining habitat and extensive restoration. Class III ecoregions are globally or regionally outstanding ecoregions that present rare opportunities to conserve large blocks of intact habitat. Class IV ecoregions are bioregionally and nationally important ecoregions requiring protection of remaining habitat and extensive restoration. Class V ecoregions are bioregionally and nationally important ecoregions requiring protection of representative habitat blocks and proper management elsewhere for biodiversity conservation.

TABLE 6.1. The Distribution of Ecoregions by Biome and Class.

Biome	Class I	Class II	Class III	Class IV	Class V	Total
Tropical and subtropical moist broadleaf forests	30	29	16	11	3	89
Tropical and subtropical dry broadleaf forests	5	5	0	6	0	16
Tropical and subtropical coniferous forests	1	3	0	0	0	4
Temperate broadleaf and mixed forests	0	1	2	0	0	3
Temperate coniferous forests	1	1	0	0	1	3
Tropical and subtropical grasslands, savannas, and shrublands	1	0	1	0	0	2
Flooded grasslands and savannas	0	0	1	0	0	1
Montane grasslands and shrublands	0	1	6	0	1	8
Deserts and xeric shrublands	0	0	0	4	3	7
Mangroves	3	4	0	0	0	7

Ecoregions Needing the Most Urgent Attention

These are the Class I ecoregions that are globally outstanding for biological distinctiveness and are critical or endangered for conservation status. Immediate action is needed to save the remaining pockets of habitat and source pools of biodiversity, especially the endemic species. This class also represents the ecoregions that are highly threatened by imminent logging concessions and human population threats to the remaining habitats (i.e., the ecoregions reclassified as critical or endangered in the threat analysis). The portfolio consists of the following (see appendix J for complete descriptions).

Indian Subcontinent

South Western Ghats Montane Rain Forests [2]. The montane and submontane forests along the Western Ghats of the Deccan Peninsula support outstanding levels of endemism; for example, the Western Ghats harbor about 75 percent of India's endemic amphibians. About 37 percent of the natural vegetation is still intact, and 17 percent of this (representing 6 percent of the ecoregion area) is included in protected areas. But habitat fragmentation and degradation threaten the ecological integrity and biodiversity of the ecoregion. Many endemic and focal species (e.g., the endemic lion-tailed macaque [*Macaca silenus*] and Malabar grey hornbill [*Ocyceros griseus*] and the tiger) that need contiguous, undisturbed habitat may not be able to survive in small forest fragments. Thus, efforts to map and link the existing protected areas using remaining forest fragments and restoration of other critical habitat where necessary to conserve representative biodiversity are urgently needed. Because substantial intact habitat remains outside of protected areas, the opportunity exists to expand the protected areas system and create linkages.

**The *shola*-grassland mosaics of the South Western Ghats montane rain forests [2] are unique to the higher elevations of this mountain range along the southwestern coast of India. It is home to the endemic Nilgiri Tahr, an important flagship species for this ecoregion.
Photo credit: Gopal Rawat**

This leaf insect found in the Sri Lanka lowland rain forests [3] uses mimicry to prevent detection from predators. Insect diversity is closely linked to plant diversity, and together comprise the majority of living species found on Earth.
Photo credit: Eric Wikramanayake

The adjacent *South Western Ghats Moist Deciduous Forests [15]* is also important for some of the larger vertebrates such as the tiger. Although this semi-deciduous forest ecoregion surrounding the montane moist forests is only regionally outstanding, conservation activities there will enhance effectiveness by creating larger landscapes and providing contiguity between the lowland and montane forests. Thus, both ecoregions should ideally be considered together, as a complex, for conservation planning purposes.

Sri Lanka Lowland Rain Forests [3]. This ecoregion in southwestern Sri Lanka harbors globally outstanding levels of endemism, including more than 250 species of Rhacophorid frogs (Pethiyagoda and Manamendra-Arachchi 1998), three endemic lizard genera (*Cerataphora*, *Cophotis*, and *Lyriocephalus*), and several endemic species of mammals, birds, and fishes; also, 98 percent of the dipterocarp trees are endemic (Dassanayake and Fosberg 1980). The ecoregion also overlaps with a Global 200 freshwater ecoregion (Olson and Dinerstein 1998). The habitat in this ecoregion is highly fragmented, and less than 500 km^2 lies within protected areas. The Sri Lanka Forest Department recently conducted a biological survey of all forest fragments greater than 80 km^2 and used a complementarity analysis to identify the patches that would contribute to biodiversity conservation. It is imperative that these proposed areas and as many of the other patches as possible are conserved.

Sri Lanka Montane Rain Forests [4]. This ecoregion in the central mountains of Sri Lanka also supports globally outstanding endemic flora and fauna. More than 50 percent of Sri Lanka's endemic vertebrates and endemic flowering plants are limited to this small montane unit. The ecoregion is also critical for watershed protection because all major rivers of Sri Lanka originate and radiate out from the central mountains. Much of the natural habitat in this ecoregion has been lost, having been converted into large tea plantations during the past two centuries. Because very little of the ecoregion is within effectively managed protected areas, the remaining habitat patches should be protected and conserved.

Eastern Highlands Moist Deciduous Forests [12]. This large ecoregion has some of the last blocks of remaining forests of the Deccan Peninsula that still harbor intact large vertebrate assemblages characteristic of the Indian Subcontinent's dry forests. These large mammals include tiger, barasingha, Asian elephant, wild dog, sloth bear (*Ursus ursinus*), leopard, gaur, chousingha (*Tetracerus quadricornis*), blackbuck, and chinkara (*Gazella bennettii*). About 75 percent of the ecoregion has been cleared or degraded through shifting cultivation, quarrying, mining, large-scale agriculture, and hydroelectric projects. Urgent conservation efforts are now needed to ensure that the remaining large habitat blocks are integrated into a conservation landscape that meets the ecological needs of large vertebrate assemblages.

Terai-Duar Savanna and Grasslands [35]. This swath of highly productive savanna grassland along the outer foothills of the Himalayan mountain range represents the world's tallest grasslands. The ecoregion harbors an assemblage of large mammals that includes Asia's largest predator—the tiger—and the two largest herbivores, the Asian elephant and greater one-horned rhinoceros (*Rhinoceros unicornis*). The pygmy hog (*Sus salvanius*) and hispid hare (*Caprolagus hispidus*) are two endemic mammals that inhabit the grasslands along the large

The Terai-Duar savannas and grasslands [35] are one of the last places in Asia that contain the greater one-horn rhinoceros, Asian elephant, and tigers. Creating large conservation landscapes where linked protected areas form the cornerstone of conservation will provide the best hope for the long-term survival of these large, wide-ranging species.
Photo credit: William Thompson

rivers and watercourses. Unfortunately, habitat fragmentation from settlements and agriculture has confined most of these species to small, isolated protected areas. However, in parts of Nepal a localized reversal of the deforestation trend has been achieved through community forestry and buffer zone ecotourism programs. The success of these restoration efforts suggests that similar efforts could be applied more widely to link existing protected areas and remaining habitat fragments, to create large conservation landscapes across the ecoregion.

Sundarbans Mangroves [33]. The Sundarbans is the largest mangrove ecosystem in the world. A large transboundary protected area, spanning across India and Bangladesh, protects about 25 percent of the remaining habitat. However, widespread woodcutting, clearing, and hunting by populous settlements have resulted in degradation and clearing of the natural habitat. Other developments, such as a planned fertilizer plant, will exacerbate the degradation of remaining fragments. The ecoregion harbors several large carnivores, such as the tiger (the only example of tigers adapted to live in mangrove ecosystems), the mugger (*Crocodylus palustris*), estuarine crocodile (*Crocodylus porosus*), gharial (*Gavialis gangeticus*), Ganges dolphin (*Platanista gangetica*), and sharks. These predators need large areas of undisturbed and undisrupted habitats, including intact, unpolluted aquatic habitats. Thus, sound protection of the large protected areas already in situ, with effort directed toward preventing further developments that threaten the ecosystem, is needed.

Indochina

Southeastern Indochina Dry Evergreen Forests [59]. The dry evergreen forests that snake along the low hills of Cambodia, Thailand, and Laos are a transition between the vast dry deciduous forests that cover the lowlands of Indochina and the montane moist forests. The ecoregion still contains several large blocks of habitat, especially in Cambodia, although these are now highly threatened by logging concessions. Because the ecoregion harbors several large vertebrates—such as tiger, Asian elephant, Javan rhinoceros (*Rhinoceros sondaicus*), gaur (*Bos gaurus*), banteng (*Bos javanicus*), clouded leopard (*Felis nebulosa*), Malayan sun bear (*Ursus malayanus*), gibbons (*Hylobates* spp.), and hornbills—that need large areas of intact habitat, a landscape conservation plan that addresses the ecological needs of these species should be implemented quickly before logging concessions foreclose conservation options.

Sunda Shelf and Philippines

Sumatra Lowland Rain Forests [82]. This large lowland moist forest ecoregion in Sumatra and the smaller Nias, Batu, and Simeuleu islands off Sumatra's western coast harbor several large vertebrates of conservation significance, including the tiger, Asian elephant, Sumatran rhinoceros, orangutan, sun bear, clouded leopard, and an incredible assemblage of ten hornbill species. The ecoregion also boasts two plants that produce the largest flowers in the world: the better-known *Rafflesia*

**Recently planted oil palm plantations are now a dominant feature of the Sumatran lowland rain forests [82]. Throughout Indonesia, the extraordinarily diverse tropical forests are being cut down, burned, and replaced with agricultural monocultures such as oil palm.
Photo credit: Dan Lebbin**

arnoldi and the lesser-known giant aroid lily *Amorphophallus titanum* (Whitten and Whitten 1992). Although the habitat, and thus the biodiversity, is under severe threat from conversion to agriculture, plantations, and logging, several large habitat blocks still remain. But land conversion has accelerated, especially since the recent economic crisis and the decentralization of administrative authority to the districts, and further loss of natural habitat will occur as the districts are compelled to rely on natural resources to fund local development. Because very little of this ecoregion is protected, it is important to secure additional areas of these rich lowland forests as large blocks within conservation landscapes.

Western Java Montane Rain Forests [93]. The endemic fauna of the wet montane forests of western Java includes the world's largest bat, *Pteropus vampyrus* (Whitten et al. 1996), the Javan gibbon (*Hylobates moloch*), the grizzled leaf monkey (*Presbytis comata*), and twenty-five species of birds, among which are trogons, nightjars, barbets, fruit doves, and pigeons. The numerous volcanic peaks strung along the central cordillera are endemism hotspots. Because of the steep slopes, this ecoregion's habitat is less threatened than this populous island's lowland forests. Nevertheless, only about one-fifth of the original habitat remains, occurring as fragments scattered across the mountain chain. Because of the beta-diversity associated with the mountain range, a critical assessment of the existing protected area system may be necessary to ascertain whether all endemism hotspots are included within it.

Borneo Lowland Rain Forests [96]. This vast ecoregion supports globally outstanding levels of species richness. Some of the mammals of conservation significance include the charismatic orangutan as well as two species of gibbons, four species of leaf monkeys, clouded leopard (*Felis nebulosa*), hairy nosed otter, and several tree shrews, all species that need intact habitat to survive. Similarly, of the almost 400 bird species in the ecoregion, the seven hornbills and a number of pheasants also need undisturbed habitats. At present, almost half of the natural forests in this ecoregion are still intact, but only about 3 percent of the ecoregion receives formal protection; thus, threats from proposed logging concessions place it in the critical list. There are several proposed protected areas that will cover more than 10 percent of the ecoregion, but a conservation assessment to ascertain whether these proposed areas will adequately represent the ecoregion's biodiversity is urgently needed before logging concessions foreclose conservation options.

Palawan Rain Forests [97]. This small, narrow ecoregion in the lowland moist forests of Palawan island harbors globally outstanding levels of endemism; almost 50 percent of the mammals are endemic, including the Palawan flying fox and *Palawanomys furvus*, a small endemic rodent that belongs to its own genus. The twenty endemic birds include the Palawan peacock-pheasant, Palawan hornbill, Palawan scops-owl, Mantanani scops-owl, and several babblers and flycatchers. Only about 13 percent of the ecoregion's natural habitat now remains, and none of this is in formal protected areas. A review of the distributions of endemic

Large-scale commercial forestry projects such as this landscape in East Kalimantan are devastating the Borneo lowland rain forests [96]. Borneo is the center of diversity for dipterocarp tree species (see essay 4.2), and plant endemicity approaches 40 percent of the vascular flora. Habitat loss and fragmentation will likely cause species extinctions of a magnitude never before experienced. Photo credit: WWF-Indonesia

species and forest fragments is needed to design a system of protected areas to conserve endemic species.

Luzon Rain Forests [98]. This ecoregion covers the lowland moist forests of Luzon Island. It harbors globally outstanding levels of endemism; the remaining rain forest fragments on Mount Makiling, a 1,109-m tall dormant volcano, alone contains more woody plant species than the continental United States (Gonzales and Rees 1988; Scanland et al. 1989). Mount Isarog in the south harbors one of two *Rafflesia* species endemic to the Philippines. Bird endemism is especially high: thirty-four birds are endemic to this ecoregion, including the Luzon bleeding-heart, Luzon hornbill, and flame-breasted fruit-dove. Other prominent birds include the Philippines' conservation icon, the great Philippine eagle. These and other specialized species need intact habitat to survive. Unfortunately, habitat loss and degradation have reached a critical state, with only about 7 percent of the natural habitat left in small, scattered fragments. Because less than 1 percent of the ecoregion receives any formal protection, as much as possible of the remaining forest fragments should be protected to conserve the irreplaceable endemic species in this ecoregion.

Mindoro Rain Forests [99]. Mindoro was isolated from Luzon and Palawan throughout the Pleistocene and retains its own unique character, including an endangered endemic buffalo species, the tamaraw (*Bubalus mindorensis*). Three critically endangered bird species, the Mindoro bleeding-heart, the Mindoro imperial pigeon, and the black-hooded coucal, are strict island endemics. Mindoro's endemic birds can be split into montane and lowland species. Although both are in urgent need of conservation, the situation for the lowland species is particularly dire because the lowland forests are almost entirely gone. The only protected population of the critically endangered Philippine crocodile (*Crocodylus mindorensis*), which was historically found throughout the Philippines, is found in Mindoro's Naujan Lake. Mindoro is now one of the most severely deforested islands in the country. Only the most rugged portions of the island's central spine have been spared from commercial logging, and the forest is still under pressure.

Greater Negros–Panay rain forests [100]. The Greater Negros-Panay ecoregion appears as a number of isolated islands, but during the last ice ages, these islands were (for the most part) part of one continuous island. The islands contain a unique mix of Sundaic and Philippine mammals and birds. Two endemic large vertebrates are especially notable: the Philippine spotted deer (*Cervus alfredi*) and the Visayan warty pig (*Sus cebifrons*). The Visayan warty pig is critically endangered, and the Philippine spotted deer is endangered (IUCN 2000). A total of twenty-two endemic or near-endemic bird species occur in the ecoregion, and an extremely high proportion of these birds are imperiled. These include the Negros bleeding-heart (*Gallicolumba keayi*), Negros fruit-dove (*Ptilonopus arcanus*), Visayan hornbill (*Penelopides panini*), writhed-billed hornbill (*Aceros waldeni*),

This red-bellied pitta (*Pitta erythrogaster*) found on the island of Sibuyan in the Greater Negros-Panay rain forests [100] is distributed throughout the Melanesia region. Bird endemism ranges between 4 and 17 percent in Philippine ecoregions, with most ecoregions exceeding 10 percent. Photo credit: D. Willard

white-throated jungle flycatcher (*Rhinomyias albigularis*), scarlet-collared flowerpecker (*Dicaeum retrocinctum*), and Cebu flowerpecker (*Dicaeum quadricolor*). Lowland evergreen dipterocarp rain forest was once the dominant vegetation of the ecoregion. But this ecoregion has suffered a disproportionate share of deforestation, and the dipterocarp forest is underrepresented in the national protected area system.

Mindanao–Eastern Visayas Rain Forests [101]. The lowland moist forests in the Visayas and the southern Philippine island of Mindanao also harbor globally outstanding levels of endemism. The endemic mammals include the Visayan spotted deer, (*Cervus alfredi*), considered one of the two most endangered deer species in the world (Heaney and Regalado 1998). More than forty bird species are endemic to the ecoregion, and many, such as the great Philippine eagle, Mindanao bleeding-heart, and the several hornbills, fruit pigeons, and doves, need undisturbed, intact habitats and are highly threatened by habitat loss. Less than 10 percent of the natural habitat now remains as scattered fragments across the ecoregion, and less than 1 percent of the ecoregion receives any formal protection. Thus, all remaining blocks of natural habitat should be protected to conserve biodiversity. Restoration should be considered (after a careful and critical assessment of the distribution of endemic species and their ecological needs) as a long-term conservation option.

Sunda Shelf Mangroves [107]. These mangrove forests rank as some of the most species-rich and diverse in the world. They support waterbirds, and Borneo's coasts provide habitat for the charismatic proboscis monkey (*Nasalis larvatus*), one of the few large mammals limited to mangrove environments. More importantly, the mangroves are critical nursery grounds and habitat for numerous species of fish, shrimp, and shellfish. More than three-fourths of the natural habitat has been lost to logging, wood-chip extraction, charcoal production, and clearing for shrimp culture. The few large patches that remain along the coasts of Brunei Darussalam are some of the best-preserved mangroves in southeast Asia (Spalding et al. 1997). Several proposed protected areas cover about 16 percent of the ecoregion, but only 1 percent of the ecoregion is within existing protected areas. All remaining patches of mangroves should be protected because of the rapid loss of this important habitat. Restoration should be undertaken, especially in areas that have been logged and extracted, and in defunct shrimp farms.

Peninsular Malaysian Peat Swamp Forests [87]. This small, disjunct ecoregion along the coast of peninsular Malaysia harbors many of the species that are found in the lowland rain forest ecoregion and several species that are specialized to this habitat. Each of the isolated habitat blocks has its endemic fishes, crabs, or shrimps (essay 4). About half the ecoregion's natural forest still remains, but it receives almost no formal protection. A conservation plan to identify important areas for protection should be conducted.

Mindanao Montane Rain Forests [103]. The montane forests of Mindanao support globally outstanding levels of endemism. Twenty-nine percent of the ecoregion's mammals are endemic, as are 23 percent of the birds. The latter include focal species such as the Philippine cockatoo (*Cacutua haematuropygia*) and

**The Sunda Shelf mangroves [107] are home to the proboscis monkey (*Nasalis larvatus*). Unlike other primates, this species is restricted to the mangrove forests, and are adapted to feed exclusively on leaves and the hard, starchy fruit of mangrove trees. Loss of the mangrove habitat threatens the survival of this charismatic primate.
Photo credit: Russel Mittermeier**

great Philippine eagle (*Pithecophagus jefferyi*). But more than 80 percent of the natural habitat has been lost to logging and shifting agriculture, and the resulting erosion on the steep slopes has degraded the habitat further. Only about 1 percent of the ecoregion is within a protected area system. Additional habitat should be brought into the system to conserve the endemic biodiversity in one of the last bastions of the famous Philippines monkey-eating eagle.

Luzon Tropical Pine Forests [106]. This ecoregion represents the small area of indigenous tropical pine forests (*Pinus kesiya*) in the northern part of Luzon Island in the Philippines. It contains sixteen species of small mammals and twenty-three species of birds that are endemic to the ecoregion. Most of the habitat has already been cleared, but about 6 percent of this ecoregion is protected.

Wallacea

Sulawesi Lowland Rain Forests [109]. The lowland rain forests of Sulawesi harbor some of the highest levels of endemism among the Indo-Pacific ecoregions; 23 percent of the mammal fauna is endemic to the ecoregion, including the charismatic babirusa (*Babyrousa babyrussa*), lowland anoa (*Bubalus depressicornis*), and three macaque species. The seventy-nine bird species endemic to the ecoregion include hornbills, owls, kingfishers, mynas, and parrots. Less than a third of the ecoregion's natural forests still remain, and only 5 percent of the ecoregion (but 9 percent of the habitat) is within protected areas. This is insufficient to protect the ecoregion's extraordinary endemism, and a conservation assessment to ensure representation of the biodiversity is urgently needed.

Sulawesi Montane Rain Forests [110]. Like the lowland forests, this montane forest ecoregion on Sulawesi supports some of the highest levels of endemism in the Indo-Pacific region. The ecoregion harbors more than thirty species of endemic mammals, among which are two species of tarsier (*Tarsius* spp.) and several murine rodents, shrews, and bats. The more than fifty species of endemic

**This species of bamboo rat (*Rhynchomys soricoides*) is but one example of a number of mammal species restricted to single mountain tops, in this case the montane forests of the Luzon tropical pine forests [106].
Photo credit: Larry Heaney**

**This male babirusa (*Babyrousa babyrussa*) in the Sulawesi lowland rain forests [109] is endemic to the island of Sulawesi. Years of isolation has resulted in high levels of endemism in these archipelago ecoregions, and Sulawesi contains one of the highest levels of mammal endemism in the entire region.
Photo credit: Riza Marlon**

birds include imperial pigeons, fruit doves, owls, whistlers, hornbills, and flycatchers. The ecoregion's natural habitat is still intact in several large and midsized blocks. However, only 8 percent of the ecoregion is within protected areas, which is probably insufficient to capture and represent the ecoregion's diversity and endemism. A conservation plan to ensure representation within protected and conservation areas is needed.

Lesser Sundas Deciduous Forests [111]. The ecoregion is spread over a series of islands, from Lombok to Flores in eastern Indonesia. The fauna is a rich mix of Asian and Australian species, but because of the long period of isolation, the ecoregion now harbors globally outstanding levels of endemism. The most famous denizen of this ecoregion is undoubtedly the Komodo dragon (*Varanus komodoensis*), the largest lizard in the world. There are also five species of small mammals and twenty-nine species of birds that are endemic to the ecoregion. Almost 60 percent of the habitat has been converted into agricultural areas, but large blocks of habitat remain on some of the islands. Because only 4 percent of the ecoregion is within protected areas, an assessment to see whether these protected areas capture the diversity and endemism of the ecoregion is recommended before habitat fragmentation results in further loss of irreplaceable biodiversity.

Timor and Wetar Deciduous Forests [112]. This ecoregion also covers several islands in eastern Indonesia, with one, Timor, making up most of the ecoregion. The ecoregion has high endemism, including five small mammals and thirty-five bird species, among which are several imperial pigeons, figbirds, lories, lorikeets, honeyeaters, and flycatchers. More than 60 percent of the habitat has been cleared for cultivation, and large areas are burned regularly for hunting. A large block of habitat exists in Timor, but the rest of the habitat is scattered as fragments throughout the ecoregion. Because only about 1 percent of the ecoregion is within protected areas, additional areas must be brought under protection to conserve the ecoregion's endemism.

**The Komodo dragon is the top predator in the Lesser Sundas deciduous forests [111]. Unlike large mammalian carnivores, the biology of these large reptiles allows them to grow to large sizes but survive on the limited prey base found on small islands.
Photo credit: Eric Wikramanayake**

New Guinea and Melanesia

Vogelkop-Aru Lowland Rain Forests [119]. This ecoregion in the "bird's-head" lowlands of New Guinea harbors globally outstanding levels of species richness, particularly for plants and birds. There are more than 360 bird species in this ecoregion, including an impressive assemblage of birds of paradise, bower birds, brush turkeys, fruit pigeons and doves, lories, cockatoos, and cassowaries. Currently almost 90 percent of the natural habitat is still intact, but the high threat from logging concessions places the ecoregion on the critical list for conservation status. Because only 6 percent of this highly diverse ecoregion receives protection and because the (currently) intact habitat presents an opportunity to do conservation right, an immediate conservation plan for the ecoregion is recommended to ensure that the ecoregion's biodiversity is adequately represented in the protected area system.

Southern New Guinea Lowland Rain Forests [128]. The lowland rain forests of New Guinea to the south of the central cordillera support a rich flora and fauna. Because about 95 percent of the ecoregion's habitat is intact, it still maintains intact vertebrate assemblages and ecological processes, phenomena that are almost extinct from fragmented lowland rain forests throughout the Indo-Pacific. However, this ecoregion is threatened by large-scale logging plans that could disrupt these processes. A conservation plan to provide additional protection above the current levels of 8 percent is urgently needed to capture the biodiversity and conserve the integrity of these processes and assemblages before the logging concessions become operational.

New Britain–New Ireland Lowland Rain Forests [133]. Although we know little about this ecoregion's biodiversity, the levels of endemism among the known fauna and flora qualify it for globally outstanding status. There are nine endemic mammals and fifty-five endemic birds in the ecoregion, including the blue-eyed cockatoo, New Britain rail, and several lories and imperial pigeons. Almost all the habitat in the ecoregion is still intact, and about 7 percent of the ecoregion is protected; however, only one of these protected areas, Whitman Mountains, is large. Most of the lowland forests on the two big islands remain unprotected. Because of the high threat from plans to log these rich forests and clear areas for oil palm plantations, the ecoregion is on the critical list for conservation status. Thus, there is an urgent need to complete inventories of the flora and fauna of these lowland forests, identify the distribution of endemism and ecological processes, and review the protected area system to ensure that the rich and unique biodiversity of this ecoregion is adequately conserved.

New Britain–New Ireland Montane Rain Forests [134]. The montane moist forests in New Britain and New Ireland are little explored, but what is known indicates that the ecoregion is globally outstanding for endemism. The Nakanai Plateau on eastern New Britain and the higher plateaus of west New Britain contain large areas of *Nothofagus* forests on limestone substrates that should be a conservation target (Johns 1993; Miller et al. 1994). Forty-one species of the known birds are near endemics, most known from the Verron and Hans Meyer Ranges in southern New Ireland. The endemic species include pied goshawk, slaty-mantled sparrowhawk, New Britain sparrowhawk, black honey-buzzard, Melanesian scrubfowl, New Britain rail, and several species of fruit doves and pigeons that need mature forests. Although the ecoregion is still clothed in intact habitat, plans for extensive logging concessions place it on the critical list for conservation status. About 11 percent of the ecoregion is now within protected areas, but given the high endemism and beta-diversity, it is essential to assess the location and configuration of the protected areas to ensure that they adequately represent the biodiversity of this ecoregion.

New Guinea Mangroves [129]. Located in the Indo-Malayan center of diversity, southern New Guinea has the highest diversity of mangroves in the world. There are two bat species recorded only from this mangrove ecoregion and eight species of endemic birds. Mangroves are a keystone habitat that provides vital ecological functions in the terrestrial-marine-freshwater interface. They accumulate sediment, stabilizing and protecting the coastline from erosion, and allow for exchange of nutrients between the terrestrial and marine environments. Mangroves also provide a nursery for many coastal fishes. Four protected areas in Indonesia and Papua New Guinea protect approximately 33 percent of the ecoregion, but the ecoregion is highly threatened by logging concessions. Additional protection may be necessary to ensure that the biodiversity and ecological processes of the mangroves are maintained.

Vanuatu Rain Forests [138]. Though faced with population pressures and regular destructive cyclones, Vanuatu's natural heritage is almost intact for now. The island's mammals include several bats, such as the Fijian blossom-bat, Vanuatu flying-fox, Banks flying-fox, Temotu flying-fox, and Vanikoro flying-fox (Flannery 1995; Flannery and Groves 1998; Bonaccorso et al. in prep.) that are near-endemic to the ecoregion. Additionally, thirty-one (39 percent) of the ecoregion's bird species are found nowhere else in the world. Because of limited distributions, these birds and large flying-foxes are vulnerable from a combination of habitat loss and hunting.

New Caledonia Rain Forests [139]. This ecoregion is an original fragment of Gondwanaland. Thus, the long period of isolation and evolution has resulted in a highly endemic flora that includes five endemic plant families (Amborellaceae, Oncothecaceae, Papracrypyiaceae, Phellinaceae, and Strasburgiaceae). Nearly 14 percent of the plant genera and 80 percent of the species are endemic. The ecoregion lacks native amphibians and has only three snake and eleven mammal species (all of which are bats). But there are sixty-eight lizard species, of which sixty are endemic, and one ancient, endemic bird family, Rynochetidae, that is represented by only a single species, the kagu (*Rhynochetos jubatus*). The forests have suffered large losses, especially from logging and mining, and only about 10 percent are undisturbed. Introduced species, especially pigs, goats, cats, dogs, and rats, prey on the indigenous fauna. Less than 3 percent of the habitat is within protected areas, which are ineffectively managed. There is a critical need for additional protection and sound management of the protected areas.

New Caledonia Dry Forests [140]. Eons of isolation, an ancient source of plant life, and an extreme geology have resulted in globally outstanding levels of endemism in this ecoregion. Of the ecoregion's 379 native plant species, 59 percent are endemic to New Caledonia, and 59 are found only in the island's dry forests. Among these are *Captaincookia noumeana* (Rubiaceae), which belongs to a mono-

An amazing diversity of plants in the New Caledonia rain forests [139] have evolved extraordinary adaptations to survive on the nutrient-poor and highly toxic soils associated with ultramafic bedrock. These maquis (ultramafic) communities support 1,844 plant species (58 percent of the island's flora), representing 440 genera and 119 families. Over 90 percent of the maquis plant species are endemic to these globally rare habitats.
Photo credit: David Olson

The New Caledonia dry forests [139] contain a unique assemblage of plants with 59 endemics. The flagship species for this ecoregion, *Captaincookia margaretae*, has declined to two small populations threatened with extinction before WWF and its collaborators stepped in to protect these magnificent plants. Photo credit: Arnaud Greth

typic genus. Now less than 2 percent of these unique dry forests remain as small, isolated patches. Most of the habitat in this ecoregion has been lost to ranches and cattle grazing, and almost all of the remaining habitat is privately owned. The protected area network of New Caledonia is inadequate, covering less than 3 percent of the ecoregion (Jaffré et al. 1998). Because the remaining patches are all likely to contain species found nowhere else, they should be protected. Introduced species, especially pigs, goats, cats, dogs, and rats, present problems for native species. There is a critical urgency for additional protection and sound conservation management to conserve this ecoregion's unique biodiversity.

High Priority Ecoregions

These are the Class III ecoregions that are globally outstanding for biological distinctiveness and are relatively intact and stable in terms of habitat and conservation status. The large, contiguous areas of natural habitat present opportunities to create conservation landscapes with linked and buffered protected areas that include the most important areas for biodiversity conservation. Despite the intactness, however, there is an urgency to act before logging concessions, plantations, and other large-scale developments fragment and degrade the landscape.

Indian Subcontinent

Eastern Himalayan Broadleaf Forests [26]. This temperate broadleaf forest ecoregion along the mid-hills of the eastern Himalayas is globally outstanding for endemism and richness. The oak- and chestnut-dominated forests harbor several vertebrate species of conservation importance, including the endemic golden langur (*Semnopithecus geei*), tiger, and clouded leopard, in the eastern extent of the ecoregion. There are six hornbill species and several pheasants in the ecoregion that need intact habitats. Approximately two-thirds of this ecoregion is intact, with most of the remaining large habitat blocks located in northeastern India and Bhutan. Six percent of the ecoregion's habitat is within protected areas, but to ensure the conservation of the many space-sensitive species, additional protected areas and conservation landscapes may be necessary.

The rhododendron forests, such as these in Phubjica Valley, Bhutan, in the Eastern Himalayan broadleaf forests [26] contain the highest levels of plant, bird, and mammal endemism in the Himalayan forests. As the center of diversity for rhododendrons, these eastern Himalayan forests contain a rich assemblage of these colorful plants.
Photo credit: Eric Wikramanayake

Rann of Kutch Seasonal Salt March [36]. This seasonal salt marsh at the edge of a desert harbors the last populations of the Asiatic wild ass (*Equus hermionus*). During the summer monsoon, the ecoregion floods to a depth of about a half meter, mostly from the runoff from several rivers that drain one of the oldest mountain ranges in the region, the Aravalli Hills. But this bleak ecoregion also harbors a surprising number of other large mammals and birds, including the chinkara (*Gazella bennettii*), nilgai (*Boselaphus tragocamelus*), wolf (*Canis lupus*), blackbuck (*Antilope cervicapra*), striped hyena (*Hyaena hyaena*), desert cat (*Felis silverstris*), and caracal (*Felis caracal*) and the globally threatened lesser florican (*Eupodotis indica*) and Houbara bustard (*Chlamydotis undulata*) (IUCN 2000) and lesser flamingo (*Phoeniconaias minor*) (WII 1993). Three protected areas cover more than three-fourths of this ecoregion, and the large Wild Ass Sanctuary in India was designated especially to protect the last population of this endangered species. Cattle grazing, salt mining, and vehicular traffic cause damage to this fragile ecosystem.

Eastern Himalayan Alpine Shrub and Meadows [39]. This ecoregion harbors one of the world's richest alpine floras, with high levels of endemism. Rhododendrons dominate the scrub vegetation and exhibit considerable species turnover along the west-east gradient, from eastern Nepal to northern Myanmar. The ecoregion supports a large vertebrate assemblage that includes predators (snow leopard, wolf, Asiatic black bear) and their ungulate prey (the endemic red goral, blue sheep, takin). The bird fauna includes pheasants, the lammergeier (*Gypaetus barbatus*), and Himalayan griffon (*Gyps himalayensis*). About 20 percent of the ecoregion is protected, but most of the protection is in Bhutan, with very little in Myanmar. Because of the high beta-diversity, the degree of protection and conservation effort should be equitably distributed to conserve the representative biodiversity in the ecoregion.

Indochina

Kayah-Karen Montane Rain Forests [51]. This montane rain forest straddling the Thailand-Myanmar border is globally outstanding for species richness. Its rich large vertebrate fauna includes an assemblage of seven cats (of which the largest is the tiger) the Asian elephant, two gibbons, Asiatic black bear, Malayan sun bear, gaur, banteng, wild water buffalo, and the world's smallest bat, Kitti's hog-nosed bat (*Craseonycteris thonglongyai*). The birds include six hornbills, among

The snow leopard (*Uncia uncia*) inhabits the high elevations of the Himalaya and Kashmir regions, including the Eastern Himalayan alpine shrub and meadows [39]. A keystone species, the snow leopard is well adapted to its role as a top predator by having long and dense hair, an enlarged nasal cavity, shortened limbs, a well-developed chest muscles for climbing and hunting in high altitude environments.
Photo credit: Bruce Bunting

which are the great hornbill (*Buceros bicornis*), wreathed hornbill (*Rhyticeros undulatus*), and rufous-necked hornbill (*Aceros nipalensis*), several pheasants, the Sarus crane (*Grus antigone*), white-winged duck (*Cairina scutulata*), and Pallas' fish-eagle (*Haliaeetus leucoryphus*). Many of these species need large habitat areas. A conservation plan for this ecoregion should strive to maintain the large conservation landscapes that are still possible. Almost 70 percent of the natural habitat is still intact, primarily because of the ruggedness of the terrain and low density of people. Although almost 20 percent of the ecoregion is now in protected areas, most is concentrated in a large protected area complex in Thailand. Protected areas in Myanmar could include a transboundary component to conserve the intact habitats that lie mostly along the boundary with Thailand.

Tenasserim–South Thailand Semi-Evergreen Rain Forests [53]. The forests along the Tenasserim Mountains along the Thailand-Myanmar border are globally outstanding for species richness. Within the dominant semi-evergreen rain forests that are represented by this ecoregion are a variety of forest types, including Myanmar's most diverse dipterocarp-dominated wet forests, and patches of limestone karst towers in the peninsula and on islands off the southwest coast of Thailand. The rich fauna includes the endangered, endemic Fea's muntjac (*Muntiacus feae*) and several large vertebrates such as the tiger (and seven other felids), Asian elephant, Malayan tapir, gaur, gibbons, and four species of leaf monkeys. The birds include an incredible assemblage of ten hornbills, several pheasants, and the endemic Gurney's pitta (*Pitta gurneyi*). Although almost 50 percent of the habitat is intact, most of the forests in Thailand have been logged and converted to agriculture and plantations. After Thailand banned timber exploitation in its forests in 1988, Myanmar granted large logging concessions to Thai companies, and timber extraction in Myanmar by Thai loggers has become common in recent years (WWF and IUCN 1995). The protected areas that cover 9 percent of the ecoregion are insufficient to conserve the representative biodiversity in this ecoregion, and there is an urgent need for more conservation efforts before the intact landscapes become fragmented.

Northern Annamites Rain Forests [54]. This ecoregion is largely unexplored scientifically; in the past five years, biodiversity surveys have yielded five new species of large mammals alone, including two in new monotypic genera: *Pseudoryx nghetinhensis* (saola) and *Megamuntiacus vuquangensis* (giant muntjac). Even the known levels of biological diversity qualify it for globally outstanding status. Among the other vertebrates of conservation importance are Douc langur (*Pygathrix nemaeus*), Francois' leaf monkey (*Semnopithecus francoisi*), Phayre's leaf monkey (*Semnopithecus phayrei*), gibbons, tiger, Asian elephant, Himalayan black bear and the Malayan sun bear, gaur, and several pheasants. Almost half of the natural habitat in this ecoregion is still intact, mostly because of the rugged terrain and low human population density. Although 16 percent of the ecoregion is in protected areas, an assessment should consider the degree to which these areas represent overall biodiversity.

Cardamom Mountains Rain Forests [55]. This ecoregion is one of the few tropical rain forest ecoregions in Asia with large, intact habitats that can sustain intact large vertebrate communities. The ecoregion harbors populations of tiger and six other felids, the Asian elephant, gibbons, wild dog, Malayan sun bear, gaur, banteng, and the endangered smooth-coated otter. The birds include several hornbills, pheasants, and forest-dwelling pigeons. A population of the rare Siamese crocodile (*Crocodylus siamensis*) was discovered during recent explorations. Because about 75 percent of the habitat is in relatively intact regions of Cambodia and about 26 percent of the ecoregion is included in large protected areas, the ecoregion provides a rare opportunity to conserve intact ecological processes in large, forested landscapes. However, conservation efforts should ensure that the existing protected areas are well managed and are not completely insularized as increased anthropogenic impacts are (inevitably) brought to bear on these forests.

Northern Triangle Subtropical Forests [73]. This ecoregion harbors globally outstanding levels of species richness, including several large vertebrates of conservation significance that need large habitat areas (e.g., tiger, Asian elephant, Hoolock gibbon, wild dog, Himalayan black bear, Malayan sun bear, clouded leopard, gaur, and tufted deer [*Elaphodus cephalophus*]) or specialized habitat (e.g., red panda). Although more than 90 percent of the habitat is still intact, only 4 percent of the ecoregion is within protected areas. Because it is likely that the ecoregion's rich timber resources will be logged or cleared for agriculture, additional conservation efforts including protected areas within larger conservation landscapes should be considered.

Northern Triangle Temperate Forests [76]. This small ecoregion harbors globally outstanding levels of species richness. Some of the larger mammals of conservation importance—tiger, clouded leopard, wild dogs, Asiatic black bear—from the adjacent subtropical forests are also found in this ecoregion. But most of the species that make up the mammal community are smaller mammals, such as murine rodents and bats. The birds include hornbills and pheasants that need mature forests for survival. More than 90 percent of the ecoregion's habitat is still intact, but only 2 percent is protected. Thus, additional protection to preempt widespread loss from logging activities is urgently needed.

Sunda Shelf and Philippines

Sumatra Montane Rain Forests [83]. This ecoregion is globally outstanding for both richness and endemism. Although most of Sumatra's lowland forests have been converted, there are large blocks of natural montane habitat along the Leuser system. These large habitat areas are especially important for some of the last populations of the Sumatran rhinoceros, Sumatran tiger, the endemic North Sumatra leaf monkey (*Presbytis thomasi*), gibbons, Asian elephants, nine hornbill species, and several pheasants. Three protected areas cover a large portion (25 percent) of the mountain range and, from the landscape perspective, additional protected forests that have been reserved for water catchment protection also serve as conservation areas. However, since the recent economic crisis, forest conversion has accelerated and has begun to intrude into these montane forests, including protected areas. Poaching of rhinoceros, elephants, songbirds, and other fauna and flora remains a major problem. Unless urgent actions are taken to stop poaching and habitat fragmentation, most of this biodiversity will be lost.

Borneo Montane Rain Forests [95]. The montane forests in central Borneo are globally outstanding for biological richness. Mammals of conservation importance include the orangutan and Sumatran rhinoceros, three species of leaf monkeys, gibbons, and several tree shrews that need intact, mature forests for survival. Of the island's thirty endemic bird species, twenty-three occur only in this montane region (MacKinnon and MacKinnon 1986). More than 90 percent of the ecoregion's natural habitat is still intact, and about 12 percent of the ecoregion is within protected areas, including the largest protected block of rain forest in Borneo, the Kayan Mentarang Nature Reserve (WWF Indonesia n.d.). Commercial logging activities, road building, and intensive extraction of valuable nontimber forest products now threaten the intactness of these montane forests. Because of the richness and high beta-diversity, an assessment to ascertain whether the protected areas adequately represent the biodiversity of the ecoregion is recommended.

Kinabalu Montane Alpine Meadows [108]. This alpine ecoregion within the tropical belt is globally outstanding for both richness and endemism. Despite its small size, the ecoregion supports more than 100 mammal species with particularly good representation of horseshoe bats (Rhinolophidae and Hipposideridae families). The twenty-two birds endemic to the ecoregion include partridges, flowerpeckers, flycatchers, and spider-hunters. The slopes and surrounding area include the greatest concentration of wild orchids on Earth, with more than seventy-seven species (WWF and IUCN 1995). Seventy-five *Ficus* species have been collected on Mount Kinabalu alone, of which thirteen are endemic. About 65 percent of the ecoregion's habitat is intact, and about 30 percent of the ecoregion is in protected areas.

Wallacea

Halmahera Rain Forests [114]. This ecoregion represents the moist forests on Halmahera, Morotai, Obi, Bacan, and the other nearby Maluku islands in the northeastern Indonesian archipelago. Overall diversity is low in this ecoregion, but overall endemism is moderate to high when compared with other ecoregions in Indo-Malaysia. This ecoregion falls within the Wallacean biogeographic zone and thus exhibits a mixture of Asian and Australian fauna. The ecoregion contains four monotypic endemic bird genera, *Habroptila, Melitorgrais, Lycocorax,* and *Semioptera.* These species include the invisible rail (*Habroptila wallacii*), white-streaked friarbird (*Melitograis gilolensis*), paradise-crow (*Lycocorax pyrrhopterus*), and standardwing (*Semioptera wallacii*). Although there is some forest exploitation by logging companies, extensive blocks of habitat still cover all the islands, and nearly 80 percent of the original forest is still intact. Most of the remaining habitat in this ecoregion is semi-evergreen rain forest. Currently, Halmahera's wet evergreen lowland forests in the northwest of the island are exploited by logging companies, primarily for the valuable damar trees (*Agathis* spp.). On Ternate, Tidore, and nearby islands the wet evergreen forests have been lost for centuries because their rich volcanic soils have been aggressively cultivated for cloves and other spices.

Seram Rain Forests [116]. This ecoregion covers the forests of the eastern Indonesian island of Seram and harbors globally outstanding levels of endemism, especially among birds. Thirty-eight (18 percent) of the birds are endemic to the ecoregion, and include the colorful lories, cryptic white-eyes, and the salmon-crested cockatoo (*Cacatua moluccensis*). The mammals consist of phalangers, small murine rodents, shrews, and several bats. More than 80 percent of the ecoregion's natural habitat is still intact, and 11 percent is within protected areas; thus the conservation status is far from dire. However, small-scale logging and illegal capture of birds for the wildlife trade pose threats to the endemic bird populations.

New Guinea and Melanesia

Huon Peninsula Montane Rain Forests [124]. This New Guinea ecoregion is globally outstanding for species richness and extends into alpine habitat. The ecoregion is especially rich in plants and birds. The mammal fauna includes three species of echidna, of which *Zaglossus bartoni* is a near endemic. The eleven endemic birds include the emperor bird-of-paradise (*Paradisaea guilielmi*) and several honeyeaters and parrots. More than 80 percent of the natural habitat still remains, and 16 percent of the ecoregion is within two large protected areas. Threats are not high enough at present to place the ecoregion on the critical list, but an assessment to ensure that the protected areas capture biodiversity is recommended.

Central Range Montane Rain Forests [125]. This large ecoregion along the rugged central cordillera of New Guinea is globally outstanding for both species richness and endemism. The array of habitat types and beta-diversity across the mountain range provides the ecoregion with an exceptionally rich biota. Forty-five mammal species are endemic to the ecoregion, as are sixty bird species, among which are several honeyeaters and birds-of-paradise. More than 80 percent of the ecoregion's natural habitat still remains, and about 15 percent is within protected areas. However, because of the high beta-diversity across the mountain range and the restricted range distributions of several species, an assessment to ascertain how well the protected areas represent the biodiversity is recommended.

Southeastern Papuan Rain Forests [126]. This ecoregion is also globally outstanding for both species richness and endemism. Many of the species in the ecoregion have very local distributions because of the dissected landscape and edaphic variations; for instance, plants such as Sericolea and Piora are recorded only from Mount Piora and Mount Amungwiwa (Johns 1993). The Owen Stanley highlands, which comprise the major mountain range in the ecoregion, have extensive alpine areas and vast tracts of pristine montane forests (Miller et al. 1994), with the higher peaks supporting ecologically fragile high alpine areas (Beehler 1994). Thus, plant diversity and endemism are high. The endemic mammals include several tree kangaroos (*Dendrolagus* spp.), cuscus (*Phalanger* spp.), flying foxes (*Pteropus* spp.), tube-nosed bats (*Nyctimene* spp.), horseshoe bats (*Hipposideros* spp.), bentwing bats (*Miniopterus* spp.), and a number of murine rodents. The forty near-endemic birds include four birds-of-paradise. About three-fourths of the habitat is still intact, but only 5 percent of the ecoregion is within protected areas, which probably is insufficient to include and represent its rich and unique biodiversity.

**The island of New Guinea has many unique Australasian species such as Doria's tree kangaroo (*Dendrolagus dorianus*) found in the Southeastern Papuan Rain Forests [126]. New Guinea represents one of the few opportunities in the Indo-Pacific to protect large swaths of intact habitat.
Photo credit: WWF**

Southern New Guinea Freshwater Swamp Forests [127]. The Southern New Guinea Freshwater Swamp Forests are a unique mosaic of swamp grassland, savanna, gallery forest, and lakes (including Papua New Guinea's largest lake, Lake Murray). Overall richness and endemism generally are low to moderate when compared with those of other ecoregions in Indo-Malaysia. Nine mammal species either are ecoregional endemics, have 75 percent of their range in the ecoregion, or are restricted-range species. They include the lowland tree-kangaroo (*Dendrolagus spadix*), Arfak long-beaked echidna (*Zaglossus bruijnii*), Fly River horseshoe bat (*Hipposideros muscinus*), Fly River trumpet-eared bat (*Kerivoula muscina*), and Papuan long-beaked echidna (*Zaglossus bartoni*). The ecoregion is also home to several restricted-range birds, which include the striated lorikeet (*Charmosyna multstiata*), olive-yellow robin (*Poecilodryas placens*), white-bellied pitohui (*Pitohui incertus*), and greater bird-of-paradise (*Paradisaea apoda*). With the exception of the largest mine in Papua New Guinea, the majority of this lightly populated ecoregion is still pristine; about 80 percent of the natural habitat in this ecoregion is still intact. Ten protected areas cover 15 percent of the ecoregion, although most of this coverage is in Indonesia.

Trans Fly Savanna and Grasslands [130]. This grassland ecoregion in southern New Guinea supports globally outstanding levels of endemism. Structurally and floristically these grasslands have some affinities with those of northern Australia (Gillison 1983 cited in Miller et al. 1994), but they also have an endemic fauna; for instance, 11 percent of the mammals are endemics. The rich bird fauna includes several birds-of-paradise, imperial pigeons, southern cassowary (*Casuarinus casuarius*), and the palm cockatoo (*Probosciger aterrimus*). The Tonda Wildlife Management Area is a globally significant wintering ground for waders and waterfowl migrating from Australia and the Palearctic (Beehler 1994). The Trans Fly region is also critical habitat for several species of endemic amphibians and reptiles and is the only location of the pitted turtle (*Carettochelys insculpta*), a unique species in its own family (Allison 1993). More than 90 percent of the habitat is still intact, and about 26 percent of the ecoregion is in protected areas.

Central Range Sub-Alpine Grasslands [131]. This high-altitude ecoregion along the central cordillera of New Guinea harbors globally outstanding levels of endemism. About 70 to 80 percent of the flora is endemic to the ecoregion. The fauna is not rich, but does include four endemic mammal species and nine endemic birds. About 70 percent of the ecoregion's natural habitat is still intact, and 26 percent of the ecoregion is in protected areas.

Priority Ecoregions

These are the ecoregions that are globally outstanding for biological distinctiveness and vulnerable for conservation status or are regionally outstanding for biological distinctiveness and relatively stable for conservation status (table 6.2).

In the former category, the globally outstanding biological distinctiveness warrants priority conservation actions, but because the conservation status is vulnerable the urgency for action is less than for the ecoregions that are critical or endangered. In the latter category, the ecoregions still have intact habitat that provides good opportunities to plan and create conservation landscapes to conserve regionally outstanding levels of endemism.

Detailed descriptions of these priority ecoregions—their biodiversity values, conservation status, and overarching threats—are provided in the individual ecoregion descriptions in appendix J.

Summary

The entire priority portfolio consists of sixty ecoregions, representing 43 percent of the Indo-Pacific ecoregions. Of these, thirty-two (23 percent of all ecoregions) are Most Urgent, twenty (14 percent) are High Priority, and eight (6 percent) are deemed Priority.

With the exception of the deserts and xeric shrublands, all Indo-Pacific biomes are represented in the priority portfolio, justifying the use of this approach to

TABLE 6.2. Portfolio of Ecoregions in the Indo-Pacific for Third-Tier Priority Conservation Actions.

Bioregion	Ecoregion	Biological Distinctiveness	Conservation Status	Class
Indian Subcontinent Bioregion	Meghalaya Subtropical Forests [11]	Globally Outstanding	Vulnerable	I
	Eastern Himalaya Subalpine Conifer Forests [28]	Globally Outstanding	Vulnerable	I
Indochina Bioregion	Mizoram-Manipur-Kachin Rain Forests [50]	Globally Outstanding	Vulnerable	I
	Southern Annamites Montane Rain Forests [57]	Globally Outstanding	Vulnerable	I
	Central Indochina Dry Forests [72]	Globally Outstanding	Vulnerable	I
	Northern Indochina Subtropical Forests [74]	Globally Outstanding	Vulnerable	I
Sunda Shelf and Philippines Bioregion	Peninsular Malaysian Rain Forests [80]	Globally Outstanding	Vulnerable	I
New Guinea and Melanesia Bioregion	Trobriand Islands Rain Forests [135]	Regionally Outstanding	Relatively Stable	III

set priorities with a goal of representation. The deserts and xeric shrubland ecoregions are only locally important for biological distinctiveness and probably represent degraded dry forests (Puri et al. 1989).

Seventy-one percent of the priority ecoregions are in the tropical and subtropical moist broadleaf forest biome. But the overall trend of representation of ecoregions by biome shows good concordance with the representation of ecoregions within the complete set of Indo-Pacific ecoregions (figure 6.2).

Analysis by bioregions shows that all biomes in the Sunda Shelf and Philippines, Wallacea, and New Guinea and Melanesia are represented in the priority portfolio, but the Indian Subcontinent and Indochina bioregions lack representation in three and two biomes, respectively (table 6.3). In the former bioregion, the tropical dry forests, subtropical conifer forests, and flooded grasslands and savannas biomes were not represented in the portfolio of Most Urgent, High Priority, and Priority ecoregions. In the latter bioregion, the subtropical pine forests and mangrove biomes were not represented.

To meet the principles of representation of all biomes at the bioregional level, we placed the following ecoregions in the Priority category.

Indian Subcontinent Bioregion

Chota-Nagpur Dry Deciduous Forests [18]. This regionally outstanding ecoregion contains large habitat blocks and was also included as a Global 200 ecoregion for representation.

Himalayan Subtropical Pine Forests [31]. The only subtropical pine forest ecoregion in the bioregion.

Northwestern Thorn Scrub Forests [24]. This ecoregion has the highest species richness among the desert thorn scrub biome ecoregions in this bioregion.

Indochina Bioregion

Northeast India–Myanmar Pine Forests [77]. This is the only tropical pine forest ecoregion in this bioregion.

Indochina Mangroves [79]. This mangrove ecoregion in the delta of the Mekong River is more extensive than the mangroves in the Irrawaddy delta and is contiguous with several other globally outstanding ecoregions.

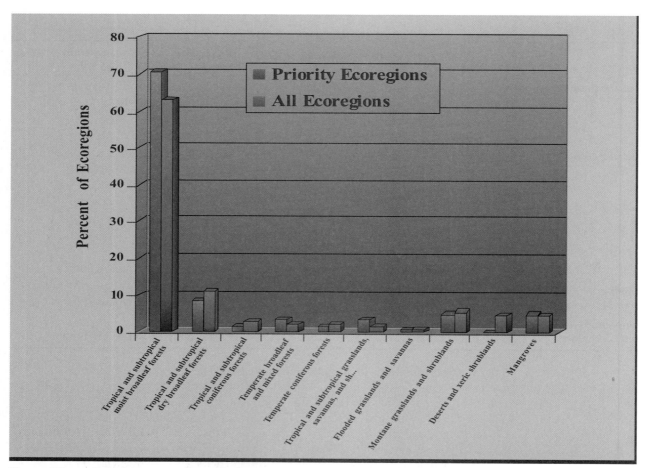

Figure 6.2. Relative proportions of priority ecoregions and all ecoregions by biome.

Taken together, these priority ecoregions contain the most important and threatened areas for biodiversity. These areas must be on the agenda of all major donors, governments, and NGOs. Mechanisms to achieve conservation victories in these priority ecoregions are the subject of chapter 7.

TABLE 6.3. Number of Priority Ecoregions in Each Bioregion and Biome.

	Most Urgent	High Priority	Priority	Representation	Total Priority Ecoregions
Indian Subcontinent					
Tropical and subtropical moist broadleaf forests	5	0	1	0	6
Tropical and subtropical dry broadleaf forests	0	0	0	1	1
Tropical and subtropical coniferous forests	0	0	0	1	1
Temperate broadleaf and mixed forests	0	1	0	0	1
Temperate coniferous forests	0	0	1	0	1
Tropical and subtropical grasslands, savannas, and shrublands	1	0	0	0	1
Flooded grasslands and savannas	0	1	0	0	1
Montane grasslands and shrublands	0	1	0	0	1
Desert and xeric shrublands	0	0	0	1	1
Mangroves	1	0	0	0	1
Total	7	3	2	3	15
Indochina					
Tropical and subtropical moist broadleaf forests	0	5	3	0	8
Tropical and subtropical dry broadleaf forests	1	0	1	0	2
Tropical and subtropical coniferous forests	0	0	0	1	1
Temperate broadleaf and mixed forests	0	1	0	0	1
Mangroves	0	0	0	1	1
Total	1	6	4	2	13
Sunda Shelf and Philippines					
Tropical and subtropical moist broadleaf forests	10	2	1	0	13
Tropical and subtropical coniferous forests	1	0	0	0	1
Montane grasslands and shrublands	0	1	0	0	1
Mangroves	1	0	0	0	1
Total	12	3	1	0	16
Wallacea					
Tropical and subtropical moist broadleaf forests	2	2	0	0	4
Tropical and subtropical dry broadleaf forests	2	0	0	0	2
Total	4	2	0	0	6
New Guinea and Melanesia					
Tropical and subtropical moist broadleaf forests	6	4	1	0	11
Tropical and subtropical dry broadleaf forests	1	0	0	0	1
Tropical and subtropical grasslands, savannas, and shrublands	0	1	0	0	1
Montane grasslands and shrublands	0	1	0	0	1
Mangroves	1	0	0	0	1
Total	8	6	1	0	15

ESSAY 18

Saving the Last Remnants of Tropical Dry Forest of New Caledonia

Arnaud Greth, Arnaud Collin, and David Olson

Scanning through binoculars on a recent visit to western New Caledonia, we searched the landscape for any sign of our quarry. A warning from local scientists (Bouchet et al. 1995) had prompted us to come to this ancient land—a fragment of the continent Gondwanaland—on a last-ditch effort to help save something precious. Unlike other conservation surveys trying to find the last populations of a critically endangered species, we were seeking the last remnants of an endangered ecosystem. No patches of green canopy came into view that morning, and we visited one of the last remaining forest fragments that afternoon with a local biologist as our guide. Sadly, the tropical dry forest of New Caledonia is almost gone. However, saving any of the last patches is a global conservation priority because of the uniqueness of the plants and animals that persist in them.

Tropical dry forests are one of the rarest terrestrial habitats on the planet, covering much less of the tropics than do rain forests. Although not as rich in species as rain forests, dry forests contain plants and animals that are highly distinctive and display unique adaptations for dealing with lengthy dry seasons, hot climates, and fire. Many domestic crops are descended from dry forest species. These extraordinary forests are globally threatened. Only in a few places, such as the Chiquitano lowlands of Bolivia and the coast of Jalisco in western Mexico, can one still find large landscapes of intact dry forests. Everywhere else in the world they have been extensively cleared and burned, especially because they offer pleasant climates and fertile soils for agriculture. The most diverse dry forests occur in Bolivia and Mexico, but the most unique occur in Madagascar and New Caledonia, both remarkable for their richness in primitive and unique higher taxa (e.g., genera, families) and very high levels of endemism (Chazeau 1993; Jaffré 1993). For example, 80 percent of the known 3,320 species of New Caledonia plants are endemic, with 14 percent of the genera and five whole families limited to the island, a level of floral uniqueness rivaled only by continents. The extraordinary endemism in both plants and animals results from the long isolation of the island and its origin as a fragment of the ancient southern supercontinent of Gondwanaland. Numerous plants also display globally rare adaptations to the prevalent, but toxic, ultramafic soils (Jaffré 1993). For example, both ultramafic and karst soils occur in the dry forest ecoregion (Jaffré and Veillon 1994). The great complexity of soils, topography, and climate of New Caledonia also contributes to the prevalence of species with limited distributions, many being found only within a few hectares (Jaffré 1993; Morat 1993).

The loss of the New Caledonia sclerophyll forests would be a tragedy of global significance. The dry forests support 329 native plant species, including at least 59 species that occur only in dry forest habitats and many more that are endemic to New Caledonia (Bouchet et al. 1995). A monotypic plant, *Captaincookia margaretae* (Rubiaceae), a treelet whose entire trunk becomes covered with fluted scarlet blooms when it flowers, is known from only two fragments. Two other sites are the only known locations for a species of wild rice, *Oryza neocalidonea*, with amazing adaptations to lengthy dry seasons, a trait with great potential for domesticated rice production. Other dry forest specialists include geckos, skinks, landsnails, beetles, and the New Caledonia fruit bat (*Pteropus vetulus*); many of these species are naturally found at only one or a few sites.

Dry forests once covered at least 4,000 km^2 along the drier western side of New Caledonia. Today, the New Caledonian dry forest has been reduced to less than fifty isolated patches, few exceeding more than 20 ha and most in the 2- to 5-ha range (collectively representing less than 2 percent of the original habitat). A single 200-ha fragment is the largest block. These forest fragments are situated in the most developed and inhabited region of the island. They are becoming increasingly degraded from clearing for pasture, intentional or uncontrolled burning during the dry season, browsing by cattle and introduced rusa deer, and the invasion of exotic plants and animals (Lowry 1996; Mittermeier et al. 1999). An introduced ant, *Wasmannia auropunctata,* is extirpating many native lizards, insects, and landsnails. Out of 117 dry forest plant species evaluated for IUCN classification, 59 (50 percent) are threatened (vulnerable or higher). The first recorded plant extinction in New Caledonia occurred within the dry forests when a fire wiped out the only known population of *Pittosporum tanianum*, a distinct species discovered in the mid-1980s (Bouchet et al. 1995; T. Jaffré, pers. comm., 1997). The severe reduction of natural dry forest and the propensity of many of its species to have limited ranges increases the probability that several plants and invertebrates went extinct before they were ever described. And the small patches of habitat that still survive are likely to contain the only known population of species found nowhere else in the world. The extinction of specialized species is all the more likely because the extensive dry forests that once occurred on basaltic substrates have been completely destroyed. Recent threat analyses, led by the Overseas Scientific Research Organization, have identified the dry forest ecosystem as one of the most endangered ecosystems of New Caledonia.

The protected area network of New Caledonia is poor both in scope, covering 2.8 percent of the land area, and in resources, reducing the protected areas to little more than paper parks (Veillon 1993). A few of the remaining dry forest fragments are in private hands, while most belong to local communities, Province Sud and Province Nord, or the French government. Jaffré et al. (1998) consider the dry forests completely unprotected. Even the traditional act of putting up fences would be insufficient to protect species from fires like the one that wiped out the only known populations of *Pittosporum tanianum*.

For these reasons, immediate action to save the last fragments of dry forest is critical, or we will witness the ongoing loss of species and the ultimate extinction of an entire ecoregion, perhaps within a decade. These remnants are necessary to prevent the total extinction of a host of species and to act as source pools for eventual restoration of larger dry forest blocks. A long-term strategy would involve protection and expansion (through restoration) of all remaining forest patches, with the goal of linking larger blocks through corridors or stepping-stone habitats. The feasibility of this approach is shown by the successful restoration efforts of dry forests in Costa Rica (Janzen 1986). Amelioration of the threats to the New Caledonian dry forest is also needed.

Building support and capacity for conservation has been an ongoing part of the effort. Some of the partners working together to save the dry forest include the local authorities of Province Nord and Province Sud; the French Government; the Overseas Scientific Research Organization, supported by the French Government; the Centre de Coopération Internationale en Recherche Agronomique pour le Développement; the Centre d'Initiation à l'Environnement, a local NGO; the new government of New Caledonia; the Institut de Recherche pour le Développement; the Institut Agronomique Néo-Calédonien; the Maruia Society; Kanak (*Ti-Va-Ouere*) tribal groups (indigenous peoples); the University of New Caledonia; the Worldwide Fund for Nature–France; the MacArthur Foundation; and local companies. If these efforts are not too late, one day the source pools will become thriving dry forest landscapes once again.

CHAPTER 7

Solutions to the Biodiversity Crisis Facing the Indo-Pacific Region

Despite the depressing trends in habitat loss and overhunting, a number of innovative conservation approaches provide glimmers of hope for biodiversity conservation in the Indo-Pacific region. In this chapter we highlight some of the most promising initiatives: ecoregion and landscape-scale conservation, carbon credits, conservation performance payments, local guardianship of natural habitats, and privately managed reserves that can stem the loss of biodiversity in many ecoregions. All these initiatives share two common features: the need to address biodiversity conservation targets and threats over larger spatial scales and longer time horizons than ever before and bold leadership and decisions to change the status quo (Dinerstein 2001). We also explain how these mechanisms can lead to a renewed commitment to establish strict protected areas and to upgrade management of the numerous paper parks strewn across the region, thereby reversing the trends identified by Terborgh and Soulé (1999). We also outline a more strategic role for core protected areas, forest certification, and integrated conservation and development projects that, nested within landscape-scale conservation, can usher in an era of Big Conservation in the region.

Big Conservation: The Next Frontier

This analysis, together with the Global 200 (Olson and Dinerstein 1998), highlights the ecoregions that deserve urgent attention within global and regional conservation strategies. But meaningful conservation rarely happens at such large scales. The next step is to set priorities at finer spatial scales within priority ecoregions, and several international NGOs, major donors, and government agencies are involved in designing, financing, and implementing such conservation activities. For example, in the Indo-Pacific region WWF is developing ecoregion-scale strategies for the Eastern Himalayas, the Lower Mekong forests, the tropical lowlands of Irian Jaya, and the Sulu-Sulawesi Sea. Other NGOs, such as The Nature Conservancy, Conservation International, IUCN, and Birdlife International, have also embraced large-scale conservation. Here, we provide a brief primer on the conservation goals of ecoregion conservation and then examine the Lower Mekong forests initiative to illustrate the process and the results.

Conservation Planning and Action at the Ecoregion Scale

Just over three years ago, WWF conservationists adopted a new paradigm in conservation planning. This shift was driven primarily by the realization that across the tropics and much of the temperate zone, conservationists were winning a few isolated battles but were losing the war to save nature. Too often conservationists were fighting rear-guard actions to head off potential threats at isolated, small sites, even though many of these sites were too small by themselves to achieve the goals of biodiversity conservation (Dinerstein et al. 2000). Clearly, conservationists had to become far more ambitious, proactive, and even visionary in their efforts to save and restore nature. Ecoregion conservation meets these criteria.

There is no precise recipe for conducting ecoregion conservation, but some basic steps guide the process. A first step involves assembling the most knowledgeable biologists available to define what success would look like if we were able to conserve the full expression of biodiversity in an ecoregion. A comprehensive assessment of the ecoregion's biodiversity culminates in a biological vision that articulates bold long-term goals for conserving the ecoregion's biodiversity. It describes the most ambitious aspirations for conserving and, where appropriate, restoring the ecoregion's biodiversity over the next fifty years or more. The vision embraces five fundamental goals:

- All distinct natural communities must be represented.
- Ecological and evolutionary processes must be maintained.
- Viable populations of species must be maintained.
- The resilience of the ecosystem must be maintained (i.e., it must remain in or be able to recover to a healthy condition if affected by large-scale periodic disturbances or long-term change).
- There must be a concerted effort to prevent the introduction of invasive species and to eradicate or control established invasive species.

The biodiversity vision adopted by those committed to nature conservation would then serve as a touchstone for conservation planning and activities. We have long recognized that if we as conservationists cannot articulate clearly what our vision of success looks like, others are ready to promote alternative forms of land and resource use that are incompatible with conservation of native species and habitats. A second guiding principle is to look beyond what exists on the map today as protected areas and to design the most ambitious conservation strategy possible, even if it might take fifty years to achieve.

Ecoregions as a Unit for Conservation Action

Ecoregions are defined in biological terms. Therefore, they are logical units for conserving biodiversity and for assessing what will be necessary to maintain the full array of species, communities, and habitats, particularly those that are most distinct and irreplaceable. Conservation action at an ecoregional scale allows an informed evaluation of trade-offs between different targets and a commitment to actions with impacts that are likely to endure over time.

Ecoregions, like other ecologically defined units, transcend political boundaries. Managers, decision-makers, and other constituents who are able to think and plan beyond their own jurisdictions can create better opportunities to conserve large-scale ecological and evolutionary processes. Furthermore, a meaningful assessment of how people relate to and alter their environment may reveal that ecoregion traits exert an influence that is as strong as the political system. People living in a given ecoregion often share a common relationship with the land and water, regardless of the country, state, or province in which they reside.

Ecoregion conservation also enables strategic assessment of threats to a particular region and identifies opportunities to mitigate them at meaningful scales. In the following section, we illustrate these points with an example of ecoregion conservation planning in forests of the Lower Mekong.

The Forests of the Lower Mekong: A Case Study in Ecoregion Conservation

Ecoregion conservation arose, in part, in recognition of the need to operate biodiversity conservation at a scale larger than site-based projects. A larger scale is necessary to achieve conservation results that are ecologically viable by conserving networks of key sites, migration corridors, and ecological processes that maintain ecosystems. A larger scale can address the broader social, economic, and policy factors that are essential to long-term success. Most ecoregions span political boundaries. Focusing conservation efforts at the ecoregion scale can stimulate transboundary cooperation and avoid redundancy of independent agendas. The goal of ecoregion conservation is to develop a suite of priority sites within an ecoregion or group of ecoregions that represents the region's outstanding or characteristic biodiversity. Future conservation investment can then be targeted at the ecoregion's most important biological areas instead of squandered on an ad hoc mechanism of prioritization.

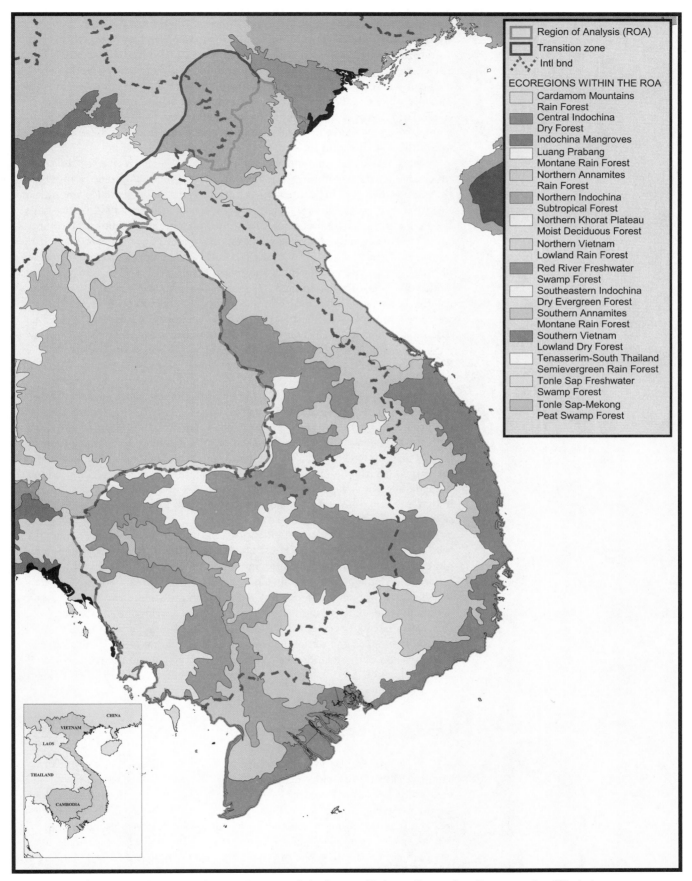

Figure 7.1. Region of analysis for the Forests of the Lower Mekong Ecoregion Conservation Workshop.

With these goals in mind, a workshop was convened in March 2000 in Phnom Penh, Cambodia to identify a suite of biological priority sites for the forests of the Lower Mekong ecoregion complex (figure 7.1; Balzer in prep.; Nguyen and Shore 2001). This priority-setting exercise brought together more than eighty biologists, based both in the region (Cambodia, Laos, Vietnam) and elsewhere, representing most of the organizations and institutions working on biological field research. The region of analysis consisted of fifteen ecoregions that included all of Cambodia, most of Laos and Vietnam, and parts of Thailand (figure 7.1). Many distinct habitat types, ranging from freshwater swamps and mangroves to extensive dry forests and lowland and montane wet evergreen forests, were included in the ecoregion complex.

Biological survey work and accessible literature for this region are limited. Much of the data and information are scattered throughout obscure journals, in the unpublished gray literature, and archived as personal knowledge among many scientists. Few comprehensive analyses of regional patterns of biodiversity have been conducted. Therefore, before the workshop four desk studies were commissioned to collect and collate the available information on mammals, birds, fish, and vegetation. Summaries of these desk studies and maps of proposed biological priority sites (birds and mammals only) were presented at the beginning of the workshop. This information base was used to build on and develop a portfolio of priority landscapes for each taxonomic group. These taxon priority landscapes were used as overlays to identify a set of biological priority sites for the region. The biological priority sites represent the landscapes where there is greatest overlap between the taxon priority sites—in other words, the most biologically diverse landscapes. Each was also ranked for importance (figure 7.2a). Throughout the workshop, scientists discussed the type and scale of threats to these landscapes and (using remotely sensed remaining habitat data, protected areas, and infrastructure information) attempted to measure the integrity and degree of threat of each biological priority site (figure 7.2b). Combining this information with the biological rankings, a biological priority map was developed for the region of analysis (figure 7.2c). To complete the landscapes, habitat linkages were recommended to promote connectivity of vital priority areas, and the need for more detailed biological surveys of remote regions was expressed. Critical and underrepresented habitat types and endangered and irreplaceable species populations were also included within landscapes.

Fifteen such landscapes were identified in the region of analysis, named the forests of the Lower Mekong. These landscapes include primarily the underexplored rain forests of the Annamite mountain range along the Laos and Vietnam border, the Cardamom Mountains in southwestern Cambodia, several vast landscapes of dry Dipterocarp forests in Cambodia, and the dry forests in northeastern Cambodia and southern Laos that extend into the low mountains of the Bolovans Plateau. Other critical areas include the Tonle Sap and surrounding flooded forests and the wetland grasslands of the Mekong Delta (figure 7.2c). Conservation of these landscapes will help conserve most of the representative and irreplaceable biodiversity of the ecoregions included in the forests of the Lower Mekong region of analysis.

Landscape-Scale Conservation

We also see the trend toward landscape-scale conservation—a scale typically smaller than the ecoregion scale—as a second major advance over ad hoc conservation of a few sites. This shift is seen as being essential to conserve Asia's large mammals, which were historically widely distributed. But over the years habitat loss and fragmentation have combined to limit these species to smaller and smaller spaces, and today they are mostly limited to small protected areas, many of which are too small to harbor populations that can be considered viable over the long term. Thus, many of these megavertebrates are becoming anthropogenic endemic species.

This is best illustrated by the five most conspicuous of Asia's charismatic megavertebrates, the tiger, elephant, and the three rhinoceros species, which are

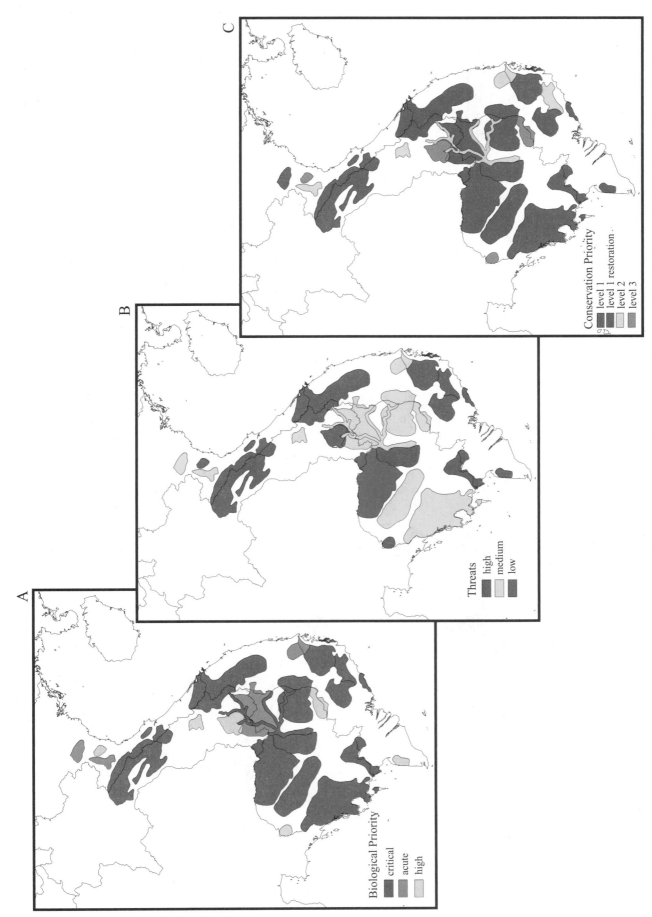

Figure 7.2. Priority sites identified in the Forests of the Lower Mekong Ecoregion Conservation Workshop. Map A shows the biological rankings. Map B shows the threat rankings. Map C shows the final conservation priority rankings, based on combining biological and threat rankings.

Solutions to the Biodiversity Crisis Facing the Indo-Pacific Region

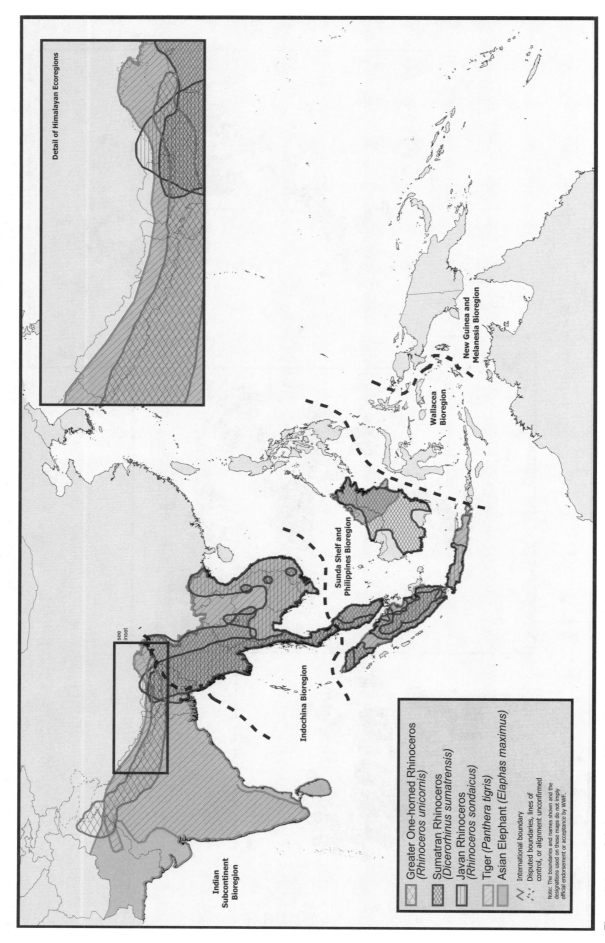

Figure 7.3. Original range of tiger, Asian elephant, and rhinoceros in the region of analysis.

TABLE 7.1. Original and Current Range Overlap of Tiger, Asian Elephant, and Three Asian Rhinoceros Species with Ecoregions.

Species	Number of Ecoregions in Original Distribution	Number of Ecoregions in Present Distribution
Tiger (*Panthera tigris*)	75	56
Asian elephant (*Elephas maximus*)	82	40
Greater one-horned rhinoceros (*Rhinoceros unicornis*)	13	7
Javan rhinoceros (*Rhinoceros sondaicus*)	41	2
Sumatran rhinoceros (*Dicerorhinus sumatrensis*)	38	10

Historic and current range data are derived from Seidensticker et al. (1999) and Dinerstein et al. (1997) for tigers, Sukumar (1999) for elephants, and Foose and van Strien (1997) for rhinoceros.

also natural candidates to serve as centerpieces for conservation planning. The original range distributions of all these species have shrunk drastically (figure 7.3). The tiger's range historically overlapped with seventy-five ecoregions, but this majestic predator has become extinct from nineteen of these ecoregions. The viable populations of the Asian elephant now overlap with forty ecoregions, whereas the historic range overlapped with eighty-two ecoregions. The three rhinoceros species have suffered the most, with the existing known populations limited to a few small protected areas. The Javan rhinoceros and Sumatran rhinoceros once were widely distributed throughout the Indochina bioregion and the Sunda Shelf and Philippines bioregion and even extended into parts of the Indian subcontinent bioregion. Today, both species have disappeared from 95 percent and 74 percent, respectively, of the ecoregions in which they used to live (table 7.1). Similarly, the range distribution of greater one-horned rhinoceros has been greatly reduced, and the current populations are limited to a few protected areas that overlap with seven ecoregions.

One effort that accelerated landscape-scale conservation in the region was the collaboration between WWF and the Wildlife Conservation Society to define and prioritize geographic areas where tigers live across the Indo-Pacific (Dinerstein et al. 1997; Wikramanayake et al. 1998). Together, we delineated 159 large landscapes within the tiger's range that we called Tiger conservation units (TCUs). A TCU is defined as "a block or cluster of blocks of existing habitats that contain or have the potential to contain interacting populations of tigers" (Dinerstein et al. 1997). The purpose of the TCU analysis was to map, for the first time, the distribution of tigers across their range; devise a hierarchical strategy to conserve the ecological role of tigers in a variety of habitats and the adaptations of tigers to living in those habitats; and systematically prioritize those landscapes within the ecological framework where tiger populations have the best chance of long-term persistence. The ranking of TCUs was based on the intactness of the habitat; poaching pressure on tigers, their prey, and the ability to mitigate those pressures; and broad trends in tiger populations. (For a detailed look at TCUs see Dinerstein et al. 1997 or Wikramanayake et al. 1998.)

The introduction of TCUs has led to some dramatic changes in landscape-scale conservation for several other large mammal species in the region. For instance, WWF created its Asian Rhinoceros and Elephant Strategies (AREAS) program to design conservation landscapes for the highest-priority rhinoceros and elephant populations in the Indo-Pacific based on this concept. The Wildlife Conservation Society has recently reorganized its field programs around the conservation of such landscape species.

A Vision for Tigerland: The Terai Arc

The TCU approach has provided the framework for one of the most ambitious landscape-scale conservation programs in the region, the Terai Arc Wildlife Corridor (formerly known as the Corbett-to-Chitwan Corridor), which we use as an example to illustrate the landscape approach and concept. The concept and conservation goal are based on management of a metapopulation, where the protected areas will harbor source populations from which animals will disperse and colonize the overall matrix (essay 19).

The southern slopes and foothill valleys of the Himalayas, known as the Siwaliks, support a spectacular assemblage of Asia's megafauna that includes the tiger, greater one-horned rhinoceros, Asian elephant, and eleven ungulate species. The ecoregion also supports migratory and breeding habitat for about 10 percent of the world's bird species and is one of the major biogeographic corridors between the Palearctic and Indo-Malayan realms.

However, the high human population density and accessible natural resources in the Terai Zone—a 5- to 40-km wide, 1,200-km long strip of forests and grasslands along the foothills and inner valleys—have resulted in degradation and fragmentation of the natural habitats. Much of the habitat loss and degradation is recent. Until the mid-1950s, the Terai was protected from human settlement and development by the widespread occurrence of malaria. With the eradication of malaria, resettlement programs in the 1950s, and construction of the East-West highway in the 1960s, land use in the Nepalese Terai changed dramatically. These tall grasslands—the tallest on Earth, with grasses reaching 7 meters by the end of the monsoon season—of the Ganges and Brahmaputra watersheds are now the most endangered habitat in south Asia (Dinerstein and Price 1991).

In the 1970s, the government of Nepal became concerned about loss of biodiversity in these habitats and established a network of protected area systems that began with creation of Royal Chitwan National Park in 1972. In the Nepalese section of the Terai this system has been expanded to include five protected areas, and there are another six in Uttar Pradesh, India. Tigers, greater one-horned rhinoceros, Asian elephants, and many of the other charismatic large mammals are now protected within the eleven reserves in the Terai Arc, but their numbers are diminishing rapidly in the forests and floodplain habitats that form the matrix.

Unfortunately, no single reserve is large enough to maintain a viable tiger population (see essay 19 by Joshi et al.). Therefore, the goal of the Terai Arc restoration program is to link these protected areas through corridors of natural habitat. Restoration efforts within the corridors will provide additional habitat in which these space-needing large mammals can live and disperse between core refuges. Successful community forestry projects in the buffer zone of Chitwan National Park in Nepal have shown that corridor restoration is feasible (essay 19).

Although the large herbivores—including elephants and rhinoceros—disperse along watercourses and over agricultural lands at night, tigers are less willing to cross open areas that are greater than 5 kilometers. Therefore, tigers will be the primary focal species when designing the landscape linkages.

The Terai Arc is an ambitious and visionary project. Implementing it will entail addressing several fundamental conservation issues and overcoming many hurdles. First, tigers share the larger landscape with a dense human population, many of whom live at a subsistence level and could perceive conservation as a threat to their livelihood. Second, the sections of the Terai floodplain still in natural habitat are highly productive and rare agricultural land waiting to be exploited. Third, the governments of Nepal and India lack the resources to implement the Terai Arc corridor, let alone provide adequate support to the existing reserve network. Finally, tigers, rhinoceros, and wild elephants can endanger human life and agriculture, and the local people could easily view their presence unfavorably.

If this is the case, what cause is there for optimism for reconnecting the Terai Arc? The answer lies largely in the successes that have been achieved in Chitwan National Park and its buffer zone during the last decade or so (Dinerstein 2002; Dinerstein et al. 1999). Essentially, changes in national legislation that allowed the diversion of 50 percent of park revenues to local communities created a powerful economic incentive for the local people to conserve wildlife and habitats. Community-based ecotourism in Chitwan's buffer zone, with tigers and rhinoceros as the main attractions, brought enough revenue to the local communities to change their attitudes from hostility to benevolence toward wildlife. Tigers, rhinoceros, and other endangered wildlife became more valuable alive than dead.

Thus, the mechanisms for habitat restoration were put in place with the legislative changes.

There are other driving forces. For instance, large cattle herds have grazed forests, resulting in extensive degradation. These forests lack undergrowth, and the local communities that depend on forest products find that the grazing practices deprive them of these important resources. The cattle have become an economic liability. In Nepal, Hindu customs and laws forbid the slaughter of cattle for meat. Milk production is also minimal. In neighboring India, cattle owners have begun to exchange cattle for chickens; the going price is four cattle for one chicken (A. S. Negi, pers. comm., 2000). But protecting community forests from grazing promotes a dramatic regeneration, increasing availability of the needed forest products. At the same time, the people have come to accept the fact that a healthy forest that yields important natural resources will also contain wildlife.

The creation of a Terai Arc Trust Fund, now in the planning stages, will help finance the Terai Arc Wildlife Corridor. The long history of conservation activities in the Terai Arc, the strict protection of core areas, and the encouraging success of local guardianship of buffer zones suggest that if we devote enough effort and resources to this project, we can safeguard one of the most spectacular wildlands on Earth. That this can be achieved in a place as impoverished and degraded as the Terai suggests that large landscapes in less populated and more prosperous areas in the Indo-Pacific are also possible.

Innovative Mechanisms to Achieve Conservation Gains

One of the principal recommendations of this and other conservation assessments for the Indo-Pacific is to greatly increase the size and degree of representation of strict protected areas. In chapter 5 we generated a scorecard of how many ecoregions fall below the threshold, accepted by some biologists, of at least 50 percent remaining habitat per ecoregion under some form of protection. How can such an important target be achieved in a short enough time frame to save biodiversity before it is lost forever?

Increasing conservation effort in montane forests is particularly important because disproportionate numbers of single-ecoregion endemic vertebrates—the group likely to be most threatened with extinction—are concentrated in upland forests. However, conservation of montane forests can be justified on purely economic grounds in a language that government planners, decision-makers, and local farmers understand, without having to invoke the cause of biodiversity conservation. Montane forest conservation is synonymous with watershed protection, flood control, and agricultural productivity. With water resources becoming a more prominent issue, major cities and towns look to upland forest conservation as essential to avoiding water problems in the future. Protection of montane areas also makes sense because they are too steep to log or because the timber found there is of lower quality than in the lowlands.

The greater challenge by far will be to maintain and restore lowland forests, especially those below 300 m (see figure 5.4). Several new mechanisms offer, at a minimum, powerful incentives to maintain forest cover and possibly even to extend the conservation effort beyond protected areas and into the adjacent landscapes.

Carbon Credits and International Conventions

All new economic mechanisms to maintain forest cover are based on the rule that an annual payment for ecological or biological services provided to the landowners or managers must exceed a perceived value of what that forest is worth if put to other uses. This would include the value of turning the forests into timber, extracting agricultural crops, planting oil palm, or grazing the land after it was cleared. The potentially disastrous effects of climate change, the loss of forest biodiversity in the tropics, and the realization that intact and maturing tropical moist forests can play a significant role in carbon budgets suggest the wisdom of transferring payments to tropical countries to maintain forest cover. To ensure that these carbon sink forests and their ecological processes are kept intact, roads should not be built into them, and wildlife hunting should be prohibited or at least carefully controlled. This approach could be covered by international treaties and is examined in detail by Joyotee Smith in essay 20.

Conservation Performance Payments

During the last decade of the twentieth century, integrated conservation and development projects (ICDPs, also called ecodevelopment projects) were widely used to link local development activities to biodiversity conservation. After a decade of experimentation and perhaps overenthusiastic implementation, this approach has come under harsh criticism by biologists as being less cost-effective than investing in strict protected areas (Terborgh 1991; Oates et al. 2000). However, a new assessment of ICDPs by Paul Ferraro (see essay 21) illustrates the complexity of these projects and how the conservation benefits often are indirect rather than explicit and easily measured. Ferraro argues convincingly for establishing a system of conservation performance payments where local landowners or communities are given a direct cash payment for maintaining forest cover on biologically important lands. The conservation performance payment approach differs greatly from the ICDP paradigm (see table 7.4 in essay 21). Ferraro's central point is that if one adds up the costs of designing, implementing, and sustaining an ICDP, it may be financially more competitive to offer an annual payment for a service that is easily monitored (essay 21).

As Ferraro points out in his essay, conservation performance payments are hardly a panacea for maintaining forest cover or biodiversity values. For example, for this program to work effectively, stable land tenure is a prerequisite, but few frontier societies enjoy clearly defined land tenure. However, one can see a critical role for conservation performance payments in areas that are adjacent to strict protected areas, are critical for dispersal linkages, or are rare habitats that are conservation imperatives. Ferraro notes that Western governments already use conservation performance payments to pay farmers to keep certain areas in natural vegetation rather than planting crops. This approach extends the concept to the tropics, where the monitoring agency would pay landowners for stewardship of the forests and wildlife resources under their tenure.

Payments for Ecological Services Provided by Protected Areas

James et al. (1999) suggest that the most equitable way to stop the rapid extinction of species and loss of habitats is to redirect the vast sums that governments spend on harmful subsidies to support and expand protected areas instead. They argue that protected areas, besides protecting biodiversity, provide essential ecological services that are not accounted for in current economic valuations. They suggest that it would be inexpensive—roughly US$16.6 billion per year for thirty years—in addition to the US$6 billion currently spent to maintain a representative network of protected areas (covering 15 percent of global land area) around the globe. For the Indo-Pacific, this figure is estimated at US$4.4 billion per year. These figures represent less than a quarter of the US$1 trillion per year governments currently spend on environmentally harmful subsidies (James et al. 1999).

Privately Managed Reserves

Private nature reserves are proliferating around the world (Langholz and Lassoie in press). These are often financed by wealthy people who are concerned about the loss of nature. The high-tech revolution and economic progress in Asia have produced many extremely wealthy individuals and companies (some as a result of exploiting the rich natural resources). The Indo-Pacific region desperately needs some of these individuals and companies to come forward and purchase failing oil palm plantations and restore the diverse forest habitats and charismatic large mammal populations—tigers, rhinos, elephants, apes—that used to live in them. They can also purchase or take out long-term leases on critical lands that have been slated for extensive exploitation or development and manage these lands for biodiversity conservation, or they can help finance the management of existing reserves. The model for such an endeavor exists on east African game ranches, where conservation-minded philanthropists have helped to recover endangered black rhinoceros populations and other wildlife on well-guarded private lands. The potential for ecotourism in such a unique set of reserves is an enormous opportunity to make such ventures self-sustainable.

Reevaluating Existing Mechanisms in a New Light

There is a tendency in any endeavor to discard existing mechanisms for achieving stated goals in favor of the newest approaches and ideas. In biodiversity conservation, protected areas, ICDPs, ecotourism, and strict enforcement of conservation laws all play vital roles. However, they should be applied rationally rather than embraced and implemented with little regard for how appropriate they are under the circumstances—a problem that has occurred repeatedly in the Indo-Pacific region.

Expanding the Network of Protected Areas

Protected areas remain the cornerstones of any conservation strategy. Much of the region's biodiversity still persists today because of strict protected areas. However, as discussed previously (see chapter 5), during the last decade there has been a decline in the establishment of protected areas, and conservation efforts have shifted toward sustainable use and natural resource management areas, supported by expensive ICDPs.

As this analysis has shown, many ecoregions still lack adequate protected area networks (table 7.2). Thus it is imperative that the pendulum shift towards establishing the time-tested strict protected areas in key locations to protect the natural biodiversity in these ecoregions. All the Class I ecoregions listed in chapter 6 should be given highest priority to develop a protected area network as rapidly as possible. The WWF–WorldBank forest alliance, which has focused on increasing protection of forested ecoregions, could be one mechanism to establish reserves in these ecoregions.

A More Strategic Role for Integrated Conservation and Development Projects

Few mechanisms that promote biodiversity conservation have received as much withering criticism as ICDPs. Many conservation biologists view the blindly enthusiastic embrace of ICDPs by major foreign donors—with no consideration of how appropriate they are to the project at hand and their near-retreat from financing strict protected areas—as a major tactical error (Terborgh 1999). Dinerstein (2002) outlines a list of steps for integrating ICDPs into large-scale conservation that gives preeminence to strictly protected core areas, buffer zones, and wildlife corridors. In many ways, the landscape-scale example of the Terai Arc grew out of an ICDP that began as a tree nursery on a single hectare of private farmland in 1987. In thirteen years, the vision has grown from increasing efforts of local farmers in Nepal to grow native tree species on their own lands to habitat restoration on a grand scale (Dinerstein 2002). However, an essential lesson from this experience is that designing an ICDP without linking the project area to a strictly protected core area is like building on quicksand. ICDPs should never again be viewed as one-off projects but should be nested within a larger landscape paradigm (essay 22, by Kathy MacKinnon).

Ecoregion- and landscape-scale conservation is in effect a zoning exercise driven by a biological vision. Once conservation biologists and planners begin to

TABLE 7.2. Degree of Protection of Class I Priority Ecoregions.

Class I Ecoregions	Total Area (km²)	Percentage Protected	Average Size of Protected Area (km²)
New Guinea Mangroves [129]	26,700	33	2,193
Mindoro Rain Forests [99]	10,300	26	893
Eastern Himalaya Subalpine Conifer Forests [28]	26,900	19	434
Southeastern Indochina Dry Evergreen Forests [59]	123,800	18	748
Sunda Shelf Mangroves [107]	37,200	18	251
Borneo Lowland Rain Forests [96]	425,400	12	826
New Britain–New Ireland Montane Rain Forests [134]	12,100	11	765
Sri Lanka Montane Rain Forests [4]	3,000	11	91
Sundarbans Mangroves [33]	20,400	10	383
Sumatra Lowland Rain Forests [82]	258,300	8	625
Meghalaya Subtropical Forests [11]	41,600	8	22
Terai-Duar Savanna and Grasslands [35]	34,500	8	267
Western Java Montane Rain Forests [93]	26,200	8	136
Southern New Guinea Lowland Rain Forests [128]	122,300	8	3,110
New Britain–New Ireland Lowland Rain Forests [133]	34,700	7	1,050
South Western Ghats Montane Rain Forests [2]	22,500	6	200
Luzon Tropical Pine Forests [106]	7,100	6	108
Mindanao Montane Rain Forests [103]	17,800	6	118
Vogelkop-Aru Lowland Rain Forests [119]	77,400	6	676
Southern Annamites Montane Rain Forests [57]	46,200	5	151
Central Indochina Dry Forests [72]	319,000	5	723
Vogelkop Montane Rain Forests [118]	21,900	5	1,137
Northern Indochina Subtropical Forests [74]	288,300	5	520
South China–Vietnam Subtropical Evergreen Forests [75]	37,500	4	91
Luzon Rain Forests [98]	94,800	4	100
Lesser Sundas Deciduous Forests [111]	39,300	4	144
Sri Lanka Lowland Rain Forests [3]	12,500	4	70
Peninsular Malaysian Rain Forests [80]	125,000	3	161
Indochina Mangroves [79]	26,800	3	117
New Caledonia Rain Forests [139]	14,500	3	19
Eastern highlands Moist Deciduous Forests [12]	340,200	3	435
Mizoram-Manipur-Kachin Rain Forests [50]	135,200	3	251
Solomon Islands Rain Forests [137]	35,700	2	930
Greater Negros–Panay Rain Forests [100]	34,200	2	93
Mindanao–Eastern Visayas Rain Forests [101]	105,000	2	130
Godavari-Krishna Mangroves [34]	6,800	1	310
Myanmar Coastal Mangroves [78]	21,900	1	42
Timor and Wetar Deciduous Forests [112]	33,300	0	0
Indus River Delta–Arabian Sea Mangroves [32]	5,800	0	0
Peninsular Malaysian Peat Swamp Forests [87]	3,600	0	0
Palawan Rain Forests [97]	13,900	0	0
Vanuatu Rain Forests [138]	13,200	0	0
New Caledonia Dry Forests [140]	4,400	0	0

design large conservation landscapes like the Terai Arc, they can create zones for ICDPs, areas slated for conservation performance payments (see essay 21), forest certification, and concentrated intensive extraction. Zoning ICDPs, forest certification programs, and plantations where they will do the least damage to biodiversity while providing other benefits will be a critical task in the years to come.

Certified Ecotourism, Local Guardianship, and Enforcement

One of the fastest-growing industries in the world today is tourism, and ecotourism is an important component of this growth spurt. However, some studies, such as Dinerstein et al.'s (1999) study of Nepal, show that ecotourism has little impact on local economies despite the high profits. Instead, ecotourism typically employs few people locally, and almost all areas where ecotourism is practiced lack legal mechanisms for returning the profits generated at a site to local communities.

The government of Nepal has led the way by enacting legislation that requires 50 percent of all revenues generated by parks and ecotourism concessions to be shared locally. This change immediately allows local people to become involved in biodiversity conservation in a meaningful way, and it encourages them to assume the role of local guardians of endangered wildlife in buffer zones surrounding core areas (Dinerstein 2002).

A market transformation in the ecotourism industry could further strengthen the involvement of local citizens in protecting wildlife and their habitats. Only operators and concession holders that meet explicit criteria and return a certain percentage of their profits to local community development should be certified. A media campaign could be launched to educate ecotourists about the choices they can make that could have profound benefits for conservation.

The rising middle classes in many Indo-Pacific countries are also a huge potential market for ecotourism. The huge Indian middle class is now discovering India's remarkable system of parks. In 1998, for the first time, Nepalese tourists became the most frequent visitors to Royal Chitwan National Park, followed by citizens from India. If homegrown ecotourism is accelerating in a country as poor as Nepal, there should be even greater opportunities for ecotourism in other countries of the region.

Bold Leadership: Who Will Be the Next Indira Gandhi of the Indo-Pacific?

Bold leadership, especially by enlightened political figures, will greatly facilitate conservation at large landscape scales. Indira Gandhi was one such figure. When the trend in tiger conservation was spiraling downward, Indira Gandhi made this a personal issue and brought about a dramatic turnaround in India. However, with her death, gains were eroded quickly because subsequent leaders did not share her passion for nature conservation. In this new millennium, we need charismatic leaders from the region to step forward and challenge their constituents to adopt a more enlightened approach to nature conservation.

Nature's End or a New Beginning?

We began this book by asking this question. We end this analysis by offering three different scenarios for the future of nature conservation in the Indo-Pacific. Depending on what goals we seek, what decisions we make, and what commitments we make, it could be the end of nature or a new beginning for conservation in the Indo-Pacific.

The worst case is the wholesale destruction of the remaining intact forests and grasslands of the region and attendant widespread extinctions. In this scenario, large-scale commercial logging and its aftermath will hasten the conversion of all remaining lowland forests to oil palm and other monoculture crops. Only archipelagos of inaccessible montane forests will remain, only the most irreplaceable habitats will be stoutly defended, and only the most unproductive lands will be spared from encroachment. Most of the region's wild vertebrate populations will end up as wildlife produce in local markets or shipped for consumption elsewhere. In some ecoregions, such as in the Philippines, publication of the flora probably will read like an obituary of extinct species. The environmental effects of the loss of forest cover will be devastating and eventually will result in widespread drought, desertification, and floods, with accompanying famine, disease outbreaks, lack of shelter, and other forms of human misery. The consequences of widespread loss of natural forests are already being manifested on local scales in many parts of Asia and the Pacific in the form of huge forest fires in Indonesia, acute droughts in northwestern India, and devastating floods throughout most of Asia.

In a seemingly less disastrous second scenario, some of the innovative mechanisms for maintaining forest cover are successful, but efforts to stop market hunting of wildlife fail. In this scenario, the Indo-Pacific will be left with a string of empty forests—forests structurally intact but without the populations of vertebrates that are integral components of forest community structure. If many of the ecological linkages that maintain ecological structure are severed, it is unlikely that the ecosystems will function over the long term. Many of the ecoregions in the Indochina bioregion currently reflect this scenario.

The most optimistic scenario is that all the conservation mechanisms described in this chapter will take root, that bold leaders will step forward, and that a heightened awareness of conservation will flourish across the region. The persistence of wildlife in India, Nepal, and Sri Lanka in the face of expanding populations and economic growth indicates that in some ecoregions, we will be able to co-exist with nature, if not live in greater harmony. Asserting that gross national happiness is more important than gross national product, the Royal Government of Bhutan, the tiny kingdom nestled on the southern slopes of the Himalayas, has pledged to proceed with economic development cautiously, primarily in deference to the consequences of environmental degradation and biodiversity loss.

We hope that the countries throughout the region will make decisions and commitments to embark on the road to the third scenario, to herald a new beginning. Twenty years from now, we hope that the logging of natural forests is no longer practiced and that the region's pulp, paper, and wood needs are met instead by well-managed plantations of fast-growing native species. We would like to envision a future where a representative network of protected areas is a reality and where these conservation landscapes transcend political boundaries. We would like to believe that some of the seeds of restoration, sprouting from the Terai Arc in the west to the tropical dry forests of New Caledonia in the east, are the harbingers of a return of the region's biological wealth. And we fervently hope that Bhutan's future generations will not be alone in benefiting from bold conservation decisions made by today's governments and their constituents.

Essay 19

The Terai Arc: Managing Tigers and Other Wildlife as Metapopulations

Anup Joshi, Eric Dinerstein, and David Smith

A critical question for conservationists and wildlife managers in the Terai parks and reserves of lowland Nepal and India is whether the protected area system is large enough to conserve wide-ranging species such as tigers. The Terai Arc consists of eleven protected areas: Royal Bardia National Park, Royal Chitwan National Park, Koshi Tappu Wildlife Reserve, Parsa Wildlife Reserve and Royal Sukla Phanta Wildlife Reserve in Nepal, and Sohelwa Wildlife Sanctuary, Katarniaghat Wildlife Sanctuary, Dudhwa National Park, Kishanpur Wildlife Sanctuary, Valmiki Wildlife Sanctuary, and Corbett National Park in India (figure 7.4). Forests connecting these protected areas are in various stages of degradation and fragmentation caused by human exploitation. The Terai Arc design calls for creating a network of corridors linking all eleven protected areas, a landscape approach that is a cornerstone to the long-range, large-scale thinking that ecoregion conservation entails.

Metapopulation Structure

Long-term tiger research in Nepal has focused on the metapopulation structure of tigers across the Terai. There are five major tiger populations, each centered in one of the region's protected areas. Estimated population sizes range from sixteen breeding animals in Sukla Phanta Wildlife Reserve to sixty-five in Chitwan (table 7.3). These population sizes are considered small by conservation biologists. Kenney et al.'s (1995) demographic model suggests that the smaller of these populations may not survive the next 100 years. If environmental and genetic stochasticity is incorporated into these demographic models, the probability of persistence is even lower.

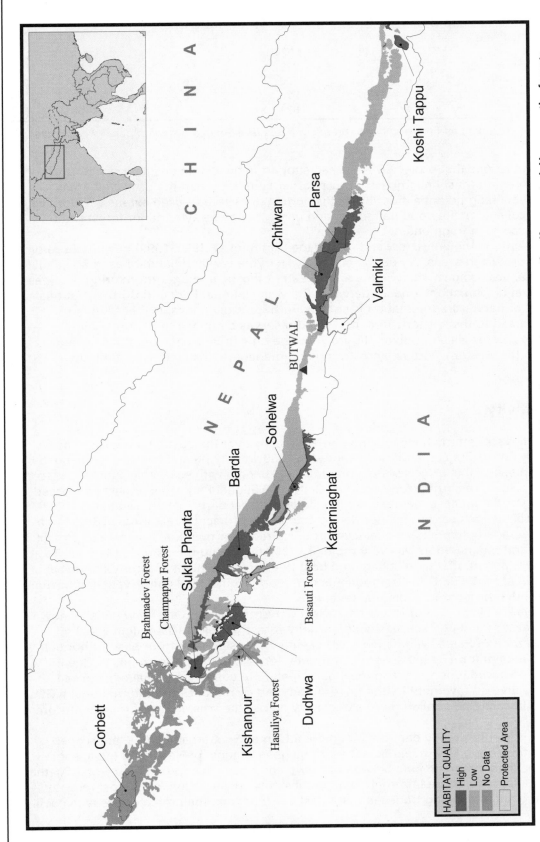

Figure 7.4. The Terai Arc. There are eleven protected areas that could contribute to tiger conservation if connected. However, the forests connecting these protected areas are in various stages of degradation and fragmentation because of human exploitation. Royal Sukla Phanta Wildlife Reserve (RSWR) and Royal Bardia National Park are no longer connected through terai forests but could be linked by developing a corridor along the foothills of the Siwalik range. RSWR is linked to the Brahmadev Forest on the foothills of Siwalik through the Champapur Forest, which is now dissected by the east-west highway. Growing settlements along the highway have become a major threat to this linkage. Dudhwa National Park and Katarniaghat Wildlife Sanctuary of India can be linked only through Hasuliya and Basauti Forests in the Kailali district of Nepal. Hasuliya Forest is highly degraded and encroached upon by local villagers and needs immediate restoration. Basauti Forest is in better shape than Hasuliya and is used by rhinos, elephants, and tigers.

Solutions to the Biodiversity Crisis Facing the Indo-Pacific Region

TABLE 7.3. Estimated Tiger Population in the Terai Protected Areas.

Populations	Total Area (km²)	Area in Reserves (km²)	Estimated Breeding Tigers	Estimated Breeding in Reserves
Royal Chitwan National Park and Parsa Wildlife Reserve, Nepal	2,543	1,921	69	49
Royal Bardia National Park, Nepal	1,840	938	50	30
Royal Sukla Phanta Wildlife Reserve, Nepal	320	320	16	15
Dudhwa National Park, India	570	570	18	18
Corbett National Park, India*		521		30

*Area outside the park is unknown, and the number of breeding tigers is estimated based on density estimates from Nepal.

The small tiger populations that protected areas can support indicate that any serious effort to conserve tigers in this ecoregion must encompass a much larger land base than the protected areas alone. However, the remaining potential tiger habitat is human dominated—used extensively for livestock grazing, fodder collection, firewood gathering, and harvest of an array of minor forest products—and is rapidly becoming fragmented.

Recent efforts to examine the entire lowland landscape (Smith et al. 1998, 1999) have focused on these forests outside protected areas. A preliminary analysis of the lowland habitat based on satellite Thematic Mapper data, discussions with wildlife and forestry officers, and tiger monitoring data has identified nine potential dispersal bottlenecks between Chitwan National Park and Corbett National Park where the threat of habitat fragmentation is critical. Perhaps most critical is that Chitwan National Park is connected to the western Terai parks and reserves by only a thin strip of highly degraded forest in the Siwalik Hills in Butwal (figure 7.4). The only hope to maintain this linkage between Chitwan and the western Terai is by restoring and managing this land as community forest at the local level.

Restoring Connectivity

In Nepal, community forestry efforts have expanded dramatically over the past fifteen years. The Baghmara Community Forest Project in Chitwan started in 1989 with 32 ha and has since expanded to about 50 km². The purpose of this project was to provide local people with economic alternatives to overexploitation of forests in the buffer zone of Royal Chitwan National Park. This project was based on an adaptive management process that resulted in increasing local participation (Dinerstein 2002). Beginning in 1995, local village user groups decided to extend the project by restoring 16.8 km² of a forest with a highly degraded understory. Local user groups hired their own forest guards to protect these forests, and natural regeneration was very rapid. By 1998 tigers and rhinos used the area. To attract tourists, villagers began offering elephant and boat rides and overnight accommodations at a watchtower (Dinerstein et al. 1999). Other components of the project included conservation education and cottage industries that brought in additional revenue.

Groups from all over the lowlands of Nepal have visited this project, and it is now widely known throughout the Terai as a model of a successful community forestry project (Dinerstein 2002). Villagers in the buffer zone of Parsa Wildlife Reserve recently requested help from park authorities to restore degraded national forest patches into community forests. In the buffer zones of Royal Bardia National Park, community forestry and local participation in conservation have increased significantly through active involvement by the government and by national and international NGOs. Similar programs could easily be applied more extensively across the western Terai to counter forest fragmentation.

Though new, this project is already demonstrating the advantages of an ambitious landscape approach. A series of transboundary meetings between the governments of Nepal and India, sponsored by WWF, has resulted in agreements between the two countries. They propose to work together to curb the illegal wildlife trade, share information on biodiversity conservation, and, most recently, collaborate in restoring and managing transnational forest corridors and the protected areas that will be linked.

One such linkage will be in the western Terai between Royal Bardia National Park and Sukla Phanta Wildlife Reserve in Nepal and Katarniaghat Wildlife Sanctuary and Dudhwa National Park in India. When fully developed, the western Terai complex would encompass a total of 6,200 km², roughly half in protected areas and half in buffer zones. The other linkage will be in one of the TCUs between Chitwan National Park and Parsa Wildlife Reserve in Nepal and Valmiki Wildlife Sanctuary in India, a total area of 4,890 km^2. Moreover, two of the transnational forest corridors encompass fragile patches of forest that are surrounded by human settlements. A major community-based reforestation effort therefore will be launched in the corridor areas of each country.

Making Tigers and Other Wildlife Worth More Alive Than Dead

The Terai Arc corridor provides an opportunity to integrate several innovative habitat restoration options. Ecotourism has already proved to be a major incentive for regenerating the buffer zone around Chitwan National Park (Dinerstein 2002). The Terai Arc restoration program could invest immediately in promoting ecotourism in Sukla Phanta Wildlife Reserve and Bardia. Investing revenues generated by ecotourism in buffer zone conservation not only provides income for local communities but also provides better protection for core areas (national parks and reserves).

For tiger corridors to be effective, they must contain nearly continuous vegetative cover and an adequate prey base of wild ungulates (deer and boar). We propose that local communities be paid a bonus for maintaining wild ungulate biomass in key corridors under their management or adjacent to their communities. For example, a breeding population of sambar (*Cervus unicolor*) or wild boar occupying a defined area of the corridor would net the local community an extra annual sum per hectare per year. The densities of these ungulates could be monitored by field staff by sampling permanent pellet plots or by using transects. Maintaining healthy populations of wild ungulates in some of the corridors entails reducing domestic cattle herds. Stall-feeding of cattle in these areas could be another incentive. Payment for conservation performance has worked successfully in several areas such as the United States, where Defenders of Wildlife pays ranchers to allow swift foxes and bears to live on their holdings (http://www.defenders.org/wildlife/grizzly/grizcomp.html).

Additional bonuses could be given to local communities if tiger sign is discovered in a section of corridor, recruitment of tigers occurs (or cub tracks are spotted), rhinoceros sign is discovered or rhinoceros are spotted, recruitment of rhinoceros occurs (tracks or sighting of young calves), or elephants move through the corridor.

At present in both India and Nepal, local communities are reimbursed for crop destruction by wildlife or loss of livestock to predators. The incentives described here offer a more proactive approach to reducing tensions between people and wildlife by offering cash incentives to promote local guardianship of the corridor areas.

A second method being implemented in the western Terai is the establishment of a network of citizen rangers to monitor biodiversity. These rangers are collecting data on the geographic distribution and rates of tiger predation on domestic livestock. This information is important for identifying habitat connectivity or areas of potential tiger-human conflict. Professional pride has developed among these rangers, and the director general of Nepal's Department of National Parks and Wildlife Conservation has publicly recognized the role these villager rangers play in collecting data needed by the department for landscape-scale planning. In the next phase of the project these village rangers will distribute conservation education materials to local schools in their areas.

The ultimate objective of community-based conservation is to ask local people to design community-based projects that help them meet their economic needs while conserving biodiversity. Villagers who have helped to explain how the system works and who have collected data about the needs and habitat use of wildlife are likely to be committed and able to design projects to meet local and conservation objectives.

Based on the widely known Baghmara community forest model and new approaches for engaging local people, there is potential to expand conservation efforts to key high-priority areas. Current efforts of many international and local NGOs and the 1995 Buffer Zone Act provide the funding needed to undertake community forestry activities. This corridor approach is visionary because it goes beyond the traditional approach of setting aside protected areas for conservation; rather, it proposes management of the entire complex as a single unit, in partnership with local communities and through transboundary collaboration between Nepal and India. Successful implementation will result in a larger land base by interconnecting protected areas through forest corridors, which will facilitate dispersal of individuals and consequently genetic exchange between different populations. This landscape-level approach will help not only tigers, rhinos, and elephants but also entire faunal and floral communities of the ecoregion, including human beings, by providing flood control, a steady supply of firewood, and other natural resources.

ESSAY 20

Forests and the Kyoto Protocol: Implications for Asia's Forestry Agenda

Joyotee Smith

Economists have long argued that payment for the environmental services of forests could contribute to forest conservation and sustainable use. The Clean Development Mechanism (CDM) of the Kyoto Protocol could make this concept a reality by allowing carbon-emitting industries in developed countries to meet a part of their emission reduction commitments by financing forestry projects that sequester carbon or protect carbon stocks in developing countries. Deforestation prevention, conversion of degraded or agricultural land to forestry plantations or agroforestry, assisted natural regeneration, biomass energy plantations, and improved forest management are all activities that could contribute to climate change mitigation. Decisions on which of these, if any, will be eligible for inclusion in the CDM have been postponed until 2001 because no agreement was reached at the meeting of the Parties to the Framework Convention on Climate Change in November 2000. Many observers believe that controversy over the inclusion of forests as a mitigation strategy was one of the crunch issues that contributed to the lack of agreement. Analysis of the potential role of the CDM in achieving sustainable forest use could therefore clarify the debate.

For the forestry community the CDM is of special interest because the Kyoto Protocol specifies that CDM projects should help host countries achieve sustainable development. The Kyoto Protocol also states that commitments under international environmental agreements and the promotion of sustainable forest practices are to be taken into account in climate change mitigation activities, although this latter clause is not specifically linked to the CDM.

Asia contributes about 38 percent of the global carbon emissions from land-use change and forestry operations. Tropical forests in Asia are subjected to a wide variety of pressures that pose environmental and social risks. To what extent could the CDM realistically contribute toward mitigating these pressures?

I argue that the CDM should not be seen as yet another silver bullet for saving tropical forests. However, there are ways to implement the CDM that could enhance the effectiveness of more conventional approaches and leverage support from stakeholders with conservation and development agendas. Proactive efforts will be needed in crafting CDM rules and designing projects.

How Large Is the Market for CDM Forestry Likely to Be?

The price of carbon and the magnitude of financial flows to forestry projects under the CDM are difficult to predict because of technical uncertainties and uncertainties about the rules under which the CDM will operate. Estimates indicate that the price of carbon may be around US$15–25/tC and that the flow of funds for forestry projects in developing countries as a whole, though sizable (around US$2–7 billion), is likely to be small compared with estimates of funding needs for forestry. As scientists have analyzed more carefully the contribution forestry can make to mitigating climate change, a consensus has begun to emerge that most forestry projects under the CDM are likely to delay the buildup of emission concentrations rather than reduce them permanently. This is primarily because the mitigation benefits of forestry projects (with the exception of biomass energy plantations) last only as long as the forest or its products exist. Therefore, the contribution of forestry projects to mitigation depends on their duration, unlike mitigation options in the energy sector. Adjustments to the cost of forestry projects to take account of project duration may increase costs of many medium-duration forestry projects (those around 20 to 30 years) by 50 percent or even severalfold. This may put the cost of some forestry projects above predicted market prices. Longer-term obligations, on the other hand, may be difficult to guarantee and in any case may be unattractive to host countries because of the loss of flexibility they imply. As a result, forestry is increasingly seen as a stopgap strategy for buying time until more permanent ways to reduce emissions become economically viable.

Surveys of pioneer investors indicate that countries with high economic or political risks may be unable to attract investments. Countries that attract even one large-scale forest protection or plantation project may reap great economic benefits, however. Large-scale pilot forest protection projects, for example, are expected to mitigate around 5 million tC over 25 to 30 years, which at indicative market prices could yield around $100 million. The challenge facing Asian countries will be to harness these potentially large sums for mitigating unsustainable pressures on their forests.

Pressures on Asian Forests

Tropical Asia can be divided very broadly into regions where most forests are in the Extensive Use stage (such as some parts of Myanmar, Cambodia, Lao PDR, and Papua province in Indonesia), regions in the Intensive Exploitation stage (such as most parts of Indonesia and Malaysia), and those in the Forest Depleted stage (such as most parts of India, Papua New Guinea, and tropical China).

Regions in the first two categories typically still contain large existing natural wealth. Regions in the Extensive Use stage are characterized by limited infrastructure and market access, which reduces pressures on their forests. Forest access and use are regulated primarily through traditional institutions of local communities. Long-fallow swidden agriculture and extraction of subsistence and high value nontimber forest products tend to be the dominant land uses. In Asia today, many of the regions in this stage are on the verge of moving into the Intensive Exploitation stage.

Forests in the Intensive Exploitation stage are subjected to high extraction pressures from export-oriented timber-processing industries. In some countries these forests are simultaneously subjected to conversion pressures from large-scale agricultural plantations. Typically, local communities benefit little from the timber wealth, with governments often superimposing large-scale concessions in areas where local communities had traditional rights to forest resources. Because of the valuable timber in these forests, the forestry sector in countries at this stage is particularly prone to corruption, illegal logging, and frenzied struggles between different stakeholders to capture the remaining forest wealth. Governments in some countries have kept a lid on these struggles through punitive measures and police or military action, but the struggles tend to erupt when controls are relaxed, as is currently occurring in Indonesia, which is in a period of transition to democracy after a history of suppression. Recently, some countries have started experimenting with alternative approaches such as providing incentives for better management (e.g., certification) or giving communities a stake in forests by recognizing traditional rights.

In countries where forests have already been depleted of most of their timber wealth, on the other hand, the scarcity of forests has induced logging bans, reforestation, and protection of existing forests. Much of the remaining forest in these countries often is under government control. However, local communities tend to have greater access to forests in these countries than in countries in the Intensive Exploitation stage, either in the form of collectivized or even private forests (as in tropical China) or through joint state-community management of forests, as in India. However, forests are still subjected to high extraction pressures because of massive gaps between domestic demand and supply.

The New Broader Forestry Agenda

Global processes, such as the International Forum on Forests (now the United Nations Forum on Forests) and the Convention on Biodiversity (CBD), have made significant advances in developing a consensus on a more environmentally and socially friendly forestry agenda, which the CBD calls the ecosystem approach. This broader approach explicitly recognizes the complexity of forest ecosystems and adaptively manages them for a wide range of goods and services, with the balance between the different management objectives determined with the full participation of all those who have a legitimate interest in the forest. Realistic and equitable approaches to biodiversity maintenance are advocated, such as the right balance and spatial distribution of protected areas, managed forests, and non-forest land, rather than vast areas of pristine forests. Creation of enabling policy and institutional environments, equitable access to forests, and empowerment and full participation of local communities in forest management decisions are key pillars of the approach.

The challenge for Asia will be to harness the CDM to increase the feasibility of implementing this broader forest agenda. If CDM projects are implemented with only carbon farming in mind, there is a danger that they will derail the delicate processes that are working to achieve an international consensus on forestry.

Harnessing the CDM to Implement the Broader Forestry Agenda

Embed CDM Projects in an Integrated Program for Sustainable Forest Management

Many of the pressures facing Asian forests cannot be solved through isolated CDM projects because they may not be able to address the underlying causes of unsustainable forest use. For example, a compartmentalized CDM project that funds the incremental costs of switching from conventional logging to reduced-impact logging (RIL) may be highly successful in enforcing better logging practices in the project area. However, where capacity in the processing sector, for example, exceeds sustainable harvests (as often occurs in countries in the Intensive Exploitation stage), timber harvesting restrictions under RIL may be compensated by increased unsustainable harvesting outside the project area or even in other countries. Real long-term improvements in logging practices would be far more likely if the project were part of an integrated package of measures. This could include the removal of subsidies that result in excess profits for the processing sector, for example. To the extent that premature reentry logging occurs to feed pulp mills, sustainable feedstock could be provided by CDM-funded fast-growing plantations. Leveraging other market-based mechanisms such as certifying CDM-funded RIL projects could be another supportive measure.

CDM projects are much more likely to result in real benefits for Asian countries if they are targeted to countries that are already taking proactive steps to support forestry activities compatible with CDM. In some forest-depleted countries (such as China and India), governments are attempting to encourage local communities to produce more biomass through various measures; these include increased participation of local communities in forest management decisions, increased stability of forest tenure, and measures to reduce fuelwood demand, such as improved stoves and biogas plants. As a result, about 11 percent of the forest area in India is under Joint Forest Management (JFM) programs, with local communities undertaking activities such as assisted forest regeneration and establishment of plantations that include species used by local artisans. In China, stabilization of forest tenure has stimulated farmers to spontaneously develop collaborative forest management institutions to capitalize on economies of scale in forest management. In these enabling environments, CDM projects could be used to leverage the proactive efforts of governments and local communities. CDM guidelines therefore should prioritize projects in which proactive efforts are being made to create enabling policy and institutional environments.

Prevent Perverse Incentives

Projects should be targeted to places where enabling policies exist if benefits are to be long lasting. But countries with the poorest policies may appear to have the most scope for CDM projects because the prospect for emission reductions is so high. For example, where deforestation occurs because of agricultural subsidies or forest fires are deliberately induced to justify conversion to agricultural plantations, there is the danger that the large sums to be earned from large-scale forest protection projects could discourage efforts to implement better policies. Therefore, CDM rules should include measures to avoid rewarding poor practices or policies. One proposed measure is that project emissions should be evaluated against standardized baseline emissions, incorporating minimum good practice standards.

Use the CDM to Leverage Financial Support for Other Environmental Services and Livelihood Benefits

The inclusion of forests in the CDM could be a powerful signal of the international community's willingness to compensate owners and managers of tropical forests for the services their forests provide. Payment for the contribution to greenhouse gas mitigation could stimulate mechanisms to support other environmental services and development goals such as poverty alleviation. The emerging consensus on the ecosystem approach, with its emphasis on environmental and social benefits, provides significant opportunities for leveraging the CDM.

Under the new paradigm of establishing biodiversity reserves through participatory land-use decisions, CDM funding could be used to develop capacity for participatory planning and for strengthening, rather than denying, local use rights. This approach would be particularly valuable in regions in the Extensive Use stage where the pressures of the Intensive Exploitation stage are beginning to build up. Many local forest uses (such as resins, medicines, and wild game) are compatible with carbon stock protection. In negotiating the trade-offs involved in establishing protected areas, project developers could include strategically located areas of nearly natural forests managed extensively for carbon-compatible uses. Where forests must be maintained in a pristine state because of exceptionally high biodiversity values, project developers could help local communities establish legal rights and sustainably manage an alternative forested area. The transaction costs of such an approach may be far higher than those of conventional forest protection projects, in which forest users are financially compensated for ceasing activities within the protected area, while parks become embattled islands of biodiversity. However, the social benefits of the new paradigm and the more realistic prospect of biodiversity conservation that it offers could attract support from stakeholders with development and environmental agendas.

Reforestation under the CDM could also be implemented to increase social and environmental benefits. For instance, CDM funds could be used to reinforce multispecies Joint Forest Management plantations in India or the spontaneous collaborative forest management institutions emerging in China. These institutions provide greater voice for local communities and tend to result in systems closer to the principles of the ecosystem approach. However, transaction costs may be significantly higher for such projects, which deal with multiple small-scale holders, than for those with one large-scale operator. Commercial CDM investors may prefer a project in India in which degraded lands are allocated to the timber industry for intensive monospecific tree planting, for example. Alternatively, in China the state may prefer to establish large-scale intensive plantations on government waste lands at low transaction costs. However, degraded lands often provide forest products important to the poorest households. "Waste lands" often are the last remaining native plant communities in the locality. Establishment of large-scale intensive plantations may make it difficult for communities to sell commercial trees that are being produced under collaborative management. Therefore, it would be essential to have mandatory social and environmental assessments of CDM projects so that the trade-offs between projects are stated explicitly, thus offering opportunities to leverage support from interested stakeholders for projects with multiple benefits. CDM rules could also provide incentives (such as exemption from CDM taxes) for projects that support other international processes and conventions.

Conclusions

Although the CDM is not a silver bullet for saving tropical forests, it could provide significant opportunities for implementing the new, broader forestry agenda that has emerged out of the last few decades of forestry experience and is now entering international processes. Implementing the CDM purely as a tool for carbon farming could derail the new forestry agenda. However, if the CDM is implemented to provide environmental and social benefits also, it could leverage support from other interested stakeholders and contribute to realistic and equitable approaches to forest conservation and sustainable use. In crafting CDM rules and projects, proactive efforts (such as described in this essay) will be needed to make this prospect a reality.

ESSAY 21

Reconciling Long-Term Biodiversity Conservation and Local Communities: A Case for Conservation Performance Payments

Paul J. Ferraro

While most of the tropical world continues to experiment with indirect development interventions to promote ecosystem conservation (e.g., ICDPs), some nations are experimenting with contracting approaches that use performance payments to achieve conservation results. Conservation contracts specify that international and national actors make periodic performance payments to local actors if a targeted ecosystem remains intact or if target levels of wildlife are found in the ecosystem. In many cases, paying individuals or groups directly for conservation performance could be a simpler and more effective approach to achieving habitat and wildlife protection in low-income nations.

Recent analyses (e.g., Ferraro 2000) have identified several attributes of development-based conservation initiatives that limit their effectiveness. Such initiatives are excessively complex, difficult to

TABLE 7.4. Development-Based Conservation Interventions versus Conservation Performance Payments.

Attribute	Development Interventions	Performance Payments
Link to conservation	*Indirect:* Incentives often are absent or counter to conservation objectives.	*Direct:* If a contract is formed, incentives are clear and unambiguous.
Institutional complexity	*Medium to high:* Long-lived institutions must allocate and enforce rights and responsibilities, resolve conflict, prevent corruption, coordinate individual and institutional behavior, and monitor ecosystem health.	*High:* The same institutions necessary or development interventions are needed, and there may be more complicated property rights issues.
Technological complexity	*High:* New technology and employment opportunities must be developed and adopted precisely; the opportunities must be tailored to economic, biological, and cultural variations.	*Low:* No new technologies are needed.
Infrastructure complexity	*High:* Rural agricultural and eco-enterprise development interventions often necessitate improvements in transportation and market infrastructure.	*Low:* With the possible exception of access for monitoring contract compliance, no new infrastructure is needed.
Targeting	*Diffuse:* Controlling the location and timing of new technology or employment adoption is difficult. Political constraints may misdirect conservation funds.	*Precise:* Payments can be targeted when and where conservation objectives are best achieved. Political constraints may misdirect conservation funds.
Monetary costs	*High:* Substantial, long-term funding is needed. Limited short-term investments will not reduce pressure on ecosystems. Changes in market conditions may necessitate expensive reinterventions.	*High:* Substantial, long-term funding is needed. Changes in market conditions necessitate adjusting payment levels and target areas over time.
Short-term impact	*Little to none:* New technologies, markets, and attitudes take many years to develop and work their way through societies.	*Potentially high:* As soon as finances and institutions are ready, payments can be made and the link between conservation and local welfare established.
Long-term impact	*Medium to none:* Conservation impacts are difficult to sustain after changes in economic or political conditions. Long-term funding and intervention are needed to sustain impact.	*High to none:* Conservation impacts are difficult to sustain after political changes. They are moderately difficult to sustain after economic changes. Long-term funding and intervention are needed to sustain impact.
Landscape impact	*Difficult:* Institutions, infrastructure, and appropriate livelihood opportunities must be tailored to each zone.	*Moderately difficult:* Supporting institutions and contract terms must be tailored to each zone.
Immigration incentives	*Salient:* Difficult to prevent entry, which can render initiatives ineffective.	*Less salient:* Easier to prevent entry because only contract holders can benefit from program.

implement at a landscape scale, and ill-suited for the immediacy of conservation objectives and donor demands. Furthermore, they often have no effect on conservation incentives, or they exacerbate the threats to habitat and wildlife. In cases in which interventions can effect desirable behavioral changes, practitioners often find that the changes can be sustained only under a narrow range of economic and political conditions.

In contrast to development interventions, conservation contracting focuses practitioner resources on a small set of critical parameters that must be affected to achieve conservation goals (namely, institutional design and property rights), permits precise targeting at a landscape scale, generates immediate conservation impacts and allows rapid adaptation over time, and forges clear, unambiguous links between individual well-being, individual actions, and ecosystem conservation.

The best-known conservation contracting initiatives are the agricultural land diversion programs of high-income nations. In Europe, fourteen nations spent an estimated US$11 billion (1993–1997) to divert more than 20 million ha into long-term set-aside and forestry contracts. In the United States, the Conservation Reserve Program spends about US$1.5 billion annually to contract for 12 to 15 million ha, an area twice the size of all national and state wildlife refuges in the lower forty-eight states. Because of their popularity among stakeholders and the opportunities they afford for flexible targeting and adjustment to local conditions, conservation contracting initiatives are among the fastest-growing payments to farmers in high-income nations (OECD 1997: 14). NGOs in high-income nations have also developed innovative direct payment approaches (e.g., Delta Waterfowl Foundation).

Among low-income nations, the best-known initiative is Costa Rica's Environmental Services Payment program, through which local, national, and international beneficiaries of ecosystem services compensate those who protect ecosystems (Castro et al. 1998; Calvo and Navarrete 1999). Funds are allocated through the National Forestry Financial Fund (FONAFIFO), which works directly with landowners and indirectly through third-party intermediaries (e.g., NGOs). FONAFIFO raises money from international donors and national sources, such as a fuel tax and payments made by hydroelectric plants. FONAFIFO then distributes the money through contractual arrangements with private individuals and groups.

FONAFIFO establishes contracts for three land-use categories: reforestation, sustainable forest management, and forest preservation. The forest preservation category is the most common contract. Each category is associated with a fixed annual payment per hectare. Regional conservation agents and third-party NGOs identify potential participants based on regional conservation priorities. They often target land buffers around protected areas. Individuals or groups who are awarded contracts receive annual payments if they comply with the contract. Similar initiatives are under way elsewhere in the tropical world (e.g., Mexico, Colombia, and El Salvador).

To implement a conservation contracting initiative in low-income nations, practitioners face substantial obstacles in matters of institutional design, property rights allocation, and strategic behavior by potential beneficiaries. However, development-based conservation interventions also require institutions that can monitor ecosystem health, resolve conflict, coordinate individual and group behavior, and allocate and enforce rights and responsibilities over time (Brown and Wycoff-Baird 1992; Wells et al. 1999). Furthermore, development interventions also require greater sophistication on the part of donors and practitioners in developing and transferring new technologies, anticipating market trends, improving market and transportation infrastructure, and predicting the conservation effects of specific investments.

The notion of paying for people to protect habitat may strike some as an expensive proposition. However, many of the regions in which conservation practitioners work are at the margins of the economy, where land uses are not very profitable. In Costa Rica, for example, annual payments of US$35 per hectare generate excess demand for conservation contracts (Calvo and Navarrete 1999). With a budget equivalent to half the 1996 payments of the U.S. Conservation Reserve Program (US$1.8 billion), practitioners could make annual payments of US$30/ha on 30 million ha. With appropriately targeted payments across the landscape, the area effectively protected could easily be four times this amount. To put this area into perspective, consider that in 1996, 126 million ha were in IUCN protected area classes I–IV in southeast Asia, south Asia, and east Asia (Green and Paine 1997: 13).

Although there is no guarantee that a direct payment initiative will succeed, the contrasts between direct and indirect conservation approaches (see table 7.4) suggest that conservation performance payments may be one of the most effective and efficient mechanisms for ensuring the survival of Asia's biodiversity through the next decades. Scholars and practitioners would do well to begin experimenting with contracting initiatives in the field.

ESSAY 22

Reconciling the Long-Term Conservation Needs of Large Mammals and Local Communities: Integrating Landscape Conservation and Development

Kathy MacKinnon

Lessons learned from a variety of conservation approaches introduced over the past two decades confirm an obvious point: protected areas remain the cornerstones of biodiversity conservation. Approximately 9 percent of the world's land surface is included in protected areas, but these protected areas often are small, and many important habitats, especially lowland tropical forests, are poorly represented within protected area systems. Throughout Asia, natural habitats and their wildlife are coming under increasing pressure from expanding human populations, logging activities, agricultural conversion, and habitat fragmentation. Logging roads and new access exacerbate the problem, opening up new wilderness areas to agricultural encroachment and hunting. As habitat is lost, many of the largest (and most charismatic) mammals—the big cats, rhinoceros, elephants, apes, monkeys, and large ungulates—are becoming increasingly rare and endangered. To survive, many of these animals need extensive home ranges or access to a range of different habitat types.

The fate of the tiger (*Panthera tigris*) is a good example of how habitat loss and fragmentation can affect endangered mammals and the other species with which they interact. Core protected areas are essential for tiger survival, but so is landscape-level planning that ensures that protected areas are large and include lowland forests, linking forest corridors, and forested buffer zones to support viable populations of tigers and their prey (Dinerstein et al. 1997).

Besides providing and managing core protected areas, a landscape-level approach entails changing human behavior and land-use patterns across the larger landscape so that humans and wildlife can co-exist. Throughout Asia, ICDPs are under way to link the conservation of biological diversity in parks and reserves with social and economic development for local communities. These ICDP projects usually provide small-scale alternative economic activities for local communities as an incentive to reduce the threat to parks, natural habitats, and wildlife from agricultural encroachment and hunting activities.

This essay describes two ambitious projects—the Kerinci-Seblat ICDP in Sumatra and the India Ecodevelopment Project—that attempt to reconcile the landscape-level conservation needs of large vertebrates and the legitimate development needs of local communities. The Kerinci-Seblat ICDP in Indonesia addresses the ecological needs of tigers but was not explicitly designed to do so, whereas the India Ecodevelopment Project focuses on tiger reserves.

Kerinci-Seblat ICDP: Linking Conservation and Regional Development in Sumatra

Kerinci-Seblat National Park is one of the largest and most important conservation areas in southeast Asia. It supports one of the last viable populations of Sumatran tigers as well as the endemic Sumatran hare (*Nesolagus netscheri*), Sumatran rhinoceros (*Dicerorhinus sumatrensis*), clouded leopard (*Neofelis nebulosa*), Malay tapir (*Tapirus indicus*), and Asian elephant (*Elephas maximus*). Extending south along the Barisan range, the park straddles four provinces and covers almost 1 million ha. It includes a range of habitats, from species-rich lowland forests, to hill forests and unique highland wetland systems, to montane forests and sub-alpine habitats on Mount Kerinci, which at 3,805 m is Sumatra's highest mountain. The park is remarkable for its species richness, with records of more than 4,000 plants (one-sixtieth of the world total), more than 350 recorded birds (one-thirtieth of all birds) including 14 of the 20 Sumatran mainland endemics, and 144 mammals (73 percent of the Sumatran mammal fauna and one-thirtieth of the world total).

Many of the habitats and species represented in the park and its immediate forest buffer zone are poorly represented or absent from other conservation areas in Sumatra and elsewhere in Asia. Unfortunately, the integrity of the park and its biodiversity values are threatened by encroachment from shifting agriculture and cinnamon plantations, mining concessions that overlap part of park lands, and logging in the lowland and hill forests. Poaching of Sumatran tigers and Sumatran rhinos has been a particular problem in recent years, exacerbated by lack of park staff and equipment to effectively patrol and protect the park.

The Kerinci-Seblat National Park project aims to secure the biodiversity of the park and to stop further habitat fragmentation by strengthening protection and management of the protected area, enlisting the support of local communities, and promoting sustainable management and the maintenance of permanent forest cover in the buffer zone concession areas (World Bank 1996a). The project will develop a model for integrating conservation with regional and district development, and this model can be applied to other parks in Indonesia and elsewhere in Asia. This integrated approach will help to stabilize the park boundary and take pressure off the rain forest flora and fauna by providing adjacent communities with alternative livelihood opportunities and encouraging rural development that is consistent with park conservation objectives. More than 130 boundary villages around the park are targeted for some kind of development assistance under the ICDP, but they will be required to enter into reciprocal conservation agreements and to respect the park boundaries and wildlife. The project involves partnerships between government agencies, local and national NGOs, and universities that deliver conservation and development services.

A particular concern is to stop further habitat loss, especially of species-rich lowland forests, and to prevent further fragmentation of the park. A proposed road development from Muara Labuh to Kambang in the province of west Sumatra was canceled because it would have bisected the park and provided further access for agricultural expansion and poaching. There will be no new road development until completion of a regional spatial development plan and park management plan, which will include recommendations for zoning and protecting areas of high biodiversity. Three mining concessions that overlap park boundaries will be allowed to continue their exploration activities, but no further concessions will be granted until park boundaries are agreed. Any proposals to mine exploitable mineral deposits will be evaluated on a case-by-case basis through a rigorous environmental impact process, with the added recommendation that any mining infrastructure would be outside park boundaries.

The project also seeks to minimize biodiversity loss in the adjacent production forests, some of Sumatra's last remaining lowland forests. As part of park boundary stabilization and rationalization of forest concessions, the project will assist the Ministry of Forestry to review and revise the boundary between the park and the seventeen neighboring concessions according to criteria based on biodiversity, land use, and biophysical and watershed values. The project will support rapid ecological assessment surveys to determine areas of high biodiversity and importance for wildlife within the concessions, and areas identified as important for biodiversity or watershed conservation will be returned to the park or maintained as protected forest within the concessions.

A particular concern is to ensure that lightly logged forest is not later cleared for agriculture, as has happened over much of lowland Sumatra. The project will also provide technical assistance, training, monitoring, and independent auditing of forest management to help concessionaires implement sustainable management.

Project personnel will also work with forestry agencies and local communities to promote improved land use, including community forestry, consistent with biodiversity conservation and maintenance of a permanent forest estate. Forestry management to integrate conservation values into forestry practice and to maintain permanent forest cover in the Kerinci forest buffer zone will effectively increase the conservation estate by maintaining natural habitat beyond park boundaries, and it could provide a model for sustainable forestry elsewhere in Indonesia. Many ungulate species prefer the more open secondary growth of logged forests and achieve higher population densities in logged areas. Therefore, it is expected that Kerinci's tigers will benefit both from the protection of large contiguous blocks of forest habitat and from increased prey numbers in logged areas if the management authority can strengthen enforcement and antipoaching activities.

India Ecodevelopment Project: An Ecosystem Approach to Conservation

More than 4.3 percent of India's land area is designated as protected areas, but many of these areas are small islands of protected habitat in an agricultural landscape. The US$67-million India Ecodevelopment Project focuses on seven priority reserves (World Bank 1996b). Six of these reserves protect valuable tiger habitat and harbor important tiger populations (Buxa, Nagarhole, Palamau, Pench, Periyar, and Ranthambore), and the seventh, Gir National Park, has the world's only population of wild Asiatic lions. Pench, Periyar, and Nagarhole are identified as top-priority TCUs (Dinerstein et al. 1997) with large blocks of suitable tiger habitat, adequate core areas, low poaching pressure, and the best chance of retaining viable tiger populations over the long term. Nagarhole is part of the Nilgiri Biosphere Reserve, one of the largest conservation areas in India. Apart from its importance for tigers and its high ungulate population densities, Nagarhole is a central link in the seasonal migrations of elephants to Bandipur National Park in the southeast and Wynad Wildlife Sanctuary in the southwest.

Buxa, Palamau, and Ranthambore are cut off from other habitat blocks with tigers and are subject to habitat fragmentation and high poaching pressure. However, these reserves harbor small populations of tigers that are important in national and state strategies for tiger conservation. Improved protection and strengthened management at these smaller reserves, including strenuous antipoaching measures to protect the tigers and their ungulate prey, should provide a favorable environment for these fast-breeding cats to recover.

At all six sites the project covers both the protected areas and some of the surrounding village areas, consistent with a landscape approach to conservation that acknowledges that reserves and their wildlife are negatively affected by surrounding land use. The village communities subsist on long-established sedentary agricultural systems. The need for land for crops and livestock grazing and the losses of cattle to tigers leaving protected areas have led to increasing conflicts between humans and wildlife and between villages and protected areas. Therefore, these external impacts must be addressed if conservation within the core areas is to succeed.

The project seeks to promote conservation by taking a broad ecoregional approach, focusing on strengthening conservation management within the protected areas and promoting more conservation-friendly development in the broader ecosystem beyond reserve boundaries. Specifically, the project will attempt to reduce the impact of local people on the protected areas and their wildlife while mitigating the impacts on the local people of protected areas and limits on natural resource use.

From the total project cost of US$67 million, US$60 million will flow directly to the seven protected areas, with a focus on three main components: improved protected area management, including boundary rationalization, strengthened management, and antipoaching activities; village ecodevelopment to take pressure off and encourage support for the conservation areas; and education, visitor management, monitoring, and research. The project will also seek to integrate protected area and conservation concerns into regional planning and regulation to ensure that regional development programs support, rather than threaten, protected area integrity and viability. As part of the management strategy there will be greater consultation with and involvement of local communities to increase local support and to develop benefit-sharing arrangements for protected area byproducts and enterprises such as ecotourism.

More than half of all project funding (US$34 million) is allocated for village ecodevelopment activities. This will fund alternative livelihoods and promote rural development that is appropriate and consistent with conservation goals, such as establishment of alternative fuel wood sources. The recipient communities will commit to conservation agreements and measurable actions to improve conservation efforts. Across the seven reserves, about 427,000 villagers will participate in the project; of these, more than 89,000 live within the protected areas, mainly in enclaves and other noncore parts of the tiger reserves where human settlement is allowed. Some 39 percent of these project beneficiaries are tribal peoples, the poorest of the poor and the most dependent on biological resources, including nontimber forest products harvested from the protected area. Activities for changing behavior and livelihoods will focus on the villages having the greatest impact on the protected area.

Ecodevelopment investments and reciprocal conservation agreements with surrounding villages can be expected to minimize human-wildlife conflicts and promote activities that emphasize the benefits of tiger conservation (e.g., ecotourism). If this integrated conservation and development strategy proves effective, the government intends to expand the model to another 100 to 200 protected areas within the national protected area system.

Linking Conservation to Development: Opportunities and Risks

Many reserves across Asia are small, fragmented, and isolated. The long-term survival of these reserves and their constituent species, especially the large, wide-ranging species, will depend on large-scale conservation planning efforts that hinge on a combination of political, social, and economic factors. The challenge of ICDPs is to reconcile the legitimate needs of local communities with the conservation objectives of protected areas and the larger conservation landscape and to show an explicit link between conservation objectives and the development and poverty alleviation efforts.

The ecodevelopment approach has evolved to support the dual goals of conservation and poverty alleviation in contexts where there is competition between wildlife and communities for land and resource use. Appropriate development can be a useful conservation tool, helping to address some of the social and economic needs that lead to biodiversity loss. Small-scale development opportunities are presented to local communities as incentives for supporting protected areas and conservation, with clear and explicit linkages between conservation and the economic investments. Strengthening protection and management of the protected areas is an integral part of the strategy, as are awareness activities to change human attitudes and behaviors in the wider landscape and create an environment compatible with biodiversity conservation. The fate of tiger populations in and around the targeted reserves will be the clearest indicator of the success of this conservation approach.

Even some of Asia's large protected areas may not be large enough to protect viable populations of tigers and other endangered species unless effective management measures are also put in place to maintain a permanent forest estate and extend conservation management into the production landscape beyond park boundaries. In doing so, we are challenged with protecting and strengthening key core conservation zones (the parks themselves) and seeking to influence development outside the park boundaries to maintain the whole forest ecosystem and encourage development that is compatible with conservation. Such an approach will not only protect biodiversity but also safeguard ecosystem services such as watershed protection, soil conservation, and climate amelioration. The parks provide important recreational, education, research, and employment opportunities for people living close to their borders. Integrating conservation with regional development offers perhaps the best chance for ensuring the survival of Asia's biodiversity through the coming decades.

APPENDIX A
Details on Ecoregion Delineation

Introduction

Some of the more recent biogeographic classifications (Dasmann 1973; Udvardy 1975; Schultz 1995; Bailey 1998) are valuable contributions to our understanding of global patterns of biodiversity but are at too coarse a scale to guide conservation planning and priority-setting in a region as biogeographically complex as the Indo-Pacific. Other classifications are based on distributions of single taxa, such as BirdLife International's Endemic Bird Areas (Stattersfield et al. 1998) or combine distributions of endemics with levels of threat (Myers et al. 2000). These classifications do not address the fundamental goal of biodiversity conservation: representation of all distinct natural communities within conservation landscapes and a network of protected areas (after Noss 1991). The classification most consistent with our approach is the delineation of biounits by MacKinnon and MacKinnon (1986) and MacKinnon (1997). But even the biounits combine vastly different habitat types, making representation difficult.

The foundation for delineating ecoregions is patterns of vegetation with similar ecological dynamics. We relied on existing vegetation maps of the Indo-Pacific region to distinguish ecoregions by biome, especially MacKinnon's (1997) reconstruction of the original forest cover data from Pakistan to New Guinea, digitized forest cover maps from Bouchet et al. (1995) for New Caledonia, and the IUCN Atlas of Tropical Forests (IUCN 1991) for the other South Pacific islands. To aid in classification of forest formations we used Blasco et al. (1996) and Whitmore (1984) and accompanying vegetation maps to help classify forest formations for the Indian subcontinent and Indochina and for Malesia, respectively. Using this baseline, we delineated or modified ecoregion boundaries based on climate and elevation (i.e., dry forest vs. wet forest, and lowland vs. montane) and analyses and reviews of published biogeographic zones (Udvardy 1975; MacKinnon and MacKinnon 1986; Rogers and Panwar 1988; MacKinnon 1997; Stattersfield et al. 1998). In keeping with Stattersfield et al. (1998) we chose 1,000 m as a universal elevation to distinguish between lowland and montane ecoregions. However, at a finer (subecoregional) scale we recommend accounting for the Massenerhebung heating effect, which may move the lowland-montane transition zone either above or below 1,000 m.

Our approach is similar to Dasmann's (1972) and Udvardy's (1975) hierarchy of biotic regions and provinces that includes both composition and process. Dasmann's (1972) system is based on the biomes of Clements and Shelford (1939). He subdivided biomes into biotic regions based on Wallace's (1876) faunal regions. Using existing biogeographic information he then subdivided the biotic regions into biotic provinces. Dasmann separated montane regions and island groups into unique biotic provinces, like ecoregions. Biogeographer Miklos Udvardy conducted a similar exercise for the world in 1969 and in 1975 incorporated the work of Dasmann (1972, 1973), Hagmeier and Stults (1964), and Hagmeier (1966). Udvardy (1975) changed Dasmann's terminology from biotic region to biogeographic realm and biotic provinces to biogeographic provinces.

Our ecoregional approach represents conservation units of much finer scale than Udvardy's (1975) biogeographic provinces. There are examples in our approach in which historical factors are given higher priority than present climatic features. In the Philippines ecoregions, the presence or absence of land bridges between island groups during the last ice age has had a more pronounced effect on the distributions of species than present rainfall gradients. Therefore, we used bathymetry and the distribution of land bridges to define ecoregions in the Philippines.

We present a set of ecological units that can be used for conservation planning and prioritizing at regional or even global scales. It is not our intention merely to introduce yet another permutation of biogeographic units. In many cases, we built on biogeographic units already recognized by regional and national biodiversity experts. The following section describes in detail how each ecoregion was delineated and how it corresponds to the major biogeographic classifications of Udvardy (1975) and MacKinnon (1997), which preceded ecoregions.

Delineating Ecoregions: Basis and Justification

Indian Subcontinent Bioregion

The Indian subcontinent, and India in particular, has been the subject of several efforts at biogeographic classification (Blanford 1901 cited in MacKinnon and MacKinnon 1997; Champion and Seth 1968; Mani 1974; Rodgers 1985; Rodgers and Panwar 1988; MacKinnon and MacKinnon 1986; and MacKinnon 1997). Although some of these were aimed at mapping vegetation zones, several were attempts to identify biomes and biounits for conservation purposes (e.g., MacKinnon and MacKinnon 1986; Rodgers and Panwar 1988; MacKinnon 1997).

MacKinnon and MacKinnon (1986) used a combination of Udvardy's (1975) biounits and, to a greater extent, Rodgers' (1985) biogeographic zones of India to identify thirty-three biounits in the Indian subcontinent bioregion. Rodgers and Panwar (1988) then reclassified India into ten biogeographic zones that contained twenty-five biotic provinces. (Biogeographic zones are zones with a distinctive set of physical and historical conditions. A biotic province is a subunit based on a further level of detail within these biogeographic zones [Rodgers and Panwar 1988]). In a recent revision of MacKinnon and MacKinnon (1986), MacKinnon (1997) used Rodgers and Panwar (1988) and Panwar's (1990) classification as a basis to revise the biounits of the Indian subcontinent (ten biounits with twenty-seven subunits).

However, based on our rationale that ecoregions are more appropriate conservation units than biounits, we used the biogeographic zone classification of Rodgers and Panwar (1988) for ecoregions that fall within India and MacKinnon's (1997) biounits for the rest of the Indian Subcontinent (including the ecoregions that extend outside the national boundary of India) as a basic framework for delineating ecoregions.

Rodgers and Panwar (1988) and MacKinnon (1997) divided the largest biogeographic unit in India, the Deccan Peninsula, into five subunits or biotic provinces. Of these the Deccan Plateau South (6A) consists of dry deciduous forests, thorn forests, and some wetlands; the Central Plateau North (6B) of subtropical forests, dry deciduous forests, moist deciduous forests, and wetlands; the Eastern Highlands (6C) of subtropical forests, moist deciduous forests, coastal plains, and wetlands; the Chota-Nagpur (6D) of dry deciduous forests, moist deciduous forests, and wetlands; and the Central Highlands (6E) of subtropical forests, dry deciduous forests, moist deciduous forests, and wetlands. In our classification, we subdivided the Deccan peninsula biogeographic zone into nine ecoregions based on forest types with a regional spatial scale. They include the Narmada Valley Dry Deciduous Forests [20], Chhota-Nagpur Dry Deciduous Forests [18], Eastern Highlands Moist Deciduous Forests [12], Northern Dry Deciduous Forests [19], Central Deccan Plateau Dry Deciduous Forests [21], Deccan Thorn Scrub Forests [23], South Deccan Plateau Dry Deciduous Forests [22], East Deccan Dry Evergreen Forests [6], and Godavari-Krishna Mangroves [34].

The semi-arid northwestern biogeographic zones in the Punjab (4A) and Gujarat-Rajwara (4B) biotic provinces of Rodgers and Panwar (1988) are represented by two ecoregions, the Kathiarbar-Gir Dry Deciduous Forests [16] and the Northwestern Thorn Scrub Forests [24], which are made up of the two widespread and dominant vegetation types. The Kathiarbar-Gir Dry Deciduous Forests [16] overlap with a portion of Udvardy's (1975) Indus-Ganges monsoon forest biogeographic province. The Northwestern Thorn Scrub Forests [24] lie in both the Indus-Ganges monsoon forest and Thar Desert biogeographic provinces.

The deserts further west, extending from northwestern India into Pakistan and classified as four subunits (I3a–d) by MacKinnon, were reclassified as eight ecoregions based on the extent of distinctive habitat of regional spatial scales. Subunit I3a, which is made up mostly of the Thar Desert [46], also included some thorn scrub habitat, which was separated out and included in the extensive Northwestern Thorn Scrub Forests [24] ecoregion that extends from MacKinnon's subunits I3a, b, and c to include most of I4a and the southern section of I4b. We delineated the Rann of Kutch Seasonal Salt Marsh [36] ecoregion from MacKinnon's subunit I3b and the Indus Valley Desert [47] ecoregion from subunit I3c. Both the Rann of Kutch Seasonal Salt Marsh [36] and the Indus Valley Desert [47] lie within the Thar Desert biogeographic province. We pulled out the mangrove systems around the delta of the Indus River and in the Gulf of Khambhat as the Indus River Delta–Arabian Sea Mangroves [32]. They also lie within the Thar Desert biogeographic province.

We identified five ecoregions—Sulaiman Range Alpine Meadows [42], South Iran Nubo-Sindian Desert and Semi-Desert [43], Baluchistan Xeric Woodlands [44], Rajasthan–North Pakistan Sandy Desert [45], and East Afghan Montane Coniferous Forests [30]—from the Baluchistan subunit (I3d). All five of these ecoregions extend westward and have a portion of their ecoregion beyond the limits of this analysis. These ecoregions overlap with numerous Udvardy biogeographic provinces outside the scope of this analysis. These include the Hindu Kush highlands to the north and the Anatolian-Iranian Desert, Iranian Desert, and the Caucaso-Iranian highlands to the south and west.

MacKinnon (1997) followed Rodgers and Panwar (1988) in his classification of the Himalayan biounits and identified two biounits: the Trans-Himalayan (I1) and the Himalayas (I2). The Trans-Himalayan biounit was in turn divided into two subunits, the Ladakh mountains (subunit I1a) and Tibetan Plateau (I1b), which includes the alpine steppe, bare rock, and glacier habitats across the Himalayan range. The Himalayan range was also divided into four subunits along the longitudinal axis: the Northwest Himalayas (I2a), West Nepal (I2b), Central Himalayas (I2c), and Eastern Himalayas (I2d). We distinguished the altitudinal bands of habitat as distinct ecoregions while retaining several of MacKinnon's subunits that are based on east-west–oriented biogeographic barriers.

We classified the high-altitude alpine shrub and meadow habitat into three ecoregions: the Northwestern Himalayan Alpine Shrub and Meadows [37], which extend from the northern Indian state of Himachal Pradesh, through Jammu and Kashmir, into Pakistan and China; the Western Himalayan Alpine Shrub and Meadows [38], which extend eastward from Himachal Pradesh, through western Nepal, to the Kali Gandaki river, considered to be a biogeographic barrier separating western and eastern flora and fauna; and the Eastern Himalayan Alpine Shrub and Meadows [39], which extend eastward from the Kali Gandaki to Arunachal Pradesh. We adopted MacKinnon's boundary to separate the Palearctic species in the northwest and differentiate between the northwestern and western ecoregions.

The Kali Gandaki River was used as a boundary to separate the broadleaf and sub-alpine conifer forests into four western and eastern ecoregions. This river flows through the world's deepest gorge and is widely recognized as a biogeographic barrier. The ecoregions are Western Himalayan Broadleaf Forests [27], Western Himalayan Sub-Alpine Conifer Forests [29], Eastern Himalayan Broadleaf Forests [26], and Eastern Himalayan Sub-Alpine Conifer Forests [28].

The subtropical broadleaf forests are distributed primarily in the eastern and central Himalayas, whereas the subtropical pine forests dominated by Chir pine (*Pinus wallachiana*) are more extensive in the west. We delineated the Himalayan Subtropical Broadleaf Forests [25] to include the subtropical broadleaf forests in the eastern and central Himalayas and the Himalayan Subtropical Pine Forests [31] that extend from Pakistan to Arunachal Pradesh in northeastern India.

The terai and duar savanna grasslands along the foothills and to the north of the Siwalik hills are represented by the Terai-Duar Savanna and Grasslands [35].

We identified two ecoregions—Karakoram–West Tibetan Plateau Alpine Steppe [40] and Central Tibetan Plateau Alpine Steppe [41]—in MacKinnon's (1997) Trans-Himalayan biounit. These ecoregions extend into China.

All the Himalayan ecoregions are part of Udvardy's Himalayan highlands biogeographic province. The Karakoram–West Tibetan Plateau Alpine Steppe [40] extends into the Tibetan biogeographic province.

For the Gangetic plains biogeographic zone, we used the upper and lower Gangetic plains biotic provinces as in Rodgers and Panwar (1988), but we included most of Bangladesh in the Lower Gangetic Plains Moist Deciduous Forests [10] following MacKinnon (1997). This ecoregion roughly corresponds to Udvardy's Bengalian rain forest biogeographic province but extends further west into the Indus-Ganges monsoon forest biogeographic province and further east into the Burma monsoon forest biogeographic province. The moister, semi-evergreen rain forests to the windward side of the Eastern Ghats in Orissa identified by MacKinnon (1997) as part of the original vegetation of the peninsula were included in a distinct ecoregion, the Orissa Semi-Evergreen Forests [5]. This ecoregion lies within Udvardy's Mahanadian biogeographic province. The Upper Gangetic Plains Moist Deciduous Forests [17] roughly corresponds to MacKinnon's Upper Gangetic plain biounit (I7a) and lies within Udvardy's Indus-Ganges monsoon forest biogeographic province.

Following MacKinnon (1997) and Rodgers and Panwar (1988), we delineated the coastal moist deciduous forests along the western coast of India as a single ecoregion: the Malabar Coast Moist Deciduous Forests [13]. Although little of the original habitat now remains, these coastal plains had distinctive floral and faunal elements (Champion and Seth 1968). However, we identified four ecoregions along the Western Ghats mountain range, whereas MacKinnon (1997) and Rogers and Panwar (1988) placed the mountain range into a single unit. We used the transition zone between the southern *Cullenia*-dominated forests and the northern drier dipterocarp forests in the Wyanad Evergreen forests of Kerala-Karnataka (Rogers and Panwar 1988) to differentiate the north-south boundary and separated the montane rain forests and moist deciduous forests within these to delineate the North Western Ghats Montane Rain Forests [1], North Western Ghats Moist Deciduous Forests [14], South Western Ghats Montane Rain Forests [2], and South Western Ghats Moist Deciduous Forests [15]. These five ecoregions all fall within Udvardy's Malabar rain forest biogeographic province.

In the northeastern zone of the Indian subcontinent bioregion, we followed MacKinnon (1997), who extended Rodgers and Panwar's (1988) units outside the national boundaries of India. Here, MacKinnon identified two biounits, Brahmaputra valley (I9a) and Northeast Hills (I9b), which we largely retain as the Brahmaputra Valley Semi-Evergreen Rain Forests [8] and the Meghalaya Subtropical Forests [11], respectively. These ecoregions lie in two of Udvardy's biogeographic provinces, the Himalayan highlands and Burma monsoon forest. We identified the extensive mangrove forests in the Sundarbans and the adjacent freshwater swamp forests as two distinct ecoregions: the Sundarbans Mangroves [33] and the Sundarbans Freshwater Swamp Forests [9], respectively. These two ecoregions fall within Udvardy's Bengalian rain forest biogeographic province.

MacKinnon (1997) identified a broad transition zone in biounit 09C, along the Indo-Myanmar border. Within this and the biounit immediately to the north (09b) we identified five ecoregions based on the distribution of the large extents of habitat and what appear to be biogeographic barriers. The ecoregions are the large Mizoram-Manipur-Kachin Rain Forests [50], Chin Hills–Arakan Yoma Montane Rain Forests [48], Northern Triangle Subtropical Forests [73], Northern Triangle Temperate Forests [76] further north in the Golden Triangle, and the Northeast India–Myanmar Pine Forests [77]. The Mizoram-Manipur-Kachin Rain Forests [50] and Chin Hills–Arakan Yoma Montane Rain Forests [48] overlap with two of Udvardy's biogeographic provinces, the Burma monsoon forest and Burman rain forest. The Northern Triangle Subtropical Forests [73], Northern Triangle Temperate Forests [76], and the Northeast India–Myanmar Pine Forests [77] lie within Udvardy's Burma monsoon forest biogeographic province.

In Sri Lanka, we delineated the dry evergreen forests throughout most of the island as the Sri Lanka Dry-Zone Dry Evergreen Forests [7]. This ecoregion corresponds to MacKinnon's biounit S13. But the tropical moist forests in the central and southwest quarter of the island that MacKinnon delineated as a single biounit (02) were divided into lowland and montane forests, represented by the Sri Lanka Lowland Rain Forests [3] and Sri Lanka Montane Rain Forests [4], respectively. Udvardy identified two biogeographic provinces in Sri Lanka. The Ceylonese rain forests covers the southwestern part of the island, and the Ceylonese monsoon forest covers the remaining area.

Indochina Bioregion

Early classifications of biogeographic units in Indochina have included efforts by Vidal (1960 cited in MacKinnon 1997), Udvardy (1975), and MacKinnon and MacKinnon (1986). More recently, MacKinnon (1997) revised and updated MacKinnon and MacKinnon (1986), which in turn was based largely on classifications of Vidal and Udvardy. Therefore, we used the biounit framework in MacKinnon (1997) to delineate ecoregions and Blasco et al. (1996) to aid classification of forests into biomes.

MacKinnon divided the Indochina bioregion into fifteen biogeographic subunits nested within seven major biounits. Each of these biounits and biogeographic subunits contains a mix of biomes. For instance, the Burmese coast biounit (04) is made up mostly of tropical wet evergreen forests but also includes tropical montane evergreen forests, semi-evergreen rain forests, dry dipterocarp forests, and large areas of freshwater swamp forests and mangroves in the Irrawaddy River delta. We assigned these distinct habitat types into separate ecoregions. Thus MacKinnon's Burmese coast biounit is represented by four ecoregions: the Myanmar Coastal Rain Forests [49], Irrawaddy Freshwater Swamp Forests [63], Myanmar Coastal Mangroves [78], and Mizoram-Manipur-Kachin Rain Forests [50]. MacKinnon's deciduous forests subunit (09a) is represented by the Irrawaddy Dry Forests [71] and the Irrawaddy Moist Deciduous Forests [67].

The Myanmar Coastal Rain Forests [49] also encompass the southern rain forests of the Burmese Monsoon Zone Biounit (09), and the Mizoram-Manipur-Kachin Rain Forests [50] include the northern rain forests. The dry, seasonal vegetation in the Irrawaddy valley is represented by the Irrawaddy Moist Deciduous Forests [67] and Irrawaddy Dry Forests [71] that comprise the teak (*Tectona grandis*)-dominated moist deciduous forests and the drier patches of *Dipterocarpus*-dominated dry deciduous forests, respectively.

MacKinnon's subunit 09b includes the tropical and subtropical moist broadleaf forests in the Kachin and upper Chindwin areas of northern Myanmar. We included the tropical moist forests within the Mizoram-Manipur-Kachin Rain Forests [50], which then extends east across northern Myanmar, and the subtropical moist broadleaf forests of the Golden Triangle to the north in the Northern Triangle Subtropical Forests [73].

This region does not correspond well to Udvardy's biogeographic provinces. The Myanmar Coastal Rain Forests [49] cover the Udvardy's Burman rain forest, the southwestern portion of the Thai monsoon forest, and the western portion of the Indochinese rain forest. The Irrawaddy Moist Deciduous Forests [67], Irrawaddy Dry Forests [71], Chin Hills–Arakan Yoma Montane Rain Forests [48], Northeast India–Myanmar Pine Forests [77], Northern Triangle Subtropical Forests [73], and Mizoram-Manipur-Kachin Rain Forests [50] roughly correspond to Udvardy's Burman rain forest and Burma monsoon forest. The Irrawaddy Freshwater Swamp Forests [63] and Myanmar Coastal Mangroves [78] are contained within Udvardy's Burman rain forest.

The westernmost subunit (09c) represents a transition zone between the Indian Subcontinent and Indochina bioregions and is made up of forest formations belonging to the tropical moist forests biome. We included the montane forests of the rugged and highly dissected Chin Hills and Arakan Yomas in a distinct ecoregion: Chin Hills–Arakan Yoma Montane Rain Forests [48]. The lower-elevation tropical evergreen moist forests were included in the Mizoram-Manipur-Kachin Rain Forests [50]. The pine forests were placed in the Northeast India–Myanmar Pine Forests [77].

The extensive Indochina biounit (10) identified by MacKinnon (1997) comprises three subunits that represent the tropical lowland plains, subtropical hills, and temperate montane areas. This biounit is a mix of tropical moist forests and tropical dry forests. We delineated seven ecoregions that have a high degree of overlap with this biounit. These are the Kayah-Karen Montane Rain Forests [51], Luang Prabang Montane Rain Forests [52], Southeastern Indochina Dry Evergreen Forests [59], Northern Thailand–Laos Moist Deciduous Forests [68], Northern Khorat Plateau Moist Deciduous Forests [69], Central Indochina Dry Forests [72], and Northern Indochina Subtropical Forests [74]. In addition, there are four other ecoregions that extend partly into this biounit: the Northern Annamites Rain Forests [54], Southern Annamites Montane Rain Forests [57], Tonle Sap–Mekong Peat Swamp Forests [62], and Tonle Sap Freshwater Swamp Forests [65]. The tropical dry forests are represented by the *Dipterocarpus*-dominated Central Indochina Dry Forests [72], and the narrow bands of dry evergreen forests along the lower elevations of the mountains are included in the Southeastern Indochina Dry Evergreen Forests [59].

The large areas of subtropical forests in MacKinnon's subunits 10b and 10c are represented by the Northern Indochina Subtropical Forests [74]. The tropical moist deciduous forests to the west and northwest of the Luang Prabang range in northwestern Thailand and Laos are an extension of the teak (*Tectona grandis*)-dominated deciduous forests (Rundel and Boonpragob 1995), but the vegetation community of the moist deciduous forests in the Mekong plains near Vientianne are dominated by Fabaceae, Lythraceae, and Rubiaceae. Therefore, these forests were placed in separate ecoregions, the former in the Northern Thailand–Laos Moist Deciduous Forests [68] and the latter in the Northern Khorat Plateau Moist Deciduous Forests [69]. The extensive dry dipterocarp forests throughout Indochina (subunit 10b) were included in the Central Indochina Dry Forests [72].

The Kayah-Karen Montane Rain Forests [51], Luang Prabang Montane Rain Forests [52], Southeastern Indochina Dry Evergreen Forests [59], Northern Thailand–Laos Moist Deciduous Forests [68], Northern Khorat Plateau Moist Deciduous Forests [69], Central Indochina Dry Forests [72], and Northern Indochina Subtropical Forests [74] and large areas of the Northern Annamites Rain Forests [54], Southern Annamites Montane Rain Forests [57], Tonle Sap–Mekong Peat Swamp Forests [62], and Tonle Sap Freshwater Swamp Forests [65] encompass Udvardy's Thai monsoon forest biogeographic province.

The remaining portion of the Northern Annamites Rain Forests [54], Southern Annamites Montane Rain Forests [57], Tonle Sap–Mekong Peat Swamp Forests [62], and Tonle Sap Freshwater Swamp Forests [65] fall into Udvardy's South Chinese rain forest biogeographic province. The South China–Vietnam Subtropical Evergreen Forests [75], Red River Freshwater Swamp Forests [66], Northern Vietnam Lowland Rain Forests [56], Southern Vietnam Lowland Dry Forests [58], Indochina Mangroves [79], Cardamom Mountains Rain Forests [55], Chao Phraya Freshwater Swamp Forests [64], Chao Phraya Lowland Moist Deciduous Forests [70], and Tenasserim–South Thailand Semi-Evergreen Rain Forests [53] also lie within Udvardy's South Chinese rain forest biogeographic province.

MacKinnon (1997) included the coastal habitats of Vietnam, Cambodia, and Thailand in the Coastal Indochina biounit (05), consisting of four subunits that have several biomes. We delineated several ecoregions that represent these biomes in the coastal and lowland areas within the biounit. MacKinnon chose the Hai Van pass to represent a transition from the tropical south and the subtropical north and recognized two subunits (South Annam 05b and North Annam 05c). Within the general bounds of these subunits, we delineated ecoregions to represent the biomes. We placed the lowland evergreen moist forests north of the Hai Van pass, up to the Red River delta, in the Northern Vietnam Lowland Rain Forests [56], and the lowland and coastal dry forests south of the Hai Van pass to the Mekong Delta in the Southern Vietnam Lowland Dry Forests [58].

The narrow band of dry evergreen forests further inland was included in the Southeastern Indochina Dry Evergreen

Forests [59]. This ecoregion winds around Indochina to include the band of semi-evergreen dry forest that extends north from the Mekong delta area in Vietnam to cross central Cambodia, then west along the low mountains of the Thailand-Cambodia border and into the Petchabun mountain range in northern Thailand.

We assigned the mangroves, freshwater swamp forests, and peat swamps in MacKinnon's subunit 05a to the Indochina Mangroves [79], Tonle Sap Freshwater Swamp Forests [65], and Tonle Sap–Mekong Peat Swamp Forests [62], respectively.

The moist forests along the Annamite mountain range were included in two ecoregions: the Northern Annamites Rain Forests [54] and Southern Annamite Montane Rain Forests [57]. MacKinnon (1997) included both lowland and montane tropical moist forests along the northern Annamite mountains within the North Annam subunit 05c; however, we chose to separate the moist montane and submontane forests from the lowland forests. Thus, the moist forests that lie along the spine of the northern Annamite range, above 1,000 m in elevation and between 19°N and the low hills between Hue and Da Nang were placed in the Northern Annamites Rain Forests [54].

The montane moist forests in the southern Annamite range were placed in the Southern Annamites Montane Rain Forests [57]. But this ecoregion has three disjunct sections and includes the forests above 1,000 m in the Kon Tum plateau to the north and the Da Lat Plateau in the south. It also includes the band of submontane moist forests on the western slopes of the Annamite range. MacKinnon (1997) recognized the Kon Tum and Da Lat plateau areas as distinct subunits: the Central Annam mountains (Ma) and the Da Lat Plateau (Mb). But both areas have granitic substrates and probably contain similar floristic communities (P. Rundel, pers. comm., 1999). Based on this recommendation, we combined the two areas into a single ecoregion.

The tropical moist forests in the Cardamom and Elephant mountain ranges in southwestern Cambodia and parts of Thailand are included within the Cardamom Mountains Rain Forests [55]. MacKinnon included these rain forests in the large Cardamom Mountains subunit (05d), which also included the freshwater swamp forests and mangroves of the Chao Phraya river and estuary. We placed these latter biomes in the Chao Phraya Freshwater Swamp Forests [64] and Indochina Mangroves [79], respectively. MacKinnon's subunit 05d also extends into the Kayah Karen and Tenasserim mountains, and we included these montane forests in the Kayah-Karen Montane Rain Forests [51] and Tenasserim–South Thailand Semi-Evergreen Rain Forests [53], respectively. The lowland moist deciduous forests along the lower reaches of the Chao Phraya river were placed in the Chao Phraya Lowland Moist Deciduous Forests [70].

We included MacKinnon's Tropical Southern China subunit (06a), with biogeographic affinities with southern China, as part of the larger South China–Vietnam Subtropical Evergreen Forests [75], which extends from southern China. However, instead of following the main tributary of the Red River as MacKinnon did, we used the upper Red River (Yuan Jiang) as the boundary to be consistent with the boundary between two zoogeographic divisions in China: the Fujian Guangdong coast and South Yunnan Mountains (Yongzu et al. 1997). These two zoogeographic divisions largely coincide with the South China–Vietnam Subtropical Evergreen Forests [75] and the Northern Indochina Subtropical Forests [74] and intersect the Vietnamese border at the upper Red River tributary.

The freshwater swamp forests and mangroves in the Red River valley and coastal areas of MacKinnon's subunit 06a were included within the Red River Freshwater Swamp Forests [66] and Indochina Mangroves [79], respectively.

Following MacKinnon (1997), we placed the Andaman Islands in a distinct ecoregion, the Andaman Islands Rain Forests [60]. However, we included the Nicobar Islands—Nicobar Islands Rain Forests [61]—in this bioregion based on recommendations by Tim Whitmore (pers. comm., 1999). Udvardy placed both island chains into the Andaman and Nicobar Islands biogeographic province.

Sunda Shelf and Philippines Bioregion

This bioregion is one of three we have identified for Malesia. Udvardy (1975) identified five biogeographic provinces within the Sunda Shelf and Philippines Bioregion, namely the Malayan rain forests (4.7.1), Sumatra (4.21.12), Java (and Bali) (4.22.12), Borneo (4.25.12), and the Philippines (4.26.12). MacKinnon (1997) placed these provinces within the Sundaic subregion, with two exceptions: he included the Nicobar islands with Sumatra (in biounit 21) and Palawan with Borneo (in biounit 25). More recently, the Philippines has been divided into fifteen biogeographic zones, based on a combined set of faunal, floral, and geological criteria (Philippine BAP 1997).

We used MacKinnon's (1997) biounits and interpretation of the distribution of the original vegetation as a general framework to guide ecoregion delineation for this bioregion. We also used Whitmore's (1984) vegetation map of Malesia as a guide to aid in vegetation classification. Because endemism is a characteristic feature in these island ecoregions, we also used Endemic Bird Areas (Stattersfield et al. 1998) as a modifier to delineate ecoregions in the smaller islands and island groups. Several ecoregions and their boundaries were also modified based on comments by regional experts, specifically Tony Whitten, Tim Whitmore, and Domingo Madulid (pers. comm., 1999).

The Peninsular Malaysia Rain Forests [80] represents the large extent of the lowland broadleaf rain forests extending south of the Kangar-Pattani line to Singapore.

This ecoregion approximates the spatial extent of MacKinnon's (1997) subunit 07a, but we also included the Lingga and Riau archipelagos. Although the FAO (1974) indicated that these islands contain heath forests, both Whitmore (1984) and MacKinnon (1997) disputed this classification, and MacKinnon identifies the vegetation as lowland moist forest.

The tropical montane evergreen moist forests above 1,000 m were placed in the Peninsular Malaysian Montane Rain Forests [81]. We also extracted the large areas of peat swamp forests along the coast of Peninsular Malaysia into the Peninsular Malaysian Peat Swamp Forests [87]. Thus, we created three ecoregions within MacKinnon's Malay Peninsular subunit (07a) and Udvardy's Malayan rain forest biogeographic province.

In keeping with our definition, we recognized six ecoregions on Sumatra. MacKinnon (1997) placed all the biomes in Sumatra within two subunits (21a and 21b), with the subunit division based on a faunal break to the south of Lake Toba. However, according to Tony Whitten (pers. comm., 1999) the differences in floral and faunal communities across MacKinnon's subunit boundary (within a habitat type) are not great enough to warrant differentiation into distinct ecoregions. Thus, we placed the lowland rain forests of Sumatra into a single ecoregion, the Sumatran Lowland Rain Forests [82]. Although FAO (1974) and MacKinnon (1997) reported that Bangka Island has extensive heath forests, Whitmore (1984) reported that the heath forests are less extensive; therefore, we included most of this island as part of the Sumatran Lowland Rain Forests [82] ecoregion.

We used the 1,000-m contour from a digital elevation model (U.S. Geological Survey 1996) to delineate the montane forests along the Bukit Barisan–Gunung Leuser mountain range as the Sumatran Montane Rain Forests [83]. These montane forests include extensive areas of forests over limestone that harbor a unique flora and fauna. Whitmore (1984) showed small, scattered patches of limestone forests in his vegetation map of Malesia, although MacKinnon (1997) and Whitten showed more extensive patches. The FAO map (1974) showed no limestone. Much of the uncertainty probably arises because these limestone substrates do not show extensive outcropping (Whitmore 1984). Based on recommendations by Tony Whitten (pers. comm., 1999), we treated these limestone forests as a distinct habitat type within the more broadly distributed montane moist forests rather than placing them in a separate ecoregion.

The *Pinus merkusii*–dominated conifer forests along the Gunung Leuser range were also shown and identified as a distinct ecoregion, the Sumatran Tropical Pine Forests [105]. These forests represent the only stands of *Pinus* found south of the equator (Whitmore 1984).

Whitmore (1984) and MacKinnon (1997) showed large extents of peat swamp forests along the northern coast of Sumatra, especially in Riau province. We delineated the Sumatran Peat Swamp Forests [85] to represent these forests but extracted the smaller patches of freshwater swamp forests into the Sumatran Freshwater Swamp Forests [88] and the mangroves in the Sunda Shelf Mangroves [107].

We placed the Mentawai Islands and Enggano Island—MacKinnon's subunit 21c—into a distinct ecoregion, the Mentawai Islands Rain Forests [84]. These small islands have ten endemic mammal species, of which four are primates (Whitten et al. 1987). The other satellite islands on the west coast, Nias, Batu, and Simeuleu, are less distinctive and were included in the Sumatran Lowland Rain Forests [82]. Although MacKinnon placed these smaller islands in separate subunits (21d and 21e), he acknowledged that they are not as biologically distinctive as the Mentawai islands.

MacKinnon's biounit 21 largely corresponds to Udvardy's Sumatra biogeographic province. However, Udvardy did not include the Nicobar Islands. There are eight ecoregions that overlap Udvardy's Sumatra biogeographic province: Sumatran Lowland Rain Forests [82], Sumatran Montane Rain Forests [83], Mentawai Islands Rain Forests [84], Sumatran Peat Swamp Forests [85], Sumatran Freshwater Swamp Forests [88], Sundaland Heath Forests [90], Sumatran Tropical Pine Forests [105], and Sunda Shelf Mangroves [107].

MacKinnon (1997) included the islands of Java and Bali in biounit 22 (with three subunits). Western Java is wetter than the eastern half of the island, and the forests are richer in species (Whitten et al. 1996; MacKinnon 1997). There are also floristic differences between the lowland and montane vegetation in Java and Bali (Whitmore 1984; Whitten et al. 1996). Therefore, using MacKinnon's subunit boundary, we delineated the Western Java Rain Forests [91] to represent the moister evergreen forests to the west and the Eastern Java–Bali Rain Forests [92] to represent the drier, less species-rich forests of eastern Java and in Bali. However, we also extracted the montane forests into distinct ecoregions—Western Java Montane Rain Forests [93] and Eastern Java–Bali Montane Rain Forests [94]—using the 1,000-m elevation contour from a digital elevation model (U.S. Geological Survey 1996). MacKinnon's Java and Bali biounit corresponds to Udvardy's Java biogeographic province, and Udvardy's province overlaps with four Java and Bali ecoregions.

The large island of Borneo was divided into seven ecoregions. Most of the island's lowland and submontane forests are dominated by dipterocarp species (MacKinnon et al. 1996). Udvardy (1975) combined all of Borneo into the Borneo biogeographic province, whereas MacKinnon and MacKinnon (1986) divided the island's lowland forests into six subunits with a central subunit representing the montane forests. MacKinnon (1997) revised the boundaries of these seven subunits but retained the same general configuration. These authors used the major rivers, the Kapuas and Barito, to represent zoogeographic barriers to a few species of mammals and based subunits largely on these barriers but also used climatic regimes for the drier eastern biounits (MacKinnon and MacKinnon 1986; MacKinnon 1997).

Because ecoregions are based on biomes, we first isolated the central montane ecoregion—the Borneo Montane Rain Forests [95]—above the 1,000-m elevation contour using a digital elevation model (U.S. Geological Survey 1996). We then assigned the large patches of peat forests, heath forests, freshwater swamp forests, and mangroves in the lowlands and along the periphery of the island into their own ecoregions, namely the Borneo Peat Swamp Forests [86], Sundaland Heath Forests [90] (which also includes Belitung Island and the heath forests in Bangka Island), Southern Borneo Freshwater Swamp Forests [89], and Sunda Shelf Mangroves [107], respectively. The alpine habitats of the Kinabalu mountain range were represented by the Kinabalu Montane Alpine Meadows [108].

The remaining lowland dipterocarp forests in Borneo fall into six of MacKinnon's subunits (25a, b, f, g, h, and i). The lowland moist forest ecoregions in Borneo were, to some extent, demarcated based on biogeographic barriers used by MacKinnon and MacKinnon (1986) and MacKinnon (1997). These units were split, in part based on Borneo's major rivers, such as the Lupara, Barita, Kayan, and Padas rivers. However, these do not represent sound dispersal barriers to most species, and we deviated from MacKinnon by combining all the lowland dipterocarp forests into a single ecoregion, the Borneo Lowland Rain Forests [96].

MacKinnon (1997) identified seven subunits in the Philippines, and the Philippine Biodiversity Action Plan (Philippine BAP 1997) demarcated fifteen biogeographic units. Udvardy (1975) identified the Philippines as a single biogeographic province. We delineated nine ecoregions in the Philippine islands, including Palawan. We deviated from Udvardy (1975), MacKinnon (1997), Stattersfield et al. (1998), and the Philippine Biodiversity Action Plan (1997) to varying degrees and based our delineation of the Philippine ecoregions on Heaney (1993).

During the most recent ice age 18,000–12,000 years ago, the sea level was 120 m below its present level. In the Philippines this opened numerous land bridges between the present-day islands. These land bridges facilitated the exchange and migration of flora and fauna. The biodiversity and vegetation that remain today are closely linked to the land bridges that formed more than 10,000 years ago. Heaney (1993) identified where the land bridges were located and which present-day islands were once connected. We followed Heaney's lead in delineating our ecoregions of the Philippines.

We placed Palawan, Calamian Islands, and Cuyo Islands into a single ecoregion, the Palawan Rain Forests [97]. Palawan has closer zoogeographic affinities to Borneo.

In Luzon we delineated three ecoregions, which correspond to MacKinnon's subunit 26a. First, we used the 1,000-m contour from the digital elevation model (U.S. Geological Survey 1996) to delineate the montane forests from the lowland forests. The Luzon Montane Rain Forests [102] are made up primarily of the montane moist evergreen forests along the Sierra Madre and Zambales mountain ranges. MacKinnon (1997) showed an area of freshwater swamp forests as part of the original vegetation of Luzon Island, which we combined with the remaining lowland forest of Luzon to form the Luzon Rain Forests [98]. These freshwater swamps in the valley to the east of the Zambales mountain range and in the Cagayan river plains have now been converted to rice fields (D. Madulid, pers. comm., 1999). The *Pinus kesiya*–dominated conifer forests in the central cordillera were designated as the Luzon Tropical Pine Forests [106].

MacKinnon (1997) designated Mindoro Island as subunit 26f and included the Lubang Islands. We delineated the island of Mindoro as the Mindoro Rain Forests [99]. Following Stattersfield et al. (1998) and Dickinson et al. (1991), we placed Lubang Islands with the Luzon Rain Forests [98].

We placed the islands of Negros, Panay, Tablas, Sibuyan, Masbate, Ticao, Siquijor, and Cebu into the Greater Negros–Panay Rain Forests [100] ecoregion. MacKinnon also grouped these islands into a single subunit (26b). However, he also included the island of Bohol in this subunit. We followed Heaney (1993) and included Bohol as part of the Mindanao–Eastern Visayas Rain Forests [101] because during the Pleistocene ice ages it was connected to Leyte Island.

The islands of Leyte, Samar, Dinagat, and Bohol were combined with the lowland rain forests of Mindanao island to form the Mindanao–Eastern Visayas Rain Forests [101]. We also included the Basilan Islands off the southwest peninsula of Mindanao in this ecoregion, based on Heaney (1993). In Mindanao we used the 1,000-m contour from the digital elevation model (U.S. Geological Survey 1996) to delineate the montane forests from the lowland forests. The montane forests of Mindanao were placed into their own ecoregion, the Mindanao Montane Rain Forests [103]. In our delineation of the Mindanao–Eastern Visayas Rain Forests [101] and Mindanao Montane Rain Forests [103] ecoregions, we deviated from MacKinnon (1997). MacKinnon placed both of Mindanao's lowland and montane forests in a single subunit (26c). The Basilan Islands were part of subunit 26d, and the islands of Leyte and Samar made up subunit 26e.

The islands of the Sulu archipelago were delineated as a separate ecoregion, the Sulu Archipelago Rain Forests [104]. This ecoregion includes the Tawitawi, Tapul, Jolo, and Samales island groups. These islands, with a lowland moist or semi-evergreen moist forest vegetation (Whitmore 1984) are also an endemic bird area (Stattersfield et al. 1998) and have been identified as a distinct biounit by MacKinnon (1997) and a biogeographic zone by the Philippine BAP (1997).

Wallacea Bioregion

This bioregion represents a transition zone between the Asian and Australian faunas, but the flora is mainly of Malesian affinity (Monk et al. 1997). There have been several attempts to divide the bioregion into biogeographic units (MacKinnon 1997; Stattersfield et al. 1998; van Balgooy 1971 cited in Monk et al. 1997; MacKinnon and Artha 1981; MacKinnon and MacKinnon 1986; MacKinnon et al. 1982; van Steenis 1950; Udvardy 1975). Because many of the islands have distinct natural faunal communities and a high degree of endemism (Monk et al. 1997), the more recent attempts have used faunal dissimilarities—especially birds—to identify distinct biogeographic units (MacKinnon 1997; Stattersfield et al. 1998; MacKinnon and Artha 1981; MacKinnon and MacKinnon 1986). Because detailed floral data are largely unavailable across most of the bioregion, we followed these authors in delineating ecoregions based on distribution of habitat types and vertebrate communities.

On Sulawesi Island we delineated two ecoregions: the Sulawesi Lowland Rain Forests [109] and Sulawesi Montane Rain Forests [110]. These represent the tropical lowland and montane tropical moist forests, respectively. We used the 1,000-m contour from the digital elevation model (U.S. Geological Survey 1996) to distinguish the montane forests from the lowland forests. The small patch of monsoon forests in the southwest peninsula of Sulawesi and on Butung Island (Whitmore 1984) were included in the Sulawesi Lowland Rain Forests [109]. We also included the island groups of Sangihe, Talaud, and Sula into the Sulawesi Lowland Rain Forests [109]. MacKinnon (1997) placed these island groups with Sulawesi into a single biounit (24). MacKinnon divided this unit into seven subunits. The Sangihe-Talaud island groups made up one, the Sula island groups a second, and the remaining five divided the island of Sulawesi. MacKinnon's biounit corresponds largely to Udvardy's Celebes biogeographic province, although Udvardy included the island of Buru as well.

The vegetation of the islands in the Maluku archipelago is primarily tropical evergreen or semi-evergreen rain forest, with patches of semi-deciduous forests in the rain shadows, limestone forests, and peat forests (Monk et al. 1997). The altitudinal limits to montane forests are also difficult to assess and delineate (see Monk et al. 1997). Therefore, we combined the lowland and montane forest types in our ecoregion delineation of this small-island archipelago and followed the biogeographic units, presented by Monk et al. (1997, after MacKinnon and Artha 1981 and MacKinnon et al. 1982).

The Sula Islands were included within the Sulawesi lowland rain forests [109], and the Aru Islands were included in the Vogelkop-Aru Lowland Rain Forests [119]. MacKinnon identified the Aru islands as subunit P3a within the New Guinea biounit. Buru Island, identified as a distinct subunit (13c) by MacKinnon (1997) and as an endemic bird area (Stattersfield et al. 1998), was delineated as a distinct ecoregion, the Buru Rain Forests [115]. Seram, the larger island to the east of Buru, was also delineated as a single ecoregion, the Seram Rain Forests [116]. MacKinnon also identified Seram as a distinct subunit within the Maluku biounit. The larger Halmahera Rain Forests [114] includes Obi island, which MacKinnon (1997) recognized as a separate subunit (13b) from Halmahera Island (subunit 13a). Stattersfield et al. (1998) also combined Obi Island with Halmahera to create a single endemic bird area. We created the Banda Sea Islands Moist Deciduous Forests [117] by combining the islands in the Kai and Tanimbar archipelagos, which were distinguished as a biogeographic unit by Monk et al. (1997). The primary vegetation on the islands in both these archipelagos is moist deciduous forests and semi-evergreen forests, whereas the vegetation in the other, nearby large islands (Seram and Aru) is evergreen rain forests (Monk et al. 1997). Stattersfield et al. (1998) also combined these archipelagos into a single endemic bird area.

The drier forests in Nusa Tenggara were placed in three ecoregions that corresponded to the biogeographic units identified in Monk et al (1997). These are Lesser Sundas Deciduous Forests [111], which includes the chain of islands extending from Lombok, Sumbawa, Komodo, Flores, and the smaller satellite islands corresponding to the Flores biogeographic unit; Timor and Wetar Deciduous Forests [112], corresponding to the Timor biogeographic unit; and the Sumba Deciduous Forests [113], corresponding to the Sumba biogeographic unit. All three ecoregions belong to the tropical dry forests biome.

Udvardy (1975) lumped the Halmahera Rain Forests [114], Seram Rain Forests [116], and Banda Sea Islands Moist Deciduous Forests [117] into his Papuan biogeographic province. The Papuan province was included in his Oceanian Realm. The remaining ecoregions in the Wallacea bioregion—Lesser Sundas Deciduous Forests [111], Timor and Wetar Deciduous Forests [112], and Sumba Deciduous Forests [113]—make up the Lesser Sunda Islands biogeographic province in the Indomalayan Realm.

New Guinea and Melanesia Bioregion

This bioregion includes the large island of New Guinea, its satellite islands, and the eastern oceanic islands, to Vanutau and New Caledonia. The bioregion line separates the Sahul shelf, with its predominantly Australian fauna, from Wallacea.

Using Whitmore's (1984) map of the vegetation of Malesia and MacKinnon's (1997) reconstruction of the original vegetation, we delineated the large areas of distinct habitat types as ecoregions.

Thus, the Vogelkop-Aru Lowland Rain Forests [119] represents the tropical lowland moist forests in the Vogelkop region of New Guinea. The ecoregion corresponds largely to subunits P3d and P3b identified by MacKinnon (1997); however, we placed the tropical montane moist forests (>1,000 m) in the Vogelkop Montane Rain Forests [118]. The tropical lowland moist and freshwater swamp forests to the north of the central cordillera were placed in the Northern New Guinea Lowland Rain and Freshwater Swamp Forests [123], and the montane forests were placed in the Northern New Guinea Montane Rain Forests [122] (based largely on recommendations by Bob Johns, pers. comm., 1999). This ecoregion corresponds to MacKinnon's (1997) biounits P3e and P3j. The tropical montane evergreen forests in the Huon peninsula were delineated as another distinct ecoregion, the Huon Peninsula Montane Rain Forests [124], and correspond to MacKinnon's (1997) biounit P3k.

The montane evergreen moist forests along the central cordillera, including the Snow Mountains, Star Mountains, Central Highlands, and Eastern Highlands, were placed in the Central Range Montane Rain Forests [125]. This ecoregion roughly corresponds to MacKinnon's subunits P3g, P3h, and P3i. The moist forests in the southeastern peninsula were distinguished as the Southeastern Papuan Rain Forests [126]. This ecoregion consists mostly of montane forests but also includes some lowland forests along the coasts and is roughly equivalent to MacKinnon's (1997) biounit P3n. We used the 1,000-m contour from a digital elevation model (U.S. Geological Survey 1996) to define the montane-lowland transition. All along the central cordillera and in the Huon Peninsula, we separated the alpine habitat into a distinct ecoregion, the Central Range Sub-Alpine Grasslands [131].

The tropical lowland moist evergreen forests to the south of the central cordillera were placed in the Southern New Guinea Lowland Rain Forests [128], and the extensive freshwater swamp forests were placed in the Southern New Guinea Freshwater Swamp Forests [127]. The freshwater swamp forests in the southern Vogelkop were also included in this ecoregion.

The savanna and grasslands in the Trans Fly region were placed in the Trans Fly Savanna and Grasslands [130], under the grasslands, savannas, and shrublands biome. This ecoregion also extends across the Arafura Sea to Australia, which was outside the region of analysis.

Yapen and Biak islands, which MacKinnon combined within biounit P3c, were delineated as separate ecoregions: Yapen Rain Forests [121] and Biak-Numfoor Rain Forests [120], respectively, based on recommendations by Bob Johns (vegetation) and the patterns of mammal distribution.

The mangroves along the coast of New Guinea were delineated as the New Guinea Mangroves [129]. We used the *World Mangrove Atlas* (Spalding et al. 1997) to determine the distribution and boundaries of mangroves.

We delineated two ecoregions to represent the montane and lowland evergreen moist forests in the New Britain and

New Ireland island complex: the New Britain–New Ireland Lowland Rain Forests [133] and the New Britain–New Ireland Montane Rain Forests [134]. The 1,000-m contour of the digital elevation model (U.S. Geological Survey 1996) was used as the transition between lowland and montane ecoregions. We placed the Admiralty Islands Lowland Rain Forests [132] into a distinct ecoregion, following Stattersfield et al. (1998). MacKinnon (1997) combined these three ecoregions into a single subunit (P3p).

The Trobriand Islands Rain Forests [135] and Louisiade Archipelago Rain Forests [136] were made distinct ecoregions based on Stattersfield et al. (1998). MacKinnon (1997) placed these two ecoregions together into subunit P3o.

The other distinct island groups were placed in their own ecoregions, namely the Solomon Islands Rain Forests [137] and the Vanuatu Rain Forests [138]. We followed Stattersfield et al. (1998) in delineating these ecoregions. MacKinnon (1997) did not extend his assessment beyond the island of Bougainville in the Solomon Islands. However, we followed Bouchet et al. (1995) and separated the distinctive dry forests in New Caledonia from the moist forests to delineate the New Caledonia Rain Forests [139] and the New Caledonia Dry Forests [140]. Stattersfield et al. (1998) did not make this distinction.

Udvardy (1975) placed all the ecoregions in the New Guinea and Melanesia bioregion, with the exception of New Caledonia, into the Papuan biogeographic province of the Oceanian Realm. New Caledonia was placed in the New Caledonian biogeographic province.

Appendix B
Terrestrial Biome Descriptions

Ecological processes, patterns of biodiversity, and responses to disturbance vary widely in their scale and importance between different habitat types. To address this variation in conservation planning, we assign each ecoregion to a biome. Biomes are not geographically defined units; rather, they refer to the dynamics of ecological systems and the broad habitat structures and patterns of species diversity that define them. Three features that are particularly relevant to conservation planning are the primary discriminators for biomes:
- broad temporal and spatial patterns of biodiversity (such as high vs. low beta-diversity, patchy resources, relative dominance of taxa)
- minimum area and condition needs of focal species and ecological processes
- the sensitivity and resilience to disturbance

Criteria used in the analyses of an ecoregion's biological distinctiveness and conservation status can be tailored to these specific characteristics of its biome. The design of conservation landscapes within ecoregions can be guided by the broad patterns and dynamics of the biome. Biomes have another important role in conservation planning at global and regional scales. By organizing ecoregions within biomes and biogeographic realms (e.g., Neotropics) and bioregions (e.g., Caribbean), one can facilitate representation of whole kinds of ecosystems in regional and global conservation strategies.

The following are descriptions of terrestrial biomes with some examples of broad patterns and dynamics that help define them.

Temperate Coniferous Forests

Temperate evergreen forests are found predominantly in areas with warm summers and cool winters and vary widely in their kinds of plant life. Temperate evergreen forests are common in the coastal areas of regions that have mild winters and heavy rainfall or inland in drier climates or montane areas. Many tree species inhabit these forests, including pine, cedar, fir, and redwood. The understory also contains a wide variety of herbaceous and shrub species. Temperate conifer forests sustain the highest levels of biomass in any terrestrial ecosystem and are notable for huge trees in temperate rain forest regions. The major threats to these forests include commercial logging, grazing, acid deposition, urban sprawl, pollution, conversion to pasture and agricultural land, suppression of natural fire regimes, and the introduction of exotic species.

Biodiversity Patterns

Most tree species and larger vertebrates have relatively widespread distributions; much local endemism and beta-diversity are seen in some ecoregions in invertebrates, understory plants, and lichens, particularly on unusual soils; invertebrate faunas or herbaceous floras may be extremely diverse; and altitudinal specialization occurs but is less pronounced than in the tropics.

Minimum Requirements

Disturbances such as fire, windthrow, and epizootics can vary widely within this biome, but the extremes typically are of sufficient size and frequency so that small patches of natural forest have only limited conservation value. Many species are highly specialized in late-successional forests, larger carnivores are very wide-ranging with large home ranges, and some species track resources that vary widely in space in time (e.g., epizootic outbreaks, fire events, cone production) and need large natural landscapes.

Sensitivity to Disturbance

Larger carnivores are highly sensitive to human activities, including low-intensity hunting; large numbers of species are highly sensitive to logging and fragmentation of natural forests, particularly late-successional species; late-successional species and features typically regenerate slowly; many temperate forests need periodic fires to maintain successional processes and many species; and exotic species can have extensive and significant impacts on natural forest communities.

Temperate Broadleaf and Mixed Forests

Temperate broadleaf forests thrive on nutrient-rich soils across western Europe, eastern Asia, and eastern North America. The Appalachians of eastern North America and the forests of southwestern China are the richest temperate forests. Temperate broadleaf forests grow in regions where summers are warm but winters cold, and in some locales, such as southeast Australia, rainfall is so great that temperate rain forests occur. Because of the agreeable climate, these regions are home to some of the most populated areas in the world. As a result, many areas of temperate broadleaf forests have been cleared for agriculture and settlement. There are many tree species in the temperate deciduous forests, but three to five species, such as oak, maple, beech, elm, and ash in eastern North America, often dominate. Characteristic mammal species of these eastern North American forests include deer, squirrels, raccoons, opossums, pumas, bears, and timber wolves. Many resident birds include chickadees, grouse, and *Accipiter* hawks, as well as migrant songbirds such as warblers and tanagers.

Biodiversity Patterns

Most dominant species have widespread distributions, but in many ecoregions there can be a large number of ecoregional and local endemics; beta-diversity can be high for plants, invertebrates, and some smaller vertebrates in some ecoregions; unusual soils can harbor many specialist plants and invertebrates; and some ecoregions can have very high alpha and gamma diversity for plants, particularly understory species and herbaceous floras. Altitudinal specialization occurs but is less pronounced than in the tropics.

Minimum Requirements

Larger native carnivores need large natural landscapes to persist, periodic large-scale disturbance events such as fire necessitate the conservation of large blocks of forest, and many species of plants, lichens, fungi, and invertebrates depend on late-successional forests.

Sensitivity to Disturbance

Certain species are highly sensitive to habitat fragmentation, such as breeding songbirds exposed to parasitism or elevated nest predation; many forest understory species are also unable to cross deforested areas; restoration potential for these forests is high; exotic species can have extensive and significant impacts on native communities; and the loss of large native predators has many cascading impacts on forest structure and ecology.

Deserts and Xeric Shrublands

Deserts are dry areas that receive less than 10 inches of rain annually and may go years with no rain at all. They usually feature humidity near zero. Many deserts, such as the Sahara, are hot year-round, but others, such as Asia's Gobi, become quite cold in winter. Temperature extremes are a characteristic of most deserts. Searing daytime heat gives way to cold nights because there is no insulation provided by humidity and cloud cover. All deserts feature sparse vegetation, although areas range from extreme desert with little or no vegetation to areas with a dense layer of shrubs. They also harbor animals that have adapted to the uniquely harsh living conditions of the region. Common wildlife includes reptiles, gerbils, ground squirrels, and larger animals such as gazelles and foxes. In the Namib Desert in Africa, elephants, lions, and black rhinos have also been found. Desert plant species are drought resistant; cacti collect and store water in stems, and others absorb dew or shed leaves to help conserve water in periods of drought. Currently, desert areas worldwide are growing because of droughts and poor farming and herding practices such as groundwater depletion and overgrazing. Dry shrub areas receive a bit more rain than deserts and therefore feature coarse thorny vegetation such as woody bushes and sometimes grasses. Human inhabitants often survive in dry shrub areas by practicing pastoral farming. However, overgrazing by sheep and goats can eliminate native vegetation and cause the area to degenerate into true desert.

Biodiversity Patterns

Deserts and xeric shrublands may have extraordinarily rich floras with very high alpha and beta-diversity, reptile faunas may also be very diverse, and local endemism may be pronounced in some regions.

Minimum Requirements

Many species track seasonally variable and patchy resources and need large natural landscapes to persist; water sources and riparian habitats are critical for the persistence of many species.

Sensitivity to Disturbance

These areas are highly sensitive to grazing, soil disturbance, burning, plowing, and other cover alteration; restoration potential can be very low and regeneration very slow; and exotic species may be a serious problem.

Montane Grasslands and Shrublands

This biome includes high-elevation (montane and alpine) grasslands and shrublands, including the puna and paramo in South America, subalpine heath in New Guinea and east Africa, steppes of the Tibetan plateaus, and similar subalpine habitats around the world. They are tropical, subtropical, and temperate. Montane grasslands and shrublands, because they are found in somewhat difficult environments, often are more sensitive than those found at lower altitudes. The paramo of Venezuela, Colombia, and Ecuador is characterized by grasses, but there are also dwarf shrubs. The puna of Peru and Bolivia is a grassland that also contains many herbaceous plants and low-growing cacti. A unique feature of many tropical paramos is the presence of giant rosette plants from a variety of plant families, such as *Lobelia* (Africa), *Puya* (South America), *Cyathea* (New Guinea), and *Argyroxiphium* (Hawaii); these plant forms can reach elevations of 4,500–4,600 m above sea level.

Biodiversity Patterns

These habitats may display high beta-diversity, particularly between isolated montane areas and along altitudinal gradients; local and regional endemism can be pronounced in some regions.

Minimum Requirements

Large natural landscapes are needed in some regions because larger vertebrates track widely distributed seasonal or patchy resources; water sources and riparian vegetation are important for wildlife in drier regions.

Sensitivity to Disturbance

These fragile habitats are highly sensitive to plowing, overgrazing, and excessive burning because of their challenging climatic and soil conditions; larger vertebrates are sensitive to even low levels of hunting.

Flooded Grasslands and Savannas

Some outstanding examples of flooded savannas and grasslands occur in the Everglades, the Pantanal of South America, and the Okavango Delta in southern Africa. These types of wetlands are found inland and along the coast near river mouths, bays, and protected coastlines. They are dominated by grasses and grass-shrub mosaics and are periodically inundated by seasonal rains or the movement of tides.

Abundant grasses shelter fish, amphibians, aquatic invertebrates, and other animals. Wading birds are common and feed on plentiful vegetation and insects. The Pantanal, for example, supports more than 260 species of fish, 700 birds, 90 mammals, 160 reptiles, 45 amphibians, 1,000 butterflies, and 1,600 plants. It is one of the largest continental wetlands on Earth, formed when the Paraguay River and its tributaries overflow their banks onto flat terrain during the rainy season. The Okavango Delta forms when the Okavango River floods and spreads over a 15,000-km^2 area in the Kalahari each year. The Everglades ecosystem, which covers saw grass marshes, sloughs, tropical hardwood hammocks, cypress heads, bayheads, and mangrove estuaries, is home to some 11,000 species of seed-bearing plants, 25 varieties of orchids, 300 bird species, and 150 fish species. Heavily affected by human actions such as conversion to agricultural lands, canal construction, and water diversion, flooded grasslands and savannas are critical habitats to preserve, both for their biodiversity value and for the ecosystem services they provide.

Biodiversity Patterns

Most terrestrial species have widespread ranges in these habitats, alpha and beta-diversity are not pronounced, and endemism in terrestrial species is low.

Minimum Requirements

Maintaining hydrographic integrity is critical to these habitats, many species track flooding patterns and seasonal abundance of resources, and riparian and gallery habitats important for many species.

Sensitivity to Disturbance

Diversion and channelization of water flow greatly decrease the integrity of these habitats, loss of riparian and gallery habitats can affect wildlife populations, these areas are sensitive to water quality changes from pollution and eutrophication, and alteration of natural fire regimes may shift the composition and structure of communities.

Tropical and Subtropical Grasslands, Savannas, and Shrublands

Vast landscapes and great herds of large mammals characterize tropical grasslands and savannas. It's here that you will find giraffes, elephants, zebras, antelopes, towering termite mounds, lions, migrating wildebeest, and many other species. More than 5 million square miles of savannas are found around the world in places such as Australia, South America, India, and Africa, where they cover one-third of the continent. The Serengeti of east Africa is one of the most renowned examples of savanna habitat because so much of the area remains in a natural state. Tall grasses dominate, dotted with scattered trees such as baobabs and acacias. Savannas serve as a transition zone between grasslands and forests. Savannas occur in warm, tropical regions where a prolonged dry season alternates with a rainy season. Grasses and trees occur in proportion to the amount and distribution of annual rainfall. The longer and more intense the dry season, the more sparse the tree growth. Savanna grasses are well adapted to these seasonal cycles. For example, in the rainy season the savanna region of the Serengeti is transformed from dusty brown to green as the new growth springs up almost overnight. Large migrations of animals that follow seasonal rainfall patterns are one of the main characteristics of this biome. After the rainy season, vegetation above the ground withers quickly, but the parts of the plants below the ground continue to absorb any available moisture. Fires, caused by lightning or people, are essential to the maintenance of the ecosystem by limiting trees and maintaining grasses as the dominant vegetation in savannas and grasslands. Some savannas have a large component of shrubs and trees.

Biodiversity Patterns

Diverse large mammal assemblages in abundant aggregations can be a characteristic feature; most vertebrates display widespread distributions; and plant alpha diversity typically is low, but in some regions beta-diversity and gamma diversity can be very high.

Minimum Requirements

Large natural landscapes are necessary to allow large grazers and their associated predators to track seasonal rainfall or to migrate to new areas during periodic droughts; large-scale fires also necessitate the conservation of larger natural landscapes; some large predators, such as wild dogs of Africa, need large natural areas because of their home range size and sensitivity to humans; and sources of water are critical for many species.

Sensitivity to Disturbance

Restoration potential in these systems is high, but plowing, overgrazing by domestic livestock, and excessive burning can quickly degrade and alter natural communities. Alteration of surface water patterns can have significant impacts on the persistence of many vertebrate species, and many species are highly sensitive to low-intensity hunting or other human activities.

Tropical and Subtropical Moist Broadleaf Forests (Includes Rain Forest)

Tropical and subtropical moist broadleaf forests are home to more species than any other terrestrial ecosystem: half the world's species may live in these forests, where a square kilometer may be home to more than 1,000 tree species. These forests are found around the world, particularly in the Indo-Malayan archipelagos, the Amazon Basin, and the African Congo. A perpetually warm, wet climate promotes more explosive plant growth than in any other environment on Earth. Annual rainfall usually exceeds 2,000 mm. A tree here may grow more than 75 feet in height in just five years. From above, the forest appears as an unending sea of green, broken only by occasional, taller emergent trees. These towering emergents are the realm of hornbills, toucans, and the harpy eagle. The canopy is home to many of the forest's animals, including apes and monkeys. Below the canopy, a lower understory hosts snakes and big cats. The forest floor, clear of undergrowth because of the thick canopy above, is prowled by other animals such as gorillas and deer. All levels of these forests contain an unparalleled diversity of invertebrate species, including New Guinea's unique stick insects and bird wing butterflies that can grow more than 1 foot long. These forests are under tremendous threat from human activity. Many forests are being cleared for farmland, and others are subject to large-scale commercial logging. An area the size of Ireland is destroyed every few years, largely by commercial logging and secondary impacts. Such activities threaten the future of these forests and are the primary contributor to the extinction of 100–200 species a day on average over the next forty years (exotics on islands and loss of island habitats are other major factors). At the current rate of deforestation, more than 17,000 species will go extinct every year, which is more than 1,000 times the rate before humans arrived on this planet.

Biodiversity Patterns

These areas are characterized by very high alpha diversity, very high beta-diversity in some regions, and very high gamma diversity in regions and by a high degree of habitat specialization, particularly on such features as karst, ultramafic soils, and white sands, and with elevation.

Minimum Requirements

Many plant and animal species with low population densities need vast natural forests to maintain viable populations; many species track patchy or low-density resources and therefore need large areas of habitat and natural corridors to survive; large, remote forests are critical for maintaining populations of species sensitive to hunting; a large percentage of forest cover may be needed to maintain natural rainfall regimes (perhaps more than 70 percent cover across whole regions); and many species need intact altitudinal gradients for seasonal movements and intact complexes of riverine and upland forests.

Sensitivity to Disturbance

Low-intensity hunting can have significant impacts on species that live at low densities; removal of forest cover and disruption of soils through logging can quickly lead to nutrient loss and erosion; restoration of natural communities can be very difficult or impossible; exotic species can have significant impacts on natural communities; and loss of large predators, herbivores, and frugivores can have cascading effects throughout the ecosystem.

Tropical and Subtropical Dry Broadleaf Forests

Tropical dry forests are found in southern Mexico, southeastern Africa, the Lesser Sundas, central India, Indochina, Madagascar, New Caledonia, eastern Bolivia and central Brazil, the Caribbean, valleys of the northern Andes, and along the coasts of Ecuador and Peru. Although these forests occur in climates that are warm year-round and may receive several hundred centimeters or rain per year, they are subject to long dry seasons that last several months and vary with geographic location. These seasonal droughts have great impact on all living things in the forest. Deciduous trees predominate in these forests, and during the drought a leafless period occurs, which varies with species type. Because trees lose moisture though their leaves, the shedding of leaves allows trees such as teak and mountain ebony to conserve water during dry periods. The newly bare trees open up the canopy layer, enabling sunlight to reach ground level and facilitate the growth of thick underbrush. Though less biologically diverse than rain forests, tropical dry forests are still home to a wide variety of wildlife including monkeys, large cats, parrots, various rodents, and ground-dwelling birds. Many of these species display extraordinary adaptations to the difficult climate. Logging, agricultural expansion, human-induced fires, and overgrazing are serious threats to the health of tropical dry forests across the globe. When trees are removed, the remaining vegetation and soil are scorched during the dry season. Vegetation then dies off, and the bare soil is very susceptible to erosion. These forests are more threatened than even tropical moist forests.

Biodiversity Patterns

Species tend to have wider ranges than moist forest species, although in some regions many species have highly restricted ranges; most dry forest species are limited to tropical dry forests, particularly plants; and beta-diversity and alpha diversity are high but typically are lower than in adjacent moist forests.

Minimum Requirements

Large natural areas are needed to maintain larger predators and other vertebrates, large areas are also needed to buffer sensitive species from hunting pressure, the persistence of riparian forests and water sources is critical for many dry forest species, and larger blocks of intact forest are needed to absorb the effects of occasional large fires.

Sensitivity to Disturbance

Dry forests are highly sensitive to excessive burning and deforestation, overgrazing and exotic species can also quickly alter natural communities, and restoration is possible but challenging, particularly if degradation has been intense and persistent.

Tropical and Subtropical Coniferous Forests

The world's richest and most complex tropical coniferous forests are found in Mexico, and others are located in the Caribbean and Indochina. Many migratory birds and butterflies winter in tropical and subtropical coniferous forests. These biomes feature a thick, closed canopy that blocks light to the floor and allows little underbrush. As a result, the ground often is covered with fungi and ferns. Shrubs and small trees compose a diverse understory. These forests occur in regions with high temperatures and very high rainfall of 380–960 cm/year. These forests harbor a wealth of plant and animal species and are still under significant threats, including overgrazing, frequent burning, logging, and agricultural expansion.

Biodiversity Patterns

Much local endemism and beta-diversity occur in some ecoregions in invertebrates, understory plants, and lichens, particularly in moister forests or on unusual soils; some larger vertebrates and dominant tree species may have widespread ranges; these areas may have extremely diverse floras; and altitudinal specialization occurs.

Minimum Requirements

Disturbance regimes such as fire, windthrow, and epizootics can vary widely within this biome, but the extremes typically are of sufficient size and frequency so that small patches of natural forest have only limited conservation value; many species are highly specialized in late-successional forests; larger carnivores are very wide-ranging with large home ranges; and some species track resources that vary widely in space in time (e.g., epizootic outbreaks, fires, cone production) and need large natural landscapes.

Sensitivity to Disturbance

Larger carnivores are highly sensitive to human activities, including low-intensity hunting; large numbers of species are highly sensitive to logging and fragmentation of natural forests, particularly late-successional species; late-successional species and features typically regenerate slowly; many temperate forests need periodic fires to maintain successional processes and many species; and exotic species can have extensive and significant impacts on natural forest communities.

Mangroves

Mangroves occur in the waterlogged, salty soils of sheltered tropical and subtropical shores. They are subject to the twice-daily ebb and flow of tides, fortnightly spring and neap tides, and seasonal weather fluctuations. They stretch from the intertidal zone up to the high-tide mark. Mangrove trees include more than sixty unrelated species. With their distinctive nest of stilt and proplike roots, mangroves can thrive in areas of soft, waterlogged, and oxygen-poor soil by using aerial and even horizontal roots to gain a foothold. The roots also absorb oxygen from the air, and the tree's leaves can excrete excess salt. The nests of roots often collect silt and create new landforms, and animals such as algae, barnacles, and mollusks cling to them. Mangroves are home to fantastic birds such as the scarlet ibis and many fish-eaters such as the anhinga and the egret and also act as nurseries for numerous fish, crustacean, insect, and other animal species. Many fish and birds are migratory visitors. Some mangrove species have seeds, which develop on the tree, then drop off as partially grown seedlings, spear-shaped to take root in the mud. Mangroves and other wetlands are being destroyed rapidly for shrimp farming, salt production, agriculture, and urban development. In some tropical countries, such as India, the Philippines, and Vietnam, more than 50 percent of mangrove ecosystems have been lost in this century. Many mangrove areas are also located at the interface of the terrestrial, freshwater, and marine environments, such as river estuaries, making them particularly vulnerable to industrial and wastewater pollution and changes in salinity. Industries such as prawn and rice farming in southeast Asia have also led to the removal of the area's mangrove swamps. Logging and charcoal production are significant threats in many areas.

Biodiversity Patterns

Most species typically have widespread distributions; diversity of floras is low, but overall alpha diversity is very high when terrestrial and aquatic species are considered; beta-diversity and ecoregional endemism are low; some highly localized species exist; there is strong zonation along gradients; and there are several distinct mangrove habitat formations.

Minimum Requirements

Mangroves need relatively intact hydrographic and salinity regimes; without these conditions, the persistence or restoration of mangroves is difficult or impossible.

Sensitivity to Disturbance

Alterations of hydrography and substrate have great impact, but restoration potential is high; mangroves are susceptible to pollution, particularly oil and other petroleum compounds; and alteration of salinity levels can have dramatic impacts on mangroves.

Appendix C

Methods for Assessing the Biological Distinctiveness of Terrestrial Ecoregions

Databases

We used the available literature, especially field guides, to derive ecoregion-by-ecoregion presence or absence and endemism data for mammals and birds. For areas in the Indo-Pacific region where distributions were not clearly documented, we requested expert review and revision.

Endemism for mammal and bird species was assessed by using the following decision rules:
1. If the total species range is more than 50,000 km^2,
    a. and a single ecoregion contains 75–100 percent of the species' range, the species was recorded as endemic to that ecoregion and recorded as present in all others containing it, but
    b. if no single ecoregion contains more than 75 percent of the species' total range, the species was recorded as present but not endemic in all ecoregions containing it.
2. If the total species range is less than 50,000 km^2,
    a. and a species occurs in five or fewer ecoregions, the species was recorded as endemic in all ecoregions containing it, but
    b. if the species occurs in more than five ecoregions (many small, disjunct areas), the species is recorded as present but not endemic in all ecoregions containing it.

If a mammal or bird was identified as endemic to multiple ecoregions, it was called *near endemic*. If a mammal or bird's entire range fell within only a single ecoregion, it was identified as a *strict endemic*.

For plants, we asked experts to make assessments (in some cases estimates) of the number of plant species and the percentage endemism for each ecoregion. In many cases, these numbers are underestimates because most of the experts opined that many species remain unnamed and await discovery and scientific description.

All references, sources, and experts are listed in table C.1.

Biodiversity Distinctiveness Index

To create the BDI, we first constructed three graphs—one for mammals, one for birds, and one for plants—for each biome. Each graph was a plot of the species richness results, for the given taxonomic group, for all ecoregions in that biome. Natural breaks in each graph were used to identify four levels of biological distinctiveness: globally outstanding (GO), regionally outstanding (RO), bioregionally outstanding (BO), and locally important (LI). These categories (table C.2) were assigned 5, 3, 2, and 1 points, respectively. For example, an

TABLE C.1. References, Sources, and Expert Opinion Used to Derive the Biological Distinctiveness Index (BDI).

References and Sources	*Expert Review*
Mammals	**Mammals**
Corbet and Hill (1992)	Dr. Frank Bonaccorso: Wallacea and New Guinea and Melanesia bioregions
Prater (1971)	
Bonaccorso (1998)	**Birds**
Flannery (1990, 1995)	Robert Tizard: Indochina bioregion
Flannery and Groves (1998)	
Lekagul and McNeely (1977)	**Plants**
Wilson and Reeder (1993)	Dr. Peter Ashton: all bioregions
Nowak (1999a,b)	Dr. Jim LaFrankie: all bioregions
	Dr. Gopal Rawat: Indian Subcontinent bioregion
Birds	Dr. Philip Rundel: Indochina bioregion
Doughty et al. (1999)	Dr. Max van Balgooy: Sunda Shelf and Philippines and Wallacea bioregions
Stattersfield et al. (1998)	Dr. Domingo Madulid: Sunda Shelf and Philippines bioregion
King et al. (1975)	Dr. Robert Johns: New Guinea and Melanesia bioregion
Robson (2000)	Dr. Tim Whitmore: Sunda Shelf and Philippines bioregion
Beehler et al. (1986)	Dr. Peter Lowry: New Guinea and Melanesia bioregion
Dickinson et al. (1991)	
Grimmett et al. (1998)	
Henry (1998)	
MacKinnon and Phillipps (1993)	
Coates and Bishop (1997)	
Coates (1985)	
Juniper and Parr (1998)	
Kennedy et al. (2000)	
Bregulla (1992)	

ecoregion that was globally outstanding in its biome for mammal richness was assigned 5 points for the mammal richness category.

This approach allowed us to tailor the evaluation of richness specifically to each biome. For instance, the globally outstanding threshold for bird richness in the tropical moist forests was twice as high as the threshold for the savannas and grasslands biomes.

We treated the endemism scores for birds and mammals similarly, using total number of endemics. However, plant endemism was evaluated by experts who placed each ecoregion into four classes based on the estimated level of endemism. The classes were 0–5 percent, 6–25 percent, 26–75 percent, and 76–100 percent. The results were then translated into point values from 1 to 4, representing the BDI classes. The scales for endemism are given in table C.3.

Next we summed the richness scores for the three taxonomic groups to create a Richness Index (RI), ranging from 3 to 15 points. The ecoregions were then classified as globally outstanding (GO), regionally outstanding (RO), bioregionally outstanding (BO), or locally important (LI) depending on the RI score, based on the point system depicted in

TABLE C.2. Species Richness Categories for Mammals, Birds, and Plants.

Biome	Mammals	Birds	Plants	Category	Points
Tropical and subtropical moist broadleaf forests	≥140	≥350	≥5,000	GO	5
	80–139	300–349	2,500–4,999	RO	3
	35–79	175–299	1,150–2,499	BO	2
	≤35	≤175	≤1,150	LI	1
Tropical and subtropical dry broadleaf forests and deserts and xeric shrublands	≥100	≥350	≥2,500	GO	5
	70–99	280–349	1,200–2,499	RO	3
	30–69	175–279	1,000–1,199	BO	2
	≤30	≤175	≤1,000	LI	1
Temperate broadleaf and mixed forests	≥100	≥350	≥2,000	GO	5
	50–99	200–349	1,500–1,999	RO	3
	11–49	100–199	1,150–1,499	BO	2
	≤10	≤100	≤1,150	LI	1
Tropical and subtropical coniferous forests and temperate coniferous forests	≥100	≥350	≥1,500	GO	5
	50–99	200–349	1,000–1,499	RO	3
	11–49	100–199	500–1,000	BO	2
	≤10	≤100	≤500	LI	1
Tropical and subtropical grasslands, savannas, and shrublands, flooded grasslands and savannas, and montane grasslands and shrublands	≥100	≥150	≥2,000	GO	5
	50–99	100–149	1,200–1,999	RO	3
	11–49	25–99	1,000–1,199	BO	2
	≤10	≤25	≤1,000	LI	1

These ranges were derived by graphing actual numbers (stratified by biome) and identifying natural breaks in the curves. The numbers under mammals, birds, and plants indicate numbers of species.

TABLE C.3. Endemism Categories for Mammals, Birds, and Plants.

Biome	Mammals	Birds	Plants	Category	Points
Tropical and subtropical moist broadleaf forests	≥8	≥30	4	GO	5
	4–7	10–29	3	RO	3
	2–3	2–9	2	BO	2
	≤1	≤1	1	LI	1
Tropical and subtropical dry broadleaf forests and deserts and xeric shrublands	≥5	≥30	4	GO	5
	3–4	9–29	3	RO	3
	2	2–8	2	BO	2
	≤1	≤1	1	LI	1
Temperate broadleaf and mixed forests	≥5	≥10	4	GO	5
	3–4	5–9	3	RO	3
	2	2–4	2	BO	2
	≤1	≤1	1	LI	1
Tropical and subtropical coniferous forests and temperate coniferous forests	≥5	≥5	4	GO	5
	3–4	3–4	3	RO	3
	2	2	2	BO	2
	≤1	≤1	1	LI	1
Tropical and subtropical grasslands, savannas, and shrublands, flooded grasslands and savanna, and montane grasslands and shrublands	≥5	≥5	4	GO	5
	3–4	3–4	3	RO	3
	2	2	2	BO	2
	≤1	≤1	1	LI	1

The ranges for mammals and birds were derived by graphing species numbers (stratified by biome) and identifying natural breaks in the results. The data for plant endemism were already indexed on a 1–4 scale where 1 = 0–5%, 2 = 6–25%, 3 = 26–75%, and 4 = 76–100% endemism.

Box C.1. Taxonomic Scores for the Richness and Endemism Indices (RI and EI).

Taxon 1	Taxon 2	Taxon 3	Total Points	RI	EI
5	5	5	15	GO	GO
5	5	3	13	GO	GO
5	5	2	12	GO	GO
5	3	3	11	GO	GO
5	3	2	10	RO	GO
5	3	1	9	RO	RO
5	2	2	9	RO	RO
5	2	1	8	RO	RO
5	1	1	7	RO	RO
3	3	3	9	RO	RO
3	3	2	8	RO	RO
3	3	1	7	BO	BO
3	2	2	7	BO	BO
3	2	1	6	BO	BO
3	1	1	5	BO	BO
2	2	2	6	BO	BO
2	2	1	5	BO	BO
2	1	1	4	LI	LI
1	1	1	3	LI	LI

GO = globally outstanding, RO = regionally outstanding, BO = bioregionally outstanding, LI = locally important.

box C.1. To become globally outstanding for the RI (11 to 15 points), an ecoregion had to be globally outstanding for at least one taxonomic group (5 points) and regionally outstanding for two taxonomic groups (3 + 3 points).

We analyzed the endemism data in a similar fashion, except that 10 points were needed for an ecoregion to be rated as globally outstanding for the Endemism Index (EI). Thus an ecoregion that was globally outstanding for one taxonomic group, regionally outstanding for a second, and bioregionally outstanding for the third group was considered globally outstanding in the EI (box C.1). We were more conservative with the cut-off for globally outstanding for endemism because of its importance for archipelagic regions that are often depauperate in vegetation.

Next, we created a Richness and Endemism Index (REI) by comparing the two indices (RI and EI) and assigning a final (REI) category to each ecoregion by taking the higher classification. Thus, an ecoregion that was bioregionally outstanding for richness and globally outstanding for endemism was considered to be globally outstanding in the final REI standing. This was done because of the differences in biodiversity values between continental and archipelagic ecoregions. Archipelagic ecoregions tended to have few species, with higher endemism, whereas continental ecoregions were more speciose but had fewer endemics.

Finally, to create the Biological Distinctiveness Index (BDI), we elevated to globally outstanding status any ecoregions that exhibit evolutionary and ecological phenomena of global significance or that represent extreme habitat rarity (figure C.1). The criteria used to determine these features are as follows.

Higher Taxonomic Uniqueness

Rationale: Represents the distinctiveness of the higher taxa (such as endemic genera and families or representatives of relict or primitive genera, families, or orders) of an ecoregion. Because genera and families are at a higher hierarchical level than species in the taxonomy of living organisms, the presence of an endemic higher taxon would contribute more to an ecoregion's biotic distinctiveness than would an endemic species. Naturally rare representatives of relict or primitive genera, families, or orders also contribute to the distinctiveness of an ecoregion's biota and the urgency of conservation action.

Criteria: An ecoregion was considered globally outstanding for this feature if

- The ecoregion contains more than four endemic plant families, or more than 30 percent of all the ecoregion's vascular plant genera are endemic to it (using the same endemism decision rules as those applied to species).
- The ecoregion harbors two or more endemic plant families and at least one endemic family of birds, mammals, invertebrates (where information was deemed adequate for regional analyses), reptiles, or amphibians. (This analysis examined only terrestrial ecoregions and therefore excluded fish in determining ecological phenomena.)

The criteria reflect levels that occur only in a few regions on Earth, primarily because of the great age, stability, or isolation of their biotas or unique environmental conditions.

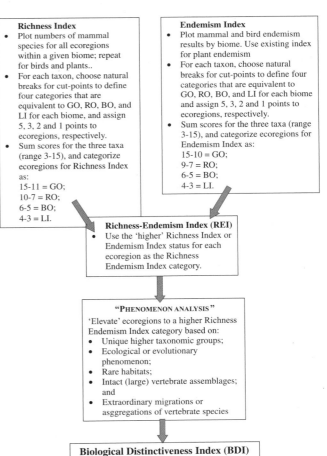

Figure C.1. Flow diagram showing the steps taken to create the Biological Distinctiveness Index (BDI). GO = globally outstanding; RO = regionally outstanding; BO = biologically outstanding; LI = locally important.

Appendix C: Methods for Assessing the Biological Distinctiveness of Terrestrial Ecoregions

Extraordinary Evolutionary Phenomena

Extraordinary Adaptive Radiations

Rationale: Emphasized assemblages that display extreme sympatric radiations.

Criteria: An ecoregion was considered globally outstanding for adaptive radiations if

- It displays extensive adaptive radiations (fifteen or more species) within one or more vertebrate families.
- It displays extensive adaptive radiations (twenty-five or more species) within one or more plant families.
- Adaptive radiations occur within two or more vertebrate families (ten or more species displaying clear adaptive radiation within each family).

Extraordinary Ecological Phenomena

Intact Large Vertebrate Assemblages

Rationale: Ecosystems that still harbor intact communities of larger vertebrates are becoming increasingly rare. We defined an intact community as one in which all large native vertebrate species are extant and their populations fluctuate within natural ranges. This criterion acts as an umbrella for intact ecological phenomena, such as predator-prey relationships.

Criteria: An ecoregion was considered globally outstanding for this feature if it contains all the largest indigenous carnivores, herbivores, frugivores, and species of other feeding guilds that were historically represented there, and these species still play their ecological roles. Where it was impossible to determine whether ecological processes are intact, we used as a proxy the occurrence of these species in populations that still fluctuate within natural ranges. Additional rules include the following:

- No more than three ecoregions per biome per biogeographic realm could be selected as globally outstanding for this feature if the large carnivores, herbivores, and frugivores have widespread distributions throughout the biogeographic realm (i.e., we emphasized the three assemblages that are deemed most intact).
- More than three ecoregions per biome per biogeographic realm could be selected only when the best available knowledge indicates that the ecoregions harbor similarly intact assemblages.
- Migrations or extraordinary aggregations of large vertebrates are present.

Rationale: Large-scale movements and aggregations of larger vertebrates are increasingly rare around the world as habitats and populations are decimated. Although it is difficult to quantify, we believe it is critical to highlight the conservation importance of remaining large-scale movements and aggregations. Migrations that are considered extinct are not used.

Criteria: An ecoregion was considered globally outstanding for this feature if it fulfilled any of the following:

- A migration of large terrestrial vertebrates occurs within the ecoregion or spans adjacent units; the migration exceeds 100 km in length and includes more than several thousand individuals and is accompanied by the full complement of native large predators.
- Enormous aggregations (millions) of breeding individuals of migratory birds (more than ten species) occur throughout an ecoregion.
- Enormous aggregations (tens of thousands) of breeding individuals of migratory nonavian vertebrates occur throughout an ecoregion.

Global Rarity of Habitat Types

Rationale: The global rarity of habitat types (habitat types occupy a hierarchical level below biomes) can be evaluated to identify critical opportunities for habitat conservation. This criterion encompasses ecological and evolutionary phenomena, but it also addresses those characteristics at the scale of whole ecosystems and biotas as well as structural features of ecosystems and habitats. We stand on the brink of losing whole habitat types such as temperate rain forests as well as whole biomes such as temperate grasslands and tropical dry forests. We counted only naturally occurring rarity for the distinctiveness analyses, although human-induced rarity is an important condition to assess when developing conservation strategies.

Criteria: Fewer than eight ecoregions occur globally that contain the habitat type in question.

Ecoregions Harboring Examples of Large, Relatively Intact Ecosystems

Rationale: Our regional and global distinctiveness assessments did not consider the current status of biodiversity features of an ecoregion, with the exception of intact biotas and extant migrations. For this reason, we did not highlight the presence of larger, relatively intact ecosystems within ecoregions as a distinctiveness feature. These wilderness or frontier areas are a critical conservation target and are often associated with intact biotas and migrations (therefore, they may be captured to some extent through the criteria described above). Other analyses have looked at remaining wilderness and larger intact habitats for forests (WRI Forest Frontiers, CI Tropical Forest Wilderness Areas), mangroves (WCMC Mangrove Atlas), and coral reefs (WRI Reefs at Risk), but few or no analyses have been done for the rest of Earth's habitat types.

Criteria: None used.

Appendix D
Methods for Assessing the Conservation Status of Terrestrial Ecoregions

The Conservation Status Index (CSI) is designed to measure degrees of habitat alteration and spatial patterns of remaining natural habitats across landscapes. The index uses landscape features as indicators for the ecological integrity of ecosystems because we know that with increasing habitat loss, degradation, and fragmentation, ecological processes cease to function naturally, and major components of biodiversity are steadily eroded. Here we assess the conservation status of ecoregions in the tradition of the IUCN *Red List* categories for threatened and endangered species (critical, endangered, and vulnerable), except that we estimate the state of whole biotas and ecological processes (Mace and Stuart 1994; Hilton-Taylor 2000). The conservation status categories we defined are *extinct*, *critical*, *endangered*, *vulnerable*, *relatively stable*, and *relatively intact*.

The categories are defined as follows:

Extinct. No natural communities remain. Some of the original species are still present, but no opportunities to restore the original natural communities exist because of permanent alteration of physical conditions, loss of source pools of native species, substantive alteration of natural ecological processes, or an inability to eradicate or control aggressive alien species.

Critical. The remaining intact habitat is limited to isolated small fragments with low probabilities of persistence over the next five to ten years without immediate and intensive protection and restoration. Many species are already extirpated or extinct because of the loss of viable habitat. Remaining habitat fragments do not meet the minimum area requirements for maintaining viable populations of many species and ecological processes. Land use in areas between remaining fragments often is incompatible with maintaining most native species and communities. Spread of alien species may be a serious ecological problem, particularly on islands. Top predators are or have almost been exterminated.

Endangered. The remaining intact habitat is limited to isolated fragments of varying size (a few larger blocks may be present), with medium to low probabilities of persistence over the next ten to fifteen years without immediate or continuing protection or restoration. Some species already are extirpated because of loss of viable habitat. Remaining habitat fragments do not meet minimum area requirements for most species' populations and large-scale ecological processes. Land use in areas between fragments is largely incompatible with maintaining most native species and communities. Top predators are almost exterminated.

Vulnerable. The remaining intact habitat occurs in habitat blocks ranging from large to small; many intact clusters probably will persist over the next fifteen to twenty years, especially if given adequate protection and moderate restoration. In many areas, some sensitive or exploited species have been extirpated or are declining, particularly top predators, larger primates, and game species. Land use in areas between remaining fragments sometimes is compatible with maintaining most native species and communities.

Relatively stable. Natural communities have been altered in certain areas, causing local declines in exploited populations and disruption of ecosystem processes. These disturbed areas can be extensive but are still patchily distributed relative to the area of intact habitats. Ecological linkages between intact habitat blocks are still largely functional. Guilds of species that are sensitive to human activities, such as top predators, the larger primates, and ground-dwelling birds, are present but at densities below the natural range of variation.

Relatively intact. Natural communities within an ecoregion are largely intact, with species, populations, and ecosystem processes occurring within their natural ranges of variation. Guilds of species that are sensitive to human activities, such as top predators, the larger primates, and ground-dwelling birds, occur at densities within the natural range of variation. Species move and disperse naturally within the ecoregion. Ecological processes fluctuate naturally throughout largely contiguous natural habitats.

We used four variables—habitat loss, size and number of larger habitat blocks, habitat fragmentation, and degree of protection—to evaluate the snapshot conservation status. The snapshot conservation status presents the current condition of the ecoregion, as reflected by the degree of protection, available habitat, and habitat integrity. The status is assessed using the most recent forest cover, habitat, and protected area data (WCMC 1997) for the region.

We believe that some landscape parameters should be given greater weight than others when determining the snapshot conservation status (as described later in this appendix). We gave greatest weight to habitat loss because it is widely recognized as a primary factor in the loss of terrestrial populations, species, and ecosystems. We designed this criterion to emphasize the rapid loss of species predicted to occur when the total area of remaining habitat falls below minimum critical levels. We gave second greatest weight to the number and extent of blocks of contiguous intact habitat. Large blocks of habitat sustain larger species populations, which are more likely to be viable over the long term. Large blocks also permit a broader range of species and ecosystem dynamics to persist. Descriptions of each of the criteria follow.

- Habitat loss   40 percent
- Size and number of larger habitat blocks   25 percent
- Habitat fragmentation   20 percent
- Degree of protection   15 percent

Habitat Loss

We measured habitat loss as a percentage of intact remaining habitat (table D.1). An ecoregion with greater habitat loss received a lesser score relative to an ecoregion with low habitat loss.

In evaluating this criterion, it was necessary to assign land cover types (identifiable from remotely sensed imagery) to habitat integrity levels. These levels were intact, degraded, or heavily altered.

Intact habitat consists of largely undisturbed areas that are characterized by the maintenance of most original ecological processes and by communities containing most of their original suite of native species. Maps of forest cover (original and remaining) were supplied by WCMC (1997).

Degraded habitat consists of areas that are more substantially affected by human disturbance but retain the potential to sustain species and processes or to recover over time through natural succession processes, moderate restoration, and colonization by nearby source pools.

The degraded habitats are categorized as follows:

Secondary scrub (SCR)

Secondary forest (SEC)

Degraded forest (DEG)

Shifting cultivation (ASW)

(The codes [SCR, SEC, etc.] are habitat codes used in the WCMC 1997 maps and database.)

Heavily altered habitat consists of areas that have been degraded to the point of retaining little or no potential value for biodiversity conservation without long-term and extensive restoration. Heavily altered habitats have been almost entirely converted, and the likelihood of natural succession or restoration is low or nonexistent. Exotic species or crops have replaced native species, and surface water patterns have been severely manipulated. Heavily altered habitats are considered areas of habitat loss.

Heavily altered habitats are categorized as follows:

Bare land (BAR)

Cleared land (CL)

Cultivated land (CUL)

Plantations (PLA)

Urban (URB)

Seasonally flooded agriculture (SFL)

Table D.1 gives the index to calculate habitat loss, using the percentages of habitat types within each ecoregion. The points assigned to each ecoregion for habitat loss consist of the sum from the "Intact Habitat" and "Degraded Habitat" columns. For example, if an ecoregion has 60 percent intact habitat and 20 percent degraded habitat, it will get a score of 30 + 3 = 33 points.

Size and Number of Remaining Habitat Blocks

Two important components of conservation status are the number and extent of blocks of intact habitat large enough for populations and ecosystem dynamics to function naturally. Large blocks of habitat have the potential to sustain viable populations, and they permit a broader range of species and ecosystem dynamics to persist. The presence of multiple large blocks has the potential to conserve an even wider range of habitats, environmental gradients, and species ranges.

The number of large habitat blocks is an important component of this criterion. Redundancy theory suggests that the presence of three or more examples of a habitat type significantly increases its probability of long-term persistence. Factors such as fire, disease, pollution, invasion by exotics, deforestation, or degradation can eliminate species or natural habitats within blocks. The presence of several habitat blocks with similar communities allows recolonization of species and the persistence of particular habitat types and species over time. Multiple blocks that are well distributed across the landscape are particularly important for conserving species and habitats in ecoregions that are characterized by a high degree of beta-diversity (species turnover with distance or along environmental gradients).

We measured the number and size of remaining habitat blocks by intersecting the remaining habitat blocks with the ecoregion map and querying the number of habitat blocks and the size of each block within each ecoregion. The analysis was done using GIS.

The threshold size requirements for blocks of habitat are tailored to each biome to reflect the different scales at which ecological processes and disturbances operate. We can consider the ecological needs of tigers as an example. Tigers are a major predator in Asian ecosystems and have a significant influence on community dynamics. Tiger home ranges are correlated with prey density (Sunquist 1981). Because prey organisms usually are more dense and abundant in subtropical and tropical moist forests than in temperate broadleaf forests (Karanth 1995; Miquelle at al. 1999), conservation management of tigers takes larger habitats in temperate broadleaf forests than in moist forests. Therefore, a large habitat block in a temperate broadleaf forest ecoregion scores more points than a similar-sized block in a tropical and subtropical moist forest ecoregion.

The ecoregions vary greatly in area, with several island ecoregions substantially smaller than the continental ecoregions. For this reason we divided ecoregions into two size categories and used a different set of habitat block thresholds for each category. This avoids drawing potentially misleading conclusions by applying continental size thresholds to very small, disjunct, or island ecoregions.

TABLE D.1. Index for Calculating Habitat Loss.

Percentage Remaining Habitat	Score for Percentage	
	Intact Habitat	Degraded Habitat
≥90%	40	10
50–90%	30	8
25–49%	20	5
10–24%	10	3
0–9%	0	0

Alpine bare rock, glaciers, and water bodies have been excluded when calculating ecoregion and remaining habitat areas. Point total = Intact habitat score + Degraded habitat score.

Table D.2 provides the decision rules for assigning a point value to each ecoregion for this criterion. To assign points, read down from the top of the column for the appropriate ecoregion size and add up the numbers of points for each true statement. The highest score for habitat blocks that an ecoregion can receive is 25 points.

Degree of Fragmentation

Habitat fragmentation is widely perceived as a major threat to the conservation of terrestrial species for two key reasons. First, the resulting diminishment and splintering of species populations places many low-density species in demographic jeopardy (Berger 1990; Laurance 1991; Newmark 1991; Wilcove et al. 1986). Second, as fragmentation increases, the amount of unaltered core habitat area decreases, and ecosystems increasingly experience edge effect degradation from hunting pressure, fires from surrounding human activity, changes in microclimates, high levels of predation or parasitism, and invasion of exotic species (Lovejoy 1980; Saunders et al. 1991; Skole and Tucker 1993).

Fragments under 100 km^2 generally are inadequate for maintaining viable populations of most large vertebrates. However, small fragments can be particularly valuable for conserving populations of other species with very localized habitat needs and small ranges, particularly in regions characterized by high levels of beta-diversity. Many invertebrates, plants, fungi, and small vertebrates can be conserved effectively in small blocks of original habitat. More vagile species such as birds and butterflies also may be able to persist as metapopulations in sets of small remaining habitat fragments (Shafer 1995).

Table D.3 describes how we assigned points for the habitat fragmentation criterion. Two people, working independently, each visually scanned a map of remaining habitat in the ecoregion and assigned points according to their evaluation of the fragmentation. There was concordance between these assessment for most of the ecoregions. Where there was

TABLE D.2. Decision Rules for Assigning Points for Size and Number of Habitat Blocks.

Points	Block Size Ecoregion ≥20,000 km^2	Ecoregion ≤20,000 km^2
I. Tropical and subtropical broadleaf forests		
12	1 block ≥5,000 km^2 or 3+ blocks each ≥2,000 km^2	1 block ≥3,000 km^2 or 3+ blocks each ≥1,000 km^2
8	1 block ≥3,000 km^2 or 3+ blocks each ≥1,000 km^2	1 block ≥1,000 km^2
4	1 block ≥2,000 km^2	1 block ≥500 km^2
1	1 block ≥1,000 km^2	1 block ≥250 km^2
0	None ≥1,000 km^2	None ≥250 km^2
II. Temperate broadleaf and conifer forests		
12	1 block ≥4,000 km^2 or 3+ blocks each ≥1,500 km^2	1 block ≥2,500 km^2 or 3+ blocks each ≥800 km^2
8	1 block ≥3,000 km^2 or 3+ blocks each ≥1,000 km^2	1 block ≥800 km^2
4	1 block ≥2,000 km^2	3 blocks each ≥500 km^2
1	1 block ≥1,000 km^2	1 block ≥250 km^2
0	None ≥1,000 km^2	None ≥250 km^2
III. Grasslands, savannas, and shrublands		
12	1 block ≥2,000 km^2 or 3+ blocks each ≥800 km^2	1 block ≥1,000 km^2 or 3+ blocks each ≥500 km^2
8	1 block ≥1,000 km^2	1 block ≥500 km^2
4	1 block ≥500 km^2	1 block ≥250 km^2
1	1 block ≥250 km^2	1 block ≥100 km^2
0	None ≥250 km^2	None ≥100 km^2
IV. Deserts and xeric shrublands		
12	2+ blocks each ≥750 km^2 or 3+ each blocks ≥500 km^2	2+ blocks each ≥500 km^2 or 3+ blocks each ≥300 km^2
8	1 block ≥750 km^2	1 block ≥500 km^2
4	1 block ≥500 km^2	1 block ≥250 km^2
1	1 block ≥250 km^2	1 block ≥100 km^2
0	None ≥250 km^2	None ≥100 km^2

TABLE D.3. Fragmentation Index.

Points	Degree of Habitat Fragmentation
20	*Relatively contiguous:* high connectivity, low fragmentation; long-distance dispersal along elevational and climatic gradients still possible.
16	*Low fragmentation:* high connectivity; more than half of all fragments clustered to some degree (i.e., have some degree of interaction with other habitat blocks).
12	*Medium fragmentation:* intermediate connectivity; fragments display moderate clustering; intervening landscape allows dispersal of many taxa through some parts of the ecoregion.
5	*Advanced fragmentation:* low connectivity; fragments highly isolated; intervening landscape precludes dispersal
1	*High fragmentation:* little to no connectivity of highly disjunct habitat blocks; most fragments small or noncircular; little core habitat because of edge effects; intervening landscapes preclude dispersal for most taxa.

disagreement, a score was arrived at through discussion and mutual agreement.

Degree of Current Protection

This index assesses how well sufficiently large blocks of intact habitat have been brought under some form of protection for biodiversity conservation. We considered only the protected areas under IUCN categories I–IV (Hilton-Taylor 2000).

Ideally, a comprehensive analysis of protected areas would consider the following:

- The degree to which remaining large blocks of habitat are adequately protected within a system of protected areas.
- The level of redundancy of habitat types within protected areas.
- The degree to which representative habitat types, communities, ecological gradients, endangered species, or critical habitats for resident or migratory species are contained within a system of protected areas.
- The degree of connectivity between reserves for the dispersal of species and contiguity of large-scale ecosystem processes (such as migrations).
- The effectiveness of management of protected areas and of the landscapes surrounding them. (Most protected areas currently are too few and too small to encompass complete ecosystems and are effective only if the surrounding landscape is managed intelligently for biodiversity conservation.)

However, at the scale of this study, we considered only the first measure. We strongly suggest that any assessments at smaller scales (i.e., within a given biome or ecoregion) should include all these measures in an analysis of the degree of protection. Because of our inability to assess protected areas fully, the degree of protection criterion is weighted low (15 percent) in the Conservation Status Index analysis.

We overlaid the protected areas coverage (WCMC 1997) on the ecoregion and remaining forest cover coverages. Then we calculated the percentage of each ecoregion's intact habitats that were included in protected areas. We scored the ecoregions using the index provided in table D.4.

TABLE D.4. Index for Evaluating the Degree of Protection.

Percentage of Intact Habitat in the Ecoregion Protected	Points
≥20%	15
10–20%	12
6–9%	9
3–5%	5
1–2%	2
0%	0

Snapshot Conservation Status

The four variables were weighted and combined to provide a Snapshot Conservation Index ranging from 1 to 100. They were then assigned to the six conservation status categories according to their score. Expert opinion determined whether an ecoregion qualified as extinct.

Extinct	expert assessment
Critical	0–20 points
Endangered	21–40 points
Vulnerable	41–70 points
Relatively stable	71–89 points
Relatively intact	≥89 points

Modifying Snapshot Conservation Status with Degree of Threat

An ecoregion that today is largely unaffected by anthropogenic disturbance may be classified as relatively stable or intact under the snapshot assessment, but impending threats might affect that ecoregion in the future. Certain activities in particular, such as large-scale logging, may create such rapid and widespread change that an intact ecoregion could quickly become highly degraded. For this reason, we examined impending threats over then next twenty to forty years to each ecoregion and used that information to modify the snapshot conservation status to produce a threat-modified (or final) conservation status. In essence, the threat-modified conservation status reflects both ongoing and future threats and the urgency for conservation action over the next decades.

Threat analyses are inherently complex because of the many synergistic factors that can impinge, either directly or indirectly, on the ecosystems. The spatial scale of this analysis precludes the evaluation of all possible threats, their interactions, and feedback loops for all ecoregions. For this analysis we considered two major types of threats on the habitat: degradation threats and conversion threats. These threats were assessed independently of each other for each ecoregion. We strongly suggest that conservation assessments at smaller spatial scales (i.e., where the region of analysis is a specific habitat type or landscape within an ecoregion) consider a more comprehensive threat assessment.

Degradation Threats

Degradation threats are chronic, low-intensity anthropogenic impacts, such as forest exploitation by communities for natural resources; small-scale clearing for agriculture, plantations, or villages; and wildlife exploitation.

We used human population density (people/km^2, according to the Center for International Earth Science Information Network's gridded population of the world, prepared by the National Center for Geographic Information and Analysis [NCGIA]) as a proxy variable to measure degradation threats.

We conducted the analysis using GIS. Because the population data were in grid format, we first buffered the remaining habitat data (WCMC 1997) by 5 km and then converted them to grid format. Five kilometers was chosen because much of local communities' fuel wood comes from within this distance. These converted habitat data were then used to capture the population data within that area. Using this coverage, we calculated the average population density within remaining habitat blocks and the 5-km buffer zones and the standard deviation for each ecoregion.

We calculated the population densities for all habitat blocks greater than 250 km^2 for ecoregions less than 20,000 km^2 in total size and for forest blocks greater than 1,000 km^2 for ecoregions greater than 20,000 km^2.

TABLE D.5. Index for Assessing Threats from Logging Concessions.

Size of Habitat Blocks Analyzed	Percentage of Habitat Blocks Affected	Level of Threat
I. Large ecoregions ≥20,000 km²		
≥500 km²	≤10%	Low
≥500 km²	10–50%	Medium
≥500 km²	≥50%	High
II. Small ecoregions ≤20,000 km²		
≥300 km²	≤10%	Low
≥300 km²	10–40%	Medium
≥300 km²	≥40%	High

We then used the population density and standard deviation results to evaluate each ecoregion for the level of degradation threat. We considered population density to reflect potential threat from the number of people using the forest and the standard deviation as a measure of concentration of the threats. A high standard deviation would suggest that the population is largely clustered in its distribution, and thus the threats would also be concentrated in a few areas, whereas a low standard deviation would indicate that the population is more evenly distributed, and so would be the threats.

Our next task was to choose cut-off points to generate levels of population density for our analysis. We considered a family unit to consist of eight people. We then considered the average size of a village to be ten houses. Thus, the average size of a village would be eighty people. Therefore, it follows that

1–8 people/km² = 1 household/km²
(sparse population density)

9–80 people/km² = >1 household to 1 village/km²
(low population density)

81–160 people/km² = >1 village to 2 villages/km²
(medium population density)

>161 people/km² = >2 villages/km²
(high population density)

We then mapped the population distributions in all habitat blocks. The standard deviations of population estimates for the ecoregions were then plotted; based on this distribution, we chose a standard deviation of fifty as being the cut-off point for clustered versus even distribution.

All ecoregions were then categorized according to the average population density and standard deviation as follows:

1 = sparse population distribution

2 = low population density, clustered distribution

3 = low population density, even distribution

4 = medium population density, clustered distribution

5 = medium population density, even distribution

6 = high population density

7 = high population density but no large blocks of intact remaining habitat

Snapshot conservation status levels of ecoregions in categories 5, 6, or 7 were elevated by one threat category (e.g., from endangered to critical); those of ecoregions in the remaining categories were not changed.

Conversion Threats

Conversion threats are high-impact threats from large-scale logging concessions, plantations, hydropower schemes, and large commercial developments. However, we analyzed only large-scale logging concessions in several countries where we were able to gather this data.

We used logging concession maps (ongoing and planned) for Cambodia, Indonesia, and New Guinea to determine threats to large habitat blocks. If these threat sources were adjacent to or within the large habitat blocks, the threat to the ecoregion was assessed using the index in table D.5.

Because large-scale logging poses catastrophic threats, any ecoregion assessed as having a high conversion threat was elevated to critical status regardless of its snapshot conservation status.

APPENDIX E

Summary of Ecoregion Scores for Species Richness, Endemism, and Biological Distinctiveness Index

Indian Subcontinent Bioregion

Ecoregion Number	Ecoregion	Ecoregion Size (km²)	Biome[a]	Mammal Richness	Mammal Endemism	Bird Richness	Bird Endemism	Plant Richness[b]	Plant Endemism[c]	Richness Index[d]	Endemism Index[d]	Richness and Endemism Index[d]	Phenomena[e]	Biological Distinctiveness Index[d]
1	North Western Ghats Montane Rain Forests	30,800	1	89	2	327	8	1,750	1	RO	BO	RO	No	RO
2	South Western Ghats Montane Rain Forests	22,500	1	96	10	309	13	1,540	2	BO	GO	GO	Intact	GO
3	Sri Lanka Lowland Rain Forests	12,500	1	75	8	229	18	2,000	3	BO	GO	GO	Radiation	GO
4	Sri Lanka Montane Rain Forests	3,000	1	76	13	179	20	2,000	4	BO	GO	GO	Radiation	GO
5	Orissa Semi-Evergreen Forests	22,200	1	59	0	215	0	900	1	BO	LI	BO	No	BO
6	East Deccan Dry Evergreen Forests	25,400	2	66	0	230	0	1,000	1	BO	LI	BO	No	RO
7	Sri Lanka Dry-Zone Dry Evergreen Forests	48,200	2	74	4	270	9	1,000	1	BO	RO	RO	No	RO
8	Brahmaputra Valley Semi-Evergreen Rain Forests	56,600	1	122	2	372	2	850	1	RO	BO	BO	No	BO
9	Sundarbans Freshwater Swamp Forests	16,400	1	55	0	189	0	1,000	1	BO	LI	RO	No	RO
10	Lower Gangetic Plains Moist Deciduous Forests	253,500	1	126	1	383	5	2,000	1	RO	BO	BO	No	RO
11	Meghalaya Subtropical Forests	41,600	1	113	0	453	1	1,100	1	GO	LI	GO	No	GO
12	Eastern Highlands Moist Deciduous Forests	340,200	1	81	1	313	1	750	1	BO	LI	BO	Intact	BO
13	Malabar Coast Moist Deciduous Forests	35,300	1	97	6	280	1	1,150	1	BO	BO	BO	No	BO
14	North Western Ghats Moist Deciduous Forests	48,000	1	88	0	346	5	1,400	2	RO	RO	RO	No	RO
15	South Western Ghats Moist Deciduous Forests	23,700	1	90	6	323	9	1,500	2	RO	BO	RO	No	RO
16	Khathiarbar-Gir Dry Deciduous Forests	266,400	2	81	0	299	1	1,200	1	RO	LI	RO	No	BO
17	Upper Gangetic Plains Moist Deciduous Forests	262,600	1	81	0	293	0	1,800	1	BO	LI	BO	No	RO
18	Chhota-Nagpur Dry Deciduous Forests	122,100	2	77	0	278	0	1,200	1	RO	LI	RO	No	BO
19	Northern Dry Deciduous Forests	58,200	2	68	0	261	0	1,000	1	BO	LI	BO	No	BO
20	Narmada Valley Dry Deciduous Forests	169,500	2	78	0	276	0	1,000	1	BO	LI	BO	No	BO
21	Central Deccan Plateau Dry Deciduous Forests	239,400	2	82	0	298	1	700	1	BO	LI	BO	No	BO
22	South Deccan Plateau Dry Deciduous Forests	81,900	2	76	1	261	2	800	1	BO	LI	BO	No	BO
23	Deccan Thorn Scrub Forests	339,000	13	96	3	345	4	1,000	1	RO	BO	RO	No	LI
24	Northwestern Thorn Scrub Forests	487,500	13	90	1	405	2	1,000	1	RO	LI	RO	No	LI
25	Himalayan Subtropical Broadleaf Forests	38,100	1	97	1	343	1	1,500	1	RO	LI	RO	No	RO
26	Eastern Himalayan Broadleaf Forests	82,900	4	128	5	490	11	2,000	2	GO	GO	GO	No	GO
27	Western Himalayan Broadleaf Forests	55,800	4	76	2	315	9	1,200	1	RO	BO	RO	No	RO
28	Eastern Himalayan Sub-Alpine Conifer Forests	26,900	5	89	3	202	5	1,500	2	GO	RO	GO	No	GO
29	Western Himalayan Sub-Alpine Conifer Forests	39,400	5	58	1	285	9	1,000	1	RO	RO	RO	No	BO
30	East Afghan Montane Coniferous Forests	7,300	5	22	0	123	0	500	1	BO	LI	BO	No	BO
31	Himalayan Subtropical Pine Forests	76,200	3	121	3	480	11	1,000	1	GO	RO	RO	No	RO
32	Indus River Delta–Arabian Sea Mangroves	5,800	14	*	*	123	*	1,000	1	*	*	RO	No	RO
33	Sundarbans Mangroves	20,400	14	*	*	174	*	900	1	*	*	GO	No	GO
34	Godavari-Krishna Mangroves	6,800	14	*	*	142	*	500	1	*	*	RO	No	RO
35	Terai-Duar Savanna and Grasslands	34,500	7	94	2	376	3	600	1	GO	BO	GO	Intact	GO
36	Rann of Kutch Seasonal Salt Marsh	27,800	9	50	1	205	0	1,200	1	GO	LI	GO	No	GO

220

#	Name												
	Northwestern Himalayan Alpine												
37	Shrub and Meadows	46,900	10	50	0	139	1,500	1	RO	LI	RO	No	RO
38	Western Himalayan Alpine Shrub and Meadows	34,600	10	37	0	127	2,000	2	RO	LI	RO	No	RO
39	Eastern Himalayan Alpine Shrub and Meadows	40,100	10	63	1	115	7,000	3	GO	BO	GO	No	GO
40	Karakoram–West Tibetan Plateau Alpine Steppe	112,600	10	36	1	122	1,500	2	RO	BO	BO	No	RO
41	Central Tibetan Plateau Alpine Steppe	13,600	10	34	0	122	1,000	2	BO	LI	BO	No	BO
42	Sulaiman Range Alpine Meadows	19,600	10	40	1	171	300	2	RO	BO	BO	No	RO
43	South Iran Nubo-Sindian Desert and Semi-Desert	40,500	13	34	0	86	1,500	1	BO	LI	BO	No	BO
44	Baluchistan Xeric Woodlands	255,200	13	67	1	348	500	1	BO	BO	BO	No	BO
45	Rajasthan–North Pakistan Sandy Desert	68,300	13	28	2	44	1,000	2	LI	LI	BO	No	BO
46	Thar Desert	238,300	13	63	0	141	1,000	1	BO	LI	BO	No	BO
47	Indus Valley Desert	19,500	13	36	0	190	1,000	1	BO	LI	RO	No	RO

Indochina Bioregion

#	Name												
48	Chin Hills–Arakan Yoma Montane Rain Forests	29,600	1	102	0	414	2,000	2	RO	BO	RO	No	RO
49	Myanmar Coastal Rain Forests	65,200	1	124	2	357	2,000	1	RO	LI	RO	No	RO
50	Mizoram-Manipur-Kachin Rain Forests	135,200	1	149	2	580	2,500	1	GO	BO	GO	No	GO
51	Kayah-Karen Montane Rain Forests	119,200	1	168	1	568	3,000	1	GO	LI	GO	No	GO
52	Luang Prabang Montane Rain Forests	71,600	1	131	0	541	2,000	3	RO	BO	RO	No	RO
	Tenasserim–South Thailand												
53	Semi-Evergreen Rain Forests	96,900	1	178	3	560	3,500	2	GO	BO	GO	No	GO
54	Northern Annamites Rain Forests	47,100	1	106	5	524	2,500	2	GO	RO	GO	No	GO
55	Cardamom Mountains Rain Forests	44,100	1	103	0	324	1,800	2	RO	BO	RO	Intact	GO
56	Northern Vietnam Lowland Rain Forests	22,500	1	110	1	306	2,500	1	RO	LI	RO	No	RO
57	Southern Annamites Montane Rain Forests	46,200	1	110	3	414	2,500	3	GO	RO	GO	No	GO
58	Southern Vietnam Lowland Dry Forests	34,900	2	93	1	320	1,500	1	RO	LI	RO	No	RO
59	Southeastern Indochina Dry Evergreen Forests	123,800	2	132	4	454	2,500	1	GO	BO	GO	No	GO
60	Andaman Islands Rain Forests	5,600	1	27	7	123	2,500	2	BO	RO	BO	No	RO
61	Nicobar Islands Rain Forests	1,700	1	21	6	82	2,000	2	LI	RO	RO	No	RO
62	Tonle Sap–Mekong Peat Swamp Forests	29,300	1	79	0	301	1,000	1	BO	LI	BO	No	BO
63	Irrawaddy Freshwater Swamp Forests	15,100	1	53	0	269	1,000	1	BO	BO	BO	No	BO
64	Chao Phraya Freshwater Swamp Forests	38,900	1	123	3	307	1,000	1	BO	LI	BO	No	BO
65	Tonle Sap Freshwater Swamp Forests	25,900	1	84	0	299	1,000	1	BO	LI	BO	No	BO
66	Red River Freshwater Swamp Forests	10,700	1	77	0	276	1,000	1	BO	LI	BO	No	RO
67	Irrawaddy Moist Deciduous Forests	137,900	1	118	0	342	1,500	1	RO	LI	RO	No	RO
68	Northern Thailand–Laos Moist Deciduous Forests	42,000	1	134	0	379	1,200	1	RO	LI	RO	No	RO
69	Northern Khorat Plateau Moist Deciduous Forests	16,800	1	103	0	390	1,500	1	RO	LI	RO	No	RO
70	Chao Phraya Lowland Moist Deciduous Forests	20,300	1	137	1	368	1,500	1	RO	LI	BO	No	RO
71	Irrawaddy Dry Forests	35,000	2	107	1	338	1,000	2	RO	BO	RO	No	RO
72	Central Indochina Dry Forests	319,000	2	165	2	491	1,200	1	GO	BO	GO	No	GO
73	Northern Triangle Subtropical Forests	55,600	1	138	8	372	2,000	2	GO	RO	GO	No	GO
74	Northern Indochina Subtropical Forests	288,300	1	183	9	707	2,000	2	GO	RO	GO	No	GO
	South China–Vietnam Subtropical												
75	Evergreen Forests	37,500	1	102	2	602	2,000	2	GO	BO	GO	No	GO
76	Northern Triangle Temperate Forests	10,700	4	90	0	365	1,500	2	GO	LI	GO	No	GO
77	Northeast India–Myanmar Pine Forests	9,700	3	71	0	286	1,000	1	RO	LI	RO	No	RO

(continues)

Ecoregion	Ecoregion Number	Ecoregion Size (km²)	Biome[a]	Mammal Richness	Mammal Endemism	Bird Richness	Bird Endemism	Plant Richness[b]	Plant Endemism[c]	Richness Index[d]	Endemism Index[d]	Richness and Endemism Index[d]	Phenomena[e]	Biological Distinctiveness Index[d]
Myanmar Coastal Mangroves	78	21,900	14	*	*	185	*	200	1	*	*	RO	No	RO
Indochina Mangroves	79	26,800	14	*	*	299	*	200	1	*	*	RO	No	RO
Sunda Shelf and Philippines Bioregion														
Peninsular Malaysian Rain Forests	80	125,000	1	193	0	452	0	6,000	2	GO	BO	RO	No	GO
Peninsular Malaysian Montane Rain Forests	81	17,100	1	175	1	258	5	3,000	3	RO	BO	GO	No	RO
Sumatran Lowland Rain Forests	82	258,300	1	159	1	463	5	6,000	2	GO	BO	RO	No	GO
Sumatran Montane Rain Forests	83	72,600	1	150	9	308	29	4,000	2	GO	GO	GO	No	GO
Mentawai Islands Rain Forests	84	6,500	1	44	17	212	3	1,500	1	BO	RO	RO	No	RO
Sumatran Peat Swamp Forests	85	87,100	1	130	0	391	0	1,000	2	RO	LI	RO	No	RO
Borneo Peat Swamp Forests	86	67,100	1	153	5	396	1	1,000	2	GO	BO	GO	No	GO
Peninsular Malaysian Peat Swamp Forests	87	3,600	1	183	0	381	0	1,000	2	GO	LI	RO	No	RO
Sumatran Freshwater Swamp Forests	88	18,000	1	130	0	391	0	1,500	2	RO	LI	RO	No	RO
Southern Borneo Freshwater Swamp Forests	89	36,600	1	125	0	361	1	1,500	2	RO	LI	RO	No	RO
Sundaland Heath Forests	90	76,200	1	105	1	353	1	1,500	2	RO	LI	RO	No	RO
Western Java Rain Forests	91	41,400	1	98	5	355	9	2,000	2	RO	BO	RO	No	RO
Eastern Java–Bali Rain Forests	92	53,500	1	99	2	346	10	2,000	2	RO	BO	GO	No	GO
Western Java Montane Rain Forests	93	26,200	1	63	14	235	30	1,000	2	BO	BO	BO	No	BO
Eastern Java–Bali Montane Rain Forests	94	15,800	1	96	1	213	18	1,000	2	BO	GO	BO	No	GO
Borneo Montane Rain Forests	95	115,100	1	147	11	256	28	5,000	3	GO	GO	GO	No	GO
Borneo Lowland Rain Forests	96	425,400	1	193	12	467	17	10,000	2	RO	GO	RO	No	RO
Palawan Rain Forests	97	13,900	1	54	15	246	20	2,600	2	RO	GO	GO	No	GO
Luzon Rain Forests	98	94,800	1	78	15	365	40	4,000	2	BO	GO	GO	No	GO
Mindoro Rain Forests	99	10,300	1	42	8	256	11	2,500	2	BO	GO	GO	No	GO
Greater Negros–Panay Rain Forests	100	34,200	1	58	13	300	23	3,000	2	RO	GO	BO	No	BO
Mindanao–Eastern Visayas Rain Forests	101	105,000	1	77	16	334	36	4,000	3	GO	GO	GO	No	GO
Luzon Montane Rain Forests	102	8,300	1	34	9	163	28	2,000	2	LI	GO	RO	No	RO
Mindanao Montane Rain Forests	103	17,800	1	42	14	167	34	2,500	3	BO	GO	GO	No	GO
Sulu Archipelago Rain Forest	104	2,000	1	18	1	184	10	1,000	2	LI	BO	BO	No	BO
Sumatran Tropical Pine Forests	105	2,700	3	100	1	166	12	1,000	1	RO	RO	RO	No	RO
Luzon Tropical Pine Forests	106	7,100	3	40	16	149	23	1,000	2	BO	GO	GO	No	GO
Sunda Shelf Mangroves	107	37,200	14	*	*	267	*	200	2	*	*	GO	No	GO
Kinabalu Montane Alpine Meadows	108	4,300	10	113	16	184	25	3,500	4	GO	GO	GO	Rare	GO
Wallacea Bioregion														
Sulawesi Lowland Rain Forests	109	114,100	1	104	28	337	70	4,000	2	RO	GO	GO	No	GO
Sulawesi Montane Rain Forests	110	75,500	1	102	33	168	44	2,000	2	BO	GO	GO	No	GO
Lesser Sundas Deciduous Forests	111	39,300	2	50	5	273	29	1,500	2	BO	GO	GO	No	GO

112 Timor and Wetar Deciduous Forests	33,300	2	38	5	229	35	1,500	1	BO	GO	GO	No	GO
113 Sumba Deciduous Forests	10,700	2	17	0	180	12	1,000	1	LI	BO	BO	No	BO
114 Halmahera Rain Forests	26,800	1	38	8	223	43	2,000	1	BO	GO	GO	No	GO
115 Buru Rain Forests	8,600	1	25	4	178	29	1,500	1	BO	BO	BO	No	BO
116 Seram Rain Forests	19,300	1	38	9	213	33	2,000	2	BO	BO	GO	No	GO
117 Banda Sea Islands Moist Deciduous Forests	7,500	1	22	3	225	43	1,500	1	BO	RO	RO	No	RO
New Guinea and Melanesia Bioregion													
118 Vogelkop Montane Rain Forests	21,900	1	42	7	354	20	5,500	3	GO	RO	GO	No	GO
119 Vogelkop-Aru Lowland Rain Forests	77,400	1	47	8	366	21	5,500	3	GO	GO	GO	No	GO
120 Biak-Numfoor Rain Forests	2,900	1	29	5	107	13	1,500	3	LI	RO	RO	No	RO
121 Yapen Rain Forests	2,300	1	37	1	147	2	1,500	3	LI	BO	BO	No	BO
122 Northern New Guinea Montane Rain Forests	23,200	1	51	6	264	12	5,500	3	RO	RO	RO	No	RO
123 Northern New Guinea Lowland Rain and Freshwater Swamp Forests	134,700	1	75	12	359	17	1,500	1	RO	RO	RO	No	RO
124 Huon Peninsula Montane Rain Forests	16,400	1	80	6	370	16	5,500	3	GO	GO	GO	No	GO
125 Central Range Montane Rain Forests	171,200	1	90	44	348	55	7,000	4	RO	RO	RO	No	GO
126 Southeastern Papuan Rain Forests	77,000	1	138	28	509	40	4,000	3	GO	GO	GO	No	GO
127 Southern New Guinea Freshwater Swamp Forests	99,400	1	50	6	339	11	2,000	3	BO	BO	RO	No	RO
128 Southern New Guinea Lowland Rain Forests	122,300	1	69	13	344	5	7,000	3	RO	GO	RO	Intact	GO
129 New Guinea Mangroves	26,700	14	*	*	306	*	300	3	*	*	GO	No	GO
130 Trans Fly Savanna and Grasslands	26,600	7	44	5	306	5	2,000	1	RO	GO	GO	No	GO
131 Central Range Sub-Alpine Grasslands	15,500	10	9	4	84	28	1,500	4	BO	GO	GO	Rare	GO
132 Admiralty Islands Lowland Rain Forests	2,100	1	18	5	64	13	2,500	3	BO	RO	RO	No	RO
133 New Britain–New Ireland Lowland Rain Forests	34,700	1	47	9	185	55	2,500	3	BO	BO	GO	No	GO
134 New Britain–New Ireland Montane Rain Forests	12,100	1	45	8	103	30	2,500	3	BO	BO	GO	No	GO
135 Trobriand Islands Rain Forests	4,200	1	38	9	136	2	3,000	3	BO	RO	RO	No	RO
136 Louisiade Archipelago Rain Forests	1,600	1	24	3	100	7	3,000	3	BO	BO	BO	No	BO
137 Solomon Islands Rain Forests	35,700	1	47	26	198	89	2,000	3	BO	BO	GO	No	GO
138 Vanuatu Rain Forests	13,200	1	12	6	79	30	1,000	2	LI	LI	GO	No	GO
139 New Caledonia Rain Forests	14,500	1	9	6	90	31	2,000	4	LI	GO	GO	Unique	GO
140 New Caledonia Dry Forests	4,400	2	8	5	79	23	1,500	4	BO	GO	GO	Unique	GO

[a] 1 = tropical and subtropical moist broadleaf forests; 2 = tropical and subtropical dry broadleaf forests; 3 = tropical and subtropical coniferous forests; 4 = temperate broadleaf and mixed forests; 5 = temperate coniferous forests; 7 = tropical and subtropical grasslands, savannas, and shrublands; 9 = flooded grasslands and savannas; 10 = montane grasslands and shrublands; 13 = deserts and xeric shrublands; 14 = mangroves.

[b] Plant richness based on expert estimates.

[c] 1 = 0–5%; 2 = 6–25%; 3 = 26–75%; 4 = 76–100%.

[d] GO = globally outstanding; RO = regionally outstanding; BO = bioregionally outstanding; LI = locally important

[e] No = no globally outstanding phenomena; Intact = intact predator-prey systems and habitat; Radiation = globally outstanding species radiations; Rare = rare biome; Unique = unique evolutionary or ecological phenomena.

*Mangrove ecoregions were not analyzed for mammals or bird endemism.

APPENDIX F

Summary of Ecoregion Scores for the Conservation Status Index and Integration Matrix

Ecoregion	Ecoregion Number	Ecoregion Size (km²)	Biome[a]	Total Habitat Loss Points[b]	Block Score[c]	Fragmentation Score[d]	Protection Score[e]	Conservation Status Score	Snapshot Conservation Status	Population Threat[f]	Logging Concessions Threat	Final Conservation Status	Biodiversity Index[g]	Conservation Class (Integration)
Indian Subcontinent Bioregion														
North Western Ghats Montane Rain Forests	1	30,800	1	20	5	12	12	49	Vulnerable	6	Not evaluated	Endangered	RO	II
South Western Ghats Montane Rain Forests	2	22,500	1	20	8	12	12	52	Vulnerable	6	Not evaluated	Endangered	GO	I
Sri Lanka Lowland Rain Forests	3	12,500	1	0	1	5	15	21	Critical	6	Not evaluated	Critical	GO	I
Sri Lanka Montane Rain Forests	4	3,000	1	10	1	5	15	31	Endangered	6	Not evaluated	Critical	GO	I
Orissa Semi-Evergreen Forests	5	22,200	1	0	0	1	0	1	Critical	6	Not evaluated	Critical	BO	IV
East Deccan Dry Evergreen Forests	6	25,400	2	0	0	1	2	3	Critical	7	Not evaluated	Critical	BO	IV
Sri Lanka Dry-Zone Dry Evergreen Forests	7	48,200	2	20	13	12	15	60	Vulnerable	3	Not evaluated	Vulnerable	RO	II
Brahmaputra Valley Semi-Evergreen Rain Forests	8	56,600	1	10	1	5	12	28	Endangered	6	Not evaluated	Critical	RO	II
Sundarbans Freshwater Swamp Forests	9	16,400	1	0	0	1	0	1	Critical	7	Not evaluated	Critical	BO	IV
Lower Gangetic Plains Moist Deciduous Forests	10	253,500	1	0	8	1	15	24	Vulnerable	6	Not evaluated	Vulnerable	RO	II
Meghalaya Subtropical Forests	11	41,600	1	20	21	12	15	68	Vulnerable	2	Not evaluated	Vulnerable	GO	I
Eastern Highlands Moist Deciduous Forests	12	340,200	1	20	24	12	12	68	Vulnerable	6	Not evaluated	Endangered	GO	I
Malabar Coast Moist Deciduous Forests	13	35,300	1	0	0	1	12	13	Critical	6	Not evaluated	Critical	BO	IV
North Western Ghats Moist Deciduous Forests	14	48,000	1	10	1	5	12	28	Endangered	6	Not evaluated	Critical	RO	II
South Western Ghats Moist Deciduous Forests	15	23,700	1	20	8	5	15	48	Vulnerable	6	Not evaluated	Endangered	RO	II
Khathiarbar-Gir Dry Deciduous Forests	16	266,400	2	10	20	5	12	47	Vulnerable	6	Not evaluated	Endangered	RO	II
Upper Gangetic Plains Moist Deciduous Forests	17	262,600	1	0	4	1	12	17	Critical	6	Not evaluated	Critical	BO	IV
Chhota-Nagpur Dry Deciduous Forests	18	122,100	2	20	24	16	9	69	Vulnerable	6	Not evaluated	Endangered	RO	II
Northern Dry Deciduous Forests	19	58,200	2	10	25	5	3	43	Vulnerable	5	Not evaluated	Endangered	BO	IV
Narmada Valley Dry Deciduous Forests	20	169,500	2	20	17	16	9	62	Vulnerable	6	Not evaluated	Endangered	BO	IV
Central Deccan Plateau Dry Deciduous Forests	21	239,400	2	10	20	12	3	45	Vulnerable	6	Not evaluated	Endangered	BO	IV
South Deccan Plateau Dry Deciduous Forests	22	81,900	2	10	16	5	3	34	Endangered	6	Not evaluated	Endangered	BO	IV
Deccan Thorn Scrub Forests	23	339,000	13	0	20	5	3	28	Endangered	6	Not evaluated	Critical	LI	IV
Northwestern Thorn Scrub Forests	24	487,500	13	0	0	1	3	4	Critical	6	Not evaluated	Critical	LI	IV
Himalayan Subtropical Broadleaf Forests	25	38,100	1	25	1	1	12	39	Endangered	4	Not evaluated	Endangered	RO	II
Eastern Himalayan Broadleaf Forests	26	82,900	4	33	20	20	12	85	Relatively Intact	2	Not evaluated	Relatively Intact	GO	III
Western Himalayan Broadleaf Forests	27	55,800	4	20	1	5	3	29	Endangered	2	Not evaluated	Endangered	RO	II
Eastern Himalayan Sub-Alpine Conifer Forests	28	26,900	5	30	1	12	15	58	Vulnerable	3	Not evaluated	Vulnerable	GO	I
Western Himalayan Sub-Alpine Conifer Forests	29	39,400	5	20	0	12	9	41	Vulnerable	3	Not evaluated	Vulnerable	RO	II
East Afghan Montane Coniferous Forests	30	7,300	5	20	5	5	12	42	Vulnerable	2	Not evaluated	Vulnerable	BO	V
Himalayan Subtropical Pine Forests	31	76,200	3	30	24	12	3	69	Vulnerable	4	Not evaluated	Vulnerable	RO	II
Indus River Delta–Arabian Sea Mangroves	32	5,800	14	10	0	5	0	15	Critical	7	Not evaluated	Critical	RO	II
Sundarbans Mangroves	33	20,400	14	20	13	16	15	64	Vulnerable	6	Not evaluated	Endangered	GO	I
Godavari-Krishna Mangroves	34	6,800	14	10	0	5	12	27	Endangered	6	Not evaluated	Critical	RO	II
Terai-Duar Savanna and Grasslands	35	34,500	7	13	0	1	12	26	Endangered	6	Not evaluated	Critical	GO	I
Rann of Kutch Seasonal Salt Marsh	36	27,800	9	40	12	12	15	79	Relatively Stable	3	Not evaluated	Relatively Stable	GO	III

Northwestern Himalayan Alpine Shrub and Meadows	37	46,900	10	30	24	16	12	82	Relatively Stable	2	Not evaluated	Relatively Stable	RO	III
Western Himalayan Alpine Shrub and Meadows	38	34,600	10	30	25	12	12	79	Relatively Stable	3	Not evaluated	Relatively Stable	RO	III
Eastern Himalayan Alpine Shrub and Meadows	39	40,100	10	40	5	16	15	76	Relatively Stable	3	Not evaluated	Relatively Stable	GO	III
Karakoram–West Tibetan Plateau Alpine Steppe	40	112,600	10	20	12	16	15	63	Vulnerable	3	Not evaluated	Vulnerable	RO	II
Central Tibetan Plateau Alpine Steppe	41	13,600	10	40	12	16	3	71	Relatively Stable	3	Not evaluated	Relatively Stable	BO	V
Sulaiman Range Alpine Meadows	42	19,600	10	40	17	20	2	79	Relatively Stable	3	Not evaluated	Relatively Stable	RO	III
South Iran Nubo-Sindian Desert and Semi-Desert	43	40,500	13	0	1	5	0	6	Critical	3	Not evaluated	Critical	BO	IV
Baluchistan Xeric Woodlands	44	255,200	13	0	1	1	9	11	Critical	4	Not evaluated	Critical	BO	IV
Rajasthan–North Pakistan Sandy Desert	45	68,300	13	20	21	5	5	51	Vulnerable	1	Not evaluated	Vulnerable	BO	V
Thar Desert	46	238,300	13	30	12	12	2	56	Vulnerable	4	Not evaluated	Vulnerable	BO	V
Indus Valley Desert	47	19,500	13	40	12	12	15	79	Relatively Stable	6	Not evaluated	Vulnerable	BO	V
Indochina Bioregion														
Chin Hills–Arakan Yoma Montane Rain Forests	48	29,600	1	35	20	20	0	75	Relatively Stable	3	Not evaluated	Relatively Stable	RO	III
Myanmar Coastal Rain Forests	49	65,200	1	23	24	16	0	63	Vulnerable	4	Not evaluated	Vulnerable	RO	II
Mizoram-Manipur-Kachin Rain Forests	50	135,200	1	25	24	12	3	64	Vulnerable	3	Not evaluated	Vulnerable	GO	I
Kayah-Karen Montane Rain Forests	51	119,200	1	33	24	20	15	92	Relatively Intact	3	Not evaluated	Relatively Intact	GO	III
Luang Prabang Montane Rain Forests	52	71,600	1	28	5	5	15	53	Vulnerable	3	Not evaluated	Vulnerable	RO	II
Tenasserim–South Thailand Semi-Evergreen Rain Forests	53	96,900	1	23	24	20	12	79	Relatively Stable	3	Not evaluated	Relatively Stable	GO	III
Northern Annamites Rain Forests	54	47,100	1	35	8	12	15	70	Relatively Stable	2	Not evaluated	Relatively Stable	GO	III
Cardamom Mountains Rain Forests	55	44,100	1	30	20	20	15	85	Relatively Intact	3	medium	Relatively Intact	GO	III
Northern Vietnam Lowland Rain Forests	56	22,500	1	0	0	1	15	16	Critical	4	Not evaluated	Critical	RO	II
Southern Annamites Montane Rain Forests	57	46,200	1	25	1	5	12	43	Vulnerable	2	Not evaluated	Vulnerable	GO	I
Southern Vietnam Lowland Dry Forests	58	34,900	1	0	0	1	3	4	Critical	2	Not evaluated	Critical	RO	II
Southeastern Indochina Dry Evergreen Forests	59	123,800	2	23	24	20	15	82	Relatively Stable	3	High	Critical	GO	I
Andaman Islands Rain Forests	60	5,600	1	30	13	16	0	59	Vulnerable	3	Not evaluated	Vulnerable	RO	II
Nicobar Islands Rain Forests	61	1,700	1	30	1	12	0	43	Vulnerable	7	Not evaluated	Endangered	RO	II
Tonle Sap–Mekong Peat Swamp Forests	62	29,300	1	10	0	1	2	13	Critical	5	Not evaluated	Critical	BO	IV
Irrawaddy Freshwater Swamp Forests	63	15,100	1	0	0	1	0	1	Critical	7	Not evaluated	Critical	BO	IV
Chao Phraya Freshwater Swamp Forests	64	38,900	1	0	0	1	0	1	Critical	5	Not evaluated	Critical	BO	IV
Tonle Sap Freshwater Swamp Forests	65	25,900	1	20	12	20	15	67	Vulnerable	4	Not evaluated	Vulnerable	BO	V
Red River Freshwater Swamp Forests	66	10,700	1	0	0	1	12	13	Critical	7	Not evaluated	Critical	BO	IV
Irrawaddy Moist Deciduous Forests	67	137,900	1	20	24	16	3	63	Vulnerable	2	Not evaluated	Vulnerable	RO	II
Northern Thailand–Laos Moist Deciduous Forests	68	42,000	1	23	24	12	9	68	Vulnerable	3	Not evaluated	Vulnerable	BO	II
Northern Indochina Moist Deciduous Forests	69	16,800	1	13	1	5	3	22	Critical	2	Not evaluated	Critical	RO	II
Chao Phraya Lowland Moist Deciduous Forests	70	20,300	1	10	1	12	15	38	Endangered	3	Not evaluated	Endangered	BO	II
Irrawaddy Dry Forests	71	35,000	2	10	1	1	0	12	Critical	2	Not evaluated	Critical	BO	II
Central Indochina Dry Forests	72	319,000	2	20	24	5	12	61	Vulnerable	3	medium	Vulnerable	GO	I
Northern Triangle Subtropical Forests	73	55,600	1	40	20	20	3	83	Relatively Stable	1	Not evaluated	Relatively Stable	GO	III
South China–Vietnam Subtropical Evergreen Forests	74	288,300	1	25	20	12	2	59	Vulnerable	2	Not evaluated	Vulnerable	GO	I
	75	37,500	1	10	1	5	12	28	Endangered	4	Not evaluated	Endangered	GO	I
Northern Triangle Temperate Forests	76	10,700	4	40	25	20	2	87	Relatively Intact	3	Not evaluated	Relatively Intact	GO	III
Northeast India–Myanmar Pine Forests	77	9,700	3	20	5	5	0	30	Endangered	3	Not evaluated	Endangered	RO	II

(continues)

Ecoregion	Ecoregion Number	Ecoregion Size (km²)	Biome[a]	Total Habitat Loss Points[b]	Block Score[c]	Fragmentation Score[d]	Protection Score[e]	Conservation Status Score	Snapshot Conservation Status	Population Threat[f]	Logging Concessions Threat	Final Conservation Status	Biodiversity Index[g]	Conservation Class (Integration)
Myanmar Coastal Mangroves	78	21,900	14	20	0	5	2	27	Endangered	2	Not evaluated	Endangered	GO	I
Indochina Mangroves	79	26,800	14	10	0	5	12	27	Endangered	3	Not evaluated	Endangered	GO	I
Sunda Shelf and Philippines Bioregion														
Peninsular Malaysian Rain Forests	80	125,000	1	13	13	12	12	50	Vulnerable	4	Not evaluated	Vulnerable	GO	I
Peninsular Malaysian Montane Rain Forests	81	17,100	1	30	17	20	12	79	Relatively Stable	4	Not evaluated	Relatively Stable	RO	III
Sumatran Lowland Rain Forests	82	258,300	1	10	24	5	9	48	Endangered	2	High	Critical	GO	I
Sumatran Montane Rain Forests	83	72,600	1	30	20	12	15	77	Relatively Stable	2	Low	Relatively Stable	GO	III
Mentawai Islands Rain Forests	84	6,500	1	30	13	5	12	60	Endangered	3	High	Critical	RO	II
Sumatran Peat Swamp Forests	85	87,100	1	20	24	16	9	69	Endangered	3	High	Critical	RO	II
Borneo Peat Swamp Forests	86	67,100	1	20	12	20	3	55	Endangered	2	High	Critical	RO	II
Peninsular Malaysian Peat Swamp Forests	87	3,600	1	30	9	12	0	51	Vulnerable	6	Not evaluated	Endangered	GO	II
Sumatran Freshwater Swamp Forests	88	18,000	1	10	5	5	0	20	Critical	4	High	Critical	RO	II
Southern Borneo Freshwater Swamp Forests	89	36,600	1	20	8	12	3	43	Vulnerable	2	High	Critical	RO	II
Sundaland Heath Forests	90	76,200	1	20	24	16	9	69	Vulnerable	3	Medium	Vulnerable	RO	II
Western Java Rain Forests	91	41,400	1	0	0	1	15	16	Critical	6	Not evaluated	Critical	RO	II
Eastern Java–Bali Rain Forests	92	53,500	1	0	0	1	15	16	Critical	6	Not evaluated	Critical	RO	II
Western Java Montane Rain Forests	93	26,200	1	10	0	5	15	30	Endangered	6	Not evaluated	Critical	GO	I
Eastern Java–Bali Montane Rain Forests	94	15,800	1	20	5	5	15	45	Vulnerable	6	Not evaluated	Endangered	BO	IV
Borneo Montane Rain Forests	95	115,100	1	40	25	20	12	97	Relatively Intact	3	Medium	Relatively Intact	GO	III
Borneo Lowland Rain Forests	96	425,400	1	20	24	12	12	68	Vulnerable	3	High	Critical	GO	I
Palawan Rain Forests	97	13,900	1	3	1	5	2	11	Critical	3	Not evaluated	Critical	GO	I
Luzon Rain Forests	98	94,800	1	0	0	1	9	10	Critical	5	Not evaluated	Critical	GO	I
Mindoro Rain Forests	99	10,300	1	0	0	1	5	6	Critical	7	Not evaluated	Critical	GO	I
Greater Negros–Panay Rain Forests	100	34,200	1	0	0	1	15	16	Critical	7	Not evaluated	Critical	GO	I
Mindanao–Eastern Visayas Rain Forests	101	105,000	1	3	1	5	9	18	Critical	5	Not evaluated	Critical	GO	I
Luzon Montane Rain Forests	102	8,300	1	23	8	12	0	43	Vulnerable	5	Not evaluated	Endangered	RO	II
Mindanao Montane Rain Forests	103	17,800	1	13	8	12	12	45	Vulnerable	5	Not evaluated	Endangered	GO	I
Sulu Archipelago Rain Forest	104	2,000	1	0	0	1	0	1	Critical	6	Not evaluated	Critical	GO	I
Sumatran Tropical Pine Forests	105	2,700	3	30	5	5	15	55	Vulnerable	3	Medium	Vulnerable	GO	I
Luzon Tropical Pine Forests	106	7,100	3	20	5	5	3	33	Endangered	4	Not evaluated	Endangered	RO	II
Sunda Shelf Mangroves	107	37,200	14	10	0	12	3	25	Critical	3	High	Critical	GO	I
Kinabalu Montane Alpine Meadows	108	4,300	10	30	12	20	15	77	Relatively Stable	3	Not evaluated	Relatively Stable	GO	III
Wallacea Bioregion														
Sulawesi Lowland Rain Forests	109	114,100	1	30	24	12	9	75	Relatively Stable	2	High	Critical	GO	I
Sulawesi Montane Rain Forests	110	75,500	1	30	17	20	12	79	Relatively Stable	3	High	Critical	GO	I
Lesser Sundas Deciduous Forests	111	39,300	2	10	8	5	9	32	Endangered	4	Medium	Endangered	GO	I

ID	Ecoregion	Area	c1	c2	c3	c4	c5	c6	Status1	c7	Status2	c8	Status3	Code1	Code2
112	Timor and Wetar Deciduous Forests	33,300	2	10	1	5	2	18	Endangered	3	Endangered	Medium	Endangered	GO	I
113	Sumba Deciduous Forests	10,700	2	10	5	5	3	23	Endangered	3	Endangered	Not evaluated	Endangered	BO	IV
114	Halmahera Rain Forests	26,800	1	30	25	20	0	75	Relatively Stable	2	Relatively Stable	Medium	Relatively Stable	GO	III
115	Buru Rain Forests	8,600	1	30	12	20	0	62	Vulnerable	3	Vulnerable	Medium	Vulnerable	BO	V
116	Seram Rain Forests	19,300	1	30	17	20	12	79	Relatively Stable	3	Relatively Stable	Medium	Relatively Stable	GO	III
117	Banda Sea Islands Moist Deciduous Forests	7,500	1	30	13	20	2	65	Vulnerable	3	Vulnerable	Not evaluated	Vulnerable	RO	II
	New Guinea and Melanesia Bioregion														
118	Vogelkop Montane Rain Forests	21,900	1	30	13	20	3	66	Vulnerable	1	Vulnerable	Medium	Vulnerable	GO	I
119	Vogelkop-Aru Lowland Rain Forests	77,400	1	30	20	20	9	79	Relatively Stable	1	Critical	High	Critical	GO	I
120	Biak-Numfoor Rain Forests	2,900	1	30	9	20	12	71	Relatively Stable	3	Critical	High	Critical	RO	II
121	Yapen Rain Forests	2,300	1	40	8	20	15	83	Relatively Stable	1	Critical	High	Critical	BO	IV
122	Northern New Guinea Lowland Rain Forests	23,200	1	40	13	20	12	85	Relatively Intact	1	Critical	High	Critical	RO	II
123	Northern New Guinea Lowland Rain and Freshwater Swamp Forests	134,700	1	30	24	20	12	86	Relatively Intact	1	Critical	High	Critical	RO	II
124	Huon Peninsula Montane Rain Forests	16,400	1	30	12	20	12	74	Relatively Stable	1	Relatively Stable	Medium	Relatively Stable	GO	III
125	Central Range Montane Rain Forests	171,200	1	30	12	20	12	74	Relatively Stable	1	Relatively Stable	Medium	Relatively Stable	GO	III
126	Southeastern Papuan Rain Forests	77,000	1	30	12	20	9	71	Relatively Stable	1	Relatively Stable	Medium	Relatively Stable	GO	III
127	Southern New Guinea Freshwater Swamp Forests	99,400	1	30	24	20	12	86	Relatively Intact	1	Relatively Intact	Medium	Relatively Intact	RO	III
128	Southern New Guinea Lowland Rain Forests	122,300	1	40	20	20	9	89	Relatively Intact	1	Critical	High	Critical	GO	I
129	New Guinea Mangroves	26,700	14	30	0	16	12	58	Vulnerable	1	Vulnerable	High	Critical	GO	I
130	Trans Fly Savanna and Grasslands	26,600	7	40	12	20	15	87	Relatively Intact	1	Relatively Intact	Not evaluated	Relatively Intact	GO	III
131	Central Range Sub-Alpine Grasslands	15,500	10	30	13	20	15	78	Relatively Stable	1	Relatively Stable	Not evaluated	Relatively Stable	GO	III
132	Admiralty Islands Lowland Rain Forests	2,100	1	40	8	20	12	80	Relatively Stable	1	Relatively Stable	High	Critical	RO	II
133	New Britain–New Ireland Lowland Rain Forests	34,700	1	40	21	20	9	90	Relatively Intact	1	Relatively Stable	High	Critical	GO	I
134	New Britain–New Ireland Montane Rain Forests	12,100	1	40	13	20	12	85	Relatively Intact	1	Relatively Stable	High	Critical	GO	I
135	Trobriand Islands Rain Forests	4,200	1	40	13	20	2	75	Relatively Stable	1	Relatively Stable	Medium	Relatively Stable	RO	III
136	Louisiade Archipelago Rain Forests	1,600	1	40	4	20	0	64	Vulnerable	1	Vulnerable	Not evaluated	Vulnerable	BO	V
137	Solomon Islands Rain Forests	35,700	1	20	12	20	9	61	Vulnerable	1	Vulnerable	Not evaluated	Vulnerable	GO	I
138	Vanuatu Rain Forests	13,200	1	30	13	12	0	55	Vulnerable	7	Vulnerable	Not evaluated	Endangered	GO	I
139	New Caledonia Rain Forests	14,500	1	20	1	12	12	45	Vulnerable	7	Endangered	Not evaluated	Endangered	GO	I
140	New Caledonia Dry Forests	4,400	2	10	0	1	5	16	Critical	7	Critical	Not evaluated	Critical	GO	I

[a] 1 = tropical and subtropical moist broadleaf forests; 2 = tropical and subtropical dry broadleaf forests; 3 = tropical and subtropical coniferous forests; 4 = temperate broadleaf and mixed forests; 5 = temperate coniferous forests; 7 = tropical and subtropical grasslands, savannas, and shrublands; 9 = flooded grasslands and savannas; 10 = montane grasslands and shrublands; 13 = deserts and xeric shrublands; 14 = mangroves.

[b] 0 = 0–9% remaining habitat intact; 10 = 10–24%; 20 = 25–49%; 30 = 50–89%; 40 = 90–100%

[c] See Table D.2 in Appendix D.

[d] 1 = high fragmentation; 5 = advanced fragmentation; 12 = medium fragmentation; 16 = low fragmentation; 20 = relatively contiguous

[e] 0 = 0% of intact habitat protected; 2 = 1–2%; 5 = 3–5%; 9 = 6–9%; 12 = 10–20%; 15 = 20+%

[f] See Table D.5 in Appendix D.

[g] GO = globally outstanding; RO = regionally outstanding; BO = bioregionally outstanding; LI = locally important

Appendix G

Habitat Types by Ecoregion

Ecoregion	Habitat Type	Area (km²)	Percentage
North Western Ghats Montane Rain Forests [1]	Cleared land	17,980	58
	Subtropical broadleaved hillforest	567	2
	Semi-evergreen rain forest	8,260	27
	Tropical wet evergreen forest	3,969	13
South Western Ghats Montane Rain Forests [2]	Cleared land	14,274	63
	Subtropical broadleaved hillforest	617	3
	Semi-evergreen rain forest	2,616	12
	Tropical wet evergreen forest	4,909	22
Sri Lanka Lowland Rain Forests [3]	Cleared land	11,349	91
	Tropical wet evergreen forest	1,049	8
Sri Lanka Montane Rain Forests [4]	Cleared land	2,470	81
	Tropical montane evergreen forest	28	1
	Tropical wet evergreen forest	567	19
Orissa Semi-Evergreen Rain Forests [5]	Cleared land	21,299	96
	Semi-evergreen rain forest	832	4
East Deccan Dry Evergreen Forests [6]	Cleared land	24,420	96
	Tropical dry evergreen forest	997	4
Sri Lanka Dry-Zone Dry Evergreen Forests [7]	Cleared land	37,078	77
	Freshwater lakes	478	1
	Semi-evergreen rain forest	1,508	3
	Tropical dry evergreen forest	9,011	19
Brahmaputra Valley Semi-Evergreen Rain Forests [8]	Cleared land	48,031	85
	Semi-evergreen rain forest	904	2
	Tropical moist deciduous forest	1,862	3
	Tropical wet evergreen forest	5,440	10
Sundarbans Freshwater Swamp Forests [9]	Cleared land	16,136	98
	Freshwater lakes	257	2
Lower Gangetic Plains Moist Deciduous Forests [10]	Cleared land	242,085	95
	Dry dipterocarp forest	4,558	2
	Degraded	2,918	1
	Tropical moist deciduous forest	2,212	1
Meghalaya Subtropical Forests [11]	Cleared land	24,882	60
	Degraded	637	2
	Subtropical broadleaved hillforest	690	2
	Semi-evergreen rain forest	3,017	7
	Tropical moist deciduous forest	12,272	29
Eastern Highlands Moist Deciduous Forests [12]	Cleared land	256,103	75
	Tropical moist deciduous forest	83,798	25
Malabar Coast Moist Deciduous Forests [13]	Cleared land	33,495	95
	Tropical moist deciduous forest	1,779	5
North Western Ghats Moist Deciduous Forests [14]	Cleared land	37,237	78
	Tropical moist deciduous forest	10,785	22
South Western Ghats Moist Deciduous Forests [15]	Cleared land	17,265	73
	Tropical moist deciduous forest	6,401	27
Khathiarbar-Gir Dry Deciduous Forests [16]	Cleared land	221,212	83
	Dry dipterocarp forest	44,445	17
Upper Gangetic Plains Moist Deciduous Forests [17]	Cleared land	252,737	96
	Dry dipterocarp forest	5,391	2
	Tropical moist deciduous forest	4,047	2
Chhota-Nagpur Dry Deciduous Forests [18]	Cleared land	71,182	58
	Dry dipterocarp forest	50,947	42
Northern Dry Deciduous Forests [19]	Cleared land	45,391	78
	Dry dipterocarp forest	12,725	22
Narmada Valley Dry Deciduous Forests [20]	Cleared land	106,125	63
	Dry dipterocarp forest	63,340	37

Ecoregion	Habitat Type	Area (km²)	Percentage
Central Deccan Plateau Dry Deciduous Forests [21]	Cleared land	191,931	80
	Dry dipterocarp forest	47,475	20
South Deccan Plateau Dry Deciduous Forests [22]	Cleared land	66,548	81
	Dry dipterocarp forest	14,835	18
	Thorn scrub forest	542	1
Deccan Thorn Scrub Forests [23]	Cleared land	313,595	93
	Thorn scrub forest	24,767	7
Northwestern Thorn Scrub Forests [24]	Cleared land	463,701	95
	Freshwater river	3,752	1
	Sand desert vegetation	6,579	1
	Thorn scrub forest	9,599	2
Himalayan Subtropical Broadleaf Forests [25]	Tropical moist deciduous forest	8,670	23
	Cleared land	13,316	35
	Dry dipterocarp forest	535	1
	Degraded	12,588	33
	Lower hardwood forest	1,412	4
	Subtropical pine forest	1,170	3
Eastern Himalayan Broadleaf Forests [26]	Montane wet temperate forest	24,979	43
	Subtropical broadleaved hillforest	2,570	4
	Upper hardwood forest	9,470	16
	Cleared land	11,591	20
	Degraded	8,285	14
Western Himalayan Broadleaf Forests [27]	Cleared land	35,249	63
	Degraded	1,154	2
	Himalayan dry temperate forest	1,152	2
	Himalayan moist temperate forest	15,803	28
	Sub-montane dry evergreen forest	897	2
	Subtropical pine forest	809	1
Eastern Himalayan Sub-Alpine Conifer Forests [28]	Alpine meadow	141	1
	Alpine scrub	829	3
	Cleared land	7,397	28
	Degraded	1,790	7
	Glacier	1,718	6
	Himalayan moist temperate forest	245	1
	Montane wet temperate forest	305	1
	Sub-alpine conifer forest	14,072	52
Western Himalayan Sub-Alpine Conifer Forests [29]	Cleared land	24,920	63
	Alpine meadow	608	2
	Alpine scrub	484	1
	Birch forest	394	1
	Blue pine forest	1,367	3
	Degraded	2,169	6
	Freshwater river	212	1
	Glacier	327	1
	Himalayan dry conifer	1,739	4
	Himalayan moist temperate forest	865	2
	Sub-alpine conifer forest	6,173	16
East Afghan Montane Coniferous Forests [30]	Cleared land	5,452	75
	Himalayan dry conifer	1,443	20
	Seasonal salt marsh	368	5
Himalayan Subtropical Pine Forests [31]	Tropical wet evergreen forest	5,138	5
	Cleared land	43,000	42
	Degraded	9,540	9
	Subtropical pine forest	42,399	42
Indus River Delta–Arabian Sea Mangroves [32]	Cleared land	5,107	88
	Mangroves	654	11
Sundarbans Mangroves [33]	Cleared land	12,660	62
	Mangroves	7,648	38
Godavari-Krishna Mangroves [34]	Cleared land	6,203	91
	Mangroves	640	9
Terai-Duar Savanna and Grasslands [35]	Cleared land	25,229	73
	Degraded	4,657	13
	Tropical moist deciduous forest	4,562	13

Ecoregion	Habitat Type	Area (km²)	Percentage
Rann of Kutch Seasonal Salt Marsh [36]	Cleared land	650	2
	Freshwater river	141	1
	Seasonal salt marsh	27,023	97
Northwestern Himalayan Alpine Shrub and Meadows [37]	Alpine dry steppe	18,687	40
	Alpine meadow	17,424	37
	Alpine scrub	3,229	7
	Cleared land	1,326	3
	Glacier	5,540	12
	Himalayan dry conifer	623	1
Western Himalayan Alpine Shrub and Meadows [38]	Alpine meadow	17,209	50
	Alpine scrub	2,228	6
	Birch forest	1,129	3
	Cleared land	3,137	9
	Degraded	301	1
	Glacier	10,111	29
	Himalayan moist temperate forest	184	1
	Sub-alpine conifer forest	271	1
Eastern Himalayan Alpine Shrub and Meadows [39]	Alpine meadow	9,441	24
	Alpine scrub	7,700	19
	Alpine bare rock	1,230	3
	Bare land	619	2
	Birch forest	474	1
	Cleared land	1,836	5
	Degraded	409	1
	Glacier	12,956	32
	Montane wet temperate forest	442	1
	Sub-alpine conifer forest	4,777	12
Karakoram–West Tibetan Plateau Alpine Steppe [40]	Alpine dry steppe	42,492	38
	Alpine meadow	47,606	42
	Cleared land	6,410	6
	Glacier	15,212	14
Central Tibetan Plateau Alpine Steppe [41]	Alpine meadow	13,319	98
	Glacier	280	2
Sulaiman Range Alpine Meadows [42]	Alpine dry steppe	18,307	93
	Cleared land	1,215	6
South Iran Nubo-Sindian Desert and Semi-Desert [43]	Bare land	1,891	5
	Cleared land	36,001	89
	Sand desert vegetation	496	1
	Seasonal salt marsh	2,031	5
Baluchistan Xeric Woodlands [44]	Alpine dry steppe	2,486	1
	Arid subtropical forest	8,956	4
	Cleared land	241,984	95
Registan–North Pakistan Sandy Desert [45]	Thorn scrub forest	1,608	2
	Cleared land	42,873	63
	Sand desert vegetation	20,917	31
	Seasonal salt marsh	2,471	4
Thar Desert [46]	Sand desert vegetation	237,307	100
Indus Valley Desert [47]	Cleared land	191	1
	Sand Desert Vegetation	19,282	99
Chin Hills–Arakan Yoma Montane Rain Forests [48]	Cleared land	566	2
	Degraded	8,913	30
	Sub-alpine conifer forest	1,605	5
	Semi-evergreen rain forest	5,022	17
	Subtropical montane forest	11,973	40
	Tropical moist deciduous forest	1,276	4
	Tropical montane evergreen forest	260	1
Myanmar Coastal Rain Forests [49]	Cleared land	35,453	54
	Degraded	5,513	8
	Mangroves	724	1
	Semi-evergreen rain forest	6,662	10
	Tropical wet evergreen forest	16,705	26

Appendix G: Habitat Types by Ecoregion

Ecoregion	Habitat Type	Area (km²)	Percentage
Mizoram-Manipur-Kachin Rain Forests [50]	Cleared land	25,270	19
	Degraded	45,314	34
	Freshwater lakes	949	1
	Montane wet temperate forest	2,942	2
	Semi-evergreen rain forest	48,046	36
	Subtropical montane forest	1,466	1
	Tropical moist deciduous forest	3,024	2
	Tropical wet evergreen forest	6,916	5
Kayah-Karen Montane Rain Forests [51]	Cleared land	13,751	12
	Dry dipterocarp forest	15,431	13
	Degraded	25,118	21
	Montane deciduous forest	2,681	2
	Semi-evergreen rain forest	21,879	18
	Subtropical pine forest	1,273	1
	Tropical moist deciduous forest	14,140	12
	Tropical montane evergreen forest	20,424	17
	Tropical wet evergreen forest	4,132	3
Luang Prabang Montane Rain Forests [52]	Tropical montane evergreen forest	3,424	5
	Cleared land	8,516	12
	Cultivated land	7,979	11
	Dry dipterocarp forest	1,334	2
	Degraded	29,487	41
	Freshwater river	761	1
	Grassland	1,272	2
	Plantations	972	1
	Secondary scrub	7,404	10
	Semi-evergreen rain forest	6,457	9
	Subtropical pine forest	683	1
	Tropical moist deciduous forest	2,915	4
Tenasserim–South Thailand Semi-Evergreen Rain Forests [53]	Shifting cultivation	14,289	15
	Cleared land	37,194	38
	Degraded	883	1
	Mangroves	938	1
	Semi-evergreen rain forest	38,373	40
	Tropical montane evergreen forest	4,471	5
Northern Annamites Rain Forests [54]	Bamboo	665	1
	Cleared land	4,555	10
	Cultivated land	2,269	5
	Dry dipterocarp forest	1,153	2
	Degraded	14,935	32
	Grassland	732	2
	Limestone forest	5,780	12
	Secondary scrub	3,473	7
	Semi-evergreen rain forest	9,085	19
	Tropical moist deciduous forest	466	1
	Tropical montane evergreen forest	2,463	5
	Tropical Pine Forest	303	1
	Tropical Wet Evergreen Forest	1,149	2
Cardamom Mountains Rain Forests [55]	Shifting cultivation	2,511	6
	Bamboo	1,175	3
	Cleared land	6,375	14
	Cultivated land	1,420	3
	Dry dipterocarp forest	5,479	12
	Grassland	495	1
	Sub-montane dry evergreen forest	622	1
	Semi-evergreen rain forest	2,226	5
	Swamp grassland	451	1
	Tropical montane evergreen forest	991	2
	Tropical Wet Evergreen Forest	21,850	50

Ecoregion	Habitat Type	Area (km²)	Percentage
Northern Vietnam Lowland Rain Forests [56]	Cleared land	6,546	29
	Cultivated land	13,105	58
	Limestone forest	147	1
	Secondary scrub	1,151	5
	Semi-evergreen rain forest	348	2
	Tropical montane evergreen forest	329	1
	Tropical semi-evergreen forest	194	1
	Tropical wet evergreen forest	665	3
Southern Annamites Montane Rain Forests [57]	Cleared land	12,161	26
	Cultivated land	1,746	4
	Degraded	8,288	18
	Secondary scrub	8,411	18
	Semi-evergreen rain forest	5,076	11
	Tropical moist deciduous forest	619	1
	Tropical montane evergreen forest	5,117	11
	Tropical pine forest	4,032	9
	Tropical semi-evergreen forest	357	1
Southern Vietnam Lowland Dry Forests [58]	Cleared land	19,940	57
	Cultivated land	9,891	28
	Secondary scrub	1,996	6
	Semi-evergreen rain forest	1,573	5
	Tropical moist deciduous forest	895	3
	Tropical semi-evergreen forest	414	1
Southeastern Indochina Dry Evergreen Forests [59]	Shifting cultivation	6,913	6
	Bamboo	1,764	1
	Cleared land	31,836	26
	Cultivated land	16,659	13
	Dry dipterocarp forest	11,106	9
	Degraded	3,613	3
	Grassland	1,091	1
	Secondary scrub	6,487	5
	Sub-montane dry evergreen forest	790	1
	Semi-evergreen rain forest	35,860	29
	Tropical moist deciduous forest	5,072	4
	Tropical semi-evergreen forest	1,261	1
Andaman Islands Rain Forests [60]	Cleared land	399	7
	Semi-evergreen rain forest	2,730	49
	Tropical wet evergreen forest	2,469	44
Nicobar Islands Rain Forests [61]	Cleared land	294	17
	Semi-evergreen rain forest	574	34
	Tropical wet evergreen forest	825	49
Tonle Sap–Mekong Peat Swamp Forests [62]	Shifting cultivation	285	1
	Cleared land	7,186	25
	Cultivated land	18,807	64
	Freshwater swamp	374	1
	Freshwater lakes	171	1
	Plantations	313	1
	Peat swamp	229	1
	Swamp grassland	1,782	6
Irrawaddy Freshwater Swamp Forests [63]	Cleared land	14,642	97
	Freshwater swamp	456	3
Chao Phraya Freshwater Swamp Forests [64]	Cleared land	38,866	100
Tonle Sap Freshwater Swamp Forests [65]	Cleared land	6,591	25
	Cultivated land	9,116	35
	Freshwater swamp	3,921	15
	Freshwater lakes	2,983	12
	Swamp grassland	3,098	12
Red River Freshwater Swamp Forests [66]	Cleared land	538	5
	Cultivated land	10,066	94
	Mangroves	67	1
Irrawaddy Moist Deciduous Forests [67]	Cleared land	70,265	51
	Tropical moist deciduous forest	67,284	49

Appendix G: Habitat Types by Ecoregion

Ecoregion	Habitat Type	Area (km²)	Percentage
Northern Thailand–Laos Moist Deciduous Forests [68]	Cleared land	13,597	32
	Cultivated land	3,554	8
	Dry dipterocarp forest	6,569	16
	Degraded	2,464	6
	Secondary scrub	2,443	6
	Semi-evergreen rain forest	250	1
	Tropical moist deciduous forest	12,375	29
	Tropical montane evergreen forest	413	1
Northern Khorat Plateau Moist Deciduous Forests [69]	Cleared land	8,478	50
	Cultivated land	1,996	12
	Dry dipterocarp forest	1,412	8
	Degraded	1,392	8
	Freshwater river	382	2
	Grassland	579	3
	Secondary scrub	1,042	6
	Semi-evergreen rain forest	229	1
	Tropical moist deciduous forest	1,284	8
Chao Phraya Lowland Moist Deciduous Forests [70]	Shifting cultivation	511	3
	Cleared land	16,029	79
	Semi-evergreen rain forest	141	1
	Tropical moist deciduous forest	3,424	17
	Tropical montane evergreen forest	127	1
Irrawaddy Dry Forests [71]	Cleared land	29,549	84
	Dry dipterocarp forest	5,381	15
Central Indochina Dry Forests [72]	Shifting cultivation	6,960	2
	Cleared land	185,553	58
	Cultivated land	20,543	6
	Dry dipterocarp forest	71,312	22
	Degraded	7,510	2
	Grassland	2,137	1
	Secondary scrub	10,006	3
	Semi-evergreen rain forest	3,462	1
	Tropical moist deciduous forest	7,063	2
Northern Triangle Subtropical Forests [73]	Alpine meadow	392	1
	Cleared land	3,225	6
	Degraded	798	1
	Montane wet temperate forest	15,134	28
	Sub-alpine conifer forest	3,868	7
	Subtropical lowland forest	30,120	56
Northern Indochina Subtropical Forests [74]	Bamboo	6,428	2
	Cleared land	72,378	25
	Cultivated land	19,538	7
	Degraded	95,752	33
	Limestone forest	4,115	1
	Montane deciduous forest	39,522	14
	Subtropical broadleaved hillforest	2,768	1
	Secondary scrub	21,265	7
	Sub-montane dry evergreen forest	4,705	2
	Semi-evergreen rain forest	4,661	2
	Subtropical montane forest	12,986	5
South China–Vietnam Subtropical Evergreen Forests [75]	Cleared land	23,302	62
	Cultivated land	5,595	15
	Limestone forest	4,738	13
	Subtropical broadleaved hillforest	253	1
	Secondary scrub	2,607	7
	Sub-montane dry evergreen forest	927	2
Northern Triangle Temperate Forests [76]	Alpine meadow	124	1
	Cleared land	635	6
	Montane wet temperate forest	7,559	71
	Sub-alpine conifer forest	1,186	11
	Subtropical lowland forest	1,191	11

Ecoregion	Habitat Type	Area (km²)	Percentage
Northeast India–Myanmar Pine Forests [77]	Cleared land	6,606	68
	Degraded	539	6
	Montane wet temperate forest	729	8
	Subtropical pine forest	1,803	19
Myanmar Coastal Mangroves [78]	Shifting cultivation	607	3
	Cleared land	14,045	64
	Freshwater swamp	110	1
	Mangroves	6,373	29
	Tropical montane evergreen forest	330	2
	Tropical wet evergreen forest	441	2
Indochina Mangroves [79]	Shifting cultivation	718	3
	Cleared land	10,134	38
	Cultivated land	12,276	46
	Mangroves	2,955	11
	Swamp grassland	145	1
	Tropical wet evergreen forest	428	2
Peninsular Malaysian Rain Forests [80]	Shifting cultivation	4,974	4
	Cleared land	85,784	69
	Degraded	6,943	6
	Tropical montane evergreen forest	2,403	2
	Tropical wet evergreen forest	24,267	19
Peninsular Malaysian Montane Rain Forests [81]	Tropical wet evergreen forest	7,374	43
	Cleared land	3,094	18
	Degraded	611	4
	Tropical montane evergreen forest	5,984	35
Sumatran Lowland Rain Forests [82]	Cleared land	187,374	73
	Ironwood forest	3,896	2
	Limestone forest	13,294	5
	Peat swamp	2,974	1
	Tropical wet evergreen forest	47,459	18
Sumatran Montane Rain Forests [83]	Cleared land	27,309	38
	Freshwater lakes	1,398	2
	Limestone forest	3,301	5
	Tropical montane evergreen forest	19,684	27
	Tropical montane limestone	12,120	17
	Tropical pine forest	452	1
	Tropical wet evergreen forest	8,120	11
Mentawai Islands Rain Forests [84]	Cleared land	1,978	31
	Freshwater swamp	122	2
	Mangroves	84	1
	Tropical wet evergreen forest	4,296	66
Sumatran Peat Swamp Forests [85]	Cleared land	47,992	55
	Peat swamp	38,865	45
Borneo Peat Swamp Forests [86]	Cleared land	22,874	34
	Cultivated land	11,720	17
	Degraded	587	1
	Heath forest	705	1
	Mangroves	1,209	2
	Peat swamp	27,513	41
	Tropical wet evergreen forest	2,410	4
Peninsular Malaysian Peat Swamp Forests [87]	Cleared land	1,456	40
	Degraded	89	2
	Freshwater swamp	2,064	57
Sumatran Freshwater Swamp Forests [88]	Cleared land	14,832	82
	Freshwater swamp	3,093	17
Southern Borneo Freshwater Swamp Forests [89]	Cleared land	19,582	54
	Freshwater swamp	16,925	46
Sundaland Heath Forests [90]	Cleared land	33,939	45
	Cultivated land	4,991	7
	Heath forest	36,072	47
	Tropical wet evergreen forest	754	1

Ecoregion	Habitat Type	Area (km²)	Percentage
Western Java Rain Forests [91]	Cleared land	39,207	95
	Freshwater swamp	255	1
	Mangroves	432	1
	Tropical wet evergreen forest	1,335	3
Eastern Java–Bali Rain Forests [92]	Cleared land	51,514	96
	Limestone forest	615	1
	Semi-evergreen rain forest	859	2
Western Java Montane Rain Forests [93]	Cleared land	20,872	80
	Tropical montane evergreen forest	3,671	14
	Tropical wet evergreen forest	1,607	6
Eastern Java–Bali Montane Rain Forests [94]	Bare land	199	1
	Cleared land	11,553	73
	Semi-evergreen rain forest	962	6
	Tropical montane evergreen forest	2,700	17
	Tropical Wet Evergreen Forest	407	3
Borneo Montane Rain Forests [95]	Cleared land	6,860	7
	Cultivated land	1,475	1
	Degraded	1,004	1
	Tropical montane evergreen forest	37,221	35
	Tropical wet evergreen forest	58,676	56
Borneo Lowland Rain Forests [96]	Cleared land	164,400	39
	Cultivated land	34,000	8
	Degraded	5,970	1
	Tropical wet evergreen forest	209,540	49
	Tropical montane evergreen forest	5,540	1
	Limestone forest	1,688	1
Palawan Rain Forests [97]	Cleared land	9,528	69
	Degraded	1,586	11
	Mangroves	315	2
	Semi-evergreen rain forest	412	3
	Montane deciduous forest	370	3
	Tropical moist deciduous forest	1,692	12
Luzon Rain Forests [98]	Cleared land	81,359	86
	Degraded	7,013	7
	Tropical moist deciduous forest	2,140	2
	Tropical wet evergreen forest	3,177	3
Mindoro Rain Forests [99]	Cleared land	9,708	94
	Degraded	275	3
	Semi-evergreen rain forest	258	3
Greater Negros–Panay Rain Forests [100]	Cleared land	33,664	97
	Degraded	433	1
	Tropical moist deciduous forest	174	1
Mindanao–Eastern Visayas Rain Forests [101]	Cleared land	86,007	83
	Degraded	12,364	12
	Freshwater swamp	799	1
	Semi-evergreen rain forest	2,758	3
	Tropical wet evergreen forest	1,201	1
Luzon Montane Rain Forests [102]	Cleared land	3,194	39
	Degraded	1,890	23
	Semi-evergreen rain forest	155	2
	Tropical montane evergreen forest	844	10
	Tropical wet evergreen forest	1,431	17
	Montane deciduous forest	284	3
	Tropical moist deciduous forest	378	5
	Tropical pine forest	92	1
Mindanao Montane Rain Forests [103]	Cleared land	10,042	55
	Degraded	3,558	20
	Semi-evergreen rain forest	1,021	6
	Tropical montane evergreen forest	3,346	18
	Montane deciduous forest	101	1
Sulu Archipelago Rain Forest [104]	Cleared land	2,020	100

Ecoregion	Habitat Type	Area (km²)	Percentage
Sumatran Tropical Pine Forests [105]	Cleared land	409	15
	Tropical montane evergreen forest	319	12
	Tropical montane limestone	445	16
	Tropical pine forest	1,547	56
	Tropical wet evergreen forest	18	1
Luzon Tropical Pine Forests [106]	Cleared land	4,379	62
	Degraded	141	2
	Tropical moist deciduous forest	330	5
	Tropical pine forest	2,202	31
Sunda Shelf Mangroves [107]	Cleared land	24,251	65
	Cultivated land	3,384	9
	Degraded	587	2
	Freshwater swamp	288	1
	Mangroves	8,279	22
	Tropical wet evergreen forest	198	1
Kinabalu Montane Alpine Meadows [108]	Cleared land	1,536	36
	Tropical montane evergreen forest	1,453	34
	Tropical wet evergreen forest	1,331	31
Sulawesi Lowland Rain Forests [109]	Cleared land	56,038	49
	Freshwater swamp	1,067	1
	Forest on ultrabasic	7,668	7
	Freshwater river	685	1
	Limestone forest	2,332	2
	Mangroves	951	1
	Semi-evergreen rain forest	30,082	26
	Tropical montane evergreen forest	2,634	2
	Tropical wet evergreen forest	12,667	11
Sulawesi Montane Rain Forests [110]	Cleared land	16,471	22
	Forest on ultrabasic	3,188	4
	Freshwater river	580	1
	Semi-evergreen rain forest	11,857	16
	Tropical montane evergreen forest	31,596	42
	Tropical wet evergreen forest	11,601	15
Lesser Sundas Deciduous Forests [111]	Cleared land	23,108	59
	Limestone forest	8,199	21
	Semi-evergreen rain forest	5,486	14
	Tropical montane evergreen forest	1,725	4
	Tropical montane limestone	795	2
Timor and Wetar Deciduous Forests [112]	Cleared land	20,070	60
	Forest on ultrabasic	643	2
	Limestone forest	8,418	25
	Semi-evergreen rain forest	1,913	6
	Tropical moist deciduous forest	300	1
	Tropical montane evergreen forest	704	2
	Tropical montane limestone	973	3
Sumba Deciduous Forests [113]	Cleared land	7,661	72
	Limestone forest	2,399	22
	Semi-evergreen rain forest	419	4
	Tropical montane limestone	74	1
	Tropical wet evergreen forest	154	1
Halmahera Rain Forests [114]	Cleared land	4,011	15
	Forest on ultrabasic	2,001	7
	Limestone forest	3,988	15
	Semi-evergreen rain forest	12,049	45
	Tropical moist deciduous forest	2,964	11
	Tropical montane evergreen forest	448	2
	Tropical wet evergreen forest	1,247	5
Buru Rain Forests [115]	Cleared land	1,941	23
	Limestone forest	54	1
	Semi-evergreen rain forest	3,230	38
	Tropical moist deciduous forest	1,597	19
	Tropical montane evergreen forest	1,632	19
	Tropical montane limestone	80	1

Appendix G: Habitat Types by Ecoregion

Ecoregion	Habitat Type	Area (km²)	Percentage
Seram Rain Forests [116]	Cleared land	3,695	19
	Limestone forest	1,485	8
	Mangroves	102	1
	Semi-evergreen rain forest	8,013	42
	Tropical montane evergreen forest	1,073	6
	Tropical wet evergreen forest	4,912	25
Banda Sea Islands Moist Deciduous Forests [117]	Cleared land	2,990	39
	Limestone forest	2,120	27
	Mangroves	84	1
	Tropical moist deciduous forest	2,500	33
Vogelkop Montane Rain Forests [118]	Cleared land	2,888	13
	Limestone forest	1,293	6
	Tropical montane evergreen forest	8,596	39
	Tropical montane limestone	1,133	5
	Tropical semi-evergreen forest	514	2
	Tropical wet evergreen forest	7,099	32
	Upper montane forest	372	2
Vogelkop-Aru Lowland Rain Forests [119]	Cleared land	7,636	10
	Degraded	800	1
	Freshwater swamp	978	1
	Forest on ultrabasic	1,043	1
	Limestone forest	9,185	12
	Mangroves	473	1
	Peat swamp	1,533	2
	Semi-evergreen rain forest	3,093	4
	Tropical montane evergreen forest	990	1
	Tropical semi-evergreen forest	2,609	3
	Tropical wet evergreen forest	48,973	63
Biak-Numfoor Rain Forests [120]	Cleared land	670	23
	Tropical montane evergreen forest	126	4
	Tropical wet evergreen forest	2,095	72
Yapen Rain Forests [121]	Tropical montane evergreen forest	578	24
	Tropical wet evergreen forest	1,804	76
Northern New Guinea Montane Rain Forests [122]	Cleared land	1,788	8
	Freshwater swamp	495	2
	Forest on ultrabasic	206	1
	Tropical montane evergreen forest	2,959	13
	Tropical wet evergreen forest	17,762	76
Northern New Guinea Lowland Rain and Freshwater Swamp Forests [123]	Cleared land	42,219	31
	Freshwater swamp	16,984	13
	Peat swamp	11,688	9
	Tropical montane evergreen forest	806	1
	Tropical wet evergreen forest	62,278	46
Huon Peninsula Montane Rain Forests [124]	Cleared land	2,857	17
	Limestone forest	498	3
	Tropical montane evergreen forest	4,205	26
	Tropical wet evergreen forest	8,885	54
Central Range Montane Rain Forests [125]	Cleared land	27,393	16
	Limestone forest	8,639	5
	Peat swamp	2,636	2
	Tropical montane evergreen forest	70,469	41
	Tropical wet evergreen forest	52,713	31
	Upper montane forest	8,107	5
Southeastern Papuan Rain Forests [126]	Cleared land	18,412	24
	Freshwater swamp	3,129	4
	Savanna	626	1
	Tropical montane evergreen forest	12,084	16
	Tropical wet evergreen forest	40,002	52
	Upper montane forest	2,407	3

Ecoregion	Habitat Type	Area (km²)	Percentage
Southern New Guinea Freshwater Swamp Forests [127]	Cleared land	18,439	19
	Freshwater swamp	32,455	33
	Grassland	2,455	2
	Mangroves	669	1
	Peat swamp	11,772	12
	Savanna	26,904	27
	Tropical dry forest	3,282	3
	Tropical wet evergreen forest	2,841	3
Southern New Guinea Lowland Rain Forests [128]	Cleared land	6,505	5
	Freshwater swamp	2,348	2
	Limestone forest	908	1
	Peat swamp	11,364	9
	Savanna	24,707	20
	Semi-evergreen rain forest	2,717	2
	Tropical wet evergreen forest	73,528	60
New Guinea Mangroves [129]	Cleared land	4,723	18
	Freshwater swamp	1,216	5
	Mangroves	17,068	64
	Peat swamp	1,455	5
	Savanna	1,773	7
	Tropical wet evergreen forest	383	1
Trans Fly Savanna and Grasslands [130]	Cleared land	1,100	4
	Grassland	15,683	59
	Savanna	7,797	29
	Tropical dry forest	2,055	8
Central Range Sub-Alpine Grasslands [131]	Alpine meadow	1,073	7
	Cleared land	4,593	30
	Limestone forest	715	5
	Tropical montane evergreen forest	5,670	37
	Upper montane forest	3,439	22
Admiralty Islands Lowland Rain Forests [132]	Cleared land	72	3
	Freshwater swamp	229	11
	Limestone forest	85	4
	Tropical wet evergreen forest	1,706	82
New Britain–New Ireland Lowland Rain Forests [133]	Cleared land	698	2
	Freshwater swamp	1,691	5
	Limestone forest	2,231	6
	Tropical wet evergreen forest	29,691	86
New Britain–New Ireland Montane Rain Forests [134]	Cleared land	129	1
	Limestone forest	2,035	17
	Tropical montane evergreen forest	2,222	18
	Tropical wet evergreen forest	7,694	64
Trobriand Islands Rain Forests [135]	Cleared land	194	5
	Limestone forest	1,098	26
	Mangroves	96	2
	Tropical montane evergreen forest	101	2
	Tropical wet evergreen forest	2,688	64
Louisiade Archipelago Rain Forests [136]	Tropical wet evergreen forest	1,573	100
Solomon Islands Rain Forests [137]	Cleared land	4,174	12
	Freshwater swamp	2,333	7
	Tropical montane evergreen forest	1,571	4
	Tropical wet evergreen forest	5,029	14
	Tropical moist upper forest	800	2
	Tropical moist lowland forest	21,800	61
Vanuatu Rain Forests [138]	Forest	6,000	45
	Altered	7,200	55
New Caledonia Rain Forests [139]	Rain forest	4,100	28
	Altered	10,400	72
New Caledonia Dry Forests [140]	Dry forest	500	11
	Altered	3,900	89

Habitat types are derived from WCMC (1997) database. Only habitat types greater than 1 percent of the ecoregion are listed.

Appendix G: Habitat Types by Ecoregion

APPENDIX H

Near-Endemic and Strict Endemic Mammals by Ecoregion

	Family	Species
North Western Ghats Montane Rain Forests [1]	Molossidae	*Otomops wroughtoni**
	Viverridae	*Viverra civettina*
South Western Ghats Montane Rain Forests [2]	Crocidurinae	*Suncus montanus*
	Pteropodinae	*Latidens salimalii*
	Cercopithecidae	*Trachypithecus johnii*
	Viverridae	*Viverra civettina*
	Viverridae	*Paradoxurus jerdoni*
	Bovidae	*Hemitragus hylocrius**
	Sciuridae	*Funambulus layardi*
	Sciuridae	*Funambulus sublineatus*
	Muridae	*Mus famulus**
	Muridae	*Vandeleuria nilagirica**
Sri Lanka Lowland Rain Forests [3]	Crocidurinae	*Suncus montanus*
	Crocidurinae	*Suncus zeylanicus**
	Colobinae	*Semnopithecus vetulus*
	Viverridae	*Paradoxurus zeylonensis*
	Sciuridae	*Funambulus layardi*
	Sciuridae	*Petinomys fuscocapillus*
	Muridae	*Mus fernandoni*
	Muridae	*Vandeleuria nolthenii*
Sri Lanka Montane Rain Forests [4]	Crocidurinae	*Suncus montanus*
	Soricidae	*Feroculus feroculus**
	Soricidae	*Crocidura miya**
	Soricidae	*Solisorex pearsoni**
	Colobinae	*Semnopithecus vetulus*
	Viverridae	*Paradoxurus zeylonensis*
	Sciuridae	*Funambulus layardi*
	Sciuridae	*Funambulus sublineatus*
	Sciuridae	*Petinomys fuscocapillus*
	Muridae	*Mus fernandoni*
	Muridae	*Vandeleuria nolthenii*
	Muridae	*Rattus montanus**
	Muridae	*Srilankamys ohiensis**
Sri Lanka Dry-Zone Dry Evergreen Forests [7]	Colobinae	*Semnopithecus vetulus*
	Viverridae	*Paradoxurus zeylonensis*
	Muridae	*Mus fernandoni*
	Muridae	*Vandeleuria nolthenii*
Brahmaputra Valley Semi-Evergreen Forests [8]	Colobinae	*Semnopithecus geei*
	Leporidae	*Caprolagus hispidus*
Lower Gangetic Plains Moist Deciduous Forests [10]	Molossidae	*Tadarida teniotis**
Eastern Highlands Moist Deciduous Forests [12]	Rhinolophidae	*Hipposideros durgadasi**
Malabar Coast Moist Deciduous Forests [13]	Crocidurinae	*Suncus dayi*
	Colobinae	*Trachypithecus johnii*
	Viverridae	*Viverra civettina*
	Viverridae	*Paradoxurus jerdoni*
	Sciuridae	*Petinomys fuscocapillus*
	Muridae	*Rattus ranjiniae**
South Western Ghats Moist Deciduous Forests [15]	Crocidurinae	*Suncus dayi*
	Pteropodidae	*Latidens salimalii*
	Colobinae	*Trachypithecus johnii*
	Viverridae	*Viverra civettina*
	Viverridae	*Paradoxurus jerdoni*
	Sciuridae	*Funambulus layardi*
South Deccan Plateau Dry Deciduous Forests [22]	Pteropodidae	*Latidens salimalii*

	Family	Species
Deccan Thorn Scrub Forests [23]	Rhinolophidae	*Hipposideros schistaceus**
	Muridae	*Millardia kondana**
	Muridae	*Cremnomys elvira**
Northwestern Thorn Scrub Forests [24]	Rhinolophidae	*Triaenops persicus**
Himalayan Subtropical Broadleaf Forests [25]	Colobinae	*Semnopithecus geei*
Eastern Himalayan Broadleaf Forests [26]	Colobinae	*Semnopithecus geei*
	Cervidae	*Muntiacus gongshanensis*
	Sciuridae	*Petaurista magnificus*
	Sciuridae	*Biswamoyopterus biswasi**
	Muridae	*Niviventer brahma*
Western Himalayan Broadleaf Forests [27]	Vespertilionidae	*Myotis longipes**
	Vespertilionidae	*Murina grisea*
Eastern Himalayan Sub-Alpine Conifer Forests [28]	Sciuridae	*Petaurista magnificus*
	Sciuridae	*Petaurista nobilis*
	Muridae	*Apodemus gurkha**
Western Himalayan Sub-Alpine Conifer Forests [29]	Muridae	*Hyperacrius wynnei**
Himalayan Subtropical Pine Forests [31]	Vespertilionidae	*Murina grisea*
	Cervidae	*Muntiacus gongshanensis*
	Sciuridae	*Petaurista magnificus*
Terai-Duar Savanna and Grasslands [35]	Suidae	*Sus salvanius**
	Leporidae	*Caprolagus hispidus*
Rann of Kutch Seasonal Salt Marsh [36]	Equidae	*Equus hemionus**
Eastern Himalayan Alpine Shrub and Meadows [39]	Sciuridae	*Petaurista nobilis*
Karakoram–West Tibetan Plateau Alpine Steppe [40]	Sciuridae	*Eupetaurus cinereus**
Sulaiman Range Alpine Meadows [42]	Dipodidae	*Salpingotus michaelis*
Baluchistan Xeric Woodlands [44]	Dipodidae	*Salpingotus michaelis*
Rajasthan–North Pakistan Sandy Desert [45]	Dipodidae	*Allactaga hotsoni**
Myanmar Coastal Rain Forests [49]	Vespertilionidae	*Eudiscopus denticulus*
	Vespertilionidae	*Pipistrellus lophurus**
Mizoram-Manipur-Kachin Rain Forests [50]	Vespertilionidae	*Pipistrellus joffrei**
	Muridae	*Hadromys humei*
Kayah-Karen Montane Rain Forests [51]	Craseonycteridae	*Craseonycteris thonglongyai**
Tenasserim–South Thailand Semi-Evergreen Rain Forests [53]	Rhinolophidae	*Hipposideros halophyllus*
	Vespertilionidae	*Eptesicus demissus**
	Cervidae	*Muntiacus feae*
Northern Annamites Rain Forests [54]	Cercopithecidae	*Pygathrix nemaeus*
	Bovidae	*Pseudoryx nghetinhensis**
	Cervidae	*Megamuntiacus vuquangensis**
	Cervidae	*Muntiacus* sp. (Buhouang muntjak)*
	Cervidae	*Muntiacus truongsonensis**
Northern Vietnam Lowland Rain Forests [56]	Rhinolophidae	*Paracoelops megalotis**
Southern Annamites Montane Rain Forests [57]	Cercopithecidae	*Pygathrix nemaeus*
	Muridae	*Rattus osgoodi**
	Muridae	*Maxomys moi*
Southern Vietnam Lowland Dry Forests [58]	Cercopithecidae	*Pygathrix nemaeus*
Southeastern Indochina dry evergreen forests [59]	Tupaiidae	*Dendrogale murina**
	Cercopithecidae	*Pygathrix nemaeus*
	Bovidae	*Bos sauveli*
	Muridae	*Maxomys moi*
Andaman Islands Rain Forests [60]	Soricidae	*Crocidura hispida**
	Soricidae	*Crocidura andamanensis**
	Soricidae	*Crocidura jenkinsi**
	Pteropodidae	*Pteropus melanotus*
	Pteropodidae	*Myotis horsfieldii*
	Rhinolophidae	*Rhinolophus cognatus**
	Muridae	*Rattus stoicus**
Nicobar Islands Rain Forests [61]	Soricidae	*Crocidura nicobarica**
	Tupaiidae	*Tupaia nicobarica**
	Pteropodidae	*Pteropus faunulus**
	Pteropodidae	*Pteropus melanotus*
	Vespertilionidae	*Pipistrellus javanicus*
	Muridae	*Rattus palmarum**

	Family	Species
Chao Phraya Freshwater Swamp Forests [64]	Rhinolophidae	*Hipposideros halophyllus*
	Muridae	*Niviventer hinpoon**
	Muridae	*Leopoldamys neilli**
Chao Phraya Lowland Moist Deciduous Forests [70]	Rhinolophidae	*Hipposideros halophyllus*
Irrawaddy Dry Forests [71]	Muridae	*Millardia kathleenae**
Central Indochina Dry Forests [72]	Vespertilionidae	*Pipistrellus pulveratus**
	Bovidae	*Bos sauveli*
Northern Triangle Subtropical Forests [73]	Talpidae	*Talpa grandis**
	Talpidae	*Scaptonyx fusicauda**
	Soricidae	*Chimarrogale styani**
	Vespertilionidae	*Pipistrellus anthonyi**
	Sciuridae	*Sciurotamias davidianus**
	Sciuridae	*Callosciurus quinquestri*
	Muridae	*Niviventer brahma*
Northern Indochina Subtropical Forests [74]	Loridae	*Nycticebus intermedius**
	Cercopithecidae	*Pygathrix avunculus*
	Hylobatidae	*Hylobates leucogenys**
	Viverridae	*Hemigalus owstoni**
	Cervidae	*Muntiacus feae*
	Cervidae	*Muntiacus rooseveltorum**
	Sciuridae	*Callosciurus quinquestri*
	Sciuridae	*Dremomys gularis*
	Muridae	*Eothenomys olitor**
South China–Vietnam Subtropical Evergreen Forests [75]	Cercopithecidae	*Pygathrix avunculus*
	Sciuridae	*Dremomys gularis*
Peninsular Malaysian Montane Rain Forests [81]	Muridae	*Maxomys inas**
Sumatran Lowland Rain Forests [82]	Molossidae	*Mormopterus doriae**
Sumatran Montane Rain Forests [83]	Soricidae	*Crocidura baluensis*
	Cercopithecidae	*Presbytis thomasi**
	Sciuridae	*Hylopetes winstoni**
	Muridae	*Mus criciduroides**
	Muridae	*Rattus korinchi**
	Muridae	*Rattus hoogerwerfi*
	Muridae	*Maxomys hylomyoides**
	Muridae	*Maxomys inflatus**
	Leporidae	*Nesolagus netscheri**
Mentawai Islands Rain Forests [84]	Rhinolophidae	*Hipposideros breviceps**
	Cercopithecidae	*Presbytis potenziani**
	Cercopithecidae	*Nasalis concolor**
	Hylobatidae	*Hylobates klossi**
	Viverridae	*Paradoxurus lignicolor**
	Sciuridae	*Callosciurus melanogaster**
	Sciuridae	*Lariscus obscurus**
	Sciuridae	*Iomys sipora**
	Sciuridae	*Hylopetes sipora**
	Sciuridae	*Petinomys lugens**
	Muridae	*Rattus lugens**
	Muridae	*Rattus adustus**
	Muridae	*Rattus hoogerwerfi*
	Muridae	*Rattus enganus**
	Muridae	*Chiropodomys karlkoopman**
	Muridae	*Leopoldamys siporanus**
	Muridae	*Maxomys pagensis**
Borneo Peat Swamp Forests [86]	Rhinolophidae	*Hipposideros doriae*
	Cercopithecidae	*Presbytis hosei*
	Sciuridae	*Callosciurus adamsi*
	Sciuridae	*Petaurillus hosei*
	Sciuridae	*Aeromys thomasi*
Sundaland Heath Forests [90]	Cercopithecidae	*Presbytis hosei*

Appendix H: Near-Endemic and Strict Endemic Mammals by Ecoregion

	Family	Species
Western Java Rain Forests [91]	Rhinolophidae	*Rhinolophus canuti*
	Molossidae	*Otomops formosus*
	Hylobatidae	*Hylobates moloch*
	Suidae	*Sus verrucosus*
	Muridae	*Sundamys maxi*
Eastern Java–Bali Rain Forests [92]	Suidae	*Sus verrucosus*
	Cervidae	*Axis kuhlii**
Western Java Montane Rain Forests [93]	Soricidae	*Crocidura orientalis**
	Soricidae	*Crocidura paradoxura*
	Pteropodidae	*Megaerops kusnotoi*
	Rhinolophidae	*Rhinolophus canuti*
	Vespertilionidae	*Glischropus javanus**
	Molossidae	*Otomops formosus*
	Cercopithecidae	*Presbytis comata*
	Hylobatidae	*Hylobates moloch*
	Sciuridae	*Hylopetes bartelsi**
	Muridae	*Mus vulcani**
	Muridae	*Maxomys bartelsii**
	Muridae	*Pithecheir melanurus**
	Muridae	*Kadarsanomys sodyi**
	Muridae	*Sundamys maxi*
Eastern Java–Bali Montane Rain Forests [94]	Pteropodidae	*Megaerops kusnotoi*
Borneo Montane Rain Forests [95]	Tupaiidae	*Tupaia montana*
	Cercopithecidae	*Presbytis hosei*
	Viverridae	*Diplogale hosei**
	Sciuridae	*Callosciurus baluensis*
	Sciuridae	*Callosciurus orestes*
	Sciuridae	*Glyphotes simus*
	Sciuridae	*Sundasciurus brookei*
	Sciuridae	*Petaurillus hosei*
	Muridae	*Maxomys alticola*
	Muridae	*Maxomys ochraceiventer*
	Muridae	*Maxomys baeodon*
Borneo Lowland Rain Forests [96]	Tupaiidae	*Tupaia picta**
	Rhinolophidae	*Hipposideros doriae*
	Vespertilionidae	*Pipistrellus kitcheneri**
	Vespertilionidae	*Pipistrellus cuprosus**
	Vespertilionidae	*Murina rozendaali**
	Cercopithecidae	*Presbytis hosei*
	Sciuridae	*Callosciurus adamsi*
	Sciuridae	*Sundasciurus brookei*
	Sciuridae	*Petaurillus hosei*
	Sciuridae	*Aeromys thomasi*
	Muridae	*Maxomys ochraceiventer*
	Muridae	*Maxomys baeodon*
Palawan Rain Forests [97]	Pteropodidae	*Acerodon leucotis**
	Cervidae	*Axis calamianensis**
	Sciuridae	*Sundasciurus steerii**
	Sciuridae	*Sundasciurus moellendorf**
	Sciuridae	*Sundasciurus rabori**
	Sciuridae	*Hylopetes nigripes**
	Muridae	*Chiropodomys calamianens*
	Muridae	*Maxomys panglima**
	Muridae	*Palawanomys furvus**
	Hystricidae	*Hystrix pumila**
	Soricidae	*Crocidura palawanensis**
	Muridae	*Haeromys* sp. A*
	Sciuridae	*Sundasciurus hoogstraali**
	Sciuridae	*Sundasciurus juvencus**
	Tupaiidae	*Tupaia palawanensis**

	Family	Species
Luzon Rain Forests [98]	Pteropodidae	*Otopteropus cartilagonodus*
	Muridae	*Abditomys latidens*
	Muridae	*Apomys abrae*
	Muridae	*Apomys datae*
	Muridae	*Apomys microdon*
	Muridae	*Apomys sacobianus*
	Muridae	*Archboldomys luzonensis**
	Muridae	*Batomys granti*
	Muridae	*Bullimus luzonicus*
	Muridae	*Chrotomys gonzalesi**
	Muridae	*Crunomys fallax**
	Muridae	*Phloeomys cumingi**
	Muridae	*Phloeomys pallidus*
	Muridae	*Rhynchomys isarogensis**
	Muridae	*Tryphomys adustus*
Mindoro Rain Forests [99]	Soricidae	*Crocidura mindorus*
	Bovidae	*Bubalus mindorensis**
	Muridae	*Rattus mindorensis**
	Muridae	*Anonymomys mindorensis**
	Muridae	*Crateromys paulus**
	Muridae	*Apomys gracilirostris**
	Muridae	*Apomys* sp. E*
	Pteropodidae	*Pteropus* sp. A*
Greater Negros–Panay Rain Forests [100]	Soricidae	*Crocidura mindorus*
	Soricidae	*Crocidura negrina**
	Muridae	*Apomys* sp. A*
	Muridae	*Apomys* sp. B* (Sibuyan only)
	Muridae	*Apomys* sp. C* (Sibuyan only)
	Muridae	*Chrotomys* sp. A*(Sibuyan only)
	Muridae	*Crateromys heaneyi**
	Muridae	*Tarsomys* sp. A* (Sibuyan only)
	Pteropodidae	*Nyctimene rabori**
	Pteropodidae	*Dobsonia chapmani**
	Pteropodidae	*Haplonycteris* sp. A* (Sibuyan only)
	Suidae	*Sus cebifrons**
	Cervidae	*Cervus alfredi**
Mindanao–Eastern Visayas Rain Forests [101]	Erinaceidae	*Podogymnura aureospinula**
	Soricidae	*Crocidura beatus*
	Tupaiidae	*Urogale everetti*
	Cynocephalidae	*Cynocephalus volans*
	Pteropodidae	*Ptenochirus minor*
	Rhinolophidae	*Hipposideros coronatus**
	Sciuridae	*Sundasciurus philippinensis*
	Muridae	*Bullimus bagobus*
	Muridae	*Batomys salomonseni*
	Muridae	*Batomys russatus** (Dinagat only)
	Muridae	*Crateromys australis** (Dinagat only)
	Muridae	*Crunomys melanius**
	Muridae	*Apomys* sp. D* (Camiguin only)
	Muridae	*Bullimus* sp. A* (Camiguin only)
	Muridae	*Tarsomys echinatus**
	Sciuridae	*Exilisciurus concinnus*
Luzon Montane Rain Forests [102]	Pteropodidae	*Otopteropus cartilagonod*
	Muridae	*Bullimus luzonicus*
	Muridae	*Tryphomys adustus*
	Muridae	*Apomys abrae*
	Muridae	*Apomys sacobianus*
	Muridae	*Apomys datae*
	Muridae	*Phloeomys pallidus*
	Muridae	*Apomys microdon*
	Muridae	*Archboldomys musseri**

Appendix H: Near-Endemic and Strict Endemic Mammals by Ecoregion

	Family	Species
Mindanao Montane Rain Forests [103]	Erinaceidae	*Podogymnura truei**
	Soricidae	*Crocidura beatus*
	Soricidae	*Crocidura grandis**
	Tupaiidae	*Urogale everetti*
	Pteropodidae	*Alionycteris paucidentata**
	Sciuridae	*Petinomys crinitus**
	Sciuridae	*Sundasciurus philippinensis*
	Sciuridae	*Exilisciurus concinnus*
	Muridae	*Bullimus bagobus*
	Muridae	*Limnomys sibuanus**
	Muridae	*Tarsomys apoensis**
	Muridae	*Batomys salomonseni*
	Muridae	*Crunomys suncoides**
	Muridae	*Limnomys sp. B**
Sulu Archipelago Rain Forests [104]	Muridae	*Rattus tawitawiensis**
Sumatran Tropical Pine Forests [105]	Soricidae	*Crocidura paradoxura*
Luzon Tropical Pine Forests [106]	Pteropodidae	*Otopteropus cartilagonod*
	Muridae	*Bullimus luzonicus*
	Muridae	*Tryphomys adustus*
	Muridae	*Abditomys latidens*
	Muridae	*Apomys abrae*
	Muridae	*Apomys sacobianus*
	Muridae	*Apomys datae*
	Muridae	*Carpomys melanurus**
	Muridae	*Carpomys phaeurus**
	Muridae	*Batomys granti*
	Muridae	*Batomys dentatus**
	Muridae	*Phloeomys pallidus*
	Muridae	*Crateromys schadenbergi**
	Muridae	*Chrotomys whiteheadi**
	Muridae	*Rhynchomys soricoides**
	Muridae	*Celaenomys silaceus**
Kinabalu Montane Alpine Meadows [108]	Crocidurinae	*Suncus ater**
	Soricidae	*Crocidura baluensis*
	Tupaiidae	*Tupaia montana*
	Cercopithecidae	*Presbytis comata*
	Mustelidae	*Melogale everetti**
	Sciuridae	*Callosciurus baluensis*
	Sciuridae	*Callosciurus orestes*
	Sciuridae	*Glyphotes simus*
	Sciuridae	*Sundasciurus brookei*
	Sciuridae	*Petaurillus hosei*
	Sciuridae	*Aeromys thomasi*
	Muridae	*Rattus baluensis**
	Muridae	*Chiropodomys muroides**
	Muridae	*Maxomys alticola*
	Muridae	*Maxomys ochraceiventer*
	Muridae	*Maxomys baeodon*
Sulawesi Lowland Rain Forests [109]	Phalangeridae	*Phalanger pelengensis**
	Soricidae	*Crocidura elongata*
	Soricidae	*Crocidura lea*
	Soricidae	*Crocidura levicula*
	Pteropodidae	*Acerodon humilis**
	Pteropodidae	*Neopteryx frosti**
	Pteropodidae	*Nyctimene minutus*
	Rhinolophidae	*Hipposideros inexpectatu*
	Vespertilionidae	*Pipistrellus minahassae*
	Vespertilionidae	*Hesperoptenus gaskelli*
	Cercopithecidae	*Macaca maura**
	Cercopithecidae	*Macaca ochreata**
	Cercopithecidae	*Macaca nigra*
	Suidae	*Babyrousa babyrussa*
	Bovidae	*Bubalus depressicornis**
	Muridae	*Rattus koopmani**

	Family	Species
	Muridae	*Rattus xanthurus*
	Muridae	*Rattus bontanus**
	Muridae	*Rattus elaphinus**
	Muridae	*Maxomys hellwaldii**
	Muridae	*Haeromys minahassae**
	Muridae	*Margaretamys beccarii**
	Muridae	*Taeromys celebensis**
	Muridae	*Taeromys punicans**
	Muridae	*Taeromys taerae**
	Muridae	*Echiothrix leucura**
	Muridae	*Melomys fulgens*
	Muridae	*Melomys caurinus**
Sulawesi Montane Rain Forests [110]	Soricidae	*Crocidura elongata*
	Soricidae	*Crocidura lea*
	Soricidae	*Crocidura levicula*
	Rhinolophidae	*Hipposideros inexpectatu*
	Vespertilionidae	*Pipistrellus minahassae*
	Vespertilionidae	*Hesperoptenus gaskelli*
	Tarsiidae	*Tarsius pumilus**
	Tarsiidae	*Tarsius dianae**
	Cercopithecidae	*Macaca nigra*
	Suidae	*Babyrousa babyrussa*
	Bovidae	*Bubalus quarlesi**
	Sciuridae	*Hyosciurus heinrichi**
	Sciuridae	*Prosciurillus weberi**
	Sciuridae	*Prosciurillus abstrusus**
	Muridae	*Rattus mollicomulus**
	Muridae	*Rattus xanthurus*
	Muridae	*Rattus marmosurus**
	Muridae	*Maxomys dollmani**
	Muridae	*Maxomys wattsi**
	Muridae	*Crunomys celebensis**
	Muridae	*Bunomys coelestis**
	Muridae	*Bunomys prolatus**
	Muridae	*Bunomys fratrorum**
	Muridae	*Bunomys heinrichi**
	Muridae	*Bunomys penitus**
	Muridae	*Eropeplus canus**
	Muridae	*Margaretamys elegans**
	Muridae	*Margaretamys parvus**
	Muridae	*Taeromys hamatus**
	Muridae	*Taeromys arcuatus**
	Muridae	*Melasmothrix naso**
	Muridae	*Melasmothrix rhinogradoi**
	Muridae	*Melasmothrix macrocercus**
Lesser Sundas Deciduous Forests [111]	Soricidae	*Suncus mertensi**
	Pteropodidae	*Pteropus lombocensis**
	Vespertilionidae	*Nyctophilus heran**
	Muridae	*Bunomys naso**
	Muridae	*Komodomys rintjanus**
Timor and Wetar Deciduous Forests [112]	Soricidae	*Crocidura tenuis**
	Pteropodidae	*Pteropus chrysoproctus*
	Rhinolophidae	*Rhinolophus canuti*
	Muridae	*Papagomys armandvillei**
	Muridae	*Rattus timorensis**
Halmahera Rain Forests [114]	Phalangeridae	*Phalanger ornatus**
	Pteropodidae	*Pteropus chrysoproctus*
	Pteropodidae	*Nyctimene minutus*
	Muridae	*Melomys obiensis**
	Pteropodidae	*Pteropus personatus**
	Phalangeridae	*Phalanger rothschildi**
	Phalangeridae	*Phalanger* sp.*
	Muridae	*Rattus* sp.*

Appendix H: Near-Endemic and Strict Endemic Mammals by Ecoregion

	Family	Species
Buru Rain Forests [115]	Pteropodidae	*Pteropus chrysoproctus*
	Pteropodidae	*Pteropus ocularis*
	Pteropodidae	*Nyctimene minutus*
	Suidae	*Babyrousa babyrussa*
Seram Rain Forests [116]	Peroryctidae	*Rhyncholemes prattorum**
	Pteropodidae	*Pteropus chrysoproctus*
	Pteropodidae	*Pteropus ocularis*
	Muridae	*Rattus feliceus**
	Muridae	*Melomys fulgens*
	Muridae	*Melomys aerosus**
	Muridae	*Melomys fraterculus**
	Pteropodidae	*Pteropus argenatatus**
	Muridae	*Stenomys ceramicus**
Banda Sea Islands Moist Deciduous Forests [117]	Macropodidae	*Thylogale bruinii*
	Vespertilionidae	*Myotis stalkeri**
	Emballonuridae	*Taphozous achates**
Vogelkop Montane Rain Forests [118]	Tachyglossidae	*Zaglossus bruijni*
	Dasyuridae	*Phascolosorex doriae*
	Peroryctidae	*Microperoryctes murina*
	Macropodidae	*Dendrolagus ursinus*
	Pseudocheiridae	*Pseudocheirus schlegeli**
	Muridae	*Leptomys elegans*
	Muridae	*Hyomys dammermani*
Vogelkop-Aru Lowland Rain Forests [119]	Tachyglossidae	*Zaglossus bruijni*
	Dasyuridae	*Phascolosorex doriae*
	Macropodidae	*Dendrolagus ursinus*
	Macropodidae	*Thylogale bruinii*
	Muridae	*Melomys lorentzi*
	Muridae	*Pogonomelomys mayeri*
	Muridae	*Pogonomelomys bruijni*
	Macropodidae	*Dorcopsis muelleri*
Biak-Numfoor Rain Forests [120]	Petauridae	*Petaurus biacensis**
	Muridae	*Rattus jobiensis*
	Muridae	*Uromys boeadii**
	Muridae	*Uromys emmae**
	Pteropodidae	*Dobsonia emersa**
Yapen Rain Forests [121]	Muridae	*Rattus jobiensis*
Northern New Guinea Montane Rain Forests [122]	Peroryctidae	*Echymipera clara*
	Petauridae	*Petaurus abidi**
	Muridae	*Paraleptomys rufilatus*
	Muridae	*Xenuromys barbatus*
	Macropodidae	*Dendrolagus scottae**
	Tachyglossidae	*Zaglossus attenboroughi**
Northern New Guinea Lowland Rain and Freshwater Swamp Forests [123]	Peroryctidae	*Echymipera clara*
	Peroryctidae	*Echymipera echinista*
	Macropodidae	*Dorcopsis hageni**
	Pteropodidae	*Nyctimene draconilla*
	Emballonuridae	*Emballonura furax*
	Rhinolophidae	*Hipposideros wollastoni*
	Vespertilionidae	*Kerivoula muscina*
	Molossidae	*Otomops secundus*
	Muridae	*Paraleptomys rufilatus*
	Muridae	*Hydromys hussoni*
	Muridae	*Pogonomelomys mayeri*
	Macropodidae	*Dorcopsis muelleri*
	Rhinolophidae	*Hipposideros edwardshill**
Huon Peninsula Montane Rain Forests [124]	Macropodidae	*Dendrolagus matschiei**
	Muridae	*Pogonomelomys mayeri*
	Muridae	*Abeomelomys sevia*
	Muridae	*Rattus novaeguineae*
	Muridae	*Leptomys ernstmayeri*
	Muridae	*Melomys gracilis*

	Family	Species
Central Range Montane Rain Forests [125]	Dasyuridae	*Neophascogale lorentzi**
	Dasyuridae	*Phascolosorex doriae*
	Dasyuridae	*Antechinus wilhelmina*
	Peroryctidae	*Microperoryctes murina*
	Peroryctidae	*Echymipera clara*
	Macropodidae	*Dorcopsulus macleayi*
	Phalangeridae	*Phalanger matanim**
	Phalangeridae	*Phalanger vestitus**
	Petauridae	*Dactylopsila megalura**
	Pseudocheiridae	*Pseudocheirus caroli**
	Pseudocheiridae	*Pseudocheirus mayeri**
	Pteropodidae	*Aproteles bulmerae*
	Pteropodidae	*Syconycteris hobbit*
	Emballonuridae	*Emballonura furax*
	Rhinolophidae	*Hipposideros corynophyll**
	Vespertilionidae	*Pipistrellus collinus*
	Vespertilionidae	*Nyctophilus microdon**
	Vespertilionidae	*Kerivoula muscina*
	Molossidae	*Otomops secundus*
	Muridae	*Leptomys elegans*
	Muridae	*Paraleptomys wilhelmina**
	Muridae	*Neohydromys fuscus*
	Muridae	*Pseudohydromys murinus*
	Muridae	*Mayermys ellermani*
	Muridae	*Hydromys hussoni*
	Muridae	*Hydromys habbema**
	Muridae	*Crossomys moncktoni*
	Muridae	*Macruromys elegans**
	Muridae	*Xenuromys barbatus*
	Muridae	*Melomys fellowsi**
	Muridae	*Melomys lorentzi*
	Muridae	*Melomys lanosus**
	Muridae	*Pogonomelomys mayeri*
	Muridae	*Pogonomelomys bruijni*
	Muridae	*Abeomelomys sevia*
	Muridae	*Coccymys albidens**
	Muridae	*Rattus giluwensis**
	Muridae	*Rattus novaeguineae*
	Rhinolophidae	*Hipposideros muscinus*
	Muridae	*Hyomys dammermani*
	Muridae	*Hydromys shawmayeri*
	Muridae	*Leptomys ernstmayeri*
	Muridae	*Melomys gracilis*
	Muridae	*Pogonomys championi**
Southeastern Papuan Rain Forests [126]	Dasyuridae	*Murexia rothschildi**
	Dasyuridae	*Planigale novaeguineae*
	Peroryctidae	*Peroryctes broadbenti**
	Peroryctidae	*Microperoryctes papuensi**
	Macropodidae	*Dorcopsulus macleayi*
	Macropodidae	*Dorcopsis luctuosa*
	Macropodidae	*Thylogale bruinii*
	Pteropodidae	*Syconycteris hobbit*
	Vespertilionidae	*Pipistrellus collinus*
	Vespertilionidae	*Pharotis imogene**
	Vespertilionidae	*Kerivoula muscina*
	Molossidae	*Otomops papuensis*
	Molossidae	*Otomops secundus*
	Muridae	*Leptomys elegans*
	Muridae	*Neohydromys fuscus*
	Muridae	*Pseudohydromys murinus*
	Muridae	*Mayermys ellermani*
	Muridae	*Crossomys moncktoni*
	Muridae	*Chiruromys forbesi*
	Muridae	*Chiruromys lamia**

(continues)

	Family	Species
Southeastern Papuan Rain Forests [126]	Muridae	Xenuromys barbatus
(continued)	Muridae	Melomys levipes*
	Muridae	Rattus novaeguineae
	Rhinolophidae	Hipposideros muscinus
	Muridae	Hydromys shawmayeri
	Muridae	Leptomys ernstmayeri
	Muridae	Melomys gracilis
	Muridae	Stenomys vandeuseni*
Southern New Guinea Freshwater Swamp Forests [127]	Macropodidae	Dorcopsis luctuosa
	Macropodidae	Dendrolagus spadix
	Macropodidae	Thylogale bruinii
	Vespertilionidae	Kerivoula muscina
	Peroryctidae	Echymipera echinista
	Muridae	Leptomys signatus*
Southern New Guinea Lowland Rain Forests [128]	Macropodidae	Dorcopsis luctuosa
	Macropodidae	Dendrolagus spadix
	Pteropodidae	Aproteles bulmerae
	Pteropodidae	Nyctimene draconilla
	Emballonuridae	Emballonura furax
	Rhinolophidae	Hipposideros wollastoni
	Vespertilionidae	Kerivoula muscina
	Molossidae	Otomops papuensis
	Muridae	Melomys lorentzi
	Muridae	Pogonomelomys bruijni
	Macropodidae	Dorcopsis muelleri
	Rhinolophidae	Hipposideros muscinus
	Muridae	Melomys gracilis
New Guinea Mangroves [129]	Emballonuridae	Emballonura furax
Trans Fly Savanna and Grasslands [130]	Dasyuridae	Planigale novaeguineae
	Dasyuridae	Dasyurus spartacus*
	Dasyuridae	Sminthopsis archeri*
	Macropodidae	Dorcopsis luctuosa
	Macropodidae	Thylogale bruinii
Central Range Sub-Alpine Grasslands [131]	Dasyuridae	Antechinus wilhelmina
	Muridae	Pseudohydromys occidenta*
	Muridae	Stenomys richardsoni*
	Muridae	Mallomys gunung*
Admiralty Islands Lowland Rain Forests [132]	Pteropodidae	Pteropus admiralitatum
	Emballonuridae	Emballonura serii
	Phalangeridae	Spilocuscus kraemeri*
	Muridae	Melomys matambuai*
	Pteropodidae	Dobsonia anderseni
New Britain–New Ireland Lowland Rain Forests [133]	Pteropodidae	Melonycteris melanops
	Pteropodidae	Nyctimene major
	Pteropodidae	Pteropus admiralitatum
	Pteropodidae	Pteropus gilliardorum
	Emballonuridae	Emballonura serii
	Muridae	Hydromys neobritannicus
	Muridae	Uromys neobritannicus
	Pteropodidae	Dobsonia anderseni
	Pteropodidae	Dobsonia praedatrix
New Britain–New Ireland Montane Rain Forests [134]	Pteropodidae	Melonycteris melanops
	Pteropodidae	Nyctimene major
	Pteropodidae	Pteropus admiralitatum
	Pteropodidae	Pteropus gilliardorum
	Muridae	Hydromys neobritannicus
	Muridae	Uromys neobritannicus
	Pteropodidae	Dobsonia anderseni
	Pteropodidae	Dobsonia praedatrix
Trobriand Islands Rain Forests [135]	Vespertilionidae	Pipistrellus collinus
	Muridae	Chiruromys forbesi
	Pteropodidae	Nyctimene major
	Macropodidae	Dorcopsis atrata*
	Vespertilionidae	Kerivoula agnella

	Family	Species
	Phalangeridae	*Phalanger lullulae**
	Petauridae	*Dactylopsila tatei**
	Peroryctidae	*Echymipera davidi**
	Pteropodidae	*Pteropus pannietensis*
Louisiade Archipelago Rain Forests [136]	Pteropodidae	*Nyctimene major*
	Vespertilionidae	*Kerivoula agnella*
	Pteropodidae	*Pteropus pannietensis*
Solomon Islands Rain Forests [137]	Pteropodidae	*Melonycteris fardoulisi**
	Pteropodidae	*Melonycteris woodfordi**
	Pteropodidae	*Nyctimene major*
	Pteropodidae	*Pteralopex anceps**
	Pteropodidae	*Pteralopex atrata**
	Pteropodidae	*Pteralopex pulchra**
	Pteropodidae	*Pteralopex* sp.*
	Pteropodidae	*Pteropus admiralitatum*
	Pteropodidae	*Pteropus howensis**
	Pteropodidae	*Pteropus mahaganus**
	Pteropodidae	*Pteropus rayneri**
	Pteropodidae	*Pteropus rennelli**
	Pteropodidae	*Pteropus woodfordi**
	Rhinolophidae	*Anthops ornatus**
	Molossidae	*Chaerephon solomonis**
	Muridae	*Melomys bougainville**
	Muridae	*Melomys spechti**
	Muridae	*Solomys ponceleti**
	Muridae	*Solomys salamonis**
	Muridae	*Solomys salebrosus**
	Muridae	*Solomys sapientis**
	Muridae	*Uromys imperator**
	Muridae	*Uromys porculus**
	Muridae	*Uromys rex**
	Pteropodidae	*Dobsonia inermis**
	Pteropodidae	*Nyctimene vizcaccia**
Vanuatu Rain Forests [138]	Pteropodidae	*Notopteris macdonaldi*
	Pteropodidae	*Nyctimene sanctacrucis**
	Pteropodidae	*Pteropus anetianus**
	Pteropodidae	*Pteropus nitendiensis**
	Pteropodidae	*Pteropus tuberculatus**
	Pteropodidae	*Pteropus fundatus**
New Caledonia Rain Forests [139]	Pteropodidae	*Pteropus vetulus*
	Pteropodidae	*Notopteris macdonaldi*
	Pteropodidae	*Pteropus ornatus*
	Vespertilionidae	*Chalinolobus neocaledonicus*
	Vespertilionidae	*Miniopterus robustior**
	Vespertilionidae	*Nyctophilus* sp.
New Caledonia Dry Forests [140]	Pteropodidae	*Pteropus vetulus*
	Pteropodidae	*Notopteris macdonaldi*
	Pteropodidae	*Pteropus ornatus*
	Vespertilionidae	*Chalinolobus neocaledonicus*
	Vespertilionidae	*Nyctophilus* sp.

*Strict endemic, a species whose range is limited to this ecoregion.

Appendix H: Near-Endemic and Strict Endemic Mammals by Ecoregion

Appendix I

Near-Endemic and Strict Endemic Birds by Ecoregion

	Family	Common Name	Species
North Western Ghats Montane Rain Forests [1]	Columbidae	Nilgiri wood-pigeon	Columba elphinstonii
	Psittacidae	Malabar parakeet	Psittacula columboides
	Bucconidae	Malabar grey hornbill	Ocyceros griseus
	Corvidae	White-bellied treepie	Dendrocitta leucogastra
	Turdidae	White-bellied shortwing	Brachypteryx major
	Pycnonotidae	Grey-headed bulbul	Pycnonotus priocephalus
	Timaliidae	Grey-breasted laughingthrush	Garrulax jerdoni
	Timaliidae	Rufous babbler	Turdoides subrufus
South Western Ghats Montane Rain Forests [2]	Columbidae	Nilgiri wood-pigeon	Columba elphinstonii
	Psittacidae	Malabar parakeet	Psittacula columboides
	Bucconidae	Malabar grey hornbill	Ocyceros griseus
	Corvidae	White-bellied treepie	Dendrocitta leucogastra
	Turdidae	White-bellied shortwing	Brachypteryx major
	Muscicapidae	Black-and-rufous flycatcher	Ficedula nigrorufa
	Muscicapidae	Nilgiri flycatcher	Eumyias albicaudata
	Pycnonotidae	Grey-headed bulbul	Pycnonotus priocephalus
	Sylviidae	Broad-tailed grassbird	Schoenicola platyura*
	Timaliidae	Rufous-breasted laughingthrush	Garrulax cachinnans*
	Timaliidae	Grey-breasted laughingthrush	Garrulax jerdoni
	Timaliidae	Rufous babbler	Turdoides subrufus
	Motacillidae	Nilgiri pipit	Anthus nilghiriensis*
Sri Lanka Lowland Rain Forests [3]	Phasianidae	Ceylon spurfowl	Galloperdix bicalcarata
	Phasianidae	Ceylon junglefowl	Gallus lafayetii
	Columbidae	Ceylon wood-pigeon	Columba torringtoni
	Psittacidae	Ceylon hanging-parrot	Loriculus beryllinus
	Psittacidae	Layard's parakeet	Psittacula calthropae
	Cuculidae	Red-faced malkoha	Phaenicophaeus pyrrhocephalus
	Cuculidae	Green-billed coucal	Centropus chlororhynchus*
	Strigidae	Chestnut-backed owlet	Glaucidium castanonotum
	Bucconidae	Ceylon grey hornbill	Ocyceros gingalensis
	Capitonidae	Yellow-fronted barbet	Megalaima flavifrons
	Corvidae	Ceylon magpie	Urocissa ornata
	Turdidae	Spot-winged thrush	Zoothera spiloptera
	Sturnidae	White-faced starling	Sturnus senex
	Sturnidae	Ceylon myna	Gracula ptilogenys
	Muscicapidae	Kashmir flycatcher	Ficedula subrubra
	Timaliidae	Brown-capped babbler	Pellorneum fuscocapillum
	Timaliidae	Orange-billed babbler	Turdoides rufescens
	Dicaeidae	White-throated flowerpecker	Dicaeum vincens*
Sri Lanka Montane Rain Forests [4]	Phasianidae	Ceylon junglefowl	Gallus lafayetii
	Columbidae	Ceylon wood-pigeon	Columba torringtoni
	Psittacidae	Ceylon hanging-parrot	Loriculus beryllinus
	Psittacidae	Layard's parakeet	Psittacula calthropae
	Cuculidae	Red-faced malkoha	Phaenicophaeus pyrrhocephalus
	Strigidae	Chestnut-backed owlet	Glaucidium castanonotum
	Bucconidae	Ceylon grey hornbill	Ocyceros gingalensis
	Capitonidae	Yellow-fronted barbet	Megalaima flavifrons
	Corvidae	Ceylon magpie	Urocissa ornata
	Turdidae	Ceylon whistling-thrush	Myiophonus blighi*
	Turdidae	Spot-winged thrush	Zoothera spiloptera

(continues)

	Family	Common Name	Species
Sri Lanka Montane Rain Forests [4] (continued)	Sturnidae	White-faced starling	Sturnus senex
	Sturnidae	Ceylon myna	Gracula ptilogenys
	Muscicapidae	Kashmir flycatcher	Ficedula subrubra
	Muscicapidae	Dull-blue flycatcher	Eumyias sordida*
	Pycnonotidae	Yellow-eared bulbul	Pycnonotus penicillatus*
	Zosteropidae	Ceylon white-eye	Zosterops ceylonensis*
	Sylviidae	Ceylon bush-warbler	Bradypterus palliseri*
	Timaliidae	Brown-capped babbler	Pellorneum fuscocapillum
	Timaliidae	Orange-billed babbler	Turdoides rufescens
Sri Lanka Dry-Zone Dry Evergreen Forests [7]	Phasianidae	Ceylon spurfowl	Galloperdix bicalcarata
	Phasianidae	Ceylon junglefowl	Gallus lafayetii
	Psittacidae	Ceylon hanging-parrot	Loriculus beryllinus
	Psittacidae	Layard's parakeet	Psittacula calthropae
	Cuculidae	Red-faced malkoha	Phaenicophaeus pyrrhocephalus
	Bucconidae	Ceylon grey hornbill	Ocyceros gingalensis
	Capitonidae	Yellow-fronted barbet	Megalaima flavifrons
	Turdidae	Spot-winged thrush	Zoothera spiloptera
	Timaliidae	Brown-capped babbler	Pellorneum fuscocapillum
Brahmaputra Valley Semi-Evergreen Rain Forests [8]	Phasianidae	Manipur bush-quail	Perdicula manipurensis
	Timaliidae	Marsh babbler	Pellorneum palustre
Meghalaya Subtropical Forests [11]	Phasianidae	Manipur bush-quail	Perdicula manipurensis
	Timaliidae	Brown-capped laughingthrush	Garrulax austeni
	Timaliidae	Marsh babbler	Pellorneum palustre
	Timaliidae	Tawny-breasted wren-babbler	Spelaeornis longicaudatus
	Timaliidae	Wedge-billed wren-babbler	Sphenocichla humei
Eastern Highlands Moist Deciduous Forests [12]	Glareolidae	Jerdon's courser	Rhinoptilus bitorquatus
Malabar Coast Moist Deciduous Forests [13]	Bucconidae	Malabar grey hornbill	Ocyceros griseus
North Western Ghats Moist Deciduous Forests [14]	Columbidae	Nilgiri wood-pigeon	Columba elphinstonii
	Psittacidae	Malabar parakeet	Psittacula columboides
	Bucconidae	Malabar grey hornbill	Ocyceros griseus
	Pycnonotidae	Grey-headed bulbul	Pycnonotus priocephalus
	Timaliidae	Rufous babbler	Turdoides subrufus
South Western Ghats Moist Deciduous Forests [15]	Columbidae	Nilgiri wood-pigeon	Columba elphinstonii
	Psittacidae	Malabar parakeet	Psittacula columboides
	Bucconidae	Malabar grey hornbill	Ocyceros griseus
	Corvidae	White-bellied treepie	Dendrocitta leucogastra
	Muscicapidae	Black-and-rufous flycatcher	Ficedula nigrorufa
	Muscicapidae	Nilgiri flycatcher	Eumyias albicaudata
	Pycnonotidae	Grey-headed bulbul	Pycnonotus priocephalus
	Pycnonotidae	Yellow-throated bulbul	Pycnonotus xantholaemus
	Timaliidae	Rufous babbler	Turdoides subrufus
Khathiarbar-Gir Dry Deciduous Forests [16]	Paridae	White-winged tit	Parus nuchalis
Central Deccan Plateau Dry Deciduous Forests [21]	Glareolidae	Jerdon's courser	Rhinoptilus bitorquatus
South Deccan plateau Dry Deciduous Forests [22]	Pycnonotidae	Yellow-throated bulbul	Pycnonotus xantholaemus
	Timaliidae	Rufous babbler	Turdoides subrufus
Deccan Thorn Scrub Forests [23]	Phasianidae	Ceylon junglefowl	Gallus lafayetii
	Glareolidae	Jerdon's courser	Rhinoptilus bitorquatus
	Capitonidae	Yellow-fronted barbet	Megalaima flavifrons
	Pycnonotidae	Yellow-throated bulbul	Pycnonotus xantholaemus
Northwestern Thorn Scrub Forests [24]	Sylviidae	Rufous-vented prinia	Prinia burnesii*
	Paridae	White-winged tit	Parus nuchalis
Himalayan Subtropical Broadleaf Forests [25]	Phasianidae	Chestnut-breasted partridge	Arborophila mandellii

	Family	Common Name	Species
Eastern Himalayan Broadleaf Forests [26]			
	Phasianidae	Chestnut-breasted partridge	*Arborophila mandellii*
	Turdidae	Rusty-bellied shortwing	*Brachypteryx hyperythra*
	Sylviidae	Grey-crowned prinia	*Prinia cinereocapilla*
	Timaliidae	Elliot's laughingthrush	*Garrulax elliotii*
	Timaliidae	Immaculate wren-babbler	*Pnoepyga immaculata*
	Timaliidae	Rufous-throated wren-babbler	*Spelaeornis caudatus**
	Timaliidae	Mishmi wren-babbler	*Spelaeornis badeigularis*
	Timaliidae	Snowy-throated babbler	*Stachyris oglei*
	Timaliidae	Spiny babbler	*Turdoides nipalensis*
	Timaliidae	Hoary-throated barwing	*Actinodura nipalensis*
	Timaliidae	Ludlow's fulvetta	*Alcippe ludlowi*
Western Himalayan Broadleaf Forests [27]			
	Phasianidae	Himalayan quail	*Ophrysia superciliosa*
	Phasianidae	Western tragopan	*Tragopan melanocephalus*
	Phasianidae	Cheer pheasant	*Catreus wallichi*
	Muscicapidae	Kashmir flycatcher	*Ficedula subrubra*
	Sittidae	Kashmir nuthatch	*Sitta cashmirensis*
	Aegithalidae	White-cheeked tit	*Aegithalos leucogenys*
	Aegithalidae	White-throated tit	*Aegithalos niveogularis*
	Fringillidae	Spectacled finch	*Callacanthis burtoni*
	Fringillidae	Orange bullfinch	*Pyrrhula aurantiaca*
Eastern Himalayan Sub-Alpine Conifer Forests [28]			
	Phasianidae	Szecheny's partridge	*Tetraophasis szechenyii**
	Phasianidae	Chestnut-breasted partridge	*Arborophila mandellii*
	Timaliidae	Immaculate wren-babbler	*Pnoepyga immaculata*
	Timaliidae	Hoary-throated barwing	*Actinodura nipalensis*
	Timaliidae	Ludlow's fulvetta	*Alcippe ludlowi*
Western Himalayan Sub-Alpine Conifer Forests [29]			
	Phasianidae	Himalayan quail	*Ophrysia superciliosa*
	Phasianidae	Western tragopan	*Tragopan melanocephalus*
	Sittidae	Kashmir nuthatch	*Sitta cashmirensis*
	Aegithalidae	White-cheeked tit	*Aegithalos leucogenys*
	Aegithalidae	White-throated tit	*Aegithalos niveogularis*
	Timaliidae	Immaculate wren-babbler	*Pnoepyga immaculata*
	Timaliidae	Hoary-throated barwing	*Actinodura nipalensis*
	Fringillidae	Spectacled finch	*Callacanthis burtoni*
	Fringillidae	Orange bullfinch	*Pyrrhula aurantiaca*
Himalayan Subtropical Pine Forests [31]	Phasianidae	Chestnut-breasted partridge	*Arborophila mandellii*
	Phasianidae	Western tragopan	*Tragopan melanocephalus*
	Phasianidae	Cheer pheasant	*Catreus wallichi*
	Turdidae	Rusty-bellied shortwing	*Brachypteryx hyperythra*
	Timaliidae	Elliot's laughingthrush	*Garrulax elliotii*
	Timaliidae	Immaculate wren-babbler	*Pnoepyga immaculata*
	Timaliidae	Mishmi wren-babbler	*Spelaeornis badeigularis*
	Timaliidae	Snowy-throated babbler	*Stachyris oglei*
	Timaliidae	Spiny babbler	*Turdoides nipalensis*
	Timaliidae	Hoary-throated barwing	*Actinodura nipalensis*
	Timaliidae	Ludlow's fulvetta	*Alcippe ludlowi*
Sunderbans Mangroves [33]	Alcedinidae	Brown-winged kingfisher	*Pelargopsis amauropterus*
Terai-Duar Savanna and Grasslands [35]	Phasianidae	Manipur bush-quail	*Perdicula manipurensis*
	Sylviidae	Grey-crowned prinia	*Prinia cinereocapilla*
	Timaliidae	Spiny babbler	*Turdoides nipalensis*
Western Himalayan Alpine Shrub and Meadows [38]	Phasianidae	Cheer pheasant	*Catreus wallichi*
Eastern Himalayan Alpine Shrub and Meadows [39]	Phasianidae	Chestnut-breasted partridge	*Arborophila mandellii*
Chin Hills–Arakan Yoma Montane Rain Forests [48]	Sittidae	White-browed nuthatch	*Sitta victoriae**
	Timaliidae	Striped laughingthrush	*Garrulax virgatus*
	Timaliidae	Brown-capped laughingthrush	*Garrulax austeni*
Mizoram-Manipur-Kachin Rain Forests [50]			
	Phasianidae	Manipur bush-quail	*Perdicula manipurensis*
	Timaliidae	Striped laughingthrush	*Garrulax virgatus*

(continues)

Appendix I: Near-Endemic and Strict Endemic Birds by Ecoregion

	Family	Common Name	Species
Mizoram-Manipur-Kachin Rain Forests [50] (continued)	Timaliidae	Brown-capped laughingthrush	Garrulax austeni
	Timaliidae	Marsh babbler	Pellorneum palustre
	Timaliidae	Tawny-breasted wren-babbler	Spelaeornis longicaudatus
	Timaliidae	Wedge-billed wren-babbler	Sphenocichla humei
Kayah-Karen Montane Rain Forests [51]	Timaliidae	Deignan's babbler	Stachyris rodolphei*
	Timaliidae	Burmese yuhina	Yuhina humilis*
Tenasserim–South Thailand Semi-Evergreen Rain Forests [53]	Phasianidae	Mountain peacock-pheasant	Polyplectron inopinatum
	Phasianidae	Malayan peacock-pheasant	Polyplectron malacense*
	Pittidae	Gurney's pitta	Pitta gurneyi*
	Pycnonotidae	Spectacled bulbul	Pycnonotus erythropthalmos
Northern Annamites Rain Forests [54]	Phasianidae	Imperial pheasant	Lophura imperialis*
	Phasianidae	Edwards's pheasant	Lophura edwardsi
	Phasianidae	Vietnamese fireback	Lophura haitensis*
	Phasianidae	Crested argus	Rheinardia ocellata
	Timaliidae	Short-tailed scimitar-babbler	Jabouilleia danjoui
	Timaliidae	Sooty babbler	Stachyris herberti*
Cardamom Mountains Rain Forests [55]	Phasianidae	Chestnut-headed partridge	Arborophila cambodiana*
	Phasianidae	Siamese partridge	Arborophila diversa*
Northern Vietnam Lowland Rain Forests [56]	Phasianidae	Annam partridge	Arborophila merlini*
	Phasianidae	Edwards's pheasant	Lophura edwardsi
	Timaliidae	Short-tailed scimitar-babbler	Jabouilleia danjoui
	Timaliidae	Grey-faced tit-babbler	Macronous kelleyi
Southern Annamites Montane Rain Forests [57]	Phasianidae	Edwards's pheasant	Lophura edwardsi
	Phasianidae	Germain's peacock-pheasant	Polyplectron germaini
	Phasianidae	Crested argus	Rheinardia ocellata
	Timaliidae	Collared laughingthrush	Garrulax yersini*
	Timaliidae	Golden-winged laughingthrush	Garrulax ngoclinhensis*
	Timaliidae	Short-tailed scimitar-babbler	Jabouilleia danjoui
	Timaliidae	Grey-faced tit-babbler	Macronous kelleyi
	Timaliidae	Black-crowned barwing	Actinodura sodangorum*
	Timaliidae	Grey-crowned crocias	Crocias langbianis*
	Fringillidae	Vietnamese greenfinch	Carduelis monguilloti
Southern Vietnam Lowland Dry Forests [58]	Phasianidae	Germain's peacock-pheasant	Polyplectron germaini
	Timaliidae	Grey-faced tit-babbler	Macronous kelleyi
Southeastern Indochina Dry Evergreen Forests [59]	Phasianidae	Orange-necked partridge	Arborophila davidi*
	Phasianidae	Germain's peacock-pheasant	Polyplectron germaini
	Timaliidae	Grey-faced tit-babbler	Macronous kelleyi
Andaman Islands Rain Forests [60]	Accipitridae	Andaman serpent-eagle	Spilornis elgini*
	Rallidae	Andaman crake	Rallina canningi*
	Columbidae	Andaman wood-pigeon	Columba palumboides
	Columbidae	Andaman cuckoo-dove	Macropygia rufipennis
	Cuculidae	Andaman coucal	Centropus andamanensis*
	Strigidae	Andaman scops-owl	Otus balli*
	Strigidae	Andaman hawk-owl	Ninox affinis
	Bucconidae	Narcondam hornbill	Aceros narcondami*
	Picidae	Andaman woodpecker	Dryocopus hodgei*
	Dicruridae	Andaman drongo	Dicrurus andamanensis*
	Corvidae	Andaman treepie	Dendrocitta bayleyi*
	Sturnidae	White-headed starling	Sturnus erythropygius
Nicobar Islands Rain Forests [61]	Accipitridae	Nicobar serpent-eagle	Spilornis minimus*
	Accipitridae	Nicobar sparrowhawk	Accipiter butleri*
	Megapodiidae	Nicobar scrubfowl	Megapodius nicobariensis
	Columbidae	Andaman wood-pigeon	Columba palumboides
	Columbidae	Andaman cuckoo-dove	Macropygia rufipennis*
	Psittacidae	Nicobar parakeet	Psittacula caniceps*
	Strigidae	Andaman hawk-owl	Ninox affinis
	Sturnidae	White-headed starling	Sturnus erythropygius
	Pycnonotidae	Nicobar bulbul	Hypsipetes nicobariensis*

	Family	Common Name	Species
Tonle Sap–Mekong Peat Swamp Forests [62]	Threskiornithidae	Giant ibis	Pseudibis gigantea
Chao Phraya Freshwater Swamp Forests [64]	Hirundinidae	White-eyed river-martin	Pseudochelidon sirintarae
Tonle Sap Freshwater Swamp Forests [65]	Threskiornithidae	Giant ibis	Pseudibis gigantea
Irrawaddy Moist Deciduous Forests [67]	Timaliidae	White-throated babbler	Turdoides gularis
Irrawaddy Dry Forests [71]	Corvidae	Hooded treepie	Crypsirina cucullata
	Timaliidae	White-throated babbler	Turdoides gularis
Central Indochina Dry Forests [72]	Hirundinidae	White-eyed river-martin	Pseudochelidon sirintarae
	Timaliidae	Grey-faced tit-babbler	Macronous kelleyi
Northern Triangle Subtropical Forests [73]	Turdidae	Rusty-bellied shortwing	Brachypteryx hyperythra
Northern Indochina Subtropical Forests [74]	Timaliidae	Short-tailed scimitar-babbler	Jabouilleia danjoui
Northern Triangle Temperate Forests [76]	Turdidae	Rusty-bellied shortwing	Brachypteryx hyperythra
Northeast India–Myanmar pine forests [77]	Timaliidae	Striped laughingthrush	Garrulax virgatus
	Timaliidae	Brown-capped laughingthrush	Garrulax austeni
Myanmar Coastal Mangroves [78]	Alcedinidae	Brown-winged kingfisher	Pelargopsis amauropterus
Indochina Mangroves [79]	Alcedinidae	Brown-winged kingfisher	Pelargopsis amauropterus
	Pittidae	Gurney's pitta	Pitta gurneyi
	Pycnonotidae	Spectacled bulbul	Pycnonotus erythropthalmos
Peninsular Malaysian Montane Rain Forests [81]	Phasianidae	Mountain peacock-pheasant	Polyplectron inopinatum
	Phasianidae	Crested argus	Rheinardia ocellata
	Turdidae	Malayan whistling-thrush	Myiophonus robinsoni
	Muscicapidae	Rufous-vented niltava	Niltava sumatrana
	Timaliidae	Marbled wren-babbler	Napothera marmorata
Sumatran Lowland Rain Forests [82]	Strigidae	Simeulue scops-owl	Otus umbra*
	Irenidae	Blue-masked leafbird	Chloropsis venusta
	Pycnonotidae	Cream-striped bulbul	Pycnonotus leucogrammicus
	Pycnonotidae	Spot-necked bulbul	Pycnonotus tympanistrigus
	Pycnonotidae	Blue-wattled bulbul	Pycnonotus nieuwenhuisii
Sumatran Montane Rain Forests [83]	Phasianidae	Red-billed partridge	Arborophila rubrirostris*
	Phasianidae	Sumatran pheasant	Lophura hoogerwerfi*
	Phasianidae	Salvadori's pheasant	Lophura inornata*
	Phasianidae	Bronze-tailed peacock-pheasant	Polyplectron chalcurum
	Columbidae	Green-spectacled pigeon	Treron oxyura
	Columbidae	Pink-headed fruit-dove	Ptilinopus porphyreus
	Cuculidae	Sumatran ground-cuckoo	Carpococcyx viridis*
	Strigidae	Rajah scops-owl	Otus brookii
	Apodidae	Waterfall swift	Hydrochous gigas
	Trogonidae	Blue-tailed trogon	Harpactes reinwardtii
	Pittidae	Schneider's pitta	Pitta schneideri*
	Pittidae	Black-crowned pitta	Pitta venusta
	Dicruridae	Sumatran drongo	Dicrurus sumatranus*
	Campephagidae	Sunda minivet	Pericrocotus miniatus
	Irenidae	Blue-masked leafbird	Chloropsis venusta
	Turdidae	Shiny whistling-thrush	Myiophonus melanurus
	Muscicapidae	Rufous-vented niltava	Niltava sumatrana
	Muscicapidae	Sunda robin	Cinclidium diana
	Muscicapidae	Sumatran cochoa	Cochoa beccarii*
	Pycnonotidae	Cream-striped bulbul	Pycnonotus leucogrammicus
	Pycnonotidae	Spot-necked bulbul	Pycnonotus tympanistrigus
	Pycnonotidae	Sunda bulbul	Hypsipetes virescens
	Zosteropidae	Black-capped white-eye	Zosterops atricapillus
	Sylviidae	Sunda warbler	Seicercus grammiceps
	Timaliidae	Sunda laughingthrush	Garrulax palliatus
	Timaliidae	Vanderbilt's babbler	Malacocincla vanderbilti*

(continues)

Appendix I: Near-Endemic and Strict Endemic Birds by Ecoregion

	Family	Common Name	Species
Sumatran Montane Rain Forests [83] (continued)	Timaliidae	Rusty-breasted wren-babbler	Napothera rufipectus
	Timaliidae	Marbled wren-babbler	Napothera marmorata
	Fringillidae	Mountain serin	Serinus estherae
Mentawai Islands Rain Forests [84]	Strigidae	Enggano scops-owl	Otus enganensis*
	Strigidae	Mentawai scops-owl	Otus mentawi*
	Zosteropidae	Enggano white-eye	Zosterops salvadorii*
Borneo Peat Swamp Forests [86]	Zosteropidae	Javan white-eye	Zosterops flavus
Southern Borneo Freshwater Swamp Forests [89]	Zosteropidae	Javan white-eye	Zosterops flavus
Sundaland Heath Forests [90]	Zosteropidae	Pygmy white-eye	Oculocincta squamifrons
Western Java Rain Forests [91]	Accipitridae	Javan hawk-eagle	Spizaetus bartelsi
	Charadriidae	Javan plover	Charadrius javanicus
	Charadriidae	Sunda lapwing	Vanellus macropterus
	Cuculidae	Sunda coucal	Centropus nigrorufus
	Sturnidae	Black-winged starling	Sturnus melanopterus
	Zosteropidae	Javan white-eye	Zosterops flavus
	Timaliidae	White-breasted babbler	Stachyris grammiceps
	Timaliidae	White-bibbed babbler	Stachyris thoracica
	Timaliidae	Crescent-chested babbler	Stachyris melanothorax
	Timaliidae	Grey-cheeked tit-babbler	Macronous flavicollis
Eastern Java–Bali Rain Forests [92]	Accipitridae	Javan hawk-eagle	Spizaetus bartelsi
	Charadriidae	Javan plover	Charadrius javanicus
	Charadriidae	Sunda lapwing	Vanellus macropterus
	Cuculidae	Sunda coucal	Centropus nigrorufus
	Sturnidae	Bali myna	Leucopsar rothschildi*
	Timaliidae	White-breasted babbler	Stachyris grammiceps
	Timaliidae	White-bibbed babbler	Stachyris thoracica
	Timaliidae	Crescent-chested babbler	Stachyris melanothorax
	Timaliidae	Grey-cheeked tit-babbler	Macronous flavicollis
Western Java Montane Rain Forests [93]	Accipitridae	Javan hawk-eagle	Spizaetus bartelsi
	Phasianidae	Chestnut-bellied partridge	Arborophila javanica*
	Columbidae	Green-spectacled pigeon	Treron oxyura
	Columbidae	Pink-headed fruit-dove	Ptilinopus porphyreus
	Columbidae	Dark-backed imperial-pigeon	Ducula lacernulata
	Strigidae	Javan scops-owl	Otus angelinae*
	Caprimulgidae	Salvadori's nightjar	Caprimulgus pulchellus
	Apodidae	Waterfall swift	Hydrochous gigas
	Apodidae	Volcano swiftlet	Aerodramus vulcanorum*
	Trogonidae	Blue-tailed trogon	Harpactes reinwardtii
	Capitonidae	Brown-throated barbet	Megalaima corvina*
	Capitonidae	Flame-fronted barbet	Megalaima armillaris
	Rhipiduridae	Rufous-tailed fantail	Rhipidura phoenicura
	Rhipiduridae	White-bellied fantail	Rhipidura euryura
	Campephagidae	Sunda minivet	Pericrocotus miniatus
	Muscicapidae	Sunda robin	Cinclidium diana
	Muscicapidae	Javan cochoa	Cochoa azurea*
	Aegithalidae	Pygmy tit	Psaltria exilis*
	Pycnonotidae	Sunda bulbul	Hypsipetes virescens
	Zosteropidae	Javan grey-throated white-eye	Lophozosterops javanicus
	Sylviidae	Javan tesia	Tesia superciliaris*
	Sylviidae	Sunda warbler	Seicercus grammiceps
	Timaliidae	Rufous-fronted laughingthrush	Garrulax rufifrons
	Timaliidae	White-bibbed babbler	Stachyris thoracica
	Timaliidae	Crescent-chested babbler	Stachyris melanothorax
	Timaliidae	Grey-cheeked tit-babbler	Macronous flavicollis
	Timaliidae	Javan fulvetta	Alcippe pyrrhoptera*
	Timaliidae	Spotted crocias	Crocias albonotatus*
	Nectariniidae	White-flanked sunbird	Aethopyga eximia
	Fringillidae	Mountain serin	Serinus estherae
Eastern Java–Bali Montane Rain Forests [94]	Accipitridae	Javan hawk-eagle	Spizaetus bartelsi
	Columbidae	Pink-headed fruit-dove	Ptilinopus porphyreus
	Columbidae	Dark-backed imperial-pigeon	Ducula lacernulata
	Strigidae	Javan owlet	Glaucidium castanopterum

(continues)

	Family	Common Name	Species
Eastern Java–Bali Montane Rain Forests [94] (continued)	Caprimulgidae	Salvadori's nightjar	*Caprimulgus pulchellus*
	Capitonidae	Flame-fronted barbet	*Megalaima armillaris*
	Rhipiduridae	Rufous-tailed fantail	*Rhipidura phoenicura*
	Rhipiduridae	White-bellied fantail	*Rhipidura euryura*
	Campephagidae	Sunda minivet	*Pericrocotus miniatus*
	Pycnonotidae	Sunda bulbul	*Hypsipetes virescens*
	Zosteropidae	Javan grey-throated white-eye	*Lophozosterops javanicus*
	Sylviidae	Sunda warbler	*Seicercus grammiceps*
	Timaliidae	Rufous-fronted laughingthrush	*Garrulax rufifrons*
	Timaliidae	White-bibbed babbler	*Stachyris thoracica*
	Timaliidae	Crescent-chested babbler	*Stachyris melanothorax*
	Timaliidae	Grey-cheeked tit-babbler	*Macronous flavicollis*
	Nectariniidae	White-flanked sunbird	*Aethopyga eximia*
	Fringillidae	Mountain serin	*Serinus estherae*
Borneo Montane Rain Forests [95]	Accipitridae	Mountain serpent-eagle	*Spilornis kinabaluensis*
	Falconidae	White-fronted falconet	*Microhierax latifrons*
	Phasianidae	Red-breasted partridge	*Arborophila hyperythra*
	Phasianidae	Crimson-headed partridge	*Haematortyx sanguiniceps*
	Strigidae	Rajah scops-owl	*Otus brookii*
	Podargidae	Dulit frogmouth	*Batrachostomus harterti*
	Trogonidae	Whitehead's trogon	*Harpactes whiteheadi*
	Capitonidae	Mountain barbet	*Megalaima monticola*
	Capitonidae	Golden-naped barbet	*Megalaima pulcherrima*
	Capitonidae	Bornean barbet	*Megalaima eximia*
	Eurylaimidae	Hose's broadbill	*Calyptomena hosii**
	Eurylaimidae	Whitehead's broadbill	*Calyptomena whiteheadi*
	Pachycephalidae	Bornean whistler	*Pachycephala hypoxantha*
	Oriolidae	Black oriole	*Oriolus hosii**
	Turdidae	Everett's thrush	*Zoothera everetti*
	Turdidae	Fruit-hunter	*Chlamydochaera jefferyi*
	Muscicapidae	Eyebrowed jungle-flycatcher	*Rhinomyias gularis*
	Zosteropidae	Black-capped white-eye	*Zosterops atricapillus*
	Zosteropidae	Pygmy white-eye	*Oculocincta squamifrons*
	Zosteropidae	Mountain black-eye	*Chlorocharis emiliae*
	Sylviidae	Bornean stubtail	*Urosphena whiteheadi*
	Sylviidae	Friendly bush-warbler	*Bradypterus accentor*
	Timaliidae	Sunda laughingthrush	*Garrulax palliatus*
	Timaliidae	Bare-headed laughingthrush	*Garrulax calvus*
	Timaliidae	Mountain wren-babbler	*Napothera crassa*
	Timaliidae	Chestnut-crested yuhina	*Yuhina everetti*
	Dicaeidae	Black-sided flowerpecker	*Dicaeum monticolum*
	Nectariniidae	Whitehead's spiderhunter	*Arachnothera juliae*
Borneo Lowland Rain Forests [96]	Falconidae	White-fronted falconet	*Microhierax latifrons*
	Phasianidae	Red-breasted partridge	*Arborophila hyperythra*
	Phasianidae	Crimson-headed partridge	*Haematortyx sanguiniceps*
	Strigidae	Mantanani scops-owl	*Otus mantananensis*
	Capitonidae	Mountain barbet	*Megalaima monticola*
	Capitonidae	Bornean barbet	*Megalaima eximia*
	Pittidae	Blue-headed pitta	*Pitta baudii*
	Artamidae	Bornean bristlehead	*Pityriasis gymnocephala*
	Muscicapidae	Long-billed blue-flycatcher	*Cyornis caerulatus*
	Muscicapidae	White-crowned shama	*Copsychus stricklandii*
	Pycnonotidae	Blue-wattled bulbul	*Pycnonotus nieuwenhuisii*
	Zosteropidae	Javan white-eye	*Zosterops flavus*
	Zosteropidae	Pygmy white-eye	*Oculocincta squamifrons*
	Timaliidae	Sunda laughingthrush	*Garrulax palliatus*
	Timaliidae	Black-browed babbler	*Malacocincla perspicillata**
	Timaliidae	Bornean wren-babbler	*Ptilocichla leucogrammica*
	Timaliidae	Chestnut-crested yuhina	*Yuhina everetti*
Palawan Rain Forests [97]	Phasianidae	Palawan peacock-pheasant	*Polyplectron emphanum**
	Columbidae	Grey imperial-pigeon	*Ducula pickeringii*
	Psittacidae	Blue-headed racquet-tail	*Prioniturus platenae**
	Strigidae	Mantanani scops-owl	*Otus mantananensis*
	Strigidae	Palawan scops-owl	*Otus fuliginosus**

(continues)

	Family	Common Name	Species
Palawan Rain Forests [97] (continued)	Apodidae	Palawan swiftlet	Aerodramus palawanensis*
	Bucconidae	Palawan hornbill	Anthracoceros marchei*
	Monarchidae	Blue paradise-flycatcher	Terpsiphone cyanescens*
	Irenidae	Yellow-throated leafbird	Chloropsis palawanensis*
	Muscicapidae	Palawan flycatcher	Ficedula platenae*
	Muscicapidae	Palawan blue-flycatcher	Cyornis lemprieri*
	Muscicapidae	White-vented shama	Copsychus niger*
	Pycnonotidae	Sulphur-bellied bulbul	Ixos palawanensis*
	Timaliidae	Ashy-headed babbler	Malacocincla cinereiceps*
	Timaliidae	Palawan babbler	Malacopteron palawanense*
	Timaliidae	Falcated wren-babbler	Ptilocichla falcata*
	Timaliidae	Palawan striped-babbler	Stachyris hypogrammica*
	Paridae	Palawan tit	Parus amabilis*
	Paridae	White-fronted tit	Parus semilarvatus
	Dicaeidae	Palawan flowerpecker	Prionochilus plateni*
Luzon Rain Forests [98]	Turnicidae	Spotted buttonquail	Turnix ocellata
	Turnicidae	Luzon buttonquail	Turnix worcesteri
	Rallidae	Brown-banded rail	Lewinia mirificus
	Columbidae	Luzon bleeding-heart	Gallicolumba luzonica
	Columbidae	Whistling green-pigeon	Treron formosae
	Columbidae	Flame-breasted fruit-dove	Ptilinopus marchei
	Columbidae	Cream-breasted fruit-dove	Ptilinopus merrilli
	Psittacidae	Luzon racquet-tail	Prioniturus montanus
	Psittacidae	Green racquet-tail	Prioniturus luconensis*
	Cuculidae	Red-crested malkoha	Phaenicophaeus superciliosus*
	Cuculidae	Scale-feathered malkoha	Phaenicophaeus cumingi
	Cuculidae	Rufous coucal	Centropus unirufus
	Strigidae	Luzon scops-owl	Otus longicornis
	Strigidae	Ryukyu scops-owl	Otus elegans
	Bucconidae	Luzon hornbill	Penelopides manilloe
	Pittidae	Whiskered pitta	Pitta kochi
	Pachycephalidae	Green-backed whistler	Pachycephala albiventris
	Monarchidae	Short-crested monarch	Hypothymis helenae
	Monarchidae	Celestial monarch	Hypothymis coelestis
	Oriolidae	White-lored oriole	Oriolus albiloris
	Oriolidae	Isabela oriole	Oriolus isabellae*
	Campephagidae	Blackish cuckoo-shrike	Coracina coerulescens
	Turdidae	Ashy thrush	Zoothera cinerea
	Muscicapidae	Rusty-flanked jungle-flycatcher	Rhinomyias insignis
	Muscicapidae	Ash-breasted flycatcher	Muscicapa randi
	Muscicapidae	Furtive flycatcher	Ficedula disposita*
	Muscicapidae	Blue-breasted flycatcher	Cyornis herioti
	Muscicapidae	Luzon redstart	Rhyacornis bicolor
	Sylviidae	Philippine bush-warbler	Cettia seebohmi
	Sylviidae	Long-tailed bush-warbler	Bradypterus caudatus
	Sylviidae	Grey-backed tailorbird	Orthotomus derbianus
	Timaliidae	Luzon wren-babbler	Napothera rabori*
	Timaliidae	Golden-crowned babbler	Stachyris dennistouni
	Timaliidae	Chestnut-faced babbler	Stachyris whiteheadi
	Timaliidae	Luzon striped-babbler	Stachyris striata*
	Rhabdornithidae	Long-billed rhabdornis	Rhabdornis grandis
	Paridae	White-fronted tit	Parus semilarvatus
	Estrildidae	Green-faced parrotfinch	Erythrura viridifacies
	Dicaeidae	Flame-crowned flowerpecker	Dicaeum anthonyi
	Zosteropidae	Lowland white-eye	Zosterops meyeni
Mindoro Rain Forests [99]	Columbidae	Mindoro bleeding-heart	Gallicolumba platenae*
	Columbidae	Mindoro imperial-pigeon	Ducula mindorensis*
	Cuculidae	Black-hooded coucal	Centropus steerii*
	Strigidae	Mindoro scops-owl	Otus mindorensis*
	Strigidae	Mantanani scops-owl	Otus mantananensis
	Bucconidae	Mindoro hornbill	Penelopides mindorensis*

(continues)

	Family	Common Name	Species
Mindoro Rain Forests [99] (*continued*)	Pachycephalidae	Green-backed whistler	*Pachycephala albiventris*
	Turdidae	Ashy thrush	*Zoothera cinerea*
	Muscicapidae	Luzon redstart	*Rhyacornis bicolor*
	Dicaeidae	Scarlet-collared flowerpecker	*Dicaeum retrocinctum*
	Laniidae	Grey-capped shrike	*Lanius validirostris*
Greater Negros–Panay Rain Forests [100]	Turnicidae	Spotted buttonquail	*Turnix ocellata*
	Columbidae	Negros bleeding-heart	*Gallicolumba keayi**
	Columbidae	Negros fruit-dove	*Ptilinopus arcanus**
	Strigidae	Mantanani scops-owl	*Otus mantananensis*
	Apodidae	Philippine needletail	*Mearnsia picina*
	Bucconidae	Tarictic hornbill	*Penelopides panini**
	Bucconidae	Writhe-billed hornbill	*Aceros waldeni**
	Monarchidae	Celestial monarch	*Hypothymis coelestis*
	Campephagidae	Blackish cuckoo-shrike	*Coracina coerulescens*
	Campephagidae	White-winged cuckoo-shrike	*Coracina ostenta**
	Irenidae	Philippine leafbird	*Chloropsis flavipennis*
	Muscicapidae	Negros jungle-flycatcher	*Rhinomyias albigularis**
	Muscicapidae	Ash-breasted flycatcher	*Muscicapa randi*
	Muscicapidae	Black shama	*Copsychus cebuensis**
	Pycnonotidae	Streak-breasted bulbul	*Ixos siquijorensis**
	Timaliidae	Flame-templed babbler	*Stachyris speciosa**
	Timaliidae	Panay striped-babbler	*Stachyris latistriata**
	Timaliidae	Negros striped-babbler	*Stachyris nigrorum**
	Estrildidae	Green-faced parrotfinch	*Erythrura viridifacies*
	Dicaeidae	Cebu flowerpecker	*Dicaeum quadricolor**
	Dicaeidae	Visayan flowerpecker	*Dicaeum haematostictum**
	Dicaeidae	Scarlet-collared flowerpecker	*Dicaeum retrocinctum*
	Laniidae	Grey-capped shrike	*Lanius validirostris*
Mindanao–Eastern Visayas Rain Forests [101]	Rallidae	Brown-banded rail	*Lewinia mirificus*
	Columbidae	Mindanao bleeding-heart	*Gallicolumba criniger*
	Columbidae	Mindanao brown-dove	*Phapitreron brunneiceps*
	Columbidae	Grey imperial-pigeon	*Ducula pickeringii*
	Cuculidae	Black-faced coucal	*Centropus melanops**
	Strigidae	Mindanao eagle-owl	*Mimizuku gurneyi*
	Apodidae	Philippine needletail	*Mearnsia picina*
	Alcedinidae	Silvery kingfisher	*Alcedo argentata**
	Alcedinidae	Blue-capped kingfisher	*Actenoides hombroni*
	Bucconidae	Mindanao hornbill	*Penelopides affinis*
	Bucconidae	Samar hornbill	*Penelopides samarensis**
	Bucconidae	Writhed hornbill	*Aceros leucocephalus*
	Pittidae	Azure-breasted pitta	*Pitta steerii**
	Eurylaimidae	Wattled broadbill	*Eurylaimus steerii**
	Eurylaimidae	Visayan wattled broadbill	*Eurylaimus samarensis**
	Rhipiduridae	Blue fantail	*Rhipidura superciliaris**
	Monarchidae	Short-crested monarch	*Hypothymis helenae*
	Monarchidae	Celestial monarch	*Hypothymis coelestis*
	Muscicapidae	Little slaty flycatcher	*Ficedula basilanica**
	Muscicapidae	Russet-tailed flycatcher	*Ficedula crypta*
	Pycnonotidae	Zamboanga bulbul	*Ixos rufigularis**
	Pycnonotidae	Yellowish bulbul	*Ixos everetti*
	Sylviidae	Long-tailed bush-warbler	*Bradypterus caudatus*
	Sylviidae	Rufous-headed tailorbird	*Orthotomus heterolaemus*
	Sylviidae	Yellow-breasted tailorbird	*Orthotomus samarensis**
	Sylviidae	White-browed tailorbird	*Orthotomus nigriceps**
	Sylviidae	White-eared tailorbird	*Orthotomus cinereiceps**
	Timaliidae	Striated wren-babbler	*Ptilocichla mindanensis**
	Timaliidae	Pygmy babbler	*Stachyris plateni*
	Timaliidae	Rusty-crowned babbler	*Stachyris capitalis*
	Timaliidae	Brown tit-babbler	*Macronous striaticeps*
	Timaliidae	Miniature tit-babbler	*Micromacronus leytensis*
	Paridae	White-fronted tit	*Parus semilarvatus*
	Dicaeidae	Whiskered flowerpecker	*Dicaeum proprium*

(*continues*)

	Family	Common Name	Species
Mindanao–Eastern Visayas Rain Forests [101] (continued)	Dicaeidae	Olive-capped flowerpecker	Dicaeum nigrilore
	Dicaeidae	Flame-crowned flowerpecker	Dicaeum anthonyi
Luzon Montane Rain Forests [102]	Turnicidae	Spotted buttonquail	Turnix ocellata
	Turnicidae	Luzon buttonquail	Turnix worcesteri
	Rallidae	Brown-banded rail	Lewinia mirificus
	Columbidae	Luzon bleeding-heart	Gallicolumba luzonica
	Columbidae	Flame-breasted fruit-dove	Ptilinopus marchei
	Columbidae	Cream-breasted fruit-dove	Ptilinopus merrilli
	Psittacidae	Luzon racquet-tail	Prioniturus montanus
	Cuculidae	Scale-feathered malkoha	Phaenicophaeus cumingi
	Cuculidae	Rufous coucal	Centropus unirufus
	Strigidae	Luzon scops-owl	Otus longicornis
	Bucconidae	Luzon hornbill	Penelopides manilloe
	Pittidae	Whiskered pitta	Pitta kochi
	Pachycephalidae	Green-backed whistler	Pachycephala albiventris
	Oriolidae	White-lored oriole	Oriolus albiloris
	Turdidae	Ashy thrush	Zoothera cinerea
	Muscicapidae	Rusty-flanked jungle-flycatcher	Rhinomyias insignis
	Muscicapidae	Ash-breasted flycatcher	Muscicapa randi
	Muscicapidae	Blue-breasted flycatcher	Cyornis herioti
	Muscicapidae	Luzon redstart	Rhyacornis bicolor
	Sylviidae	Philippine bush-warbler	Cettia seebohmi
	Sylviidae	Long-tailed bush-warbler	Bradypterus caudatus
	Timaliidae	Golden-crowned babbler	Stachyris dennistouni
	Timaliidae	Chestnut-faced babbler	Stachyris whiteheadi
	Rhabdornithidae	Long-billed rhabdornis	Rhabdornis grandis
	Estrildidae	Green-faced parrotfinch	Erythrura viridifacies
	Dicaeidae	Flame-crowned flowerpecker	Dicaeum anthonyi
	Fringillidae	White-cheeked bullfinch	Pyrrhula leucogenis
	Laniidae	Grey-capped shrike	Lanius validirostris
Mindanao Montane Rain Forests [103]	Columbidae	Mindanao brown-dove	Phapitreron brunneiceps
	Psittacidae	Mindanao racquet-tail	Prioniturus waterstradti*
	Loriidae	Mindanao lorikeet	Trichoglossus johnstoniae*
	Strigidae	Mindanao scops-owl	Otus mirus*
	Strigidae	Mindanao eagle-owl	Mimizuku gurneyi
	Apodidae	Whitehead's swiftlet	Aerodramus whiteheadi
	Apodidae	Philippine needletail	Mearnsia picina
	Alcedinidae	Blue-capped kingfisher	Actenoides hombroni
	Bucconidae	Mindanao hornbill	Penelopides affinis
	Bucconidae	Writhed hornbill	Aceros leucocephalus
	Rhipiduridae	Black-and-cinnamon fantail	Rhipidura nigrocinnamomea*
	Campephagidae	McGregor's cuckoo-shrike	Coracina mcgregori*
	Sturnidae	Apo myna	Basilornis miranda*
	Muscicapidae	Mindanao jungle-flycatcher	Rhinomyias goodfellowi*
	Muscicapidae	Russet-tailed flycatcher	Ficedula crypta
	Zosteropidae	Mindanao white-eye	Lophozosterops goodfellowi*
	Zosteropidae	Cinnamon white-eye	Hypocryptadius cinnamomeus*
	Sylviidae	Long-tailed bush-warbler	Bradypterus caudatus
	Sylviidae	Rufous-headed tailorbird	Orthotomus heterolaemus
	Timaliidae	Bagobo babbler	Trichastoma woodi*
	Timaliidae	Pygmy babbler	Stachyris plateni
	Timaliidae	Rusty-crowned babbler	Stachyris capitalis
	Timaliidae	Brown tit-babbler	Macronous striaticeps
	Timaliidae	Miniature tit-babbler	Micromacronus leytensis
	Estrildidae	Red-eared parrotfinch	Erythrura coloria*
	Dicaeidae	Whiskered flowerpecker	Dicaeum proprium
	Dicaeidae	Olive-capped flowerpecker	Dicaeum nigrilore
	Dicaeidae	Flame-crowned flowerpecker	Dicaeum anthonyi
	Nectariniidae	Grey-hooded sunbird	Aethopyga primigenius*
	Nectariniidae	Mount Apo sunbird	Aethopyga boltoni*
	Nectariniidae	Linas sunbird	Aethopyga linaraborae*

(continues)

	Family	Common Name	Species
Mindanao Montane Rain Forests [103] (*continued*)	Fringillidae	Mountain serin	*Serinus estherae*
	Fringillidae	White-cheeked bullfinch	*Pyrrhula leucogenis*
	Laniidae	Grey-capped shrike	*Lanius validirostris*
Sulu Archipelago Rain Forests [104]	Columbidae	Sulu bleeding-heart	*Gallicolumba menagei**
	Columbidae	Dark-eared dove	*Phapitreron cinereiceps**
	Columbidae	Grey imperial-pigeon	*Ducula pickeringii*
	Psittacidae	Blue-winged racquet-tail	*Prioniturus verticalis**
	Strigidae	Mantanani scops-owl	*Otus mantananensis*
	Apodidae	Philippine needletail	*Mearnsia picina*
	Bucconidae	Sulu hornbill	*Anthracoceros montani**
	Monarchidae	Celestial monarch	*Hypothymis coelestis*
	Pycnonotidae	Yellowish bulbul	*Ixos everetti*
	Timaliidae	Brown tit-babbler	*Macronous striaticeps*
Sumatran Tropical Pine Forests [105]	Phasianidae	Bronze-tailed peacock-pheasant	*Polyplectron chalcurum*
	Campephagidae	Sunda minivet	*Pericrocotus miniatus*
	Irenidae	Blue-masked leafbird	*Chloropsis venusta*
	Turdidae	Shiny whistling-thrush	*Myiophonus melanurus*
	Muscicapidae	Rufous-vented niltava	*Niltava sumatrana*
	Muscicapidae	Sunda robin	*Cinclidium diana*
	Pycnonotidae	Cream-striped bulbul	*Pycnonotus leucogrammicus*
	Pycnonotidae	Spot-necked bulbul	*Pycnonotus tympanistrigus*
	Pycnonotidae	Sunda bulbul	*Hypsipetes virescens*
	Zosteropidae	Black-capped white-eye	*Zosterops atricapillus*
	Timaliidae	Sunda laughingthrush	*Garrulax palliatus*
	Timaliidae	Rusty-breasted wren-babbler	*Napothera rufipectus*
Luzon Tropical Pine Forests [106]	Turnicidae	Luzon buttonquail	*Turnix worcesteri*
	Rallidae	Brown-banded rail	*Lewinia mirificus*
	Columbidae	Luzon bleeding-heart	*Gallicolumba luzonica*
	Columbidae	Flame-breasted fruit-dove	*Ptilinopus marchei*
	Psittacidae	Luzon racquet-tail	*Prioniturus montanus*
	Cuculidae	Scale-feathered malkoha	*Phaenicophaeus cumingi*
	Strigidae	Luzon scops-owl	*Otus longicornis*
	Apodidae	Whitehead's swiftlet	*Aerodramus whiteheadi*
	Pittidae	Whiskered pitta	*Pitta kochi*
	Pachycephalidae	Green-backed whistler	*Pachycephala albiventris*
	Oriolidae	White-lored oriole	*Oriolus albiloris*
	Turdidae	Ashy thrush	*Zoothera cinerea*
	Muscicapidae	Rusty-flanked jungle-flycatcher	*Rhinomyias insignis*
	Muscicapidae	Ash-breasted flycatcher	*Muscicapa randi*
	Muscicapidae	Luzon redstart	*Rhyacornis bicolor*
	Sylviidae	Philippine bush-warbler	*Cettia seebohmi*
	Sylviidae	Long-tailed bush-warbler	*Bradypterus caudatus*
	Timaliidae	Golden-crowned babbler	*Stachyris dennistouni*
	Timaliidae	Chestnut-faced babbler	*Stachyris whiteheadi*
	Rhabdornithidae	Long-billed rhabdornis	*Rhabdornis grandis*
	Dicaeidae	Flame-crowned flowerpecker	*Dicaeum anthonyi*
	Fringillidae	White-cheeked bullfinch	*Pyrrhula leucogenis*
	Laniidae	Grey-capped shrike	*Lanius validirostris*
Sunda Shelf Mangroves [107]	Muscicapidae	White-crowned shama	*Copsychus stricklandii*
Kinabalu Montane Alpine Meadows [108]	Accipitridae	Mountain serpent-eagle	*Spilornis kinabaluensis*
	Phasianidae	Red-breasted partridge	*Arborophila hyperythra*
	Phasianidae	Crimson-headed partridge	*Haematortyx sanguiniceps*
	Podargidae	Dulit frogmouth	*Batrachostomus harterti*
	Trogonidae	Whitehead's trogon	*Harpactes whiteheadi*
	Capitonidae	Mountain barbet	*Megalaima monticola*
	Capitonidae	Golden-naped barbet	*Megalaima pulcherrima*
	Capitonidae	Bornean barbet	*Megalaima eximia*
	Eurylaimidae	Whitehead's broadbill	*Calyptomena whiteheadi*
	Pachycephalidae	Bornean whistler	*Pachycephala hypoxantha*
	Turdidae	Everett's thrush	*Zoothera everetti*
	Turdidae	Fruit-hunter	*Chlamydochaera jefferyi*
	Muscicapidae	Eyebrowed jungle-flycatcher	*Rhinomyias gularis*

(continues)

	Family	Common Name	Species
Kinabalu Montane Alpine Meadows [108] (*continued*)	Zosteropidae	Black-capped white-eye	*Zosterops atricapillus*
	Zosteropidae	Pygmy white-eye	*Oculocincta squamifrons*
	Zosteropidae	Mountain black-eye	*Chlorocharis emiliae*
	Sylviidae	Bornean stubtail	*Urosphena whiteheadi*
	Sylviidae	Friendly bush-warbler	*Bradypterus accentor*
	Timaliidae	Sunda laughingthrush	*Garrulax palliatus*
	Timaliidae	Bare-headed laughingthrush	*Garrulax calvus*
	Timaliidae	Mountain wren-babbler	*Napothera crassa*
	Timaliidae	Chestnut-crested yuhina	*Yuhina everetti*
	Dicaeidae	Black-sided flowerpecker	*Dicaeum monticolum*
	Nectariniidae	Bornean spiderhunter	*Arachnothera everetti**
	Nectariniidae	Whitehead's spiderhunter	*Arachnothera juliae*
Sulawesi Lowland Rain Forests [109]	Accipitridae	Small sparrowhawk	*Accipiter nanus*
	Megapodiidae	Sula scrubfowl	*Megapodius bernsteinii**
	Megapodiidae	Maleo	*Macrocephalon maleo*
	Rallidae	Platen's rail	*Aramidopsis plateni*
	Rallidae	Bare-faced rail	*Gymnocrex rosenbergii*
	Rallidae	Isabelline waterhen	*Amaurornis isabellinus*
	Columbidae	Dusky cuckoo-dove	*Macropygia magna*
	Columbidae	Sulawesi ground-dove	*Gallicolumba tristigmata*
	Columbidae	Maroon-chinned fruit-dove	*Ptilinopus subgularis**
	Columbidae	White-bellied imperial-pigeon	*Ducula forsteni*
	Columbidae	Grey-headed imperial-pigeon	*Ducula radiata*
	Columbidae	Pink-headed imperial-pigeon	*Ducula rosacae*
	Columbidae	Grey imperial-pigeon	*Ducula pickeringii*
	Columbidae	Silver-tipped imperial-pigeon	*Ducula luctuosa**
	Psittacidae	Yellowish-breasted racquet-tail	*Prioniturus flavicans**
	Psittacidae	Moluccan hanging-parrot	*Loriculus amabilis*
	Psittacidae	Sangihe hanging-parrot	*Loriculus catamene**
	Psittacidae	Pygmy hanging-parrot	*Loriculus exilis**
	Loriidae	Red-and-blue lory	*Eos histrio**
	Loriidae	Yellow-and-green lorikeet	*Trichoglossus flavoviridis*
	Cuculidae	Sulawesi hawk-cuckoo	*Cuculus crassirostris*
	Cuculidae	Bay coucal	*Centropus celebensis**
	Strigidae	Ochre-bellied hawk-owl	*Ninox ochracea*
	Strigidae	Speckled hawk-owl	*Ninox punctulata*
	Tytonidae	Minahassa owl	*Tyto inexspectata*
	Tytonidae	Taliabu owl	*Tyto nigrobrunnea**
	Tytonidae	Sulawesi owl	*Tyto rosenbergii*
	Caprimulgidae	Diabolical nightjar	*Eurostopodus diabolicus*
	Caprimulgidae	Sulawesi nightjar	*Caprimulgus celebensis**
	Alcedinidae	Sulawesi kingfisher	*Ceyx fallax**
	Alcedinidae	Lilac kingfisher	*Cittura cyanotis**
	Alcedinidae	Black-billed kingfisher	*Pelargopsis melanorhyncha**
	Alcedinidae	Talaud kingfisher	*Todirhamphus enigma**
	Alcedinidae	Green-backed kingfisher	*Actenoides monachus**
	Alcedinidae	Scaly kingfisher	*Actenoides princeps*
	Meropidae	Purple-bearded bee-eater	*Meropogon forsteni*
	Coraciidae	Purple-winged roller	*Coracias temminckii*
	Bucconidae	Sulawesi hornbill	*Penelopides exarhatus**
	Bucconidae	Knobbed hornbill	*Aceros cassidix*
	Acanthizidae	Rufous-sided gerygone	*Gerygone dorsalis*
	Pachycephalidae	Sulphur-bellied whistler	*Pachycephala sulfuriventer*
	Pachycephalidae	Drab whistler	*Pachycephala griseonota*
	Rhipiduridae	Rusty-flanked fantail	*Rhipidura teysmanni*
	Monarchidae	Cerulean paradise-flycatcher	*Eutrichomyias rowleyi**
	Monarchidae	White tipped monarch	*Monarcha everetti**
	Dicruridae	Sulawesi drongo	*Dicrurus montanus*
	Corvidae	Banggai crow	*Corvus unicolor**
	Campephagidae	Cerulean cuckoo-shrike	*Coracina temminckii*
	Campephagidae	Pied cuckoo-shrike	*Coracina bicolor**
	Campephagidae	White-rumped cuckoo-shrike	*Coracina leucopygia*
	Campephagidae	Sula cuckoo-shrike	*Coracina sula**

(*continues*)

	Family	Common Name	Species
Sulawesi Lowland Rain Forests [109] (*continued*)	Campephagidae	Slaty cuckoo-shrike	*Coracina schistacea**
	Campephagidae	White-rumped triller	*Lalage leucopygialis**
	Turdidae	Rusty-backed thrush	*Zoothera erythronota**
	Sturnidae	Pale-bellied myna	*Acridotheres cinereus*
	Sturnidae	Sulawesi myna	*Basilornis celebensis*
	Sturnidae	Helmeted myna	*Basilornis galeatus**
	Sturnidae	White-necked myna	*Streptocitta albicollis*
	Sturnidae	Bare-eyed myna	*Streptocitta albertinae**
	Sturnidae	Fiery-browed myna	*Enodes erythrophris*
	Sturnidae	Finch-billed myna	*Scissirostrum dubium**
	Muscicapidae	Henna-tailed jungle-flycatcher	*Rhinomyias colonus**
	Muscicapidae	Rufous-throated flycatcher	*Ficedula rufigula**
	Zosteropidae	Sulawesi white-eye	*Zosterops consobrinorum**
	Zosteropidae	Black-ringed white-eye	*Zosterops anomalus*
	Sylviidae	Sulawesi leaf-warbler	*Phylloscopus sarasinorum*
	Dicaeidae	Crimson-crowned flowerpecker	*Dicaeum nehrkorni*
	Dicaeidae	Red-chested flowerpecker	*Dicaeum maugei*
	Dicaeidae	Grey-sided flowerpecker	*Dicaeum celebicum**
	Nectariniidae	Elegant sunbird	*Aethopyga duyvenbodei**
Sulawesi Montane Rain Forests [110]	Accipitridae	Small sparrowhawk	*Accipiter nanus*
	Megapodiidae	Maleo	*Macrocephalon maleo*
	Rallidae	Platen's rail	*Aramidopsis plateni*
	Rallidae	Bare-faced rail	*Gymnocrex rosenbergii*
	Rallidae	Isabelline waterhen	*Amaurornis isabellinus*
	Scolopacidae	Sulawesi woodcock	*Scolopax celebensis**
	Columbidae	Sulawesi ground-dove	*Gallicolumba tristigmata*
	Columbidae	Red-eared fruit-dove	*Ptilinopus fischeri**
	Columbidae	White-bellied imperial-pigeon	*Ducula forsteni*
	Columbidae	Grey-headed imperial-pigeon	*Ducula radiata*
	Columbidae	Sombre pigeon	*Cryptophaps poecilorrhoa**
	Loriidae	Yellow-and-green lorikeet	*Trichoglossus flavoviridis*
	Cuculidae	Sulawesi hawk-cuckoo	*Cuculus crassirostris*
	Strigidae	Ochre-bellied hawk-owl	*Ninox ochracea*
	Tytonidae	Minahassa owl	*Tyto inexspectata*
	Caprimulgidae	Diabolical nightjar	*Eurostopodus diabolicus*
	Alcedinidae	Scaly kingfisher	*Actenoides princeps*
	Meropidae	Purple-bearded bee-eater	*Meropogon forsteni*
	Coraciidae	Purple-winged roller	*Coracias temminckii*
	Meliphagidae	Dark-eared honeyeater	*Myza celebensis**
	Meliphagidae	Greater streaked honeyeater	*Myza sarasinorum**
	Pachycephalidae	Olive-flanked whistler	*Hylocitrea bonensis**
	Pachycephalidae	Maroon-backed whistler	*Coracornis raveni**
	Pachycephalidae	Sulphur-bellied whistler	*Pachycephala sulfuriventer*
	Rhipiduridae	Rusty-flanked fantail	*Rhipidura teysmanni*
	Dicruridae	Sulawesi drongo	*Dicrurus montanus*
	Campephagidae	Cerulean cuckoo-shrike	*Coracina temminckii*
	Campephagidae	Pygmy cuckoo-shrike	*Coracina abbotti**
	Turdidae	Geomalia	*Geomalia heinrichi**
	Turdidae	Sulawesi thrush	*Cataponera turdoides**
	Turdidae	Great shortwing	*Heinrichia calligyna**
	Sturnidae	Pale-bellied myna	*Acridotheres cinereus*
	Sturnidae	Sulawesi myna	*Basilornis celebensis*
	Sturnidae	Fiery-browed myna	*Enodes erythrophris*
	Muscicapidae	Lompobattang flycatcher	*Ficedula bonthaina**
	Muscicapidae	Matinan flycatcher	*Cyornis sanfordi**
	Muscicapidae	Blue-fronted flycatcher	*Cyornis hoevelli**
	Zosteropidae	Black-ringed white-eye	*Zosterops anomalus*
	Zosteropidae	Streak-headed white-eye	*Lophozosterops squamiceps**
	Sylviidae	Chestnut-backed bush-warbler	*Bradypterus castaneus*
	Sylviidae	Sulawesi leaf-warbler	*Phylloscopus sarasinorum*
	Timaliidae	Malia	*Malia grata**
	Dicaeidae	Crimson-crowned flowerpecker	*Dicaeum nehrkorni*
	Fringillidae	Mountain serin	*Serinus estherae*

	Family	Common Name	Species
Lesser Sundas Deciduous Forests [111]	Columbidae	Dusky cuckoo-dove	*Macropygia magna*
	Columbidae	Flores green-pigeon	*Treron floris**
	Columbidae	Pink-headed imperial-pigeon	*Ducula rosacea*
	Columbidae	Dark-backed imperial-pigeon	*Ducula lacernulata*
	Psittacidae	Wallace's hanging-parrot	*Loriculus flosculus**
	Loriidae	Olive-headed lorikeet	*Trichoglossus euteles*
	Strigidae	Flores scops-owl	*Otus alfredi**
	Strigidae	Wallace's scops-owl	*Otus silvicola**
	Alcedinidae	Cinnamon-backed kingfisher	*Todirhamphus australasia*
	Alcedinidae	White-rumped kingfisher	*Caridonax fulgidus**
	Meliphagidae	Sunda honeyeater	*Lichmera lombokia**
	Pachycephalidae	Bare-throated whistler	*Pachycephala nudigula**
	Rhipiduridae	Brown-capped fantail	*Rhipidura diluta**
	Monarchidae	Flores monarch	*Monarcha sacerdotum**
	Corvidae	Flores crow	*Corvus florensis**
	Campephagidae	Sumba cuckoo-shrike	*Coracina dohertyi*
	Campephagidae	Flores minivet	*Pericrocotus lansbergei**
	Turdidae	Chestnut-backed thrush	*Zoothera dohertyi*
	Muscicapidae	Flores jungle-flycatcher	*Rhinomyias oscillans*
	Zosteropidae	Yellow-spectacled white-eye	*Zosterops wallacei*
	Zosteropidae	White-browed white-eye	*Lophozosterops superciliaris**
	Zosteropidae	Dark-crowned white-eye	*Lophozosterops dohertyi**
	Zosteropidae	Flores white-eye	*Heleia crassirostris**
	Sylviidae	Russet-capped Tesia	*Tesia everetti**
	Sylviidae	Timor leaf-warbler	*Phylloscopus presbytes*
	Dicaeidae	Golden-rumped flowerpecker	*Dicaeum annae**
	Dicaeidae	Black-fronted flowerpecker	*Dicaeum igniferum**
	Dicaeidae	Red-chested flowerpecker	*Dicaeum maugei*
	Nectariniidae	Flame-breasted sunbird	*Nectarinia solaris*
Timor and Wetar Deciduous Forests [112]	Columbidae	Dusky cuckoo-dove	*Macropygia magna*
	Columbidae	Slaty cuckoo-dove	*Turacoena modesta**
	Columbidae	Wetar ground-dove	*Gallicolumba hoedtii**
	Columbidae	Timor green-pigeon	*Treron psittacea**
	Columbidae	Pink-headed imperial-pigeon	*Ducula rosacea*
	Columbidae	Timor imperial-pigeon	*Ducula cineracea**
	Psittacidae	Olive-shouldered parrot	*Aprosmictus jonquillaceus**
	Loriidae	Olive-headed lorikeet	*Trichoglossus euteles*
	Loriidae	Iris lorikeet	*Psitteuteles iris**
	Alcedinidae	Cinnamon-backed kingfisher	*Todirhamphus australasia*
	Acanthizidae	Plain gerygone	*Gerygone inornata**
	Meliphagidae	White-tufted honeyeater	*Lichmera squamata*
	Meliphagidae	Yellow-eared honeyeater	*Lichmera flavicans**
	Meliphagidae	Black-chested honeyeater	*Lichmera notabilis**
	Meliphagidae	Crimson-hooded myzomela	*Myzomela kuehni**
	Meliphagidae	Black-breasted myzomela	*Myzomela vulnerata**
	Meliphagidae	Streak-breasted honeyeater	*Meliphaga reticulata**
	Meliphagidae	Timor friarbird	*Philemon inornatus**
	Pachycephalidae	Fawn-breasted whistler	*Pachycephala orpheus**
	Oriolidae	Timor oriole	*Oriolus melanotis**
	Oriolidae	Timor figbird	*Sphecotheres viridis**
	Oriolidae	Wetar figbird	*Sphecotheres hypoleucus**
	Turdidae	Chestnut-backed thrush	*Zoothera dohertyi*
	Turdidae	Orange-banded thrush	*Zoothera peronii*
	Muscicapidae	Black-banded flycatcher	*Ficedula timorensis**
	Muscicapidae	Timor blue-flycatcher	*Cyornis hyacinthinus**
	Muscicapidae	Timor bushchat	*Saxicola gutturalis**
	Zosteropidae	Timor white-eye	*Heleia muelleri**
	Sylviidae	Timor stubtail	*Urosphena subulata*
	Sylviidae	Timor leaf-warbler	*Phylloscopus presbytes*
	Sylviidae	Buff-banded bushbird	*Buettikoferella bivittata**
	Estrildidae	Tricolored parrotfinch	*Erythrura tricolor*

(continues)

	Family	Common Name	Species
Timor and Wetar Deciduous Forests [112] (continued)	Estrildidae	Timor sparrow	*Padda fuscata**
	Dicaeidae	Red-chested flowerpecker	*Dicaeum maugei*
	Nectariniidae	Flame-breasted sunbird	*Nectarinia solaris*
Sumba Deciduous Forests [113]	Turnicidae	Sumba buttonquail	*Turnix everetti**
	Columbidae	Sumba green-pigeon	*Treron teysmannii**
	Columbidae	Red-naped fruit-dove	*Ptilinopus dohertyi**
	Strigidae	Sumba boobook	*Ninox rudolfi**
	Alcedinidae	Cinnamon-backed kingfisher	*Todirhamphus australasia*
	Bucconidae	Sumba hornbill	*Aceros everetti**
	Campephagidae	Sumba cuckoo-shrike	*Coracina dohertyi*
	Turdidae	Chestnut-backed thrush	*Zoothera dohertyi*
	Muscicapidae	Flores jungle-flycatcher	*Rhinomyias oscillans*
	Muscicapidae	Sumba flycatcher	*Ficedula harterti**
	Zosteropidae	Yellow-spectacled white-eye	*Zosterops wallacei*
	Nectariniidae	Apricot-breasted sunbird	*Nectarinia buettikoferi**
Halmahera Rain Forests [114]	Accipitridae	Moluccan goshawk	*Accipiter henicogrammus**
	Accipitridae	Rufous-necked sparrowhawk	*Accipiter erythrauchen*
	Megapodiidae	Moluccan scrubfowl	*Megapodius wallacei*
	Megapodiidae	Dusky scrubfowl	*Megapodius freycinet*
	Rallidae	Invisible rail	*Habroptila wallacii**
	Scolopacidae	Moluccan woodcock	*Scolopax rochussenii**
	Columbidae	Scarlet-breasted fruit-dove	*Ptilinopus bernsteinii**
	Columbidae	Blue-capped fruit-dove	*Ptilinopus monacha**
	Columbidae	Grey-headed fruit-dove	*Ptilinopus hyogastra**
	Columbidae	Carunculated fruit-dove	*Ptilinopus granulifrons**
	Columbidae	White-eyed imperial-pigeon	*Ducula perspicillata*
	Columbidae	Spice imperial-pigeon	*Ducula myristicivora*
	Columbidae	Pink-headed imperial-pigeon	*Ducula rosacea*
	Columbidae	Cinnamon-bellied imperial-pigeon	*Ducula basilica**
	Psittacidae	Moluccan hanging-parrot	*Loriculus amabilis*
	Cacatuidae	White cockatoo	*Cacatua alba**
	Loriidae	Violet-necked lory	*Eos squamata*
	Loriidae	Chattering lory	*Lorius garrulus**
	Cuculidae	Moluccan cuckoo	*Cacomantis heinrichi**
	Cuculidae	Pied bronze-cuckoo	*Chrysococcyx crassirostris*
	Cuculidae	Goliath coucal	*Centropus goliath**
	Strigidae	Moluccan hawk-owl	*Ninox squamipila*
	Aegothelidae	Moluccan owlet-nightjar	*Aegotheles crinifrons**
	Alcedinidae	Blue-and-white kingfisher	*Todirhamphus diops**
	Alcedinidae	Sombre kingfisher	*Todirhamphus funebris**
	Coraciidae	Purple roller	*Eurystomus azureus**
	Pittidae	Ivory-breasted pitta	*Pitta maxima**
	Meliphagidae	Olive honeyeater	*Lichmera argentauris*
	Meliphagidae	White-streaked friarbird	*Melitograis gilolensis**
	Meliphagidae	Dusky friarbird	*Philemon fuscicapillus**
	Pachycephalidae	Drab whistler	*Pachycephala griseonota*
	Monarchidae	White-naped monarch	*Monarcha pileatus*
	Monarchidae	Moluccan flycatcher	*Myiagra galeata*
	Corvidae	Long-billed crow	*Corvus validus**
	Paradisaeidae	Paradise-crow	*Lycocorax pyrrhopterus**
	Paradisaeidae	Wallace's standardwing	*Semioptera wallacii**
	Oriolidae	Halmahera oriole	*Oriolus phaeochromus**
	Campephagidae	Moluccan cuckoo-shrike	*Coracina atriceps*
	Campephagidae	Halmahera cuckoo-shrike	*Coracina parvula**
	Campephagidae	Pale-grey cuckoo-shrike	*Coracina ceramensis*
	Campephagidae	Rufous-bellied triller	*Lalage aurea**
	Zosteropidae	Cream-throated white-eye	*Zosterops atriceps**
	Dicaeidae	Flame-breasted flowerpecker	*Dicaeum erythrothorax*
Buru Rain Forests [115]	Accipitridae	Rufous-necked sparrowhawk	*Accipiter erythrauchen*
	Megapodiidae	Forsten's scrubfowl	*Megapodius forstenii*
	Megapodiidae	Moluccan scrubfowl	*Megapodius wallacei*
	Columbidae	White-eyed imperial-pigeon	*Ducula perspicillata*
	Columbidae	Long-tailed mountain-pigeon	*Gymnophaps mada*
	Psittacidae	Buru racquet-tail	*Prioniturus mada**

(continues)

	Family	Common Name	Species
Buru Rain Forests [115] (continued)	Psittacidae	Black-lored parrot	Tanygnathus gramineus*
	Loriidae	Red lory	Eos bornea
	Loriidae	Blue-fronted lorikeet	Charmosyna toxopei*
	Strigidae	Moluccan hawk-owl	Ninox squamipila
	Tytonidae	Lesser masked-owl	Tyto sororcula
	Meliphagidae	Buru honeyeater	Lichmera deningeri*
	Meliphagidae	Wakolo myzomela	Myzomela wakoloensis
	Meliphagidae	Black-faced friarbird	Philemon moluccensis
	Pachycephalidae	Drab whistler	Pachycephala griseonota
	Rhipiduridae	Cinnamon-backed fantail	Rhipidura superflua*
	Monarchidae	White-naped monarch	Monarcha pileatus
	Monarchidae	Black-tipped monarch	Monarcha loricatus*
	Monarchidae	Moluccan flycatcher	Myiagra galeata
	Oriolidae	Buru oriole	Oriolus bouroensis
	Campephagidae	Buru cuckoo-shrike	Coracina fortis*
	Campephagidae	Pale-grey cuckoo-shrike	Coracina ceramensis
	Turdidae	Moluccan thrush	Zoothera dumasi
	Muscicapidae	Buru jungle-flycatcher	Rhinomyias addita*
	Muscicapidae	Cinnamon-chested flycatcher	Ficedula buruensis
	Zosteropidae	Buru white-eye	Zosterops buruensis*
	Zosteropidae	Rufous-throated white-eye	Madanga ruficollis*
	Sylviidae	Chestnut-backed bush-warbler	Bradypterus castaneus
	Dicaeidae	Flame-breasted flowerpecker	Dicaeum erythrothorax
Seram Rain Forests [116]	Accipitridae	Rufous-necked sparrowhawk	Accipiter erythrauchen
	Megapodiidae	Forsten's scrubfowl	Megapodius forstenii
	Megapodiidae	Moluccan scrubfowl	Megapodius wallacei
	Columbidae	White-eyed imperial-pigeon	Ducula perspicillata
	Columbidae	Long-tailed mountain-pigeon	Gymnophaps mada
	Cacatuidae	Salmon-crested cockatoo	Cacatua moluccensis*
	Loriidae	Red lory	Eos bornea
	Loriidae	Blue-eared lory	Eos semilarvata*
	Loriidae	Purple-naped lory	Lorius domicella*
	Cuculidae	Pied bronze-cuckoo	Chrysococcyx crassirostris
	Strigidae	Moluccan hawk-owl	Ninox squamipila
	Tytonidae	Lesser masked-owl	Tyto sororcula
	Alcedinidae	Lazuli kingfisher	Todirhamphus lazuli*
	Meliphagidae	Olive honeyeater	Lichmera argentauris
	Meliphagidae	Seram honeyeater	Lichmera monticola*
	Meliphagidae	Seram myzomela	Myzomela blasii*
	Meliphagidae	Wakolo myzomela	Myzomela wakoloensis
	Meliphagidae	Seram friarbird	Philemon subcorniculatus*
	Pachycephalidae	Drab whistler	Pachycephala griseonota
	Rhipiduridae	Streaky-breasted fantail	Rhipidura dedemi*
	Monarchidae	Black-chinned monarch	Monarcha boanensis*
	Monarchidae	Moluccan flycatcher	Myiagra galeata
	Oriolidae	Seram oriole	Oriolus forsteni*
	Campephagidae	Moluccan cuckoo-shrike	Coracina atriceps
	Campephagidae	Pale-grey cuckoo-shrike	Coracina ceramensis
	Turdidae	Moluccan thrush	Zoothera dumasi
	Sturnidae	Long-crested myna	Basilornis corythaix*
	Muscicapidae	Cinnamon-chested flycatcher	Ficedula buruensis
	Zosteropidae	Ambon white-eye	Zosterops kuehni*
	Zosteropidae	Rufescent white-eye	Tephrozosterops stalkeri*
	Zosteropidae	Grey-hooded white-eye	Lophozosterops pinaiae*
	Sylviidae	Chestnut-backed bush-warbler	Bradypterus castaneus
	Dicaeidae	Ashy flowerpecker	Dicaeum vulneratum
Banda Sea Islands Moist Deciduous Forests [117]	Megapodiidae	Tenimbar megapode	Megapodius tenimberensis*
	Megapodiidae	Forsten's scrubfowl	Megapodius forstenii
	Columbidae	Dusky cuckoo-dove	Macropygia magna
	Columbidae	Wallace's fruit-dove	Ptilinopus wallacii
	Columbidae	Pink-headed imperial-pigeon	Ducula rosacea
	Cacatuidae	Tanimbar cockatoo	Cacatua goffini*

(continues)

	Family	Common Name	Species
Banda Sea Islands Moist Deciduous Forests [117] (*continued*)	Loriidae	Red lory	*Eos bornea*
	Loriidae	Blue-streaked lory	*Eos reticulata**
	Loriidae	Olive-headed lorikeet	*Trichoglossus euteles*
	Cuculidae	Green-cheeked bronze-cuckoo	*Chrysococcyx rufomerus**
	Cuculidae	Pied bronze-cuckoo	*Chrysococcyx crassirostris*
	Cuculidae	Kai coucal	*Centropus spilopterus**
	Strigidae	Moluccan hawk-owl	*Ninox squamipila*
	Tytonidae	Lesser masked-owl	*Tyto sororcula*
	Alcedinidae	Cinnamon-backed kingfisher	*Todirhamphus australasia*
	Acanthizidae	Rufous-sided gerygone	*Gerygone dorsalis*
	Meliphagidae	White-tufted honeyeater	*Lichmera squamata*
	Meliphagidae	Banda myzomela	*Myzomela boiei**
	Meliphagidae	Black-faced friarbird	*Philemon moluccensis*
	Eopsaltriidae	Golden-bellied flyrobin	*Microeca hemixantha**
	Pachycephalidae	Drab whistler	*Pachycephala griseonota*
	Pachycephalidae	Wallacean whistler	*Pachycephala arctitorquis**
	Rhipiduridae	Cinnamon-tailed fantail	*Rhipidura fuscorufa**
	Rhipiduridae	Long-tailed fantail	*Rhipidura opistherythra**
	Monarchidae	White-naped monarch	*Monarcha pileatus*
	Monarchidae	Black-bibbed monarch	*Monarcha mundus**
	Monarchidae	White-tailed monarch	*Monarcha leucurus**
	Monarchidae	Moluccan flycatcher	*Myiagra galeata*
	Oriolidae	Buru oriole	*Oriolus bouroensis*
	Campephagidae	Kai cuckoo-shrike	*Coracina dispar**
	Turdidae	Slaty-backed thrush	*Zoothera schistacea**
	Turdidae	Orange-banded thrush	*Zoothera peronii*
	Turdidae	Fawn-breasted thrush	*Zoothera machiki**
	Sturnidae	Tanimbar starling	*Aplonis crassa**
	Muscicapidae	Cinnamon-chested flycatcher	*Ficedula buruensis*
	Muscicapidae	Damar flycatcher	*Ficedula henrici**
	Zosteropidae	Great Kai white-eye	*Zosterops grayi**
	Zosteropidae	Little Kai white-eye	*Zosterops uropygialis**
	Sylviidae	Timor stubtail	*Urosphena subulata*
	Sylviidae	Tanimbar bush-warbler	*Cettia carolinae**
	Estrildidae	Tricolored parrotfinch	*Erythrura tricolor*
	Dicaeidae	Ashy flowerpecker	*Dicaeum vulneratum*
	Dicaeidae	Red-chested flowerpecker	*Dicaeum maugei*
Vogelkop Montane Rain Forests [118]	Rallidae	White-striped forest-rail	*Rallina leucospila**
	Ptilonorhynchidae	Vogelkop bowerbird	*Amblyornis inornatus**
	Acanthizidae	Vogelkop scrubwren	*Sericornis rufescens**
	Meliphagidae	Vogelkop honeyeater	*Melidectes leucostephes**
	Meliphagidae	Arfak honeyeater	*Melipotes gymnops**
	Pachycephalidae	Vogelkop whistler	*Pachycephala meyeri**
	Paradisaeidae	Western parotia	*Parotia sefilata**
	Paradisaeidae	Arfak astrapia	*Astrapia nigra**
	Estrildidae	Grey-banded munia	*Lonchura vana**
	Rallidae	Chestnut forest-rail	*Rallina rubra*
	Psittacidae	Modest tiger-parrot	*Psittacella modesta*
	Meliphagidae	Rufous-sided honeyeater	*Ptiloprora erythropleura*
	Paradisaeidae	Long-tailed paradigalla	*Paradigalla carunculata*
	Melanocharitidae	Obscure berrypecker	*Melanocharis arfakiana*
	Eopsaltriidae	Smoky robin	*Peneothello cryptoleucus*
	Caprimulgidae	Mountain eared-nightjar	*Eurostopodus archboldi*
	Meliphagidae	Black-backed honeyeater	*Ptiloprora perstriata*
	Meliphagidae	Cinnamon-browed honeyeater	*Melidectes ochromelas*
	Eopsaltriidae	Green-backed robin	*Pachycephalopsis hattamensis*
	Paradisaeidae	Greater melampitta	*Melampitta gigantea*
Vogelkop-Aru Lowland Rain Forests [119]	Megapodiidae	Bruijn's brush-turkey	*Aepypodius bruijnii**
	Alcedinidae	Kofiau paradise-kingfisher	*Tanysiptera ellioti**
	Monarchidae	Black-backed monarch	*Monarcha julianae**
	Paradisaeidae	Wilson's bird-of-paradise	*Cicinnurus respublica**
	Paradisaeidae	Red bird-of-paradise	*Paradisaea rubra**

(*continues*)

	Family	Common Name	Species
Vogelkop-Aru Lowland Rain Forests [119] (continued)	Megapodiidae	Dusky scrubfowl	Megapodius freycinet
	Loriidae	Violet-necked lory	Eos squamata
	Alcedinidae	Spangled kookaburra	Dacelo tyro
	Corvidae	Brown-headed crow	Corvus fuscicapillus
	Columbidae	Western crowned-pigeon	Goura cristata
	Loriidae	Black lory	Chalcopsitta atra
	Alcedinidae	Red-breasted paradise-kingfisher	Tanysiptera nympha
	Meliphagidae	Olive honeyeater	Lichmera argentauris
	Meliphagidae	Silver-eared honeyeater	Lichmera alboauricularis
	Paradisaeidae	Greater bird-of-paradise	Paradisaea apoda
	Dicaeidae	Olive-crowned flowerpecker	Dicaeum pectorale
	Megapodiidae	Moluccan scrubfowl	Megapodius wallacei
	Megapodiidae	Red-billed brush-turkey	Talegalla cuvieri
	Columbidae	Wallace's fruit-dove	Ptilinopus wallacii
	Columbidae	Spice imperial-pigeon	Ducula myristicivora
	Alcedinidae	Little paradise-kingfisher	Tanysiptera hydrocharis
Biak-Numfoor Rain Forests [120]	Megapodiidae	Geelvink scrubfowl	Megapodius geelvinkianus*
	Psittacidae	Geelvink pygmy-parrot	Micropsitta geelvinkiana*
	Loriidae	Black-winged lory	Eos cyanogenia*
	Cuculidae	Biak coucal	Centropus chalybeus*
	Alcedinidae	Biak paradise-kingfisher	Tanysiptera riedelii*
	Alcedinidae	Numfor paradise-kingfisher	Tanysiptera carolinae*
	Acanthizidae	Biak gerygone	Gerygone hypoxantha*
	Monarchidae	Biak monarch	Monarcha brehmii*
	Monarchidae	Biak flycatcher	Myiagra atra*
	Sturnidae	Long-tailed starling	Aplonis magna*
	Zosteropidae	Biak white-eye	Zosterops mysorensis*
	Columbidae	Spice imperial-pigeon	Ducula myristicivora
	Columbidae	Yellow-bibbed fruit-dove	Ptilinopus solomonensis
Yapen Rain Forests [121]	Columbidae	Spice imperial-pigeon	Ducula myristicivora
	Eopsaltriidae	Green-backed robin	Pachycephalopsis hattamensis
Northern New Guinea Montane Rain Forests [122]	Rallidae	Mayr's forest-rail	Rallina mayri*
	Ptilonorhynchidae	Golden-fronted bowerbird	Amblyornis flavifrons*
	Ptilonorhynchidae	Fire-maned bowerbird	Sericulus bakeri*
	Meliphagidae	Mayr's honeyeater	Ptiloprora mayri*
	Meliphagidae	Rufous-backed honeyeater	Ptiloprora guisei
	Meliphagidae	Cinnamon-browed honeyeater	Melidectes ochromelas
	Eopsaltriidae	Smoky robin	Peneothello cryptoleucus
	Eopsaltriidae	Green-backed robin	Pachycephalopsis hattamensis
	Cinclosomatidae	Brown-capped jewel-babbler	Ptilorrhoa geislerorum
	Paradisaeidae	Greater melampitta	Melampitta gigantea
	Paradisaeidae	Carola's parotia	Parotia carolae
	Paradisaeidae	Wahnes's parotia	Parotia wahnesi
Northern New Guinea Lowland Rain and Freshwater Swamp Forests [123]	Meliphagidae	Brass's friarbird	Philemon brassi*
	Monarchidae	Rufous monarch	Monarcha rubiensis
	Monarchidae	Rufous-collared monarch	Arses insularis
	Paradisaeidae	Jobi manucode	Manucodia jobiensis
	Paradisaeidae	Pale-billed sicklebill	Epimachus bruijnii*
	Psittacidae	Edwards's fig-parrot	Psittaculirostris edwardsii
	Psittacidae	Salvadori's fig-parrot	Psittaculirostris salvadorii
	Corvidae	Brown-headed crow	Corvus fuscicapillus
	Loriidae	Brown lory	Chalcopsitta duivenbodei
	Alcedinidae	Red-breasted paradise-kingfisher	Tanysiptera nympha
	Meliphagidae	Silver-eared honeyeater	Lichmera alboauricularis
	Pachycephalidae	White-bellied whistler	Pachycephala leucogastra
	Apodidae	Papuan swiftlet	Aerodramus papuensis
	Cinclosomatidae	Brown-capped jewel-babbler	Ptilorrhoa geislerorum
	Eopsaltriidae	Green-backed robin	Pachycephalopsis hattamensis
	Paradisaeidae	Greater melampitta	Melampitta gigantea

	Family	Common Name	Species
Huon Peninsula Montane Rain Forests [124]			
	Psittacidae	Edwards's fig-parrot	*Psittaculirostris edwardsii*
	Psittacidae	Madarasz's tiger-parrot	*Psittacella madaraszi*
	Loriidae	Brown lory	*Chalcopsitta duivenbodei*
	Caprimulgidae	Mountain eared-nightjar	*Eurostopodus archboldi*
	Apodidae	Papuan swiftlet	*Aerodramus papuensis*
	Meliphagidae	Olive-streaked honeyeater	*Ptiloprora meekiana*
	Meliphagidae	Rufous-backed honeyeater	*Ptiloprora guisei*
	Meliphagidae	Cinnamon-browed honeyeater	*Melidectes ochromelas*
	Meliphagidae	Huon wattled honeyeater	*Melidectes foersteri*
	Meliphagidae	Spangled honeyeater	*Melipotes ater*
	Cinclosomatidae	Brown-capped jewel-babbler	*Ptilorrhoa geislerorum*
	Cinclosomatidae	Blue-capped ifrita	*Ifrita kowaldi*
	Paradisaeidae	Wahnes's parotia	*Parotia wahnesi*
	Paradisaeidae	Huon astrapia	*Astrapia rothschildi*
	Paradisaeidae	Emperor bird-of-paradise	*Paradisaea guilielmi**
	Motacillidae	Alpine pipit	*Anthus gutturalis*
Central Range Montane Rain Forests [125]			
	Loriidae	Orange-billed lorikeet	*Neopsittacus pullicauda*
	Ptilonorhynchidae	Archbold's bowerbird	*Archboldia papuensis**
	Eopsaltriidae	White-winged robin	*Peneothello sigillatus*
	Pachycephalidae	Sooty shrike-thrush	*Colluricincla umbrina**
	Cinclosomatidae	Papuan whipbird	*Androphobus viridis**
	Paradisaeidae	Short-tailed paradigalla	*Paradigalla brevicauda**
	Paradisaeidae	King-of-Saxony bird-of-paradise	*Pteridophora alberti**
	Campephagidae	Hooded cuckoo-shrike	*Coracina longicauda*
	Estrildidae	Snow mountain munia	*Lonchura montana**
	Estrildidae	Black-breasted munia	*Lonchura teerinki**
	Melanocharitidae	Streaked berrypecker	*Melanocharis striativentris*
	Rallidae	Chestnut forest-rail	*Rallina rubra*
	Psittacidae	Modest tiger-parrot	*Psittacella modesta*
	Loriidae	Striated lorikeet	*Charmosyna multistriata*
	Aegothelidae	Archbold's owlet-nightjar	*Aegotheles archboldi*
	Apodidae	Bare-legged swiftlet	*Aerodramus nuditarsus*
	Climacteridae	Papuan treecreeper	*Cormobates placens*
	Meliphagidae	Orange-cheeked honeyeater	*Oreornis chrysogenys*
	Meliphagidae	Leaden honeyeater	*Ptiloprora plumbea*
	Meliphagidae	Rufous-sided honeyeater	*Ptiloprora erythropleura*
	Meliphagidae	Yellow-browed honeyeater	*Melidectes rufocrissalis*
	Pachycephalidae	Lorentz's whistler	*Pachycephala lorentzi*
	Pachycephalidae	Wattled ploughbill	*Eulacestoma nigropectus*
	Paradisaeidae	Yellow-breasted bird-of-paradise	*Loboparadisea sericea*
	Paradisaeidae	Loria's bird-of-paradise	*Cnemophilus loriae*
	Paradisaeidae	Long-tailed paradigalla	*Paradigalla carunculata*
	Paradisaeidae	Carola's parotia	*Parotia carolae*
	Paradisaeidae	Lawes's parotia	*Parotia lawesii*
	Paradisaeidae	Splendid astrapia	*Astrapia splendidissima*
	Paradisaeidae	Ribbon-tailed astrapia	*Astrapia mayeri*
	Paradisaeidae	Blue bird-of-paradise	*Paradisaea rudolphi*
	Psittacidae	Painted tiger-parrot	*Psittacella picta*
	Psittacidae	Madarasz's tiger-parrot	*Psittacella madaraszi*
	Acanthizidae	Papuan thornbill	*Acanthiza murina*
	Meliphagidae	Spot-breasted meliphaga	*Meliphaga mimikae*
	Meliphagidae	Olive-streaked honeyeater	*Ptiloprora meekiana*
	Meliphagidae	Sooty honeyeater	*Melidectes fuscus*
	Meliphagidae	Belford's honeyeater	*Melidectes belfordi*
	Eopsaltriidae	Greater ground-robin	*Amalocichla sclateriana*
	Eopsaltriidae	Alpine robin	*Petroica bivittata*
	Eopsaltriidae	Smoky robin	*Peneothello cryptoleucus*
	Pachycephalidae	Black sittella	*Daphoenositta miranda*
	Paradisaeidae	Crested bird-of-paradise	*Cnemophilus macgregorii*
	Paradisaeidae	MacGregor's bird-of-paradise	*Macgregoria pulchra*
	Paradisaeidae	Brown sicklebill	*Epimachus meyeri*
	Paradisaeidae	Princess Stephanie's astrapia	*Astrapia stephaniae*

(continues)

	Family	Common Name	Species
Central Range Montane Rain Forests [125] (continued)	Estrildidae	Mountain firetail	Oreostruthus fuliginosus
	Caprimulgidae	Mountain eared-nightjar	Eurostopodus archboldi
	Apodidae	Papuan swiftlet	Aerodramus papuensis
	Meliphagidae	Black-backed honeyeater	Ptiloprora perstriata
	Cinclosomatidae	Blue-capped ifrita	Ifrita kowaldi
	Meliphagidae	Rufous-backed honeyeater	Ptiloprora guisei
	Meliphagidae	Cinnamon-browed honeyeater	Melidectes ochromelas
	Eopsaltriidae	Green-backed robin	Pachycephalopsis hattamensis
	Paradisaeidae	Greater melampitta	Melampitta gigantea
Southeastern Papuan Rain Forests [126]	Alcedinidae	Brown-headed paradise-kingfisher	Tanysiptera danae*
	Ptilonorhynchidae	Streaked bowerbird	Amblyornis subalaris*
	Paradisaeidae	Eastern parotia	Parotia helenae*
	Estrildidae	Alpine munia	Lonchura monticola*
	Estrildidae	Grey-headed munia	Lonchura caniceps*
	Apodidae	Bare-legged swiftlet	Aerodramus nuditarsus
	Climacteridae	Papuan treecreeper	Cormobates placens
	Meliphagidae	Leaden honeyeater	Ptiloprora plumbea
	Meliphagidae	Yellow-browed honeyeater	Melidectes rufocrissalis
	Pachycephalidae	Wattled ploughbill	Eulacestoma nigropectus
	Paradisaeidae	Yellow-breasted bird-of-paradise	Loboparadisea sericea
	Paradisaeidae	Loria's bird-of-paradise	Cnemophilus loriae
	Paradisaeidae	Lawes's parotia	Parotia lawesii
	Paradisaeidae	Blue bird-of-paradise	Paradisaea rudolphi
	Melanocharitidae	Obscure berrypecker	Melanocharis arfakiana
	Psittacidae	Painted tiger-parrot	Psittacella picta
	Psittacidae	Madarasz's tiger-parrot	Psittacella madaraszi
	Acanthizidae	Papuan thornbill	Acanthiza murina
	Meliphagidae	Silver-eared honeyeater	Lichmera alboauricularis
	Meliphagidae	Spot-breasted meliphaga	Meliphaga mimikae
	Meliphagidae	Olive-streaked honeyeater	Ptiloprora meekiana
	Meliphagidae	Sooty honeyeater	Melidectes fuscus
	Meliphagidae	Belford's honeyeater	Melidectes belfordi
	Eopsaltriidae	Greater ground-robin	Amalocichla sclateriana
	Eopsaltriidae	Alpine robin	Petroica bivittata
	Pachycephalidae	White-bellied whistler	Pachycephala leucogastra
	Pachycephalidae	Black sittella	Daphoenositta miranda
	Paradisaeidae	Crested bird-of-paradise	Cnemophilus macgregorii
	Paradisaeidae	MacGregor's bird-of-paradise	Macgregoria pulchra
	Paradisaeidae	Brown sicklebill	Epimachus meyeri
	Paradisaeidae	Princess Stephanie's astrapia	Astrapia stephaniae
	Estrildidae	Mountain firetail	Oreostruthus fuliginosus
	Motacillidae	Alpine pipit	Anthus gutturalis
	Caprimulgidae	Mountain eared-nightjar	Eurostopodus archboldi
	Meliphagidae	Black-backed honeyeater	Ptiloprora perstriata
	Cinclosomatidae	Brown-capped jewel-babbler	Ptilorrhoa geislerorum
	Cinclosomatidae	Blue-capped ifrita	Ifrita kowaldi
	Meliphagidae	Rufous-backed honeyeater	Ptiloprora guisei
	Meliphagidae	Cinnamon-browed honeyeater	Melidectes ochromelas
	Paradisaeidae	Greater melampitta	Melampitta gigantea
Southern New Guinea Freshwater Swamp Forests [127]	Megapodiidae	Red-billed brush-turkey	Talegalla cuvieri
	Columbidae	Wallace's fruit-dove	Ptilinopus wallacii
	Columbidae	Western crowned-pigeon	Goura cristata
	Loriidae	Black lory	Chalcopsitta atra
	Alcedinidae	Little paradise-kingfisher	Tanysiptera hydrocharis
	Pachycephalidae	White-bellied pitohui	Pitohui incertus
	Paradisaeidae	Greater bird-of-paradise	Paradisaea apoda
	Sylviidae	Fly River grassbird	Megalurus albolimbatus
	Estrildidae	Grey-crowned munia	Lonchura nevermanni
	Estrildidae	Black munia	Lonchura stygia
	Dicaeidae	Olive-crowned flowerpecker	Dicaeum pectorale

	Family	Common Name	Species
Southern New Guinea Lowland Rain Forests [128]	Megapodiidae	Red-billed brush-turkey	*Talegalla cuvieri*
	Loriidae	Striated lorikeet	*Charmosyna multistriata*
	Alcedinidae	Little paradise-kingfisher	*Tanysiptera hydrocharis*
	Meliphagidae	Spot-breasted meliphaga	*Meliphaga mimikae*
	Paradisaeidae	Greater bird-of-paradise	*Paradisaea apoda*
New Guinea Mangroves [129]	Megapodiidae	Red-billed brush-turkey	*Talegalla cuvieri*
	Columbidae	Wallace's fruit-dove	*Ptilinopus wallacii*
	Columbidae	Western crowned-pigeon	*Goura cristata*
	Psittacidae	Salvadori's fig-parrot	*Psittaculirostris salvadorii*
	Loriidae	Black lory	*Chalcopsitta atra*
	Loriidae	Brown lory	*Chalcopsitta duivenbodei*
	Apodidae	Papuan swiftlet	*Aerodramus papuensis*
	Alcedinidae	Red-breasted paradise-kingfisher	*Tanysiptera nympha*
	Pachycephalidae	White-bellied pitohui	*Pitohui incertus*
	Dicaeidae	Olive-crowned flowerpecker	*Dicaeum pectorale*
Trans Fly Savanna and Grasslands [130]	Alcedinidae	Spangled kookaburra	*Dacelo tyro*
	Alcedinidae	Little paradise-kingfisher	*Tanysiptera hydrocharis*
	Sylviidae	Fly River grassbird	*Megalurus albolimbatus*
	Estrildidae	Grey-crowned munia	*Lonchura nevermanni*
	Estrildidae	Black munia	*Lonchura stygia*
Central Range Sub-Alpine Grasslands [131]	Phasianidae	Snow mountain quail	*Anurophasis monorthonyx**
	Eopsaltriidae	Snow mountain robin	*Petroica archboldi**
	Aegothelidae	Archbold's owlet-nightjar	*Aegotheles archboldi*
	Meliphagidae	Orange-cheeked honeyeater	*Oreornis chrysogenys*
	Meliphagidae	Short-bearded honeyeater	*Melidectes nouhuysi**
	Meliphagidae	Long-bearded honeyeater	*Melidectes princeps**
	Meliphagidae	Huon wattled honeyeater	*Melidectes foersteri*
	Meliphagidae	Spangled honeyeater	*Melipotes ater*
	Pachycephalidae	Lorentz's whistler	*Pachycephala lorentzi*
	Paradisaeidae	Splendid astrapia	*Astrapia splendidissima*
	Paradisaeidae	Ribbon-tailed astrapia	*Astrapia mayeri*
	Paradisaeidae	Huon astrapia	*Astrapia rothschildi*
	Psittacidae	Painted tiger-parrot	*Psittacella picta*
	Acanthizidae	Papuan thornbill	*Acanthiza murina*
	Meliphagidae	Sooty honeyeater	*Melidectes fuscus*
	Meliphagidae	Belford's honeyeater	*Melidectes belfordi*
	Eopsaltriidae	Greater ground-robin	*Amalocichla sclateriana*
	Eopsaltriidae	Alpine robin	*Petroica bivittata*
	Pachycephalidae	Black sittella	*Daphoenositta miranda*
	Paradisaeidae	Crested bird-of-paradise	*Cnemophilus macgregorii*
	Paradisaeidae	MacGregor's bird-of-paradise	*Macgregoria pulchra*
	Paradisaeidae	Brown sicklebill	*Epimachus meyeri*
	Paradisaeidae	Princess Stephanie's astrapia	*Astrapia stephaniae*
	Estrildidae	Mountain firetail	*Oreostruthus fuliginosus*
	Motacillidae	Alpine pipit	*Anthus gutturalis*
	Meliphagidae	Black-backed honeyeater	*Ptiloprora perstriata*
	Cinclosomatidae	Blue-capped ifrita	*Ifrita kowaldi*
	Meliphagidae	Rufous-backed honeyeater	*Ptiloprora guisei*
Admiralty Islands Lowland Rain Forests [132]	Strigidae	Manus hawk-owl	*Ninox meeki**
	Tytonidae	Manus owl	*Tyto manusi**
	Pittidae	Black-headed pitta	*Pitta superba**
	Meliphagidae	White-naped friarbird	*Philemon albitorques**
	Rhipiduridae	Manus fantail	*Rhipidura semirubra**
	Monarchidae	Manus monarch	*Monarcha infelix**
	Psittacidae	Meek's pygmy-parrot	*Micropsitta meeki*
	Meliphagidae	Ebony myzomela	*Myzomela pammelaena*
	Columbidae	Pied cuckoo-dove	*Reinwardtoena browni*
	Columbidae	Yellow-tinted imperial-pigeon	*Ducula subflavescens*
	Zosteropidae	Black-headed white-eye	*Zosterops hypoxanthus*
	Megapodiidae	Melanesian scrubfowl	*Megapodius eremita*
	Columbidae	Yellow-bibbed fruit-dove	*Ptilinopus solomonensis*

Appendix I: Near-Endemic and Strict Endemic Birds by Ecoregion

	Family	Common Name	Species
New Britain–New Ireland Lowland Rain Forests [133]	Psittacidae	Green-fronted hanging-parrot	*Loriculus tener**
	Cacatuidae	Blue-eyed cockatoo	*Cacatua ophthalmica**
	Alcedinidae	Bismarck kingfisher	*Alcedo websteri**
	Alcedinidae	New Britain kingfisher	*Todirhamphus albonotatus**
	Meliphagidae	Scarlet-bibbed myzomela	*Myzomela sclateri**
	Rhipiduridae	Matthias fantail	*Rhipidura matthiae**
	Monarchidae	White-breasted monarch	*Monarcha menckei**
	Sylviidae	Rusty thicketbird	*Megalurulus rubiginosus**
	Estrildidae	Mottled munia	*Lonchura hunsteini**
	Estrildidae	New Ireland munia	*Lonchura forbesi**
	Estrildidae	New Hanover munia	*Lonchura nigerrima**
	Accipitridae	Slaty-mantled sparrowhawk	*Accipiter luteoschistaceus**
	Accipitridae	New Britain sparrowhawk	*Accipiter brachyurus**
	Accipitridae	Black honey-buzzard	*Henicopernis infuscatus**
	Rallidae	New Britain rail	*Gallirallus insignis*
	Columbidae	New Britain bronzewing	*Henicophaps foersteri**
	Columbidae	Knob-billed fruit-dove	*Ptilinopus insolitus*
	Columbidae	Finsch's imperial-pigeon	*Ducula finschii*
	Columbidae	Bismarck imperial-pigeon	*Ducula melanochroa*
	Psittacidae	Meek's pygmy-parrot	*Micropsitta meeki*
	Psittacidae	Finsch's pygmy-parrot	*Micropsitta finschii*
	Loriidae	Cardinal lory	*Chalcopsitta cardinalis*
	Loriidae	White-naped lory	*Lorius albidinuchus*
	Loriidae	Red-chinned lorikeet	*Charmosyna rubrigularis*
	Cuculidae	Violaceous coucal	*Centropus violaceus*
	Cuculidae	Pied coucal	*Centropus ateralbus*
	Strigidae	Bismarck hawk-owl	*Ninox variegata**
	Strigidae	Russet hawk-owl	*Ninox odiosa*
	Tytonidae	Bismarck owl	*Tyto aurantia*
	Meliphagidae	New Ireland myzomela	*Myzomela pulchella*
	Meliphagidae	Ebony myzomela	*Myzomela pammelaena*
	Meliphagidae	Black-bellied myzomela	*Myzomela erythromelas**
	Meliphagidae	New Britain friarbird	*Philemon cockerelli*
	Meliphagidae	New Ireland friarbird	*Philemon eichhorni*
	Meliphagidae	Ashy myzomela	*Myzomela cineracea**
	Rhipiduridae	Bismarck fantail	*Rhipidura dahli*
	Monarchidae	Black-tailed monarch	*Monarcha verticalis*
	Monarchidae	Dull flycatcher	*Myiagra hebetior*
	Dicruridae	Ribbon-tailed drongo	*Dicrurus megarhynchus*
	Artamidae	Bismarck woodswallow	*Artamus insignis**
	Sturnidae	Atoll starling	*Aplonis feadensis*
	Dicaeidae	Red-banded flowerpecker	*Dicaeum eximium*
	Accipitridae	Pied goshawk	*Accipiter albogularis*
	Columbidae	Yellow-legged pigeon	*Columba pallidiceps*
	Columbidae	Pied cuckoo-dove	*Reinwardtoena browni*
	Columbidae	Red-knobbed imperial-pigeon	*Ducula rubricera*
	Columbidae	Yellow-tinted imperial-pigeon	*Ducula subflavescens*
	Psittacidae	Singing parrot	*Geoffroyus heteroclitus*
	Apodidae	Mayr's swiftlet	*Aerodramus orientalis*
	Turdidae	New Britain thrush	*Zoothera talaseae*
	Zosteropidae	Black-headed white-eye	*Zosterops hypoxanthus*
	Estrildidae	Bismarck munia	*Lonchura melaena*
	Megapodiidae	Melanesian scrubfowl	*Megapodius eremita*
	Columbidae	Yellow-bibbed fruit-dove	*Ptilinopus solomonensis*
	Zosteropidae	Louisiade white-eye	*Zosterops griseotinctus*
New Britain–New Ireland Montane Rain Forests [134]	Accipitridae	New Britain goshawk	*Accipiter princeps**
	Meliphagidae	Bismarck honeyeater	*Melidectes whitemanensis**
	Sylviidae	Bismarck thicketbird	*Megalurulus grosvenori**
	Accipitridae	Pied goshawk	*Accipiter albogularis*
	Rallidae	New Britain rail	*Gallirallus insignis*

(continues)

	Family	Common Name	Species
New Britain–New Ireland Montane Rain Forests [134] (continued)	Columbidae	Knob-billed fruit-dove	Ptilinopus insolitus
	Columbidae	Finsch's imperial-pigeon	Ducula finschii
	Columbidae	Bismarck imperial-pigeon	Ducula melanochroa
	Loriidae	White-naped lory	Lorius albidinuchus
	Loriidae	Red-chinned lorikeet	Charmosyna rubrigularis
	Cuculidae	Violaceous coucal	Centropus violaceus
	Cuculidae	Pied coucal	Centropus ateralbus
	Strigidae	Russet hawk-owl	Ninox odiosa
	Tytonidae	Bismarck owl	Tyto aurantia
	Meliphagidae	New Ireland myzomela	Myzomela pulchella
	Meliphagidae	New Britain friarbird	Philemon cockerelli
	Meliphagidae	New Ireland friarbird	Philemon eichhorni
	Rhipiduridae	Bismarck fantail	Rhipidura dahli
	Monarchidae	Black-tailed monarch	Monarcha verticalis
	Monarchidae	Dull flycatcher	Myiagra hebetior
	Dicruridae	Ribbon-tailed drongo	Dicrurus megarhynchus
	Dicaeidae	Red-banded flowerpecker	Dicaeum eximium
	Columbidae	Yellow-legged pigeon	Columba pallidiceps
	Columbidae	Red-knobbed imperial-pigeon	Ducula rubricera
	Psittacidae	Singing parrot	Geoffroyus heteroclitus
	Apodidae	Mayr's swiftlet	Aerodramus orientalis
	Turdidae	New Britain thrush	Zoothera talaseae
	Zosteropidae	Black-headed white-eye	Zosterops hypoxanthus
	Estrildidae	Bismarck munia	Lonchura melaena
	Megapodiidae	Melanesian scrubfowl	Megapodius eremita
	Columbidae	Yellow-bibbed fruit-dove	Ptilinopus solomonensis
Trobriand Islands Rain Forests [135]	Paradisaeidae	Curl-crested manucode	Manucodia comrii*
	Paradisaeidae	Goldie's bird-of-paradise	Paradisaea decora*
Louisiade Archipelago Rain Forests [136]	Meliphagidae	White-chinned myzomela	Myzomela albigula*
	Meliphagidae	Tagula honeyeater	Meliphaga vicina*
	Pachycephalidae	White-bellied whistler	Pachycephala leucogastra
	Cracticidae	Tagula butcherbird	Cracticus louisiadensis*
	Zosteropidae	White-throated white-eye	Zosterops meeki*
	Zosteropidae	Louisiade white-eye	Zosterops griseotinctus
	Dicaeidae	Louisiade flowerpecker	Dicaeum nitidum*
Solomon Islands Rain Forests [137]	Accipitridae	Imitator sparrowhawk	Accipiter imitator*
	Accipitridae	Solomon sea-eagle	Haliaeetus sanfordi*
	Rallidae	Woodford's rail	Nesoclopeus woodfordi*
	Rallidae	Roviana rail	Gallirallus rovianae*
	Rallidae	San Cristobal moorhen	Gallinula silvestris*
	Columbidae	Crested cuckoo-dove	Reinwardtoena crassirostris*
	Columbidae	Thick-billed ground-dove	Gallicolumba salamonis*
	Columbidae	Choiseul pigeon	Microgoura meeki*
	Columbidae	Silver-capped fruit-dove	Ptilinopus richardsii*
	Columbidae	White-headed fruit-dove	Ptilinopus eugeniae*
	Columbidae	Chestnut-bellied imperial-pigeon	Ducula brenchleyi*
	Columbidae	Pale mountain-pigeon	Gymnophaps solomonensis*
	Cacatuidae	Ducorps's cockatoo	Cacatua ducorpsii*
	Loriidae	Yellow-bibbed lory	Lorius chlorocercus*
	Loriidae	Meek's lorikeet	Charmosyna meeki*
	Loriidae	Duchess lorikeet	Charmosyna margarethae*
	Cuculidae	Buff-headed coucal	Centropus milo*
	Strigidae	Solomon hawk-owl	Ninox jacquinoti*
	Strigidae	Fearful owl	Nesasio solomonensis*
	Alcedinidae	Ultramarine kingfisher	Todirhamphus leucopygius*
	Alcedinidae	Moustached kingfisher	Actenoides bougainvillei*
	Pittidae	Black-faced pitta	Pitta anerythra*
	Meliphagidae	Bougainville honeyeater	Stresemannia bougainvillei*
	Meliphagidae	Scarlet-naped myzomela	Myzomela lafargei*
	Meliphagidae	Yellow-vented myzomela	Myzomela eichhorni*

(continues)

	Family	Common Name	Species
Solomon Islands Rain Forests [137] (*continued*)	Meliphagidae	Red-bellied myzomela	*Myzomela malaitae**
	Meliphagidae	Black-headed myzomela	*Myzomela melanocephala**
	Meliphagidae	Sooty myzomela	*Myzomela tristrami**
	Meliphagidae	Guadalcanal honeyeater	*Guadalcanaria inexpectata**
	Meliphagidae	San Cristobal honeyeater	*Melidectes sclateri**
	Pachycephalidae	Mountain whistler	*Pachycephala implicata**
	Rhipiduridae	White-winged fantail	*Rhipidura cockerelli**
	Rhipiduridae	Brown fantail	*Rhipidura drownei**
	Rhipiduridae	Dusky fantail	*Rhipidura tenebrosa**
	Rhipiduridae	Rennell fantail	*Rhipidura rennelliana**
	Rhipiduridae	Malaita fantail	*Rhipidura malaitae**
	Monarchidae	Rennell shrikebill	*Clytorhynchus hamlini**
	Monarchidae	Bougainville monarch	*Monarcha erythrostictus**
	Monarchidae	Chestnut-bellied monarch	*Monarcha castaneiventris**
	Monarchidae	White-capped monarch	*Monarcha richardsii**
	Monarchidae	Black-and-white monarch	*Monarcha barbatus**
	Monarchidae	Kulambangra monarch	*Monarcha browni**
	Monarchidae	White-collared monarch	*Monarcha viduus**
	Monarchidae	Steel-blue flycatcher	*Myiagra ferrocyanea**
	Monarchidae	Ochre-headed flycatcher	*Myiagra cervinicauda**
	Dicruridae	Solomon Islands drongo	*Dicrurus solomenensis**
	Corvidae	White-billed crow	*Corvus woodfordi**
	Corvidae	Bougainville crow	*Corvus meeki**
	Campephagidae	Solomon cuckoo-shrike	*Coracina holopolia**
	Turdidae	Olive-tailed thrush	*Zoothera lunulata**
	Turdidae	San Cristobal thrush	*Zoothera margaretae**
	Sturnidae	Rennell starling	*Aplonis insularis**
	Sturnidae	Brown-winged starling	*Aplonis grandis**
	Sturnidae	San Cristobal starling	*Aplonis dichroa**
	Sturnidae	White-eyed starling	*Aplonis brunneicapilla**
	Zosteropidae	Rennell white-eye	*Zosterops rennellianus**
	Zosteropidae	Banded white-eye	*Zosterops vellalavella**
	Zosteropidae	Ganongga white-eye	*Zosterops splendidus**
	Zosteropidae	Splendid white-eye	*Zosterops luteirostris**
	Zosteropidae	Solomon Islands white-eye	*Zosterops kulambangrae**
	Zosteropidae	Kulambangra white-eye	*Zosterops murphyi**
	Zosteropidae	Yellow-throated white-eye	*Zosterops metcalfii**
	Zosteropidae	Grey-throated white-eye	*Zosterops rendovae**
	Zosteropidae	Malaita white-eye	*Zosterops stresemanni**
	Zosteropidae	Bare-eyed white-eye	*Woodfordia superciliosa**
	Sylviidae	Shade warbler	*Cettia parens*
	Sylviidae	San Cristobal leaf-warbler	*Phylloscopus makirensis**
	Sylviidae	Kulambangra leaf-warbler	*Phylloscopus amoenus**
	Sylviidae	Bougainville thicketbird	*Megalurulus llaneae*
	Dicaeidae	Midget flowerpecker	*Dicaeum aeneum**
	Dicaeidae	Mottled flowerpecker	*Dicaeum tristrami**
	Psittacidae	Finsch's pygmy-parrot	*Micropsitta finschii*
	Loriidae	Cardinal lory	*Chalcopsitta cardinalis*
	Sturnidae	Atoll starling	*Aplonis feadensis*
	Accipitridae	Pied goshawk	*Accipiter albogularis*
	Columbidae	Yellow-legged pigeon	*Columba pallidiceps*
	Columbidae	Red-knobbed imperial-pigeon	*Ducula rubricera*
	Psittacidae	Singing parrot	*Geoffroyus heteroclitus*
	Apodidae	Mayr's swiftlet	*Aerodramus orientalis*
	Meliphagidae	Cardinal myzomela	*Myzomela cardinalis*
	Turdidae	New Britain thrush	*Zoothera talaseae*
	Sylviidae	Guadalcanal thicketbird	*Megalurulus whitneyi*
	Estrildidae	Bismarck munia	*Lonchura melaena*
	Megapodiidae	Melanesian scrubfowl	*Megapodius eremita*
	Acanthizidae	Fan-tailed gerygone	*Gerygone flavolateralis*
	Monarchidae	New Caledonian flycatcher	*Myiagra caledonica*
	Campephagidae	Melanesian cuckoo-shrike	*Coracina caledonica*
	Campephagidae	Long-tailed triller	*Lalage leucopyga*
	Columbidae	Yellow-bibbed fruit-dove	*Ptilinopus solomonensis*

Ecoregion	Family	Common Name	Species
Vanuatu Rain Forests [138]	Megapodiidae	New Hebrides scrubfowl	*Megapodius layardi**
	Columbidae	Santa Cruz ground-dove	*Gallicolumba sanctaecrucis**
	Columbidae	Tanna ground-dove	*Gallicolumba ferruginea**
	Columbidae	Tanna fruit-dove	*Ptilinopus tannensis**
	Columbidae	Baker's imperial-pigeon	*Ducula bakeri**
	Loriidae	Palm lorikeet	*Charmosyna palmarum**
	Alcedinidae	Chestnut-bellied kingfisher	*Todirhamphus farquhari**
	Meliphagidae	New Hebrides honeyeater	*Phylidonyris notabilis**
	Monarchidae	Vanikoro monarch	*Mayrornis schistaceus**
	Monarchidae	Buff-bellied monarch	*Neolalage banksiana**
	Monarchidae	Black-throated shrikebill	*Clytorhynchus nigrogularis**
	Monarchidae	Vanikoro flycatcher	*Myiagra vanikorensis**
	Sturnidae	Polynesian starling	*Aplonis tabuensis**
	Sturnidae	Rusty-winged starling	*Aplonis zelandica**
	Sturnidae	Mountain starling	*Aplonis santovestris**
	Zosteropidae	Santa Cruz white-eye	*Zosterops santaecrucis**
	Zosteropidae	Yellow-fronted white-eye	*Zosterops flavifrons**
	Zosteropidae	Sanford's white-eye	*Woodfordia lacertosa**
	Estrildidae	Royal parrotfinch	*Erythrura regia**
	Accipitridae	Pied goshawk	*Accipiter albogularis*
	Columbidae	Red-bellied fruit-dove	*Ptilinopus greyii*
	Meliphagidae	Dark-brown honeyeater	*Lichmera incana*
	Meliphagidae	Cardinal myzomela	*Myzomela cardinalis*
	Rhipiduridae	Streaked fantail	*Rhipidura spilodera*
	Monarchidae	Southern shrikebill	*Clytorhynchus pachycephaloides*
	Sylviidae	Guadalcanal thicketbird	*Megalurulus whitneyi*
	Acanthizidae	Fan-tailed gerygone	*Gerygone flavolateralis*
	Monarchidae	New Caledonian flycatcher	*Myiagra caledonica*
	Campephagidae	Melanesian cuckoo-shrike	*Coracina caledonica*
	Campephagidae	Long-tailed triller	*Lalage leucopyga*
New Caledonia Rain Forests [139]	Rallidae	New Caledonian rail	*Gallirallus lafresnayanus**
	Rhynochetidae	Kagu	*Rhynochetos jubatus**
	Columbidae	Cloven-feathered dove	*Drepanoptila holosericea**
	Loriidae	New Caledonian lorikeet	*Charmosyna diadema**
	Meliphagidae	Crow honeyeater	*Gymnomyza aubryana**
	Zosteropidae	Large lifou white-eye	*Zosterops inornatus**
	Zosteropidae	Small lifou white-eye	*Zosterops minutus**
	Accipitridae	White-bellied goshawk	*Accipiter haplochrous*
	Columbidae	New Caledonian imperial-pigeon	*Ducula goliath*
	Psittacidae	Horned parakeet	*Eunymphicus cornutus*
	Aegothelidae	New Caledonian owlet-nightjar	*Aegotheles savesi*
	Meliphagidae	New Caledonian myzomela	*Myzomela caledonica*
	Meliphagidae	New Caledonian friarbird	*Philemon diemenensis*
	Meliphagidae	Barred honeyeater	*Phylidonyris undulata*
	Eopsaltriidae	Yellow-bellied robin	*Eopsaltria flaviventris*
	Pachycephalidae	New Caledonian whistler	*Pachycephala caledonica*
	Corvidae	New Caledonian crow	*Corvus moneduloides*
	Campephagidae	New Caledonian cuckoo-shrike	*Coracina analis*
	Sturnidae	Striated starling	*Aplonis striata*
	Zosteropidae	Green-backed white-eye	*Zosterops xanthochrous*
	Sylviidae	New Caledonian grassbird	*Megalurulus mariei*
	Estrildidae	Red-throated parrotfinch	*Erythrura psittacea*
	Columbidae	Red-bellied fruit-dove	*Ptilinopus greyii*
	Meliphagidae	Dark-brown honeyeater	*Lichmera incana*
	Meliphagidae	Cardinal myzomela	*myzomela cardinalis*
	Rhipiduridae	Streaked fantail	*Rhipidura spilodera*
	Monarchidae	Southern shrikebill	*Clytorhynchus pachycephaloides*
	Acanthizidae	Fan-tailed gerygone	*Gerygone flavolateralis*
	Monarchidae	New Caledonian flycatcher	*Myiagra caledonica*
	Campephagidae	Melanesian cuckoo-shrike	*Coracina caledonica*
	Campephagidae	Long-tailed triller	*Lalage leucopyga*

Appendix I: Near-Endemic and Strict Endemic Birds by Ecoregion

	Family	Common Name	Species
New Caledonia Dry Forests [140]	Accipitridae	White-bellied goshawk	Accipiter haplochrous
	Columbidae	Red-bellied fruit-dove	Ptilinopus greyii
	Columbidae	New Caledonian imperial-pigeon	Ducula goliath
	Psittacidae	Horned parakeet	Eunymphicus cornutus
	Aegothelidae	New Caledonian owlet-nightjar	Aegotheles savesi
	Acanthizidae	Fan-tailed gerygone	Gerygone flavolateralis
	Meliphagidae	Dark-brown honeyeater	Lichmera incana
	Meliphagidae	New Caledonian myzomela	Myzomela caledonica
	Meliphagidae	New Caledonian friarbird	Philemon diemenensis
	Meliphagidae	Barred honeyeater	Phylidonyris undulata
	Eopsaltriidae	Yellow-bellied robin	Eopsaltria flaviventris
	Pachycephalidae	New Caledonian whistler	Pachycephala caledonica
	Rhipiduridae	Streaked fantail	Rhipidura spilodera
	Monarchidae	Southern shrikebill	Clytorhynchus pachycephaloides
	Monarchidae	New Caledonian flycatcher	Myiagra caledonica
	Corvidae	New Caledonian crow	Corvus moneduloides
	Campephagidae	Melanesian cuckoo-shrike	Coracina caledonica
	Campephagidae	New Caledonian cuckoo-shrike	Coracina analis
	Campephagidae	Long-tailed triller	Lalage leucopyga
	Sturnidae	Striated starling	Aplonis striata
	Zosteropidae	Green-backed white-eye	Zosterops xanthochrous
	Sylviidae	New Caledonian grassbird	Megalurulus mariei
	Estrildidae	Red-throated parrotfinch	Erythrura psittacea

*Strict endemic, a species whose range is restricted to this ecoregion

Appendix J

Ecoregion Descriptions

We present detailed descriptions of the ecoregions of the Indo-Pacific region in this appendix. These descriptions were written in collaboration with local experts (the authors and contributors to each ecoregion are listed at the end of each description). Each description introduces and summarizes the outstanding environmental and biological features of the ecoregion that contribute to its conservation ranking in this broad-scale analysis. Before the ecoregion descriptions of the Philippines and New Guinea we include a brief summary highlighting their evolutionary history, biodiversity, conservation status, and threats.

For each ecoregion, we list the near and strict endemic mammals and birds, derived from databases compiled by the Conservation Science Program, WWF-US. These databases are freely available on request. We also list the protected areas in each ecoregion, derived from the WCMC (1997) database of protected areas of the world. The extent of the protected areas as listed in the tables represent the areas that fall within the respective ecoregion boundaries (calculated using an ArcInfo GIS analysis), not the actual spatial extent of an entire protected area (the difference is reflected only when a protected area extends across two or more ecoregions). The descriptions also provide lists of important or focal species for conservation in each ecoregion, lists of several local conservation organizations and contacts, and nonexhaustive lists of several priority conservation actions. We end each description with a section detailing the basis and rationale for delineating the ecoregion boundaries. There are three ecoregions for which we do not include descriptions because more than 75 percent of their range is outside of our region of analysis. These ecoregions are the Central Tibetan Plateau Alpine Steppe [41], South Iran Nubo-Sindian Desert and Semi-Desert [43], and Registan–North Pakistan Sandy Desert [45]. Short descriptions of their biodiversity and conservation status, in addition to the rest of the 867 terrestrial ecoregions of the world, can be found at the following Web site: www.worldwildlife.org/ecoregions.

Ecoregion Number:	**1**
Ecoregion Name:	**North Western Ghats Montane Rain Forests**
Bioregion:	**Indian Subcontinent**
Biome:	**Tropical and Subtropical Moist Broadleaf Forests**
Political Units:	**India**
Ecoregion Size:	**30,800 km²**
Biological Distinctiveness:	**Regionally Outstanding**
Conservation Status:	**Endangered**
Conservation Assessment:	**II**

The North Western Ghats Montane Rain Forests [1] are one of two montane rain forest ecoregions (the other is the South Western Ghats Montane Rain Forests [2]) along the Western Ghats Mountain Range, renowned for its large number of species, particularly endemics. For instance, about a third of the plants, almost half the reptiles, and more than three-fourths of the amphibians known in India are found in this narrow strip of rain forest near the west coast of India. Even among the nonendemic fauna, species such as tigers, elephants, langurs, hornbills, and king cobras that conjure up romantic visions of the subcontinent's jungles inhabit these forests. The Wyanad evergreen forests of Kerala-Karnataka represent a transition zone from the moist *Cullenia*-dominated forests in the south Western Ghats to the northern drier dipterocarp forests (Rodgers and Panwar 1988).

Location and General Description

The ecoregion extends through the states of Tamil Nadu, Kerala, and Karnataka, on to Maharashtra State in western India. The ecoregion includes the middle- and upper-elevation biomes of the North Western Ghats Mountain Range.

As part of the Deccan Plate, the ecoregion has Gondwanaland origins, although the Western Ghats Mountains were created by post-Cretaceous uplift after the northward drifting Deccan Plate collided with the northern Eurasian continent. The resulting mountains intercepted the June to September southwest monsoon, resulting in the deposition of more than 2,500 mm of rainfall along their windward western slopes. The range's rainshadow creates drier conditions on the eastern slopes. Three large rivers, the Godavari, Krishna, and Cauvery, carry the rainfall from the monsoon rains eastward, all the way across the vast Deccan Plateau. The mountain range ascends abruptly on the western side from near sea level to 2,700 m and descends just as abruptly to 500 m onto the Deccan Plateau. The deeply dissected terrain produces localized variations in rainfall and habitat types and creates hotspots of endemism by limiting species distributions.

The multistoried rain forests with tall, buttressed trees exceeding 45 m are replete with climbers, lianas and epiphytes. Orchids in particular are plentiful. The forests are dominated by species of Dipterocarpaceae, Clusiaceae, Anacardiaceae, Sapotaceae, and Meliaceae (Puri et al. 1989). Bamboos, palms, and canes form a dense and impenetrable understory. The ground vegetation forms luxuriant carpets of species of *Strobilanthes*, *Ixora*, *Canthium*, *Selaginella*, *Arisaema*, Acanthaceae, Annonaceae, and various ferns (WWF and IUCN 1995). Several threatened and endemic plants in these rain forests include *Actinodaphne lanata*, *Meteoromyrtus wynaadensis*, *Cryptocoryne tortuosa*, *Cyathea nilgirensis*, *Ceropegia beddomei*, *Impatiens anaimudica*, and *Paphiopedilum druryi*.

At elevations above 900 m, Lauraceae are prominent (Pascal 1988; Puri et al. 1989). In Tamil Nadu, Karnataka, and Kerala, patches of evergreen forests above 1,500 m contain both tropical and temperate flowering elements. At even higher elevations in the southern parts of the ecoregion are patches of stunted forests contained within extensive grasslands, known locally as *shola* forests. The characteristic species in these *shola* forests include *Syzigium* spp., *Rhododendron nilgiricum*, *Mahonia nepalensis*, *Eleocarpus recurvatus*, *Ilex denticulata*, *Michaelia nilagirica*, and *Actinodaphne bourdellonii* (Karunakaran et al. 1998). The grasslands that surround the *shola* forests are characterized by several fire- and frost-resistant grasses: *Chrysopogon zeylanicus*, *Cymbopogon flexuosus*, *Arundinella ciliata*, *Arundinella mesophylla*, *Arundinella tuberculata*, *Themeda tremula*, and *Sehima nervosum* (Karunakaran et al. 1998).

Biodiversity Features

The rich forests of the Western Ghats harbor a large portion of India's biological diversity and include most of the endemic species. Although the forests in the southern part of the Western Ghats Mountain Range are richer than those in the northern sections, the latter also harbor important elements of diversity. These moist forests also extend into the dry forests along the rivers and act as riparian corridors for many mammals, birds, and reptiles that are typically found in the moist forests. For instance, giant squirrels (*Ratufa* spp.) inhabit the moist riparian forests within the dry deciduous and thorn forest ecoregions that provide poor habitat for them. The montane rain forests of the North Western Ghats also represent the northern limit for many evergreen and endemic trees, including *Myristica malabarica* and *Diospyros sylvatica*. These rain forests are also rich in orchids, which as a group are more diverse here than in any other ecoregion with the exception of the Himalayan broadleaf forest. These orchids have specialized pollination systems and often depend on a single pollinator species. Many orchids are epiphytes that need closed-canopy forests with mature trees.

The ecoregion harbors almost ninety mammal species, including two endemic species (table J1.1). The Malabar civet (*Viverra civettina*) is a near-endemic species that also inhabits the southern Western Ghats ecoregion. But the entire known population of the Wroughton's free-tailed bat (*Otomops wroughtoni*), estimated at a little more than forty animals, is limited to a single cave.

Because of its small population and extremely limited range, the Wroughton's free-tailed bat is considered critically endangered (Hilton-Taylor 2000). Other threatened species in the ecoregion include the tiger (*Panthera tigris*), Asian elephant (*Elephas maximus*), gaur (*Bos gaurus*), sloth bear (*Ursus ursinus*), and Malabar squirrel (*Ratufa indica*) (Hilton-Taylor 2000). The southern part of the ecoregion is included within a high-priority tiger conservation unit (TCU) that extends across most of the Western Ghats Mountain Range, with the northern boundary located close to the large city of Pune (Dinerstein et al. 1997). Some of India's most important tiger reserves, such as Bandipur and Dandeli, are included in this TCU. India's largest elephant population, in the Nilgiri Hills, also wanders into the extreme southern areas of the ecoregion.

More than 325 bird species are known to inhabit the ecoregion. These include eight near-endemic species (table J1.2) that are shared with the South Western Ghats Montane Rain Forests [2]. The distributions of the Nilgiri wood-pigeon, Malabar grey hornbill, grey-headed bulbul, rufous babbler, and the Malabar parakeet also extend down to the surrounding moist deciduous forests. The white-bellied treepie, white-bellied shortwing, and grey-breasted laughingthrush are primarily montane species (Grimmet et al. 1998).

The white-bellied shortwing is also a threatened species (Hilton-Taylor 2000). The ecoregion overlaps with the Western Ghats endemic bird area (EBA 123) (Stattersfield et al. 1998).

The montane forests are also critically important for maintaining watershed integrity in the Tamil Nadu plains and along the Malabar Coast. The large rivers that collect rainfall along the mountain range flow right across the vast, arid plateau.

Status and Threats

Current Status

More than half (58 percent) of the natural habitat in this ecoregion has now been cleared. Habitat loss and fragmentation are especially heavy in the northern reaches of the ecoregion, close to the large cities of Mumbai and Pune. Satellite images indicate the presence of one large block of intact habitat still evident along the central part of the ecoregion. The thirteen protected areas in the ecoregion cover almost 4,000 km^2 (13 percent) of the ecoregion (table J1.3). One protected area, Dandeli, is more than 1,000 km^2, but the average size of the protected areas is only 307 km^2. Eight other protected areas overlap across this ecoregion and the North Western Ghats Moist Deciduous Forests [14] (table

TABLE J1.1. Endemic and Near-Endemic Mammal Species.

Family	Species
Molossidae	*Otomops wroughtoni**
Viverridae	*Viverra civettina*

An asterisk signifies that the species' range is limited to this ecoregion.

TABLE J1.2. Endemic and Near-Endemic Bird Species.

Family	Common Name	Species
Columbidae	Nilgiri wood-pigeon	*Columba elphinstonii*
Bucconidae	Malabar grey hornbill	*Ocyceros griseus*
Corvidae	White-bellied treepie	*Dendrocitta leucogastra*
Turdidae	White-bellied shortwing	*Brachypteryx major*
Pycnonotidae	Grey-headed bulbul	*Pycnonotus priocephalus*
Timaliidae	Grey-breasted laughingthrush	*Garrulax jerdoni*
Timaliidae	Rufous babbler	*Turdoides subrufus*
Psittacidae	Malabar parakeet	*Psittacula columboides*

An asterisk signifies that the species' range is limited to this ecoregion.

TABLE J1.3. WCMC (1997) Protected Areas That Overlap with the Ecoregion.

Protected Area	Area (km²)	IUCN Category
Tansa [14]	130	IV
Koyna [14]	160	PRO
Chandoli [14]	80	IV
Radhanagari	350	IV
Kudremukh	820	II
Someswara	40	IV
Dandeli	1,060	UA
Anshi [14]	80	II
Bhadra [14]	200	IV
Shettihally [14]	470	IV
Sharavathi Valley [14]	370	IV
Mookambika [14]	160	IV
Pushpagiri [2]	70	IV
Total	3,990	

Ecoregion numbers of protected areas that overlap with additional ecoregions are listed in brackets.

J1.3). Three of these protected areas, Bhadra, Shettihally, and Sharavathi Valley, are each larger than 500 km².

Types and Severity of Threats

The major threats to this ecoregion stem from agriculture, mining, hydroelectric projects, and urban expansion. All of these overarching threats are widespread throughout the bioregion. Most of the commercially valuable trees in this ecoregion have already been harvested (IUCN 1991), and ironically, logging is not a significant threat. The paper pulp, plywood, and fiber industries and sawmills were the major consumers of timber and bamboo in the past. Mining for iron and manganese ore are now large contributors to habitat destruction.

Many of the valleys that supported large stands of species-rich forests have been submerged by reservoirs created by the construction of hydroelectric dams. In addition to this inundation of large areas, the secondary activities associated with dam construction, such as road building, access and encroachment into the intact forests, settlements, and fuelwood collection, have exacerbated habitat loss and degradation. The important riparian habitat is the first to be lost during these development enterprises. Many of the remaining forest patches that harbor endemic species are being converted to rubber, areca, and coffee plantations.

Fuelwood and fodder collection, grazing, and collection of nonwood forest products are intensifying as rural populations grow. The grasslands of this ecoregion are highly vulnerable to fire, and frequent fires retard the growth and regeneration of *shola* forests. The degraded habitat is then colonized by the exotic *Lantana camera* and *Eupatorium odorata*, which inhibit regeneration of native vegetation.

The prevalence of guns, ostensibly for crop protection, encourages widespread poaching.

Priority Conservation Actions

Rodgers and Panwar (1988) have identified several conservation gaps and needs in a comprehensive analysis. Some of the short and longer term conservation actions are as follows.

Short-Term Conservation Actions (1–5 Years)

As habitat fragmentation and loss continue to isolate existing protected areas, they become increasingly vulnerable to external influences. Thus, protected areas should be sufficiently buffered.

- Protect the riparian habitat and maintain connectivity.
- Survey remaining forest fragments, especially for the smaller, cryptic taxonomic groups that can harbor high levels of endemism. Pay particular attention to the forests in the Puttur division of South Kanara (Subramanya Hills) and adjoining areas.
- Encourage conservation and protection of sacred groves that lie outside existing protected areas.
- Manage habitat to provide an adequate prey base for tigers in the high-priority TCU.
- Mitigate human-wildlife conflicts, with emphasis on elephant, tiger, and sloth bear conflicts.
- Adequately protect the riparian habitat and the river systems because they provide water to the vast Deccan Plateau.

Longer-Term Conservation Actions (5–20 Years)

- Review land-use practices and patterns and sources of habitat loss. Develop a plan to create a conservation landscape that includes habitat linkages between protected areas in this and the adjacent moist deciduous forest ecoregion (Rodgers and Panwar 1988). If necessary, critical habitat should be restored.
- Implement conservation actions in the Level I TCU to create a conservation landscape that includes adequate core protected areas and dispersal habitat.

Focal Species for Conservation Action

Habitat specialists: hornbills (Malabar grey hornbill, Malabar pied hornbill, great hornbill) as indicators of mature, intact forests

Endangered endemic species: white-bellied treepie and white-bellied shortwing

Area sensitive species: Asian elephant, tiger

Top predator species: tiger

Other: endemic trees, orchids, sloth bear, gaur

Selected Conservation Contacts

Bombay Natural History Society
Centre for Ecological Sciences, Bangalore
Chief Wildlife Warden, Karnataka State
Indian Institute of Sciences, Bangalore
Pulney Hills Conservation Society
State Forest Departments
Western Ghats Development Authority
Western Ghats Conservation Board
WWF-India

Justification for Ecoregion Delineation

In an earlier analysis, Rodgers and Panwar (1988) placed the Western Ghats Mountain Range into a single biogeographic unit. But Rodgers and Panwar acknowledged that the Western Ghats Mountain Range is too large to represent a single unit for conservation planning. They divided the range into northern and southern areas, using the Wyanad as the boundary. Here, the southern *Cullenia*-dominated forests see a transition into drier northern dipterocarp forests. Following Rodgers and Panwar, we used this transition to make a more explicit division of the Western Ghats into northern and southern ecoregions. But in keeping with our rules for representing distinct habitat types within separate ecoregions and for separating the lowland from the montane forests, we placed the montane rain forests north of the Wyanad into the North Western Ghats Montane Rain Forests [1]. We used MacKinnon's (1997) map of original vegetation types in the Indo-Malayan region to identify the extent of the habitat. And we used the 1,000-m contour from a digital elevation model (DEM) to separate the montane rain forests from the surrounding moist deciduous forests. This ecoregion falls within Udvardy's Malabar rain forest biogeographic province.

Prepared by Gopal S. Rawat, Ajay Desai, Hema Somanathan, and Eric D. Wikramanayake.

Ecoregion Number:	**2**
Ecoregion Name:	**South Western Ghats Montane Rain Forests**
Bioregion:	**Indian Subcontinent**
Biome:	**Tropical and Subtropical Moist Broadleaf Forests**
Political Units:	**India**
Ecoregion Size:	**22,500 km²**
Biological Distinctiveness:	**Globally Outstanding**
Conservation Status:	**Endangered**
Conservation Assessment:	**I**

The South Western Ghats Montane Rain Forests [2] are the most species-rich ecoregion in the Deccan Peninsula. They also harbor the highest levels of endemics. Consider the numbers: 35 percent of the plants, 42 percent of the fishes, 48 percent of the reptiles, and 75 percent of the amphibians that live in these rain forests are endemic species (GOI 1997). Ten mammals and thirteen birds are endemic or near endemic to the ecoregion. More than 80 percent of the flowering plants characteristic of this mountain range are in the species-rich forests of the south. Large, charismatic mammals such as the tiger (*Panthera tigris*), Asian elephant (*Elephas maximus*), sloth bear (*Melursus ursinus*), gaur (*Bos gaurus*), and raucous hornbill inhabit the forests. These species evoke images of wild jungles of the Indian subcontinent. Large expanses of high-elevation, undulating grasslands interspersed with patches of stunted *shola* forests harbor the endemic Nilgiri tahr and India's largest elephant population. Every twelve years, the blue flowers of Neelakurunji (*Phlebophyllum kunthianum*) impart a blue hue to these grassland-*shola* mountains.

Location and General Description

This ecoregion represents the montane rain forests above 1,000 m along the southern half of the Western Ghats. It extends as a long and narrow unit through the Indian states of Kerala and Tamil Nadu. The northern boundary of this ecoregion is the Wyanad, where the habitat makes a transition from the drier forest in the north to the more moist forest in the south. From here the ecoregion runs along the Western Ghats Mountain Range, about 35 km inland and parallel to the Malabar Coast, all the way to the southern end of the range.

The Deccan Plateau itself was once part of Gondwanaland, evident from relicts of the ancient southern flora and fauna. After becoming detached from this southern continent during the Cretaceous, it drifted northward to finally crash into the northern Laurasian continent. After this initial collision, a series of geological uplifts created the Western Ghats Mountain Range, with several peaks higher than 2,000 m. The highest of these is Anaimudi, which rises to 2,695 m.

As the moisture-laden southwest monsoon winds sweep in from the Malabar Coast and rise above the mountain range, they release more than 2,500 mm of rainfall. The northeast monsoon from October to November supplements the June to September southwest monsoon rainfall, for an average annual precipitation that exceeds 2,800 mm. But because of the deeply dissected topography, some areas can receive more than 8,000 mm of rainfall throughout the year (Mishra and Johnsingh 1998). This produces local variations in habitat types and localized centers of endemism (Rodgers and Panwar 1988; Kendrick 1989). The Periyar River, which originates from Periyar Tiger Reserve, is one of the larger rivers that carry the monsoon rains east across the peninsula.

The habitat types include the wet montane evergreen forests and *shola*-grassland complexes in the higher elevations. The montane evergreen forests are diverse, multistoried, and rich in epiphytes, with a low canopy at 15 to 20 m (Puri et al. 1989; WWF and IUCN 1995; Ganesh et al. 1996). The forest communities are characterized by *Cullenia exarillata, Mesua ferrea, Palaquium ellipticum, Gluta travancorica,* and *Podocarpus wallichiana* (Ramesh et al. 1999). *Podocarpus* represents a Gondwanaland relict carried across during the long northward journey. Other evergreen species in these montane forests include *Calophyllum austroindicum, Garcinia rubro-echinata, Garcinia travancorica, Diospyros barberi, Memecylon subramanii, Memecylon gracile, Goniothalamus rhyncantherus,* and *Vernonia travancorica* (Ramesh 1999).

The montane *shola*-grassland complexes occur between 1,900 and 2,220 m. These consist of stunted montane forests surrounded by undulating grasslands. The upper story of the forests is characterized by *Pygeum gardneri, Schefflera racemosa, Linnociera ramiflora, Syzigium* spp., *Rhododenron nilgiricum, Mahonia nepalensis, Eleocarpus recurvatus, Ilex denticulata, Michaelia nilagirica, Actinodaphne bourdellonii,* and *Litsea wightiana* (Karunakaran et al. 1998). A low, twisted second story of *Ilex wightiana, Rapanaea wightiana, Ternstroemia gymnanthera, Symplocos* spp., and *Microtropis* spp. and a dense shrub layer of saplings of *Strobilanthes,*

Psychotria, and *Lasianthus* spp. usually is present. The grasslands that surround the *shola* forests consist of several fire- and frost-resistant grasses: *Chrysopogon zeylanicus*, *Cymbopogon flexuosus*, *Arundinella ciliata*, *Arundinella mesophylla*, *Arundinella tuberculata*, *Themeda tremula*, and *Sehima nervosum* (Karunakaran et al. 1998).

Biodiversity Features

The levels of endemism in these montane forests are truly astounding. More than half the tree species are endemic, especially among the Dipterocarpaceae and Ebenaceae. The majority of the fifty endemic plant genera are also monotypic. But the distribution of richness and endemism is not uniform. There are localized areas that harbor exceptional levels of diversity and endemism. For instance, the Agasthyamalai and Nilgiri hills, recognized as centers of plant diversity, are exceptional for plant richness (WWF and IUCN 1995).

A high proportion of the trees in these forests have dioecious breeding systems. In the *shola* forests in particular, dioecy is quite prevalent among the Lauraceae and Moraceae. Therefore, many of the tree populations are especially vulnerable to deforestation and other disturbances because of mate isolation and reduction of numbers of mates. In the *Cullenia*-dominated stands, for instance, a large proportion of trees have less than five individuals in a 10-ha area, making them locally rare.

This ecoregion harbors almost 20 percent of India's mammal fauna. The seventy-nine mammal species attributed to the ecoregion include ten endemics (table J2.1). Three of these are strict endemic species (i.e., limited to this ecoregion), and the others are considered to be near-endemic species (i.e., shared with adjacent ecoregions).

The Salim Ali fruit bat (*Latidens salimalii*) and the rodent *Platacanthomys lasiurus* represent monotypic, endemic genera. The rare and endemic Nilgiri tahr (*Hemitragus hylocrius*) is currently limited to a narrow, 400-km stretch of *shola*-grassland mosaic, from the Nilgiri Hills to the Ashambu Hills (Mishra and Johnsingh 1998). At lower altitudes the tahr also use patches of grassland that grow on rock-sheeted substrates. Tigers, leopards (*Panthera pardus*), and wild dogs (*Cuon alpinus*) are the natural predators of Nilgiri tahr. However, the population decline of the tahr has been caused by heavy hunting pressure and habitat conversion (Mishra and Johnsingh 1998). The largest population of tahr, estimated at 250–300 animals, now resides in the Grass Hills of the Anamalai Sanctuary (Mishra and Johnsingh 1998).

The charismatic endangered lion-tailed macaque (*Macaca silenus*) and Nilgiri macaque (*Semnopithecus johnii*) are other endemic species that need intact habitat and are highly threatened by habitat conversion. This ecoregion also harbors India's largest Asian elephant population (Sukumar 1989) and contains critical habitat for tigers. It overlaps with two Level I TCUs (Dinerstein et al. 1997). The survival of the tiger and elephant is threatened by habitat fragmentation.

Among the other threatened species are the sloth bear (*Melursus ursinus*), gaur (*Bos gaurus*), and wild dog (Hilton-Taylor 2000).

The bird fauna in the ecoregion is estimated at 309 species. Ten species are near-endemics and three are strict endemics (table J2.2).

The broad-tailed grassbird (*Schoenicola platyura*) and Nilgiri pipit (*Anthus nilghiriensis*) are high-elevation grassland species (Grimmet et al. 1998).

The ecoregion is included within the Western Ghats EBA (123) (Stattersfield et al. 1998).

The high endemism levels seen among the mammals and birds extend to other taxonomic groups as well. About 90 of India's 484 reptile species are endemic to these forests, including eight endemic genera (*Brachyophidium*, *Dravidogecko*, *Melanophidium*, *Plectrurus*, *Ristella*, *Salea*, *Teretrurus*, and *Xylophis*). The amphibian fauna exhibits even greater levels of endemism: almost 50 percent of India's 206 amphibian species are endemic to this ecoregion, among

TABLE J2.1. Endemic and Near-Endemic Mammal Species.

Family	Species
Soricidae	*Suncus montanus*
Pteropodidae	*Latidens salimalii*
Cercopithecidae	*Trachypithecus johnii*
Viverridae	*Viverra civettina*
Viverridae	*Paradoxurus jerdoni*
Bovidae	*Hemitragus hylocrius**
Sciuridae	*Funambulus layardi*
Sciuridae	*Funambulus sublineatus*
Muridae	*Mus famulus**
Muridae	*Vandeleuria nilagirica**

An asterisk signifies that the species' range is limited to this ecoregion.

TABLE J2.2. Endemic and Near-Endemic Bird Species.

Family	Common Name	Species
Columbidae	Nilgiri wood-pigeon	*Columba elphinstonii*
Bucconidae	Malabar grey hornbill	*Ocyceros griseus*
Corvidae	White-bellied treepie	*Dendrocitta leucogastra*
Turdidae	White-bellied shortwing	*Brachypteryx major*
Pycnonotidae	Grey-headed bulbul	*Pycnonotus priocephalus*
Timaliidae	Grey-breasted laughingthrush	*Garrulax jerdoni*
Timaliidae	Rufous babbler	*Turdoides subrufus*
Muscicapidae	Black-and-rufous flycatcher	*Ficedula nigrorufa*
Muscicapidae	Nilgiri flycatcher	*Eumyias albicaudata*
Sylviidae	Broad-tailed grassbird	*Schoenicola platyura**
Timaliidae	Nilgiri laughingthrush	*Garrulax cachinnans**
Motacillidae	Nilgiri pipit	*Anthus nilghiriensis**
Psittacidae	Malabar parakeet	*Psittacula columboides*

An asterisk signifies that the species' range is limited to this ecoregion.

which are six endemic genera (*Indotyphlus*, *Melanobatrachus*, *Nannobatrachus*, *Nyctibatrachus*, *Ranixalus*, and *Uraeotyphlus*).

Status and Threats

Current Status

Nearly two-thirds of the natural forests in this ecoregion have already been cleared. The remaining habitat is fragmented, except for one large intact habitat block in the southern area of the ecoregion. About 3,200 km^2 of the ecoregion is already included within sixteen protected areas (table J2.3).

Among these, Periyar, Anamalai, and Kalakad-Mundanthurai represent three important reserves. Parambikulam, Anamalai, and Eravikulam lie adjacent to each other and form a large protected area complex that harbors important Nilgiri tahr populations (Mishra and Johnsingh 1998). Four reserves—Periyar, Anamalai, Eravikulam, and Megamalai (proposed)—extend into the adjacent South Western Ghats Moist Deciduous Forests [15].

Despite 15 percent of the intact habitat being within the protected area system, the management status and threats to these protected areas vary enormously. For example, the Mukurty National Park in Tamil Nadu has no human inhabitants but contains small abandoned plantations, whereas Kerala's Parambikulam Wildlife Sanctuary has several large commercial plantations within it (IUCN 1991).

Types and Severity of Threats

The threats to this ecoregion's natural habitats and biodiversity are manifold. Some of the major threats include conversion of forests into tea, coffee, potato, teak, *Eucalyptus*, and cardamom plantations, as well as road construction, tourism pressures, and livestock grazing (Rodgers and Panwar 1988; WII 1999). Illegal taking of timber is high and is considered a major threat to the remaining forests. Many people own guns for crop protection that they also use for poaching.

The Nilgiri and Cardamom hills in particular harbor high levels of richness and endemism as well as some of the most important populations of elephants and tigers. These areas are especially affected by the tea, coffee, and rubber plantations. Shifting cultivation has begun to clear patches of old-growth forest (Kendrick 1989). In the Anamalais, fig trees from the *sholas* are felled or lopped to feed camp elephants. Because many of these species are dioecious, this affects the sexual selection among trees and causes reproductive isolation. Fig trees are also keystone food resources for several species (from giant squirrels to hornbills), and their removal results in cascading ecological effects on the frugivore community.

Hydroelectric power development along the rivers in this ecoregion is also a serious threat. In addition to inundating critical habitat, dam construction also causes tremendous habitat destruction and disturbances.

The mountains, especially in the south, are mineral-rich. Therefore, mining is a potential future threat that should be addressed with preemptive measures.

Priority Conservation Actions

In their comprehensive conservation analysis for India, Rodgers and Panwar (1988) identified several conservation gaps and actions for the southern Western Ghats. These are still relevant for this ecoregion. In this section, we also present several short-term and long term conservation actions.

Short-Term Conservation Actions (1–5 Years)

- Review the land-use practices and patterns and sources of habitat loss. Develop a plan to create conservation landscapes that includes habitat linkages between key protected areas.
- Manage the habitat to provide an adequate prey base for tigers in the Level I TCUs.
- The Nilgiri tahr populations, under threat from habitat loss and hunting, should be protected. Consider reintroducing tahr into parts of its former range. Such reintroductions will also help renew interest in conservation of additional *shola* forests, especially since these forests have been overlooked for conservation because the charismatic large vertebrates are now missing.
- Protect the habitat and populations of lion-tailed macaque and Nilgiri langur.
- Plan conservation landscapes for the elephant populations in the Nilgiri Hills and in Mudumalai, including additions to the Nagarhole-Bandipur-Mudumalai protected areas complex to include seasonal habitat needs for the elephants (Rodgers and Panwar 1988).
- Restore to natural habitats the cardamom and tea estates that have been acquired by the Forest Department.
- Survey the remaining forests, with a focus on the smaller, secretive taxonomic groups that usually harbor high levels of endemism.
- Encourage and continue conservation of the sacred groves that have received traditional protection.
- Mitigate human-wildlife conflicts, with a focus on elephant, tiger, and sloth bear conflicts.

TABLE J2.3. WCMC (1997) Protected Areas That Overlap with the Ecoregion.

Protected Area	Area (km^2)	IUCN Category
Pushpagiri [1]	60	IV
Talakaveri	250	PRO
Brahmagiri	190	PRO
Aralam	50	IV
Karimpuzha	230	PRO
Mukurty National Park	60	PRO
Silent Valley	110	II
Megamalai [15]	120	PRO
Periyar [15]	540	IV
Anamalai [15]	600	IV
Eravikulam [15]	70	II
Parambikulam	260	IV
Idukki	80	IV
Shenduruny	300	IV
Kalakad-Mundanthurai	290	IV
Peppara [15]	40	IV
Total	3,250	

Ecoregion numbers of protected areas that overlap with additional ecoregions are listed in brackets.

Longer-Term Conservation Actions (5–20 Years)

- Restore and recreate the habitat between the Periyar Tiger Reserve and the Ashambu Hills and between the southern and northern areas of the Nilgiri Biosphere Reserve in Tamil Nadu to allow elephant movements. Proposals to create these corridors have not been implemented and should be acted on (Johnsingh and Williams 1999).
- Implement conservation actions in the two Level I TCUs to create conservation landscapes that include adequate core protected areas and dispersal habitat.

Focal Species for Conservation Action

Habitat specialists: hornbills (Malabar grey hornbill, Malabar pied hornbill, great hornbill) as indicators of mature and intact forests

Endangered endemic species: white-bellied shortwing, Nilgiri marten, Nilgiri pipit, Nilgiri laughing thrush, lion-tailed macaque, Nilgiri macaque, Nilgiri tahr

Area-sensitive species: Asian elephant, tiger

Top predator species: tiger, wild dog

Other: Nilgiri langur, sloth bear, gaur, Salim Ali fruit bat

Selected Conservation Contacts

Bombay Natural History Society

Centre for Ecological Sciences, Bangalore

Chief Wildlife Wardens, Kerala and Tamil Nadu States

Director, Anamalai Wildlife Sanctuary, Tamil Nadu

Field director, Kalakad-Mundanthurai Tiger Reserve, Tamil Nadu

French Institute of Pondicherry

Gardens and Research Institute, Thiruvananthapuram

High Range Wildlife Preservation Society, Munnar, Kerala

Kerala Forest Research Institute, Pichi, Kerala

Salim Ali Institute of Ornithology and Natural History, Coimbatore Tropical Botanic

Wildlife Institute of India

WWF-India

Justification for Ecoregion Delineation

In an earlier analysis of conservation units of India, Rodgers and Panwar (1988) placed the Western Ghats Mountain Range into a single biogeographic unit. But they acknowledged that the entire mountain range was too large to represent a single unit for conservation planning and divided the mountain range into northern and southern areas, using the Wyanad as a transition zone. Here the moister southern *Cullenia*-dominated forests grade into the drier northern dipterocarp forests. In our analysis, we also used this transition to make a more explicit division of the northern and southern ecoregions in the Western Ghats. But in keeping with our definition of ecoregions (i.e., to represent distinct habitat types and lowland and montane forests in separate ecoregions), we placed the montane forests to the south of the Wyanad area in the South Western Ghats Montane Rain Forests [2]. We used the 1,000-m contour from a DEM and MacKinnon's (1997) map of original vegetation to define the boundary between the montane rain forests and moist deciduous forests. This ecoregion falls within Udvardy's Malabar rain forest biogeographic province.

Prepared by Gopal S. Rawat, Ajay Desai, Hema Somanathan, and Eric D. Wikramanayake.

Ecoregion Number:	**3**
Ecoregion Name:	**Sri Lanka Lowland Rain Forests**
Bioregion:	**Indian Subcontinent**
Biome:	**Tropical and Subtropical Moist Broadleaf Forests**
Political Units:	**Sri Lanka**
Ecoregion Size:	**12,500 km^2**
Biological Distinctiveness:	**Globally Outstanding**
Conservation Status:	**Critical**
Conservation Assessment:	**I**

In 1996, Dutta and Manamendra-Arachchi listed fifty-three amphibian species from Sri Lanka. This was a substantial increase from the thirty-five species-groups described by Kirtisinghe (1957) in what was the definitive work on the Amphibia of Sri Lanka for almost half a century. But dedicated amphibian surveys have since increased this number to more than 250 frog species, most of them endemic to the rain forests in the southwestern quarter of the island (Pethiyagoda and Manamendra-Arachchi 1998). The globally outstanding levels of endemism in the Sri Lanka Lowland Rain Forests [3] are epitomized by this assemblage of frogs.

Location and General Description

The Sri Lanka Lowland Rain Forests [3] represents the tropical rain forests below 1,000 m in elevation in the southwestern quarter of Sri Lanka.

A continental island, Sri Lanka is separated from the Indian peninsula by the shallow Palk Strait. The island was part of Gondwanaland until the Cretaceous, when as part of the Deccan Plate it became detached and drifted northward. The Deccan Plate collided with the Asian mainland—the Southern Laurasian coastline—about 55 million years later (Audley-Charles et al. 1981). Therefore, the ecoregion harbors several ancient Gondwana taxonomic groups. The island first became separated from the mainland Indian subcontinent during the late Miocene. Since then, climatic changes have interposed drier conditions between the moist forests in southwest Sri Lanka and the Western Ghats in India, the closest other moist forests. Despite the several land bridge connections with the Indian Peninsula since the initial separation, the moist forest and its wet forest–adapted biota have been ecologically isolated (Deraniyagala 1992).

The May to September southwest monsoon, extended by the intermonsoonal season before and after the true monsoon, brings more than 5,000 mm of rainfall to the ecoregion. Temperatures remain nearly constant, 27–30°C, year-round. Convectional winds from the ocean also ameliorate daily temperatures, especially along the coastal areas. Relative humidity is high (80–85 percent). The long years of isolation combined with these warm, moist conditions have resulted in the evolution of specialized species found nowhere else on Earth.

The ecoregion partially encircles the central massif, which rises to more than 2,500 m, and the detached Knuckles Mountain Range to the northeast. These mountains are placed in their own ecoregion, the Sri Lanka Montane Rain Forests [4]. The ecoregion's topography is characterized by deep valleys of the major rivers that radiate out of the central mountains. The soils in this lowland wet-zone ecoregion are red-yellow podzolic soils (Survey Department 1988).

The vegetation is influenced primarily by climate, with topography and edaphic conditions contributing secondarily (de Rosayro 1950). The lowland wet evergreen forests are characterized by two floral communities: the *Dipterocarpus*-dominated community and the *Mesua-Shorea* community (Ashton and Gunatilleke 1987; Singhakumara 1995). The former is dominated by *Dipterocarpus zeylanicus* and *D. hispidus*, with *Vitex altissima, Chaetocarpus castanocarpus, Dillenia retusa, D. triquetra, Myristica dactyloides,* and *Semecarpus gardneri* (de Rosayro 1942). The canopy of the *Mesua-Shorea* community consists of *Anisophyllea cinnamomoides, Cullenia rosayroana, Mesua ferrea, M. nagassarium, Myristica dactyloides, Palaquium petiolare, Shorea affinis, S. congestiflora, S. disticha, S. megistophylla, S. trapezifolia, S. worthingtoni, Syzygium rubicundum,* and a subcanopy of *Chaetocarpus castanocarpus, Garcinia hermonii, Syzygium neesianum,* and *Xylopia championi* (Ashton et al. 1992; Gunatilleke and Gunatilleke 1981; Singhakumara 1995). Undisturbed forests have four strata, with a main canopy at 30–40 m, a sub-canopy at 15–30 m, a 5–15 m understory, and a sparse shrub layer (Singhakumara 1995). Emergent species rise above the upper canopy to 45 m (Abeywickrama 1980).

Among the distinct habitat types included within the ecoregion are patches of swamp forests closer to the coastlines, although most of these have long been converted to agriculture. Small patches of *Avicennia-Rhizophora-Sonneratia*–dominated riverine and fringing mangroves line the coastlines, especially near the mouths of the major rivers (Arulchelvam 1969; Spalding et al. 1997).

Biodiversity Features

Almost all of Sri Lanka's endemic flora and fauna are confined to the rain forests in the southwest quarter of the island, where the warm, moist climatic conditions and the longer period of isolation of the wet forest–adapted species have promoted the evolution of endemism and specialization (Pethiyagoda and Manamendra-Arachchi 1998; Phillips 1980; Sri Lanka National Report 1991; Wikramanayake 1990).

For instance, more than 60 percent of the 306 tree species that are endemic to Sri Lanka are found only in the lowland rain forests represented by this ecoregion, and another 61 species are shared with the montane rain forests and dry forests. Of the twelve endemic floral genera on the island, eleven are confined to the rain forests (Kendrick 1989). Ninety-eight percent of the fifty-eight species in the family Dipterocarpaceae—the dominant tree family in Asian rain forests—are endemic to the rain forests (Dassanayake and Fosberg 1980; Ashton and Gunatilleke 1987), and these include two endemic genera, *Doona* and *Stemonoporus* (Kostermans 1992). An endemic ground orchid, *Anoectochilus setaceus*, commonly known as the king of the forest or *wanaraja*, is found only in undisturbed portions of these rain forests. Several other plants have highly localized distributions. Examples include *Diospyros oppositifolia*, which is confined to the top of a small but species-rich peak known as Hinidumkanda, *Stemonoporus moonii*, and *Mesua stylosa*, which are known only from an inland marsh forest in Bulathsinhala, and an aquatic Cyperaceae, *Mappania immersa*, which is limited to some streams at Sinharaja. Floristically, the lowland and lower hill forests are the richest in Sri Lanka (Erdelen 1988; Gunatilleke and Gunatilleke 1990; Singhakumara 1995) and of all south Asia (Ashton and Gunatilleke 1987).

As a small island, Sri Lanka lacks the space to support many of the megavertebrates found on the mainland, although the fossil record indicates that ancestral forms of rhinoceroses, hippopotamuses, and lions once roamed here (Deraniyagala 1992). Despite the small number of species, the ecoregion contains several near-endemic mammals, including one strict endemic (table J3.1).

The two endemic shrews, *Suncus zeylanicus* and *Suncus montanus*, are endangered and vulnerable, respectively (Hilton-Taylor 2000). The Sri Lankan leopard (*Panthera pardus kotiya*) is considered a threatened genotype (Hilton-Taylor 2000). The ecoregion also contains a small population of the endangered Asian elephant (*Elephas maximus*). Unlike the larger elephant populations in the dry-zone ecoregion, this small rain forest population is greatly threatened by habitat loss and fragmentation.

The ecoregion is completely contained within an EBA, Sri Lanka (124) (Stattersfield et al. 1998). Sixteen birds are considered near-endemic species, and two are strict endemics (table J3.2). The endemic green-billed coucal and Sri Lanka whistling-thrush (*Myiophonus blighi*) are considered threatened (Hilton-Taylor 2000).

TABLE J3.1. Endemic and Near-Endemic Mammal Species.

Family	Species
Soricidae	*Suncus montanus*
Soricidae	*Suncus zeylanicus**
Rhinolophidae	*Hipposideros halophyllus*
Cercopithecidae	*Semnopithecus vetulus*
Viverridae	*Paradoxurus zeylonensis*
Sciuridae	*Funambulus layardi*
Muridae	*Petinomys fuscocapillus*
Muridae	*Mus fernandoni*
Muridae	*Vandeleuria nolthenii*

An asterisk signifies that the species' range is limited to this ecoregion.

TABLE J3.2. Endemic and Near-Endemic Bird Species.

Family	Common Name	Species
Columbidae	Ceylon wood-pigeon	Columba torringtoni
Bucconidae	Ceylon grey hornbill	Ocyceros gingalensis
Cuculidae	Red-faced malkoha	Phaenicophaeus pyrrhocephalus
Cuculidae	Green-billed coucal	Centropus chlororhynchus*
Phasianidae	Ceylon spurfowl	Galloperdix bicalcarata
Phasianidae	Ceylon junglefowl	Gallus lafayetii
Corvidae	Ceylon magpie	Urocissa ornata
Turdidae	Spot-winged thrush	Zoothera spiloptera
Sturnidae	White-faced starling	Sturnus senex
Sturnidae	Ceylon myna	Gracula ptilogenys
Muscicapidae	Kashmir flycatcher	Ficedula subrubra
Timaliidae	Brown-capped babbler	Pellorneum fuscocapillum
Timaliidae	Orange-billed babbler	Turdoides rufescens
Dicaeidae	White-throated flowerpecker	Dicaeum vincens*
Capitonidae	Yellow-fronted barbet	Megalaima flavifrons
Psittacidae	Ceylon hanging-parrot	Loriculus beryllinus
Psittacidae	Layard's parakeet	Psittacula calthropae
Strigidae	Chestnut-backed owlet	Glaucidium castanonotum

An asterisk signifies that the species' range is limited to this ecoregion.

Among the other vertebrate groups, two reptiles, the mugger crocodile (Crocodylus palustris) and the spineless forest lizard (Calotes liocephalus), are listed as endangered, as are eight freshwater fish species (Hilton-Taylor 2000).

Several other taxonomic groups also exhibit high levels of endemism. The rhacophorid frogs in particular have undergone a remarkable radiation, and this richness places these rain forests at the apex in terms of amphibian species numbers per unit area (Pethiyagoda and Manamendra-Arachchi 1998). Many of these species have only limited range distributions, often less than 0.5 km^2, and are now limited to the undisturbed habitat fragments. Thus, extinctions surely must have accompanied the widespread habitat loss in this ecoregion.

Status and Threats

Current Status

During the past two centuries, nearly all the natural forests in this ecoregion have been cleared for tea, rubber, and coconut plantations, rice paddies, and human settlements. Only about 8 percent of the lowland wet forests now remain, as several small, isolated patches in a highly fragmented landscape. Nevertheless, because many of the endemic species in the ecoregion have small habitat needs, these patches can provide adequate refuge if effectively protected and managed to ameliorate edge effects and other external threats.

TABLE J3.3. WCMC (1997) Protected Areas That Overlap with the Ecoregion.

Protected Area	Area (km^2)	IUCN Category
Sri Jayewardenepura Bird Sanctuary	30	IV
Sinharaja	100	IV
Telwatte	20	IV
Attidiya Marsh	10	IV
Peak Wilderness [4]	100	IV
Total	260	

Ecoregion numbers of protected areas that overlap with additional ecoregions are listed in brackets.

Currently, just about 2 percent of the ecoregion's intact habitat is protected within five protected areas (table J3.3). The most important of these are undoubtedly the Sinharaja Natural Heritage Wilderness Area and Peak Wilderness Sanctuary. The latter extends into the montane rain forest ecoregion. Together, these two protected areas represent the two largest forest patches. However, because most of Sri Lanka's endemic species and species richness lie within the rain forest ecoregions, the current level of protection is inadequate. The Forest Department has also set aside several conservation areas from its portfolio of forest reserves to address conservation gaps (IUCN 1992). The Sinharaja reserve is contiguous with several forest reserves, namely Morapitiya, Runakanda, Panagala, and Delgoda, that together form the largest forest block, accounting for about 43 percent of the remaining wet-zone forests (Zoysa and Raheem 1990). Although these reserves will contribute significantly to the overall protected areas system, it is imperative that all remaining habitat patches be conserved to safeguard the remaining beta-diversity. Unless additional protection with effective management is provided, Sri Lanka's most important elements of biodiversity will be lost forever.

Types and Severity of Threats

Fifty-five percent of Sri Lanka's human population lives in this small ecoregion, which represents less than 25 percent of the land area. Clearing land for agricultural expansion and settlements to support this population and illegal logging, albeit small-scale, in the remaining forests are the most serious threats to the survival of the endemic species (Gunatilleke and Gunatilleke 1990).

Priority Conservation Actions

Priority actions have been identified in a national biodiversity action plan.

Short-Term Conservation Actions (1–5 Years)

- Implement without delay the management plans that have been prepared for Sinharaja, Peak Wilderness, and the Attidiya Marsh.
- Prepare and implement management plans for the conservation areas identified by the Forest Department. A moratorium on clearing and logging in intact forests is urgently needed.

- Conduct biodiversity surveys, especially of the secretive fauna, in all intact forests.
- Protect additional areas that include habitats of the majority of Sri Lanka's endemic flora and fauna.

Longer-Term Conservation Actions (5–20 Years)

- Restore habitat to add to existing conservation areas and to link existing protected areas, especially with the protected areas that extend into the montane region. Because all rivers originate in these mountains and radiate outward, the altitudinal connectivity will also provide more effective watershed protection.
- Monitor ecological integrity of the remaining forests and protected areas, using endemic species and assemblages.

Focal Species for Conservation Action

Endangered endemic species: green-billed coucal, white-throated flowerpecker, *Suncus zeylanicus*, *Suncus montanus*

Area-sensitive species: Asian elephant, leopard

Top predator species: leopard

Other: purple-faced leaf monkey; golden palm civet; endemic lizards *Calotes liolepis*, *Ceratophora aspera*, and *Lyriocephalus scutatus* (need mature and undisturbed forests); assemblage of endemic rhacophorid frogs; endemic freshwater fishes; mugger crocodile (*Crocodylus palustris*); spineless forest lizard (*Calotes liocephalus*)

Selected Conservation Contacts

Biodiversity and Elephant Conservation Trust

Bird Club, Sri Lanka

Department of Wildlife Conservation, Government of Sri Lanka

Field Ornithological Group

Forest Department, Government of Sri Lanka

IUCN Sri Lanka

March for Conservation

University of Colombo

University of Peradeniya

Wildlife Heritage Trust

Wildlife and Nature Conservation Society, Sri Lanka

Young Zoologists Association

Justification for Ecoregion Delineation

Sri Lanka's forests have been divided into two broad climatic sub-regions, the wet zone and the dry zone. In a previous analysis of conservation units of the Indo-Malayan realm, MacKinnon (1997) placed the wet-zone forests into a single biounit, Ceylon Wet Zone 02. We used MacKinnon's regional classification as a guiding framework in delineating ecoregions across the Indo-Pacific region. But because we differentiated between lowland and montane forests in delineating ecoregions, we used the 1,000-m contour to extract and place the lowland rain forests into a distinct ecoregion, the Sri Lanka Lowland Rain Forests [3]. The ecoregion also overlaps with floristic zones 5, 6, 7, and 11 identified by Ashton and Gunatilleke (1987).

Prepared by Eric D. Wikramanayake and Savithri Gunatilleke

Ecoregion Number:	**4**
Ecoregion Name:	**Sri Lanka Montane Rain Forests**
Bioregion:	**Indian Subcontinent**
Biome:	**Tropical and Subtropical Moist Broadleaf Forests**
Political Units:	**Sri Lanka**
Ecoregion Size:	**3,000 km²**
Biological Distinctiveness:	**Globally Outstanding**
Conservation Status:	**Critical**
Conservation Assessment:	**I**

The number of known floral and faunal endemic species of the Sri Lanka Montane Rain Forests [4] is higher than in the lowland rain forests (Phillips 1980; Singhakumara 1995), which are also renowned for high levels of endemism. But new ongoing surveys that have focused on the hitherto neglected secretive fauna reveal a plethora of additional new species limited to the montane forests (Pethiyagoda and Manamendra-Arachichi 1998; Ng 1995; Bahir 1998). These surveys demonstrate that this area's biodiversity is even more important globally than previously thought. The known levels of endemism apparently are only the tip of the iceberg. In light of the emerging information, the Sri Lanka Montane Rain Forests [4] can be considered a super-hotspot within the endemism hotspot of global importance recognized by Myers et al. (2000).

Location and General Description

The Sri Lanka Montane Rain Forests [4] represents the montane and submontane moist forests above 1,000 m in the central massif and in the Knuckles Mountain Range to the northeast.

Sri Lanka first separated from the Deccan Peninsula during the late Miocene, but geologically it has Gondwanaland origins. The biogeographic patterns indicate that the wet forests in the southwestern quarter became isolated from the nearest other wet forests in the mainland soon after the first separation when climatic changes created drier, warmer conditions throughout most of the lowland areas (Deraniyagala 1992). During subsequent Pleistocene-period land-bridge connections, the intervening drier habitat prevented the exchange of wet forest–adapted species with the wet forests in India. Therefore, the species in the southwestern quarter evolved and speciated in isolation, giving rise to the high levels of endemism that are evident today (Senanayake 1993; Wikramanayake 1990).

The post-Miocene uplift that created three peneplains, or steep plains of erosion (Deraniyagala 1958), have also served as biogeographic barriers, isolating taxonomic groups (Senanayake 1993; Wikramanayake 1990). This is best exemplified in the distribution of the freshwater fish fauna, where the lowland streams have the highest species richness, the second peneplain fewer, and the upper peneplain streams support a limited fish fauna (Wikramanayake 1990). Floral surveys also show that plant diversity is greater in the lowland forests but that the montane forests support greater numbers of endemic plants (Singhakumara 1995).

The higher peaks of the central massif attain heights of more than 2,500 m, but the average height is about 1,800 m. The isolated Knuckles Range to the northwest of the central massif rises to more than 1,800 m, but the main range is about 1,500 m. All of Sri Lanka's major rivers originate in the central mountains and radiate outward. The predominant soil type is red-yellow podzolic soils (Survey Department 1988).

There is a general trend of a 0.5°C decrease in air temperature for every 100-m rise in elevation (Ashton et al. 1997); therefore, the climate is cooler than in the lowlands. Ground frost is not uncommon in the higher elevations during the nights from December to February. Many of the high peaks are also enveloped in mist and fog, creating cloud forests.

The ecoregion receives 2,500–5,000 mm of annual rainfall. Most of the rainfall is from the May to September southwestern monsoon, which expends its moisture on the western, windward side. The northeastern monsoon, which sweeps in from the northeastern coast and ascends the gentler eastern slopes, brings additional rainfall to the ecoregion from December to March.

The vegetation is influenced primarily by climate (de Rosayro 1950) and ranges from *Dipterocarpus*-dominated montane moist forests to montane savanna and cloud forests with *Rhododendron* species. The submontane forests in the central massif are dominated by a *Shorea-Calophyllum-Syzygium* community with a canopy of *Shorea gardneri*, *S. trapezifolia*, *Palaquium* spp., *Homalium zelanicum*, *Calophyllum calaba*, *C. tomentosum*, *C. pulcherrimum*, *Syzygium* spp., *Cullenia* spp., *Myristica dactyloides*, *Cryptocarya wightiana*, and *Neolitsea involucrata* (Ashton and Gunatilleke 1987; Singhakumara 1995). The forests above 1,500 m are composed of twisted, stunted *Syzygium-Gordonia-Michelia-Elaeocarpus* formations, draped in a rich epiphytic community of orchids, mosses, and filmy ferns, and a *Strobilanthes*-dominated understory (Singhakumara 1995). Other prominent genera include *Cinnamomum*, *Litsea*, *Neolitsea*, *Symplocus*, *Semecarpus*, *Carallia*, *Garcinia*, and *Mastixia*. The montane forests of the Peak Wilderness area have forests that are locally dominated by the endemic Dipterocarpaceae genus *Stemnoporus*, possibly representing the only area of a dipterocarp-dominated montane forest (Singhakumara 1995), but in general the montane forests are dominated by Lauraceae, Myrtaceae, Clusiaceae, and Symplocaceae. *Rhododendron* species occur at the highest elevations—where cloud forests prevail—and in the transition zones with the wet montane grasslands, locally known as wet *pathanas*. These grasslands are thought to be fire-maintained and support a rich herb community with both temperate and tropical elements. Several animal species have also adapted their life history and behavior (e.g., nesting patterns) to these fire regimes. Not too long ago, Asian elephants (*Elephas maximus*) roamed these savannas (Phillips 1980), although they have now become locally extinct.

The flora of the detached Knuckles Range is different from the central massif flora and is characterized by a *Myristica-Cullenia-Aglaia-Litsea* community with subtropical affinities and a tropical *Callophyllum* zone (Gunatilleke and Gunatilleke 1990).

Biodiversity Features

The montane forests harbor higher levels of endemism than the lowland ecoregion. Half of Sri Lanka's endemic flowering plants and 51 percent of the endemic vertebrates are limited to this ecoregion. The smaller, isolated Knuckles Range has several species of relict, endemic flora and fauna that are distinct from the montane forests of the central massif (Senanayake et al. 1977).

More than 34 percent of Sri Lanka's endemic trees, shrubs, and herbs are limited to the montane rain forests (Peeris 1975 in Gunatilleke and Gunatilleke 1990) represented by this ecoregion.

The ecoregion harbors eight near-endemic mammals, and five are strict endemics (table J4.1). Two genera, *Srilankamys* and *Feroculus*, are monotypic and are strictly endemic to this ecoregion.

This ecoregion also supports a population of the Sri Lankan leopard (*Panthera pardus kotiya*) recognized by IUCN (1996) as a threatened genotype. The leopard is also the largest carnivore in Sri Lanka. Other endangered species include the endemic *Rattus montanus* and five shrews: *Crocidura miya*, *Feroculus feroculus*, *Solisorex pearsoni*, *Suncus fellowsgordoni*, and *Suncus zeylanicus* (Hilton-Taylor 2000).

The Sri Lanka Montane Rain Forests [4] ecoregion is included within an EBA, Sri Lanka (124) (Stattersfield et al. 1998). The ecoregion harbors twenty near-endemic bird species, of which five are strict endemics (table J4.2).

TABLE J4.1. Endemic and Near-Endemic Mammal Species.

Family	Species
Soricidae	*Suncus montanus*
Soricidae	*Feroculus feroculus**
Soricidae	*Crocidura miya**
Soricidae	*Solisorex pearsoni**
Cercopithecidae	*Semnopithecus vetulus*
Viverridae	*Paradoxurus zeylonensis*
Sciuridae	*Funambulus layardi*
Sciuridae	*Funambulus sublineatus*
Muridae	*Petinomys fuscocapillus*
Muridae	*Mus fernandoni*
Muridae	*Vandeleuria nolthenii*
Muridae	*Rattus montanus**
Muridae	*Srilankamys ohiensis**

An asterisk signifies that the species' range is limited to this ecoregion.

TABLE J4.2. Endemic and Near-Endemic Bird Species.

Family	Common Name	Species
Columbidae	Ceylon wood-pigeon	Columba torringtoni
Bucconidae	Ceylon grey hornbill	Ocyceros gingalensis
Cuculidae	Red-faced malkoha	Phaenicophaeus pyrrhocephalus
Phasianidae	Ceylon junglefowl	Gallus lafayetii
Corvidae	Ceylon magpie	Urocissa ornata
Sturnidae	White-faced starling	Sturnus senex
Sturnidae	Ceylon myna	Gracula ptilogenys
Timaliidae	Orange-billed babbler	Turdoides rufescens
Turdidae	Ceylon whistling-thrush	Myiophonus blighi*
Muscicapidae	Dull-blue flycatcher	Eumyias sordida*
Pycnonotidae	Yellow-eared bulbul	Pycnonotus penicillatus*
Zosteropidae	Ceylon white-eye	Zosterops ceylonensis*
Sylviidae	Ceylon bush-warbler	Bradypterus palliseri*
Turdidae	Spot-winged thrush	Zoothera spiloptera
Muscicapidae	Kashmir flycatcher	Ficedula subrubra
Timaliidae	Brown-capped babbler	Pellorneum fuscocapillum
Capitonidae	Yellow-fronted barbet	Megalaima flavifrons
Psittacidae	Ceylon hanging-parrot	Loriculus beryllinus
Psittacidae	Layard's parakeet	Psittacula calthropae
Strigidae	Chestnut-backed owlet	Glaucidium castanonotum

An asterisk signifies that the species' range is limited to this ecoregion.

TABLE J4.3. WCMC (1997) Protected Areas That Overlap with the Ecoregion.

Protected Area	Area (km²)	IUCN Category
Pidurutalagala National Park	80	VIII
Hakgala	20	I
Knuckles	217	IV
Peak Wilderness [3]	120	IV
Horton Plains	20	II
Total	457	

Ecoregion numbers of protected areas that overlap with additional ecoregions are listed in brackets.

Despite years of biological exploration, new species are still being discovered in this ecoregion, especially within the cryptic fauna, such as the frogs, lizards, fishes, and crabs. In addition to the large number of Rhacophorid frogs (Pethiyagoda and Manamendra-Arachchi 1998a), new species have been added to the endemic agamid lizard genera, *Lyriocephalus*, *Ceratophora*, and *Cophotis*, over the past few years (Pethiyagoda and Manamendra-Arachchi 1998b). Dedicated surveys of the crab fauna have increased the number of described species from seven in 1993 to thirty-five, some of which represent endemic genera (WHT 2000). Undoubtedly many more of these secretive yet fascinating species await discovery in this ecoregion.

Status and Threats

Current Status

Over the past 200 years, most of the montane rain forests have been cleared to establish large tea (*Camellia sinensis*) plantations (Ashton et al. 1997). Therefore, we will never know the full extent of the endemic biodiversity that has been lost forever, although the localized distribution patterns of species provide some indication of what this ecoregion's forests may have held.

Because many of the endemic species are habitat specialists and have small spatial needs, even small habitat fragments can provide viable habitat for many of the ecoregion's species that are key to its biodiversity. Therefore, protecting the remaining forest patches and their natural biodiversity is imperative. Currently, the ecoregion has five protected areas—of which one (Peak Wilderness) extends partially into the ecoregion from the lowland rain forests ecoregion—that cover less than 500 km² (table J4.3). None have good protection measures or conservation plans in place.

In addition to instituting good management in existing reserves, an attempt should be made to link the forest patches that are contiguous with the lowland and submontane forests to create more effective conservation landscapes across the altitudinal gradient. Because the major rivers originate in the mountains and radiate outward, the altitudinal connectivity will protect the watersheds and stabilize the substrate. The frequent landslides that result in loss of life and property are a manifestation of the loss of ground cover in these montane forests. Failure to act soon will result in the loss of Sri Lanka's most valuable contribution to global conservation.

Types and Severity of Threats

The primary threats are from land clearing for plantations, agriculture, and illegal logging (Gunatilleke and Gunatilleke 1990), even within protected areas. Most of the natural habitat has already been cleared for large-scale tea plantations, but the few remaining patches are still being cleared for other agricultural crops. Mining for precious stones have devastated large areas of forests.

The cloud forests in the high montane areas, especially in the Horton Plains National Park, are experiencing diebacks. Investigations are under way to determine the causes.

In the Knuckles Range the undergrowth in large forest areas is being cleared for cardamom (*Elettaria cardamomum*) plantations (IUCN 1994). Therefore, the regeneration capacity of these forests—though seemingly intact because of the intact canopy—is highly compromised.

Priority Conservation Actions

Priority actions have been identified in a national biodiversity action plan.

Short-Term Conservation Actions (1–5 Years)

- Institute effective conservation management for Peak Wilderness National Park, Horton Plains National Park, and Knuckles National Park.
- Put into effect a moratorium on clearing and logging in intact forests.
- Conduct biodiversity surveys, especially of the cryptic fauna, in all intact forests. The recent surveys by the Wildlife Heritage Trust indicate that the taxonomic groups that have been neglected during surveys to date have very high levels of endemism.
- Protect additional areas to include the important habitats that harbor most of Sri Lanka's endemic flora and fauna.

Longer-Term Conservation Actions (5–20 Years)

- Restore and recreate habitat linkages between the montane and lowland forests, with possible extensions into the dry zone. This will recreate landscapes with original connectivity between habitats.
- Monitor ecological integrity of the remaining forests and protected areas, using endemic species and assemblages.

Focal Species for Conservation Action

Endangered endemic species: Rattus montanus, Crocidura miya, Feroculus feroculus, Solisorex pearsoni, Suncus fellowsgordoni, Suncus zeylanicus, Ceylon whistling-thrush, dull-blue flycatcher, yellow-eared bulbul, Ceylon white-eye, and Ceylon bush-warbler

Area-sensitive species: Asian elephant, leopard

Top predator species: leopard

Other: purple-faced leaf monkey, golden palm civet, endemic lizards (*Ceratophora* and *Lyriocephalus*), and assemblage of endemic rhacophorid frogs.

The montane streams lack a rich fish community, but the endemic crabs and shrimps can be used as indicators of the quality of aquatic systems.

Selected Conservation Contacts

Biodiversity and Elephant Conservation Trust

Bird Club, Sri Lanka

Department of Wildlife Conservation, Government of Sri Lanka

Field Ornithological Group

Forest Department, Government of Sri Lanka

IUCN Sri Lanka

March for Conservation

University of Colombo

University of Peradeniya

Wildlife Heritage Trust

Wildlife and Nature Conservation Society, Sri Lanka

Young Zoologists Association

Justification for Ecoregion Delineation

Sri Lanka's forests have been divided into two broad climatic sub-regions: the wet zone and the dry zone. In a previous analysis of conservation units of the Indo-Malayan realm, MacKinnon (1997) placed the wet-zone forests into a single biounit, Ceylon Wet Zone 02. We used MacKinnon's regional classification as a guiding framework in delineating ecoregions across the Indo-Pacific region. But because we differentiated between lowland and montane forests in delineating ecoregions, we used the 1,000-m contour to extract and place the montane rain forests into a distinct ecoregion: the Sri Lanka Montane Rain Forests [4]. This ecoregion includes the montane floristic zones (9, 10, 12, 13, 14, and 15) identified by Ashton and Gunatilleke (1987).

Prepared by Eric D. Wikramanayake and Savithri Gunatilleke

Ecoregion Number:	**5**
Ecoregion Name:	**Orissa Semi-Evergreen Rain Forests**
Bioregion:	**Indian Subcontinent**
Biome:	**Tropical and Subtropical Moist Broadleaf Forests**
Political Units:	**India**
Ecoregion Size:	**22,200 km^2**
Biological Distinctiveness:	**Bioregionally Important**
Conservation Status:	**Critical**
Conservation Assessment:	**IV**

The Orissa Semi-Evergreen Rain Forests [5] are neither exceptionally species-rich nor high in endemism. But this ecoregion does harbor several of the Indian Subcontinent bioregion's charismatic large vertebrates. Some of these include the tiger (*Panthera tigris*), which is the region's largest predator, and the Asian elephant (*Elephas maximus*), large herds of gaur (*Bos gaurus*), and one of the most dangerous mammals in the region, the sloth bear (*Melursus ursinus*).

Location and General Description

The ecoregion's position on the low hills in the northeastern Indian state of Orissa makes it vulnerable to the full force of the southwestern monsoon winds that sweep in from the Bay of Bengal. The rainfall from this monsoon and the ameliorating year-round oceanic influences create moister climatic conditions here than in the rest of the Deccan Peninsula. Therefore, the ecoregion represents the original extent of distinctly moister semi-evergreen forests that once existed to the east of the Eastern Ghats Mountains (MacKinnon 1997).

The ecoregion has an ancient geological lineage of Gondwanaland origins, and it still harbors relicts of an ancient biota.

Within the predominant semi-evergreen forests of this ecoregion are patches of several other habitat types, such as canebrakes, wet bamboo brakes, moist bamboo brakes, lateritic semi-evergreen forests, and secondary moist bamboo breaks (Champion and Seth 1968). But most of these habitat types may not be identifiable in the field because of excessive deforestation and changes in land-use practices. Gaussen et al. (1973) report the following five series of natural vegetation from this ecoregion: *Shorea-Buchanania-Cleistanthus*, *Shorea-Cleistanthus-Croton*, *Shorea-Terminalia-Adina*, *Shorea-Syzygium operculatum-Toona*, and *Shorea-Dillenia-Pterospermum*.

The typical floral community of these forests includes *Artocarpus lakoocha*, *Michelia champaca*, *Celtis tetrandra*, *Bridelia tomentosa*, *B. verrucosa*, *Dillenia pentagyna*, *Saraca indica*, *Ficus* spp., *Mangifera indica*, and *Firmiana colorata* in the upper story (Champion and Seth 1968; Puri et al. 1989; Saxena and Dutta 1975). The second story is characterized by *Aphanamixis polystachya*, *Mesua ferrea*, *Phoebe lanceolata*, *Polyalthia* spp., *Macaranga peltata*, *Glochidion* spp., and *Litsea nitida* (Saxena and Dutta 1975). An understory of evergreen shrubs, canes, and herbs is also present. In the hilly areas with lateritic soils (the residual product of rock decay), the vegetation is characterized by *Xylia xylocarpa*, *Pterocarpus marsupium*, *Anogeissus latifolia*, *Grewia tiliaefolia*, *Terminalia tomentosa*, and *Terminalia bellirica* (Champion and Seth 1968).

Biodiversity Features

The ecoregion harbors fifty-nine mammal species. Although none are endemic to the ecoregion, several threatened species warrant conservation attention: the tiger, Asian elephant, gaur, wild dog (*Cuon alpinus*), sloth bear, and chousingha (*Tetracerus quadricornis*). The forests of this ecoregion—especially along the higher elevations—may provide dispersal habitat for tigers and leopards from Simlipal in the north to Andhra Pradesh to the south of Orissa.

Bird richness is higher, with more than 215 species, although none are endemic to the ecoregion. The ecoregion harbors a globally threatened species, the lesser florican (*Eupodotis indica*). Other birds that warrant conservation attention include the Oriental darter (*Anhinga melanogaster*), greater flamingo (*Phoenicopterus ruber*), and white-bellied sea eagle (*Haliaeetus leucogaster*).

Status and Threats

Current Status

More than 95 percent of this ecoregion's habitat has been cleared. The remaining forest consists of scattered fragments of *Xylia xylocarpa*, *Pterocarpus marsupium*, *Anogeissus latifolia*, *Grewia tiliaefolia*, *Terminalia tomentosa*, and *Terminalia bellirica*. Even thirty years ago, Champion and Seth (1968) alluded to the sub-climax forests throughout most of the ecoregion when they said that "the undisturbed climatic climax no longer exists and we can only surmise that the rapid invasion of the damp deciduous forests and coastal sal by evergreen species after successful fire protection will lead to a climax . . . included as tropical fully evergreen forests."

There are three protected areas that cover 1,102 km^2 (5 percent) of the ecoregion, but with the exception of the proposed 970-km^2 Chilka reserve, the other protected areas are small, at less than 100 km^2 (table J5.1).

Types and Severity of Threats

Cultivation, settlements, and grazing pressure from domestic livestock are the primary threats to the small areas of remaining habitat in the ecoregion.

Priority Conservation Actions

Short-Term Conservation Actions (1–5 Years)

- Restore native species in degraded areas, especially in areas that are critical for elephant and tiger movements. The degraded hill slopes should be rehabilitated.
- Because poverty is a primary driver of deforestation in the ecoregion, provide economic benefits that are linked to forest conservation efforts. Create community forests that will be a source of fuelwood.

Longer-Term Conservation Actions (5–20 Years)

- Strengthen the existing protected areas network.

Focal Species for Conservation Action

Area-sensitive species: tiger, Asian elephant

Top predator species: tiger

Other: gaur, wild dog, sloth bear, lesser florican

Selected Conservation Contacts

Chief Wildlife Warden, Orissa

Field director, Simlipal Tiger Reserve

WWF-India

Justification for Ecoregion Delineation

In an earlier analysis of biogeography and conservation gaps of India, Rodgers and Panwar (1988) divided the Gangetic plains into the upper and lower Gangetic plains biotic provinces. In a subsequent regional revision of conservation units, MacKinnon (1997) included most of Bangladesh in the lower Gangetic plain biounit. However, MacKinnon also included the patch of coastal semi-evergreen rain forest along the eastern side of the Eastern Ghats with the moist deciduous forests along the Gangetic plains. But in keeping with our rules for defining distinct habitat types of regional extent as separate ecoregions, we extracted these semi-evergreen forests that are distinct from the surrounding moist deciduous forests into the Orissa Semi-Evergreen Rain Forests [5]. This ecoregion lies within Udvardy's Mahanadian biogeographic province.

Prepared by Gopal S. Rawat, Ajay Desai, Hema Somanathan, and Eric D. Wikramanayake.

TABLE J5.1. WCMC (1997) Protected Areas That Overlap with the Ecoregion.

Protected Area	Area (km^2)	IUCN Category
Nandankanan	50	UA
Chilka	980	PRO
Balukhand Konark	70	IV
Total	1,100	

Ecoregion numbers of protected areas that overlap with additional ecoregions are listed in brackets.

Ecoregion Number:	**6**
Ecoregion Name:	**East Deccan Dry Evergreen Forests**
Bioregion:	**Indian Subcontinent**
Biome:	**Tropical and Subtropical Dry Broadleaf Forests**
Political Units:	**India**
Ecoregion Size:	**25,400 km^2**
Biological Distinctiveness:	**Bioregionally Outstanding**
Conservation Status:	**Critical**
Conservation Assessment:	**IV**

The vegetation in the East Deccan Dry Evergreen Forests [6] has an unusual physiognomy that makes it distinctive from most of the other dry forests. Unlike other dry forests of the Indo-Pacific region that lose their leaves for part of the year (i.e., during the nonmonsoonal season), these forests stay green during the long dry season.

Location and General Description

The ecoregion extends along the southern coastal areas of Andhra Pradesh and Tamil Nadu states in India, where it represents the narrow strip of dry evergreen forest formation.

Geologically, this ecoregion has Gondwanaland origins. The average rainfall is 800 mm. Most of the precipitation occurs during the brief northeast monsoon between October and December. Maximum temperatures during the long dry season can reach a stifling 44°C.

The original vegetation had emergent species of deciduous elements, such as *Albizzia amara* and *Chloroxylon* spp. But the original canopy-forming deciduous species have succumbed to human pressures, and the shrubby evergreen species now form a closed evergreen understory. Therefore, the existing vegetation type represents a low forest (up to 10 m) with a complete, closed canopy consisting of mostly small leathery-leaved evergreen trees with short trunks and spreading crowns. A large number of climbers are present, but bamboos are completely absent (Champion and Seth 1968).

The characteristic floristic elements are *Manilkara hexandra*, *Mimusops elengi*, *Diospyros ebernum*, *Strychnos nux-vomica*, *Eugenia* spp., *Drypetes sepiaria*, and *Flacourtia indica* (Champion and Seth 1968). The degraded stages of this forest have been categorized as tropical dry evergreen scrub by Puri et al. (1989) and are typically dominated by thorny species such as *Zyzyphus glabarrima*, *Dicrostachys cinerea*, *Randia dumetorum*, and *Carissa spinarum*.

Biodiversity Features

The ecoregion does not contain any endemic mammals or birds. But the sixty-six known mammal species include two threatened species: the wild dog (*Cuon alpinus*) and sloth bear (*Melursus ursinus*). Other species that deserve conservation attention in this ecoregion include its largest predator, the common leopard (*Panthera pardus*), and some of the smaller predators such as the jungle cat (*Felis chaus*) and leopard cat (*Prionailurus bengalensis*). The mammal community includes several ungulates of conservation importance such as the blackbuck (*Antilope cervicapra*), the chinkara (*Gazella bennettii*), and the small Indian chevrotain or mouse deer (*Moschiola meminna*). The Sriviliputhur (Kamarajar District in Tamil Nadu) and Amaravathy Nagar (Coimbatore District) have the largest remaining populations of the threatened grizzled giant squirrel (*Ratufa macroura*).

Bird richness is greater, with 230 species. The Jerdon's courser (*Rhinoptilus bitorquatus*) is endangered, and the spot-billed pelican (*Pelecanus philippensis*) and lesser florican (*Eupodotis indica*) are globally threatened (Hilton-Taylor 2000). Some other birds that deserve conservation attention as focal species because of their need for relatively intact habitat and low tolerances of disturbance include the woolly-necked stork (*Ciconia episcopus*), white-bellied sea-eagle (*Haliaeetus leucogaster*), and Indian grey hornbill (*Ocyceros birostris*).

Status and Threats

Current Status

Hundreds of years of human impact have taken a heavy toll on the natural habitat of this ecoregion, and more than 95 percent of the ecoregion is deforested. The remaining forests are scattered small fragments. The two small, protected areas cover less than 200 km^2 (table J6.1), which is less than 1 percent of the ecoregion's area. Marakanam, a sacred grove near Pondicherry, is fairly well protected and represents an important example of this vegetation type.

Types and Severity of Threats

Like most of the other ecoregions in the Indian Subcontinent, this ecoregion is also subjected to heavy deforestation and grazing pressure from domestic livestock. The stunted scrub vegetation present throughout most of the ecoregion is indicative of long years of grazing practices. The remnant sal (*Shorea robusta*) forests are being rapidly lost to *podu*, or shifting cultivation.

Invasion by *Prosopsis*, a thorny exotic plant that is unpalatable to domestic livestock, is being used extensively in reforestation programs and will certainly usurp the preferred habitat of the Jerdon's courser (Rawat and Babu 1995).

Poaching is rampant in areas with Naxalite conflicts, especially in the Satmala Hills, Pakla Wildlife Sanctuary, and Etunagaram. The ground situation makes patrolling by the forest department staff nearly impossible.

Priority Conservation Actions

Short-Term Conservation Actions (1–5 Years)

- Monitor the population of Jerdon's courser and manage its habitat.
- Stop the use of *Prosopsis* in reforestation programs; use indigenous species instead.

TABLE J6.1. WCMC (1997) Protected Areas That Overlap with the Ecoregion.

Protected Area	Area (km^2)	IUCN Category
Vettangudi [23]	30	IV
Nelapattu	130	IV
Total	160	

Ecoregion numbers of protected areas that overlap with additional ecoregions are listed in brackets.

- Protect two large plots of *sal* forests that still remain in the Narasimha Sanctuary and protect a 100-m-wide strip of riparian forests along the streams.

Longer-Term Conservation Actions (5–20 Years)

- Establish long-term vegetation-monitoring plots to determine effects of grazing on the vegetation.

Focal Species for Conservation Action

Area-sensitive species: leopard, wild dog

Top predator species: leopard, wild dog

Other: Jerdon's courser, lesser florican, spot-billed pelican, woolly-necked stork, white-bellied sea eagle, Indian grey hornbill

Selected Conservation Contacts

Chief Wildlife Warden, Andhra Pradesh

Dr. K. Narayana Rao, Professor of Botany, Shri Venkateshwara University, Tirupati, Andhra Pradesh

French Institute, Pondicherry

WWF-India

Justification for Ecoregion Delineation

During previous analyses of conservation units, Rodgers and Panwar (1988) and MacKinnon (1997) divided the largest biogeographic unit in India, the Deccan Peninsula, into five subunits or biotic provinces. Each of these units contained several distinct habitat types. In keeping with our rules for defining ecoregions as conservation units that represent distinct habitat types, we identified the distinct dry evergreen forests along the southeastern coast of the Deccan Peninsula as the East Deccan Dry Evergreen Forests [6].

Prepared by Gopal S. Rawat, Ajay Desai, Hema Somanathan, and Eric D. Wikramanayake

Ecoregion Number:	**7**
Ecoregion Name:	**Sri Lanka Dry-Zone Dry Evergreen Forests**
Bioregion:	**Indian Subcontinent**
Biome:	**Tropical and Subtropical Dry Broadleaf Forests**
Political Units:	**Sri Lanka**
Ecoregion Size:	**48,200 km²**
Biological Distinctiveness:	**Regionally Outstanding**
Conservation Status:	**Vulnerable**
Conservation Assessment:	**II**

The Sri Lanka Dry-Zone Dry Evergreen Forests [7] harbor one of Asia's largest and most viable Asian elephant (*Elephas maximus*) populations and a large protected area system designed specifically for elephant conservation. Therefore, the ecoregion provides one of the best opportunities to conserve Asia's largest vertebrate over the long-term. Unlike most other dry forests, the trees in the Sri Lanka Dry-Zone Dry Evergreen Forests [7] retain their leaves during the dry season. Only two other ecoregions exhibit this phenology: the small East Deccan Dry Evergreen Forests [6] and the Southeastern Indochina Dry Evergreen Forests [59] in the Indochina bioregion. Because the evergreen dry forest ecoregion in the east Deccan Plateau is small and has lost most of its intact habitat, the Sri Lanka Dry-Zone Dry Evergreen Forests [7] are the only viable example of evergreen dry forests in the bioregion.

Location and General Description

The Sri Lanka Dry-Zone Dry Evergreen Forests [7] represent the tropical dry forests throughout most of the island of Sri Lanka, except for the southwestern quarter, the central mountain range, and the Jaffna Peninsula in the extreme north.

The island's geological roots date back to the Cretaceous period when, as part of the Deccan Plateau, it detached from Gondwanaland and drifted north to collide with the northern Eurasian continent about 50 million years later. Therefore, the ecoregion harbors elements of the ancient Gondwana biota. The island then became separated from the Deccan Plateau during the late Miocene about 20 million years later. Since that time, there have been several land bridges that allowed species exchanges between the island and the mainland until the final separation during the Pleistocene (Deraniyagala 1992).

The ecoregion receives about 1,500–2,000 mm of annual rainfall in the December to March northeast monsoon but is mostly dry the rest of the year. Topographically, the ecoregion is flat, except for scattered inselbergs and isolated low hills. Ritigala, a 766-m isolated peak in the central part of Sri Lanka, is the highest point between the central massif and the Western Ghats of India (Jayasuriya 1984).

Tropical dry forests cover most of the ecoregion, but within the ecoregion there are a number of distinct habitat types (of sub-regional extent). These include patches of sub-montane savanna and grassland—known locally as *talawa*—especially along the eastern and southeastern slopes of the central massif. In the northeast, lowland grasslands, locally known as *villus*, are associated with the floodplains of the river systems. These grasslands provide critical water and fodder for herbivores during the dry season.

Most of this ecoregion was settled and cultivated until about 500 years ago; therefore, the forest is secondary. However, several patches of old-growth forests remain and are included within protected areas (e.g., Wasgomuwa National Park and parts of Ruhuna National Park).

The evergreen dry forests are dominated by *Manilkara hexandra*, *Chloroxylon sweitenia*, *Drypetes sepiaria*, *Feronia limonia*, *Vitex altissima*, *Syzygium* spp., *Drypetes sepiaria*, and *Chukrasia tabularis*, with the scrub and regenerating forests characterized by *Bauhinia racemosa*, *Pterospermum suberifolium*, *Cassia fistula*, and *Dichrostachys cineria* (Gunatilleke and Gunatilleke 1990). *Acacia* thorn scrub grows in disturbed areas.

The *talawa* savannas are characterized by *Terminalia chebula*, *T. belerica*, *Pterocarpus marsupium*, *Butea monosperma*, *Careya arborea*, *Anogeissus latifolia*, *Phyllanthus embilica*, and *Zizyphus* spp. (McKay 1973). The dominant grasses in the villus include *Cymbopogon* spp., *Eragrostis* spp., *Themeda* spp., and *Imperata* spp. (McKay 1973).

Ritigala, the isolated hill in central Sri Lanka, is a hotspot of endemic species within this ecoregion with several endemic plants such as *Madhuca clavata* (Jayasuriya 1984).

Biodiversity Features

Compared with Sri Lanka's rain forests, this ecoregion does not contain very high levels of endemism. Nevertheless, it harbors one of Asia's largest elephant populations, estimated at 2,500 to 4,000 animals. The large protected area system that was designed using the elephant as a focal species provides an excellent chance for the long-term conservation of this endangered species.

The seventy-four mammal species known from the ecoregion include four near-endemic species (table J7.1).

Several of the ecoregion's mammals are also listed as threatened: the endangered Asian elephant, the Sri Lankan genotype of the common leopard (*Panthera pardus kotiya*), and the vulnerable sloth bear (*Melursus ursinus*), purple-faced leaf monkey (*Semnopithecus vetulus*), and slender loris (*Loris tardigradus*) (Hilton-Taylor 2000).

Bird richness is greater, with 270 species, of which nine species are near endemic (table J7.2).

The spot-billed pelican (*Pelecanus philippensis*) is a globally threatened species. BirdLife International has included this entire ecoregion within an EBA, Sri Lanka (124) (Stattersfield et al. 1998).

Viable populations of the freshwater and mugger crocodiles (*Crocodylus porosus* and *C. palustris*, respectively) are found along the rivers and estuaries and have also adapted to the numerous ancient reservoirs that formed part of the extensive irrigation system. Two other large lizards, the water monitor (*Varanus salvator*)—the second largest lizard in the world—and the Bengal monitor (*V. bengalensis*), which are persecuted and hunted throughout most of their ranges, have found a safe haven in this ecoregion.

Status and Threats

Current Status

About three-quarters of this ecoregion has been deforested; however, extensive areas of contiguous, intact forest remain in the north and north central area. Thirty-eight protected areas cover 7,842 km², or 17 percent of the ecoregion area (table J7.3). This represents the largest proportion of intact

TABLE J7.1. Endemic and Near-Endemic Mammal Species.

Family	Species
Cercopithecidae	*Semnopithecus vetulus*
Viverridae	*Paradoxurus zeylonensis*
Muridae	*Mus fernandoni*
Muridae	*Vandeleuria nolthenii*

An asterisk signifies that the species' range is limited to this ecoregion.

TABLE J7.2. Endemic and Near-Endemic Bird Species.

Family	Common Name	Species
Bucconidae	Ceylon grey hornbill	*Ocyceros gingalensis*
Cuculidae	Red-faced malkoha	*Phaenicophaeus pyrrhocephalus*
Phasianidae	Ceylon junglefowl	*Gallus lafayetii*
Phasianidae	Ceylon spurfowl	*Galloperdix bicalcarata*
Turdidae	Spot-winged thrush	*Zoothera spiloptera*
Timaliidae	Brown-capped babbler	*Pellorneum fuscocapillum*
Capitonidae	Yellow-fronted barbet	*Megalaima flavifrons*
Psittacidae	Ceylon hanging-parrot	*Loriculus beryllinus*
Psittacidae	Layard's parakeet	*Psittacula calthropae*

An asterisk signifies that the species' range is limited to this ecoregion.

TABLE J7.3. WCMC (1997) Protected Areas That Overlap with the Ecoregion.

Protected Area	Area (km²)	IUCN Category
Ruhuna (Yala) Block 1	1,030	II
Kudumbigala	40	IV
Yala East Block 1	230	II
Uda Walawe	320	II
Yala	230	I
Kataragama	10	IV
Katagamuwa	10	IV
Wirawila-Tissa	50	IV
Bundala	70	IV
Kalametiya Kalapuwa	10	IV
Maduru Oya Block 1	570	II
Nuwaragala	400	VIII
Nelugala	90	UA
Gal Oya Valley NE	120	IV
Gal Oya Valley	280	II
Sneanayake Samudra	80	IV
Gal Oya Valley SW	120	IV
Sagamam	2	IV
Lahugala Kitulana	170	II
Vavunikulam	20	IV
Madhu Road	250	IV
Kokilai Lagoon	50	IV
Giant's Tank	50	IV
Padaviya Tank	30	IV
Wilpattu Block 1	130	II
Wilpattu North	10	IV
Anaulandewa	310	VIII
Trincomalee Naval Headworks	170	IV
Great Sober Island	1	IV
Mahakandarawewa	10	IV
Anuradhapura	40	IV
Mihintale	7	IV
Hurulu	250	VIII
Seruwila-Allai	140	IV
Somawathiya Block 1	370	II
Tirikonamadu	240	IV
Ritigala	30	I
Minneriya-Giritale	60	IV
Victoria-Randenigala-Rantambe	140	IV
Floodplains	130	II
Polonnaruwa	10	IV
Minneriya-Giritale Block 1	140	IV
Wasgomuwa Lot 1	370	II
Total	6,790	

Ecoregion numbers of protected areas that overlap with additional ecoregions are listed in brackets.

forests that are included within the protected area systems of the dry forest ecoregions in the Indo-Pacific region.

Two of these protected areas complexes—Yala/Ruhuna and Wilpattu national parks—exceed 1,000 km^2 in area, and several other reserves form contiguous complexes that exceed 500 km^2. Many of these larger reserves and reserve complexes contain elephant populations that can contribute significantly to a regional conservation program.

Types and Severity of Threats

The primary overarching threats are from deforestation caused by agriculture, resettlements, and small-scale logging. Encroachment into protected areas, several of which lack adequate protection and management, also poses important threats that warrant attention. Overall, however, the protected area system is extensive enough to conserve the ecoregion's biodiversity, given effective management. The cultural and religious taboos against killing and hunting wildlife—most of the country is Buddhist—have served well to protect wildlife relative to the rest of Asia.

Priority Conservation Actions

Priority actions have been identified in a national biodiversity action plan.

Short-Term Conservation Actions (1–5 Years)

- Require effective conservation management for the larger reserves such as Ruhuna, Gal Oya, Wasgomuwa, and Uda Walawe.
- Restore effective management to Wilpattu National Park, which has been closed because of the ongoing armed conflict in the north and parts of the northeast. It is one of the most important protected areas in this ecoregion.
- Institute effective management of the protected areas that were created in the Mahaweli River plains for elephant conservation. At present, most of these protected areas remain paper parks.
- Protect Ritigala Strict Nature Reserve. This isolated mountain contains many endemic species but is being degraded because of poor management.
- Create conservation landscapes with core areas and well-defined buffer zones to minimize human-elephant conflicts in this ecoregion, as indicated by recent elephant surveys (Fernando et al., in prep.). These landscapes should be designated and appropriate land uses planned and specified.

Longer-Term Conservation Actions (5–20 Years)

- Create and manage conservation landscapes in this ecoregion using the Asian elephant as a focal species.
- Protect the patches of old-growth dry forests.
- Train wildlife managers. Although Sri Lanka was at the forefront of wildlife conservation in the region three decades ago, subsequent lack of leadership and trained staff has led to a deterioration of management capabilities, especially to handle the emerging threats and issues caused by new irrigation and resettlement programs.

Focal Species for Conservation Action

Area-sensitive species: Asian elephant, leopard

Top predator species: leopard, mugger crocodile, monitor lizards

Other: sloth bear, spot-billed pelican, purple-faced leaf monkey, golden palm civet, rusty-spotted cat (*Felis rubiginosa*), Sri Lanka junglefowl, Sri Lanka spurfowl, Sri Lanka grey hornbill, red-faced malkoha, mugger crocodile

Selected Conservation Contacts

Biodiversity and Elephant Conservation Trust
Bird Club, Sri Lanka
Department of Wildlife Conservation, Government of Sri Lanka
Field Ornithological Group
Forest Department, Government of Sri Lanka
IUCN Sri Lanka
March for Conservation
University of Colombo
University of Peradeniya
Wildlife Heritage Trust
Wildlife and Nature Conservation Society, Sri Lanka
Young Zoologists Association

Justification for Ecoregion Delineation

The dry zone has been commonly used as a distinct bioclimatic zone in Sri Lanka. We used MacKinnon's (1997) map of original forest cover to separate the dry evergreen forests of Sri Lanka from the wet-zone moist forests and represented the former in the Sri Lanka Dry-Zone Dry Evergreen Forests [7]. This ecoregion corresponds to MacKinnon's (1997) biounit S13 and overlaps with floristic zones 1 and 2 identified by Ashton and Gunatilleke (1987).

Prepared by Eric D. Wikramanayake and Savithri Gunatilleke

Ecoregion Number:	**8**
Ecoregion Name:	**Brahmaputra Valley Semi-Evergreen Rain Forests**
Bioregion:	**Indian Subcontinent**
Biome:	**Tropical and Subtropical Moist Broadleaf Forests**
Political Units:	**India and Bhutan**
Ecoregion Size:	**56,600 km^2**
Biological Distinctiveness:	**Regionally Outstanding**
Conservation Status:	**Critical**
Conservation Assessment:	**II**

The Brahmaputra Valley Semi-Evergreen Rain Forests [8] used to be some of the most productive areas in the Indian Subcontinent bioregion. The ecoregion lies along the alluvial plains of the Brahmaputra River, which flows through Assam and West Bengal before it becomes confluent with the Ganges River and heads south to the Bay of Bengal. The ecoregion's high productivity has been the primary cause for its widespread loss of habitat and natural biodiversity. The valley has been densely settled by humans and cultivated for thousands of years. Yet despite the long history of habitat loss and degradation, the ecoregion still harbors an impressive biological diversity in the small fragments of habitat that lie scattered throughout. For instance, some of India's viable populations of Asian elephants (*Elephas maximus*) and the world's largest population of the greater one-horned rhinoceros (*Rhinoceros unicornis*) still live here.

Location and General Description

The ecoregion represents the swath of semi-evergreen forests along the upper Brahmaputra River plains. Most of the ecoregion lies within the eastern Indian state of Assam, but small sections extend into the neighboring states of Arunachal Pradesh and Nagaland and also into the southern lowlands of Bhutan.

This northeastern region represents the area where the northward-migrating Deccan Peninsula first made contact with the Eurasian continent during the early Tertiary period and is a gateway for species exchanges between the typically Indian and Malayan' faunas (Rodgers and Panwar 1988). The wide Brahmaputra River is also a biogeographic barrier for several species. For instance, the golden langur (*Semnopithecus geei*), hispid hare (*Caprolagus hispidus*), and pygmy hog (*Sus salvanius*) are limited to the north bank of the river, whereas the hoolock gibbon (*Hylobates hoolock*) and stump-tailed macaque (*Macaca arctoides*) are limited to the south bank (Rodgers and Panwar 1988).

The June to September southwest monsoon is funneled through the Gangetic River plains, flanked by the Himalayas to the north and the Mizo Hills to the south, deluging the ecoregion with 1,500–3,000 mm of rainfall, depending on the topographic variation. The substrate consists of deep alluvial deposits, washed down over the centuries by the Brahmaputra and other rivers such as the Manas and Subansiri, which drain southern slopes of the Eastern Himalaya. The ecoregion's vegetation therefore is thus influenced by the rich alluvial soils and the monsoon rains.

Champion and Seth (1968) recognize the following forest types in this ecoregion: Assam Valley semi-evergreen forest, Assam alluvial plains semi-evergreen forest, eastern submontane semi-evergreen forest, sub-Himalayan light alluvial semi-evergreen forest, eastern alluvial secondary semi-evergreen forest, sub-Himalayan secondary wet mixed forest, and Cachar semi-evergreen forest. But most of the ecoregion's original semi-evergreen forests have been converted to grasslands by centuries of fire and other human influences. Only small patches of forests now remain, scattered along the Indo-Bhutan border and along the border of Assam and Meghalaya. Many of these forest patches are confined to protected areas.

According to Champion and Seth (1968), the typical evergreen tree species in these forests are *Syzygium*, *Cinnamomum*, *Artocarpus*, and Magnoliacea, and the common deciduous species include *Terminalia myriocarpa*, *Terminalia citrina*, *Terminalia tomentosa*, *Tetrameles* spp., and *Stereospermum* spp. *Shorea robusta* is present in disturbed habitats, especially in areas that have been subjected to fire, and represents a sub-climax community. Other *Dipterocarpus* species are considered to be indicative of a forest in retrogression from the tropical evergreen or as a preclimax stage. Typically, the canopy trees are 20–30 m high.

The understory is of Lauraceae (mostly *Phoebe* spp., *Machilus* spp., and *Actinodaphne* spp.), Anonaceae (*Polyalthia* spp.), Meliaceae (*Aphanamixis* spp.), *Mesua ferrea*, *Tetrameles* spp., *Stereospermum* spp., and species of Meliaceae, Anacardiaceae, Myristicaceae, Lauraceae, and Magnoliaceae, with several bamboos such as *Bambusa arundinaria*, *Dendrocalamus hamilitonii*, and *Melocanna bambusoides* (Champion and Seth 1968; Puri et al. 1989).

The riparian areas along the Brahmaputra River that have been cleared are characterized by wet grasslands with similar communities and dynamics as described under the Terai-Duar Savanna and Grasslands [35] description. Species such as the hispid hare and pygmy hog are found in these grasslands, especially in the soft soils and muddy areas along the river courses.

Biodiversity Features

This ecoregion harbors India's largest elephant population (Rodgers and Panwar 1988), the world's largest population of the greater one-horned rhinoceros, tigers (*Panthera tigris*), and wild water buffalo (*Bubalus arnee*) (WII 1997). The ecoregion overlaps with a high-priority (Level I) TCU that extends north to include the subtropical and temperate forests of the Himalayan midhills (Wikramanayake et al. 1999).

The known mammal fauna consists of 122 species, including 2 near-endemic species (table J8.1). Of these, the pygmy hog and the hispid hare are confined to the grassland habitats.

There are several threatened mammal species in this ecoregion, including the tiger, Asian elephant, greater one-horned rhinoceros, swamp deer (*Cervus duvaucelii*), gaur (*Bos gaurus*), clouded leopard (*Pardofelis nebulosa*), hispid hare, pygmy hog, capped leaf monkey (*Semnopithecus pileatus*), Asiatic black bear (*Ursus thibetanus*), and sloth bear (*Melursus ursinus*) (Hilton-Taylor 2000). Although it is not strictly a part of the terrestrial ecoregion, the Brahmaputra River also harbors the freshwater dolphin (*Platanista gangetica*), which is also a threatened species (Hilton-Taylor 2000).

The bird fauna is richer, with more than 370 species and 2 species that are near endemic (table J8.2). The ecoregion overlaps with BirdLife International's EBA, Assam Plains (131) (Stattersfield et al. 1998), which contains three restricted-range bird species.

TABLE J8.1. Endemic and Near-Endemic Mammal Species.

Family	Species
Cercopithecidae	*Semnopithecus geei*
Bovidae	*Caprolagus hispidus*

An asterisk signifies that the species' range is limited to this ecoregion.

Table J8.2. Endemic and Near-Endemic Bird Species.

Family	Common Name	Species
Phasianidae	Manipur bush-quail	Perdicula manipurensis
Timaliidae	Marsh babbler	Pellorneum palustre

An asterisk signifies that the species' range is limited to this ecoregion.

The Bengal florican is one of the most endangered species here and is largely limited to the protected areas. Conversion of grasslands and other human disturbances can make this species locally extinct.

Status and Threats

Current Status

Nearly three-quarters of the habitat in this ecoregion has been cleared or degraded; however, some large blocks of intact habitat—including sub-regional–scale swampland, grasslands, and fringing woodland habitats that are critical for the waterbirds and the large herbivores—still remain. The larger forest blocks are in central Assam.

At present, twelve protected areas cover about 2,500 km^2 of intact habitat, or 5 percent of the ecoregion (table J8.3). Of these, Manas, Dibru-Saikowa, Kaziranga, and Mehao are the larger and more important reserves. Mehao extends over two other ecoregions and is only partially within this ecoregion (table J8.3). Kaziranga has the world's largest population of the greater one-horned rhinoceros, estimated at 1,100 individuals (Foose and van Strien 1997). Because of the large number of wide-ranging large vertebrates in this ecoregion, additional protection is urgently needed. Specifically, habitat connectivity should be provided within the Buxa-Manas complex and the Barail-Intanki-Kaziranga complex to allow elephants to disperse and migrate (Rodgers and Panwar 1988).

Types and Severity of Threats

The overarching threats to the ecoregion's natural habitat stem from forest clearing and livestock grazing. But the settlements and agriculture go back thousands of years and have already taken a heavy toll on the natural habitat and biodiversity. Vast areas of original habitat were taken over by large tea plantations.

The Naxalite insurgency prevents effective government administration of conservation areas (protected areas and reserve forests). The movement is also funded to some extent by poaching and the wildlife and timber trade.

Priority Conservation Actions

Short-Term Conservation Actions (1–5 Years)

- Institute better management of domestic buffalo to prevent contact and breeding with wild buffalo. (Domestic livestock have begun to breed with some of the last populations of wild water buffalo in this ecoregion.)
- Stop clearing and encroachment into the remaining patches of grasslands and forests by local people.
- Prevent the spread of exotic weeds, especially *Eupatorium odoratum*, *Mikania cordata*, and *Lantana camara*. Eradication programs should begin as part of the protected areas management prescriptions.
- Stop poaching and hunting of rhinoceros and tigers.
- Maintain intact habitat and habitat linkages, especially where it would allow elephant migrations and tiger dispersal. Restore critical linkages where necessary.
- Study and manage the population of feral horses in the grasslands of this ecoregion as part of a tiger conservation program because these horses form an important prey base for tigers.

Longer-Term Conservation Actions (5–20 Years)

- Prevent the spread of exotic weeds, especially *Eupatorium odoratum*, *Mikania cordata*, and *Lantana camara*. Eradication programs should be included as part of the protected areas management prescriptions.
- Prevent poaching and hunting of rhinoceroses and tigers.
- Maintain intact habitat and habitat linkages, especially where this would benefit elephant migration and tiger dispersal. Critical linkages should be restored where necessary.

Focal Species for Conservation Action

Area-sensitive species: tiger, clouded leopard, leopard, elephant

Top predator species: tiger, clouded leopard, leopard, wild dog

Other: swamp deer, hispid hare, pygmy hog, Asian elephant, greater one-horned rhinoceros, tiger, clouded leopard, sloth bear, wild water buffalo, gaur, endemic birds

Selected Conservation Contacts

Chief Wildlife Warden, Assam, India

Director, Buxa Tiger Reserve, Assam, India

Director, Kaziranga National Park, India

Director, Manas National Park, Assam, India

Director, Royal Manas National Park, Bhutan

Nature Conservation Section, Ministry of Agriculture, Royal Government of Bhutan

Table J8.3. WCMC (1997) Protected Areas That Overlap with the Ecoregion.

Protected Area	Area (km^2)	IUCN Category
Mehao [26, 31]	190	IV
Manas	560	IV
Barnadi	90	IV
Kaziranga	320	II
Orang	110	IV
Laokhowa	170	IV
Pobitora	80	IV
Sonai Rupai	160	IV
Nameri	90	IV
Dibru-Saikowa	490	PRO
D'Ering Memorial	190	IV
Pabha	110	IV
Total	2,560	

Ecoregion numbers of protected areas that overlap with additional ecoregions are listed in brackets.

Royal Society for Conservation of Nature, Bhutan
WWF-Bhutan
WWF-India

Justification for Ecoregion Delineation

In previous analyses of conservation units in the region, Rodgers and Panwar (1988) and MacKinnon identified bio-units—Brahmaputra Valley 8A and Brahmaputra valley (I9a), respectively—that cover the semi-evergreen forests along the Brahmaputra River plains. We retained the boundary of their units as the Brahmaputra Valley Semi-Evergreen Rain Forests [8]. This ecoregion overlaps with two of Udvardy's biogeographic provinces, the Himalayan highlands and Burma monsoon forest.

Prepared by Gopal S. Rawat, Ajay Desai, Hema Somanathan, and Eric D. Wikramanayake

Ecoregion Number:	**9**
Ecoregion Name:	**Sundarbans Freshwater Swamp Forests**
Bioregion:	**Indian Subcontinent**
Biome:	**Tropical and Subtropical Moist Broadleaf Forests**
Political Units:	**India, Bangladesh**
Ecoregion Size:	**16,400 km^2**
Biological Distinctiveness:	**Bioregionally Outstanding**
Conservation Status:	**Critical**
Conservation Assessment:	**IV**

The Sundarbans Freshwater Swamp Forests [9] ecoregion is nearly extinct. Hundreds of years of habitation and exploitation by one of the world's densest human populations have exacted a heavy toll of this ecoregion's habitat and biodiversity. Because it sits in the vast, productive delta of the Ganges and Brahmaputra rivers, the annual alluvial deposits make the ecoregion exceptionally productive. Therefore, most of the natural habitat has long been converted to agriculture, making it almost impossible to even surmise the original composition of the ecoregion's biodiversity.

Location and General Description

This ecoregion represents the brackish swamp forests that lie behind the Sundarbans Mangroves [33] where the salinity is more pronounced. The freshwater ecoregion is an area where the water is only slightly brackish and becomes quite fresh during the rainy season, when the freshwater plumes from the Ganges and Brahmaputra rivers push the intruding salt water out and also bring a deposit of silt (Champion and Seth 1968). Like the vast mangrove ecoregion, the freshwater swamp forest ecoregion also straddles the boundary between Bangladesh and India's state of West Bengal.

The June to September southwest monsoon brings heavy rains. Frequent, devastating cyclones also sweep in from the Bay of Bengal, causing widespread destruction. Annual rainfall can exceed 3,500 mm, and the daytime temperatures can rise above 48°C during these monsoon months, which when coupled with the humidity can be unbearable.

Habitat loss is so extensive, and the remaining habitat is so fragmented, that it is difficult to ascertain the composition of the original vegetation of this ecoregion. According to Champion and Seth (1968), the freshwater swamp forests are characterized by *Heritiera minor*, *Xylocarpus molluccensis*, *Bruguiera conjugata*, *Sonneratia apetala*, *Avicennia officinalis*, and *Sonneratia caseolaris*, with *Pandanus tectorius*, *Hibiscus tiliaceus*, and *Nipa fruticans* along the fringing banks (Champion and Seth 1968).

Biodiversity Features

This ecoregion, together with the mangrove ecoregion, is an important refuge for the tiger (*Panthera tigris*). It is especially important to conserve tigers that are behaviorally and ecologically adapted to mangrove and swamp habitats. The known mammal fauna includes fifty-five species, but none are endemic to the ecoregion. In addition to the endangered tiger, there are several other threatened mammal species, such as the capped langur (*Semnopithecus pileatus*), smooth-coated otter (*Lutrogale perspicillata*), Oriental small-clawed otter (*Aonyx cinerea*), and great Indian civet (*Viverra zibetha*). The ecoregion also contains the leopard (*Panthera pardus*) and several smaller predators such as the jungle cat (*Felis chaus*), fishing cat (*Prionailurus viverrinus*), and leopard cat (*Prionailurus bengalensis*).

About 190 bird species have been recorded from the ecoregion. None are endemic, but the osprey (*Pandion haliaetus*) and grey-headed fish-eagle (*Ichthyophaga ichthyaetus*) should be focal species because of their role as predators in this ecosystem that includes a significant aquatic component.

Conservation activities in the aquatic habitat should also include the Gangetic dolphin (*Platanista gangetica*) and the three crocodile species that co-occur here: the marsh crocodile or mugger (*Crocodylus palustris*), estuarine crocodile (*Crocodylus porosus*), and gharial (*Ghavialis gangeticus*).

Status and Threats

Current Status

This ecoregion is nearly extinct, the victim of large-scale clearing and settlement to support one of the densest human populations in Asia. There are two protected areas that cover a mere 130 km^2 of the ecoregion (table J9.1).

Types and Severity of Threats

The large human population that has settled in this ecoregion and used its resources for hundreds of years has exacted a heavy toll on the natural habitat. Very little of the natural habitat of this ecoregion now remains, and even the remaining fragments are degraded.

TABLE J9.1. WCMC (1997) Protected Areas That Overlap with the Ecoregion.

Protected Area	Area (km^2)	IUCN Category
Narendrapur	110	IV
Ata Danga Baor WS	20	
Total	130	

Ecoregion numbers of protected areas that overlap with additional ecoregions are listed in brackets.

Illegal hunting and habitat loss have caused several local extinctions of key species such as swamp deer (*Cervus duvaucelii*). Others, such as the gharial, are on the brink of extinction. The tiger is severely threatened.

Oil spills from the ships that travel up the river to the large port city of Calcutta and pollution from large settlements and developments—a proposed fertilizer plant is one example—rank among some of the other significant threats.

The diversion of more than a third of the Ganges dry-season flow through the Farraka Barrage in India will also affect water flows. Changes to the water flows will be especially critical to this ecoregion because the vegetation characteristics here depend on salinity levels.

Priority Conservation Actions

Conservation actions in this ecoregion are essentially salvage operations to try to identify and save what remains. Critical habitat may have to be restored. Here we identify some short-term and long-term conservation actions for this ecoregion.

Short-Term Conservation Actions (1–5 Years)

- Conserve and monitor some of the species that might be considered to be typical of this habitat, including the swamp francolin (*Francolinus gularis*), which is now confined to the Terai along the upper Gangetic Plains, and freshwater turtles such as *Milanochelys tricarinata* and *Milanochelys trijuga*.
- Conduct status surveys of the tiger, crocodile, gharial, and swamp francolin.
- Survey and assess remaining intact forests for conservation and protection.
- Address threats such as pollution, habitat loss, hunting, and unsustainable natural resource use.
- Initiate transboundary cooperation in conservation efforts between India and Bangladesh.

Longer-Term Conservation Actions (5–20 Years)

- Develop an ecoregion-based conservation plan for this ecoregion and the adjoining Sundarbans Mangroves [33]. The latter ecoregion is included in a Level I TCU (Dinerstein et al. 1997). Therefore, the tiger should be used as a focal species in identifying core protected areas across this conservation landscape.

Focal Species for Conservation Action

Area-sensitive species: tiger, leopard

Top predator species: tiger, leopard

Other: capped langur, smooth-coated otter, Oriental small-clawed otter, great Indian civet, Gangetic dolphin, fishing cat, marsh crocodile, estuarine crocodile, gharial, osprey, grey-headed fish eagle

Selected Conservation Contacts

Chief Wildlife Warden, Government of West Bengal, Calcutta, India

Department of Wildlife Conservation, Government of Bangladesh, Dacca

WWF-India

Justification for Ecoregion Delineation

We used MacKinnon's (1997) digital map of the original habitat to identify the boundaries of this ecoregion. In keeping with our rules for defining ecoregions, we separated the freshwater swamp forests from the mangroves in the Ganges River delta and placed them in their own ecoregion, the Sundarbans Freshwater Swamp Forests [9]. This ecoregion falls within Udvardy's Bengalian rain forest biogeographic province.

Prepared by Gopal S. Rawat and Eric D. Wikramanayake

Ecoregion Number:	**10**
Ecoregion Name:	**Lower Gangetic Plains Moist Deciduous Forests**
Bioregion:	**Indian Subcontinent**
Biome:	**Tropical and Subtropical Moist Broadleaf Forests**
Political Units:	**India, Bangladesh**
Ecoregion Size:	**253,500 km^2**
Biological Distinctiveness:	**Regionally Outstanding**
Conservation Status:	**Critical**
Conservation Assessment:	**II**

The Lower Gangetic Plains Moist Deciduous Forests [10] lie along the confluence of two of Asia's largest rivers, the Ganges and Brahmaputra rivers, which run the length of the Himalayan foothills and drain its breadth. It once harbored impressive populations of tiger (*Panthera tigris*), greater one-horned rhinoceros (*Rhinoceros unicornis*), Asian elephant (*Elephas maximus*), gaur (*Bos gaurus*), swamp deer (*Cervus duvaucelli*), and Bengal florican (*Eupodotis bengalensis*). Today, the ecoregion supports one of the densest human populations on Earth, and the fertile alluvial plains have been cleared and intensely cultivated. The human activities that date back thousands of years have taken a very heavy toll on the natural biodiversity of the ecoregion, and many of these species have disappeared from the ecoregion.

Location and General Description

The tropical moist deciduous forests represented by this ecoregion once stretched along the lower reaches of the Ganges and Brahmaputra river plains across the Indian states of Bihar, West Bengal, Assam, Uttar Pradesh, and Orissa, and most of Bangladesh. The southwest monsoon that sweeps in from the Bay of Bengal deluges the ecoregion with more than 3,500 mm of rainfall during the four months from June to September. Devastating cyclones sweep in from the Bay of Bengal, causing widespread flooding.

The alluvial substrate deposited by the rivers is clayey and drains poorly, but on the more stable but flood-prone riverine flats the soil tends to be more sandy, with only local clay patches.

The vegetation is semi-deciduous; the upper canopy contains the deciduous species, and the second story is dominated by evergreen species (Puri et al. 1989). Open forests are dominated by *Bombax ceiba* in association with *Albizzia procera*, *Duabanga sonneratioides*, and *Sterculia vilosa*. These are early seral communities that will eventually become sal (*Shorea ro-*

TABLE J10.1. Endemic and Near-Endemic Mammal Species.

Family	Species
Molossidae	Tadarida teniotis

An asterisk signifies that the species' range is limited to this ecoregion.

busta) forest if succession is allowed to proceed. But in most places the forests fail to reach a climax stage for various reasons—many human-caused—perpetuating the sub-climax community (Champion and Seth 1968).

The riparian forests are characterized by an *Acacia-Dalbergia* association that includes *Acacia catechu*, *Albizzia procera*, *Bombax ceiba*, *Sterculia villosa*, and *Dalbergia sisso*. In the upper Assam areas of the ecoregion, the forests are made up of *Duabanga-Pterospermum-Terminalia* in association with *Bombax ceiba*, *Pterspermum acerifolium*, *Laportea crenulata*, *Duabanga sonneratioides*, *Terminalia myriocarpa*, and *Calamus tenuis* (Puri et al. 1989).

In permanently wet or moist areas with fine, clayey soils and a rich humus layer, impenetrable canebrakes grow profusely in the understory. Annual fires are common, so fire-hardy species such as *Zizyphus mauritiana*, *Madhuca latifolia*, *Aegle marmelos*, *Butea monosperma*, *Terminalia tomentosa*, *Ochna pumila*, and several others also occur in these fire-prone areas (Puri et al. 1989).

Biodiversity Features

The ecoregion does not exhibit high levels of endemism but does harbor several threatened species. The mammal fauna consists of 126 known species, including one near-endemic species, a bat, *Tadarida teniotis* (table J10.1). The mammal assemblage also includes several threatened species such as the tiger, Asian elephant, gaur, sloth bear (*Melursus ursinus*), smooth-coated otter (*Lutrogale perspicillata*), great Indian civet (*Viverra zibetha*), and four-horned antelope or chousingha (*Tetracerus quadricornis*). The ecoregion's large habitat blocks are an important contribution to conservation of the large, wide-ranging mammals of northeastern India and overlaps with three Level I TCUs (Dinerstein et al. 1997). The elephant population in Midnapur district is considered to be of high conservation importance (WII 1999).

There are more than 380 known bird species, although none are endemic. However, the ecoregion harbors two globally threatened species, the Bengal florican and the lesser florican (*Eupodotis indica*), as well as the threatened Pallas's fish-eagle (*Haliaeetus leucoryphus*) and swamp francolin (*Francolinus gularis*). The Indian grey hornbill (*Ocyceros birostris*) and Oriental pied-hornbill (*Anthracoceros albirostris*) are indicators of intact forests and deserve conservation attention.

Status and Threats

Current Status

Despite hundreds of years of human settlement, much forest still remained until the early twentieth century. Since then deforestation has accelerated, and now the ecoregion's natural habitat borders on the verge of extinction. Only about 3 percent of the ecoregion is now under natural forest, and only one large block of intact habitat (south of Varanasi) remains in this ecoregion. Although more than forty protected areas are represented in the ecoregion, they cover only about 3 percent of the ecoregion, and more than half of these protected areas are small, being less than 100 km^2 in area (table J10.2).

Types and Severity of Threats

The dense human population is still growing rapidly. The urbanization, industrialization, and agriculture associated with this growing population and its resource and economic needs pose serious threats to the remaining forest fragments. The small, protected areas are vulnerable to this tidal wave of

TABLE J10.2. WCMC (1997) Protected Areas That Overlap with the Ecoregion.

Protected Area	Area (km2)	IUCN Category
Gautam Budha [18]	140	IV
Udaipur	50	IV
Kaimur	2,370	IV
Chandra Prabha	80	IV
Rajgir	290	IV
Kaimur	120	IV
Valmikinagar [25]	230	IV
Raiganj	40	IV
Kabartal	30	PRO
Ramsagar	10	UA
Ganga Dolphin	160	PRO
Nakti Dam	200	IV
Narendrapur	90	IV
Jaldapara	70	IV
Chalan Beel WS	170	?
Bil Bhatia WS	30	?
Madhupur	110	V
Hail Haor	100	PRO
Roa	10	IV
Bhawal	90	V
Rema-Kalenga	50	IV
Gumti	420	IV
Sepahijala	20	IV
Trishna	270	IV
Aila Beel WS	30	VIII
Kawadighi WS	10	?
Unnamed	20	?
Dakhar Haor WS	40	?
Dubriar Haor WS	20	?
Erali Beel WS	10	?
Hakaluki Haor WS	160	?
Kawadighi WS	20	?
Kuri Beel WS	30	?
Meda Beel WS	20	?
Rajkandi WS	50	?
Tangua Haor WS	160	?
West Bhanugach WS	50	?
Hazarikhil	30	PRO
Rampahar-Sitapahar	10	PRO
Chunati	90	IV
Chimbuk WS	30	?
Himchari	20	UA
Teknaf	100	VIII
Sangu Matamuhari	960	?
Total	7,010	

Ecoregion numbers of protected areas that overlap with additional ecoregions are listed in brackets.

human growth and are inadequate to conserve the ecoregion's biodiversity. Finding additional habitat for protection will be challenging. Therefore, the existing protected areas should be effectively managed and protected, and restoring critical habitat should be considered where necessary.

Priority Conservation Actions

Short-Term Conservation Actions (1–5 Years)

- Identify sizable, intact habitat blocks. Reintroduce the swamp deer and the wild water buffalo, which is now extinct from this region (Rodgers and Panwar 1988).
- Conserve the small patches of natural habitats (Rodgers and Panwar 1988).
- Protect small patches of wetlands that are waterfowl habitat along the central and south of West Bengal (Rodgers and Panwar 1988).

Longer-Term Conservation Actions (5–20 Years)

- Identify and restore critical habitat areas (especially riparian habitat) to create conservation landscapes that can support a metapopulation of elephants and tigers.

Focal Species for Conservation Action

Area-sensitive species: tiger, Asian elephant

Top predator species: tiger, clouded leopard, wild dog

Other: gaur, swamp deer (now locally extinct but can be reintroduced), wild water buffalo (now locally extinct but can be reintroduced), Gangetic dolphin, Bengal florican, swamp francolin, lesser florican

Selected Conservation Contacts

Chief Wildlife Warden, Government of West Bengal, Calcutta

Department of Wildlife Conservation, Government of Bangladesh, Dacca

Wildlife Institute of India, Dehra Dun, India

WWF-India

Justification for Ecoregion Delineation

In previous analyses of conservation units in the region, Rodgers and Panwar (1988) and then MacKinnon (1997) placed these forests in one conservation unit but also included the freshwater swamp forests and mangroves with the moist deciduous forests. In keeping with our rules for defining ecoregions, we separated the freshwater swamp forests and mangroves into their own ecoregions and defined the moist forests as the Lower Gangetic Plains Moist Deciduous Forests [10]. We used MacKinnon's (1997) map of original vegetation as a guide to delineate the spatial extents of the vegetation types. This ecoregion roughly corresponds to Udvardy's Bengalian rain forest biogeographic province but extends further westward into the Indus-Ganges monsoon forest biogeographic province and further east into the Burma monsoon forest biogeographic province.

Prepared by Gopal S. Rawat and Eric D. Wikramanayake

Ecoregion Number:	**11**
Ecoregion Name:	**Meghalaya Subtropical Forests**
Bioregion:	**Indian Subcontinent**
Biome:	**Tropical and Subtropical Moist Broadleaf Forests**
Political Units:	**India**
Ecoregion Size:	**41,600 km²**
Biological Distinctiveness:	**Globally Outstanding**
Conservation Status:	**Vulnerable**
Conservation Assessment:	**I**

The Meghalaya Subtropical Forests [11] ecoregion is one of the wettest ecoregions in the Indo-Pacific region, with parts receiving more than 11 *meters* of rainfall over the course of a year. More than a century ago, these Khasi-Jaintia Hills of Meghalaya were described as one of the richest botanical habitats of Asia (in Rodgers and Panwar 1988). Not surprisingly, even today the ecoregion is considered one of the most species-rich in the bioregion for mammals, birds, and plants. For instance, the Khasi Hills alone are endowed with seventy-five orchid genera, represented by 265 species (GOI 1997). These hills are also considered the center of diversity for several primitive tree genera such as *Magnolia* and *Michelia* and for families such as Elaeocarpaceae and Elaeagnaceae (Davis et al. 1995). The Meghalaya Subtropical Forests [11] are also the gateway to the Malayan fauna. For instance, the haunting gibbon calls that echo through tall forests from Burma to Sumatra are also part of the early morning chorus here but not further west. The tiger (*Panthera tigris*), one of India's premier flagship species, is also considered to have entered the subcontinent from this region (Johnsingh et al. 1991).

Location and General Description

The ecoregion represents the subtropical forests of the Khasi and Garo hills that rise to more than 1,800 m in the eastern Indian states of Meghalaya and Assam. These hills were geologically part of the Deccan Plateau. Therefore, the ecoregion has ancient Gondwanaland origins.

The dominant substrate is a low-nutrient acrisoil, but the high rainfall—parts of the Khasi Hill can receive more than 11 m of annual rainfall—and local climatic variation as a result of the complex landform have promoted high biological diversity and centers of endemism here (Rodgers and Panwar 1988). The endemic pitcher plant *Nepenthes khasiana*, which is confined to the Khasi Hills, is but one example of the many plant species with very limited range distributions.

The major forest types in this ecoregion, according to Champion and Seth (1968), are Assam subtropical hill savanna, Khasi subtropical hill forest, Assam subtropical pine forest, and Assam subtropical pine savanna. More recently, Haridasan and Rao (1985) described the forest vegetation in this area as consisting of tropical evergreen forests in low-lying areas with high rainfall and characterized by species such as *Bischofia javanica, Fermiana colorata, Pterygota alata, Mesua ferrea, Castonopsis indica, Talauma hodgsonii, Pterospermum acerifolium*, and *Acrocarpus fracinifolius*; tropical semi-evergreen forest (up to elevations of 1,200 m) where annual rainfall is 1,500–2,000 mm and typical species include

Eleocarpus floribundus, Dillenia pentagyna, Dillenia indica, Hovenia acerba, and *Lithocarpus fenestratus*; tropical moist and deciduous forests in areas with less than 1,500 mm rainfall and characterized by *Shorea robusta, Tectona grandis* (most likely from plantations), *Terminalia myriocarpa, Tetrameles nudiflora, Schima wallichii,* and *Albizzia lebbeck*; grasslands and savanna of *Saccarum spontaneum, Neyraudia reynaudiana, Chrysopogon aciculatus,* and *Setaria glauca* on the tops of the Khasi, Jaintia, and Garo hills; isolated patches of temperate forests of *Lithocarpus fenestratus, Castanopsis kurzii, Quercus griffithii, Quercus semiserrata, Schima khasiana,* and *Myrica esculenta* along the southern slopes of the Khasi and Jaintia hills; and subtropical pine forests with pure stands of *Pinus khesia* confined to the higher reaches of the Shillong Plateau.

The mature forests have a dense undergrowth of patches of bamboos and canes (Rao 1991), with *Dendrocalamus hamiltonii* being especially common (Puri et al. 1989). The trees are draped in climbers and epiphytes (Puri et al. 1989; WWF and IUCN 1995).

Biodiversity Features

More than 110 mammal species are known from the Meghalaya Subtropical Forests [11], but none are endemic to this ecoregion. Some of the species of conservation importance represented here include the tiger (*Panthera tigris*), clouded leopard (*Pardofelis nebulosa*), Asian elephant (*Elephas maximus*), wild dog (*Cuon alpinus*), Malayan sun bear (*Ursus malayanus*), sloth bear (*Melursus ursinus*), smooth-coated otter (*Lutrogale perspicillata*), large Indian civet (*Viverra zibetha*), Chinese pangolin (*Manis pentadactyla*), Indian pangolin (*Manis crassicaudata*), Assamese macaque (*Macaca assamensis*), bear macaque (*Macaca arctoides*), capped leaf monkey (*Semnopithecus pileatus*), and hoolock gibbon (*Hylobates hoolock*). The tiger, clouded leopard, Asian elephant, Assamese macaque, bear macaque, capped leaf monkey, wild dog, sloth bear, and smooth-coated otter are threatened species (Hilton-Taylor 2000).

Despite the sparse large mammal fauna, the small mammals are well represented, especially the bats and the small carnivore community, which is unrivaled (Rodgers and Panwar 1988) across this bioregion. The ecoregion's position at the interface between different regional faunas is apparent with the overlap between similar species with different range distributions (e.g., the Indian and Chinese pangolins and the Malayan sun bear and sloth bear).

The bird fauna is richer, with more than 450 species. These include five species that are endemic to the ecoregion (table J11.1). All these species are near endemic (i.e., shared with adjacent ecoregions).

The ecoregion also harbors the globally threatened rufous-necked hornbill (*Aceros nipalensis*) and several other threatened species, including the white-winged duck (*Cairina scutulata*), ferruginous pochard (*Aythya nyroca*), Pallas's fish-eagle (*Haliaeetus leucoryphus*), marsh babbler (*Pellorneum palustre*), tawny-breasted wren-babbler (*Spelaeornis longicaudatus*), Manipur bush-quail (*Perdicula manipurensis*), bristled grassbird (*Chaetornis striatus*), Blyth's kingfisher (*Alcedo hercules*), greater spotted eagle (*Aquila clanga*), black-breasted parrotbill (*Paradoxornis flavirostris*), dark-rumped swift (*Apus acuticauda*), and beautiful nuthatch (*Sitta formosa*).

Several of the other hornbill species—wreathed hornbill (*Aceros undulatus*), brown hornbill (*Anorrhinus tickelli*), and great hornbill (*Buceros bicornis*)—are indicators of intact and mature habitat; these birds depend on tall, mature trees for nesting, without which the populations will decline.

The ecoregion also overlaps with two of BirdLife International's EBAs, Assam Hills (131) and Eastern Himalayas (130) (Stattersfield et al. 1998), attesting to the rich bird fauna with limited breeding ranges.

Status and Threats

Current Status

More than two-thirds of this ecoregion has been cleared or degraded; however, extensive stretches of intact habitat can still be found in the northeastern parts. The protected area system in this ecoregion amounts to a mere 154 km^2, which is less than 1 percent of the ecoregion's area (table J11.2). All seven protected areas are extremely small, with the largest being only 60 km^2. However, several patches of habitat are included within sacred forests that have been traditionally protected (Rao and Haridasan 1982).

Two protected areas are of special interest in this ecoregion: Baghmara pitcher plant sanctuary (about 0.1 km^2 for the protection of *Nepenthes khasiana*) and the small Nokrek National Park in the Garo Hills, known for its rich evergreen forests containing wild *Citrus* spp. Among the glaring protection gaps are large habitat areas that will allow the

TABLE J11.1. Endemic and Near-Endemic Bird Species.

Family	Common Name	Species
Phasianidae	Manipur bush-quail	*Perdicula manipurensis*
Timaliidae	Marsh babbler	*Pellorneum palustre*
Timaliidae	Brown-capped laughingthrush	*Garrulax austeni*
Timaliidae	Tawny-breasted wren-babbler	*Spelaeornis longicaudatus*
Timaliidae	Wedge-billed wren-babbler	*Sphenocichla humei*

An asterisk signifies that the species' range is limited to this ecoregion.

TABLE J11.2. WCMC (1997) Protected Areas That Overlap with the Ecoregion.

Protected Area	Area (km^2)	IUCN Category
Nokrek	50	II
Siju Wildlife	10	IV
Balphakram	10	II
Baghmara Pitcher Plant	10	IV
Nongkhyllem	10	IV
Garampani	60	IV
Rangapahar	4	IV
Total	154	

Ecoregion numbers of protected areas that overlap with additional ecoregions are listed in brackets.

important elephant populations in Assam to undertake unhindered seasonal migrations (Rodgers and Panwar 1988).

Types and Severity of Threats

The primary threats to this ecoregion's biodiversity stem from the following. Deforestation is a huge factor, especially that which results from shifting cultivation in and around the remaining blocks of forests. Hunting for tigers and elephants is rife. Mining for coal and limestone along the migratory routes of elephants will threaten their traditional movements (Johnsingh and Williams 1998) and result in escalating human-elephant conflicts that are already intense. These lands have been leased by the tribal landholders to private mining companies.

Priority Conservation Actions

More than 90 percent of the land in this ecoregion belongs to local communities. Therefore, community participation in conservation activities is essential. Some of the short-term and long-term activities to conserve the ecoregion's biodiversity include the following.

Short-Term Conservation Actions (1–5 Years)

- Conduct surveys to identify the distribution patterns of the ecoregion's characteristic biodiversity.
- Promote the establishment of additional sacred groves. If each council establishes its own sacred grove, this will add substantially to the protected area system.
- Conduct a survey of the ecoregion's tiger population, its habitat, and its prey base. The ecoregion overlaps with a TCU that has been identified as a priority unit for surveys.
- Maintain the traditional migratory routes of elephants in this region through proper land-use planning and zoning.
- Create an awareness program to increase the involvement of local communities in conservation activities.
- Reduce the dependence on shifting cultivation, especially on steep slopes and areas critical for biodiversity conservation.

Longer-Term Conservation Actions (5–20 Years)

- Identify and develop conservation landscapes that can support viable populations of elephants and tigers over the long term.

Focal Species for Conservation Action

Habitat specialists: hornbill assemblage (wreathed hornbill, brown hornbill, great hornbill) as indictors of intact mature forests

Area-sensitive species: tiger, Asian elephant

Top predator species: tiger, clouded leopard

Other: Nepenthes khasiana, wild *Citrus* spp., Assamese macaque, capped leaf monkey, large Indian civet, hoolock gibbon, sun bear, sloth bear, smooth-coated otter, bear macaque, white-winged duck, ferruginous pochard, marsh babbler, tawny-breasted wren-babbler, Manipur bush-quail, bristled grassbird, Blyth's kingfisher, greater spotted eagle, black-breasted parrotbill, dark-rumped swift, beautiful nuthatch

Selected Conservation Contacts

Chief Wildlife Warden, Meghalaya State, India

Department of Environmental Sciences, Northeastern Hill University, Shillong, India

Wildlife Institute of India, Dehra Dun, India

WWF-India

Justification for Ecoregion Delineation

Rodgers and Panwar (1988) delineated the northeastern region of India into two units, of which one, Assam Hills (8B), included the Manipur Hills to the west. Subsequently, MacKinnon (1997) split unit 8B into two biounits, the Northeast Hills (I9b) and the Burma transition (09c). The former represents a distinct vegetation type in the hills of Meghalaya, as depicted in MacKinnon's reconstruction of the original vegetation in the region. Therefore, we retained this biounit, with very slight modification, as the Meghalaya Subtropical Forests [11]. We used MacKinnon's (1997) digital map of the original vegetation, aided by a DEM, to draw the boundaries of the ecoregion to include the subtropical forests above 1,000 m in elevation. This ecoregion overlaps with two of Udvardy's biogeographic provinces: the Himalayan highlands and Burma monsoon forest.

Prepared by Gopal S. Rawat and Eric D. Wikramanayake

Ecoregion Number:	**12**
Ecoregion Name:	**Eastern Highlands Moist Deciduous Forests**
Bioregion:	**Indian Subcontinent**
Biome:	**Tropical and Subtropical Moist Broadleaf Forests**
Political Units:	**India**
Ecoregion Size:	**340,200 km²**
Biological Distinctiveness:	**Globally Outstanding**
Conservation Status:	**Endangered**
Conservation Assessment:	**I**

The Eastern Highlands Moist Deciduous Forests [12] are considered globally outstanding for the large vertebrate assemblages and intact ecological processes they still support. The ecoregion still retains large blocks of habitat that are more than 5,000 km². These are essential for the long-term conservation of the megavertebrates of Asia and are now a limited ecological resource.

The complex landform coupled with the monsoon rains that sweep in from the Bay of Bengal also make this ecoregion an ice age refuge for elements of the moist forest flora from the faraway Western Ghats mountains and from the Eastern Himalayas (Rawat 1997). It is also a present-day refuge for many of the bioregion's large vertebrates, such as tigers (*Panthera tigris*), wolves (*Canis lupus*), gaur (*Bos gaurus*), and sloth bear (*Melursus ursinus*), that are increasingly confined to fragments of the natural habitat that once covered vast expanses of the Indian Subcontinent. Unfortunately Asia's largest terrestrial vertebrate, the Asian elephant (*Elephas maximus*), has already been extirpated from this ecoregion.

Location and General Description

This expansive ecoregion represents the moist deciduous forests along the low hills of the northern sections of the Eastern Ghats Range of Andhra Pradesh and the eastern parts of the Satpura Range in Madhya Pradesh, India. These hills rise above 1,300 m and mark the northern boundary of the Indian Peninsula (Kendrick 1989). They also provide some relief to the otherwise featureless, flat Deccan Plateau. The ecoregion also extends across several other Indian states, including Bihar and Orissa. Geologically, it dates back to the Cretaceous and has Gondwanaland origins.

The low Eastern Ghats Mountains along the peninsula's eastern coast are not as effective in trapping the moisture from the monsoon winds that sweep in from the Bay of Bengal. Thus, this ecoregion receives about 1,000 mm of rainfall, which is about four times as much precipitation as falls in the drier areas to the west, lying in the rainshadow of the much higher and steeper Western Ghats Mountains.

The Eastern Ghats have biogeographic affinities with the eastern Himalayas and with the Western Ghats. The typical Himalayan species include *Arisaema tortuosum*, *Elatostemma cuneatum*, *Didymocarpus pygmaea*, *Rubus ellipticus*, *Thalictrum foliolosum*, *Boehmeria macrophylla*, and *Zanthoxylum rhetsa*. Species with Western Ghats affinities include *Schefflera venulosa*, *Artocarpus integrifolia*, *Gnetum ula*, and *Calamus rotang* (Rawat and Babu 1995). These relict formations along with their floral and faunal communities suggest that this ecoregion once formed an important dispersal route for species between the Western Ghats and the Eastern Himalayas (Abdulali 1949). This is contrary to the Satpura hypothesis put forth by Hora (1949), who suggested that the dispersal bridge was the Satpura Range rather than the Eastern Ghats.

The hill forest vegetation is influenced by topography, soil, and precipitation, including microclimatic moisture regimes manifested by the complex landform. Gaussen et al. (1969) identified four vegetation types in this region: *Shorea-Terminalia-Adina* series, *Shorea–Syzygium operculatum–Toona* series, *Shorea-Buchanania-Cleistanthus* series, and *Terminalia-Anogeissus-Cleistanthus* series. The forests growing on the lateritic and crystalline rock substrates along the hills are dominated by *Shorea robusta* in association with *Syzygium cumini*, *Dendrocalamus strictus*, *Phoenix acaulis*, and *Themeda quadrivalvis* (Champion and Seth 1968). On undulating hills in Madhya Pradesh, *Terminalia tomentosa*, *Anogeissus latifolia*, *Pterocarpus marsupium*, *Madhuca indica*, and *Lagerstroemia parviflora* become part of the association. Bamboos, especially *Dendrocalamus strictus*, grow in the undergrowth. The riverbanks are lined with *Bambusa arundinaria*.

Biodiversity Features

The Eastern Highlands Moist Deciduous Forests [12] harbor intact assemblages of the large vertebrates of the Indian Subcontinent bioregion's tropical moist forests. Because the ecoregion still retains about 84,000 km^2 of intact habitat (representing about 25 percent of its area) in several large habitat blocks, it represents a rare opportunity to create large conservation landscapes that can support viable populations of species that need large spaces, such as tigers. Wikramanayake et al. (1999) recognized these forests as a high-priority TCU.

Although its intact habitat makes an important contribution to conservation, the ecoregion does not harbor high levels of endemism. The only ecoregional endemic mammal is a threatened Rhinolophidae bat, *Hipposideros durgadasi* (table J12.1).

But the mammal fauna includes a long list of threatened large vertebrates that includes the tiger, sloth bear (*Melursus ursinus*), wild dog (*Cuon alpinus*), gaur (*Bos gaurus*), chousingha (*Tetracerus quadricornis*), blackbuck (*Antilope cervicapra*), and chinkara (*Gazella bennettii*) (Hilton-Taylor 2000). A population of the threatened wild water buffalo (*Bubalus arnee*) may still be living in Bastar District, Madhya Pradesh, near the junction of the Eastern Ghats Mountains and the eastern highlands (Rawat and Babu 1995).

The 313 bird species known from the ecoregion include two threatened species: the green avadavat (*Amandava formosa*) and Pallas's fish-eagle (*Haliaeetus leucoryphus*) (Hilton-Taylor 2000).

Botanically, this ecoregion contains several interesting and threatened species of conservation value, such as *Arisaema tortuosum*, *Cyathea gigantea*, *Entada rheedii*, *Gnetum ula*, *Rauvolfia serpentina*, *Musa rosacea*, *Linociera ramiflora*, *Dioscorea anguina*, *Litsea monopetala*, and *Didymoplexis pallens* and the endemic *Leucas mukerjiana* and *Phlebophyllum jeyporensis* (Rao et al. 1986; Puri et al. 1989).

Status and Threats

Current Status

About three-fourths of the natural forests of this ecoregion have been cleared. But unlike most of the other forested ecoregions in this bioregion, the remaining forests are in several large blocks that exceed 5,000 km^2. These extensive habitats are important to conserve the large vertebrates in the region (Wikramanayake et al. 1998), and conservation actions should be directed to secure them from further fragmentation and degradation. The thirty-one protected areas in the ecoregion cover more than 13,500 km^2, or about 4 percent of the ecoregion's intact habitat (table J12.2).

Despite the large number of protected areas, few are large enough to meet the ecological needs of one of the ecoregion's important focal species, the tiger. The largest protected area, Simlipal, is just over 2,000 km^2, and Kawal and Indravati are just over 1,000 km^2. Kanha, one of the most important tiger reserves in India, is just under 1,000 km^2. But twenty-three of the thirty-one reserves are less than 500 km^2 in extent.

TABLE J12.1. Endemic and Near-Endemic Mammal Species.

Family	Species
Rhinolophidae	*Hipposideros durgadasi**

An asterisk signifies that the species' range is limited to this ecoregion.

Types and Severity of Threats

The primary threats to these remaining habitat blocks stem from quarries, coal mines, large-scale clearing for agriculture, and hydroelectric projects. Shifting (*podu*) cultivation and excessive dependence of the local communities on the non-timber forest products, especially in the northern parts of Andhra Pradesh, have also begun to degrade the ecological integrity of the forests. Organized timber thieves also operate in this ecoregion, illegally felling trees in remaining forests. The Naxalite movements and ongoing conflicts prevent effective conservation and management actions.

Priority Conservation Actions

The ecoregion's large habitat blocks have been included within a high-priority TCU (Wikramanayake at al. 1999), and a landscape conservation plan for these forests using the tiger and the associated predator-prey relationships as foci is needed.

Short-Term Conservation Actions (1–5 Years)

- Assess tiger population, habitat condition, and prey base in the TCU.
- Evaluate the protected areas to see how well they can harbor viable populations of the area-sensitive species, contain ecological processes, and sustain long-term viability of the habitat itself.
- Promote more efficient agricultural systems to replace *podu* (shifting cultivation).
- Promote protection of sacred groves that have been historically effective conservation areas.
- Control and prevent illegal logging operations.

Longer-Term Conservation Actions (5–20 Years)

- Strengthen the existing protected area network, using the tiger as a focal species. Conservation efforts in the high-priority TCU should focus on a landscape conservation strategy that includes establishing additional core protected areas, strengthening protection and management of existing protected areas, and promoting conservation compatible land uses outside strict protected areas.
- Monitor the tiger populations and their prey base.

Focal Species for Conservation Action

Endangered endemic species: Hipposideros durgadasi

Area-sensitive species: tiger

Top predator species: tiger

Other: sloth bear, wild dog, gaur, chousingha, blackbuck, chinkara, wild water buffalo, green avadavat, ferruginous pochard, and Pallas's fish-eagle

Selected Conservation Contacts

Bombay Natural History Society

Chief Wildlife Wardens of Andhra Pradesh and Madhya Pradesh

Conservator of Forests, Northern Circle, Andhra Pradesh

Divisional Forest Officer, Bastar Division, Madhya Pradesh

Wildlife Institute of India

WWF-India

Justification for Ecoregion Delineation

In previous analyses of conservation units of the Indian subcontinent, Rodgers and Panwar (1988) and MacKinnon (1997) divided the Deccan Peninsula into five biotic provinces, with each unit including several different vegetation types ranging from moist deciduous forests to dry forests to thorn scrub. The moist deciduous forests of the Eastern Ghats and Satpura Range were included in the Central Plateau North (6B), Eastern Highlands (6C), and Central Highlands (6E) biotic provinces by Rodgers and Panwar.

In keeping with our definition of an ecoregion (i.e., an ecosystem of regional extent) and following our rules for ecoregion delineation (representing distinct vegetation types of regional extent in separate ecoregions), we placed the these moist deciduous forests in the Eastern Highlands Moist Deciduous Forests [12]. We used MacKinnon's (1997) map of the original habitat to identify the ecoregion boundaries.

Prepared by Gopal S. Rawat, Ajay Desai, Hema Somanathan, and Eric D. Wikramanayake

TABLE J12.2. WCMC (1997) Protected Areas That Overlap with the Ecoregion.

Protected Area	Area (km²)	IUCN Category
Kolleru [21]	480	IV
Kawal	1,080	IV
Indravati	1,150	II
Pranahita	130	IV
Bhairamgarh	160	IV
Lanjamadugu	80	DE
Eturnagaram	120	IV
Pamed	60	IV
Pakhal	120	IV
Balimela	130	DE
Kondakameru	400	DE
Kinnerasani	290	IV
Papikonda	530	IV
Achanakmar	550	IV
Kanha	900	II
Phen	100	IV
Sitanadi	670	IV
Udanti	340	IV
Pachmarhi	500	IV
Satpura	490	II
Bori	460	IV
Barnawapara	240	IV
Mahuadaur	60	IV
Badalkhol	120	IV
Simlipal	2,550	IV
Hadgarh	140	IV
Satkosia Gorge	790	IV
Baisipalli	170	IV
Kotgarh	400	IV
Karlapat	150	IV
Kakhari Valley	180	IV
Total	13,540	

Ecoregion numbers of protected areas that overlap with additional ecoregions are listed in brackets.

Ecoregion Number: **13**
Ecoregion Name: **Malabar Coast Moist Deciduous Forests**
Bioregion: **Indian Subcontinent**
Biome: **Tropical and Subtropical Moist Broadleaf Forests**
Political Units: **India**
Ecoregion Size: **35,300 km²**
Biological Distinctiveness: **Bioregionally Outstanding**
Conservation Status: **Critical**
Conservation Assessment: **IV**

Many years ago, the Malabar Coast Moist Deciduous Forests [13] ecoregion was a swath of lush tropical evergreen forest that extended along the western coast of the Deccan Peninsula between the Western Ghats Mountains and the Indian Ocean (Champion and Seth 1968). These forests were once inhabited by tigers (*Panthera tigris*), Asian elephants (*Elephas maximus*), wild dogs (*Cuon alpinus*), sloth bear (*Melursus ursinus*), and a host of hornbills. But today very little of the natural habitat is left (Rodgers and Panwar 1988), the result of years of forest clearing to establish teak (*Tectona grandis*) plantations, human settlements, and other human activities such as fires set to clear forests for agriculture and promote grazing lands for livestock. Therefore, the original evergreen character of the forests has changed from the evergreen vegetation to a semi-evergreen condition (Champion and Seth 1968). Many of the large, space-dependent species have disappeared from the ecoregion, victims of habitat loss and fragmentation. This is another ecoregion on the verge of extinction.

Location and General Description

The ecoregion represents the semi-evergreen forests along India's Malabar Coast, a narrow strip of land lying between the Indian Ocean to the west and extending up to the 250-m contour of the steep Western Ghats Mountains to the east (after Rodgers and Panwar 1988). It extends through the Indian states of Kerala, Karnataka, and Maharashtra.

Geologically, the Deccan Plateau—and thus the ecoregion—has Gondwanaland origins. After the Deccan Plate became detached and drifted northward to attach itself to the northern Eurasian continent, the Western Ghats Mountains were created by geological uplift. The mountains then began to intercept the southwestern monsoon rains, creating moister conditions on the western slopes of the range and drier conditions on the eastern side.

The southwestern monsoon brings more than 2,500 mm of annual rainfall to the ecoregion, influencing the vegetation. The southern parts of the ecoregion, in Kerala State, receive more rainfall, and the vegetation has responded by being tropical wet evergreen in the south with a gradual trend tending to drier conditions to the north.

The original vegetation along the west coast of the Deccan Peninsula was tropical evergreen (Champion and Seth 1968). But the forests have been largely replaced or interspersed with teak, giving the vegetation a semi-deciduous character; the teak is now considered indicative of a secondary successional stage or presence of plantations.

Champion and Seth (1968) characterized these southern moist mixed deciduous forests with the following species: *Tetrameles nudiflora*, *Stereospermum personatum*, *Dysoxylum binectariferum*, *Ficus nervosa*, *Ficus glomerata*, *Pterocarpus marsupium*, *Salmalia malabarica*, *Terminalia bellerica*, *Terminalia tomentosa*, *Anogeissus latifolia*, *Dalbergia latifolia*, *Lannea coromandelica*, *Madhuca indica*, *Garuga pinnata*, *Syzygium cumini*, *Olea dioica*, *Pouteria tomentosa*, *Bridelia retusa*, *Mangifera* spp., and *Actinodaphne angustifolia*. There is generally a second story of *Erythrina variegata*, *Butea monosperma*, *Wrightia tinctoria*, *Bauhinia racemosa*, and *Zizyphus rugosa* and a shrub layer of *Flacourtia* spp., *Woodfordia fruticosa*, *Meyna laxiflora*, and *Carissa congesta* (Puri et al. 1989). Along the northern coast of Karnataka State several patches of moist deciduous forests are represented by an association of *Lagerstroemia microcarpa*, *Tectona grandis*, and *Dillenia pentagyna* (Pascal et al. 1982), representing the drier climatic conditions. The *Myristica* swamps and the inland lagoons represent distinct habitat types within this ecoregion (Rodgers and Panwar 1988) that are now endangered.

Biodiversity Features

The ecoregion's mammal fauna, estimated at ninety-seven mammal species, includes five near-endemic species and a single small rodent species that is strictly endemic to this ecoregion (table J13.1). The near-endemic species are shared with the ecoregions along the Western Ghats Mountain Range. Many of the other species, especially the larger mammals, are widespread throughout the Indian subcontinent but are threatened by shrinking habitat. The threatened species include the Asian elephant, gaur (*Bos gaurus*), slender loris (*Loris tardigradus*), wild dog, sloth bear, Jerdon's palm civet, and grizzled giant squirrel (*Ratufa macroura*). The habitat in this ecoregion is too fragmented to harbor viable populations of wide-ranging species such as tigers (Wikramanayake et al. 1999). Elephants no longer roam the forests of this ecoregion, although one of India's most important elephant populations ranges along the Nelliampathi Hills (Sukumar 1989) adjacent to the southern parts of this ecoregion.

The ecoregion has an estimated 280 bird species, including one near-endemic species, the Malabar grey hornbill (*Ocyceros griseus*), which is shared with the submontane and montane ecoregions of the Western Ghats Mountains (table J13.2).

TABLE J13.1. Endemic and Near-Endemic Mammal Species.

Family	Species
Soricidae	*Suncus dayi*
Cercopithecidae	*Semnopithecus johnii*
Viverridae	*Viverra civettina*
Viverridae	*Paradoxurus jerdoni*
Muridae	*Petinomys fuscocapillus*
Muridae	*Rattus ranjiniae**

An asterisk signifies that the species' range is limited to this ecoregion.

TABLE J13.2. Endemic and Near-Endemic Bird Species.

Family	Common Name	Species
Bucconidae	Malabar grey hornbill	Ocyceros griseus

An asterisk signifies that the species' range is limited to this ecoregion.

The ecoregion also harbors several other species of conservation importance, including the globally threatened lesser florican (*Eupodotis indica*), a grassland species whose range fringes on this ecoregion. Other species such as the greater flamingo (*Phoenicopterus ruber*), Indian grey hornbill (*Ocyceros birostris*), and great hornbill (*Buceros bicornis*), though not threatened, are nevertheless sensitive to disturbances and can be used as focal species for conservation.

Status and Threats

Current Status

More than 95 percent of the ecoregion's natural habitat has been cleared or converted. The moist southern forests have been converted into coconut plantations and rice paddies and the northern forests into teak, rosewood, and rubber plantations. No large blocks of intact forest habitat now exist, although several smaller forest fragments have been preserved by local people as sacred groves.

The three protected areas in this ecoregion cover a mere 300 km^2 (<1 percent of its area). All are small, with an average of only 100 km^2 (table J13.3). In their assessment, more than a decade ago, Rodgers and Panwar (1988) indicated that the national target of a 5 percent threshold for protected areas would not be possible in these coastal forests.

Types and Severity of Threats

Most of the natural forest in this ecoregion has been converted to agriculture and plantations. Extensive teak plantations have created a semi-deciduous vegetation where there used to be evergreen forests. The unique *Myristica* swamps, limited to parts of northern Karnataka (e.g., Kumta) and Kerala and close to the coast, are being converted to areca and coconut plantations.

Continuing threats to this ecoregion include habitat damage from livestock grazing and trampling. The pastoralists themselves also overexploit the fodder trees and burn large areas to create grasslands. In Karnataka State, huge herds of domestic livestock have become uncontrollable. Meanwhile, Sanjay National Park in Maharashtra is subject to more than a hundred fires every year, and the park staff is not equipped to fight them.

TABLE J13.3. WCMC (1997) Protected Areas That Overlap with the Ecoregion.

Protected Area	Area (km^2)	IUCN Category
Sanjay	50	II
Bhagwan Mahavir	150	IV
Peechi Vazhani	100	IV
Total	300	

Ecoregion numbers of protected areas that overlap with additional ecoregions are listed in brackets.

Hydroelectric projects have also flooded large areas of forests; for instance, eleven dams have been constructed along the Periyar River, one of the largest waterways in the state of Kerala. Secondary effects of the dam projects, such as roads, have also allowed more human encroachment, which is causing serious fragmentation of the remaining forests (IUCN 1991).

Priority Conservation Actions

Years of human impacts and dense settlements—this ecoregion has one of the highest human population densities in the region—have resulted in large-scale habitat loss. Rodgers and Panwar (1988) proposed a suite of conservation actions more than a decade ago, which include expanding the existing protected areas and protecting even the small, remaining habitat blocks in an effort to conserve the biodiversity of these forests. Conservation actions have the most potential in areas where the protected areas extend into adjacent ecoregions, namely the North Western Ghats Moist Deciduous Forests [14] and South Western Ghats Moist Deciduous Forests [15].

Short-Term Conservation Actions (1–5 Years)

- Promote conservation of traditional sacred groves.
- Conserve the coastal mangroves and inland lagoons that represent important habitat types in this ecoregion. These habitats, although not extensive, play a crucial role in coast protection and fisheries.
- Identify and conserve the relict patches of *Myristica* swamps (Rodgers and Panwar 1988; Rodgers et al. 2000).
- Increase the protected area system by expanding the Sanjay Gandhi National Park and extending the Phansad Wildlife Sanctuary (Rodgers and Panwar 1988).
- Protect remaining habitat patches in Honavar and Pilarkhan in Karnataka, Chirikala in Kerala, and Keenparai in Tamil Nadu (Rodgers and Panwar 1988).

Longer-Term Conservation Actions (5–20 Years)

- Restore habitat along the foothills, especially adjacent to the larger forest patches and protected areas of the South Western Ghats Moist Deciduous Forests [15] so that there will be contiguous habitat and conservation landscapes for the wide-ranging species.

Focal Species for Conservation Action

Habitat specialists: Indian grey hornbill, Malabar pied hornbill, great hornbill

Top predator species: leopard, wild dog

Other: gaur, slender loris, sloth bear, Jerdon's palm civet, grizzled giant squirrel

Selected Conservation Contacts

Bombay Natural History Society
Centre for Ecological Sciences, Bangalore
Chief Wildlife Wardens of Kerala and Karnataka States
Kerala Forest Research Institute
Wildlife Institute of India
WWF-India

Justification for Ecoregion Delineation

In an earlier analysis of conservation units of India, Rodgers and Panwar (1988) delineated the coastal moist deciduous forests along the western coast of India as a single conservation unit, the Coastal Plain (5A). In a subsequent regional analysis, MacKinnon (1997) also recognized this unit as a distinct biounit. We retained their classification and placed the lowland moist deciduous forests in the Malabar Coast Moist Deciduous Forests [13]. In doing so, we used MacKinnon's (1997) map of the original vegetation to delineate the ecoregion boundaries. This ecoregion falls within Udvardy's Malabar rain forest biogeographic province.

Prepared by Gopal S. Rawat, Ajay Desai, Hema Somanathan, and Eric D. Wikramanayake.

Ecoregion Number:	**14**
Ecoregion Name:	**North Western Ghats Moist Deciduous Forests**
Bioregion:	**Indian Subcontinent**
Biome:	**Tropical and Subtropical Moist Broadleaf Forests**
Political Units:	**India**
Ecoregion Size:	**48,000 km^2**
Biological Distinctiveness:	**Regionally Outstanding**
Conservation Status:	**Critical**
Conservation Assessment:	**II**

The North Western Ghats Moist Deciduous Forests [14] surround the montane rain forest ecoregion of the northern areas of the Western Ghats Range and create a conservation landscape from the rich montane rain forests to the lowland moist deciduous forests. Together the two ecoregions vastly increase the conservation potential and importance of the Western Ghats ecosystem. Many of the large herbivores have larger populations in the drier moist deciduous forests, relative to the montane rain forests (although the rain forests have higher endemicity). The populations of predators such as the tigers, leopards, and wild dogs are also correspondingly larger, a response to the greater prey availability.

Location and General Description

The ecoregion represents the swath of moist deciduous forests around the montane rain forests in the northern section of the Western Ghats Mountains. It extends through Maharashtra and Karnataka States in western India.

The ecoregion has Gondwana origins. During the Cretaceous, the Deccan Plate detached from Gondwanaland and drifted northward to reattach with the northern Eurasian continent. Millions of years later, geological uplift gave rise to the Western Ghats Mountains, which began to intercept the southwestern monsoon rains, creating a moisture gradient from wet conditions in the windward side of the mountains to increasingly dry habitats on the leeward side. Rainfall ranges from 1,500 to 2,000 mm annually, with a 4- to 5-month dry season. Mean annual temperature ranges from 24 to 27°C, but maximum temperatures can exceed a stifling 40°C inland.

The vegetation, influenced by the southwestern monsoon, corresponds to Champion and Seth's (1968) southern Indian Moist Deciduous Forest type, and consists of moist, teak-bearing forests, moist mixed deciduous forest without teak, and secondary moist mixed deciduous forests. Characteristic species include *Tectona grandis*, *Grewia tiliaefolia*, *Lagerstroemia lanceolata*, *Dillenia pentagyna*, *Kydia calycina*, *Bambusa arundinacea*, *Dalbergia latifolia*, *Adina cordifolia*, *Pterocarpus marsupium*, *Xylia xylocarpa*, *Wrightia tinctoria*, and *Schleichera oleosa* (Champion and Seth 1968). The teak forests on lateritic soils have typical understory species represented by *Cleistanthus collinus*, *Holarrhena antidysenterica*, *Bauhinia racemosa*, and *Kydia calycina*. Important climbers include *Ichnocarpus frutescens*, *Dioscorea* spp., *Butea superba*, *Bauhinia vahlii*, and *Smilax macrophylla* (Puri et al. 1989).

Patches of evergreen forests extend into these moist deciduous forests, with rain forest species such as *Cinnamomum malabaricum*, *Syzygium gardneri*, *Mrystica dactyloides*, *Kneema attenuata*, and *Diospyros sylvatica*.

Biodiversity Features

The ecoregion harbors eighty-seven mammal species, but unlike in the montane rain forests, none are endemic to this ecoregion. However, several of the mammals in the ecoregion are threatened (Hilton-Taylor 2000), including the tiger (*Panthera tigris*), Asian elephant (*Elephas maximus*), gaur (*Bos gaurus*), slender loris (*Loris tardigradus*), sloth bear (*Ursus ursinus*), chousinga (*Tetracerus quadricornis*), and squirrel (*Ratufa indica*). Tigers used to roam the forests of this habitat, but because of extensive habitat loss and hunting, populations have been severely depleted. The southern section of the ecoregion overlaps with a high-priority TCU (Dinerstein et al. 1997) and also borders on the Nilgiri Hills, which harbor one of Asia's most important elephant populations (Sukumar 1989).

The bird fauna is much richer, with more than 345 species, including 5 near-endemic species (table J14.1) that are shared with the adjacent montane forest ecoregion.

The globally threatened spot-billed pelican (*Pelecanus philippensis*) and lesser florican (*Eupodotis indica*) are also part of the ecoregion's bird community. Several hornbills—Malabar grey hornbill (*Ocyceros griseus*), Indian grey hornbill (*Ocyceros birostris*), and great hornbill (*Buceros bicornis*)—that need tall, mature forests for nesting can be used as focal species for conservation management. The ecoregion overlaps with an EBA, Western Ghats (123), identified by BirdLife International (Stattersfield et al. 1998).

TABLE J14.1 Endemic and Near-Endemic Bird Species.

Family	Common Name	Species
Columbidae	Nilgiri wood-pigeon	*Columba elphinstonii*
Bucconidae	Malabar grey hornbill	*Ocyceros griseus*
Pycnonotidae	Grey-headed bulbul	*Pycnonotus priocephalus*
Timaliidae	Rufous babbler	*Turdoides subrufus*
Psittacidae	Malabar parakeet	*Psittacula columboides*

An asterisk signifies that the species' range is limited to this ecoregion.

However, relative to the south Western Ghats, the northern parts of the Western Ghats have been poorly surveyed.

Status and Threats

Current Status

More than three-fourths of the natural habitat in this ecoregion has been cleared or converted, and only scattered fragments remain. However, even these fragments represent important habitat for the Indian megafauna, especially as an extension of the montane forest ecoregion and in the context of connectivity with lowland habitats. The thirteen protected areas cover about 2,200 km^2, or 5 percent of the ecoregion (table J14.2), but most of these protected areas are small, with the average size being only 173 km^2. Three of the eight protected areas—Bhadra, Shettihally, Sharavathi Valley—that overlap with the North Western Ghats Montane Moist Forests [1] are more than 500 km^2.

Types and Severity of Threats

Forest clearing and human encroachment remain the biggest threats to this ecoregion's natural habitat and biodiversity. Large areas have already been converted into rubber, tea, and coffee plantations. Because of extensive forest fragmentation, it may not be possible to create additional large protected areas without extensive restoration. Poaching is a serious problem in the ecoregion.

The dense human population led to a change in conservation status from endangered to critical.

Priority Conservation Actions

Rodgers and Panwar (1988) identified several conservation gaps and needs in a comprehensive analysis. Here we present some short- and longer-term conservation actions.

Short-Term Conservation Actions (1–5 Years)

- Buffer existing protected areas.
- Stop poaching and illegal timber removal.
- Survey habitat status in protected areas. Restore degraded habitat as deemed necessary.
- Manage habitat to provide an adequate prey base for tigers in the high-priority TCU.
- Mitigate human-wildlife conflicts, with a focus on elephant, tiger, and sloth bear conflicts.

Longer-Term Conservation Actions (5–20 Years)

- Review land-use practices and develop a plan to create a conservation landscape that includes habitat linkages between protected areas and incorporates the adjacent montane rain forest ecoregion and this moist deciduous ecoregion (Rodgers and Panwar 1988).
- Implement conservation actions in the Level I TCU to create a conservation landscape that includes adequate core protected areas and dispersal habitat.

Focal Species for Conservation Action

Habitat specialists: hornbills (Malabar grey hornbill, Malabar pied hornbill, great hornbill) as indicators of mature, intact forests

Area-sensitive species: Asian elephant, tiger

Top predator species: tiger

Other: sloth bear, gaur, chousingha, spot-billed pelican, Nilgiri wood-pigeon

Selected Conservation Contacts

Bombay Natural History Society
Centre for Ecological Sciences, Bangalore
Chief Wildlife Warden, Karnataka State
Indian Institute of Sciences, Bangalore
Pulney Hills Conservation Society
State Forest Departments
Western Ghats Conservation Board
Western Ghats Development Authority
WWF-India

Justification for Ecoregion Delineation

In an earlier analysis Rodgers and Panwar (1988) placed the Western Ghats Mountain Range into a single biogeographic unit. But Rodgers and Panwar (1988) acknowledged that the Western Ghats Mountain Range is too large to represent a single unit for conservation planning and divided the range into northern and southern areas. They used the Wyanad as the boundary, where the southern *Cullenia*-dominated forests transition into drier northern dipterocarp forests. We also used this transition to make a more explicit division of the Western Ghats into northern and southern ecoregions. But in keeping with our rules for representing distinct habitat types within separate ecoregions, we placed the belt of moist deciduous forests that surround the montane rain forests into the North Western Ghats Moist Deciduous Forests [14].

We used the 1,000-m contour from a DEM and MacKinnon's (1997) map of original vegetation to define the boundary between the montane rain forests and moist deciduous forests. The outer boundaries, between the dry forests in the Deccan Plateau to the east and the distinctly different lowland moist deciduous forests along the Malabar coast to

TABLE J14.2 WCMC (1997) Protected Areas That Overlap with the Ecoregion.

Protected Area	Area (km^2)	IUCN Category
Bhimashankar [20]	30	IV
Schollpaneshwar	330	IV
Purna	150	IV
Karnala	50	IV
Tansa [1]	80	IV
Koyna [1]	90	PRO
Chandoli [1]	100	IV
Cotigao	170	IV
Anshi [1]	280	II
Bhadra [1]	330	IV
Shettihally [1]	280	IV
Sharavathi Valley [1]	220	IV
Mookambika [1]	140	IV
Total	2,250	

Ecoregion numbers of protected areas that overlap with additional ecoregions are listed in brackets.

the west, were defined using MacKinnon's (1997) reconstruction of the distribution of the original vegetation types. This ecoregion falls within Udvardy's Malabar rain forest biogeographic province.

Prepared by Gopal S. Rawat, Ajay Desai, Hema Somanathan, and Eric D. Wikramanayake

Ecoregion Number:	**15**
Ecoregion Name:	**South Western Ghats Moist Deciduous Forests**
Bioregion:	**Indian Subcontinent**
Biome:	**Tropical and Subtropical Moist Broadleaf Forests**
Political Units:	India
Ecoregion Size:	23,700 km^2
Biological Distinctiveness:	Regionally Outstanding
Conservation Status:	Endangered
Conservation Assessment:	II

The South Western Ghats Moist Deciduous Forests [15] lie adjacent to the montane rain forest ecoregion in the southern extent of the Western Ghats Mountain Range. This ecoregion creates a landscape that extends from the lowlands to the highest peaks of one of the bioregion's richest and most diverse ecosystems. The ecoregion is wider in the drier, leeward side of the mountain range, where it drops down to the dry Deccan Plateau to encompass prime habitat where some of the most important populations of tiger (*Panthera tigris*), Asian elephant (*Elephas maximus*), and gaur (*Bos gaurus*) live. The ecoregion represents a transition area between the South Western Ghats Montane Rain Forests [2] and the South Deccan Plateau Dry Deciduous Forests [22], and includes species from both. Therefore, species richness is high. But more importantly, the ecoregion provides continuity of ecological processes between the lowland and montane ecosystems. The moister forests of this ecoregion support many of the larger vertebrates also found in the dry forest ecoregion but at higher densities.

Location and General Description

The ecoregion represents the band of moist deciduous forests that surround the montane evergreen rain forests in the southern part of the Western Ghats Mountains. It extends across the southern Indian states of Kerala and Tamil Nadu. Being part of the Deccan Plateau, the ecoregion has Gondwanaland origins. But the moist deciduous nature of the vegetation is a manifestation of the geological uplift that occurred during the upper Tertiary period and created the Western Ghats Mountain Range, which then began to intercept the southwestern monsoon and create a complex rainshadow. The swath of moist deciduous forests is very narrow on the steeper, windward side of the mountain range, where the southwest monsoon rains promote a wet evergreen forest. But on the shallower leeward side, the drier conditions caused by the rainshadow result in a broader, uneven swath of moist deciduous forests that extends further, into the Deccan Plateau. Rainfall on the leeward side is influenced by the complex landform, with some areas receiving less than a fifth of the 3,000 mm or more of annual precipitation that is deposited higher in the mountains.

Champion and Seth (1968) classify these forests as a Southern Indian Moist Deciduous Forest. The vegetation is characterized by *Adina cordifolia*, *Albizzia odoratissima*, *Albizzia procera*, *Alstonia scholaris*, *Bombax ceiba*, *Toona ciliata*, *Dalbergia latifolia*, *Grewia tiliaefolia*, *Holoptelea integrifolia*, *Hymenodictyon excelsum*, *Lagerstroemia lanceolata*, *Lagerstroemia speciosa*, *Lannea coromandelica*, *Miliusa velutina*, *Pterocarpus marsupium*, *Schleichera oleosa*, *Spondias pinnata*, *Radermachera xylocarpa*, *Tectona grandis*, *Terminalia bellerica*, *Terminalia paniculata*, *Terminalia tomentosa*, *Vitex altissima*, *Xylia xylocarpa*, and *Machilus macrantha* (Champion and Seth 1968).

Biodiversity Features

Although it does not serve as home to as many endemic species as the montane rain forests, the moist deciduous forest ecoregion is actually richer in species, harboring more mammal and bird species. Among the eighty-nine mammal species known from here, six are near-endemic species (table J15.1). But none of the species are strict endemics, entirely limited to this ecoregion. Three mammal species, Nilgiri langur (*Semnopithecus johnii*), Malabar large-spotted civet (*Viverra civettina*), and Jerdon's civet (*Paradoxurus jerdoni*), also extend into the montane rain forests and the semi-deciduous coastal forests on either side, whereas the shrew *Suncus dayi* is also found in the coastal forests. Layard's striped squirrel (*Funambulus layardi*) is also found in the adjacent montane rain forest ecoregion.

Among the larger threatened mammal species in this ecoregion are the tiger, Asian elephant, gaur, Nilgiri langur, wild dog (*Cuon alpinus*), and sloth bear (Hilton-Taylor 2000). Several of these large species need large spaces. Two of India's most important elephant populations—the Nilgiri–Eastern Ghats population, estimated at more than 6,300 animals, and the Anaimalais-Nelliampathis population, estimated at 1,200 to 2,000 animals (Sukumar 1989)—venture into this ecoregion. The ecoregion is also part of a high-priority (Level I) TCU (Wikramanayake et al. 1999) that also includes the montane rain forests. Some of the smaller threatened species include Jerdon's palm civet (*Paradoxurus jerdoni*), slender loris (*Loris tardigradus*), grizzled giant squirrel (*Ratufa macroura*), and Indian giant squirrel (*Ratufa indica*) (Hilton-Taylor 2000).

TABLE J15.1 Endemic and Near-Endemic Mammal Species.

Family	Species
Soricidae	*Suncus dayi*
Pteropodidae	*Latidens salimalii*
Cercopithecidae	*Semnopithecus johnii*
Viverridae	*Viverra civettina*
Sciuridae	*Paradoxurus jerdoni*
Sciuridae	*Funambulus layardi*

An asterisk signifies that the species' range is limited to this ecoregion.

TABLE J15.2. Endemic and Near-Endemic Bird Species.

Family	Common Name	Species
Columbidae	Nilgiri wood-pigeon	Columba elphinstonii
Bucconidae	Malabar grey hornbill	Ocyceros griseus
Pycnonotidae	Grey-headed bulbul	Pycnonotus priocephalus
Timaliidae	Rufous babbler	Turdoides subrufus
Corvidae	White-bellied treepie	Dendrocitta leucogastra
Muscicapidae	Black-and-rufous flycatcher	Ficedula nigrorufa
Muscicapidae	Nilgiri flycatcher	Eumyias albicaudata
Pycnonotidae	Yellow-throated bulbul	Pycnonotus xantholaemus
Psittacidae	Malabar parakeet	Psittacula columboides

An asterisk signifies that the species' range is limited to this ecoregion.

The 322 bird species known from this ecoregion include nine near-endemic species (table J15.2).

These species are shared with the South Western Ghats Montane Rain Forests [2], but five species that are endemic to the montane ecoregion are absent from this lower, moist deciduous forest ecoregion.

The globally threatened lesser florican (*Eupodotis indica*) can be found in patches of grassland habitats in this ecoregion. BirdLife International has included these moist deciduous forests within an EBA, Western Ghats (123) (Stattersfield et al. 1998).

Status and Threats

Current Status

Nearly three-fourths of the natural vegetation in the ecoregion has been cleared or converted, and the remaining forests are severely fragmented. However, there are fourteen protected areas that cover almost 5,000 km² (21 percent), of the ecoregion's intact habitat (table J15.3).

One protected area, Bandipur, exceeds 1,000 km² and is contiguous with Nagarhole, Mudumalai, and Wyanad. Together this protected area complex includes the largest protected elephant population, estimated at more than 2,500 elephants in India. Two other protected areas that exceed 1,000 km², Periyar and Anamalai, straddle this and the adjacent South Western Ghats Montane Rain Forests [2]. But eleven of the other reserves are small, less than 500 km², and four are less than 100 km². Therefore, many of the protected areas cannot support viable populations of the focal large mammals in this ecoregion.

Types and Severity of Threats

The most significant threat to the remaining habitat and existing reserves is from livestock grazing and associated impacts such as trampling, excessive fodder collection, and burning to create grasslands for livestock (IUCN 1991). These threats are especially severe in Karnataka State. Dam construction for hydroelectricity and irrigation has also caused habitat fragmentation, concentrating elephant populations in forest fragments and exacerbating human-elephant conflicts. Habitat is also being fragmented by forest clearing for plantations.

Priority Conservation Actions

In a comprehensive analysis, Rodgers and Panwar (1988) identified several conservation gaps and needs. Here we present some short- and longer-term conservation actions.

Short-Term Conservation Actions (1–5 Years)

- Manage habitat to provide prey base for tigers, one of the focal species for conservation in this ecoregion.
- Mitigate human-wildlife conflicts. Especially address elephant depredation issues.
- Make additions to the Nagarhole-Bandipur-Mudumalai protected area complex to include seasonal habitat for elephants (Rodgers and Panwar 1988).
- Protect the habitat linkages between the Nagarhole-Bandipur-Mudumalai-Wyanad complex with conservation areas to the south and in the Eastern Ghats. These linkages now exist and will maintain connections and landscape integrity between the montane and dry forest ecoregions.
- Reforest the tea and cardamom plantations recently acquired by the forest department.
- Review land-use practices and develop a plan to conserve representative habitat and create a conservation landscape that includes habitat linkages between protected areas (Rodgers and Panwar 1988).
- Provide alternative cooking facilities and regulate natural resource use to reduce damage to the environment during pilgrimages that currently result in excessive fuel wood collection in several areas of the ecoregion.

TABLE J15.3. WCMC (1997) Protected Areas That Overlap with the Ecoregion.

Protected Area	Area (km²)	IUCN Category
Nagarahole	620	II
Bandipur	1,110	II
Wynad	430	IV
Mudumalai	400	IV
Bilgiri Ranga Swamy Temple [22]	370	PRO
Megamalai [2]	310	PRO
Periyar [2]	470	IV
Anamalai [2]	620	IV
Eravikulam [2]	90	II
Chinnar	50	IV
Parambikulam (extension)	60	PRO
Peppara [2]	40	IV
Neyyar	150	IV
Kalakad	240	IV
Total	4,960	

Ecoregion numbers of protected areas that overlap with additional ecoregions are listed in brackets.

Longer-Term Conservation Actions (5–20 Years)

- Restore and recreate the habitat between the Periyar Tiger Reserve and the Ashambu Hills and between the southern and northern areas of the Nilgiri Biosphere Reserve in Tamil Nadu to allow elephant movements. Proposals to create these corridors have not been implemented (Johnsingh and Williams 1999).
- Implement conservation actions in the Level I TCU to create a conservation landscape that includes adequate core protected areas and dispersal habitat.

Focal Species for Conservation Action

Habitat specialists: Malabar grey hornbill

Area-sensitive species: Asian elephant, tiger

Top predator species: tiger

Other: wild dog, Jerdon's palm civet, sloth bear, gaur, Salim Ali fruit bat, Nilgiri langur, Nilgiri wood-pigeon, grey-headed bulbul, Malabar parakeet

Selected Conservation Contacts

Bombay Natural History Society

Centre for Ecological Sciences, Bangalore

Chief Wildlife Wardens, Kerala, Tamil Nadu States

Director, Anamalai Wildlife Sanctuary, Tamil Nadu

Field director, Kalakad-Mundanthurai Tiger Reserve, Tamil Nadu

French Institute of Pondicherry

High Range Wildlife Preservation Society, Munnar, Kerala

Kerala Forest Research Institute, Pichi, Kerala

Salim Ali Institute of Ornithology and Natural History, Coimbatore

Tropical Botanic Gardens and Research Institute, Thiruvananthapuram

Wildlife Institute of India

WWF-India

Justification for Ecoregion Delineation

In earlier analyses, MacKinnon (1997), and Rodgers and Panwar (1988) placed the Western Ghats Mountain Range into a single biogeographic unit. But Rodgers and Panwar (1988) acknowledged that the Western Ghats Range is too large to represent a single unit for conservation planning and implicitly divided the mountain range into northern and southern areas. They used the Wyanad as a transition zone from the southern *Cullenia*-dominated forests and the northern drier dipterocarp forests in this division. We also used this transition to make a more explicit division of the northern and southern ecoregions in the Western Ghats. But in keeping with our definition of ecoregions, we also separated the distinct belt of moist deciduous forests that surround the montane rain forests and placed them in different ecoregions. Thus the moist deciduous forests to the south of the Wyanad area were placed in the South Western Ghats Moist Deciduous Forests [15]. This ecoregion falls within Udvardy's Malabar rain forest biogeographic province.

We used the 1,000-m contour from a DEM and MacKinnon's (1997) map of original vegetation to define the boundary between the montane rain forests and moist deciduous forests. The outer boundaries, between the dry forests in the Deccan Plateau to the east and the distinctly different lowland moist deciduous forests along the Malabar coast to the west, were defined using MacKinnon's (1997) reconstruction of the distribution of the original vegetation types.

Prepared by Gopal S. Rawat, Ajay Desai, Hema Somanathan, and Eric D. Wikramanayake

Ecoregion Number:	**16**
Ecoregion Name:	**Kathiarbar-Gir Dry Deciduous Forests**
Bioregion:	**Indian Subcontinent**
Biome:	**Tropical and Subtropical Dry Broadleaf Forests**
Political Units:	**India**
Ecoregion Size:	**266,400 km²**
Biological Distinctiveness:	**Regionally Outstanding**
Conservation Status:	**Endangered**
Conservation Assessment:	**II**

The Kathiarbar-Gir Dry Deciduous Forests [16] support the only population of the Asiatic lion (*Panthera leo*), which is now limited to the small Gir National Park in northwestern India. One of the world's oldest mountain ranges, the Aravalli Hill Range, dating back to the Precambrian, runs through this ecoregion, rising to 1,721 m at its highest point, Mt. Abu. This region is also the meeting point of the Afrotropical and South Indian floras.

Location and General Description

This large ecoregion extends across the India's northwestern states of Gujarat, Rajasthan, and Madhya Pradesh. A small, disjunct portion of this ecoregion is completely included within the Northwestern Thorn Scrub Forests [24] and represents the Gir forests.

The region is arid for more than eight months of the year. Annual rainfall ranges from about 550 to 700 mm, and ambient temperatures regularly exceed 40°C.

The dry deciduous forests of this ecoregion are composed of three stories, with the upper canopy reaching from 15 to 25 m (Puri et al. 1989). Even within this dry area, relative differences in moisture levels determine forest composition. Teak (*Tectona grandis*) forests grow in areas that are less dry, in association with *Aegle marmelos, Boswellia serrata, Ougeinia oojeinensis, Diospyros* spp., *Bombax ceiba, Sterculia urens, Emblica officinalis, Dalbergia paniculata,* and *Terminalia tomentosa*. But in drier areas, the teak-dominated forests are replaced by almost pure stands of *Anogeissus pendula* that grow predominantly on the quartzite ridges and gneiss hillocks of the Aravalli system, in association with *Acacia catechu*. The rocky hillocks are characterized by *Euphorbia caducifolia, Maytenus emarginata, Acacia senegal, Commiphora mukul, Wrightia tinctoria, Securinega leucopyrus, Grewia tenax,* and *Grewia villosa*.

Mt. Abu is covered with tropical dry deciduous forests, and thorn scrub dominates the rest of the Aravalli Hills in south-

TABLE J16.1 Endemic and Near-Endemic Bird Species.

Family	Common Name	Species
Paridae	White-naped tit	Parus nuchalis

An asterisk signifies that the species' range is limited to this ecoregion.

east Rajasthan. Most of the ecoregion's endemic flora occurs in these hills and includes *Dicliptera abuensis*, *Strobilanthes halbergii*, and *Veronica anagallis* (WWF and IUCN 1995).

The riparian habitats have a distinctive flora characterized by *Phoenix sylvestris* and *Ficus racemosa*.

Biodiversity Features

Although most of the original natural habitat has been degraded or converted, the protected habitats harbor high densities of ungulates that in turn support the large carnivore assemblage, which includes the Asiatic lion, leopard (*Panthera pardus*), wolf (*Canis lupus*), and hyena (*Hyena hyena*).

There are eighty known mammal species in this ecoregion. The only population of the Asiatic lion is now limited to the Gir National Park. In addition to these lions, several other threatened mammal species deserve conservation attention, namely the threatened chousingha (*Tetracerus quadricornis*), blackbuck (*Antilope cervicapra*), and chinkara (*Gazella bennettii*) (Hilton-Taylor 2000).

The bird fauna approaches 300 species, a high number for these dry forests. These include the white-naped tit, a near-endemic species (table J16.1) that prefers thorn scrub habitat, shared with the adjacent Northwestern Thorn Scrub Forests [24].

The ecoregion also harbors two globally threatened bird species, the Indian bustard (*Ardeotis nigriceps*) and lesser florican (*Eupodotis indica*), which should be focal species for conservation management.

Status and Threats

Current Status

The ecoregion is highly deforested and degraded; however, a few large blocks of habitat still remain and are worth conserving for possible translocations of the Asiatic lion to establish additional populations. Currently, the twenty-one protected areas (average size 427 km^2), cover a little over 8,900 km^2, or 3 percent of the ecoregion. Of these, Gir National Park is the largest and most probably the best known because of its lion population. Three other reserves, Balaram-Ambaji, Sariska, and Gandhi Sagar, are more than 900 km^2, and three others exceed 500 km^2 (table J16.2).

Types and Severity of Threats

The numerous threats to the habitat and biodiversity in the ecoregion stem from the resource needs of a fast-growing human population. The most severe threats include livestock grazing, burning, and woodcutting by the indigenous people. However, large-scale mining for marble and minerals have begun to degrade the habitat further. Powerful industrialists with mining interests have lobbied for denotification of protected areas so they can be mined.

Priority Conservation Actions

An important conservation target in this ecoregion is the only population of the Asiatic lion, now limited to the Gir National Park. Sharma and Johnsingh (1996) provided a comprehensive analysis of the conservation needs for this park and the lions. Here we present some of the short- and longer-term conservation actions for the ecoregion.

Short-Term Conservation Actions (1–5 Years)

- Conduct biological surveys to identify the distribution patterns of the ecoregion's characteristic biodiversity.
- Institute controls on mining activities. Several sanctuaries have been denotified because of political pressure and influence by powerful mine owners and industrialists. These actions should be countered through political action.
- Implement management prescriptions for Gir National Park provided by Sharma and Johnsingh (1996).

Longer-Term Conservation Actions (5–20 Years)

- Identify alternative habitat areas for Asiatic lions and translocate lions to create a second population.
- Conduct voluntary relocation of pastoral communities (Maldharis) from lion habitat in the Gir forest in a phased manner, as proposed in Sharma and Johnsingh (1996).

Focal Species for Conservation Action

Top predator species: Asiatic lion, leopard, wolf

Other: hyena, chousingha, blackbuck, chinkara, Indian bustard, lesser florican

TABLE J16.2 WCMC (1997) Protected Areas That Overlap with the Ecoregion.

Protected Area	Area (km^2)	IUCN Category
Gir	1,720	IV
Mt. Abu	490	IV
Balaram-Ambaji	980	IV
Sariska	980	IV
Nahargarh	120	IV
National Chambal [17]	30	PRO
Ranthambore	550	II
Ghatigaon Great Indian Bustard	520	IV
Palpur	330	IV
Ramgarh	330	PRO
Madhav	170	II
Madhav extension	170	PRO
Jawahar Sagar	130	IV
Darrah	150	IV
Gandhi Sagar	900	IV
Kumbhalgarh	290	IV
Jaisamand	40	IV
Sita Mata	600	IV
Sailana	10	IV
Ratanmahal	300	IV
Jambughoda	170	IV
Total	8,980	

Ecoregion numbers of protected areas that overlap with additional ecoregions are listed in brackets.

Selected Conservation Contacts

Chief Wildlife Warden, Gujarat, India
Director, Gir National Park
The Gujarat Ecological Education and Research Foundation
Wildlife Institute of India
WWF-India

Justification for Ecoregion Delineation

In a previous analysis of conservation units, Rodgers and Panwar (1988) identified two units in this region: the Punjab (4A) and Gujarat-Rajwara (4B). Then MacKinnon (1997) reconstructed the original extent of the vegetation in the region, which indicated the presence of several distinct habitat types (dry deciduous forests, thorn scrub, and desert) within each of the units identified by Rodgers and Panwar (1988). In keeping with our rules for representing large areas of distinct habitat types within separate ecoregions, we used MacKinnon's (1997) digital map of the original vegetation to separate the thorn scrub and desert habitats from the dry deciduous forests and placed the latter in the Kathiarbar-Gir Dry Deciduous Forests [16]. The Kathiarbar-Gir Dry Deciduous Forests [16] overlap with a portion of Udvardy's (1975) Indus-Ganges monsoon forest biogeographic province.

Prepared by Gopal S. Rawat and Eric D. Wikramanayake

Ecoregion Number: 17

Ecoregion Name:	**Upper Gangetic Plains Moist Deciduous Forests**
Bioregion:	**Indian Subcontinent**
Biome:	**Tropical and Subtropical Moist Broadleaf Forests**
Political Units:	**India**
Ecoregion Size:	**262,600 km²**
Biological Distinctiveness:	**Bioregionally Outstanding**
Conservation Status:	**Critical**
Conservation Assessment:	**IV**

Many years ago, the Upper Gangetic Plains Moist Deciduous Forests [17] harbored impressive populations of tiger (*Panthera tigris*), greater-one horned rhinoceros (*Rhinoceros unicornis*), Asian elephants (*Elephas maximus*), gaur (*Bos gaurus*), and swamp deer (*Cervus duvaucelli*), to name but a few of the large vertebrates that used to roam here. Large hornbills made daily migrations from their feeding areas to mature groves, where they nested in cavities of tall trees and roosted for the night. Today, this natural biodiversity has been replaced by one of the densest human populations on Earth, and the fertile alluvial plains have been cleared and intensely cultivated. There is so little natural forest left, it is difficult even to assign a particular vegetation type to it with any certainty. The small patches of forests that are left suggest that much of the upper Gangetic Plains may have supported a tropical moist deciduous forest with sal (*Shorea robusta*) as a climax species.

Location and General Description

This ecoregion lies along one of Asia's largest rivers, the Ganges. Originating in the western Himalayas, the river flows east along most of the length of the long mountain range to eventually join with the Brahmaputra River and head south to the Bay of Bengal. This ecoregion extends through the upper reaches of the Ganges River, across the northern Indian states of Uttar Pradesh, Haryana, and Bihar.

The southwest monsoon originating in the Bay of Bengal provides the ecoregion's precipitation. Therefore, the eastern reaches receive more moisture, with a gradual trend towards drier conditions as one goes west. Annual rainfall averages less than 500 mm. Topographically, the ecoregion has little relief except for floodplain scarps and ravines carved out by gully erosion (Rodgers and Panwar 1988). The substrate consists of deep alluvial soils deposited over the eons by the Ganges River.

The original moist deciduous habitat probably was dominated by *Shorea robusta* that formed a top canopy reaching 25–35 m. Other associated species included *Terminalia tomentosa, Terminalia belerica, Lagerstroemia parviflora, Adina cordifolia, Dillenia pentagyna, Stereospermum suaveolens,* and *Ficus* spp. (Puri et al. 1989). Patches of mesophilous grasslands, or savanna ecosystems, with *Saccharum spontaneum, Saccharum narenga, Saccharum benghalense,* and *Vetiveria zizanioides* intersperse the forest lands (Puri et al. 1989), representing early seral stages maintained by fire, flood, and grazing by domestic livestock.

Biodiversity Features

Several centuries ago, when the habitat was intact, this ecoregion harbored a rich wildlife community that included tiger, greater one-horned rhinoceros, Asian elephant, wild water buffalo, swamp deer, sloth bear (*Melursus ursinus*), and several hornbill species (Rodgers and Panwar 1988). Many of these space-dependent species are gone from much of the ecoregion, victims of the relentless wave of habitat destruction that has swept through most of the ecoregion.

The seventy-nine known mammal species do not include any ecoregional endemic species. But there are some threatened mammals, including the tiger, Asian elephant, sloth bear, chousingha (*Tetracerus quadricornis*), and possibly small refuge populations of swamp deer in some habitat patches. With the exception of a few forested landscapes along the Himalayan foothills, habitat fragmentation is too far advanced throughout much of the ecoregion to support viable populations of the larger predators and herbivores such as tigers and Asian elephants that used to be common here. Nevertheless, the intact habitat landscapes and the protected areas within these forest fragments have been included in a high-priority (Level I) TCU (Dinerstein et al. 1997) that extends into the adjacent ecoregions to the north.

The bird fauna consists of about 290 species. There are no ecoregional endemic species, although the Indian bustard (*Ardeotis nigriceps*) and lesser florican (*Eupodotis indica*) are globally threatened species. The Indian grey hornbill (*Ocyceros birostris*) and Oriental pied-hornbill (*Anthracoceros albirostris*) need mature habitat for nesting and can be used as focal species for conservation management.

The Ganges River supports a population of freshwater dolphins (*Platanista gangetica*), and the associated wetlands support a rich and diverse waterfowl community that includes many migrant bird species as well as the mugger crocodile (*Crocodylus palustris*) and Gangetic gharial (*Gavialis gangeticus*).

Status and Threats

Current Status

More than 95 percent of this vast ecoregion has been degraded or converted into agriculture and settlement areas by the dense human population that has settled here for thousands of years. Only one large block of habitat now remains, running along the Himalayan foothills in Uttar Pradesh and including Rajaji, Corbett, and Dudwa national parks.

The nine protected areas cover only 2,050 km^2, amounting to less than 1 percent of the ecoregion area (table J17.1). Among these, Rajaji and Corbett are important tiger reserves. Whereas Rajaji is the largest reserve completely within the ecoregion, Corbett (1,300 km^2) is a larger reserve that also extends into the adjacent ecoregion. None of the other reserves are more than 300 km^2 in extent, and the average size of the protected areas represented within this ecoregion is only 227 km^2.

Nevertheless, the landscape that extends from Rajaji National Park to Corbett National Park is one of the most important conservation areas for tigers and elephants. This area is believed to have about 750 elephants and about 140 tigers (Singh 1986, in Johnsingh et al. 1990). With the inclusion of Corbett National Park and Sonanadi Wildlife Sanctuary in Rajaji National Park, this potential conservation landscape can be as large as 2,500 km^2 (Johnsingh et al. 1990).

Types and Severity of Threats

Impacts from human activities have devastated the natural habitat in this river plain for many years and continue to threaten the small patches of remaining forests and the biodiversity within them. Rodgers and Panwar (1988), subsequently updated by Rodgers et al. (2000), have highlighted the conservation gaps in this region with specific proposals for additional protection.

But habitat fragmentation continues unabated, severing important habitat links that permit movements of large mammals such as elephants and tigers (Johnsingh et al. 1990; Johnsingh and Williams 1999) across the landscape and between protected areas. Road construction erodes the steep, fragile slopes and is invariably followed by human settlements because of the easier access. Grazing of cattle and other livestock is widespread. The threats to the Chila-Motichur corridor, critical for maintaining genetic exchange among elephants and tigers within their range in northwestern India, exemplifies the prevalent widespread threats. The habitat within this corridor has been degraded by settlements, overgrazing by domestic livestock, and invasions by exotic weeds (Johnsingh et al. 1990).

Priority Conservation Actions

Because most of the ecoregion is densely settled and cultivated, conservation actions here are essentially salvage operations to identify and protect whatever habitat is left. However, because this ecoregion contains large, important populations of two species that epitomize space-dependent megavertebrates, every effort should be made to protect and restore some of the critical corridors that link existing protected areas, as identified by Rodgers and Panwar (1988) and Johnsingh et al. (1990).

Short-Term Conservation Actions (1–5 Years)

- Stop encroachment by agricultural expansion and clearing of remaining forests.
- Protect and maintain the forests between the protected areas (Corbett, Rajaji, and possibly Dudwa) to serve as a dispersal corridor for large mammals, including tigers and elephants (see Johnsingh et al. 1990). Acquire land from the evacuees of the Tehri Dam and army on a priority basis. Totally protect from forestry operations and development the strip where the Malin River and Kotdwar-Amsod corridors are situated.
- Stop large-scale conversions of remaining forests into agricultural lands by new settlers.
- Prevent overgrazing by domestic livestock, especially within protected areas and their buffer zones. Ban cattle grazing and other human activities from the islands on the Ganges River. Restore the degraded habitat with *Acacia catechu* and *Dalbergia sissoo* (Johnsingh et al. 1990).

Longer-Term Conservation Actions (5–20 Years)

- Identify and develop conservation landscapes that can support viable populations of elephants and tigers over the long term.

Focal Species for Conservation Action

Habitat specialists: hornbills (as indicators of forest quality)

Area-sensitive species: tiger, Asian elephant

Top predator species: tiger

Other: Sloth bear, chousingha, Indian bustard, lesser florican, waterfowl community, mugger crocodile, gharial

Selected Conservation Contacts

Chief Wildlife Warden, Uttar Pradesh

Director, Corbett Tiger Reserve

Table J17.1 WCMC (1997) Protected Areas That Overlap with the Ecoregion.

Protected Area	Area (km^2)	IUCN Category
Corbett [31]	260	II
National Chambal [16]	130	PRO
Rajaji	790	II
Hastinapur	20	IV
Karera Great Indian Bustard	200	IV
Ranipur	230	IV
Ken Gharial	80	IV
Kishanpur	70	IV
Sohagabarwa [25]	270	IV
Total	2,050	

Ecoregion numbers of protected areas that overlap with additional ecoregions are listed in brackets.

Directors of Dudwa and Rajaji National Parks
Forest Research Institute, Dehra Dun, India
Wildlife Institute of India, Dehra Dun, India
WWF-India

Justification for Ecoregion Delineation

In a previous analysis, Rodgers and Panwar (1988) identified conservation units in India and placed the moist deciduous forests along the upper Gangetic valley in the Upper Gangetic Plains biotic province (7A). We retained this unit and represented it with the Upper Gangetic Plains Moist Deciduous Forests [17]. The Upper Gangetic Plains Moist Deciduous Forests [17] lie within Udvardy's Indus-Ganges monsoon forest biogeographic province.

Prepared by Gopal S. Rawat and Eric D. Wikramanayake

Ecoregion Number:	**18**
Ecoregion Name:	**Chhota-Nagpur Dry Deciduous Forests**
Bioregion:	**Indian Subcontinent**
Biome:	**Tropical and Subtropical Dry Broadleaf Forests**
Political Units:	**India**
Ecoregion Size:	**122,100 km²**
Biological Distinctiveness:	**Regionally Outstanding**
Conservation Status:	**Endangered**
Conservation Assessment:	**II**

The Chhota-Nagpur Dry Deciduous Forests [18] still harbor large populations of Asia's largest predator and largest herbivore, the tiger (*Panthera tigris*) and the Asian elephant (*Elephas maximus*), respectively. Both are still able to roam and live within the large habitat blocks of this ecoregion, a rare phenomenon in this bioregion. The Chhota-Nagpur Plateau also has a flora and fauna that are distinct from the adjacent areas (Rodgers and Panwar 1988), with several pockets of endemic plants. In the geological past, this plateau formed a link between Satpura Hill Ranges and eastern Himalaya that allowed species exchanges between these ranges (Hora 1949).

Location and General Description

The ecoregion represents the dry deciduous forests on the Chhota-Nagpur Plateau and lies between the moist deciduous forests of the Eastern Ghats and Satpura Range and of the lower reaches of the Gangetic Plains. It extends across the eastern Indian states of Bihar, Madhya Pradesh, and West Bengal.

The Gondwana substrates attest to the plateau's ancient origin. It is part of the Deccan Plate, which broke free from the southern continent during the Cretaceous to embark on a 50-million-year journey that was violently interrupted by the northern Eurasian continent. The northeastern part of the Deccan Plateau, where this ecoregion sits, was the first area of contact with Eurasia.

The plateau receives less rainfall than the adjacent ecoregions that support moist deciduous forests. Therefore, the vegetation is drier than in the adjacent ecoregions. The dry deciduous forests typically are composed of three stories, with an upper canopy reaching 15–25 m, a high understory at 10–15 m, and an undergrowth at about 3–5 m. The vegetation is characterized by *Shorea robusta*, usually in association with *Anogeissus latifolia*, *Terminalia alata*, *Lagerstroemia parviflora*, *Pterocarpus marsupium*, *Aegle marmelos*, *Syzygium operculatum*, *Symplocus racemosa*, and *Croton oblongifolius* (Puri et al. 1989; Davis et al. 1995). Lianas are common in denser forests.

A dry deciduous scrub that grows to about 3–6 m in height is a common habitat type in this ecoregion. This scrub includes bamboo and shrubs such as *Holarrhena* and *Dodonaea* (Puri et al. 1989). At higher altitudes, there are shola-type forests characterized by *Phoenix robusta*, *Pterospermum acerifolium*, and *Clematis nutans* (WWF and IUCN 1995). The ecoregion also includes patches of moist deciduous forests and swampy areas, with several interesting plant species (*Syzygium cumini*, *Manilkara hexandra*, *Ficus* spp., *Mallotus philippinensis*) that are typical of moist deciduous forests.

Biodiversity Features

Although *Aglaia haselettiana*, *Carum villosum*, and *Pycnocyclea glauca* are endemic, endangered plant species in this ecoregion, *Diospyros melanoxylon*, *Madhuca longifolia*, *Butea monosperma*, and *Shleichera oleosa*, are more economically useful species.

The ecoregion's fauna is neither exceptionally species-rich nor distinctive. The ecoregion's mammal fauna includes seventy-seven species, but none are endemic to it. Nevertheless, several of India's large charismatic vertebrates, including several that are threatened such as the tiger, Asian elephant, wild dog (*Cuon alpinus*), sloth bear (*Melursus ursinus*), chousingha (*Tetracerus quadricornis*), blackbuck (*Antilope cervicapra*), and chinkara (*Gazella bennettii*) (Hilton-Taylor 2000), are found here. Importantly, unlike in many other ecoregions in this bioregion, there still are large areas of habitat capable of supporting viable populations of these species. These habitat blocks are included in two Level I TCUs (Dinerstein et al. 1997) that extend into the ecoregion.

There are almost 280 bird species, but none are endemic to the ecoregion. However, it harbors several species that are of conservation importance, including the globally threatened lesser florican (*Eupodotis indica*). Some of the hornbills, notably the Indian grey hornbill (*Ocyceros birostris*) and Oriental pied-hornbill (*Anthracoceros albirostris*), which need mature trees for nesting, should be focal species for conservation management.

Status and Threats

Current Status

More than half of the natural forests in this ecoregion have been cleared; however, extensive areas of intact forest remain in the northern areas. The thirteen protected areas in the ecoregion cover more than 6,700 km², representing about 6 percent of the ecoregion (table J18.1). The average size of a protected area exceeds 500 km², and two—Sanjay and Palamau—are larger than 1,000 km². These large protected areas provide the essential core areas for the large vertebrates in this ecoregion. However, these reserves must be included in larger conservation landscapes for more effective long-term conservation of these species (Wikramanayake et al. 1999).

Types and Severity of Threats

Like many of the other ecoregions in this bioregion, livestock grazing has greatly affected the vegetation (Puri et al. 1989). However, extensive areas that are being mined for iron ore and open-pit coal mines pose very significant threats to the habitat, especially to critical migration paths of elephants and dispersal routes for tigers.

Priority Conservation Actions

The ecoregion's large habitat blocks have been included in a high-priority TCU (Wikramanayake et al. 1999). A landscape conservation plan for these units based on the tiger population and the associated predator-prey relationships is needed. The migration routes of elephant herds are being disrupted by habitat fragmentation and degradation; therefore, landscape conservation efforts are necessary to maintain these habitat links. Rodgers and Panwar (1988) identified conservation gaps and needs for this area, which have been updated by Rodgers et al. (2000). Here we offer some short and longer-term conservation actions.

Short-Term Conservation Actions (1–5 Years)

- Assess tiger population, habitat condition, and prey base in the TCUs.
- Identify and rehabilitate the degraded mine sites that are critical for maintaining habitat linkages between core protected areas, are in important migration and dispersal paths of elephants and tigers, contribute to the ecological integrity of conservation landscapes, and harbor endemic flora.
- Adopt rotational grazing practices for domestic livestock to prevent excessive habitat degradation.
- Survey to identify botanical hotspots, including areas of fossil-bearing rocks. Develop site-specific conservation plans for these hotspots.

Longer-Term Conservation Actions (5–20 Years)

- Strengthen the existing protected area network, using the tiger and elephant as focal species. The TCUs that overlap with this ecoregion's intact habitat should be assessed for adequacy of the protected area system for long-term tiger conservation. Increase effectiveness of management and protection within the existing and any additional protected areas. Plan and promote conservation-compatible land use outside the protected areas. Monitor tiger populations and their prey base.
- Address elephant conservation needs, including maintaining—and where necessary restoring—migration paths.
- Address human-elephant conflicts.

Focal Species for Conservation Action

Habitat specialists: Indian grey hornbill, Oriental pied hornbill

Wide-ranging species: tiger, Asian elephant

Top predator species: tiger

Other: wild dog, sloth bear, chousingha, blackbuck, chinkara, lesser florican

Selected Conservation Contacts

Chief Wildlife Warden, Bihar
District Forest Officer, Porahat Forest Division, India
District Forest Officer, Saranda Forest Division, India
Wildlife Institute of India, Dehra Dun, India
WWF-India

Justification for Ecoregion Delineation

In a previous analysis of conservation units, Rodgers and Panwar (1988), and subsequently MacKinnon (1997), divided the Deccan Peninsula into five biotic provinces and sub-biounits, respectively. But each of these units contains several vegetation types that extend over large areas. In this analysis, we represent these large ecosystems in separate ecoregions (in keeping with our definition of an ecoregion as an ecosystem of regional extent). Therefore, we modified the boundaries of Rodgers and Panwar's biotic province 6D (which is the same as MacKinnon's sub-biounit 16b) to include only the large expanse of dry deciduous forest in the Chhota-Nagpur Dry Deciduous Forests [18]. We used MacKinnon's (1997) digital map of the original vegetation to define the boundaries of this ecoregion.

Prepared by Gopal S. Rawat and Eric D. Wikramanayake

TABLE J18.1 WCMC (1997) Protected Areas That Overlap with the Ecoregion.

Protected Area	Area (km²)	IUCN Category
Sanjay [20]	1,020	II
Lawalang	410	IV
Koderma	180	IV
Hazaribagh	450	IV
Palamau	1,330	II
Topchanchi	40	IV
Ramnabagan	150	IV
Dalma	630	IV
North Simlipal	420	II
Bhimbandh	910	IV
Gautam Budha [10]	110	IV
Tamor Pingla	600	IV
Semarsot	470	IV
Total	6,720	

Ecoregion numbers of protected areas that overlap with additional ecoregions are listed in brackets.

Ecoregion Number:	**19**
Ecoregion Name:	**Northern Dry Deciduous Forests**
Bioregion:	**Indian Subcontinent**
Biome:	**Tropical and Subtropical Dry Broadleaf Forests**
Political Units:	**India**
Ecoregion Size:	**58,200 km²**
Biological Distinctiveness:	**Bioregionally Outstanding**
Conservation Status:	**Endangered**
Conservation Assessment:	**IV**

The Northern Dry Deciduous Forests [19] ecoregion is neither exceptionally species-rich nor high in numbers of endemic species. But it does harbor several large vertebrates, including Asia's largest and most charismatic carnivore, the tiger (*Panthera tigris*). This ecoregion also represents the northernmost extent of dry deciduous forests in India.

Location and General Description

The ecoregion extends across the Indian states of Bihar, Orissa, and Madhya Pradesh. It represents a north-south-directed island of dry deciduous forests in the rainshadow of the Eastern Ghats Mountain Range and is completely surrounded by the Eastern Highlands Moist Deciduous Forests [12]. As with the rest of the Deccan Plateau, the ecoregion's geological history dates back to the Cretaceous, when it was part of southern Gondwanaland.

The forest type in the ecoregion corresponds to the *Shorea-Buchanania-Cleistanthus* and *Shorea-Cleistanthus-Croton* vegetation mapped by Gaussen et al. (1973). Most of these forests are open scrub influenced by human activities. The original *sal* (*Shorea robusta*)–dominated, multistoried vegetation has been replaced by teak (*Tectona grandis*), which favors drier conditions. In many areas, intensive livestock grazing, fire, and nontimber forest product harvest have converted the habitat to scrub and savanna woodland.

The vegetation is made up of associations of *Anogeissus latifolia*, *Dalbergia latifolia*, *Pterocarpus marsupium*, *Stereospermum suaveolens*, *Spondias pinnata*, *Cleistanthus collinus*, *Acacia lenticularis*, *Flacourtia indica*, *Boswellia serrata*, *Butea monosperma*, *Sterculia urens*, *Cochlospermum religiosum*, and *Euphorbia nivulia* (Puri et al. 1989; WWF and IUCN 1995). Gregarious patches of *Dendrocalamus strictus* tend to occur in moister areas. *Cleistanthus collinus* can release substances toxic to other species; therefore, monospecific stands can occur in places.

Biodiversity Features

Like many of the Deccan Plateau dry forest ecoregions, this region does not harbor large numbers of endemic species, nor is it exceptionally rich in biodiversity. The known mammal fauna consists of sixty-eight species. There are no ecoregional endemic species, but the threatened species include the tiger, wild dog (*Cuon alpinus*), sloth bear (*Melursus ursinus*), and chousingha (*Tetracerus quadricornis*) (Hilton-Taylor 2000).

The 261 bird species in the ecoregion do not include endemic species. The Indian grey hornbill (*Ocyceros birostris*) and Oriental pied-hornbill (*Anthracoceros albirostris*) need tall, mature trees for nesting and can be used as focal species for conservation management.

Status and Threats

Current Status

More than three-fourths of the ecoregion's natural habitat has been cleared or degraded. But a few mid-sized (i.e., more than 2,000 km²) blocks of intact habitat still remain. The four protected areas in the ecoregion (table J19.1) amount to just over 1,400 km², representing about 2.5 percent of the ecoregion area. But of these, only one exceeds 500 km².

Types and Severity of Threats

Land clearing and degradation remain the primary threat to the remaining habitat. Fires are regularly set to encourage grazing lands for livestock. Extensive poaching and collection of nontimber forest products by the tribal communities are also a serious concern. But degradation threats from industry and timber companies are also high.

The large tribal populations in these areas are also shifting from subsistence to more materially demanding lifestyles. The growing populations and economic aspirations and shrinking resources result in conflicts with conservation interests and authorities. Until these problems are addressed effectively, the conflicts will increase.

The Naxalite conflicts in Andhra Pradesh and at the junction of Madhya Pradesh–Maharashtra–Andhra Pradesh prevent effective government management and protection of conservation areas (especially protected areas and reserve forests). The conflicts are funded by poaching of rhinoceroses, tigers, and elephants.

Priority Conservation Actions

Although most of the habitat is degraded or converted, the remaining patches of natural habitat contain nontimber forest products that are used by the tribal groups. Conservation actions should focus on sustainable use of these products. Here we offer some short and longer-term conservation actions.

Short-Term Conservation Actions (1–5 Years)

- Carry out biodiversity surveys of Kanger Valley on which to base a conservation plan for one of the few remaining protected forest patches in the ecoregion.
- Assess cultural natural resource use to help evaluate their impacts on natural biodiversity.
- Monitor and study the tribal collection of nontimber forest products. These include bamboo shoots, *Bauhinia* leaves,

TABLE J19.1 WCMC (1997) Protected Areas That Overlap with the Ecoregion.

Protected Area	Area (km²)	IUCN Category
Sunabeda	600	IV
Kanger Ghati	230	II
Debrigarh	340	IV
Gomarda	290	IV
Total	1,460	

Ecoregion numbers of protected areas that overlap with additional ecoregions are listed in brackets.

silk cocoon, wild mango, *Terminalia chebula*, tendu (*Diospyros melanoxylon*), mahua (*Madhuca latifolia*), char fruits (*Buchanania lanzan*), mushrooms, honey, sal seeds, roots, and tubers.

Longer-Term Conservation Actions (5–20 Years)

- Monitor programs for sustainable natural resource use.
- Restore critical habitat to create linkages and conservation landscapes with the surrounding Eastern Highlands Moist Deciduous Forests [12]. Tigers and hornbills can be focal species for this conservation plan.
- Address resource use conflicts by tribal peoples through creation of long-term land-use and natural resource management plans.

Focal Species for Conservation Action

Habitat specialists: Indian grey hornbill, Oriental pied-hornbill

Area-sensitive species: tiger

Top predator species: tiger, leopard

Other: sloth bear, chousingha, lesser florican, nontimber forest products (NTFPs) (*Bauhinia* spp., *Mangifera* spp., *Terminalia chebula*, *Diospyros melanoxylon*, *Madhuca latifolia*, *Buchanania lanzan*)

Selected Conservation Contacts

Bombay Natural History Society

Chief Wildlife Warden, Madhya Pradesh

District Forest Officer, Kanger Valley Madhya Pradesh

Forest Department, Bihar

Forest Department, Orissa

Wildlife Institute of India

WWF-India

Justification for Ecoregion Delineation

In a previous analysis of conservation units, Rodgers and Panwar (1988) divided the Deccan Peninsula into five biotic provinces. The Eastern Highlands (6C) and Chhota-Nagpur (6D) biotic provinces of Rodgers and Panwar contain this large patch of dry deciduous forests, which is surrounded by an extensive area of moist deciduous forests (MacKinnon 1997). In keeping with our definition of an ecoregion (i.e., an ecosystem of regional extent) and following our rules for ecoregion delineation (represent distinct vegetation types of regional extent in separate ecoregions), we extracted the dry deciduous forests (as mapped by MacKinnon 1997) from the surrounding moist deciduous forests and represented them with the Northern Dry Deciduous Forests [19].

Prepared by Gopal S. Rawat, Ajay Desai, Hema Somanathan, and Eric D. Wikramanayake

Ecoregion Number:	**20**
Ecoregion Name:	**Narmada Valley Dry Deciduous Forests**
Bioregion:	**Indian Subcontinent**
Biome:	**Tropical and Subtropical Dry Broadleaf Forests**
Political Units:	**India**
Ecoregion Size:	**169,500 km^2**
Biological Distinctiveness:	**Bioregionally Outstanding**
Conservation Status:	**Endangered**
Conservation Assessment:	**IV**

The Narmada Valley Dry Deciduous Forests [20] are neither exceptionally species-rich nor high in numbers of endemic species. But this ecoregion still retains more than a third of its natural habitat in several large blocks that exceed 5,000 km^2 in area. In this bioregion, such large dry forest habitats present rare and important opportunities to conserve Asia's largest carnivore, the tiger (*Panthera tigris*).

Location and General Description

The ecoregion represents the dry deciduous forests along the Narmada River Valley and the flanking Vindhya Mountain Range and the western part of the Satpura Mountain Range in the central Indian states of Madhya Pradesh and Maharashtra. These hill ranges, rising to more than 1,300 m, mark the northern boundary of the Indian Peninsula (Kendrick 1989). The Deccan Plateau itself—and thus the ecoregion—traces its geological roots back to the ancient circumpolar continent Gondwanaland. Hora (1949) hypothesized that the Satpura Range was a dispersal bridge that allowed species exchanges between the eastern Himalayas and the Western Ghats. The presence of fossils of species such as *Anisopteris*, *Cynometra*, *Dipterocarpus*, *Dryobalanops*, *Gluta*, *Hopea*, and *Mesua* suggests that evergreen moist forests covered this area during the Miocene (Meher-Homji 1989).

The seven- to eight-month dry season is relieved by the southwest monsoon, which brings 1,200–1,500 mm of annual rainfall. The vegetation is influenced by this seasonality. Three stories—an upper canopy at 15–25 m, a 10- to 15-m understory, and 3–4 m undergrowth—characterize the forests. Teak (*Tectona grandis*) dominates the vegetation and is associated with *Diospyros melanoxylon*, *Anogeissus latifolia*, *Lagerstroemia parviflora*, *Terminalia tomentosa*, *Lannea coromandelica*, *Hardwickia binata*, and *Boswellia serata* (Champion and Seth 1968). Riparian habitats with species such as *Terminalia arjuna*, *Syzygium cumini*, *Syzygium heyneanum*, *Salix tetrasperma*, *Homonoia riparia*, and *Vitex negundo* create moist forest corridors.

Biodiversity Features

Although it is not exceptional in terms of endemism or diversity, the ecoregion still retains important habitat for many of the Indian Subcontinent's large animals, such as the tiger, gaur (*Bos gaurus*), wild dog (*Cuon alpinus*), sloth bear (*Melursus ursinus*), chousingha (*Tetracerus quadricornis*), and blackbuck (*Antilope cervicapra*). Throughout most of their ranges, these larger vertebrate species are being increasingly confined to small forest fragments that do not offer much hope for long-term survival of declining populations. Therefore, it is important to safeguard large habitat areas where they exist. Most of the large blocks of remaining habitat have been included in TCUs (Dinerstein et al. 1997). These habitat landscapes present the best opportunities for long-term conservation of viable tiger populations. However, very little is known of the habitat integrity and the status of the prey populations in these forests. Surveys have been recommended to determine their importance and potential contribution toward a regional tiger conservation strategy.

The mammal fauna in the ecoregion includes seventy-six species. Although none of them are endemic, there are several threatened species, including the tiger, gaur, wild dog, sloth bear, chousingha, and blackbuck (Hilton-Taylor 2000).

None of the 276 bird species in this ecoregion are endemic. But the bird fauna includes the globally threatened lesser florican (*Eupodotis indica*) and the endangered Indian bustard (*Ardeotis nigriceps*) (Hilton-Taylor 2000).

Status and Threats

Current Status

Nearly two-thirds of the natural forests of this ecoregion have been cleared, but the remaining habitat includes several large blocks that cover extensive areas, especially along the Satpura and Vindhya Ranges. The seventeen protected areas cover more than 7,500 km^2, or almost 5 percent of the ecoregion's area (table J20.1). Two of the protected areas, Melghat and Noradehi, exceed 1,300 km^2.

Types and Severity of Threats

The large patches of forests are still greatly threatened by ongoing forest clearing and conversion. But the threats from a series of dams on the Narmada River are even more serious than the small-scale degradation threats. These dams will flood critical habitat, and they will also displace a large number of tribal and local communities into adjacent intact forests (Anonymous 1994).

The conservation status of this ecoregion was changed from vulnerable to endangered because of threats from the dense human population to the intact forest blocks. However, impending threats from the hydropower schemes may warrant elevation to critical status.

As the tribal populations shift from a subsistence lifestyle to a more material one and as the populations continue to increase rapidly, conflicts are beginning to occur with conservation interests and authorities. These problems must be addressed in a timely manner.

Priority Conservation Actions

A landscape conservation plan for the remaining large habitat areas should be the focus of conservation actions in this ecoregion. These forests have been identified as important tiger conservation areas (Dinerstein et al. 1997); therefore, tigers and their ecological needs should be the focus of such a conservation plan.

Short-Term Conservation Actions (1–5 Years)

- Implement the environmental mitigation plans for the ongoing hydroelectric projects, as suggested by WII (1994).
- Protect remaining habitat patches from possible settlement by people displaced by the hydroelectric project.
- Survey the large habitat blocks for tigers and their ecological needs (e.g., prey and seasonal water availability, disturbance levels, poaching levels, core refugia).

Longer-Term Conservation Actions (5–20 Years)

- Strengthen the existing protected area network, using the tiger as a focal species. Focus conservation efforts in the high-priority TCU on a landscape conservation strategy that includes establishing additional core protected areas, strengthening protection and management of existing protected areas, and promoting conservation-compatible land uses outside strict protected areas.
- Create long-term land-use and natural resource use plans to address the needs of the tribal people.
- Monitor the tiger populations and their prey base.

Focal Species for Conservation Action

Area-sensitive species: tiger

Top predator species: tiger

Other: wild dog, hyena, sloth bear, nilgai, smooth Indian otter, chousingha, gaur, blackbuck, chinkara, Indian bustard, lesser florican, *Crocodylus palustris*

Selected Conservation Contacts

Bombay Natural History Society

Forest Department, Madhya Pradesh

Medha Patekar, leader of Narmada Bachao Andolan

Narmada Development Authority, Bhopal

Wildlife Institute of India

WWF-India

TABLE J20.1 WCMC (1997) Protected Areas That Overlap with the Ecoregion.

Protected Area	Area (km^2)	IUCN Category
Panna	820	II
Noradehi	1,380	IV
Singhori	220	IV
Ratapani	490	IV
Kheoni	80	IV
Son Gharial	210	IV
Bagdara	540	IV
Sanjay [18]	690	II
Sanjay (Dubri)	350	IV
Bandhavgarh	360	II
Panpatha	300	IV
Sardarpur	120	IV
Aner Dam	70	IV
Melghat	1,490	IV
Yawal	100	IV
Gugamal	350	II
Bhimashankar [14]	30	IV
Total	7,600	

Ecoregion numbers of protected areas that overlap with additional ecoregions are listed in brackets.

Justification for Ecoregion Delineation

In a previous analysis of conservation units, Rodgers and Panwar (1988) divided the Deccan Peninsula into five biotic provinces. This ecoregion largely corresponds to the Rodgers and Panwar biotic province 6E, the Central Highlands, but we redrew the boundary to exclude the small area of moist deciduous forest represented in the biotic province as shown in MacKinnon's (1997) map of the original vegetation. This modification of Rodgers and Panwar (1988) is to conform to our rules for delineating ecoregions, that is, to represent distinct habitat types of regional extent. The dry deciduous forests therefore were represented by the Narmada Valley Dry Deciduous Forests [20].

Prepared by Gopal S. Rawat and Eric D. Wikramanayake

Ecoregion Number:	**21**
Ecoregion Name:	**Central Deccan Plateau Dry Deciduous Forests**
Bioregion:	**Indian Subcontinent**
Biome:	**Tropical and Subtropical Dry Broadleaf Forests**
Political Units:	**India**
Ecoregion Size:	**239,400 km^2**
Biological Distinctiveness:	**Bioregionally Outstanding**
Conservation Status:	**Endangered**
Conservation Assessment:	**IV**

The Central Deccan Plateau Dry Deciduous Forests [21] make up a large ecoregion that is neither exceptionally species-rich nor high in numbers of endemic species. But the ecoregion still contains about 20 percent of its natural habitat as several large blocks, some of which exceed 5,000 km^2. In a region characterized by a high human population density and the presence of several large vertebrate species, the presence of large natural habitat areas is both unusual and important, respectively. These large habitat blocks have been recognized as high-priority landscapes for a long-term tiger conservation strategy (Wikramanayake et al. 1999); therefore, the ecoregion makes an important contribution to a regional tiger conservation strategy.

Location and General Description

This ecoregion represents the *Hardwickia*-dominated woodlands in the central Deccan Plateau, a vegetation type that is distinct from the teak (*Tectona grandis*) or sal (*Shorea robusta*) dominated dry forests that cover most of the Deccan Plateau. The ecoregion extends across the central Indian states of Madhya Pradesh, Maharashtra, Karnataka, and Andhra Pradesh. It includes the western extent of the Satpura Range of hills within the northern extent. The Godavari River, which originates in the Western Ghats, crosses the ecoregion as it traverses the Deccan Plateau.

The Deccan Plateau itself traces its geological history back to Gondwanaland. A Miocene-era fossil flora of evergreen rain forest genera including *Anisopteris*, *Cynometra*, *Dipterocarpus*, *Dryobalanops*, *Gluta*, *Hopea*, and *Mesua* reveals the past moist climatic history (Meher-Homji 1989).

TABLE J21.1. Endemic and Near-Endemic Bird Species.

Family	Common Name	Species Name
Glareolidae	Jerdon's courser	*Rhinoptilus bitorquatus*

An asterisk signifies that the species' range is limited to this ecoregion.

Structurally, the dry forests in this ecoregion have an upper canopy at 15–25 m and an understory at 10–15 m. Lianas drape the trees in mature forests, but the undergrowth is sparse (Puri at al. 1989). The characteristic tree association is *Hardwickia binata–Albizia amara* woodland with *Tectona grandis*, *Boswellia serrata*, *Lannea coromandelica*, *Anogeissus latifolia*, *Albizia lebbek*, *Lagerstroemia parvifolia*, *Diospyros tomentosa*, and *Acacia catechu* in the northern areas and *Pterocarpus santalinus*, *P. marsupium*, *Chyloroxylon swietenia*, *Terminalia chebula*, *T. tomentosa*, *Albizia lebbek*, and *Dalbergia latifolia* in the southern areas of the ecoregion (Puri et al. 1989). The Nagarjunasagar-Srisailam Tiger Reserve in Andhra Pradesh represents the typical habitat in this ecoregion. Several sacred groves containing evergreen forests in the state of Andhra Pradesh make an important contribution to conservation and to the ecoregion's diversity.

Biodiversity Features

Although not exceptional in terms of endemism and diversity, this ecoregion harbors several of India's large, threatened vertebrates such as the tiger (*Panthera tigris*), wild buffalo (*Bubalus arnee*), wild dog (*Cuon alpinus*), sloth bear (*Melursus ursinus*), chousingha (*Tetracerus quadricornis*), gaur (*Bos gaurus*), blackbuck (*Antilope cervicapra*), and chinkara (*Gazella bennettii*) (Hilton-Taylor 2000). Significantly, the large habitat blocks provide an opportunity to conserve these wide-ranging species whose populations are fast declining throughout most of their ranges because of habitat loss and hunting.

The known mammal fauna in the ecoregion includes eighty-two species, but none are endemic to the ecoregion. In addition to the larger, threatened mammal species mentioned earlier, the ecoregion also harbors the threatened Malabar squirrel (*Ratufa indica*) (Hilton-Taylor 2000).

The bird fauna is richer, with almost 300 species, of which one, the globally threatened Jerdon's courser (*Rhinoptilus bitorquatus*), is a near-endemic species (table J21.1). This species was rediscovered in 1986, after the last record in 1900 (Grimmet et al. 1998). The small distribution range of the only known population of this rare species extends across this ecoregion and the neighboring Deccan Thorn Scrub Forests [23].

Status and Threats

Current Status

About 80 percent of the natural habitat in this ecoregion has already been lost, but the remaining habitat includes several blocks that exceed 5,000 km^2. These large habitat blocks along the southern and eastern boundaries of the ecoregion have been the basis for the high-priority TCU (Wikramanayake at al. 1999).

TABLE J21.2 WCMC (1997) Protected Areas That Overlap with the Ecoregion.

Protected Area	Area (km²)	IUCN Category
Katepurna	420	IV
Bor	80	IV
Navegaon	170	II
Gautala Autramghat	260	IV
Tadoba	110	II
Andhari	480	IV
Tipeshwar	160	PRO
Painganga	130	IV
Chaprala	130	IV
Pocharam	120	IV
Manjira	30	IV
Nelapattu	200	IV
Nagarjunasagar	3,568	PRO
Gundla Brahmeswaram	120	IV
Nagzira	150	IV
Pench	440	II
Kolleru [12]	220	IV
Total	6,788	

Ecoregion numbers of protected areas that overlap with additional ecoregions are listed in brackets.

The sixteen protected areas in this ecoregion cover almost 3 percent of its area. The Nagarjunasagar Tiger Reserve, at more than 3,500 km², is one of the largest and most important protected areas in this bioregion. Recent ecodevelopment and habitat restoration efforts in this tiger reserve have proved successful in bringing back the dwindling tiger population.

Types and Severity of Threats

The remaining natural habitat is now under severe threat from conversion to cash crop plantations, excessive fuelwood collection, and overgrazing by cattle. The habitat loss results in a decreasing prey base for tigers that then turn to the livestock that are grazed around and within the forests. Retaliation by the local people against these depredations has affected tiger populations. But although these degradation threats nibble away at the intact forests, large hydroelectric projects present more severe, high-impact threats (Davis et al. 1995).

As the tribal people in the area shift from a subsistence lifestyle to more material lifestyles and the populations grow, there are inevitable conflicts with diminishing resources. These conflicts must be addressed early with resource and land-use plans.

The ecoregion's conservation status was changed from vulnerable to endangered because of the projected impacts of the human population on the remaining forest blocks

The Naxalite conflict in northeast India is also being funded by proceeds from rhinoceros, tiger, and elephant poaching.

Priority Conservation Actions

One of the most important conservation actions in this ecoregion is to ensure that the recovering tiger population is well protected and conserved over the long term. The large habitat blocks have been included in a high-priority TCU (Wikramanayake at al. 1999), and the large Nagarjunasagar-Srisailam Tiger Reserve has been established to conserve the tiger population. These should be important foci for conservation actions. The population of Jerdon's courser should also be protected. Rodgers and Panwar (1988) presented an analysis of conservation gaps and recommendations. Here we offer some short- and longer-term conservation actions.

Short-Term Conservation Actions (1–5 Years)

- Assess tiger population, habitat condition, and prey base in the TCU.
- Increase protection and management of the Nagarjunasagar-Srisailam Tiger Reserve.
- Conduct ecological studies and status surveys on the Jerdon's courser.
- Mitigate human-wildlife conflicts. Livestock depredations by tigers have resulted in escalation of retaliation against tigers. Sloth bear-human conflicts have also greatly increased in recent years.
- Control forested land conversion into other land uses.
- Give special conservation status to the fossil beds of Nawegaon area in Madhya Pradesh to preserve the fossils (Rodgers and Panwar 1988)

Longer-Term Conservation Actions (5–20 Years)

- Strengthen the existing protected area network, using the tiger as a focal species. Conservation efforts in the high-priority TCU should focus on a landscape conservation strategy that includes establishing additional core protected areas, strengthening protection and management of existing protected areas, and promoting conservation compatible land uses outside strict protected areas.
- Create long-term land and resource use plans to address the needs of tribal people.
- Monitor the tiger populations and their prey base.

Focal Species for Conservation Action

Endangered endemic species: Jerdon's courser

Area-sensitive species: Asian elephant, tiger

Top predator species: tiger

Other: wild dog, wild buffalo, chousingha, gaur, blackbuck, chinkara, Malabar squirrel

Selected Conservation Contacts

Bombay Natural History Society
Centre for Ecological Sciences, Bangalore
Chief Wildlife Wardens, Madhya Pradesh, Maharashtra
Field director, Kanha National Park
Field director, Nagarjunasagar-Srisailam Tiger Reserve
French Institute of Pondicherry
WWF-India

Justification for Ecoregion Delineation

In a previous analysis of conservation units, Rodgers and Panwar (1988), and subsequently MacKinnon (1997), divided the Deccan Peninsula into five biotic provinces.

The *Hardwickia*-dominated woodlands represented by this ecoregion extend across the Deccan Plateau South (6A) and Central Plateau (6B) biotic provinces identified by Rodgers and Panwar (1988). However, both provinces also include other vegetation types. In keeping with our definition of an ecoregion (i.e., an ecosystem of regional extent) and following our rules for ecoregion delineation (representing distinct vegetation types of regional extent in separate ecoregions), we placed the *Hardwickia* dry forests into the Central Deccan Plateau Dry Deciduous Forests [21]. We used MacKinnon's (1997) map of the original vegetation to delineate the ecoregion boundaries.

Prepared by Gopal S. Rawat, Ajay Desai, Hema Somanathan, and Eric D. Wikramanayake

Ecoregion Number:	**22**
Ecoregion Name:	**South Deccan Plateau Dry Deciduous Forests**
Bioregion:	**Indian Subcontinent**
Biome:	**Tropical and Subtropical Dry Broadleaf Forests**
Political Units:	**India**
Ecoregion Size:	**81,900 km^2**
Biological Distinctiveness:	**Bioregionally Outstanding**
Conservation Status:	**Critical**
Conservation Assessment:	**IV**

Although not exceptionally outstanding for biological richness or endemism by itself, the South Deccan Plateau Dry Deciduous Forests [22] ecoregion is contiguous with the moist deciduous forests that lie along the foothills of the southern extent of the Western Ghats Mountains. Two of India's most important elephant conservation areas, the Nilgiris–Eastern Ghats and the Anamalais-Nelliampathis (Sukumar 1989), and two of the most important TCUs (Wikramanayake et al. 1999) extend across three ecoregions: the South Western Ghats Montane Rain Forests [2], South Western Ghats Moist Deciduous Forests [15], and South Deccan Plateau Dry Deciduous Forests [22]. Together, these three ecoregions provide an important, contiguous habitat landscape for conservation of Asia's largest terrestrial herbivore and predator.

Location and General Description

The ecoregion represents a large area of tall, tropical dry forests in the southern Deccan Plateau, on the leeward side of the Western Ghats Mountain Range. It extends across the southern Indian states of Karnataka and Tamil Nadu. The ecoregion's links to the ancient, southern circumpolar continent, Gondwanaland, are evident in the biotic links to Africa and Madagascar (Meher-Homji 1989).

The ecoregion's vegetation is highly influenced by climate. The tall Western Ghats Mountain Range intercepts the moisture from the southwest monsoon. Therefore, the eastern slopes and the Deccan Plateau receive very little rainfall; annual rainfall ranges from 900 to 1,500 mm. The undulating hillsides have very shallow soils.

The dry deciduous forests of this ecoregion are flanked by the moist deciduous forests along the lower elevations and foothills of the Western Ghats to the west and by thorn scrub to the east. Therefore, this ecoregion probably represents a transition zone between the moister western vegetation and the drier vegetation to the east.

Champion and Seth (1968) classified these forests as Southern Dry Mixed Deciduous Forests, where teak (*Tectona grandis*) is not conspicuous. Thorny plants become more common in areas where grazing pressure is high.

Structurally, these dry forests have a three-storied forest, with an upper canopy at 15–25 m, an understory at 10–15 m, and undergrowth at 3–5 m (Puri et al. 1989). Lianas drape the trees in denser, mature forests. The vegetation is characterized by *Boswellia serrata, Anogeissus latifolia, Acacia catechu, Terminalia tomentosa, Terminalia paniculata, Terminalia belirica, Chloroxylon swietenia, Albizzia amara, Cassia fistula, Hardwickia binata, Dalbergia latifolia, Sterospermum personatum, Pterocarpus marsupium, Diospyros montana,* and *Shorea talura,* among others (Champion and Seth 1968). One of the important species of this forest, sandalwood (*Santalum album*), has been selectively removed from most of the forests in this ecoregion.

Biodiversity Features

Although not high in endemism or diversity, the ecoregion harbors several important populations of India's large threatened vertebrates whose populations are fast declining throughout their ranges because of habitat loss and hunting pressure. For instance, the elephant population that ranges from the Nilgiri Hills to the Eastern Ghats, estimated at more than 6,000 animals, is considered to be the largest single elephant population in India (Sukumar 1989). A second important population ranges along the Anaimalai and Nelliampathi Hills.

Two high-priority TCUs (Periyar-Kalakad and Dandeli-Bandipur) overlap with parts of this ecoregion (Wikramanayake et al. 1999).

The ecoregion's mammal fauna includes seventy-five species. One, the critically endangered Salim Ali fruit bat (Hilton-Taylor 2000), is a near-endemic species (table J22.1). Other threatened mammals include the Asian elephant (*Elephas maximus*), wild dog (*Cuon alpinus*), sloth bear (*Melursus ursinus*), chousingha (*Tetracerus quadricornis*), gaur (*Bos gaurus*), and grizzled giant squirrel (*Ratufa macruora*).

The bird fauna consists of about 260 species, of which two are near-endemic species (table J22.2).

TABLE J22.1 Endemic and Near-Endemic Mammal Species.

Family	Species
Pteropodidae	*Latidens salimalii*

An asterisk signifies that the species' range is limited to this ecoregion.

TABLE J22.2. Endemic and Near-Endemic Mammal Species.

Family	Common Name	Species
Timaliidae	Rufous babbler	*Turdoides subrufus*
Pycnonotidae	Yellow-throated bulbul	*Pycnonotus xantholaemus*

An asterisk signifies that the species' range is limited to this ecoregion.

Two species in this ecoregion, the Indian bustard (*Ardeotis nigriceps*) and lesser florican (*Eupodotis indica*), are globally threatened and warrant conservation attention.

Status and Threats

Current Status

More than 80 percent of the ecoregion's forests have been cleared, but two large blocks of contiguous habitat remain in the north. There are seven protected areas that cover about 1,100 km^2, or a little more than 1 percent of the ecoregion. But none of these exceed 500 km^2; therefore, the ecoregion's protected area system is woefully inadequate to conserve the several large vertebrates in this ecoregion.

Types and Severity of Threats

Most of the dry deciduous forests represented by this ecoregion have already been degraded to thorn scrub (Champion and Seth 1968). The remaining forests are highly threatened with conversion to cash crop plantations, excessive fuelwood collection, and overgrazing by large herds of cattle. The habitat contiguity with the moist deciduous forests along the foothill of the Western Ghats, in particular, has been severed in several areas by conversion to agricultural lands. Large areas of forested land are lost to development projects such as dams, mining, and resettlement of displaced people. Many of the habitat areas that will be lost to large-scale hydroelectric projects now serve as refuges for the ecoregion's biodiversity. The conservation status of the ecoregion was changed from endangered to critical because of the threats from human population to the remaining forest blocks.

Poaching is a major threat in this ecoregion. The capacity to protect and manage the ecoregion's biodiversity is lacking, especially because these habitats are perceived as being unimportant for biodiversity conservation. Where conservation efforts have been made, poor planning has resulted in the introduction of exotic species and conversion of grassland and open scrub habitat that are important for many focal species such as bustards and floricans.

A major threat in the future will be the development of these areas for mining and industrialization. Licenses have been granted to exploit several areas in the ecoregion, and there is an urgent need to counter these initiatives by setting aside key areas for conservation of the ecoregion's representative biodiversity.

Priority Conservation Actions

Immediate and longer-term actions are necessary to ensure that conservation landscapes for the important elephant and tiger populations are maintained. Other actions are also necessary to conserve some distinct habitat types within this ecoregion.

Short-Term Conservation Actions (1–5 Years)

- Protect more effectively the low hills in the central plains of Tamil Nadu (Rodgers and Panwar 1988).
- Evaluate the protected areas to determine the long-term viability of the habitat itself and how well it can harbor viable populations of the area-sensitive species and contain ecological processes.
- Create a sanctuary in the Dhamapuri district to protect sandalwood, as proposed by Rodgers and Panwar (1988).
- Formulate land-use plans to prevent unplanned and unregulated forest clearing.
- Define areas for social forestry to address fuelwood and fodder needs for local people and livestock.
- Define and manage conservation areas (not necessarily strict protected areas) and reserve forests to restore the original flora and fauna.
- Remove exotic species.
- Review the protected area system with regard to elephant and tiger conservation needs; revise as deemed necessary.
- Manage prey populations for tigers.
- Manage habitat for the important elephant populations and address human-elephant conflicts.
- Maintain traditional water harvesting and storage systems that provide important habitat for migratory waterfowl (Rodgers and Panwar 1988).
- Conduct a status survey of the ecoregion's biodiversity.
- Create an awareness and education campaign about the conservation importance of these dry forests.

Longer-Term Conservation Actions (5–20 Years)

- Restore habitat, especially to recreate the habitat linkages that will allow elephant movements and tiger dispersal.
- Formulate conservation and land-use plans to accommodate mining operations and industrialization.

Focal Species for Conservation Action

Endangered endemic species: Rufous babbler, yellow-throated bulbul

Area-sensitive species: Asian elephant, tiger

Top predator species: tiger, leopard

TABLE J22.3. WCMC (1997) Protected Areas That Overlap with the Ecoregion.

Protected Area	Area (km^2)	IUCN Category
Bilgiri Ranga Swamy Temple [15]	290	PRO
Bannerghatta	140	II
Melkote Temple	50	IV
Ranganthittu	20	IV
Arabithittu	40	IV
Cauvery	500	IV
Vedanthangal	70	IV
Total	1,110	

Ecoregion numbers of protected areas that overlap with additional ecoregions are listed in brackets.

Other: wild dog, sloth bear, chousingha, gaur, Salim Ali fruit bat, grizzled giant squirrel, sandalwood, Indian bustard, lesser florican

Selected Conservation Contacts

Bombay Natural History Society

Centre for Ecological Sciences, Bangalore

District Forest Officer, Kanyakumari District

Field Director, Bandipur Tiger Reserve

French Institute of Pondicherry

Kerala Forest Research Institute, Pichi, Kerala

Pulney Hills Conservation Society

Salim Ali Institute of Ornithology and Natural History, Coimbatore

Tropical Botanic Gardens and Research Institute, Thiruvananthapuram

Wildlife Institute of India

WWF-India

Justification for Ecoregion Delineation

In a previous analysis of conservation units, Rodgers and Panwar (1988), and subsequently MacKinnon (1997), divided the Deccan Peninsula into five biotic provinces. This ecoregion lies within Rodgers and Panwar's Deccan Plateau South (6A) biotic province. But this biotic province contains several vegetation types. Therefore, in keeping with our definition of an ecoregion (i.e., an ecosystem of regional extent) and following our rules for ecoregion delineation (representing distinct vegetation types of regional extent in separate ecoregions), we placed the tropical dry deciduous forests in this biotic province in the South Deccan Plateau Dry Deciduous Forests [22]. We used MacKinnon's (1997) map of original vegetation to identify the extent of the dry deciduous forests and exclude the large areas of thorn forests.

Prepared by Gopal S. Rawat, Ajay Desai, Hema Somanathan, and Eric D. Wikramanayake

Ecoregion Number:	**23**
Ecoregion Name:	**Deccan Thorn Scrub Forests**
Bioregion:	**Indian Subcontinent**
Biome:	**Deserts and Xeric Shrublands**
Political Units:	**India, Sri Lanka**
Ecoregion Size:	**339,000 km²**
Biological Distinctiveness:	**Locally Important**
Conservation Status:	**Critical**
Conservation Assessment:	**IV**

The Deccan Thorn Scrub Forests [23] harbor the last populations of the globally threatened Jerdon's courser (*Rhinoptilus bitorquatus*), rediscovered recently, eighty-six years since it was last recorded in 1900. Otherwise, the ecoregion is neither exceptionally species-rich nor high in endemism. Many ecologists believe that the thorn scrub vegetation represents a degraded stage of the tropical dry forests, modified by human and livestock use over hundreds of years (Puri et al. 1989).

Location and General Description

The ecoregion represents the thorn scrub vegetation in the arid parts of the Deccan Plateau. It sprawls across the Indian states of Tamil Nadu, Andhra Pradesh, Karnataka, and Maharashtra and also includes part of northern Sri Lanka.

The Deccan Plateau itself was part of the ancient southern continent, Gondwanaland, that disintegrated during the Cretaceous to give rise to the Indian Subcontinent as well as Africa, Madagascar, Australia, South America, and New Guinea and some of the smaller islands such as New Caledonia and Tasmania. After the Deccan Plateau drifted northward to collide with the Eurasian continent about 50 million years ago, geological uplift gave rise to the Western Ghats Mountains along the western coast of the peninsula. This mountain range then intercepted the moisture-laden southwest monsoons and created a dry rainshadow in the vast plateau, affecting its vegetation. But in the more recent past, human influences have altered the vegetation to create vast areas of thorn scrub from what was believed to be tropical dry forests.

Annual rainfall in the ecoregion is less than 750 mm. All rain is received during the brief wet season, and there is practically no rainfall from November to April. Ambient temperatures can exceed a sweltering 40°C during the hotter months of the year.

The forest type in this ecoregion is mostly southern tropical thorn scrub, as defined by Champion and Seth (1968), but includes patches of tropical dry deciduous forests, which are believed to be the original vegetation. The former consists of open, low vegetation characterized by thorny trees with short trunks and low, branching crowns that rarely meet to form a closed canopy. The trees attain heights of 6–9 m. The second story is poorly developed and consists of spiny and xerophytic species, mostly shrubs. During the brief wet season an ill-defined lower story can be discerned. The dominant vegetation is *Acacia* species, with *Balanites roxburghii*, *Cordia myxa*, *Capparis* spp., *Prosopis* spp., *Azadirachta indica*, *Cassia fistula*, *Diospyros chloroxylon*, *Carrisa carandas*, and *Phoenix sylvestris*.

Champion and Seth (1968) have also identified several habitat types within this vast thorn scrub ecoregion. In areas of particularly low rainfall and rocky soils, the thorn scrub transitions into an *Euphorbia*-dominated scrub (i.e., the southern Euphorbia scrub). Here the soil usually is bare, although some grassy growth may appear during the short monsoon season.

In parts of Tamil Nadu, where rainfall is even less, the vegetation is made up of open thorny forests with scattered *Acacia planifrons* that are characterized by umbrella-shaped crowns. This vegetation is described as Carnatic umbrella thorn forests by Champion and Seth (1968).

Scattered amid the thorn scrub are patches of dry grasslands that provide habitat for the native fauna. For example, the grasslands of southern Andhra Pradesh support a good population of the Indian bustard (*Ardeotis nigriceps*) and blackbuck (*Antilope cervicapra*). The typical grasses in this habitat include *Chrysopogon fulvus*, *Heteropogon contortus*,

Eremopogon foveolatus, *Aristida setacea*, and *Dactyloctenium* spp. (Rawat and Babu 1995).

Patches of dry deciduous forests, especially along the Tirupathi Hill Ranges, are known for a large number of medicinal plants and various other species of botanical interest, among which are the rare endemic cycad (*Cycas beddomei*) and *Psilotum nudum*. The latter usually is found along steep escarpments. A small patch of the dipterocarp *Shorea talura* exists within the Chittoor forest division, part of which is being maintained as a preservation plot by the Forest Department of Andhra Pradesh.

The Srilankamalleswara Sanctuary between the Nallamalais and Sechachalam hill ranges is known for a rare, endemic tree species, red sanders (*Pterocarpus santalinus*). This area is also the southern distributional limit of the nilgai (*Boselaphus tragocamelus*) in the Indian Peninsula.

Biodiversity Features

Until the recent past, this ecoregion provided important habitat for the tiger (*Panthera tigris*) (in the Indian sector) and Asian elephant (*Elephas maximus*). But over the years, their populations have dwindled and even become locally extinct because of the adverse influences from the dense human population.

The mammal fauna in the ecoregion includes ninety-six species, including two endemic rodents and an endemic bat (table J23.1).

The endemic rodents are threatened (Hilton-Taylor 2000). Other threatened species in the ecoregion include tiger, gaur (*Bos gaurus*), wild dog (*Cuon alpinus*), sloth bear (*Melursus ursinus*), chousingha (*Tetracerus quadricornis*), and blackbuck (*Antilope cervicapra*) (Hilton-Taylor 2000). The important elephant populations are now only marginally included within this ecoregion. Small wolf populations may still be left, although most have been eliminated by a combination of loss of prey and poisoning by people as retribution for livestock predation.

The ecoregion's bird fauna consists of almost 350 species, of which three are near-endemics (table J23.2).

The Jerdon's courser (*Rhinoptilus bitorquatus*) is a globally threatened species that was rediscovered in this ecoregion in 1986, after the last record in 1900 (Grimmet et al. 1998). The known population of this species is limited to a small area in this and the neighboring Central Deccan Plateau Dry Deciduous Forests [21]. The Ceylon junglefowl (*Gallus lafayetii*) is limited to the ecoregion's area in northern Sri Lanka. The globally threatened lesser florican (*Eupodotis indica*) and Indian bustard are other birds of conservation importance in this ecoregion.

Status and Threats

Current Status

More than 90 percent of the ecoregion's natural habitat has been degraded or cleared, but one large block of habitat remains in southern Andhra Pradesh. The eleven protected areas cover more than 4,000 km^2, but this represents just about 1 percent of the ecoregion area (table J23.3). The Great Indian Bustard reserve accounts for most of the protected areas system.

Types and Severity of Threats

The forests in this ecoregion have been degraded to thorn scrub solely as a result of these human activities (Puri et al. 1989). Among the more serious sources of degradation is pastoralism, both from heavy cattle grazing and from forest produce extracted by the pastoralists. Several village pastures have been taken over by an exotic thorny shrub, *Prosopis juliflora*, resulting in the loss of grazing areas for the cattle and encroachment into the reserved forests or protected areas for grazing (Rawat and Babu 1995). The conservation status of the ecoregion was changed from endangered to critical after the analysis of projected threats from the human population. There is a common perception that these dry forests are not important for conservation. Therefore, grazing and forest clearing, especially for fuelwood, are rampant.

Priority Conservation Actions

Although the thorn scrub represented by this ecoregion is a degraded vegetation type, there are still scattered patches of the original habitat and other distinct habitat types that harbor several threatened species. Conservation actions should

TABLE J23.1. Endemic Mammal Species.

Family	Species
Rhinolophidae	*Hipposideros schistaceus**
Muridae	*Millardia kondana**
Muridae	*Cremnomys elvira**

An asterisk signifies that the species' range is limited to this ecoregion.

TABLE J23.2. Endemic and Near-Endemic Bird Species.

Family	Common Name	Species
Glareolidae	Jerdon's courser	*Rhinoptilus bitorquatus*
Phasianidae	Ceylon junglefowl	*Gallus lafayetii*
Capitonidae	Yellow-fronted barbet	*Megalaima flavifrons*

An asterisk signifies that the species' range is limited to this ecoregion.

TABLE J23.3. WCMC (1997) Protected Areas That Overlap with the Ecoregion.

Protected Area	Area (km^2)	IUCN Category
Chandikulam	120	IV
Vettangudi [6]	20	IV
Srivenkateswara	500	IV
Nandur Madmesh War	80	IV
Jaikwadi	230	IV
Great Indian Bustard	2,600	IV
Great Indian Bustart (extension)	250	PRO
Sagareshwar	50	IV
Ghataprapha	110	IV
Tungabadra	90	DE
Ranebennur	60	IV
Total	4,110	

Ecoregion numbers of protected areas that overlap with additional ecoregions are listed in brackets.

target these species and habitats. Here we offer some short- and longer-term conservation actions.

Short-Term Conservation Actions (1–5 Years)

- Conduct experimental studies to control the invasion of exotic species, especially *Prosopis juliflora* and its impact in wildlife habitat (Rawat and Babu 1995). This exotic species seems to threaten the habitat of Jerdon's courser, Indian bustard, and blackbuck, which favor grasslands.
- Conduct ecological studies and status surveys on the Jerdon's courser, golden gecko (*Calaodactylodes aureus*), *Cycas beddomei*, and red sanders, especially with respect to habitat management. Complete protection of the open scrub may result in a dense scrub that is unsuitable for species such as bustards, floricans, and blackbuck.
- Control land conversion into agriculture and industrial lands.
- Create awareness and educate people, even the forest department staff, of the importance of these dry forests for conservation.

Longer-Term Conservation Actions (5–20 Years)

- Monitor key wildlife areas, especially those brought under recent afforestation programs, where important grasslands and open scrub are being converted into plantations and "forests" using exotics.
- Monitor programs for the Jerdon's courser, Indian bustard, golden gecko, *Cycas beddomei*, and red sanders.
- Reintroduce nilgai into its past range within the ecoregion.

Focal Species for Conservation Action

Area-sensitive species: tiger

Top predator species: tiger, wolf, leopard

Other: wild dog, gaur, sloth bear, chousingha, blackbuck, *Millardia kondana, Cremnomys elvira,* red sanders, golden gecko, endemic bat, and rodents

Selected Conservation Contacts

Bombay Natural History Society

Centre for Ecological Sciences, Bangalore

Chief Wildlife Wardens, Tamil Nadu, Andhra Pradesh, Karnataka States

Conservator of Forests, Southern Circle, Andhra Pradesh

French Institute of Pondicherry

Wildlife Institute of India

WWF-India

Justification for Ecoregion Delineation

In a previous analysis of conservation units, Rodgers and Panwar (1988), and subsequently MacKinnon (1997), divided the Deccan Peninsula into five biotic provinces.

This ecoregion includes Rodgers and Panwar's Central Plateau North (6B) biotic province and partially includes the Deccan Plateau South (6A) biotic province.

In keeping with our definition of an ecoregion (i.e., an ecosystem of regional extent) and following our rules for ecoregion delineation (represent distinct vegetation types of regional extent in separate ecoregions), we placed the thorn scrub, as mapped by MacKinnon (1997), that extends across these two biotic provinces within the Deccan Thorn Scrub Forests [23].

Prepared by Gopal S. Rawat, Ajay Desai, Hema Somanathan, and Eric D. Wikramanayake

Ecoregion Number:	**24**
Ecoregion Name:	**Northwestern Thorn Scrub Forests**
Bioregion:	**Indian Subcontinent**
Biome:	**Deserts and Xeric Shrublands**
Political Units:	**India, Pakistan**
Ecoregion Size:	**487,500 km^2**
Biological Distinctiveness:	**Locally Important**
Conservation Status:	**Critical**
Conservation Assessment:	**IV**

The Northwestern Thorn Scrub Forests [24] ecoregion represents a large expanse of degraded dry forest surrounding the Thar Desert. Neither exceptionally species-rich nor high in endemism, the ecoregion nevertheless harbors viable populations of chinkara (*Gazella bennettii*), chousingha (*Tetracerus quadricornis*), and blackbuck (*Antilope cervicapra*).

Location and General Description

The ecoregion represents the thorn scrub forests in the northwestern region of the Indian subcontinent. Many ecologists consider this thorn scrub to represent a degraded state of tropical dry forests (e.g., Champion and Seth 1968; Puri et al. 1989). The ecoregion stretches across the border between India and Pakistan and covers parts of Gujarat, Rajasthan, Haryana, and Punjab states of India and the lower parts of Jammu and Kashmir. The average annual rainfall in this arid ecoregion is less than 750 mm. The temperature can exceed 45°C during the hottest months, and in winter temperatures can drop to below freezing.

The flat alluvial lowlands extend into the low hills. Local variations in soil salinity affect the distribution of vegetation; patches of highly saline soil usually are bare of vegetation. The vegetation is stunted and open, dominated by *Acacia* species such as *A. senegal* and *A. leucophloea* that rarely exceed 6 m in height. Other characteristic species that make up the vegetation are *Prosopis spicigera, Capparis zeylanica, Salvadora* spp., *Carissa* spp., *Gymnosporia* spp., *Grewia* spp., and *Gardenia* spp. and xerophytic climbers such as species of *Tragia, Rivea, Tinospora, Vitis,* and *Peristrophe* (Champion and Seth 1968; Puri et al. 1989). In drier areas, the thorn forest transitions into xerophytic shrubland and semiarid vegetation, usually dominated by *Euphorbia* species. Intermingled with the *Euphorbia* scrub is a *Zizyphus* scrub (Champion and Seth 1968) that is characterized by *Zizyphus nummularia* with *Acacia leucocephala, Acacia senegal, Anogeissus pendula,* and *Dicrostachys cinerea*. The poor soils along rocky tracts promote a *Cassia-Butea* community. Closer to the coast, where the soils are more saline, the community includes *Salvadora* and *Tamarix* (Puri et al. 1989). The southwestern part of the

TABLE J24.1 Endemic and Near-Endemic Mammal Species.

Family	Species
Rhinolophidae	Triaenops persicus*
Rhinopomatidae	Rhinopoma muscatellum*

An asterisk signifies that the species' range is limited to this ecoregion.

TABLE J24.2. Endemic and Near-Endemic Bird Species.

Family	Common Name	Species
Paridae	White-winged tit	Parus nuchalis
Sylviidae	Rufous-vented prinia	Prinia burnesii*

An asterisk signifies that the species' range is limited to this ecoregion.

Aravalli Range supports a distinct deciduous forest characterized by *Anogeissus pendula, Aegle marmelos, Boswellia serreta, Cassia fistula, Mitragyna parviflora, Diospyros melanxylon,* and *Wrightia tinctoria.*

Biodiversity Features

The ecoregion is not exceptionally rich or high in endemism but does harbor several large mammals of conservation importance, including the leopard (*Panthera pardus*), caracal (*Felis caracal*), chinkara (*Gazella bennettii*), chousingha (*Tetracerus quadricornis*), and blackbuck (*Antilope cervicapra*). Overall, the mammal fauna consists of about ninety species, including two bats that are endemic to the ecoregion (table J24.1).

Both of these bats are strict endemics, their known range being limited to this ecoregion (Corbet and Hill 1986). The chousingha and blackbuck are threatened species (Hilton-Taylor 2000) and should be focal species for conservation actions.

The Aravalli Hill Ranges and the surrounding flat areas support two very distinct rodent communities that are influenced by the geology, soil, and vegetation conditions. Detailed studies are under way to determine the species composition of these communities (I. Prakash, pers. comm., 2000).

The bird fauna consists of an impressive list of more than 400 species, among the highest for ecoregions in this bioregion. The list includes two endemic species (table J24.2).

The rufous-vented prinia is a strict endemic limited to this ecoregion, and the white-winged tit is a near endemic shared with the adjacent Khathiarbar-Gir Dry Deciduous Forests [16]. The ecoregion also harbors the globally threatened Indian bustard (*Ardeotis nigriceps*) and lesser florican (*Eupodotis indica*) (Hilton-Taylor 2000).

Status and Threats

Current Status

More than 90 percent of this ecoregion's natural habitat has been converted, and only small, scattered fragments remain. There are an astounding sixty protected areas in the ecoregion (table J24.3), but they cover only about 11,000 km², or just over 2 percent of the ecoregion area. Many of the protected areas are small, at an average size of just over 175 km².

TABLE J24.3. WCMC (1997) Protected Areas That Overlap with the Ecoregion.

Protected Area	Area (km²)	IUCN Category
Dhrun [44]	310	II
Kachau	220	IV
Khurkhera	180	IV
Kirthar	2,980	II
Mahal Kohistan	650	IV
Hab Dam	460	IV
Kinjhar Lake	180	IV
Hadero Lake	10	IV
Haleji Lake	20	IV
Bijoro Chach	1	IV
Norange	2	IV
Deh Jangisar	3	UA
Keb Bunder North	90	IV
Pai	20	UA
Lakhi	1	IV
Goleen Gol	10	UA
Deh Sahib Saman	3	UA
Khipro	40	UA
Narayan Sarovar	830	IV
Barda	210	IV
Velavadar	50	II
Nal Sarovar	50	IV
Jessore	260	IV
Doli Closed Area	700	VII
Indus River #1	1,370	UA
Dhoung Block	20	IV
Dosu Forest	20	UA
Drigh Lake	1	IV
Mando Dero	140	UA
Resi	50	UA
Sheikh Buddin	240	IV
Kalabagh Game Reserve	10	
Thanadarwala	40	UA
Nemal Lake	4	IV
Chashma Lake	220	IV
Taunsa Barrage	70	IV
Kot Zabzai	100	UA
Sodhi	50	IV
Daphar	30	IV
Head Qadirabad	30	UA
Bhon Fazil	30	UA
Gat Wala	60	UA
Bhono	20	UA
Kharar Lake	2	IV
Kamalia Plantation	40	UA
Chichawatni Plantation	50	UA
Abohar	190	IV
Indo-Pak Border	30	UA
Chaupalia	100	UA
Head Islam/Chak Kotora	30	UA
Daulana	20	UA
Bahwaalpur Plantation	5	UA
Harike Lake	1	IV
Bir Motibagh	30	IV
Bir Bunerheri	10	IV
Chautala	110	IV
Bir Gundial Pura	30	IV
Simbalbara	160	IV
Sultanpur	20	IV
Renuka	40	IV
Total	10,653	

Ecoregion numbers of protected areas that overlap with additional ecoregions are listed in brackets.

The local people, especially the Bishnoi communities, have a cultural reverence for wild animals, particularly blackbuck and the tree *Prosopis cinerea*. For this reason, the wildlife and habitat have been protected over the years.

Types and Severity of Threats

This ecoregion is subject to intensive degradation threats from livestock grazing. Associated human impacts such as lopping and cutting of vegetation for fuelwood, setting fires to create grazing lands, and settlements also contribute to habitat degradation. The marble beds of the ancient Aravalli Range are being heavily mined. Rodgers and Panwar (1988) present an analysis of conservation gaps and needs for the Indian part of the ecoregion.

Priority Conservation Actions

Here we identify some short and long-term conservation actions in this regard.

Short-Term Conservation Actions (1–5 Years)

- Control mining activities within 10–15 km of the protected area boundaries.
- Ensure proper livestock management and animal health care to avoid transmission of diseases to wildlife.
- Promote the traditional methods of natural resource use and harness the religious sentiments of the local people for conservation.
- Assess and revise the protected area system.
- Establish a conservation and monitoring program for the Indian bustard.

Longer-Term Conservation Actions (5–20 Years)

- Develop a long-term conservation plan, using the conservation needs of focal species in this ecoregion.

Focal Species for Conservation Action

Endangered endemic species: rufous-vented prinia

Wide-ranging species: leopard

Top predator species: leopard

Other: caracal, chinkara, chousingha, blackbuck, white-winged tit, Indian bustard, lesser florican

Selected Conservation Contacts

Central Arid Zone Research Institute, Jodhpur

Forest Departments of Rajasthan and Gujarat States, India

Wildlife Institute of India, Dehra Dun, India

WWF-India

Justification for Ecoregion Delineation

In a previous analysis of conservation units, MacKinnon (1997) placed these desert and thorn scrub forests of northwestern India and Pakistan in Biounit (I3) along with other habitat types that included dry deciduous forests and deserts. In keeping with our rules for representing distinct habitat types of regional extent in separate ecoregions, we used MacKinnon's (1997) digital map of the original vegetation to delineate and separate the thorn scrub forests from the desert and dry deciduous forests. The thorn scrub around the Thar Desert therefore was placed in the Northwestern Thorn Scrub Forests [24]. The Northwestern Thorn Scrub Forests [24] lie in both the Indus-Ganges monsoon forest and Thar Desert biogeographic provinces.

Prepared by Gopal S. Rawat and Eric D. Wikramanayake

Ecoregion Number:	**25**
Ecoregion Name:	**Himalayan Subtropical Broadleaf Forests**
Bioregion:	**Indian Subcontinent**
Biome:	**Tropical and Subtropical Moist Broadleaf Forests**
Political Units:	**Bhutan, India, and Nepal**
Ecoregion Size:	**38,000 km²**
Biological Distinctiveness:	**Regionally Outstanding**
Conservation Status:	**Endangered**
Conservation Assessment:	**II**

The Himalayan Subtropical Broadleaf Forests [25] ecoregion includes several forest types along its length as it traverses an east to west moisture gradient. The forest types include *Dodonea* scrub, subtropical dry evergreen forests of *Olea cuspidata*, northern dry mixed deciduous forests, dry Siwalik *sal* (*Shorea robusta*) forests, moist mixed deciduous forests, subtropical broadleaf wet hill forests, northern tropical semi-evergreen forests, and northern tropical wet evergreen forests (Champion and Seth 1968).

The ecoregion also forms a critical link in the chain of interconnected Himalayan ecosystems that extend from the Terai and Duar grasslands along the foothills to the high alpine meadows at the top of the world's highest mountain range. For instance, several Himalayan birds and mammals exhibit seasonal altitudinal migrations and depend on contiguous habitat to permit these movements. Therefore, conservation actions in the Himalayas must pay due attention to habitat connectivity because degradation or loss of a habitat type along this chain will disrupt these important ecological processes.

Location and General Description

This ecoregion represents the east-west–directed band of Himalayan subtropical broadleaf forests along the Siwaliks or Outer Himalayan Range, lying between 500 and 1,000 m. The ecoregion achieves its greatest coverage in the middle hills of central Nepal, but the long, narrow ecoregion extends through Darjeeling into Bhutan and also into the Indian State of Uttar Pradesh. The Kali Gandaki River, which has gouged the world's deepest river valley through the Himalayan Range, bisects the ecoregion.

The Himalayas rose from beneath the ancient Tethys Sea when the Deccan Plateau collided with the Eurasian continent about 50 million years ago and forced the latter upward to form the highest mountain range in the world (Wadia 1966). The Himalayas now consist of three east-west–directed parallel zones: the southernmost outer Himalayas, which represent the Siwaliks; the Middle Himalayas, representing a series of ridges and valleys that rise to about 5,000 m; and the Inner Himalayas, which include the tallest mountain peaks in the world. The Siwalik Hills, where this ecoregion lies, are composed of alluvium deposited over the ages by the rivers that drain this young mountain range.

Rainfall varies from east to west, but annual rainfall can be as much as 2,000 mm. The Himalayas capture moisture from the monsoons that sweep in from the Bay of Bengal, and most of this rainfall is expended in the eastern Himalayas. Therefore, the western Himalayas are drier, a trend reflected in the timberline that declines from 4,000 m in the east to about 3,500 m in the west (Kendrick 1989).

The forests in this ecoregion are very rich in biodiversity. The forest types are varied because of the subtropical climate, complex topography, rich alluvial soils, moisture gradient, and intermingling of taxa from the Indo-Malayan and Palearctic regions (Rawat and Mukherjee 1999). These forest types consist of *Dodonea* scrub, subtropical dry evergreen forests of *Olea cuspidata*, northern dry mixed deciduous forests, dry Siwalik *sal* forests, moist mixed deciduous forests, subtropical broadleaf wet hill forests, northern tropical semi-evergreen forests, and northern tropical wet evergreen forests (Champion and Seth 1968).

The forests generally reach 30 m, although in favorable areas the canopy can reach as high as 50 m. The top canopy is less dense than the tropical evergreen forests, and a midcanopy and shrubby undergrowth are recognizable. Grasses are absent, but there is a well-developed herb cover. Climbers and epiphytes are common.

The diversity and richness of woody species decrease from east to west. In the western foothills of Himachal Pradesh and Uttar Pradesh, characteristic tree communities are represented by *Shorea robusta*, *Terminalia tomentosa*, *Anogeissus latifolia*, *Mallotus philippinensis*, *Olea cuspidata*, *Bauhinia restusa*, and *Bauhinia variegata*. In the eastern foothills, the characteristic species include *Schima wallichii*, *Castanopsis tribuloides*, *C. indica*, *Terminalia crenulata*, *Terminalia bellerica*, *Engelhardtia spicata*, *Betula* spp., and *Anogeissus* spp. (Champion and Seth 1968). In eastern Nepal, *Engelhardtia spicata*, *Erythrina* spp., and *Albizia* spp. are important components of the subtropical forest associations. *Alnus nepalensis* is an early-successional species that invades landslide areas and forms monospecific stands. Many of the trees in these broadleaf forests such as *Gnetum montanus*, *Cycas pectinata*, *Cyathea spinulosa* (tree ferns), *Rauwolfia serpentina*, *Pandanus nepalensis*, *Calamus lalifolius*, *C. leptospadix*, *Phoenix humilis*, and *Phoenix sylvestris* have become very rare in Nepal.

Biodiversity Features

This ecoregion is a critical link in the Himalayan ecosystem, where altitudinal connectivity between the habitat types (each represented by a different ecoregion) is important for ecosystem function. In addition to its importance in maintaining ecosystem dynamics, the ecoregion also harbors several threatened species that warrant conservation attention. The large areas of intact habitat have been included within two high-priority (Level I) TCUs (Dinerstein et al. 1997). They extend across adjacent ecoregions representing the broadleaf and subtropical conifer forests and the savanna grasslands along the foothills.

TABLE J25.1. Endemic and Near-Endemic Mammal Species.

Family	Species
Cercopithecidae	*Semnopithecus geei*

An asterisk signifies that the species' range is limited to this ecoregion.

TABLE J25.2. Endemic and Near-Endemic Bird Species.

Family	Common Name	Species
Phasianidae	Chestnut-breasted partridge	*Arborophila mandellii*

An asterisk signifies that the species' range is limited to this ecoregion.

The mammal fauna consists of ninety-seven species including one that is endemic to this ecoregion (table J25.1).

This charismatic, endemic golden langur (*Semnopithecus geei*) has a small range distribution, being limited to the broadleaf forests north of the Brahmaputra River. It is shared between this and the adjacent Eastern Himalayan Broadleaf Forests [26].

Several of the ecoregion's mammals are threatened species. These include the tiger (*Panthera tigris*), Asian elephant (*Elephas maximus*), golden langur, smooth-coated otter (*Lutrogale perspicillata*), clouded leopard (*Pardofelis nebulosa*), gaur (*Bos gaurus*), serow (*Capricornis sumatraensis*), Irrawaddy squirrel (*Callosciurus pygerythrus*), and particoloured squirrel (*Hylopetes alboniger*) (Hilton-Taylor 2000).

The bird fauna is very rich; there are more than 340 birds in the ecoregion. One species is endemic to the ecoregion (table J25.2).

The chestnut-breasted partridge (*Arborophila mandellii*) is shared with several of the other eastern Himalayan ecoregions (Eastern Himalayan Broadleaf Forests [26], Eastern Himalayan Sub-Alpine Conifer Forests [28], and Himalayan Subtropical Pine Forests [31]) but has a very limited range within these ecoregions. The globally threatened whitewinged wood duck (*Carina scutulata*) and five hornbill species are found here. The latter in particular need mature forests for nesting and are good indicators of habitat quality. BirdLife International's EBA, Eastern Himalayas (130) (Stattersfield et al. 1998), overlaps with this ecoregion.

Status and Threats

Current Status

More than 70 percent of the natural forests in the ecoregion have been cleared or degraded. Cultivation is especially extensive in the fertile valleys of large rivers such as the Karnali, Babai, and Rapti and in the flat lands between the Trisuli River and Kali Gandaki (Shrestha and Joshi 1997). But most of the hill forests above 1,000 m still remain uncut because the shallow, erosion-prone soils are unsuitable for cultivation.

The eight protected areas that extend into this ecoregion (table J25.3) cover a little over 2,700 km^2, representing about 7 percent of the ecoregion's area. Several of these protected areas—especially Royal Manas, Royal Chitwan, Royal Bardia, and Valmikigar—are important for the large vertebrates that can be considered umbrella species (the tiger, Asian elephant, clouded leopard, and hornbills are candidates) for overall biodiversity.

Although almost all the protected areas are smaller than 500 km² within the ecoregion (the exception is Royal Chitwan National Park); Royal Bardia National Park and Royal Manas National Park extend into the adjacent ecoregions. Both parks are more than 800 km². Three of the other protected areas also overlap across adjacent ecoregions (table J25.3).

Types and Severity of Threats

The primary threats to the ecoregion's natural habitat stem from fuelwood collection, intensive livestock grazing (FAO 1981), and annual burning by pastoralists to encourage the growth of new shoots for livestock. Heavy grazing even within intact forests has destroyed the undergrowth, including the saplings that should eventually replace the mature canopy trees. Therefore, the long-term viability of these forests is compromised. However, farmers have also begun to plant and maintain fodder trees on their land to feed livestock, and these agroforestry practices have begun to ameliorate the degradation of natural habitats.

Priority Conservation Actions

Conservation actions in this ecoregion should be identified within the context of the larger Himalayan ecosystem. This ecoregion's biodiversity should be represented, and its contribution to the overall functioning of the Himalayan ecosystem should be considered in a regional-scale conservation plan. Some of the important conservation actions that should be addressed include the following.

Short-Term Conservation Actions (1–5 Years)

- Identify botanical hotspots and characteristic habitats of the threatened fauna (the distribution of the white-wing wood duck, for instance, is not well known in this ecoregion).
- Stop livestock grazing in protected areas and regulate grazing immediately around protected areas.
- Assess forested landscapes to identify protected areas, corridors, and habitat linkages for elephant conservation. Allow for mitigation of human-elephant conflicts. The forested landscapes that are critical for elephant conservation are becoming fragmented as people settle in areas that are heavily used by elephant herds.
- Implement a conservation management and monitoring program for arboreal species, especially the endemic golden langur, that are losing habitat because of fragmentation and loss of canopy trees.

Longer-Term Conservation Actions (5–20 Years)

- Develop an ecoregion-based conservation plan in conjunction with the other Himalayan ecoregions. The eastern extent of the ecoregion falls within a Level I TCU (Dinerstein et al. 1997). Therefore, the conservation plan should consider the ecological needs and long-term viability of the tiger population. An ecoregion-based conservation plan is being developed by WWF and ICIMOD (2001) for the eastern Himalayas and includes this ecoregion. The actions needed to realize this long-term conservation vision for the eastern Himalayas should be implemented.
- Promote transfrontier peace parks between India, Nepal, and Bhutan to improve regional conservation.

Focal Species for Conservation Action

Habitat specialists: hornbills (especially Oriental pied, great, rufous-necked, and wreathed hornbills) as indicators of mature, intact forests

Endangered endemic species: chestnut-breasted partridge

Threatened species: tiger, Asian elephant, golden langur, wild dog, back-striped weasel, smooth-coated otter, clouded leopard, gaur, southern serow, Irrawaddy squirrel, particoloured squirrel, great Indian civet, white-winged wood duck

Area-sensitive species: tiger, Asian elephant

Top predator species: tiger, clouded leopard, common leopard

Other: golden langur, wild dogs, gaur, white-wing wood duck

Selected Conservation Contacts

Department of National Parks and Wildlife Conservation, Government of Nepal

Forest Departments of Uttar Pradesh, West Bengal, Assam, and Arunachal Pradesh

King Mahendra Trust for Nature Conservation, Royal Government of Nepal

Nature Conservation Division, Royal Government of Bhutan.

Royal Society for Nature Conservation, Bhutan

WWF-Bhutan

WWF-India

WWF-Nepal

Justification for Ecoregion Delineation

In a previous conservation assessment, MacKinnon (1997) identified four biounits along the Himalayas. But these units were based on longitudinal boundaries, and each included the

TABLE J25.3. WCMC (1997) Protected Areas That Overlap with the Ecoregion.

Protected Area	Area (km²)	IUCN Category
Sohagabarwa [17]	150	IV
Valmikinagar [10]	160	IV
Royal Bardia National Park [35]	510	II
Parsa Wildlife Reserve [35]	400	IV
Royal Chitwan National Park [35]	560	II
Khaling/Neoli	260	IV
Phibsoo	240	IV
Royal Manas [26]	430	II
Total	2,710	

Ecoregion numbers of protected areas that overlap with additional ecoregions are listed in brackets.

range of habitat types represented from the lowlands to the alpine habitats. In our analysis, we sought to represent distinct ecosystems of regional extent in separate ecoregions. Therefore, we used MacKinnon's (1997) digital map of the distribution of original vegetation to delineate the boundaries of the subtropical forests that run the length of the eastern and central Himalayas, flanked by the Terai and Duar savanna grasslands and the temperate broadleaf forests. These subtropical forests were then defined as the Himalayan Subtropical Broadleaf Forests [25]. All the Himalayan ecoregions are part of Udvardy's Himalayan highlands biogeographic province.

Prepared by Gopal S. Rawat and Eric D. Wikramanayake, with contributions by Pralad Yonzon

Ecoregion Number:	**26**
Ecoregion Name:	**Eastern Himalayan Broadleaf Forests**
Bioregion:	**Indian Subcontinent**
Biome:	**Temperate Broadleaf and Mixed Forests**
Political Units:	**Bhutan, India, and Nepal**
Ecoregion Size:	**82,900 km²**
Biological Distinctiveness:	**Globally Outstanding**
Conservation Status:	**Relatively Intact**
Conservation Assessment:	**III**

The Eastern Himalayan Broadleaf Forests [26] is one of the few Indo-Pacific ecoregions that is globally outstanding for both species richness and levels of endemism. The eastern Himalayas are a crossroads of the Indo-Malayan, Indo-Chinese, Sino-Himalayan, and East Asiatic floras as well as several ancient Gondwana relicts that have taken refuge here. Overall, this ecoregion is a biodiversity hotspot for rhododendrons and oaks; for instance, Sikkim has more than fifty rhododendron species, and there are more than sixty species in Bhutan.

In addition to the outstanding levels of species diversity and endemism, the ecoregion also plays an important role in maintaining altitudinal connectivity between the habitat types that make up the larger Himalayan ecosystem. Several birds and mammals exhibit altitudinal seasonal migrations and depend on contiguous habitat up and down the steep Himalayan slopes for unhindered movements. Habitat continuity and intactness are also essential to maintain the integrity of watersheds along these steep slopes. If any of the habitat layers, from the Terai and Duar grasslands along the foothills through the broadleaf forests and conifers to the alpine meadows in the high mountains, are lost or degraded, these processes will be disrupted. For instance, several bird species are found in the temperate broadleaf forests of Bhutan where the habitat is more intact and continuous with the subtropical broadleaf forests lower down, but in Nepal where the habitat continuity has been disrupted, these same birds have limited ranges (C. Inskipp, pers. comm., 2000).

Location and General Description

This ecoregion represents the band of temperate broadleaf forest between 2,000 and 3,000 m, stretching from the deep Kali Gandaki River gorge in central Nepal, eastward through Bhutan, into India's eastern states of Arunachal Pradesh and Nagaland. Champion and Seth (1968) identified a number of broadleaf forest types across the midelevations (1,500–3,000 m) of the Himalayas: the east Himalayan moist mixed deciduous forests, east Himalayan wet temperate forests, Naga Hills wet temperate forests, Alder forests, east Himalayan oak-rhododendron forests, and Himalayan temperate parkland.

The Himalayas themselves trace their origin to the collision between the Eurasian continent and the northward-drifting Deccan Plateau more than 50 million years ago. The northern edge of the Deccan Plateau pushed beneath the Eurasian continent to raise the latter from beneath the Tethys Sea to create what is now the Tibetan Plateau. During this and three subsequent periods of geologic upheaval and uplift, the Himalayas were thrust upward to form the highest mountain range in the world (Wadia 1966).

The Himalayan Mountain Range comprises three east-west–directed parallel zones. The southernmost outer Himalayas, or Siwaliks, lie alongside the Indo-Gangetic Plain and is composed of alluvial deposits that have washed down from the north. It is more recent in origin than the other ranges. The next is the Middle Himalayas, a highly folded system of ridges and valleys that rise to about 5,000 m. The third is the Inner Himalayas, which contain the tallest mountains in the world: Everest, Makalu, Dhaulagiri, and Jomalhari.

The monsoon rains provide about 2,000 mm of precipitation from May to September. Because these monsoons are funneled in from the Bay of Bengal, the eastern Himalayas receive the greatest rainfall, with a progressively drier trend toward the west. Thus, precipitation, topography, and temperature combine to influence the vegetation across this ecoregion.

Two distinct ecological formations of broadleaf forests can be distinguished in this ecoregion depending on the geology and slope (moisture regime): the temperate evergreen forests of oaks (*Quercus* spp.), especially *Quercus lamellosa* in association with *Lithocarpus pachyphylla, Rhododendron arboreum, Rhododendron falconeri, Rhododendron thomsonii, Michelia excelsa, Michelia cathcartii, Bucklandia populnea, Symplocos cochinchinensis*, and other species of *Magnolia, Cinnamomum,* and *Machilus*; and temperate deciduous forests dominated by *Acer campbellii, Juglans regia, Alnus nepalensis, Betula alnoides, Betula utilis,* and *Echinocarpus dasycarpus* (Rao 1994; Puri et al. 1989; Shrestha and Joshi 1997).

In the wetter parts of eastern Nepal, the forests are composed of a *Magnolia-Acer-Osmanthus* association, characterized by *Magnolia campbellii, Acer campbellii, Osmanthus suavis, Schefflera impressa,* and *Corylus ferox* (Chaudhary 1998). Common herbs the include *Luculia gratissima, Lilium wallichianum, Pipanthus nepalensis,* and *Aster himalaicus,* which form a ground cover. Shrestha and Joshi (1997) have reported that an understory of bamboo (*Arundinaria* spp. and *Bambusa* spp.) or species of *Symplocos, Eurya, Rhododendron, Acer, Alnus, Carpinus,* and *Prunus* can also be present in some areas in these elevations in eastern Nepal. In mature forests, the trees are draped with mosses, ferns, and other epiphytes.

Biodiversity Features

The Eastern Himalayan Broadleaf Forests [26] is globally outstanding for both species richness and endemism, especially for its flora. It contains several localized areas of floral richness and endemism—floral hotspots—which are especially rich in rhododendrons and oaks.

The 125 mammal species known to occur in here include four species that are endemic to the ecoregion (table J26.1).

Three of these species are shared with adjacent ecoregions, but the Namdapha flying squirrel, *Biswamoyopterus biswasi*, is a strict endemic whose range distribution is limited to the Eastern Himalayan Broadleaf Forests [26] (from Corbet and Hill 1992). The golden langur is limited to the broadleaf forests to the north of the Brahmaputra River, which flows along the foothills and between the Sankosh and Manas rivers, which flow south from the mountains. Therefore, despite being shared between this ecoregion and the adjacent Himalayan Subtropical Broadleaf Forests [25], the golden langur has a very limited range distribution.

The ecoregion also harbors several threatened species, including the endangered tiger (*Panthera tigris*), red panda (*Ailurus fulgens*), takin (*Budorcas taxicolor*), and serow (*Capricornis sumatraensis*) and the vulnerable Vespertilionidae bat (*Myotis sicarius*), Assamese macaque (*Macaca assamensis*), stump-tailed macaque (*Macaca arctoides*), wild dog (*Cuon alpinus*), back-striped weasel (*Mustela strigidorsa*), clouded leopard (*Pardofelis nebulosa*), and Irrawaddy squirrel (*Callosciurus pygerythrus*) (Hilton-Taylor 2000).

The ecoregion overlaps with a high priority (Level I) TCU (Dinerstein et al. 1997), a landscape with the ecological resources to support a viable tiger population over the long term. It also represents the only opportunity to conserve a tiger population that is adapted to survive in these Himalayan temperate broadleaf forests, where, unlike the productive habitats in the Terai and Duar grasslands and tropical dry forests, the tiger's prey densities are much lower. Under these conditions, the tigers are considered to have adapted their territory size, reproductive strategies, and hunting strategies to these ecological conditions (Wikramanayake et al. 1998).

The red panda is more characteristic of the sub-alpine fir forest ecoregion adjacent to and above this temperate broadleaf forest ecoregion. In this broadleaf forest ecoregion, the red panda is limited to patches of mature fir (*Abies*) forests with a bamboo understory (Yonzon and Hunter 1991).

There are almost 500 bird species, among the highest across the ecoregions in this bioregion. Twelve species are endemic to the ecoregion (table J26.2). Of these, eleven are near-endemic species, and one, the rufous-throated wren-babbler, is a strict endemic that is restricted to the Eastern Himalayan Broadleaf Forests [26].

The bird assemblage also includes several threatened species of pheasants, tragopans, and hornbills that need mature forests and have low tolerances for disturbance. These species, namely the globally threatened rufous-necked hornbill (*Aceros nipalensis*) and Sclater's monal (*Lophophorus sclateri*), and the threatened white-bellied heron (*Ardea insignis*), Blyth's tragopan (*Tragopan blythii*) and Ward's trogon (*Harpactes wardi*) (Hilton-Taylor 2000), therefore can be considered indicators of habitat integrity and deserve conservation attention. The ecoregion also overlaps with BirdLife International's EBA, Eastern Himalayas (130) (Stattersfield et al. 1998).

Status and Threats

Current Status

About two-thirds of this ecoregion's natural habitat is still intact, with most of the remaining large habitat blocks located in northeastern India and Bhutan. Even though the species-rich *Quercus lamellosa* forests are quite well represented within protected areas, the *Quercus lanata* forests in the lower elevations are not. Many of the small areas that remain are so badly degraded that restoration should be considered as a conservation option. Mt. Phulchowki, in the Kathmandu valley, represents an important area of habitat that is currently unprotected and highly threatened (Wikramanayake et al. 1998).

The fifteen protected areas that extend into the ecoregion cover about 5,800 km^2 (7 percent) of the ecoregion (table J26.3). With the exception of Namdapha, none exceed 1,000 km^2. However, there are several very large reserves that over-

TABLE J26.1. Endemic and Near-Endemic Mammal Species.

Family	Species
Cercopithecidae	Semnopithecus geei
Sciuridae	Petaurista magnificus
Sciuridae	Biswamoyopterus biswasi*
Muridae	Niviventer brahma

An asterisk signifies that the species' range is limited to this ecoregion.

TABLE J26.2. Endemic and Near-Endemic Bird Species.

Family	Common Name	Species
Phasianidae	Chestnut-breasted partridge	Arborophila mandellii
Timaliidae	Hoary-throated barwing	Actinodura nipalensis
Timaliidae	Ludlow's fulvetta	Alcippe ludlowi
Turdidae	Rusty-bellied shortwing	Brachypteryx hyperythra
Timaliidae	Elliot's laughingthrush	Garrulax elliotii
Paradoxornithidae	Grey-headed parrotbill	Paradoxornis gularis
Timaliidae	Immaculate wren-babbler	Pnoepyga immaculata
Sylviidae	Grey-crowned prinia	Prinia cinereocapilla
Timaliidae	Mishmi wren-babbler	Spelaeornis badeigularis
Timaliidae	Rufous-throated wren-babbler	Spelaeornis caudatus*
Timaliidae	Snowy-throated babbler	Stachyris oglei
Timaliidae	Spiny babbler	Turdoides nipalensis

An asterisk signifies that the species' range is limited to this ecoregion.

lap across several ecoregions, although only parts of the reserves are represented in this one. Examples include Thrumsing La, Jigme Dorji, and Black Mountains national parks in Bhutan. The Jigme Dorji National Park exceeds 4,000 km^2 and sprawls across three ecoregions to include the alpine meadows, sub-alpine conifer forests, and temperate broadleaf forests represented by this ecoregion. Two others, Kulong Chu and Black Mountains, exceed 1,000 km^2, and Cha Yu, Makalu-Barun, Mehao, and Thrumsing La are more than 500 km^2.

Bhutan recently revised its protected area system to link the existing reserves (Sherpa and Norbu 1999). Plans to develop similar linkages through conservation landscapes have been proposed in the other areas in the eastern Himalayas (WWF and ICIMOD 2001). These plans use ecoregions as basic conservation units for representation of biodiversity.

Types and Severity of Threats

The primary threat to this ecoregion's natural biodiversity is forest clearing for agriculture, plantations, and settlements. The upper elevation limit of cultivation in the eastern Himalayas is about 2,100 m, but the land above the agricultural areas is used for livestock grazing, especially during the summer, and exploited for wood and foliage. Pastoralists often clear and burn these forests to create grazing lands for livestock. Even in the well-represented *Quercus lamellosa*–dominated forests, the lower areas have been extensively cleared.

Because Bhutan is less densely populated than Nepal, habitat loss and degradation are not severe. Therefore, most of the large, intact habitat blocks are in Bhutan.

Chakma and Tibetan refugees and lamas from Bhutan have settled the area surrounding Namdapha National Park in northern India. These settlers have begun to clear forests for shifting cultivation, cut trees for timber and fuelwood, and illegally extract *agar*, *dhoop*, and other forest products (Davis et al. 1995).

A proposal to build a dam across the Noa-Dihing River at Burma Nala will result in inundation of land and increased human settlements, presenting serious threats to the natural biodiversity in this area of the ecoregion (Davis et al. 1995).

Priority Conservation Actions

Conservation actions in this ecoregion should be planned and implemented within the context of the larger Himalayan ecosystem. The long-term biological vision developed for the eastern Himalayas (WWF and ICIMOD 2001) should be the foundation for these conservation actions. The ecoregion's biodiversity and conservation needs will be represented in this plan. Some of the important conservation actions that should be addressed include the following.

Short-Term Conservation Actions (1–5 Years)

- Impose strict controls on illegal hunting of threatened species.
- Reforest and restore critical habitats, especially in the broadleaf forests in the lower elevations of this ecoregion.
- Stop livestock grazing within protected areas and regulate activities in buffer zones.
- Stop deforestation and agricultural expansion on steeper slopes. This is especially relevant in Nepal, where higher and steeper slopes are being cleared.
- Conduct biological inventories of many of the protected areas and conservation landscapes. These inventories should be designed with a conservation management focus and not with the intent of merely compiling species lists.

Longer-Term Conservation Actions (5–20 Years)

- Develop a long-term ecoregion-based conservation plan for the eastern Himalayas (WWF and ICIMOD 2001). Implement the priority actions identified in this plan.

Focal Species for Conservation Action

Habitat specialists: rufous-necked hornbill

Endangered endemic species: rufous-throated wren babbler, *Myotis sicarius*

Area-sensitive species: tiger, altitudinal migrant birds

Top predator species: tiger, leopard

Other: red panda, takin, southern serow, great Indian civet, stump-tailed macaque, wild dog, back-striped weasel, clouded leopard, Irrawaddy squirrel, Sclater's monal, white-bellied heron, Blyth's tragopan, Ward's trogon

Selected Conservation Contacts

Department of Forestry, Royal Government of Bhutan

Department of Wildlife Conservation, Government of Nepal

Forest Department, Government of Sikkim, India

Forest Departments of Uttar Pradesh, West Bengal, Assam, and Arunachal Pradesh, India

ICIMOD

King Mahendra Trust for Nature Conservation, Nepal

Resources Himalaya, Kathmandu, Nepal

TABLE J26.3. WCMC (1997) Protected Areas That Overlap with the Ecoregion.

Protected Area	Area (km²)	IUCN Category
Eagle Nest [31]	120	IV
Mouling [31]	200	II
Mehao [31, 8]	110	IV
Royal Manas [25]	410	II
Shivapuri Waters & Wildlife Reserve	90	IV
Makalu-Barun Conservation Area [28, 39]	70	II
Neora Valley	40	II
Torsa [28, 39]	240	I
Jigme Dorji [28, 39]	560	II
Thrumsing La [39]	310	II
Kulung Chhu WS [28]	390	IV
Black Mountain [28, 39]	600	II
Namdapha	1,580	PRO
Kamlang	780	IV
Cha Yu [28]	320	
Total	5,820	

Ecoregion numbers of protected areas that overlap with additional ecoregions are listed in brackets.

Royal Society for Nature Conservation, Bhutan
WWF-Bhutan
WWF-India
WWF-Nepal

Justification for Ecoregion Delineation

MacKinnon (1997) identified four biounits along the Himalayas. Because these were based on longitudinal boundaries, each included the range of habitat types represented along the north-south axis of the Himalayas, from the lowlands to the alpine habitats. Because we define an ecoregion as an ecosystem of regional extent, in our analysis we sought to represent these distinct ecosystems in separate ecoregions.

We used the deep Kali Gandaki River gorge, an acknowledged biogeographic barrier, as a boundary to separate the band of temperate forests that run along the east-west–directed length of the Himalayas into eastern and western broadleaf forest ecoregions. We then used MacKinnon's (1997) digital map of the distribution of original vegetation to separate the temperate forests from the broadleaf subtropical forests to the south and the sub-alpine conifer forests to the north. Therefore, the temperate broadleaf forests to the east of the Kali Gandaki River are represented within the Eastern Himalayan Broadleaf Forests [26]. All the Himalayan ecoregions are part of Udvardy's Himalayan highlands biogeographic province.

Prepared by Gopal S. Rawat and Eric D. Wikramanayake, with contributions by Pralad Yonzon

Ecoregion Number:	**27**
Ecoregion Name:	**Western Himalayan Broadleaf Forests**
Bioregion:	**Indian Subcontinent**
Biome:	**Temperate Broadleaf and Mixed Forests**
Political Units:	**Nepal, India, Pakistan**
Ecoregion Size:	**55,800 km²**
Biological Distinctiveness:	**Regionally Outstanding**
Conservation Status:	**Endangered**
Conservation Assessment:	**II**

Because the western Himalayas are drier than the eastern extent, the Western Himalayan Broadleaf Forests [27] are less species-rich than their eastern counterpart. This ecoregion is nevertheless of regional conservation importance for its biodiversity and for its role as a critical link in the chain of Himalayan ecosystems that are layered along the steep south-facing slopes. Several of the Himalayan birds and mammals that exhibit altitudinal seasonal migrations depend on contiguous habitat up and down the steep Himalayan slopes for unrestricted movements. If any of the habitat layers—which range from the alluvial savanna and grasslands in the Terai, through the broadleaf forests and conifers, to the alpine meadows in the high mountains—are lost or degraded, these processes will be disrupted. The fact that several bird species known from the subtropical broadleaf forests to the south are expected to occur in this ecoregion but do not may result from the more advanced state of habitat fragmentation in these western Himalayan broadleaf forests, relative to the eastern broadleaf forests (C. Inskipp, pers. comm., 2000). Habitat loss in these steep slopes will also compromise the ecological integrity and hydrology of the watersheds, with far-reaching consequences that will be felt in the Ganges delta thousands of kilometers away.

Location and General Description

This ecoregion represents the temperate broadleaf forests of the western Himalayas, to the west of the Kali Gandaki River gorge in Nepal, through northern India's states of Uttar Pradesh and Himachal Pradesh, and into Jammu and Kashmir, with small sections extending into Pakistan. These temperate forests form a narrow east-west–directed band between 1,500 and 2,600 m.

The Himalayan Mountain Range dates back 50 million years, to the period when the northward-drifting Deccan Plateau collided with the Eurasian continent and inexorably pushed the southern edge of Eurasia upward, raising it from beneath the Tethys Sea to create the Tibetan Plateau, which now lies at more than 4,000 m above sea level. During this and three subsequent periods of geologic upheaval and uplift, the Himalayan Mountains were thrust upward to form what is now the highest mountain range in the world (Wadia 1966). The Himalayas consist of three parallel ranges. The Outer Himalayas lie alongside the Indo-Gangetic Plain. The Middle Himalayas, a highly folded system of ridges and valleys, rise to about 5,000 m. The Inner Himalayas contain the highest mountain peaks in the world. This ecoregion lies along the middle Himalayas.

The Himalayan Range receives most of its moisture from the southwestern monsoon that originates in the Bay of Bengal. The moisture-laden monsoon winds are funneled through the Gangetic Plains toward the mountain range, where most of the precipitation is intercepted by the eastern Himalayas. The western extent therefore receives less precipitation. The drier climate in the west influences the vegetation. For instance, the treeline declines from 4,000 m in the east to about 3,500 m in the west (Kendrick 1989).

Two distinct forest types can be recognized in this ecoregion: evergreen broad-leaved forests and deciduous broad-leaved forests. The former, dominated by *Quercus semecarpifolia*, *Quercus dilatata*, *Quercus lamellosa*, and *Quercus incana*, usually is on the moister southern slopes, which are more influenced by the monsoon (Puri et al. 1989). These forests often are associated with species of Lauraceae (e.g., *Machilus odoratissima*, *Litsea umbrosa*, *Litsea lanuginosa*, *Phoebe pulcherrima*). They also have a dense understory with mosses, ferns, and several epiphytes on the trees, typical conditions expected in moist forests in the lower elevations. The drier forests, especially on the north-facing slopes and along the higher elevations, are characterized by *Quercus ilex*, sometimes mixed with conifers such as *Abies*, *Picea*, *Cedrus*, and *Pinus* spp., with an *Arundinaria*-dominated understory (Puri et al. 1989).

Deciduous broadleaf forests are distributed along riverbanks to the west of the Kali Gandaki River. These forests are composed of *Aesculus indica*, *Juglans regia*, *Carpinus viminea*, *Alnus nepalensis*, and several *Acer* spp. In drier places such as the Upper Karnali, these species are associated with *Populus ciliata*, *Ulmus wallichiana*, and *Corylus colurna*. Whereas *Alnus nepalensis* is the common species along most of the riverine forests, the forests along the upper Karnali River are dominated by *Alnus nitida*, which has not been recorded elsewhere in Nepal (Shrestha and Joshi 1997).

Biodiversity Features

This western Himalayan ecoregion is less species-rich than the eastern temperate forests but nevertheless harbors several large, focal vertebrates of conservation importance. Some of these species include the Asiatic black bear (*Ursus thibetanus*), leopard (*Panthera pardus*), and, in open, steeper hills and woodland habitats, the Himalayan tahr (*Hemitragus jemlahicus*). As important as the species is the ecoregion's role as an integral part of the overall Himalayan ecosystem, which relies on altitudinal connectivity for ecosystem function.

The mammal fauna consists of seventy-six species, of which two vespertilionid bats are endemic to this ecoregion (table J27.1). Whereas *Murina grisea* is a near-endemic species that is shared with the Himalayan Subtropical Pine Forests [31], the known range of *Myotis longipes* (Corbet and Hill 1992) is limited to this ecoregion.

The ecoregion's mammals also include several threatened species, including *Murina grisea*, and the serow (*Capricornis sumatraensis*) (Hilton-Taylor 2000).

The bird fauna is rich, with about 315 species. These include ten species that are near endemic to the ecoregion (table J27.2). All ten are shared with adjacent ecoregions. However, the Himalayan quail is now presumed to be extinct (Grimmet et al. 1998). The Tytler's leaf-warbler and the Kashmir flycatcher breed in the northwestern Himalaya, including in this ecoregion, but migrate to the Western Ghats during the winter.

Several of the pheasants in this ecoregion are candidates as focal species for conservation management. Some of these include the western tragopan (*Tragopan melanocephalus*), Satyr tragopan (*Tragopan satyra*), Koklass pheasant (*Pucrasia macrolopha*), Himalayan monal (*Lophophorus impejanus*), and Cheer pheasant (*Catreus wallichi*). The ecoregion overlaps with the EBA, Western Himalayas (128), identified by BirdLife International (Stattersfield et al. 1998).

Status and Threats

Current Status

Nearly two-thirds of this ecoregion has been cleared or degraded. But several large patches of forests still remain in the extreme western part of the ecoregion.

The twenty protected areas cover 2,770 km^2, or about 5 percent of the ecoregion's area (table J27.3). Most ecoregions are small (average size 139 km^2). One of these, Kistar, exceeds 1,000 km^2, but only about half the ecoregion is represented in this ecoregion. The entire protected area extends into the adjacent Western Himalayan Sub-Alpine Conifer Forests [29] and Northwestern Himalayan Alpine Shrub and Meadows [37]. Two other protected areas, Rupti Bhabha and Govind Pashu Vihar, are also large (more than 800 and 500 km^2, respectively), but only a small proportion of each is represented in this ecoregion.

Types and Severity of Threats

The threats to the remaining blocks of natural habitat from logging and agricultural clearing continue unabated. In the

TABLE J27.1. Endemic and Near-Endemic Mammal Species.

Family	Species
Vespertilionidae	*Myotis longipes**
Vespertilionidae	*Murina grisea*

An asterisk signifies that the species' range is limited to this ecoregion.

TABLE J27.2. Endemic and Near-Endemic Bird Species.

Family	Common Name	Species
Phasianidae	Cheer pheasant	*Catreus wallichi*
Phasianidae	Himalayan quail	*Ophrysia superciliosa*
Phasianidae	Western tragopan	*Tragopan melanocephalus*
Aegithalidae	White-cheeked tit	*Aegithalos leucogenys*
Aegithalidae	White-throated tit	*Aegithalos niveogularis*
Fringillidae	Spectacled finch	*Callacanthis burtoni*
Muscicapidae	Kashmir flycatcher	*Ficedula subrubra*
Sylviidae	Tytler's leaf-warbler	*Phylloscopus tytleri*
Fringillidae	Orange bullfinch	*Pyrrhula aurantiaca*
Sittidae	Kashmir nuthatch	*Sitta cashmirensis*

An asterisk signifies that the species' range is limited to this ecoregion.

TABLE J27.3. WCMC (1997) Protected Areas That Overlap with the Ecoregion.

Protected Area	Area (km^2)	IUCN Category
Kistar National Park [37, 29]	540	II
Overa-Aru [37, 29]	110	IV
Rupti Bhabha [37, 29]	370	IV
Hirapora [31]	50	IV
Manshi	20	IV
Salkhala	10	IV
Ghamot	110	
Machayara	90	
Gulmarg	130	V
Lachipora	80	IV
Limber	10	IV
Dachigam	330	II
Gamgul Siahbehi	90	IV
Kais	10	IV
Khokhan	20	IV
Nargu [31]	110	IV
Pong Dam	250	IV
Daranghati	30	IV
Askot [31]	140	IV
Govind Pashu Vihar [38]	270	IV
Total	2,770	

Ecoregion numbers of protected areas that overlap with additional ecoregions are listed in brackets.

wetter areas where *Quercus lamellosa* dominates, the upper limit of cultivation is about 2,100 m; therefore, most of the forest below this elevation has been extensively cleared for farming. But the higher regions are used for grazing, especially during the summer, and regularly set fires to promote new plant growth for cattle destroy the understory (FAO 1981). Therefore, forest regeneration is retarded, and the long-term viability of the forest is compromised. Erosion from a combination of road building, overgrazing, and excessive fuelwood collection is of serious concern in this steep-sloped ecoregion.

Priority Conservation Actions

Because of the close links this ecoregion has with the other ecoregions along the southern face of the western Himalayas, conservation actions should be planned and implemented within the context of the larger western Himalayan ecosystem. However, some conservation actions that should be addressed include the following.

Short-Term Conservation Actions (1–5 Years)

- Control forest clearing and agricultural expansion, especially on steeper slopes.
- Assess protected areas for their size, representation of habitat types, and biodiversity hotspots. Assess them also for their relationships to existing habitat blocks and other protected areas.
- Stop illegal hunting.
- Halt livestock grazing in protected areas and regulate it in buffer zones and steep slopes that are vulnerable to erosion.
- Explore alternatives to excessive dependence on forests for fuelwood by an increasing human population.
- Institute better planning and more effective regulations to ensure that large pilgrimages and unregulated tourism practices do not result in ill-planned and unsustainable development, such as roads, trails, hotels, and other infrastructure, that has a heavy impact on the environment and biodiversity.
- Implement more effective planning and regulations to stop the rapid conversion of forests to orchards.
- Control agriculture's heavy use of pesticides to decrease the pollution of water courses and the loss of biodiversity.

Longer-Term Conservation Actions (5–20 Years)

- Develop a long-term ecoregion-based conservation plan for the entire western Himalayas with the biological vision that is contained within the plan developed for the eastern Himalayas (WWF and ICIMOD 2001).

Focal Species for Conservation Action

Endangered endemic species: Myotis longipes, Murina grisea

Top predator species: leopard

Other: altitudinal migrant birds, Asiatic black bear, southern serow, wild dog, Himalayan tahr, western tragopan, Satyr tragopan, Koklass pheasant, Himalayan monal, Kalij pheasant, Cheer pheasant

Selected Conservation Contacts

Chief Conservator of Forests, Uttarakhand Region, Nainital, India

Department of National Parks and Wildlife Conservation, Government of Nepal

Forest Departments of Himachal Pradesh and Jammu and Kashmir

Forest Research Institute, Dehra Dun, India

G. B. Pant Institute of Himalayan Environment and Development, Almora, India

ICIMOD

King Mahendra Trust for Nature Conservation, Nepal

Wildlife Institute of India, Dehra Dun, India

WWF-India

WWF-Nepal

Justification for Ecoregion Delineation

In a previous analysis, MacKinnon (1997) identified four biounits along the Himalayas. These units, with longitudinal boundaries, encompass the range of habitat types along the north-south axis of the Himalayas, from the lowlands to the alpine habitats. In a conservation analysis of India, Rodgers and Panwar (1988) used a similar concept to identify conservation units for the Indian Himalaya.

Our analysis is based on ecoregions, defined as ecosystems of regional extent. Therefore, we delineated ecoregions so that each will represent the distinct ecosystems. We used the deep Kali Gandaki River gorge, an acknowledged biogeographic barrier, to separate the band of temperate forests that run along the length of the Himalayas into eastern and western broadleaf forest ecoregions. We then used MacKinnon's (1997) digital map of the distribution of original vegetation to separate and extract the temperate forests from the broadleaf subtropical forests to the south and the sub-alpine conifer forests to the north. Therefore, the temperate broadleaf forests to the west of the Kali Gandaki River are represented in the Western Himalayan Broadleaf Forests [27]. All the Himalayan ecoregions are part of Udvardy's Himalayan highlands biogeographic province.

Prepared by Gopal S. Rawat and Eric D. Wikramanayake, with contributions by Pralad Yonzon

Ecoregion Number:	**28**
Ecoregion Name:	**Eastern Himalayan Sub-Alpine Conifer Forests**
Bioregion:	**Indian Subcontinent**
Biome:	**Temperate Coniferous Forests**
Political Units:	**Bhutan, India, and Nepal**
Ecoregion Size:	**26,900 km²**
Biological Distinctiveness:	**Globally Outstanding**
Conservation Status:	**Vulnerable**
Conservation Assessment:	**I**

The Eastern Himalayan Sub-Alpine Conifer Forests [28] represent the transition from the forested ecoregions of the Himalayas to treeless alpine meadows and boulder-strewn alpine screes. Their ecological role within the interconnected Himalayan ecosystem, which extends from the alluvial grasslands along the foothills to the high alpine meadows, makes the forests of this ecoregion a conservation priority. Conservation of the Himalayan biodiversity is contingent on protecting the interconnected processes among the Himalayan ecosystems. For instance, several Himalayan birds and mammals exhibit altitudinal seasonal migrations and depend on contiguous habitats that permit these movements. The integrity of the watersheds of the rivers that originate in the high mountains of this majestic range depends on the intactness of habitat, from the high elevations to the lowlands. If any of the habitat layers are lost or degraded, these processes will also be disrupted.

The ecoregion also straddles the transition from the southern Indo-Malayan to the northern Palearctic fauna. Here tigers yield to snow leopards, and sambar are replaced by blue sheep. But the ecoregion also has its own specialized flora and fauna, such as the musk deer and red panda, which are limited to these mature temperate conifer forests.

Location and General Description

The ecoregion represents the belt of conifer forest between 3,000 and 4,000 m, from east of the Kali Gandaki River in Nepal through Bhutan and into the state of Arunachal Pradesh, in India. These forests usually are confined to the steeper, rocky, north-facing slopes and therefore are inaccessible to human habitation and cultivation.

The ecoregion has recent origins. The Himalayas were born of the tumultuous collision between the northward-drifting Deccan Plate and the Eurasian continent during the Cretaceous. The Himalayas were created during this and three subsequent periods of geologic upheaval and uplift (Wadia 1966).

Today, the Himalayan Mountain Range comprises three east-west–directed parallel zones. The southernmost is the outer Himalayas, also known as the Siwaliks, which lies adjacent to the Indo-Gangetic Plain. The Middle Himalayas represent a series of ridges and valleys that rise to about 5,000 m and contain this ecoregion. The Inner Himalayas are where the imposing, giant peaks such as Everest, Makalu, and Dhaulagiri tower far above the Deccan Plateau.

The Himalayas capture moisture from the monsoons that sweep in from the Bay of Bengal. Therefore, the eastern Himalayas, being closer to the source, receive more precipitation than the western Himalayas. This trend is reflected in the timberline, which declines from 4,500 m in the east to about 3,600 m in the west (Kendrick 1989).

These temperate conifer forests are dominated by fir (*Abies spectabilis*), larch (*Larix griffithii*), hemlock (*Tsuga dumosa*), *Juniperus recurva*, and *Juniperus indica*. Several colorful rhododendrons (*Rhododendron hodgsonii*, *R. barbatum*, *R. campylocarpum*, *R. campanulatum*, *R. fulgens*, *R. thomsonii*) grow profusely in the understory, along with *Viburnum grandiflorum*, *Lonicera angustifolia*, *Betula utilis*, *Acer* spp., *Sorbus* spp., *Juniperus indica*, and *J. recurva*. There is considerable species turnover among the rhododendrons along the east-west axis. Species common in central Nepal can be absent from the community in eastern Bhutan. This contributes to the overall diversity of the ecoregion.

Between 2,500 and 3,000 m, the wetter areas are dominated by *Tsuga dumosa*, which also forms mixed stands with fir. Blue pine (*Pinus wallichiana*) occurs in the drier inner valleys closer to Tibet. Blue pine is especially common in the Khumbu region, where it forms a distinct community. Patches of pure birch (*Betula utilis*) represent a pioneer stage that climaxes as a *Betula-Abies* mixed forest (Puri et al. 1989). *Taxus baccata* is an important, uncommon species in this forest community.

Juniper woodland is another distinct habitat type within this sub-alpine conifer zone. It characteristically grows along the flat, inner river valleys with willow (*Salix* spp.) and *Prunus* spp. The juniper woodlands in Tsarijathang, in Jigme Dorji National Park, Bhutan, in particular are used by takin (*Budorcas taxicolor*) as summer habitat.

Biodiversity Features

Overall, the mammal fauna consists of eighty-nine species. The ecoregion lies within the ecotone between the Indo-Malayan and Palearctic zoogeographic zones. Therefore, the mammal fauna from both zones, such as civets, martens, Himalayan tahr, and muntjac are included within the ecoregion. Three species—two squirrels and a murid rodent—with limited distributions are considered to be endemic to the ecoregion (table J28.1). Whereas the squirrels are near-endemic species that are shared with neighboring ecoregions, *Apodemus gurkha* is a strict endemic that has been recorded only from this ecoregion (Corbet and Hill 1992).

Although not endemic species, the red panda and the Himalayan musk deer (*Moschus chrysogaster*) are characteristic of the mature fir forests represented by this ecoregion. The red pandas usually are limited to *Abies-ringal* bamboo forests between 3,000 and 4,000 m, where precipitation is high (Yonzon and Hunter 1991). The musk deer is widely hunted for its musk glands, which allegedly have pharmaceutical properties. The waxy, brown musk (a single musk gland yields about 25 grams of musk) can fetch as much as US$45,000 per kilogram on the international market.

The ecoregion harbors several threatened species, including the endangered red panda (*Ailurus fulgens*), takin (*Budorcas taxicolor*), serow (*Capricornis sumatraensis*), and particolored squirrel (*Hylopetes alboniger*) (Hilton-Taylor 2000). The endangered tiger (*Panthera tigris*) roams the broadleaf forests lower down and occasionally ventures into these conifer forests. But the prey base in the conifer forests is insufficient to harbor a viable resident population within this ecoregion. In addition to these endangered species, the following species are considered vulnerable (Hilton-Taylor 2000) and should also receive conservation attention: the Vespertilionidae bat *Myotis sicarius*, wild dog (*Cuon alpi-*

TABLE J28.1. Endemic and Near-Endemic Mammal Species.

Family	Species
Sciuridae	*Petaurista magnificus*
Sciuridae	*Petaurista nobilis*
Apodemus	*Apodemus gurkha**

An asterisk signifies that the species' range is limited to this ecoregion.

TABLE J28.2. Endemic and Near-Endemic Bird Species.

Family	Common Name	Species
Phasianidae	Chestnut-breasted partridge	Arborophila mandellii
Phasianidae	Buff-throated partridge	Tetraophasis szechenyii*
Timaliidae	Hoary-throated barwing	Actinodura nipalensis
Timaliidae	Ludlow's fulvetta	Alcippe ludlowi
Timaliidae	Immaculate wren-babbler	Pnoepyga immaculata
Psittacidae	Derbyan parakeet	Psittacula derbiana*

An asterisk signifies that the species' range is limited to this ecoregion.

nus), Asiatic black bear (*Ursus thibetanus*), and Himalayan tahr (*Hemitragus jemlahicus*).

There are more than 200 species of birds known from this ecoregion, of which six are considered endemic to it (table J28.2). Four species are shared with the adjacent ecoregions, and two are strict endemics limited to this ecoregion. These two species—buff-throated partridge and Derbyan parakeet—are known only from the sub-alpine conifer forests in the northeastern part of Arunachal Pradesh (Grimmet et al. 1998).

The ecoregion's other birds also include the globally threatened Tibetan eared-pheasant (*Crossoptilon harmani*) and Sclater's monal (*Lophophorus sclateri*) (Hilton-Taylor 2000). Several other species, such as some of the pheasants, tragopans, and partridges—blood pheasant (*Ithaginis cruentus*), Blyth's tragopan (*Tragopan blythii*), Satyr tragopan (*Tragopan satyra*), Ward's trogon (*Harpactes wardi*), and chestnut-breasted partridge (*Arborophila mandellii*)—are characteristic of these sub-alpine Himalayan forests and have low disturbance thresholds. Therefore, they should receive conservation attention and can be used as focal species to monitor habitat integrity. Two large birds of prey, the lammergeier (*Gypaetus barbatus*) and the Himalayan griffon (*Gyps himalayensis*), which soar high above the mountains in these alpine regions and embody the aura of the large open spaces of the high Himalayas, are space-dependent species that should be conservation targets.

The ecoregion overlaps with two EBAs, Eastern Himalayas (130) and Southern Tibet (133), which have been identified by BirdLife International as important areas that harbor birds with limited breeding ranges (Stattersfield et al. 1998). Between them, these two EBAs contain twenty-four such restricted-range bird species.

Status and Threats

Current Status

Only about a third of the ecoregion's natural habitat has been cleared or degraded. Most of the remaining habitat occurs in large, contiguous blocks. The fourteen protected areas in the ecoregion cover more than 6,000 km², or about 22 percent of the ecoregion's area (table J28.3).

Some of these protected areas—Langtang, Makalu-Barun, Black Mountain, and Sakteng—extend across several ecoregions and exceed 1,000 km² in area. Annapurna and Jigme Dorji also extend across multiple ecoregions but are much larger, exceeding 4,000 km². More importantly, several of these protected areas will be linked by natural habitat and included within conservation landscapes designed to maintain the altitudinal habitat continuity (WWF and ICIMOD 2001).

Types and Severity of Threats

Threats to the ecoregion's remaining natural habitats stem mostly from cutting trees for fuelwood by the local people and by tourist trekkers and mountaineers. In some parts of the ecoregion, fires are set to clear juniper forests and create grasslands for livestock grazing. Musk deer and Asiatic black bear are hunted for their musk and gall bladder, respectively.

Priority Conservation Actions

Because the ecoregion is largely inaccessible and unsuitable for agriculture and settlement, the threats are not severe. Conservation actions, especially long-term actions, in this Himalayan ecoregion should be developed within the context of the entire Himalayan ecosystem. Some of the important conservation actions that are needed include the following.

Short-Term Conservation Actions (1–5 Years)

- Assess the current protective status to see whether key habitats in the ecoregion are included in the protected area system.
- Institute proper management in protected areas, especially to guard against cutting, burning, and grazing in juniper woodlands and mature fir forests.
- Survey status of pheasants and black-necked cranes (including in Arunachal Pradesh).
- Survey, manage, and protect musk deer and red pandas.
- Survey status *Taxus baccata* var. *wallichiana* and control its illegal collection.

TABLE J28.3. WCMC (1997) Protected Areas That Overlap with the Ecoregion.

Protected Area	Area (km²)	IUCN Category
Annapurna Conservation Area [38, 39]	980	
Langtang National Park [39]	260	II
Makalu-Barun Conservation Area [26, 39]	350	II
Singalila	120	II
Singba WS	90	?
Kyongnosla WS	50	IV
Torsa [26, 39]	300	I
Jigme Dorji [26, 39]	1,110	II
Sakteng WS [39]	390	IV
Kulung Chhu WS [26]	850	IV
Black Mountain [26, 39]	870	II
Cha Yu [26]	260	?
Mo Tuo [39]	380	?
Dibang Valley [39]	150	PRO
Total	6,160	

Ecoregion numbers of protected areas that overlap with additional ecoregions are listed in brackets.

Longer-Term Conservation Actions (5–20 Years)

- Implement and manage conservation landscapes that have been proposed (WWF and ICIMOD 2001).

Focal Species for Conservation Action

Endangered endemic species: Derbyan parakeet, buff-throated partridge

Area-sensitive species: lammergeier, Himalayan griffon

Migratory species: black-necked crane, other altitudinal migratory birds

Other: takin, Juniper woodlands habitat, red panda, southern serow, particoloured squirrel, great Indian civet, *Myotis sicarius*, Himalayan tahr, musk deer, Himalayan black bear, blood pheasant, Blyth's tragopan, Satyr tragopan, Ward's trogon, chestnut-breasted partridge, Szecheny's partridge

Selected Conservation Contacts

Chief Wildlife Warden, Itanagar, Arunachal Pradesh

Department of National Parks and Wildlife Conservation, Government of Nepal

Forest Department, Government of Sikkim

Forest Department, Royal Government of Nepal

G. B. Pant Institute of Himalayan Environment and Development, Sikkim Unit

ICIMOD

Mountain Institute, Kathmandu, Nepal

Nature Conservation Section, Royal Government of Bhutan

Resources Himalaya, Kathmandu, Nepal

WWF-Bhutan

WWF-India

WWF-Nepal

Justification for Ecoregion Delineation

We used the Kali Gandaki River, widely considered a biogeographic barrier that defines the eastern and western Himalayan biotas, as a boundary to separate the sub-alpine conifer forests that extend across the length of the Himalayas into western and eastern ecoregions. We then used digital forest cover maps from MacKinnon (1997) to identify the distribution of the conifer forests, bordered by alpine meadows and broadleaf forests to the north and south, respectively. The belt of sub-alpine conifer forest to the east of the Kali Gandaki River was then placed in the Eastern Himalayan Sub-Alpine Conifer Forests [28]. All the Himalayan ecoregions are part of Udvardy's Himalayan highlands biogeographic province.

Prepared by Gopal S. Rawat and Eric D. Wikramanayake, with contributions from Pralad Yonzon

Ecoregion Number:	**29**
Ecoregion Name:	**Western Himalayan Sub-Alpine Conifer Forests**
Bioregion:	**Indian Subcontinent**
Biome:	**Temperate Coniferous Forests**
Political Units:	**India, Nepal, Pakistan**
Ecoregion Size:	**39,400 km²**
Biological Distinctiveness:	**Regionally Outstanding**
Conservation Status:	**Vulnerable**
Conservation Assessment:	**II**

The Western Himalayan Sub-Alpine Conifer Forests [29] represent the frontline of the forested ecoregions in the western Himalayan region, standing against the treeless alpine meadows to the north. This ecoregion plays a critical ecological role as part of the Himalayan ecosystem, with interconnected processes that extend from the Terai and Duar grasslands along the foothills to the high alpine meadows and boulder-strewn scree that lie above the treeline. Several Himalayan birds and mammals exhibit seasonal migrations up and down the steep mountain slopes and depend on contiguous habitat for these movements. If any of the habitat layers are lost or degraded, these movements can be disrupted. Therefore, conservation of this ecoregion is critical to maintain the biodiversity—species and processes—of this youngest and tallest mountain range on Earth.

Location and General Description

The ecoregion represents the sub-alpine conifer forests between 3,000 and 3,500 m in the Himalayan Mountain Range to the west of the Kali Gandaki River in central Nepal. It extends from western Nepal through the northern Indian states of Uttar Pradesh and Himachal Pradesh into Jammu and Kashmir and into eastern Pakistan. The western extents of the Himalayas have more extensive conifer forests of blue pine (*Pinus wallichiana*), chilgoza pine (*Pinus gerardiana*), fir (*Abies spectabilis*), silver fir (*Abies pindrow*), and spruce (*Picea smithiana*) than the moister eastern part of the mountain range.

The Middle Himalayas within which the ecoregion lies rise to about 5,000 m. The Middle Himalayas are flanked by the Outer Himalayas, also known as the Siwaliks to the south and the Inner Himalayas to the north. The former are composed of alluvial deposits that have washed down from the north for thousands of years and rise to about 1,000 m. The latter contain the tallest peaks on Earth, such as Everest, Makalu, and Dhaulagiri.

The Himalayas are a young mountain range that dates back to a period more than 50 million years ago, when the north-drifting Deccan Plateau collided with the northern Eurasian continent. This collision and subsequent uplift gave rise to the Himalayan Range that now stretches for more than 3,000 km, from Pakistan to Myanmar and beyond.

The Himalayas receive rainfall from the southwestern monsoon that sweeps in from the Bay of Bengal. Most of the monsoon rains are intercepted and expended in the eastern Himalayas; therefore, the western extent receives less precipitation and is drier. This moisture gradient influences the vegetation. For instance, the treeline descends from 4,000 m in the east to about 3,300 m in the west (Kendrick 1989).

The ecoregion has several recognizable forest types based on floral associations. These include pure fir forest (*Abies spectabilis*), mixed oak-fir forest (*Quercus semecarpifolia* and *Abies spectabilis*), mixed rhododendron, fir, and birch forest (*Rhododendron campanulatum*, *Abies spectabilis*, and *Betula utilis*), and mixed coniferous forest (*Abies spectabilis*, *Pinus wallichiana*, and *Picea smithiana*) (Shrestha and Joshi 1997). Cypress (*Cupressus torulosa*) and deodar (*Cedrus deodara*) are common above 2,400 m (FAO 1981). Fir (*Abies spectabilis*) usually forms a continuous belt between 3,000 and 3,500 m on the southern side of the main ranges in central Nepal and can be mixed with *Quercus semecarpifolia*, *Betula utilis*, and a rhododendron understory. These sub-alpine areas have a number of economically important species such as *Daphne bholua*, *Arundinaria* spp., *Betula utilis*, and a large number of medicinal plants (Shrestha and Joshi 1997).

Biodiversity Features

This belt of conifer forest sitting immediately below the alpine meadows in the western Himalayas does not have a spectacularly rich fauna or flora but does harbor several focal species of large mammals of conservation importance, including the brown bear (*Ursus arctos*).

The ecoregion's mammal fauna of fifty-eight species includes one strict endemic rodent (table J29.1) that has been recorded only from this ecoregion (Corbet and Hill 1992).

The mammals include several threatened species, including the southern serow (*Naemorhedus sumatraensis*), Himalayan tahr (*Hemitragus jemlahicus*), and markhor (*Capra falconeri*) (IUCN 1996) that warrant conservation attention.

The ecoregion's bird fauna consists of 285 species, of which 9 are endemic to the ecoregion (table J29.2). However, none of these are strict endemics (i.e., limited to this ecoregion).

Other species such as pheasants, and tragopans—e.g., Koklass pheasant (*Pucrasia macrolopha*), western tragopan (*Tragopan melanocephalus*), and Himalayan monal (*Lophophorus impejanus*)—are characteristic of these subalpine western Himalayan forests and have low disturbance thresholds. Therefore, they should receive conservation attention and can be used as focal species to monitor habitat integrity and as focal species for conservation management. The Himalayan griffon (*Gyps himalayensis*), a large bird of prey that soars high above the mountains in these alpine regions and embodies the sense of space in the high Himalayas, can be another focal species. The ecoregion overlaps with an EBA, Western Himalaya (128), identified by BirdLife International (Stattersfield et al. 1998).

Status and Threats

Current Status

Although the ecoregion is less populated than some of the other Himalayan ecoregions (especially those in the lower elevations), more than 70 percent of the natural habitat has been cleared or degraded. Nevertheless, this ecoregion contains some of the least disturbed forests in the western Himalayas. The eleven protected areas cover 2,400 km², or about 6 percent of the ecoregion (table J29.3). Most of the protected areas that fall within the ecoregion are small (<500 km²). However, some of the protected areas are large and overlap across several ecoregions (table J29.3). Notable among these are Kistwar, Royal Dhorpatan, and Rupti Bhabha.

Types and Severity of Threats

The steep slopes of some of the high mountains have been deforested for intensive cultivation, although the practice of terracing has greatly reduced erosion. Large-scale collection of

TABLE J29.1. Endemic and Near-Endemic Mammal Species.

Family	Species
Muridae	Hyperacrius wynnei

An asterisk signifies that the species' range is limited to this ecoregion.

TABLE J29.2. Endemic and Near-Endemic Bird Species.

Family	Common Name	Species
Phasianidae	Himalayan quail	Ophrysia superciliosa
Phasianidae	Western tragopan	Tragopan melanocephalus
Timaliidae	Hoary-throated barwing	Actinodura nipalensis
Aegithalidae	White-cheeked tit	Aegithalos leucogenys
Aegithalidae	White-throated tit	Aegithalos niveogularis
Fringillidae	Spectacled finch	Callacanthis burtoni
Timaliidae	Immaculate wren-babbler	Pnoepyga immaculata
Fringillidae	Orange bullfinch	Pyrrhula aurantiaca
Sittidae	Kashmir nuthatch	Sitta cashmirensis

An asterisk signifies that the species' range is limited to this ecoregion.

TABLE J29.3. WCMC (1997) Protected Areas That Overlap with the Ecoregion.

Protected Area	Area (km²)	IUCN Category
Great Himalayan (west) [37]	370	II
Kugti WS [37]	110	IV
Tundah WS [37]	270	IV
Kistar National Park [27, 37]	400	II
Overa-Aru [27, 37]	160	IV
Rupti Bhabha [27, 37]	130	IV
Ayub	10	V
Tangir GR	160	
Rara National Park	120	II
Khaptad National Park	250	II
Royal Dhorpatan Hunting Reserve [38]	420	
Total	2,400	

Ecoregion numbers of protected areas that overlap with additional ecoregions are listed in brackets.

the morel mushroom (*Morchella esculenta*) from this ecoregion by the local people for export coincides with the breeding season of several pheasants and high-altitude mammals. Collection of wood by the local people for their own use and for sale to tourist trekkers and mountaineering parties is also a substantial threat, especially because the high-altitude forests are very slow to regenerate.

Priority Conservation Actions

Conservation actions, especially any long-term actions, should be developed within the context of the entire Himalayan ecosystem. Some of the important conservation actions that should be addressed include the following.

Short-Term Conservation Actions (1–5 Years)

- Identify and protect all remaining old-growth conifer forests in this ecoregion.
- Monitor the collection of wild morel mushrooms and implement regulations that will minimize disturbances to breeding birds and mammals.
- Strictly control the collection of *Taxus baccata*, if necessary, through special legal provisions.

Longer-Term Conservation Actions (5–20 Years)

- Identify conservation landscapes within this and other Himalayan ecoregions that include altitudinal connectivity, adequate protected areas, and representation of the biodiversity of each ecoregion.
- Implement conservation activities in these conservation landscapes.

Focal Species for Conservation Action

Area-sensitive species: lammergeier, Himalayan griffon

Migratory species: altitudinal migratory birds

Other: Himalayan tahr, markhor, southern serow, endemic birds, pheasants (Koklass pheasant, western tragopan, Himalayan monal) as indicators of habitat quality, brown bear, southern serow

Selected Conservation Contacts

Chief Conservator of forests, Uttarakhand Region, Nainital, India

Conifer Research Institute, Shimla, Himachal Pradesh, India

Department of National Parks and Wildlife Conservation, Government of Nepal

Forest Departments of Himachal Pradesh, Jammu, and Kashmir

Forest Research Institute, Dehra Dun, India

G. B. Pant Institute of Himalayan Environment and Development, Himachal Unit, India

ICIMOD

King Mahendra Trust for Conservation of Nature, Kathmandu, Nepal

Wildlife Institute of India, Dehra Dun, India

WWF-India

WWF-Nepal

Justification for Ecoregion Delineation

We used the Kali Gandaki River as the eastern boundary of the ecoregion. This deep gorge is widely considered to be a biogeographic barrier that defines the eastern and western Himalayan biotas. We used digital forest cover maps from MacKinnon (1997) to identify the distribution of the Himalayan temperate conifer forests to the west of the Kali Gandaki River, bordered by alpine meadows and broadleaf forests to the north and south, respectively. This belt of sub-alpine conifer forest was then placed in the Western Himalayan Sub-Alpine Conifer Forests [29]. All the Himalayan ecoregions are part of Udvardy's Himalayan highlands biogeographic province.

Prepared by Gopal S. Rawat and Eric D. Wikramanayake, with contributions by Pralad Yonzon

Ecoregion Number:	**30**
Ecoregion Name:	**East Afghan Montane Coniferous Forest**
Bioregion:	**Indian Subcontinent**
Biome:	**Temperate Coniferous Forests**
Political Units:	**Pakistan and Afghanistan**
Ecoregion Size:	**7,300 km²**
Biological Distinctiveness:	**Bioregionally Outstanding**
Conservation Status:	**Vulnerable**
Conservation Assessment:	**V**

The East Afghan Montane Coniferous Forests [30] are found between 2,000 and 3,300 m. These temperate coniferous forests of western Pakistan and northeastern Afghanistan support a variety of avifauna and harbor the largest remaining populations of Chiltan markhor (*Capra falconeri chiltanensis*).

Location and General Description

The East Afghan Montane Coniferous Forest [30] spans from Eastern Hindu Kush at Jalalbad Valley, Kunar Range, and Ghazni Province in the north. In the south, it extends to lower Kohistan of Indus, Safed Koh, Takhat-i-Suleiman, and Koh-i-Maran ranges and Quetta Pass of western Pakistan. The average precipitation is 200–400 mm. Temperatures vary from −12°C to 40°C. The soil in the north is made up of clay substrate covered with coarse gravel, small stones or rocks, and organic detritus. South, the ecoregion has a bedrock of sedimentary limestone, and soils on the flat lands are sandy loam. The Alingar and Kunar rivers in northeastern Afghanistan join the Kabul River and flow southeast into the Indus River of West Pakistan. Hashmat Khan, Sarobi, and Duronta are three lakes significant as staging areas for breeding waterfowl species that are situated in northeast Afghanistan and designated as Important Bird Areas by BirdLife International.

Two types of forests make up this ecoregion because of the influence of the monsoon. Areas between 2,100 and 2,500 m receive less monsoon rain and have dry coniferous species. Pinaceae forest creates a belt between the Fagaceae and the Cedrus community. These include *Pinus gerardiana* and

Quercus baloot species. The understory vegetation of this elevation includes *Indigofera gerardiana*, *Sambucus ebulus*, and *Plecanthrus rugosus*. As one moves to higher elevations (2,500–3,100 m) where there is a continuous rain from the monsoon, a temperate deciduous species mixes with the conifers. *Picea smithiana*, *Pinus wallichiana*, *Quercus semecarpifolia*, and *Cedrus deodara* develop into dense forest cover at this elevation with trees 15–30 m high. *Cedrus deodara* is one of the most important timber species in both Pakistan and Afghanistan. At 3,100–3,300 m the precipitation decreases and the cedar forest is replaced by junipers (*Juniperus seravschanica*). The dominant herbaceous ground vegetation at higher elevations is composed of hemicryptophytes such as *Polygonum amplexicaule*, *Rumex nepalense*, *Fragaria nubicola*, and *Berberis* spp. and geophytes such as *Polygonatum geminiflorum*, *Lilium polyphyllum*, and *Habenaria aitchisonii* (Hassinger 1968; Freitag 1971).

Biodiversity Features

This ecoregion is a prime location for migrant bird species and breeding waterfowl. There are approximately 30,000 migratory birds in the northern part of this ecoregion alone, found mostly near lakes. Waterfowl species of this area include pochards, coots, moorhens, black-necked grebe, greater flamingos, spoonbills, shelducks, and marbled teals. The largest number of the highly threatened western tragopan (*Tragopan melanocephalus*) and long-billed bush-warbler (*Bradypterus major*) are found here. These birds have a specialized habitat need: pockets of pristine forests, which have been severely fragmented and thus led to their dwindling population. Brooks's leaf-warbler (*Phylloscopus subviridis*), Kashmir nuthatch (*Sitta cashmirensis*), orange bullfinch (*Pyrrhula aurantiaca*), and Tytler's leaf-warbler (*Phylloscopus tytleri*) are some of the restricted-range species in the open coniferous and deciduous forest throughout the ecoregion (Stattersfield et al. 1998). There are no endemic bird species.

Mammals such as Chiltan wild goat (*Capra falconeri chiltanensis*), Himalayan black bear (*Ursus thibetanus langier*), leopard cat (*Prionailurus bengalensis*), grey langur (*Semnopithecus entellus*), rhesus macaque (*Macaca mulatta*), jackal (*Canis aureus*), and red fox (*Vulpes vulpes*) are found throughout the ecoregion. Royle's high mountain voles (*Alticola royeli*), Afghan vole (*Microtus afghans*), marbled polecat (*Vormela peregusna*), and Euphrates jerboa (*Allactaga euphratica*) are found mainly in the northern part of this ecoregion. Other small mammals include two bat species (whiskered bat and grey long-eared bat), two squirrel species (giant red flying squirrel and Kashmir flying squirrel), and yellow throated marten, Turkestan rat, and long-tailed field mouse. The Chiltan wild goat is listed as critical, and the Himalayan bear, Kashmir flycatcher, and Himalayan musk deer are listed on the Red Data List as vulnerable (Hilton-Taylor 2000). There are no endemic mammal species.

Status and Threats

Current Status

Only 8 percent of the ecoregion is covered by the protected area systems (table J30.1). Hazar Ganji–Chiltan National Park has helped to recover the Chiltan markhor population, which was nearing extinction. In the 1950s the population of Chiltan markhor used to exceed 1,200. Studies in 1970 indicated that the number fell to less than 150. Since the area was designated as a park, this number has grown to about 300. A total area of 2.2 million ha (34 percent) of well-preserved coniferous forest located in Kunar, Paktia, and Nangarhar provinces of east and north Afghanistan are thinning. Some no longer produce sustainable products because of the widespread extraction of timber without any consideration for future environmental degradation. Of the three forest regions, Kunar Province has the highest diversity content and the biggest area.

Types and Severity of Threats

One of the main threats of this area is logging of deodar, fir, blue pine, and oak to fulfill the ever-increasing demand for timber, fuel, and fodder. The abundant and dense forests of this ecoregion have been replaced by construction sites for industrial developments. Freshwater lakes and wetlands have been severely affected by irrigation channels. This has caused low water levels during the dry months, which disturb breeding birds. There are also introduced amphibian species such as a toad (*Bufo virdis*) and frogs of *Rana* spp. in the wetlands of northeast Afghanistan. The rivers and lakes have been polluted by domestic sewage, agricultural effluents, industrial waste, and garbage dumps. The use of chemical pesticides in irrigation channels has contributed to the decrease in the number of mammals such as jackals. Although the red fox is still widespread, it is ruthlessly hunted for its valuable pelt.

Priority Conservation Actions

Short-Term Conservation Actions (1–5 Years)

- Protect areas of rich biodiversity, such as Kunar Province, with a complete ban on timber extraction.
- Legislate and implement the draft forest law of Afghanistan.
- Designate for protection freshwater lakes and rivers that are biologically outstanding to minimize human activities.

Longer-Term Conservation Actions (5–20 Years)

- Begin wide-scale conservation assessment to identify areas of outstanding biodiversity contents.
- Clear land mines from forest areas in Afghanistan to make possible research and management of resources.

TABLE J30.1. WCMC (1997) Protected Areas That Overlap with the Ecoregion.

Protected Area	Area (km²)	IUCN Category
Surkhab	40	DE
Hazar Ganji-Chiltan	620	V
Wam	100	VI
Sasnamana	60	IV
Gogi	80	VI
Zawarkhan	40	UA
Total	940	

Ecoregion numbers of protected areas that overlap with additional ecoregions are listed in brackets.

Focal Species for Conservation Action

Area-sensitive species: rhesus macaque, jackal, red fox, Euphrates jerboa, yellow-throated marten, Royle's high mountain voles, Afghan vole

Migratory species: waterfowl species

Other: Chiltan wild goat, Himalayan black bear, Himalayan musk deer, Kashmir flycatcher

Justification for Ecoregion Delineation

We identified five ecoregions—Sulaiman Range Alpine Meadows [42], South Iran Nubo-Sindian Desert and Semi-Desert [43], Baluchistan Xeric Woodlands [44], Rajasthan–North Pakistan Sandy Desert [45], and East Afghan Montane Coniferous Forests [30]—from the Baluchistan sub-unit (I3d). All five of these ecoregions extend westward and have a portion of their ecoregion beyond the limits of this analysis. These ecoregions overlap with numerous Udvardy biogeographic provinces outside the scope of this analysis. These include the Hindu Kush highlands to the north and the Anatolian-Iranian Desert, Iranian Desert, and Caucaso-Iranian highlands to the south and west.

Prepared by Meseret Taye

Ecoregion Number:	**31**
Ecoregion Name:	**Himalayan Subtropical Pine Forests**
Bioregion:	**Indian Subcontinent**
Biome:	**Tropical and Subtropical Coniferous Forests**
Political Units:	**Bhutan, India, Nepal, and Pakistan**
Ecoregion Size:	**76,200 km²**
Biological Distinctiveness:	**Regionally Outstanding**
Conservation Status:	**Vulnerable**
Conservation Assessment:	**II**

The Himalayan Subtropical Pine Forests [31] are the largest in the Indo-Pacific region. They stretch throughout most of the 3,000-km length of this the world's youngest and highest mountain range. Some scientists believe that climate change and human disturbance are causing the lower-elevation oak forests to be gradually degraded and invaded by the drought-resistant Chir pine (*Pinus roxburghii*), the dominant species in these subtropical pine forests. Biologically, the ecoregion does not harbor exceptionally high levels of species richness or endemism, but it is a distinct facet of the region's biodiversity that should be represented in a comprehensive conservation portfolio.

Location and General Description

The subtropical pine forests represented by this ecoregion extend as a long, disjunct strip from Pakistan in the west, through the states of Jammu and Kashmir, Himachal Pradesh, and Uttar Pradesh in northern India, into Nepal and Bhutan. Although Champion and Seth (1968) indicate the presence of large areas of Chir pine in Arunachal Pradesh, the eastern-most extent of large areas of Chir pine is in Bhutan.

The world's deepest river valley, the Kali Gandaki, bisects the ecoregion in Nepal, dividing it into a drier, western conifer forest and a wetter and richer eastern conifer forest. However, the species assemblages, community structure, and ecosystem dynamics are not sufficiently different to separate this pine forest into eastern and western ecoregions, as was done for the broadleaf forest ecoregions.

The Himalayan Mountain Range was formed about 50 million years ago, when the northward drifting Deccan Plateau collided with the northern Eurasian continent.

The mountain range is now made up of three east-west–directed parallel zones, with the southernmost Outer Himalayas, also known as the Siwaliks, lying adjacent to the Indo-Gangetic Plain. The next is the Middle Himalayas, representing a series of ridges and valleys that rise to about 5,000 m, and the third is the Inner Himalayas with imposing high peaks such as Everest, Makalu, and Dhaulagiri.

Most of the rainfall is brought by the southwestern monsoon from the Bay of Bengal. The monsoon rains are intercepted and expended in the eastern Himalayas, which are closer to the Bay of Bengal; therefore, the western region receives less precipitation. The climatic gradient influences the vegetation in the Himalayas. For instance, the treeline in the western Himalayas is more than 500 m lower than in the east (Kendrick 1989).

The dominant species in this belt of subtropical pine forest is Chir pine. Because of frequent fires, pine forests do not have a well-developed understory. However, frequently burnt slopes support a rich growth of grasses including *Arundinella setosa*, *Imperata cylindrica*, *Themeda anathera*, and *Cymbopogon distans* and a number of shrubs such as species of *Berberis*, *Rubus*, and other thorny bushes (Shrestha and Joshi 1997).

In Himachal Pradesh, extensive patches of Chir pine grow on the lower parts of Kangra and Una Districts, but toward the eastern parts of Himachal Pradesh and in the lower areas of the Uttar Pradesh hills, Chir pine grows in scattered patches commonly in association with *Shorea robusta*, *Anogeissus latifolia*, and *Cordia vestita* (Rawat and Mukherjee 1999). Extensive Chir pine plantations are present in Himachal Pradesh and in northwestern Uttar Pradesh.

In western Nepal, Chir pine forests predominate on all aspects of the slope, in contrast to areas further west, where they develop mainly on the southern slopes. In central and eastern Nepal, this ecoregion covers parts of the Siwalik and Mahabharat ranges on the south-facing slopes between 1,000 and 2,000 m (Shrestha and Joshi 1997).

Biodiversity Features

Compared with the adjacent broadleaf forests, this ecoregion is neither exceptionally rich in species nor high in endemic species. However, it does provide habitat to several endemic bird species that are found in adjacent ecoregions.

The ecoregion's mammal fauna consists of about 120 species. Most of the mammals are not specialized or limited to this ecosystem. Because of the lack of undergrowth and browse, the ecoregion does not support a significant herbivore community. Therefore, the carnivore community that is dependent on the herbivore (i.e., prey) density is also depressed. Some of the characteristic mammals that can be used as focal

TABLE J31.1. Endemic and Near-Endemic Bird Species.

Family	Common Name	Species
Phasianidae	Chestnut-breasted partridge	Arborophila mandellii
Phasianidae	Cheer pheasant	Catreus wallichi
Timaliidae	Ludlow's fulvetta	Alcippe ludlowi
Turdidae	Rusty-bellied shortwing	Brachypteryx hyperythra
Timaliidae	Elliot's laughingthrush	Garrulax elliotii
Timaliidae	Immaculate wren-babbler	Pnoepyga immaculata
Timaliidae	Snowy-throated babbler	Stachyris oglei
Timaliidae	Hoary-throated barwing	Actinodura nipalensis
Timaliidae	Spiny babbler	Turdoides nipalensis
Timaliidae	Mishmi wren-babbler	Spelaeornis badeigularis
Phasianidae	Western tragopan	Tragopan melanocephalus

An asterisk signifies that the species' range is limited to this ecoregion.

species in this ecoregion include goral (*Nemorhaedus goral*), barking deer (*Muntiacus muntjak*), and yellow-throated marten (*Martes flavigula*).

The bird fauna consists of 480 species, including 11 species that are endemic to this ecoregion (table J31.1). However, none of these species are strict endemics (i.e., limited to this ecoregion), being shared with adjacent ecoregions.

This ecoregion overlaps with two of BirdLife International's EBAs: Eastern Himalayas (128) and Western Himalayas (130) (Stattersfield et al. 1998).

Status and Threats

Current Status

More than half of this ecoregion's natural habitat has been cleared or degraded. In central and eastern Nepal, terraced agriculture plots, especially between 1,000 and 2,000 m, have replaced nearly all the natural forest. Other than in the less populated western regions, little natural forest remains in Nepal. Similarly, habitat loss is widespread in Pakistan, Jammu and Kashmir, and Himachal Pradesh and Uttar Pradesh states in India. The few larger blocks of remaining habitat blocks are now found in Bhutan.

There are twenty-six protected areas in the ecoregion that cover a little over 3,000 km^2, representing about 4 percent of its area (table J31.2). Most of these protected areas are small (average 119 km^2). Four protected areas extend into adjacent ecoregions, but with the exception of Corbett National Park that exceeds 500 km^2, these are also small reserves.

Types and Severity of Threats

Threats to the forests in this densely populated ecoregion include overgrazing, overexploitation for fuelwood and fodder, and shifting cultivation. Since 1975, extensive quarrying on the lower slopes of some areas has destroyed extensive areas of forest. As a result of loss of forest cover and undergrowth, erosion has become a serious problem. Especially in areas where road construction is taking place, there is overgrazing by livestock, and excessive fuelwood is collected. However, Chir pine forests are resilient and can tolerate considerable human pressure.

Priority Conservation Actions

Because this ecoregion is an integral part of the Himalayan ecosystem, any conservation actions here should be planned within the context of the overall Himalayan ecosystem. Some of the important conservation actions that should be addressed include the following.

Short-Term Conservation Actions (1–5 Years)

- Assess and map Chir pine and oak zones for long-term monitoring of succession.
- Manage grass and fuelwood harvesting from this ecoregion.
- Prepare and implement fire control and management regimes.

Longer-Term Conservation Actions (5–20 Years)

- Monitor changes in oak and pine zones.
- Implement the conservation actions in the eastern extent of the ecoregion, as identified in the long-term ecoregion-based conservation plan being developed for the eastern Himalayas (WWF and ICIMOD 2001).
- Develop and implement a similar conservation plan for the western extent of the Himalayas.

TABLE J31.2. WCMC (1997) Protected Areas That Overlap with the Ecoregion.

Protected Area	Area (km^2)	IUCN Category
Ayubia	140	V
Margalla Hills	100	V
Kazinag GR	70	II
Moji WS	80	
Hirapora [27]	70	IV
Mori Said Ali GR	50	
Killan GR	80	
Nargu [27]	170	IV
Askot [27]	80	IV
Shikari Devi	80	IV
Bandli	10	IV
Gobind Sagar	180	IV
Naina Devi	40	IV
Majathal	20	IV
Darlaghat	120	IV
Simla Water Catchment	40	Ia
Chail	130	IV
Shilli	2	IV
Talra	40	IV
Corbett [17]	310	II
Fambong Lho	50	IV
Maenam	50	IV
Kane	50	IV
Pakhui	860	IV
Itanagar	190	IV
Sessa Orchid	100	IV
Total	3,112	

Ecoregion numbers of protected areas that overlap with additional ecoregions are listed in brackets.

Focal Species for Conservation Action

Endangered endemic species: endemic birds, especially the Cheer pheasant

Other: goral

Selected Conservation Contacts

Chief Conservator of Forests, Uttarakhand Region, Nainital, India

Chief Wildlife Warden, Itanagar, Arunachal Pradesh, India

Conifer Research Institute, Shimla, Himachal Pradesh, India

Department of Forestry, Royal Government of Bhutan

Department of National Parks and Wildlife Conservation, Government of Nepal

Forest Department, Government of Sikkim

Forest Departments, Himachal Pradesh and Jammu and Kashmir

Forest Department, Nepal

Forest Research Institute, Dehra Dun, India

G. B. Pant Institute of Himalayan Environment and Development, Himachal Unit, India

G. B. Pant Institute of Himalayan Environment and Development, Sikkim Unit, India

King Mahendra Trust for Nature Conservation, Nepal

Nature Conservation Division, Royal Government of Bhutan

Wildlife Institute of India, Dehra Dun, India

WWF-Bhutan

WWF-India

WWF-Nepal

Justification for Ecoregion Delineation

Previous analyses of conservation units in the Indo-Malayan region by MacKinnon (1997) and for the Indian Himalaya by Rodgers and Panwar (1988) placed the Himalayan subtropical pine forests in conservation units that included the range of habitat types within the south-facing slopes of the Himalayas. In this analysis, we sought to represent the distinct ecosystems of regional extent in separate ecoregions. Therefore, we used MacKinnon's (1997) digital map of the distribution of original vegetation to delineate the boundaries of the subtropical pine forests that run the length of the Himalayan Mountain Range and place them in a separate ecoregion, the Himalayan Subtropical Pine Forests [31]. All the Himalayan ecoregions are part of Udvardy's Himalayan highlands biogeographic province.

Prepared by Gopal S. Rawat and Eric D. Wikramanayake, with contributions from Pralad Yonzon

Ecoregion Number:	**32**
Ecoregion Name:	**Indus River Delta–Arabian Sea Mangroves**
Bioregion:	**Indian Subcontinent**
Biome:	**Mangroves**
Political Units:	**India, Pakistan**
Ecoregion Size:	**5,800 km^2**
Biological Distinctiveness:	**Regionally Outstanding**
Conservation Status:	**Critical**
Conservation Assessment:	**II**

The Indus River Delta–Arabian Sea Mangroves [32] represent a mangrove habitat that is adapted to some of the most extreme temperatures and salinity conditions in the Indo-Pacific region. As a transition from the marine to freshwater and terrestrial systems, mangroves provide critical habitat for numerous species of fishes and crustaceans that are adapted to live among the tangled mass of pneumatophores, the roots that reach up from the muddy, anaerobic substrate to get the mangroves' supply of oxygen.

Location and General Description

This ecoregion lies at the delta of the Indus River, which originates in the Tibetan Plateau and flows through the northwestern part of India and into the arid Thar Desert in Pakistan before finally emptying into the Arabian Sea. The mangrove sloughs are exceptionally saline because of the high evaporation rates and from the salts that are washed down by the river which flows through a highly saline area. Climatic conditions are extreme. Ambient temperatures range from near-freezing temperatures in the winter to higher than 50°C during the summer. All rainfall is associated with the July to September southwest monsoon, which brings a mere 100–500 mm of precipitation.

Mangroves in general are not diverse compared with most other terrestrial ecosystems, and undisturbed mangrove forests have a dense canopy with little stratification and an undergrowth made up of seedlings and saplings from the canopy trees. The Indus River mangroves are even less diverse, being composed of nearly monospecific stands of *Avicennia marina*, a species that is highly resistant to high salinity levels and capable of surviving the region's extreme conditions (Spalding et al. 1997). Other species that are sometimes associated with the *Avicennia* include *Rhizophora apiculata* and *Acanthus ilicifolius*, with occasional smaller patches of *Rhizophora mucronata* and *Ceriops tagal* scattered throughout. The former usually is closer to creeks.

Biodiversity Features

This mangrove ecoregion provides important habitat for fish and invertebrates and serves as critically important spawning grounds and nurseries for fishes and aquatic crustaceans.

There are 123 bird species known from this ecoregion, although none are considered endemic to the ecoregion.

Status and Threats

Current Status

Although this area has been heavily degraded and more than 85 percent of the habitat has been lost, several patches of intact habitat still exist. There are three protected areas that cover about 820 km², or 16 percent of the ecoregion (table J32.1), but even these protected areas are degraded.

Types and Severity of Threats

Threats to this mangrove ecoregion stem from several sources. Mangroves are cut for fuelwood, the major source of firewood for the more than 10,000 people who live in the area, and for fodder to feed the domestic livestock, which include camels (Spalding et al. 1997).

Other threats stem from industrial pollution from the city of Karachi, oil spills and discharges from the ships that anchor in the city's harbor, and increased salinity levels caused by flow diversions in the Indus River. Seasonal fishers live in the mangrove forests and cut and clear the forests while setting up camps.

Priority Conservation Actions

It is imperative that the remaining larger patches of mangroves be protected. Habitat should be restored where possible, especially within and around the protected areas. Here we offer some short and longer-term conservation actions.

Short-Term Conservation Actions (1–5 Years)

- Identify the larger blocks of habitat for protection.
- Conduct biodiversity surveys in this little-studied ecosystem.
- Place controls to prevent or minimize pollution threats that emanate from ships and from the city of Karachi.
- Control the currently unsustainable wood and fodder harvest levels.
- Strengthen protected area management.

Longer-Term Conservation Actions (5–20 Years)

- Monitor the integrity of the mangrove habitat, including its role as a nursery for fishes and invertebrates, pollution levels, and habitat regeneration.

Focal Species for Conservation Action

Fishes and selected invertebrates (especially the juvenile stages of fishes and crustaceans), *Avicennia marina*

TABLE J32.1. WCMC (1997) Protected Areas That Overlap with the Ecoregion.

Protected Area	Area (km²)	IUCN Category
Marho Kotri	810	IV
Cut Munarki Chach	3	UA
Mirpur Sakro	10	IV
Total	823	

Ecoregion numbers of protected areas that overlap with additional ecoregions are listed in brackets.

Justification for Ecoregion Delineation

We used MacKinnon's (1997) digital map of original vegetation and Spalding et al. (1997) to identify the extent of the mangroves represented in this ecoregion. This ecoregion lies within Udvardy's Thar Desert biogeographic province.

Prepared by Eric D. Wikramanayake

Ecoregion Number:	**33**
Ecoregion Name:	**Sundarbans Mangroves**
Bioregion:	**Indian Subcontinent**
Biome:	**Mangroves**
Political Units:	**Bangladesh, India**
Ecoregion Size:	**20,400 km²**
Biological Distinctiveness:	**Globally Outstanding**
Conservation Status:	**Endangered**
Conservation Assessment:	**I**

The Sundarbans Mangroves [33] ecoregion is the world's largest mangrove ecosystem. Named after the dominant mangrove species *Heritiera fomes*, locally known as *sundri*, this is the only mangrove ecoregion that harbors the Indo-Pacific region's largest predator, the tiger (*Panthera tigris*). Unlike in other habitats, here tigers live and swim among the mangrove islands, where they hunt scarce prey such as chital deer (*Cervus axis*), barking deer (*Muntiacus muntjak*), wild pig (*Sus scrofa*), and even macaques (*Macaca mulatta*). Quite frequently, the people who venture into these impenetrable forests to gather honey, to fish, and to cut mangrove trees to make charcoal also fall victim to the tigers.

But the ecoregion's importance is not based solely on its role as a priority tiger conservation area (Dinerstein et al. 1997). Mangroves are a transition from the marine to freshwater and terrestrial systems. They provide critical habitat for numerous species of fishes and crustaceans that are adapted to live, reproduce, and spend their juvenile lives among the tangled mass of roots, known as pneumatophores, that grow upward from the anaerobic mud to get the trees' supply of oxygen.

Location and General Description

The ecoregion lies in the vast delta formed by the confluence of the Ganges, Brahmaputra, and Meghna rivers. The maze of mangrove channels extends across southern Bangladesh and India's West Bengal State. The June to September monsoon brings heavy rains and frequent, devastating cyclones that cause widespread destruction. Annual rainfall can exceed 3,500 mm, and daytime temperatures can exceed a stifling 48°C during these monsoon months.

Mangroves are not diverse compared with most other terrestrial ecosystems. The undisturbed forests have an unstratified, dense canopy and an undergrowth made up of seedlings and saplings of the canopy trees (Tomlinson 1986). In the Sundarbans, the mangrove forests are characterized by *Heritiera fomes*, a species valued for its timber. Other species that make up the forest assemblage include *Avicennia* spp., *Xylocarpus mekongensis*, *X. granatum*, *Sonneratia apetala*, *Bruguiera gymnorrhiza*, *Cereops decandra*, *Aegiceras corniculatum*, *Rhizophora mucronata*, and the palm *Nypa fruticans* (Puri et al. 1989).

TABLE J33.1. Endemic and Near-Endemic Bird Species.

Family	Common Name	Species
Alcedinidae	Brown-winged kingfisher	Pelargopsis amauropterus*

An asterisk signifies that the species' range is limited to this ecoregion.

Biodiversity Features

This vast mangrove ecosystem along the marine, freshwater, and terrestrial interfaces provides critical ecosystem functions. The tangled mass of roots from mangrove trees provides safe havens for juvenile stages of a gamut of species, from fish fry to shrimp nauplii.

The ecoregion harbors several mammals, but the most charismatic undoubtedly is the majestic Bengal tiger that swims from island to mangrove island searching for and hunting scarce prey. Because this ecoregion represents the only example of tigers ecologically adapted to a life in the mangroves, it has been designated a Level I TCU (Dinerstein et al. 1997). The tiger's reputation as a human-eater is greater here than anywhere else in its range; the people who venture into the Sundarbans must take great precautions to avoid being attacked. One fascinating deterrent is wearing a mask in the back of the head because tigers are believed to be less liable to attack a human looking directly at them.

Several other predators dwell in this labyrinth of channels. Two species of crocodiles (*Crocodylus porosus* and *C. palustris*), the Gangetic gavial (*Gavialis gangeticus*), and the water monitor lizard (*Varanus salvator*) use both land and water to hunt and bask in. Sharks and the Gangetic freshwater dolphins (*Platanista gangetica*) inhabit the waterways. And several birds of prey patrol the sky overhead. More cryptic but equally fascinating are the mudskippers, a gobioid fish that climbs out of the water into mudflats and even climbs trees. An abundance of crabs, hermit crabs, and shrimp scavenge among the roots.

More than 170 bird species are known to inhabit these mangrove forests, including a single endemic species (table J33.1). This brown-winged kingfisher is limited to the coastal habitats in this ecoregion.

The bird assemblage also includes the globally threatened lesser adjutant (*Leptoptilos javanicus*) and threatened masked finfoot (*Heliopais personata*) (Hilton-Taylor 2000). There are twelve birds of prey that coexist here, including the osprey (*Pandion haliaetus*), white-bellied sea eagle (*Haliaeetus leucogaster*), and grey-headed fish-eagle (*Ichthyophaga ichthyaetus*). The mangrove ecosystem also is an important staging and wintering area for migratory birds, which include several species of shorebirds, gulls, and terns.

Status and Threats

Current Status

Bangladesh supports one of the world's highest human population densities. About half of this ecoregion's mangrove forests have been cut down to supply the fuelwood and other natural resources extracted from these forests by this large population. Despite the intense and large-scale exploitation, the ecoregion still is one of the largest contiguous areas of mangroves in the world.

There are seven protected areas that cover almost 2,700 km^2, or 15 percent of the ecoregion (table J33.2). Despite the high proportion of the ecoregion being within the protected area system, only one of these, Sajnakhali, is large enough to support a space-dependent species such as the tiger. Many of the protected areas also lack trained and dedicated personnel and infrastructure to adequately manage them.

Types and Severity of Threats

The conservation status of the ecoregion was changed from vulnerable to endangered because of the projected human threats. The human population in the Sundarbans, now estimated at more than 2 million (Spalding et al. 1997), continues to increase very rapidly. Hunting and trapping wildlife, cutting and lopping trees for fuelwood and to make charcoal, and overexploiting the trees for timber by the forestry industry are some of the most severe current threats. Shrimp fry are being collected at unsustainable levels to supply the shrimp grow-out industry, and the mangrove forests are cut and cleared to build shrimp grow-out ponds, contributing to degradation and habitat loss. There is also the potential for harmful effluents to enter the mangrove waterways from a proposed fertilizer plant.

But some of the greatest threats to this ecoregion's biodiversity emanate from thousands of kilometers away. The rivers that feed and flush the mangroves bring down heavy silt loads as a result of deforestation and erosion in the Himalayan Range. This silt and water turbidity have profound effects on the sensitive mangrove ecosystem and its flora and fauna, especially on the juvenile stages that the mangroves support. The diversion of more than 30 percent of the Ganges River's dry season flow through the Farraka Barrage in India to provide irrigation for agriculture has drastically increased salinity levels and disrupted fish migration and breeding patterns. The delicately balanced community composition in these mangroves is determined to a large part by salinity levels.

Priority Conservation Actions

The ecoregion has been included within a high-priority TCU (Dinerstein et al. 1997), and a landscape conservation plan for the ecoregion using the tiger and its associated ecological needs as a management focus is recommended. Seidensticker

TABLE J33.2. WCMC (1997) Protected Areas That Overlap with the Ecoregion.

Protected Area	Area (km^2)	IUCN Category
Sajnakhali	2,090	IV
Sundarbans East	210	IV
Char Kukri-Mukri	30	IV
Sundarbans South	200	IV
Sundarbans West	130	IV
Halliday Island	4	IV
Lothian Island	20	IV
Total	2,684	

Ecoregion numbers of protected areas that overlap with additional ecoregions are listed in brackets.

and Hai (1983), Santiapillai (1999), and Moudud (1999) provide recommendations for short-term and long-term conservation actions for the Sundarbans in Bangladesh, using the tiger as a focal species.

Short-Term Conservation Actions (1–5 Years)

- Make a clear plan to demarcate human use and core protected zones.
- Conduct a status survey for tigers and prey species both within and outside protected areas.
- Conduct mass education and awareness campaigns for tiger conservation and to mark the UNESCO World Heritage Site declaration.
- Develop EIA guidelines for gas and oil exploration.
- Develop and implement stronger legal measures to curb hunting.
- Extend the protected areas to provide larger core habitat that can support tiger populations.
- Prohibit commercial fishing in the vicinity of the protected areas.
- Provide alternative means of livelihood for people who are dependent on honey and other nontimber forests products.

Longer-Term Conservation Actions (5–20 Years)

- Strengthen the existing protected area network and develop land- and resource-use plans for the landscape, using the tiger as a focal species.
- Over the longer term, maintain hydrologic and flow regimes to ensure that the mangrove system is flushed and salinity regimes are maintained at natural levels.

Focal Species for Conservation Action

Area-sensitive species: tiger, Gangetic dolphin

Top predator species: tiger

Migratory species: migratory birds

Other: Gangetic ghavial

Selected Conservation Contacts

Chief Wildlife Warden, Calcutta, West Bengal

Department of Forests and Wildlife Conservation, Bangladesh

Field Director, Sundarbans Tiger Reserve

WWF-India

Justification for Ecoregion Delineation

In keeping with our definition of representing habitat types of regional extent in separate ecoregions, we placed the large area of mangrove forests into the Sundarbans Mangroves [33]. We used MacKinnon's (1997) digital map of the original vegetation, aided by Spalding et al. (1997) to define the ecoregion boundaries. This ecoregion falls within Udvardy's Bengalian rain forest biogeographic province.

Prepared by Gopal S. Rawat and Eric D. Wikramanayake

Ecoregion Number:	**34**
Ecoregion Name:	**Godavari-Krishna Mangroves**
Bioregion:	**Indian Subcontinent**
Biome:	**Mangroves**
Political Units:	**India**
Ecoregion Size:	**6,800 km²**
Biological Distinctiveness:	**Regionally Outstanding**
Conservation Status:	**Critical**
Conservation Assessment:	**II**

The Godavari-Krishna Mangroves [34] provide a critical buffer between the marine and terrestrial ecosystems along the eastern coast of the Indian subcontinent, especially at the estuaries formed by the Godavari and Krishna rivers, which originate in the Western Ghats and run right across the vast Deccan Plateau.

Location and General Description

The ecoregion extends along the coastline as narrow disjunct patches but forms larger habitat blocks in the estuaries of Godavari and Krishna rivers in the states of Orissa and Andhra Pradesh, respectively.

Mangrove ecoregions are not exceptionally rich, but their importance should not be based solely on their species richness. Mangroves play a critical ecological role as a transition habitat from the marine to freshwater and terrestrial systems. They also provide important habitat for numerous species of fishes and crustaceans that are adapted to live and reproduce among the tangled mass of mangrove roots. The juvenile stages of many of these species depend on these mangroves for refuge and survival.

The mangrove vegetation is influenced by tidal fluctuations and salinity patterns. They are not diverse compared with most of the other terrestrial ecosystems (Tomlinson 1986). Undisturbed forests have a dense, unstratified canopy. The undergrowth is made up of seedlings and saplings from the canopy trees. The characteristic flora of these mangroves includes *Avicennia marina*, *Suaeda* spp., *Rhizophora* spp., and *Bruguiera* spp. Other species such as *Avicennia officinalis*, *Aegiceras corniculatum*, *Ceriops*, *Lumnitzera racemosa*, and *Excoecaria agallocha* are less common. Climbers such as *Derris trifoliata* and *Dalbergia spinosa*, undershrubs of *Suaeda* spp. and *Acanthus ilicifolius* and, very rarely, *Sonneratia apetala*, *Xylocarpus mekongensis*, *Salicornia brachiata*, *Arthrocnemum indicum*, and *Sesuvium portulacastrum* provide additional structure to these forests (Puri 1989).

Biodiversity Features

This mangrove ecoregion provides critical habitat for many species of vertebrates and invertebrates, and is an important spawning ground and nursery for fish fry, shrimp, crabs, and other invertebrates. Crocodiles (*Crocodylus* spp.), monitor lizards (*Varanus* spp.) (Pandav and Choudhury 1996), and various snake species forage for smaller lizards and other prey. Hermit crabs, fiddler crabs, and mudskippers take refuge in the spaces among the tangle of mangrove roots from egrets and other wading birds that hunt them.

The ecoregion harbors more than 140 bird species, including the globally threatened lesser florican (*Eupodotis indica*) and a large community of aquatic birds that includes flamingoes (*Phoenicoptreus* spp.), spot-billed pelicans (*Pelecanus philippensis*), spoonbills (*Platalea* spp.), and painted storks (*Mycteria leucocephala*). In the Krishna River mangroves, Prasad (1989) has reported heron rookeries that are 2–3 ha.

Status and Threats

Current Status

More than 90 percent of this ecoregion's natural habitat has been destroyed. Three small protected areas cover a mere 930 km^2 (table J34.1). Although this represents about 14 percent of the ecoregion's area, most of the protected areas are degraded.

Types and Severity of Threats

Because of continuing threats from human activities, the conservation status of this ecoregion was changed from endangered to critical. Most of the threats stem from clearing the forests for shrimp culture, agriculture, plantations, and urban development. Pollution from urban and agricultural runoff exerts great stresses on this delicate ecosystem and on the juvenile stages of many species of fishes and invertebrates that use the mangroves as nurseries and are highly susceptible to changes in environmental quality (Spalding et al. 1997).

Other threats include freshwater diversion for agriculture that prevents or reduces the regular flushing within the system, causing the mangroves to become stagnant. Construction of harbors and channels also reduces freshwater flows, thus increasing the salinity beyond the tolerance levels of the floral and faunal communities.

Priority Conservation Actions

Most of the natural habitats in these mangrove habitats have been lost, and conservation actions will necessarily be salvage actions—attempts to save the few patches that are left.

Short-Term Conservation Actions (1–5 Years)

- Prohibit clearing of additional mangrove habitats for prawn farms.
- Stop further wood cutting and implement an ecodevelopment project to provide alternate sources of fuelwood.
- Stop cattle grazing and provide cattle check posts at key points.
- Stop additional road construction (see Prasad 1989).
- Strengthen management of the existing protected area network.

Longer-Term Conservation Actions (5–20 Years)

- Monitor the habitat quality of the ecoregion.
- Monitor key indicators of ecosystem integrity, such as salinity, presence of fish fry and shrimp larvae, migrant bird populations, and wading bird populations.

Focal Species for Conservation Action

Aquatic community (especially selected estuarine fishes and invertebrates and their juvenile stages), crocodiles, lesser floricans, assemblage of wading birds, the assemblage of migratory birds that indicate intact large-scale processes.

Selected Conservation Contacts

Chief Wildlife Wardens, Orissa and Andhra Pradesh
French Institute, Pondicherry
Wildlife Institute of India
WWF-India

Justification for Ecoregion Delineation

We used MacKinnon's (1997) digital map of original vegetation and Spalding et al. (1997) to delineate the mangrove habitat along the coast of the Deccan Peninsula. These mangrove habitats were represented by this Godavari-Krishna Mangroves [34].

Prepared by Gopal S. Rawat and Eric D. Wikramanayake

Ecoregion Number:	**35**
Ecoregion Name:	**Terai-Duar Savanna and Grasslands**
Bioregion:	**Indian Subcontinent**
Biome:	**Tropical and Subtropical Grasslands, Savannas, and Shrublands**
Political Units:	**India, Nepal, Bhutan**
Ecoregion Size:	**34,500 km^2**
Biological Distinctiveness:	**Globally Outstanding**
Conservation Status:	**Critical**
Conservation Assessment:	**I**

This ecoregion contains the highest densities of tigers, rhinos, and ungulates in Asia. One of the features that elevates it to the Global 200 is the diversity of ungulate species and extremely high levels of ungulate biomass recorded in riverine grasslands and grassland-forest mosaics (Seidensticker 1976; Dinerstein 1980). The world's tallest grasslands, found in this ecoregion, are the analogue of the world's tallest forests and are a phenomenon unto themselves. Very tall grasslands are rare worldwide in comparison with short grasslands and are the most threatened. Tall grasslands are indicators of mesic or wet conditions and nutrient-rich soils. Most have been converted to agricultural use.

Location and General Description

The Terai-Duar Savanna and Grasslands [35] ecoregion sits at the base of the Himalayas, the world's youngest and tallest mountain range. About 25 km wide, this narrow lowland ecoregion is a continuation of the Gangetic Plain. The

TABLE J34.1. WCMC (1997) Protected Areas That Overlap with the Ecoregion.

Protected Area	Area (km^2)	IUCN Category
Point Calimere	250	IV
Pulicat Lake	640	IV
Bhitar Kanika	40	IV
Total	930	

Ecoregion numbers of protected areas that overlap with additional ecoregions are listed in brackets.

ecoregion stretches from southern Nepal's Terai, Bhabar, and Dun Valleys eastward to Banke and covers the Dang and Deokhuri Valleys along the Rapti River. A small portion reaches into Bhutan, and each end crosses the border into India's states of Uttar Pradesh and Bihar.

This ecoregion covers a wide range of habitat types—savanna grasslands, evergreen and deciduous forests, thorn forest, and steppe—corresponding to different moisture conditions (Shrestha and Joshi 1997). Low in elevation, this ecoregion is hot and humid in the summer, and during the late dry season temperatures commonly reach 40°C (FAO 1981). Annual monsoon floods deposit silt from the rivers that meander across the grasslands (Mishra and Jeffries 1991). Areas buried in silt return to tall grasslands by the end of the following monsoon, and low-lying areas inundated for a few days only are recharged with an annual load of nutrients (Dinerstein 2002).

In the Nepal Terai, which include the tallest grasslands in the world, characteristic species include *Saccharum spontaneum*, *Saccharum benghalesis*, *Phragmitis kharka*, *Arundo donax*, *Narenga porphyracoma*, *Themeda villosa*, *Themeda arundinacea*, and *Erianthus ravennae* and shorter species such as *Imperata cylindrica*, *Andropogon* spp., and *Aristida ascensionis* (Shrestha and Joshi 1997). The grasses are fire and flood resistant and spread rapidly under favorable conditions.

Saccharum spontaneum (*kans*) grasslands are dominated by this tallgrass species, which is the first to colonize the exposed silt plains after the retreat of monsoon floods. This often occurs in almost pure stands, forming a thin strip on the first terrace of the floodplain. This is the keystone habitat for rhinoceroses and other large mammals. *Saccharum benghalesis* (*baruwa*) grasslands dominate the next terrace above the *S. spontaneum* band along the river's edge. Grazing lawns (*chaurs*) are very short mixes of grasses maintained by intense grazing by greater one-horned rhinoceroses and other large ungulates. *Imperata cylindrica*, *Chrysopogon aciculatus* (*kuro*), *Eragrostis* spp., and many short grasses typically dominate them. *Cymbopogon* spp. (*ganaune gans*) is another short grass species that occurs in distinct associations on the floodplain and is eaten by greater one-horned rhinoceroses and elephants. The tall grasses *Arundo donax* and *Phragmites karka* (*narkot*) surround oxbows and lakes (Dinerstein 2002).

The alluvial terrain blends into the forested hills, where sal (*Shorea robusta*) forests are common. These forests average between 25 and 40 m in height but may reach 45 m under favorable conditions (FAO 1981). Moist sal forest is found in eastern and central Nepal, whereas western Nepal's sal forests are drier (Shrestha and Joshi 1997). Common associates include *Terminalia tomentosa*, *Syzygium cuminii*, *Anogeissus latifolia*, *Terminalia alata*, *T. belerica*, *T. chebula*, *Lagerstromia parviflora*, *Dillenia pentagyna*, *Syzygium operculata*, *Carya arborea*, and *Buchanania latifolia*, as well as chir pine (*Pinus roxburghii*) on higher reaches (Dinerstein 2002; FAO 1981).

This ecoregion also contains small patches of tropical deciduous riverine forest dominated by *Mallotus philippinensis*, *Syzygium cuminii*, *Bombax ceiba*, *Trewia nudiflora*, and *Garuga pinnata* (Shrestha and Joshi 1997). Another variant, tropical evergreen forest, is made up of *Michelia champaca*, *Syzygium*, *Cedrela toona*, *Garuga pinnata*, and *Duabanga* (Shrestha and Joshi 1997). Around the Koshi-Tappu Wildlife Reserve, Ghodaghodi Tal, Bishajaari Tal, oxbow lakes and wetlands provide an additional habitat type.

Biodiversity Features

This ecoregion contains the highest densities of tigers, rhinos, and ungulates in Asia. Royal Chitwan National Park in southern Nepal contains more than 500 of the world's 1,000 endangered greater one-horned rhinoceroses (*Rhinoceros unicornis*) and about seventy breeding tigers (*Panthera tigris*) (King Mahendra Trust for Nature Conservation, pers. comm.). Royal Bardia National Park, Royal Shukla Phanta Wildlife Reserve, and Dudwa National Park contain approximately eighty-five more breeding tigers (WWF and ICIMOD 2001). This ecoregion overlaps with three Level I TCUs and one Level II TCU (Dinerstein et al. 1997). The Chitwan-Parsa-Valmiki is a Level 1 TCU that spans southern Nepal and northern India. These three reserves form an important transboundary area for wildlife conservation in general and for tigers in particular. This TCU was evaluated as the most important among all the alluvial grassland units containing tigers on the Indian subcontinent (Dinerstein et al. 1997; Wikramanayake et al. 1998). The TCU supports a healthy leopard population and at least a small population of the rare clouded leopard, the last new large mammal to be reported for Chitwan, representing a range extension of more than 250 km to the west (Dinerstein and Mehta 1989).

Many of the *terai* grasslands and floodplain forests support five deer species (swamp deer, sambar, axis deer, hog deer, and barking deer), an unusually diverse assemblage of cervids. Four large herbivores, the Asian elephant, greater one-horned rhinoceros, gaur (seasonal occupant), and nilgai or blue bull (in drier grasslands) also co-exist (Dinerstein 2002). Several endangered mammalian herbivores are also present, including the Asiatic wild buffalo and the near-endemic hispid hare (*Caprolagus hispidus*) (Bell et al. 1990) (table J35.1). The pygmy hog is another highly endangered ungulate of the very tall grasslands and a strict endemic to this ecoregion (Oliver 1980) (table J35.1).

TABLE J35.1. Endemic and Near-Endemic Mammal Species.

Family	Species
Suidae	*Sus salvanius**
Leporidae	*Caprolagus hispidus*

An asterisk signifies that the species' range is limited to this ecoregion.

TABLE J35.2. Endemic and Near-Endemic Bird Species.

Family	Common Name	Species
Phasianidae	Manipur bush-quail	*Perdicula manipurensis*
Sylviidae	Grey-crowned prinia	*Prinia cinereocapilla*
Timaliidae	Spiny babbler	*Turdoides nipalensis*

An asterisk signifies that the species' range is limited to this ecoregion.

Riverine grasslands in the Terai also provide critical habitat for now-endangered reptiles, including the rare, primitive crocodilian, the gharial (Maskey 1979), mugger crocodile, and soft-shelled turtles (Zug and Mitchell 1995).

This ecoregion overlaps with small portions of two EBAs. It overlaps with the Central Himalayas EBA (129) in western Nepal and the far western portion of the Assam Plains EBA (131) south of Bhutan (Stattersfield et al. 1998). Three near-endemic bird species are found in the Terai (table J35.2). The Manipur bush-quail (*Perdicula manipurensis*) is considered vulnerable (Hilton-Taylor 2000). The Terai's diverse grasslands, riparian woodlands, hill forests, and scrub forests provide a diverse set of habitats for many bird species. More than 375 bird species are found in this ecoregion.

Nepal's list of threatened birds, numbering 130 breeding and wintering species, includes 44 species that are found in grasslands or wetlands and 14 species that are grassland specialists. The grassland-associated birds include the Bengal florican (*Houbaropsis bengalensis*), lesser florican (*Sypheotides indica*), sarus crane (*Grus antigone*), and large grass warbler (*Graminicola bengalensis*). Most are declining in numbers (Dinerstein 2002).

Status and Threats

Current Status

The alluvial grassland fragments of this ecoregion now represent remnants of a once-extensive ecosystem. The extremely productive alluvial grasslands, which provide important habitats to endangered large animals such as tigers and elephants, are also good arable land, and most of the grasslands have been converted to agriculture. Perhaps no more than 2.0 percent of the alluvial grasslands of the Gangetic floodplain remains intact, and the best-conserved examples of floodplain grasslands are in Royal Chitwan National Park, Royal Shukla Phanta Wildlife Reserve, Dudhwa National Park, and to a lesser extent Royal Bardia National Park (table J35.3) (Dinerstein 2002). An extensive network of reserves has been established in the Terai; the challenge now is to connect these reserves to allow wide-ranging species, such as tigers, elephants, and rhinoceros, to move among reserves.

Table J35.3. WCMC (1997) Protected Areas That Overlap with the Ecoregion.

Protected Area	Area (km²)	IUCN Category
Dudwha	570	II
Katarniaghat	530	IV
Royal Shukla Phanta	320	IV
Royal Bardia National Park [25]	1,840	II
Parsa Wildlife Reserve [25]	90	IV
Royal Chitwan National Park [25]	932	II
Mahananda	60	IV
Buxa	100	IV
Garumara	10	IV
Koshi Tappu Wildlife Reserve	140	IV
Total	4,592	

Ecoregion numbers of protected areas that overlap with additional ecoregions are listed in brackets.

Almost all the forest remnants are found in the dry *bhabar* region, which consists of gravelly soil that has eroded from the foothills and is unsuitable for agriculture (Inskipp 1989). One of the most important forest remnants is protected by Royal Chitwan National Park in southern Nepal (Mishra and Jeffries 1991).

For a long time, malaria kept the human population density low in the Terai, allowing some of the habitat to be set aside, first as hunting reserves for royalty and later as wildlife sanctuaries. More recently, landmark legislation that allowed park revenues and buffer zone projects to make possible viable ecotourism and community forestry projects has enhanced conservation efforts and promoted regeneration of wildlife habitats (Dinerstein 2002).

In 1993, a major reform in Nepal's national policy allowed legal buffer zones to be created around existing protected areas. Management of these zones was taken on by local User Group Committees (UGCs), providing that they developed effective management plans based on sustainable resource use. Additional landmark legislation came in 1995, when Nepal's parliament ratified a series of bylaws requiring that 50 percent of the revenue generated by protected areas be allocated to local development programs in these buffer zones instead of returning to the Ministry of Finance. Now operational, these two initiatives paved the way for establishing legal economic incentives to reduce pressures on core reserves and to conserve wildlife habitats outside parks. More importantly, they allowed villagers to become partners in the recovery of the buffer zones and to serve as guardians of endangered wildlife and habitat (Dinerstein 2002).

Types and Severity of Threats

The Terai is Nepal's major area for logging and wood industries (FAO 1981). Sawmilling is the largest wood-based industry, with private sawmills spread over most of the Terai districts. Fuelwood production is also important, with most consumed within the country and the rest exported to India (FAO 1981). Main species harvested from these forests are *Shorea robusta, Terminalia tomentosa, Dalbergia sissoo, Bombax ceiba,* and *Adina cordifolia* (FAO 1981). In addition to the recorded log production, some unauthorized cutting for export is known to have occurred (FAO 1981).

Growing population pressure in the hills has led to migration to and settlement in the Terai, both spontaneously and through government-sponsored resettlement programs (FAO 1981). The southern parts of the Terai therefore are densely populated, and most of the area is under cultivation, although northern regions have a lower population density (FAO 1981). Water diversion, especially for irrigation projects, poses another significant threat. Poaching and overgrazing are also problems here. Much of the savanna grasslands may be created by burning by pastoralists and other human intervention (Shrestha and Joshi 1997).

Priority Conservation Actions

Short-Term Conservation Actions (1–5 Years)

- Maintain and strengthen effective antipoaching teams and information networks.

- Expand successful buffer zone restoration program to all twelve reserves in the Nepalese and Indian sections of the Terai.
- Develop and implement the Terai Arc Wildlife Corridor stretching from Corbett National Park in India to Chitwan National Park in Nepal (see essay 19, chapter 7).
- Develop Trust Fund to finance Terai Arc restoration program.
- Expand restoration program in reserves through translocation of large mammal populations that have been extirpated. These include translocation of swamp deer and wild water buffalo back into Chitwan from Shukla Phanta and Kosi Tappu reserves, respectively.

Longer-Term Conservation Actions (5–20 Years)
- Continue successful translocation of greater one-horned rhinoceros populations from Chitwan to Bardia, Shukla Phanta, Dudhwa, and Corbett.
- Reestablish all of the linkages among the twelve reserves of the Terai Arc (see essay 19).

Focal Species for Conservation Action

Endangered endemic species: hispid hare (Bell et al. 1990), pygmy hog (Oliver 1980), Manipur bush-quail, grey-crowned prinia, spiny babbler

Area-sensitive species: greater one-horned rhinoceros (Dinerstein 2002; Mishra and Jeffries 1991), tiger (Dinerstein 2002; Mishra and Jeffries 1991; Franklin et al. 1999; Tilson et al. 1994; Dinerstein et al. 1997; Seidensticker et al. 1999), Asian elephant (Santiapillai and Jackson 1990; Santiapillai 1997)

Top predator species: tiger (Dinerstein 2002; Mishra and Jeffries 1991; Franklin et al. 1999; Tilson et al. 1994; Dinerstein et al. 1997; Seidensticker et al. 1999), clouded leopard (Dinerstein and Mehta 1989), leopard

Other: gharial (Maskey 1979), mugger crocodile, soft-shelled turtles (Zug and Mitchell 1995), gaur, nilgai, swamp deer, sambar, axis deer, hog deer, barking deer, Bengal florican, lesser florican, sarus crane, large grass warbler

Selected Conservation Contacts

King Mahendra Trust for Nature Conservation
WWF-Bhutan
WWF-India
WWF-Nepal

Justification for Ecoregion Delineation

MacKinnon (1997) follows Rodgers and Panwar (1988) in his classification of the Himalayan biounits and identified two biounits: the Trans-Himalayan (I1) and the Himalayas (I2). The Himalayan Range is divided into four subunits along the longitudinal axis: the Northwest Himalayas (I2a), West Nepal (I2b), Central Himalayas (I2c), and Eastern Himalayas (I2d). We distinguished the altitudinal bands of habitat as distinct ecoregions while retaining several of MacKinnon's subunits that are based on east-west–oriented biogeographic barriers. The Terai-Duar savannas and grasslands are one of these habitats, and they are located along the foothills and to the north of the Siwalik Hills. All the Himalayan ecoregions are part of Udvardy's Himalayan highlands biogeographic province.

Prepared by Eric Dinerstein and Colby Loucks

Ecoregion Number:	**36**
Ecoregion Name:	**Rann of Kutch Seasonal Salt Marsh**
Bioregion:	**Indian Subcontinent**
Biome:	**Flooded Grasslands and Savannas**
Political Units:	**India, Pakistan**
Ecoregion Size:	**27,800 km²**
Biological Distinctiveness:	**Globally Outstanding**
Conservation Status:	**Relatively Stable**
Conservation Assessment:	**III**

Perhaps the bleakest, dustiest, and hottest region in India is the Great Rann of Kutch. It stretches for hundreds of square kilometers in the State of Gujarat, from the frontier with Pakistan's Sind Desert, southward to the Little Rann and the Gulf of Kutch. The Rann of Kutch is described as "a desolate area of unrelieved, sun-baked saline clay desert, shimmering with the images of a perpetual mirage" (Cubitt and Mountfort 1991). Despite this bleak description, the Rann of Kutch seasonal salt marsh provides refuge for the last population of the endangered Asiatic wild ass (*Equus hermionus*) and supports the one of the world's largest breeding colonies of the greater and lesser flamingos (*Phoenicopterus ruber* and *P. minor*).

Location and General Description

The Rann of Kutch Seasonal Salt Marsh [36] lies at the end of the Luni River, which drains the Aravalli Hills and flows southward only to dissipate into the dry, arid salt flats represented by this ecoregion. Geographically, the ecoregion extends across the northwestern Indian State of Gujarat and southern Pakistan's Sind Desert region, sitting along the Tropic of Cancer.

Since the Mesozoic, the Little and Great Ranns were extensions of the shallow Arabian Sea until geological uplift closed off the connection with the sea, creating a vast lake that was still navigable during the time of Alexander the Great (WII 1993). But over the centuries, silting has turned the lake into a vast, saline mudflat. During the brief wet season, the mudflat becomes flooded. Then it becomes parched under the relentless, searing heat of the long dry season; the ecoregion has one of the highest annual evaporation rates in the region (WII 1993). Average summer temperatures hover around 44°C but can reach highs of 50°C, and the minimum winter temperatures approach or even drop below freezing (WII 1993).

The July to September monsoon rains flood the vast, flat area to a depth of about 0.5 m. Several rivers—the Bhambhan, Kankavati, Godhra, and Umai from the south, the Rupen and Saraswati from the east, and Banas from the northeast—drain into the Ranns during the monsoon. Several sandy, salt-free areas of higher ground that rise to 2–3 m lie above the flood level and provide wet-season

TABLE J36.1. Endemic Mammal Species.

Family	Species
Equidae	*Equus hemionus*

An asterisk signifies that the species' range is limited to this ecoregion.

refuges for the ecoregion's wildlife. These refuges are known as bets, and three (Pung-bet, Vacha-bet, and the Jhilandan-bet) in particular are important refuges for wild asses.

The vegetation consists of grasses and dry thorny scrub such as *Apluda aristata*, *Cenchrus* spp., *Pennisetum* spp., *Cymbopogon* spp., *Eragrostis* spp., and *Elionurus* spp. (Puri et al. 1989). There are hardly any large trees except in the bets, where the exotic *Prosopis juliflora* has begun to invade. Seedpods from the *Prosopis* provide year-round food for the wild asses. In the Little Rann the vegetation is classified into *Salvadora* scrub and tropical *Euphorbia* scrub (Champion and Seth 1968).

Biodiversity Features

Despite the inhospitable conditions, this ecoregion harbors several charismatic mammal species of conservation importance, including the Asiatic wild ass, chinkara (*Gazella bennettii*), nilgai (*Boselaphus tragocamelus*), wolf (*Canis lupus*), blackbuck (*Antilope cervicapra*), striped hyena (*Hyaena hyaena*), desert cat (*Felis silverstris*), and caracal (*Felis caracal*).

The overall mammal fauna is estimated at about fifty species, including a single strict endemic species (table J36.1). The Asiatic wild ass and blackbuck also represent threatened species (Hilton-Taylor 2000).

The ecoregion also harbors more than 200 bird species. Although none are endemic, the ecoregion does harbor the globally threatened lesser florican (*Eupodotis indica*) and houbara bustard (*Chlamydotis undulata*) (Hilton-Taylor 2000). Other birds of conservation importance include the demoiselle crane (*Anthropoides virgo*) and lesser flamingo (*Phoeniconaias minor*) (WII 1993).

Status and Threats

Current Status

Three protected areas cover more than three-fourths of this ecoregion (table J36.2). The large Wild Ass Sanctuary in India was designated especially to protect the last population of this endangered species.

TABLE J36.2. WCMC (1997) Protected Areas That Overlap with the Ecoregion.

Protected Area	Area (km²)	IUCN Category
Keti Bunder South	980	IV
Kachchh Desert	13,540	IV
Wild Ass Sanctuary	7,170	IV
Total	21,690	

Ecoregion numbers of protected areas that overlap with additional ecoregions are listed in brackets.

Types and Severity of Threats

The primary threats to this ecoregion's habitat are from cattle grazing even within the protected areas, vehicular traffic that damages the fragile ecosystem, and cutting trees to make charcoal. The proposed expansion of the commercial salt extraction operations will result in disturbances to wildlife, especially to the wild ass population and the floricans, bustards, flamingoes, and pelicans (WII 1993). Feral pigs around the fringes of the sanctuary carry disease, degrade habitat, and disrupt reproduction of ground-nesting birds.

Priority Conservation Actions

Short-Term Conservation Actions (1–5 Years)

- Deal with expansion of the salt concessions and encroachment into the protected areas by salt manufacturers.
- Prohibit vehicular traffic through the fragile ecosystem.
- Regulate livestock grazing. Stop grazing within protected areas.
- Institute a monitoring program for the Asiatic wild ass population and its habitat (WII 1993).
- Set up a monitoring program for the lesser florican, houbara bustard, dalmatian pelican, and nesting flamingo flocks.

Longer-Term Conservation Actions (5–20 Years)

- Develop a long-term conservation plan to conserve and monitor the Asiatic wild ass population and its habitat. Consider translocations to create additional populations as a hedge against diseases.

Focal Species for Conservation Action

Endangered endemic species: Asiatic wild ass

Top predator species: wolf, striped hyena, desert cat

Other: blackbuck, chinkara, lesser florican, houbara bustard, dalmatian pelican, nesting flamingo flocks

Selected Conservation Contacts

Forest Department, Gujarat, India
Sanctuary Warden, Little Rann of Kutch
Wildlife Institute of India
WWF-India
WWF-Pakistan

Justification for Ecoregion Delineation

We used MacKinnon's (1997) digital map of original habitat that depicts the extent of the Rann of Kutch salt marshes to define the boundaries of this ecoregion. Both the Rann of Kutch Seasonal Salt Marsh [36] and the Indus Valley Desert [47] lie within the Thar Desert biogeographic province.

Prepared by Gopal S. Rawat and Eric D. Wikramanayake

Ecoregion Number: **37**
Ecoregion Name: **Northwestern Himalayan Alpine Shrub and Meadows**
Bioregion: **Indian Subcontinent**
Biome: **Montane Grasslands and Shrublands**
Political Units: **China, India, and Pakistan**
Ecoregion Size: **46,900 km²**
Biological Distinctiveness: **Regionally Outstanding**
Conservation Status: **Relatively Stable**
Conservation Assessment: **III**

The remote Northwestern Himalayan Alpine Shrub and Meadows [37] ecoregion is some of the most intact and undisturbed habitat for the snow leopard (*Uncia uncia*), the Himalayan high-altitude carnivore that hunts large mountain ungulates such as ibex (*Capra ibex*), markhor (*Capra falconeri*), blue sheep (*Pseudois nayur*), and tahr (*Hemitragus jemlahicus*). However, the snow leopard is not the lone predator here, for the ecoregion harbors the Tibetan wolf (*Canis lupus*) and large avian predators such as the lammergeier (*Gypaetus barbatus*) and golden eagle (*Aquila chrysaetos*), which soar high above the mountain peaks searching for colonial marmots (*Marmota himalayana*).

For a short time in late spring and summer, the meadows become resplendent with a colorful tapestry of delphiniums, gentians, poppies, roseroots, louseworts, anemones, and asters, to name but a few. No less bright are the flowers of the rhododendrons that characterize the alpine scrub habitat closer to the treeline. Within this species-rich landscape are hotspots of endemism, created by the varied topography, which results in very localized climatic variations, promoting the evolution of specialized plant communities.

Location and General Description

This ecoregion represents the alpine scrub and meadows between 3,300 and 3,600 m in the northwestern reaches of the Himalayan Range. The ecoregion extends across northwestern India and northern Pakistan, mostly in the Vale of Kashmir and associated mountains. The vegetation consists of stunted, twisted *krummholz*, dominated by rhododendron and juniper scrub, interspersed with patches of birch forests. It also includes patches of moist scrub vegetation within some parts of the trans-Himalaya, such as Lahul Valley, the northwestern parts of the Pir Panjal Range, and adjacent areas in Pakistan.

The Himalayas trace their origin to the recent geological past when the collision between the northward drifting Deccan Plateau and the northern Eurasian continent took place. During this, and several subsequent periods of geological upheaval and uplift, the Himalayan Mountains were thrust upward and above the Tibetan Plateau, which represented the bottom of the Tethys Sea before the collision (Wadia 1966). Evidence suggests that this northwest region of the Himalaya is younger than the eastern extent of the mountain range (Gupta 1994).

Because the southwestern monsoon that sweeps in from the Bay of Bengal expends most of its precipitation in the eastern Himalayas, the western Himalayas are drier (Gupta 1994). Therefore, the ecoregion is arid to semi-arid, with little precipitation falling in the winter and spring. Ambient temperature approaches freezing during the winter, but summers are hot. In response to these dry and harsh climatic conditions, the treeline is almost 1,000 m lower in the western Himalayas than in the east.

Aspect is another important determinant of local climatic variation; the north-facing slopes are less exposed to sunlight and therefore are cooler and retain more moisture. But within this general trend, the complex topography also creates rainshadows, resulting in very localized climatic variations.

There are several vegetation types in this ecoregion. Closer to the timberline the vegetation consists of low near-evergreen scrub made up of a stunted birch-rhododendron community. Only five rhododendron species are reported from this belt (Rao 1994), compared to more than sixty species known from the eastern Himalaya. Above this scrub is the alpine meadow, comprising a diverse herb community that includes species of *Doronicum*, *Delphinium*, *Gentiana*, *Meconopsis*, *Pedicularis*, *Anemone*, *Aster*, *Polygonum*, *Primula*, and *Mertensia*, among others. Scree habitat contains cushion-forming plants such as *Caragana*, *Saxifraga*, *Draba*, and *Gypsophila*. A high proportion of alpine and sub-alpine herbs in Fumariaceae, Primulaceae, Saxifragaceae, and Scrophulariaceae are endemic and several are endangered (WWF and IUCN 1995).

Biodiversity Features

Little information is available on the biodiversity of this remote and scientifically unexplored area. Although most of the faunal representatives in the high altitudes are small (Mani 1994), the ecoregion contains several large mammals that are important for conservation and can be used as focal species. The snow leopard takes prominence, but a successful and comprehensive conservation program should also include the larger ungulates such as the markhor, Asiatic ibex, blue sheep, and argali that make up this predator's prey species. The ecoregion also represents the southernmost limit for the Asiatic ibex, which prefers drier scrub vegetation on steeper slopes adjacent to the trans-Himalaya. There is a distinct ecological separation between the Asiatic ibex and the Himalayan tahr, which prefer moister scrub and meadow vegetation in the southern slopes of the Greater Himalaya.

The known mammal fauna includes forty-nine species. None of the mammals are considered endemic to the ecoregion, but there are several threatened species, including the snow leopard, serow (*Capricornis sumatraensis*), markhor, Himalayan tahr, argali, Himalayan musk deer (*Moschus chrysogaster*), and brown bear (*Ursus arctos*) (Hilton-Taylor 2000).

The bird fauna consists of 139 species. None are endemic, but there are several characteristic Himalayan species such as the lammergeier, golden eagle, Himalayan griffon, snow partridge (*Lerwa lerwa*), Tibetan snowcock (*Tetraogallus tibetanus*), and Himalayan snowcock (*Tetraogallus himalayensis*), which should be focal species for conservation efforts.

Status and Threats

Current Status

Because of its high elevation and harsh climate, as well as its lack of commercially valuable timber trees, the ecoregion's habitat is relatively stable. There are fifteen protected areas that cover 4,500 km^2 (10 percent) of the ecoregion (table J37.1). Two of these—Siachu Tuan Nalla Wildlife Sanctuary and Pin Valley—are large, the latter exceeding 1,000 km^2. Three other protected areas—Great Himalayan (690 km^2), Kistar (1,260 km^2), and Rupti Bhabha (850 km^2)—that overlap with adjacent ecoregions are also large. But most of the others are small, with the average area of the protected areas being only 300 km^2.

Types and Severity of Threats

The threats to the natural habitat in this ecoregion stem largely from overgrazing by domestic livestock and fuelwood collection. Already, all lands that can support agriculture (i.e., alluvial fans and valleys) have been cleared and cultivated (Schaller 1977). Commercial harvest of wild medicinal herbs in many areas of western Himalayas also is a major threat.

Priority Conservation Actions

Little is known about the status and conservation needs of this remote ecoregion. Therefore, an immediate action would be to conduct a conservation assessment. Some of the other important conservation actions that should be addressed include the following.

Short-Term Conservation Actions (1–5 Years)

- Protect the relict patches of riverine vegetation, patches of *Juniperus macropoda* (in Lahul Valley), and willow scrub in Kashmir and parts of Pakistan (Chundawat and Rawat 1994).
- Conduct a status survey of brown bear, markhor, and pheasants.
- Ensure protection of snow leopards and their prey species.
- Control commercial exploitation of medicinal plants.
- Strengthen protection and management of existing protected areas.

Longer-Term Conservation Actions (5–20 Years)

- Develop a long-term conservation plan based on a conservation assessment.

Focal Species for Conservation Action

Area-sensitive species: snow leopard (Fox 1989; Jackson 1992), lammergeier

Top predator species: snow leopard (Fox 1989; Jackson 1992), golden eagle

Other: Tibetan wolf, serow, markhor, Himalayan tahr, argali (see Schaller 1977 for montane goats and sheep), Himalayan musk deer, brown bear, snow partridge, Tibetan snowcock, Himalayan snowcock, and the endemic alpine plants

Selected Conservation Contacts

Chief Wildlife Warden, Jammu and Kashmir
ICIMOD
Kashmir University, Srinagar (especially Botany Department)
Pakistan Museum of Natural History
Wildlife Institute of India
WWF-India
WWF-Pakistan
Zoological Society of Pakistan

Justification for Ecoregion Delineation

In a previous analysis of conservation units in the Indo-Malayan region, MacKinnon (1997) divided the Himalayan Range into four subunits along the longitudinal axis: the Northwest Himalayas (I2a), West Nepal (I2b), Central Himalayas (I2c), and Eastern Himalayas (I2d). We used MacKinnon's (1997) vegetation map to extract the alpine scrub and meadows in biounit 12d into a separate ecoregion—Northwestern Himalayan Alpine Shrub and Meadows [37]—as per our rules for defining ecoregions (i.e., represent ecosystems of regional extent within separate ecoregions). We used a digital map of the vegetation provided by MacKinnon (1997), aided by a DEM (USGS) to define the boundaries of the ecoregion. All the Himalayan ecoregions are part of Udvardy's Himalayan highlands biogeographic province.

Prepared by Gopal S. Rawat and Eric D. Wikramanayake, with contributions from Pralad Yonzon

TABLE J37.1. WCMC (1997) Protected Areas That Overlap with the Ecoregion.

Protected Area	Area (km^2)	IUCN Category
Chitral Gol National Park	110	II
Gahriat Gol GR	140	
Parit Gol GR	140	
Baltal	150	IV
Siachu Tuan Nalla WS	880	
Manali WS	100	IV
Pin Valley	1,190	II
Kanawar	80	IV
Great Himalayan (east)	200	II
Great Himalayan (west) [29]	320	II
Kugti WS [29]	160	IV
Tundah WS [29]	180	IV
Kistar National Park [27, 29]	320	II
Overa-Aru [27, 29]	180	IV
Rupti Bhabha [27, 29]	350	IV
Total	4,500	

Ecoregion numbers of protected areas that overlap with additional ecoregions are listed in brackets.

Ecoregion Number:	**38**
Ecoregion Name:	**Western Himalayan Alpine Shrub and Meadows**
Bioregion:	**Indian Subcontinent**
Biome:	**Montane Grasslands and Shrublands**
Political Units:	**India, Nepal**
Ecoregion Size:	**34,600 km²**
Biological Distinctiveness:	**Regionally Outstanding**
Conservation Status:	**Relatively Stable**
Conservation Assessment:	**III**

The Western Himalayan Alpine Shrub and Meadows [38] ecoregion has large areas of habitat suitable for conserving viable populations of the high-altitude Himalayan predator, the snow leopard (*Uncia uncia*), and the large montane ungulates such as blue sheep (*Pseudois nayur*), Himalayan tahr (*Hemitragus jemlahicus*), Himalayan musk deer (*Moschus chrysogaster*), and serow (*Capricornis sumatraensis*), which form its prey. Wild yak (*Bos grunniens*) used to inhabit the innermost ranges of the western Himalayan alpine meadows adjacent to the Tibetan Plateau. These large beasts are now considered locally extinct throughout most of their former distribution and have not been recorded in Indian territory recently. Most of the ecoregion's mammals are small species, such as the Himalayan palm civet, pale weasel, Himalayan weasel, pikas, and voles, that scurry about the undergrowth and among the boulders.

In spring and summer the meadows, known locally as the *bughiyals*, become a resplendent tapestry of contrasting blue, purple, yellow, pink, and red flowers of delphiniums, gentians, poppies, roseroots, louseworts, anemones, and asters. The shrub layer closer to the treeline is bright with the flowers of *Rhododendron* shrubs.

Location and General Description

This ecoregion represents the Himalayan alpine meadows and shrublands between about 3,000 and 5,000 m, and to the west of the deep Kali Gandaki River gorge in central Nepal and on to the northern Indian State, Uttar Pradesh. About a fourth of this high-elevation ecoregion is bare rock and ice.

This alpine area was formed during the early Pleistocene, after a long geological process that thrust up the Inner Himalayas (Ram and Singh 1994). The beginning of the process dates back to the Cretaceous, when the drifting Deccan Plateau collided with the northern Eurasian continent, causing the latter to buckle upward (Wadia 1966). The Himalayan Mountain Range now comprises three parallel east-west–directed zones: the Outer, Middle, and the imposing Inner Himalaya, which contains the tallest peaks in the world.

There is a general trend of decreasing moisture levels from east to west along the length of the Himalayan Mountain Range. The southwestern monsoon that provides most of the moisture originates from the Bay of Bengal and expends most of its precipitation in the eastern Himalayas. Therefore, the west receives less rainfall, usually ranging from 1,500 to almost 1,900 mm (Ram and Singh 1994). The treeline responds to the drier conditions in the west by being almost 1,000 m lower than in the east.

The ecoregion experiences cold winters during which snowfall blankets the slopes. Snowmelt during April and May results in additional moisture. The summers are mild. Cloudy days and fog are common, with clear skies limited to a few hours during the day. Climate and aspect influence vegetation. The north-facing slopes, which are less exposed to sunlight, are cooler and retain more moisture. These harbor a more typical Himalayan flora. But within this general trend, the complex topography also creates rainshadows, resulting in very localized climatic variations.

The early vegetation in this region was dominated by *Quercus semecarpifolia* and *Betula utilis*, but periodic climatic changes and glaciation events have given rise to the present alpine vegetation (Ram and Sing 1994). The forest-alpine transition is now marked by *krummholz*-type vegetation of *Rhododendron campanulatum*, *Rhododendron barbatum*, *Salix* spp., and *Syringa emodia* (Ram and Singh 1994). The alpine scrub flora is dominated by several colorful, dwarf rhododendron species, among which are other shrubby species such as *Hippophae rhamnoides* and *Cotoneaster microphyllus* (Shrestha and Joshi 1997). Along the transition area with the alpine conifers lower down, junipers (*Juniperus* spp.) contribute to the floral diversity. Many of the rhododendron species in these western Himalayan habitats are different from those found in the eastern extent of this long mountain range.

The rich meadow flora is dominated by herbaceous plants, especially species of *Anaphalis*, *Aster*, *Cynanthus*, *Jurinea*, *Morina*, *Potentilla*, *Gentiana*, *Delphinium*, *Gentiana*, *Meconopsis*, *Pedicularis*, *Anemone*, *Aster*, *Polygonum*, *Primula*, and *Saussurea* (WWF and IUCN 1995; Shrestha and Joshi 1996; Polunin and Stainton 1997). Willows (*Salix* spp.) line the sides of shallow, montane streams. At higher elevations, species of *Saxifraga*, *Allium*, *Corydalis*, *Eriophyton*, *Stellaria*, *Soroseris*, and *Cremanthodium* grow on the alpine scree with scattered rocks and boulders. A steppe-type vegetation of *Caragana pygma*, *C. gerardiana*, *Lonicera spinosa*, *Juniperus squamata*, *J. indica*, *Ephedra gerardiana*, *Hippophae tibetana*, *Myricaria rosea*, *Lonicera spinulosa*, and *Berberis* spp. is found in the northern extents of the ecoregion.

Biodiversity Features

The ecoregion contains several localized hotspots of floral diversity and endemism. Notable areas are the Valley of Flowers, Nanda Devi (India's second-highest mountain), Tinker Valley, Rara-Shey-Dolpho, Dhorpatan-Annapurna, Pokhara, and Gorkha-Himalchuli.

The snow leopard (*Uncia uncia*) and the Tibetan wolf (*Canis lupus*) roam these high-altitude landscapes, hunting large ungulates such as the goral (*Nemorhaedus goral*), serow (*Naemorhedus sumatraensis*), Himalayan tahr (*Hemitragus jemlahicus*), argali (*Ovis ammon*), and blue sheep (*Pseudois*

TABLE J38.1. Endemic and Near-Endemic Bird Species.

Family	Common Name	Species
Phasianidae	Cheer pheasant	*Catreus wallichi*

An asterisk signifies that the species' range is limited to this ecoregion.

nayur). This ecoregion probably represents the eastern limit of the distribution of the brown bear (*Ursus arctos*); there have been unconfirmed reports of this species from western Nepal.

The forty mammal species known in the ecoregion do not include any endemics. But several species are threatened, including the snow leopard, serow, Himalayan tahr, argali, and Himalayan goral (Hilton-Taylor 2000).

The bird fauna is richer, with almost 130 species, including one that is endemic to the ecoregion (table J38.1). This near-endemic species is shared with the other western Himalayan ecoregions (Western Himalayan Broadleaf Forest [27] and Himalayan Subtropical Pine Forests [31]).

Other birds that are typical of these high-altitude ecosystems and can be used as focal species for conservation management planning include the blood pheasant (*Ithaginis cruentus*), western tragopan (*Tragopan melanoephalus*), Satyr tragopan (*Tragopan satyra*), and Himalayan monal (*Lophophorus impejanus*), which inhabit the shrubby ground cover, and the large avian predators, lammergeier (*Gypaetus barbatus*), golden eagle (*Aquila chrysaetos*), and Himalayan griffon (*Gyps himalayensis*).

Status and Threats

Current Status

Because of its remoteness and inaccessibility, much of the ecoregion's habitat is still intact. There are eleven protected areas that cover more than 8,500 km^2 representing about 25 percent of the ecoregion area (table J38.2). These include several large protected areas (e.g., Shey-Phuksundo, Annapurna, Kedarnath, Nanda Devi, and Dhorpatan). Three large protected areas also overlap across adjacent ecoregions (table J38.2).

Types and Severity of Threats

Grazing and trampling by large herds of domestic livestock, including cows, buffalos, horses, sheep, goats, and yak, are beginning to severely degrade the natural habitat (Ram and Singh 1994; WWF and IUCN 1995; Rawat 1998). As livestock herds increase in number and size and exceed the carrying capacity in the lower habitats, pastoralists have begun to drive the herds into these alpine meadows, increasing the degradation threats here (Ram and Singh 1994; Rawat 1998). An overnight campsite for pastoralists and their livestock as they migrate up and down the mountains can lay bare an area up to a hectare (Singh 1991, cited in Ram and Singh 1994).

Overexploitation of rare medicinal herbs is another conservation threat in the area (Rawat 1998). Most of the collection sites are already under heavy grazing pressure, and the harvest of medicinal plants places additional stresses on this fragile ecosystem.

Adding to the localized degradation threats are several historical trade routes between India, Nepal, and Tibet that are still heavily used in this region. The accessibility brings with it associated degradation threats such as fuelwood collection and heavy use by people and pack animals. Because the ecoregion includes national borders, concentrations of defense personnel exploit its natural resources, especially timber and fuelwood.

Priority Conservation Actions

Despite its remoteness, the ecoregion's biodiversity is under threat from various sources. A large proportion of the ecoregion is already within protected areas, but these reserves need effective protection and management. An important conservation target in Himalaya is to ensure altitudinal connectivity to permit the ecological processes to function. Here we identify some short- and long-term conservation actions in this regard.

Short-Term Conservation Actions (1–5 Years)

- Equip and train field staff in protected areas. Strengthen protection and management of existing protected areas.
- Protect and monitor snow leopard populations and their prey species.
- Involve defense personnel posted in the border posts in conservation efforts and raise conservation awareness among them.
- Control grazing and set aside core areas within the biosphere reserves and other protected areas where livestock grazing is prohibited.
- Control commercial exploitation of medicinal plants. Promote cultivation of highly useful and commercially important medicinal plants in the high-altitude villages.
- Restrict collection of fuelwood from trailside forests by tourists and pilgrimage parties.

Longer-Term Conservation Actions (5–20 Years)

- Develop a long-term conservation plan based on an ecoregion-based conservation assessment of the western Himalayan ecoregions.
- Assess adequacy of existing protected areas for each ecoregion.
- Link core protected areas with natural habitat.
- Maintain north-south connectivity and restore it where necessary.

TABLE J38.2. WCMC (1997) Protected Areas That Overlap with the Ecoregion.

Protected Area	Area (km^2)	IUCN Category
Royal Dhorpatan Hunting Reserve [29]	870	
Govind Pashu Vihar [27]	240	IV
Govind	490	II
Lippa-Asrang WS	60	IV
Raksham Chitkul	30	IV
Gangotri	270	II
Kedarnath	970	IV
Valley of Flowers	70	II
Nanda Devi	810	Ia
Shey-Phoksumdo National Park	3,590	IV
Annapurna Conservation Area [28, 39]	1,260	
Total	8,660	

Ecoregion numbers of protected areas that overlap with additional ecoregions are listed in brackets.

Focal Species for Conservation Action

Area-sensitive species: snow leopard, lammergeier, Himalayan griffon

Top predator species: snow leopard, golden eagle

Other: Tibetan wolf, serow, Himalayan tahr, blue sheep, argali, Himalayan goral, Asiatic black bear, Cheer pheasant, musk deer, blood pheasant, western tragopan, Satyr tragopan, Himalayan monal, Himalayan blue poppy (*Mecanopsis aculeata*), endemic alpine plants, several medicinal and rare plants including *Saussurea obvallata*, *Nardostachys jatamansi*, *Picrorhiza kurrooa*, *Aconitum heterophyllum*, *Parnassia nubicola*, *Boschniackia himalaica*, *Falconeria himalaica*, and *Circaeaster agrestis*

Selected Conservation Contacts

Department of National Parks and Wildlife Conservation, Government of Nepal

Director, Nanda Devi Biosphere Reserve

G. B. Pant Institute of Himalayan Environment and Development, Almora, India

ICIMOD

Principal Conservator of Forests, Uttarakhand, Naini Tal, India

Wildlife Institute of India

WWF-India

WWF-Nepal

Justification for Ecoregion Delineation

In a previous analysis of conservation units, MacKinnon (1997) divided the Himalayan Range into four subunits along the longitudinal axis, namely the Northwest Himalayas (I2a), West Nepal (I2b), Central Himalayas (I2c), and Eastern Himalayas (I2d). In our analysis, we combined the alpine scrub and meadows in units 12b and 12c. All the Himalayan ecoregions are part of Udvardy's Himalayan highlands biogeographic province.

We considered the Kali Gandaki River to be a biogeographic barrier, as widely acknowledged, and took this as the eastern extent of the ecoregion. We used MacKinnon's (1997) digital map of the original vegetation (and aided by a DEM) to identify the extent of the alpine meadows shown in biounits I2b and 12c, which were then defined as the Western Himalayan Alpine Shrub and Meadows [38].

Prepared by Gopal S. Rawat and Eric D. Wikramanayake, with contributions from Pralad Yonzon

Ecoregion Number:	**39**
Ecoregion Name:	Eastern Himalayan Alpine Shrub and Meadows
Bioregion:	Indian Subcontinent
Biome:	Montane Grasslands and Shrublands
Political Units:	Bhutan, India, Myanmar, and Nepal
Ecoregion Size:	40,100 km²
Biological Distinctiveness:	Globally Outstanding
Conservation Status:	Relatively Stable
Conservation Assessment:	Class III

The Eastern Himalayan Alpine Shrub and Meadows [39] ecoregion supports one of the world's richest alpine floral displays that becomes vividly apparent during the spring and summer when the meadows explode into a riot of color from the contrasting blue, purple, yellow, pink, and red flowers of alpine herbs. Rhododendrons characterize the alpine scrub habitat closer to treeline. The tall, bright-yellow flower stalk of the noble rhubarb, *Rheum nobile* (Polygonaceae), stands above all the low herbs and shrubs like a beacon, visible from across the valleys of the high Himalayan slopes.

The plant richness in this ecoregion sitting at the top of the world is estimated at more than 7,000 species, a number that is three times what is estimated for the other alpine meadows in the Himalayas. In fact, from among the Indo-Pacific ecoregions, only the famous rain forests of Borneo are estimated to have a richer flora. Within the species-rich landscape are hotspots of endemism, created by the varied topography, which results in very localized climatic variations and high rainfall, enhancing the ability of specialized plant communities to evolve. Therefore, the ecoregion boasts the record for a plant growing at the highest elevation in the world: *Arenaria bryophylla*, a small, dense, tufted cushion-forming plant with small, stalkless flowers, was recorded at an astonishing 6,180 m by A. F. R. Wollaston (Wollaston 1921, in Polunin and Stainton 1997).

Location and General Description

The Eastern Himalayan Alpine Shrub and Meadows [39] represent the alpine scrub and meadow habitat along the Inner Himalayas to the east of the Kali Gandaki River in central Nepal. Within it are the tallest mountains in the world—Everest, Makalu, Dhaulagiri, and Jomalhari—which tower high above the Gangetic Plains. The alpine scrub and meadows in the eastern Himalayas are nested between the treeline at 4,000 m and the snowline at about 5,500 m and extend from the deep Kali Gandaki gorge through Bhutan and India's northeastern state of Arunachal Pradesh, to northern Myanmar.

In addition to being the world's tallest mountain range, the Himalayas are also one of the youngest. Their origin has been traced back to the collision between the northward-drifting Deccan Plateau and the northern Eurasian continent. During this collision the northern edge of the Deccan Plateau pushed beneath Eurasia and began to raise the northern continent from beneath the Tethys Sea to create what is now the

Tibetan Plateau, 4,000 m above sea level. The Himalayan mountains were thrust upward during subsequent geologic uplifts and upheavals to form the highest mountain range in the world (Wadia 1966).

The eastern Himalayas are wetter than the western extents of the mountain range because of precipitation from the May–September southwest monsoon. The water it brings from the Bay of Bengal is first intercepted and expended here. But within this general trend, the complex topography creates rainshadows, resulting in very localized climatic variations. For instance, Pokhara, in the southern Annapurna Range, faces the brunt of the monsoonal rains and receives more than 3,500 mm of annual rainfall, but Jomsom, just 65 km north and in Pokhara's rainshadow, gets only 300 mm of rainfall (Polunin and Stainton 1997). Aspect is another important criterion that determines local climatic variation. The north-facing slopes are less exposed to sunlight and are thus cooler and retain more moisture. Therefore, they are more likely to harbor a specialized Himalayan flora adapted to these moist, microclimatic conditions.

The scrub vegetation of this ecoregion is dominated by colorful *Rhododendron* species that exhibit high species turnover along the west-east gradient from eastern Nepal to northern Myanmar. For instance, some common species in the Nepal alpine scrublands, *Rhododendron campanulatum*, *R. wallichi*, *R. campylocarpum*, *R. thomsonii*, and *R. wightii* (Shrestha and Joshi 1997) drop out of the assemblage in Bhutan, where *R. bhutanense*, *R. aeruginosum*, *R. succothii*, *R. fragariiflorum*, *R. pumilum*, *R. baileyi*, and *R. pogonophyllum* (Pradhan et al., in prep) are added to the assemblage. And further east in northern Myanmar, the characteristic assemblage consisting of *R. calciphila*, *R. crebriflorum*, *R. chryseum*, *R. riparium*, *R. sanguineum*, and *R. saluenense* (WWF and IUCN 1995) is entirely different.

The herbs that lend springtime color to the alpine meadows include hundreds of species from genera such as *Alchemilla*, *Androsace*, *Primula*, *Diapensia*, *Impatiens*, *Draba*, *Anemone*, *Gentiana*, *Leontopodium*, *Meconopsis*, *Saxifraga*, *Sedum*, *Saussurea*, *Rhododendron*, *Potentilla*, *Pedicularis*, and *Viola* (Puri et al. 1989; WWF and IUCN 1995; Shrestha and Joshi 1996; Polunin and Stainton 1997). Several of these (e.g., *Picrorhiza*, *Rheum*, *Aconitum*, and *Paris*) are prized as medicinal herbs (Shrestha and Joshi 1996). The splendor and richness of these meadow communities in full bloom are difficult to describe.

The upper elevations of the ecoregion transition into rock screes, where shallow soils support grasses, cushion-forming plants, and rhododendron scrub among large boulders and craggy rock faces.

Biodiversity Features

The fauna of this ecoregion is surprisingly rich in large vertebrates. Important mammal species include the snow leopard (*Uncia uncia*), which roams the high-altitude meadows; blue sheep (*Pseudois nayur*); Himalayan tahr (*Hemitragus jemlahicus*); and the formidable takin (*Budorcas taxicolor*). Avian predators such as the lammergeier (*Gypaetus barbatus*), Himalayan griffon (*Gyps himalayensis*), black eagle (*Ictinaetus malayensis*), and northern goshawk (*Accipiter gentilis*) soar high among the peaks searching for colonial

TABLE J39.1. Endemic and Near-Endemic Mammal Species.

Family	Species
Vespertilionidae	Eptesicus gobiensis

An asterisk signifies that the species' range is limited to this ecoregion.

TABLE J39.2. Endemic and Near-Endemic Bird Species.

Family	Common Name	Species
Phasianidae	Chestnut-breasted partridge	Arborophila mandellii

An asterisk signifies that the species' range is limited to this ecoregion.

marmots (*Marmota himalayana*), which build extensive burrows in which they gain refuge.

The ecoregion harbors about 100 mammal species, but because the small and mid-sized mammals are poorly studied in this inaccessible habitat, this number probably is only a conservative estimate. Only one species is considered near endemic to this ecoregion (table J39.1). The Vespertilionid bat is also found in the Karakoram–West Tibetan Plateau Alpine Steppe [40] ecoregion, much further west, and probably inhabits several other ecoregions.

There are several threatened species such as the endangered snow leopard, takin, and Himalayan goral (*Naemorhedus baileyi*), and the vulnerable serow (*Capricornis sumatraensis*) and Himalayan tahr (*Hemitragus jemlahicus*) (in eastern Nepal and Sikkim) (Hilton-Taylor 2000) in this ecoregion. A recent record of a single takin bull was reported from Sikkim, far from the nearest known takin population in Jigme Dorji National Park in Bhutan.

The 115 bird species known in this ecoregion include one near-endemic species (table J39.2). This partridge is limited to the eastern Himalayas and is also found in the adjacent Himalayan Subtropical Broadleaf Forests [25], Eastern Himalayan Broadleaf Forests [26], and Eastern Himalayan Sub-Alpine Conifer Forests [28].

There are several other high-elevation specialists, such as the Himalayan snowcock (*Tetraogallus himalayensis*), Tibetan partridge (*Perdix hodgsoniae*), snow partridge (*Lerwa lerwa*), Satyr tragopan (*Tragopan satyra*), lammergeier, and the Himalayan griffon, that also need conservation attention.

Status and Threats

Current Status

The ecoregion has fourteen protected areas that cover more than 11,680 km^2, including several—such as Annapurna, Makalu Barun, Sagarmatha, Jigme Dorgi, and Sakteng—that exceed 1,000 km^2 (or, as in the case of Annapurna and Jigme Dorji, 2,500 km^2) (table J39.3). Although the total area protected represents about 30 percent of the ecoregion's area, the reserves are inequitably distributed. Most of the protected areas are in Nepal and Bhutan, whereas the eastern section of the ecoregion, especially in Myanmar, receives little or no formal protection. Because of the high species turnover along the east-west axis, more equitable protection is necessary for better representation of the ecoregion's biodiversity.

Moreover, about half of the areas that lie within the existing protected areas represent bare rock and areas covered with permanent ice, not very important habitat for biodiversity conservation.

Types and Severity of Threats

The assessment of the habitat based on satellite imagery indicates that there is very little habitat loss, but ground surveys indicate widespread habitat degradation. Overgrazing and trampling by large herds of livestock (especially yak) and the unregulated commercial harvest of medicinal plants are the main threats to biodiversity in this ecoregion. These threats extend to the protected areas, so these fragile ecosystems lack effective core protection.

Snow leopards are hunted for their pelts but are also killed by herders, who consider them to be livestock predators. Herders also claim that large herbivores compete with livestock for food.

Priority Conservation Actions

Conservation actions, especially any long-term actions, should be developed within the context of the entire Himalayan ecosystem. Some of the important conservation actions that should be addressed include the following.

Short-Term Conservation Actions (1–5 Years)

- Train and equip the field staff of several of the protected areas in this ecoregion.
- Make defense personnel posted in the high-elevation border stations aware of conservation needs and involve them in conservation efforts.
- Control and regulate domestic livestock grazing. This grazing has been a traditional practice, but herd sizes have increased tremendously. Core areas should be set aside within biosphere reserves and other protected areas where grazing is not allowed.
- Regulate commercial exploitation of medicinal and rare plants. Promote cultivation of some of the commercially important medicinal plants in the high-altitude villages.
- Mitigate impacts on the natural habitats from tourism and pilgrimages.
- Formulate management plans for protected areas and include a zoning plan to for multiple-use and core areas.
- Control musk deer and snow leopard poaching.
- Conduct status survey for the Sikkim stag (*Cervus elaphus wallichi*).

Longer-Term Conservation Actions (5–20 Years)

- Create transboundary peace parks between Sikkim, Tibet, and Bhutan to provide additional protection to takin and their seasonal ranges.
- Implement the landscape-level conservation plans under the ecoregion-based conservation program.

Focal Species for Conservation Action

Area-sensitive species: snow leopard (Fox 1989; Jackson 1992)

Top predator species: snow leopard (Fox 1989; Jackson 1992)

Migratory species: takin, blue sheep, altitudinal migratory birds

Other: juniper scrub habitat, Himalayan goral, Southern serow, Himalayan tahr, Asiatic black bear, takin, chesnut-breasted partridge, lammergeier, Himalayan griffon, Satyr tragopan, Himalayan snowcock, Tibetan partridge, snow partridge, Himalayan blue poppy (*Mecanopsis* spp.), and several medicinal and rare plants such as *Saussurea obvallata, Nardostachys jatamansi, Picrorhiza kurrooa, Aconitum heterophyllum, Parnassia nubicola, Rheum nobile,* and *Panax pseudo-ginseng*

Selected Conservation Contacts

Department of Wildlife and Nature Conservation, Royal Government of Nepal

Forest Department, Arunachal Pradesh, India

Forest Department, Sikkim, India

ICIMOD

King Mahendra Trust for Conservation of Nature, Nepal

Mountain Institute, Kathmandu, Nepal

Nature Conservation Section, Royal Government of Bhutan

Royal Society for Protection of Nature, Bhutan

Wildlife Institute of India

WWF-Bhutan

WWF-India

WWF-Nepal

Justification for Ecoregion Delineation

In a previous analysis of conservation units, MacKinnon divided the Himalayan Range into four subunits along the longitudinal axis: the Northwest Himalayas (I2a), West

TABLE J39.3. WCMC (1997) Protected Areas That Overlap with the Ecoregion.

Protected Area	Area (km²)	IUCN Category
Langtang National Park [28]	790	II
Makalu-Barun National Park	1,500	II
Sagarmatha National Park	1,110	II
Makalu-Barun Conservation Area [26, 28]	450	II
Torsa [26, 28]	110	I
Jigme Dorji [26, 28]	2,510	II
Sakteng WS [28]	1,050	IV
Black Mountain [26, 28]	70	II
Thrumsing La [26]	530	II
Walong National Park/WS	180	?
Dong Jiu	220	?
Mo Tuo [28]	120	?
Dibang Valley [28]	380	PRO
Annapurna Conservation Area [28, 31]	2,660	
Total	11,680	

Ecoregion numbers of protected areas that overlap with additional ecoregions are listed in brackets.

Nepal (I2b), Central Himalayas (I2c), and Eastern Himalayas (I2d). In doing so, he included all the habitat types, from subtropical broadleaf forests in the lowlands to the alpine habitats, within each subunit. In defining ecoregions, we strove to delineate distinct habitats of regional extent as separate ecoregions. Therefore, we distinguished between the bands of habitat types along the latitudinal axis of the Himalayan Range and placed them within individual ecoregions. We used the Kali Gandaki River, widely considered a biogeographic barrier that defines the eastern and western Himalayan biotas, as a boundary to separate the eastern Himalayan alpine meadows and shrublands from the western alpine meadows and shrublands. We used the digital map of the habitat types of the Himalayas from MacKinnon (1997), aided by a DEM, to delineate the northern and southern boundaries of the alpine meadows and scrub, and placed them in the Eastern Himalayan Alpine Shrub and Meadows [39]. All the Himalayan ecoregions are part of Udvardy's Himalayan highlands biogeographic province.

Prepared by Gopal S. Rawat and Eric D. Wikramanayake, with contributions from Pralad Yonzon.

Ecoregion Number:	**40**
Ecoregion Name:	**Karakoram–West Tibetan Plateau Alpine Steppe**
Bioregion:	**Indian Subcontinent**
Biome:	**Montane Grasslands and Shrublands**
Political Units:	**Pakistan, China, Afghanistan, India**
Ecoregion Size:	**112,600 km²**
Biological Distinctiveness:	**Regionally Outstanding**
Conservation Status:	**Vulnerable**
Conservation Assessment:	**II**

The Karakoram–West Tibetan Plateau Alpine Steppe [40] contains some of the highest densities of ungulates in the region, including the endangered Marco Polo sheep. The alpine vegetation supports numerous mountain sheep and goats, which in turn provide a substantial prey base for the endangered snow leopard. A majority of this ecoregion is prime habitat for the snow leopard and, like its ungulate prey, this large predator often comes into conflict with the region's domestic animals that use the same rangelands. A resolution to this conflict will help ensure habitat for the region's native flora and fauna.

Location and General Description

This ecoregion makes up a majority of the Karakoram high mountain region to the west of the Himalayas in Kashmir. It includes snow and glaciers of some of the world's highest mountains (such as K2) as well as the lower-elevation alpine, sub-alpine, and interdispersed coniferous vegetation. The predominant mountain ranges are the Karakoram Range, Ladakh Range, Chang Chenmo Range (China), and Deosai Mountains.

Mountain slopes support mainly unstable, excessively drained shallow to moderately deep gravelly, loamy soils on bedrock and are subject to severe sheet, rill, and gully erosion. The mean annual precipitation varies in the ecoregion but ranges from 200 to 900 mm, 90 percent in the form of snow (Shengji 1996; IUCN 1993).

Within this broad mountain ecosystem, small distances result in large changes based on altitude, aspect, geology, and soils, giving rise to a wide variety of microclimates and biodiversity. The predominant vegetation is characterized by sparse grasslands and herbaceous vegetation on mountainous slopes. On the alpine slopes or in sheltered ravines, *Salix denticulata*, *Mertensia tibetica*, *Potentilla desertorum*, *Juniperus polycarpus*, *Polygonum viviparum*, *Berberis pachyacantha*, *Rosa webbiana*, and *Spiraea lycoides* dominate. In the highest elevations, above 4,500 m, the vegetation thins out. Common species found at these altitudes include *Delphinium cashmerianum*, *Glechoma tibetica*, *Silene longicarpophora*, *Potentilla fruticosa*, and *Nepeta* spp. (Shengji 1996; IUCN 1993).

Shrublands and patchy forests are found in the valley bottoms. The primary plant species include *Hippophae rhamnoides*, *Myricaria elegans*, *Salix viminalis*, *Capparis spinosa*, *Tribulus terrestris*, *Pegamum harmala*, *Sophora alopecuroides*, and *Lycium ruthenicum* (Shengji 1996). A steppe juniper forest, once common to most of central Asia, remains in relict populations on cliffs and sloped land. These forest fragments are dominated by *Juniperus macropoda* and *J. indica* (IUCN 1993).

Biodiversity Features

The unique microclimatic features and harsh climatic conditions force plants to adapt to survive. This gives rise to numerous endemic plant species. In Pakistan, an estimated sixty-six plant species are endemic to the Kashmir region, and montane plant species make up 90 percent of Pakistan's endemic flora (Ali 1978).

Most of the species found in this ecoregion are wide-ranging species found throughout most of the high mountains or Tibetan Plateau of the Karakorams, Hindu-Kush, and Himalayas. However, a single endemic mammal is found in this ecoregion: the woolly flying squirrel (table J40.1). This squirrel is found at high elevations in sparse *Pinus* or *Picea* forests. Brown bear is found in the lower elevations, especially in the forest of the Deosai plains of Pakistan.

Ungulates are the most diverse set of species in this region and include the Marco Polo sheep (*Ovis ammon poli*), the largest of its genus. Most of the ungulate species in this ecoregion are endangered or vulnerable (Hilton-Taylor 2000). They include markhor (*Capra falconeri*), ibex (*Capra ibex*), and urial (*Ovis orientalis*). An additional ungulate found in this ecoregion is the Tibetan argali (*Ovis ammon hodgsoni*).

The snow leopard (*Uncia uncia*) inhabits the high elevations of the Himalaya and Kashmir regions. This ecoregion includes prime habitat for the snow leopard and is therefore

TABLE J40.1. Endemic and Near-Endemic Mammal Species.

Family	Species
Sciuridae	*Eupetaurus cinereus**

An asterisk signifies that the species' range is limited to this ecoregion.

vital to its survival. The snow leopard is adapted to the high altitudes by having an enlarged nasal cavity, shortened limbs, well-developed chest muscles for climbing, long and dense hair, and a tail up to 1 m long (75–90 percent of head-body length) (Hemmer 1972; Fox 1989; Jackson 1992). In general, their most common prey consists of wild sheep and goats but also includes pikas (*Ochotona* spp.), hares (*Lepus* spp.), and gamebirds (chukar partridge and snowcocks) (Hemmer 1972; Schaller 1977; Jackson 1979; Mallon 1984; Fox 1989).

Additional mammals found in this ecoregion include the Altai weasel (*Mustela altaica*), stone marten (*Martes foina*), brown bear (*Ursos arctos*), Himalayan black bear (*Selenarctos thibetanus*), lynx (*Felis lynx*), fox (*Vulpes vulpes*), and wolf (*Canis lupus*).

There are no endemic bird species, and bird richness is low. Common bird species of the montane region are found here and include rosefinches (*Carpodacus* spp.), Guldenstadt's redstart (*Phoenicurus erythrogaster*), Himalayan monal (*Lophophorus impejanus*), raptors, and vultures. Passerines such as the black-throated thrush (*Turdus ruficollis*) or robin accentor (*Prunella rubeuloides*) winter in this ecoregion (IUCN 1993).

There are no amphibians, but at least three lizard species are found in this ecoregion: *Agama himalayan* (common), *Scincella ladacensis* (sparse), and *Phrynocephalus theobaldi* (stone deserts) (IUCN 1993).

Status and Threats

Current Status

Protected areas cover large swaths of montane habitat in this ecoregion, but most of these protected areas do not conserve the most important ecological areas (table J40.2). Governments have haphazardly designated protected areas, and nature conservation has been given a low priority.

There is no consistency in nature conservation policy in this region, and most protected areas have no management plans because personnel lack the expertise and training to develop them. Population and grazing pressures exert enormous pressure on the region, and people still use the land within protected areas for grazing livestock, collecting firewood, cutting trees, and hunting illegally (Shengji 1996).

Only two protected areas within the snow leopard's range can support viable populations. A viable population is defined as more than fifty breeding adults. These areas are Khunjerab II in Pakistan and Taxkorgan IV in China. Protected areas account for only about 10 to 15 percent of the snow leopard's total range (Fox 1994).

The effects of war can be seen in this region as well. On the positive side, natural pastures and natural vegetative cover have had a chance to recover because of decreased grazing pressures, but the forests have been decimated.

Types and Severity of Threats

Trophy hunting for markhor, ibex, snow leopard, and game birds (such as falcons) is prevalent in this ecoregion and has decimated their populations. Ibex and snow leopard face extinction in this ecoregion because of hunting pressures. There is a demand from the Chinese medicinal trade for snow leopard bones to use as substitutes for tiger bone (Liao and Tan 1988). The furs from snow leopards have been commonly used for coats, and furs have been seen on sale throughout China and Taiwan (Low 1991; Jackson 1992; Fox 1994).

Livestock rely on rangelands for forage, and overgrazing of natural vegetation is a common. Domestic grazing competes directly with native ungulates for precious resources, and grazing is a greater threat than hunting to this ecoregion's native species. In elevations up to about 1,500 m, the pastures are grazed throughout the entire year. The higher elevations, between 1,500 and 3,300 m, are grazed only in the summer.

Priority Conservation Actions

Conservation actions, especially any long-term actions, should be developed within the context of the entire Karakoram ecosystem. Shengji (1996) presents a complete list of action by country and should be consulted when working in this ecoregion. Some of the more important conservation actions that should be addressed include the following.

Short-Term Conservation Actions (1–5 Years)

- Strengthen and reorganize the National Council for the Conservation of Wildlife in Pakistan to form an effective coordinating body.
- Train and equip the field staff of several of the protected areas in this ecoregion.
- Make the defense personnel posted in the high elevation border stations aware of conservation needs and involve them in conservation efforts.
- Control and regulate domestic livestock grazing. This grazing has been a traditional practice, but herd sizes have

TABLE J40.2. WCMC (1997) Protected Areas That Overlap with the Ecoregion.

Protected Area	Area (km²)	IUCN Category
Karakoram	1,790	IV
Hemis National Park/WS	1,400	II
Hemis National Park/WS	3,890	II
Khunjerab II	2,269	II
Gya-Meru WS	80	?
Tingri WS	50	?
Lungnag WS	940	?
Rupshu WS	190	?
Naltar WS	210	IV
Kargah WS	240	IV
Chassi/Bowshdar GR	480	?
Naz Bar GR	410	?
Sherqillah GR	130	?
Danyor Nullah GR	160	?
Pakura Nullah GR	140	?
K2 National Park	2,330	?
Nar Nullah GR	160	?
Astore WS	750	IV
Baltistan	330	IV
Askor Nullah GR	150	?
Kanji WS	90	IV
Boodkharbu WS	100	?
Rangdum WS	280	?
Agram Basti	270	IV
Total	16,839	

Ecoregion numbers of protected areas that overlap with additional ecoregions are listed in brackets.

increased tremendously. Core areas should be set aside within biosphere reserves and other protected areas where grazing is not allowed.
- Regulate commercial exploitation of medicinal and rare plants. Promote cultivation of some of the commercially important medicinal plants in the high-altitude villages.
- Formulate management plans for protected areas and include a zoning plan to allow multiple use and core areas.
- Control snow leopard poaching.
- Establish a database on biological resources of the region. Establish a phytogeographic and ecological database on the flora of the region.

Longer-Term Conservation Actions (5–20 Years)

- Implement the landscape-level conservation plans under the ecoregion-based conservation program.
- Establish management plans for all protected areas in the region.
- Encourage member countries to cooperate to solve transboundary conservation issues such as common habitats and migratory species.

Focal Species for Conservation Action

Area-sensitive species: snow leopard (Fox 1989; Jackson 1992), Marco Polo sheep (Schaller 1977)

Top predator species: snow leopard (Fox 1989; Jackson 1992)

Migratory species: blue sheep, altitudinal migratory birds

Other: ibex, argali, urial (Schaller 1977 for wild sheep and goats)

Selected Conservation Contacts

ICIMOD
International Council for Bird Preservation, Pakistan
National Council for the Protection of Wildlife, Pakistan
Pakistan Environmental Protection Council
Pakistan provincial wildlife departments
Wildlife Institute of India
WWF-India
WWF-Pakistan

Justification for Ecoregion Delineation

We identified two ecoregions—Karakoram–West Tibetan Plateau Alpine Steppe [40] and Central Tibetan Plateau Alpine Steppe [41]—in MacKinnon's (1997) Trans-Himalayan biounit. These ecoregions extend into China. All the Himalayan ecoregions are part of Udvardy's Himalayan highlands biogeographic province.

Prepared by Colby Loucks

Ecoregion Number:	**42**
Ecoregion Name:	**Sulaiman Range Alpine Meadows**
Bioregion:	**Indian Subcontinent**
Biome:	**Montane Grasslands and Shrublands**
Political Units:	**Pakistan, Afghanistan**
Ecoregion Size:	**19,600 km^2**
Biological Distinctiveness:	**Regionally Outstanding**
Conservation Status:	**Relatively Stable**
Conservation Assessment:	**III**

The magnificent biotic variation and composition of the Sulaiman Range Alpine Meadows [42], a central site for plant endemism, is the result of the ecoregion's role as a transitional zone between Palearctic and Indo-Malayan zoogeographic regions. The Toba Kakar highlands in the south have thirty-one plant endemics, whereas the Chitral Valley has at least thirty-nine. The only surviving but vulnerable patches of *Chilghoza* pine (*P. geraridiana*) forest of the world are located in a 200-km^2 area on Sulaiman Hills. The oldest woodlands of *Juniperus macropoda* are also found at higher peaks south of this ecoregion.

Location and General Description

This ecoregion traverses the side valleys of Chitral and Swat in the north and cuts through parts of the ranges of Safed Koh and Waziristan. It then descends to the high-elevation areas of Chialtan, Toba Kakar, and Takht-I-Suleiman ranges of southwestern Pakistan and northeastern Afghanistan. Most of the area has average elevation of 1,600–3,500 m. This ecoregion is characterized by extreme aridity and severe temperature extremes. Mean annual precipitation is no more than 225 mm. The temperature rises to a maximum of 40°C and falls to a minimum of –12°C in the winter. Much of the precipitation occurs in the form of rain or snow during the winter, and the snow remains on higher peaks from December until March. The foothills of Safed Koh and Waziristan hills and most of the south receive similar amounts of rainfall because of the effect of the monsoon. Approximately 25 percent of the rainfall occurs in the winter, but rain occurs in every month of the year.

This ecoregion is made up of gravel and scree slopes with widely scattered isolated tufts of bunch grasses, thorny hassock-shaped clumps of plants such as *Onobrychis* and *Acantholimon* spp. (T. Roberts, pers. comm., 2000). Forest cover is sparse and highly concentrated in gullies. In the lower part of Chitral, Safed Koh, Nuristan, and Waziristan Hills, where the moisture level is higher because of the influence of the monsoon, western Himalayan evergreen sclerophyllous forests and woodlands that resemble the flora of the Mediterranean region are common. These include species of Fagaceae, in particular *Quercus ilex*, which is found only in the Northern Waziristan, Koh-i-Safed, and Chitral foothills, as well as east oleander (*Nerium*), tropical adhatoda, and *Fraxinus xanthoxyloides*. The lower slopes of Takhatu, Zarghun, Wam-Pilghar, and Toba Kakar, between 2,000 and 3,300 m, have dry alpine steppe of Irano-Turanian affinities. These are made up of Cupressaceae, Anacardiaceae, Pinaceae, Oleaceae, and Fagaceae families. *Juniperus macropoda*,

J. polycarpos, *Pinus gerardiana*, and *P. wallichiana* are typical species of this elevation. Ephedraceae, Labiatae, Gramineae, Compositae, and Leguminosae make up the understory shrubs and perennial grass of this ecoregion. Some of these are *Artemisia* spp., *Astragalu* spp., *Cotoneaster persica*, *Berberis baluchistani*, and *Ephedra intermedia*. *Ferrula oopoda*, *Eremurus stenophylla*, and *Saliva* spp. are some of the flowering plant species.

Biodiversity Features

This ecoregion is predominantly Palearctic in origin but does contain some Indo-Malay affinities. Approximately fifty mammal species are scattered throughout this ecoregion. There is one near-endemic mammal species (table J42.1).

Chiltan markhor (*Capra falconeri chiltanensis*), Sind ibex (*Capra hircus*), Ladahk urial (*Ovis vignei*), brown bear (*Ursus arctos*), grey wolf (*Canis lupus*), Asiatic black bear (*Ursus thibetanus*), and leopard (*Panthera pardus*) are located at upper north and mid-latitudes of the ecoregion. Carnivores such as two species of fox (*V. vulpes* and *V. lupus*), striped hyena (*Hyaena hyaena*), and Asiatic jackal (*C. aureus*) are found south of the ecoregion. Small mammals and rodents such as collard pika (*Occhotona rufescens*), Royle's pika (*Ochotona roylei*), stone marten (*Martes foina*), migratory hamster (*Cricetulus migratorius*), and three sand rat species are spread widely. The Suleiman markhor, Afghan urial, and Ladahk urial are listed as endangered, and the Chiltan wild goat is indicated as critically endangered on IUCN's Red List of threatened animals database (Hilton-Taylor 2000). The flare horned markhor, sarmantier (*Vormela peregusna*), hyena, wolf, and leopard are also threatened in this ecoregion. The Asiatic black bear may have been extirpated from this ecoregion; as recent WWF-Pakistan surveys have not found any evidence in this ecoregion (T. Roberts, pers. comm., 2000).

More than 150 bird species have been recorded in this ecoregion. Although there are no endemic birds, the white-cheeked tit (*Aegithalos leucogenys*) and the Kashmir nuthatch (*Sitta cashmirensis*) are limited to this ecoregion (Stattersfield et al. 1998). Other noteworthy bird species include black-rumped flameback (*Dinopium benghalense*), Eurasian cuckoo (*Cuculus canorus*), alpine swift (*Apus apus*), European nightjar (*Caprimulgus europaeus*), blue rock thrush (*Monticola solitarius*), and two owl species (*Otus scops* and *O. brucei*). Conspicuous species include redstarts (*Phoenicurus* spp.), finches, buntings (*Caprodacus* and *Emberiza* spp.), serins, and tits (Paridae).

Important reptile species include Afghan tortoise (*Testudo horsefieldi*), rock Agsama lizard (*Agama caucasica*), and the Perisan horned viper (*Pseudocerastes persicus*).

Table J42.1. Endemic and Near-Endemic Mammal Species.

Family	Species
Dipodidae	Salpingotus michaelis

An asterisk signifies that the species' range is limited to this ecoregion.

Table J42.2. WCMC (1997) Protected Areas That Overlap with the Ecoregion.

Protected Area	Area (km²)	IUCN Category
Koh-e-Geish	240	IV

Ecoregion numbers of protected areas that overlap with additional ecoregions are listed in brackets.

Status and Threats

Current Status

Higher-elevation areas are highly intact and suitable for wildlife because of the sparsity of human settlement. However, at lower elevations threats such as deforestation, shifting cultivation, urban and industrial development, and effect of introduced species are imminent. Only 240 km² of the total area of 19,600 km², or 0.01 percent of this ecoregion, is under protection (table J42.2).

Types and Severity of Threats

The *Chilghoza* forests of Sulaiman Ranges are at risk from local timber extraction. In the early 1990s, close to 40,000 trees were harvested annually. WWF-Pakistan has worked with the local communities on implementing a conservation agreement with alternative economic incentives for forest landowners (Shengji et al. 1996).

Local gypsy tribes catch the young cubs of brown bears and leopards for village shows and baiting. In 1993, a survey by WWF-Pakistan revealed that 205 bear cubs were captured by local villagers for display (Roberts 1997). The adult bears are also hunted by sport hunters and locals for their medicinal value. Another significant threat is poaching of wild animals to protect domestic livestock. Habitat loss caused by high fragmentation is the main threat for the avifauna of this ecoregion. High competition for fodder with domestic animals has caused a severe decline of wild ungulates. The heavy human settlement, possession of firearms, and disruption of the food chain caused by scarcity of ungulates have also led carnivores, especially leopards, to near extinction. Unless preventive measures are taken, extirpation is likely.

There is severe depletion of subterranean aquifers, caused by the installation of tube wells and increased irrigation. The groundwater is not being replenished, and the *Kharezes* originally used for crop irrigation are mostly now dry and inoperative. In the case of the *Chilghoza*, because it grows so slowly there has never been a single attempt to produce nursery stock. The Pakistan Forestry Department is trying to introduce alien species such as *Pinus halipensis* instead.

Overgrazing by goats prevents any natural regeneration and is a more serious threat to the *Chilghoza* than timber felling.

Priority Conservation Actions

Short-Term Conservation Actions (1–5 Years)

- Secure for preservation the entire areas of the chilghosa pine and juniper forests; portions of them are already in protected area systems.
- Support legislation that would stop the use of threatened animals for public shows.

- Implement environmental education in school curricula and other local community-based programs.

Longer-Term Conservation Actions (5–20 Years)

- Designate for protection areas such as the Toba Kakar and Safed Koh Ranges that have high plant and animal biodiversity.
- Conduct extensive research of natural resources of the ecoregion to make possible biodiversity priority setting.

Focal Species for Conservation Action

Top predator: leopard

Area-sensitive species: collard pika, forest dormouse, stone marten, migratory hamster, hyena

Other: Sind ibex, brown bear, Asiatic black bear, Sulaiman markhor, Chiltan wild goat, Marco Polo sheep, Ladahk urial, juniper scrub, and chilghosa pine forest habitat

Selected Conservation Contacts

Forestry and Wildlife Department, Government of Baluchistan

ICIMOD

IUCN

Peshawar Forest Department

WWF-Pakistan

Justification for Ecoregion Delineation

We identified five ecoregions—Sulaiman Range Alpine Meadows [42], South Iran Nubo-Sindian Desert and Semi-Desert [43], Baluchistan Xeric Woodlands [44], Rajasthan–North Pakistan Sandy Desert [45], and East Afghan Montane Coniferous Forests [30]—from MacKinnon's Baluchistan sub-unit (I3d). All five of these ecoregions extend westward and have a portion of their ecoregion beyond the limits of this analysis. These ecoregions overlap with numerous Udvardy biogeographic provinces outside the scope of this analysis. These include the Hindu Kush highlands to the north, and the Anatolian-Iranian Desert, Iranian Desert, and the Caucaso-Iranian highlands to the south and west.

Prepared by Meseret Taye, with contributions from Tom Roberts

Ecoregion Number:	**44**
Ecoregion Name:	**Baluchistan Xeric Woodlands**
Bioregion:	**Indian Subcontinent**
Biome:	**Deserts and Xeric Shrublands**
Political Units:	**Pakistan, Afghanistan**
Ecoregion Size:	**255,200 km²**
Biological Distinctiveness:	**Bioregionally Outstanding**
Conservation Status:	**Critical**
Conservation Assessment:	**IV**

Climatic and slope variations have given this ecoregion many of world's biomes. The juniper forest of north central Baluchistan is believed to be the most extensive remaining in the world and is home to the distinctive and highly threatened Baluchistan bear and straight-horned markhor. Some individual *Juniperus macropoda* have been in existence for 2,500 years. Tiger (*Panthera tigris virgata*), Asiatic cheetah (*Aciononyx jubartus venaticus*), and Asiatic wild ass (*Equus hermionus*) have all been extirpated from this ecoregion within the past 400 years.

Location and General Description

This Baluchistan Xeric Woodlands [44] ecoregion spans from the Las Bela Valley and the high barren plateau of Baluchistan from Southwest Pakistan to eastern Afghanistan. It extends to the north, cutting through the Trans-Indus Plains of the North-West Frontier Province and through Peshawar, Kohat, and Bannu, ending at the border of Eastern Hindu Kush and the Himalayan Mountains. A maze of ranges, hills, and mountains lie within and around this ecoregion. These include the Sulaiman, Kirthar, Safed Koh, and Pub ranges, Torghar and Kaliphat Waziristan hills, and the Tobakakar, Takhatu, and Zarghun mountains with elevations from 1,000 to 3,000 m. Large passes such as Quetta and Khyber cut through these ranges. Short rivers that originate from the hills of Baluchistan Plateau drain into shallow lakes or are absorbed in the sandy deserts. The submontane plateaus of the north are fed mostly by the tributaries of the Indus and Kabul rivers.

This area is characterized by medium-altitude arid to semi-arid scrub forest, and it experiences low rainfall and severe temperature fluctuations. The average annual rainfall is less than 150 mm. Temperature is generally high in the summer and can reach up to 40 C. The Baluchistan Plateau reaches temperatures of 45°C. Northerly hot winds called *loo* blow across the south during the day in the summer, causing dust storms with wind velocities between 60 and 110 mph. The period for the southwest monsoon is from June to September. Winters generally are cold. Soils are classified as gypsum and pedocals that are characterized by high calcium carbonate and low organic matter content. Many of the ridges in the north are made up of uneven limestone filled with lacustrine clays, gravel, or boulder.

The area supports tropical steppe flora (below 1,500 m) and open xeric woodlands (1,500 to 2,000 m) (Hassinger 1968). The montane vegetation covers areas of the Baluchistan and Kurram valleys that consist of open woodlands of pistachio (*Pistachia atlantica, P. khinjuk*), almond (*Prunus rosaceae, P. eburnea*), barberry (*Berberis*), honeysuckle bush (*Lonicera caprifoliaceae, L. hypoleuca*), lycium (Solanaceae), sage or wormwood (*Artemesia* spp.), and juniper (*Juniperus macropoda, J. semiglobosa*, and *J. seravschanica*). The transitional woodland between the subtropical woodlands and the alpine vegetation of sclerophyllus forest supports the olive (*Olea cuspidata*) accompanied by the shrub varnish leaf (*Dodonaea viscosa*, Sapindaceae) found on the foothills in Pakistan. The ground layer that dominates areas within the woodlands south of the ecoregion is composed of perennial grasses, tropical shrubs, and *Acacia*. These include *Amygdalus communis, A. kuramica*, and *Fraxinus xanthoxyloides*. The Indus Plain of the immediate vicinity east of Indus River and north of the ecoregion is heavily degraded because of heavy logging and overgrazing. It supports Tamaricaceae,

Gramineae, Leguminosae, and Rhamnaceae communities. These include species such as *Tamarix* spp., *Saccharum spontaneum*, *Acacia arabica*, *Savadora oleoides*, *S. persica*, and *Ziziphus mauritiana*. Thorny small trees and *Acacia* such as *Olea ferruginea*, *Acacia modesta*, and *Artemisia maritima*, perennial grasses of *Poa* and *Bromus* spp., and bulbous plants such as *Iris*, *Tulipa*, and *Allium* spp. are also found on these slopes.

Biodiversity Features

This ecoregion is known for its richness rather than its endemism. It has more than 300 bird species. The number of migratory bird species is three times higher than that of restricted-range birds. The majority of migratory bird species are passerines. These include different redstarts (Muscicapidae), swallows (Hirundinidae), larks (Alaudidae), sparrows (Passeridae), common and European goldfinch (Fringillidae), and buntings. Other bird species include bee eaters (*Merops* spp.) and wheat eaters (*Oenanthse* spp.), the Egyptian vulture (*Neophron percnopterus*), Eurasian sparrowhawk (*Accipiter nisus*), and greater spotted eagle (*Aquila clanga*), pheasants, doves, and three types of owls. White-cheeked tit (*Aegithalos leucogenys*) and Brooks's leaf-warbler (*Phylloscopus subviridis*) are limited to the juniper woodlands. The Indus valley is rich with migratory and resident waterfowl. The migratory species include ducks (mallard, pintail, shoveler, pochard, and teals), egrets, kingfishers, coots, plovers, sandpipers, and snipes. The marbled teal (*Marmaronetta angustirostris*) and great Indian bustard (*Otis bengalensis*) are listed on IUCN's Red List as vulnerable because of fragmentation.

Wild cats such as common leopard (*Panthera pardus*) and caracal (*Caracal caracal*) are found on the hills of Baluchistan, whereas the jungle cat (*Felis chaus*) and leopard cat (*Prionailurus bengalensis*) are in the northern plains. Antelopes chinkara or Indian gazelle (*Gazella bennettii*) are found in western Baluchistan, and hog deer (*Axis porcinus*) are common to the Indus plain and southwest Quetta. The stripped hyena (*Hyaena hyaena*) and the red fox (*Vulpes vulpes*) are found in the major hills of Baluchistan.

The major mountain ranges around the Sulaiman Range, northeastern Baluchistan, southern Northwest Frontier Province, and small areas in northeastern Afghanistan support the straight-horned or Sulaiman markhor (*Capra falconeri jerdoni* or *C. f. megaceros*), flare-horned markhor (*C. f. falconeri* or *C. f. falconeri cashmirensis*), Baluchistan black bear (*Ursus thibetanus gedrosianus*), Afghan urial (*Ovis vignei cycloceros*), Sind ibex (*Capra hircus aegargus*), Eurasian wild boar (*Sus scrofa*), Baluchistan gerbel (*Gerbillus nanus*), and Hotson's long-tailed hamster (*Calomyscus hotsoni*). The freshwater Indus River dolphin (*Platanista minor*) and the terrestrial mammal listed on table J44.1 are near-endemic species of this ecoregion.

TABLE J44.1. Endemic and Near-Endemic Mammal Species.

Family	Species
Dipodidae	Salpingotus michaelis

An asterisk signifies that the species' range is limited to this ecoregion.

The endangered mugger or marsh crocodiles (*Crocodylus palustris*) are found around the Indus River as well as the High and Dasht rivers of Baluchistan. The crocodile recovery that is being carried by the Sind wildlife department from 1983 to 2000 has helped to recover the population successfully. Other reptiles in the area are leopard gecko (*Eublepharis macularius*), which is common throughout the ecoregion; the Indian cobra (*Naja naja*), found in Baluchistan; and the Central Asian cobra (*Naja oxiana*) from the northwest plains. Other important reptile species include Afghan tortoise (*Testudo horsefieldi*), rock Agsama lizard (*Agama caucasica*), and Perisan horned viper (*Pseudocerastes persicus*). Amphibians such as three species of toads and frogs such as ant frog (*Rana cyanphylyctic*), *R. rudibunda*, *R. sternostignata*, and *R. khuli* are common in Baluchistan, and the tiger frog (*Rana tigrina*) is common in the northern plains.

Three species of goitered gazelles, the Chiltan wild goat, and the Afghan urial are considered vulnerable. The Indus River dolphin, the Baluchistan bear, the Suleiman markhor, Hotson's long-tailed hamster, and the Central Asian cobra are classified as endangered (Hilton-Taylor 2000).

Status and Threats

Current Status

This ecoregion has lost much of its woodlands because of intensive logging that has occurred over hundreds of years. The Indus Plains to the northeast have been cleared for cultivation. The remaining forest in this area exists in small patches and is highly fragmented. There are four national

TABLE J44.2. WCMC (1997) Protected Areas That Overlap with the Ecoregion.

Protected Areas	Area (km²)	IUCN Category
Raghai Rakhshan	1,250	IV
Chorani	190	IV
Shashan	290	IV
Bund Khush Dil Khan	10	UA
Maslakh	460	IV
Ziarat Juniper	370	IV
Dhrun [24]	840	II
Surjan, Sumbak, Eri, and Hothiano	390	UA
Dureji	1,770	IV
Bilyamin	40	UA
Manglot	10	IV
Nizampur	10	UA
Shina-Wari Chapri	1,160	UA
Kala Chitta	360	UA
Borraka	20	IV
Khari Murat	60	UA
Islamabad	70	IV
Chinji	40	II
Diljabba-Domeli	190	UA
Chumbi Surla	560	IV
Rasool Barrage	10	IV
Bajwat	50	IV
Total	8,150	

Ecoregion numbers of protected areas that overlap with additional ecoregions are listed in brackets.

parks, twelve wildlife sanctuaries, ten game reserves, and one waterfowl sanctuary in this ecoregion. However, the 22 areas listed in table J44.2 are strictly protected from human impact.

Out of 255,200 km² total area of the ecoregion, 8,150 km², or 3.2 percent, is in a protected area system. Less than 0.2 percent of the remaining 4 percent arid subtropical habitat and 0.1 percent of the 3 percent thorn scrub forest of the northern plains is within the protected areas (Mackinnon and Mackinnon 1997). The protected area systems support a number of all the restricted species of avian species, but few of them have large enough area to support viable populations (Stattersfield et al. 1998).

Types and Severity of Threats

Deforestation for fuel, fodder, charcoal, building materials, commercial logging, and food by natives and the 3 million refugees in the bordering areas of Pakistan and Afghanistan is one of the main threats to the area. The assignment of large portions of land as rangelands, where some pastures support livestock populations that are three times greater than their carrying capacity, has caused a serious overgrazing problem that led to biodiversity degradation, soil erosion, and desertification. The construction of dams and barrages on the Indus and Kabul rivers to control floods and store water for irrigation has benefited the wintering waterfowl by expanding the amount of wetland habitat. However, it has interfered with the movement of migratory fish species, severely affecting the habitat of the Indus River dolphin and other fish fauna.

Surface irrigation and seepage from unlined irrigation canals and channels have resulted in the rise of the water table, which in turn is causing land loss through waterlogging and salinization. The conversion of land for agriculture has caused high fragmentation and destruction of the ground vegetation, which in turn affects the fauna that depend on it. Hunting is controlled in Pakistan. However, permits in game reserves often are given on the basis of influence rather than ecological considerations. Political instability and the presence of extensive mine fields in Afghanistan have hindered the conservation efforts by causing uncontrolled timber and wood use and hunting of wildlife. No activities have been undertaken in Afghanistan in the field of conservation and protected areas since 1979 (WCMC 1992).

Priority Conservation Actions

Short-Term Conservation Actions (1–5 Years)

- Strictly enforce the law for sustainable harvest in commercially valuable forests, where most wildlife sanctuaries are found, to safeguard the habitat and to make clear the purpose of wildlife sanctuaries.
- Enact policies to properly manage game reserves.
- Write management plans for all national parks that consider the needs of the local people.
- Encourage enactment of the draft forest law in Afghanistan with provisions for establishing and managing much wider coverage of protected areas than now exist.

Longer-Term Conservation Actions (5–20 Years)

- Carry out on a wider scale a comprehensive system review of protected areas.
- Modify management categories so nature reserves and parks can replace wildlife sanctuaries.

Focal Species for Conservation Action

Area-sensitive species: stone marten, migratory hamsters, Afghan hedgehog, Baluchistan gerbil, and greater horseshoe bat

Migratory species: Chiltan wild goat, migratory land and waterfowl

Other: Baluchistan black bear, Chiltan wild goat, straight-horned or Suleiman markhor, Afghan urial, goitered gazelle, Sind ibex, hyena, red fox (Schaller 1977; Roberts 1997), juniper scrub habitat

Selected Conservation Contacts

Forestry and Wildlife Department Government of Baluchistan

ICIMOD

IUCN

Peshawar Forest Department

WWF-Pakistan

Justification for Ecoregion Delineation

We identified five ecoregions—Sulaiman Range Alpine Meadows [42], South Iran Nubo-Sindian Desert and Semi-Desert [43], Baluchistan Xeric Woodlands [44], Rajasthan–North Pakistan Sandy Desert [45], and East Afghan Montane Coniferous Forests [30]—from MacKinnon's Baluchistan subunit (I3d). All five of these ecoregions extend westward and have a portion of their ecoregion beyond the limits of this analysis. These ecoregions overlap with numerous Udvardy biogeographic provinces outside the scope of this analysis. These include the Hindu Kush highlands to the north and the Anatolian-Iranian Desert, Iranian Desert, and Caucaso-Iranian highlands to the south and west.

Prepared by Meseret Taye, with contributions from Tom Roberts

Ecoregion Number:	**46**
Ecoregion Name:	**Thar Desert**
Bioregion:	**Indian Subcontinent**
Biome:	**Deserts and Xeric Shrublands**
Political Units:	**India, Pakistan**
Ecoregion Size:	**238,300 km²**
Biological Distinctiveness:	**Bioregionally Outstanding**
Conservation Status:	**Vulnerable**
Conservation Assessment:	**IV**

The arid Thar Desert [46] is the world's seventh largest desert and is without doubt the most inhospitable ecoregion in the Indo-Pacific region. However, 4,000 to 5,000 years ago this area supported what is considered to be one of the world's oldest civilizations, the Mohenjo Daro and Harappa (Chaudhry et al. 1997).

Location and General Description

This large ecoregion lies to the west of the Aravalli Mountain Range in northwestern India and includes the deserts that cover portions of the Indian states of Gujarat, Rajasthan, and Punjab, as well as the Punjab and Sind in Pakistan.

The climate is extreme; annual temperatures can range from near-freezing in the winter to more than 50°C during the summer. All rainfall is associated with the short July–September southwest monsoon that brings a mere 100–500 mm of precipitation (Hawkins 1986). About 10 percent of this ecoregion is composed of sand dunes, and the other 90 percent of craggy rock forms, compacted salt-lake bottoms, and interdunal and fixed dune areas (Grewal 1992).

The habitat is greatly influenced by the extreme climate. The sparse vegetation consists of xerophilious grasslands of *Eragrostis* spp. *Aristida adscensionis, Cenchrus biflorus, Cympogon* spp., *Cyperus* spp., *Eleusine* spp., *Panicum* spp., *Lasiurus scindicus, Aeluropus lagopoides,* and *Sporobolus* spp. (Mares 1999). Scrub vegetation consists of low trees such as *Acacia nilotica, Prosopis cineraria, P. juliflora, Tamrix aphylla, Zizyphus mauritiana, Capparis decidua,* and shrubs such as *Calligonum polygonoides, Calotropis* spp., *Aerva* spp., *Crotalaria* spp., and *Haloxylon salicornicum. Haloxylon recurvum* is also present (Puri et al. 1989; Mares 1999).

Biodiversity Features

Despite the climate, several species have evolved to survive the extreme conditions here. Among the mammal fauna, the blackbuck (*Antilope cervicapra*), chinkara (*Gazella bennettii*), caracal (*Felis caracal*), and desert fox (*Vulpes bengalensis*) inhabit the open plains, grasslands, and saline depressions known as *chappar* or *rann* in the core area of the desert (Grewal 1992; Rodgers and Panwar 1988). The overall mammal fauna consists of forty-one species. None are endemic to the ecoregion, but the blackbuck is a threatened species (Hilton-Taylor 2000) whose populations take refuge in this harsh environment.

Among the 141 birds known in this ecoregion, the great Indian bustard (*Chirotis nigricaps*) is a globally threatened species (Hilton-Taylor 2000) whose populations in this ecoregion have rebounded in recent years. A migration flyway used by cranes (*Grus grus, Anthropoides virgo*) and flamingos (*Phoenicopterus* spp.) on their way to the Rann of Kutch (Grewal 1992) further south crosses this ecoregion.

Chaudhry et al. (1997) reported eleven reptile species from ten genera from the Cholistan desert in the western Thar.

Status and Threats

Current Status

There are eleven protected areas that cover almost 44,000 km^2, representing about 18 percent of the ecoregion's area (table J46.1). These include several reserves that are quite large, with one (Cholistan) exceeding 20,000 km^2, and two (Nara Desert, Rann of Kutch) being over 5,000 km^2.

Types and Severity of Threats

The Thar probably is the world's most densely populated desert. Grazing of livestock, mostly sheep and goats, is intensive, affecting soil fertility and destroying native vegetation. Many palatable perennial species are being replaced with inedible annual species (Hawkins 1986), thus changing the vegetation composition and the ecosystem dynamics. Availability of water since the completion of the Indira Gandhi Canal Scheme has provided irrigation water to the once nonarable desert, thus attracting farmers to the area (Allan and Warren 1993). Salt pans for commercial salt production will have serious impacts on Sambhar Lake in Rajasthan. Together with recent climatic changes, these pressures combine to degrade and destroy the fragile desert ecosystems.

Priority Conservation Actions

This desert ecoregion already has large protected areas for which adequate management and protection are needed to mitigate the widespread threats, many of which occur from originate outside the protected areas. Here we present some short- and long-term conservation actions.

Short-Term Conservation Actions (1–5 Years)

- Conserve populations of blackbuck, chinkara, and great Indian bustard.
- Conserve the Houbara bustard and falcons, including protection from Royal Arabian sheiks who congregate in the Cholistan area of the Thar desert to practice falconry (Chaudhry et al. 1997).
- Institute land-use planning and zoning to mitigate ad hoc and uncontrolled agricultural practices, especially with the availability of irrigation water. Failure to do so will result in agricultural practices that are unsustainable over the long term, in which case the people who have been attracted to the area will begin to exploit the natural resources for short-term gain.
- Control livestock grazing.

Longer-Term Conservation Actions (5–20 Years)

- Strengthen the existing protected area network.
- Monitor populations of key species such as the great Indian bustard, blackbuck, chinkara, houbara bustard, and falcons.
- Monitor status and extent of desert vegetation regarding habitat degradation.

TABLE J46.1. WCMC (1997) Protected Areas That Overlap with the Ecoregion.

Protected Area	Area (km^2)	IUCN Category
Nara Desert	7,290	IV
Desert	2,710	II
Tando Mitha Khan	220	UA
Rann of Kutch	10,540	IV
Nara	600	UA
Cholistan	21,830	UA
Rahri Bungalow	50	UA
Abbasia	100	UA
Ramgarh Bundi	300	IV
Lal Suhanra	70	V
Tal Chapar	80	IV
Total	43,790	

Ecoregion numbers of protected areas that overlap with additional ecoregions are listed in brackets.

Focal Species for Conservation Action

Area-sensitive species: blackbuck

Top predator species: fox

Other: caracal, great Indian bustard, houbara bustard, falcons, chinkara

Selected Conservation Contacts

Central Arid Zone Research Institute, Jodhpur, India

Director, Desert National Park, Rajasthan

Gujarat Institute of Desert Ecology, Bhuj

Pakistan Museum of Natural History

Wildlife Institute of India, Dehra Dun, India

WWF-India

WWF-Pakistan

Zoological Society of Pakistan

Justification for Ecoregion Delineation

In a previous assessment of conservation units, MacKinnon (1997) assigned the deserts in northwestern India and Pakistan into four subunits (I3a–d). We used MacKinnon's (1997) digital map of original habitat to reclassify these biounits into eight ecoregions based on the extent of distinctive habitat of regional spatial scales. Under this schema, we assigned the Thar deserts into its own ecoregion, the Thar Desert [46]. This ecoregion falls within Udvardy's Thar Desert biogeographic province.

Prepared by Gopal S. Rawat and Eric D. Wikramanayake

Ecoregion Number:	**47**
Ecoregion Name:	**Indus Valley Desert**
Bioregion:	**Indian Subcontinent**
Biome:	**Deserts and Xeric Shrublands**
Political Units:	**Pakistan**
Ecoregion Size:	**19,500 km²**
Biological Distinctiveness:	**Bioregionally Outstanding**
Conservation Status:	**Vulnerable**
Conservation Assessment:	**IV**

The Indus Valley Desert [47], like the larger Thar Desert, is one of the most inhospitable ecoregions in the Indo-Pacific region. Biodiversity conservation should focus on the large mammals and birds of the region.

Location and General Description

This arid ecoregion is located in Pakistan's Indus Valley. The foothills of the Glaiman Range and the Chenab River define its western and eastern limits, respectively.

The extreme annual temperature variations can range from near-freezing in the winter to highs of more than 45°C during the summer. Annual rainfall averages from 640 to 760 mm (Grewal 1992), which is slightly more than in the Thar Desert.

The vegetation is greatly influenced by the extreme climatic regime. The desert thorn scrub vegetation is characterized by isolated clumps of *Prosopis* spp., *Salvadora oleoides* and *Caparis* spp., and taller thorn-scrub forests of *Acacia* spp., *Tamarix* spp., *Albizzia lebbek*, and *Morus alba* (Grewal 1992).

Biodiversity Features

This desert ecoregion is not high in richness or endemism, but it does harbor a few large vertebrates that can serve as focal species for conservation. These include the wolf (*Canis lupus*), hyena (*Hyaena hyaena*), caracal (*Felis caracal*), leopard (*Panthera pardus*), and Punjab urial (*Ovis orientalis punjabensis*). The overall mammal fauna consists of thirty-two species, but none are endemic to the ecoregion.

Bird richness is higher, with 190 species, but none are considered endemic species.

Status and Threats

Current Status

The single protected area covers more than 13,000 km², or almost 70 percent of the ecoregion area (table J47.1).

Types and Severity of Threats

Because the harsh climate is unsuited for settlement, agriculture, and livestock grazing, direct human threats are not as significant as in other ecoregions.

Priority Conservation Actions

Most of this desert ecoregion is already within a large protected area. Conservation actions should focus on strengthening management of this protected area and long-term monitoring of key species and habitat degradation.

Short-Term Conservation Actions (1–5 Years)

- Strengthen management of the protected area.

Longer-Term Conservation Actions (5–20 Years)

- Monitor populations of key species such as the wolf, Punjab urial, hyena, and caracal.
- Monitor the status and extent of desert vegetation for habitat degradation.

Focal Species for Conservation Action

Top predator species: wolf, leopard

Other: hyena, caracal, Punjab urial

Selected Conservation Contacts

Pakistan Museum of Natural History

WWF-Pakistan

Zoological Society of Pakistan

TABLE J47.1. WCMC (1997) Protected Areas That Overlap with the Ecoregion.

Protected Area	Area (km²)	IUCN Category
Thal	13,290	UA

Ecoregion numbers of protected areas that overlap with additional ecoregions are listed in brackets.

Justification for Ecoregion Delineation

MacKinnon (1997) placed the deserts of northwestern India and Pakistan into four subunits (I3a–d). We reclassified these subunits into eight ecoregions based on the extent of distinctive habitat of regional spatial scales. Using MacKinnon's biounit framework and his digital map of original habitat, we delineated the desert habitat in subunit I3c as the Indus Valley Desert [47]. Both the Rann of Kutch Seasonal Salt Marsh [36] and the Indus Valley Desert [47] lie within Udvardy's Thar Desert biogeographic province.

Prepared by Gopal S. Rawat and Eric D. Wikramanayake

Ecoregion Number:	**48**
Ecoregion Name:	**Chin Hills–Arakan Yoma Montane Rain Forests**
Bioregion:	**Indochina**
Biome:	**Tropical and Subtropical Moist Broadleaf Forests**
Political Units:	**Myanmar**
Ecoregion Size:	**29,600 km²**
Biological Distinctiveness:	**Regionally Outstanding**
Conservation Status:	**Relatively Stable**
Conservation Assessment:	**III**

The Chin Hills–Arakan Yoma Montane Rain Forests [48] are globally outstanding for bird richness, partly because they acted as a refugia during recent glaciation events. This ecoregion still harbors many taxa characteristic of the Palearctic realm and a diverse assemblage of subtropical species distributed across its elevational gradients. Much of the southern Chin Hills remains biologically unexplored.

Location and General Description

This ecoregion represents the montane moist forests along the length of the Chin Hills and Arakan Yomas mountain ranges along the west coast of Myanmar. The Köppen climate zone classifies this ecoregion in the tropical wet climate zone (National Geographic Society 1999).

Below 1,000 m the vegetation is characterized by several canopy dominants, such as *Bauhinia variegata*, *Lagerstroemia speciosa*, *Derris robusta*, *Ficus* spp. *Hibiscus*, and *Strobilanthes* (Davis et al. 1995). Mature forests are richly draped with lianas, especially *Congea tomentosa* and *Mucuna pruriens* (Davis et al. 1995). Between 1,000 and 2,100 m, the mixed evergreen broadleaf forest is composed of *Quercus* spp., *Castanopsis* spp., *Eugenia*, *Saurauia*, *Eriobotrya*, and *Schima*. Above 2,000 m, the forest is dominated by Himalayan tree taxa such as *Alnus nepalensis*, *Betula alnoides*, *Carpinus*, *Prunus*, *Pyrus*, and *Torreya*, and these transition into *Castanea*, *Cornus*, *Eriobotrya*, *Laurus*, and *Taxus* at higher elevations (Davis et al. 1995). Epiphytes are numerous in this montane cloud forest environment and include *Aeschynanthus*, *Agapetes*, *Rhododendron cuffeanum*, various aroids, and the orchid taxa *Dendrobium* and *Pleione* (WWF and IUCN 1995). *Quercus xylocarpa* is the dominant oak between 2,400 and 2,750 m, whereas *Rhododendron arboreum* and *Q. semecarpifolia* are dominant above 2,750 m (WWF and IUCN 1995). At the highest altitudes (c. 3,000 m) the forests become shrubby, with the vegetation consisting of *Hypericum patulum* and *Rhododendron burmanicum*, and the herbaceous plants include *Aconitum*, *Lactuca*, *Pedicularis*, and *Veronica* (WWF and IUCN 1995). In some areas, the summit ridges are covered by a temperate savanna composed of various shrubs such as *Rhododendron* spp., *Buddleja* spp., *Daphne* spp., *Leycesteria* spp., and *Lonicera* spp., as well as tall grasses and herbs such as *Aconitum*, *Delphinium*, *Geranium*, and *Thalictrum* (WWF and IUCN 1995).

TABLE J48.1. Endemic and Near-Endemic Bird Species.

Family	Common Name	Species
Sittidae	White-browed nuthatch	*Sitta victoriae**
Timaliidae	Striped laughingthrush	*Garrulax virgatus*
Timaliidae	Brown-capped laughingthrush	*Garrulax austeni*

An asterisk signifies that the species' range is limited to this ecoregion.

Biodiversity Features

The summit of Natma Taung (Mt. Victoria), in northwest Myanmar, is a southern enclave of the Palearctic temperate flora and a refugium for several species such as *Geranium* spp. (WWF and IUCN 1995). Lilies such as *Lilium wallichianum* also occur on these slopes, and large areas are covered by communities of alpine forbs (WWF and IUCN 1995). Besides presenting an island of Himalayan alpine vegetation at tropical latitudes, this high-altitude, isolated mountain contains many endemics above 1,800 m. Examples include *Rhododendron burmanicum*, *R. cuffeanum*, *Agapetes unwinii*, *Viola unwinii*, *Mantisia wardii*, *Geranium wardii*, *Roscoea australis*, *Agrostis burmanica*, *A. mackliniae*, *Apocopis anomalus*, and *Tripogon wardii* (WWF and IUCN 1995). *Tetracentron sinense*, a CITES-listed species, is one of the species that remained in this important refuge (WWF and IUCN 1995). Although the southern portion of this ecoregion is poorly known botanically, the Rongklang Range is known to be the only site in Myanmar for plants such as *Trichodesma khasianum* and *Pyrus khasiana* (WWF and IUCN 1995).

This ecoregion overlaps with a Level I TCU (Dinerstein et al. 1997) and the Eastern Himalayas EBA (130) (Stattersfield et al. 1998). Mammal species of conservation significance in this ecoregion include the threatened and endangered Hoolock gibbon (*Hylobates hoolock*), tiger (*Panthera tigris*), clouded leopard (*Pardofelis nebulosa*), leopard (*Panthera pardus*), sun bear (*Ursus malayanus*), thamin (*Cervus eldii*), and gaur (*Bos gaurus*). There are no endemic mammals.

This ecoregion contains three endemic bird species: the striped laughingthrush (*Garrulax virgatus*), brown-capped laughingthrush (*Garrulax austeni*), and white-browed nuthatch (*Sitta victoriae*) (table J48.1).

Status and Threats

Current Status

This ecoregion is in good condition, with about two-thirds of the forest intact. Most of the clearing has taken place in northern and central Myanmar, where the primary climax vegetation, below about 2,500 m, has been almost eliminated

(Davis et al. 1995). One proposed protected area covers about 1 percent of the ecoregion.

Types and Severity of Threats

Shifting cultivation, locally known as *taunggya*, is the most serious threat to the remaining undisturbed natural vegetation (Davis et al. 1995). Although this system once caused minimal damage, increasing population pressure has forced farmers deeper into the forest and reduced the time allowed for the cycle (IUCN 1991). When the fallow period is too short to permit forest regrowth, secondary bamboo scrub results (Davis et al. 1995). Hunting and habitat loss have led to the local extinctions of several mammals in recent times, including the gaur (*Bos gaurus*), elephant (*Elephas maximus*), and rhinoceros (Davis et al. 1995).

Priority Conservation Actions

This ecoregion still retains large amounts of its natural vegetation. However, increasing populations have begun to threaten both the flora and fauna.

Short-Term Conservation Actions (1–5 Years)

- Increase protected area system in the ecoregion.
- Place all protected areas under effective conservation and management plans.

Longer-Term Conservation Actions (5–20 Years)

- The protected area system mandates that 5 percent of land area be strictly dedicated to conservation. Implement and enforce this mandate.
- The Nature and Wildlife Directorate should be the sole authority to manage the protected areas, with sufficient personnel and funding.

Focal Species for Conservation Action

Endangered endemic species: white-browed nuthatch, striped laughingthrush, brown-capped laughingthrush, *Rhododendron burmanicum*, *R. cuffeanum*, *Agapetes unwinii*, *Viola unwinii*, *Mantisia wardii*, *Geranium wardii*, *Roscoea australis*, *Agrostis burmanica*, *A. mackliniae*, *Apocopis anomalus*, and *Tripogon wardii*

Area-sensitive species: tiger (Franklin et al. 1999; Tilson et al. 1994; Dinerstein et al. 1997; Seidensticker et al. 1999)

Top predator species: tiger (Franklin et al. 1999; Tilson et al. 1994; Dinerstein et al. 1997; Seidensticker et al. 1999), leopard

Other: Hoolock gibbon, clouded leopard, sun bear, thamin, and gaur

TABLE J48.2. WCMC (1997) Protected Areas That Overlap with the Ecoregion.

Protected Area	Area (km²)	IUCN Category
Natma Taung	300	PRO

Ecoregion numbers of protected areas that overlap with additional ecoregions are listed in brackets.

Selected Conservation Contacts

Myanmar Forest Department
Myanmar Wildlife Conservation Society

Justification for Ecoregion Delineation

MacKinnon's westernmost subunit (09c) represents a transition zone between the Indian Subcontinent and Indochina bioregions and is essentially made up of forest formations belonging to the tropical moist forests biome. We included the montane forests of the rugged and highly dissected Chin Hills and Arakan Yomas in a distinct ecoregion: Chin Hills–Arakan Yoma Montane Rain Forests [48]. This ecoregion overlaps with a Level I TCU (Dinerstein et al. 1997), the Eastern Himalayas (130) EBA (Stattersfield et al. 1998).

Prepared by Eric Wikramanayake

Ecoregion Number:	**49**
Ecoregion Name:	**Myanmar Coastal Rain Forests**
Bioregion:	**Indochina**
Biome:	**Tropical and Subtropical Moist Broadleaf Forests**
Political Units:	**Myanmar, Bangladesh**
Ecoregion Size:	**65,200 km²**
Biological Distinctiveness:	**Regionally Outstanding**
Conservation Status:	**Vulnerable**
Conservation Assessment:	**II**

The Myanmar Coastal Rain Forests [49] are a diverse set of climatic niches and habitats that include flora and fauna from the Indian, Indochina, and Sundaic regions. Though low in endemism, this ecoregion has a tremendous species diversity. However, the forests have been increasingly destroyed to make way for agriculture, and poaching has become the dominant threat to the remaining wildlife populations.

Location and General Description

This ecoregion represents the lowland evergreen and semi-evergreen rain forests of the western side of Arakan Yoma and Tenasserim ranges along the west coast of Myanmar. A small area extends into southeast Bangladesh. It falls within the tropical wet climate zone of the Köppen climate system (National Geographic Society 1999).

The copious monsoonal rainfall results in very lush vegetation (MacKinnon 1997). Most of the dominant tree species are characterized by dipterocarp species such as *Dipterocarpus alatus*, *D. turbinatus*, *D. obtusifolius*, *D. pilosa*, *Anisoptera glabra*, *Hopea odorata*, *Lagerstroemia calyculata*, *L. floribunda*, *L. speciosa*, *Parashorea stellata*, *Pentace birmanica*, and *Swintonia floribunda*, with undergrowth of *Calamus* palms and the creeping bamboo *Temostachyon helferi* (WWF and IUCN 1995). In the southern part (Tenasserim portion) of the ecoregion, there are some *Heritiera*-dominated brackish and fresh-water habitats along the Tenasserim river. The former is characterized by *Bruguiera parviflora*, *Aquilaria agallocah*, *Sonneratia griffithii*, and *Cynometra mimosoides* and the latter by *Amoora cucullata*, *Dysophyllum cochinchinensis*, *D. turbinatus*, *Intsia bijuga*, *Barringtonia acutangula*, and

Table J49.1. Endemic and Near-Endemic Mammal Species.

Family	Species
Vespertilionidae	Eudiscopus denticulus
Vespertilionidae	Pipistrellus lophurus*

An asterisk signifies that the species' range is limited to this ecoregion.

Combretum trifoliatum. There are also large tracts of mixed delta scrub and low forest (*ken-byauk*), where the principal tree species are *Elaeocarpus hygrophilus, Calophyllum amoenum, Litsea nitida, Eugenia* spp., and *Diospyros burmanica,* with a dense undergrowth of *Calamus erectus* and *Pinanga gracilis.*

Biodiversity Features

Although low in endemism, it has a rich fauna and flora, largely as a result of its lush vegetation, extensive range of habitats from mangroves to mountains, and position as a corridor between the Sundaic, Indochinese, and Indian subregions (MacKinnon 1997). Among the several threatened and endangered mammals of conservation significance are the tiger (*Panthera tigris*), Asian elephant (*Elephas maximus*), Malayan tapir (*Tapirus indicus*), wild dog (*Cuon alpinus*), sun bear (*Ursus malayanus*), clouded leopard (*Pardofelis nebulosa*), leopard (*Panthera pardus*), Asiatic golden cat (*Felis temmincki*), and gaur (*Bos gaurus*).

The ecoregion overlaps with three Level I TCUs (Dinerstein et al. 1997). One near-endemic mammal occurs here, the disc-footed bat (*Eudiscopus denticulus*), and an endemic pipistrelle (*Pipistrellus lophurus*) (table J49.1). More than 350 bird species are found in the diverse habitats of this ecoregion, although none are considered ecoregional endemics.

The Sumatran rhinoceros was once part of this diverse forest ecosystem, but they have been extirpated from this ecoregion for more than fifteen years. The last Sumatran rhinoceros in this ecoregion was caught in 1984 in Arakan Yoma and its horn sold to a traditional Chinese medicine shop in Yangon.

Status and Threats

Current Status

Most of the seasonal evergreen forest and almost all the freshwater swamp of this ecoregion has been cleared for agriculture (MacKinnon 1997), especially along the fertile, densely populated plains of the Irrawaddy (IUCN 1991). Heavy degradation is evident around Myeik (Mergui) and Dawei (Tavoy) (WWF and IUCN 1995). Further north, large tracts of forest have been cut, including the gorges of the Thanlwin (Salween) River where it enters the Andaman Sea at Mawlamyine, an area that once harbored many local or endemic species of orchids, begonias, and other herbs (WWF and IUCN 1995). This ecoregion is inadequately protected; there are five proposed protected areas that cover about 2,700 km^2 (4 percent) of the ecoregion area (table J49.2). Of these one, Pegu Yomas, shared with Irrawaddy Moist Deciduous Forests [67], accounts for almost 2,500 km^2.

Along Myanmar's western coast, extensive areas of forest remain that are worthy of conservation and should be brought under protection and managed effectively to increase representation in this diverse ecoregion.

Types and Severity of Threats

The continued development of flat, lowland areas for irrigated paddy rice and subsistence crops such as hill rice, cassava, yams, and vegetables on hilly ground will be a major threat in the future (MacKinnon 1997). Forests are being exploited extensively for timber because the country is hungry for foreign currency.

Wildlife trade and poaching are a major threat to the rapidly declining large mammals and medicinal plants in both regions of Arakan and Tenasserim coasts. Tigers are almost extinct in the northern part of the ecoregion along the Arakan Yoma because of intense demand in China and Thailand.

Priority Conservation Actions

Short-Term Conservation Actions (1–5 Years)

- The northwestern part of the Arakan Yoma should be protected because various wild species still exist there, including tigers, elephants, and migratory birds.
- It is essential to stop all land appropriation from within protected areas.
- All protected areas should be placed under effective conservation and management plans.

Longer-Term Conservation Actions (5–20 Years)

- The status of all protected areas should be thoroughly reevaluated, and those that have been wholly converted to other land uses should be traded for more pristine areas.
- The protected area system mandates that 5 percent of land area be strictly dedicated to conservation. Implement and enforce this mandate.
- The Nature and Wildlife Directorate should be the sole authority to manage the protected areas, with sufficient personnel and funding.

Focal Species for Conservation Action

Endangered endemic species: disc-footed bat, *Pipistrellus lophurus*

Area-sensitive species: tiger (Franklin et al. 1999; Tilson et al. 1994; Dinerstein et al. 1997; Seidensticker et al. 1999),

Table J49.2. WCMC (1997) Protected Areas That Overlap with the Ecoregion.

Protected Area	Area (km^2)	IUCN Category
Pegu Yomas [67]	2,490	PRO
Naaf River WS	140	?
Mohingyi	130	PRO
Dipayon	20	PRO
Letkokkon	4	PRO
Total	2,784	

Ecoregion numbers of protected areas that overlap with additional ecoregions are listed in brackets.

Asian elephant (Santiapillai and Jackson 1990; Santiapillai 1997)

Top predator species: tiger (Franklin et al. 1999; Tilson et al. 1994; Dinerstein et al. 1997; Seidensticker et al. 1999), leopard, Asiatic golden cat, clouded leopard

Other: Malayan tapir, gaur, banteng

Selected Conservation Contacts

Myanmar Forest Department

Myanmar Wildlife Conservation Society

Justification for Ecoregion Delineation

MacKinnon's Burmese coast biounit is represented by four ecoregions: the Myanmar Coastal Rain Forests [49] (lowland rain forests), Irrawaddy Freshwater Swamp Forests [63], Myanmar Coastal Mangroves [78], and Mizoram-Manipur-Kachin Rain Forests [50] (montane semi-evergreen forests). The Myanmar Coastal Rain Forests [49] encompass the lowland rain forests in the southern extent of the Burmese Monsoon Zone Biounit (09).

Prepared by U Tin Than and Eric Wikramanayake

Ecoregion Number:	**50**
Ecoregion Name:	**Mizoram-Manipur-Kachin Rain Forests**
Bioregion:	**Indochina**
Biome:	**Tropical and Subtropical Moist Broadleaf Forests**
Political Units:	**Myanmar, India, Bangladesh**
Ecoregion Size:	**135,200 km²**
Biological Distinctiveness:	**Globally Outstanding**
Conservation Status:	**Vulnerable**
Conservation Assessment:	**I**

The Mizoram-Manipur-Kachin Rain Forests [50] has the highest bird species richness of all ecoregions that are completely within the Indo-Pacific region. (The only ecoregions that have more birds are the Northern Indochina Subtropical Forests [74] and South China–Vietnam Subtropical Evergreen Forests [75] that extend into China.) Except the pioneering explorations of Kingdon-Ward (1921, 1930, 1952) and Burma Wildlife Survey made by Oliver Milton and Richard D. Estes (1963), few scientific surveys have been made in this ecoregion. Recent surveys have been limited to explorations by the Wildlife Conservation Society (WCS) and Smithsonian Institution's reptile survey in northwestern Myanmar. Therefore these rugged mountains' biodiversity remains largely unknown.

Location and General Description

This large ecoregion represents the semi-evergreen submontane rain forests that extend from the midranges of the Arakan Yoma and Chin Hills north into the Chittagong Hills of Bangladesh, the Mizo and Naga hills along the Myanmar-Indian border, and into the northern hills of Myanmar. It divides the Brahmaputra and Irrawaddy valleys, through which two of Asia's largest rivers flow. Some areas in this ecoregion receive more than 2,000 mm of rainfall annually from the monsoons that sweep in from the Bay of Bengal.

An area of deep gorges and dissected landscapes, this mountain range was created 40 to 50 million years ago when the Laurasian mainland crumpled under the inexorable force of the northward-drifting Deccan Plateau; this northeastern corner bore the brunt of the initial contact. These mountains now represent a biogeographic crossroads for the Indian, Indo-Malayan, and Indo-Chinese biotas (Rodgers and Panwar 1988). The ecoregion's position along this ecotone endows it with high biological diversity. These semi-evergreen forests are characterized by several species of Dipterocarpaceae that include *Dipterocarpus alatus*, *D. turbinatus*, and *D. griffithii*, and *Parashorea stellata*, *Hopea odorata*, *Shorea burmanica*, *Swintonia floribunda*, *Anisoptera scaphula*, *Eugenia grandis*, *Xylia xylocarpa*, *Gmelina arborea*, *Bombax insignis*, *B. ceiba*, *Albizia procera*, and *Castanopsis* spp. (FAO 1981). The dense understory includes some evergreen trees and a dense growth of bamboo, such as *Cephalostachyum pergracile* (tin-wa), *Gigantochloa nigrociliata*, and *Dendrocalamus hamiltonii*. In the Arakan area, *Melocanna baccifera* grows almost monotypic stands and may inhibit forest regeneration. Along stream banks and in low-lying areas, *Hopea odorata* and *Lagerstroemia speciosa* dominate.

Biodiversity Features

The Mizoran-Manipur-Kachin Rain Forests [50] ecoregion still retains almost half of its natural habitat. The 149 mammal species known from the ecoregion include two near endemic species: the bat *Pipistrellus joffrei* and the murid rodent *Hadromys humei* (table J50.1). The lower forests in Nagaland harbor two primates: the stump-tailed macaque (*Macaca arctoides*) and the pig-tailed macaque (*Macaca nemestrina*) (Rodgers and Panwar 1988). The forests of Manipur may still harbor the critically endangered Eld's deer or Thamin (*Cervus eldii*) (Corbet and Hill 1992). And these large, intact forests have been identified as priority areas for the long-term conservation of the region's largest predator, the majestic tiger *Panthera tigris* (Dinerstein et al. 1997; Wikramanayake et al. 1998). There are also several threatened species, including the red panda (*Ailurus fulgens*), Asian elephant (*Elephas maximus*), clouded leopard (*Pardofelis nebulosa*), gaur (*Bos gaurus*), goral (*Nemorhaedus goral*), great Indian civet (*Viverra zibetha*), Assamese macaque (*Macaca assamensis*), capped langur (*Semnopithecus pileatus*), hoolock gibbon (*Hylobates hoolock*), back-striped weasel (*Mustela strigidorsa*), and smooth-coated otter (*Lutrogale perspicillata*).

The ecoregion's large, contiguous habitat areas are included within two high-priority (Level I) TCUs (Dinerstein

TABLE J50.1. Endemic and Near-Endemic Mammal Species.

Family	Species
Vespertilionidae	*Pipistrellus joffrei**
Muridae	*Hadromys humei*

An asterisk signifies that the species' range is limited to this ecoregion.

TABLE J50.2. Endemic and Near-Endemic Bird Species.

Family	Common Name	Species
Phasianidae	Manipur bush-quail	Perdicula manipurensis
Timaliidae	Striped laughingthrush	Garrulax virgatus
Timaliidae	Brown-capped laughingthrush	Garrulax austeni
Timaliidae	Marsh babbler	Pellorneum palustre
Timaliidae	Tawny-breasted wren-babbler	Spelaeornis longicaudatus
Timaliidae	Wedge-billed wren-babbler	Sphenocichla humei

An asterisk signifies that the species' range is limited to this ecoregion.

et al. 1997). Maintaining landscapes with large habitat blocks is important to conserve these large predators, and this ecoregion provides one of few opportunities in the region to do so.

The ecoregion harbors 580 bird species, with 6 being near endemic (table J50.2). Two EBAs—Eastern Himalayas (130) and Assam Plains (131)—that contain thirteen restricted-range birds them (Stattersfield et al. 1998) overlap with this ecoregion.

Complementing the endemic bird species are four pheasant species (Blyth's tragopan [*Tragopan blythii*], grey peacock-pheasant [*Polyplectron bicalcaratum*], green peafowl [*Pavo muticus*], and Kalij pheasant [*Lophura leucomelanos*]) and three hornbills (great hornbill [*Buceros bicornis*], wreathed hornbill [*Aceros undulatus*], and oriental pied-hornbill [*Anthracoceros albirostris*]), which are good indicators of intact forests because of their need for mature trees and low thresholds for disturbances.

Status and Threats

Current Status

Almost half of this ecoregion's natural habitat is still intact, especially in the eastern areas within Myanmar. There

TABLE J50.3. WCMC (1997) Protected Areas That Overlap with the Ecoregion.

Protected Area	Area (km²)	IUCN Category
Intanki	310	IV
Tamanthi	1,250	UA
Barail	370	PRO
Keibul Lamjao	40	II
Yagoupokpi Lokchao	190	IV
Murlen	200	DE
Khawnglung	40	IV
Phawngpui	50	II
Unnamed	60	?
Puliebadze	120	IV
Dampa	600	IV
Pablakhali	240	IV
Ngengpui	150	IV
Bogakine Lake WS	10	?
Kyaukpandaung	140	PRO
Total	3,770	

Ecoregion numbers of protected areas that overlap with additional ecoregions are listed in brackets.

are fifteen protected areas that cover about 3,700 km² (3 percent) of the ecoregion (table J50.3). Nonetheless, several other intact habitats should be incorporated to create a more comprehensive and representative protected area network that includes the diverse habitats and biodiversity contained within this ecoregion.

Types and Severity of Threats

In the past, these forests were logged heavily for their timber. Currently, the primary causes of deforestation are shifting cultivation, although illegal logging still occurs. Cutting trees for fuelwood and fodder, regular burning to encourage new growth for livestock, and overgrazing and trampling by livestock are other overarching threats (FAO 1981). Demand for wildlife and wildlife products from China market is a serious threat to the biodiversity of these areas. Lack of enforcement encourages the poachers and wildlife traders. Rarity of tiger in this ecoregions is the result of the tiger trade over the last two decades.

Priority Conservation Actions

Short-Term Conservation Actions (1–5 Years)

- The northwestern part of the Arakan Yoma should be protected because various wildlife still exist there, including tigers and elephants as well as migratory birds.
- It is essential to stop all land appropriation from within protected areas.
- All protected areas should be placed under effective conservation and management plans.

Longer-Term Conservation Actions (5–20 Years)

- The status of all protected areas should be thoroughly reevaluated, and those that have been wholly converted to other land uses should be traded for more pristine areas.
- The protected area system mandates that 5 percent of land area be strictly dedicated to conservation. Implement and enforce this mandate.
- The Nature and Wildlife Directorate should be the sole authority to manage the protected areas, with sufficient personnel and funding.

Focal Species for Conservation Action

Habitat specialists: hornbills, pheasants

Area-sensitive species: tiger (Franklin et al. 1999; Tilson et al. 1994; Dinerstein et al. 1997; Seidensticker et al. 1999), Asian elephant (Santiapillai and Jackson 1990; Santiapillai 1997), Asiatic black bear

Top predator species: tiger (Franklin et al. 1999; Tilson et al. 1994; Dinerstein et al. 1997; Seidensticker et al. 1999)

Other: red panda, clouded leopard, hoolock gibbon, gaur

Selected Conservation Contacts

Myanmar Forest Department

Myanmar Wildlife Conservation Society

Justification for Ecoregion Delineation

In a previous analysis of conservation units in the Indo-Malayan realm, MacKinnon (1997) identified a sub-biounit (09b) that covers the Kachin and upper Chindwin areas of northern Myanmar. But this subunit includes several vegetation types in the tropical and subtropical moist broadleaf forest biome in the region. We extracted the semi-evergreen, submontane moist forests into the Mizoram-Manipur-Kachin Rain Forests [50], which then extends east across northern Myanmar to include the full extent of the distribution of these forests. In keeping with our rules for first defining ecoregions based on distinct habitat types and then separating the lowland and montane forests, we used the boundary between the moist deciduous and semi-evergreen forests to define the lowland boundary. We then used the 1,000-m contour from a DEM to define the upper boundary. MacKinnon's westernmost subunit (09c) represents a transition zone between the Indian Subcontinent and Indochina bioregions and is essentially made up of forest formations belonging to the tropical and subtropical moist broadleaf forests biome. We extended this ecoregion to include the lower-elevation tropical moist forests in this subunit.

This region does not correspond well to Udvardy's biogeographic provinces. The Myanmar Coastal Rain Forests [49] cover the Udvardy's Burman rain forest, southwestern portion of the Thai monsoon forest, and western portion of the Indochinese rain forest. The Irrawaddy Moist Deciduous Forests [67], Irrawaddy Dry Forests [71], Chin Hills–Arakan Yoma Montane Rain Forests [48], Northeast India–Myanmar Pine Forests [77], Northern Triangle Subtropical Forests [73], and Mizoram-Manipur-Kachin Rain Forests [50] correspond roughly to Udvardy's Burman rain forest and Burma monsoon forest.

Prepared by Eric Wikramanayake and U Tin Than

Ecoregion Number:	**51**
Ecoregion Name:	**Kayah-Karen Montane Rain Forests**
Bioregion:	**Indochina**
Biome:	**Tropical and Subtropical Moist Broadleaf Forests**
Political Units:	**Myanmar, Thailand**
Ecoregion Size:	**119,200 km²**
Biological Distinctiveness:	**Globally Outstanding**
Conservation Status:	**Relatively Intact**
Conservation Assessment:	**III**

The Kayah-Karen Montane Rain Forests [51] ecoregion harbors globally outstanding levels of species richness. Among the ecoregions of Indochina, it ranks second for bird species richness and fourth for mammal species richness. The world's smallest mammal, Kitti's hog-nosed bat (*Craseonycteris thonglongyai*), equal in mass to a large bumblebee, resides in the limestone caves of this ecoregion. Because the ecoregion remains unexplored scientifically, especially the parts that lie in Myanmar, it probably will yield more biological surprises.

Location and General Description

This ecoregion includes the northern part of the Tenasserim Mountain Range, which forms the border between Thailand and Myanmar. Much of the region consists of hills of Paleozoic limestone that have been dissected by chemical weathering. The overhanging cliffs, sinkholes, and caverns characteristic of tropical karst landscapes are all present in this ecoregion. Large patches of limestone forest are associated with the tropical karst. The flora and fauna here is distinct and includes several endemic species. Because complex habitats are little explored, it is likely that they contain undescribed endemic species.

Terrain throughout much of this ecoregion is rugged and intricately folded. Hillsides tend to be steep, and ridges exceed 2,000 m elevation. Valley bottoms are narrow but fertile and tend to lie at about 300 m elevation. The western slopes drain into the Salween River, which flows through Myanmar and into the Gulf of Martaban in the Indian Ocean. The eastern slopes drain into the Chao Phraya River, which drains into the gulf of Thailand.

Although the limestone that makes up much of this mountainous ecoregion was deposited in a shallow marine environment more than 300 million years ago, the mountains themselves are much younger and owe their existence to the collision between the Indian and Eurasian continental plates that produced the Himalaya about 50 million years ago.

The entire region has a monsoonal climate with warm, moist summers and mild winters that tend to be dry. Overall annual rainfall averages 1,500 to 2,000 mm. Although this ecoregion lies within the Tropic of Cancer, winter temperatures can be cool, especially at the higher elevations, where frost has been recorded from the northern part of the ecoregion. West-facing slopes (on the Myanmar side) face the Bay of Bengal and receive more precipitation. East-facing slopes (on the Thailand side) lie within a partial rainshadow and tend to be drier. This climatic difference is clearly reflected in the vegetation. Forests to the east are dominated, especially at the lower elevations, by trees that have a drought-deciduous phenology, whereas the west-facing slopes are a mixture of deciduous and evergreen species.

Edaphic factors also affect the vegetation: forests on granite, with a higher water-holding capacity, tend to support a higher proportion of evergreen broadleaf species, whereas forests on limestone are mostly drought-deciduous. Plants on bare limestone crags have a distinctive physiognomy, with fleshy stems and small, sometimes ephemeral leaves.

At low elevation (below about 1,000 m) on the east side of the Tenasserim Hills, potential vegetation consists of drought deciduous forest or savanna woodland. Although fire is common today, there is little consensus as to the historical frequency of fire or its importance in this ecosystem. An important unresolved question is whether fire (mostly anthropogenic) or premonsoon drought stress (nonanthropogenic) is primarily responsible for limiting species diversity in these places. Forests of teak *Tectona grandis* represent the climax vegetation at low elevation in the absence of fire, but today the teak forests are nearly extirpated in Thailand and declining rapidly in Myanmar. Places at low elevation where fires are more common support a savanna woodland. Doi Suthep National Park near Chiangmai, Thailand provides an example. Here the woodlands are dominated by dipterocarps such as *Dipterocarpus tuberculatus* and *D. obtusifolia* and oaks such as *Quercus kerrii*. Trees grow up through a grassy ground cover that dries in the winter and often burns in spring. Lack of closed-canopy forests encourages the growth of annual grasses, which in turn promote fire. Rare, distinctive plants include the drought-adapted epiphyte *Discidia major*, which grows in mutualistic symbiosis with ants that live inside specially modified leaves, and *Phoenix acaulis*, a small, fire-adapted palm.

Higher elevations support much richer broad-leaved forest communities with a mixture of evergreen and deciduous species. At 800–1,200 m, a well-developed shrub understory grows beneath a tall, closed forest canopy that includes some very large, buttressed trees that share an affinity with tropical Asia together with temperate tree taxa in the families Magnoliaceae and Lauraceae (IUCN 1991). Shady conditions at ground level favor woody climbers that include strangler figs (*Ficus* spp.) and the cablelike *Gnetum*, an unusual climbing gymnosperm. *Sapria himalaica* is a rare component of the forest understory here. This root parasite, with its large, red flowers and underground stems, is related to the giant *Rafflesia* of Borneo and Sumatra.

Ridgetops include a plethora of Himalayan plant taxa including members of the oak family (*Castanopsis*, *Quercus*, and *Lithocarpus*), *Schima wallichii*, and members of the birch and alder families. *Rhododendron* occurs at the highest elevations throughout this ecoregion.

Species diversity in Doi Suthep National Park, near Chiangmai in northern Thailand, rivals that of the most diverse seasonal forests. On Thailand's sandy northern hills, a fire-maintained climax of open stands of the three-needle pine *Pinus kesiya* and the two-needle pine *P. merkusii* occur at 550–1,800 m elevation (IUCN 1991).

Biodiversity Features

The ecoregion is the fourth richest in the Indo-Pacific region for mammals, with 168 known species. These include one ecoregional endemic species, the tiny Kitti's hog-nosed bat, *Craseonycteris thonglongyai* (table J51.1). This bat, weighing a mere 2 g and with an 8-cm wingspan, is confined to the limestone caves in western Thailand. This bat is now threatened because collectors catch large numbers to make taxidermy mounts that are sold to tourists. The forests around the caves have also been cleared and probably will lead to a change in the cave microclimate.

Some of the other mammals of conservation importance include several threatened species such as the tiger (*Panthera tigris*), Asian elephant (*Elephas maximus*), gaur (*Bos gaurus*), banteng (*Bos javanicus*), wild water buffalo (*Bubalus arnee*), southern serow (*Naemorhedus sumatraensis*), clouded leopard (*Pardofelis nebulosa*), Malayan tapir (*Tapirus indicus*), wild dog (*Cuon alpinus*), Asiatic black bear (*Ursus thibetanus*), Assamese macaque (*Macaca assamensis*), stump-tailed macaque (*Macaca arctoides*), smooth-coated otter (*Lutrogale perspicillata*), great Indian civet (*Viverra zibetha*), and particoloured flying squirrel (*Hylopetes alboniger*).

Sumatran rhinoceros is believed to have inhabited remote regions of the Tenasserim Hills in recent years, but this critically endangered species is now thought to have been extirpated from this ecoregion.

The relatively intact, contiguous habitat has potential to conserve large landscapes that will provide adequate habitat to maintain a viable population of Asia's largest carnivore, the tiger, as well as other species of critical conservation significance. Therefore, the ecoregion lies within a high-priority (Level I) TCU (Dinerstein et al. 1997). Several of Thailand's largest and most intact wildlife reserves lie within this ecoregion, including Huai Kha Khaeng Wildlife Sanctuary (2,575 km^2) and several other protected areas with which it forms a contiguous network. Huai Kha Khaeng is prized for the high diversity of cat species it supports and its relatively intact vertebrate communities and intact lowland dipterocarp forests. Moister habitats on the Myanmar side of the Tenasserim Range also include significant amounts of intact habitat, probably still in better condition overall than the forest on the eastern (Thai) side of the range. However, it is difficult to assess ecological conditions in the forests of eastern Myanmar at this time.

The ecoregion's 568 bird species make it the second highest in terms of species richness. These include two ecoregional endemic species (table J51.2).

But there are several other birds that are indicators of habitat integrity and are thus of conservation importance, such as Lady Amherst's pheasant (*Chrysolophus amherstiae*), Hume's pheasant (*Syrmaticus humiae*), silver pheasant (*Lophura nycthemera*), grey peacock-pheasant (*Polyplectron bicalcaratum*), green peafowl (*Pavo muticus*), Kalij pheasant (*Lophura leucomelanos*), brown hornbill (*Anorrhinus tickelli*), plain-pouched hornbill (*Aceros subruficollis*), rufous hornbill (*Buceros hydrocorax*), great hornbill (*Buceros bicornis*), wreathed hornbill (*Aceros undulatus*), and white-winged duck (*Cairina scutulata*). The great slaty woodpecker (*Muellerpicus pulverulentus*), the largest Old World woodpecker, is an uncommon resident of lowland forests and low hills in this ecoregion. It is particularly sensitive to degradation of late-successional forests because it needs large, dead trees. Some of these are threatened species, and the white-winged duck and plain-pouched hornbill in particular are globally threatened.

TABLE J51.1. Endemic and Near-Endemic Mammal Species.

Family	Species
Craseonycteridae	*Craseonycteris thonglongyai**

An asterisk signifies that the species' range is limited to this ecoregion.

TABLE J51.2. Endemic and Near-Endemic Bird Species.

Family	Common Name	Species
Timaliidae	Deignan's babbler	*Stachyris rodolphei**
Timaliidae	Burmese yuhina	*Yuhina humilis**

An asterisk signifies that the species' range is limited to this ecoregion.

Status and Threats

Current Status

About a third of this ecoregion has been cleared or degraded; however, the twenty-eight protected areas cover almost 23,500 km² (20 percent) of its area (table J51.3). The average size of the protected areas is 725 km². But the system includes several protected area complexes, such as Thailand's Huay Kha Khaeng-Thung Yai Naresuan Reserve complex (with eleven protected areas) and the Omgoy–Mae Ping–Mae Tuen reserve complex, which cover much larger areas.

Types and Severity of Threats

Shifting cultivation is the main cause of deforestation throughout the region. But in areas such as northern Thailand, where previously nomadic tribal peoples have been settled, pressure exists to convert forest into more intensive agricultural land devoted to cash crops such as cabbage, coffee, and lychee. Opium replacement efforts (in Thailand but not Myanmar) have compelled local people to grow alternative crops that need more cultivated land area and higher pesticide inputs. The land needs of an increasing population have forced itinerant farmers to reduce the cycle of cultivation-fallow periods and have pushed them deeper into the forest and into more marginal areas (IUCN 1991).

Remote hill areas occupied by ethnic minorities have suffered considerable degradation (IUCN 1991). However, the places occupied by the majority culture are more likely to have been completely converted to rice cultivation.

Hunting has decimated most of the large mammal populations, such as elephant, banteng, gaur, and tiger (IUCN 1991). Gibbons and hornbills, species important for dispersing seeds of many forest tree species, have also been severely reduced in many areas.

Priority Conservation Actions

Short-Term Conservation Actions (1–5 Years)

- Stop illegal logging in and around protected areas.
- Revoke forest concessions in areas of high biodiversity outside protected areas.
- Stop wildlife poaching, especially in key parks and reserves.
- Urge Myanmar to join CITES.

Longer-Term Conservation Actions (5–20 Years)

- Reassess representation of habitat in protected areas system in the ecoregion.
- Ensure that the protected area system includes linkages between lowland and montane habitats.
- Restore integrity of buffer zones of protected areas.
- Implement rigorous conservation actions in the Level I and Level II TCUs in the ecoregion.

Focal Species for Conservation Action

Habitat specialists: hornbills

Endangered endemic species: Kitti's hog-nosed bat, Burmese yuhina, Deignan's babbler

Area-sensitive species: Asian elephant (Santiapillai and Jackson 1990; Santiapillai 1997), tiger (Franklin et al. 1999; Tilson et al. 1994; Dinerstein et al. 1997; Seidensticker et al. 1999)

Top predator species: tiger (Franklin et al. 1999; Tilson et al. 1994; Dinerstein et al. 1997; Seidensticker et al. 1999), clouded leopard

Other: great slaty woodpecker, Assamese macaque

Selected Conservation Contacts

WWF-Thailand

Justification for Ecoregion Delineation

The extensive Indochina biounit (10) identified by MacKinnon (1997) in his conservation analysis of the Indo-Malayan realm comprises three subunits that represent the tropical lowland plains, subtropical hills, and temperate montane areas. The largest of these subunits, Central Indochina (10a), is a mix of tropical moist forests and tropical dry forests. We delineated eight ecoregions that overlap with this biounit by assigning the different vegetation types of ecoregional extent into each. This ecoregion represents the semi-evergreen forests of the Kayah-Karen mountains in the broad transition zone between the subtropical broadleaf evergreen forests in the north and the southern dry deciduous forests.

Prepared by Eric Wikramanayake and Chris Carpenter

TABLE J51.3. WCMC (1997) Protected Areas That Overlap with the Ecoregion.

Protected Area	Area (km²)	IUCN Category
Mae Yuam Fang Khwa	1,510	IV
Doi Chiang Dao	510	IV
Namtok Mae Surin	400	II
Doi Suthep-Pui	190	II
Doi Inthanon	480	II
Salawin	930	IV
Doi Khuntan	250	II
Doi Pha Muang	220	IV
Omgoy	1,060	IV
Kahilu	110	UA
Mae Tuen	1,400	IV
Ton Krabak Yai	250	II
Lansang	110	II
Klong Wang Chao	700	II
Umphang	2,440	IV
Khlong Lan	430	II
Mulayit	70	UA
Mae Wong	930	II
Thung Yai Naresuan	3,930	IV
Huai Kha Khaeng	2,480	IV
Si Nakarin	1,830	II
Chaloem Rattanakosin	60	II
Salak-Phra	830	UA
Erawan	530	II
Mae Ping [68, 72]	100	II
Khao Laem [53, 70]	380	II
Unnamed [72]	1,080	?
Doi Pha Muang [72]	270	IV
Total	23,480	

Ecoregion numbers of protected areas that overlap with additional ecoregions are listed in brackets.

Ecoregion Number: **52**
Ecoregion Name: Luang Prabang Montane Rain Forests
Bioregion: Indochina
Biome: Tropical and Subtropical Moist Broadleaf Forests
Political Units: Laos, Vietnam, Thailand
Ecoregion Size: 71,600 km^2
Biological Distinctiveness: Regionally Outstanding
Conservation Status: Vulnerable
Conservation Assessment: II

The Luang Prabang Montane Rain Forests [52] are globally outstanding for bird richness. This ecoregion has had limited biological exploration. Although more than 70 percent of this ecoregion's original montane vegetation has been converted to scrub or degraded forest, the remaining area presents some of the best opportunities for large mammal conservation in Indochina.

Location and General Description

The Luang Prabang Montane Rain Forests [52] ecoregion comprises areas largely above 800 m in north central Laos. This region grades to the north very gradually into the Northern Indochina Subtropical Forest Ecoregion and to the east into the Northern Annamites Rain Forests Ecoregion. Precise boundary lines between these ecoregions are somewhat arbitrary. The eastern gradient involves a gradual change in the amount of rainfall and length of the dry season.

Rather than a single community type, these forests include a variety of forest associations. Included within these forest habitats are montane hardwood forests of Fagaceae and Lauraceae, mixed conifer-hardwood forests, open montane forests, and open conifer forests. Montane habitats in northern Laos typically have 2,000–3,000 mm of annual rainfall, and a long dry season.

Transitional communities of humid evergreen forest occur at about 800 m elevation, with *Dipterocarpus turbinatus* and *Toxicodendron succedanea* as the dominant canopy trees. Also characteristic are a variety of palms including *Arenga saccharifera*, *Caryota*, and *Calamus* spp. Montane habitats at 1,000–1,500 m may be dominated by a variety of forest communities representing both substrate differences and history of human disturbance. Evergreen forest originally represented by a variety of Fagaceae and other hardwoods is now dominated by *Castanopsis hystrix*, with *Phyllanthus emblica* as a codominant and *Antidesma collettii*, *Helicia terminalis*, and *Wendlandia paniculata* as important associates. The low stature of trees in this community and open understory with an abundance of broad-leaved monocots and grasses suggest severe past impacts from burning and clearance. Open conifer hardwood forests are also present, with open *Quercus griffithii*, *Q. serrata*, and the conifer *Keteleeria evelyniana* as the dominant tree species. Thin, granitic soils support an edaphic climax overwhelmingly dominated by *Engelhardtia spicata* with an understory cover of bracken fern, *Pteridium aquilinum*. Another edaphic climax occurs with open stands dominated by *Pinus kesiya* on skeletal soils of clay schist or sandstone.

Biodiversity Features

This ecoregion lacks known endemic mammals or birds, but that may be because of its inaccessibility and lack of biological surveys. However, the ecoregion harbors several threatened and endangered mammals of conservation significance, including the Francois's leaf monkey (*Semnopithecus francoisi*), silvered leaf monkey (*S. cristatus*), douc langur (*Pygathrix nemaeus*), tiger (*Panthera tigris*), Asian elephant (*Elephas maximus*), wild dog (*Cuon alpinus*), Himalayan black bear (*Ursus thibetanus*), sun bear (*Ursus malayanus*), clouded leopard (*Pardofelis nebulosa*), common leopard (*Panthera pardus*), thamin (*Cervus eldii*), and gaur (*Bos gaurus*). This ecoregion overlaps with two Level I TCUs and two Level II TCUs (Dinerstein et al. 1997). There are more than 540 bird species whose ranges are shown to extend into this ecoregion, including the green peafowl (*Pavo muticus*).

Status and Threats

Current Status

Today many of the montane broadleaf forests in this ecoregion have been converted to scrub or degraded habitat, primarily as a result of widespread shifting cultivation, and less than 30 percent of the original habitat now remains. In northeastern Thailand, the Khorat Plateau is mostly deforested, although extensive forest still exists on the mountain ranges (IUCN 1991), especially in the steeper, more inaccessible slopes. The twelve protected areas cover 9,300 km^2 (13 percent) of the ecoregion (table J52.1). There are three protected areas—Nam Poui, Phou Khao Khoay, and Phu Khieo—that are greater than 1,000 km^2. Thailand's Nam Nao National Park and Phu Khieo Reserves are encompassed by Level I TCUs, and Laos's Nam Poui is included within a Level II TCU (Dinerstein et al. 1997).

Types and Severity of Threats

The primary threats to biological diversity in this ecoregion are from shifting cultivation, especially in the steeper slopes that people (especially in Vietnam and Thailand) are beginning to clear because of lack of land to support increasing populations, illegal hunting for a huge commercial trade that

TABLE J52.1. WCMC (1997) Protected Areas That Overlap with the Ecoregion.

Protected Area	Area (km2)	IUCN Category
Nam Poui	1,900	VIII
Phou Khao Khoay	1,890	VIII
Phu Miang–Phu Thong	500	IV
Phu Rua	140	II
Unnamed	360	?
Phu Luang	850	IV
Phu Hin Rong Kla	310	II
Ton Krabak Yai	270	II
Nam Nao	920	II
Phu Khieo	1,590	IV
Phu Pha Man	350	?
Chatrakhan [68]	230	?
Total	9,310	

Ecoregion numbers of protected areas that overlap with additional ecoregions are listed in brackets.

supplies China, and loss of habitat caused by large hydro projects (especially in Laos and Vietnam), illegal logging, gahuru-wood and widlife (elephant and gaur) poaching, and road construction in Nam Poui (Boonratana 1997, 1998b).

Priority Conservation Actions

- Initiate a survey to determine the distributional range and density of the Asian elephant in the Nam Poui National Biodiversity Conservation Area (NBCA).
- Initiate a study looking into the behavior and ecology of the Asian elephant in Nam Poui NBCA.
- Seek donor support to strengthen management and activities in Nam Poui NBCA. (LSFP, the current donor, will pull out its support for Phou Khao Khouay and Nam Poui NBCAs in September 2000. Phou Khao Khouay NBCA has other projects that will support it, but Nam Poui NBCA is severely threatened by this void in financial support.)
- Address the issue of transborder log and wildlife poaching (from Laos into Thailand and Vietnam).
- Manage tourism in Thailand's protected areas.
- Reforestation in the degraded areas of Thailand's protected areas.
- Promote conservation awareness (plus awareness of forestry, wildlife, protected area laws, rules, and regulations) to members of the security force (e.g., military, paramilitary, militia, border police) working in and adjacent to these protected areas.

Longer-Term Conservation Actions (5–20 Years)

- Establish a long-term research, monitoring, and training program (for protected-area staff and potential Laotian biologists) for the Asian elephant in Nam Poui NBCA.
- Assess current and future development plans for the region to avoid conflicts with conservation areas, programs, and efforts.
- Complete the exercise of land allocation and land use planning for settlements in and around the protected areas (for Laotian protected areas) and simultaneously establish village rules and regulations for the concerned villages.

Focal Species for Conservation Action

Habitat specialists: hornbills

Wide-ranging species: Asian elephant (Boonratana 1997, 1998a; Payne et al. 1995), gaur (Boonratana 1997, 1998a), tiger (Boonratana 1997, 1998a)

Top predator species: tiger (Boonratana 1997, 1998a)

Other: silvered leaf monkey (Boonratana 1998a), douc langur (Payne et al. 1995), green peafowl, Francois's langur

Selected Conservation Contacts

Bangkok Bird Club

BirdLife-Indochina

Center for Conservation Biology, Mahidol University, Bangkok, Thailand

Department of Forest Resources and Conservation, Lao PDR

FFI-Indochina

IUCN Lao PDR

IUCN Vietnam

WCS-Lao PDR

Wildlife Fund Thailand

WWF-Indochina

WWF-Lao PDR

WWF-Thailand

Justification for Ecoregion Delineation

The extensive Indochina biounit (10) identified by MacKinnon (1997) comprises three subunits that represent the tropical lowland plains, subtropical hills, and temperate montane areas. The largest of these subunits, Central Indochina (10a), is a mix of tropical moist forest and tropical dry forest biomes, and we delineated eight ecoregions that overlap with this biounit but represent the distinct biomes. This ecoregion represents the large area of montane evergreen forests in the Luang-Prabang Range.

Prepared by Eric Wikramanayake, Ramesh Boonratana, Philip Rundel, and Nantiya Aggimarangsee

Ecoregion Number:	**53**
Ecoregion Name:	**Tenasserim–South Thailand Semi-Evergreen Rain Forests**
Bioregion:	**Indochina**
Biome:	**Tropical and Subtropical Moist Broadleaf Forests**
Political Units:	**Myanmar, Thailand, Malaysia**
Ecoregion Size:	**96,900 km²**
Biological Distinctiveness:	**Globally Outstanding**
Conservation Status:	**Relatively Stable**
Conservation Assessment:	**III**

The Tenasserim–South Thailand Semi-Evergreen Rain Forests [53] cover the transition zone from continental dry evergreen forests common in the north to semi-evergreen rain forests to the south. As a consequence, this ecoregion contains some of the highest diversity of both bird and mammal species found in the Indo-Pacific region. The relatively intact hill and montane forests form some of the best remaining habitat essential to the survival of Asian elephants and tigers in the Indo-Pacific region. However, the lowland forests are heavily degraded, and many lowland specialists such as the endemic Gurney's pitta survive in a few isolated reserves.

Location and General Description

This ecoregion encompasses the mountainous, semi-evergreen rain forests of the southern portion of the Tenasserim Range, which separates Thailand and Myanmar, and the numerous small ranges of peninsular Thailand. This ecoregion also includes the extensive lowland plains that lie between the peninsular mountains and until recent decades supported extensive lowland forest. The southern margin of this ecoregion is defined by the Kangar-Pattani floristic boundary

(Whitmore 1984), which separates Indochina from the Malesia.

Annual precipitation increases southward as the length of the dry season and the magnitude of premonsoon drought stress declines. The southern mountain ranges receive rain from both the northeast and southwest monsoons so that, unlike in mountain ranges further north, there is no significant rainshadow. The Köppen climate system places this ecoregion in the tropical wet climate zone (National Geographic Society 1999).

The vegetation of this ecoregion includes both lowland and montane forests. It is transitional between the drought-deciduous forests of Central Thailand at 12° N latitude, where climax species include teak *Tectona grandis* and *Xylia dolabriformis* (WWF and IUCN 1995), and seasonal evergreen rain forests that occur south of about 6° N. Tropical hardwood trees in the family Dipterocarpaceae dominate forests throughout the ecoregion, but species turn over with both elevation and latitude. The diverse dipterocarp forests that occur in the southern portion of this ecoregion include numerous species such as *Dipterocarpus alatus*, *D. griffithii*, *D. laevis*, *D. turbinatus*, *Shorea* spp., *Hopea odorata*, *Fagraea fragrans*, *Bassia longifolia*, *Mesua ferrea*, *Delina sarmentosa*, *Tetracera assa*, *Dillenia aurea*, and *Talauma mutabilis* (WWF and IUCN 1995). The mature forest trees are buttressed and draped with numerous lianas and epiphytes, including *Drynaria* basket ferns, and more than 700 orchid species. Distinctive, thorny climbing palms known as rattans are common in undisturbed sites, although their economic value means that they are greatly reduced in most accessible, unprotected forest locations. Lichens and algae are also speciose components of the epiphyte community.

Karst limestone towers are evident in many locations throughout peninsular Thailand, including islands off Thailand's southwest coast. Extensive karst limestone vegetation is also found in Ao Phangnga National Park (IUCN 1991). In general, forests on limestone are more easily drought stressed and tend to be deciduous, whereas forests on granite benefit from the soil's increased water-holding capacity and tend to support a higher proportion of evergreen species.

Forests of this ecoregion support innumerable plant species that have distinctive and fascinating life histories. The insectivorous pitcher plants *Nepenthes* grow in high-elevation bogs and other nitrogen-deficient habitats. Forests of this ecoregion also support *Rafflesia*, a curious root parasite specific to vines of the genus *Tetrastigma*. This plant occurs throughout Malesia. The species found here, *R. kerrii*, is by no means the largest-flowered member of the genus, but the flowers do attain a diameter of 70 cm, which makes them the largest flowers in Thailand or Myanmar. *Rafflesia* is completely devoid of leaves and possesses very little vascular tissue. The entire plant is underground, except the large but ephemeral, rank-smelling flowers. It is thought that the smell of carrion attracts flies that serve as pollinators (MacKinnon et al. 1996).

Biodiversity Features

This ecoregion contains one of the most intact vertebrate faunas of Indochina, including one of the richest mammal assemblages in Asia. The fauna is also distinctive, with characteristics of the islands of the Malay Archipelago as well as the mountains of China and India. The relatively intact and contiguous hill and montane habitat has potential to conserve large landscapes that will provide adequate habitat to maintain viable populations of Asia's largest carnivore, the tiger (*Panthera tigris*), and the Asian elephant (*Elephas maximus*). This ecoregion lies within a high-priority (Level I) TCU (Dinerstein et al. 1997). This range of forests in conjunction with the Kayah-Karen mountains represents some of the best landscapes for Asian elephant conservation in Indochina.

Numerous other mammals are of conservation significance, primarily the elusive and endemic Fea's muntjac (*Muntiacus feae*) (table J53.1). The population of the Malayan tapir (*Tapirus indicus*), the only Old World tapir representative, has been drastically reduced. It survives in the hill and montane protected areas of this ecoregion and scattered pockets throughout peninsular Malaysia and Sumatra. Several primate species are found in these forests and include the threatened banded langur (*Trachypithecus melalophus*) and slow loris (*Loris nycticebus*), a small, nocturnal prosimian. Other species of conservation concern include the Dyak fruit bat (*Dyacopterus spadiceus*), the endangered clouded leopard (*Pardofelis nebulosa*), common leopard (*Panthera pardus*), sun bear (*Helarctos malayanus*), binturong (*Arctictis binturong*), gaur (*Bos gaurus*), and banteng (*Bos javanicus*) (Stewart-Cox 1995).

The diverse habitats within this ecoregion, from deciduous forests in the north to seasonal evergreen forests in the south, lowland to montane, make it one of the richest in bird species for the entire Indo-Pacific. A total of 560 bird species have been recorded here. However, with rapid habitat loss in the lowlands, many of these forest birds are threatened. One of the most critically endangered, Gurney's pitta (*Pitta gurneyi*), is endemic to this ecoregion (table J53.2). This species, once thought to be extinct, survives in a few locations in lowland

TABLE J53.1. Endemic and Near-Endemic Mammal Species.

Family	Species
Rhinolophidae	*Hipposideros halophyllus*
Vespertilionidae	*Eptesicus demissus**
Cervidae	*Muntiacus feae*

An asterisk signifies that the species' range is limited to this ecoregion.

TABLE J53.2. Endemic and Near-Endemic Bird Species.

Family	Common Name	Species
Phasianidae	Mountain peacock-pheasant	*Polyplectron inopinatum*
Phasianidae	Malayan peacock-pheasant*	*Polyplectron malacense**
Pittidae	Gurney's pitta*	*Pitta gurneyi**
Pycnonotidae	Spectacled bulbul	*Pycnonotus erythropthalmos*

An asterisk signifies that the species' range is limited to this ecoregion.

forest in the central part of peninsular Thailand. More than twenty-five pairs have been found in some of the last remaining forests, now contained within Bang Kram wildlife sanctuary (Stewart-Cox 1995). The Malayan peacock-pheasant (*Polyplectron malacense*) is endemic to the hill and montane forests of this ecoregion (table J53.2). The lowland, alluvial, and wetland forests also support a wide variety of waterfowl. Species range from the large and colorful purple swamphen (*Porphyrio porphyrio*) to several majestic egret species (Stewart-Cox 1995). The lowland, hill, and montane dipterocarp forests are home to several species of the vivacious hornbills. Hornbills prefer to nest in tall trees (usually dipterocarps) in primary forests. At least nine hornbill species are found in these forests and include the nearly extinct wrinkled hornbill (*Aceros corrugatus*).

Status and Threats

Current Status

The existing protected areas system includes twenty-two reserves that cover 11,530 km^2, or 12 percent of the ecoregion's area (table J53.3). Most of these protected areas are located in Thailand. Large blocks of intact seasonal evergreen forest habitat remain in Myanmar, but most of these are not protected. Some protected areas have been designated in the portion of this ecoregion that lies within Myanmar, but their effectiveness is difficult to assess at this time because of the political instability of the region.

Overall, more than 50 percent of the ecoregion's habitat has been converted to agriculture. Despite a logging ban in the late 1980s, the extensive lowland forests of peninsular Thailand have been nearly extirpated. Only one flat, lowland area in peninsular Thailand—Bang Kram—still supports a significant amount of late-successional forest. Other areas support vast tracts of rubber plantation, monocultures of a fast-growing, short-lived tree species native to South America, and plantations of oil palm. Pineapple may be grown here for a few years as a rotation crop after the removal of senescent rubber trees. Paddy rice is also grown in some lowland areas. Unfortunately, none of these crops provide significant support for natural biological diversity.

Hill slopes support more native forest than the lowland areas, and the hill forests of southern Thailand are relatively intact, although swidden (slash-and-burn) agriculture is still practiced in some hill areas in the northern part of the ecoregion. Mature forest cut for swidden agriculture generally is succeeded in this ecoregion by a grassy subclimax that supports far fewer species than the mature forest (IUCN 1991). Kaeng Krachan National Park (2,910 km^2) provides important protection to a variety of moist forest habitats and is exceptionally rich for birds.

Khao Sok National Park (645 km^2) is another important protected area in southern Thailand, although both of these either contain or adjoin large artificial reservoirs from which protrude the skeleton trunks of climax forest trees that have been inundated. Tarutao Marine National Park includes several large islands that support extensive stands of late-successional evergreen forest. Terrestrial mammals that occur on these islands include flying lemur and mouse deer.

Types and Severity of Threats

After Thailand banned timber exploitation in its forests in 1988, Myanmar granted large logging concessions to Thai companies, and illegal timber extraction in Myanmar by Thai loggers has become common in recent years (WWF and IUCN 1995). The soils of this ecoregion are very vulnerable to erosion once exposed, and the long-term ecological effects of large-scale clear felling would be catastrophic. This area is also subject to heavy pressure from development, especially in Kanchanaburi Province, Thailand, where dams and highways are being constructed (IUCN 1991), and in certain areas of peninsular Thailand such as Phuket and Krabi, where coastal resort development has been proceeding with an imprudent degree of urgency. The lowland and alluvial forests are more endangered than the montane region. Large tracts of these forests are being converted to rubber and oil palm plantations. Less than 5 percent of the level lowlands still retain their forest cover, and the degradation threats are slowly moving upslope to hill and montane forests (Stewart-Cox 1995).

Priority Conservation Actions

Short-Term Conservation Actions (1–5 Years)

- Stop illegal logging in and around protected areas.
- Revoke forest concessions in areas of high biodiversity outside protected areas.
- Place under protection all remaining lowland forests.
- Stop wildlife poaching, especially in key parks and reserves.
- Urge Myanmar to join CITES.

TABLE J53.3. WCMC (1997) Protected Areas That Overlap with the Ecoregion.

Protected Area	Area (km^2)	IUCN Category
Khao Laem [51, 70]	530	II
Sai Yok [70]	500	II
Kaeng Krachan	2,560	II
Pakchan	940	PRO
Unnamed	440	?
Unnamed	130	?
Unnamed	490	?
Khlong Nakha	460	IV
Khlong Saeng	1,070	IV
Khao Sok	690	II
Unnamed	260	?
Khao Luang	640	II
Khlong Phraya	60	IV
Unnamed	20	?
Unnamed	80	?
Khao Phanom Bencha	60	UA
Khao Pu–Khao Ya	710	II
Khao Banthat	1,160	IV
Hat Chao Mai	260	II
Ton Nga Chang	190	IV
Thaleban	100	II
Unnamed	180	?
Total	11,530	

Ecoregion numbers of protected areas that overlap with additional ecoregions are listed in brackets.

Longer-Term Conservation Actions (5–20 Years)
- Reassess representation of habitat in the protected area system in the ecoregion.
- Ensure that the protected area system includes linkages between lowland and montane habitats.
- Restore integrity of buffer zones of protected areas.
- Implement rigorous conservation actions in the Level I and Level II TCUs in the ecoregion

Focal Species for Conservation Action

Habitat specialists: hornbills (need mature trees for nests), especially the wrinkled hornbill

Endangered endemic species: Fea's muntjac, Gurney's pitta

Area-sensitive species: Asian elephant (Santiapillai and Jackson 1990; Santiapillai 1997), tiger (Franklin et al. 1999; Tilson et al. 1994; Dinerstein et al. 1997; Seidensticker et al. 1999)

Top predator species: tiger (Franklin et al. 1999; Tilson et al. 1994; Dinerstein et al. 1997; Seidensticker et al. 1999)

Other: Dyak fruit bat, banded langur, clouded leopard, gaur, banteng

Selected Conservation Contacts

Wildlife Fund Thailand
WWF-Thailand
WWF-Malaysia

Justification for Ecoregion Delineation

MacKinnon's subunit 05d extends into the Kayah Karen and Tenasserim mountain ranges. We extracted the montane forests along the Tenasserims to form this ecoregion.

Prepared by Colby Loucks, Chris Carpenter, and J. F. Maxwell

Ecoregion Number:	**54**
Ecoregion Name:	**Northern Annamites Rain Forests**
Bioregion:	**Indochina**
Biome:	**Tropical and Subtropical Moist Broadleaf Forests**
Political Units:	**Laos and Vietnam**
Ecoregion Size:	**47,100 km^2**
Biological Distinctiveness:	**Globally Outstanding**
Conservation Status:	**Relatively Stable**
Conservation Assessment:	**III**

The discovery of a new vertebrate species usually is a newsworthy achievement. But to discover five large or midsized mammals, including two that belong to new, monotypic genera within the space of five years, is truly astounding. It also epitomizes the globally outstanding biological diversity of the Northern Annamites Rain Forests [54]. It was just under a decade ago that the Saola (*Pseudoryx nghetinhensis*) was discovered during biological surveys of the intact forests along the rugged and little-explored mountains of the northern Annamite Range (Dung et al. 1993). Soon thereafter the giant muntjac (*Megamuntiacus vuquangensis*) was discovered in the same area (Tuoc et al. 1994; Schaller and Vrba 1996). Since then there has been a spate of discoveries and rediscoveries of species that have not been seen for years since their first description (Giao et al. 1998; Groves et al. 1998).

Location and General Description

The Northern Annamites Rain Forests [54] ecoregion lies largely in Laos but with a significant area across the crest of the Annamite Range in Vietnam. Geological substrates are varied over this ecoregion, but there are notable large areas of limestone karst topography. Climatic conditions often change abruptly along the mountain crest. Lower moisture input from the northeast monsoon winds produces mean annual rainfall in montane habitats of this ecoregion in Laos that are lower at 1,500–2,500 mm and more seasonal than those of the Northern Vietnam Lowland Rain Forests [56]. However, higher elevations produce cooler temperatures and thus distinct conditions for plant growth. This ecoregion shows strong floristic relationships with the mountains of northern Vietnam and southern China and with more southern areas of the Annamite Range. Vascular plant levels of endemism are high.

Mesic lower montane forests at 800–1,200 m in the northern Annamite Range generally consist of a two-tiered forest canopy reaching to about 15–25 m in height. The dominant floristic elements in this forest are the Myrtaceae, Fagaceae, Elaeocarpaceae, and Lauraceae, with high endemism. Also notable in this forest community is the presence of species of Podocarpaceae. Epiphytes, particularly orchids, are abundant and rich in endemic species. The dense canopy structure of undisturbed humid montane forest allows little light to penetrate to ground level, so understory vegetation is sparse. However, the understory in open areas of forest often supports thick stands of woody bamboo (*Chimonobambusa* and *Dendrocalamus*). It has been suggested that typhoons that regularly reach this area provide a natural disturbance regime that continually opens the canopy and promotes bamboo growth.

Less mesic sites with thin, acid soils at elevations up to 1,500 m often support pine forests with *Pinus kesiya* as the dominant species. Under reasonable growing conditions, *P. kesiya* reaches 20–30 m in height. The most common tree associates of *P. kesiya* are *Keteleeria evelyniana*, *Pieris ovalifolia*, *Schima wallichii*, and *Tristania merguensis*. Frequent fire seems to promote the establishment and maintenance of these pine forests at the expense of hardwoods. A second community of open conifer forest in montane areas of northern Laos is dominated by *Keteleeria evelyniana*, which may occur in pure stands or mixed with *Pinus kesiya* and species of Fagaceae. *Keteleeria* may reach up to 35 m in height and occurs from subtropical areas of southern China southward into Laos and Vietnam.

Humid montane forests at about 1,200–1,800 m may support a rich forest dominated by *Fokienia hodginsii*, *Podocarpus imbricatus*, and *Cunninghamia lanceolata*. *Fokienia* can reach 40–50 m in height and 1.5 m in diameter in this community. The cool and moist conditions of these *Fokienia* forests often promote lush growth of cryptogamic epiphytes and an accumulation of organic litter on the forest floor.

Long histories of human impact in the northern Annamite Range of Laos and Vietnam have led to the degradation of large areas of what were once montane evergreen forest. Such forests have been replaced by thickets or savannas with diverse structure depending on the degree of degradation and the time for successional recovery since major disturbance. Thickets may include diverse assemblages of woody species or be dominated by pure stands of a single colonizing species.

Further floristic and faunal surveys are needed to assess the biological potential of this ecoregion, especially in Laos. However, the Vietnamese portion of this ecoregion contains high species diversity and is notable for its significant endemism.

Biodiversity Features

This treasure trove of biological diversity is still divulging its true riches. Each new survey turns up new species of mammals, birds, fishes, herptiles, fishes, butterflies, and plants. Of the 134 mammals now known from the ecoregion, three are near-endemic species and four are endemic (table J54.1). More than half of these are recently described species; therefore, it is very likely that several species could be added to this list as surveys continue.

The mammal assemblage also includes several threatened species, including the endangered Douc langur, tiger (*Panthera tigris*), Asian elephant (*Elephas maximus*), gaur (*Bos gaurus*), banteng (*Bos javanicus*), serow (*Capricornis sumatraensis*), and particoloured flying squirrel (*Hylopetes alboniger*) (Hilton-Taylor 2000).

TABLE J54.1. Endemic and Near-Endemic Mammal Species.

Family	Species
Cercopithecidae	Semnopithecus francoisi
Hylobatidae	Hylobates gabriellae
Bovidae*	Pseudoryx nghetinhensis*
Cervidae*	Megamuntiacus vuquangensis*
Cervidae*	Muntiacus truongsonensis*
Cervidae*	Muntiacus (Buhouang muntjak)*
Cercopithecidae	Pygathrix nemaeus

An asterisk signifies that the species' range is limited to this ecoregion.

TABLE J54.2. Endemic and Near-Endemic Bird Species.

Family	Common Name	Species
Phasianidae	Imperial pheasant*	Lophura imperialis*
Phasianidae	Edwards's pheasant	Lophura edwardsi
Phasianidae	Vietnamese fireback*	Lophura haitensis*
Phasianidae	Crested argus	Rheinardia ocellata
Timaliidae	Short-tailed scimitar-babbler	Jabouilleia danjoui
Timaliidae	Sooty babbler*	Stachyris herberti*

An asterisk signifies that the species' range is limited to this ecoregion.

The ecoregion overlaps with a Level I TCU and two Level II TCUs (Dinerstein et al. 1997), but the prey base necessary to maintain tigers is declining rapidly because of hunting pressure. Most of the habitat in the ecoregion is too steep for elephants, which are limited to the lower elevations. But as these more accessible midmontane forests are cleared for agriculture, elephants are becoming pocketed in small forest patches.

The bird fauna in the ecoregion is estimated at more than 525 species, including 3 near-endemic and 3 endemic species (table J54.2). The imperial pheasant (*Lophura imperialis*) and Edwards's pheasant (*L. edwardsi*) are critically endangered, and the Vietnamese fireback (*L. haitensis*) and white-winged duck (*Cairina scutulata*) are endangered species found in this ecoregion. Not surprisingly, the ecoregion overlaps with an EBA, Annamese Lowlands (143), which contains nine restricted-range bird species (Stattersfield et al. 1998).

Status and Threats

Current Status

More than half of this ecoregion has been cleared or degraded. However, the forests in the northwest part of the ecoregion are still in fairly good condition. A large block of montane forest straddles the Laos-Vietnam border. There are nine protected areas that cover about 12,200 km^2 (26 percent) of the ecoregion (table J54.3). Of these, five—Pu Mat in Vietnam and Nam Kading, Phou Xang He, Nakai-Nam Theun, and Khammouane Limestone NBCAs in Laos—exceed 1,000 km^2.

Because of the high elevations and steep slopes of this ecoregion, human population density is moderate (MacKinnon 1997). However, shifting cultivation and logging in the submontane forests are major sources of deforestation (IUCN 1991). The resultant erosion in the steep slopes exacerbates degradation and drives shifting cultivation into higher and more remote areas (IUCN 1991). Cleared forests usually regrow as grasslands of *Imperata cylindrica* or bamboo forests (*Bambusa*, *Thyrsostachys*, and *Oxytenanthera*) (FAO 1981).

TABLE J54.3. WCMC (1997) Protected Areas That Overlap with the Ecoregion.

Protected Area	Area (km^2)	IUCN Category
Ho Ke Go [56]	180	PRO
Nam Kading	1,600	VIII
Pu Mat	1,400	?
Dong Phong Nha	440	IV
Phou Xang He	1,170	VIII
Nakai-Nam Theun	3,710**	VIII
Hin Namno	865	VIII
Vu Quang	850	IV
Khammouane Limestone NBCA*	2,000	VIII
Total	12,215	

Ecoregion numbers of protected areas that overlap with additional ecoregions are listed in brackets.

*Now commonly called Phou Hin Poun NBCA (Phou Hin Poun = limestone mountain).

**Currently undergoing redelineation of boundaries, so the area will differ.

Types and Severity of Threats

Both Laos and Vietnam have begun or are planning to develop several major hydropower schemes that will inundate large areas of habitat and provide ready access to intact forest areas, thus increasing the probabilities of further habitat degradation. Major illegal and legal logging and local and transboundary wildlife poaching and trade (Laos into Vietnam) still occur. The presence of unexploded ordnances similarly poses a severe threat to wildlife, researchers, and protected area staff.

Priority Conservation Actions

Short-Term Conservation Actions (1–5 Years)

- Implement the conservation action plan for saola in Lao PDR (for specific recommendations see Robichaud 1999).
- Gazette the proposed northern extension of Nakai–Nam Theun NBCA.
- Gazette the proposed corridor between Nakai-Nam Theun NBCA and Phou Hin Poun NBCA.
- Initiate a species conservation research and management study on the Asian elephant, covering Nakai-Nam Theun NBCA, Phou Hin Poun NBCA, and the proposed corridor.
- Initiate a survey to determine the distribution, density, and movement of the Asian elephant in the ecoregion.
- Seek donors to initiate management and related activities in Nam Kading NBCA.
- Address the issue of transborder wildlife poaching and trade (from Laos into Vietnam).
- Initiate a wildlife and habitat survey in the Phou Hai-Phou Mon area (a proposed extension to Phou Hin Poun NBCA.)
- Implement activities as specified in the management plans for Phou Hin Poun NBCA.
- Provide training and capacity building for protected area staff and their associates.
- Clear unexploded ordnance.

Longer-Term Conservation Actions (5–20 Years)

- Carry forward the implementation of the management plan for Phou Hin Poun NBCA.
- Assess current and future development plans for the region to avoid conflicts with conservation areas, programs, and efforts.
- Complete the exercise of land allocation and land use planning for settlements in and around the protected areas (for Lao PDR protected areas) and simultaneously establish village rules and regulations for the concerned villages.
- Stamp out transboundary wildlife poaching and trade.
- Initiate a program to study and monitor the movements of the Asian elephant in the Nam Theun area after inundation (with the construction of Nam Theun II).

Focal Species for Conservation Action

Habitat specialists: Francois's langur (Boonratana 1998c; Steinmetz 1998; Evans et al., in prep.; Timmins and Khounboline 1996), southern serow (Boonratana 1998c; Steinmetz 1998; Tobias 1997; Timmins and Khounboline 1996), hornbill spp. (Boonratana 2000a; Tobias 1997; Timmins and Khounboline 1996)

Endangered endemic species: saola (Schaller and Rabinowitz 1995; Tobias 1997; Robichaud 1999), giant muntjac (Boonratana 1998a, 1998b, 2000a; Schaller 1995; Tobias 1997; Timmins and Khounboline 1996), Annamite muntjac, *Buhouang muntjak,* imperial pheasant, Vietnamese fireback, sooty babbler, Annamite striped rabbit

Area-sensitive species: elephant (Boonratana 1998a, 1998b, 2000a; Steinmetz 1998; Evans et al., in prep.; Tobias 1997), gaur (Boonratana 1998a, 1998b, 2000a; Steinmetz 1998; Evans et al., in prep.; Timmins and Khounboline 1996), banteng (Boonratana 1998b; Tobias 1997), tiger (Boonratana 1998a, 1998b, 2000a; Steinmetz 1998; Tobias 1997), hornbill spp. (Boonratana 2000b; Tobias 1997)

Top predator species: tiger (Boonratana 1998a, 1998b, 2000b; Steinmetz 1998; Tobias 1997; Timmins and Khounboline 1996), leopard (Boonratana 1998b), clouded leopard (Tobias 1997), dhole (Boonratana 1998a, 1998b, 2000b; Steinmetz 1998; Evans et al., in prep.; Tobias 1997)

Other: Phayre's langur (Evans et al., in prep.), douc langur (Boonratana 1998a, 1998b, 2000b; Evans et al., in prep.; Tobias 1997; Timmins and Khounboline 1996), white-cheeked and yellow-cheeked gibbon (Boonratana 1998a, 2000b; Steinmetz 1998; Evans et al., in prep.; Tobias 1997; Timmins and Khounboline 1996), Asiatic black bear (Boonratana 2000b; Steinmetz 1998; Tobias 1997), Malayan sun bear (Boonratana 2000b; Tobias 1997), bear spp. (Boonratana 1998a, 1998b; Steinmetz 1998; Evans et al., in prep.; Tobias 1997; Timmins and Khounboline 1996), serow (Boonratana 1998b; Tobias 1997), golden cat (Boonratana 1998b; Tobias 1997), green peafowl (Boonratana 1998a), crested argus (Tobias 1997)

Selected Conservation Contacts

BirdLife-Indochina

Department of Forest Resources and Conservation, Lao PDR

FFI-Indochina

IUCN Lao PDR

IUCN Vietnam

WCS-Lao PDR

WWF-Indochina

WWF-Lao PDR

Justification for Ecoregion Delineation

In a previous analysis of conservation units across the Indo-Malayan realm, MacKinnon (1997) included both lowland and montane tropical moist forests along the northern Annamite mountains within a single sub-biounit, the North Annam subunit 05c. In keeping with our rules for delineating ecoregions, we separated the montane and submontane forests from the lowland forests using the 1,000-m contour from a DEM. The southern boundary to separate the northern from the southern Annamites was based on the Hai Van Pass, widely recognized as a boundary between the northern and southern flora. Thus, the moist forests that lie along the spine of the northern Annamite Range, above 1,000 m in elevation and between 19°N and the low hills between Hue and Da Nang were placed in the Northern Annamites Rain Forests [54].

Prepared by Ramesh Boonratana, Philip Rundel, and Eric Wikramanayake

Ecoregion Number:	**55**
Ecoregion Name:	**Cardamom Mountains Rain Forests**
Bioregion:	**Indochina**
Biome:	**Tropical and Subtropical Moist Broadleaf Forests**
Political Units:	**Cambodia, Thailand, and Vietnam**
Ecoregion Size:	**44,100 km²**
Biological Distinctiveness:	**Globally Outstanding**
Conservation Status:	**Relatively Intact**
Conservation Assessment:	**III**

The Cardamom Mountains Rain Forests [55] ecoregion sits astride the Cardamom Mountains (locally known as *Kravanh*) and the Elephant Range (locally known as *Dom rei*) in southwestern Cambodia and extends slightly across the border into southeastern Thailand. It is separated from the nearest other rain forest by the vast, dry Khorat Plateau in central Thailand to the north and east and by the Gulf of Thailand in the west. The Cardamom Mountain rain forests are considered by some to be one of the most species-rich and intact natural habitats in the region, but they are also one of the least explored. Although scientific explorations have begun to reveal the unique biological riches in other intact and hitherto unexplored rain forests such as in the Annamite Mountain Range, the Cardamom Mountains have been neglected until recently (Fauna & Flora International 2000). It would not be surprising if this ecoregion, which contains the essential ingredients for speciation (i.e., isolation, moist and stable conditions, intact and undisturbed habitat, and rugged terrain), yields a number of new species that will add to its biological diversity.

Location and General Description

The ecoregion represents the original extent (see MacKinnon 1997) of the wet evergreen forests that cover the Cardamom and Elephant mountains in southwest Cambodia and along the mountains east of Bangkok, in Thailand. The small Dao Phu Quoc Island, which belongs to Vietnam but is just off the southern coast of Thailand, is also included in this ecoregion.

The mountain range rises from sea level to more than 1,500 m to intercept and extract the moisture from the monsoon winds. The orientation of their topography along the Gulf of Thailand produces unusually wet conditions of 3,000–4,000 mm annual rainfall on the southwestern slopes of these ranges, and only a short dry season occurs. The mean annual rainfall total exceeds 5,000 mm in the Emerald Valley near Bokor in the Elephant Range, whereas Kirirom, more distant from the coast in this range, receives about 2,000 mm annually. These ranges are largely Mesozoic sandstone, with localized areas of limestone and volcanic rock. Younger basalts in these ranges have produced rich supplies of gemstones (rubies, sapphires, and zircons). Also included in the ecoregion is the northeastern part of this mountain unit, composed of granite ridges that reach a maximum elevation of 1,813 m at Phnom Aural, the highest point in Cambodia. These ranges rise rapidly from the coast, leaving only a narrow coastal plain. They gently grade down into the interior lowlands to the north and northeast.

The ecological and floristic composition of these wet evergreen forest communities is poorly studied, and there would undoubtedly be recognizable communities within this association were it better known and investigated. From limited available data, local endemism appears to be significant. One of the most abundant canopy species in wet evergreen forests is *Hopea pierrei*, a small tree of limited distribution outside of this area. Other dipterocarps once formed the dominant canopy elements of a tall evergreen forest in coastal area, with *Shorea hypochra*, *Anisoptera costata*, *Dipterocarpus costatus*, and *Hopea odorata* all abundant. Also important as canopy trees in this area were *Parkia streptocarpa*, *Heritiera javanica*, *Swintonia pierrei*, and *Syzygium cinereum*. These forests have largely disappeared today.

In addition to typical lowland evergreen rain forests, the southern slopes of the Elephant Mountains support an unusual dwarf rain forest community reaching no more than 12 m height in areas of poorly drained depressions. The dominant species in these waterlogged sites typically are *Dacrydium elatum* and *Podocarpus neriifolius*, with a scattered distribution of *P. (Nageia) fleuryi* and *P. (Dacrycarpus) imbricatus*.

Upper elevation areas above about 700 m in the Cardamom and Elephant Mountains contain a distinct montane forest community. These forests are structured with dense evergreen tree canopies reaching up to 30 m in height. The Fagaceae are notably dominant, including *Lithocarpus cambodienseis*, *L. guinieri*, *L. farinulenta*, *L. harmandii*, and *Castanopsis cambodiana*. Also important are species of Lauraceae (*Cinnamomum* and *Litsea*) and Myrtaceae (*Syzyngium* and *Tristania*), whereas legumes are less common. There is a rich understory in these habitats, with shrubs of Rubiaceae and Euphorbiaceae, palms (*Arenga pinnata* and *Pinanga cochinchinensis*), arborescent ferns (*Cibotium*, *Cyathea*, and *Oleandra*), Pandanus, and Araliaceae. Epiphytes, most notably orchids, may often be abundant, particularly in sites that receive frequent fogs or mists.

A distinctive dwarf forest 5–10 m in height is present on the acid and skeletal soils on the sandstone plateau of the southern Elephant Mountains. This community typically is dominated by *Dacrydium elatum*, with another conifer, *Podocarpus (Dacrycarpus) imbricatus*, also commonly present. Other important associates in this dwarf forest are a variety of Fagaceae and Myrtaceae, *Vaccinium viscifolium*, and *Schima crenata*. On ridgelines or other areas habitually exposed to strong winds, this community reaches no more than 5 m height. Sphagnum bogs are also present.

A prominent area of *Pinus merkusii* occurs on the Kirirom Plateau in the Elephant Range, where *P. merkusii* grows with *Dipterocarpus obtusifolius, Rhodomyrtus tomentosa, Phyllanthus officinalis,* and a variety of Melastomataceae and Rubiaceae.

Biodiversity Features

The Cardamom Mountains Rain Forests [55] ecoregion is one of the few ecoregions in Indochina with intact rain forests that still have potential for landscape-level conservation actions. Natural disturbance regimes and large predator-prey interactions still occur. Much of the ecoregion's biota probably remains intact. This combination of factors makes the ecoregion unique in Indochina.

The ecoregion is considered to harbor more than 100 mammal species. There are no known ecoregional endemics. Nevertheless, there are several threatened species, including the endangered tiger (*Panthera tigris*), Asian elephant (*Elephas maximus*), clouded leopard (*Pardofelis nebulosa*), dhole (*Cuon alpinus*), gaur (*Bos gaurus*), banteng (*Bos javanicus*), khting vor (*Pseudonovibos spiralis*), pileated gibbon (*Hylobates pileatus*), and serow (*Capricornis sumatraensis*). It is possible that several large mammals that have disappeared from Indochina's forests—the Sumatran (*Dicerorhinus sumatrensis*) and Javan rhinoceros (*Rhinoceros sondaicus*) are examples—may still find safe haven in these forests.

The elephant population in the Cardamom and Elephant ranges is widely considered to be the most important in Cambodia and among the largest in Indochina, although surveys are necessary to confirm this assertion. These intact forests were also recognized as a Level I TCU (Dinerstein et al. 1997), where the large habitat areas allow the predator-prey dynamics associated with tigers to occur under undisturbed conditions. The Cardamom Range probably harbors the highest density of pileated gibbons throughout the species' distributional range (Boonratana 1999a).

The bird fauna is estimated at more than 450 species and includes 2 strict endemic species (table J55.1). However, in all probability many more endemic species will be added to this list after more comprehensive surveys.

Recent surveys conducted by the Flora and Fauna International (FFI) have found a population of the critically endangered Siamese crocodile (*Crocodylus siamensis*) and several other small mammals and amphibians (Fauna & Flora International 2000).

Status and Threats

Current Status

Because of the low human population of this ecoregion, the forests in Cambodia are relatively intact; however, the areas in southeastern Thailand have been greatly reduced and now exist only in a few protected areas in hilly regions (IUCN 1991). Sixteen protected areas cover about 14,500 km² (33 percent) of the ecoregion (table J55.2). Six of these protected areas—Aural, Phnom Bokor, Botum-Sakor, Roniem Daun Sam, Khao Ang Ru Nai, and Phnom Samkos—are larger than 1,000 km². Phnom Samkos National Park exceeds 3,000 km².

Types and Severity of Threats

Despite this high level of formal protection, very few reserves have effective management and workforces; they are paper parks. Several are now under threat from illegal logging operations and from adjacent concessions that encroach on the unprotected protected areas. The wildlife trade has also resulted in widespread hunting throughout Cambodia and Thailand, exacting a heavy toll from endangered wildlife populations. The widespread presence of antipersonnel landmines pose severe threat to both wildlife and humans (including researchers).

Priority Conservation Actions

Short-Term Conservation Actions (1–5 Years)

- Carry out more wildlife and botanical surveys in the area.
- Initiate a study on the density, distribution, and ecology of megaherbivores, focusing on khting vor.

Longer-Term Conservation Actions (5–20 Years)

- Streamline and coordinate the activities of the Cambodia's Ministry of Environment (which has overall responsibility for development and management of protected areas) and the Ministry of Agriculture, Forestry and Fisheries (which is responsible wildlife protection).
- Proactive law enforcement and international cooperation are needed to stamp out wildlife poaching and trade, focusing on restaurants, markets, and international border crossings.

TABLE J55.1. Endemic and Near-Endemic Mammal Species.

Family	Common Name	Species
Phasianidae	Chestnut-headed partridge*	*Arborophila cambodiana**
Phasianidae	Siamese partridge*	*Arborophila diversa**

An asterisk signifies that the species' range is limited to this ecoregion.

TABLE J55.2. WCMC (1997) Protected Areas That Overlap with the Ecoregion.

Protected Area	Area (km²)	IUCN Category
Aural	2,420	IV
Kirirom	250	II
Phnom Bokor	1,580	II
Ream	140	II
Botum-Sakor [79]	1,520	II
Dong Peng [79]	150	VIII
Peam Krasop [79]	80	IV
Roniem Daun Sam	1,980	IV
Khao Ang Ru Nai	1,030	IV
Khao Soi Dao	720	IV
Khao Chamao-Khao Wong	80	II
Khao Khitchakut	70	II
Samlaut	630	VIII
Namtok Phlui	100	II
Phnom Samkos	3,170	IV
Mu Ko Chang	660	II
Total	14,580	

Ecoregion numbers of protected areas that overlap with additional ecoregions are listed in brackets.

- A long-term personnel and institutional capacity-building and strengthening program is needed for the relevant government agencies dealing with the conservation of natural resources. The program's components should include field surveys, patrolling, monitoring, law enforcement, protected area management, species management, organizational matters, and conservation education and awareness.

Focal Species for Conservation Action

Habitat specialists: hornbills (Boonratana 1999a)

Endangered endemic species: chestnut-headed partridge (Boonratana 1999a), Siamese partridge, Siamese crocodile

Wide-ranging species: Elephant (Boonratana 1999a), gaur (Boonratana 1999a), banteng (Boonratana 1999a), khting vor (Boonratana 1999a), tiger (Boonratana 1999a)

Top predator species: tiger (Boonratana 1999a), leopard (Boonratana 1999a), clouded leopard (Boonratana 1999a), dhole (Boonratana 1999a), Asiatic jackal (Boonratana 1999a)

Other: golden cat (Boonratana 1999a), Malayan sun bear (Boonratana 1999a), Asiatic black bear (Boonratana 1999a), large Indian civet (Boonratana 1999a), pileated gibbon (Boonratana 1999a), silvered leaf monkey (Boonratana 1999a), fishing cat (Fauna & Flora International 2000)

Selected Conservation Contacts

Bangkok Bird Club

BirdLife-Indochina

Center for Conservation Biology, Mahidol University, Bangkok, Thailand

FFI-Indochina

IUCN Vietnam

Ministry of Environment, Cambodia

WCS-Cambodia

Wildlife Fund Thailand

Wildlife Protection Office, Cambodia

WWF-Cambodia

WWF-Indochina

WWF-Thailand

Justification for Ecoregion Delineation

The large Indochina biounit (10) identified by MacKinnon (1997) in his analysis of conservation units in the Indo-Malayan realm comprises three subunits that represent the vegetation in the tropical lowland plains, subtropical hills, and temperate montane areas. The largest of these subunits, Central Indochina (10a), is a mix of tropical moist and tropical dry forests. Adjacent to this is the Cardamom Mountains subunit (05d), which also includes the freshwater swamp forests and mangroves of the Chao Phraya river and estuary (MacKinnon 1997). We extracted the tropical wet evergreen vegetation that covers the Cardamom and Elephant mountain ranges (included within MacKinnon's subunits 10a and 05d) and represented it with the Cardamom Mountains Rain Forests [55]. The swamp forests, dry forests, and mangroves were assigned to separate ecoregions.

Prepared by Ramesh Boonratana, Philip Rundel, and Eric Wikramanayake

Ecoregion Number:	**56**
Ecoregion Name:	**Northern Vietnam Lowland Rain Forests**
Bioregion:	**Indochina**
Biome:	**Tropical and Subtropical Moist Broadleaf Forests**
Political Units:	**Vietnam**
Ecoregion Size:	**22,500 km^2**
Biological Distinctiveness:	**Regionally Outstanding**
Conservation Status:	**Critical**
Conservation Assessment:	**II**

The Northern Vietnam Lowland Rain Forests [56] are in a dismal state. Less than 10 percent of the native vegetation of the ecoregion remains, and very little of that is protected. The remaining patches of habitat are small and scattered throughout the ecoregion, so that any natural ecological processes that once occurred have been lost. This ecoregion has lost most of its outstanding biodiversity, although the white-cheeked gibbon and Francois's leaf monkey can still be found locally in these forests.

Location and General Description

The Northern Vietnam Lowland Rain Forests [56] ecoregion extends from the freshwater swamp forests of the Red River Valley south along the north-central coast of Vietnam to the region south of Tam Ky. Geological formations are varied, but there are extensive limestone substrates. The north-central coastal area of Vietnam typifies the tropical monsoon climate, with high temperatures and abundant precipitation, but rainfall peaks later in the year than in other parts of southeast Asia, and the dry season is less extreme. Mean annual rainfall at Hanoi and Vinh is about 1,800 mm, with the peak of rainfall in September and October. Every month receives at least 50 mm of rainfall. Peak rainfall is pushed back even further at Quang Tri, along the central coast, where peak rainfall comes in October and November, and mean annual total is higher at about 2,500 mm. The pronounced wet season here extends from September through January, and no month of the year has less than 60 mm of rainfall. Total rainfall continues to increase moving southward down the coast, with about 3,000 mm at Hue, but drops sharply and becomes more seasonal in south-central coastal areas. Moving inland, the boundary between the Northern Vietnam Lowland Rain Forests [56] ecoregion and the Northern Annamites Rain Forests [54] ecoregion is poorly defined as lowland wet evergreen forests grade into montane forest communities.

The high rainfall and short dry season characterizing the coastal habitats of this ecoregion produce conditions that once supported diverse wet evergreen forests. Such forest habitats have largely been cleared and exist only in isolated patches today. This ecoregion is best preserved in Cuc Phuong and Pu Mat National Parks. At Cuc Phuong,

1,800 vascular plant species have been described for a small area with limited topographic diversity. Overall, the flora of these wet evergreen forests shows a stronger affinity to that of northern Vietnam and southern China than to that of southern Vietnam. Although the Dipterocarpaceae is an ecologically significant element of the lower-elevation wet evergreen forests, the species richness in this family is lower than that of similar habitats in the southern Annamite Range.

Primary wet evergreen forest consists of a dense, three-tiered canopy reaching 25–35 m and occasionally 45 m height in undisturbed sites, with large emergent trees extending above this level to give a rough upper surface to the canopy. The upper canopy is dominated by a species of *Hopea*, *Castanopsis hystrix*, and *Madhuca pasquieri*. The fan palm *Livistona saribus* is a common subcanopy species in small gaps and reaches 20 m in height. Wet evergreen forest stands disturbed by logging show a characteristic presence of *Knema erratica*, a fast-growing colonizer, and an increased dominance of *Livistona saribus* in the upper canopy.

Biodiversity Features

Most of this ecoregion's biodiversity has been lost because of the extensive habitat loss. Nevertheless, it still harbors several mammals and birds of conservation significance, including the Owston's banded civet (*Hemigalus owstoni*), white-cheeked gibbon (*Hylobates leucogenys*), red-shanked douc langur (*Pygathrix nemaeus*), and Francois's leaf monkey (*Semnopithecus francoisi*). One endemic bat species is found here (*Paracoelops megalotis* table J56.1). The ecoregion overlaps with a Level II TCU (Dinerstein et al. 1997). There are more than 300 bird species in this ecoregion, including three near-endemic and one endemic species (table J56.2). The Annamese Lowlands (143) EBA (Stattersfield et al. 1998) overlaps with this ecoregion.

Status and Threats

More than 90 percent of the natural habitat in this ecoregion has been converted, and the remaining habitat is scattered as small fragments. The nine protected areas in the ecoregion cover less than 900 km^2 (4 percent) of the ecoregion (table J56.3). And many of these small protected areas (average size 99 km^2) are degraded.

Types and Severity of Threats

Vietnam's high human population density has taken a heavy toll on its natural habitats. The coastal and alluvial forests in particular have been hardest hit because the human population densities are highest in these lowland habitats. The rampant illegal wildlife trade has also exacerbated the threats to wildlife and other natural resources.

Priority Conservation Actions

Short-Term Conservation Actions (1–5 Years)

- Protect remaining fragments of habitat.
- Identify and restore degraded habitats wherever possible to increase the available habitat, in line with the Forest Protection and Development Act of 1991.
- Prohibit and prevent hunting, especially of endangered species for the commercial trade.
- Implement the National Biodiversity Action Plan.

Longer-Term Conservation Actions (5–20 Years)

- Strive for effective buffer-zone management and increased protected areas through habitat restoration. Existing programs that grant stewardship to local people to manage reforested state lands (under Decision No 327/CT) provides ways to increase forested areas and protected area coverage.

Focal Species for Conservation Action

Habitat specialists: Owston's banded civet, white-cheeked gibbon, red-shanked douc langur, Francois's leaf monkey

Endemic species: Paracoelops megalotis, Annam partridge, Edwards's pheasant, short-tailed scimitar-babbler, grey-faced tit-babbler

Area-sensitive species: tiger, white-cheeked gibbon, red-shanked douc langur, Francois's leaf monkey

Top predator species: tiger

TABLE J56.1. Endemic and Near-Endemic Mammal Species.

Family	Species
Rhinolophidae*	Paracoelops megalotis*

An asterisk signifies that the species' range is limited to this ecoregion.

TABLE J56.2. Endemic and Near-Endemic Bird Species.

Family	Common Name	Species
Phasianidae	Annam partridge*	Arborophila merlini*
Phasianidae	Edwards's pheasant	Lophura edwardsi
Timaliidae	Short-tailed scimitar-babbler	Jabouilleia danjoui
Timaliidae	Grey-faced tit-babbler	Macronous kelleyi

An asterisk signifies that the species' range is limited to this ecoregion.

TABLE J56.3. WCMC (1997) Protected Areas That Overlap with the Ecoregion.

Protected Area	Area (km^2)	IUCN Category
Cuc Phuong [74]	200	II
Ngoc Trao [74]	7	UA
Chua Huong Tich	40	UA
Tam Quy	40	IV
Den Ba Trieu	10	UA
Ben En	130	II
Ho Ke Go [54]	140	PRO
Bac Hai Van	40	?
Bach Ma [57]	280	II
Total	887	

Ecoregion numbers of protected areas that overlap with additional ecoregions are listed in brackets.

Selected Conservation Contacts

BirdLife International, Indochina
FFI-Indochina
Forest Planning and Inventory Unit, Hanoi, Vietnam
Forest Protection Department, Hanoi, Vietnam
Institute for Ecology and Biological Resources, Hanoi, Vietnam
IUCN Vietnam
Provincial forest departments
WWF-Indochina

Justification for Ecoregion Delineation

MacKinnon (1997) included the coastal habitats of Vietnam, Cambodia, and Thailand in the Coastal Indochina biounit (05), consisting of four subunits. Each of these subunits includes several biomes. We delineated several ecoregions based on the distribution of these biomes. Like MacKinnon, we chose the Hai Van pass to represent a transition from the tropical south and the subtropical north and placed the lowland evergreen moist forests between the Hai Van pass and the Red River delta, in the Northern Vietnam Lowland Rain Forests [56].

Prepared by Eric Wikramanayake and Philip Rundel

Ecoregion Number:	**57**
Ecoregion Name:	**Southern Annamites Montane Rain Forests**
Bioregion:	**Indochina**
Biome:	**Tropical and Subtropical Moist Broadleaf Forests**
Political Units:	**Laos, Vietnam, and Cambodia**
Ecoregion Size:	**46,200 km²**
Biological Distinctiveness:	**Globally Outstanding**
Conservation Status:	**Vulnerable**
Conservation Assessment:	**I**

The Southern Annamites Montane Rain Forests [57] ecoregion in the remote montane forests of Kontuey Neak, or "the dragon's tail"—in the extreme northwest of Cambodia, where the boundaries of Cambodia, Laos, and Vietnam meet—is globally outstanding for its biodiversity. The intact forests of the ecoregion are little explored; it takes two weeks of intense walking and braving hazards such as mines and bombs that lie scattered throughout the landscape to get to some of the remote areas of the ecoregion.

But the known flora and fauna attest to the region's biological diversity, which includes some of Asia's charismatic fauna. Among the larger vertebrates, the tiger (*Panthera tigris*), Asian elephant (*Elephas maximus*), douc langur (*Pygathrix nemaeus*), gibbon (*Hylobates gabriellae*), wild dog (*Cuon alpinus*), sun bear (*Ursus malayanus*), clouded leopard (*Pardofelis nebulosa*), gaur (*Bos gaurus*), banteng (*Bos javanicus*), and Eld's deer (*Cervus eldii*) are better known.

Location and General Description

The Southern Annamites Montane Rain Forests [57] ecoregion extends along the greater Annamite Range from central Vietnam south to the Bolovans Plateau of Laos and the Central Highlands of Vietnam. It includes a broad topographic range from lowlands with wet evergreen forests to montane habitats with evergreen hardwood and conifer forests.

The geology of this ecoregion is also extremely diverse. The Kontum Massif of exposed granitic basement rock extends over an area extending 250 km from north to south and inland for 200 km. Ngoc Linh (Ngoc Pan), at the northwestern margin of the Kontum Massif, is the highest point of the Annamite Range in central Vietnam at 2,598 m. To the south the Annamite Range includes a complex mosaic of volcanic basalts, granites, and sedimentary substrates. Chu Yang Sinh, located 80 km inland from Nha Trang, forms the highest peak in the southern Annamite Range at 2,410 m. The Dac Lac (Darlac), Pleiku, Haute Cochinchine, Djirling, and Bolovans plateaus of southern Vietnam and Laos are volcanic remnants that reach maximum elevations as high as 2,200 m. Weathered basalts in these areas produce highly fertile soils with good agricultural potential. The heavily eroded Dalat Plateau of southern Vietnam, including the peak of Lang Bian, which reaches an elevation of 2,163 m, has soils formed from schists and quartzites and extensive areas of granites and volcanic rhyolites, andesites, and dactites.

Strong climatic gradients of rainfall and temperature are present within the ecoregion. The higher Kontum Massif and Dalat plateaus of central and southern Vietnam, and to a lesser degree the plateaus of Pleiku, Haut Chhlong, Djiring, and Haut Cochinchine, all rise to elevations or have exposures sufficient to give them more humid climatic zones. The semi-humid submontane zone at elevations of about 800–1,000 m has moderate levels of rainfall, with 1,500–2,200 mm annually. The peak of this rainfall occurs in September or October under the influence of the southward return of the monsoon with northeastern winds. Mean annual temperatures are about 20–21°C. The humid submontane climate zone comprises upland areas at middle elevations below about 1,100 m that are exposed to humid winds. Mean annual rainfall generally is above 2,500 mm and quite regular, with brief dry seasons lasting three months or less. Peak months for rainfall are typically August through October. Dew is abundant in the brief dry season, and foggy mists are common. The climatic zone covers much of the Kontum Massif, Haut Chhlong, and Haut Cochinchine and small areas east of the Dac Lac (Darlac) Plateau. The final climatic zone is the lower montane climate of southern Vietnam. This zone is present in the Kontum Massif and around the Dalat Plateau, where rainfall is about 1,800–2,000 mm, and there are 3–4 dry months of less than 50 mm rainfall. Very heavy rains of more than 3,850 mm annually occur on the eastern margin of the Dalat Plateau, with no dry month. These rains peak very late in November under the influence of the northeastern monsoon winds. Morning dew and foggy mists are common throughout this zone.

As expected from the complex geological, topographic, and climatic gradients present in the Southern Annamites Montane Rain Forests [57] ecoregion, forest structure and composition are highly variable. Reconstruction of pristine forest structure and composition has been made very difficult by the high degree of landscape degradation that has taken place, much of it as the result of swidden agricultural practices. Wet evergreen forests at 600–900 m elevation are dominated by species of Fagaceae, Myrtaceae, and Lauraceae, with high overall species richness. For the Fagaceae, as many as twenty species of *Lithocarpus*, five species of *Castanopsis*, and three species of *Quercus* may be present in this formation. Very large emergent trees are also present among the Anacardiaceae, Burseraceae, Dipterocarpaceae, and *Schima crenata*. *Hopea pierrei* may be the most abundant of the large trees, forming up to 90 percent of all trunks reaching more than 1 m in diameter. Lianas form an important component of this forest community. Lower-elevation areas are dominated by wet evergreen forest in mesic sites and semi-evergreen forest in drier sites.

Montane hardwood forests above 900 m elevation in this ecoregion vary in structure and composition depending on geological substrate and moisture availability. Evergreen hardwood forests generally have an upper canopy reaching to about 30 m in height. Canopy heights decline with increasing elevation and decreasing soil depth.

The family best represented in the upper canopy of these habitats is the Fagaceae, with important contributions from the Magnoliaceae, Aceraceae, Podocarpaceae, Lauraceae, and Theaceae. The diversity of conifers is high in montane forests, with five genera present (*Podocarpus sensu latu.*, *Calocedrus*, *Fokienia*, *Cephalotaxus*, and *Taxus*). A number of significant endemic species are present, including *Pinus dalatensis* and *P. krempfii*. Epiphytes form a notable part of the biodiversity of these montane forests. Particularly diverse are the orchids in the upper canopy and ferns in the middle and lower canopy.

Pinus kesiya extends over a fairly broad range of montane areas of the Dalat Plateau and other uplands in southern Vietnam at elevations up to 1,800 m. These are less mesic habitats than those of the high-elevation humid montane forests, where more diverse conifer forests are present. Human impacts on montane landscapes have almost certainly promoted the expansion of the range of *P. kesiya* at the expense of what were once montane evergreen forests. *Keteleeria evelyniana* may be present in denser gallery forests along the margins of the pine stands but seldom as an associate of the pure pine formations. Fire is a common factor in these forests.

The highest elevations and the slopes most exposed to humid winds in the montane zone often are bathed in moist clouds. The abundant dew and fog in these areas compensate for the lack of rainfall in the short dry season, producing lush conditions of mossy forest. Conifers (*Fokienia hodginsii*, *Podocarpus* spp.), Fagaceae (*Quercus* spp., *Lithocarpus* spp.), Theaceae (*Gordonia*, *Pyrenaria*, *Ternstroemia*), and Ericaceae form the dominant elements of these forests.

Biodiversity Features

Of the 122 mammal species known from the ecoregion, three are near-endemic species, and two are endemic (table J57.1).

Some of the threatened species in this assemblage include the tiger, Asian elephant, douc langur, gaur, banteng, Eld's deer, serow, clouded leopard, pygmy loris (*Nycticebus pygmaeus*), pig-tailed macaque (*Macaca nemestrina*), wild dog, Malayan sun bear, and smooth-coated otter (*Lutrogale perspicillata*).

The large habitat blocks that remain have been included within a high-priority (Level I) TCU (Dinerstein et al. 1997). But because of rampant hunting to supply the wildlife trade (WWF-Indochina 1998), tiger and prey populations have been depleted.

More than 410 bird species are known from this ecoregion. Five of these species are near endemic, and five are endemic (table J57.2). Among the other bird species that need conservation attention are the globally threatened white-winged duck (*Cairina scutulata*), the critically endangered Edwards's pheasant (*Lophura edwardsi*), and the threatened Siamese fireback (*Lophura diardi*), green peafowl (*Pavo muticus*), and Germain's peacock-pheasant (*Polyplectron germaini*). Several other species such as the great hornbill (*Buceros bicornis*), Austen's brown hornbill (*Anorrhinus austeni*), wreathed hornbill (*Aceros undulatus*), and crested argus (*Rheinardia ocellata*) are indicators of low disturbance levels and relatively intact forests. The ecoregion also overlaps with two EBAs identified by BirdLife International, the Da Lat Plateau (145) and South Vietnamese Lowlands (144), which have eight and

TABLE J57.1. Endemic and Near-Endemic Mammal Species.

Family	Species
Hylobatidae	*Hylobates gabriellae*
Muridae*	*Rattus hoxaensis**
Cercopithecidae	*Pygathrix nemaeus*
Muridae*	*Rattus osgoodi**
Muridae	*Maxomys moi*

An asterisk signifies that the species' range is limited to this ecoregion.

TABLE J57.2. Endemic and Near-Endemic Bird Species.

Family	Common Name	Species
Phasianidae	Edwards's pheasant	*Lophura edwardsi*
Phasianidae	Germain's peacock-pheasant	*Polyplectron germaini*
Phasianidae	Crested argus	*Rheinardia ocellata*
Timaliidae	Collared laughingthrush*	*Garrulax yersini**
Timaliidae	Golden-winged laughingthrush*	*Garrulax ngoclinhensis**
Timaliidae	Short-tailed scimitar-babbler	*Jabouilleia danjoui*
Timaliidae	Grey-faced tit-babbler	*Macronous kelleyi*
Timaliidae	Black-crowned barwing*	*Actinodura sodangorum**
Timaliidae	Grey-crowned crocias*	*Crocias langbianis**
Fringillidae	Vietnamese greenfinch*	*Carduelis monguilloti**

An asterisk signifies that the species' range is limited to this ecoregion.

three restricted-range bird species, respectively (Stattersfield et al. 1998).

Status and Threats

Current Status

More than 75 percent of this ecoregion's natural habitat has been converted or degraded. The remaining forest is distributed in small, isolated fragments. There are sixteen protected areas that average 150 km² and cover almost 2,500 km², or 5 percent of the ecoregion area (table J57.3). Many of these protected areas are paper parks that lack good management and effective protection.

Types and Severity of Threats

Because of the high elevations and steep slopes of this ecoregion, the human population density here is moderate (MacKinnon 1997). However, anthropogenic influences are pervasive throughout the ecoregion in the form of regular burning to create open woodlands and shifting cultivation. Large areas in the Da Lat Plateau in Vietnam have been cleared for settlements, plantations, and agriculture. Shifting cultivation is prevalent in the upper slopes. Bolovans Plateau is severely threatened by clearing of the forested area for coffee plantations. Wildlife poaching and excessive harvesting of NTFPs are severely threatening the integrity of Dong Hua Sao.

Priority Conservation Actions

Short-Term Conservation Actions (1–5 Years)

- Stop further encroachment (for coffee plantations) into Dong Hua Sao (Bolovans Plateau).
- Monitor waterbird migration and use of wetlands in Dong Hua Sao.

Longer-Term Conservation Actions (5–20 Years)

- Continue monitoring waterbird migration and use of wetlands in Dong Hua Sao.

TABLE J57.3. WCMC (1997) Protected Areas That Overlap with the Ecoregion.

Protected Area	Area (km²)	IUCN Category
Bach Ma [56]	160	II
Dong Hua Sao [72]	400	VIII
Ngoc Linh	200	IV
Kong Cha Rang	210	IV
Kon Kai Kinh	200	IV
Bana-Nui Chua	60	IV
Nui Thanh [58]	20	UA
Rung Kho Phan Rang [58]	30	IV
Deo Ngoau Muc [58]	30	?
Kalon Song Mao [58]	30	IV
Chu Yang Sinh	190	IV
Thuong Da Nhim	390	IV
Mt. Lang Bian	110	IV
Nam Lung	120	IV
Nui Dai Binh	100	IV
Phnom Nam Lyr [59]	160	IV
Total	2,410	

Ecoregion numbers of protected areas that overlap with additional ecoregions are listed in brackets.

Focal Species for Conservation Action

Habitat specialists: hornbill (Boonratana 1998a; Duckworth et al. 1994; Evans et al. 1996)

Wide-ranging species: tiger (Boonratana 1998a; Duckworth et al. 1994; Evans et al. 1996), gaur (Boonratana 1998a; Duckworth et al. 1994; Evans et al. 1996), hornbill (Boonratana 1998a; Duckworth et al. 1994; Evans et al. 1996)

Top predator species: tiger (Boonratana 1998a; Duckworth et al. 1994; Evans et al. 1996), leopard (Boonratana 1998a), dhole (Boonratana 1998a; Duckworth et al. 1994; Evans et al. 1996)

Other: Pangolin (Boonratana 1998a; Duckworth et al. 1994; Evans et al. 1996), silvered langur (Boonratana 1998a; Duckworth et al. 1994; Evans et al. 1996), douc langur (Boonratana 1998a; Duckworth et al. 1994; Evans et al. 1996), white-cheeked and yellow-cheeked gibbon (Boonratana 1998a; Duckworth et al. 1994; Evans et al. 1996), dhole (Boonratana 1998a; Duckworth et al. 1994; Evans et al. 1996), sun bear (Boonratana 1998a; Duckworth et al. 1994; Evans et al. 1996), tiger (Boonratana 1998a; Duckworth et al. 1994; Evans et al. 1996), gaur (Boonratana 1998a; Duckworth et al. 1994; Evans et al. 1996), serow (Boonratana 1998a; Duckworth et al. 1994; Evans et al. 1996), green peafowl (Boonratana 1998a; Duckworth et al. 1994; Evans et al. 1996).

Selected Conservation Contacts

BirdLife-Indochina

Cambodia Ministry of Environment

FFI-Indochina

IUCN Lao PDR

IUCN Vietnam

Lao PDR Department of Forest Resources and Conservation

WCS-Cambodia

WCS-Lao PDR

Wildlife Protection Office, Cambodia

WWF-Cambodia

WWF-Indochina

WWF-Lao PDR

Justification for Ecoregion Delineation

In a previous analysis of conservation units across the Indo-Malayan realm, MacKinnon (1997) included the coastal habitats of Vietnam, Cambodia, and Thailand in the Coastal Indochina biounit (05). This biounit consisted of four sub-units, and each includes large areas of different vegetation types. In keeping with our rules for delineating ecoregions, we extracted the wet evergreen forests that cover the mountain range and plateaus (to the south of the Hai Van Pass) from the adjacent dry forests and placed them in the Southern Annamites Montane Rain Forests [57].

Prepared by Eric Wikramanayake, Philip Rundel, and Ramesh Boonratana

Ecoregion Number: **58**
Ecoregion Name: **Southern Vietnam Lowland Dry Forests**
Bioregion: **Indochina**
Biome: **Tropical and Subtropical Dry Broadleaf Forests**
Political Units: **Vietnam**
Ecoregion Size: **34,900 km^2**
Biological Distinctiveness: **Regionally Outstanding**
Conservation Status: **Critical**
Conservation Assessment: **II**

The Southern Vietnam Lowland Dry Forests [58] ecoregion is the single most degraded or otherwise converted dry forest ecoregion outside of India. Only India's East Deccan Dry Evergreen Forests [6] are in worse shape. This ecoregion retains less than 10 percent of its forests, and only 2 percent are protected. The protected areas are surrounded by a burgeoning human population, and many of the "protected" forests are devoid of their fauna species because of intensive hunting pressures.

Location and General Description

The semi-arid coastal areas of southern Vietnam are the most arid in Vietnam because of the rainshadow effects of the plateaus of the southern Annamite Range, which restrict the flow of humid air in the early monsoon season. Mean annual rainfall is less than 1,500 mm in the coastal belt south of Nha Trang and below 800 m at Cape Padaran south of Phan Rang. These areas have been heavily affected by human activities and greatly altered.

The coastal region around Nha Trang, Cam Ranh Bay, and Phan Rang in southern coastal Vietnam is formed as a succession of small alluvial plains to the east of the Annamite Range and separated by low hills up to 1,000 m elevation formed of rhyolite or granite. The first woody community encountered away from the beach in southern Vietnam is generally a thicket community that gives way to a low scrubby forest further inland.

Areas with extensive coverage of unstabilized dunes support a distinctive community of shrubby trees with extensive root systems. On low dunes these species have a shrubby growth form. However, where a significant phreatophytic zone of freshwater develops at the base of these dunes, a true forest may develop with a variety of tree species reaching up to 12 m in height. Moist depressions within these dunes are dominated by *Baeckia frutescens* (Myrtaceae), *Melaleuca leucadendron*, and diverse graminoids.

Extensive areas of coastal dunes formed of stabilized red sands are a characteristic component of this coastal region around Cam Ranh Bay on granitic parent material. Two endemic species of Dipterocarpaceae, *Hopea cordata* and *Shorea falcata*, are known only from dune forests on these sands. These are two of only six endemic Dipterocarpaceae for all of Indochina, suggesting that many other endemic species may be found in this habitat if it is more carefully studied. Evergreen and semi-evergreen forest cover may be present on the coastal hills that reach higher elevations. A unique low forest or thicket community occurring on semi-arid slopes along the coast of southern Vietnam and notably rich in endemic species has been described near Phan Rang, Ba Ngoi, and Nha Trang, but it is heavily degraded today.

The low rolling hills of Binh Chau–Phouc Buu Nature Reserve, southwest of Phan Thiet, contain a variety of dry forest communities including notable tree endemics such as *Lithocarpus dinhensis*, *Dalbergia bariensis*, and the recently discovered *Dipterocarpus caudatus*. The forest is dominated by varying combinations of *Dipterocarpus caudatus*, *D. intricatus*, *Shorea siamensis*, and *S. roxburghii*.

Biodiversity Features

There are several large mammals of conservation significance in this ecoregion, including the endangered douc langur (*Pygathrix nemaeus*), red-cheeked gibbon (*Hylobates gabriellae*), and pileated gibbon (*Hylobates pileatus*) and potentially the tiger (*Panthera tigris*). The ecoregion also overlaps with a portion of the South Vietnamese Lowlands (144) EBA (Stattersfield et al. 1998) and contains two near-endemic bird species (table J58.1).

Status and Threats

Current Status

More than 90 percent of the habitat in this ecoregion has been cleared. There are ten very small protected areas (the average size is only 69 km^2) in the ecoregion that cover less than 700 km^2 (about 2 percent) of the ecoregion (table J58.2). Even these small protected areas do not have effective protection and are now insularized by forest clearing throughout the larger landscape, increasing the edge effects.

Types and Severity of Threats

The main threats to this habitat are agriculture, exploitation of valuable hardwood trees and other plant resources, and the

TABLE J58.1. Endemic and Near-Endemic Bird Species.

Family	Common Name	Species
Phasianidae	Germain's peacock-pheasant	*Polyplectron germaini*
Timaliidae	Grey-faced tit-babbler	*Macronous kelleyi*

An asterisk signifies that the species' range is limited to this ecoregion.

TABLE J58.2. WCMC (1997) Protected Areas That Overlap with the Ecoregion.

Protected Area	Area (km^2)	IUCN Category
Ban Dao Son Tra	50	IV
Cu Lao Cham	20	IV
Ba To	20	UA
Nam Hai Van	100	?
Nui Thanh [57]	40	UA
Deo Ca Hon Ron	210	UA
Rung Kho Phan Rang [57]	50	IV
Deo Ngoau Muc [57]	30	?
Kalon Song Mao [57]	70	IV
Binh Chan Phuoc Buu	100	IV
Total	690	

Ecoregion numbers of protected areas that overlap with additional ecoregions are listed in brackets.

rampant hunting primarily to supply the huge commercial market in both Vietnam and China (WWF and IUCN 1995; Compton 1998).

Priority Conservation Actions

Short-Term Conservation Actions (1–5 Years)

- Protect the remaining habitat fragments and restore adjacent degraded habitats to increase the areas of natural habitat. The Forest Protection and Development Act of 1991 provides for such restoration efforts.
- Prohibit and prevent hunting, especially of endangered species for the commercial trade.
- Implement the prescriptions under the National Biodiversity Action Plan.

Longer-Term Conservation Actions (5–20 Years)

- Increase the extent of protected areas through habitat restoration. Existing programs that grant stewardship to local people to manage reforested state lands (under Decision No. 327/CT) provide ways to increase forested areas and protected area coverage.

Focal Species for Conservation Action

Habitat specialists: douc langur, red-cheeked gibbon, pileated gibbon

Endemic species: Germain's peacock-pheasant, grey-faced tit-babbler

Area-sensitive species: tiger, banteng, douc langur, red-cheeked gibbon, pileated gibbon

Top predator species: tiger

Selected Conservation Contacts

BirdLife International, Indochina

FFI-Indochina

Forest Planning and Inventory Unit, Hanoi, Vietnam

Forest Protection Department, Hanoi, Vietnam

Institute for Ecology and Biological Resources, Hanoi, Vietnam

IUCN Vietnam

Provincial forest departments

WWF-Indochina

Justification for Ecoregion Delineation

MacKinnon (1997) included the coastal habitats of Vietnam, Cambodia, and Thailand in the Coastal Indochina biounit (05), consisting of four subunits. Each of these subunits includes several biomes. We delineated several ecoregions based on the distribution of these biomes. Like MacKinnon, we chose the Hai Van pass to represent a transition from the tropical south and the subtropical north and placed the lowland dry coastal forests in the Southern Vietnam Lowland Dry Forests [58].

Prepared by Eric Wikramanayake and Philip Rundel

Ecoregion Number:	**59**
Ecoregion Name:	**Southeastern Indochina Dry Evergreen Forests**
Bioregion:	**Indochina**
Biome:	**Tropical and Subtropical Dry Broadleaf Forests**
Political Units:	**Vietnam, Cambodia, Laos, and Thailand**
Ecoregion Size:	**123,800 km^2**
Biological Distinctiveness:	**Globally Outstanding**
Conservation Status:	**Critical**
Conservation Assessment:	**I**

The Southeastern Indochina Dry Evergreen Forests [59] ecoregion is globally outstanding for the large vertebrate fauna it harbors within large intact landscapes. Among the impressive large vertebrates are the Indo-Pacific region's largest herbivore, the Asian elephant (*Elephas maximus*), and largest carnivore, the tiger (*Panthera tigris*). The list includes the second known population of the critically endangered Javan rhinoceros (*Rhinoceros sondaicus*)—comprising a handful of animals in Vietnam's Cat Loc reserve—Eld's deer (*Cervus eldi*), banteng (*Bos javanicus*), gaur (*Bos gaurus*), clouded leopard (*Pardofelis nebulosa*), common leopard (*Panthera pardus*), Malayan sun bear (*Ursus malayanus*), and the enigmatic khting-vor (*Pseudonovibos spiralis*), known to science only by a few horns. But the ecoregion's conservation priority does not rest merely on its charismatic biodiversity. Importantly, it also represents a rare instance of a nonmontane ecoregion with large expanses of intact habitat that can allow viable populations of these species to survive over the long term. Unfortunately, all is not well in this haven, for plans to log Cambodia's forests, where most of the large habitat blocks lie, will result in large-scale habitat loss and fragmentation. Therefore, the ecoregion has been placed on the critical list.

Location and General Description

The Southeastern Indochina Dry Evergreen Forests [59] ecoregion occurs in a broad band across northern and central Thailand into Laos, Cambodia, and Vietnam. Dry evergreen forest is more appropriately called semi-evergreen forest because a significant proportion of canopy tree species are deciduous at the height of the dry season. Extensive areas of this ecoregion in southern Laos and northeastern Cambodia occur with a largely deciduous forest canopy. The Southeastern Indochina Dry Evergreen Forests [59] ecoregion occurs in humid and subhumid climatic regions where mean annual rainfall is generally between 1,200 and 2,000 mm, and a significant dry period of 3–6 months occurs each year. Semi-evergreen forest is the predominant forest cover in this ecoregion, but it often occurs in mosaics of deciduous dipterocarp or mixed deciduous forest communities. The distribution of dry evergreen forest habitats across a landscape catena is largely a function of gradients of soil moisture availability, with soil parent material unimportant. Therefore, it is common to see dry evergreen forest on both calcareous and crystalline rock substrates but with a deep soil profile and good capacity for soil moisture storage as a common characteristic.

Semi-evergreen forests represent a highly heterogeneous group of communities whose classification together derives from the parallel climatic and edaphic regimes that promote their formation. These communities have a characteristic tall and multilayered forest structure similar to that of lowland evergreen rain forest but grow in areas of lower and more seasonal rainfall regimes. In contrast to evergreen rain forests, species diversity is lower and the canopies and understory more open. Although dry evergreen forests have floristic relationships to lowland Indo-Malaysian forest communities, they exhibit a unique floristic structure that is rich in species endemic to mainland southeast Asia. With the broad distribution of these types of communities across Thailand and Laos, however, there are few local endemics.

The canopy of semi-evergreen forests generally is multi-layered and reaches about 30–40 m, with an open structure. Dipterocarps are a major component of the forest structure and form emergent tree canopies with such species as *Dipterocarpus alatus*, *D. costatus*, *Hopea odorata*, *Shorea guiso*, *S. hypochra*, and *Anisoptera costata*. Other giant emergents are species of *Ficus*, *Tetrameles nudiflora*, and *Heritiera javanica*, each forming large buttresses. In broad valley areas, semi-evergreen forest often occurs as a fringing gallery forest along streams, grading out into deciduous dipterocarp forest on drier sites with shallow soils. Unusually large trees such as *Dipterocarpus turbinatus*, *D. alatus*, and *D. costatus* (at the head of valleys) once formed dense stands in these forests, with a variety of other dipterocarps in ravines and scattered in such habitats. These species begin to drop out at about 700 m elevation and are replaced by montane species.

Bamboos are common in the forest, particularly as colonizers of open gaps after disturbance. Palms are present, most notably along watercourses, but less abundant and diverse than in lowland tropical rain forest habitats. Lianas are abundant, but understory formations are less complex than in lowland rain forest, and species richness is also lower.

Semi-evergreen forests generally are intolerant of fire. In comparison to species in adjacent dry deciduous dipterocarp communities, woody species in dry evergreen forest resprout poorly after fire. Similarly, dry evergreen forest species are sensitive to drought, apparently because of their less well-developed root systems.

Biodiversity Features

The ecoregion is globally outstanding for species richness, especially for the large vertebrate assemblage and associated ecological processes. The known mammal fauna of 160 species includes tiger, Asian elephant, douc langur (*Pygathrix nemaeus*), red-cheeked gibbon (*Hylobates gabriellae*), pileated gibbon (*Hylobates pileatus*), wild dog, Malayan sun bear, clouded leopard, common leopard, gaur, banteng, Javan rhinoceros, Eld's deer, and southern serow (*Naemorhedus sumatraensis*). The douc langur and the murid *Maxomys moi* are near endemics, whereas the northern smooth-tailed tree shrew (*Dendrogale murina*) is endemic (table J59.1). Many of these species are threatened.

Asia's largest predator, the tiger, was once widespread throughout the Indochina bioregion. But hunting and loss of the prey base and habitat have fragmented its distribution, confining the viable populations to the few areas where large

TABLE J59.1. Endemic and Near-Endemic Mammal Species.

Family	Species
Tupaiidae*	Dendrogale murina*
Muridae	Maxomys moi
Cercopithecidae	Pygathrix nemaeus

An asterisk signifies that the species' range is limited to this ecoregion.

TABLE J59.2. Endemic and Near-Endemic Bird Species.

Family	Common Name	Species
Phasianidae	Orange-necked partridge*	Arborophila davidi*
Phasianidae	Germain's peacock-pheasant	Polyplectron germaini
Timaliidae	Grey-faced tit-babbler	Macronous kelleyi

An asterisk signifies that the species' range is limited to this ecoregion.

areas of intact habitat still remain, known as TCUs (Dinerstein et al. 1997). These forests overlap with three high-priority (Level I) TCUs. Even within these TCUs, hunting and habitat loss continue to take their toll on this endangered species, which evokes images of the wild spaces in Asia that used to exist not very long ago. This ecoregion is one of the few in the Indochina bioregion that still has intact habitat and prey populations that provides hope for the tiger's long-term survival.

The ecoregion also harbors the critically endangered Javan rhinoceros, only the second population of this species known on Earth, the first being in the Ujung Kulon peninsula in western Java, Indonesia. The Indochinese population, discovered by scientists in the early 1990s, is limited to a handful of individuals confined to the small Cat Loc reserve in southern Vietnam. Their long-term survival is doubtful. However, most of this ecoregion's intact forests have not been scientifically explored, and a rhinoceros was allegedly captured from southern Laos, near the border with Cambodia, in the late 1980s. Therefore, discovery of additional populations is not improbable, especially because other species are known from body parts or reports.

The 455 bird species known from the ecoregion include two near-endemic species and one endemic species, the endangered orange-necked partridge (*Arborophila davidi*) (table J59.2). Other species of conservation importance include the critically endangered giant ibis (*Pseudibis gigantea*) and the endangered white-winged duck (*Cairina scutulata*). The ecoregion overlaps slightly with BirdLife International's EBA, South Vietnamese Lowlands (144) (Stattersfield et al. 1998), near the Vietnam-Cambodia border.

Status and Threats

Current Status

About two-thirds of the original forest in this ecoregion has been cleared or seriously degraded, especially in Vietnam and Thailand, but the habitat is relatively intact in Cambodia. A few large forest blocks also remain in Thailand and Laos. The thirty-one protected areas in this ecoregion cover 22,230 km^2 (18 percent) of the ecoregion (table J59.3). These include seven reserves—Kulen Promtep, Thap Lan, Khao Yai, Xe Piane, Phou Xiang Thong, Beng Per, and Virachey—which are larger than 1,000 km^2. Overall, the protected areas in this ecoregion are large, with an average size of almost 750 km^2.

Types and Severity of Threats

Most of the forests in Vietnam have already been replaced by plantations (FAO 1981). Shifting agriculture has further degraded some areas of this ecoregion (IUCN 1991). But the greatest threats now are from large-scale logging concessions that have been granted to multinational companies by the Cambodia government; therefore, the conservation status has been changed from relatively stable to critical.

Hunting to supply the huge wildlife trade has created empty forests throughout most of the ecoregion. From small, homemade crossbows used to kill small mammals for local consumption to bombs hidden in baited traps to kill tigers and pitfall traps for elephants, hunting has taken a very heavy toll on wildlife. The ravages of war and conflict have also had lasting effects; mines and bombs scattered across the landscape and the easy availability of automatic weapons that have replaced the crossbows have had deadly consequences.

TABLE J59.3. WCMC (1997) Protected Areas That Overlap with the Ecoregion.

Protected Area	Area (km^2)	IUCN Category
Xe Bang Nouane [72]	680	VIII
Xe Piane [72]	1,250	VIII
Phnom Prich [72]	520	IV
Kulen Promtep [72]	1,630	IV
Thap Lan	2,170	II
Khao Yai	2,270	II
Huai Sa La	400	IV
Preah Vihear	90	V
Banteay Chmar	790	V
Pang Sida	900	II
Phnom Kulen	340	II
Phou Xiang Thong	1,070	VIII
Kaeng Tana	70	II
Yod Dom	160	IV
Beng Per	2,730	IV
Virachey	3,325	II
Mom Ray	940	IV
Suoi Trai	230	IV
Ho Lac	50	UA
Phnom Nam Lyr [57]	230	IV
Bien Lac–Nui Ong	300	IV
Bu Gia Map	270	IV
Snoul	710	IV
Nui Ba Ra	20	UA
Lo Go–Sa Mat	140	IV
Cat Loc	200	PRO
Nam Bai Cat Tien	560	II
Nui Ba Den	40	UA
Duong Minh Chau	60	DE
Khao Sam Lan National Park	40	?
Khao Sam Lan	45	?
Total	22,230	

Ecoregion numbers of protected areas that overlap with additional ecoregions are listed in brackets.

Priority Conservation Actions

Short-Term Conservation Actions (1–5 Years)

- Biological surveys are needed in most protected areas and other forested landscapes.
- Protected areas need management; many are still paper parks that need staff. The staff, once recruited, should be trained.
- Hunting and poaching should be stopped. Many endangered species, such as tigers, elephants, gaur, and sun bears, are being hunted.

Longer-Term Conservation Actions (5–20 Years)

- Develop a conservation landscape, complete with habitat linkages that allow dispersal of wildlife between the protected areas.
- Establish a biodiversity trust fund to provide for sustainable conservation financing for the conservation landscape.

Focal Species for Conservation Action

Habitat specialists: Francois' langur (R. Boonratana, pers. obs., 1999), serow (Boonratana 1998a, 1999b; Evans et al. 1996; Timmins and Bleisch 1995), hornbill (Boonratana 1998a, 1999b; Duckworth et al. 1994)

Area-sensitive species: Asian elephant (Boonratana 1999b; Duckworth et al. 1994; Evans et al. 1996; Timmins and Bleisch 1995), tiger (Boonratana 1998a, 1999b; Duckworth et al. 1994; Timmins and Bleisch, 1995), gaur (Duckworth et al. 1994; Timmins and Bleisch 1995), banteng (Boonratana 1998a; Duckworth et al. 1994; Evans et al. 1996)

Top predator species: tiger (Boonratana 1998a, 1999b; Duckworth et al. 1994; Evans et al. 1996; Timmins and Bleisch 1995), leopard (Boonratana 1998a), clouded leopard (Boonratana 1999b), dhole (Boonratana 1998a; Duckworth et al. 1994), golden jackal (Duckworth et al. 1994)

Other: large-antlered muntjac (Timmins and Bleisch 1995), pangolin (Boonratana 1998a, 1999b; Duckworth et al. 1994; Evans et al. 1996), silvered langur (Duckworth et al. 1994; Timmins and Bleisch 1995), douc langur (Timmins and Bleisch 1995), white-cheeked gibbon (Boonratana 1999b), white-cheeked and yellow-cheeked gibbon (Duckworth et al. 1994; Evans et al. 1996; Timmins and Bleisch 1995), sun bear (Duckworth et al. 1994), bear spp. (Boonratana 1998a, 1999b; Duckworth et al. 1994; Evans et al. 1996; Timmins and Bleisch 1995), Irrawaddy dolphin (Duckworth et al. 1994), green peafowl (Duckworth et al. 1994; Evans et al. 1996), white-winged duck (Duckworth et al. 1994), black ibis (Duckworth et al. 1994), lesser adjutant (Duckworth et al. 1994)

Selected Conservation Contacts

Bangkok Bird Club

BirdLife-Indochina

Center for Conservation Biology, Bangkok, Thailand

Department of Forest Resources and Conservation, Vientiane, Lao PDR
FFI-Indochina
IUCN Lao PDR
IUCN Vietnam
Ministry of Environment, Phnom Penh, Cambodia
WCS-Cambodia
WCS-Lao PDR
Wildlife Fund Thailand
Wildlife Protection Office, Cambodia
WWF-Cambodia
WWF-Indochina
WWF-Lao PDR
WWF-Thailand

Justification for Ecoregion Delineation

The extensive Indochina biounit (10) identified by MacKinnon (1997) in his analysis of conservation units across the Indo-Pacific region comprises three subunits that represent the tropical lowland plains, subtropical hills, and temperate montane areas. The largest of these subunits, Central Indochina (10a), is a mix of tropical moist forests and tropical dry forests. In keeping with our rules for defining ecoregions, we separated the distinct habitats of regional extent to delineate eight ecoregions using the overall framework of MacKinnon's biounit. The distinct narrow bands of dry evergreen forests along the low hills that wind through Indochina represent one of these ecoregions, the Southeastern Indochina Dry Evergreen Forests [59].

Prepared by Eric Wikramanayake, Philip Rundel, Ramesh Boonratana, and Nantiya Aggimarangsee

Ecoregion Number:	**60**
Ecoregion Name:	**Andaman Islands Rain Forests**
Bioregion:	**Indochina**
Biome:	**Tropical and Subtropical Moist Broadleaf Forests**
Political Units:	**India, Myanmar**
Ecoregion Size:	**5,600 km²**
Biological Distinctiveness:	**Regionally Outstanding**
Conservation Status:	**Vulnerable**
Conservation Assessment:	**II**

The Andaman Islands were connected to mainland Myanmar during the Pleistocene. However, the islands still contain a lot of endemic animals and species shared only with the Nicobar Islands. Interestingly, the Andamans are floristically more like the mainland than the Nicobars, with less overlap of plant species than one might expect. Much of the islands are part of protected areas, but the protected areas often are not in the best location for conserving terrestrial species. There are also increasing threats to the biodiversity of the islands from a growing immigration of mainland people.

Location and General Description

The Andamans are made up of 204 islands of varying size and are located in the eastern Indian Ocean as part of the Bay of Bengal. Politically almost all the islands belong to India, although a few small islands in the northernmost end of the archipelago belong to Myanmar (e.g., Table, Great Coco, and Little Coco islands).

North Andaman lies 285 km south of Myanmar (a few smaller islands are closer), and the 150-km-wide Ten Degree Channel separates the Andamans from the Nicobars. The Andamans are warm tropical, with temperatures ranging from 22 to 30°C and 3,000–3,800 mm annual average rainfall. Rainfall is heavily influenced by monsoons, which come from the southwest (May to September) and from the northeast (October to December).

The Andamans are geologically part of long island arch that runs from Arakan Yoma in Myanmar to the Mentawai Islands off Sumatra and include the Nicobar Islands Rain Forests [61] and many underwater sea mounts. The arch was formed as the uplift along the subduction of the Indian-Australian plate in the late Eocene or early Oligocene. The opening of the Andaman Sea in the middle Miocene (about 10.8 m.y. ago) marked the first isolation of the Andamans from the mainland. Falling sea levels of the Pleistocene reconnected the islands to the mainland Myanmar for a period about 10,000 years ago (Das 1999). The highest point in the Andamans is Saddle Peak, at 720 m. The higher elevations of the Andamans often contain serpentine and gabbro formations, and at lower elevations Eocene sediments (sandstones, shales, and siltstones) with ultrabasic igneous intrusions predominate (Rao 1996).

The main categories of natural vegetation of the Andamans are coastal and mangrove forests and the interior evergreen and deciduous forests. Mangroves are extensive in the Andamans and make up about 15 percent of the total land area. All the most common trees belong to the family Rhizophoraceae and tend to reach heights of 6–24 m (Balakrishnan 1989). Evergreen forests form on clayey loam soils with poor humus content on top of micaceous sandstones. Dominant tree species reach heights of 40–60 m, including *Dipterocarpus griffithii*, *D. turbinatus*, *Sideroxylon longipetiolatum*, *Hopea odorata*, *Endospermum malaccense*, and *Planchonia andamanica*. Deciduous forests exist mainly on North Andaman, Middle Andaman, and Baratang Island and parts of South Andaman. They shed their leaves (either fully or partially) during the dry season and often are composed of tall trees reaching 40–50 m, including *Terminalia procera*, *T. bialata*, *T. manii*, *Canarium euphyllum*, *Ailanthes kurzii*, *Parishia insignis*, *Diploknema butyracea*, *Albizia lebbek*, *Tetrameles nudiflora*, and *Pterocymbium tinctorium* (Balakrishnan 1989; Rao 1996).

Biodiversity Features

The Andaman Islands have five mammal species that are strictly endemic to the ecoregion (table J60.1). All five species are listed as threatened (categories vulnerable and above) (Hilton-Taylor 2000).

There are eight strictly endemic bird species and four near endemics in the Andamans (table J60.2). Additionally, one

TABLE J60.1. Endemic and Near-Endemic Mammal Species.

Family	Species
Sorcidae	*Crocidura hispida**
Sorcidae	*Crocidura andamanensis**
Sorcidae	*Crocidura jenkinsi**
Rhinolophidae	*Rhinolophus cognatus**
Muridae	*Rattus stoicus**

An asterisk signifies that the species' range is limited to this ecoregion.

TABLE J60.2. Endemic and Near-Endemic Bird Species.

Family	Common Name	Species
Accipitridae	Andaman serpent-eagle	*Spilornis elgini**
Rallidae	Andaman crake	*Rallina canningi**
Columbidae	Andaman wood-pigeon	*Columba palumboides*
Columbidae	ndaman cuckoo-dove	*Macropygia rufipennis*
Cuculidae	Andaman coucal	*Centropus andamanensis**
Strigidae	Andaman scops-owl	*Otus balli**
Strigidae	Andaman hawk-owl	*Ninox affinis*
Bucconidae	Narcondam hornbill	*Aceros narcondami**
Picidae	Andaman woodpecker	*Dryocopus hodgei**
Dicruridae	Andaman drongo	*Dicrurus andamanensis**
Corvidae	Andaman treepie	*Dendrocitta bayleyi**
Sturnidae	White-headed starling	*Sturnus erythropygius*

An asterisk signifies that the species' range is limited to this ecoregion.

species not listed in table J60.2 is Nicobar scrubfowl (*Megapodius nicobariensis*), which used to live in both the Andamans and Nicobars but is now found only in the latter. One of the strict endemics, *Aceros narcondami*, is considered threatened (IUCN categories VU and above) and is found only on the small volcanic island of Narcondam.

The Andaman Islands have forty-five reptile species, thirteen of which are endemic. Twelve amphibian species (all frogs and toads) are in the Andamans, seven of which are endemic (Das 1999).

Floristically, the Andamans have much more in common with northeast India, Myanmar, and Thailand than with the Nicobars, which have affinities with Malaysia and Indonesia. In fact, the Andamans and Nicobars share only 28 percent of angiosperm species (Rao 1996). The plants that are not shared between the two island groups are just as revealing. The genera *Dipterocarpus* and *Pterocarpus* are both common in the Andamans but are absent from the Nicobars. *Otanthera*, *Astronia*, *Cyrtandra*, *Stemonurus*, *Bentinckia*, *Rhopaloblaste*, and *Spathoglottis* all occur in the Nicobars but not in the Andamans (Balakrishnan 1989; Rao 1996).

Status and Threats

Current Status

The Andaman Islands remain largely forested, and several protected areas exist (table J60.3). However, the protected areas cover mainly marine habitats and not terrestrial ones (Das 1999). The Narcondum Island protected area was created for the island's endemic hornbill, *Aceros narcondami*.

Types and Severity of Threats

The main threats to the Andamans are the result of an influx of people from the mainland. Population growth has put greater demands on the natural resources of the islands. Logging rates are high, as are agricultural encroachment (Das 1999). Indigenous people of the Andamans are permitted to exploit wildlife within the parks (unlike in mainland India), but increasing wildlife exploitation is caused mostly by recent immigrants and foreigners.

Introduced species are a problem in the Andamans. Typical island introductions such as rats, dogs, and cats may be harming the endemic Andaman crake (*Rallina canningi*) (Stattersfield et al. 1998). Spotted deer (*Axis axis*) are now widespread throughout the Andamans, as is the African giant snail (*Achatina fulica*). Elephants (*Elephas maximus*) have been introduced to Interview Island and North Andaman.

The islands have increasingly become a tourist destination, but the industry is not properly regulated (Sinha 1992).

Priority Conservation Actions

A major conservation plan for the Andamans and Nicobars has been proposed by Saldanha (1989), and even some years later, these recommendations appear crucial.

Short-Term Conservation Actions (1–5 Years)

- Saldanha (1989: 31) pointed out the need to "check immigration into the islands by instituting an entry permit for a specified stay even for mainland Indian citizens."
- He recommended allocating more staff to enforce the protection of forests and reducing the amount of legal logging.
- He put forward a list of areas that should be protected, and many of these have since become protected areas.
- Saldanha stated that tourism should be encouraged but should be selective and not destroy the environment.
- Saldanha (1989: 31) believes that the tribal lands should not be encroached upon regardless of "pressures for land and timber."
- He also recommended that islanders be educated about the environmental significance of where they live

TABLE J60.3. WCMC (1997) Protected Areas That Overlap with the Ecoregion.

Protected Area	Area (km^2)	IUCN Category
Little Andaman Island	300	?
Narcondum Island	7	?
North Andaman Island	180	?
South Andaman Island	110	?
Total	597	

Ecoregion numbers of protected areas that overlap with additional ecoregions are listed in brackets.

(Saldanha 1989). This task is being pursued in schools by the Andaman and Nicobar Islands Environmental Team using a teacher's manual written by Rao (1995) and Kalpvriksh, an NGO in New Delhi (Das 1999).
- Sinha (1992), like Saldanha (1989), recommended controlling the very recent population growth of the islands and the number of tourists (both foreign and domestic) who are increasingly making the Nicobars their destination.

Longer-Term Conservation Actions (5–20 Years)

- Continue to follow the short-term recommendations by Saldanha (1989), Sinha (1992), Rao (1995), and Das (1999).

Focal Species for Conservation Action

Habitat specialists: edible-nest swiftlet (*Collocalia fucifaga*), Narcondam hornbill (*Aceros narcondami*)

Endangered endemic species: Andaman spiny shrew (*Crocidura hispida*), Andaman shrew (*C. andamanensis*), Jenkin's shrew (*C. jenkinsi*), Andaman horseshoe bat (*Rhinolophus cognatus*), Andaman rat (*Rattus stoicus*), Andaman wood-pigeon (*Columba palumboides*), Narcondam hornbill (*Aceros narcondami*)

Area-sensitive species: wild pig (*Sus scrofa*), *Pteropus melanotus*, *P. giganteus*, *P. vampyrus*, *Cynopterus brachyotis*, *C. sphinx*, *Eonycteris spelaea*, crested serpent-eagle (*Spilornis cheela*), Andaman serpent-eagle (*Spilornis elgini*), Narcondam hornbill (*Aceros narcondami*)

Top predator species: crested serpent-eagle (*Spilornis cheela*), Andaman serpent-eagle (*Spilornis elgini*)

Selected Conservation Contacts

Kalpvriksh, an NGO based in New Delhi

Two local NGOs interested in biodiversity are listed by Ellis et al. (2000: 78); the following are direct quotes: "Society for Andaman Nicobar Ecology (SANE) based at Port Blair, with about 25 members, mainly consisting of scientists, carrying out several biodiversity related programmes particularly in the North, Middle, and South Andamans." "Andaman Prakriti Samsad, based at Port Blair, concentrates its work on educational programmes relating to biodiversity."

Justification for Ecoregion Delineation

Following MacKinnon (1997), we placed the Andaman Islands in a distinct ecoregion, the Andaman Islands Rain Forests [60]. However, we included the Nicobar Islands Rain Forests [61] in this bioregion based on recommendations by Tim Whitmore (pers. comm., 1999). Udvardy (1975) placed both island chains into the Andaman and Nicobar Islands biogeographic province.

Prepared by John Lamoreux

Ecoregion Number:	**61**
Ecoregion Name:	**Nicobar Islands Rain Forests**
Bioregion:	**Indochina**
Biome:	**Tropical and Subtropical Moist Broadleaf Forests**
Political Units:	**India**
Ecoregion Size:	**1,700 km²**
Biological Distinctiveness:	**Regionally Outstanding**
Conservation Status:	**Endangered**
Conservation Assessment:	**II**

The isolation of the Nicobar Islands Rain Forests [61] has given rise to endemic plant and animal species. The rain forests are in good shape and are afforded a high level of protection, but the future biodiversity of the ecoregion is not yet secure.

Location and General Description

The Nicobar Islands consist of twenty-two islands of varying size and are located in the eastern Indian Ocean as part of the Bay of Bengal. The Nicobars are separated from the Andamans in the north by a 150-km-wide channel and are 189 km from Sumatra to the southeast. The climate of the Nicobar Islands is warm tropical, with temperatures ranging from 22 to 30°C and 3,000–3,800 mm annual average rainfall. Rainfall is heavily influenced by monsoons, which come from the southwest (May to September) and the northeast (October to December). The only perennial rivers are found on Great Nicobar.

The islands are geologically part of a long island arc that runs from Arakan Yoma in Myanmar to the Mentawai Islands off Sumatra and includes the Andamans and many underwater sea mounts. The arc was formed as the uplift along the subduction of the Indian-Australian plate in the late Eocene or early Oligocene. Isolation of the Nicobars from the mainland resulted from the opening of the Andaman Sea in the middle Miocene about 10.8 m.y. ago. Unlike the Andamans, which are thought to have been connected to mainland Myanmar during periods of falling sea levels of the Pleistocene, the Nicobars remained as islands. Falling sea levels during this time joined many of the islands into three distinct groups, each of which has its own biological character today: Great Nicobar in the south (including Great Nicobar, Little Nicobar, Meroe, and satellites), Nancowry and the middle Nicobars (Nancowry, Katchall, Camorta, Teressa Chaura, and Tillanchong), and Car Nicobar to the north (including Car Nicobar and Batti Malv). The highest point in the Nicobars is Mt. Thullier, at 670 m. The higher elevations of the Nicobars often contain serpentine and gabbro formations, whereas at lower elevations Eocene sediments (sandstones, shales, and siltstones) with ultrabasic igneous intrusions predominate (Rao 1996). Great Nicobar contains younger substrates from the Tertiary that are more like the soils of parts of Sumatra than the other islands.

The vegetation of the Nicobars typically is divided into the coastal and mangrove forests and the interior evergreen and deciduous forests. Additionally, Kamorta, Katchall, Nancowry, and Car Nicobar all contain extensive interior grasslands, but these are thought to be anthropogenic in origin (Rao 1996; Daniels 1996). The grasslands are composed mainly of *Imperata cylindrica, Saccharum spontaneum, Heteropogon contortus, Chloris barbata, Chrysopogon aciculatus,* and *Scleria cochinchinensis,* along with many herbs and shrubs. Evergreen forests of Great Nicobar, Kamorta, and Katchall are dominated by *Calophyllum soulattri, Sideroxylon longipetiolatum, Garcinia xanthochymus, Pisonia excelsa,* and *Mangifera sylvatica.* Other important species of Kamorta and Katchall are *Artocarpus peduncularis, Radermachera lobbi, Symplocos leiostachya,* and *Bentinckia nicobarica.* Deciduous forests occur at lower elevations on Great Nicobar and include *Terminalia procera* and *T. bialata.*

Biodiversity Features

The isolation of the Nicobars has given rise to a number of endemic plant and animal species. The Nicobar Islands Rain Forests [61] contain twenty-five native mammal species, four of which are strict endemics (table J61.1). The majority of species are bats, and rodents (all rats) are the second most numerous order. Several larger species exist, however, including wild pig (*Sus scrofa*) and Nicobar macaque (*Macaca fascicularis umbra*). There is also a unique tree-shrew (*Tupaia nicobarica*), which is the most arboreal member of its genus. All four endemics in table J61.1 are considered threatened by IUCN (categories VU and above) (Hilton-Taylor 2000).

There are eighty-two native, nonpelagic bird species in the Nicobars, and nine of them are endemic (table J61.2).

TABLE J61.1. Endemic and Near-Endemic Mammal Species.

Family	Species
Sorcidae	*Crocidura nicobarica**
Tupaiidae	*Tupaia nicobarica**
Pteropodidae	*Pteropus faunulus**
Muridae	*Rattus palmarum**

An asterisk signifies that the species' range is limited to this ecoregion.

TABLE J61.2. Endemic and Near-Endemic Bird Species.

Family	Common Name	Species
Accipitridae	Nicobar serpent-eagle	*Spilornis minimus**
Accipitridae	Nicobar sparrowhawk	*Accipiter butleri**
Megapodiidae	Nicobar scrubfowl	*Megapodius nicobariensis***
Columbidae	Andaman wood-pigeon	*Columba palumboides*
Columbidae	Andaman cuckoo-dove	*Macropygia rufipennis*
Psittacidae	Nicobar parakeet	*Psittacula caniceps**
Strigidae	Andaman hawk-owl	*Ninox affinis*
Sturnidae	White-headed starling	*Sturnus erythropygius*
Pycnonotidae	Nicobar bulbul	*Hypsipetes nicobariensis**

An asterisk signifies that the species' range is limited to this ecoregion.

**Megapodius nicobariensis is currently limited to the ecoregion but was formerly found in the Andamans as well.

The list of endemics matches the list of nine restricted-range species from BirdLife International's Nicobar Islands EBA (Stattersfield et al. 1998). There are also many species and subspecies shared only with the Andamans (Sankaran 1997, 1998). Two of the endemics are threatened (VU or above) (*Megapodius nicobariensis, Hypsipetes nicobariensis*) (Hilton-Taylor 2000).

The types of birds present in the Nicobars are particularly interesting. There are very high numbers of some groups, including eight heron species, seven hawk species, six kingfisher species, and six pigeon species. However, there are no babblers, only two sylvine warblers, and just one bulbul species.

The Nicobar Islands have forty-three reptile species, of which eleven are endemic. Eleven amphibian species (all frogs and toads) are in the Nicobars; two are endemic.

The Nicobars are more similar to Sumatra and Malaysia botanically than to Burma, Thailand, or even the Andamans. In fact, the Nicobars and Andamans share only 28 percent of angiosperm species. The Nicobars contain more than 580 flowering plant species. A rate of endemism for the angiosperms is available only for the Nicobars and Andamans jointly and is about 14 percent of all species.

Status and Threats

Current Status

Protected areas cover about 30 percent of the Nicobar islands, and only 14 percent of the ecoregion's native forest has been lost (table J61.3). Unfortunately, the protected areas, as they currently stand, are not situated with regard to the distribution of endemic species (Das 1999; Sankaran 1997). Sankaran (1997) describes several key unprotected sites for endemic birds that are important for all taxa. This includes the southern tip of Great Nicobar, which not only contains the greatest number of endemic birds in all the Nicobars but also is the largest uninhabited lowland forest in the ecoregion. The southern tip is also one of the most susceptible to future development. A large portion of Great Nicobar is dominated by the Great Nicobar Biosphere Reserve, which consists of two national parks: Campbell Bay National Park and Galathea National Park. The two parks are separated by a 12-km buffer zone that is uninhabited primary forest. Sankaran (1997) argues that the buffer zone should be designated as national park to preserve what is currently a large contiguous forest, thus preventing roads or

TABLE J61.3. Protected Areas That Overlap with the Ecoregion, Derived from Sankaram (1997).

Protected Area	Area (km²)	IUCN Category
Batti Malv Wildlife Sanctuary	2	?
Tillanchong Wildlife Sanctuary	17	?
Megapod Island Wildlife Sanctuary	0.13	?
Galathea National Park	110	?
Campbell Bay National Park	426	?
Total	555.13	

Ecoregion numbers of protected areas that overlap with additional ecoregions are listed in brackets.

other disturbance from bisecting the biosphere. The Nancowry group of islands is also in need of expanded protected areas. It has levels of bird endemism matching the other two island groups but a much higher percentage of threatened species (Sankaran 1997).

Types and Severity of Threats

Habitat conversion poses the greatest threat to the ecoregion. Aboriginal peoples have inhabited the Nicobars for at least 2,000 years and are estimated to have converted about 10 percent of the forest cover in that time. Settlement programs brought mainlanders to the Nicobars starting in the late 1960s; they now make up 36 percent of the population. Just in the past twenty-five years, 4 percent of the Nicobars' original forest cover has been lost to mainlanders. The settlement program no longer exists, but the Nicobars are still at a critical juncture where decisions about how to control development and conserve its resources must be made. There are proposals to make the Nicobars a major tourist destination, make Great Nicobar a free trade port, and increase the military presence on the islands. Road development and cash crop promotion (particularly rubber and cashews) are also future threats. Wildlife exploitation threatens the edible-nest swiftlet in the Nicobars, the Nicobar megapode, crocodiles, and sea turtles (Das 1999; Sankaran 1997).

Priority Conservation Actions

Saldanha (1989) has proposed a major conservation plan for the Andamans and Nicobars. All of these recommendations are crucial. However, Sankaran (1997) in a conservation assessment and series of recommendations appeals mainly to the Indian government to expand and reconfigure the national parks in the Nicobars and to stop or regulate trade in some species under CITES.

Short-Term Conservation Actions (1–5 Years)

- Saldanha (1997: 31) raises the need to "check immigration into the islands by instituting an entry permit for a specified stay even for mainland Indian citizens."
- He recommended allocating more staff to enforce the protection of forests and reducing the amount of legal logging.
- He put forward a list of areas that should be protected, and many of these have since become protected areas.
- Saldanha stated that tourism should be encouraged but should be selective and not destroy the environment.
- Saldanha (1989: 31) believes that the tribal lands should not be encroached upon regardless of "pressures for land and timber."
- He also recommended that islanders be educated about the environmental significance of where they live (Saldanha 1989). This task is being pursued in schools by the Andaman and Nicobar Islands Environmental Team using a teacher's manual written by Rao (1995) and Kalpvriksh, an NGO in New Delhi (Das 1999).
- Sinha (1992), like Saldanha (1989), recommended controlling the very recent population growth of the islands and the number of tourists (both foreign and domestic) who are increasingly making the Nicobars their destination.

Longer-Term Conservation Actions (5–20 Years)

- Continue to follow the short-term recommendations by Saldanha (1989), Sankaran (1997), Sinha (1992), Rao (1995), and Das (1999).

Focal Species for Conservation Action

Habitat specialists: edible-nest swiftlet (*Collocalia fucifaga*) (Sankaran 1995)

Endangered endemic species: Nicobar megapode (*Megapodius nicobariensis*), Nicobar bulbul (*Hypsipetes nicobariensis*)

Area-sensitive species: wild pig (*Sus scrofa*), Nicobar flying-fox (*Pteropus faunulus*), black-eared flying-fox (*P. melanotus*), Indian flying-fox (*P. giganteus*), large flying-fox (*P. vampyrus*), lesser short-nosed fruit-bat (*Cynopterus brachyotis*), greater short-nosed fruit-bat (*C. sphinx*) (for all bats see Mickleburgh et al. 1992); crested serpent-eagle (*Spilornis cheela*); Nicobar serpent-eagle (*Spilornis minimus*)

Top predator species: crested serpent-eagle (*Spilornis cheela*), Nicobar serpent-eagle (*Spilornis minimus*)

Selected Conservation Contacts

Parirakshak, "based at Campbell Bay, Great Nicobar, carries out educational and biodiversity preservation work in the Nicobar region" (listed as an NGO interested in biodiversity, Ellis et al. 2000: 78).

Justification for Ecoregion Delineation

Following MacKinnon (1997), we placed the Andaman Islands in a distinct ecoregion, the Andaman Islands Rain Forests [60]. However, we included the Nicobar Islands Rain Forests [61] in this bioregion based on recommendations by Tim Whitmore (pers. comm., 1999). Udvardy (1975) placed both island chains into the Andaman and Nicobar Islands biogeographic province.

Prepared by John Lamoreux

Ecoregion Number:	**62**
Ecoregion Name:	**Tonle Sap–Mekong Peat Swamp Forests**
Bioregion:	**Indochina**
Biome:	**Tropical and Subtropical Moist Broadleaf Forests**
Political Units:	**Vietnam, Cambodia**
Ecoregion Size:	**29,300 km^2**
Biological Distinctiveness:	**Bioregionally Outstanding**
Conservation Status:	**Critical**
Conservation Assessment:	**IV**

The Tonle Sap–Mekong Peat Swamp Forests [62] are only a small vestige of their former range and function. More than 90 percent of this ecoregion has been converted to scrub or degraded forests. Intensive agriculture and the alteration of the hydrodynamics of the river systems in the region have altered the natural river fluctuations, adversely affecting the remaining native vegetation.

Location and General Description

The Tonle Sap–Mekong Peat Swamp Forests [62] ecoregion extends over areas permanently inundated with shallow freshwater, although the region as mapped includes mosaics of swamp forest and herbaceous wetland interposed with upland areas of dry forest. However, care must be given in separating permanently flooded swamp forests of southeast Asia from seasonal swamp forests that characterize extensive areas of the Tonle Sap Basin and the floodplain of major Cambodian rivers. Conditions of permanent flooding compared with flooding for 6–8 months produce differential selective factors and thus a distinctive floristic assemblage. Such conditions produce several notable characteristics of these floodplain habitats. Native palms, often a characteristic component of typical swamp forests, generally are absent from the Tonle Sap floodplain with the exception of the local occurrences of rattans in some gallery forests.

Large areas of swamp forest are present in poorly drained landscapes of the Haut Chhlong and Blao regions of Vietnam where the water table reaches to the surface. These soils are saturated, with an upper horizon very rich in organic matter. Several formations of swamp forest have been described. The most widespread of these occurs at an elevation of 600–900 m and is characterized by a dominance of *Livistona cochinchinensis* (Arecaceae). This palm reaches up to 30 m in height. Many associated dicot canopy species have stilt roots or pneumatophores. These include species of *Eugenia* (Myrtaceae), *Elaeocarpus* (Elaeocarpaceae), and *Calophyllum* (Guttiferae).

Extensive areas of grass and sedge wetlands are also included in this ecoregion. Tall grasses such as *Phragmites karka*, *Saccharum arundinaceum*, and *Coix gigantea* generally dominate fertile alluvial clay sediments. Less fertile soils typically are characterized by a diverse assemblage of Poaceae and Cyperaceae and other herbaceous species.

Brackish and freshwater wetlands behind mangrove areas in Cambodia and the Mekong delta area of Vietnam typically are dominated by dense stands of *Meleuca leucadendron* called the paperbark swamps or rear mangrove communities. Paperbark swamps once covered large areas of the Mekong delta region with acid sulfate soils subjected to seasonal inundation. Today the area of *Melaleuca* forests has been greatly diminished, although successful reforestation efforts have occurred. The largest remaining stands of *Melaleuca* forest occur on peat soils of the U Minh area of Minh Hai Province and on the acidic soils of the Plain of Reeds and Ha Tien plain in Vietnam.

Paperbark swamps are low in plant diversity but have a great significance in maintaining natural ecosystem function. These swamps reduce water flow in the wet season and thus minimize flooding, store fresh water, reduce soil acidification, promote biodiversity of many aquatic organisms, and provide a sustainable source of wood for construction and fuel.

Biodiversity Features

Mammal species of conservation significance include the possibly extinct wild water buffalo (*Bubalus arnee*), Eld's deer (*Cervus eldii siamensis*), Indochinese hog deer (*Axis porcinus annamiticus*), and banteng (*Bos javanicus*). The reed beds are important sites for waterfowl, and the habitats provide feeding grounds for the eastern sarus crane (*Grus antigone*), the near-endemic giant ibis (*Pseudibis gigantea*), white-shouldered ibis (*P. davisoni*), glossy ibis (*Plegadis falcinellus*), black-headed ibis (*Threskiornis melanocephalus*), Asian openbill (*Anastomus oscitans*), and possibly the lesser adjutant (*Leptoptilos javanicus*) (IUCN 1991) (table J62.1).

TABLE J62.1. Endemic and Near-Endemic Bird Species.

Family	Common Name	Species
Threskiornithidae	Giant ibis	Pseudibis gigantea

An asterisk signifies that the species' range is limited to this ecoregion.

Status and Threats

Current Status

This ecoregion is heavily altered, and very little (<10 percent) of the original habitat is left. Excessive forest exploitation has reduced many areas to scrub or secondary forest comprising nonindigenous species. (FAO 1981). Very little (<1 percent) of the ecoregion is protected, although the Tonle Sap Great Lake reserve, which extends into the ecoregion, is large (2,725 km^2) (table J62.2).

Types and Severity of Threats

Large areas of acid sulfate peat soils have been drained, cultivated, and subsequently abandoned, and these areas have become extensive reed beds (MacKinnon 1997). The mangroves and *Melaleuca* forests of the Mekong delta were severely affected by military activities in the Vietnamese wars; however, they have partially recovered through replanting programs (MacKinnon 1997). The vast floodplains of the Mekong River and Tonle Sap are extensively cultivated during the dry season, and local fishing communities have greatly altered the Tonle Sap swamp forests (IUCN 1991). The area of the Mekong delta that lies in southern Vietnam has also been severely affected by deforestation in water catchments in Laos, Thailand, and southern China. The river is now prone to flood more frequently and violently in the wet season but to reach very low levels in the dry season; this altered flow is creating increasing problems for local agriculture (IUCN 1991).

Priority Conservation Actions

Short-Term Conservation Actions (1–5 Years)

- Urgent protection and effective management of remaining larger habitat blocks are needed.
- Biological surveys are needed.

TABLE J62.2. WCMC (1997) Protected Areas That Overlap with the Ecoregion.

Protected Area	Area (km^2)	IUCN Category
Hon Chong	40	UA
Vo Doi	20	IV
Total	60	

Ecoregion numbers of protected areas that overlap with additional ecoregions are listed in brackets.

Longer-Term Conservation Actions (5–20 Years)

- Protection and management of remaining habitat should be monitored.

Focal Species for Conservation Action

Habitat specialists: wild water buffalo, eastern sarus crane, giant ibis, white-shouldered ibis, glossy ibis, black-headed ibis

Area-sensitive species: wild water buffalo

Top predator species: tiger, wild dog, leopard

Selected Conservation Contacts

BirdLife-Indochina
Cambodia Ministry of Environment
FFI-Indochina
IUCN Vietnam
WCS-Cambodia
Wildlife Protection Office, Cambodia
WWF-Cambodia
WWF-Indochina
WWF-Lao PDR

Justification for Ecoregion Delineation

We separated out the mangroves, freshwater swamp forests, and peat swamps in MacKinnon's (1997) subunit 05a and represented the peat swamps with the Tonle Sap–Mekong Peat Swamp Forests [62].

Prepared by Eric Wikramanayake and Philip Rundel

Ecoregion Number:	**63**
Ecoregion Name:	**Irrawaddy Freshwater Swamp Forests**
Bioregion:	**Indochina**
Biome:	**Tropical and Subtropical Moist Broadleaf Forests**
Political Units:	**Myanmar**
Ecoregion Size:	**15,100 km^2**
Biological Distinctiveness:	**Bioregionally Outstanding**
Conservation Status:	**Critical**
Conservation Assessment:	**IV**

In 1929 the *Burma Game Manual* stated its guiding principle: "A countryside devoid of wildlife is uninteresting and unnatural, and life under such conditions can adversely affect the national character." Therefore, invaluable natural and national assets had to be saved from destruction. This has not happened in large portions of Myanmar, especially the fertile lands of the Irrawaddy freshwater swamp forest. Most of the ecoregion's original forests and wildlife such as Asian elephants and tigers have been destroyed. Protection of the last remaining bits of habitat and restoration ecology will be key elements of returning this ecoregion to its natural state.

Location and General Description

The Irrawaddy River flows into the Bay of Bengal, and its delta is made up of mangroves and freshwater swamp forests of this ecoregion. This ecoregion is an extremely fertile area because of the riverborne silt deposited in the delta. The southern portion of the ecoregion transitions into the Myanmar Coastal Mangroves [78] and is made up of fanlike marshes with oxbow lakes, islands, and meandering rivulets and streams.

Topographically the region is primarily flatlands. The western part of the region is bounded by the Rakhine (Arakan) Yomas, with the highest elevation at about 1,287 m to the north, tapering down to the south to 428 m.

The geological formation in the west is composed of shale and sedimentary rocks. The soil is yellow brown to gley soil, and to the east it is sandy loam and calcareous. To the north the soils are lateritic. The region is rich in limestone. Along riverbanks there are loamy sands deposited by the rivers. Toward river mouths the soil is more loamy as the result of tidal actions and marshes.

The region is made up of a mix of deciduous forests. The forests to the north support teak, *Xylia kerri*, *Salmalia malabrica*, *S. insigni*, *Milletia pendula*, *Dalbergia kurzii*, *Spondias pinnata*, *Lannea grandis*, *Terminalia balerica*, *Anogeissus accuminata*, *Eugenia* spp., *Terminalia tomentosa*, *T. chebula*, *Vitex* spp., *Schleichera oleosa*, and *Manglietia insignis*. The forests to the south are made up of *Xylia keri*, *Salmalia malabrica*, *S. insigni*, *Dalbergia kurzii*, *Lannea grandis*, *Teminalia balerica*, *T. chebula*, *Eugenia* spp., *Anogeissus acuminata*, *Terminalia* spp., *Vitex pubescens*, *Adina cordifolia*, and *Spondias pinnata*.

Bamboo breaks of *Melocanna bambusoides* also are prevalent. *Melocanna bambusoides* is found extensively on the eastern aspect of Arakan Yomas. Other species are *Dinochloa m'clellandi*, *Oxytenanthera albo-ciliata*, *Bambusa tulda*, *Dendrocalamus longispathus*, *Cephalostachyum pergracile*, and *Bambusa polymorpha*. Bamboo associates to the north are *Melocanna bambusoides*, *Bambusa polymorpha*, *Cephalostachyum pergracile*, and *Bambusa tulda*.

Biodiversity Features

Large mammals have been largely extirpated from this ecoregion. In a 1982 survey, Richard Salter found only fifteen Asian elephants in this ecoregion, but species such as sambar (*Cervus unicolor*), hog deer (*C. porcinus*), and wild boar (*Sus scrofa*) appeared to be numerous. Tigers and leopards were once abundant as well. Today, however, this ecoregion harbors very little of its original biodiversity. There are no endemic mammals in this ecoregion.

This ecoregion is an important wetland for migratory birds. These birds arrive by the thousands every year and include plovers (Mongolian plover [*Charadrius mongolus*]), sandpipers (spoon-billed sandpiper [*Eurynorhynchus pygmeus*]), black-tailed godwit (*Limosa limosa*), Eurasian curlew (*N. arquata*), Temminck's stint (*C. temminckii*), and Asian openbill stork (*Anastomus oscitans*). Unfortunately, their populations have been steadily declining. There are also numerous resident and wintering waterbirds. These include common waterbirds such as bitterns (cinnamon bittern [*Ixobrychus cinnamomeus*]), herons and egrets (Indian pond-heron [*Ardeola grayii*], Pacific reef-egret [*Egretta sacra*]), storks (woolly-necked stork [*Ciconia episcopus*]), ibis (black-headed ibis [*Threskiornis melanocephalus*]), ducks (spot-billed duck [*Anas poecilorhyncha*]), jacanas (pheasant-tailed jacana [*Hydrophasianus chirurgus*]), pratinoles (oriental pratincole [*Glareola maldivarum*]), and terns (black-bellied tern [*Sterna acuticauda*]). There are no endemic birds in this ecoregion.

Reptiles such as the estuarine crocodile (*Crocodylus porosus*) may still exist, but the population probably is low. In 1982 the population was estimated to be about 4,000, but they have been exploited since 1978 by the Pearl and Fishery Corporation, which reared hatchlings at a farm in Rangoon. The hatchery is still in existence.

Status and Threats

Current Status

The Irrawaddy is one of the most heavily silted rivers in the world. The sedimentation rate was 299 million tons/year, and it ranked fifth behind the Yellow, Ganges, Amazon, and Mississippi in silt deposition. Today the sedimentation rate is worsening as deforestation and agricultural erosion continue at a phenomenal rate.

The future survival of wildlife in this ecoregion is bleak. There are no large mammals of sizable populations because of habitat fragmentation. The situation for birds is no better. Birds do not receive much attention from the staff of the Forest Department or the Nature and Wildlife Conservation Division. There are no protected areas in this ecoregion.

Types and Severity of Threats

The ecoregion has experienced agricultural expansion, firewood extraction, commercial logging, fish and prawn culture, and other development works. It is highly degraded, and the future threats remain the same as those that have destroyed the habitat to this point.

Priority Conservation Actions

Short-Term Conservation Actions (1–5 Years)

- Establish protected areas.
- Provide sufficient staffing and funding to maintain the protected areas.

Longer-Term Conservation Actions (5–20 Years)

- Manage protected areas under an appropriate action plan.
- Provide sufficient staffing and funding to maintain the protected areas.

Focal Species for Conservation Action

Migratory species: migratory birds

Other: resident bird species, deer species

Selected Conservation Contacts

Myanmar Forest Department

Myanmar Wildlife Conservation Society

Myanmar Wildlife and Sanctuaries Division

Justification for Ecoregion Delineation

MacKinnon divided the Indochina bioregion into fifteen biogeographic subunits nested within seven major biounits. Each of these biounits and biogeographic subunits contains a mix of biomes. For instance, the Burmese coast biounit (04) is made up mostly of tropical wet evergreen forests but also includes tropical montane evergreen forests, semi-evergreen rain forests, dry dipterocarp forests, and large areas of freshwater swamp forests and mangroves in the Irrawaddy River delta. We assigned these distinct habitat types into separate ecoregions. Thus MacKinnon's Burmese coast biounit is represented by four ecoregions: the Myanmar Coastal Rain Forests [49], Irrawaddy Freshwater Swamp Forests [63], Myanmar Coastal Mangroves [78], and Mizoram-Manipur-Kachin Rain Forests [50].

This region does not correspond well to Udvardy's biogeographic provinces. The Irrawaddy Freshwater Swamp Forests [63] and Myanmar Coastal Mangroves [78] are contained within Udvardy's Burman rain forest.

Prepared by U Saw Han and Saw Tun Khaing

Ecoregion Number: **64**
Ecoregion Name: **Chao Phraya Freshwater Swamp Forests**
Bioregion: **Indochina**
Biome: **Tropical and Subtropical Moist Broadleaf Forests**
Political Units: **Thailand**
Ecoregion Size: **38,900 km²**
Biological Distinctiveness: **Bioregionally Outstanding**
Conservation Status: **Critical**
Conservation Assessment: **IV**

Like most lowland habitats on alluvial floodplains elsewhere in Asia, the area has been severely altered. Almost none of the original vegetation remains. Descriptions of the flora and fauna must be inferred from similar habitats in surrounding countries. Because this is one of the most densely populated regions of Asia, supporting one of its larger cities, Bangkok (estimated population 8 million), there is little hope that any extensive protected areas can be set up or that any significant original vegetation remains. Nonetheless, the area retains important conservation attributes. Appropriate land-use planning may enable the area to retain some of them.

Location and General Description

This ecoregion consists of the freshwater swamp forests in the lowland alluvial plains of the Chao Phraya River in central Thailand and extends north up the valleys of its major tributaries, the River Ping and River Nan. It is roughly 400 km north to south and, at its widest, about 180 km from east to west.

The area can be roughly divided into two parts. The Lower Central Plain, which extends north as far as the province of Ang Thong (ca. 15° north), represents an area of Quaternary deposits of silt, of 15–30 m depth, overtopping the soft marine clays laid down when the area was once a huge bay of the South China Sea, about 6,000 to 8,000 years B.P., when sea levels were approximately 4 m higher than at present. The area is flat and low lying. The Lower Central Plain has an average elevation of about 2 m above mean sea level. Above this, the Upper Central Plain extends north up the Chao Phraya River and lower parts of the valleys of the Ping and Nan rivers and lies at >20 m above sea level. This plain was never subject to significant tidal flooding (Sinsakul 1997).

The area has a moist monsoonal climate, receiving approximately 1,400 mm rainfall per year. Mean maximum and mean minimum temperatures are around 33°C and 24°C, respectively, for Bangkok (Meteorological Department 1987).

Four major rivers enter the Central Plain. These are the Me Klong (Kwae) River system on the west; the Chao Phraya and its major branch, the Tachin River, in roughly the center of the region; the Pasak River, a tributary of the Chao Phraya; and the Bang Pakong River, entering from the east.

As alluvium built up, brackish water swamps were gradually superseded by freshwater habitats in which extensive open swamps were occupied by permanent or seasonally inundated vegetation. Successional vegetation, including open water, mats of floating vegetation, *Typha*, *Phragmites*, and scrub alternated with stands of towering *Dipterocarpus alatus* trees, and other species on higher ground, such as along river banks; *Pandanus* was common in many swampy forest areas.

These forests and swamp vegetation formerly transitioned into mangroves toward the coast and along the lower reaches of major rivers.

Phragmites and other marsh grasses once were widespread but have largely disappeared and been replaced by *Typha angustifolia*. A remnant of *Phragmites* marsh is present at Khao Sam Roi Yot National Park (shared with ecoregion [70]), which to some extent represents a microcosm of central plains swamp habitat.

The northern part of the area has long been settled and to some extent cultivated. It included the Kingdoms of Sukkhothai (ca. twelfth to thirteenth century) and Ayutthaya (mid-fourteenth century to mid-eighteenth century.) Sukkhothai was based on an older, less intensive Mon-Khmer style of cultivation, which allowed wildlife to coexist.

Although Nakhon Pathom on the western margin of the lower central plain was a major center of civilization as early as the fifth century A.D., much of the lower Central Plain remained a little disturbed swampland until the Siamese kings moved their capital to what is now Bangkok in 1782. In the early nineteenth century, the area began to be canalized, the banks beng claimed for agriculture by the aristocracy. With Siam's increased integration into the world economy after the signing of the Bowring treaty in 1855, further canals were dug and the export of rice on a significant scale began. The emergence of a free peasant class and a merchant class helped speed the conversion of remaining swamp to paddy land so that the fate of the megafauna, including Schomburgk's deer, was already sealed by the beginning of the twentieth century. Older, broadcast rice systems in the more deeply flooded areas were superseded by transplanted rice systems as control over irrigation and drainage increased, especially with a slew of irrigation projects that were implemented in the first half of the twentieth century, culminating in the completion of the Greater Chao Phraya Project in 1962.

Biodiversity Features

More than a quarter of all the threatened birds in Thailand, as well as many mammals, live in wetlands (IUCN 1991). Most of the remaining colonies of large waterbirds breeding in Thailand are situated in this ecoregion.

Some of the mammals of conservation significance found in these freshwater swamps until at least the mid-nineteenth century included tiger (*Panthera tigris*), elephant (*Elephas maximus*), and Javan rhinoceros (Bradley 1876). Small populations of some of these species were still present in northern parts of the region during the early twentieth century.

There are two ecoregional endemic mammals and one near-endemic mammal in this ecoregion (table J64.1). One further ecoregional endemic, Schomburgk's deer (*Cervus schomburki*), was extirpated by the early twentieth century. The area still supports nationally important populations of Lyle's flying-fox (*Pteropus lylei*).

Many larger birds probably disappeared simultaneously with the mammals, so there are no historical nest records of spot-billed pelicans or adjutant storks, although it is likely that the ecoregion once included the huge waterbird rookeries described by Oates (in Smythies 1986) for the Sittang plain in Burma.

At the time Madoc (1950) wrote, a colony of Oriental darters still nested on trees in the middle of the city. He also found what may have been the last nest of sarus crane (*Grus antigone*) for the ecoregion, near Kamphaengphet toward its northern boundary.

Some commensals of humans, such as vultures, were still present into the mid-1960s (B. King, pers. comm., 1999) but have since vanished. Wintering populations of black kite (*Milvus migrans*) have collapsed within the past decade, and the breeding population numbers only a few pairs.

The white-eyed river martin (*Pseudochelidon sirintarae*) may have been endemic to this ecoregion, but its life cycle and habitat needs remain unknown (table J64.2). However, it is likely that it nested on riverine sandbars on the northern part of the region and was already nearing extinction at the time of its discovery in 1968. There have been no records

TABLE J64.1. Endemic and Near-Endemic Mammal Species.

Family	Species
Muridae*	*Niviventer hinpoon**
Muridae*	*Leopoldamys neilli**
Rhinolophidae	*Hipposideros halophyllus*
Cervidae	*Cervus schomburgki* (extinct)

An asterisk signifies that the species' range is limited to this ecoregion.

TABLE J64.2. Endemic and Near-Endemic Bird Species.

Family	Common Name	Species
Hirundinidae	White-eyed river-martin	*Pseudochelidon sirintarae*

An asterisk signifies that the species' range is limited to this ecoregion.

apart from ten birds collected in 1968 and a sight record of five or six in 1977 (King and Kanwanich 1977).

The area continues to support possibly the largest known concentration of the globally near-threatened Asian openbill (*Anastomus oscitans*), currently scattered in four or five colonies in the ecoregion and thought to number more than 10,000 pairs. Other large waterbirds including painted stork (*Mycteria leucocephala*), spot-billed pelican (*Pelecanus philippensis*), and black-headed ibis (*Threskiornis melanocephalus*), which probably formerly bred, are annual visitors, probably entering the country from Cambodia. Rarely adjutants (*Leptotilos* spp.) also occur as nonbreeding visitors. The area supports significant concentrations of the globally near-threatened grey-headed lapwing (*Vanellus cinereus*) and significant wintering and breeding populations of waterbirds including egrets (*Egretta* spp.).

Status and Threats

Current Status

Almost all of the natural habitat has been cleared for agriculture and settlements. The Central Plain of the Chao Phraya River is now almost completely under rice cultivation (IUCN 1991), with smaller areas occupied by sugar cane, bananas, fruit orchards, and other crops. Creeping urbanization and land speculation have also greatly affected the area within the past couple of decades. Most large wildlife has disappeared, and only grassland, scrub, or commensal species remain (IUCN 1991). Few wetlands are represented in Thailand's protected area system, and this ecoregion receives no protection. Such protected areas exist as "nonhunting areas" and are too small and isolated to have any relevance to the larger conservation situation. Tyson Roberts (pers. comm., 1999) pointed out that a major Holocene extinction has already taken place among fishes in river deltas throughout Asia, including the Chao Phraya Basin, with the conversion of swamps to paddy land.

There are no protected areas in this ecoregion with an IUCN designation I–IV (WCMC and ABC 1997).

Types and Severity of Threats

Almost all the wetlands have been drained or severely disturbed (IUCN 1991). Rice paddies have replaced the original freshwater swamp and monsoon forest of this ecoregion, none of which remains intact (IUCN 1991). Rice paddies have considerable importance as a modified form of wetland habitat (sustaining populations of wintering and resident waterfowl and other species). However, the area of rice is declining gradually with a switch to more profitable and destructive agriculture (e.g., vegetables), which requires more intensive management and a lesser volume of water. Increased use of land for housing estates and factories takes place without reference to any zoning plan. The Industrial Estates Authority of Thailand is aggressively promoting industrialization throughout the lower Chao Phraya Basin without reference to any zoning plan that takes account of its environmental attributes. Other inappropriate forms of land use include prawn farming, in which brackish or saline water is trucked or piped into former freshwater areas and the wastewater is disposed of inappropriately, leading to gradual salinization.

There is a gradually increasing pesticide and herbicide load. Although rice farming per se entails less intensive herbicide use than other forms of agriculture, in some instances pesticides (including persistent organochlorines such as endosulfan) are dumped into rice fields by farmers to kill the introduced snail *Pomacea*, which is a pest of rice crops.

Priority Conservation Actions

Although it has ratified the Ramsar Convention, the Thai government has the dubious distinction of being the Asian government to have done least in real terms to meet the requirements of the convention. (Thailand's only Ramsar site constitutes only a small proportion of a single swamp in southern Thailand.) Wetlands remain almost entirely outside the protected area system (apart from a few nonhunting areas). The only freshwater wetland to receive national park status, Khao Sam Roi Yot, has been badly encroached upon by prawn farmers and land speculators. A wetland inventory for Thailand's central region is in preparation by Office of Environmental Policy and Planning.

Short-Term Conservation Actions (1–5 Years)

- Pressure should be placed on the Thai government to declare the freshwater marsh in and adjacent to Khao Sam Roi Yot National Park, Prachuap Khiri Khan Province, and Bung Boraphet Non-Hunting Area as Ramsar sites.
- Use of persistent organochlorine pesticides should be banned.
- Assays of pesticide residues among terrestrial vertebrates should be conducted.
- The lower central plain should be surveyed to determine the distribution and location of patches of significant habitat. These should include larger, better-quality water bodies; roosting and breeding colonies for large waterbirds and fruit-bats; and significant patches of regenerating seminatural scrub vegetation. Management options for land surrounding such patches could be considered, especially the possibility of acquiring privately owned nature wetland reserves in collaboration with Thai NGOs because private ownership is the only viable option.
- Regular basinwide waterfowl censuses should be conducted.
- Surveys for smaller, more secretive mammals (e.g., fishing cat [*Prionailurus viverrinus*] and otters) should be conducted.

Longer-Term Conservation Actions (5–20 Years)

- Ensure that options for wetland conservation are fully taken into account in any review of biodiversity issues or representativeness of nature reserves in Thailand.
- Ensure that a system of appropriate land-use zoning is developed and applied for agricultural and settled areas to maximize their contribution to biodiversity conservation.
- Establish a functioning network of small wetland nature reserves throughout the basin through acquisition by private NGOs.
- Develop appropriate legislation to prohibit inappropriate or damaging land use and encourage more traditional systems in agricultural areas. An appropriate system of financial compensation should be considered.

Focal Species for Conservation Action

Habitat specialists: fishing cat, Asian openbill
Endangered endemic species: white-eyed river martin
Top predator species: leopard, wild dog
Migratory species: migratory birds

Selected Conservation Contacts

Bangkok Bird Club
Bird Conservation Society of Thailand
Green World Foundation
Office of Environmental Policy and Planning
Thailand Center for Conservation Biology
WCS-Lao PDR
WCS-Thailand
Wildlife Conservation Division, Royal Forest Department
Wildlife Fund Thailand
WWF-Thailand

Justification for Ecoregion Delineation

The tropical moist forests in the Cardamom and Elephant mountain ranges in southwestern Cambodia and parts of Thailand are included within the Cardamom Mountains Rain Forests [55]. MacKinnon included these rain forests in the large Cardamom Mountains subunit (05d), which also included the freshwater swamp forests and mangroves of the Chao Phraya river and estuary. We placed the latter biomes in the Chao Phraya Freshwater Swamp Forests [64] and Indochina Mangroves [79], respectively.

Prepared by Philip Round, Sompoad Srikosamatara, Nantiya Aggimarangsee, and Eric Wikramanayake.

Ecoregion Number:	**65**
Ecoregion Name:	**Tonle Sap Freshwater Swamp Forests**
Bioregion:	**Indochina**
Biome:	**Tropical and Subtropical Moist Broadleaf Forests**
Political Units:	**Vietnam, Cambodia**
Ecoregion Size:	**25,900 km²**
Biological Distinctiveness:	**Bioregionally Outstanding**
Conservation Status:	**Vulnerable**
Conservation Assessment:	**V**

This ecoregion comprises the seasonally inundated forests that surround southeast Asia's largest lake, the Tonle Sap. Although most of the ecoregion, including the lake, was declared a protected area recently, it was too little too late. The protected area is a paper park with no protection or management, and it was declared protected after most of the habitat had been cleared for agriculture. This is prime rice-growing habitat.

Location and General Description

The swamp shrublands and forest of the Tonle Sap Freshwater Swamp Forests [65] ecoregion include two forest associations that have been described for the extensive floodplain area of Tonle Sap, a short tree shrubland covering the majority of the area and a stunted swamp forest around the lake itself. Similar swamp forests are also present along floodplains of the Mekong and other major rivers in Cambodia. The structure and composition of woody vegetation on the floodplain appear to be largely a function of the microheterogeneity of soil moisture conditions and seasonal flood dynamics. Much of this ecoregion is flooded for at least a six-month period extending from August to January or February.

In general, the dominant woody species of the short tree shrubland form a nearly continuous canopy of deciduous species reaching no more than 4 m in height. The height reached by individual species appears to be related to soil moisture conditions, with the tallest individuals occurring closer to the permanent lake basin and smaller individuals present at the periphery of the floodplain area. Several characteristic species with shrubby growth forms in this community are capable of reaching tree size in swamp forest habitats. The flora of these short-tree shrublands is dominated by species of Euphorbiaceae, Fabaceae, and Combretaceae, together with *Barringtonia acutangula*. *Terminalia cambodiana* is an important local endemic.

A band of stunted swamp forest, 7–15 m in height, originally dominated the dry-season shoreline of Tonle Sap, covering about 10 percent of the floodplain, and a similar community once occurred as a gallery forest along the seasonal floodplains of the Mekong, Bassac, and other major rivers in southern Cambodia and Vietnam. This community is covered by water for six to eight months each year, at which time the majority of species lose their leaves. Rather than forming a continuous forest, this community is broken into a mosaic of stands of large trees and open areas with floating aquatic herbs typical of the lake itself. Two tree species, *Barringtonia acutangula* and *Diospyros cambodiana*, are the primary dominants of this community.

The giant mimosa (*Mimosa pigra*) presents a serious invasive species problem. This aggressive species becomes established in fallow fields and disturbed shrubland and swamp forest area after clearance or burning. Once established, giant mimosa forms dense, impenetrable thickets of spiny growth that choke out other native species and have little value as wildlife habitat.

The strong seasonal cycle of flooding around the floodplain of Tonle Sap has made the great majority of woody species deciduous. Rather than lose their leaves in the dry season, however, these species lose their leaves when submerged as the lake deepens and the plants are submerged. However, there are several woody species that remain evergreen despite submergence for six to eight months each year. With only a few exceptions, flowering and fruit production in the floodplain trees and shrubs are delayed for several months after the flush of new leaves. The majority of woody species are laden with fruits and seeds at the time of submergence, suggesting that fish may serve as important dispersal agents.

The shrublands and swamp forests of the Tonle Sap floodplain have been heavily affected by human activities, and this impact has accelerated over the past decade. Very little of the original forest cover remains in pristine condition today.

In addition to areas of woody vegetation, the Tonle Sap Freshwater Swamp Forests [65] ecoregion includes extensive areas of seasonally inundated grasslands growing in a mosaic of scattered individuals of *Barringtonia acutangula*. These hydromorphic savannas, called *veal* in the French literature, are commonly saturated for at least six months of the year.

Biodiversity Features

Mammals of conservation significance include the endangered pileated gibbon (*Hylobates pileatus*), tiger (*Panthera tigris*), and several threatened species, including wild dog (*Cuon alpinus*), sun bear (*Ursus malayanus*), clouded leopard (*Pardofelis nebulosa*), common leopard (*Panthera pardus*), and banteng (*Bos javanicus*). There is one near-endemic bat species (table J65.1). This ecoregion overlaps with a Level I TCU (Dinerstein et al. 1997).

Although unsuitable for agriculture, areas that have been degraded to reed beds nonetheless are still important sites for waterfowl, providing feeding grounds for the eastern sarus crane (*Grus antigone*), white-shouldered ibis (*P. davisoni*), and near-endemic giant ibis (*Pseudibis gigantea*) (IUCN 1991) (table J65.2).

Conservation Status

Current Status

This ecoregion once consisted primarily of permanent and seasonal freshwater swamp forests, but much of the natural habitat has been cleared. Excessive forest exploitation has reduced many areas to scrub or secondary forest invaded by nonindigenous species, and the natural regeneration of large species is very slow (FAO 1981). Most of the remaining habitat is found in northern Cambodia, around the Tonle Sap and Tonle rivers. The large reserve surrounding the Tonle Sap Lake in Cambodia accounts for most of the 5,490 km² of protected areas in this ecoregion (table J65.3). However, none of the protected areas have effective management or protection.

Types and Severity of Threats

The mangroves and *Melaleuca* forests of the Mekong delta were severely affected by military activities in the Vietnamese wars; however, these have partially recovered through replanting programs (MacKinnon 1997). The vast floodplains of the Mekong River and Tonle Sap are extensively cultivated during the dry season (IUCN 1991). Local fishing communities have greatly altered the Tonle Sap swamp forests (IUCN 1991). The area of the Mekong delta that lies in southern Vietnam has been severely affected by deforestation in water catchments in Laos, Thailand, and southern China. As a result, there are dramatic fluctuations in the water level of the Mekong river—with frequent floods and low water levels during the dry season—that create increasing problems for local agriculture (IUCN 1991).

Priority Conservation Actions

Short-Term Conservation Actions (1–5 Years)

- Recruit a staff, provide training in conservation management, and implement effective protected area management in the existing reserve.
- Stop hunting, especially of the endangered and keystone species.
- Allow regeneration in critical areas, especially along the riverine forests.

Longer-Term Conservation Actions (5–20 Years)

- Monitor protected area management and conservation in the Tonle Sap Great Lake reserve.

Focal Species for Conservation Action

Habitat specialists: eastern sarus crane, white-shouldered ibis, giant ibis
Endangered species: pileated gibbon, wild dog, sun bear
Area-sensitive species: tiger
Top predator species: tiger, clouded leopard, common leopard, wild dog

Selected Conservation Contacts

BirdLife-Indochina
FFI-Indochina
IUCN Vietnam
Ministry of Environment, Phnom Penh, Cambodia
WCS-Cambodia
Wildlife Protection Office, Cambodia
WWF-Cambodia
WWF-Indochina

TABLE J65.1. Endemic and Near-Endemic Mammal Species.

Family	Species
Rhinolophidae	Hipposideros halophyllus

An asterisk signifies that the species' range is limited to this ecoregion.

TABLE J65.2. Endemic and Near-Endemic Bird Species.

Family	Common Name	Species
Threskiornithidae	Giant ibis	Pseudibis gigantea

An asterisk signifies that the species' range is limited to this ecoregion.

TABLE J65.3. WCMC (1997) Protected Areas That Overlap with the Ecoregion.

Protected Area	Area (km²)	IUCN Category
Tam Nong	70	IV
Tonle Sap Great Lake	5,420	VIII
Total	5,490	

Ecoregion numbers of protected areas that overlap with additional ecoregions are listed in brackets.

Justification for Ecoregion Delineation

We assigned the mangroves, freshwater swamp forests, and peat swamps in MacKinnon's subunit 05a to the Indochina Mangroves [79], Tonle Sap Freshwater Swamp Forests [65], and Tonle Sap–Mekong Peat Swamp Forests [62], respectively.

Prepared by Eric Wikramanayake and Philip Rundel

Ecoregion Number: **66**
Ecoregion Name: **Red River Freshwater Swamp Forests**
Bioregion: **Indochina**
Biome: **Tropical and Subtropical Moist Broadleaf Forests**
Political Units: **Vietnam**
Ecoregion Size: **10,700 km^2**
Biological Distinctiveness: **Bioregionally Outstanding**
Conservation Status: **Critical**
Conservation Assessment: **IV**

This ecoregion is nearly extinct. Representing the swamp forests of the Red River in northern Vietnam, the natural habitat has long been cleared for agriculture. Today it is nearly impossible to even ascertain the natural biodiversity that used to occur in this ecoregion.

Location and General Description

This ecoregion comprises the freshwater swamp forests along the lower Red River in northern Vietnam.

Swamp forest occurs inland of the mangroves in freshwater conditions. Freshwater swamp forest occurs on permanently or seasonally flooded mineral soils, often in zones up to 5 km wide, along rivers or around freshwater lakes. The original vegetation in these freshwater swamp forests may have been *Melaleuca*-dominated (BAP 1995). This ecoregion used to provide habitat for several endangered species of mammals and birds.

Biodiversity Features

Freshwater swamps support a wide variety of plant species, many limited to this specialized habitat, and they support a great diversity of freshwater fish, birds, and mammals. There are no endemic birds or mammals in this ecoregion.

Status and Threats

Current Status

This freshwater swamp forest ecoregion along the lower Red River in northern Vietnam has been almost totally cleared of its original habitat. There are no protected areas in this ecoregion, and little freshwater swamp forest remains.

Types and Severity of Threats

Most of these forests have been converted for agricultural use and settlements.

Priority Conservation Actions

Short-Term Conservation Actions (1–5 Years)

- Protect whatever small patches of natural habitats remain.
- Reforest and restore any degraded habitat wherever possible.
- Prohibit and prevent hunting in protected areas.

Longer-Term Conservation Actions (5–20 Years)

- Monitor conservation and management of any protected areas that are established.

Focal Species for Conservation Action

Top predator species: Owston palm civet (Boonratana and Le 1994), Tay Nguyen civet (Boonratana 1999b)

Other: pygmy loris (Boonratana 1999b), Francois' leaf monkey (Boonratana 1999b), Chinese pangolin (Boonratana 1999b)

Selected Conservation Contacts

BirdLife International, Indochina

FFI-Indochina

Forest Planning and Inventory Unit, Hanoi, Vietnam

Forest Protection Department, Hanoi, Vietnam

Institute for Ecology and Biological Resources, Hanoi, Vietnam

IUCN Vietnam

Provincial forest departments

WWF-Indochina

Justification for Ecoregion Delineation

The freshwater swamp forests and mangroves in the Red River Valley and coastal areas of MacKinnon's subunit 06a were included within the Red River Freshwater Swamp Forests [66] and Indochina Mangroves [79], respectively.

Prepared by Eric Wikramanayake and Ramesh Boonratana

Ecoregion Number: **67**
Ecoregion Name: **Irrawaddy Moist Deciduous Forests**
Bioregion: **Indochina**
Biome: **Tropical and Subtropical Moist Broadleaf Forests**
Political Units: **Myanmar**
Ecoregion Size: **137,900 km^2**
Biological Distinctiveness: **Regionally Outstanding**
Conservation Status: **Vulnerable**
Conservation Assessment: **II**

Like many of the region's lowland forests, the Irrawaddy Moist Deciduous Forests [67] ecoregion has been intensively cultivated and its forests converted over hundreds of years. As a consequence, most of the region's biodiversity has been extirpated, and because of political forces over the past few decades very little current information on the biodiversity status of this ecoregion is known.

Location and General Description

This ecoregion is located within the Irrawaddy River Basin, the catchments of Bago Yoma, and the foothills of Rakhine Yoma. The soils belong to the Irrawaddian series, which consists of the fluvial sands with terrestrial and aquatic vertebrate fossils. Silicified wood fossils are found among ferruginous, calcareous, and siliceous concretions, with quartz pebbles. The Irrawaddian rocks are distinct from other Tertiary rock groups. Their occurrence reaches up to the Kachin State in the north and in Chindwin districts in Sagaing division. The southern distribution is down to Rangoon.

Moist deciduous forests dominate this ecoregion but are found not only in the Irrawaddy River Basin but also throughout the country along the Chindwin, Sittang, and Salween rivers. Topographically the forests are found on well-drained hilly or undulating land up to 1,000 m. The region is best characterized by more than 1,500 mm of rainfall interrupted by dry spells. According to Champion's classification, it is moist upper mixed deciduous forest.

The forests are closed high forest. They cover large areas in Pegu Yoma. The westward extension is across Irrawaddy River onto Yakhine Yoma foothills, and the northern extension is up to the Kachin State. Trees reach a height of more than 30 m. The dominant species are teak (*Tectona grandis*) and Pyinkado or ironwood (*Xylia kerri*). Species composition is varied and intimately mixed with bamboo groves. In the matrix of deciduous species, some evergreen dominants appear in places.

Common tree species are teak, *Xylia kerri*, *Terminalia tomentosa*, *T. belerica*, *T. pyrifolia*, *Homalium tomentosum*, *Salmalia insigni*, *Ginelina arborea*, *Lannea grandis*, *Odina wodia*, *Pterocarpus macrocarpus*, *Millettia pendula*, *Berrya ammonilla*, *Mitravgyna rotundifdlia*, and *Vitex* spp.

Bambusa polymorpha and *Cephalostachyum pergracile* are the most common bamboos in lower Burma. In the north *Dendrocalamus hamiltonii*, *D. membranaceus*, and *Cephalostachyum pergracile* are common bamboos.

The undergrowth often consists of *Leea* spp., *Barleria strigosa*, and other Acanthaceae. *Eupatorium odoratum*, a noxious weed, colonized the areas when timber extraction left gaps.

Biodiversity Features

Wildlife is intensively exploited outside the protected areas. Most knowledge about the remaining wildlife populations comes from work inside selected national parks. The general trend has been the extirpation of the large charismatic megafauna and the persistent survival of birds and medium-small mammals.

Asian elephants (*Elephas maximus*) survive in this ecoregion, but their numbers have slowly decreased as the habitat has been fragmented. Tigers (*Panthera tigris*) were common in this ecoregion only fifteen years ago, but severe hunting pressure has nearly extirpated the population. A recent WCS survey using camera traps found no signs of pugmarks or tigers. Eld's deer (*Cervus eldi*) was thought to survive approximately fifteen years ago at Yinmabin. Today there are no recent signs of Eld's deer in the ecoregion.

Wild species that persist in the ecoregion's national parks include the gaur (*Bos gaurus*), sambar (*Cervus unicolor*), serow (*Capricornis sumatrensis*), golden cat (*Felis temmincki*), masked palm civet (*Paguma larvata*), marbled cat (*Felis marmorata*), leopard cat (*Felis bengalensis*), spotted linsang (*Prionodon pardicolor*), Himalayan black-bear (*Selenarctos thibetanus*), Himalayan sun-bear (*Helarctos malayanus*), binturong (*Arctictis binturong*), dhole (*Cuon alpinus*), and capped langur (*Presbytis pileatus*). There are no endemic mammals.

Almost 350 birds are found in this ecoregion, and they include both water and forest birds found in a wide variety of habitats. The forest birds includes several species of woodpecker, laughing thrushes, babblers, orioles, drongos, parakeets, barbets, pigeons, doves, and magpies. Common waterbirds include the red-wattled lapwing (*Vanellus indicus*), wagtails, sandpipers, forktails, and river chat (*Thamnolacea leucocephala*). A single near-endemic bird species is found in this ecoregion (table J67.1).

This ecoregion also corresponds to the Irrawaddy Plains EBA (132), which contains two restricted-range bird species (Stattersfield et al. 1998). In addition to the white-throated babbler, it also identifies the hooded treepie (*Crypsirina cucullata*) as vulnerable.

Status and Threats

Current Status

Today the remaining intact forests are found mainly on Pegu Yoma and the northern extension of the ecoregion. There are five protected areas in the ecoregion (with IUCN categories I–IV or unknown) (table J67.2). Master plans have been drafted for several of these reserves. Unfortunately, they were never approved or implemented.

There is an ongoing attempt to establish more protected areas to reach the Forest Policy's mandated target of 5 to 10 percent, but very few effective measures have been taken to manage, protect, and staff the areas that have been established. Therefore, cutting of valuable trees and poaching are rife.

The Moyingyi Bird Sanctuary (104 km^2) was established in 1988. It is the water impoundage constructed to irrigate the surrounding paddy fields. Annually thousands of winter-

TABLE J67.1. Endemic and Near-Endemic Bird Species.

Family	Common Name	Species
Timaliidae	White-throated babbler	*Turdoides gularis*

An asterisk signifies that the species' range is limited to this ecoregion.

TABLE J67.2. WCMC (1997) Protected Areas That Overlap with the Ecoregion.

Protected Area	Area (km^2)	IUCN Category
Alaungdaw Kathapa	1,490	II
Kyatthin	230	UA
Minwun Taung	140	UA
Shwesettaw	340	UA
Pegu Yomas [49]	2,220	PRO
Total	4,420	

Ecoregion numbers of protected areas that overlap with additional ecoregions are listed in brackets.

ing birds arrive in December through February. Therefore, it is imperative to conserve the wetland for the migratory birds. However, the bird population is now being reduced because of poaching. The protected area, though in place, is not effective.

Types and Severity of Threats

This ecoregion remains under various threats. Conversion of forests to agriculture and shifting cultivation is prevalent. Illegal cutting of timber, firewood, and nonforest timber products is equally common. Intense poaching of protected animals, such as tigers, for their parts is reducing the population in the remaining forests.

The lack of political will to conserve this and other ecoregions in Myanmar essentially permits all these activities to occur without thought of retribution. Hunters and poachers are almost never arrested or punished, and the wildlife protection law exists only on paper.

Priority Conservation Actions

Short-Term Conservation Actions (1–5 Years)

- Prepare short-term management plans for current protected areas and stop the current development projects.
- Enforce existing wildlife protection laws.
- Adequately fund protected areas.
- Survey remaining forests to determine status of wildlife.

Longer-Term Conservation Actions (5–20 Years)

- Develop political will for conservation.
- Develop management plans for protected areas.
- Adequately fund protected areas.

Focal Species for Conservation Action

Endangered endemic species: white-throated babbler
Area-sensitive species: tiger (Franklin et al. 1999; Tilson et al. 1994; Dinerstein et al. 1997; Seidensticker et al. 1999), Asian elephant (Santiapillai and Jackson 1990; Santiapillai 1997)
Top predator species: tiger (Franklin et al. 1999; Tilson et al. 1994; Dinerstein et al. 1997; Seidensticker et al. 1999), leopard, Asiatic golden cat, and clouded leopard
Migratory species: migratory bird species
Other: Eld's deer, Asiatic black bear, gaur, banteng

Selected Conservation Contacts

Myanmar Forest Department
Myanmar Wildlife Conservation Society
Myanmar Wildlife and Sanctuaries Division

Justification for Ecoregion Delineation

MacKinnon's Burmese coast biounit is represented by four ecoregions: the Myanmar Coastal Rain Forests [49], Irrawaddy Freshwater Swamp Forests [63], Myanmar Coastal Mangroves [78], and Mizoram-Manipur-Kachin Rain Forests [50]. MacKinnon's deciduous forests subunit (09a) are represented by the Irrawaddy Dry Forests [71] and the Irrawaddy Moist Deciduous Forests [67]. The dry, seasonal vegetation in the Irrawaddy Valley is represented by the Irrawaddy Moist Deciduous Forests [67] and Irrawaddy Dry Forests [71], which comprise the teak (*Tectona grandis*)-dominated moist deciduous forests and the drier patches of *Dipterocarpus*-dominated dry deciduous forests, respectively. The Irrawaddy Moist Deciduous Forests [67], Irrawaddy Dry Forests [71], Chin Hills–Arakan Yoma Montane Rain Forests [48], Northeast India–Myanmar Pine Forests [77], Northern Triangle Sub-tropical Forests [73], and Mizoram-Manipur-Kachin Rain Forests [50] roughly correspond to Udvardy's Burman rain forest and Burma monsoon forest.

Prepared by U Saw Han and Saw Tun Khaing

Ecoregion Number:	**68**
Ecoregion Name:	**Northern Thailand–Laos Moist Deciduous Forests**
Bioregion:	**Indochina**
Biome:	**Tropical and Subtropical Moist Broadleaf Forests**
Political Units:	**Laos, Thailand**
Ecoregion Size:	**42,000 km²**
Biological Distinctiveness:	**Regionally Outstanding**
Conservation Status:	**Vulnerable**
Conservation Assessment:	**II**

The Northern Thailand–Laos Moist Deciduous Forests [68] still retain large blocks of teak-dominated forests. However, despite good coverage of protected areas, these forests are devoid of wildlife. Most of the larger animals, such as tigers, have been driven to low populations or extirpation because of pervasive illegal hunting over the past fifty years. New biodiversity surveys are needed to reassess the status of wildlife in this ecoregion and to plan future conservation actions.

Location and General Description

This ecoregion is situated on the upper reaches of the Ping, Wang, Yom, and Nan rivers, tributaries of the Chao Phraya, and the upper Pasak River. Part of the area also lies in the Mekong Drainage. Annual rainfall throughout the region is around 1,000–1,200 m, and mean minimum and maximum temperatures are around 20°C and 32°C, respectively (Meteorological Department 1987). Most of the area is on steep hill slopes, interspersed with narrow, north-south–aligned valleys carrying alternately swift-flowing and slow-flowing rivers.

Teak (*Tectona grandis*) is a co-dominant of the moist mixed deciduous forest characterizing this zone. A detailed description of vegetation at one site in Mae Yom National Park, Phrae Province, Thailand is given in Center for Conservation Biology (1992). Teak contributed 27.2 percent of all tree individuals, *Xylia xylocarpa* 11.4 percent, and *Pterocarpus macrocarpus* 10.0 percent. Other important tree species in the mixed deciduous community included *Millettia brandisiana* (6.7 percent), *Lagestroemia cochinchinensis* (4.1 percent), *Bombax kerrii* (3.0 percent), and *Afzelia xylocarpa* (2.3 percent). Bamboo is common and is an indicator of high human disturbance. This vegetation type extends from the valley floors (which range from less than 200 m elevation to about 400 m elevation, depending on site) at varying distances from the montane transition (800–1,000 m).

Biodiversity Features

Most large mammals were extirpated from this ecoregion by the 1960s or early 1970s, but small populations of elephants, banteng, and gaur still remain at a few sites. There are at least five banteng individuals in Sri Satchanalai National Park and a few gaur in Thung Saleng Luang National Park (Srikosamatara and Suteethorn 1995). There is very little information on carnivores. It is conceivable that tiger may be already extirpated from this region. The ecoregion overlaps with two Level I TCUs and one Level II TCU (Dinerstein et al. 1997).

The Ping River once supported important riparian bird communities, including such species as Brahminy kite, river lapwing, and river tern, associated with rapids, sand bars, and islands. Possibly the largest such area of rapids in the region, on the Ping River (McClure and Lekagul 1961), was submerged by the Bhumibol Dam in the early 1960s, so this rich river community has now disappeared from most areas (Murray 1995). Green peafowl (*Pavo muticus*) was once widespread. It is still represented by a few populations in this ecoregion, the largest of which are found on the northern border of Mae Yom National Park. Small populations have been found persisting at three other sites in Chiang Mai Province (Wina Meckvichai, pers. comm., 2000).

Some of the forest bird species would have been shared with those found in the Central Indochina Dry Forests [72], but bird communities have been affected greatly by the gradual loss of large trees. A few green imperial pigeons (*Ducula aenea*) were still present in Mae Yom National Park in 1991 (Center for Conservation Biology 1992). Yellow-footed pigeon (*Treron phoenicoptera*) also still occurs in the region but has been reduced to near extirpation. Huge flocks of parakeets, mainly *Psittacula alexandri*, were recorded by Deignan (1945), and Alexandrine parakeet (*P. eupatria*) was once widespread. The former is still present in small numbers, along with both *P. roseata* and *P. finschii*, but *P. eupatria* probably is close to being extirpated.

The only non–breeding-season record of the globally threatened blackthroat (*Luscinia obscura*), which breeds in west and central China, was from lowland forest or secondary growth in this region (Lekagul and Round 1991). There is a subsequent sight record (March 2000) of two birds in lowland forest remnants in the newly established Mae Jarim National Park, Nan Province.

Overall, because the area supports the highest proportion of woody cover of any region of Thailand, it continues to be of great conservation importance. The continued presence of such large and sensitive species as green peafowl raises the possibility that some species of birds and perhaps large mammals may be able to recolonize regenerating secondary forest, given a likely future reduction in hunting pressure. Even in Thailand, much of this area has not been surveyed and assessed for its biodiversity attributes.

Status and Threats

Current Status

The valleys of this ecoregion have long been cultivated, and shifting cultivation has been responsible for more recent destruction of upland regions (IUCN 1991). In Thailand, previously itinerant hill tribes such as the Hmong and Yao, who cultivate upland rice and opium replacement crops such as coffee, cabbage, and ornamental flowers, have settled illegally in many areas that are ostensibly protected. The subsistence and, increasingly, commercial agricultural activities of both upland shifting cultivators and especially lowland Thai farmers have been responsible for much of the recent degradation that has occurred inside and outside Thailand's protected areas. Large areas of forest land are now occupied by cotton fields and fruit orchards. Almost all remaining forest has been selectively logged and is significantly degraded, and has been damaged by burning and other disturbances. The predominance of teak probably has been increased by burning, although most larger trees are gone. In addition, much of this area is also occupied by teak plantations rather than native forest. In Laos, teak forests have been mostly destroyed; meanwhile, shifting cultivation, regular fires, and continual erosion of the hills turn these areas into scrubland of bamboo or other grass species, preventing reestablishment of the original forest (FAO 1981). However, overall, the original habitat has been heavily altered.

Small-scale hunting is widely practiced and has greatly reduced populations of mammals and larger birds. There is a large amount of trade in wildlife and wildlife parts throughout the region. One such market specializing in wildlife is situated near the provincial forestry office in Lampang. Sale of wildlife, though illegal, continues with little or no official interference.

Riverine habitat has been greatly altered. River valleys were first settled and agriculture practiced along their banks. In silty areas, dry season cultivation follows the retreating water so that almost no undisturbed riparian habitat remains. Three of the four major rivers of northern Thailand now have hydroelectric dams. In addition, the spread of *Mimosa pigra* into disturbed areas has replaced areas of native riparian and floodplain scrub and low herbage. Construction of new roads and enlargement of existing roads under government development programs seem likely to further fragment the larger habitat patches.

There are fourteen protected areas in this ecoregion (table J68.1). For some ecoregions, we were not able to get accurate area information, so the total area reflects only those listed.

Types and Severity of Threats

The construction of a combined hydroelectric and irrigation dam on the Yom River inside Mae Yom National Park, which supports one of the best and most extensive remnants of teak-dominated forest, has been approved by the Thai government.

Hunting is widespread and has reduced or locally extirpated populations of mammals and larger birds. Illegal logging carried out by influential businesses in conjunction with corrupt officials and police officers is a major threat. A Thai government body, the Forest Industries Organization, is empowered to remove and sell illegally felled timber from the forest, including inside protected areas. Therefore, there is no incentive to suppress illegal logging. In addition, many villages around Mae Yom National Park and other areas also continue a thriving, small-scale, illegal timber trade, assembling teakwood houses, which can be readily dismantled and sold. Fires set by farmers and hunters sweep through the region annually.

Priority Conservation Actions

Much of this ecoregion remains little known, but because of the high proportion of woody cover and the wide distribution of protected areas, it continues to have great potential for conservation.

Short-Term Conservation Actions (1–5 Years)

- Develop improved capacity among protected area staff in national parks and wildlife sanctuaries through training.
- Survey biodiversity attributes of remaining habitat patches inside and outside protected areas.
- Survey rivers to determine which areas have the richest riparian communities.
- Prioritize remaining habitat patches for establishment of new protected area or incorporation into existing protected areas.
- Ban removal of illegally felled timber from the forest by government agencies.
- Ban use of persistent organochlorine pesticides for cotton and other crops.
- Reforest.

Longer-Term Conservation Actions (5–20 Years)

- Develop an integrated conservation and management plan for the entire region. This should also cover ecotourism because this ecoregion contains many historical sites and antiquities. Better mechanisms to distribute income from ecotourism could provide a powerful incentive to help conserve biodiversity.

TABLE J68.1. WCMC (1997) Protected Areas That Overlap with the Ecoregion.

Protected Area	Area (km²)	IUCN Category
Namtok Chatrakarn National Park [52]	310	?
Doi Chiang Dao		IV
Doi Pha Chang [72, 74]	180	IV
Doi Luang Wildlife Sanctuary	100	IV
Doi Luang National Park		II
Doi Khun Tan	255	II
Wiang Kosai	440	II
Mae Yom		
Si Satchanalai	170	II
Thung Saleng Luang	1,260	II
Mae Ping [51, 72]	480	II
Khun Jae		II
Jae Sorn	592	II
Mae Jarim		II
Total	3,787	

Ecoregion numbers of protected areas that overlap with additional ecoregions are listed in brackets.

Focal Species for Conservation Action

Area-sensitive species: green peafowl, Asian elephant (Santiapillai and Jackson 1990; Santiapillai 1997), banteng

Top predator species: tiger (Franklin et al. 1999; Tilson et al. 1994; Dinerstein et al. 1997; Seidensticker et al. 1999), leopard

Migratory species: blackthroat

Selected Conservation Contacts

Bangkok Bird Club
Bird Conservation Society of Thailand
Center for Conservation Biology, Thailand
Department of Forest Resources and Conservation, Lao PDR
Green World Foundation
IUCN Laos
Office of Environmental Policy and Planning
WCS-Lao PDR
WCS-Thailand
Wildlife Conservation Division, Royal Forest Department
Wildlife Fund Thailand
WWF-Thailand

Justification for Ecoregion Delineation

The tropical moist deciduous forests to the west and northwest of the Luang Prabang Range in northwestern Thailand and Laos are an extension of the teak (*Tectona grandis*)-dominated deciduous forests (Rundel and Boonpragob 1995), but the vegetation community of the moist deciduous forests in the Mekong plains near Vientiane are dominated by Fabaceae, Lythraceae, and Rubiaceae. Therefore, these forests were placed in separate ecoregions, the former in the Northern Thailand–Laos Moist Deciduous Forests [68] and the latter in the Northern Khorat Plateau Moist Deciduous Forests [69].

Prepared by Philip D. Round, Sompoad Srikosamatara, Nantiya Aggimarangsee, and Eric Wikramanayake

Ecoregion Number:	**69**
Ecoregion Name:	**Northern Khorat Plateau Moist Deciduous Forests**
Bioregion:	**Indochina**
Biome:	**Tropical and Subtropical Moist Broadleaf Forests**
Political Units:	**Laos, Thailand**
Ecoregion Size:	**16,800 km²**
Biological Distinctiveness:	**Regionally Outstanding**
Conservation Status:	**Critical**
Conservation Assessment:	**II**

This small ecoregion along the northern reaches of the Mekong River represents a transition from the dry forests of the Khorat Plateau to the moister forests of the Annamite Mountains. Like most ecotone habitats, the ecoregion contains a mix of species from dry and mesic habitats, increasing the overall biodiversity. But again, like many productive habitats that lie along river plains, much of the natural habitat has been cleared for agriculture.

Location and General Description

The Northern Khorat Plateau Moist Deciduous Forests [69] are located in the middle Mekong River Valley, along the border of Thailand and Laos. The average annual rainfall for Nong Khai Province for the period 1956–1985 was 1,629 mm, and mean minimum and maximum temperatures were 21.6 and 31.8, respectively (Meteorological Department 1987).

The floristic structure of mixed deciduous forests in the Mekong Valley of southern Laos and adjacent areas of northeastern Cambodia suggests an intermediate ecological habitat between more mesic semi-evergreen forests and more xeric deciduous dipterocarp forests. Lowlands in this area commonly receive 2,000–3,000 mm of precipitation annually, with five to six months of dry season.

In many respects these forests represent a deciduous form of semi-evergreen forest and therefore are similar to the Central Indochina Dry Forests [72]. The most common canopy species are *Lagerstroemia angustifolia*, *Afzelia xylocarpa*, *Xylia xylocarpa*, *Peltophorum dasyrrachis*, and *Pterocarpus macrocarpus*. In some areas, *Lagerstroemia* may form almost single-species dominance. In northern Cambodia, there is a transition between mixed deciduous forest and forms of semi-evergreen forest where deciduous species exhibit a strong co-dominance with evergreen species. Any separation of such types along this transition is somewhat arbitrary. This separation is complicated by the fact that many taxa such as *Xylia*, *Pterocarpus*, *Lagerstroemia*, and *Irvingia* have a wide range of ecological tolerance and are found in deciduous dipterocarp woodlands and mixed deciduous and semi-evergreen forest communities.

Biodiversity Features

Within this ecoregion, dry evergreen forest contains much greater mammalian diversity than deciduous forest because no arboreal mammals are known to feed on the fruit or foliage of any dipterocarp except for *Shorea platyclados* (Lekagul and McNeely n.d.). Several endangered species live here, including the pileated gibbon (*Hylobates pileatus*) and Asian elephant (*Elephas maximusas*), and several threatened species such as wild dog (*Cuon alpinus*), Himalayan black bear (*Ursus thibetanus*), sun bear (*Ursus malayanus*), clouded leopard (*Pardofelis nebulosa*), and common leopard (*Panthera pardus*). Tiger (*Panthera tigris*), gaur (*Bos gaurus*), and banteng (*Bos javanicus*) probably have already disappeared from the Thai portions of this area (Srikosamatara and Suteethorn 1995). The ecoregion overlaps with a Level II TCU (Dinerstein et al. 1997).

There appear to be significant differences in the forest bird community on the Lao and Thai banks of the Mekong River. The Indochinese endemic red-collared woodpecker (*Picus rabieri*) has been found at Sangthong, Lao PDR (Duckworth 1996) but not as far on the Thai bank.

Some sections of the Mekong River in this ecoregion are outstanding examples of upper perennial riverine habitat, with exposed bedrock, sand, and shingle bars and rapids. Extensive areas are covered by *Homonoia riparia* scrub (Euphorbiaceae), which support what is probably the largest population of the scarce Jerdon's bushchat (*Saxicola jerdoni*) known anywhere in southeast Asia (Duckworth 1996). A few great thick-knees (*Esacus recurvirostris*) are also present, as are river lapwings (*Vanellus duvaucelii*) and plain martins (*Riparia paludicola*). All these species are nationally at risk in Thailand (Round 2000) and Laos (Thewlis et al. 1998).

Status and Threats

Current Status

The five protected areas cover 1,965 km^2 (12 percent) of the ecoregion (table J69.1). The IUCN category on three of the protected areas is unknown, and so is the extent to which the ecoregion's biodiversity is protected.

Types and Severity of Threats

Most larger bird and mammal species have been greatly reduced or extirpated. A few elephants remain inside Phu Wua Wildlife Sanctuary. Hunting by ethnic Lao people is even more ubiquitous than in northern Thailand, for example, so that even common, open country birds are scarcer here than in many other ecoregions. Use of fire by hunters and farmers is widespread and continues to degrade remaining forest. Human use of fisheries on the Mekong River is almost certainly unsustainable, and other forms of human use (e.g., recreation, ferry traffic) place added pressure on riparian habitats.

Priority Conservation Actions

Much of this ecoregion remains little known, but it continues to have great potential for conservation.

Short-Term Conservation Actions (1–5 Years)

- Designate cross-border nature reserves along key sections of the Mekong River.
- Survey all remaining forest patches in the ecoregion for possible establishment of nature reserves.
- Identify possible habitat corridors and forest regeneration corridors linking nature reserves.

Longer-Term Conservation Actions (5–20 Years)

- Develop an integrated conservation and management plan for the entire region.

TABLE J69.1. WCMC (1997) Protected Areas That Overlap with the Ecoregion.

Protected Area	Area (km^2)	IUCN Category
Phu Phan National Park	645	?
Huai Huat National Park	830	?
Phu Sri Than WS	250	?
Phu Wua	190	IV
Phu Kao-Phu Phan Kham [72]	50	II
Total	1,965	

Ecoregion numbers of protected areas that overlap with additional ecoregions are listed in brackets.

Focal Species for Conservation Action

Area-sensitive species: Asian elephant (Santiapillai and Jackson 1990; Santiapillai 1997)

Top predator species: leopard

Selected Conservation Contacts

Bangkok Bird Club

Bird Conservation Society of Thailand

Center for Conservation Biology, Thailand

Department of Forest Resources and Conservation, Lao PDR

Green World Foundation

IUCN Lao PDR

Office of Environmental Policy and Planning

WCS-Lao PDR

WCS-Thailand

Wildlife Conservation Division, Royal Forest Department

Wildlife Fund Thailand

WWF-Lao PDR

WWF-Thailand

Justification for Ecoregion Delineation

The tropical moist deciduous forests to the west and northwest of the Luang Prabang Range in northwestern Thailand and Laos are an extension of the teak (*Tectona grandis*)-dominated deciduous forests (Rundel and Boonpragob 1995), but the vegetation community of the moist deciduous forests in the Meking plains, near Vientianne, are dominated by Fabaceae, Lythraceae, and Rubiaceae. Therefore, these forests were placed in separate ecoregions, the former in the Northern Thailand–Laos Moist Deciduous Forests [68] and the latter in the Northern Khorat Plateau Moist Deciduous Forests [69].

Prepared by Philip D. Round, Sompoad Sriksomatara, Nantiya Aggimarangsee, and Eric Wikramanayake

Ecoregion Number:	**70**
Ecoregion Name:	**Chao Phraya Lowland Moist Deciduous Forests**
Bioregion:	**Indochina**
Biome:	**Tropical and Subtropical Moist Broadleaf Forests**
Political Units:	**Thailand**
Ecoregion Size:	**20,300 km²**
Biological Distinctiveness:	**Regionally Outstanding**
Conservation Status:	**Endangered**
Conservation Assessment:	**II**

The Chao Phraya Lowland Moist Deciduous Forests [70] have had intense anthropogenic influence over time. A majority of the forests that remain are degraded, and most of the larger wildlife no longer is found in these forests. However, these forests still abut intact forests to the west along the Tenasserim mountains and, if allowed to regenerate, might support viable populations of Asian elephants and tigers in the future.

Location and General Description

This ecoregion is not a homogeneous unit but contains forest patches having affinities with other ecoregions. Forest on the west of the Chao Phraya, in the drainage of the Khwae River system, grades into Tenasserim–South Thailand Semi-Evergreen Rain Forests [53] in wetter areas and Central Indochina Dry Forests [72] in the more seasonal or drier areas. As in most of northern and western Thailand, the gibbon present is the wide-ranging *Hylobates lar*.

The area to the east of the Chao Phraya has more affinities with the Southeastern Indochina Dry Evergreen Forests [59], and is characterized by the presence of some mammals (e.g., *Hylobates pileatus*) and species or subspecies of birds that are not found west of the Thai Lower Central Plain.

Average annual rainfall is in the region of 1,000–1,100 mm per year in the west to 1,300 mm per year in the east, roughly 80 percent of which falls during the southwest monsoon, May to October. Mean maximum and minimum temperatures are around 34°C to 23°C (Meteorological Department 1987).

In the western area of the region, some areas contain fragments of a lush, moist rain forest formation that has more in common with Tenasserim–South Thailand Semi-Evergreen Rain Forests [53] and supports a few characteristic Sundaic bird species. Such habitats may have been more widespread than formally realized but have been largely lost.

Huge areas have been converted to agricultural land, principally growing tapioca and sugar cane.

Limestone karst formations are found in the plains around Ratchaburi and Phetchaburi and still support some drier forest. One of the most significant such formations lies inside Khao Sam Roi Yot National Park, Prachuap Khiri Khan Province.

Biodiversity Features

Biodiversity information that is known from this ecoregion has been gathered in a few key protected areas. For example, the Khao Ang Ru Nai Wildlife Sanctuary lies on the margins of this ecoregion. Larger mammals present include banteng and pileated gibbon. The area also supports a few freshwater crocodiles (*Crocodylus siamensis*). Some typical Indochinese birds found here, which are not found west of the Lower Central Plain, include Siamese fireback (*Lophura diardi*), Indochinese magpie (*Cissa hypoleuca*), and scaly-crowned babbler (*Malacopteron cinereum*). Most of these species were once found slightly further west, in the plains of Chonburi Province, before these areas were so completely deforested. A few larger birds are still present in Khao Ang Ru Nai,

TABLE J70.1. Endemic and Near-Endemic Mammal Species.

Family	Species
Vespertilionidae*	*Myotis rosseti**

An asterisk signifies that the species' range is limited to this ecoregion.

including, at least until the early 1990s, up to five pairs of woolly-necked stork (*Ciconia episcopus*, the last such remaining in Thailand), pompadour pigeon (*Treron pompadora*), and green imperial pigeon (*Ducula aenea*).

Khao Sam Roi Yot National Park supports a few serow (*Capricornis sumatrensis*). A Tenasserimese pheasant, previously placed as a race of Kalij pheasant (*L. leucomelana crawfurdii*) but treated by McGowan and Panchen as a race of silver pheasant (*L. nycthemera*), occurs here and in other protected areas west of the Chao Phraya plain.

Most of the larger wildlife species, such as Asian elephant and tiger, no longer survive in the wild in this ecoregion. This is because of the heavy human influence on the forests. A single endemic bat is found in this ecoregion (table J70.1).

Status and Threats

Current Status

Forest in the western area, though largely degraded, is still contiguous with the 17,000 km^2 contiguous block of protected habitat in Thailand's western forest complex. The eastern portion is contiguous, or nearly so, with Khao Ang Ru Nai and Khao Soi Dap wildlife sanctuaries, which constitute part of the Southeast Indochina Dry Evergreen Forests [59] and Cardamom Mountains Rain Forests [55], respectively. Though degraded, such fragments still remain and therefore have the potential to be recolonized by large mammals and some birds.

The four protected areas cover more than 1,400 km^2 (7 percent) of the ecoregion, but none are greater than 1,000 km^2 (table J70.2).

Types and Severity of Threats

Very little of the original forest cover remains, and most of what is left has been selectively logged and degraded by fire. New roads and the spread of industry and new housing from Bangkok, including proposed new towns and a new Bangkok International Airport, will place pressure on remaining land, especially through land speculation. Poor villagers continue to produce charcoal from forest trees in degraded forest areas, leading to further degradation of forest cover.

The Thai government recently completed a gas pipeline from the gulf of Marteban, which has created a new corridor through previously intact forest between this zone and the Tenasserim South Thailand rain forest zone.

The Industrial Estates Authority of Thailand is aggressively promoting industrial development throughout the more rural areas of this ecoregion.

TABLE J70.2. WCMC (1997) Protected Areas That Overlap with the Ecoregion.

Protected Area	Area (km^2)	IUCN Category
Khao Laem [51, 53]	840	II
Sai Yok [53]	440	II
Khao Sam Roi Yot	90	II
Chalerm Ratanakosin National Park	60	?
Total	1,430	

Ecoregion numbers of protected areas that overlap with additional ecoregions are listed in brackets.

Priority Conservation Actions

Though degraded, this ecoregion can contribute to an overall conservation strategy through the restoration of its degraded forests, which are contiguous with more intact or protected regions.

Short-Term Conservation Actions (1–5 Years)

- Zone the area to facilitate regeneration of forest on degraded public lands, where feasible, and restrict or limit spread of industry and new housing.
- Identify possible corridors, or greenbelts, that could link remaining patches of forest and other nonurban or suburban habitat.

Longer-Term Conservation Actions (5–20 Years)

- Develop an integrated conservation and management plan for the entire region.

Focal Species for Conservation Action

Habitat specialists: plain-pouched hornbill

Area-sensitive species: banteng, serow, woolly-necked stork

Top predator species: tiger (Franklin et al. 1999; Tilson et al. 1994; Dinerstein et al. 1997; Seidensticker et al. 1999), leopard

Other: freshwater crocodile

Selected Conservation Contacts

Bangkok Bird Club
Bird Conservation Society of Thailand
Center for Conservation Biology, Thailand
Green World Foundation
Office of Environmental Policy and Planning
WCS-Lao PDR
WCS-Thailand
Wildlife Conservation Division, Royal Forest Department
Wildlife Fund Thailand
WWF-Thailand

Justification for Ecoregion Delineation

MacKinnon's subunit 05d also extends into the Kayah Karen and Tenasserim mountains, and we included these montane forests in the Kayah-Karen Montane Rain Forests [51] and Tenasserim–South Thailand Semi-Evergreen Rain Forests [53], respectively. The lowland moist deciduous forests along the lower reaches of the Chao Phraya River were placed in the Chao Phraya Lowland Moist Deciduous Forests [70].

Prepared by Sompoad Srikosmatara, Philip D. Round, and Eric Wikramanayake

Ecoregion Number: **71**
Ecoregion Name: **Irrawaddy Dry Forests**
Bioregion: **Indochina**
Biome: **Tropical and Subtropical Dry Broadleaf Forests**
Political Units: **Myanmar**
Ecoregion Size: **35,000 km^2**
Biological Distinctness: **Regionally Outstanding**
Conservation Status: **Critical**
Conservation Assessment: **II**

The Irrawaddy Dry Forests [71], like the surrounding moist deciduous forests, have been under intensive conversion pressure for hundreds of years. However, until recently most of its large mammal fauna, such as the tiger, still persisted in the degraded forests. Only recently has the larger mammal fauna been hunted to the brink of extinction in this ecoregion. The little protection afforded this ecoregion has hindered conservation efforts.

Location and General Description

This ecoregion falls in the dry zone of central Myanmar. The region has a harsh climate and is extremely dry. The average rainfall is about 650 mm per year. Rains start in mid-July and last until October. There are rarely more than fifteen days of rain per year. When rainfall does occur, it falls in torrential showers. In addition to rains, the dry zone is subject to southerly winds during the summer, resulting in wind erosion of the topsoil.

The soil substrate is primarily soft sandstone and clays and belong to the Irrawaddian series. The wide diurnal temperature range weathers the soft rocks. The weathered rocks and soil are then washed away during seasonal showers. Therefore, the landscape of the dry zone is marked with exposed rocks. Besides the harsh climate, the region is extremely short of potable water. There is no underground water recharge because of uncontrolled surface runoff and the low moisture retention of the subsoil. This causes an extreme water shortage.

The forests in this ecoregion are dominated by dry forests where rainfall is less than 800 mm. The stands are low in stature, with thorny trees. Common species include *Terminalia oliveri* and *Tectona homiltoniana*. Trees associated with them are *Acacia catechu* and *Bauhinia racemosa*. Pests such as *Euphobia neriifolia*, *E. nirculia*, *E. tirucalli*, and the introduced *Prosopsis juliflora* are hard to eradicate. They characterize the dry zone low-growing degraded forests.

There are also two types of dipterocarp forests: high *Indaing* and low *Indaing*. The former produces big timber, whereas the growth of the latter is stunted by repeated cutting and burning of high *Indaing*. Both types are widespread, but the low type is more common in the dry zone. There are stands of pure *Dipterocarpus tuberculatus* among the typical *Indaing* forests of *Shorea oblongifolia* and *Pentacme siamensis*.

A dry mixed deciduous forest also occurs in this ecoregion. Teak dominates these forests. Bamboo such as *Dendrocalamus strictus*, *Bambusa polymorpha*, and *B. tulda* are also common in these forests.

TABLE J71.1. Endemic and Near-Endemic Mammal Species.

Family	Species
Muridae	*Millardia kathleenae**

An asterisk signifies that the species' range is limited to this ecoregion.

TABLE J71.2. Endemic and Near-Endemic Bird Species.

Family	Common Name	Species
Corvidae	Hooded treepie	*Crypsirina cucullata*
Timaliidae	White-throated babbler	*Turdoides gularis*

An asterisk signifies that the species' range is limited to this ecoregion.

Biodiversity Features

The larger mammals historically found in this ecoregion have been extirpated. Wildlife found in the protected areas includes medium and small mammals. Numerous species of deer such as barking deer (*Muntiacus muntjak*), Eld's deer (*Cervus eldi*), or sambar deer (*Cervus unicolor*) persist. Primate species include the rhesus macaque (*Macaca mulatta*) and Hoolock gibbon (*Hylobates hoolock*). Mid-sized predators such as the jungle cat (*Felis chaus*), Asiatic jackal (*Canis aureus indicus*), and perhaps leopard (*Panthera pardus fuscus*) also persist. A single endemic mouse also is found in this ecoregion (table J71.1).

The birds include more than 300 species, including two rare wetland birds: the black stork (*Ciconia nigra*) and the woolly-necked stock (*Ciconia bicolor*). Two near-endemic species are found in this ecoregion (table J71.2). This ecoregion also overlaps with the Irrawaddy Plains EBA (132), which contains two restricted-range bird species, corresponding to our near-endemic species (Stattersfield et al. 1998).

Among reptiles, the dry zone is famous for vipers, including the Burmese python (*Python molurus bivittatus*). Turtles such as the yellow tortoise (*Testudo elongata*) and soft-shelled turtle (*Trionyx* spp.) were once common but now are intensively collected for food.

Status and Threats

Current Status

Most of the forests in this ecoregion have been converted to agriculture. The forests that do exist are degraded and lack the full complement of species. Those that did survive have been under intense hunting pressure. In the past twenty years the increased demand and population have driven many of the larger species to the brink of extinction within this ecoregion.

Only one nature reserve exists in this ecoregion that meets our IUCN criteria (table J71.3). This ecoregion is poorly protected, and the protected areas currently gazetted provide little real protection to its wildlife and forests.

Types and Severity of Threats

The threats to this ecoregion are similar to those facing all of Myanmar and its neighboring countries. Conversion of forests to agricultural land causes habitat loss and fragmentation of the remaining habitat. A large percentage of this conversion is illegally done. Poaching and the illegal trade of animals and their parts are rampant. The lack of protected areas and the political will to prosecute poachers leave the remaining forests devoid of wildlife.

Priority Conservation Actions

Short-Term Conservation Actions (1–5 Years)

- Establish additional protected areas in this ecoregion.
- Develop management plans for these protected areas.

Longer-Term Conservation Actions (5–20 Years)

- It is essential to stop all land appropriation in the protected areas.
- The status of all protected areas should be thoroughly reevaluated, and those that have been wholly converted to other land uses should be traded for more pristine areas.
- The protected area system mandates that 5 percent of land area be strictly dedicated to conservation. Implement and enforce this mandate.
- All protected areas should be placed under effective conservation and management plans.
- The Nature and Wildlife Directorate should be the sole authority to manage the protected areas, with sufficient personnel and funding.

Focal Species for Conservation Action

Top predator species: leopard

Migratory species: black stork, woolly-necked stork

Other: Burmese python, yellow tortoise, Eld's deer, hoolock gibbon

Selected Conservation Contacts

Myanmar Forest Department

WCS-Myanmar

Justification for Ecoregion Delineation

MacKinnon divided the Indochina bioregion into fifteen biogeographic subunits nested within seven major biounits. Each of these biounits and biogeographic subunits contains a mix of biomes. MacKinnon's deciduous forests subunit (09a) is represented by the Irrawaddy Dry Forests [71] and the Irrawaddy Moist Deciduous Forests [67].

This region does not correspond well to Udvardy's biogeographic provinces. The Myanmar Coastal Rain Forests [49] cover the Udvardy's Burman rain forest, southwestern portion of the Thai monsoon forest, and western portion of the Indochinese rain forest. The Irrawaddy Moist Deciduous Forests [67], Irrawaddy Dry Forests [71], Chin Hills–Arakan Yoma Montane Rain Forests [48], Northeast India–Myanmar Pine Forests [77], and Mizoram-Manipur-Kachin Rain Forests [50] roughly correspond to Udvardy's Burman rain forest and Burma monsoon forest.

Prepared by U Saw Han and Saw Tun Khaing

Ecoregion Number:	**72**
Ecoregion Name:	**Central Indochina Dry Forests**
Bioregion:	**Indochina**
Biome:	**Tropical and Subtropical Dry Broadleaf Forests**
Political Units:	**Thailand, Cambodia, Laos, and Vietnam**
Ecoregion Size:	**319,100 km²**
Biological Distinctiveness:	**Globally Outstanding**
Conservation Status:	**Vulnerable**
Conservation Assessment:	**I**

The Central Indochina Dry Forests [72] ecoregion covers most of central Indochina and harbors an outstanding assemblage of threatened large vertebrates that characterize the mammal fauna of the Indo-Pacific region. Just half a century ago large populations of megaherbivores such as Asian elephants (*Elephas maximus*), banteng (*Bos javanicus*), kouprey (*Bos sauveli*), gaur (*Bos gaurus*), wild water buffalo (*Bubalus arnee*), and Eld's deer (*Cervus eldii*) roamed and grazed in these dry woodlands (Wharton 1957). Where human densities were still low, the landscapes were dominated by large herds of wildlife reminiscent of the savannas of east Africa. Large carnivores such as tigers (*Panthera tigris*), clouded leopards (*Felis nebulosa*), leopards (*Panthera pardus*), and packs of wild dogs (*Cuon alpinus*) hunted these herbivores. Unfortunately, throughout the ensuing years habitat loss and hunting for trade have exacted a devastating toll on these species. Some species have even become extinct. The two rhinoceros species, the Javan (*Rhinoceros sondaicus*) and the Sumatran (*Dicerorhinus sumatrensis*) are now extinct in this ecoregion, as is Schomburgk's deer (*Cervus duvaucelli schomburgki*). The kouprey probably is globally extinct, although intermittent reports from remote areas of northern and eastern Cambodia keep hopes alive. Among the other species, the tiger, Asian elephant, Eld's deer, banteng, and gaur are endangered.

Location and General Description

The Central Indochina Dry Forests [72] ecoregion covers more area in mainland southeast Asia than any other forest type. It extends widely in Thailand, from dry lower slopes in northern Thailand and the foothills of the Tenasserim Range to uplands around the Chao Phraya Basin and then across the Khorat Plateau. It remains extensive in coverage along the broad valley of the Mekong and its tributaries in central and southern Laos and has a widespread distribution in the arid plains of northern, eastern, and south-central Cambodia. A small area of the Central Indochina Dry Forests [72] ecoregion reaches into Vietnam within the upper watersheds of the Se San and Srepok rivers. Over this range the ecoregion characteristically occurs in areas

TABLE J71.3. WCMC (1997) Protected Areas That Overlap with the Ecoregion.

Protected Area	Area (km²)	IUCN Category
Wethtigan	4	UA

Ecoregion numbers of protected areas that overlap with additional ecoregions are listed in brackets.

with 1,000–1,500 mm rainfall and five to seven months of drought. Potential evapotranspiration may exceed rainfall for up to nine months per year.

Deciduous dipterocarp forest, the name commonly used for the characteristic forest association of the Central Indochina Dry Forests [72], forms an open forest or woodland community dominated by deciduous trees. This forest formation has been called *idaing* in Burma and *forêt claire* in Laos and Cambodia. Community structure may range from a nearly closed canopy forest of low trees 5–8 m in height, to more typical woodland structure with 50 to 80 percent canopy cover and an open understory dominated by grasses. Occasional emergent trees may reach to heights of 10–12 m. Along gradients of increasing environmental stress, whether from natural drought or human intervention, dry deciduous dipterocarp communities become increasingly open in structure and lower in stature, grading eventually into savanna woodlands with decreasing woody cover.

Deciduous species of Dipterocarpaceae form the dominant element of deciduous dipterocarp forests. Only six species of the approximately 550 dipterocarps in the world are deciduous, and all of these occur in this formation. Four of these, *Shorea siamensis*, *S. obtusa*, *Dipterocarpus obtusifolius*, and *D. tuberculatus*, generally form the dominant biomass and cover. The community often is moderately rich in other small trees, particularly legumes such as *Pterocarpus macrocarpus*, *Sindora siamensis*, and *Xylia xylocarpa*. *Terminalia alata* and *Pinus merkusii* may be a co-dominant species. Cycads are common in the grassy understory. Epiphytic vascular plants and lichens are few in number and low in diversity in lowland habitats of this ecoregion but increase in abundance with higher elevations and more humid conditions.

Ground fires burning through the herbaceous understory of deciduous dipterocarp forests are a regular aspect of the environment, so this association is sometimes called a fire climax community. The question, therefore, is how much of this community has been formed by a history of human activities that have greatly increased the frequency of such fires. Some researchers believe that a significant portion of the modern coverage of this habitat represents a type conversion of what was once semi-evergreen forest into deciduous dipterocarp forest under the influence of repeated burning. Other researchers believe that such conversions have been limited. Most fires occur between December and early March, when forest conditions are driest. Dominant tree species in this formation exhibit adaptations to fire in the form of thick, corky bark to protect cambium tissues and root crowns, which readily resprout.

Biodiversity Features

The ecoregion's 167 mammal species include an impressive assemblage of threatened large vertebrates such as the critically endangered kouprey and Eld's deer, the endangered tiger, Asian elephant, gaur, banteng, wild water buffalo, serow, and other species such as the pileated gibbon, two leaf monkey species (*Semnopithecus cristatus* and *S. phayrei*), wild dog (dhole), Malayan sun bear, clouded leopard, and common leopard. It also harbors two endemic species of Vespertilionidae bats (*Myotis altarium* and *Pipistrellus pulveratus*) (table J72.1).

TABLE J72.1. Endemic and Near-Endemic Mammal Species.

Family	Species
Vespertilionidae*	*Myotis altarium*ature*
Vespertilionidae*	*Pipistrellus pulveratus*

An asterisk signifies that the species' range is limited to this ecoregion.

TABLE J72.2 Endemic and Near-Endemic Bird Species.

Family	Common Name	Species
Hirundinidae	White-eyed river-martin	*Pseudochelidon sirintarae*
Timaliidae	Grey-faced tit-babbler	*Macronous kelleyi*

An asterisk signifies that the species' range is limited to this ecoregion.

The remaining large habitat blocks overlap with six high-priority (Level I) and two Level II TCUs (Dinerstein et al. 1997). But the long-term conservation success of tigers and the ecological integrity in these conservation landscapes will be compromised if the habitat becomes fragmented.

Many of the 500 or so bird species include several that are of conservation importance for their role as focal species for conservation management and for their threatened status. The latter include the critically endangered white-eyed river-martin (*Pseudochelidon sirintarae*), the globally threatened Bengal florican (*Eupodotis bengalensis*), and the endangered greater adjutant (*Leptoptilos dubius*) and white-shouldered ibis (*Pseudibis davisoni*).

Some of the other species that are indicators of habitat integrity, and therefore of conservation importance as focal species, include the silver pheasant (*Lophura nycthemera*), Siamese fireback (*Lophura diardi*), Hume's pheasant (*Syrmaticus humiae*), grey peacock-pheasant (*Polyplectron bicalcaratum*), sarus crane (*Grus antigone*), great hornbill (*Buceros bicornis*), Austen's brown hornbill (*Anorrhinus austeni*), and wreathed hornbill (*Aceros undulatus*). There are two near-endemic birds (table J72.2).

The reptile and amphibian faunas are not well known. But the limited information suggests that the critically endangered turtle *Pelochelys cantorii*, the geckos *Gehyra lacerata* and *Gekko petricolus*, the agamid lizard *Ptyctolaemus phuwuanensis*, the two skinks *Isopachys borealis* and *Lygosoma koratense*, the earth snake *Typhlops khoratensis*, and the colubrid snake *Oligodon hamptoni* are ecoregional endemics. According to local reports, critically endangered Siamese crocodiles may persist at some isolated permanent lakes.

Status and Threats

Current Status

Although large blocks of forests remain in northeastern Cambodia, most of the natural habitat has been extensively cleared in Thailand, Laos, and Vietnam. There are twenty-one protected areas that cover more than 15,000 km² (6 percent) of the ecoregion. These include four (Phu Kao-Phu Phan Kham, Xe Piane, Phnom Prich, and Yok Don) that exceed 1,000 km² and two (Kulen Promtep and Lomphat) that exceed 2,000 km² (table J72.3). Many of these large protected areas, in Laos and Cambodia, were established recently and represent parts of the ecoregion where habitat still remains. But large areas of the ecoregion now in northeastern Thailand are deforested and lack protected areas.

Types and Severity of Threats

Most of the ecoregion lies in densely populated areas, where the natural habitat has long been converted to agriculture and settlements. Some fires are natural, but many are set to clear land for shifting agriculture and to promote new growth of grasses for livestock, to make traveling easier, and to attract wildlife, making hunting easier. In Thailand the Hmong and Yao hill tribes, who have traditionally cultivated upland rice and opium poppy, have been moving progressively further south and into lower-elevation dry forests and have begun clearing large areas of habitat (IUCN 1991). Areas of this ecoregion that are sparsely populated may be threatened by paddy rice development (encroachment) and the associated human population increase. Parts of Cambodia, Laos, and Thailand are threatened by dam development. In Cambodia, large reservoirs are proposed for the lower Se San and Srepok rivers. Some Thailand protected areas suffer from tourism. The land is being degraded, and rubbish is commonly discarded. In Laos, protected areas suffer from shifting cultivation, excessive NTFP harvesting, wildlife poaching (for domestic use and local and international trade), and illegal logging.

Priority Conservation Actions

Short-Term Conservation Actions (1–5 Years)

- Control poaching and wildlife trade (develop effective laws and alternative incomes for rural communities and implement trade control).
- Promote public education and community involvement in wildlife monitoring and conservation.
- Decommission the Pak Mun dam.
- Study seasonal migration of wildlife and implications for protection of megaherbivores.
- Study the role of fire in ecosystem function.
- Protect Dong Hua Sao's wetlands (key feature of Dong Hua Sao and important for waterbirds) and allow no human activities at or near the wetlands.
- Manage NTFP harvesting in Dong Hua Sao (current practices are depleting the NTFPs).
- Strengthen management capacity of protected-area staff and reinforce field activities (particularly patrolling and monitoring).

Longer-Term Conservation Actions (5–20 Years)

- Control poaching and wildlife trade (enforce national laws and international regulations).
- Identify or restore corridors and linkages between protected areas to conserve habitat for wide-ranging open-forest obligates.
- Promote appropriate land-use planning and management outside PAs to reduce rural dependence on wildlife trade.
- Educate policymakers on role of fire in ecosystem function.

Focal Species for Conservation Action

Habitat specialists: Cervus eldii, hornbills, serow (Boonratana 1998a)

Endangered endemic species: Bos sauveli

Area-sensitive species: Elephas maximus, gaur (Boonratana 1998c), Bos javanicus

Top predator species: tiger (Boonratana 1998a), dhole (Boonratana 1998a)

Other: Crocodylus siamensis, pangolin (Boonratana 1998a), silvered langur (Boonratana 1998a), douc langur (Boonratana 1998a), white-cheeked and yellow-cheeked gibbon (Boonratana 1998a), bear spp. (Boonratana 1998a), green peafowl (Boonratana 1998a)

Selected Conservation Contacts

Bangkok Bird Club
BirdLife-Indochina
Cambodia Ministry of Environment
Cambodia Wildlife Protection Office
Center for Conservation Biology, Thailand
Department of Forest Resources and Conservation, Lao PDR
FFI-Indochina
IUCN Lao PDR
IUCN Vietnam
WCS-Cambodia
WCS-Lao PDR
Wildlife Fund Thailand
WWF-Cambodia
WWF-Indochina
WWF-Lao PDR
WWF-Thailand

TABLE J72.3. WCMC (1997) Protected Areas That Overlap with the Ecoregion.

Protected Area	Area (km²)	IUCN Category
Phu Kao-Phu Phan Kham [69]	1,030	II
Doi Pha Chang [68, 74]	270	IV
Mae Ping [51, 68]	310	II
Mae Yom	500	II
Doi Pha Muang [51]	200	IV
Ramkamhaeng	270	II
Khao Sanam Phriang	130	IV
Klong Wang Chao	700	II
Khao Sam Lan [59]	50	II
Phu Kao-Phu Phan Kham	170	II
Tat Ton	240	II
Unnamed	330	?
Phu Kao-Phu Phan Kham	750	II
Xe Bang Nouane [59]	680	VIII
Xe Piane [59]	1,130	VIII
Phnom Prich [59]	1,710	IV
Kulen Promtep [59]	2,520	IV
Dong Hua Sao [57]	550	VIII
Lomphat	2,500	IV
Yok Don	1,050	II
Angkor Wat	90	V
Total	15,180	

Ecoregion numbers of protected areas that overlap with additional ecoregions are listed in brackets.

Justification for Ecoregion Delineation

In his analysis of the conservation units of the Indo-Pacific realm, MacKinnon (1997) included the dry deciduous forests in the single Indochina biounit (10), along with other vegetation types. In keeping with our rules for defining ecoregions, we extracted the dry deciduous forests and placed them in the Central Indochina Dry Forests [72].

Prepared by Eric Wikramanayake, Ramesh Boonratana, Philip Rundel, and Nantiya Aggimarangsee

Ecoregion Number:	**73**
Ecoregion Name:	**Northern Triangle Subtropical Forests**
Bioregion:	**Indochina**
Biome:	**Tropical and Subtropical Moist Broadleaf Forests**
Political Units:	**Myanmar**
Ecoregion Size:	**55,600 km^2**
Biological Distinctiveness:	**Globally Outstanding**
Conservation Status:	**Relatively Stable**
Conservation Assessment:	**III**

The Northern Triangle Subtropical Forests [73] are one of the least explored and scientifically known places in the world. The region's remote location, limited access, and rugged landscape have kept scientific exploration at a minimum. Yet what is known about these forests still ranks them as globally outstanding for their biological diversity. There are at least sixty-five endemic mammals known from this ecoregion, but more probably await discovery. In 1997 a new species of small deer, the leaf muntjac, was discovered high in the mountains. This ecoregion remains one of the few places in the Indo-Pacific region where conservation action can be done on a proactive rather than reactive basis.

Location and General Description

Floristically, Kachin State in northern Myanmar is one of the most diverse regions in continental Asia (WWF and IUCN 1995), but it is also one of the least explored. In 1997 a WCS team went into the region, the first in more than fifty years, since the early explorations of Kingdon-Ward (1921, 1930, 1952). Therefore, our assessment of the biodiversity in this region probably is highly underestimated; it probably harbors many more species than are now attributed to it.

The mountains trace their origins to the geological period when the collision between the Deccan Plateau and the Laurasian mainland created the Himalayas. The mountains extend as offshoots from the eastern Himalayas in four parallel ranges. The westernmost Sangpang Bum Range forms the Indo-Myanmar boundary, and the easternmost Goligong (Gaoligong) Shan demarcates the Myanmar-China border. In general, the elevation exceeds 1,500 m, but the peaks rise steeply to more than 3,000 m. The Chindwin, Mali Hka, and Mai Hka rivers originate in these mountains and converge in the lower reaches to form the Irrawaddy River.

The varied topography and biogeographic setting at the crossroads of the Assam-Indian, Eastern Himalayan, Indo-Malayan, and Chinese flora, as well as the ancient Gondwana relicts that have taken refuge here, give the ecoregion a high floral diversity. However, the Pleistocene glaciation has influenced the distributions of these floras. The Indo-Malayan elements are now limited to the river valleys, below 2,400 m, with the Indo-Himalayan flora stratified above (WWF and IUCN 1995).

This ecoregion consists primarily of the large area of subtropical broadleaf forest but includes small, sub-regional–scale patches of temperate broadleaf forests and sub-alpine conifer forests. The subtropical forests are distributed roughly between 500 and 1,600 m. Magnoliaceae, Lauraceae, and Dipterocarpaceae species make up the associations below 915 m, and species of Fagaceae, Meliaceae, tree ferns, and climbing palms make up the upper-elevation associations (WWF and IUCN 1995). Characteristic trees in these forests include *Acer pinnatinervium, Aesculus assamicus, Betula alnoides, Carpinus viminea, Castanopsis argentea, Magnolia pterocarpa, Persea* spp., *Litsea* spp., and *Lindera* spp. (WWF and IUCN 1995; Blower n.d.). In mature forests the trees are draped with lianas (*Jasminum duclouxii, J. pericallianthum, Lonicera hildebrandii, Bauhinia* spp., *Clematis* spp., *Mussaenda* spp., and *Rubus* spp.). Along the upper limits (above 1,525 m) the forest is dominated by *Bucklandia populnea* (WWF and IUCN 1995).

The Ngawchang valley, at 1,000 and 1,980 m between Htawgaw and Gangfang in the western part of the ecoregion, has a pine-oak association that is characterized by *Pinus kesiya, Quercus incana, Q. serrata,* and *Q. griffithii* (WWF and IUCN 1995). A rich, unique herb flora of *Anemone begoniifolia, Gentiana cephalantha, Gerbera piloselloides, Inula cappa, Lilium bakerianum, L. ochraceum, Primula denticulata,* and *Senecio densiflora* grows in these open forests, considered a fire-maintained preclimax community.

Several endemic species are associated with these forests, including the terrestrial orchid *Paphiopedilum wardii* and other endemics such as *Agapetes adenobotrys, A. pubiflora, Brachytome wardii, Lactuca gracilipetiolata, Lasianthus wardii, Paphiopedilum wardii,* and *Strobilanthes stramineus.*

Biodiversity Features

The ecoregion harbors almost 140 mammal species, including three near-endemic species and six endemic species (figure 73.1). One of the endemic species, *Muntiacus putaoensis*, was discovered in 1997 during the most recent scientific trek into the region (Amato et al. 1999).

TABLE J73.1. Endemic and Near-Endemic Mammal Species.

Family	Species
Talpidae	*Talpa grandis**
Talpidae	*Scaptonyx fusicauda**
Soricidae	*Chimarrogale styani**
Vespertilionidae	*Pipistrellus anthonyi**
Sciuridae	*Sciurotamias davidianus**
Sciuridae	*Callosciurus quinquestri*
Muridae	*Niviventer brahma*
Cervidae	*Muntiacus putaoensis**

An asterisk signifies that the species' range is limited to this ecoregion.

TABLE J73.2. Endemic and Near-Endemic Bird Species.

Family	Common Name	Species
Turdidae	Rusty-bellied shortwing	Brachypteryx hyperythra

An asterisk signifies that the species' range is limited to this ecoregion.

Several threatened species that make up part of this ecoregion's mammal fauna are also of conservation importance. These species include the tiger (*Panthera tigris*), red panda (*Ailurus fulgens*), Asian elephant (*Elephas maximus*), takin (*Budorcas taxicolor*), southern serow (*Naemorhedus sumatraensis*), pig-tailed macaque (*Macaca nemestrina*), Assamese macaque (*Macaca assamensis*), stump-tailed macaque (*Macaca arctoides*), capped leaf monkey (*Semnopithecus pileatus*), hoolock gibbon (*Hylobates hoolock*), Asiatic black bear (*Ursus thibetanus*), great Indian civet (*Viverra zibetha*), clouded leopard (*Pardofelis nebulosa*), red goral (*Naemorhedus baileyi*), Irrawaddy squirrel (*Callosciurus pygerythrus*), and particolored squirrel (*Hylopetes alboniger*).

The ecoregion's large, contiguous habitat areas are included within a TCU identified for immediate survey (Dinerstein et al. 1997). Maintaining landscapes with large habitat blocks is important to conserve these large predators, and this TCU provides one of few opportunities in the region to do so.

Wolf (*Canis lupus*) and musk deer (*Moschus moschiferous*) are considered to be recently extinct in this area.

The bird fauna exceeds 370 species. There is one near-endemic species, the rusty-bellied shortwing (*Brachypteryx hyperythra*) (table J73.2). However, there are several birds that can be considered focal species because of their need for mature forests and low thresholds for disturbance. Some of these species are Blyth's tragopan (*Tragopan blythii*), great hornbill (*Buceros bicornis*), wreathed hornbill (*Aceros undulatus*), and rufous-necked hornbill (*Aceros nipalensis*). The ecoregion overlaps with the far eastern portion of the Eastern Himalayas (130) EBA (Stattersfield et al. 1998).

With Assam, this region in northern Myanmar is considered to be one of the richest in the world for Lepidoptera (Mani 1974).

Status and Threats

Current Status

Because of its remoteness and inaccessibility, very little of this ecoregion has been substantially altered by human activity. More than 90 percent of the habitat is still intact in large habitat blocks, but there is little formal protection. Most of the areas covered by dense forest are demarcated as reserved forests but include the Piodaung Wildlife Sanctuary (table J73.3).

TABLE J73.3. WCMC (1997) Protected Areas That Overlap with the Ecoregion.

Protected Area	Area (km²)	IUCN Category
Piodaung	250	UA

Ecoregion numbers of protected areas that overlap with additional ecoregions are listed in brackets.

However, the forests on hill slopes are being rapidly cleared for shifting cultivation. The shifting cultivation cycle has also been reduced from twelve to twenty years to five to eight years, resulting in the perpetuation of a bamboo sub-climax that has begun to replace the broadleaf forests (WWF and IUCN 1995).

Types and Severity of Threats

The continual extraction of timber in this ecoregion and the increasing demand on the black market for timber to be transported into Yunnan Province, China is a serious threat to the future biodiversity of the region. This ecoregion has also become increasingly populated since 1988 as private companies have been allowed to extract gems, minerals, and timber.

Priority Conservation Actions

Short-Term Conservation Actions (1–5 Years)

- Establish additional protected areas in this ecoregion.
- Develop management plans for these protected areas.
- Conduct flora and fauna surveys of the region and disseminate results to conservation organizations.
- Stop illegal logging and wildlife trade.

Longer-Term Conservation Actions (5–20 Years)

- The protected area system mandates that 5 percent of land area be strictly dedicated to conservation. Implement and enforce this mandate.
- Place all protected areas under effective conservation and management plans.
- The Nature and Wildlife Directorate should be the sole authority to manage the protected areas, with sufficient personnel and funding.

Focal Species for Conservation Action

Habitat specialists: hornbills and pheasants

Endangered endemic species: Muntiacus putaoensis, Pere David's rock squirrel

Area-sensitive species: tiger (Franklin et al. 1999; Tilson et al. 1994; Dinerstein et al. 1997; Seidensticker et al. 1999), Asian elephant (Santiapillai and Jackson 1990; Santiapillai 1997)

Top predator species: tiger (Franklin et al. 1999; Tilson et al. 1994; Dinerstein et al. 1997; Seidensticker et al. 1999), leopard

Other: hoolock gibbon, red panda, takin

Selected Conservation Contacts

Myanmar Forest Department

WCS-Myanmar

Justification for Ecoregion Delineation

MacKinnon's subunit 09b includes the tropical and subtropical moist broadleaf forests in the Kachin and upper Chindwin areas of northern Myanmar. We included the tropical moist forests within the Mizoram-Manipur-Kachin Rain Forests [50], which then extends east across northern

Myanmar, and the subtropical moist broadleaf forests of the Golden Triangle to the north in the Northern Triangle Subtropical Forests [73].

This region does not correspond well to Udvardy's biogeographic provinces. The Myanmar Coastal Rain Forests [49] cover Udvardy's Burman rain forest, the southwestern portion of the Thai monsoon forest, and the western portion of the Indochinese rain forest. The Irrawaddy Moist Deciduous Forests [67], Irrawaddy Dry Forests [71], Chin Hills–Arakan Yoma Montane Rain Forests [48], Northeast India–Myanmar Pine Forests [77], Northern Triangle Subtropical Forests [73], and Mizoram-Manipur-Kachin Rain Forests [50] roughly correspond to Udvardy's Burman rain forest and Burma monsoon forest.

Prepared by U Tin Than, Tin Aung Moe, and Eric Wikramanayake

Ecoregion Number:	**74**
Ecoregion Name:	**Northern Indochina Subtropical Forests**
Bioregion:	**Indochina**
Biome:	**Tropical and Subtropical Moist Broadleaf Forests**
Political Units:	**Myanmar, Thailand, Laos, Vietnam, China**
Ecoregion Size:	**288,300 km²**
Biological Distinctiveness:	**Globally Outstanding**
Conservation Status:	**Vulnerable**
Conservation Assessment:	**I**

The Northern Indochina Subtropical Forests [74] are globally outstanding for their biological diversity; this ecoregion has the highest species richness for birds among all ecoregions in the Indo-Pacific region and ranks third for mammal richness. The ecoregion sits astride a major zoogeographic ecotone, where the northern Palearctic and the southern Indo-Malayan faunas mix, allowing langurs to mix with red pandas and muntjac and musk deer to mingle. It is also a crossroads for the south Asian and east Asian floras as well as some ancient relicts that have found refuge here during the turbulent and variable geological past.

Location and General Description

This large ecoregion extends across the highlands of northern Myanmar, Laos, and Vietnam and also includes most of southern Yunnan Province. A complex network of hills and river valleys extends south of the Yunnan Plateau into northern Indochina to include the middle catchments of the Red, Mekong, and Salween rivers. Mountains in this area are composed of intrusive igneous rocks or Paleozoic limestone and approach but seldom exceed 2,000 m, and major river valleys lie at 200–400 m elevation.

The climate throughout northern Indochina is summer monsoonal. Precipitation averages 1,200 to 2,500 mm per year, depending on location, and almost all of this consists of summer rain fetched from the Bay of Bengal and the South China Sea during April to October. From November to March, westerly subtropical winds drawn from continental Asia create dry conditions. These are moderated by easterly rain in southern Vietnam (WWF and IUCN 1995). Mean temperatures vary with elevation, but the spring premonsoon period is the hottest time of the year, and January is the coldest. Frost is known from the higher elevations, but it is infrequent.

The cool winter temperatures and high elevation here at the edge of the tropics promote a montane flora with a distinct Himalayan component. Some of the important tree taxa belong to families such as Theaceae (*Schima* spp.), Magnoliaceae (*Michelia* spp., *Magnolia* spp.), and Fagaceae (*Quercus* spp., *Castanopsis* spp., *Lithocarpus* spp.). Open-canopy pine forests occur in the higher elevations, and patches of tropical forests grow in the moist valleys. Montane deciduous forests are especially well developed in the Shan Plateau area of northern Myanmar.

Tropical forests of Xishuangbanna in the southern part of Yunnan Province, China are typical of this ecoregion. They have been classified by Cao and Zhang (1997) according to five different categories. Three of these (tropical seasonal rain forest, tropical montane rain forest, and evergreen broad-leaved forest) form a replacement sequence driven by declining drought stress with increased elevation. The other two are associated with anomalous moisture conditions: monsoon forest over limestone is determined by the fact that limestone loses water quickly, and the fifth, monsoon forest on river banks, is determined by perennial high water availability.

Subtropical broadleaf evergreen forest once covered much of the area of this ecoregion. The lower-elevation broadleaf forests up to 600–800 m appear to represent a transition from wet evergreen lowland forest to montane forests. The forest canopy is three-layered, with upper crowns reaching about 30 m in height. This transition zone contains clear elements of wet evergreen forests to the south but also subtropical Chinese elements from the north. Several distinctive forest communities can be distinguished in this region, although quantitative data on forest structure and floristics are highly limited. Above this zone, at elevations of 800–2,000 m, a clear montane evergreen forest takes over, with low-temperature conditions eliminating most of the tropical lowland forest elements. These forest communities are poorly studied in Laos and Myanmar but are better described for southern Yunnan Province in China. Little of these broadleaf evergreen forests remains intact today. Much of the former area of broadleaf evergreen forest has been converted to grasslands or savannas under the heavy impact of human overexploitation and swidden agriculture.

The characteristic families for the dominant forest trees at lower elevations are the Betulaceae, Fagaceae, Hamamelidaceae, Lauraceae, Magnoliaceae, Sapotaceae, Elaeocarpaceae, and Theaceae. These are all families characteristic of montane evergreen forests to the north in China. The montane broadleaf evergreen forests show a number of floristic elements, indicating that this is a relictual Tertiary flora. The family Rhoipteleaceae, with a single species, is endemic to this ecoregion.

Elevations above about 2,000 m in the Fan Si Pan massif of northern Vietnam support a distinct *Tsuga dumosa-Abies delavayi* var. *nukiangensis* forest unlike that present anywhere else in southeast Asia. Associated with the hemlock and fir are a moderate diversity of broadleaf trees, with the most species belonging to the Aceraceae, Hippocastanaceae, Fagaceae, Magnoliaceae, Lauraceae, Cupressaceae, Podocarpaceae, and Taxaceae.

Forests growing on patches of limestone are distinctive and are dominated by *Tetrameles nudiflora, Antiaris toxicaria, Celtis cinnamomifolia, C. wightii, Cleistanthus sumatranus, Garuga floribunda, Pterospermum menglunense, Ulmus lanceaefolia,* and *Xantolis stenopetala.*

Many basic questions about plant biodiversity remain unresolved for this ecoregion. For example, Cao and Zhang (1997) report that tropical montane forest in Xishuangbanna is species-poor, whereas studies in Doi Suthep National Park, Thailand, near the southern limit of this ecoregion, show that this forest type is the most species-rich. Another source of controversy is the role of fire in this ecosystem. Some evidence suggests that species in the seasonal forests at low elevation are adapted to a regime of frequent fire. *Phoenix* palms, for example, are well adapted to survive a ground fire. Yet most of the fires that occur now are lit by people, and it is thought that in the absence of fire, the composition of the forest would change.

Biodiversity Features

More than 183 mammal species are known from this ecoregion, of which 4 are endemic and 5 near endemic (table J74.1). Several other species are threatened, including the critically endangered Tonkin snub-nosed monkey (*Rhino-pithecus avunculus*) and the tiger (*Panthera tigris*). Other mammals of conservation significance include Asian elephant (*Elephas maximus*), gaur (*Bos gaurus*), southern serow (*Capricornis sumatraensis*), banteng (*Bos javanicus*), red panda (*Ailurus fulgens*), particoloured flying squirrel (*Hylopetes alboniger*), pygmy loris (*Nycticebus pygmaeus*), three macaque species (*Macaca nemestrina, Macaca assamensis,* and *Macaca arctoides*), wild dog (*Cuon alpinus*), Asian black bear (*Ursus thibetanus*), back-striped weasel (*Mustela strigidorsa*), inornate squirrel (*Callosciurus inornatus*), smooth-coated otter (*Lutrogale perspicillata*), Lowe's otter civet (*Cynogale lowei*), and clouded leopard (*Pardofelis nebulosa*) (Hilton-Taylor 2000).

The bird fauna here is very rich, 707 species, and includes species such as Ward's trogon (*Harpactes wardi*) (table J74.2). Among these is one near endemic, the short-tailed scimitar babbler *Jabouilleia danjoui*. This ecoregion also supports several pheasant and hornbill species that need intact, mature forests and are intolerant of human disturbance. Moreover, hornbills, as large frugivores, play an important role in dispersing the large seeds of some late-successional forest species, including trees and large lianas. The pheasants are Blyth's tragopan (*Tragopan blythii*), Temminck's tragopan (*Tragopan temminckii*), ring-necked pheasant (*Phasianus colchicus*), Lady Amherst's pheasant (*Chrysolophus amherstiae*), blood pheasant (*Ithaginis cruentus*), silver pheasant (*Lophura nycthemera*), and Siamese fireback (*Lophura diardi*). The hornbills are the plain-pouched hornbill (*Aceros subruficollis*), rufous-necked hornbill (*Aceros nipalensis*), great hornbill (*Buceros bicornis*), and brown hornbill (*Anorrhinus tickelli*). Because of their status as vulnerable indicator species and because of the important role they play in the dynamics of this ecosystem, these birds should receive special conservation attention.

Status and Threats

Current Status

Almost all forests in Vietnam have been cleared, and only very small fragments remain in this heavily populated area (MacKinnon 1997). Large areas of forest in the eastern Shan State of Myanmar, along the border with Laos and China, also have been cleared (IUCN 1991), although a few large blocks of habitat still remain in this part of the ecoregion. Forests in the hills of southern Yunnan are similarly degraded, although large tracts of monsoon forest remain in the Xishuangbanna Nature Reserve (2,418 km^2), a complex of several smaller reserves that supports small populations of Asian elephant and tiger.

There are twenty-seven protected areas that cover 15,948 km^2 (5 percent) of this vast ecoregion (table J74.3). But many of these are small (less than 100 km^2) reserves that lie in northern Vietnam and do not contribute significantly to conservation of the many space-sensitive, larger mammals and birds. The average size of protected areas is 590 km^2, but nineteen reserves (73 percent) are less than 500 km^2, and nine (35 percent) are less than 100 km^2. The large average is driven by the few large protected areas, many in Laos.

Types and Severity of Threats

The threats to biodiversity in this ecoregion stem from two main sources: land clearing for shifting cultivation, poppy cultivation, and logging; and hunting for food and income. Even though human population density is low in these northern highlands, shifting cultivation in steep and unproductive land accounts for large-scale forest clearing. Fallow lands take more than thirty years to regenerate into forest, although some ecosystem function as wildlife habitat is restored in a much shorter time. If degradation is extensive, then forest

TABLE J74.1. Endemic and Near-Endemic Mammal Species.

Family	Species
Loridae*	*Nycticebus intermedius**
Cercopithecidae	*Rhinopithecus avunculus*
Hylobatidae*	*Hylobates leucogenys**
Viverridae*	*Hemigalus owstoni**
Cervidae	*Muntiacus feae*
Cervidae*	*Muntiacus rooseveltorum**
Sciuridae	*Callosciurus quinquestriatus*
Muridae	*Dremomys gularis*
Muridae*	*Eothenomys olitor**

An asterisk signifies that the species' range is limited to this ecoregion.

TABLE J74.2. Endemic and Near-Endemic Bird Species.

Family	Common Name	Species
Timaliidae	Short-tailed scimitar-babbler	*Jabouilleia danjoui*

An asterisk signifies that the species' range is limited to this ecoregion.

TABLE J74.3. WCMC (1997) Protected Areas That Overlap with the Ecoregion.

Protected Area	Area (km²)	IUCN Category
Shwe u Daung	240	UA
Maymyo	110	UA
Taunggyi	20	UA
Nam Ha	790	VIII
Inle Lake	180	UA
Mong Pai Lake	170	PRO
Tay Con Linh	70	PRO
Muong Nhe	3,720	IV
Nui Hoang Lien	400	IV
Phou Dene Dinh	950	VIII
Nam Don	200	IV
Muong Phang	30	UA
Xuan Son	40	IV
Nam Et	2,700	VIII
Dao Ho Song Da	10	UA
Xuan Nha	580	IV
Sop Cop	60	IV
Den Hung	30	UA
Ba Vi	100	II
Thuong Tien	20	IV
Phou Loeuy	1,700	VIII
Nam Xam	850	VIII
Bu Huong	230	IV
Cuc Phuong [56]	110	II
Ngoc Trao [56]	10	UA
Xishuangbanna	2,418	UA
Doi Pha Chang [68, 72]	210	IV
Total	15,948	

Ecoregion numbers of protected areas that overlap with additional ecoregions are listed in brackets.

regeneration will not occur at all, and the vegetation consists of a subclimax of *Imperata* grasslands or the exotic weed *Eupatorium adenophorum* (MacKinnon 1987).

Hunting at unsustainable levels to supply a thriving market in China is a severe threat. And hunting is indiscriminate, targeting all species, from the large mammals to small birds, reptiles, amphibians, and invertebrates.

Priority Conservation Actions

Short-Term Conservation Actions (1–5 Years)

- Initiate a concerted transborder effort to stop the flow of wildlife through their common borders (this include joint patrols along the borders, effective law enforcement by agencies of each country, and training for border guards and customs officials).
- Conduct an intensive survey to locate remnant populations of *Rhinopithecus avunculus* and draw up appropriate species-specific conservation management plans.
- Reforest degraded areas.
- Seek ways to improve agricultural yield in permanent cultivated plots to put a stop to shifting cultivation.
- Reforest nonviable areas (with respect to biodiversity) with economic species (e.g., *Bambusa* spp.) to generate income for local communities, who otherwise need to exploit the remaining natural resources.

Longer-Term Conservation Actions (5–20 Years)

- Identify potential land corridors between fragmented Indo-forest areas and initiate a reforestation program.
- Conduct a long-term study and conservation program for *Rhinopithecus avunculus*. Training for local tertiary-level students and primatologists should be incorporated into this program.
- Reforest and manage degraded areas.
- Manage *Bambusa* spp. plantations, establishing small industries to produce paper, handicrafts, and canned bamboo shoots.

Focal Species for Conservation Action

Habitat specialists: serow (Boonratana and Le 1994; Boonratana 1999b; Davidson 1998; Showler et al. 1998), hornbills (Boonratana and Le 1994, Boonratana 1999b)

Endangered endemic species: Tonkin snub-nosed monkey (Boonratana and Le 1994, 1998a, 1998b; Boonratana 1999b)

Area-sensitive species: tiger (Davidson 1998), Asian elephant (Davidson 1998; Showler et al. 1998), gaur (Davidson 1998; Showler et al. 1998)

Top predator species: tiger (Davidson 1998), wild dog, clouded leopard

Other: red panda, particoloured squirrel, Indian civet, Owston palm civet (Boonratana and Le 1994), Tay Nguyen civet (Boonratana 1999b), pygmy loris (Boonratana 1999b), Francois' leaf monkey (Boonratana 1999b), Chinese pangolin (Boonratana 1999b; Davidson 1998), white-cheeked and yellow-cheeked gibbon (Davidson 1998; Showler et al. 1998), bear sp. (Davidson 1998)

Selected Conservation Contacts

Bangkok Bird Club
BirdLife-Indochina
Center for Conservation Biology, Thailand
FFI-Indochina
IUCN Lao PDRIUCN Vietnam
Wildlife Fund Thailand
WWF-Indochina
WWF-Lao PDR
WWF-Thailand

Justification for Ecoregion Delineation

In a previous analysis, MacKinnon (1997) included this region in three subunits: Indochina Biounit 10a, b, and c. The subtropical forests across the northern areas of Thailand, Laos, and Vietnam and in southern China extend across all three subunits. In keeping with our rules for defining ecoregions, we used the distribution of these subtropical broadleaf forests as depicted in MacKinnon's (1997) map of the original extent of vegetation within the Northern Indochina Subtropical Forests [74].

Prepared by Chris Carpenter, Ramesh Boonratana, and Philip Rundel

Ecoregion Number:	**75**
Ecoregion Name:	**South China–Vietnam Subtropical Evergreen Forests**
Bioregion:	**Indochina**
Biome:	**Tropical and Subtropical Moist Broadleaf Forests**
Political Units:	**Vietnam, China**
Ecoregion Size:	**37,500 km²**
Biological Distinctiveness:	**Globally Outstanding**
Conservation Status:	**Endangered**
Conservation Assessment:	**I**

This large ecoregion extends only partially into the region of analysis considered here, with the majority of the area extending into China. Because the ecoregion lies within two of the most densely populated countries in Asia, very little natural habitat now remains, having been cleared throughout the last several centuries for agriculture.

Location and General Description

This large ecoregion extends from northern Vietnam into southeastern China, including Hainan Island. Only the Vietnamese portion of this ecoregion is considered here.

Vietnam northeast of the Red River forms a part of the South China Platform and is geologically separate from the remainder of southeast Asia. The sediments here consist of a combination of exposed ancient metamorphic basement rock and marine sediments deposited in the late Paleozoic and early Triassic. Many of these sediments are limestones that have been uplifted and weathered to form an extensive karst landscape with steep topography. The highest peak in this region of Vietnam is Tsi Con Ling at 2,531 m, but most of the area is of only moderate relief.

Annual rainfall in this ecoregion varies from a low of about 1,800 mm in the Red River Basin to a high of about 2,850 mm near the Chinese border, where there are only two months with less than 50 mm rainfall and no totally dry months. Temperature regimes in the north have a strong seasonality, with cool winter conditions and hot and humid summers.

Very little pristine forest remains within this ecoregion. Lowland forests growing on limestone substrates commonly reach only 15–20 m in height but grow to 30–35 m on favorable sites. Few emergent trees in most stands suggest past cutting of larger trees. Dominance is mixed among a variety of tree species, with the Lauraceae, Fagaceae, and Meliaceae particularly important. This region shows strong floristic relationships with areas to the north in China, and regional endemism is high.

Biodiversity Features

A complete biodiversity database is being compiled for this ecoregion as part of a large analysis that includes China. However, the area of the ecoregion that extends into northern Vietnam, and thus into region of analysis, contains four near-endemic mammals (table J75.1), an indicator of the biodiversity and endemism levels that can be expected from this ecoregion.

TABLE J75.1. Endemic and Near-Endemic Mammal Species.

Family	Species
Cercopithecidae	Trachypithecus francoisi
Cercopithecidae	Pygathrix avunculus
Viverridae	Cynogale lowei
Muridae	Dremomys gularis

An asterisk signifies that the species' range is limited to this ecoregion.

Status and Threats

Current Status

Almost all of the lowland forests of the northern part of this ecoregion, which falls within a heavily populated section of Vietnam, have been cleared (MacKinnon 1997). In Myanmar, in eastern Shan State along the border with Laos and China, very large areas of forest have been cleared (IUCN 1991), but a few large blocks of habitat remain.

There are sixteen protected areas (in Vietnam only) that cover 1,450 km² of this ecoregion (table J75.2). Most of these protected areas are smaller than 100 km², and the largest is only 260 km².

Types and Severity of Threats

Much of the natural habitat has been altered by shifting cultivation and logging. Hunting for the wildlife trade is also a significant threat to conservation of biological diversity.

Priority Conservation Actions

Short-Term Conservation Actions (1–5 Years)

- Conduct an intensive survey to locate remnant populations of *Rhinopithecus avunculus* and draw up appropriate species-specific conservation management plans.
- Implement effective law enforcement with regard to wildlife trade, especially along major border crossings from

TABLE J75.2. WCMC (1997) Protected Areas That Overlap with the Ecoregion.

Protected Area	Area (km²)	IUCN Category
Pac Bo	50	UA
Trung Khanh	170	IV
Nui Pia Oac	110	IV
Ba Be	170	II
Bac Son	20	IV
Tan Trao	20	UA
Huu Lien	20	IV
Ai Chi Lang	30	UA
Nui Tam Dao	260	IV
Cam Son	120	?
Unnamed	80	?
Unnamed	70	?
Con Son–Kiep Bac	20	UA
Unnamed	70	?
Bai Chay	60	UA
category Ba	180	II
Total	1,450	

Ecoregion numbers of protected areas that overlap with additional ecoregions are listed in brackets.

Vietnam into China. (This will also entail training border guards and custom officials.)
- Reforest degraded areas.
- Seek ways to improve agricultural yield in permanent cultivated plots to put a stop to shifting cultivation.
- Reforest nonviable areas (with respect to biodiversity) with economic species (e.g., *Bambusa* spp.) to generate income for local communities, which otherwise need to exploit the remaining natural resources.

Longer-Term Conservation Actions (5–20 Years)
- Implement a long-term study and conservation program for *Rhinopithecus avunculus*. Training for local tertiary-level students and primatologists should be incorporated into this program.
- Reforest and manage degraded areas.
- Manage *Bambusa* spp. plantations, establishing small industries to produce paper, handicrafts, and canned bamboo shoots.

Focal Species for Conservation Action

Habitat specialists: serow (Boonratana and Le 1994; Boonratana 1999b), hornbills

Endangered endemic species: Hainan black-crested gibbon, Tonkin snub-nosed monkey (Boonratana and Le 1994, 1998a, 1998b; Boonratana 1999b)

Other: Owston palm civet (Boonratana and Le 1994), Tay Nguyen civet (Boonratana 1999b), pygmy loris (Boonratana 1999b), Francois' leaf monkey (Boonratana 1999b), Chinese pangolin (Boonratana 1999b)

Selected Conservation Contacts

BirdLife-Indochina
FFI-Indochina
IUCN Vietnam
WWF-China
WWF-Indochina

Justification for Ecoregion Delineation

We included MacKinnon's Tropical Southern China subunit (06a), with biogeographic affinities with southern China, as part of the larger South China–Vietnam Subtropical Evergreen Forests [75], which extends from southern China. However, instead of following the main tributary of the Red River, as MacKinnon does, we used the upper Red River (Yuan Jiang) as the boundary to be consistent with the boundary between two zoogeographic divisions in China: the Fujian Guangdong coast and South Yunnan Mountains (Yongzu et al. 1997). These two zoogeographic divisions largely coincide with the South China–Vietnam Subtropical Evergreen Forests [75] and the Northern Triangle Subtropical Forests [73] and intersect the Vietnamese border at the upper Red River tributary.

Prepared by Eric Wikramanayake, Ramesh Boonratana, and Philip Rundel

Ecoregion Number:	**76**
Ecoregion Name:	**Northern Triangle Temperate Forests**
Bioregion:	**Indochina**
Biome:	**Temperate Broadleaf and Mixed Forests**
Political Units:	**Myanmar**
Ecoregion Size:	**10,700 km²**
Biological Distinctiveness:	**Globally Outstanding**
Conservation Status:	**Relatively Intact**
Conservation Assessment:	**III**

The Northern Triangle Temperate Forests [76] ecoregion lies in the extreme northern area of the Golden Triangle of Myanmar. The region is scientifically unexplored, and the biological information, especially of its flora, is still based on the early, pioneering exploration done by Kingdon-Ward (1921, 1930, 1952). There have been no detailed scientific surveys in this area since then, and current assessments of its biodiversity probably are underestimated. In all probability this ecoregion harbors many more species than are now attributed to it. Satellite imagery indicates that the ecoregion is still largely clothed in intact forests and presents a rare opportunity to conserve large landscapes that will support the ecological processes and the biodiversity within this eastern Himalayan ecosystem.

Location and General Description

The mountains originated more than 40 million years ago, when the collision between the drifting Deccan Plateau and the northern Laurasian mainland created the Himalayan Mountains, including these mountains in the Golden Triangle. Therefore, the mountains are young and unweathered; the terrain is rugged and dissected, with north-south–oriented ranges that reach south, toward the central plains of Myanmar. The peaks along this range rise steeply to attain heights of more than 3,000 m. The Chindwin, Mali Hka, and N'mai Hka rivers originate in these mountains and flow south to converge in the lower reaches to form the Irrawaddy River.

Biogeographically, the mountains are an ecotone of the Assam-Indian, Eastern Himalayan, Indo-Malayan, and Chinese floras. Gondwana-era relicts that rafted north on the Deccan Plateau have also taken refuge here. Therefore, floristically, the ecoregion is extremely diverse. And the complex topography and moist conditions produced by the southwestern monsoon funneled in from the Bay of Bengal have provided the localized climatic variations that promote endemism.

Floristically, the ecoregion is very similar to the middle mountain forests of the Eastern Himalaya. The temperate forests lie between 1,830 and 2,700 m. At lower elevations the forest transitions into the subtropical forests and, in the upper elevations, into the sub-alpine conifer forests. The temperate forests are characterized by *Alnus nepalensis, Betula cylindrostachya, Castanopsis* spp., *Schima* spp., *Callophylus* spp., *Michelia* spp., and *Bucklandia populnea* (WWF and IUCN 1995; Lwin 1995). Rich epiphytic rhododendron shrub vegetation is common. At higher elevations rhododendrons, especially *Rhododendron decorum, R. magnificum, R.*

bullatum, *R. crinitum*, and *R. neriiflorum* become a dominant component in the vegetation.

At elevations above 2,100 m, the broadleaf forests transition into a mixed forest, where species of *Quercus*, *Magnolia*, *Acer*, *Prunus*, *Ilex*, and *Rhododendron* are mixed with *Picea brachytyla*, *Tsuga dumosa*, *Larix griffithiana*, and *Taiwania flousiana*. The rich, diverse shrub flora persists here but is characterized by species of *Acer*, *Berberis*, *Clethra*, *Enkianthus*, *Euonymus*, *Hydrangea*, *Photinia*, *Rhododendron*, *Rubus*, *Betula*, and *Sorbus* (WWF and IUCN 1995; Blower n.d.).

Biodiversity Features

Ninety-one mammal species are known from this ecoregion, but further surveys may well reveal the presence of additional species. However, there are several other threatened species in the ecoregion's assemblage that deserve conservation attention, namely the tiger (*Panthera tigris*), takin (*Budorcas taxicolor*), clouded leopard (*Pardofelis nebulosa*), red panda (*Ailurus fulgens*), wild dog (*Cuon alpinus*), Asiatic black bear (*Ursus thibetanus*), stump-tailed macaque (*Macaca arctoides*), capped leaf monkey (*Semnopithecus pileatus*), red goral (*Naemorhedus baileyi*), great Indian civet (*Viverra zibetha*), back-striped weasel (*Mustela strigidorsa*), Irrawaddy squirrel (*Callosciurus pygerythrus*), and particolored squirrel (*Hylopetes alboniger*).

Of the 365 birds known from this ecoregion, the rusty-bellied shortwing (*Brachypteryx hyperythra*) is the only ecoregional endemic (table J76.1). But there are several species that need mature habitats, have low tolerances for disturbances that are indicators of habitat integrity, and deserve conservation attention. These species include the Oriental pied-hornbill (*Anthracoceros albirostris*), wreathed hornbill (*Aceros undulatus*), Blyth's tragopan (*Tragopan blythii*), Himalayan flameback (*Dinopium shorii*), and Sclater's monal (*Lophophorus sclateri*). The ecoregion also overlaps with the Eastern Himalayas EBA (130), which contains seven restricted-range species (Stattersfield et al. 1998).

But the conservation priority of this ecoregion goes beyond the species lists. A critical component of the Himalayan ecosystems is the dependence of ecological processes—from seasonal bird and mammal migrations to watershed protection—on altitudinal connectivity and the intactness of ecosystems that are stratified by elevation. Loss of habitat or even large-scale degradation can have far-reaching consequences. Birds that migrate up and down the steep slopes will lose staging and feeding habitat, thus disrupting their migrations. Loss of ground cover in the high elevations can result in erosion on steep slopes, with the consequences being manifested as floods and silting far away, in lowland plains and mangroves in the river deltas.

TABLE J76.1. Endemic and Near-Endemic Bird Species.

Family	Common Name	Species
Turdidae	Rusty-bellied shortwing	*Brachypteryx hyperythra*

An asterisk signifies that the species' range is limited to this ecoregion.

Status and Threats

Current Status

Because of the remoteness and inaccessibility of this high-elevation ecoregion that sits on steep, rugged mountains, most of the habitat in this ecoregion is still intact. However, it receives no formal protection.

Types and Severity of Threats

Throughout Myanmar, the hill tribes who still practice shifting cultivation are being pushed further and further into marginal, steeper lands for a variety of reasons, including logging in the lower elevations that were the traditional agricultural areas. The consequent clearing of new forests for shifting cultivation and poppy cultivation will result in severe erosion and loss of habitat and biological diversity.

Priority Conservation Actions

Short-Term Conservation Actions (1–5 Years)

- Establish protected areas in this ecoregion.
- Develop management plans for these protected areas.
- Conduct flora and fauna surveys of the region and disseminate results to conservation organizations.
- Stop illegal logging and wildlife trade.

Longer-Term Conservation Actions (5–20 Years)

- The protected area system mandates that 5 percent of land area be strictly dedicated to conservation. Implement and enforce this mandate.
- Place all protected areas under effective conservation and management plans.
- The Nature and Wildlife Directorate should be the sole authority to manage the protected areas, with sufficient personnel and funding.

Focal Species for Conservation Action

Habitat specialists: wreathed hornbill

Endangered endemic species: Gongshan muntjac, rusty-bellied shortwing

Area-sensitive species: tiger (Franklin et al. 1999; Tilson et al. 1994; Dinerstein et al. 1997; Seidensticker et al. 1999)

Top predator species: tiger (Franklin et al. 1999; Tilson et al. 1994; Dinerstein et al. 1997; Seidensticker et al. 1999)

Other: Fea's muntjac, red panda, takin, red goral, Asiatic black bear, Blyth's tragopan, Himalayan flameback, Sclater's monal

Selected Conservation Contacts

Myanmar Forest Department
WCS-Myanmar

Justification for Ecoregion Delineation

In a previous analysis of conservation units across the Indo-Malayan region, MacKinnon (1997) identified a subunit (09b) that includes the tropical and subtropical moist broadleaf forests in the Kachin and upper Chindwin areas of northern Myanmar. In keeping with our rules for defining ecoregions, we represented the temperate broadleaf forests in this subunit with the Northern Triangle Temperate Forests [76].

Prepared by Eric Wikramanayake

Ecoregion Number:	**77**
Ecoregion Name:	**Northeast India–Myanmar Pine Forests**
Bioregion:	**Indochina**
Biome:	**Tropical and Subtropical Coniferous Forests**
Political Units:	**India, Myanmar**
Ecoregion Size:	**9,700 km²**
Biological Distinctiveness:	**Regionally Outstanding**
Conservation Status:	**Endangered**
Conservation Assessment:	**II**

The Northeast India–Myanmar Pine Forests [77] ecoregion is one of only four tropical or subtropical conifer forest ecoregions in the Indo-Pacific region. All of these ecoregions contain less biodiversity than the forests that surround them. However, they contain processes and species unique to these ecosystems. This ecoregion contains moderate levels of biodiversity but remains largely intact, providing opportunities to conserve and protect this ecoregion's biodiversity into the future.

Location and General Description

These forests are found in the north-south Burmese-Java Arc. The Arc is formed by the parallel folded mountain ranges that culminate in the Himalayas in the north. Moving south are the mountain ranges of Patkoi, Lushai Hills, Naga Hills, Manipur, and the Chin Hills. The outer southwestern fringe of mountain ranges forming the Arc is the Arakan Yomas, the southern continuation of the folded mountain ranges branching off from the Himalayas. Geologically the ecoregion has gley and black slates. Dark-colored serpentine and gabbro also are found interstratified within the shales.

The pine forest is found between 1,500 and 2,500 m. Several pine species occur. In the lower elevations *P. merkusii* is the dominant species, occasionally associated with dipterocarps. *Pinus insularis* and *Pinus excelsa* are found at higher altitudes. These species often are associated with numerous broadleaf species such as *Tsuga*, *Picea*, *Acer*, and *Quercus*. *Rhododendron*, *Ilex*, *Prunus*, and *Arundinaria* bamboo occur in the understory.

Biodiversity Features

In the late 1950s the Burma Wildlife Survey team found few species except sambar (*Cervus unicolor*), barking deer (*Muntiacus muntjac*), wild boar (*Sus scrofa*), and Asiatic black bear (*Selenarctos thibetanus*) in this region. Serow (*Capricornis sumatrensis*) was plentiful. Large squirrel (*Rafuta species*), brown squirrel, small chipmunk, and civet also occur there. The dearth of wildlife was caused by deforestation from extensive shifting cultivation resulting in the barren hillsides. There are no endemic mammals in this ecoregion.

TABLE J77.1. Endemic and Near-Endemic Bird Species.

Family	Common Name	Species
Timaliidae	Striped laughingthrush	*Garrulax virgatus*
Timaliidae	Brown-capped laughingthrush	*Garrulax austeni*

An asterisk signifies that the species' range is limited to this ecoregion.

The avifauna is also less diverse than the surrounding ecoregions. According to the Burma Wildlife Survey, conducted in the late 1950s, several distinctive birds were found in the region. These include the silver-breasted broadbill (*Serilophus lunatus*), white-naped yuhina (*Yuhina bakeri*), rufous-vented tit (*Parus rubidiventris saramatii*), stripe-throated yuhina (*Yuhina gularis*), babblers (*Timaliidae*), grey-sided laughingthrush (*Garrulax caerulatus*), rufous-chinned laughingthrush (*Garrulax rufogularis*), striated laughingthrush (*Garrulax striatus*), cochoa (*Cochoa spp.*), beautiful nuthatch (*Sitta formosa*), sultan tit (*Melanochlora sultana*), leafbird (*Chloropsis spp.*), and white-browed fulvetta (*Alcippe vinipectus*). Shelduck and bare-headed geese were seen in tens and hundreds along the Chindwin River. Two additional laughingthrush species are considered near endemics (table J77.1). More recent information on the status of the region's wildlife is not known.

Status and Threats

Current Status

Until now this ecoregion has remained largely intact because there has been little commercial forestry. Its wildlife has also been spared from commercial wildlife hunting or poaching. Local people use the forest for cooking and building materials and hunt wildlife for food. However, only one small protected area exists in this ecoregion, and its contribution to wildlife protection is minimal (table J77.2).

Types and Severity of Threats

The primary threat to the integrity of this habitat is from shifting cultivation. This practice denudes the hillsides of tress, increases erosion and sedimentation of rivers, and deprives wildlife of their habitats.

Priority Conservation Actions

Short-Term Conservation Actions (1–5 Years)

- Establish additional protected areas in this ecoregion.
- Develop management plans for these protected areas.

TABLE J77.2. WCMC (1997) Protected Areas That Overlap with the Ecoregion.

Protected Area	Area (km²)	IUCN Category
Fakim	4	IV

Ecoregion numbers of protected areas that overlap with additional ecoregions are listed in brackets.

Longer-Term Conservation Actions (5–20 Years)

- It is essential to stop all land appropriation from within the protected areas.
- The status of all protected areas should be thoroughly reevaluated, and those that have been wholly converted to other land uses should be traded for more pristine areas.
- The protected area system mandates that 5 percent of land area be strictly dedicated to conservation. Implement and enforce this mandate.
- All protected areas should be placed under effective conservation and management plans.
- The Nature and Wildlife Directorate should be the sole authority to manage the protected areas, with sufficient personnel and funding.

Focal Species for Conservation Action

Endangered endemic species: striped laughingthrush, brown-capped laughingthrush

Other: Asiatic black bear, serow

Selected Conservation Contacts

Myanmar Forest Department

WCS-Myanmar

Justification for Ecoregion Delineation

The westernmost subunit (09c) represents a transition zone between the Indian Subcontinent and Indochina bioregions and is made up of forest formations belonging to the tropical moist forests biome. We included the montane forests of the rugged and highly dissected Chin Hills and Arakan Yomas in a distinct ecoregion: Chin Hills–Arakan Yoma Montane Rain Forests [48]. The lower-elevation tropical evergreen moist forests were included in the Mizoram-Manipur-Kachin Rain Forests [50]. The pine forests were placed in the Northeast India–Myanmar Pine Forests [77].

This region does not correspond well to Udvardy's biogeographic provinces. The Myanmar Coastal Rain Forests [49] cover Udvardy's Burman rain forest, southwestern portion of the Thai monsoon forest, and western portion of the Indochinese rain forest. The Irrawaddy Moist Deciduous Forests [67], Irrawaddy Dry Forests [71], Chin Hills–Arakan Yoma Montane Rain Forests [48], Northeast India–Myanmar Pine Forests [77], and Mizoram-Manipur-Kachin Rain Forests [50] roughly correspond to Udvardy's Burman rain forest and Burma monsoon forest.

Prepared by U Saw Han and Saw Tun Khaing

Ecoregion Number:	**78**
Ecoregion Name:	**Myanmar Coastal Mangroves**
Bioregion:	**Indochina**
Biome:	**Mangroves**
Political Units:	**India, Myanmar, Thailand, Malaysia**
Ecoregion Size:	**21,900 km^2**
Biological Distinctiveness:	**Regionally Outstanding**
Conservation Status:	**Endangered**
Conservation Assessment:	**II**

The Myanmar Coastal Mangroves [78] are some of the most degraded or destroyed mangrove systems in the Indo-Pacific. The sedimentation rate of the Irrawaddy River is the fifth highest in the world. This is largely because of the deforestation that has occurred throughout central Myanmar. The mangroves have also been overexploited from forestry, agriculture, aquaculture, and development projects. The wild species have been severely reduced but hang on in isolated pockets.

Location and General Description

Myanmar Coastal Mangroves [78] ecoregion is found in the Irrawaddy delta. The mouth of the Irrawaddy River was some 170 miles inland near Prome 300,000 years ago. On the islands of Twante, Myaungmya, and Bassein, lateritic ridges stood above the water. The delta is composed largely of alluvium, and a large area is occupied by volcanic rocks.

The mangrove flora consists of three separate regions: the Rakhine mangroves, Irrawaddy mangroves, and Taninthayi mangroves. The Rakhine mangroves are made up primarily of *Rhizophora mucronata, R. candelria, Sonneratia* spp., *Kandelia rheedeii, Bruguiera* spp., *Xylocarpus granatum, X. moluccensis, Nipa fruticans,* and *Phoenix paludosa*. The Irrawaddy mangroves consist of *Rhizophora mucronata, R. conjugata, Bruguiera parviflora, B. gymnorhiza, B. cylindrica, Heritiera formes, Sonneratia apetala, S. griffithii, S. caseolaris, Xylocarpus granatum, X. molluccensis, Ceiops roxburghiana, C. mimosoides, Avicennia officinalis, Kanddelia rheedii,* and *Excoecaria agallocha*. Finally, the Taninthayi mangroves contain *Rhizophora* spp., *Sonneratia caseolaris, Ceriops tegal, Xyloxarpus granatum, Avicennia officinalis,* and *Bruguiera* spp.

Biodiversity Features

Several mammal species occur in this ecoregion but in small and scattered populations. Wild elephants (*Elephas maximus*) are represented by a small population of approximately 150 animals. Rakhine is one of the last regions where Asian elephants still roam wild. During the dry summer elephants come down from the mountains to the mangroves to drink salt water (Thet Tun, pers. comm., 1999). The population estimate for Rakhine State in 1990–1991 was between 750 and 1,100, with the country total between 4,115 and 4,639 (U Ga 2000)

The tiger used to be plentiful all over the country some forty years ago but has been persecuted to the point of extirpation. There are at most 150 tigers remaining in Myanmar, and it is unknown how many, if any, use the dwindling mangroves (Saw Tun Khaing, pers. comm., 2000). Tiger (*Panthera tigris*), leopard (*Panthera pardus*), wild dog (*Cuon apinus*), and otter (*Lutra* spp.) were reported by Salter in 1982 but probably have been extirpated or survive only in low numbers. As a result, sambar (*Cervus unicolor*), hog deer (*Cervus porcinus*), mouse deer (*Tragulus javanicus*), barking deer (*Muntiacus muntjak*), tapir (*Tapirus malayanus*), and wild boar (*Sus scrofa*), prey species for many predators, are abundant in the reserved forests. According to Salter (1982), the Malayan tapir (*Tapirus indicus*) was rapidly decreasing to the point of extinction in Myanmar. However, they still survive in parts of the Taninthayi mangroves.

Bird life in the mangroves is rich in migrants and resident waterbirds. A few resident waterbirds include the oriental darter (*Anhinga melanogaster*), little cormorant (*Phalacrocorax nigers*), reef heron (*Egretta sacra*), dusky gray heron (*Ardea sumatrana*), ruddy shelduck (*Tadorna ferruginea*), bronze-winged jacana (*Metopidius indicus*), lesser sand plover (*Charadrius mongolus*), great stone plover (*Esacus magnirostris*), black-winged stilt (*Himantopus himantopus*), spotted greenshank (*Tringa guttifer*), lesser black-back gull (*Larus fuscus*), and common moorhen (*Gallinula chloropus*). One of the most highly threatened residents is the edible-nest swiftlet (*Collocalia fuciphaga*), found inhabiting rocky limestone caves. The nests, which are an expensive delicacy, are highly valued and collected.

The southern part of the delta contains the last population of crocodiles (*Crocodylus porosus*) in the ecoregion. The river terrapin (*Batagur baska*) has also been reduced to a few small populations on offshore islands as a result of egg overcollection.

Status and Threats

Current Status

Mangrove forests are subject to severe degradation because there is no clear-cut land-use system. Forestlands have been converted to agriculture and other development activities, not only in this ecoregion but throughout Myanmar. Irrawaddy is one of the most heavily silted rivers in the world. The sedimentation rate was 299 million tons per year, and it ranked fifth behind the Yellow, Ganges, Amazon, and Mississippi rivers in silt deposition. Today the sedimentation rate in getting worse as deforestation and agricultural erosion continue. If the situation between 1977 and 1986 is maintained, it is estimated that all the mangrove forests will disappear in fifty years.

A new Forest Policy, Forest Act, and Wildlife Protection Act are in force today. However, implementation of conservation and protection activities is poor, with a shortage of staff to police and monitor the few protected areas proposed (table J78.1). This is primarily because of an inadequate budget.

Types and Severity of Threats

The entire ecoregion is under severe threat of conversion and illegal settlements. Rice yields decline on reclaimed mangroves within a few years, and new areas are cut. Fertilizer on reclaimed mangroves causes more harm than good because of the accumulation of acid sulfate over time. Mangroves are being intensively cut for firewood, charcoal burning, and nontimber produce. Mangroves are increasingly converted to fish and prawn aquaculture to meet the demand of an increasing population.

Wildlife poaching is rampant. Wild elephant, tiger, sun bear, monkey species, wild boar, banteng, and sambar are being intensively hunted. Tigers are often killed for the bones and skins for sale to wildlife traders in Thailand and China. There is no control over the extraction of forest resources despite forest reserves having legal protected status. Illegal logging is more prevalent than legal logging concessions.

Priority Conservation Actions

Short-Term Conservation Actions (1–5 Years)

- Proper demarcation of the protected areas and data collection on land and resources is a top priority.
- Develop protected area management plans.
- Obtain larger financing of the Wildlife Department to enable effective protection of the nature reserves through monitoring and patrolling.
- Stop illegal logging and wildlife poaching.

Longer-Term Conservation Actions (5–20 Years)

- Restore highly degraded forests.
- Stem the tide of illegal cutting and poaching of wildlife and key species.
- Develop management plans for a ten-year duration, like the Forest Management Plans, for the Elephant Range (Rakhine State), Lampi Marine National Park (Taninthayi Division), and Meinmahla Kyun Wildlife Sanctuary (Irrawaddy Division).

Focal Species for Conservation Action

Wide-ranging species: tiger (Franklin et al. 1999; Tilson et al. 1994; Dinerstein et al. 1997; Seidensticker et al. 1999), Asian elephant (Santiapillai and Jackson 1990; Santiapillai 1997)

Top predator species: tiger (Franklin et al. 1999; Tilson et al. 1994; Dinerstein et al. 1997; Seidensticker et al. 1999), leopard

Migratory species: waterbirds

Other: crocodiles, Malayan tapir, deer species, river terrapin

Selected Conservation Contacts

Myanmar Forest Department

WCS-Myanmar

TABLE J78.1. WCMC (1997) Protected Areas That Overlap with the Ecoregion.

Protected Area	Area (km²)	IUCN Category
Letkokkon	4	PRO
Meinmahla Kyun	120	PRO
Kadonlay Kyun	1	PRO
Total	125	

Ecoregion numbers of protected areas that overlap with additional ecoregions are listed in brackets.

Justification for Ecoregion Delineation

MacKinnon divided the Indochina bioregion into fifteen biogeographic subunits nested within seven major biounits. Each of these biounits and biogeographic subunits contains a mix of biomes. For instance, the Burmese coast biounit (04) is made up mostly of tropical wet evergreen forests but also includes tropical montane evergreen forests, semi-evergreen rain forests, dry dipterocarp forests, and large areas of freshwater swamp forests and mangroves in the Irrawaddy River delta. We assigned these distinct habitat types to separate ecoregions. Thus MacKinnon's Burmese coast biounit is represented by four ecoregions: the Myanmar Coastal Rain Forests [49], Irrawaddy Freshwater Swamp Forests [63], Myanmar Coastal Mangroves [78], and Mizoram-Manipur-Kachin Rain Forests [50].

This region does not correspond well to Udvardy's biogeographic provinces. The Irrawaddy Freshwater Swamp Forests [63] and Myanmar Coastal Mangroves [78] are contained within Udvardy's Burman rain forest.

Prepared by U Saw Han and Saw Tun Khaing

Ecoregion Number:	**79**
Ecoregion Name:	**Indochina Mangroves**
Bioregion:	**Indochina**
Biome:	**Mangroves**
Political Units:	**Vietnam, Cambodia, Thailand**
Ecoregion Size:	**26,800 km²**
Biological Distinctiveness:	**Regionally Outstanding**
Conservation Status:	**Endangered**
Conservation Assessment:	**II**

Among the most diverse and extensive mangrove ecosystems in the world, this ecoregion provides extremely important habitat for some of the world's rarest waterbirds. The largest block of Indochina Mangroves [79] in the Mekong River delta suffered large-scale habitat loss from defoliants sprayed during the Vietnam War.

Location and General Description

Mangrove forests occur in coastal areas of regular flooding by tidal or brackish water and develop on saline gleysols. The extent of mangroves in coastal areas of Thailand, Cambodia, and Vietnam was once high, but much of this area has been destroyed. Extensive mangrove forests once occurred around Pattaya in Thailand and in the areas of Veal Renh and Kompong Som Bays in Cambodia. The absence of more extensive mangrove stands in Cambodia is strongly related to the rocky coastline and lack of major estuaries or river deltas. In Vietnam, the largest area of remaining mangroves is around Camou Point at the southern tip of Vietnam, with smaller areas in the Mekong delta region, in south central Vietnam around Cam Ranh Bay, and in northern Vietnam in the Red River delta area. The central coast of Vietnam is largely free of mangroves because of the exposed coastline, absence of major river deltas, and low tidal fluctuations in this area. Far more extensive stands of mangroves once occurred around the Red River delta in northern Vietnam. The extensive military use of defoliants and napalm during the Vietnam War (1962–1972) destroyed a major part of mangrove forests in southern Vietnam, but these areas are slowly recovering under active reforestation programs today.

Mangrove diversity in the Indochina Mangroves [79] ecoregion is high, with the presence of approximately 60 percent of the mangrove species known from anywhere in south and southeast Asia and Indonesia. The most diverse mangrove communities occur in areas that are inundated at high tide but are otherwise influenced by freshwater flows. Mangrove forests in the Red River delta and associated estuaries and mud flats have lower diversity than mangrove habitats in the south. This low mangrove diversity in the Red River delta area is the result of a combination of cooler growing conditions and a longer and more intense period of human impact.

Mangrove forests typically exhibit strong patterns of zonation. The pioneer species along the open coastline is typically *Avicennia alba*. Next along a gradient of decreasing exposure and submergence by sea water are *Rhizophora apiculata* and *Brugiera parviflora*, which become established after five or six years and grow to replace *Avicennia* after about twenty years. Higher ground subject to conditions of brackish water rather than seawater is dominated by *Avicennia officinalis*, *Sonneratia caseolaris*, *Nypa fruticans*, and *Phoenix paludosa*.

Biodiversity Features

There are no endemic mammals in the ecoregion, but many species are known to use mangroves, including the tiger (*Panthera tigris*), tapir (*Tapirus indicus*), and siamang (*Hylobates syndactylus*). This ecoregion overlaps with a Level I TCU (Dinerstein et al. 1997).

Numerous waterbirds use the remaining parts of these mangroves, and many of them are endangered. Included in this assemblage are the lesser adjutant (*Leptoptilos javanicus*), Storm's stork (*Ciconia stormi*), white-winged wood duck (*Cairina scutulata*), and spot-billed pelican (*Pelicanus philippensis*).

There are several reptile species of conservation significance in this ecoregion, including the monitor lizard (*Varanus salvator*), the false gavial (*Tomistoma schlegeli*), and the estuarine crocodile (*Crocodylus porosus*). The Mekong delta supports a valuable fishery, especially for shrimp (MacKinnon 1997).

Status and Threats

Current Status

This ecoregion is highly threatened in nearly every site where it occurs. About half of the mangroves in southern Vietnam were destroyed by Agent Orange, tank movements, and bombing during the war. Since then, however, the government has launched a large-scale reforestation program. Although protected areas have been created to conserve these mangroves—seven small protected areas (average size of only 117 km²) cover a mere 820 km² (3 percent) of the ecoregion—the majority of the ecoregion is threatened by a multitude of human activities (IUCN 1991) (table J79.1).

Mangrove forests often are treated as wasteland to be cleared for development (Spalding et al. 1997; Cubitt and

Stewart-Cox 1995). In Thailand, large areas of the ecoregion have been logged, primarily to produce charcoal to supply the domestic market and markets in Malaysia, Singapore, and Hong Kong (Spalding et al. 1997). Fifty percent of the mangrove habitat in Thailand was lost between 1975 and 1991. The species most heavily exploited for charcoal are *Rhizophora apiculata*, *R. mucronata*, *Avicennia marina*, and *Xylocarpus* spp. (FAO 1981). Thailand's mangroves are also severely affected by prawn farming.

Types and Severity of Threats

In addition to exploitation for the domestic and international commercial markets, trees are cut or lopped for domestic consumption as fuelwood (Spalding et al. 1997). In Vietnam and in Thailand, large areas are cleared for aquaculture, salt ponds, and agriculture (Spalding et al. 1997). Poaching and illegal trade of animal products are another important threat, especially to estuarine crocodiles and monitor lizards. Fishing with explosives and trawlers with drag-nets has also caused extensive damage to this sensitive ecosystem (IUCN 1991).

Priority Conservation Actions

With the rapid pace of mangrove destruction, quick conservation action is urgently needed to protect the remaining fragments of habitat. However, steps should be taken to lay the groundwork for longer-term conservation as well.

Short-Term Conservation Actions (1–5 Years)

- Stop further destruction of sensitive mangrove habitats.
- Reforest degraded areas.
- Restrict the expansion aquaculture and salt ponds and manage existing ones.
- Curb the use of unsustainable fishing practices in the mangrove areas (e.g., trawling nets).

Longer-Term Conservation Actions (5–20 Years)

- Reforest degraded areas.
- Manage reforested and forested areas.

Focal Species for Conservation Action

Top predator species: Estuarine crocodile

Other: Silvered leaf monkey

TABLE J79.1. WCMC (1997) Protected Areas That Overlap with the Ecoregion.

Protected Area	Area (km²)	IUCN Category
Do Son	30	UA
Unnamed	60	?
Xuan Chuy	40	?
Con Dao	190	II
Botum-Sakor [55]	250	II
Dong Peng [55]	200	VIII
Peam Krasop [55]	50	IV
Total	820	

Ecoregion numbers of protected areas that overlap with additional ecoregions are listed in brackets.

Selected Conservation Contacts

Bangkok Bird Club
BirdLife-Indochina
Cambodia Ministry of Environment
Cambodia Wildlife Protection Office
FFI-Indochina
IUCN Vietnam
Malaysian Nature Society
Sahabat Alam Malaysia
Thailand Center for Conservation Biology
WCS-Cambodia
Wildlife Fund Thailand
WWF-Cambodia
WWF-Indochina
WWF-Malaysia
WWF-Thailand

Justification for Ecoregion Delineation

The ecoregion was delineated on the basis of the distribution of mangrove forests in MacKinnon's (1997) biounit 05. Spalding et al. (1997) was used to validate and supplement the information in MacKinnon.

Prepared by Eric Wikramanayake, Ramesh Boonratana, Philip Rundel, and Nantiya Aggimarangsee

Ecoregion Number:	**80**
Ecoregion Name:	**Peninsular Malaysian Rain Forests**
Bioregion:	**Sunda Shelf and Philippines**
Biome:	**Tropical and Subtropical Moist Broadleaf Forests**
Political Units:	**Thailand, Malaysia**
Ecoregion Size:	**125,000 km²**
Biological Distinctiveness:	**Globally Outstanding**
Conservation Status:	**Vulnerable**
Conservation Assessment:	**I**

The Peninsular Malaysian Rain Forests [80] ecoregion, with 195 mammal species, has the second most mammal species in the Indo-Pacific, behind the Borneo Lowland Rain Forests [96]. Yet most of the wide-ranging or top carnivore species lead a tenuous existence in these biologically outstanding forests. The tiger, Asian elephant, Sumatran rhinoceros, Malayan tapir, gaur, and clouded leopard all fall into this category. As in many other tropical forests in this region, habitat loss and poaching are the two primary reasons for the decline in these and other species.

Location and General Description

This ecoregion is made up of the lowland moist forests of peninsular Malaysia and the extreme southern part of Thailand. There are no clear seasons in peninsular Malaysia, and rainfall is plentiful year-round. Two monsoons punctuate the region. From October to March a northeastern monsoon brings extra rain to the eastern side of peninsular Malaysia. The southwest monsoon, which is more powerful, bathe the western side of peninsular Malaysia with rain from April to August. Based on the Köppen climate zone system, this ecoregion falls in the tropical wet climate zone (National Geographic Society 1999).

These majestic forests are dominated by the Dipterocarpaceae tree family, notably *Anisoptera* spp., *Dipterocarpus* spp., *Dryobalanops* spp., *Hopea* spp., and *Shorea* spp. (IUCN 1991). These forests contain a high diversity of tree species (an estimated 6,000 species), and dominant species are uncommon. The general characteristics of these forests are canopies of 24–36 m, with emergents reaching more than 45 m. Dipterocarpaceae is a dominant family in the emergent stratum (Whitmore 1984). In the richest forests, up to 80 percent of the emergent trees are dipterocarps. Of the dipterocarps, *Dipterocarpus*, *Dryobalanops*, and *Shorea* are the emergents, and *Hopea* and *Vatica* usually are found in the main canopy. Berseraceae and Sapotaceae are other common main canopy families (MacKinnon et al. 1996). The tallest tree species in these forests is not a dipterocarp but rather a legume. The *Koompassia excelsa* (known as *tualang*) can grow as tall as 76 m (250 feet) (Richards 1952). This tree is rarely felled when land is cleared because of its large size, hard wood, and huge buttresses. They are often laden with the wild honey-producing bees (*Apis dorsata*) and are therefore worth more money standing than being logged. Riverine forests are bordered by neram trees (*Dipterocarpus oblongifolius*), figs (*Ficus racemosa*), and kelat (*Eugenia grandis*) and are draped with numerous epiphytic orchids and ferns (Payne and Cubitt 1990). Below the canopy a layer of shade-tolerant species thrives. This layer includes many species from the Euphorbiaceae, Rubiaceae, Annonaceae, Lauraceae, and Myristicaceae families. Ground vegetation usually is sparse, mainly small trees, and herbs are uncommon.

This ecoregion also includes some karst limestone areas that are floristically rich. The limestone hills harbor more than 1,200 species of vascular plants, of which at least 129 are endemic to this habitat (IUCN 1991), most limited to one or a few isolated hills.

Biodiversity Features

Peninsular Malaysia shares many of its tremendous species diversity with Sumatra and Borneo. All three landmasses were once part of a single, larger landmass during the last ice age, when the sea level was more than 100 m lower than it is today. However, peninsular Malaysia retained a connection to Asian mainland, and the lowland forests have the greatest diversity of mammal species found in the Indo-Pacific region. Unfortunately for many of the wide-ranging or carnivorous species, survival in peninsular Malaysia is in doubt.

The tiger (*Felis tigris*) is peninsular Malaysia's largest predator. They exact estimate of the number of remaining tigers in peninsular Malaysia is not known, but estimates have ranged from 300 to 650 over the past decade. What is not in doubt is that Malaysia's tiger population is in danger of drifting toward extinction. As recent as the 1950s more than 3,000 tigers were estimated to live in Malaysia. However, poaching, habitat loss, and inadequate protection have drastically reduced the population. Taman Negara Nature Reserve remains one of the last refuges for the tiger and several other endangered species (Rabinowitz 1999). There are one Level I and two Level II TCUs in peninsular Malaysia that overlap this ecoregion (Dinerstein et al. 1997).

The Malayan tapir (*Tapirus indicus*) is the largest of the four living tapir species and only Old World representative. The population of the Malayan tapir has been drastically reduced, but they survive in protected areas. Tapirs can be seen in the Ampang Forest Reserve, and the largest surviving population probably lives around Taman Negara National Park (McClung 1997; Payne and Cubitt 1990). Peninsular Malaysia is also home to the world's smallest rhinoceros, the two-horned Sumatran rhinoceros (*Didermocerus sumatrensis*). It once ranged through much of southeast Asia, but today the entire population numbers about 500 individuals scattered in isolated populations from peninsular Malaysia to Sumatra and Borneo (McClung 1997). Populations of the rhinoceros survive at Sungai Dusun, Ulu Selama, and Endau-Rompin parks in peninsular Malaysia (Payne and Cubitt 1990).

Peninsular Malaysia's largest land animal, the Asian elephant (*Elephas maximus*), is also one of the most endangered. They range over vast tracts of land throughout peninsular Malaysia in search of food, and as habitat is destroyed the population has become fragmented and poached. The estimated population has varied widely from 600 to 6,000 elephants, demonstrating the poor knowledge of the actual status of this species in peninsular Malaysia (Sukumar 1989). The second largest mammal in these forests is wild cattle, the gaur (*Bos gaurus*), or *seladang* as it is known in Malaysia. Widespread poaching and habitat loss have drastically reduced gaur populations outside protected areas (Payne and Cubitt 1990).

Numerous other mammal species live in these forests and include squirrels, deer, otter, civet, primates, the rare sun bear (*Ursus malayanus*), the clouded leopard (*Pardofelis nebulosa*), and many bat species.

Birds species richness is also high, with more than 450 species attributed to the ecoregion. These include several pheasants: the crestless fireback pheasant (*Lophura erythrophthalma*), crested fireback (*Lophura ignita*), Malay peacock pheasant (*Polyplectron malacense*), great argus pheasant (*Argusianus argus*), and crested argus pheasant (*Rheinardia ocellata*). Hornbills, barbets, bulbuls, woodpeckers, pigeons, broadbills, and babblers are other highly speciose or characteristic groups.

Status and Threats

Current Status

Most of these lowland forests have been converted to rice fields, rubber and oil palm plantations, and orchards, with the northwestern region being the most intensively cultivated (Cubitt and Payne and Cubitt 1990). Only about a fifth of the original forest cover now remains, scattered in fragments throughout the ecoregion. There are numerous protected areas that cover 3,875 km^2 (3 percent) of this ecoregion (table J80.1). With the exception of Taman Negara, all are less than 500 km^2, many less than 10 km^2. None of the alluvial swamp forest in the ecoregion is protected in peninsular Malaysia (WWF and IUCN 1995).

Types and Severity of Threats

Logging began in the 1950s, and by the 1970s, 3,660 km² of forests were being logged annually (IUCN 1991). But as the forest has been depleted of valuable timber trees, these rates have slowed. Logging is selective and therefore not a direct cause of forest loss but is often a preliminary to other forms of land use. The major cause of forest conversion in peninsular Malaysia is large-scale clearing for plantations and agriculture, especially of rubber, coconut, and oil palm for export production (IUCN 1991). Coffee and cocoa are minor crops. The threat of extensive plantations of pine trees for paper production (MacKinnon and MacKinnon 1986) has not materialized, but other quick-growing softwoods such as *Acacia mangium* are being favored. Hydropower and irrigation projects are a second major cause of forest loss, with mining and its associated road building and quarrying around limestone areas such as Perak and Selangor being the third (IUCN 1991). The losses in terms of forest area are slight but tend to have high impact at specific sites. In comparison to these high-impact activities, the forest conversion and degradation from shifting cultivation by indigenous people is slight (IUCN 1991). Some of the last remnants of coastal dipterocarp forest in the state of Perak, peninsular Malaysia, are at high risk from the expansion of nearby urban areas (WWF and IUCN 1995). Because more than 80 percent of the mammal species of peninsular Malaysia live only in the lowland forests of this ecoregion and 53 percent are confined to primary forest, the degradation in this ecoregion severely reduces species diversity and long-term population viability in peninsular Malaysia as a whole (IUCN 1991).

Priority Conservation Actions

Short-Term Conservation Actions (1–5 Years)

- Stop illegal logging in and around protected areas.
- Revoke forest concessions in areas of high biodiversity outside protected areas.
- Place under protection all remaining lowland forests.
- Stop wildlife poaching, especially in key parks and reserves.
- Establish a trust fund to offset opportunity costs of maintaining forest cover and avoiding conversion to oil palm plantations.

Longer-Term Conservation Actions (5–20 Years)

- Reassess representativeness of the protected area system in the ecoregion.
- Ensure that redesign of the protected area system includes linkages between lowland and montane habitats.
- Restore integrity of buffer zones of protected areas.
- Implement rigorous conservation actions in the Level I and Level II TCUs in the ecoregion.

Focal Species for Conservation Action

Habitat specialists: hornbills (mature trees needed for nests)
Area-sensitive species: tiger (Franklin et al. 1999; Tilson et al. 1994; Dinerstein et al. 1997; Seidensticker et al. 1999), Asian elephant (Santiapillai and Jackson 1990; Santiapillai 1997), Sumatran rhinoceros (Foose and van Strien 1997)
Top predator species: tiger (Franklin et al. 1999; Tilson et al. 1994; Dinerstein et al. 1997; Seidensticker et al. 1999)
Other: Malayan tapir, gaur, pheasant species

Selected Conservation Contacts

WWF-Malaysia

Justification for Ecoregion Delineation

The Peninsular Malaysian Rain Forests [80] ecoregion represents the large extent of the lowland broadleaf rain forests extending south of the Kangar-Pattani line to Singapore.

This ecoregion approximates the spatial extent of MacKinnon's (1997) subunit 07a, but we also included the Lingga and Riau archipelagos. Although FAO (1974) indicates that these islands contain heath forests, both Whitmore (1984) and MacKinnon (1997) dispute this classification, and MacKinnon identified the vegetation as lowland moist forest. The tropical montane evergreen moist forests above 1,000 m were placed in the Peninsular Malaysian Montane Rain Forests [81]. We also extracted the large areas of peat swamp forests along the coast of peninsular Malaysia into the Peninsular Malaysian Peat Swamp Forests [87]. Thus, we created three ecoregions within MacKinnon's Malay Peninsular subunit (07a) and Udvardy's Malayan rain forests biogeographic province.

Prepared by Colby Loucks

TABLE J80.1. WCMC (1997) Protected Areas That Overlap with the Ecoregion.

Protected Area	Area (km²)	IUCN Category
Unnamed	210	?
Unnamed	160	?
Pulau Redang	20	PRO
Taman Negara [81]	2,770	II
Tasek Bera	270	PRO
Pulau Tioman	130	VIII
Mersing	110	PRO
Endau-Kota Tinggi (East)	50	IV
Sungei Buloh	1	V
Senko (Sungei Sembawang)	1	REC
Sungei Khatib Bongsu Heronry	1	REC
Kranji Heronry and Marshes	2	REC
Bukit Timah and Central catchment NR	30	IV
Nee Soon	1	IV
Bukit Timah extension	1	IV
Pasir Ris	1	V
Botanic Gardens	1	UA
Fort Canning	1	UA
Kent Ridge	1	REC
Labrador	1	V
Sentosa	1	REC
Pulau Bulan	100	PRO
Pulau Pasir Panjang	2	PRO
Pulau Penyengat	10	PRO
Total	3,875	

Ecoregion numbers of protected areas that overlap with additional ecoregions are listed in brackets.

Ecoregion Number: **81**
Ecoregion Name: **Peninsular Malaysian Montane Rain Forests**
Bioregion: **Sunda Shelf and Philippines**
Biome: **Tropical and Subtropical Moist Broadleaf Forests**
Political Units: **Malaysia, Thailand**
Ecoregion Size: **17,100 km²**
Biological Distinctiveness: **Regionally Outstanding**
Conservation Status: **Relatively Stable**
Conservation Assessment: **III**

The Peninsular Malaysian Montane Rain Forests [81] are one of the last refuges for several of Asia's large characteristic species. These forests still support tigers, Asian elephants, gaur, tapirs, Sumatran rhinoceros, and the spectacular and endemic crested argus. Taman Negara Nature Reserve in central peninsular Malaysia is one of the largest reserves in the Indo-Pacific and provides an essential montane refuge and linkage area to the lowland forests.

Location and General Description

This ecoregion is made up of the montane moist forests in peninsular Malaysia and southernmost Thailand. There are no clear seasons in peninsular Malaysia, and rainfall is plentiful year-round. Two monsoons inundate the region. From October to March a northeastern monsoon brings extra rain to the eastern side of peninsular Malaysia. The southwest monsoon, which is more powerful, bathes the western side of peninsular Malaysia with rain from April to August. Based on the Köppen climate zone system, this ecoregion falls in the tropical wet climate zone (National Geographic Society 1999).

This ecoregion contains several distinct montane habitat blocks. The most extensive is the Main Range, which encompasses Malaysia's largest remaining area of pristine montane rain forest, reaching about 2,180 m around Cameron Highlands. Together with the Main Range, Fraser's Hill and Genting Highlands are botanically well known, but many other areas remain to be explored. On the east of the peninsula the outlying peak of Mt. Tahan, 2,187 m, in Taman Negara National Park has various endemic plants and a slightly impoverished fauna. Other key sites are the Taiping hills (Bukit Larut, Gunung Bubu) and Gunung Benom.

In the lower elevations of montane forest (about 1,000–1,200 m) the dominance of dipterocarp species ends. These are replaced with highly speciose oak (*Quercus* and *Lithocarpus* spp.) and chestnut (*Castanopsis* spp.) forests. Myrtaceae are also important in the lower montane forests. Common genera also include *Agathis* spp., *Dacrydium* spp., *Baeckea* spp., *Leptospermum* spp., *Podocarpus* spp., and *Styphelia* spp. (Payne and Cubitt 1990). Three major and parallel changes occur in montane forests with increasing altitude. First, there is a decrease in forest height. Montane forest do not have giant emergent trees, and their overall height is much lower. Malaysia's tallest tree species, *Koompassia excelsa* (known as *tualang*), does not occur in the montane region. The canopy typically is 10–20 m high. Second, the size and shape of the leaves change. Trees with buttress usually are absent. Lowland forests are dominated by tree species with medium-large leaves. Montane forests are dominated by slender trees with small leaves and a flattish crown surface. Rhododendrons are characteristic of upper montane flora. Rhododendrons are found on acidic and peat soils and have adapted to the harsh upper montane environments (MacKinnon 1997). The third difference is the increased presence of epiphytes. Orchids, ferns, moss, lichen, and liverworts are more abundant in montane forests than in lowland rain forests (MacKinnon 1997). Orchids are found at all levels of the forest but are common epiphytes in the upper montane forests. For growth, orchids need light and moisture as well as a mycorrhiza relationship with a tree or other plant to derive their nutrients. The well-lit and moist conditions of the moss forest in the upper montane zone provide ideal growing conditions for many orchid species (Lamb and Chan 1978).

Biodiversity Features

The noticeable differences in vegetation structure and species composition also affect faunal communities found in montane forests. Few of them are intrinsically montane, but montane forest provides an important habitat reservoir and connecting links between forest patches. There is a single endemic mammal, the rodent *Maxomys inas* (table J81.1). Other primarily montane species include the gymnure (*Hylomys suillus*), siamang (*Symphalangus syndactylus*), and red-cheeked squirrel (*Dremomys rufigenys*). Numerous other mammal species that were once common in the lowland rain forest are now finding their last remaining refuges in the montane region. Many of these species are wide-ranging or top carnivore species.

The tiger (*Panthera tigris*) is peninsular Malaysia's largest predator. The exact number of remaining tigers in peninsular Malaysia is not known, but estimates range from 300 to 650 over the past decade. Taman Negara remains one of the last refuges for the tiger. As Malaysia's tiger population drifts toward extinction, montane reserves may provide some of the best places for conservation (Rabinowitz 1999). There is one Level I TCUs in peninsular Malaysia that overlaps this ecoregion and extends into the lowlands (Dinerstein et al. 1997).

Peninsular Malaysia is also home to the world's smallest rhinoceros, the two-horned Sumatran rhinoceros (*Didermocerus sumatrensis*). It once ranged through much of southeast Asia, but today the entire population numbers about 500 individuals distributed in isolated populations from peninsular Malaysia to Sumatra and Borneo (McClung 1997). The Malayan tapir (*Tapirus indicus*) is another endangered species. It is the largest of the four living tapir species and the only Old World representative. The population of the Malayan tapir has been drastically reduced, but they hang on in protected areas. Tapirs can be seen in the

TABLE J81.1. Endemic and Near-Endemic Mammal Species.

Family	Species
Muridae	*Maxomys inas**

An asterisk signifies that the species' range is limited to this ecoregion.

TABLE J81.2. Endemic and Near-Endemic Bird Species.

Family	Common Name	Species
Phasianidae	Mountain peacock-pheasant	Polyplectron inopinatum
Phasianidae	Crested argus	Rheinardia ocellata
Turdidae	Malayan whistling-thrush	Myiophonus robinsoni
Muscicapidae	Rufous-vented niltava	Niltava sumatrana
Timaliidae	Marbled wren-babbler	Napothera marmorata

An asterisk signifies that the species' range is limited to this ecoregion.

Ampang Forest Reserve, and the largest surviving population probably occurs around Taman Negara National Park (McClung 1997; Payne and Cubitt 1990).

Peninsular Malaysia's largest land animal, the Asian elephant (Elephas maximus), is also one of the most endangered. They range over vast tracts of land throughout peninsular Malaysia in search of food, and as habitat is destroyed the population has become fragmented and poached. The estimated population has ranged from 600 to 6,000 individuals, demonstrating the poor knowledge of the actual status of this species in peninsular Malaysia (Sukumar 1989). The second-largest mammal in these forests is the wild cattle, the gaur (Bos gaurus), or seladang as it is known in Malaysia. Widespread poaching and habitat loss have decimated gaur populations outside protected areas (Payne and Cubitt 1990).

Numerous other mammal species live in these forests. Nearly one-half of Malaysia's mammal species are bats. The forests also teem with squirrels, deer, otter, civet, and primate species. They include the rare sun bear (Ursus malayanus) and clouded leopard (Pardofelis nebulosa).

More than 250 bird species are known to live in this ecoregion, including five near-endemic species (table J81.2). More than seventy-five birds are montane specialists, and BirdLife International considers two of these species threatened: the mountain peacock-pheasant (Polyplectron inopinatum) and the crested argus (Rheinardia ocellata). The ecoregion also overlaps with the peninsular Malaysian portion of the Sumatra and peninsular Malaysia EBA (158) (Stattersfield et al. 1998). Eight restricted-range bird species live in this ecoregion.

Status and Threats

Current Status

Most of peninsular Malaysia's remaining forests are limited to the high, steep areas of this ecoregion. Approximately two-thirds of these forests remain intact, mainly in two large blocks of primary forest that cover the Main Range (the Titi Wangsa Mountains) and the East Coast Range in the states of Kelantan and Terengganu (IUCN 1991). Taman Negara National Park in peninsular Malaysia is one of the largest protected areas in southeast Asia. However, this park harbors only about 3 percent of the endemic tree species and about 30 percent of the palm species known in peninsular Malaysia (Soepadmo 1995). On paper there are four protected areas that cover 5,120 km^2 (30 percent) of the ecoregion area (table J81.3). However, Belum is still only a proposed nature reserve. Furthermore, in 1996 Malaysia's Department of Wildlife and National Parks discovered that the Cameron Highlands Wildlife Reserve had been degazetted in 1962.

Types and Severity of Threats

Despite the rugged terrain of this ecoregion, logging is intensive on the lower slopes. Resort development has also caused environmental degradation in some popular montane areas (WWF and IUCN 1995). A new road is planned in the Main Range, where it would cause extensive damage to this sensitive habitat (IUCN 1991).

Priority Conservation Actions

Short-Term Conservation Actions (1–5 Years)

- Stop illegal logging in and around protected areas.
- Revoke forest concessions in areas of high biodiversity outside protected areas.
- Stop wildlife poaching, especially in key parks and reserves.

Longer-Term Conservation Actions (5–20 Years)

- Ensure that the protected area system includes linkages between lowland and montane habitats.

Focal Species for Conservation Action

Area-sensitive species: tiger (Franklin et al. 1999; Tilson et al. 1994; Dinerstein et al. 1997; Seidensticker et al. 1999), Asian elephant (Santiapillai and Jackson 1990; Santiapillai 1997); Sumatran rhinoceros (Foose and van Strien 1997)

Top predator species: tiger (Franklin et al. 1999; Tilson et al. 1994; Dinerstein et al. 1997; Seidensticker et al. 1999)

Other: gaur, Malayan tapir, Malaysian peacock-pheasant, crested argus

Selected Conservation Contacts

WWF-Malaysia
WWF-Thailand

Justification for Ecoregion Delineation

The tropical montane evergreen moist forests above 1,000 m were placed in the Peninsular Malaysian Montane Rain Forests [81]. The lowland rain forests (below 1,000 m) approximates the spatial extent of MacKinnon's (1997) sub-unit 07a. A DEM was used to derive the 1,000-m contour to delineate this ecoregion's boundary. Therefore, we created three ecoregions within MacKinnon's Malay Peninsular subunit (07a) and Udvardy's Malayan rain forests biogeo-

TABLE J81.3. WCMC (1997) Protected Areas That Overlap with the Ecoregion.

Protected Area	Area (km^2)	IUCN Category
Belum	2,120	PRO
Cameron Highlands	680	IV
Krau	640	VIII
Taman Negara [80]	1,680	II
Total	5,120	

Ecoregion numbers of protected areas that overlap with additional ecoregions are listed in brackets.

graphic province. We also extracted the large areas of peat swamp forests along the coast of peninsular Malaysia into the Peninsular Malaysian Peat Swamp Forests [87]. Thus, we created three ecoregions within MacKinnon's Malay Peninsular subunit (07a) and Udvardy's Malayan rain forests biogeographic province.

Prepared by Colby Loucks

Ecoregion Number:	**82**
Ecoregion Name:	**Sumatran Lowland Rain Forests**
Bioregion:	**Sunda Shelf and Philippines**
Biome:	**Tropical and Subtropical Moist Broadleaf Forests**
Political Units:	**Indonesia**
Ecoregion Size:	**258,300 km^2**
Biological Distinctiveness:	**Globally Outstanding**
Conservation Status:	**Critical**
Conservation Assessment:	**I**

The Sumatran Lowland Rain Forests [82] are one of the most diverse forests on Earth and also one of the most threatened. These forests contain comparable levels of species diversity as the richest forests in Borneo and New Guinea. The Sumatra rain forests are home to some of the world's most charismatic flowering plants: *Rafflesia arnoldii*, which produces the largest flower in the world (up to 1 m wide), and *Amorphophallus titanum*, which stands more than 2 m tall and produces aroid flowers. The avifauna is also exceptionally rich. More than 450 bird species are found here, more than in any other ecoregion in the Sunda Shelf and Philippines bioregion except the Borneo Lowland Rain Forests [96]. In the past fifteen years rampant logging, hunting, fires, and habitat loss in the lowlands have pushed many of this ecoregion's already endangered species to the edge of extinction. These include the Sumatran rhinoceros, Malayan tapir, tiger, Asian elephant, and orangutan. Illegal logging and pervasive corruption are contributing to the more than 3,000 km^2 of forest lost every year in this ecoregion. At the current, rate no mappable natural forests will remain beyond 2005.

Location and General Description

This ecoregion represents the lowland moist forests of Sumatra, including the small islands of Simeulue, Nias, and most of Bangka. The geologic history of Sumatra provides insights into the origins of Sumatra's biodiversity. About 150 million years ago Borneo, Sumatra, and western Sulawesi split off from Gondwanaland and drifted north. Around 70 million years ago India slammed into the Asian landmass, forming the Himalayas, and an associated thrust formed Sumatra's Barisan Mountains, which run the length of Sumatra. As the Barisan Range buckled upward, it formed a deep-water channel to the west of Sumatra. During this time the islands of Simeulue and Enggano were formed. Today, to the east of the Barisan Range low hills and plains exist as a result of tectonic and volcanic events. Continued mountain building, volcanic activity, and sedimentation in the lowland occurred over the past 25 million years. Podzolic soils associated with altosols or litosols are the predominant lowland soils. Large limestone areas occur in northern Sumatra, and they are associated with brown podzolic and renzina soils (Whitten et al. 2000).

Based on the Köppen climate zone system, Sumatra falls in the tropical wet climate zone (National Geographic Society 1999). The lowland rain forests to the west of the Barisan Range receive more rainfall (~6,000 mm/year) than the lowland rain forests to the east (~2,500+ mm/year). The Barisan Range blocks much of this rainfall. However, most of Sumatra experiences less than three consecutive months of dry weather (less than 100 mm rainfall/month) (Whitten et al. 2000).

Sumatra's rain forests are quite diverse and contain levels of species diversity comparable to those of the richest forests in Borneo and New Guinea and are much richer than Java, Sulawesi, and other islands in the Indonesian Archipelago. Large, buttressed trees dominated by the Dipterocarpaceae family characterize Sumatra's lowland rain forests. Woody climbers and epiphytes are also abundant (Whitten et al. 2000). The lowland rain forests of Sumatra support 111 dipterocarp species, including 6 endemics. The emergent trees, which can reach 70 m tall, are also dipterocarps (*Dipterocarpus* spp., *Parashorea* spp., *Shorea* spp., *Dryobalanops* spp.) and, to a lesser extent, species in the Caesalpiniaceae family (*Koompasia* spp., *Sindora* spp., and *Dialium* spp.). Dipterocarps dominate the canopy layer as well. Other canopy and understory tree families that are common include Burseraceae, Sapotaceae, Euphorbiacae, Rubiaceae, Annonaceae, Lauraceae, and Myristicaceae (Whitten et al. 2000). Ground vegetation usually is sparse—mainly small trees and saplings of canopy species—and herbs are uncommon.

Figs (Moraceae) are also common in the lowland rain forest. There are more than 100 fig species in Sumatra, and each species usually is pollinated exclusively by a single fig-wasp (Agaonidae) species. Figs may produce (mast) from 500 to a million fruits twice a year and are important food sources for many forest animals (MacKinnon 1986). Dipterocarps also use mast fruiting, perhaps to escape seed predation, by satiating the appetites of seed-predators and leaving the remaining seeds to germinate (Whitten et al. 2000). Sumatra once contained pure stands of rot- and insect-resisting ironwood (*Eusideroxylon zwageri*) forests. Ironwood is a member of the laurel family and is distributed throughout southern Sumatra, Kalimantan, and the Philippines. Ironwood forests are dominated by *Eusideroxylon zwageri* but may have also contained *Shorea*, *Koompasia*, or *Intsia* species as emergents (Whitten et al. 2000).

Biodiversity Features

Sumatra shares many of its species with peninsular Malaysia and Borneo. All three landmasses were once part of a single, larger landmass during the last ice age, when sea level was more than 100 m lower than it is today. Consequently, these forests share many of the same flora and fauna. One of the most distinctive plant species in the region is *Rafflesia*. Five of the sixteen species of the parasitic *Rafflesia* plant are found in Sumatra and occur mainly in lowland forest, although they have been recorded as high as 1,800 m on Mt. Lembuh, Aceh. *Rafflesia arnoldii* is found in this ecoregion and produces the largest flower in the world. Its large brown-orange and white flowers span almost 1 m in width. *Rafflesia* have no leaves, instead deriving all their energy from the tissues of their host, the vine *Tetrastigma*. Large buds emerge from the vine and have five large, flowery petals surrounding plates, which smell like rotting meat and attract pollinating insects (Whitten et al. 2000; MacKinnon 1997). Sumatra's lowland rain forests also are home to one of the world's tallest flower, *Amorphophallus titanum*, belonging to the Arum family. The flower often grows on top of a 2-m stalk and appears every few years. These plants have blotchy stems, unusual leaves, and a fetid odor, which attracts small stingless bees that act as pollinators (Stone 1994; Whitten et al. 2000).

Other plants common to these forests are epiphytes. Common epiphyte families found in Sumatra include Orchidaceae, Gesneriaceae, Melastomaceae, Asclepidiaceae, and Rubiaceae. Rubiaceae includes the "ant-plants," *Myrmecodia* and *Hydnophytum*. These plants harbor ant colonies in their stems, and the ants protect the plant's leaves from caterpillars and other arthropod herbivores (Whitten et al. 2000).

The fauna of Sumatra can be split into two regions, one to the north of Lake Toba and the other to the south. Lake Toba formed 75,000 years ago as part of a volcanic eruption that had a devastating impact on Sumatra (Stone 1994). Seventeen bird species are found only north of Lake Toba, and ten are limited to the south. The white-handed gibbon (*Hylobates lar*) occurs only north of Lake Toba, and the dark-handed gibbon (*Hylobates agilis*) is found only to the south. The tarsier (*Tarsius bancansus*), banded leaf-monkey (*Presbytis melalophus*), and endangered Malayan tapir (*Tapirus indicus*) are found south of Lake Toba (Whitten et al. 2000). The Malayan tapir is the largest of the four living tapir species and the only Old World representative. The Sumatran population of the Malayan tapir is close to extinction, with no more than fifty animals left in the wild (McClung 1997).

The Malayan tapir is but one of the many endangered mammals living in Sumatra's rain forests. The two-horned Sumatran rhinoceros (*Dicerorhinus sumatrensis*) once ranged through much of southeast Asia. Today the entire population numbers about 300–500 individuals scattered in several populations in Sumatra, Borneo, and peninsular Malaysia. The Asian elephant (*Elephas maximus*) is found in several small populations throughout Sumatra. Only five populations number more than 200 individuals, and with rapid habitat loss the survival of many of these populations is uncertain, despite the fact that they can exploit secondary forests (Sukumar 1989). The Sumatran tiger (*Panthera tigris*) is Indonesia's largest terrestrial predator and is critically endangered (Hilton-Taylor 2000). The Sumatran tiger lives in lowland and montane rain forest and in freshwater swamp forests throughout Sumatra. There are an estimated 500 Sumatran tigers remaining in Sumatra, with approximately 100 found in Gunung Leuser National Park in northern Sumatra (Franklin et al. 1999). However, the tiger is intensively hunted for skins and to supply the traditional medicine markets. There are two Level I TCUs in Sumatra that overlap this ecoregion (Dinerstein et al. 1997).

There is only one endemic mammal in this ecoregion (table J82.1). However, these forests contain numerous primate species such as several leaf-monkey species, slow loris (*Nycticebus coucang*), long-tailed macaque (*Macaca fascicularis*), pig-tailed macaque (*M. nemestrina*), and siamang (*Hylobates syndactylus*), the region's largest gibbon and found only in Malaya and Sumatra's lowland forests. Other species include the Sunda otter-civet (*Cynogale bennettii*), wild dog (*Cuon alpinus*), sun bear (*Ursus malayanus*), and clouded leopard (*Pardofelis nebulosa*) (Whitten et al. 2000; Stone 1994).

The bird fauna consists of more than 450 species, including four near-endemic species and one endemic species, the Simeulue scops-owl (*Otus umbra*) (table J82.2). The Sumatran lowland forests harbor a remarkable ten hornbill species, including the great hornbill (*Buceros bicornis*), absent on other Indonesian islands. These rain forests are also home to the great argus pheasant (*Argusianus argus*). The male argus pheasant clears patches in the forest to perform dramatic dances, flaunting his 1.5-m tail and emitting distinctive "ki-an" calls to attract females for breeding.

Conservation Status

Current Status

The conservation status of this ecoregion's forests is critical. Before 1985 only about one-third of this ecoregion's natural forests remained. Most of this habitat had been lost to agricultural expansion and logging. However, in the past fifteen years more than 60 percent of these forests have been destroyed. The remaining areas of intact habitat are found primarily in central Sumatra. There are several protected areas in this ecoregion, which include about 9 percent of the ecoregion area (table J82.3). However, encroachment, widespread illegal logging, and fires in these protected areas are severe (Holmes, essay 1).

TABLE J82.1. Endemic and Near-Endemic Mammal Species.

Family	Species
Molossidae	*Mormopterus doriae**

An asterisk signifies that the species' range is limited to this ecoregion.

TABLE J82.2. Endemic and Near-Endemic Bird Species.

Family	Common Name	Species
Strigidae	Simeulue scops-owl	*Otus umbra**
Irenidae	Blue-masked leafbird	*Chloropsis venusta*
Pycnonotidae	Cream-striped bulbul	*Pycnonotus leucogrammicus*
Pycnonotidae	Spot-necked bulbul	*Pycnonotus tympanistrigus*
Pycnonotidae	Blue-wattled bulbul	*Pycnonotus nieuwenhuisii*

An asterisk signifies that the species' range is limited to this ecoregion.

Although almost 9 percent of this ecoregion is in protected areas, the habitat is not guaranteed protection. Since the fall of the Suharto government, many of these protected areas have been invaded and their natural resources mined, despite the efforts of conservation agencies and assistance from many outside conservation organizations. With the fall of the Suharto government in 1998 and the economic crises of 1998–1999, several groups exploited forests inside existing protected areas. Local people, feeling betrayed by the Suharto government, took control of the natural resources around them. A second group of people who worked for forest and oil palm concessionaires and sawmill owners also began to encroach on protected areas. Common thugs and corrupt local and military officials made up a third group and provided protection to all groups involved in illegal activities. One of numerous examples of illegal logging inside Indonesia's protected areas is the unabated logging that has occurred in and around the biologically valuable Gunung Leuser National Park since April 1999. The destruction includes prime habitat for the orangutan, siamang, and white-handed gibbon. This area also contains the only known population of tool-using orangutans (Colijn, http://www.bart.nl/~edcolijn/).

To ensure the survival of many of the top predators and wide-ranging species such as the tiger, the protected areas must be protected. Another problem is that the current protected area system is not large enough in size or distribution to protect many of the top predators and wide-ranging species over the long term. Many of the protected areas have no effective management. However, many of the largest reserves are reservoirs for biodiversity and retain the best chance for conservation in the future. These reserves include Gunung Leuser, Way Kambas, and Bukit Barisan Selatan.

Types and Severity of Threats

The remaining natural forests in this ecoregion will be completely gone within the next five years unless drastic actions are taken to halt rampant logging. The approximate forest loss in Sumatra from 1985 to 1997 has been 67,000 km^2, most of it in the rich lowland forests. However, the annual rate of forest loss has been increasing across Indonesia. Countrywide, the deforestation rate in the 1980s was 8,000 km^2/year. In the early 1990s this rate had increased to around 12,000 km^2/year. From about 1996 to the present the rate has almost doubled to more than 20,000 km^2/year. In Sumatra's lowland forests from 1985 to 1997 the average annual forest loss was about 2,800 km^2/year. If the current deforestation trend continues, this ecoregion's natural forests will be gone by 2005 (Holmes, essay 1).

Logging and clearing for plantations and agriculture have been especially heavy in the lowlands east of the Barisan Range, especially since the 1998–1999 economic crisis. Extensive stands of ironwood (*Eusideroxylon zwageri*), a species of great commercial importance producing an exceptionally durable timber, have been almost entirely destroyed in southern Sumatra. Deforestation has been extensive in Bengkulu Province, especially along the main road running north from Bengkulu to Muko-Muko, where international involvement has been responsible for replacing tropical rain forest with oil palm and cocoa monocultures. In the absence of these forests, heavy rains in 1988 caused massive landslides and floods, which affected thousands of people.

The island of Nias presents a similar case as Sumatra's mainland forests. Almost all the natural forests are gone and exist as extensive areas of secondary forest. They still support many of the region's endemic birds, but the situation is deteriorating. Simeulue had most of its forests in the 1980s but is rapidly losing them to clove and coffee plantations.

Priority Conservation Actions

With the rapid pace of forest destruction currently under way throughout Indonesia, quick conservation action is urgently needed to stave off the complete collapse of these forests. However, steps should be taken to lay the groundwork for longer-term conservation as well.

TABLE J82.3. WCMC (1997) Protected Areas That Overlap with the Ecoregion.

Protected Area	Area (km^2)	IUCN Category
Gunung Salawah Agam	50	PRO
Hutan Pinus/Janthoi	60	I
Gunung Leuser [83]	3,220	II
Dolok Sembelin [83]	110	VI
Pulau Simeulue	570	PRO
Singkil Barat	800	
Pulau Bengkaru	60	
Laut Tapus	40	PRO
Sibolga	240	PRO
Pulau Nias (I–IV)	710	PRO
Padang Lawas	680	PRO
Rimbo Panti	20	DE
Malampah Alahan Panjang [83]	250	PRO
Bukit Rimbang–Baling–Baling	1,220	PRO
Air Sawan	880	PRO
Seberida	2,170	
Peranap	960	PRO
Bukit Sebelah and Batang Pangean	460	VI
Bukit Besar	1,330	PRO
Way Kambas [88]	810	II
Dangku	750	IV
Bentayan	310	IV
Benakat	570	DE
Subanjeriji	1,100	DE
Rebang	120	PRO
Semidang Bukit Kabu	240	VI
Bukit Balal [83]	100	VI
Bukit Raja Mandara [83]	80	VIII
Gumai Pasemah [83]	320	IV
Isau-Isau Pasemah [83]	100	IV
Bukit Barisan Selatan [83]	1,750	II
Bukit Nantiogan Hulu/ Nanti Komerung Hulu [83]	720	VI
Paraduan Gistana and Surroundings	730	VI
Way Waya	330	VIII
Pulau Sangiang	20	DE
Total	21,880	

Ecoregion numbers of protected areas that overlap with additional ecoregions are listed in brackets.

Short-Term Conservation Actions (1–5 Years)

- Initiate a major public awareness campaign on forest values and forest loss to drive political will.
- Stop illegal logging in and around protected areas by challenging the authorities and judiciary to act on existing laws and regulations.
- Revoke forest concessions in areas of high biodiversity outside protected areas and institute a moratorium on new concessions for oil palm, timber, and other plantations, using the vast areas of deforested and cleared land instead.
- Place under active protection all remaining lowland forests.
- Mobilize major assistance for the more outstanding protected areas.
- Stop wildlife poaching, especially in key parks and reserves, by tightening surveillance around ports or airports and making the purchase of endangered species socially unacceptable.
- Establish a trust fund to offset opportunity costs of maintaining forest cover and avoiding conversion to oil palm plantations.
- Purchase through Indonesian NGOs several large tracts of lowland forest, create model ecotourism reserves with elephant and foot safaris to view wildlife and attract foreign currency, and share profits from such ventures with communities.
- Determine whether Sumatran provincial governments can be provided funding from carbon credits for maintaining forest cover and wildlife protection.
- Initiate major biodiversity surveys and collecting expeditions in doomed areas to document the biodiversity that will soon be lost.

Longer-Term Conservation Actions (5–20 Years)

- Establish a long-term fund for conservation of lowland forests in Indonesia that would benefit Sumatra.
- Ensure that redesign of the protected area system includes linkages between lowland and montane habitats.
- Devise ways to maintain the integrity of buffer zones of protected areas.
- Implement rigorous conservation actions in the Level I and Level II TCUs in the ecoregion.

Focal Species for Conservation Action

Habitat specialists: hornbills (mature trees needed for nests)

Area-sensitive species: tiger (Franklin et al. 1999; Tilson et al. 1994; Dinerstein et al. 1997; Seidensticker et al. 1999), Asian elephant (Santiapillai and Jackson 1990; Santiapillai 1997), Sumatran rhinoceros (Foose and van Strien 1997), orangutan (Rijksen and Meijaard 1999; Dijuwantoko 1991; Tilson et al. 1993)

Top predator species: tiger (Dinerstein et al. 1997; Seidensticker et al. 1999)

Other: Malayan tapir, siamang, *Rafflesia* spp., Simeulue scops-owl, *Mormopterus doriae*

Selected Conservation Contacts

ARSG
EU-Leuser
Lembaga Alam Tropika Indonesia (LATIN)
Sumatran Tiger Project
WAHLI
WARSI
WCS
Worldbank Kerinci, Telapak
WWF-Indonesia
Yayasan Keanekaragaman Hayati
Yayasan Konsorsium Pembaruan Agraria (KPA)

Justification for Ecoregion Delineation

We recognized six ecoregions in Sumatra. MacKinnon (1997) placed all the biomes in Sumatra within two subunits (21a and 21b), with the subunit division based on a faunal break to the south of Lake Toba. However, according to Tony Whitten (pers. comm., 1999) the differences in floral and faunal communities across MacKinnon's subunit boundary (within a habitat type) are not great enough to warrant differentiation into distinct ecoregions. Therefore, we placed the lowland rain forests of Sumatra into a single ecoregion, the Sumatran Lowland Rain Forests [82]. Although FAO (1974) and MacKinnon (1997) showed that Bangka Island has extensive heath forests, Whitmore (1984) and Whitten et al. (1984a) reported that the heath forests are less extensive, and its main forests are not heath forests; therefore, we included this island as part of the Sumatran Lowland Rain Forests [82] ecoregion.

MacKinnon's biounit 21 largely corresponds to Udvardy's Sumatra biogeographic province. However, Udvardy did not include the Nicobar Islands. Eight ecoregions overlap Udvardy's Sumatra biogeographic province: Sumatran Lowland Rain Forests [82], Sumatran Montane Rain Forests [83], Mentawai Islands Rain Forests [84], Sumatran Peat Swamp Forests [85], Sumatran Freshwater Swamp Forests [88], Sundaland Heath Forests [90], Sumatran Tropical Pine Forests [105], and Sunda Shelf Mangroves [107].

Prepared by Colby Loucks and Tony Whitten

Ecoregion Number: 83

Ecoregion Name:	Sumatran Montane Rain Forests
Bioregion:	Sunda Shelf and Philippines
Biome:	Tropical and Subtropical Moist Broadleaf Forests
Political Units:	Indonesia
Ecoregion Size:	72,600 km²
Biological Distinctiveness:	Globally Outstanding
Conservation Status:	Relatively Stable
Conservation Assessment:	III

The Sumatra Montane Rain Forests [83] are home to a wide variety of species and are one of the most outstanding examples of montane rain forests west of Wallace's Line.

The Sumatran montane rain forests are home to both *Amorphophallus titanum*, which grows on a stalk that measures more than 2 m tall, and the parasitic *Rafflesia arnoldii*, which produces the largest flower in the world (up to 1 m wide). Sumatra's montane forests are also home to some of the most endangered species in the Indo-Pacific region. The Sumatran rhinoceros, tiger, and Sumatran rabbit all inhabit these forests.

Location and General Description

This ecoregion represents the montane forests (>1,000 m) along the Barisan Mountain Range of Sumatra. The geologic history of Sumatra provides insights into the origins and amount of Sumatra's biodiversity. About 150 million years ago Borneo, Sumatra, and western Sulawesi split off from the Gondwanaland and drifted north. About 70 million years ago India slammed into the Asian landmass, forming the Himalayas, and an associated thrust formed Sumatra's Barisan Mountains, which run the length of Sumatra. As the Barisan Range buckled upward, it formed a deep water channel to the west of Sumatra. Today, to the east of the Barisan Range, low hills and plains exist as a result of tectonic and volcanic events. Continued mountain building, volcanic activity, and sedimentation in the lowland have occurred over the past 25 million years. Podzolic soils associated with altosols or litosols are the predominant soils found in this ecoregion. Large limestone areas occur in northern Sumatra, and they are associated with brown podzolic and renzina soils (Whitten et al. 2000).

Based on the Köppen climate zone system, Sumatra falls in the tropical wet climate zone (National Geographic Society 1999). The montane rain forests of the Barisan Range receive more rainfall on their western slopes than their eastern slopes, which are in a rainshadow. However, most of Sumatra experiences less than three consecutive months of dry weather (less than 100 mm rainfall/month), and rainfall in the montane rain forests averages more than 2,500 mm/year (Whitten et al. 2000).

Sumatra's montane rain forests can be separated into three major forest zones: lower montane forest, upper montane forest, and sub-alpine forest. Temperature and cloud level are the major factors determining these forest zones. The lower montane zone forests are similar to lowland rain forests but begin to get smaller. The canopy height typically is no more than 35 m high. Emergents may extend to 45 m, but buttresses are rare. Lianas usually are absent, and epiphytes such as orchid begin to increase in abundance. The upper montane zone sharply changes from lowland rain forests. The canopy becomes even and rarely exceeds 20 m. Emergents may extend to 25 m, but buttresses usually are absent. Trees rarely have compound leaves or lianas. Orchids and other epiphytes such as moss, lichen, and liverworts are very common. Beyond this forest lies the sub-alpine forest, a complex of grass, heath, and bog areas. Small, stunted trees may reach 10 m high, orchids become very rare, but moss, lichen, and liverworts are very abundant (Whitten et al. 2000).

The montane flora of Sumatra originates from two sources: local sources (autochthonous) and areas that have a center of origin outside of Sumatra (allochthonous). The local source can be divided into two categories: species that are characteristic of lowland rain forest, such as Dipterocarpaceae, Bombacaceae, and the genus *Ficus* (figs), and those that have a large global latitudinal distribution such as pines, Cruciferae (e.g., mustard), Theaceae (e.g., tea), and tree ferns. The allochthonous flora belong to genera whose species are found only in cold climates, not near equatorial rain forests. These species in the tropics are never found below 1,000 m and usually dominate the sub-alpine flora. Genera include *Rhododendron*, the pretty herbs *Gentiana*, and grass *Deschampsia*. Most of these species dispersed from Asia or Australia during cooler glacial periods when the Sunda region was a single landmass. Forest zones were all 350–400 m lower than their present height, providing numerous stepping stones (Steenis 1950).

The characteristic vegetation in lower montane forests changes from Dipterocarpaceae, the dominant lowland family, to Fagaceae (oaks) and Lauraceae (laurels). *Lithocarpus*, *Quercus*, and *Castanea* are common genera in the Fagaceae family, and *Cinnamomum burmansea*, *Persea americana*, and *Litsea* spp. are common Lauraceae species. Other families common to the lower montane region include Cunoniaceae, Monimiaceae, Magnoliaceae, and Hamamelidaceae (FAO 1981; Whitten et al. 2000). Tree ferns in the genus *Cyathea* are also common in the lower montane forests. The upper montane forest is characterized by conifers (pines and related trees), particularly by the Ericaceae (*Rhododendron*, *Vaccinium*) and Myrtaceae (*Eucalyptus*, *Melaleuca*) families. *Dacrycarpus imbricatus* and *Leptospermum flavescens* are also abundant in these forests, which because of their smaller stature are called elfin forests. Lichens are common to the drier parts of this zone, whereas mosses and liverworts are common in the moister parts of this zone that coincide with where clouds form and are commonly called cloud or moss forests. The sub-alpine zone is characterized by smaller specimens of the montane forest. There is also an increased abundance of grasses (*Agrostis* and *Festuca*), rushes and sedges (*Juncus*, *Carex*, *Scirpus*, and *Cyperus*), and small, colorful herbs (Whitten et al. 2000).

Five of the sixteen species of the parasitic *Rafflesia* plant are found in Sumatra and have been recorded as high as 1,800 m on Mt. Lembuh, Aceh. *Rafflesia arnoldii*, which produces the largest flower in the world, is found in this ecoregion. Its large brown-orange and white flowers can reach 1 m in diameter. *Rafflesia* have no leaves, instead deriving all their energy from the tissues of its host, the ground vine *Tetrastigma*. Large buds emerge from the vine and have five large, flowery petals surrounding spikes, which smell like rotting meat and attract pollinating insects (Whitten et al. 2000; MacKinnon 1986).

Biodiversity Features

Sumatra's montane forests contain far higher levels of mammal and bird endemism than the lowland forests, in part because of their longer periods of isolation and distinctive forest types. Seven mammal and eight bird species are endemic (table J83.1, table J83.2), whereas the lowland Sumatran Lowland Rain Forests [82] have only one mammal and one endemic bird species. The mammal species includes Thomas's leaf-monkey (*Presbytis thomasi*), one of Sumatra's four leaf-monkeys, Sumatran rabbit (*Nesolagus netscheri*), and Sumatran shrew-mouse (*Mus crociduroides*). This ecoregion also contains two near-endemic mammal species and twenty-one near-endemic bird species.

This ecoregion overlaps with a portion of the Sumatra and peninsular Malaysia EBA (158) (Stattersfield et al. 1998). Thirty-five restricted-range bird species are found in this ecoregion and include two threatened species that are also endemic to this ecoregion: the Sumatran cochoa (*Cochoa beccarii*) and Sumatran ground-cuckoo (*Carpococcyx viridis*).

Another distinctive feature of Sumatra's fauna is that it can be split into two regions, one to the north of Lake Toba and the other to the south. Lake Toba formed 75,000 years ago as part of a volcanic eruption that had a devastating impact on Sumatra (Stone 1994). Seventeen bird species are found only north of Lake Toba, and ten are found only to the south. The white-handed gibbon (*Hylobates lar*) occurs only north of Lake Toba, and the dark-handed gibbon (*Hylobates agilis*) is found to the south. The tarsier (*Tarsius bancansus*), banded leaf-monkey (*Presbytis melalophus*), and endangered Malayan tapir (*Tapirus indicus*) are all found to the south of Lake Toba (Whitten et al. 2000). The Malayan tapir is the largest of the four living tapir species and the only Old World representative. The Sumatran population of the Malayan tapir is close to extinction, with no more than fifty animals left in the wild, mostly in the lowland forests (McClung 1997).

The Sumatran tiger (*Panthera tigris*), Indonesia's largest terrestrial predator, lives in lowland and montane rain forest, as well as freshwater swamp forests throughout Sumatra. There are an estimated 500 Sumatran tigers remaining in Sumatra, with approximately 100 found in Gunung Leuser National Park in northern Sumatra (Franklin et al. 1999). There are three Level I TCUs in Sumatra that overlap this ecoregion (Dinerstein et al. 1997). The two-horned Sumatran rhinoceros (*Didermocerus sumatrensis*) once ranged through much of southeast Asia. Today the entire population numbers about 500 individuals scattered in several populations in Sumatra, Borneo, and peninsular Malaysia (McClung 1997). Another distinctive species of Sumatra's montane forests is the serow (*Capricornis sumatraensis*). The serow lives from 200 m to the vegetated summits of Sumatra's highest peaks. It can also be found on forested limestone hills (Whitten et al. 2000).

Several other mammal species are found in this ecoregion, including numerous primate species such as several leaf-monkeys, siamang (*Hylobates syndactylus*), the region's largest gibbon, wild dog (*Cuon alpinus*), sun bear (*Ursus malayanus*), and clouded leopard (*Pardofelis nebulosa*) (Whitten et al. 2000; Stone 1994).

Status and Threats

Current Status

Despite Sumatra's dense human population, this montane ecoregion contains several large blocks of intact forest, stretching along the Barisan Mountain Range, running the length of the island. Numerous protected areas are scattered

TABLE J83.1. Endemic and Near-Endemic Mammal Species.

Family	Species
Sorcidae	Crocidura baluensis
Cercopithecidae	Presbytis thomasi*
Sciuridae	Hylopetes winstoni*
Muridae	Mus crociduroides*
Muridae	Rattus korinchi*
Muridae	Rattus hoogerwerfi
Muridae	Maxomys hylomyoides*
Muridae	Maxomys inflatus*
Leporidae	Nesolagus netscheri*

An asterisk signifies that the species' range is limited to this ecoregion.

TABLE J83.2. Endemic and Near-Endemic Bird Species.

Family	Common Name	Species
Phasianidae	Red-billed partridge	Arborophila rubrirostris*
Phasianidae	Sumatran pheasant	Lophura hoogerwerfi*
Phasianidae	Salvadori's pheasant	Lophura inornata*
Phasianidae	Bronze-tailed peacock-pheasant	Polyplectron chalcurum
Columbidae	Green-spectacled pigeon	Treron oxyura
Columbidae	Pink-headed fruit-dove	Ptilinopus porphyreus
Cuculidae	Sumatran ground-cuckoo	Carpococcyx viridis*
Strigidae	Rajah scops-owl	Otus brookii
Apodidae	Waterfall swift	Hydrochous gigas
Trogonidae	Blue-tailed trogon	Harpactes reinwardtii
Pittidae	Schneider's pitta	Pitta schneideri*
Pittidae	Black-crowned pitta	Pitta venusta
Dicruridae	Sumatran drongo	Dicrurus sumatranus*
Campephagidae	Sunda minivet	Pericrocotus miniatus
Irenidae	Blue-masked leafbird	Chloropsis venusta
Turdidae	Shiny whistling-thrush	Myiophonus melanurus
Muscicapidae	Rufous-vented niltava	Niltava sumatrana
Muscicapidae	Sunda robin	Cinclidium diana
Muscicapidae	Sumatran cochoa	Cochoa beccarii*
Pycnonotidae	Cream-striped bulbul	Pycnonotus leucogrammicus
Pycnonotidae	Spot-necked bulbul	Pycnonotus tympanistrigus
Pycnonotidae	Sunda bulbul	Hypsipetes virescens
Zosteropidae	Black-capped white-eye	Zosterops atricapillus
Sylviidae	Sunda warbler	Seicercus grammiceps
Timaliidae	Sunda laughingthrush	Garrulax palliatus
Timaliidae	Vanderbilt's babbler	Malacocincla vanderbilti*
Timaliidae	Rusty-breasted wren-babbler	Napothera rufipectus
Timaliidae	Marbled wren-babbler	Napothera marmorata
Fringillidae	Mountain serin	Serinus estherae

An asterisk signifies that the species' range is limited to this ecoregion.

along the range and cover 40 percent of the ecoregion's area (table J83.3). The large Gunung Leuser national park extends into the northern part of the ecoregion. In the middle, the Kerinci-Seblat National Park—the largest reserve in Sumatra—protects the watersheds of two of Sumatra's most important rivers: the Musi and Batang Hari (IUCN 1991). To the south, Bukit Barisan Selatan, another of Sumatra's large reserves, also extends into this ecoregion, covering more than 2,000 km².

Types and Severity of Threats

At the current rate of deforestation, Sumatra's remaining lowland rain forest will be completely gone within the next ten years unless drastic actions are taken to halt the rampant logging (Holmes, Essay 1). Given this ominous prediction, the only remaining natural forests in Sumatra will be the hill and montane forests of this ecoregion. This ecoregion is extremely fragile and sensitive to disturbance, especially in the upper montane and sub-alpine zones (Whitten et al. 2000). They probably will be targeted for intense logging activities, especially in light of the rampant illegal logging currently taking place throughout Indonesia. From 1985 to 1997, 15,000 km² of montane forest was destroyed, more than 1,000 km²/year. Since 1997 this annual rate of forest loss has increased gradually, and after the fall of the Suharto government and economic collapse of 1998, even the large protected areas such as Gunung Leuser, Kerinci Seblat, and Bukit Barisan Selatan national parks are threatened by encroachment and poaching.

One example of illegal logging inside Indonesia's protected areas is the unabated illegal logging that has occurred in and around the biologically valuable Gunung Leuser National Park since April 1999. The destruction includes prime habitat for the orangutan, siamang, and white-handed gibbon. In Gunung Kerinci Seblat National Park, Indonesia's first fully gazetted national park, illegal logging is increasing, and more than 400 families are staking claims along the road bordering the park. Kerinci-Seblat National Park is an important site for one of the last remaining populations of the Sumatran rhinoceros and Sumatran rabbit, as well as siamang and agile gibbon (Colijn, http://www.bart.nl/~edcolijn/).

Poaching is another threat to the great diversity of life in these forests. From 1990 to 1996 the number of Sumatran rhinoceros in Kerinci-Seblat National Park fell from 300 to about 30 (Mittermeier et al. 1999).

Priority Conservation Actions

With the rapid pace of forest destruction under way in Indonesia, quick conservation action is urgently needed to stave off the complete collapse of the remaining forests. However, steps should be taken to lay the groundwork for longer-term conservation as well.

Short-Term Conservation Actions (1–5 Years)

- Initiate a major public awareness campaign on forest values and forest loss to drive political will.
- Stop illegal logging in and around protected areas by challenging the authorities and judiciary to act on existing laws and regulations.
- Revoke forest concessions in areas of high biodiversity outside protected areas and institute a moratorium on new concessions for oil palm, timber, and other plantations, using the vast areas of deforested and cleared land instead.
- Mobilize major assistance for the more outstanding protected areas.
- Stop wildlife poaching, especially in key parks and reserves, by tightening surveillance around ports or airports and making the purchase of endangered species socially unacceptable.
- Establish a trust fund to offset the opportunity costs of maintaining forest cover and avoiding conversion to oil palm plantations.
- Determine whether Sumatran provincial governments can be provided funding from carbon credits for maintaining forest cover and wildlife protection.
- Conduct major biodiversity surveys and collecting expeditions in doomed areas to document the biodiversity that will soon be lost.

Longer-Term Conservation Actions (5–20 Years)

- Reassess representativeness of the protected area system in the ecoregion.
- Ensure that redesign of the protected area system includes linkages between lowland and montane habitats.
- Devise ways to maintain the integrity of buffer zones of protected areas.

TABLE J83.3. WCMC (1997) Protected Areas That Overlap with the Ecoregion.

Protected Area	Area (km²)	IUCN Category
Lingga Isaq [108]	790	VI
Gunung Leuser [82]	6,000	II
Dolok Sembelin [82]	110	VI
Dolok Surungan	320	IV
Dolok Sipirok	70	I
Malampah Alahan Panjang [82]	250	PRO
Lembah Harau	220	VI
Maninjau	230	VI
Gunung Sago/Malintang/Karas	80	VI
Gunung Singgalang	240	VI
Gunung Merapi	100	VIII
Kerinci Seblat [108]	7,960	II
Punguk Bingin	100	VI
Bukit Dingin/Gunung Dempo	530	VI
Bukit Hitam	790	VIII
Bukit Balal [82]	150	VI
Bukit Raja Mandara [82]	100	VIII
Gumai Pasemah [82]	320	IV
Isau-Isau Pasemah [82]	100	IV
Gunung Patah/Bepagut/Muara Duakisim	580	VI
Bukit Balai Rejang	710	VIII
Bukit Barisan Selatan [82]	1,620	II
Bukit Nantiogan Hulu/Nanti Komerung Hulu [82]	720	VI
Gunung Raya	750	IV
Tanggamus	50	VI
Gunung Betung	60	VI
Total	22,950	

Ecoregion numbers of protected areas that overlap with additional ecoregions are listed in brackets.

- Implement rigorous conservation actions in the Level I and Level II TCUs in the ecoregion.

Focal Species for Conservation Action

Endemic species: Sumatran cochoa, Sumatran ground-cuckoo, red-billed partridge, Sumatran pheasant, Salvadori's pheasant, Sumatran drongo, Schneider's pitta, Vanderbilt's babbler, *Presbytis thomasi*, *Hylopetes winstoni*, *Mus crociduroides*, *Rattus korinchi*, *Maxomys hylomyoides*, *M. inflatus*, and *Nesolagus Netscheri*

Area-sensitive species: tiger (Franklin et al. 1999; Tilson et al. 1994; Dinerstein et al. 1997; Seidensticker et al. 1999), Sumatran rhinoceros (Foose and van Strien 1997), orangutan (Rijksen and Meijaard 1999; Dijuwantoko 1991; Tilson et al. 1993)

Top predator species: tiger (Franklin et al. 1999; Tilson et al. 1994; Dinerstein et al. 1997; Seidensticker et al. 1999)

Other: *Rafflesia* species

Selected Conservation Contacts

ARSG

EU-Leuser

Lembaga Alam Tropika Indonesia (LATIN)

Sumatran Tiger Project

WAHLI

WARSI

WCS

Worldbank Kerinci, Telapak

WWF-Indonesia

Yayasan Keanekaragaman Hayati

Yayasan Konsorsium Pembaruan Agraria (KPA)

Justification for Ecoregion Delineation

We recognized six ecoregions on Sumatra. MacKinnon (1997) placed all the biomes in Sumatra within two subunits (21a and 21b), with the subunit division based on a faunal break to the south of Lake Toba. We used the 1,000-m contour from a DEM (USGS 1996) to delineate the montane forests along the Bukit Barisan–Gunung Leuser Mountain Range as the Sumatran Montane Rain Forests [83]. However, these montane forests include extensive forest areas over limestone. Whitmore (1984a) shows small, scattered patches of limestone forests in his vegetation map of Malesia. The FAO map (1974) shows no limestone. Much of the uncertainty probably arises because these limestone substrates do not show extensive outcropping (Whitmore 1984a). Based on recommendations by Tony Whitten (pers. comm., 1999), we treated these limestone forests as a distinct habitat type within the more broadly distributed montane moist forests rather than placing them in a separate ecoregion.

MacKinnon's biounit 21 largely corresponds to Udvardy's Sumatra biogeographic province. However, Udvardy did not include the Nicobar Islands. Eight ecoregions overlap Udvardy's Sumatra biogeographic province: Sumatran Lowland Rain Forests [82], Sumatran Montane Rain Forests [83], Mentawai Islands Rain Forests [84], Sumatran Peat Swamp Forests [85], Sumatran Freshwater Swamp Forests [88], Sundaland Heath Forests [90], Sumatran Tropical Pine Forests [105], and Sunda Shelf Mangroves [107].

Prepared by Colby Loucks and Tony Whitten

Ecoregion Number:	**84**
Ecoregion Name:	**Mentawai Islands Rain Forests**
Bioregion:	**Sunda Shelf and Philippines**
Biome:	**Tropical and Subtropical Moist Broadleaf Forests**
Political Units:	**Indonesia**
Ecoregion Size:	**6,500 km^2**
Biological Distinctiveness:	**Regionally Outstanding**
Conservation Status:	**Critical**
Conservation Assessment:	**II**

The Mentawai Islands Rain Forests [84] have had a long geographic isolation that has resulted in numerous endemic mammal species, including four primates. There are seventeen endemic mammal species (39 percent), which on a per-unit area ranks it with Madagascar in endemic mammal species, notably primates. Of the four endemic primate species, these forests have the world's only exclusively monogamous leaf-monkey, the Mentawai leaf-monkey.

Location and General Description

This ecoregion covers the moist forests of Mentawai Islands and Enganno Island, off the west coast of central Sumatra, Indonesia. Approximately 70 million years ago, the Indian subcontinent collided with the Asian landmass, forming the Himalayas. An associated thrust formed Sumatra's Barisan Mountains, and as the Barisan Range buckled upward, it formed a deep water channel to the west of Sumatra. During this time the islands of Simeulue and Enggano were formed. The Mentawai Islands separated from the Sumatran mainland via the Batu Islands more than half a million years ago. The rainfall on these islands is approximately 4,500 mm/year (Whitten et al. 2000). Based on the Köppen climate zone system, this ecoregion falls in the tropical wet climate zone (National Geographic Society 1999).

The natural vegetation of these islands is tropical lowland rain forest and is similar to but not as diverse as Sumatra's lowland rain forests. These forests are characterized by large buttressed trees, dominated by the Dipterocarpaceae family. There is an abundant presence of woody climbers and epiphytes. The general characteristics of these forests are canopies 24 to 36 m high, with emergents reaching more than 45 m. The emergent trees are dominated by dipterocarp species (*Dipterocarpus* spp., *Parashorea* spp., *Shorea* spp., and *Dryobalanops* spp.) and, to a lesser extent, species in the Caesalpiniaceae family (*Koompasia* spp., *Sindora* spp., and *Dialium* spp.). Dipterocarps also dominate much of the canopy layer, but there are many other tree families such as Burseraceae, Sapotaceae, Euphorbiaceae, Rubiaceae, Annonaceae, Lauraceae, and Myristicaceae (Whitten et al. 2000). Ground vegetation is limited to small-diameter trees, and an herbaceous layer is uncommon.

Biodiversity Features

The Mentawai Islands have been separated from the mainland for more than half a million years, and the long isolation has allowed the survival of relics of an early Indo-Malayan fauna as well as the evolution of many endemics (Whitten et al. 2000). Although this small ecoregion has a depauperate mammal fauna, consisting of just forty-four species, seventeen species are endemic and one is a near endemic (table J84.1).

Four endemic primates are found on the Mentawai Islands: the Mentawai gibbon (*Hylobates klossii*), Mentawai macaque (*Macaca pagensis*), Mentawai leaf-monkey (*Presbytis potenziani*), and snub-nosed monkey (*Simias concolor*). Almost all monkeys live in single- or multi-adult male groups with more females than adult males. There is only one exception: the Mentawai leaf-monkey. Like gibbons, it lives in permanent monogamous groups (one male and one female) within home ranges. The snub-nosed monkey is in an endemic primate genus but is related to Borneo's proboscis monkey (*Nasalis larvatus*). The primary difference is that the snub-nosed monkey has a shorter tail (Whitten et al. 2000). The Mentawai gibbon has one of the most splendid songs of any land mammal.

Numerous other mammals found on the Mentawai islands are also endemic, and these include the Mentawai palm civet (*Paradoxurus lignicolor*), Mentawai squirrel (*Callosciurus melanogaster*), Mentawai three-striped squirrel (*Lariscus obscurus*), and Mentawai flying squirrel (*Hylopetes sipora*) (Whitten et al. 2000).

Enggano Island probably was never connected to Sumatra and has extremely impoverished mammal faunas. However, Enggano is an EBA (159) containing two restricted-range bird species (Stattersfield et al. 1998). There are three endemic species (table J84.2).

Status and Threats

Current Status

Although the forest cover data indicate that over 60 percent of the habitat is intact, in recent years larger-scale forestry operations have taken hold. There are three protected areas that cover 1,090 km^2 (18 percent) of the ecoregion (table J84.3). Two of the three protected areas in the ecoregion are small, but one is almost 1,000 km^2, although it is unknown how well protected these reserves are in light of the recent logging operations.

Types and Severity of Threats

The primary forests of the Mentawai Islands remained essentially intact until the influx of settlers from mainland Sumatra created population pressure and disrupted traditional management practices (WWF and IUCN 1995). Extensive areas of land have been cleared for cash crops, and logging has become a serious threat (WWF and IUCN 1995), especially to the islands' four endemic primates (IUCN 1991).

Priority Conservation Actions

With the rapid pace of forest destruction under way in Indonesia, quick conservation action is urgently needed to stave off the complete collapse of these forests. However, steps should be taken to lay the groundwork for longer-term conservation as well.

Short-Term Conservation Actions (1–5 Years)

- Initiate a major public awareness campaign on forest values and forest loss to drive political will.
- Stop illegal logging in and around protected areas by challenging the authorities and judiciary to act on existing laws and regulations.
- Revoke forest concessions in areas of high biodiversity outside protected areas and institute a moratorium on new concessions for oil palm, timber, and other plantations, using the vast areas of deforested and cleared land instead.
- Place under active protection all remaining forests.
- Mobilize major assistance for the more outstanding protected areas.
- Stop wildlife poaching, especially in key parks and reserves, by tightening surveillance around ports or airports and

TABLE J84.1. Endemic and Near-Endemic Mammal Species.

Family	Species
Rhinolophidae	Hipposideros breviceps*
Cercopithecidae	Presbytis potenziani*
Cercopithecidae	Nasalis concolor*
Cercopithecidae	Macaca pagensis*
Hylobatidae	Hylobates klossi*
Viverridae	Paradoxurus lignicolor*
Sciuridae	Callosciurus melanogaster*
Sciuridae	Lariscus obscurus*
Muridae	Iomys sipora*
Sciuridae	Hylopetes sipora*
Muridae	Petinomys lugens*
Muridae	Rattus lugens*
Muridae	Rattus adustus*
Muridae	Rattus enganus*
Muridae	Chiropodomys karlkoopmani*
Muridae	Leopoldamys siporanus*
Muridae	Maxomys pagensis*
Muridae	Rattus hoogerwerfi

An asterisk signifies that the species' range is limited to this ecoregion.

TABLE J84.2. Endemic and Near-Endemic Bird Species.

Family	Common Name	Species
Cuculidae	Enggano scops-owl*	Otus enganensis*
Strigidae	Mentawai scops-owl*	Otus mentawi*
Zosteropidae	Enggano white-eye*	Zosterops salvadorii*

An asterisk signifies that the species' range is limited to this ecoregion.

TABLE J84.3. WCMC (1997) Protected Areas That Overlap with the Ecoregion.

Protected Area	Area (km^2)	IUCN Category
Tai-tai Batti	930	IV
Muara Siberut	80	PRO
Gunung Nanu'ua	80	VI
Total	1,090	

Ecoregion numbers of protected areas that overlap with additional ecoregions are listed in brackets.

making the purchase of endangered species socially unacceptable.
- Establish a trust fund to offset the opportunity costs of maintaining forest cover and avoiding conversion to oil palm plantations.
- Determine whether Sumatran provincial governments can be provided funding from carbon credits for maintaining forest cover and wildlife protection.
- Conduct major biodiversity surveys and collecting expeditions in doomed areas to document the biodiversity that will soon be lost.

Longer-Term Conservation Actions (5–20 Years)
- Establish a long-term fund for conservation of lowland forests in Indonesia that would benefit Sumatra.
- Devise ways to maintain the integrity of buffer zones of protected areas.

Focal Species for Conservation Action

Selected endemic species: Mentawai macaque, Mentawai gibbon, Mentawai leaf-monkey, snub-nosed monkey, Mentawai palm civet, Mentawai squirrel, Mentawai three-striped squirrel, Sipura flying squirrel, Mentawai scops-owl, Enggano scops-owl, and Enggano white-eye

Selected Conservation Contacts

Lembaga Alam Tropika Indonesia (LATIN)
UNESCO
WWF-Indonesia
Yayasan Keanekaragaman Hayati
Yayasan Konsorsium Pembaruan Agraria (KPA)

Justification for Ecoregion Delineation

We placed the Mentawai Islands and Enggano Island—MacKinnon's subunit 21c—into a distinct ecoregion, the Mentawai Islands Rain Forests [84].

MacKinnon's biounit 21 largely corresponds to Udvardy's Sumatra biogeographic province. However, Udvardy did not include the Nicobar Islands. Eight ecoregions overlap Udvardy's Sumatra biogeographic province: Sumatran Lowland Rain Forests [82], Sumatran Montane Rain Forests [83], Mentawai Islands Rain Forests [84], Sumatran Peat Swamp Forests [85], Sumatran Freshwater Swamp Forests [88], Sundaland Heath Forests [90], Sumatran Tropical Pine Forests [105], and Sunda Shelf Mangroves [107].

Prepared by Colby Loucks and Tony Whitten

Ecoregion Number:	**85**
Ecoregion Name:	**Sumatran Peat Swamp Forests**
Bioregion:	**Sunda Shelf and Philippines**
Biome:	**Tropical and Subtropical Moist Broadleaf Forests**
Political Units:	**Indonesia**
Ecoregion Size:	**87,100 km^2**
Biological Distinctiveness:	**Regionally Outstanding**
Conservation Status:	**Critical**
Conservation Assessment:	**II**

The Sumatran Peat Swamp Forests [85] are a distinctive forest type, and their biodiversity is characteristic of the habitat. The peat swamp forests in Indonesia are less threatened than the freshwater swamp forests. This is partly because of their low nutrient levels, which limit the productivity of their vegetation, including agricultural crops. However, despite their poor productivity in the past five years, significant areas of peat swamp forests have been burned in Indonesia, and less than one-half of these forests remain.

Location and General Description

This ecoregion represents the peat swamp forests along the eastern coast of the island of Sumatra, Indonesia. Based on the Köppen climate zone system, this ecoregion falls in the tropical wet climate zone (National Geographic Society 1999).

The peat swamp forests of Sumatra have similar characteristics to those in Borneo and peninsular Malaysia. Peat soil is composed of more than 65 percent organic matter (Driessen 1978). Most peat deposits found behind mangroves along the coast are ombrogenous, or rain-fed peat swamps (Driessen 1977; Morley 1981). Peat swamp forests are formed when rivers drain into the inland edge of a mangrove and the sediments are trapped behind the tangle of mangrove roots. These areas begin to build up and flood less often as the coastline extends outward. The peat deposits usually are at least 50 cm thick but can extend up to 20 m. Peat swamps are domed and are rarely flooded. Because peat swamps are not drained by flooding, they are nutrient deficient and acidic, with a pH usually less than 4. Compared with other moist forest ecoregions, peat forests—at least in the lowlands—are not as species-rich and are not high in endemism (WWF and IUCN 1991).

There is not a single type of peat swamp forest but rather a gradation of forests types along a nutrient gradient. The edges of peat swamp forests are relatively nutrient-rich, whereas the center is nutrient-poor (Whitten et al. 2000). Likewise, the forest becomes smaller with an even canopy, moving from the edges to the center. Whereas Borneo has up to six types of peat swamp forest, Sumatra retains only two: a mixed peat swamp forest and a pole forest.

Both mixed swamp forest and pole forest have few tree species, but a pole forest may have a higher density. Mixed swamp forests tend to have a larger average diameter and basal area. Some characteristic species from these forests include *Tristania obovata, Ploiarium alternifolium, Polyalthia glauca, Stemonurus secundiflorus, Radermachera gigantea,*

Salacca conferta, *Livistona hasseltii*, and *Cyrtostachys lakka* (Whitten et al. 2000). Palms are not common, but several species generally are confined to these forests. The emergent *Livistona hasseltii* is characteristic, as is the bright-red sealing wax palm *Cyrtostachys lakka*.

Biodiversity Features

Peat swamp forests do not support an abundance of terrestrial wildlife, and none of the mammals are considered endemic. The Sumatran tiger (*Panthera tigris*) is Indonesia's largest terrestrial predator and is critically endangered (Hilton-Taylor 2000). The Sumatran tiger lives in lowland and montane rain forest and frequent peat swamp forests throughout Sumatra. An estimated 500 Sumatran tigers remain in Sumatra (Franklin et al. 1999). There are three Level II TCUs in Sumatra that overlap this ecoregion (Dinerstein et al. 1997). The Asian elephant (*Elephas maximus*) is found in numerous populations throughout Sumatra. However, only five populations have more than 200 individuals, but habitat loss has placed the survival of many of these populations at risk (Sukumar 1989). The number of bird species tends to be lower in peat swamp forest than in the surrounding lowland rain forests, and there are no endemic or near-endemic species.

Status and Threats

Current Status

More than half of the habitat in this ecoregion has been cleared, especially in the southern portion, where only a few blocks of habitat remain. Large areas of swamp have been drained, mainly for transmigration settlements and large-scale development projects, making this a highly vulnerable ecoregion. There are thirteen protected areas that extend into the ecoregion to cover 4,730 km² (5 percent) of the area (table J85.2). However, many of the protected areas are proposed, and the official status is still uncertain. Of the gazetted protected areas, only Berbak is greater than 1,000 km².

TABLE J85.1. WCMC (1997) Protected Areas That Overlap with the Ecoregion.

Protected Area	Area (km²)	IUCN Category
Bakau Selat Dumai [110]	320	PRO
Bukit Batu	230	PRO
Siak Kecil	940	PRO
Danau Tanjung Padang	110	PRO
Giam Duri [88]	200	PRO
Pulau Burung [110]	260	I
Berbak [110]	1,310	II
Istana Sultan Siak	240	PRO
Danau Belat/Besar Serkap	80	PRO
Sarang Barung	40	PRO
Kerumutan	110	IV
Kerumutan Lama	40	PRO
Padang Sugihan	850	IV
Total	4,730	

Ecoregion numbers of protected areas that overlap with additional ecoregions are listed in brackets.

Types and Severity of Threats

In some areas of southern Sumatra, the peat swamp has been drained for transmigration and other major development projects. The drainage of one area dries neighboring areas. Therefore, fires are common, preventing natural succession and promoting the development of extensive, nearly monospecific stands of paperbark (*Melaleuca cajuputih*) (Whitten et al. 2000). In areas where the peat itself is burned, small, shallow lakes form and become covered with floating islands of grasses and herbs (Whitten et al. 2000). Large-scale plantations, illegal logging, and timber enterprises have also led to increasing deforestation with resultant erosion and sedimentation of nearby rivers (WWF-Indonesia n.d.). Coconuts are grown along the coast, and drained swamps are used for pineapple plantations (MacKinnon and MacKinnon 1986). Logging concessions cover almost 80 percent of the ecoregion's remaining habitat and pose a serious threat to habitat integrity and conservation.

Priority Conservation Actions

With the rapid pace of forest destruction under way in Indonesia, quick conservation action is urgently needed to stave off the complete collapse of these forests. However, steps should be taken to lay the groundwork for longer-term conservation as well.

Short-Term Conservation Actions (1–5 Years)

- Initiate a major public awareness campaign on forest values and forest loss to drive political will.
- Stop illegal logging in and around protected areas by challenging the authorities and judiciary to act on existing laws and regulations.
- Revoke forest concessions in areas of high biodiversity outside protected areas and institute a moratorium on new concessions for oil palm, timber, and other plantations, using the vast areas of deforested and cleared land instead.
- Place all remaining forests under active protection.
- Mobilize major assistance for the more outstanding protected areas.
- Stop wildlife poaching, especially in key parks and reserves, by tightening surveillance around ports or airports and making the purchase of endangered species socially unacceptable.
- Establish a trust fund to offset the opportunity costs of maintaining forest cover and avoiding conversion to oil palm plantations.
- Purchase through Indonesian NGOs several large tracts of lowland forest, create model ecotourism reserves with elephant and foot safaris to view wildlife and attract foreign currency, and share profits from such ventures with communities.
- Determine whether Sumatran provincial governments can be provided funding from carbon credits for maintaining forest cover and wildlife protection.
- Conduct major biodiversity surveys and collecting expeditions in doomed areas to document the biodiversity that will soon be lost.

Longer-Term Conservation Actions (5–20 Years)

- Establish a long-term fund for conservation of lowland forests in Indonesia that would benefit Sumatra.
- Ensure that redesign of the protected area system includes linkages between lowland and montane habitats.
- Devise ways to maintain the integrity of buffer zones of protected areas.
- Implement rigorous conservation actions in TCUs in the ecoregion.

Focal Species for Conservation Action

Area-sensitive species: tiger (Franklin et al. 1999; Tilson et al. 1994; Dinerstein et al. 1997; Seidensticker et al. 1999), Asian elephant (Santiapillai and Jackson 1990; Santiapillai 1997), Sumatran rhinoceros (Foose and van Strien 1997)

Top predator species: tiger (Franklin et al. 1999; Tilson et al. 1994; Dinerstein et al. 1997; Seidensticker et al. 1999)

Selected Conservation Contacts

ARSG

EU-Leuser

Lembaga Alam Tropika Indonesia (LATIN)

Sumatran Tiger Project

WAHLI

WARSI

WCS

Worldbank Kerinci, Telapak

WWF-Indonesia

Yayasan Keanekaragaman Hayati

Yayasan Konsorsium Pembaruan Agraria (KPA)

Justification for Ecoregion Delineation

Whitmore (1984a) and MacKinnon (1997) showed large extents of peat swamp forests along the northern coast of Sumatra, especially in Riau Province. We delineated the Sumatran Peat Swamp Forests [85] to represent these forests but extracted the smaller patches of freshwater swamp forests into the Sumatran Freshwater Swamp Forests [88] and the mangroves in the Sunda Shelf Mangroves [107].

MacKinnon's biounit 21 largely corresponds to Udvardy's Sumatra biogeographic province. However, Udvardy did not include the Nicobar Islands. Eight ecoregions overlap Udvardy's Sumatra biogeographic province: Sumatran Lowland Rain Forests [82], Sumatran Montane Rain Forests [83], Mentawai Islands Rain Forests [84], Sumatran Peat Swamp Forests [85], Sumatran Freshwater Swamp Forests [88], Sundaland Heath Forests [90], Sumatran Tropical Pine Forests [105], and Sunda Shelf Mangroves [107].

Prepared by Colby Loucks and Tony Whitten

Ecoregion Number:	**86**
Ecoregion Name:	**Borneo Peat Swamp Forests**
Bioregion:	**Sunda Shelf and Philippines**
Biome:	**Tropical and Subtropical Moist Broadleaf Forests**
Political Units:	**Indonesia, Malaysia, Brunei**
Ecoregion Size:	**67,100 km^2**
Biological Distinctiveness:	**Regionally Outstanding**
Conservation Status:	**Critical**
Conservation Assessment:	**II**

Although peat swamp forests are not as biodiverse as neighboring lowland rain forests, the Borneo Peat Swamp Forests [86] are some of the most speciose peat swamp forests in the region. Peat swamp forests are a key habitat for the endangered Borneo endemic and highly unique proboscis monkey (*Nasalis larvatus*). They are also home to the world's most desirable aquarium fish, the arowana (*Scleropages formosus*).

Location and General Description

This ecoregion is made up of the peat swamp forests along the western coasts of the island of Borneo, within the Malaysian state of Sarawak and Indonesian Kalimantan. Most of the peat swamp forests are associated with coastal areas, but two large areas of peat swamp forests occur around Lake Mahakam and Lake Kapuas. Based on the Köppen climate zone system, this ecoregion falls in the tropical wet climate zone (National Geographic Society 1999).

The peat swamp forests of Borneo have vegetative and edaphic characteristics similar to those in Sumatra and peninsular Malaysia. The peat soil is predominantly organic matter, which built up behind mangrove swamps. They are ombrogenous, or rain fed, and are recent in origin (Driessen 1978; Morley 1981). Peat swamp forests form when sediments build up behind mangroves as rivers drain toward the coast. Over time, these areas build up and eventually form domes that are rarely flooded. Organic matter builds up, and peat deposits can extend up to 20 m. Because peat swamps are not drained by flooding, they are nutrient deficient and acidic (pH usually is less than 4). Compared with other moist forest ecoregions, peat swamp forests are not as species-rich or high in endemism (IUCN 1991).

Peat swamp forests encompass a sequence of forest types distributed from the perimeter to the center of each swamp. Six forest communities that have a distinct structure, physiognomy, and flora are discernible (Anderson 1983; Whitmore 1984b). The first type of forest is similar to yet less rich than lowland dipterocarp evergreen rain forests that occur on mineral soils. These forests are dominated by *Gonystylus bancanus* (the single most valuable timber species), *Dactylocladus stenostachys*, *Copaifera palustris*, and four *Shorea* species (Anderson 1983). *Shorea albida* plays a major role in the swamp forest communities (Whitmore 1984b; IUCN 1991) and dominates forest types two through four (Anderson 1983). Forest type four is also characterized by *Calophyllum obliquinervum*, *Cratoxylum glaucum*, and *Combretocarpus rotundatus* (Anderson 1983). The principal species in forest type five are *Tristania obovata, Palaquium cochleariifolium,* and *Parastemon spicatum,* and type six resembles open savanna woodland, with the most abundant species being *Dactylocladus stenostachys, Garcinia cuneifolia, Litsea crassifolia,* and *Parastemon spicatum* (Anderson 1983). Other genera of trees often found in Sarawak's peat swamp forests include *Dryobalanops* and *Melanorrhea* (IUCN 1991).

Most of the tree families of lowland dipterocarp forests are also found in peat swamp forests. Exceptions include Proteaceae, Lythraceae, Combretaceae, and Styraceae (Whitmore 1984b). Few plant species are endemic to peat swamp forests, mainly because of their recent formation. Many species found in the most acidic central portion of peat swamp forests also occur in heath forests (Sundaland Heath Forests [90]). Brünig (1973) found 146 species common to both forest types. More than thirty palm species are found in peat swamp forest, including the red-stemmed sealing wax palm, *Cyrtostachys lakka*.

Biodiversity Features

Many animal species occur in peat swamp forests, but only the bat, *Hipposideros doriae*, and the Javan white-eye (*Zosterops flavus*) are considered near endemic (table J86.1, table J86.2). With only two exceptions, monkeys, gibbons, and orangutans are all found in Borneo's peat swamp forests, but at lower densities. Long-tailed macaques (*Macaca fascicularis*) and silvered langurs (*Presbytis cristata*) have higher densities in peat swamp forests than in lowland rain forests, but only along rivers (Wilson and Wilson 1975; Marsh and Wilson 1981; Davies and Payne 1982; MacKinnon 1983). Forest productivity is higher at the river's edge, with additional nutrient and light inputs.

Peat swamp forests are key habitats for the unique proboscis monkey (*Nasalis larvatus*). Proboscis monkeys are found only in coastal and riverine habitats in Borneo. Proboscis monkeys are good swimmers and will swim across rivers, despite the presence of crocodilian predators. In the afternoon they are often found along the rivers, where they sleep in tall lookout trees. They eat primarily young leaves and the seeds of unripe fruit (Bennett and Sebastian 1988; Yeager 1989). Like other colobines, they have developed highly complex sacculated stomachs with specialized bacteria to digest this diet (Bauchop 1978; Bauchop and Martucci 1968).

Bird species diversity tends to be lower in peat swamp forests than in the surrounding lowland rain forests. However, in Tanjung Puting National Park, a freshwater and peat swamp reserve in Kalimantan, more than 200 bird species were recorded.

One of the most desirable and rare aquarium fish, the arowana (*Scleropages formosus*), is found in deep pools in peat swamp rivers. These rivers also support other typical riverine fauna such as otters, waterbirds, false gavials, crocodiles, and monitor lizards (Giesen 1987).

Status and Threats

Current Status

Peat swamp forests used to be extensive in Sarawak and also occurred in southwestern Sabah (IUCN 1991). But today about half of the area has been cleared. Brunei's peat swamp forests probably are less disturbed than those elsewhere in the region. Some of the forests in the Belait River are still in pristine condition and will soon represent the only undisturbed forests of this type in the region (WWF and IUCN 1995). The eleven protected areas cover 4,300 km^2 (6 percent) of the ecoregion. Tanjung Puting, Muara Sebuku, and Kelompok Hutan Kahayan all protect more than 500 km^2 of contiguous forest (table J86.3). The peat swamp forests are represented in Brunei's protected area network even though these forests are most intact there.

In 1997–1998, drought-driven fires intentionally set to clear forests for commercial agriculture and forestry companies destroyed vast tracts of Borneo's lowland forests. More than 7,500 km^2 of peat swamp forests were burned in this two-year period. Kutai and Muara Kaman suffered extensive, severe damage during the fires, with almost no area left untouched (IFFM/GTZ 1998; Yeager 1999a). Peat swamp forests are particularly vulnerable to fire and produce the most carcinogenic haze of any forest type when they are burned because of the release of large amounts of fine particulate matter. The widespread burning of these forests contributed a significant portion of the haze, which covered most of Indonesia and Malaysia and extended north to Thailand and

TABLE J86.1. Endemic and Near-Endemic Mammal Species.

Family	Species
Rhinolophidae	Hipposideros doriae

An asterisk signifies that the species' range is limited to this ecoregion.

TABLE J86.2. Endemic and Near-Endemic Bird Species.

Family	Common Name	Species
Pycnonotidae	Hook-billed bulbul	Setornis criniger

An asterisk signifies that the species' range is limited to this ecoregion.

TABLE J86.3. WCMC (1997) Protected Areas That Overlap with the Ecoregion.

Protected Area	Area (km^2)	IUCN Category
Muara Sebuku	570	PRO
Muara Kayan [110]	400	PRO
Unnamed	80	?
Gunung Palung [96]	250	II
Mandor	50	I
Muara Kendawangan [89, 110]	370	PRO
Tanjung Puting [89, 90, 110]	1,120	II
Kelompok Hutan Kahayan [110]	680	PRO
Danau Semayang Sungay Mahakam [89]	330	PRO
Kutai (extension) [89, 90, 97]	160	PRO
Muara Kaman Sedulang [89, 90]	290	I
Total	4,300	

Ecoregion numbers of protected areas that overlap with additional ecoregions are listed in brackets.

Bangkok. Peat fires typically burn underground as well, eliminating the seedbank and destroying soil, which may take thousands of years to replace (Yeager 1999a).

The fires also had adverse effects on the wildlife populations. Unknown numbers of birds, reptiles, amphibians, primates, and other mammals died in the fires or shortly after because of the scarcity of food. Hundreds of orangutans were killed by villagers for meat, and their orphaned babies were sold to the international pet trade as they fled into villages to escape the fires. Fires are a major threat to the continued existence of the endangered orangutan (Yeager 1999b). The charismatic proboscis monkey was the primate species that lost the greatest percentage of habitat to the fires because large areas of riverine and coastal habitats were destroyed in the fires (Yeager and Russon 1998; Yeager and Fredriksson 1999).

Types and Severity of Threats

Peat swamp forests were the first formations to be logged on a commercial scale in Sarawak and for many years were the main source of timber (IUCN 1991). In Brunei, peat swamp forests are still intact but are threatened by the planned expansion of forestry operations, which may result in overexploitation of forests rich in *Shorea albida* (WWF and IUCN 1995). The continual threat of fires clearing forest for oil palm and other commercial agriculture crops will loom over Indonesia until stricter forest policies and protection are taken.

Priority Conservation Actions

With the rapid pace of forest destruction under way in all of Indonesia, quick conservation action is urgently needed to stave off the complete collapse of these forests. However, steps should be taken to lay the groundwork for longer-term conservation as well.

Short-Term Conservation Actions (1–5 Years)

- Call for an international conference to highlight the impending biodiversity crisis and loss of forests in Indonesia (perhaps using World Bank as an organizer).
- Stop illegal logging in and around protected areas.
- Revoke forest concessions in areas of high biodiversity outside protected areas and institute a moratorium on new concessions for oil palm, timber, and other plantations.
- Place under protection all remaining forests.
- Stop wildlife poaching, especially in key parks and reserves.
- Establish a trust fund to offset the opportunity costs of maintaining forest cover and avoiding conversion to oil palm plantations.
- Purchase through Indonesian NGOs several large tracts of lowland forest, create model ecotourism reserves with elephant and foot safaris to view wildlife and attract foreign currency, and share profits from such ventures with communities.
- Determine whether Kalimantan and Sarawak provincial governments can be provided funding from carbon credits for maintaining forest cover and protecting wildlife.

Longer-Term Conservation Actions (5–20 Years)

- Reassess representativeness of protected area system in the ecoregion.
- Establish a long-term fund for conservation of lowland forests in Indonesia.
- Restore integrity of buffer zones of protected areas.

Focal Species for Conservation Action

Habitat specialists: proboscis monkey (Salter and MacKenzie 1985; Bennet and Gombek 1993; Yeager 1989, 1990, 1991; Yeager and Blondal 1992)

Area-sensitive species: orangutan (Galdikas 1971–1977; Dijuwantoko 1991; Tilson et al. 1993), hornbills

Top predator species: python, black eagle, crested sea eagle, brahminy kite, clouded leopard, golden bay cat, Malay civet, false gavial

Other: primates, Storm's stork, Malaysian sun bear, mouse deer, barking deer

Selected Conservation Contacts

CI-Indonesia
European Union Forest Management Projects
GTZ Sustainable Forest Management Projects
Lembaga Alam Tropika Indonesia (LATIN)
TNC-Indonesia
WCS
WWF-Indonesia
WWF-Malaysia
Yayasan Keanekaragaman Hayati
Yayasan Konsorsium Pembaruan Agraria (KPA)

Justification for Ecoregion Delineation

The large island of Borneo was divided into nine ecoregions. Most of the island's lowland and submontane forests are dominated by dipterocarp species (MacKinnon et al. 1996). MacKinnon and MacKinnon (1986) divided the island's lowland forests into six subunits, with a central subunit representing the montane forests. MacKinnon (1997) revised the boundaries of these seven subunits but retained the same general configuration. These authors used the major rivers, the Kapuas and Barito, to represent zoogeographic barriers to a few species of mammals and based subunits largely on these barriers but also used climatic regimes for the drier eastern biounits (MacKinnon and MacKinnon 1986; MacKinnon 1997).

Because ecoregions are based on biomes, we first isolated the central montane ecoregion—the Borneo Montane Rain Forests [95]—above the 1,000-m elevation contour using the DEM (USGS 1996). We then assigned the large patches of peat forests, heath forests, freshwater swamp forests, and mangroves, in the lowlands and along the periphery of the island, into their own ecoregions: the Borneo Peat Swamp Forests [86], Sundaland Heath Forests [90] (which also includes Belitung Island and the heath forests in Bangka island), Southern Borneo Freshwater Swamp Forests [89], and Sunda Shelf Mangroves [107], respectively.

Prepared by Colby Loucks

Ecoregion Number:	**87**
Ecoregion Name:	**Peninsular Malaysian Peat Swamp Forests**
Bioregion:	**Sunda Shelf and Philippines**
Biome:	**Tropical and Subtropical Moist Broadleaf Forests**
Political Units:	**Malaysia**
Ecoregion Size:	**3,600 km^2**
Biological Distinctiveness:	**Globally Outstanding**
Conservation Status:	**Endangered**
Conservation Assessment:	**I**

The Peninsular Malaysian Peat Swamp Forests, though not as diverse in species as the surrounding lowland rain forests, are home to many of Malaysia's endangered species. Asian elephants, Sumatran rhinoceros, tigers, clouded leopards, and Malayan tapir all inhabit these rapidly shrinking forests.

Location and General Description

This ecoregion represents the disjunct peat swamp forests in peninsular Malaysia and southern Thailand. Based on the Köppen climate zone system, this ecoregion falls in the tropical wet climate zone (National Geographic Society 1999).

The peat swamp forests of peninsular Malaysia have edaphic and vegetative characteristics similar to those in Sumatra and Borneo. The soil is infertile and primarily organic matter (Driessen 1978). Peat deposits found behind mangroves are recent in origin. They are formed when rivers drain into the inland edge of a mangrove and trap the sediments within their tangle of roots. As these areas build up, they flood less frequently. The peat deposits can extend between 50 cm and 20 m (Driessen 1977; Morley 1981). Because peat swamps are not drained by flooding, they are acidic (pH usually is less than 4) and nutrient deficient. Compared with other lowland rain forest ecoregions, peat forests are not as species-rich and have fewer endemic species (IUCN 1991).

The ecoregion plays a significant role in acting as a sink for water from surrounding habitat. The edge of this forest is characterized by strangler figs (*Ficus* spp.), whose fruits provide an important source of food for many of the mammal, bird, and fish species (Payne et al. 1994). The vegetation is not dominated by a single dipterocarp (such as *Shorea albida* in Borneo), but *Shorea macrophylla* (bintangor) is an important timber tree. Pandan and the sealingwax palm (*Cyrtostachys lakka*) give the understory a characteristic appearance.

Biodiversity Features

The freshwater swamp forest, fauna is much more diverse than the fauna of peat swamp forests, being more similar to lowland rain forests. Many of the characteristic species of lowland rain forests are also found here. There are five large endangered mammals found in this ecoregion: the tiger (*Panthera tigris*), Asian elephant (*Elephas maximus*), Sumatran rhinoceros (*Dicerorhinus sumatrensis*), Malayan tapir (*Tapirus indicus*), and clouded leopard (*Pardofelis nebulosa*). The avifauna is not as diverse as that of the Malaysian lowland rain forests and does not include any endemic species, but the grey-headed fish eagle *Ichthyophaga ichthyaetus* and brown fish owl *Ketupa zeylonensis* are thought to be specialists in this habitat. The importance of peat swamp forest for freshwater biodiversity, and in particular small fish diversity (e.g., species of *Betta*), has recently been demonstrated by surveys by Wetlands International.

Status and Threats

Current Status

More than half of this small ecoregion has been cleared or degraded. Most of the remaining habitat is found along the eastern coast of peninsular Malaysia. The ecoregion does not receive any formal protection. These forests did not escape the pervasive forest fires that raged through Malaysia and Indonesia in 1997–1998.

Types and Severity of Threats

These forests are threatened by logging, tin mining, and clearance for agriculture including rice, rubber, coconut, and oil palm. Coastal swamp forests continue to be cleared for development, and drainage of neighboring land can draw down the water table. Until recently water extraction has been an unlicensed activity; attention has been drawn to this problem by water use for eel farming in Pahang. The combination of disturbance, lowering of the water table, and surrounding activities has increased risks of fire; significant areas near the Pahang coast were burnt in 1997–1998.

Priority Conservation Actions

With the rapid pace of forest destruction, quick conservation action is urgently needed to stave off the complete collapse of these forests. However, steps should be taken to lay the groundwork for longer-term conservation as well.

Short-Term Conservation Actions (1–5 Years)

- Establish formal protection of the remaining habitat in this ecoregion.
- Stop illegal logging in and around this ecoregion.
- Revoke forest concessions in areas of high biodiversity and institute a moratorium on new concessions for oil palm, timber, and other plantations.
- Stop wildlife poaching, especially in key parks and reserves.

Longer-Term Conservation Actions (5–20 Years)

- Establish an effective protected area system in the ecoregion.

Focal Species for Conservation Action

Habitat specialists: grey-headed fish eagle, brown fish owl

Area-sensitive species: tiger (Franklin et al. 1999; Tilson et al. 1994; Dinerstein et al. 1997; Seidensticker et al. 1999), Asian elephant (Santiapillai and Jackson 1990; Santiapillai 1997), Sumatran rhinoceros (Foose and van Strien 1997)

Top predator species: tiger (Franklin et al. 1999; Tilson et al. 1994; Dinerstein et al. 1997; Seidensticker et al. 1999)

Selected Conservation Contacts

WCS

WWF-Malaysia

Justification for Ecoregion Delineation

The Peninsular Malaysian Rain Forests [80] ecoregion represents the large extent of the lowland broadleaf rain forests extending south of the Kangar-Pattani line to Singapore. The tropical montane evergreen moist forests above 1,000 m were placed in the Peninsular Malaysian Montane Rain Forests [81]. We also extracted the large areas of peat swamp forests along the coast of peninsular Malaysia into the Peninsular Malaysian Peat Swamp Forests [87]. Thus, we created three ecoregions within MacKinnon's Malay Peninsular subunit (07a). Udvardy (1975) combined all of Borneo into the Borneo biogeographic province.

Prepared by Colby Loucks

Ecoregion Number:	**88**
Ecoregion Name:	**Sumatran Freshwater Swamp Forests**
Bioregion:	**Sunda Shelf and Philippines**
Biome:	**Tropical and Subtropical Moist Broadleaf Forests**
Political Units:	**Indonesia**
Ecoregion Size:	**18,000 km²**
Biological Distinctiveness:	**Regionally Outstanding**
Conservation Status:	**Critical**
Conservation Assessment:	**II**

The Sumatran Freshwater Swamp Forests [88] contain many of the endangered and characteristic Sumatran species found in the lowland rain forests. The endangered tiger, Asian elephant, estuarine crocodile, false gharial, clouded leopard, several primate species, and a multitude of waterbirds all live in freshwater swamp forests. All will be in danger of local extinction within ten years. The freshwater swamp forests have been decimated in Sumatra. The forests are highly productive and have been cleared by illegal logging and slash-and-burn agriculture to establish plantations and agricultural fields.

Location and General Description

This ecoregion represents the disjunct patches of freshwater swamp forests along the eastern alluvial plain on the Indonesian island of Sumatra. Based on the Köppen climate zone system, this ecoregion falls in the tropical wet climate zone (National Geographic Society 1999).

The major difference between freshwater swamp forests and peat swamp forests is the lack of deep peat, and the source of water is riverine and rainwater. Freshwater swamp forests grow on fertile alluvial soils, and the wide variety of soils is reflected in a diversity of vegetation types that ranges from grassy marshes to palm or *Pandanus*-dominated forest and forests similar in structure and composition to lowland rain forests. Trees with buttresses, stilt roots, and pneumatophores are common in some areas (Whitten et al. 1987). Trees in freshwater swamp forests endure prolonged periods of flooding, causing the soils to become anaerobic. Pneumatophores, specialized respiratory structures on the roots, are common on many tree species and assist in respiration during oxygen-poor periods. Emergent trees attain heights of 50–60 m (FAO 1981). These forests are floristically very variable; the dominant species include *Adina*, *Alstonia*, *Campnosperma*, *Coccoceras*, *Dillenia*, *Dyera*, *Erythrina*, *Eugenia*, *Ficus*, *Gluta*, *Lophopetalum*, *Memecylon*, *Metroxylon*, *Pandanus*, *Pentaspadon*, *Shorea*, and *Vatica* spp. (FAO 1981). *Melaleuca leucadendron* covers extensive areas in the south of this ecoregion (FAO 1981).

Biodiversity Features

The freshwater swamp forest fauna is much more diverse than the fauna of peat swamp forests, being more similar to that of lowland rain forests. Many of the characteristic species of lowland rain forests are also found here, although these forests contain no endemic mammal or bird species. However, they are home to Asia's largest terrestrial mammal, the Asian elephant (*Elephas maximus*), which is found in numerous populations throughout Sumatra. However, only five populations number more than 200 individuals, and with the rapid habitat loss the survival of many of these populations is uncertain. Way Kambas, a freshwater swamp forest reserve in southern Sumatra, contains one of these populations (Sukumar 1989). The endangered Malayan tapir (*Tapirus indicus*), the largest of the four living tapir species and only Old World representative, is also a resident of these forests. The Sumatran population of the Malayan tapir is close to extinction, with no more than fifty animals left in the wild (McClung 1997). Another endangered species, the Sumatran tiger (*Panthera tigris*), Indonesia's largest terrestrial predator, lives in lowland, montane, and freshwater swamp forests throughout Sumatra. There are an estimated 500 Sumatran tigers remaining in Sumatra, with approximately 100 found in Gunung Leuser National Park in northern Sumatra (Franklin et al. 1999).

Numerous primate species live in these forests, including the long-tailed macaque (*Macaca fascicularis*), pig-tailed macaque (*M. nemestrina*), and siamang (*Hylobates syndactylus*), Sumatra's largest gibbon. Other species common to these forests include squirrels, monitor lizards, estuarine crocodile (*Crocodylus porosus*), false gharial (*Tomistomus schlegeli*), and the endangered clouded leopard (*Pardofelis nebulosa*) (Whitten et al. 1987; Stone 1994).

The swampy grasslands and forests provide important habitat for many waterbirds, including herons, egrets, bitterns, pond herons, whistling ducks, pygmy geese, lesser adjutant (*Leptoptilos javanicus*), milky stork (*Mycteris cinerea*), and the rare white-winged duck (*Cairina scutulata*) (Stone 1994).

Status and Threats

Current Status

Less than a fifth of the original extent of natural habitat remains in this severely threatened ecoregion. There are three protected areas that extend into the ecoregion and cover 620 km² (3 percent) of the ecoregion, but two of these are less than 200 km² (table J88.1). Freshwater swamp forests have very fertile soils that are suitable for agriculture; therefore, this ecoregion has been intensively converted and exploited (Whitten et al. 1987). Very little of the remaining habitat is in an undisturbed state, including the areas inside nature reserves.

Types and Severity of Threats

Way Kambas is one of the oldest reserves in Indonesia, declared in 1937, and was upgraded to a national park in 1989. It was established to protect important freshwater and lowland rain forest flora and fauna in southern Sumatra. Despite its long standing as a nature reserve and increased protection in 1989, most of its land has been degraded by human interference (Scott 1994).

The estuarine crocodile and the false gharial, once numerous in this ecoregion, have been decimated by hunting, both from fear of these animals and for their skins, although the estuarine crocodile is protected by law in Indonesia (Whitten et al. 1987). Large-scale logging has also occurred throughout this ecoregion, especially from 1968 to 1974, and concessions cover 17 percent of the remaining habitat. Several large fires (1972, 1974, 1976, 1982, 1998, and 2000) have also swept through this ecoregion, destroying large tracts of forest. Paperbark, a common secondary growth tree in freshwater swamps, has a bark that peels off to form a highly flammable soil cover, encouraging further burning (Whitten et al. 1987). Climax communities of this forest, which include 100-year-old dipterocarp trees, are extremely slow to regenerate (Whitten et al. 1987). Widespread illegal logging, regardless of logging concessions, has occurred throughout Indonesia since the economic crises and fall of the Suharto government in 1998. It is doubtful whether any pristine freshwater swamp forests still exist in Sumatra. If any do exist, they will be under intensive pressure in the coming years and probably will be gone within 10 years (Holmes, Essay 1; Holmes, pers. comm., 2000).

Priority Conservation Actions

With the rapid pace of forest destruction under way in Indonesia, quick conservation action is urgently needed to stave off the complete collapse of these forests. However, steps should be taken to lay the groundwork for longer-term conservation as well.

Short-Term Conservation Actions (1–5 Years)

- Initiate a major public awareness campaign on forest values and forest loss to drive political will.
- Stop illegal logging in and around protected areas by challenging the authorities and judiciary to act on existing laws and regulations.
- Revoke forest concessions in areas of high biodiversity outside protected areas and institute a moratorium on new concessions for oil palm, timber, and other plantations, using the vast areas of deforested and cleared land instead.
- Place all remaining freshwater swamp forests under active protection.
- Mobilize major assistance for the more outstanding protected areas.
- Stop wildlife poaching, especially in key parks and reserves, by tightening surveillance around ports or airports and making the purchase of endangered species socially unacceptable.
- Establish a trust fund to offset the opportunity costs of maintaining forest cover and avoiding conversion to oil palm plantations.
- Purchase through Indonesian NGOs several large tracts of lowland forest, create model ecotourism reserves with elephant and foot safaris to view wildlife and attract foreign currency, and share profits from such ventures with communities.
- Determine whether Sumatran provincial governments can be provided funding from carbon credits for maintaining forest cover and wildlife protection.
- Instigate major biodiversity surveys and collecting expeditions in doomed areas to document the biodiversity that will soon be lost.

Longer-Term Conservation Actions (5–20 Years)

- Establish a long-term fund for conservation of lowland forests in Indonesia that would benefit Sumatra.
- Devise ways to maintain the integrity of buffer zones of protected areas.

Focal Species for Conservation Action

Area-sensitive species: tiger (Franklin et al. 1999; Tilson et al. 1994; Dinerstein et al. 1997; Seidensticker et al. 1999), Asian elephant (Santiapillai and Jackson 1990; Santiapillai 1997), Sumatran rhinoceros (Foose and van Strien 1997)

Top predator species: tiger (Franklin et al. 1999; Tilson et al. 1994; Dinerstein et al. 1997; Seidensticker et al. 1999)

Other: waterbird species

Selected Conservation Contacts

ARSG
EU-Leuser
Lembaga Alam Tropika Indonesia (LATIN)
Sumatran Tiger Project
WAHLI
WARSI
WCS
Worldbank Kerinci, Telapak
WWF-Indonesia
Yayasan Keanekaragaman Hayati
Yayasan Konsorsium Pembaruan Agraria (KPA)

TABLE J88.1. WCMC (1997) Protected Areas That Overlap with the Ecoregion.

Protected Area	Area (km²)	IUCN Category
Sei Prapat Simandulang	60	PRO
Giam Duri [85]	140	PRO
Way Kambas [82]	420	II
Total	620	

Ecoregion numbers of protected areas that overlap with additional ecoregions are listed in brackets.

Justification for Ecoregion Delineation

Whitmore (1984) and MacKinnon (1997) showed large extents of peat swamp forests along the northern coast of Sumatra, especially in Riau Province. We delineated the Sumatran Peat Swamp Forests [85] to represent these forests but extracted the smaller patches of freshwater swamp forests into the Sumatran Freshwater Swamp Forests [88] and the mangroves in the Sunda Shelf Mangroves [107].

MacKinnon's biounit 21 largely corresponds to Udvardy's Sumatra biogeographic province. However, Udvardy did not include the Nicobar Islands. Eight ecoregions overlap Udvardy's Sumatra biogeographic province: Sumatran Lowland Rain Forests [82], Sumatran Montane Rain Forests [83], Mentawai Islands Rain Forests [84], Sumatran Peat Swamp Forests [85], Sumatran Freshwater Swamp Forests [88], Sundaland Heath Forests [90], Sumatran Tropical Pine Forests [105], and Sunda Shelf Mangroves [107].

Prepared by Colby Loucks and Tony Whitten

Ecoregion Number:	**89**
Ecoregion Name:	**Southern Borneo Freshwater Swamp Forests**
Bioregion:	**Sunda Shelf and Philippines**
Biome:	**Tropical and Subtropical Moist Broadleaf Forests**
Political Units:	**Indonesia**
Ecoregion Size:	**36,600 km²**
Biological Distinctiveness:	**Regionally Outstanding**
Conservation Status:	**Critical**
Conservation Assessment:	**II**

The Southern Borneo Freshwater Swamp Forests [89] can range in species diversity from numbers rivaling those of neighboring lowland rain forests to single-species forest stands of the *Mallotus* tree. Many of the animals that use lowland rain forest also use freshwater swamp forest, including all monkey and ape species. Like all other freshwater swamps found in the Indo-Pacific region, this ecoregion has been intensively converted to agricultural and plantation lands. Further protection is urgently needed to stem the loss of this ecoregion's native vegetation.

Location and General Description

This ecoregion is made up of the freshwater swamp forests in Kalimantan. These forests are located just inland from the southwestern coast, with a few small areas towards the center of the island. They are associated with coastal swamps, inland lakes, and low-lying river basins. Based on the Köppen climate zone system, this ecoregion falls in the tropical wet climate zone (National Geographic Society 1999).

Freshwater swamp forests exist where rivers meander through flat, low-lying alluvial floodplains before encountering mangrove forests. They are periodically flooded or waterlogged by mineral-rich fresh water, have a high pH (above 6), and do not contain substantial amounts of peat (Payne et al. 1994; FAO 1981). These factors combine to produce taller, more species-rich and more productive forests in comparison with peat swamp forests. The floristic composition of these forests is quite varied. They may include floating grass mats, the spiny pandan and palm vegetation, marshes, scrub, and forests. Mature freshwater swamp forest has an average tree height of 35 m, some lianas, and numerous epiphytes (Payne et al. 1994; MacKinnon et al. 1997).

Freshwater swamp forests can range from species-poor stands dominated by the genus *Mallotus* to forests that are floristically diverse. The flora is not distinct but may contain many species common in the surrounding lowland rain forest. The commonly occurring genera include *Adina, Alstonia, Campnosperma, Coccoceras, Dillenia, Dyera, Erythrina, Eugenia, Ficus, Gluta, Lophopetalum, Memecylon, Metroxylon, Pandanus, Pentaspadon, Shorea,* and *Vatica* (FAO 1981). The tall legumes *Koompassia, Calophyllum,* and *Melanorrhoea* and the swamp sago *Metroxylon sagu* also thrive in this habitat (MacKinnon 1997).

Biodiversity Features

The faunal diversity and abundance vary with the structure and diversity of the vegetation in the forest but are usually higher than in peat swamp forests. Primate densities in freshwater swamp forests can be as high as in lowland rain forests. Like peat swamp forests, they are usually the highest around rivers. Long-tailed macaques (*Macaca fascicularis*) are the most common primates in freshwater swamp forests. Long-tailed macaques are the most adaptive primate found in Borneo and can live almost anywhere. They are found in rain forests, mangrove forests, freshwater forests, disturbed or logged forest, or even human-dominated landscapes such as villages. These monkeys live in troops of twenty or more and are commonly seen in coastal areas wandering over the beach searching for food. This behavior has earned them an alternative name: the crab-eating macaque (MacKinnon et al. 1996). In inland riverine swamp forests, they are primarily frugivorous, with leaves, flowers, insects, and bark providing the remainder of the diet (Yeager 1996).

These forests are also home to the endangered orangutan (*Pongo pygmaeus*). The orangutan is a CITES I species and was added to the U.S. Endangered Species list in 1970. Despite these efforts, the populations of orangutans in both Borneo and Sumatra have continued to diminish, primarily through habitat loss and poaching (for food or the pet trade) (Yeager 1999). Orangutans have adapted to both tropical forest and freshwater swamp forests. Orangutans' primary food is fruit, but they also feed on a complex mix of nuts, leaves, insects, bark, honey, and sap. Orangutans often visit numerous trees as their fruit becomes ripe over wide areas, keeping track of dozens of fruiting trees. As the trees drop their last fruit, they move on to other tree species (notably figs). In this way orangutans act as seed dispersers for numerous tree species (Galdikas and Briggs 1999).

TABLE J89.1. Endemic and Near-Endemic Bird Species.

Family	Common Name	Species
Zosteropidae	Javan white-eye	Zosterops flavus

An asterisk signifies that the species' range is limited to this ecoregion.

The bird fauna numbers more than 360 species and includes a wide variety of hornbills and a single near-endemic species the Javan white-eye (*Zosterops flavus*) (table J89.1).

Status and Threats

Current Status

Most of the original vegetation of the freshwater swamp forests has been cleared or modified by human activity. Only about 1.4 percent of the original land area is still forested; the rest has been cleared away for agriculture, especially *sawah* rice (MacKinnon et al. 1996). The existing protected area system, consisting of nine reserves, covers 3,520 km^2 (10 percent) of the ecoregion area (table J89.2). There are several reserves (including those that extend into the ecoregion) that are more than 500 km^2. The Tanjung Puting National Park provides some of the best freshwater habitat for orangutans and other wildlife in Borneo. However, this park, like the majority of parks and reserves in Indonesia, has been subject to severe encroachment by illegal loggers and illegal gold mining in the past few years (Yeager, pers. comm., 2000).

In 1997 and 1998 El Niño–driven fires burned through large portions of Sumatra and Borneo. Most of these fires were intentionally set to clear forest for large commercial oil palm plantations. Preliminary estimates indicate that more than 7,500 km^2 of both freshwater and peat swamp forests were lost during these fires. The subsequent years of 1999 and 2000 saw further forest loss. The effects on the forest's wildlife populations were profound. The fires of 1997–1998 exceeded many species' ability to adapt to the loss of habitat and stress of lost food sources. Hundreds of adult orangutans were killed as the fled the fires and encountered human populations. Many of the orphaned juveniles were then sold into the international pet trade (Barber and Schweithelm 2000).

TABLE J89.2. WCMC (1997) Protected Areas That Overlap with the Ecoregion.

Protected Area	Area (km^2)	IUCN Category
Muara Kendawangan [86, 110]	720	PRO
Unnamed [110]	90	?
Tanjung Penghujan NR [110]	110	PRO
Tanjung Puting [86, 90, 110]	660	II
Bukit Tangkiling	50	I
Danau Semayang Sungay Mahakam [87]	900	PRO
Kutai [90, 97]	110	II
Kutai (extension) [86, 90, 97]	430	PRO
Muara Kaman Sedulang [86, 90]	450	I
Total	3,520	

Ecoregion numbers of protected areas that overlap with additional ecoregions are listed in brackets.

Types and Severity of Threats

The primary threat to this habitat is deforestation caused by conversion of forests to plantations, slash-and-burn agricultural practices, uncontrolled fires, mining, and exotic species. Fires are set in freshwater swamp forests to make room for plantation crops such as oil palm and pepper. This trend is not likely to abate given the current economic and political situation in Indonesia.

Freshwater swamp forests are highly desirable to commercial timber operations because of their high stocking level of commercially valuable species. The soil in most of the deep alluvial terraces is deep, fertile, and well watered, promoting rapid forest growth. However, it is also ideal for agriculture, and this ecoregion has been intensively cleared for cultivation (MacKinnon et al. 1996). In south and southeast Kalimantan, original freshwater swamp forests were converted to single-species stands of paperbark (*gelam*) (*Melaleuca cajuputi*), sedge and swamp grass, or wet rice fields (Whitmore 1984). Paperbark is a fire-adapted species with thick, fire-resistant bark. During repeated burns or uncontrolled fires paperbark regenerates quickly, outcompeting and eradicating most native species. Paperbark is used extensively by local people for cajeput oil, insect repellant, soap, caulking for boats, firewood, and construction (Brinkman and Xuan 1986; Klepper et al. 1990). The wide variety of uses derived from this tree promote its wider distribution into additional freshwater swamp forests.

Another increasing threat to the forest integrity is the lure of gold. Gold accumulates in the alluvial fans, and Indonesia is rapidly becoming one of the world's leading gold exporters. Most of the mining operations are illegal and use water pumps and mercury to separate out the gold. This causes increased sedimentation in the rivers and bioaccumulation of heavy metals in numerous riverine species (Galdikas and Briggs 1999).

Priority Conservation Actions

With the rapid pace of forest destruction currently under way in Indonesia, quick conservation action is urgently needed to stave off the complete collapse of these forests. However, steps should be taken to lay the groundwork for longer-term conservation as well.

Short-Term Conservation Actions (1–5 Years)

- Call for an international conference to highlight the impending biodiversity crisis and loss of forests in Indonesia (perhaps using World Bank as an organizer).
- Stop illegal logging in and around protected areas.
- Revoke forest concessions in areas of high biodiversity outside protected areas and institute a moratorium on new concessions for oil palm, timber, and other plantations.
- Place under protection all remaining freshwater forests.
- Stop wildlife poaching, especially in key parks and reserves.
- Establish a trust fund to offset the opportunity costs of maintaining forest cover and avoiding conversion to oil palm plantations.
- Purchase through Indonesian NGOs several large tracts of lowland forest, create model ecotourism reserves with

elephant and foot safaris to view wildlife and attract foreign currency, and share profits from such ventures with communities.
- Determine whether Kalimantan provincial governments can be provided funding from carbon credits for maintaining forest cover and protection of wildlife.

Longer-Term Conservation Actions (5–20 Years)
- Reassess representativeness of protected area system in the ecoregion.
- Establish a long-term fund for conservation of lowland forests in Indonesia.
- Restore integrity of buffer zones of protected areas.

Focal Species for Conservation Action

Habitat specialists: proboscis monkey (Salter and MacKenzie 1985; Bennet and Gombek 1993; Yeager, 1989, 1990, 1991; Yeager and Blondal, 1992), stork-billed kingfisher

Area-sensitive species: orangutan (Galdikas 1971–1977; Dijuwantoko 1991; Tilson et al. 1993), hornbills

Top predator species: python, black eagle, crested sea eagle, brahminy kite, clouded leopard, golden bay cat, Malay civet, false gavial

Other: primates, Storm's stork, Malaysian sun bear, mouse deer, barking deer

Selected Conservation Contacts

CI-Indonesia

European Union Forest Management Projects

GTZ Sustainable Forest Management Projects

Konsorsium SHK Kalimantan Timur

Lembaga Alam Tropika Indonesia (LATIN)

TNC-Indonesia

WWF-Indonesia

Yayasan Keanekaragaman Hayati

Yayasan Konsorsium Pembaruan Agraria (KPA)

Justification for Ecoregion Delineation

The large island of Borneo was divided into nine ecoregions. Most of the island's lowland and submontane forests are dominated by dipterocarp species (MacKinnon et al. 1996). MacKinnon and MacKinnon (1986) divided the island's lowland forests into six subunits, with a central subunit representing the montane forests. MacKinnon (1997) revised the boundaries of these seven subunits but retained the same general configuration. These authors used the major rivers, the Kapuas and Barito, to represent zoogeographic barriers to a few species of mammals and based subunits largely on these barriers but also used climatic regimes for the drier eastern biounits (MacKinnon and MacKinnon 1986; MacKinnon 1997).

Because ecoregions are based on biomes, we first isolated the central montane ecoregion—the Borneo Montane Rain Forests [95]—above the 1,000-m elevation contour using the DEM (USGS 1996). We then assigned the large patches of peat forests, heath forests, freshwater swamp forests, and mangroves, in the lowlands and along the periphery of the island, into their own ecoregions: the Borneo Peat Swamp Forests [86], Sundaland Heath Forests [90] (which also includes Belitung Island and the heath forests in Bangka island), Southern Borneo Freshwater Swamp Forests [89], and Sunda Shelf Mangroves [107], respectively. The alpine habitats of the Kinabalu Mountain Range were represented by the Kinabalu Montane Alpine Meadows [108].

Prepared by Colby Loucks

Ecoregion Number: 90

Ecoregion Name:	**Sundaland Heath Forests**
Bioregion:	**Sunda Shelf and Philippines**
Biome:	**Tropical and Subtropical Moist Broadleaf Forests**
Political Units:	**Indonesia, Malaysia**
Ecoregion Size:	**76,200 km²**
Biological Distinctiveness:	**Regionally Outstanding**
Conservation Status:	**Vulnerable**
Conservation Assessment:	**II**

The Sundaland Heath Forests [90], known in Indonesia as *kerangas* (or land too poor for rice growing once cleared, in the Iban language) foster the growth of specialist plants such as the carnivorous pitcher plant *Nepenthes*, sundews *Drosera*, and bladderwort *Utricularia*. Like the rest of Borneo and Sumatra's lowland forest types, the heath forests are not immune to intentionally set fires for commercial logging and agriculture. This quest for land is exploiting the nutritionally impoverished soils of the heath forests.

Location and General Description

This ecoregion is made up of heath forests scattered throughout Borneo on raised beaches, sandstone plateaus, and ridges. Heath forest is found on well-drained acidic soils (pH less than 4) with a low clay content, derived from siliceous rocks under ever-wet conditions. These soils are commonly called white-sand soils. These soils usually originate from old, eroded sandstone beaches that isolated during the mid-Pleistocene (Burnham 1984). A layer of peat or humus often covers these soils but is lost once the natural vegetation is cleared. If the soils become waterlogged (lose their drainage capabilities), they develop into *kerapah* forests. These forests still remain heath forests but are more swampy in character (Whitmore 1984). Based on the Köppen climate zone system, this ecoregion falls in the tropical wet climate zone (National Geographic Society 1999).

Heath forest soils degrade very quickly to bleached sand once the forest cover is removed, making this type of forest extremely fragile (Whitmore 1989). Periodic water stress and lack of available nutrients may be important in the formation of this forest, which is notoriously poor for agriculture (Whitmore 1989). Heath forests are vastly different from lowland dipterocarp forests in structure, texture, and color. Heath forests have a low, uniform single-layered canopy. Leaf size is smaller, and trees often are densely packed and difficult to penetrate. Trees reach up to 20 m in height. Large trees are rare, buttresses are smaller, and epiphytes are common (Whitmore 1984). Under favorable conditions heath forests contain many plant species found in lowland evergreen rain forest. Dipterocarps are prominent in the canopy, and palms are common. Under the worst conditions, no dipterocarps may exist, and palms may be rare. Species of the Australian Myrtaceae and Casuarinaceae families predominate, and conifers such as *Agathis*, *Podocarpus*, and *Dacrydium* (WWF and IUCN 1995) are abundant. In Kalimantan, the dominant trees are Dipterocarpaceae (*Shorea* and *Hopea* spp.), Myrtaceae, *Gonystlus* spp., *Agathis* spp., *Dacriydium elatum*, *Styphelia* spp., and *Bachea* spp. (FAO 1981). It is estimated that the heath forests of Sarawak and Brunei contain 849 tree species, and, along with the Nabawan heath forests of Sabah, these forests are richer in plant species and endemics than elsewhere in the ecoregion (WWF and IUCN 1995).

Heath forests generally are less species-rich than comparable dipterocarp forests. They share many features in common with moss forests in the upper montane zones, such as a dense undergrowth, abundant bryophytes, presence of conifers, and the presence of *Casuarina nobilis* (a nitrogen-fixing plant) (Richards 1936). They share at least 146 tree species with freshwater and peat swamp forests (Brünig 1973), including the dipterocarps *Shorea albida*, *S. pachyphylla*, and *S. scabrida*.

Ground vegetation in heath forests generally is sparse, primarily composed of mosses and liverworts, with a host of insectivorous plants. The presence of insectivorous plants may be an evolutionary response to growing in nitrogen-poor conditions. The sundews *Drosera* have leaves covered with long red hairs that entrap insects. A bladderwort *Utricularia* has a hollow bag on the end of a stalk, with the entrance guarded by hairs. If an insect touches these hairs, a rush of water is released, dragging the insect inside to be digested (Mabberley 1987). Six species of pitcher plants are common in heath forests, including *Nepenthes bicalcarata*, found exclusively in heath and peat swamp forests (Smythies 1965). In other cases, a symbiotic relationship exists between plants and insects. This is the case with *Myrmecodia* and *Hydnophytum* species. *Myrmecodia* harbors ants in its thickened stem, and the ants in turn provide the plant with much-needed nutrients in the form of dead insects and other food they bring into their colony (Payne et al. 1994).

Biodiversity Features

Animals in heath forests are confronted with many of the same problems as those in peat swamp forests. Poor soils cause low productivity, and plants defend themselves from predators with toxic or unpalatable compounds (Brünig 1974). Because of these unfavorable conditions, heath forests are less species-rich, and animal communities are reduced in diversity and abundance.

Orangutans (*Pongo pygmaeus*) may frequent this forest types, but less often than other forest types. Heath forests have no turtles and less than one-half of the number of frog, lizard, and snake species found in other Bornean forests (Lloyd et al. 1968). There is a noticeable lack of small vertebrates, helping explain why heath forests contain only one-third of the snakes found in dipterocarp forests. With a

TABLE J90.1. Endemic and Near-Endemic Mammal Species.

Family	Species
Cercopithecidae	*Presbytis hosei*

An asterisk signifies that the species' range is limited to this ecoregion.

TABLE J90.2. Endemic and Near-Endemic Bird Species.

Family	Common Name	Species
Zosteropidae	Pygmy white-eye	*Oculocincta squamifrons*

An asterisk signifies that the species' range is limited to this ecoregion.

lack of prey, snakes become less diverse, and this effect cascades up through the food chain. The area supports only a single near-endemic mammal and one near-endemic bird species (table J90.1, table J90.2).

Status and Threats

Current Status

The Sundaland Heath Forests [90] ecoregion includes two blocks of intact habitat larger than 5,000 km^2, and the most intact part is in south Kalimantan. Although more than half of the ecoregion has been cleared, heath forests cannot sustain agriculture, probably accounting for the good condition of this vegetation type. There are seven protected areas that cover 4,440 km^2 (6 percent) of the ecoregion (table J90.3). Tanjung Puting is the largest of the protected areas, Kutai, which extends into the ecoregion and is also greater than 1,000 km^2.

Types and Severity of Threats

Heath forests are easily degraded by logging or burning activities. Once degraded, they develop into an open savanna of shrubs and trees over sparse grass and sedge. This formation is called a *padang*. Heath forest can recover from a *padang*, but this takes a very long time to do. The replanting or reestablishment of native vegetation has proved ineffective (Mitchell 1963). Extensive areas of this ecoregion were severely damaged in the great forest fire of 1982–1983, which consumed a total area of 33,000 km^2, mostly in East

TABLE J90.3. WCMC (1997) Protected Areas That Overlap with the Ecoregion.

Protected Area	Area (km^2)	IUCN Category
Tanjung Putting [86, 89, 110]	1,780	II
Pararawen Baru [96]	300	PRO
Pantai Samarinda	180	V
Bukit Soeharto	840	V
Kutai [89, 97]	610	II
Kutai (extension) [86, 89, 97]	610	PRO
Muara Kaman Sedulang [86, 89]	120	I
Total	4,440	

Ecoregion numbers of protected areas that overlap with additional ecoregions are listed in brackets.

Kalimantan (IUCN 1991), and more recently during the 1997–1998 fires.

Priority Conservation Actions

With the rapid pace of forest destruction under way in Indonesia, quick conservation action is urgently needed to stave off the complete collapse of these forests. However, steps should be taken to lay the groundwork for longer-term conservation as well.

Short-Term Conservation Actions (1–5 Years)

- Call for an international conference to highlight the impending biodiversity crisis and loss of forests in Indonesia (perhaps using World Bank as an organizer).
- Stop illegal logging in and around protected areas.
- Revoke forest concessions in areas of high biodiversity outside protected areas and institute a moratorium on new concessions for oil palm, timber, and other plantations.
- Place all remaining heath forests under protection.
- Stop wildlife poaching, especially in key parks and reserves.
- Establish a trust fund to offset the opportunity costs of maintaining forest cover and avoiding conversion to oil palm plantations.
- Determine whether Kalimantan provincial governments can be provided funding from carbon credits for maintaining forest cover and protection of wildlife.

Longer-Term Conservation Actions (5–20 Years)

- Reassess representativeness of the protected area system in the ecoregion.
- Establish a long-term fund for conservation of lowland and heath forests in Indonesia.
- Restore integrity of buffer zones of protected areas.

Focal Species for Conservation Action

Endemic species: pygmy white-eye, *Presbytis hosei*

Area-sensitive species: orangutan (Galdikas 1971–1977; Dijuwantoko 1991; Tilson et al. 1993)

Selected Conservation Contacts

Konsorsium SHK Kalimantan Timur

Lembaga Alam Tropika Indonesia (LATIN)

WWF-Indonesia

Yayasan Keanekaragaman Hayati

Yayasan Konsorsium Pembaruan Agraria (KPA)

Justification for Ecoregion Delineation

The large island of Borneo was divided into nine ecoregions. Most of the island's lowland and submontane forests are dominated by dipterocarp species (MacKinnon et al. 1996). MacKinnon and MacKinnon (1986) divided the island's lowland forests into six subunits, with a central subunit representing the montane forests. MacKinnon (1997) revised the boundaries of these seven subunits but retained the same general configuration. These authors used the major rivers, the Kapuas and Barito, to represent zoogeographic barriers to a few mammal species and based subunits largely on these barriers but also used climatic regimes for the drier eastern biounits (MacKinnon and MacKinnon 1986; MacKinnon 1997).

Because ecoregions are based on biomes, we first isolated the central montane ecoregion—the Borneo Montane Rain Forests [95]—above the 1,000-m elevation contour using the DEM (USGS 1996). We then assigned the large patches of peat forests, heath forests, freshwater swamp forests, and mangroves, in the lowlands and along the periphery of the island, into their own ecoregions: the Borneo Peat Swamp Forests [86], Sundaland Heath Forests [90] (which also includes Belitung Island and the heath forests in Bangka island), Southern Borneo Freshwater Swamp Forests [89], and Sunda Shelf Mangroves [107], respectively. The alpine habitats of the Kinabalu Mountain Range were represented by the Kinabalu Montane Alpine Meadows [108].

Prepared by Colby Loucks

Ecoregion Number:	**91**
Ecoregion Name:	**Western Java Rain Forests**
Bioregion:	**Sunda Shelf and Philippines**
Biome:	**Tropical and Subtropical Moist Broadleaf Forests**
Political Units:	**Indonesia**
Ecoregion Size:	**41,400 km²**
Biological Distinctiveness:	**Regionally Outstanding**
Conservation Status:	**Critical**
Conservation Assessment:	**II**

The Western Java Rain Forests [91] are found on one of the most actively volcanic islands in the world. Once the home of the extinct Javan tiger (*Panthera tigris sondaicus*), these forests still contain one of the two remaining populations of one of the world's most threatened mammal species, the critically endangered Javan rhinoceros (*Rhinoceros sondaicus*). Unfortunately, only about 5 percent of the original habitat of this ecoregion remains.

Location and General Description

This ecoregion represents the lowland moist forests (less than 1,000 m) of western Java, Indonesia. Based on the Köppen climate zone system, this ecoregion falls in the tropical wet climate zone (National Geographic Society 1999), although as one moves eastward on Java there is increasing seasonality of precipitation. Java probably did not exist before the Miocene (24 m.y.). Truly born of fire, the island of Java is the result of the subduction and remelting of the Australian–Indian Ocean tectonic plate beneath the Eurasian tectonic plate at the Java trench. The melted crust has risen as volcanoes and, along with subsequent sedimentation, created Java. Therefore, the surface geology consists of Tertiary and Quaternary volcanics, alluvial sediments, and areas of uplifted coral limestone. Twenty of the volcanoes on Java and Bali have been active in historic times, and they are among the most active volcanic islands in the world. During previous ice ages, when sea levels were much lower, Java was connected to Sumatra, Borneo, and the rest of the Asian mainland (Whitten et al. 1996).

The natural forests in the lowlands of western Java once included several forest subtypes, including extensive evergreen rain forest, semi-evergreen rain forest, moist deciduous forest along the northern coast, and dry deciduous forest, also along the northern coast of the island. The differences are mostly related to the seasonality of rainfall. There are also small areas of azonal limestone and freshwater swamp forests. No single tree family dominates the forests of Java, as is the case with the dipterocarps in Sumatra and Borneo (Whitten et al. 1996).

The most common species in the rain forests of Java are *Artocarpus elasticus* (Moraceae), *Dysoxylum caulostachyum* (Meliaceae), langsat *Lansium domesticum* (Meliaceae), and *Planchonia valida* (Lecythidaceae). Semi-evergreen rain forest differs from evergreen rain forest by being slightly more seasonal, with two to four dry months each year (Whitten et al. 1996).

Java's deciduous forests generally are lightly closed, with few trees exceeding 25 m. *Borassus* and *Corypha* palms are good indicators of the seasonal climates that generate deciduous forests in the region. Moist deciduous forests have 1,500 to 4,000 mm of rainfall annually, with a four- to six-month dry season. Dry deciduous forests have less than 1,500 mm of annual rainfall and more than six dry months. Common lowland deciduous trees found in eastern Java and Bali are *Homalium tomentosum*, *Albizia lebbekoides*, *Acacia leucophloea*, *A. tomentosa*, *Bauhinia malabarica*, *Cassia fistula*, *Dillenia pentagyna*, *Tetrameles nudiflora*, *Ailanthus integrifolia*, and *Phyllanthus emblica*. Many herbaceous plants are confined to the deciduous forests (Whitten et al. 1996).

Limestone forests on Java have basal areas similar to those of other lowland forest types and apparently contain no plant endemics, but because they often grow on steep slopes of shallow soils, their growth pattern is affected. Limestone forests are found on Mt. Cibodas, Nusu Barung, Padalarang, and Nusa Penida (Whitten et al. 1996).

Patches of freshwater swamp forest found throughout the ecoregion are relatively poor in species (Whitten et al. 1996). Rawa Danau, Banten in west Java is the largest remaining area of swamp forest in Java and Bali, and it contains many tree species now nearly extinct elsewhere in Java, such as *Elaeocarpus macrocerus*, *Alstonia spathulata*, wild mango (*Mangifera gedebe*), and *Stemonurus secundiflora*. Other rare plants include the sedge *Machaerina rubiginosa*, the aroid *Cyrtosperma merkusii*, and floating water plants such as *Hydrocharis dubia* and water chestnut (*Trapa maximoviscii*) (Whitten et al. 1996).

Biodiversity Features

The overall richness and endemism of this ecoregion are moderate compared with those of other ecoregions in Indo-Malaysia.

The ecoregion harbors 101 mammal species, including five endemics and near endemics (table J91.1). The larger of the two known populations of the critically endangered Javan rhinoceros (*Rhinoceros sondaicus*) is found in the extreme western part of the ecoregion, in Ujung Kulon National Park. The Javan gibbon (*Hylobates moloch*) is also critically endangered (Hilton-Taylor 2000). Other species of conservation significance include the globally threatened surili (or Java) leaf monkey (*Presbytis comata*), fishing cat (*Felis viverrina*), wild dog (*Cuon alpinus*), Javan warty pig (*Sus verrucosus*), banteng (*Bos javanicus*), and slow loris (*Nycticebus coucang*) (MacKinnon and MacKinnon 1986; Whitten et al. 1996; Hilton-Taylor 2000). The Javan subspecies of the yellow-throated marten (*Martes flavigula robinsoni*) and leopard on Java (*Pantera pardus melas*) are also considered endangered (Hilton-Taylor 2000).

More than 350 bird species are known to occur in the ecoregion, including 9 endemics and near endemics (table J91.2). This ecoregion overlaps with two EBAs (Stattersfield et al. 1998): Java and Bali forests and Javan coastal zone. These two EBAs contain a total of thirty-seven restricted-range birds. Nine of these restricted-range birds are found in this ecoregion of lowland west Java. Of all the birds in the ecoregion, twelve are threatened, including the critically endangered Javanese lapwing (*Vanellus macropterus*) and the endangered Javan hawk-eagle (*Spizaetus bartelsi*) and black-winged starling (*Sturnus melanopterus*) (Stattersfield et al. 1998; MacKinnon and Phillipps 1993; BirdLife International 2000).

The forests of west Java are more species-rich for plants than the rest of the island, with more than 3,800 species, including two endemic genera (MacKinnon and MacKinnon 1986; Whitten et al. 1996). These forests harbor two species of the giant insectivorous *Rafflesia* (*R. rochussenii* and *R. patma*) (Whitten et al. 1996).

TABLE J91.1. Endemic and Near-Endemic Mammal Species.

Family	Species
Rhinolophidae	*Rhinolophus canuti*
Molossidae	*Otomops formosus*
Hylobatidae	*Hylobates moloch*
Suidae	*Sus verrucosus*
Muridae	*Sundamys maxi*

An asterisk signifies that the species' range is limited to this ecoregion.

TABLE J91.2. Endemic and Near-Endemic Bird Species.

Family	Common Name	Species
Accipitridae	Javan hawk-eagle	*Spizaetus bartelsi*
Charadriidae	Javan plover	*Charadrius javanicus*
Charadriidae	Sunda lapwing	*Vanellus macropterus*
Cuculidae	Sunda coucal	*Centropus nigrorufus*
Zosteropidae	Javan white-eye	*Zosterops flavus*
Timaliidae	White-breasted babbler	*Stachyris grammiceps*
Timaliidae	White-bibbed babbler	*Stachyris thoracica*
Timaliidae	Grey-cheeked tit-babbler	*Macronous flavicollis*
Timaliidae	Crescent-chested babbler	*Stachyris melanothorax*

An asterisk signifies that the species' range is limited to this ecoregion.

Status and Threats

Current Status

Only about 5 percent of the original habitat of this ecoregion remains. There are thirty-three protected areas that cover 3,045 km² (7 percent), but most of the protected areas (twenty-eight) are small (less than 100 km²) (table J91.3). The largest, Ujung Kulon National Park, is significant for its Javan rhinoceros population.

Java is one of the most densely populated islands in the world, so it is not surprising that very little natural habitat remains here (MacKinnon and MacKinnon 1986). Anthropogenic fires are common, and over the centuries burning has resulted in monospecific stands of fire-resistant species, usually *Tectona grandis* (FAO 1981). In many annual cropping systems, soils are left exposed during critical periods, resulting in extensive erosion (IUCN 1991). Illegal farming and felling even within protected areas are widespread, and an important timber tree *Altingia excelsa* has been nearly eliminated from the lowland forests (Whitten et al. 1996). In freshwater swamp forests, the exotic *Mimosa pigra* has the potential to become a very serious pest because it is fire-resistant and capable of forming impenetrable thickets (Whitten et al. 1996). Plans to construct a dam at the outlet of the Cidanau will destroy Rawa Danau, the only remaining extensive area of freshwater swamp in Java (Whitten et al. 1996).

Types and Severity of Threats

The threats that have degraded or destroyed most of this ecoregion in the past still threaten the remaining forest fragments. Political instability will continue to contribute to rampant destruction of these forests as existing environmental laws are routinely ignored and not implemented.

Priority Conservation Actions

With the rapid pace of forest destruction under way in Indonesia, quick conservation action is urgently needed to stave off the complete collapse of these forests. However, steps should be taken to lay the groundwork for longer-term conservation as well.

Short-Term Conservation Actions (1–5 Years)

- Initiate a major public awareness campaign on forest values and forest loss to drive political will.
- Stop illegal logging in and around protected areas by challenging the authorities and judiciary to act on existing laws and regulations.
- Revoke forest concessions in areas of high biodiversity outside protected areas and institute a moratorium on new concessions for oil palm, timber, and other plantations, using the vast areas of deforested and cleared land instead.
- Place all remaining lowland forests under active protection.
- Mobilize major assistance for the more outstanding protected areas.
- Stop wildlife poaching, especially in key parks and reserves.

Longer-Term Conservation Actions (5–20 Years)

- Determine whether any habitat remains to develop linkages between lowland and montane habitats in the protected area system.
- Devise ways to maintain the integrity of buffer zones of protected areas.

Focal Species for Conservation Action

Endangered endemic species: Javan rhinoceros (*Rhinoceros sondaicus*) (Foose and van Strien 1997), Javan gibbon (*Hylobates moloch*) (Eudey 1987), Javan or surili leaf monkey (*Presbytis comata*) (Eudey 1987), yellow-throated marten (*Martes flavigula robinsoni*), Javan leopard (*Panthera pardus melas*) (Nowell and Jackson 1996), Javan hawk-eagle (*Spizaetus bartelsi*) (BirdLife International 2000), Javanese lapwing (*Vanellus macropterus*) (BirdLife International 2000)

Area-sensitive species: banteng (*Bos javanicus*) (Hedges, in press), Javan leopard (*Panthera pardus melas*) (Nowell and Jackson 1996), wild dog (*Cuon alpinus*) (Ginsburg and Macdonald 1990), Javan warty pig (*Sus verrucosus*) (Oliver 1993), Sunda sambar (*Cervus timorensis*) (Wemmer 1998), barking deer (*Muntiacus muntjak*) (Wemmer 1998), hornbills (Bucerotidae) (rhinoceros hornbill (*Buceros rhinoceros*), Oriental pied-hornbill (*Anthracoceros albirostris*), wreathed hornbill (*Aceros undulatus*) (Kemp 1995)

TABLE J91.3. WCMC (1997) Protected Areas That Overlap with the Ecoregion.

Protected Area	Area (km²)	IUCN Category
Pulau Sangiang	20	DE
Gunung Tukung Gede	50	I
Muara Angke	30	I
Rawa Danau	70	I
Carita	30	V
Yanlapa	10	I
Depok	20	I
Tanjung Pasir	20	?
Mawuk	20	?
Gunung Karang	50	?
Jayabaya	40	?
Muara Gembong	30	?
Muara Bobos	90	?
Tanjung Sedari	90	?
Muara Cimanuk	80	?
Ujung Kulon	1,120	II
Cikepuh	140	IV
Peson Subah I, II	10	I
Ulolanang Kecubung	30	I
Unnamed	40	?
Gunung Unggaran	70	?
Ciogong	40	?
Salatri	30	?
Bojong Larang Jayanti	20	I
Gunung Selok	30	V
Leuwang Sancang	110	I
Pasir Salam	70	?
Nusakambangan	200	?
Gunung Kendeng	30	?
Cikencreng	30	?
Telaga Bodas	5	?
Telogo Ranjeng	20	I
Gunung Pangasaman [92]	400	?
Total	3,045	

Ecoregion numbers of protected areas that overlap with additional ecoregions are listed in brackets.

Top predator species: Javan leopard (*Panthera pardus melas*) (Nowell and Jackson 1996), wild dog (*Cuon alpinus*) (Ginsburg and Macdonald 1990)

Selected Conservation Contacts

Lembaga Alam Tropika Indonesia (LATIN)

WWF-Indonesia

Yayasan Keanekaragaman Hayati

Yayasan Konsorsium Pembaruan Agraria (KPA)

Justification for Ecoregion Delineation

MacKinnon (1997) included the islands of Java and Bali in biounit 22 (with three subunits). Western Java is wetter than the eastern half of the island, and the forests are richer in species (Whitten et al. 1996; MacKinnon 1997). There are also floristic differences between the lowland and montane vegetation in Java and Bali (Whitmore 1984; Whitten et al. 1996). Therefore, using MacKinnon's subunit boundary, we delineated the Western Java Rain Forests [91] to represent the moister evergreen forests to the west and the Eastern Java–Bali Rain Forests [92] to represent the drier, less species-rich forests of eastern Java and Bali. However, we also extracted the montane forests into distinct ecoregions—Western Java Montane Rain Forests [93] and Borneo Montane Rain Forests [95]—using the 1,000-m elevation contour of a DEM (USGS 1996).

Prepared by John Morrison

Ecoregion Number:	**92**
Ecoregion Name:	**Eastern Java–Bali Rain Forests**
Bioregion:	**Sunda Shelf and Philippines**
Biome:	**Tropical and Subtropical Moist Broadleaf Forests**
Political Units:	**Indonesia**
Ecoregion Size:	**53,500 km²**
Biological Distinctiveness:	**Regionally Outstanding**
Conservation Status:	**Critical**
Conservation Assessment:	**II**

The Eastern Java–Bali Rain Forests [92] are found on one of the most actively volcanic islands in the world. Once the home of the extinct Javan and Balinese tigers (*Panthera tigris sondaicus* and *Panthera tigris balica*, respectively), these forests still contain one of the most endangered and high-profile songbirds in the world, the Bali starling (*Leucopsar rothschildi*). Almost all of this ecoregion's natural habitat was cleared long ago by logging interests and for agriculture and settlements to provide for a rapidly expanding, dense human population. Only tiny fragments of natural forests remain, and these are also disturbed.

Location and General Description

This ecoregion represents the lowland moist forests of eastern Java and Bali in the Indonesian Archipelago. Based on the Köppen climate zone system, this ecoregion falls in the tropical wet and dry climate zones (National Geographic Society 1999), although as one moves east on Java there is increasing seasonality of precipitation. Java probably did not exist before the Miocene (24 m.y. ago). Bali did not emerge until as recently as 3 million years ago. Truly born of fire, the islands of Java and Bali are the result of the subduction and remelting of the Australian–Indian Ocean tectonic plate beneath the Eurasian tectonic plate at the Java trench. The melted crust has risen as volcanoes and, along with subsequent sedimentation, created Java. Therefore, the surface geology consists of Tertiary and Quaternary volcanics, alluvial sediments, and areas of uplifted coral limestone. Twenty of the volcanoes on Java and Bali have been active in historic times, and they are among the most active volcanic islands in the world. During previous ice ages, when sea levels were much lower, Java and Bali were connected to Sumatra, Borneo, and the rest of the Asian mainland (Whitten et al. 1996).

The climate in eastern Java and Bali is drier than in the western part of Java; therefore, the lowland forests are predominantly moist deciduous forests, with semi-evergreen rain forest along the south coast and dry deciduous forest along the north coast. No one family dominates the forests of Java, as is the case with the dipterocarps in Sumatra and Borneo.

Java and Bali's deciduous forests generally are lightly closed, with few trees exceeding 25 m. *Borassus* and *Corypha* palms are good indicators of the seasonal climates that generate deciduous forests in the region. Moist deciduous forests have 1,500–4,000 mm of rainfall annually, with a four- to six-month dry season. Dry deciduous forests have less than 1,500 mm of annual rainfall and more than six dry months. Common lowland deciduous trees in eastern Java and Bali are *Homalium tomentosum*, *Albizia lebbekoides*, *Acacia leucophloea*, *A. tomentosa*, *Bauhinia malabarica*, *Cassia fistula*, *Dillenia pentagyna*, *Tetrameles nudiflora*, *Ailanthus integrifolia*, and *Phyllanthus emblica*. Many herbaceous plants are confined to the deciduous forests (Whitten et al. 1996).

Semi-evergreen rain forest is slightly more seasonal than the evergreen rain forest found in western Java, with two to four dry months each year. The most common species in the rain forests of Java are *Artocarpus elasticus* (Moraceae), *Dysoxylum caulostachyum* (Meliaceae), langsat *Lansium domesticum* (Meliaceae), and *Planchonia valida* (Lecythidaceae). Little forest remains below 500 m in Bali, so the natural forest composition is somewhat unclear, but trees commonly found in the remnants of Bali's lowland rain forests include *Dipterocarpus hasseltii*, *Planchonia valida*, *Palaquium javense*, *Duabanga moluccana*, *Meliosma ferruginosa*, *Pterospermum javanicum*, and *Tabernaemontana* (Whitten et al. 1996).

Limestone forests on Java and Bali have basal areas similar to those of other lowland forest types and apparently contain no plant endemics, but because they often grow on steep slopes of shallow soils, their growth pattern is affected. Limestone forests are found on Nusu Barung and Nusa Penida (Whitten et al. 1996).

Biodiversity Features

The overall richness and endemism for this ecoregion are low to moderate when compared with those of other ecoregions in Indo-Malaysia.

The ecoregion harbors 103 mammal species. The Kangean Islands, off the northern coast of Java, have uplifted limestone deposits with numerous caves that harbor fourteen of these islands' fifteen bat species (Whitten et al. 1996). Fruit bats play an exceptionally important ecological role by pollinating and dispersing plants. Of special significance is the cave fruit bat (*Eonycteris spelaea*), which is a pollinator of the economically important durian tree (*Durio zibethinus*) (Whitten et al. 1996). The endangered Bawean (or Kuhl's) deer (*Axis kuhlii*), confined to the small island of Bawea (MacKinnon and MacKinnon 1986), is a strict endemic of this ecoregion (table J92.1). The other endemic mammal is the vulnerable Javan warty pig (*Sus verrucosus*). Additional mammal species of conservation significance are the endangered Javan yellow-throated marten (*Martes flavigula robinsoni*) and banteng (*Bos javanicus*). The Javan subspecies of leopard on Java (*Pantera pardus melas*) is also considered endangered (Hilton-Taylor 2000). Two tiger subspecies, the Javan and Balinese tigers (*Panthera tigris sundaica* and *Panthera tigris balica*, respectively), once inhabited these islands but are now extinct. The Balinese tiger disappeared by the end of World War II, but reports of Javan tigers persisted into the 1970s in the southeast (Nowell and Jackson 1996).

This ecoregion harbors almost 350 bird species, including ten endemic and near-endemic species (table J92.2). This ecoregion overlaps partially with the Java and Bali forests and Javan coastal zone EBAs. Together these two EBAs contain thirty-seven restricted-range bird species, only some of which are found in the lowlands of this ecoregion. Included are the critically endangered Bali starling (*Leucopsar rothschildi*), Javanese lapwing (*Vanellus macropterus*), and the endangered Javan hawk eagle (*Spizaetus bartelsi*) (BirdLife International 2000). Seven other non-endemic bird species are threatened in this ecoregion.

Status and Threats

Current Status

Almost all of this ecoregion's natural habitat has long been cleared by logging interests and for agriculture and settlements to provide for a rapidly expanding, dense human population. Only tiny fragments of natural forests remain, but they are also disturbed to some degree. The largest remaining blocks of forest in this ecoregion are found at Lebakharjo and Bantur, along the coast south of Malang (Whitten et al. 1996). There are eighteen protected areas that cover 2,330 km^2 (4 percent), although the majority are small (less than 100 km^2) (table J92.3). The largest is Meru Betiri National Park, although the total size of Gunung Raung that extends into the ecoregion exceeds 1,000 km^2.

Types and Severity of Threats

Anthropogenic fires are common here, and centuries of burning have resulted in monospecific stands of fire-resistant species, usually *Tectona grandis* (FAO 1981). Most of the larger trees usually are felled by villagers (Whitten et al. 1996) or have succumbed to logging operations. Hunting is rife. The larger fruit bats, known as flying-foxes, are hunted extensively in Indonesia for food and medicine, but generally all fruit bats are killed by orchard owners who think they damage crops, even though these species play an important role in flower pollination and seed dispersal (Whitten et al. 1996). Shifting cultivation by large and rapidly expanding populations has led to extensive erosion (IUCN 1991).

TABLE J92.1. Endemic and Near-Endemic Mammal Species.

Family	Species
Suidae	Sus verrucosus
Cervidae	Axis kuhlii*

An asterisk signifies that the species' range is limited to this ecoregion.

TABLE J92.2. Endemic and Near-Endemic Bird Species.

Family	Common Name	Species
Accipitridae	Javan hawk-eagle	Spizaetus bartelsi
Charadriidae	Javan plover	Charadrius javanicus
Charadriidae	Sunda lapwing	Vanellus macropterus
Cuculidae	Sunda coucal	Centropus nigrorufus
Strigidae	Javan owlet	Glaucidium castanopterum
Sturnidae	Bali myna	Leucopsar rothschildi*
Timaliidae	White-breasted babbler	Stachyris grammiceps
Timaliidae	White-bibbed babbler	Stachyris thoracica
Timaliidae	Crescent-chested babbler	Stachyris melanothorax
Timaliidae	Grey-cheeked tit-babbler	Macronous flavicollis

An asterisk signifies that the species' range is limited to this ecoregion.

TABLE J92.3. WCMC (1997) Protected Areas That Overlap with the Ecoregion.

Protected Area	Area (km^2)	IUCN Category
Bawean	190	IV
Gunung Celering	30	I
Gunung Butak	10	I
Bekutuk	20	I
Gunung Maria	160	?
Pulau Saobi (Kangean Islands)	60	?
Sumber Semen	30	?
P. Sembu	10	I
Nusu Barung	80	I
Eko Boyo	20	?
Teluk Lenggasana	130	?
Gunung Ringgit	60	?
Gunung Beser	60	?
Bali Barat [93]	310	II
Meru Betiri [93]	430	II
Gunung Raung [93]	100	?
Baluran	250	II
Banyuwangi	380	IV
Total	2,330	

Ecoregion numbers of protected areas that overlap with additional ecoregions are listed in brackets.

Priority Conservation Actions

With the rapid pace of forest destruction under way in Indonesia, quick conservation action is urgently needed to stave off the complete collapse of these forests. However, steps should be taken to lay the groundwork for longer-term conservation as well.

Short-Term Conservation Actions (1–5 Years)

- Initiate a major public awareness campaign on forest values and forest loss to drive political will.
- Place under active protection all remaining lowland forests.
- Stop wildlife poaching, especially in key parks and reserves, by tightening surveillance around ports or airports and making the purchase of endangered species socially unacceptable.

Longer-Term Conservation Actions (5–20 Years)

- Attempt to restore and protect linkages between lowland and montane habitats.

Focal Species for Conservation Action

Endangered endemic species: Bawean deer (*Axis kuhlii*) (Wemmer 1998), Javan warty pig (*Sus verrucosus*) (Oliver 1993), Javan hawk-eagle (*Spizaetus bartelsi*) (BirdLife International 2000), Javanese lapwing (*Vanellus macropterus*) (BirdLife International 2000), Bali starling (*Leucopsar rothschildi*) (BirdLife International 2000), Javan leopard (*Panthera pardus melas*) (Nowell and Jackson 1996).

Area-sensitive species: banteng (*Bos javanicus*) (Hedges, in press), Javan leopard (*Panthera pardus melas*) (Nowell and Jackson 1996), wild dog (*Cuon alpinus*) (Eudey 1987), Javan warty pig (*Sus verrucosus*) (Oliver 1993), Sunda sambar (*Cervus timorensis*) (Wemmer 1998), barking deer (*Muntiacus muntjak*) (Wemmer 1998), hornbills (Bucerotidae) (rhinoceros hornbill [*Buceros rhinoceros*], Oriental pied-hornbill [*Anthracoceros albirostris*], wreathed hornbill [*Aceros undulatus*]) (Kemp 1995)

Top predator species: leopard (*Panthera pardus melas*) (Nowell and Jackson 1996), wild dog (*Cuon alpinus*) (Eudey 1987)

Selected Conservation Contacts

Lembaga Alam Tropika Indonesia (LATIN)

WWF-Indonesia

Yayasan Keanekaragaman Hayati

Yayasan Konsorsium Pembaruan Agraria (KPA)

Justification for Ecoregion Delineation

MacKinnon (1997) included the islands of Java and Bali in biounit 22 (with three subunits). Western Java is wetter than the eastern half of the island, and the forests are richer in species (Whitten et al. 1996; MacKinnon 1997). There are also floristic differences between the lowland and montane vegetation in Java and Bali (Whitmore 1984; Whitten et al. 1996). Therefore, using MacKinnon's subunit boundary, we delineated the Western Java Rain Forests [91] to represent the moister evergreen forests to the west and the Eastern Java–Bali Rain Forests [92] to represent the drier, less species-rich forests of eastern Java and in Bali. However, we also extracted the montane forests into distinct ecoregions—Western Java Montane Rain Forests [93] and Borneo Montane Rain Forests [95]—using the 1,000-m elevation contour of a DEM (USGS 1996).

Prepared by John Morrison

Ecoregion Number:	**93**
Ecoregion Name:	**Western Java Montane Rain Forests**
Bioregion:	**Sunda Shelf and Philippines**
Biome:	**Tropical and Subtropical Moist Broadleaf Forests**
Political Units:	**Indonesia**
Ecoregion Size:	**26,200 km^2**
Biological Distinctiveness:	**Globally Outstanding**
Conservation Status:	**Critical**
Conservation Assessment:	**I**

The Western Java Montane Rain Forests [93] are found on one of the most actively volcanic islands in the world. Several mammals and nine bird species are found nowhere else on Earth. Once the home of the extinct Javan tiger (*Panthera tigris sundaicus*), only a fifth of the original habitat remains in this ecoregion, and these forests are scattered in fragments throughout the mountains.

Location and General Description

This ecoregion represents the montane forests of west Java. Based on the Köppen climate zone system, this ecoregion falls in the tropical wet climate zone (National Geographic Society 1999). Java probably did not exist before the Miocene (24 m.y.). Truly born of fire, the island of Java is the result of the subduction and remelting of the Australian–Indian Ocean tectonic plate beneath the Eurasian tectonic plate at the Java trench. The melted crust has risen as volcanoes and, along with subsequent sedimentation, created Java. Therefore, the surface geology consists of Tertiary and Quaternary volcanics, alluvial sediments, and areas of uplifted coral limestone. Twenty of the volcanoes on Java and Bali have been active in historic times, and they are among the most active volcanic islands in the world. During previous ice ages, when sea levels were much lower, Java was connected to Sumatra, Borneo, and the rest of the Asian mainland (Whitten et al. 1996).

The main forest types in this ecoregion are evergreen rain forest, semi-evergreen rain forest, and aseasonal montane forest. The evergreen and semi-evergreen rain forests generally are found in the lower portions of the ecoregion. Evergreen rain forests of Java contain *Artocarpus elasticus* (Moraceae), *Dysoxylum caulostachyum* (Meliaceae), langsat *Lansium domesticum* (Meliaceae), and *Planchonia valida* (Lecythidaceae) (Whitten et al. 1996). Semi-evergreen rain forest differs from evergreen rain forest by being slightly more seasonal, with two to four dry months each year (Whitten et al. 1996).

The transition between lowland and montane forests is a floristic one, and some plant families and genera are found only on one or the other side of this transition. Above 1,000 m, genera begin to include *Anemone, Aster, Berberis, Galium, Gaultheria, Lonicera, Primula, Ranunculus, Rhododendron, Veronica,* and *Viola*. Some lowland tree species transition out up to 1,200 m. The most abundant montane tree species in the lower montane zone are *Lithocarpus, Quercus, Castanopsis,* and laurels (Fagaceae and Lauraceae). Magnoliaceae, Hamamelidaceae, and Pococarpaceae are also well represented. Few emergents, primarily *Atingia excelsa* and *Podocarpus* spp., are found in the lower montane zone, but tree ferns are common (Whitten et al. 1996).

In the gradual transition from lower to upper montane forest, which begins at approximately 1,800 m, enormous quantities of *Aerobryum* moss begin to become prevalent on all surfaces. *Dacrycarpus* (*Podocarpus*) continues up from the lower montane. Ericaceae shrubs are very characteristic of the upper montane zone, including *Rhododendron, Vaccinium,* and *Gaultheria*. Sub-alpine forest, found above 3,000 m, contains one species-poor layer of trees, including *Rhododendron* and *Vaccinium*. Edelweiss (*Anaphais javanica*) is characteristic of the sub-alpine zone (Whitten et al. 1996)

Biodiversity Features

Overall richness and endemism for this ecoregion are moderate compared with those of other ecoregions in Indo-Malaysia.

The ecoregion harbors sixty-four mammal species, of which fourteen are endemics or near endemics (table J93.1). Of the latter, the Javan or surili leaf monkey (*Presbytis comata*) and the Javan gibbon (*Hylobates moloch*) are the most endangered primates in Indonesia (Whitten et al. 1996). Other ecoregional endemic mammals include the Javan mastiff bat (*Otomops formosus*), Javan shrew-mouse (*Mus vulcani*), and the red tree rat (*Pithecheir melanurus*). The Javan subspecies of the yellow-throated marten (*Martes flavigula robinsoni*) and leopard on Java (*Pantera pardus melas*) are also considered endangered (Hilton-Taylor 2000).

TABLE J93.1. Endemic and Near-Endemic Mammal Species.

Family	Species
Sorcidae	*Crocidura orientalis**
Sorcidae	*Crocidura paradoxura*
Pteropodidae	*Megaerops kusnotoi*
Rhinolophidae	*Rhinolophus canuti*
Vespertilionidae	*Glischropus javanus**
Molossidae	*Otomops formosus*
Cercopithecidae	*Presbytis comata*
Hylobatidae	*Hylobates moloch*
Sciuridae	*Hylopetes bartelsi**
Muridae	*Mus vulcani**
Muridae	*Maxomys bartelsii**
Muridae	*Pithecheir melanurus**
Muridae	*Kadarsanomys sodyi**
Muridae	*Sundamys maxi*

An asterisk signifies that the species' range is limited to this ecoregion.

More than 230 bird species are known to occur in the ecoregion, of which 30 are endemic or near endemic (table J93.2). The ecoregion overlaps with the western portion of the Java and Bali forests EBA. There are thirty-four restricted-range bird species in this EBA, of which thirty are found in this ecoregion. Of these, nine bird species are found nowhere else on Earth and four are threatened, including the endangered Javan hawk-eagle (*Spizaetus bartelsi*) and the vul-

TABLE J93.2. Endemic and Near-Endemic Bird Species.

Family	Common Name	Species
Accipitridae	Javan hawk-eagle	*Spizaetus bartelsi*
Phasianidae	Chestnut-bellied partridge	*Arborophila javanica**
Columbidae	Green-spectacled pigeon	*Treron oxyura*
Columbidae	Pink-headed fruit-dove	*Ptilinopus porphyreus*
Columbidae	Dark-backed imperial-pigeon	*Ducula lacernulata*
Strigidae	Javan scops-owl	*Otus angelinae**
Caprimulgidae	Salvadori's nightjar	*Caprimulgus pulchellus*
Apodidae	Waterfall swift	*Hydrochous gigas*
Apodidae	Volcano swiftlet	*Aerodramus vulcanorum**
Trogonidae	Blue-tailed trogon	*Harpactes reinwardtii*
Capitonidae	Brown-throated barbet	*Megalaima corvina**
Capitonidae	Flame-fronted barbet	*Megalaima armillaris*
Rhipiduridae	Rufous-tailed fantail	*Rhipidura phoenicura*
Rhipiduridae	White-bellied fantail	*Rhipidura euryura*
Campephagidae	Sunda minivet	*Pericrocotus miniatus*
Muscicapidae	Sunda robin	*Cinclidium diana*
Muscicapidae	Javan cochoa	*Cochoa azurea**
Aegithalidae	Pygmy tit	*Psaltria exilis**
Pycnonotidae	Sunda bulbul	*Hypsipetes virescens*
Zosteropidae	Javan grey-throated white-eye	*Lophozosterops javanicus*
Sylviidae	Javan tesia	*Tesia superciliaris**
Sylviidae	Sunda warbler	*Seicercus grammiceps*
Timaliidae	Rufous-fronted laughingthrush	*Garrulax rufifrons*
Timaliidae	White-bibbed babbler	*Stachyris thoracica*
Timaliidae	Crescent-chested babbler	*Stachyris melanothorax*
Timaliidae	Grey-cheeked tit-babbler	*Macronous flavicollis*
Timaliidae	Javan fulvetta	*Alcippe pyrrhoptera**
Timaliidae	Spotted crocias	*Crocias albonotatus**
Nectariniidae	White-flanked sunbird	*Aethopyga eximia*
Fringillidae	Mountain serin	*Serinus estherae*

An asterisk signifies that the species' range is limited to this ecoregion.

nerable Javan cochoa (*Cochoa azurea*), and Javan scops-owl (*Otus angelinae*) (BirdLife International 2000).

Status and Threats

Current Status

Only a fifth of the original habitat remains. There are twenty-five protected areas that cover 3,410 km^2 (13 percent) of the ecoregion (table J93.3). Although there are several that are larger than 100 km^2, none exceed 500 km^2; thus, the protected habitats represent isolated mountains (usually volcanic peaks) that are scattered throughout the mountain chains.

Types and Severity of Threats

Because of this ecoregion's steep terrain, it is less threatened by human activity than the island's lowlands. Population pressure is becoming more intense. Farmers are continually being forced into steeper lands in the upper watersheds and into more marginal environments. The resulting forest clearing has significant destructive effects on nutrient outflow, total water yield, peak storm flows, and stream sedimentation (IUCN 1991).

Priority Conservation Actions

With the rapid pace of forest destruction under way in Indonesia, quick conservation action is urgently needed to stave off the complete collapse of these forests. However, steps should be taken to lay the groundwork for longer-term conservation as well.

Short-Term Conservation Actions (1–5 Years)

- Initiate a major public awareness campaign on forest values and forest loss to drive political will.
- Stop illegal logging in and around protected areas by challenging the authorities and judiciary to act on existing laws and regulations.
- Place all remaining lowland forests under active protection.
- Mobilize major assistance for the more outstanding protected areas.
- Stop wildlife poaching, especially in key parks and reserves, by tightening surveillance around ports or airports and making the purchase of endangered species socially unacceptable.

Longer-Term Conservation Actions (5–20 Years)

- Attempt to develop linkages between lowland and montane habitats.
- Devise ways to maintain integrity of buffer zones of protected areas.

Focal Species for Conservation Action

Endangered endemic species: Javan gibbon (*Hylobates moloch*) (Eudey 1987), Javan or surili leaf monkey (*Presbytis comata*) (Eudey 1987), yellow-throated marten (*Martes flavigula robinsoni*), Javan leopard (*Pantera pardus melas*) (Nowell and Jackson 1996), Javan hawk-eagle (*Spizaetus bartelsi*) (BirdLife International 2000)

Area-sensitive species: Javan leopard (*Panthera pardus melas*) (Nowell and Jackson 1996), wild dog (*Cuon alpinus*) (Ginsburg and Macdonald 1990), hornbills (Bucerotidae) (rhinoceros hornbill [*Buceros rhinoceros*], Oriental pied-hornbill [*Anthracoceros albirostris*], wreathed hornbill [*Aceros undulatus*]) (Kemp 1995)

Top predator species: Javan leopard (*Panthera pardus melas*) (Nowell and Jackson 1996), wild dog (*Cuon alpinus*) (Ginsburg and Macdonald 1990)

Selected Conservation Contacts

Lembaga Alam Tropika Indonesia (LATIN)

WWF-Indonesia

Yayasan Keanekaragaman Hayati

Yayasan Konsorsium Pembaruan Agraria (KPA)

Justification for Ecoregion Delineation

MacKinnon (1997) included the islands of Java and Bali in biounit 22 (with three subunits). Western Java is wetter than the eastern half of the island, and the forests are richer in species (Whitten et al. 1996; MacKinnon 1997). There are also floristic differences between the lowland and montane vegetation in Java and Bali (Whitmore 1984; Whitten et al. 1996). Therefore, using MacKinnon's subunit boundary, we delineated the Western Java Rain Forests [91] to represent the moister evergreen forests to the west and the Eastern Java–Bali Rain Forests [92] to represent the drier, less species-

TABLE J93.3. WCMC (1997) Protected Areas That Overlap with the Ecoregion.

Protected Area	Area (km^2)	IUCN Category
Gunung Pangasaman [90]	110	?
Telaga Warna	20	I
Gunung Halimun	480	I
Gunung Gede Pangrango	220	II
Gunung Burangrang	50	I
Gunung Jagat	6	I
Gunung Tilu	90	I
Nusu Gede Pandjalu	8	I
Gunung Sawai	50	IV
Gunung Papandayan	70	V
Gunung Simpang	150	I
Kawah Gunung Tangkuban Perahu	20	V
Gunung Ciremai	160	?
Waduk Gede/Jati Gede	120	?
Masigit Kareumbi	130	?
Gunung Masigit	290	?
Kawah Kamojang	90	?
Pegunungan Pembarisan	120	?
Gunung Liman Wilis	230	?
Gunung Limbung	200	?
Gunung Perahu	330	?
Gunung Slamet	260	?
Gunung Sumbing	60	?
Pringombo I, II	6	?
Tuk Songo	140	V
Total	3,410	

Ecoregion numbers of protected areas that overlap with additional ecoregions are listed in brackets.

rich forests of eastern Java and in Bali. However, we also extracted the montane forests into distinct ecoregions—Western Java Montane Rain Forests [93] and Borneo Montane Rain Forests [95]—using the 1,000-m elevation contour of a DEM (USGS 1996).

Prepared by John Morrison

Ecoregion Number:	**94**
Ecoregion Name:	**Eastern Java–Bali Montane Rain Forests**
Bioregion:	**Sunda Shelf and Philippines**
Biome:	**Tropical and Subtropical Moist Broadleaf Forests**
Political Units:	**Indonesia**
Ecoregion Size:	**15,800 km^2**
Biological Distinctiveness:	**Bioregionally Outstanding**
Conservation Status:	**Endangered**
Conservation Assessment:	**IV**

The Eastern Java–Bali Montane Rain Forests [94] are found on one of the most actively volcanic islands in the world. Once the home of the extinct Javan and Balinese tigers (*Panthera tigris sundaica* and *Panthera tigris balica*, respectively), they contain fifteen bird species found nowhere else on Earth and more than 100 mammal species. Nearly three-quarters of the ecoregion's natural habitat has been cleared by a rapidly expanding population that is increasingly forced into these marginal lands.

Location and General Description

This ecoregion represents the montane forests of eastern Java and Bali, Indonesia. Based on the Köppen climate zone system, this ecoregion falls in the tropical wet and dry climate zones (National Geographic Society 1999). Java probably did not exist before the Miocene (24 m.y.). Bali did not emerge until 3 million years ago. Truly born of fire, the islands of Java and Bali are the result of the subduction and remelting of the Australian–Indian Ocean tectonic plate beneath the Eurasian tectonic plate at the Java trench. The melted crust has risen as volcanoes and, along with subsequent sedimentation, created Java. Therefore, the surface geology consists of Tertiary and Quaternary volcanics, alluvial sediments, and areas of uplifted coral limestone. Twenty of the volcanoes on Java and Bali have been active in historic times, and they are among the most active volcanic islands in the world. During previous ice ages, when sea levels were much lower, Java and Bali were connected to Sumatra, Borneo, and the rest of the Asian mainland (Whitten et al. 1996).

The predominant forest types found in the ecoregion include evergreen rain forest, moist deciduous forest, and seasonal and aseasonal montane forest. The forest type found at any particular location depends on a variety of factors including elevation, soils, aspect, and longitude; the ecoregion tends to become drier as one moves east. The transition between the evergreen rain forest and moist deciduous forest, which is typical of the lowlands, and montane forest generally is gradual in Java and Bali (Whitten et al. 1996). .

Evergreen rain forests of Java contain *Artocarpus elasticus* (Moraceae), *Dysoxylum caulostachyum* (Meliaceae), langsat *Lansium domesticum* (Meliaceae), and *Planchonia valida* (Lecythidaceae) (Whitten et al. 1996). Moist deciduous forests are drier than the evergreen rain forests and have 1,500–4,000 mm of rainfall annually, with a four- to six-month dry season. Common lowland deciduous trees in eastern Java and Bali are *Homalium tomentosum*, *Albizia lebbekoides*, *Acacia leucophloea*, *A. tomentosa*, *Bauhinia malabarica*, *Cassia fistula*, *Dillenia pentagyna*, *Tetrameles nudiflora*, *Ailanthus integrifolia*, and *Phyllanthus emblica*. Many herbaceous plants are confined to the deciduous forests (Whitten et al. 1996).

The transition between lowland and montane forests is a floristic one, and some plant families and genera are found only on one or the other side of this transition. Above 1,000 m, genera begin to include *Anemone*, *Aster*, *Berberis*, *Galium*, *Gaultheria*, *Lonicera*, *Primula*, *Ranunculus*, *Rhododendron*, *Veronica*, and *Viola*. Some lowland tree species transition out up to 1,200 m. The most abundant montane tree species in the lower montane zone are *Lithocarpus*, *Quercus*, *Castanopsis*, *Fagaceae*, and laurels (Lauraceae). Magnoliaceae, Hammamelidaceae, and Podocarpaceae are also well represented. Few emergents, primarily *Atingia excelsa* and *Podocarpus* spp., are found in the lower montane zone, but tree ferns are common (Whitten et al. 1996).

In the gradual transition from lower to upper montane forest, which begins at approximately 1,800 m, enormous quantities of *Aerobryum* moss begin to become prevalent on all surfaces. *Dacrycarpus* (*Podocarpus*) continues up from the lower montane. Ericaceae shrubs are very characteristic of the upper montane zone, including *Rhododendron*, *Vaccinium*, and *Gaultheria*. Sub-alpine forest, found above 3,000 m, contains one species-poor layer of trees, including *Rhododendron* and *Vaccinium*. Edelweiss (*Anaphalis javanica*) is characteristic of the sub-alpine zone (Whitten et al. 1996).

Biodiversity Features

Overall richness and endemism in this ecoregion are low to moderate compared with those of other ecoregions. The ecoregion provides habitat for approximately 100 mammal species, including one—*Megaerops kusnotoi*—that is an ecoregional near-endemic species (table J94.1). Among the mammals of conservation significance are the wild dog (*Cuon alpinus*) and endangered Javan leopard (*Panthera pardus melas*) (Nowell and Jackson 1996; Hilton-Taylor 2000). The Javan tiger (*Panthera tigris*) once inhabited this ecoregion but is now extinct (Nowell and Jackson 1996).

The bird fauna exceeds 215 species and includes 18 endemics and near endemics (table J94.2). The ecoregion overlaps with the eastern portion of the Java and Bali forests

TABLE J94.1. Endemic and Near-Endemic Mammal Species.

Family	Species
Pteropodidae	*Megaerops kusnotoi*

An asterisk signifies that the species' range is limited to this ecoregion.

TABLE J94.2. Endemic and Near-Endemic Bird Species.

Family	Common Name	Species
Accipitridae	Javan hawk-eagle	Spizaetus bartelsi
Columbidae	Pink-headed fruit-dove	Ptilinopus porphyreus
Columbidae	Dark-backed imperial-pigeon	Ducula lacernulata
Strigidae	Javan owlet	Glaucidium castanopterum
Caprimulgidae	Salvadori's nightjar	Caprimulgus pulchellus
Capitonidae	Flame-fronted barbet	Megalaima armillaris
Rhipiduridae	Rufous-tailed fantail	Rhipidura phoenicura
Rhipiduridae	White-bellied fantail	Rhipidura euryura
Campephagidae	Sunda minivet	Pericrocotus miniatus
Pycnonotidae	Sunda bulbul	Hypsipetes virescens
Zosteropidae	Javan grey-throated white-eye	Lophozosterops javanicus
Sylviidae	Sunda warbler	Seicercus grammiceps
Timaliidae	Rufous-fronted laughingthrush	Garrulax rufifrons
Timaliidae	White-bibbed babbler	Stachyris thoracica
Timaliidae	Crescent-chested babbler	Stachyris melanothorax
Timaliidae	Grey-cheeked tit-babbler	Macronous flavicollis
Nectariniidae	White-flanked sunbird	Aethopyga eximia
Fringillidae	Mountain serin	Serinus estherae

An asterisk signifies that the species' range is limited to this ecoregion.

EBA. Thirty-four bird restricted-range bird species are found in the EBA, of which thirty are found in the montane habitats of this ecoregion. Two bird species in the ecoregion are considered threatened: the Javan hawk-eagle (*Spizaetus bartelsi*) (endangered), and the Javan scops-owl (*Otus angelinae*) (vulnerable) (BirdLife International 2000).

Status and Threats

Current Status

Nearly three-quarters of the ecoregion's natural habitat has been cleared by a rapidly expanding population. The remaining forest is scattered throughout the landscape as small patches, mainly limited to the steep slopes of the volcanoes. There are twelve protected areas that cover 3,690 km^2 (23 percent) of the ecoregion (table J94.3). At 307 km^2, the average size of the protected areas in the ecoregion is large, with Gunung Raung being the largest. This mountain range has more active volcanoes than are found anywhere else in the world. The vegetation of this ecoregion has been disturbed by repeated volcanic activity (MacKinnon and MacKinnon 1986). Fires are common, and *Casuarina junghuhniana*, a secondary forest species, occurs gregariously in burned areas (FAO 1981).

Types and Severity of Threats

Java also has one of the highest human population densities in Asia, and these populations are being continually forced into steep, upper watersheds and more marginal environments, where they have had significant destructive effects on nutrient outflow, total water yield, peak storm flows, and stream sedimentation (IUCN 1991). Visitors to the mountains often collect plants, and over the years this collection has taken a toll on many of the plant populations (Whitten et al. 1996).

Priority Conservation Actions

With the rapid pace of forest destruction under way in Indonesia, quick conservation action is urgently needed to stave off the complete collapse of these forests. However, steps should be taken to lay the groundwork for longer-term conservation as well.

Short-Term Conservation Actions (1–5 Years)

- Initiate a major public awareness campaign on forest values and forest loss to drive political will.
- Place all remaining lowland forests under active protection.
- Mobilize major assistance for the more outstanding protected areas.
- Stop wildlife poaching, especially in key parks and reserves, by tightening surveillance around ports or airports and making the purchase of endangered species socially unacceptable.

Longer-Term Conservation Actions (5–20 Years)

- Attempt to design a protected area system that includes linkages between lowland and montane habitats.
- Devise ways to maintain the integrity of buffer zones of protected areas.

Focal Species for Conservation Action

Endangered endemic species: Javan hawk-eagle (*Spizaetus bartelsi*) (BirdLife International 2000), Javan leopard (*Panthera pardus melas*) (Nowell and Jackson 1996)

TABLE J94.3. WCMC (1997) Protected Areas That Overlap with the Ecoregion.

Protected Area	Area (km^2)	IUCN Category
Gunung Picis	20	I
Bromo Tengger Semeru	610	II
Gunung Lawu	290	?
Arjuno Lalijiwo	60	?
Gunung Liman Wilis	230	?
Gunung Kawi/Kelud	410	?
Sungai Kolbu Lyang Pintean	180	?
Bali Barat [91]	470	II
Meru Betiri [91]	180	II
Gunung Raung [91]	1,040	?
Batukau I/II/III NR	160	I
Gunung Abang GR	40	?
Total	3,690	

Ecoregion numbers of protected areas that overlap with additional ecoregions are listed in brackets.

Area-sensitive species: Javan leopard (*Panthera pardus melas*) (Nowell and Jackson 1996), wild dog (*Cuon alpinus*) (Ginsburg and Macdonald 1990), hornbills (Bucerotidae) (rhinoceros hornbill [*Buceros rhinoceros*], wreathed hornbill [*Aceros undulatus*]) (Kemp 1995)

Top predator species: Javan leopard (*Panthera pardus melas*) (Nowell and Jackson 1996), wild dog (*Cuon alpinus*) (Ginsburg and Macdonald 1990)

Selected Conservation Contacts

Lembaga Alam Tropika Indonesia (LATIN)

WWF-Indonesia

Yayasan Keanekaragaman Hayati

Yayasan Konsorsium Pembaruan Agraria (KPA)

Justification for Ecoregion Delineation

MacKinnon (1997) included the islands of Java and Bali in biounit 22 (with three subunits). Western Java is wetter than the eastern half of the island, and the forests are richer in species (Whitten et al. 1996; MacKinnon 1997). There are also floristic differences between the lowland and montane vegetation in Java and Bali (Whitmore 1984; Whitten et al. 1996). Therefore, using MacKinnon's subunit boundary, we delineated the Western Java Rain Forests [91] to represent the moister evergreen forests to the west and the Eastern Java–Bali Rain Forests [92] to represent the drier, less species-rich forests of eastern Java and Bali. However, we also extracted the montane forests into distinct ecoregions—Western Java Montane Rain Forests [93] and Borneo Montane Rain Forests [95]—using the 1,000-m elevation contour of a DEM (USGS 1996).

Prepared by John Morrison

Ecoregion Number: 95

Ecoregion Name:	**Borneo Montane Rain Forests**
Bioregion:	**Sunda Shelf and Philippines**
Biome:	**Tropical and Subtropical Moist Broadleaf Forests**
Political Units:	**Indonesia, Malaysia, Brunei**
Ecoregion Size:	**115,100 km²**
Biological Distinctiveness:	**Globally Outstanding**
Conservation Status:	**Relatively Intact**
Conservation Assessment:	**III**

The Borneo Montane Rain Forests [95] can be likened to montane islands in a sea of lowland dipterocarp forests. This isolation has produced a unique and diverse set of montane species. Of Borneo's endemic bird species, twenty-three (73 percent) are montane. There are more than 150 mammal species in montane forests, making this ecoregion globally outstanding for mammal richness, and it is the most speciose montane rain forest found in the Indo-Pacific region. Despite this wealth of diversity, large tracts of Borneo's montane forests have not been explored to catalog the flora and fauna.

Location and General Description

This ecoregion represents the montane forests in the central region of the island of Borneo and falls within the boundaries of all three nations with territory in Borneo: Malaysia, Indonesia, and Brunei. Montane forests are much cooler and moister than lowland forests. For every 1,000 m above sea level there is an average 5°C drop in temperature, equivalent to a 10° shift in latitude. This partly explains why montane regions contain plants normally found in temperate regions. Rainfall is higher than in the lowlands, and many forests also derive moisture from clouds that bathe the region (MacKinnon 1986). The geology of the Borneo Montane Rain Forests [95] is primarily old volcanic rocks and melange (rock fragments in clay) that is continental in origin. Montane soils also change with altitude, generally becoming more acidic and nutrient-poor. Montane soils are primarily inceptisols and ultisols (RePPProT 1990). Based on the Köppen climate zone system, this ecoregion falls in the highland climate zone (National Geographic Society 1999).

The montane flora of Borneo is derived from both Asian and Australian families, making it one of the most diverse montane habitats on Earth. Araucariaceae, Clethraceae, Ericaceae, Fagaceae, Lauraceae, Myrtaceae, Podocarpaceae, Symplocaceae, and Theaceae are all families commonly found in montane forests. In the lower elevations of montane forest (about 1,000–1,200 m) the dominance of dipterocarp species ends. These are replaced with highly speciose oak (*Quercus* and *Lithocarpus* spp.) and chestnut (Castanopsis spp.) forests. Myrtaceae are also important in the lower montane forests. Above 1,500 m this forest grades into a montane ericaceous belt followed by an alpine meadow on the very highest peaks. Three major and parallel changes occur in montane forests with increasing altitude. First, there is a decrease in forest height. Montane forests do not have giant emergent trees, and their overall height is much lower. The canopy typically is 10–20 m high. Second, the size and shape of the leaves change. Trees with buttresses usually are absent. Lowland forests are dominated by tree species with medium-large (mesophyll) and billowing canopy trees. Montane forests are dominated by slender trees, small leaves (microphyll), and a flattish crown surface. The third difference is the increased presence of epiphytes. Orchids, ferns, moss, lichen, and liverworts are more abundant in montane forests than in lowland rain forests (MacKinnon et al. 1996). Upper montane forests share many common species and features of structure and appearance with heath forests (Richards 1936).

Pitcher plants, rhododendron, and orchids are especially diverse in Borneo's montane habitats. More than one-half of Borneo's thirty pitcher plant (*Nepenthes*) species are found in this ecoregion's montane habitats. Rhododendrons are characteristic of upper montane flora, and more than twenty *Vireya* species are found in this ecoregion. Rhododendrons are found on acidic and peat soils and have adapted to the harsh upper montane environments (MacKinnon et al. 1996). Orchids are found at all levels of the forest but are common epiphytes in the upper montane forests. For growth, orchids need light and moisture as well as a mycorrhiza relationship with a tree or other plant to derive their nutrients. The well-lit and moist conditions of the moss forest in the upper montane zone provide ideal growing conditions for many orchid species (Lamb and Chan 1978).

Although in southeast Asia most limestone occurs in the lowlands, Borneo has important limestone forests in the montane zone. Gunung Api is highly diverse botanically, with fourteen of Borneo's fifteen *Monophyllea* species, disrupted altitudinal zonation compared with forest over other substrates, and montane birds occurring at atypically low altitudes. The few high-altitude swamp forests (which help regulate water supply to downstream areas) are found in the Ulu Meligan/Ulu Long Pasia montane area and the Usun Apau Plateau (WWF and IUCN 1995).

Biodiversity Features

The noticeable difference in vegetation structure and species composition also affect faunal communities found in montane forests. Animals in montane regions must face adverse climate, lack of shelter, and food shortages. For example, on Gunung Mulu in Sarawak, of the 171 bird species found on its lowland slopes, the range of most species does not exceed 900 m. By the boundary of the upper montane region at 1,300 m, only twelve species are still found. Similarly, mammal richness tends to decrease with altitude. Most primate species prefer lowland habitats, and orangutans (*Pongo pygmaeus*), gibbons, and langurs all show significant decrease in density between 500 m and 1,500 m in altitude. The macaque species density shows no large change between lowland and montane regions, perhaps being attributed to a greater dietary versatility (Caldecott 1980). Smaller mammals such as civets, tree shrews, squirrels, and rats dominate the montane region. Indeed, all near-endemic species fall into one of these categories (table J95.1). The vast tracts of montane forest still remain undisturbed and therefore still support some of the larger megafauna such as orangutan and Sumatran rhinoceros (*Dicerorhinus sumatrensis*).

More than 250 bird species are attributed to this ecoregion. This ecoregion overlaps with a large portion of the Bornean Mountains EBA (157) (Stattersfield et al. 1998). Many of the mountains of Borneo are ornithologically unexplored and poorly known, so the habitat needs and distributions of many species are incomplete (Stattersfield et al. 1998). Twenty-one near-endemic and two endemic bird species are found in these forests (table J95.2).

Status and Threats

Current Status

The ecoregion is largely intact, and only about 8 percent of the area has been cleared or converted. There are seventeen protected areas that cover 26,380 km² (about 25 percent) of the ecoregion (table J95.3). Several very large (more than 5,000 km²) reserves account for most of this protected area system. The ecoregion includes the largest protected block of rain forest in Borneo, the Kayan Mentarang National Park, which covers 140,000 km² of lowland dipterocarp forest as well as mountain forests; this park and Gunung Bentuang both exceed 5,000 km² (WWF-Indonesia n.d.). The Kayan Mentarang National Park and surrounding area is inhabited by several thousand indigenous people who depend on forest resources (WWF-Indonesia n.d.). However, commercial logging activities, road building, and intensive extraction of commercially valuable nontimber forest products now threaten the natural integrity of the reserve and the livelihoods of local people (WWF-Indonesia n.d.). The montane forests of Borneo have largely escaped the fires of 1997–1998,

TABLE J95.1. Endemic and Near-Endemic Mammal Species.

Family	Species
Viverridae	*Diplogale hosei*
Sciuridae	*Callosciurus baluensis*
Sciuridae	*Callosciurus orestes*
Sciuridae	*Sundasciurus brookei*

An asterisk signifies that the species' range is limited to this ecoregion.

TABLE J95.2. Endemic and Near-Endemic Bird Species.

Family	Common Name	Species
Accipitridae	Mountain serpent-eagle	*Spilornis kinabaluensis*
Phasianidae	Red-breasted partridge	*Arborophila hyperythra*
Phasianidae	Crimson-headed partridge	*Haematortyx sanguiniceps*
Podargidae	Dulit frogmouth	*Batrachostomus harterti*
Trogonidae	Whitehead's trogon	*Harpactes whiteheadi*
Capitonidae	Mountain barbet	*Megalaima monticola*
Capitonidae	Golden-naped barbet	*Megalaima pulcherrima*
Eurylaimidae	Hose's broadbill*	*Calyptomena hosii**
Eurylaimidae	Whitehead's broadbill	*Calyptomena whiteheadi*
Pachycephalidae	Bornean whistler	*Pachycephala hypoxantha*
Oriolidae	Black oriole*	*Oriolus hosii**
Turdidae	Everett's thrush	*Zoothera everetti*
Turdidae	Fruit-hunter	*Chlamydochaera jefferyi*
Muscicapidae	Eyebrowed jungle-flycatcher	*Rhinomyias gularis*
Zosteropidae	Pygmy white-eye	*Oculocincta squamifrons*
Zosteropidae	Mountain black-eye	*Chlorocharis emiliae*
Sylviidae	Bornean stubtail	*Urosphena whiteheadi*
Sylviidae	Friendly bush-warbler	*Bradypterus accentor*
Timaliidae	Bare-headed laughingthrush	*Garrulax calvus*
Timaliidae	Mountain wren-babbler	*Napothera crassa*
Timaliidae	Chestnut-crested yuhina	*Yuhina everetti*
Dicaeidae	Black-sided flowerpecker	*Dicaeum monticolum*
Nectariniidae	Whitehead's spiderhunter	*Arachnothera juliae*

An asterisk signifies that the species' range is limited to this ecoregion.

which were intentionally set throughout most of the lowland forests in Borneo. With the rapid pace of habitat loss in the lowland forests, Borneo's montane forests may serve as a final refuge for many of Borneo's species.

Types and Severity of Threats

Although little of the ecoregion has been cleared or degraded, significant threats from planned mining operations, large dams, and conversion to agriculture and high-altitude timber plantations are now increasing (WWF and IUCN 1995). Illegal collection of species for the commercial trade and shifting cultivation are also increasing, threatening the integrity of Borneo's highly distinctive montane biodiversity.

Priority Conservation Actions

With the rapid pace of forest destruction under way in Indonesia, quick conservation action is urgently needed in both lowland and montane forests. However, steps should be taken to lay the groundwork for longer-term conservation as well.

Short-Term Conservation Actions (1–5 Years)

- Call for an international conference to highlight the impending biodiversity crisis and loss of forests in Indonesia (perhaps using World Bank as an organizer).
- Stop illegal logging in and around protected areas.
- Revoke forest concessions in areas of high biodiversity outside protected areas and institute a moratorium on new concessions for oil palm, timber, and other plantations.
- Stop wildlife poaching, especially in key parks and reserves.

Longer-Term Conservation Actions (5–20 Years)

- Reassess representativeness of protected area system in the ecoregion.
- Ensure that redesign of the protected area system includes linkages between lowland and montane habitats.
- Restore integrity of buffer zones of protected areas.

Focal Species for Conservation Action

Endemic species: Hose's broadbill, black oriole

Area-sensitive species: Sumatran rhinoceros (Foose and van Strien 1997), orangutan (Galdikas 1971–1977; Dijuwantoko 1991; Tilson et al. 1993)

Selected Conservation Contacts

CI-Indonesia
European Union Forest Management Projects
GTZ Sustainable Forest Management Projects
Konsorsium SHK Kalimantan Timur
Lembaga Alam Tropika Indonesia (LATIN)
WCS
WWF-Indonesia
WWF-Malaysia
TNC-Indonesia
Yayasan Keanekaragaman Hayati
Yayasan Konsorsium Pembaruan Agraria (KPA)

Justification for Ecoregion Delineation

The large island of Borneo was divided into nine ecoregions. Most of the island's lowland and submontane forests are dominated by dipterocarp species (MacKinnon et al. 1996). MacKinnon and MacKinnon (1986) divided the island's lowland forests into six subunits, with a central subunit representing the montane forests. MacKinnon (1997) revised the boundaries of these seven subunits but retained the same general configuration. These authors used the major rivers, the Kapuas and Barito, to represent zoogeographic barriers to a few mammal species and based subunits largely on these barriers but also used climatic regimes for the drier eastern biounits (MacKinnon and MacKinnon 1986; MacKinnon 1997). Because ecoregions are based on biomes, we first isolated the central montane ecoregion—the Borneo Montane Rain Forests [95]—above the 1,000-m elevation contour using the DEM (USGS 1996).

Prepared by Colby Loucks

TABLE J95.3. WCMC (1997) Protected Areas That Overlap with the Ecoregion.

Protected Area	Area (km²)	IUCN Category
Bukit Tawau	290	II
Maliau Basing	450	VIII
Danum Valley [98]	90	VIII
S. Kayan S. Mentarang [97, 98]	5,290	?
Apo Kayan [97]	480	PRO
Long Bangun [97]	1,840	PRO
Batu Kristal	40	PRO
SAR (Sanctuary Reserve)	670	PRO
SAR (Sanctuary Reserve)	530	II
SAR (Sanctuary Reserve)	190	PRO
SAR (Sanctuary Reserve) [100]	350	PRO
SAR (Sanctuary Reserve) [100]	920	PRO
Gunung Bentuang [100]	5,660	II
Bukit Batutenobang	4,650	PRO
Bukit Batikap I, II, III [96]	3,100	PRO
Bukit Baka–Bukit Raya [96]	1,060	II
Gunung Penrisen/Gunung Niut [96]	770	IV
Total	26,380	

Ecoregion numbers of protected areas that overlap with additional ecoregions are listed in brackets.

Ecoregion Number:	**96**
Ecoregion Name:	**Borneo Lowland Rain Forests**
Bioregion:	**Sunda Shelf and Philippines**
Biome:	**Tropical and Subtropical Moist Broadleaf Forests**
Political Units:	**Indonesia, Malaysia**
Ecoregion Size:	**165,200 km^2**
Biological Distinctiveness:	**Globally Outstanding**
Conservation Status:	**Critical**
Conservation Assessment:	**I**

Borneo Lowland Rain Forests [96] are the richest rain forests in the world and rival the diversity of New Guinea and the Amazon. With 267 Dipterocarpaceae species (155 endemic to Borneo), Borneo is the center of the world's diversity for dipterocarps. These forests are home to the world's smallest squirrel, the 11-cm pygmy squirrel, and the endangered orangutan. In northeast Borneo, populations of Sumatran rhinoceros and Asia's largest terrestrial mammal, the Asian elephant, still tenuously survive in the last pockets of forest. These forests contain the parasitic plant *Rafflesia arnoldii*, which produces the world's largest flower (up to 1 m in diameter). These forests are globally outstanding for both bird and plant richness, with more than 380 birds and an estimated 10,000 plant species found within its boundaries. Unfortunately, these forests have been rapidly converted to oil palm plantations or commercially logged at unprecedented rates over the past ten years. In 1997–1998 fires intentionally set to clear the forest for commercial agriculture such as oil palm ravaged Kalimantan. If the current trend of habitat destruction continues, there will be no remaining lowland forests in Borneo by 2010.

Location and General Description

This ecoregion is made up of the lowland dipterocarp forests of Borneo. All of Borneo, Java, Sumatra, and mainland Malaysia and Indochina were once part of the same landmass during the Pleistocene glacial period. Land bridges connected all of these islands, fostering waves of migrations of animals, plants, and humans (MacKinnon 1997). Today Borneo is separate but shares similarities in flora and fauna with these other landmasses. The geology of the Borneo Lowland Rain Forests [96] consists of limestones, volcanic rocks, schistgneiss complexes, and sedimentary rocks. Soils are primarily ultisols and inceptisols, generally older, infertile soils (RePPProT 1990). Based on the Köppen climate zone system, this ecoregion falls in the tropical wet climate zone (National Geographic Society 1999). Monthly rainfall exceeds 200 mm year-round, and the temperature rarely fluctuates more than 10°C.

The natural vegetation of this ecoregion is tropical lowland rain forest. The region's stable climatic conditions have given rise to some of the world's richest assemblage of flowering plants. Forests contain a high diversity of tree species, and dominant species are uncommon. As many as 240 different tree species can grow within 1 ha, with another 120 in the hectare adjacent (Kartawinata et al. 1981; Ashton 1989). The general characteristics of these forests are canopies 24–36 m high, with emergents reaching up to 65 m. Dipterocarpaceae is a dominant family in the emergent stratum (Whitmore 1984). In the richest forests, up to 80 percent of the emergent trees are dipterocarps, although the white-barked *Koompassia* tree (Laurelaceae) is a readily identifiable emergent as well. Of the dipterocarp genera, *Dipterocarpus*, *Dryobalanops*, and *Shorea* are the emergents, and *Hopea* and *Vatica* usually are found in the main canopy. Berseraceae and Sapotaceae are other common main canopy families (MacKinnon et al. 1996). Bornean ironwood (*Eusideroxylon zageri*), or Belian, is a common and commercially valuable species. A third layer occurs below the canopy of shade-tolerant species, adorned with lianas, orchids, and epiphytic ferns. This layer includes many species from the Euphorbiaceae, Rubiaceae, Annonaceae, Lauraceae, and Myristicaceae families. In some cases Euphorbiaceae is more common than dipterocarps, being the second most common family in Borneo (Newberry et al. 1992). Another unique feature of this layer is the prevalence of cauliflory. Cauliflory is a phenomenon in which trees bear their flowers and fruits on their trunks. The forest durian (*Durio testudinarum*) is a common example of cauliflory. On the forest floor, herbs, seedlings, and shade-tolerant palms exploit the few places that receive sunlight.

Limestone formations are found throughout Borneo and include extensive areas in the Sangkulirang Peninsula and the limestone hills of Sarawak. Detailed botanical surveys of the limestone areas have not been completed, but preliminary studies suggest that this habitat supports a tremendous number of flora species, many of them probably endemic (Whitmore 1989). Although limestone forest has little commercially valuable timber, herbs such as Balsaminaceae, Begoniaceae, and Gesneriaceae (which are well represented in this ecoregion) have horticultural potential (Whitmore 1989). The Labi Hills of the Brunei-Sarawak border support a diverse mosaic of forests on podzols and sandy yellow soils, and these are rich in endemics, including two palms, *Livistona exigua* and *Pinanga yassinii*, known only from podzolized ridges in the Ulu Ingei area (WWF and IUCN 1995). There seems to be no special vertebrate fauna associated with limestone in this ecoregion, although banteng (*Bos javanicus*), orangutan (*Pongo pygmaeus*), Bornean gibbon (*Hylobates muelleri*), sambar (*Cervus unicolor*), muntjac (*Muntiacus muntjak*), and mousedeer (*Tragulus* spp.) all use limestone outcroppings (MacKinnon et al. 1996).

Biodiversity Features

The flora of Borneo alone consists of about 10,000–15,000 species, more than on any other island or region in Malesia and richer than all of Africa, which is forty times larger (Burley 1991; Jacobs 1988). Borneo has more than 3,000 tree species and 2,000 orchid species and is the center of distribution of dipterocarps, with 267 species, 60 percent of them endemic (Ashton 1982; Whitmore and Tantra 1987). Some of the most unique plants in the world are the five or six species of the parasitic *Rafflesia* plant found in Borneo. All *Rafflesia* produce large brown, red, orange, and white flowers. The largest in the world, *Rafflesia arnoldii*, is found in this ecoregion and produces flowers more than 3 feet wide. *Rafflesia* have no leaves, instead deriving all their energy from the tissues of the ground vine *Tetrastigma*, which it parasitizes. Large buds emerge from the vine and have five petals that surround spikes that smell like rotting meat. This smell attracts insects, which act as pollinators (MacKinnon et al. 1996).

Bornean rain forests share much of their fauna with the Asian mainland and other Sunda islands but few with Sulawesi and the eastern islands. The wildlife in the Bornean lowland rain forests is not characterized great spectacles of migrating or large charismatic species but by an enormous diversity of forest animals. As a whole, Borneo has a higher number of endemic mammals (forty-four) than the other islands in the Sunda Shelf and Philippines Bioregion, but fifteen are limited to lowland forests, and many of these are small rodents and bats (MacKinnon et al. 1996). Twelve of these are endemic, and one is near endemic to these forests (table J96.1).

Borneo also lacks some of the larger predators found in Asian mainland ecoregions, such as the tiger (*Pantera tigris*) and the leopard (*P. pardus*). In their absence, several small-medium carnivores dominate these forests, including the endangered clouded leopard (*Neofelis nebulosa*), sun bear (*Helarctos malayanus*), Sunda otter-civet (*Cynogale bennettii*), and other mustelids (MacKinnon 1997). The Asian elephant (*Elephas maximus*) and critically endangered Sumatran rhinoceros (*Dicerorhinus sumatrensis*) are found in northeastern Borneo. The Asian elephant and Sumatran rhinoceros are the largest forest herbivores, and they need a large amount of forest to survive. The intense human pressure on the natural forests will challenge their ability to survive in the long term.

Borneo's lowland forests are home to the globally recognized orangutan. The prehistoric race of orangutan distribution reached mainland Asia through Indochina and Thailand to southeastern China. Today the orangutan is limited to northern Sumatra and Borneo (Galdikas and Briggs 1999). Unlike other apes, orangutans are solitary and arboreal. They feed primarily on fruit but also feed on leaves, flowers, insects, and, during times of food stress, specifically bark (Knott 1998). The orangutans move throughout the forest, following the fruiting of numerous trees. They have the ability to catalog the location and degree of a fruit's ripeness for a large number of trees and species (Daws and Fujita 1999; Galdikas and Briggs 1999).

The orangutan is not the only primate in Borneo's lowland forests. They are home to thirteen primate species: three apes (the orangutan and two gibbon species), five langurs, two macaques, the tarsier (*Tarsius bancanus*), the slow loris (*Nycticebus coucang*), and the endangered proboscis monkey (*Nasalis larvatus*). Most of these species have overlapping ranges, but they vary with respect to dietary content and foraging strategy (MacKinnon 1997).

The fauna communities are divided into those that are active during the day and those that are nocturnal. Orangutans, pig-tailed macaques (*Macaca nemestrina*), and the white-rumped shama (*Copyschus malabaricus*), with its melodious calls, can be seen or heard during the early part of the day. Many birds, reptiles, and amphibians browse throughout the day. As night approaches, bats such as the large flying-fox (*Pteropus vampyrus*) and flying squirrels begin to appear. Pangolins (*Manis javanicus*) search for ants on the forest floor, while civets and the clouded leopard (*Neofilis nebulosa*) hunt for food. Bearded pigs, though occasionally active during the day, are predominantly nocturnal animals. The tarsier (*Tarsius bancanus*), one of Borneo's strangest animals, with its large eyes searches out smaller animals and leaves for food.

The bird fauna, like the mammal fauna, is similar to that of peninsular Malaysia and Sumatra. In Borneo there are eight hornbill species, eighteen woodpecker species, and thirteen pitta species. The 385 bird species attributed to the ecoregion include nine near-endemic species and two endemic species: the black-browed babbler (*Malacocincla perspicillata*) and the white-crowned shama (*Copsychus stricklandii*) (table J96.2). Hornbills such as the bushy-crested (*Anorrhinus galeritus*), helmeted hornbill (*Rhinoplax vigil*), and great rhinoceros (*Buceros rhinoceros*) are important seed

TABLE J96.1. Endemic and Near-Endemic Mammal Species.

Family	Species
Vespertilionidae	*Pipistrellus kitcheneri**
Tupaiidae	*Tupaia montana**
Tupaiidae	*Tupaia picta**
Vespertilionidae	*Pipistrellus cuprosus**
Vespertilionidae	*Murina rozendaali**
Cercopithecidae	*Presbytis hosei**
Sciuridae	*Callosciurus adamsi**
Sciuridae	*Petaurillus hosei**
Muridae	*Aeromys thomasi**
Muridae	*Maxomys ochraceiventer**
Muridae	*Maxomys baeodon**
Muridae	*Haeromys pusillus**
Rhinolophidae	*Hipposideros doriae*

An asterisk signifies that the species' range is limited to this ecoregion.

TABLE J96.2. Endemic and Near-Endemic Bird Species.

Family	Common Name	Species
Zosteropidae	Javan white-eye	*Zosterops flavus*
Timaliidae	Black-browed babbler*	*Malacocincla perspicillata**
Pycnonotidae	Blue-wattled bulbul	*Pycnonotus nieuwenhuisii*
Phasianidae	Red-breasted partridge	*Arborophila hyperythra*
Phasianidae	Crimson-headed partridge	*Haematortyx sanguiniceps*
Strigidae	Mantanani scops-owl	*Otus mantananensis*
Capitonidae	Mountain barbet	*Megalaima monticola*
Pachycephalidae	White-vented whistler	*Pachycephala homeyeri*
Muscicapidae	White-crowned shama*	*Copsychus stricklandii**
Zosteropidae	Pygmy white-eye	*Oculocincta squamifrons*
Timaliidae	Chestnut-crested yuhina	*Yuhina everetti*

An asterisk signifies that the species' range is limited to this ecoregion.

disperses of many fig (*Ficus* spp.) trees. Figs act as a keystone resource, providing food for many animals such as monkeys, arboreal mammals, squirrels, civets, and birds. The high diversity of fig trees in the rain forest assures food during fruiting throughout the year (MacKinnon et al. 1996).

Borneo is probably the richest island in the Sunda Shelf for reptile and amphibian diversity. Because detailed studies of the entire island have not been done, an incomplete picture emerges. For example, the earless monitor lizard (*Lanthonotus borneensis*) spends most of its life in underground caves and has been recorded in only a few places in Borneo. However, it probably has a widespread distribution throughout the island. Many populations of amphibians and reptiles are also hunted toward the brink of local extinctions.

Status and Threats

Current Status

This ecoregion is the second largest ecoregion in the Indo-Pacific region, yet more than half of the natural habitat in this ecoregion has been cleared or degraded. In recent years there has been extensive habitat loss, and current assessments predict further widespread destruction of forests. There are sixty reserves on paper that protect almost 12 percent of the region, but many are small (less than 100 km^2) or are still proposed (table J96.3). Only six reserves have been gazetted that protect more than 1,000 km^2 of contiguous lowland rain forest. The remaining reserves are proposed. These include Gunung Bentuang, Kutai, Pleihari Martapura, S. Kayan S. Mentarang, Tabin, and one SAR (Sanctuary Reserve). The protection of these reserves and the gazetting of the proposed reserves are essential to the conservation of biodiversity in this ecoregion.

This area has been heavily logged, with most logging occurring in the last thirty years. Commercial logging destroyed the forests while providing access to them. Massive agricultural projects, such as oil palm, rubber, and industrial timber for pulp and plywood, soon followed. Smaller agricultural shareholders also filled in these areas by cutting and burning patches of forests. By the late 1980s Indonesia was the world's largest plywood producer. Private and state-owned forestry companies have stripped the land clear for pulp and palm oil plantations, destroying vast tracts of forest. They are replanting with a limited number of fast-growing exotic species (Galdikas and Briggs 1999; Barber and Schweithelm 2000).

In 1982–1983 27,000 km^2 of tropical rain forest was burned in Kalimantan. The cause of what was at the time the largest forest fire ever recorded was widespread logging activities over the previous decade. Logging had transformed the fire-resistant primary forest into a degraded and fire-prone landscape. The El Niño–driven drought of that year set that stage for catastrophe when small agricultural fires escaped and ignited the forest (Grip 1986).

Indonesia did not learn from its past lessons. In 1997–1998, another intense El Niño year, the same events unfolded. In 1997–1998 fires raged across Kalimantan and Sumatra as forests were being cleared to make way for agriculture and plantation lands. More than 23,750 km^2 of lowland rain forest in Kalimantan was burned in these years alone. Primary tropical rain forests do not burn naturally, so few native plant or animal species are adapted to fire. Fires in

TABLE J96.3. WCMC (1997) Protected Areas That Overlap with the Ecoregion.

Protected Area	Area (km^2)	IUCN Category
Apar Besar [107]	1,370	PRO
Apo Kayan [95]	470	PRO
Bukit Baka–Bukit Raya [95]	800	II
Bukit Batikap I, II, III [95]	5,350	PRO
Bukit Perai	1,830	PRO
Bukit Raya (Perluasan)	3,300	PRO
Bukit Rongga	1,380	PRO
Crocker Range [108]	570	II
Danum Valley [95]	270	VIII
Gunung Bentuang [95]	2,210	II
Gunung Berau	1,560	PRO
Gunung Lumut	500	PRO
Gunung Palung [86]	780	II
Gunung Penrisen/Gunung Niut [95]	640	IV
Gunung Raya Pasi	50	I
Gunung Tunggal	640	PRO
Hutan Kapur Sangkulirang	2,150	PRO
Hutan Sambas [107]	220	PRO
Kaya Kuku NR	70	PRO
Kinabalu [108]	190	II
Kuala Kayan	610	PRO
Kutai (extension) [86, 89, 90]	320	PRO
Kutai [89, 90]	1,370	II
Long Bangun [95]	1,720	PRO
Meratus Hulu Barabai	980	PRO
Pararawen Baru [90]	520	PRO
Pararawen I, II	40	I
Pleihari Martapura	1,350	IV
S. Kayan S. Mentarang	1,660	PRO
S. Kayan S. Mentarang [95]	6,400	?
Sangalaki	1	IV
SAR (Sanctuary Reserve)	80	II
SAR (Sanctuary Reserve)	90	II
SAR (Sanctuary Reserve)	40	II
SAR (Sanctuary Reserve)	160	UA
SAR (Sanctuary Reserve)	30	UA
SAR (Sanctuary Reserve)	3	PRO
SAR (Sanctuary Reserve)	3	PRO
SAR (Sanctuary Reserve)	30	UA
SAR (Sanctuary Reserve)	540	PRO
SAR (Sanctuary Reserve)	1,730	IV
SAR (Sanctuary Reserve)	50	PRO
SAR (Sanctuary Reserve)	100	PRO
SAR (Sanctuary Reserve)	30	PRO
SAR (Sanctuary Reserve)	220	II
SAR (Sanctuary Reserve)	3	PRO
SAR (Sanctuary Reserve)	3	IV
SAR (Sanctuary Reserve)	60	II
SAR (Sanctuary Reserve)	40	II
SAR (Sanctuary Reserve)	6	UA
SAR (Sanctuary Reserve)	4	UA
SAR (Sanctuary Reserve)	110	PRO
SAR (Sanctuary Reserve) [107]	120	IV
SAR (Sanctuary Reserve) [95]	1,560	PRO
SAR (Sanctuary Reserve) [95]	1,560	PRO
Sesulu	120	PRO
Sungai Berambai	780	PRO
Tabin	1,230	IV
Unnamed	1,230	?
Unnamed	310	?
Total	49,563	

Ecoregion numbers of protected areas that overlap with additional ecoregions are listed in brackets.

pristine forests rarely escape the ground vegetation because of the high humidity and moisture. However, tropical rain forests that have been previously logged are fire-prone because large amounts of wood are left on the forest floor, and the forest canopy is opened, drying out the ground vegetation. These forests rarely escape fires such as those set throughout Indonesia in 1997 (Holmes, Essay 1; Barber and Schweithelm 2000; Yeager 1999a).

Types and Severity of Threats

The prognosis for Borneo's lowland forests is bleak. With little to no law enforcement in Indonesia, large tracts of land will be susceptible to burning every year by commercial logging agricultural projects such as oil palm, rubber, and industrial timber for pulp and plywood. Smaller agricultural shareholders threaten the forests through continued slash-and-burn agricultural practices. Little to no regard is given to the long-term environmental effects these actions will cause. Without immediate intervention to stem the loss of these forests, and given the current rate of destruction, there will be no primary lowland dipterocarp forests remaining in Indonesian Borneo within ten years.

The current system of protected areas underrepresents the habitat, and, based on the actions from 1997–1998, even protected areas will become susceptible to illegal and rampant logging and burning activities. The loss and fragmentation of the forest habitat will have drastic effects on the wildlife populations. As the forests shrink, more and more primates will interact with humans, and many will be killed or sold into the international pet trade. Vast numbers of species will become extinct before they have even been described and their role in the ecosystem determined. This collapse of the ecosystem will occur long before the last dipterocarp is cut down or set afire.

Exploration for oil and coal is a potential major threat to Gunung Lotung and the Maliau Basin in south central Sabah (WWF and IUCN 1995). On the Klias Peninsula, proposed development schemes will drain the wetlands for agriculture and pose the most serious threat (WWF and IUCN 1995).

Priority Conservation Actions

With the rapid pace of forest destruction under way in Indonesia, quick conservation action is urgently needed to stave off the complete collapse of these forests. However, steps should be taken to lay the groundwork for longer-term conservation as well.

Short-Term Conservation Actions (1–5 Years)

- Call for an international conference to highlight the impending biodiversity crisis and loss of forests in Indonesia (perhaps using World Bank as an organizer).
- Stop illegal logging in and around protected areas.
- Revoke forest concessions in areas of high biodiversity outside protected areas and institute a moratorium on new concessions for oil palm, timber, and other plantations.
- Place all remaining lowland forests under protection.
- Stop wildlife poaching, especially in key parks and reserves.
- Establish a trust fund to offset the opportunity costs of maintaining forest cover and avoiding conversion to oil palm plantations.
- Purchase through Indonesian NGOs several large tracts of lowland forest, create model ecotourism reserves, with foot safaris to view wildlife and attract foreign currency, and share profits from such ventures with communities.
- Determine whether Kalimantan provincial governments can be provided funding from carbon credits for maintaining forest cover and protection of wildlife.

Longer-Term Conservation Actions (5–20 Years)

- Reassess representativeness of protected area system in the ecoregion.
- Establish a long-term fund for conservation of lowland forests in Indonesia.
- Ensure that redesign of protected area system includes linkages between lowland and montane habitats.
- Restore integrity of buffer zones of protected areas.

Focal Species for Conservation Action

Habitat specialists: hornbills (mature trees needed for nests)

Area-sensitive species: orangutan (Galdikas 1971–1977; Dijuwantoko 1991; Tilson et al. 1993), Asian elephant (Santiapillai and Jackson 1990; Santiapillai 1997), Sumatran rhinoceros (Foose and van Strien 2000)

Top predator species: clouded leopard

Other: tarsier, primates, *Rafflesia*, banteng, Malaysian sun bear, mouse deer

Selected Conservation Contacts

CI-Indonesia
European Union Forest Management Projects
GTZ Sustainable Forest Management Projects
Konsorsium SHK Kalimantan Timur
Lembaga Alam Tropika Indonesia (LATIN)
TNC-Indonesia
WCS
WWF-Indonesia
WWF-Malaysia
Yayasan Keanekaragaman Hayati
Yayasan Konsorsium Pembaruan Agraria (KPA)

Justification for Ecoregion Delineation

The large island of Borneo was divided into seven ecoregions. Most of the island's lowland and submontane forests are dominated by dipterocarp species (MacKinnon et al. 1996). MacKinnon and MacKinnon (1986) divided the island's lowland forests into six subunits, with a central subunit representing the montane forests. MacKinnon (1997) revised the boundaries of these seven subunits but retained the same general configuration. These authors used the major rivers, the Kapuas and Barito, to represent zoogeographic barriers to a few mammal species and based subunits largely on these barriers but also used climatic regimes for the drier eastern biounits (MacKinnon and MacKinnon 1986; MacKinnon 1997).

Because ecoregions are based on biomes, we first isolated the central montane ecoregion—the Borneo Montane Rain

Forests [95]—above the 1,000-m elevation contour using the DEM (USGS 1996). We then assigned the large patches of peat forests, heath forests, freshwater swamp forests, and mangroves, in the lowlands and along the periphery of the island, into their own ecoregions: the Borneo Peat Swamp Forests [86], Sundaland Heath Forests [90] (which also includes Belitung Island and the heath forests in Bangka island), Southern Borneo Freshwater Swamp Forests [89], and Sunda Shelf Mangroves [107], respectively. The alpine habitats of the Kinabalu Mountain Range were represented by the Kinabalu Montane Alpine Meadows [108]. The remaining lowland dipterocarp forests in Borneo were combined into one ecoregion, Borneo Lowland Rain Forests [96]. This deviates from MacKinnon's use of subunits (25a, b, f, g, h, i) to divide Borneo.

Prepared by Colby Loucks

General Philippines Description

Introduction

As storehouses of biological diversity, the Philippines have been called "the Galapagos Islands multiplied tenfold." The Philippines are the site of an evolutionary radiation every bit as spectacular as that of the Galapagos and Hawaii. A volcanic archipelago, the Philippines are part of the Pacific rim's Ring of Fire, with all of the destruction and fertile rebirth associated with volcanism. The Philippines have a natural resiliency that developed in the face of constant natural disturbance: the Philippines are subject to an average of fifteen to twenty-five and as many as thirty-three typhoons in a single year. With the exception of the Palawan group, these are truly oceanic islands born in isolation and have come close to Asia only in recent geologic time. The Philippines contain approximately 8,900 vascular plant species, of which approximately 39 percent are endemic (Davis et al. 1995). Fifty-seven percent, or approximately 510 of its terrestrial vertebrates, are endemic to the Philippines, including 85 percent of the nonvolant mammals (Heaney and Regalado 1998) and 30 percent of the bird species (Kennedy et al. 2000). Sixteen new mammal species have been discovered in recent years (Heaney et al. 1998). The Philippines contains 252 known reptile species, of which 159 (63 percent) are endemic (Myers et al. 2000). The amphibian fauna is small (Inger 1999) but with a high rate of endemism; with eighty-four amphibian species total, sixty-five (77 percent) are endemic (Myers et al. 2000). The herpetofauna of the Philippines is poorly surveyed, and the species list will surely grow over time (R. Crombie, pers. comm., 2000).

Included in the high levels of endemism for all vertebrate taxa are some truly special animals, including the second largest eagle in the world, the Philippine eagle, or monkey-eating eagle; and the tamaraw, a unique dwarf water buffalo that lives only on the small island of Mindoro. Bats with nearly six-foot wingspans and flying lemurs glide through the trees while primitive primates, tarsiers, inhabit the islands connected to Mindanao during the ice age.

Almost as spectacular as the Philippines' natural heritage, however, is its rapid and nearly complete destruction since the end of World War II. Fifty-seven percent of the Philippines was intact in 1937, but only 6 percent remains in its natural state. Conservationists have identified the Philippines as a red-hot hotspot, in danger of losing many unique species, and humankind's will to avert an imminent extinction crisis will be tested in the coming years. The Philippines contain more threatened mammals (fifty) per unit area than any other country (Hilton-Taylor 2000; Heaney et al. 1998).

Geologic History

The Philippines began as island arcs approximately 50 million years ago, the result of complex interactions between the Philippine, Eurasian, Australian, and Pacific tectonic plates. The Philippines are essentially accreted terrains—an accretion of previously isolated island archipelagos that were brought together during the collision and partial subduction of large oceanic tectonic plates. The exceptions are the northern section of Palawan, the Calamianes, and the southwestern portions of Mindoro, which were rifted from the Asian mainland (below sea level) approximately 32 million years ago, transported through seafloor spreading across the growing South China Sea, and added to the growing Philippine Archipelago approximately 17 million years ago (Dickinson et al. 1991). Luzon was first to remain substantially above sea level about 15 million years ago. Only Palawan has its origins on the Asian mainland (Heaney et al. 1998). Although the components of the Philippines were essentially in place by the end of the Pliocene (Dickinson et al. 1991), much insight can be gathered by studying the Philippines' more recent ice age history: the combined effects of changing sea level and the complex Philippine Archipelago (currently more than 7,000 islands) separated by short stretches of ocean. When sea levels rise and fall, land bridges connect and disappear. During the last ice age, sea level was approximately 120 m below current levels, and the Philippines took on a completely different configuration. At that time there were fewer, larger islands, yet the Philippines were always separated from continental Asia, in contrast with Sumatra, Borneo, and Java (Heaney and Regalado 1998). Much of the Philippines biogeography can be explained by this ice age configuration.

Biogeography

The Philippines' fauna reflects influences from the southwest (Malaysia and Borneo), the west (across the South China Sea), and the southeast (Sulawesi and New Guinea). The influences from Malaysia and Borneo are strongest because the connections have been closer and more persistent. For example, of the 395 bird species that breed in the Philippines, 35 percent also breed in Borneo. There are two primary pathways from Borneo to the Philippines that were enhanced by lower sea levels: Palawan and the Sulu Islands. Palawan was the main pathway from Borneo; there are twenty-three species common to Borneo and Palawan that are absent in the remainder of the Philippines. The Philippines, being neither completely isolated nor completely connected to the Sunda Shelf, thus contains a fauna that is intermediate between an oceanic island and the mainland. It contains several of the regional groups of birds that are generally reluctant to cross water and mostly small mammals that are more likely to effect water crossings. The Philippines also contains almost no primary freshwater fish (Diamond and Gilpin 1983).

There are also two pieces of animal evidence of a connection from the Palearctic. The mammalian evidence consists of the endemic tamaraw (*Bos minorensis*), a dwarf water buffalo that is limited to the island of Mindoro and is most closely related to *Bos arnee* from southwest China. Additionally, the discoglossid frog *Barbourula busuangensis* is limited to the island of Busuanga in the Calamianes, and its closest relative inhabits southwest China also. It is also thought that one of the Philippines endemic tits, *Parus amabilis*, which is found only on Palawan, is an earlier derivative of China's *Parus venustulus* than the other Philippine endemic *Parus elegans*. There are several bird species whose ancestors presumably flew to the Philippines and are primarily associated with mountain habitats that would have been forest refugia during past glacial periods. Four essentially Australasian bird genera, *Prioniturus* (Psittacidae), *Gallicolumba* (Columbidae), *Trichoglossus* (Psittacidae), and *Basilornus* (Sturnidae), inhabit the Philippines. It is thought that these groups were rafted to the Philippines (Mindanao) by continental drift from areas in close contact with Australasia (Dickinson et al. 1991).

Because the Philippines are effectively buffered from new immigration by stretches of ocean, the islands have developed much endemism over time. A total of 510 terrestrial vertebrates are endemic to the Philippines, including 64 percent of 174 land mammals, 44 percent of 395 breeding land birds, 68 percent of 244 reptiles, and 78 percent of its approximately 85 amphibians. Eighty-five percent of the Philippines nonvolant mammals are endemic (Heaney and Regalado 1998). Many of these species are the result of adaptive radiations limited to the Philippines.

Floristically, the Philippines are predominantly Asiatic in origin, but there is a substantial Australasian influence as well. A total of 181 Asian plant genera reach their eastern limit in the Philippines, and 43 Australasian genera reach their western limit there. The Philippines contain more Australasian floral elements than Borneo. Of the approximately 9,000 vascular plant species in the Philippines, 70 percent are endemic. There is a low level of vascular plant differentiation within the islands (Dickinson et al. 1991).

Twelve natural groupings of islands are supported by clusters of endemic species: the Palawan group, the central Philippines (Negros, Panay, Guimaras, Masbate, Ticao, and Bantayan), Mindoro and its satellites, Luzon and its satellites, Samar-Leyte-Bohol, Mindanao and its satellites, Basilan, the Sulu Archipelago, Tablas-Romblon-Sibuyan (north of Panay), Cebu, Babuyanes (small islands north of Luzon), and a twelfth heterogeneous group of islands (Siquijor, Cagayancillo, Cresta de Gallo) (Dickinson et al. 1991). Samar-Leyte-Bohol, Mindanao, and Basilan often are grouped together.

Physiography

The physiography of the Philippines Archipelago, which consists of more than 7,000 islands, is a result of tectonic collisions and subduction, volcanism, and subsequent sedimentation. Only 470 of the islands are larger than a square mile (2.6 km^2). The Philippines often are simply grouped into Luzon, Mindanao, Palawan, and the Visayas, reflecting population and ethnic groups (Dickinson et al. 1991). Luzon and Mindanao are by far the largest two islands and together constitute 68 percent of the land area of the Philippines (Collins et al. 1991). Many of the upland areas are isolated volcanoes, but the Central Cordillera of northern Luzon is an uplifted area with large areas above 1,500 m. Several additional smaller mountain ranges exist in the western and northeastern sections of Luzon. A large area of swampland was originally found in north central Luzon. Mindanao is less affected by typhoons that regularly hit the more northerly islands. It contains several mountain ranges that extend to 2,804 and 2,865 m, respectively. Large rivers exist only on Luzon and Mindanao (Dickinson et al. 1991).

Palawan is long and narrow, consisting of a steep mountain range whose highest point is 2,085 m. The eastern half of the island is in a rain shadow and contains moist semi-deciduous forests. Mindoro is also mountainous and is subject to extremely high rainfall. The Visayas (Panay, Negros, Masbate, Samar, Leyte, Cebu, and Bohol) are a diverse group of mostly mountainous islands between Luzon and Mindanao.

General Vegetation

In prehistory the Philippines were completely covered in tropical forest, but as early as 1911 it was noted that original forest probably would not be encountered during a visit to the Philippines. In the early 1930s it was estimated that the islands contained approximately 45 percent rain forest, 2 percent mossy forest, 10 percent secondary forest, almost 20 percent grasslands, a bit more than 20 percent cultivated land, and another 2 percent mangroves. Much less forest remains now, but there are still eight major types of forest distributed through the Philippines: beach forest, dipterocarp or mixed dipterocarp forest, molave forest, pine forest, lower montane forest, mossy or upper montane forest, mangroves, and ultrabasic forest. Nonforest habitats include cultivated areas, grasslands, and swamp and marshlands (Dickinson et al. 1991; Heaney and Regalado 1998; L. Heaney, pers. comm., 2000).

The stunted beach forest contains *Casuarina* and *Barringtonia* mixed with other lowland species. Palms, vines, bamboo, and *Pterocarpus indicus* are present only in rare back-beach swamps. This habitat type is extremely rare because of coastal habitation. The dominant forest type in the Philippines was dipterocarp forest. This group of trees is known as Philippine Mahogany in the timber trade. This forest type occurred from sea level to elevations of 400 m or higher. Individual dipterocarps are found to 1,500 m. Philippine dipterocarp forest is quite tall (45–65 m) and dense, with three canopy layers. Lianas and bamboo are rare in mature forest but common in poorly developed evergreen forest. Ferns, orchids, and other epiphytic plants are found on the larger trees. At higher elevations there are only two canopy layers, tree stature is lower, and there are more epiphytes. Upper hill dipterocarp forest is found at elevations of 650–1,000 m and contains dominant *Shorea polysperma* and oaks, chestnuts, and elaeocarps. At approximately 1,000 m the montane forest contains oaks and laurels (Heaney and Regalado 1998). The mossy upper montane forest is generally found at elevations over 1,200 m, where humidity is constantly high. This stunted, single-story, moss- and epiphyte-covered forest contains tree ferns up to 10 m high. (Dickinson et al. 1991).

Molave forest 25–30 m high is found on limestone or sandy substrate at elevations less than 200 m, especially in areas with a pronounced dry season. This is an open forest type that is semideciduous in seasonally dry areas and contains abundant bamboo. The dominant tree species is *Vitex parviflora*, also known as molave, a teak. Consequently, this is an endangered forest type (Dickinson et al. 1991). Ultrabasic forest grows in areas containing ultrabasic rocks, a stressful habitat type (Heaney and Regalado 1998). Pine forest is limited to Luzon and Mindoro at elevations over 1,000 m (Dickinson et al. 1991) in an area with a distinct dry season and periodic fires (Heaney and Regalado 1998). The two native pine species are Benguet pine (*Pinus insularis*) and Mindoro pine (*Pinus merkusii*). There is evidence that this forest type was not always predominant in the areas it now grows in (Heaney and Regalado 1998). Mangroves are found along low-energy coastlines. Trees may reach 30 m in height and contain extensive stands of *Nipa* palm. Mangroves are another endangered forest type because the space they occupy often is cleared for charcoal or fishfarming (Dickinson et al. 1991).

Ecological Processes

Several important ecological processes affect the Philippines. Perhaps the most important is a high degree of natural disturbance. Because of their position in the Pacific's Ring of Fire, the islands are subject to volcanoes, landslides, and earthquakes that result from the constant grinding of the tectonic plates beneath the islands. Additionally, the Philippines' position in the Pacific puts it in the path of tropical typhoons that originate near the Marianas and move west. Some locations in the Philippines receive as much as one-third of their total precipitation during cyclones, which generally occur from July to November, and mostly to the north of the northern Visayas (Dickinson et al. 1991).

Seasonal fire is a significant factor in areas of Luzon that have a long dry season. Here fire-tolerant pine forest dominates where other vegetation cannot survive the frequent fires (Heaney and Regalado 1998).

Dipterocarp forest trees have poor dispersal capabilities and rely on animal dispersal. As forest clearings used for subsistence agriculture grow larger and the forest becomes reduced, the animals that normally disperse seeds and are limited to the forest disappear, reducing seed dispersal and forest regeneration (Dickinson et al. 1991).

The Philippines has been home to five species of hornbills (Bucerotidae), which are important frugivores (*Ficus*) and seed dispersers in moist tropical forests (Heaney and Regalado 1998).

Status and Threats

Only 1.3 percent of the Philippines is covered by protected areas, and two-thirds of these national parks contain human settlement, with associated destruction of natural forests (Collins et al. 1991). The single most important factor that has led to environmental degradation in the Philippines is a very high population density of 202 people per km^2 (Wood et al. 2000). Once completely forested, the area of forest was reduced from 57 percent of the country in 1934 to 24 percent in 1988. The annual rate has not been constant over time, varying from less than 2 percent per year before 1969 to approximately 3.5 percent from 1976 to 1980. Accessibility, or the distance of forest patches to roads, is the most important factor in deforestation (Liu et al. 1993).

Logging of the Philippines' natural capital has provided important income for the Philippines economy this past century. Combined with a significant population increase and the need for land for slash-and-burn cultivation, logging has destroyed the natural forests of the Philippines and provided access for a growing population of shifting cultivators so that the forests are mostly gone. Limited reforestation efforts have been largely unsuccessful (Dickinson et al. 1991). In 1990 the state of the Philippines forests was "arguably the worst in tropical Asia" (Collins et al. 1991, p. 192). Only 40 percent of once almost completely forested Philippines was considered forest land in 1990 (Collins et al. 1991), and little of that could be considered natural forest. By 1994, forest cover had decreased to 18 percent (6 percent old growth and 12 percent secondary forest; Heaney, pers. comm., 2000), and only 27 percent of keystone mangrove forest remains (Wood et al. 2000).

With so little forest land left, the Philippines forestry industry is greatly reduced. Agricultural expansion, accelerated by the logging roads, has taken over much of the formerly forested lands. With a shortage of mature forest and an absence of effective forest management, the Philippines is now heavily dependent on timber imports (Collins et al. 1991). The few remaining timber companies are doing their best to self-regulate and provide a sustainable base for their industry (Dickinson et al. 1991), but illegal commercial logging is very common (Heaney and Regalado 1998). Shifting agriculture is a traditional practice in the Philippines, but combined with high population growth it has become the main cause of deforestation. Shifting agriculture and deforestation are so intense that more than 90,000 km^2 of the country can no longer support farming because of erosion (Collins et al. 1991).

Most forest outside of protected areas is leased to private timber companies intent on logging it. The prescribed selective logging practices often are not followed, resulting in forest loss, erosion, and siltation (Collins et al. 1991). Fuel wood gathering and livestock grazing, significant threats in some parts of the world, are not considered important threats in the Philippines (Dickinson et al. 1991).

Entire habitat types are in danger of being completely eradicated. Little beach forest remains because of heavy settlement along the coasts. Dipterocarp forest has been significantly reduced from logging and agricultural expansion, and secondary forests that are more susceptible to cyclone damage are also prone to drying out more easily; this is part of a ratcheting climate effect that makes it more difficult for natural forest to regenerate. Mangroves have been heavily exploited for charcoal and for construction of fishponds in their place. Extensive areas of swamp and marshland on Luzon and Mindanao are being reclaimed for land and flood control (Dickinson et al. 1991).

As a result of this almost unprecedented habitat destruction, the Philippines has 194 species that are considered critically endangered, endangered, or vulnerable. The only other countries that contain comparable numbers of critically endangered species are much larger: Australia, Mexico, and the United States. Two species extinctions have been recorded—the Panay giant fruit bat (*Acerodon lucifer*) and the Philippine bare-backed fruit bat (*Dobsonia chapmani*) (Hilton-Taylor 2000)—although *A. lucifer* is no longer considered distinct from *A. jubatus* (Heaney, pers. comm., 2000). The Philippines fauna is poorly studied in general, and many more species are likely to be encountered (R. Crombie, pers. comm., 2000). There is a great need for additional biological survey work throughout the country and for more well-trained field biologists (L. Heaney, pers. comm., 2000)

Habitat destruction is the main threat to species in the Philippines, but hunting or capture for the pet trade is important for several species, especially the Philippine cockatoo (*Cacatua haematuropygia*) and the Palawan peacock-pheasant (*Polypectron emphanum*) (Dickinson et al. 1991).

Prepared by John Morrison

Ecoregion Number:	**97**
Ecoregion Name:	**Palawan Rain Forests**
Bioregion:	**Sunda Shelf and Philippines**
Biome:	**Tropical and Subtropical Moist Broadleaf Forests**
Political Units:	**Philippines**
Ecoregion Size:	**13,900 km^2**
Biological Distinctiveness:	**Globally Outstanding**
Conservation Status:	**Critical**
Conservation Assessment:	**I**

Palawan represents a bridge between the Sunda Shelf and Philippine bioregions and contains faunal elements from both, as well as it own unique elements. This ecoregion, though more intact than any other region in the Philippines, is under great pressure from logging interests.

Location and General Description

This ecoregion includes the island Palawan plus Balabac, Ursula Island, and the Calamian Group. Palawan itself is the sixth largest of the Philippine Islands. The climate of the ecoregion is tropical wet (National Geographic Society 1999). In northwest Palawan, a dry season lasts from November to May while the wet season lasts from June to October; the rest of the island experiences a short, one- to three-month dry season. The east coast becomes progressively drier than the west coast from north to south (Davis et al. 1995).

Palawan (along with the Calamianes and the island of Mindoro) was rifted (below water) from the Asian mainland approximately 32 million years ago, transported through seafloor spreading across the growing South China Sea, added to the growing Philippine Archipelago approximately 17 million years ago, and uplifted above water approximately 5–10 million years ago (Hall and Holloway 1998; Dickinson, Kennedy, and Parkes 1991). Metamorphic rocks are found in the northern portion of the island north of Mt. St. Paul. Volcanic rocks are found in the vicinity of Cleopatra's Needle, just south of Mt. St. Paul. Mt. St. Paul itself and the El Nido Cliffs are karst landscapes. The southern third of the island, south of the Quezon-Aboabo Gap, is dominated by ultramafics mixed with volcanic rocks and Tertiary limestone. Tertiary sandstones and shales occur along the southwest coast (Davis et al. 1995).

The channel between Palawan and Borneo is about 145 m deep. During the middle Pleistocene, sea levels were 160 m lower than today, and the islands were connected. During the last ice age (late Pleistocene), sea level was approximately 120 m below current levels, and Palawan was separated from ice age Borneo by a narrow channel. Palawan has always remained separated from the rest of the Philippines. Palawan is long and narrow, consisting of a steep mountain range whose highest point is 2,085 m (Mt. Mantalingajan). More than 45 percent of Palawan consists of mountains with slopes greater than 30 percent (Davis et al. 1995).

Vegetation types on Palawan are diverse and include beach forest, tropical lowland evergreen dipterocarp rain forest, lowland semi-deciduous forest, montane forest, and ultramafic and limestone forest. Beach forest merges with other forest types away from the coast and includes *Calophyllum inophyllum*, *Canarium asperum* var. *asperum*, *Pometia pinnata*, *Palaquium dubardii*, and *Ficus* spp. (Davis et al. 1995).

The lowland evergreen dipterocarp rain forest, which naturally occupies 31 percent of the island, is dominated by *Agalai* spp., *Dipterocarpus gracilis*, *D. grandiflorus*, *Ficus* spp., *Tristania* spp., *Exocarpus latifolius*, and *Swintonia foxworthyi*. *Sygium* spp., *Dracontomelon dao*, and *Pongamia pinnata* are emergent. Lianas and cycads are common. In southern Palawan, a *Casuarina* sp. dominates in the lowland forests (Davis et al. 1995).

The eastern half of the island is in a rain shadow and contains moist semi-deciduous forests. Soils are thin on the steeper slopes and support medium-sized trees (up to 15 m tall), which shed their leaves during the March–May dry season. The rainy season is June–July. Common tree species include *Pterocymbium tinctorium*, *Pterospermum diversifolium*, *Hymenodictyon* spp., and *Garuga floribunda* (Davis et al. 1995).

Montane forests, found between 800 and 1,500 m, are dominated by *Tristania* spp., *Casuarina* spp., *Swietenia foxworthyi*, and *Litsea* spp. in the lower elevations. Upper montane forest trees include *Agathis philippinensis*, *Dacrydium pectinatum*, *Podocarpus polystachyus*, *Gnetum latifolium*, *Cycas wadei*, *Cinnamomum rupestre*, *Nepenthes philippinensis*, and *Angiopteris* spp. (Davis et al. 1995).

Limestone forests are found on the islets surrounding Palawan and over large areas in the southern portions of the island. Represented are *Euphorbia trigona*, *Aglaia argentea*, and *Antidesma*, *Drypetes*, *Gomphandra*, *Sterculia*, *Pleomele*, and *Begonia* spp. (Davis et al. 1995).

Victoria Peak, in south-central Palawan, contains the largest region of ultramafic forest on the island. Although many of the ultramafic tree species are shared with semi-deciduous forest, several species, including *Scaevola micrantha, Brackenridgea palustris* var. *foxworthi, Exocarpus latifolius,* and *Phyllanthus lamprophyllus* are believed to be heavy metal indicators (Davis et al. 1995).

Biodiversity Features

Relative to the size of Palawan, the ecoregion contains a rich fauna, including several groups that are not found in the rest of the Philippines (carnivores, pangolins, porcupines, and some insectivores) (Heaney 1986).

There are many endemic mammals in Palawan, but nearly all the genera (96 percent) are also found in Borneo. Of twenty-five indigenous nonvolant mammal species, eleven (44 percent) are endemic to Palawan, and the remainder are shared with Borneo. Therefore, the greater Palawan region is rightly considered part of the Sunda Shelf bioregion rather than that of the Philippines. The large number of endemic species but few endemic genera of Palawan are consistent with a separation of Borneo and Palawan of approximately 160,000 (since the middle Pleistocene) (Heaney 1986). There are fifteen endemic or near-endemic mammals in greater Palawan (table J97.1).

The Calamian deer (*Axis calamianensis*) is found only in the Calamian Islands, where it survives in low densities on Busuanga, Calauit, and Culion Islands. The only protected area for this species was established to protect free-ranging African ungulates on Calauit Island (Wemmer 1998).

Balabac, Palawan, and the Calamian Islands also provide habitat for an endemic subspecies of the bearded pig (*Sus barbatus ahoenobarbus*), another subspecies of which is widely distributed in the Greater Sundas. The IUCN considers this species to be rare and declining. This species naturally inhabits tropical evergreen rain forest but is able to use a wide variety of habitats within forests. They are quite dependent on fruit supplies but consume a wide variety of foods. Directional large-scale population movements in scattered or condensed herds lasting days, weeks, or even months are reported for other subspecies in Borneo and Sumatra; this is generally associated with the mast fruiting of dipterocarps. Such movements have not been reported from the Philippines (Oliver 1993).

Several of Palawan's endemic mammals are considered threatened. Three endemic mammal species are considered endangered, including the Calamian deer, a Sunda tree squirrel (*Sundasciurus juvencus*) (recommended for delisting; Heaney et al. 1998), and the Palawan rat (*Palawanomys furvus*), which was collected only four times in 1962. A subspecies of mouse deer, the Balabac chevrotain (*Tragulus napu nigricans*), which is confined to Balabac Island, is also considered endangered. Five endemic mammal species are considered vulnerable, including *Acerodon leucotis*, the Palawan treeshrew (*Tupaia palawanensis*), the Palawan stink badger (*Mydaus marchei*), the Palawan binturong (*Arctictis binturong whitei*), and a Sunda tree squirrel (*Sundasciurus rabori*) (Hilton-Taylor 2000).

As with mammals, Philippine birds in general show a strong Bornean affinity, and it is clear that the main pathway of Asian immigration to the Philippines was through Palawan; of 395 Philippine breeding species, 137 (35 percent) also breed in Borneo. Palawan birds exhibit strong differentiation at the subspecific level when compared with its nearest Philippine neighbor, Mindoro. This is in contrast to the other partial land bridge between Borneo and the Philippines, the

TABLE J97.1. Endemic and Near-Endemic Mammal Species.

Family	Species
Pteropodidae	*Acerodon leucotis**
Cervidae	*Axis calamianensis**
Sciuridae	*Sundasciurus steerii**
Sciuridae	*Sundasciurus moellendorfi**
Sciuridae	*Sundasciurus rabori**
Sciuridae	*Hylopetes nigripes**
Muridae	*Chiropodomys calamianensis**
Muridae	*Maxomys panglima**
Muridae	*Palawanomys furvus**
Hystricidae	*Hystrix pumila**
Sorcidae	*Crocidura palawanensis**
Muridae	*Haeromys sp. A**
Sciuridae	*Sundasciurus hoogstraali**
Sciuridae	*Sundasciurus juvencus**
Tupaiidae	*Tupaia palawanensis**

An asterisk signifies that the species' range is limited to this ecoregion.

TABLE J97.2. Endemic and Near-Endemic Bird Species.

Family	Common Name	Species
Phasianidae	Palawan peacock-pheasant	*Polyplectron emphanum**
Columbidae	Grey imperial-pigeon	*Ducula pickeringii*
Psittacidae	Blue-headed racquet-tail	*Prioniturus platenae**
Strigidae	Mantanani scops-owl	*Otus mantananensis*
Strigidae	Palawan scops-owl	*Otus fuliginosus**
Apodidae	Palawan swiftlet	*Aerodramus palawanensis**
Bucconidae	Palawan hornbill	*Anthracoceros marchei**
Monarchidae	Blue paradise-flycatcher	*Terpsiphone cyanescens**
Irenidae	Yellow-throated leafbird	*Chloropsis palawanensis**
Muscicapidae	Palawan flycatcher	*Ficedula platenae**
Muscicapidae	Palawan blue-flycatcher	*Cyornis lemprieri**
Muscicapidae	White-vented shama	*Copsychus niger**
Pycnonotidae	Sulphur-bellied bulbul	*Ixos palawanensis**
Timaliidae	Ashy-headed babbler	*Malacocincla cinereiceps**
Timaliidae	Palawan babbler	*Malacopteron palawanense**
Timaliidae	Falcated wren-babbler	*Ptilocichla falcata**
Timaliidae	Palawan striped-babbler	*Stachyris hypogrammica**
Paridae	Palawan tit	*Parus amabilis**
Paridae	White-fronted tit	*Parus semilarvatus*
Dicaeidae	Palawan flowerpecker	*Prionochilus plateni**

An asterisk signifies that the species' range is limited to this ecoregion.

Sulu Islands, which have not differentiated significantly from Mindanao. Borneo and Palawan share twenty-three bird species that are not found in the rest of the Philippines. The Asian genera *Polyplectron, Malacocincla, Malacopteron, Dinopium, Aegithina, Criniger, Seicercus,* and *Gracula* are found only in Palawan within the Philippines (Dickinson et al. 1991). The island forms an important bird migration route between Borneo and the rest of the Philippines for southern migrants (Davis et al. 1995).

This ecoregion corresponds exactly with the Palawan EBA (Stattersfield et al. 1998). The EBA contains twenty restricted-range birds, seventeen of which are found nowhere else on Earth and six of which (Palawan peacock-pheasant [*Polyplectron emphanum*], grey imperial-pigeon [*Ducula pickeringii*], blue-headed racquet-tail [*Prioniturus platenae*], Palawan hornbill [*Anthracoceros marchei*], falcated wren-babbler [*Ptilocichla falcata*], and Palawan flycatcher [*Ficedula platenae*]) are considered vulnerable (Collar 1999). All these vulnerable birds are dependent on lowland and hill forest (Collar et al. 1999; Stattersfield et al. 1998). There are twenty endemic or near-endemic bird species in the Palawan ecoregion (Kennedy et al. 2000; table J97.2).

The critically endangered Philippine crocodile (*Crocodylus mindorensis*) was historically found on the islands of Luzon, Mindoro, Masbate, Samar, Jolo, Negros, Busuanga, and Mindanao. Busuanga contains one of the only remaining populations (others are found on Mindoro, Negros, and Mindanao). Whereas the decline of the species was initially driven by overexploitation, habitat loss and human persecution are now the principal threats to the Philippine crocodile. Surveys in 1980–1982 revealed a total wild population of approximately 500–1,000 individuals, but current wild populations may be approximately 100 non-hatchlings. Captive breeding efforts are being led by the Crocodile Farming Institute, an entity of the Philippine government (Ross 1998).

A total of 1,522 (Davis et al. 1995) to 1,672 (Quinnell and Balmford 1988) vascular plants have been identified on Palawan, and it is estimated that more than 2,000 species are present on the island. As detailed earlier, Palawan has an extremely diverse range of vegetation types for the Philippines. A small number of dipterocarps, an important timber tree group, are present on the island, as well as a variety of medicinal plants used by ethnic tribes and plants used in ceremony and as ornamentals (Davis et al. 1995).

Status and Threats

Current Status

Almost all of the Philippines was once completely forested (Dickinson et al. 1991). As of 1988, Palawan contained 7,410 km^2 (54 percent) of total forest remaining (SSC 1988). At the time this was the highest percentage of any of the Philippines' large islands.

Later aerial surveys (Development Alternatives 1992) indicated that significant reductions in closed-canopy forest cover had occurred since 1988 as a result of recent logging. As seen from the air, the lowlands and hillsides consist of slash-and-burn agriculture up to the edges of natural forest in the highlands. Closed-canopy forest caps only the highest areas on the island.

Palawan's forests are of low commercial value because of the small number of dipterocarps, and until the last twenty years Palawan's forests were ignored in favor of the more valuable forests of Luzon and Mindanao. Government logging regulations setting guidelines for minimum diameter, minimum rotation length, and replanting have been largely ignored (Quinnell and Balmford 1988).

Because of a generally high population density in other parts of the Philippines, large numbers of shifting cultivators (*kaingineros*) are attracted to Palawan to eke out a living on the hillsides of the island, and their cumulative impact is enormous (Quinnell and Balmford 1988).

All of Palawan was declared a Fauna and Flora Watershed Reserve, and this includes a variety of protected areas, including national parks, wilderness areas, experimental forests, forest research reserves, game refuges, wildlife sanctuaries, museum reservations and research sites, tourist zones, and marine reserves.

Recent reports in the international press indicate (and have been confirmed, L. Heaney, pers. comm., 2000) that the situation in Palawan has stabilized, that large-scale logging has been halted, and that a balance is being achieved between economic development and conservation; future monitoring will determine whether this is remains true.

Types and Severity of Threats

Habitat destruction is the main threat to biodiversity in the Philippines, and Palawan, though currently in better condition, is no different. Logging and shifting cultivation (*kaingin*) are cited as the primary forces of habitat conversion. Logging takes many forms, from industrial scale to smaller-scale operations that use water buffalo to haul logs out of the forest. Mangroves are used locally for firewood, dyes, and tannins (Davis et al. 1995), and they are sometimes removed to make way for fishponds (Quinnell and Balmford 1988).

Hunting and the wild pet trade are also significant threats in Palawan. Leopard cats have been hunted for their pelts and are sold when kittens as pets (Heaney and Regalado 1998). The Palawan binturong is hunted for meat and as pets, and the pangolin is hunted for its hide (Quinnell and Balmford 1988). The Palawan peacock-pheasant (Dickinson et al. 1991; Collar et al. 1999), blue-headed racquet-tail (Collar et al. 1999), Philippine cockatoos (*Cacatua haematuropygia*), and blue-naped parrots (*Tanygnathus lucionensis*) (Quinnell and Balmford 1988) apparently are suffering greatly from the pet trade. The final destination for these birds often is the United States (Quinnell and Balmford 1988).

Ornamental plant collecting, especially for the orchids (*Phalaenopsis amabilis* and *Paphiopedilum argus*), pitcher plants (*Nepenthes* spp.), palms (*Veitchia merrillii*), and aroids (*Amorphophallus* spp. and *Alocasia* spp.) threatens some plant populations (Davis et al. 1995).

A valuable resin, known as Manila copal, is collected from *Agathis dammara* trees. This collection weakens the trees, and slackening production and disease combined with overexploitation are threatening the species (Davis et al. 1995; Quinnell and Balmford 1988).

Currently, Palawan's mineral wealth (chromite, copper, iron, manganese, mercury, and nickel) has not been extensively exploited, but the possible future extraction of these minerals represents a potential threat (Quinnell and Balmford 1988).

Priority Conservation Actions

Short-Term Conservation Actions (1–5 Years)

- Protect all remaining old-growth forests.
- Expand existing protected areas to include more lowland rain forest and entire watersheds.
- A number of conservation actions are recommended for the Calamian deer (*Axis calamianensis*) in the IUCN Deer Status Survey and Conservation Action Plan. These actions include monitoring the species' current status, strengthening existing protected area management, incorporating buffer zones in the Calauit Game Preserve, establishing new protected areas on Culion and Busuanga, undertaking behavioral and ecological research on Calauit deer, and initiating conservation education using the Calamian deer as a flagship species (Wemmer 1998).
- Field status surveys are considered a priority to assess the condition of populations of the Palawan bearded pig (*Sus barbatus ahoenobarbus*) on Balabac and Palawan to match a similar survey begun in the Calamians. A properly structured captive breeding program is also recommended in the Pigs, Peccaries, and Hippos Status Survey and Conservation Action Plan (Oliver 1993).
- The Parrot Action Plan (Snyder et al. 2000) considers the proposed extension of St. Paul's Subterranean River National Park on Palawan to be a priority project for protecting the Philippine cockatoo (*Cacatua haematuropygia*). A program to raise awareness of the plight of the cockatoo and an export ban of wild-caught birds is also suggested. Recommended action for the endemic blue-headed racquet-tail includes protection of roost sites from trappers, a public awareness campaign, and a survey of all remaining forest areas in the ecoregion.

Longer-Term Conservation Actions (5–20 Years)

- Educate the public about watershed protection (L. Heaney, pers. comm., 2000).
- Collar et al. (1999) identified (incomplete and preliminary) four key sites that are known to support several threatened bird species and retain sufficient areas of natural habitat for their survival. Some of these key sites are protected to some degree; others are not. A future effort will develop a comprehensive set of Important Bird Areas that would be sufficient to conserve the full range of bird species. For Palawan, the key sites identified thus far are San Vicente/Taytay/Roxas, St. Paul's Subterranean Reserve National Park, Victoria/Anapalan ranges, and Mt. Mantalingahan.

Focal Species for Conservation Action

Threatened endemic species: Calamian deer (*Axis calamianensis*) (Wemmer 1998), Palawan rat (*Palawanomys furvus*) (Heaney et al. 1998), Palawan peacock pheasant (*Polypectron emphanum*) (Collar et al. 1999), Balabac chevrotain (*Tragulus napu nigricans*) (Wemmer 1998), grey imperial-pigeon (*Ducula pickeringii*) (Collar et al. 1999), blue-headed racquet-tail (*Prioniturus platenae*) (Collar et al. 1999), falcated wren-babbler (*Ptilocichla falcata*) (Collar et al. 1999), Palawan flycatcher (*Ficedula platenae*) (Collar et al. 1999), Northern Palawan tree squirrel (*Sundsciurus juvencus*) (Heaney et al. 1998), Palawan hornbill (*Anthracoceros marchei*) (Collar et al. 1999), Palawan fruit bat (*Acerodon leucotis*) (Heaney et al. 1998), Palawan tree shrew (*Tupaia palawanensis*) (Heaney et al. 1998), Palawan shrew (*Crocidura palawanensis*) (Heaney et al. 1998), Palawan montane squirrel (*Sundasciurus rabori*) (Heaney et al. 1998)

Area-sensitive species: bearded pig (*Sus barbatus ahoenobarbus*) (Oliver 1993), Calamian deer (*Axis calamianensis*) (Wemmer 1998), Balabac chevrotain (*Tragulus napu nigricans*) (Wemmer 1998), fruit bats (*Acerodon leucotis, Pteropus vampyrus, Eonycteris spelaea*) (Mickleburgh et al. 1992), crested serpent-eagle (*Spilornis cheela*), rufous-bellied eagle (*Hieraaetus kienerii*), changeable hawk-eagle (*Spizaetus cirrhatus*), Philippine hawk-eagle (*Spizaetus philippensis*), peregrine falcon (*Falco peregrinus*)

Top predator species: Asian leopard cat (*Prionailurus bengalensis*) (Nowell and Jackson 1996), binturong (*Arctitis binturong*) (Heaney et al. 1998), palm civet (*Paradoxurus hermaphroditus*) (Heaney et al. 1998), Malay civet (*Viverra tangalunga*) (Heaney et al. 1998), Oriental small-clawed otter (*Amblonyx cinereus*) (Heaney et al. 1998), crested serpent-eagle (*Spilornis cheela*), rufous-bellied eagle (*Hieraaetus kienerii*), changeable hawk-eagle (*Spizaetus cirrhatus*), Philippine hawk-eagle (*Spizaetus philippensis*), peregrine falcon (*Falco peregrinus*), Philippine crocodile (*Crocodylus mindorensis*) (Ross 1998)

Selected Conservation Contacts

Conservation International
Haribon Foundation
Palawan NGO Network Incorporated
Palawan Wildlife Rescue and Conservation Center
Philippine Council for Sustainable Development
Protected Areas & Wildlife Bureau, Philippine Department of Environment and Natural Resources
WWF-Philippines

Justification for Ecoregion Delineation

MacKinnon (1997) identified seven subunits in the Philippines, and the Philippine Biodiversity Action Plan (Philippine BAP 1997) demarcated fifteen biogeographic units. Udvardy (1975) identified the Philippines as a single biogeographic province. We delineated nine ecoregions in the Philippine islands, including Palawan. We deviated from Udvardy (1975), MacKinnon (1997), Stattersfield et al. (1998), and the Philippine BAP (1997) to varying degrees and based our delineation of the Philippine ecoregions on Heaney (1993).

We placed Palawan, Calamian Islands, and Cuyo Islands into a single ecoregion, the Palawan Rain Forests [97]. Palawan has closer zoogeographic affinities to Borneo.

Prepared by John Morrison

Ecoregion Number: **98**
Ecoregion Name: **Luzon Rain Forests**
Bioregion: **Sunda Shelf and Philippines**
Biome: **Tropical and Subtropical Moist Broadleaf Forests**
Political Units: **Philippines**
Ecoregion Size: **94,800 km²**
Biological Distinctiveness: **Globally Outstanding**
Conservation Status: **Critical**
Conservation Assessment: **I**

The Luzon Rain Forests [98] ecoregion is rich in endemic species and also contains one of the largest populations of the Philippine eagle *Pithecophaga jefferyi*. A high proportion of the ecoregion was originally forested, but now very little of this forest remains. However, the ecoregion has managed to retain one of the largest remaining tracts of primary forest in the Philippines.

Location and General Description

Luzon is located in the western Pacific Ocean. It is the largest island in the Philippines and lies at the northern end of the island group. The lowland rain forests ecoregion comprises all the areas below 1,000 m on Luzon and a few isolated volcanic mountains in the south of the island that exceed 1,000 m: Mt. Maquiling, Mt. Banashaw, Mt. Isarog, Mayon Volcano, and Bulusan Volcano. The broad Cagayan River valley to north is sheltered from typhoons lying between the two north-south mountain ranges: the Cordillera Central in the west and the Sierra Madre to the east. The fertile soil of the Cagayan Valley is the biggest rice-growing region in the country. Southern Luzon is also agricultural but is subject to typhoons and comprises less area as Luzon narrows southward. Several neighboring island groups are also part of the ecoregion, including the Batanes and Babuyan Islands to the north (rather isolated but placed here for convenience), Polillo and Catanduanes to the east, and Marinduque to southwest.

The geologic history of the Philippines is very complex and has had tremendous influence on the biota currently found there. Luzon has developed many unique plant and animal species as a result of its long-standing isolation from other landmasses. Parts of the Luzon highlands were established as a result of volcanic activity and the friction of the Australian and Asian plates at least 15 million years ago. The highlands began to take their current form over the next 10 million years. Luzon therefore is oceanic in character, having never been connected to mainland Asia. Even during the Pleistocene, as world sea levels fell 120 m, Luzon expanded to become a larger island including the modern islands of Polillo, Marinduque, and Catanduanes but never connecting to other regions of the Philippines or to mainland Asia (Heaney and Regalado 1998).

Temperatures in Luzon vary greatly with elevation, but within the lowlands temperatures are fairly uniform at about 25–28°C. Rainfall in the lowlands is seasonally variable, with four distinct types. Southwestern Luzon and Marinduque Island receive rain uniformly throughout the year. The Cagayan valley and the eastern portion of the Bataan Peninsula do not have pronounced seasons but are dry from November to April. The southeastern portions of Luzon and Polillo and Catanduanes Islands have no dry season but do have a period of increased rainfall from May to January. Northwestern Luzon has two distinct seasons, being wet from May to October and dry November to April.

Lowland vegetation of Luzon is dominated by dipterocarp trees with wide buttresses at the base. These massive trees are 1–2 m in diameter and up to 60 m high. The canopy height of mature lowland forests tends to be uneven. In areas of disturbance, rattans and lianas receive the light they need to flourish in the understory. There tends to be an abundant herbaceous undergrowth, and ferns and orchids are prevalent on large branches of tall trees. Other natural habitats in the ecoregion include mangrove forests and beach forests (consisting of *Casuarinas* and *Barrintonia*) near the coasts. There also were natural grasslands in valley bottoms and on plateaus, as evidenced by the presence of several endemic buttonquail taxa needing grasslands (Collar et al. 1999).

Biodiversity Features

In terms of mammalian endemism, perhaps the most significant area of endemism in the ecoregion is Mt. Isarog, but this has only recently become apparent. Mt. Isarog is an extinct volcano and the second highest peak in southern Luzon at 1,966 m (Mt. Mayon is higher at 2,462 m). The unique character and geographic isolation of Isarog make it difficult to lump with the other two montane ecoregions of Luzon [102, 106], so it is considered here as part of the Luzon rain forests. The Luzon rain forest ecoregion as a whole has ten species of near-endemic mammals and five strictly endemic species (table J98.1). Three of the five strict endemics are found only on Mt. Isarog, and none was been described before 1981: Isarog shrew-mouse (*Archboldomys luzonensis*), Isarog striped shrew-rat (*Chrotomys gonzalesi*), and Isarog shrew-rat (*Rhynchomys isarogensis*). All three consume earthworms, and the latter two are strongly vermivorous. The southern Luzon giant cloud rat (*Phloemys cumingi*) is another of the ecoregion's strict endemics found on Isarog, but it is also

TABLE J98.1. Endemic and Near-Endemic Mammal Species.

Family	Species
Pteropodidae	*Otopteropus cartilagonodus*
Muridae	*Abditomys latidens*
Muridae	*Apomys abrae*
Muridae	*Apomys datae*
Muridae	*Apomys microdon*
Muridae	*Apomys sacobianus*
Muridae	*Archboldomys luzonensis**
Muridae	*Batomys granti*
Muridae	*Bullimus luzonicus*
Muridae	*Chrotomys gonzalesi**
Muridae	*Crunomys fallax**
Muridae	*Phloemys cumingi**
Muridae	*Phloemys pallidus*
Muridae	*Rhynchomys isarogensis**
Muridae	*Tryphomys adustus*

An asterisk signifies that the species' range is limited to this ecoregion.

found at other locations. The fifth strict endemic is the northern Luzon shrew-mouse (*Crunomys fallax*), known from a single specimen collected at about 300 m in the northern Sierra Madres.

The ecoregion has thirteen mammal species that are listed by IUCN as threatened (categories VU and above) (Hilton-Taylor 2000). One of these species is the golden-crowned flying-fox (*Acerodon jubatus*). It is probably the largest bat in the world (at more than 1.2 kg, perhaps reaching 1.5 kg) and is widespread in the Philippines but has undergone a precipitous decline because of heavy hunting and habitat destruction (Heaney and Regalado 1998).

Five large mammals inhabit the ecoregion: long-tailed macaque (*Macaca fascicularis*), Philippine warty pig (*Sus philippensis*), Philippine brown deer (*Cervus mariannus*), Malay civet (*Viverra tangalunga*), and common palm civet (*Paradoxurus hermaphroditus*). All are fairly widespread, and none are listed as threatened by IUCN (Hilton-Taylor 2000). However, habitat destruction affects all these species, and hunting affects all but the civets. Additionally, the Philippine brown deer is said to be declining and is listed as data deficient by IUCN, although the species is not uncommon in appropriate habitat (Heaney et al. 1998; Wemmer 1998).

The lowland forests of Luzon contain thirty-four near-endemic bird species and six strict endemics (table J98.2). Only two of the strict endemics are threatened (IUCN categories VU and above): green racquet-tail *Prioniturus luconensis* and Isabela oriole *Oriolus isabellae* (Collar et al. 1999). Researchers in Luzon feared that Isabela oriole was extinct (Mallari and Jensen 1993; Poulsen 1995) until two recent reports of its existence in northern Luzon (Gamauf and Tebbich 1995; van der Linde 1995). Although these observations are encouraging, some doubt has been raised about the level of scrutiny the records were subject to (Collar et al. 1999). The green racquet-tail's decline is thought to be similar to that of many other parrots of the Philippines: it was common several decades ago but has become very rare recently as a result of deforestation and collection for the pet trade (Poulsen 1995; Snyder et al. 2000). However, Kennedy et al. (2000) stated that the decline of green racquet-tail may be less straightforward because it appears not to be heavily subject to the pet trade or to deforestation. The parrot pet trade in the Philippines has had a tremendous negative impact on certain species. The most notable example is the Philippine cockatoo (*Cacatua haematuropygia*), which at one time was widespread throughout the Philippines but is now scarce (if not extinct) in Luzon, an enormous decline caused almost entirely by the pet trade (Snyder et al. 2000).

The Philippine eagle is the national bird of the Philippines and the most famous animal in the country. Unfortunately, the eagle is critically endangered, existing in primary lowland forests of Samar, Leyte, Mindanao, and Luzon. The large area needs of the eagle coupled with the bird's low reproductive rate have made it highly susceptible to deforestation. The two largest remaining populations are in Luzon and Mindanao, although precise numbers of individuals are still speculative. Population numbers are estimated mainly by assumptions of remaining forest cover, range size, percentage occupancy, and the number of immature birds that territories include. Luzon is thought to have between 52 and 104 eagles, but probably

TABLE J98.2. Endemic and Near-Endemic Bird Species.

Family	Common Name	Species
Turnicidae	Spotted buttonquail	*Turnix ocellata*
Turnicidae	Luzon buttonquail	*Turnix worcesteri*
Rallidae	Brown-banded rail	*Lewina mirificus*
Columbidae	Luzon bleeding-heart	*Gallicolumba luzonica*
Columbidae	Whistling green-pigeon	*Treron formosae*
Columbidae	Flame-breasted fruit-dove	*Ptilinopus marchei*
Columbidae	Cream-breasted fruit-dove	*Ptilinopus merrilli*
Psittacidae	Luzon racquet-tail	*Prioniturus montanus*
Psittacidae	Green racquet-tail	*Prioniturus luconensis**
Cuculidae	Red-crested malkoha	*Phaenicophaeus superciliosus**
Cuculidae	Scale-feathered malkoha	*Phaenicophaeus cumingi*
Cuculidae	Rufous coucal	*Centropus unirufus*
Strigidae	Luzon scops-owl	*Otus longicornis*
Strigidae	Ryukyu scops-owl	*Otus elegans*
Bucconidae	Luzon hornbill	*Penelopides manilloe*
Pittidae	Whiskered pitta	*Pitta kochi*
Campephagidae	Blackish cuckoo-shrike	*Coracina coerulescens*
Turdidae	Ashy thrush	*Zoothera cinerea*
Timaliidae	Luzon wren-babbler	*Napothera rabori**
Timaliidae	Golden-crowned babbler	*Stachyris dennistouni*
Timaliidae	Chestnut-faced babbler	*Stachyris whiteheadi*
Timaliidae	Luzon striped-babbler	*Stachyris striata**
Sylviidae	Philippine bush-warbler	*Cettia seebohmi*
Sylviidae	Long-tailed bush-warbler	*Bradypterus caudatus*
Sylviidae	Grey-backed tailorbird	*Orthotomus derbianus*
Muscicapidae	Rusty-flanked jungle-flycatcher	*Rhinomyias insignis*
Muscicapidae	Ash-breasted flycatcher	*Muscicapa randi*
Muscicapidae	Furtive flycatcher	*Ficedula disposita**
Muscicapidae	Blue-breasted flycatcher	*Cyornis herioti*
Muscicapidae	Luzon redstart	*Rhyacornis bicolor*
Monarchidae	Short-crested monarch	*Hypothymis helenae*
Monarchidae	Celestial monarch	*Hypothymis coelestis*
Pachycephalidae	Green-backed whistler	*Pachycephala albiventris*
Paridae	White-fronted tit	*Parus semilarvatus*
Rhabdornithidae	Long-billed rhabdornis	*Rhabdornis grandis*
Dicaeidae	Flame-crowned flowerpecker	*Dicaeum anthonyi*
Zosteropidae	Lowland white-eye	*Zosterops meyeni*
Estrildidae	Green-faced parrotfinch	*Erythrura viridifacies*
Oriolidae	White-lored oriole	*Oriolus albiloris*
Oriolidae	Isabela oriole	*Oriolus isabellae**

An asterisk signifies that the species' range is limited to this ecoregion.

around 78 (Collar et al. 1999). The survival of the Philippine eagle is being watched as a benchmark of the health of the Philippines environment as a whole. The survival of the species is largely tied to the protection of the few remaining large tracts of forest. Such protection would benefit many other forest-dwelling species.

Conservation Status

Current Status

The largest remaining forested area in the ecoregion is in the lowlands of the northern Sierra Madres, which have remained inaccessible. The Northern Sierra Madre Natural Park (also know as the Palanan complex or wilderness) has recorded most of the ecoregion's endemic bird species and is a stronghold for the long-tailed macaque, Philippine warty pig, and Philippine brown deer. Luzon's population of Philippine eagles is joined by thirteen other threatened bird species. The park receives funding and attention from the Department of Environment and Natural Resources (DENR), several conservation organizations, the Global Environment Facility, and the European Commission. However, the park is threatened by plans to construct roads that would transect the park and is subject to high levels of encroachment (several towns are within the park boundary). Other sites, which include lowlands in the northern Sierra Madres and have been identified as important areas for biodiversity, include Mt. Cetaceo and Mt. Los Dos Cuernos. Neither of these sites receives any formal protection, and both are being cleared (Collar et al. 1999).

As noted earlier, Mt. Isarog National Park (table J98.3) is of major importance for endemic mammals. Mt. Isarog also contains four threatened bird species. The national park status of the Mt. Isarog has not effectively protected it thus far. Encroachment on the park has led to a population of several hundred people who live in the park. Also, deforestation continues at the hands of "well-financed commercial ventures" (Collar et al. 1999, p. 57; see also Heaney et al. 1999 and Heaney and Regalado 1998). However, the Haribon Foundation does have an active conservation program at Mt. Isarog that is again gaining strength.

The smaller islands in the ecoregion have not fared well. Marinduque contained only 317 ha of primary forest in 1992, compared with 813 ha in 1984, for a loss of 61 percent (Development Alternatives 1992). The current amount of primary forest on the island is unknown. Early this century, Polillo was so forested that McGregor wrote that "he had never seen an island with 'so large a proportion of the area covered with trees,'" but today no forest remains (McGregor 1910 quoted in Collar et al. 1999, p. 55). Forest cover maps from 1992 showed that Catanduanes contained some forested areas. Central Catanduanes is listed as an important area for biodiversity, but it currently receives no protection (although it is a proposed watershed reserve) (Collar et al. 1999).

Types and Severity of Threats

There are no recent estimates of primary forest cover for the Philippines. Originally 95 percent of the Philippines was covered by rain forest (Heaney and Regalado 1998). Intact lowland forest made up less than 6 percent of the total land area of the country about a decade ago, with another 12 percent being classified as degraded (Collins et al. 1991). From these figures, the state of the forest in the Philippines could be the poorest in all of tropical Asia (Poulsen 1995). In the intervening decade, more forest has been cleared as a result of unsustainable shifting agricultural practices, legal logging, and illegal logging. These threats are still present and are the main future threats to biodiversity, followed by unsustainable hunting and collection for trade.

The Luzon Rain Forests [98] ecoregion has been greatly modified by human activities. This is probably unavoidable given the high rate of population increase and the size of the population as a whole. Three of the country's largest cities are in this ecoregion: Manila, Quezon City, and Caloocan City. Many of the lowland areas were converted into agriculture long ago, but recent forest clearing is a tremendous problem. Still, Luzon's estimated 24 percent total forest cover was and probably is better than that of the Philippines as a whole. This is probably because Luzon has smaller trees than the rest of the country because of increasingly frequent typhoons in

TABLE J98.3. WCMC (1997) Protected Areas That Overlap with the Ecoregion.

Protected Area	Area (km²)	IUCN Category
Paoay Lake	7	III
Cassamata Hill	40	V
Northern Luzon Heroes Hill	110	III
Lake Malimanga	6	IV
Subic-Bataan Extension	90	PRO
Bataan	310	II
Roosevelt	20	V
Biak-na-Bato	60	V
Hinulugang Taktak	2	III
Quezon National Park	30	III
Mts. Palay–Palay–Mataas Na Gulod	40	II
Fugo Island	140	V
Palaui	90	V
Magapit	90	IV
Penablanca	120	V
Callao Cave	8	V
Northern Sierra Madre	260	II
Fuyot Spring	30	III
PNOC 1636 [102]	1,060	?
Mts. Banahaw San Cristobal	30	II
Taal Volcano	20	III
Bicol	30	II
Quezon National Park	10	III
Libmanan Cave	20	III
Catanduanes	270	PRO
Caramoan	30	III
Bulusan Volcano	50	II
Manlelung Spring	5	III
Minalungao	1	V
Capas Death March Monument	1	III
Mt. Arayat	80	V
Aurora Memorial	120	V
Mt. Isarog	150	II
Mayon Volcano	80	III
Total	3,410	

Ecoregion numbers of protected areas that overlap with additional ecoregions are listed in brackets.

the northern Philippines (Collar et al. 1999), rugged mountains, and independent ethnic communities.

Priority Conservation Actions

The conservation assessment and plan for the northern Sierra Madres (Danielsen et al. 1994) are excellent. Much of the work straddles two ecoregions (Luzon Rain Forests and Luzon Montane Rain Forests [98, 102]) but is equally important to both. Sponsored by DENR, BirdLife International, the Zoological Museum of Copenhagen University, and the Danish Ornithological Society, the project documents the importance of the biodiversity of the region through careful field surveys. It also puts forward useful recommendations. Conservation International–Philippines also has an active program in the area.

Short-Term Conservation Actions (1–5 Years)

- Establish biological surveys of the region.
- Protect remaining old-growth forests.
- Expand the existing protected area system to include additional lowland rain forest and to include entire watersheds.
- Increase public education about watershed protection (Heaney, pers. comm., 2000).

Longer-Term Conservation Actions (5–20 Years)

- Expand the existing protected area system to include additional lowland rain forest and to include entire watersheds.
- Increase public education about watershed protection (Heaney, pers. comm., 2000).
- Develop sound management plans for protected areas.

Focal Species for Conservation Action

Endangered endemic species: Luzon pygmy fruit bat (Mickleburgh et al. 1992), Luzon short-nosed rat, long-nosed Luzon forest mouse, southern Luzon giant cloud rat, northern Luzon shrew-mouse, Isarog shrew-mouse, Isarog shrew-rat, Isarog striped shrew-rat, flame-breasted fruit-dove, green racquet-tail, whiskered pitta (Lambert and Woodcock 1996), ashy thrush, Luzon redstart, rusty-flanked jungle-flycatcher, ash-breasted flycatcher, celestial monarch, green-faced parrotfinch, Isabela oriole

Area-sensitive species: Fourteen Pteropidae species (Mickleburgh et al. 1992), long-tailed macaque, Philippine warty pig (Oliver 1993), Philippine brown deer (Wemmer 1998), Jerdon's baza, oriental honey-buzzard, barred honey-buzzard, white-bellied sea-eagle, grey-headed fish-eagle, crested serpent-eagle, crested goshawk, grey-faced buzzard, Philippine eagle (Collar et al. 1999), rufous-bellied eagle, Philippine hawk-eagle, Philippine eagle-owl, rufous hornbill (Kemp 1995), Luzon hornbill (Kemp 1995)

Top predator species: palm civet (*Paradoxurus hermaphroditus*) (Heaney et al. 1998), Malay civet (*Viverra tangalunga*) (Heaney et al. 1998), Jerdon's baza (*Aviceda jerdoni*), oriental honey-buzzard (*Pernis ptilorhynchus*), barred honey-buzzard (*Pernis celebensis*), white-bellied sea-eagle (*Haliaeetus leucogaster*), grey-headed fish-eagle (*Ichthyophaga ichthyaetus*), crested serpent-eagle (*Spilornis cheela*), crested goshawk (*Accipiter trivirgatus*), Philippine eagle (*Pithecophaga jefferyi*) (Collar et al. 1999), rufous-bellied eagle (*Hieraaetus kienerii*), Philippine hawk-eagle (*Spizaetus philippensis*), Philippine eagle-owl (*Bubo philippensis*)

Selected Conservation Contacts

Conservation International
Flora and Fauna International
Haribon Foundation
Plan International
Protected Areas & Wildlife Bureau, Philippine DENR
WWF-Philippines

Justification for Ecoregion Delineation

MacKinnon (1997) identified seven subunits in the Philippines, and the Philippine Biodiversity Action Plan (Philippine BAP 1997) demarcated fifteen biogeographic units. Udvardy (1975) identified the Philippines as a single biogeographic province. We delineated nine ecoregions in the Philippine islands, including Palawan. We deviated from Udvardy (1975), MacKinnon (1997), Stattersfield et al. (1998), and the Philippine BAP (1997) to varying degrees and based our delineation of the Philippine ecoregions primarily on Heaney (1993).

In Luzon we delineated three ecoregions, which correspond to MacKinnon's subunit 26a. First, we used the 1,000-m contour from the DEM (USGS 1996) to delineate the montane forests from the lowland forests. The Luzon Montane Rain Forests [102] are made up primarily of the montane moist evergreen forests along the Sierra Madre, northern Central Cordillera, and Zambales mountain ranges. MacKinnon (1997) shows an area of freshwater swamp forests as part of the original vegetation of Luzon Island, which we combined with the remaining lowland forest of Luzon to form the Luzon Rain Forests [98]. These freshwater swamps, in the valley to the east of the Zambales Mountain Range and in the Cagayan river plains, have been converted to rice fields (D. Madulid, pers. comm., 1999). Following Stattersfield et al. (1998) and Dickinson et al. (1991), we placed the Lubang Islands and Batanes and Babuyan groups with the Luzon Rain Forests [98]. The Banguet pine (*Pinus insularis*, also known as *P. kesiya*)–dominated conifer forests in the Central Cordillera were designated as the Luzon Tropical Pine Forests [106].

Prepared by John Lamoreux

Ecoregion Number: **99**
Ecoregion Name: **Mindoro Rain Forests**
Bioregion: **Sunda Shelf and Philippines**
Biome: **Tropical and Subtropical Moist Broadleaf Forests**
Political Units: **Philippines**
Ecoregion Size: **10,300 km²**
Biological Distinctiveness: **Globally Outstanding**
Conservation Status: **Critical**
Conservation Assessment: **I**

Called the dark island by outsiders because of a virulent strain of malaria, Mindoro is located between the large islands of Luzon and the Sunda-affiliated Palawan, and it shares faunal attributes of both islands. However, Mindoro was isolated from Luzon and Palawan throughout the Pleistocene and retains its own unique character, including an endemic water buffalo species (Heaney 1986). Unfortunately, Mindoro is one of the most severely deforested islands in the country (Heaney and Mittermeir 1997). Only the most rugged portions of the island's central spine has been spared from commercial logging, and the forest is still under pressure.

Location and General Description

This ecoregion includes the island of Mindoro and the Semirara Islands. The climate of the ecoregion is tropical wet (National Geographic Society 1999). The western coast of Mindoro experiences a wet season during the southwest monsoon of June to October and a dry season during the November to February northeast monsoon because of the central mountains (High Rolling Mountains) (Collins et al. 1991). The High Rolling Mountains dominate the central portions of the island and rise to a maximum elevation of approximately 2,500 m at Mt. Halcon and Mt. Baco.

Mindoro (along with Palawan and the Calamianes) was rifted (below water) from the Asian mainland approximately 32 million years ago, transported through seafloor spreading across the growing South China Sea, added to the growing Philippine Archipelago approximately 17 million years ago, and uplifted above water approximately 5–10 million years ago (Hall and Holloway 1998; Dickinson, Kennedy, and Parkes 1991). Mindoro is separated from Palawan to the south and Luzon to the north by deepwater channels and has not been connected to those islands during the recent past (Pleistocene) (Heaney 1986).

Vegetation types on Mindoro include lowland evergreen rain forest to approximately 400 m or higher, open forest from about 650 to 1,000 m, and mossy forest above. Only small patches remain of the lowland evergreen dipterocarp rain forest that would have dominated the lowland eastern portions of the island. Semideciduous forest would have predominated on the western half of the island. Limited stands of Mindoro pine (*Pinus merkusii*) are found at elevations of 600 m or less in the northern portions of the island (Stattersfield et al. 1998; Development Alternatives 1992).

Biodiversity Features

Of the forty-two indigenous mammal species found on Mindoro, close to 20 percent endemic or near endemic (table J99.1). The nonendemic mammals are also found on Luzon. An endemic rat (*Rattus mindorenis*) is closely related to *Rattus tiomanicus*, and the endemic genus *Anonomomys* is most closely related to the genus *Haeromys*, from Palawan and some of its satellite islands. Thus colonization of Mindoro has occurred from both Luzon and Palawan (Heaney 1986).

The most unique animal feature of Mindoro must be the tamaraw (*Bubalus mindorensis*), or dwarf water buffalo. There were perhaps 10,000 living at all elevations on the island at the turn of the century. The tamaraw, like the anoas of Sulawesi (*Anoa* spp.), are wary forest animals, just over 1 m tall at the shoulder (Heaney and Regalado 1998; Nowak 1999a). Tamaraws sometimes are placed in the same genus as anoas. They should not be confused with the carabao of the Philippines, which is a small variety of domesticated Asian water buffalo (*B. bubalus*). The tamaraw needs both dense vegetation for resting and open grazing land. It is unclear whether tamaraws need wallows. Unfortunately, they are confined to areas of grassland that have taken the place of the native forest. Adult bulls are largely solitary and aggressive toward each other. Young males (up to five years) form bachelor groups. Females are found alone, with a bull, or with up to three young of differing ages. Young are born during Mindoro's June–November rainy season and stay with their mother until the age of 1.5 to 4.5 years (Nowak 1999a).

Mindoro also supports a population of the Philippine warty pig (*Sus philippensis*), which the IUCN considers rare and declining (Hilton-Taylor 2000). The Philippine warty pig is widely distributed in the still-forested areas of Luzon, Mindoro, Samar, Leyte, Mindanao, and some of the smaller satellite islands. Many of these forested areas are found in existing national parks. The Philippine warty pig is closely related to *Sus barbatus* of the Greater Sundas and was once thought to be a subspecies, analogous to the Palawan bearded pig (*Sus barbatus ahoenobarbus*). This species is still threatened by hunting and habitat loss (Oliver 1993).

An endemic subspecies of the Philippine deer (*Cervus mariannus barandanus*) is found on Mindoro. Although Philippine deer are native to Luzon, Mindoro, Samar, Leyte, Mindanao, and the Basilan Islands, *C. m. barandanus* is found only on Mindoro. The population of this subspecies is considered to be at risk over its limited range on the island (Wemmer 1998).

Greater Mindoro is home to the critically endangered Illin hairy-tailed cloud rat (*Crateromys paulus*), the endangered Mindoro shrew (*Crocidura mindorus*), and the more wide-

TABLE J99.1. Endemic and Near-Endemic Mammal Species.

Family	Species
Sorcidae	*Crocidura mindorus*
Bovidae	*Bubalus mindorensis**
Muridae	*Rattus mindorensis**
Muridae	*Anonymomys mindorensis**
Muridae	*Crateromys paulus**
Muridae	*Apomys gracilirostris**
Muridae	*Apomys sp. E**
Pteropodidae	*Pteropus sp. A**

An asterisk signifies that the species' range is limited to this ecoregion.

spread (within the Philippines) but endangered golden-crowned fruit bat (*Acerodon jubatus*) (Hilton-Taylor 2000).

This ecoregion corresponds exactly with the Mindoro EBA. The Mindoro EBA contains ten restricted-range birds, six of which are threatened. The Mindoro ecoregion contains eleven endemic or near-endemic bird species (Kennedy et al. 2000; table J99.2). Two bird species, the Mindoro bleeding-heart (*Gallicolumba platenae*) and the black-hooded coucal (*Centropus steerii*), are considered critically endangered, the Mindoro tarictic (*Penelopides mindorensis*) is endangered, and four species are considered vulnerable: Mindoro imperial-pigeon (*Ducula mindorensis*), ashy thrush (*Zoothera cinerea*), Luzon water-redstart (*Rhyacornis albiventris*), and scarlet-collared flowerpecker (*Dicaeum retrocinctum*). Mindoro's endemic birds can be split into montane and lowland species. Although both are in urgent need of conservation, the situation for the lowland species is particularly dire because the lowland forests are almost entirely gone (Dutson et al. 1992). Mindoro is also an important wintering and staging area for ducks and other waterbirds (Bagarinao 1998).

The type specimen of the critically endangered Philippine crocodile (*Crocodylus mindorensis*) was collected in Mindoro's Naujan Lake (Bagarinao 1998). They were historically found on the islands of Luzon, Mindoro, Masbate, Samar, Jolo, Negros, Busuanga, and Mindanao, but the only remaining populations are found on Mindoro, Negros, Mindanao, and Busuanga. The only protected population of Philippine crocodiles is in Lake Naujan National Park on Mindoro. Whereas the decline of the species initially was driven by overexploitation, habitat loss and human persecution are now the principal threats to the Philippine crocodile. Surveys in 1980–1982 revealed a total wild population of approximately 500–1,000 individuals, but current wild populations may be approximately 100 nonhatchlings. Captive breeding efforts are being led by the Crocodile Farming Institute, an entity of the Philippine government (Ross 1998; Hilton-Taylor 2000).

Lubang Island, near Mindoro and Luzon, is a poorly known island surrounded by deep water channels; it may well represent a small but distinct center of endemism (L. Heaney, pers. comm., 2000).

Status and Threats

Current Status

The only remaining intact forests in Mindoro are found along the top of the mountain ridge that divides the island. On the eastern side of the ridge commercial logging ended long enough ago that the remaining intact forests are buffered by secondary forests that have reestablished a closed condition, yet these same forests are again under threat from poaching and *kaingin* (slash-and-burn) agriculture. On the western side of the ridges, however, perennial fires in adjacent grasslands used for pasture are eating into the forest (Development Alternatives 1992). Only 8.5 percent of Mindoro was forested in 1988 (SSC 1988).

Several reserves have been established in Mindoro, beginning in 1936, but these have proved to be less than effective (Heaney and Regalado 1998). Table J99.3 details the existing protected areas on the island.

The largest protected area on Mindoro is Mounts Iglit-Baco National Park, which is one of two ASEAN Natural Heritage Sites in the Philippines (the other is Mount Apo National Park on Mindanao). The park covers the east-west divide and includes several physiographic regions and an important tamaraw population. Small patches of dipterocarp and mossy forest can be found in the park. The park is inhabited by the Mangyan tribal people, and much of the reserve consists of fire-maintained grassland with *Imperata cylindrica* and *Sacchareum spontaneum*. The combination of burning to maintain pasture for domestic cattle, ranching, and uncontrolled hunting activities leaves this protected area substantially altered, and the park is an insecure refuge for the endangered tamaraw (Collins et al. 1991).

Lake Naujan National Park is the only protected area in the Philippines that protects the critically endangered Philippine crocodile (Ross 1998).

The tamaraw numbered approximately 10,000 animals at the turn of the century and approximately 1,000 by 1949, and today estimates range from 100 to 200 animals (Collins et al. 1991; Heaney and Regalado 1998).

A well-funded conservation program aimed at captive breeding of tamaraws has been an expensive failure (Heaney and Regalado 1998).

TABLE J99.2. Endemic and Near-Endemic Bird Species.

Family	Common Name	Species
Columbidae	Mindoro bleeding-heart	*Gallicolumba platenae**
Columbidae	Mindoro imperial-pigeon	*Ducula mindorensis**
Cuculidae	Black-hooded coucal	*Centropus steerii**
Strigidae	Mindoro scops-owl	*Otus mindorensis**
Strigidae	Mantanani scops-owl	*Otus mantananensis*
Bucconidae	Mindoro hornbill	*Penelopides mindorensis**
Pachycephalida	Green-backed whistler	*Pachycephala albiventris*
Laniidae	Mountain shrike	*Lanius validrostris**
Turdidae	Ashy thrush	*Zoothera cinerea*
Muscicapidae	Luzon redstart	*Rhyacornis bicolor*
Dicaeidae	Scarlet-collared flowerpecker	*Dicaeum retrocinctum*

An asterisk signifies that the species' range is limited to this ecoregion.

TABLE J99.3. WCMC (1997) Protected Areas That Overlap with the Ecoregion.

Protected Area	Area (km²)	IUCN Category
Lake Naujan	90	IV
Mts. Iglit-Baco	790	II
F.B. Harrison	1,800	IV
Total	2,680	

Ecoregion numbers of protected areas that overlap with additional ecoregions are listed in brackets.

Types and Severity of Threats

Hunting by local people is a threat to all large mammals in the ecoregion, including the tamaraw, Philippine deer, and Philippine warty pig (Hedges, in press). Forestry activities and *kaingin* (slash-and-burn) agriculture continue to fragment and destroy the remaining habitat.

Although it is smaller and not as rich as some of the larger Philippine islands, Mindoro faces high levels of faunal endangerment because a larger proportion of its fauna is endangered; this level of endangerment is well-correlated with the degree of deforestation on the respective islands (Heaney 1993).

Priority Conservation Actions

Short-Term Conservation Actions (1–5 Years)

- Protect all remaining old-growth forests.
- Expand existing protected areas to include more lowland rain forest and entire watersheds.

Hedges (in press) lists recommended actions for the conservation of tamaraw. Many of these measures will benefit the protection of other taxa and communities, including the following:

- Create a dedicated tamaraw protection force to protect the known tamaraw population in Mounts Iglit-Baco National Park.
- Develop appropriate survey methods, conduct training for, and carry out an island-wide status survey and regular monitoring of all tamaraw populations.
- Enforce existing legislation protecting the tamaraw.
- Initiate a three- to five-year study of the ecology and behavior of the tamaraw.

Conservation of Mindoro's montane endemic bird species (*Ducula mindorensis* and *Otus mindorensis*) should focus in the Halcon Range, with an aim toward reducing logging and slash-and-burn agriculture. Surveys and conservation of Iglit-Baco National Park for these same species is also recommended. Three of the four lowland bird endemics are threatened with extinction (*Gallicolumba platenae*, *Centropus steerii*, *Penelopides mindorensis*, and *Dicaeum retrocinctum*), and immediate conservation action in west-central Mindoro is recommended, along with further field surveys in the far south and west to identify forest remnants that might support populations of these lowland species. Specifically recommended field survey areas are Bongabong, Central Occidental Mindoro, South Mindoro, San Vicente, and Iglit-Baco (Dutson et al. 1992).

The development of a national crocodile management program, for both the Philippine crocodile and the saltwater crocodile (*Crocodylous porosus*), is listed as a high priority in the Crocodile Status Survey and Conservation Action Plan (Ross 1998). Worldwide coordination of the Philippine crocodile captive-breeding program is considered a medium priority (Ross 1998).

Longer-Term Conservation Actions (5–20 Years)

- Educate the public about watershed protection (L. Heaney, pers. comm., 2000).

Hedges (in press) lists recommended actions for the conservation of the tamaraw. Many of these measures will benefit the protection of other taxa and communities, including the following:

- Improve management of the captive tamaraw population.
- Investigate the potential for increasing the Mounts Iglit-Baco National Park's tamaraw carrying capacity by means of habitat management.
- Survey and demarcate the Mounts Iglit-Baco National Park boundaries.
- Gradually exclude domestic cattle from the Mounts Iglit-Baco National Park.
- Implement livestock quarantine measures to reduce transmission of domestic cattle diseases to the tamaraw.
- Create a buffer zone around Mounts Iglit-Baco National Park.
- Develop a management plan for Mounts Iglit-Baco National Park.
- Conduct participatory rural appraisals around the Mounts Iglit-Baco National Park.

Collar et al. (1999) identified (incomplete and preliminary) five key sites that are known to support several threatened bird species and retain sufficient areas of natural habitat for their survival. Some of these key sites are protected to some degree; others are not. A future effort will develop a comprehensive set of Important Bird Areas that would be sufficient to conserve the full range of bird species. For Mindoro, the key sites identified thus far are Mt. Halcon, Lake Naujan, Mt. Iglit-Baco, Siburan, and Malpalom.

Focal Species for Conservation Action

Endangered endemic species: tamaraw (*Bubalus mindorensis*) (Hedges, in press), Philippine deer (*Cervus mariannus barandanus*) (Wemmer 1998), Ilin hairy-tailed cloud rat (*Crateromys paulus*) (Heaney et al. 1998), Mindoro shrew (*Crocidura mindorus*) (Heaney et al. 1998), Mindoro bleeding-heart (*Gallicolumba platenae*) (Collar et al. 1999), black-hooded coucal (*Centropus steerii*) (Collar et al. 1999), Mindoro tarictic (*Penelopides mindorensis*) (Collar et al. 1999)

Area-sensitive species: Philippine warty pig (*Sus philippensis*) (Oliver 1993), Philippine deer (*Cervus mariannus barandanus*) (Wemmer 1998), fruit bats (*Acerodon jubatus, Haplonycteris fischeri, Harpyionycteris whiteheadi, Ptenochirus jagori, Pteropus pumilus, Pteropus vampyrus, Eonycteris major, Eonycteris spelaea*) (Mickleburgh et al. 1992), crested serpent-eagle (*Spilornis cheela*), rufous-bellied eagle (*Hieraaetus kienerii*), changeable hawk-eagle (*Spizaetus cirrhatus*), Philippine hawk-eagle (*Spizaetus philippensis*), peregrine falcon (*Falco peregrinus*)

Top predator species: Malay civet (*Viverra tangalunga*) (Heaney et al. 1998), crested serpent-eagle (*Spilornis cheela*), rufous-bellied eagle (*Hieraaetus kienerii*), changeable hawk-eagle (*Spizaetus cirrhatus*), Philippine hawk-eagle (*Spizaetus philippensis*), peregrine falcon (*Falco peregrinus*), Philippine crocodile (*Crocodylus mindorensis*) (Ross 1998)

Selected Conservation Contacts

Conservation International
Flora & Fauna International
Kalikasan Mindoro Foundation, Inc.
Protected Areas & Wildlife Bureau, Philippine DENR
Tamaraw Conservation Program
WWF-Philippines

Justification for Ecoregion Delineation

MacKinnon (1997) identified seven subunits in the Philippines, and the Philippine Biodiversity Action Plan (Philippine BAP 1997) demarcated fifteen biogeographic units. Udvardy (1975) identified the Philippines as a single biogeographic province. We delineated nine ecoregions in the Philippine islands, including Palawan. We deviated from Udvardy (1975), MacKinnon (1997), Stattersfield (1998), and the Philippine BAP (1997) in varying degrees and based our delineation of the Philippine ecoregions on Heaney (1993). MacKinnon (1997) designates Mindoro island as subunit 26f and includes the Lubang Islands. We delineated the island of Mindoro as the Mindoro Rain Forests [99].

Prepared by John Morrison

Ecoregion Number:	**100**
Ecoregion Name:	**Greater Negros–Panay Rain Forests**
Bioregion:	**Sunda Shelf and Philippines**
Biome:	**Tropical and Subtropical Moist Broadleaf Forests**
Political Units:	**Philippines**
Ecoregion Size:	**34,200 km²**
Biological Distinctiveness:	**Globally Outstanding**
Conservation Status:	**Critical**
Conservation Assessment:	**I**

The Greater Negros–Panay Rain Forests [100] ecoregion, including the Western Visayas and parts of additional political regions, appears as a number of isolated islands, but during the last ice ages these islands were (for the most part) part of one continuous island. The islands contain a unique mix of Sundaic and Philippine mammals and birds, including leopard cats and endemic pigs and deer species. Sibuyan, a small mountainous island surrounded by deep water, contains five endemic mammals and several restricted-range birds and nearly qualifies as an ecoregion in itself.

Location and General Description

The ecoregion includes the large island of Negros, Panay, and Cebu and the smaller islands of Masbate, Ticao, and Guimaras; Sibuyan, Romblon, Tablas, and Siquijor are moderately isolated and distinctive (Heaney and Regalado 1998). The climate of the ecoregion is tropical wet (National Geographic Society 1999). The Visayas receive approximately 2,419 mm of rainfall annually. July and August are the wettest months. The west coasts of Panay and Negros experience a dry season between November and February (Davis et al. 1995).

Most of these islands have been uplifted above water in the last 6 million years or less (Hall and Holloway 1998). Sibuyan Island contains pre-Tertiary schists, marble, volcanics, and ultramafics (Davis et al. 1995). Cebu contains a rugged mountain spine, and 73 percent of the island contains slopes greater than 18 percent. The geology of Cebu includes limestones, marls, and karst topography (Wood et al. 2000). Negros contains a string of volcanic mountains (Wood et al. 2000).

The channel between Greater Negros–Panay and the surrounding islands of Luzon and Greater Mindanao is more than 120 m deep. During the last ice age (late Pleistocene), sea level was approximately 120 m below current levels and Greater Negros–Panay was separated from ice age Luzon and the Eastern Visayas by a narrow channel (Heaney 1986). Except for the eastern portion of Panay and much of Masbate, the islands generally are fairly rugged, with the highest elevation being 2,465 m Mt. Canlaon in north-central Negros.

Vegetation types in the ecoregion are diverse and include beach vegetation, mangroves, tropical lowland rain forest, montane forests, and grasslands and heath forests. Beach forest merges with other forest types away from the coast and includes *Casuarina* and *Barringtonia* mixed with other lowland species. Palms, vines, bamboo, and *Pterocarpus indicus* are present only in rare back-beach swamps. This habitat type is extremely rare because of coastal habitation (Davis et al. 1995; Dickinson et al. 1991).

The lowland evergreen dipterocarp rain forest was once the dominant vegetation of the ecoregion and contained *Dipterocarpus* spp., *Shorea* spp., *Hopea* spp., *Pterocarpus indicus*, and pandans (Davis et al. 1995). Philippine dipterocarp forest is quite tall (45–65 m) and dense, with three layers of canopy. Lianas and bamboo are rare in mature forest but common in poorly developed evergreen forest. Ferns, orchids, and other epiphytic plants are found on the larger trees. At higher elevations there are only two canopy layers, tree stature is lower, and there are more epiphytes. Upper hill dipterocarp forest is found at elevations of 650–1,000 m and contains dominant *Shorea polysperma* and oaks, chestnuts, and elaeocarps. At approximately 1,000 m the montane forest contains oaks and laurels (Heaney and Regalado 1998). The mossy upper montane forest generally is found at elevations over 1,200 m, where humidity is constantly high. This stunted, single-story moss- and epiphyte-covered forest contains tree ferns up to 10 m high. (Dickinson et al. 1991).

Biodiversity Features

Greater Negros–Panay contains a unique mix of Sundaic and Philippine mammals and birds, including leopard cats and endemic pig and deer species. Fifty-eight mammals inhabit the ecoregion, and thirteen mammal species are endemic or near endemic (table J100.1). Six of these species (*Crocidura mindorus*, two *Apomys* spp., and *Chrotomys*, *Tarsomys*, and *Haplonyteris* spp.) are limited to Sibuyan Island within the ecoregion (*Crocidura mindorus* is also found on Mindoro); most of these species have yet to be named. Two endemic large vertebrates are especially notable: the Philippine spotted deer (*Cervus alfredi*) and the Visayan warty pig (*Sus cebifrons*). The Visayan warty pig is critically endangered, and the Philippine spotted deer is endangered (Hilton-Taylor 2000). One found only on Negros and Cebu, the Philippine bare-backed fruit bat (*Dobsonia chapmani*), is now believed to be extinct because of guano mining, deforestation, and hunting (Heaney et al. 1998).

Sibuyan, a small mountainous (to 2,052 m) island surrounded by deep water within the ecoregion, is outstanding in itself. Sibuyan contains six endemic mammals (five strict endemics; Heaney et al. 1998; Heaney and Regalado 1998; Goodman and Ingle 1993) and is considered a Secondary Area by BirdLife International (Stattersfield et al. 1998) because of the existence of three restricted-range birds. These restricted-range birds are also found elsewhere but have no clear affinities. One hundred thirty-one bird species have been recorded on the island, a high number for an isolated island of only 463 km^2. Sibuyan still contains about half of its original forest cover (100–150 km^2), including some extremely rare (in the Philippines) lowland forest; unfortunately, logging threatens the remainder (Goodman and Ingle 1993; Stattersfield et al. 1998). Sibuyan Island is also a Centre of Plant Diversity (CPD) (Davis et al. 1995), containing beach vegetation, mangroves, lowland rain forest, montane forests, and grasslands and heath forests. There are an estimated 700 vascular plant species on the island, including 54 endemic species. The flora is closely related to that of Luzon. Sibuyan contains valuable timber trees, almaciga (*Agathis philippensis*) resins, and ornamental plants and has good potential for tourism (Davis et al. 1995).

The Asian leopard cat (*Prionailurus bengalensis*), an Asian species, is found on Negros, Panay, and Cebu within the ecoregion. Also found on Palawan, the Negros leopard cat is a subspecies (Heaney et al. 1998). Threatened endemic or near-endemic mammals include the critically endangered Negros shrew (*Crocidura negrina*) and Philippine tube-nosed fruit bat (*Nyctimene rabori*), the endangered Mindoro shrew (*Crocidura mindorus*) and Panay bushy-tailed cloud rat (*Crateromys heaneyi*), and the widespread but endangered golden-crowned fruit bat (*Acerodon jubatus*) (Hilton-Taylor 2000).

The Visayan warty pig has been extirpated from Masbate, Guimaras, Cebu, and Siquijor and now lives only in isolated areas of Negros and Panay, where hunting of them is still intense. There are no effective protected areas in the fragmented range of *S. cebifrons*. The Visayan warty pig is smaller than its closer relative, the Philippine warty pig (*Sus philippensis*), which the IUCN considers rare and declining. The Visayan warty pig once inhabited both lowland and montane rain forests and may have thrived in disturbed areas as well (Oliver 1993).

The Philippine spotted deer (*Cervus alfredi*) has been extirpated from Guimaras, Cebu, and Siquijor and is now limited to the Mt. Madja–Mt. Baloy area of west Panay and some scattered forest fragments on Negros. A very small

TABLE J100.1. Endemic and Near-Endemic Mammal Species.

Family	Species
Sorcidae	*Crocidura mindorus*
Sorcidae	*Crocidura negrina**
Muridae	*Apomys* sp. A*
Muridae	*Apomys* sp. B* (Sibuyan only)
Muridae	*Apomys* sp. C* (Sibuyan only)
Muridae	*Chrotomys* sp. A* (Sibuyan only)
Muridae	*Crateromys heaneyi**
Muridae	*Tarsomys* sp. A* (Sibuyan only)
Pteropodidae	*Nyctimene rabori**
Pteropodidae	*Dobsonia chapmani**
Pteropodidae	*Haplonycteris* sp. A* (Sibuyan only)
Suidae	*Sus cebifrons**
Cervidae	*Cervus alfredi**

An asterisk signifies that the species' range is limited to this ecoregion.

TABLE J100.2. Endemic and Near-Endemic Bird Species.

Family	Common Name	Species
Turnicidae	Spotted buttonquail	*Turnix ocellata*
Columbidae	Negros bleeding-heart	*Gallicolumba keayi**
Columbidae	Negros fruit-dove	*Ptilinopus arcanus**
Strigidae	Mantanani scops-owl	*Otus mantananensis*
Apodidae	Philippine needletail	*Mearnsia picina*
Bucconidae	Tarictic hornbill	*Penelopides panini**
Bucconidae	Writhed-billed hornbill	*Aceros waldeni**
Monarchidae	Celestial monarch	*Hypothymis coelestis*
Campephagidae	Blackish cuckoo-shrike	*Coracina coerulescens*
Campephagidae	White-winged cuckoo-shrike	*Coracina ostenta**
Irenidae	Philippine leafbird	*Chloropsis flavipennis*
Muscicapidae	Negros jungle-flycatcher	*Rhinomyias albigularis**
Muscicapidae	Ash-breasted flycatcher	*Muscicapa randi*
Muscicapidae	Black shama	*Copsychus cebuensis**
Pycnonotidae	Streak-breasted bulbul	*Ixos siquijorensis**
Timaliidae	Flame-templed babbler	*Stachyris speciosa**
Timaliidae	Panay striped-babbler	*Stachyris latistriata**
Timaliidae	Negros striped-babbler	*Stachyris nigrorum**
Estrildidae	Green-faced parrotfinch	*Erythrura viridifacies*
Dicaeidae	Cebu flowerpecker	*Dicaeum quadricolor**
Dicaeidae	Visayan flowerpecker	*Dicaeum haematostictum**
Dicaeidae	Scarlet-collared flowerpecker	*Dicaeum retrocinctum*
Lanidae	Grey-capped shrike	*Lanius validirostris*

An asterisk signifies that the species' range is limited to this ecoregion.

population apparently was located on Masbate in 1991. Small populations are protected in Mt. Canlaon National Park, North Negros Forest Reserve, and the Mount Talinis/Lake Balinsasayao Reserve. The Philippine spotted deer prefers dipterocarp rain forests but also frequents open grassland patches and secondary forest. The population has been reduced by habitat loss and hunting, which are still the main threats (Wemmer 1998).

The Greater Negros–Panay ecoregion encompasses two EBAs (Negros and Panay; Cebu) and two Secondary Areas (Tablas, Romblon, and Sibuyan; Siquijor). There are seventeen restricted-range bird species in the Negros and Panay EBA, six restricted-range species in Cebu (five additional species), and several others in the adjacent Secondary Areas. A total of twenty-three endemic or near-endemic species are found in the ecoregion (Kennedy et al. 2000; table J100.2). An extremely high proportion of these birds are threatened. These include the critically endangered Negros bleeding-heart (*Gallicolumba keayi*), Negros fruit-dove (*Ptilonopus arcanus*), writhed-billed hornbill (*Aceros waldeni*), Cebu flowerpecker (*Dicaeum quadricolor*), six endangered species and seven vulnerable species (Collar et al. 1999). Almost no forest remains on Cebu, and the species with these habitats on Cebu are in a critical situation.

The critically endangered Philippine crocodile (*Crocodylus mindorensis*) was historically found on the islands of Luzon, Mindoro, Masbate, Samar, Jolo, Negros, Busuanga, and Mindanao, but the only remaining populations are found on Mindoro, Negros, Mindanao, and Busuanga. Whereas the decline of the species initially was driven by overexploitation, habitat loss and human persecution are now the principal threats to the Philippine crocodile. Surveys in 1980–1982 revealed a total wild population of approximately 500–1,000 individuals, but current wild populations may be approximately 100 nonhatchlings. Captive breeding efforts are being led by the Crocodile Farming Institute, an entity of the Philippine government (Ross 1998; Hilton-Taylor 2000).

Status and Threats

Current Status

"Nowhere in the Philippines is environmental degradation quite so acute, and the need for immediate conservation action quite so pressing, as in the West Visayas, or Negros Faunal Region" (Oliver 1993, p. 151). This ecoregion harbors the some of the highest levels of endemism but has suffered a disproportionate share of deforestation, and to compound the problem the area is underrepresented in the national protected area system (Oliver 1993).

The Philippines were once almost completely forested (Dickinson et al. 1991), but forests have been significantly reduced in this ecoregion and the rest of the Philippines. Aerial surveys indicate that Sibuyan Island still contains substantial areas of closed-canopy forest. Forests on Masbate Island have been devastated by improper farming and pasture practices. Remaining closed-canopy forests on Panay straddle the main mountain range. Open-canopy dipterocarp forests can still be found on Panay in steep areas at elevations between 690 and 2,150 m. Closed-canopy forests can also be found on Negros. Mt. Canlaon contains broadleaf and mossy forest on steep slopes at elevations of 490 to 1,870 m; there was little to no buffer zone between this forest and cultivation in 1992. On Mt. Mandalagan, remaining mossy forest is surrounded by a large buffer of residual open-canopy forest. A patch of closed canopy forest also remains above Dumaguete City in Negros (Development Alternatives 1992). In 1988, Negros was only 4 percent forested and Panay was 8 percent forested (SSC 1988).

Cebu contains approximately 15 km^2, or 0.3 percent, of its original dipterocarp forest cover (SSC 1988; Brooks et al. 1995); this has led to the near extinction of the island's endemic species. Little of the ecoregion is protected (table J100.3).

Types and Severity of Threats

Habitat destruction is the main threat to biodiversity in the Philippines, and Greater Negros–Panay is no different. Logging and shifting cultivation (*kaingin*) are cited as the primary forces of habitat conversion. Logging takes many forms, from industrial-scale to smaller-scale operations that use water buffalo to haul logs out of the forest. Mangroves are used locally for firewood, dyes, and tannins (Davis et al. 1995).

Hunting and the wild pet trade are also significant threats in Greater Negros–Panay. Leopard cats have been hunted for their pelts, and kittens are sold as pets (Heaney and Regalado 1998).

Priority Conservation Actions

Short-Term Conservation Actions (1–5 Years)

- Protect all remaining old-growth forests.
- Expand existing protected areas to include more lowland rain forest and entire watersheds.
- Place remaining lowland forests of Sibuyan should be placed within the protected area system of the Philippine Department of Environment and Natural Resources, and enforce existing laws to prevent illegal logging (Goodman and Ingle 1993).

Longer-Term Conservation Actions (5–20 Years)

- Educate the public about watershed protection (L. Heaney, pers. comm., 2000).

TABLE J100.3. WCMC (1997) Protected Areas That Overlap with the Ecoregion.

Protected Area	Area (km^2)	IUCN Category
Sampunong-Bolo Bird Sanctuary	3	I
Bulabog-Putian	30	III
Central Cebu	200	VI
Mt. Canlaon	320	II
Sudlon	30	VI
Olango Island Complex	50	I
Guadalupe Mabugnao–Mainit Hot Spring	20	PRO
Total	653	

Ecoregion numbers of protected areas that overlap with additional ecoregions are listed in brackets.

A number of conservation actions are recommended for the Philippine spotted deer (Cervus alfredi) in the IUCN Deer Status Survey and Conservation Action Plan. These actions include implementation of existing management plans in existing protected areas, an assessment of the genetic differences between populations in Negros and Panay, the continued development of a public education campaign, the monitoring and control of illegal captures and movement of the deer, and the integration of logging techniques with the deer's needs (Wemmer 1998).

The Visayan warty pig is one of the most threatened of all suids, and the IUCN Pigs, Peccaries, and Hippos Status Survey and Conservation Action Plan includes recommendations for its conservation. Recommended measures include promoting proposals to create effective protected areas in the ecoregion; conducting field surveys in Masbate, Bohol, Sibuyan, and Negros to ascertain the current population status; developing a cooperative breeding program; promoting local conservation projects; and encouraging field studies relevant to the future management of the species (Oliver 1993).

Brooks et al. (1992) identified critical conservation priorities for Negros and, by extension, for the Western Visayas because Negros and Panay hold much of the remaining natural forest cover in the ecoregion. The fate of nine endemic bird species and twenty-six forest-dependent endemic bird subspecies is at stake. The highest priority was to allocate funding to preserve remaining forest on Negros and include Mt. Canlaon and Curenos de Negros in NIPAS. Protecting remaining forest is a higher priority than reforestation. Other recommended actions include allocating more official personnel to the field at remaining forest sites to discourage slash-and-burn agriculture, adopting a flagship species (Walden's hornbill [Aceros waldeni] was suggested), and reforesting with native species on previously forested sites. Other recommendations, specific to sites, include fencing remaining forest at Ban-ban (for Negros jungle-flycatcher [Rhinomyias albigularis]), enforcing watershed protection by the company operating in Cuernos de Negros (for Negros striped-babbler [Stachyris nigrorum]), evaluating invertebrate collecting occurring on Mt. Canlaon, and conducting field surveys at Mt. Mandalagan and Mt. Patag, Hinob-an in southwest Negros, and especially in Cuernos de Negros.

Collar et al. (1999) identified (incomplete and preliminary) twelve key sites that are known to support several threatened bird species and to retain sufficient areas of natural habitat for their survival. Some of these key sites are protected to some degree; others are not. A future effort will develop a comprehensive set of Important Bird Areas that would be sufficient to conserve the full range of bird species. For Greater Negros–Panay, the key sites identified thus far are Mt. Guiting-guiting National Park on Sibuyan; the North-west Panay peninsula, Mts. Makja-as and Hantotubig, and Mt. Baloy on Panay; Mts. Silay and Mandalagan, Mt. Canlaon National Park, Hinoba-an, Lake Balinsasayao, and Eastern Cuernos de Negros on Negros; and Mt. Bandila-an on Siquijor, Tabunan, and Olango on Cebu.

Focal Species for Conservation Action

Endangered endemic species: Philippine spotted deer (Cervus alfredi) (Wemmer 1998), Visayan warty pig (Sus cebifrons) (Oliver 1993), Philippine bare-backed fruit bat (Dobsonia chapmani) (Heaney et al. 1998), Negros shrew (Crocidura negrina) (Heaney et al. 1998), Philippine tube-nosed fruit bat (Nyctimene rabori) (Heaney et al. 1998), Mindoro shrew (Crocidura mindorus) (Heaney et al. 1998), Panay bushy-tailed cloud rat (Crateromys heaneyi) (Heaney et al. 1998), Chrotomys spp. (Heaney et al. 1998), Haplonycteris spp. (Heaney et al. 1998), Negros bleeding-heart (Gallicolumba keayi) (Collar et al. 1999), Negros fruit-dove (Ptilinopus arcanus) (Collar et al. 1999), Visayan hornbill (Penelopides panini) (Collar et al. 1999), writhed-billed hornbill (Aceros waldeni) (Collar et al. 1999), white-throated jungle flycatcher (Rhinomyias albigularis) (Collar et al. 1999), Cebu flowerpecker (Dicaeum quadricolor) (Collar et al. 1999), black shama (Copsychus cebuensis) (Collar et al. 1999), streak-breasted bulbul (Ixos siquijorensis) (Collar et al. 1999), flame-templed babbler (Stachyris (Dasychrotapha) speciosa) (Collar et al. 1999), Negros striped babbler (Stachyris nigrorum) (Collar et al. 1999)

Area-sensitive species: Philippine spotted deer (Cervus alfredi) (Wemmer 1998), Visayan warty pig (Sus cebifrons) (Oliver 1993), fruit bats (Acerodon jubatus, Dobsonia chapmani, Haplonycteris fischeri, Harpyionycteris whiteheadi, Nyctimene rabori, Ptenochirus jagori, Pteropus hypomelanus, Pteropus pumilus, Pteropus vampyrus, Eonycteris major, Eonycteris spelaea) (Mickleburgh et al. 1992), Oriental honey-buzzard (Pernis ptilorhynchus), barred honey-buzzard (Pernis celebensis), crested serpent-eagle (Spilornis cheela), rufous-bellied eagle (Hieraaetus kienerii), changeable hawk-eagle (Spizaetus cirrhatus), Philippine hawk-eagle (Spizaetus philippensis), peregrine falcon (Falco peregrinus)

Top predator species: Asian leopard cat (Prionailuru bengalensis) (Nowell and Jackson 1996), palm civet (Paradoxurus hermaphroditus) (Heaney et al. 1998), Malay civet (Viverra tangalunga) (Heaney et al. 1998), Oriental honey-buzzard (Pernis ptilorhynchus), barred honey-buzzard (Pernis celebensis), crested serpent-eagle (Spilornis cheela), rufous-bellied eagle (Hieraaetus kienerii), changeable hawk-eagle (Spizaetus cirrhatus), Philippine hawk-eagle (Spizaetus philippensis), peregrine falcon (Falco peregrinus), Philippine crocodile (Crocodylus mindorensis) (Ross 1998)

Selected Conservation Contacts

Center for Tropical Conservation Studies, Silliman University

Flora and Fauna International

Green Forum Philippines

Haribon Foundation

Negros Forest & Ecological Foundation, Inc.

Philippine Endemic Species Conservation Project

Philippine Wetland Society

Protected Areas & Wildlife Bureau, Philippine DENR

WWF-Philippines

Justification for Ecoregion Delineation

MacKinnon (1997) identified seven subunits in the Philippines, and the Philippine Biodiversity Action Plan (Philippine BAP 1997) demarcated fifteen biogeographic units. Udvardy (1975) identified the Philippines as a single biogeographic province. We delineated nine ecoregions in the Philippine islands, including Palawan. We deviated from Udvardy (1975), MacKinnon (1997), Stattersfield et al. (1998), and the Philippine BAP (1997) in varying degrees and based our delineation of the Philippine ecoregions on Heaney (1993), except that Heaney placed Sibuya, Tablas, and Romblon in a separate region.

We placed the islands of Negros, Panay, Tablas, Sibuyan, Masbate, Ticao, Siquijor, and Cebu into the Greater Negros–Panay Rain Forests [100] ecoregion. MacKinnon also grouped these islands into a single subunit (26b). However, he also included the island of Bohol in this subunit. We followed Heaney (1993) and included Bohol as part of the Mindanao–Eastern Visayas Rain Forests [101] because during the Pleistocene ice ages it was connected to Leyte Island.

Prepared by John Morrison

Ecoregion Number:	**101**
Ecoregion Name:	**Mindanao–Eastern Visayas Rain Forests**
Bioregion:	**Sunda Shelf and Philippines**
Biome:	**Tropical and Subtropical Moist Broadleaf Forests**
Political Units:	**Philippines**
Ecoregion Size:	**105,000 km^2**
Biological Distinctiveness:	**Globally Outstanding**
Conservation Status:	**Critical**
Conservation Assessment:	**I**

This ecoregion features lowland and hill forests on a number of large Philippine islands that, though currently disconnected, were part of one island during the height of the last ice age. These islands harbor Philippine warty pigs, Philippine deer, the Philippine tarsier, flying lemurs, and some of the last strongholds of the charismatic Philippine eagle. Most lowland forest has been cleared from the islands, but some large patches of hill and montane forest remain (the montane forest is a separate ecoregion). Tiny Camiguin Island, with two strictly endemic mammals of its own, is a unique feature of this ecoregion.

Location and General Description

This ecoregion includes the lowland (less than 1,000 m elevation) on the main islands of Mindanao, Samar, Leyte, Bohol, and numerous smaller satellite islands, including Biliran and Basilan. The climate of the ecoregion is tropical wet (National Geographic Society 1999). The northern Visayas (northern portions of Samar and Leyte) are in the main typhoon track that so strongly influences the more northerly Philippine islands. These typhoons typically occur from July to November: As much as one-third of an island's total annual precipitation may be collected during typhoon events. Mindanao is south of the main typhoon track (Dickinson et al. 1991).

Mindanao and the Visayas were transported across the western Pacific to their present location during the last 25 million years. Most of these islands have been uplifted above water only in the last 15 million years or less (Hall and Holloway 1998). During the Pleistocene, Mindanao, Samar, Leyte, and Bohol were all one island—Greater Mindanao—and their faunal affinities to each other persist to this day (Heaney 1986; Heaney and Regalado 1998).

Vegetation types on Mindanao and in the Eastern Visayas originally included beach forest, mangroves, lowland rain forest, and more open forest at higher elevations up to 1,000 m (Stattersfield et al. 1998).

The stunted beach forest contains *Casuarina* and *Barringtonia* mixed with other lowland species. Palms, vines, bamboo, and *Pterocarpus indicus* are present only in rare back-beach swamps. This habitat type is extremely rare because of coastal habitation (Heaney and Regalado 1998).

The dominant forest type in the Mindanao lowlands and the rest of the Philippines was dipterocarp forest. This group of trees is known as Philippine mahogany in the timber trade. This forest type occurred from sea level to elevations of 400 m or higher. Individual dipterocarps occur to 1,500 m. Philippine dipterocarp forest is quite tall (45–65 m) and dense, with three canopy layers. Lianas and bamboo are rare in mature forest but common in poorly developed evergreen forest. Ferns, orchids, and other epiphytic plants are found on the larger trees. At higher elevations there are only two canopy layers, tree stature is lower, and there are more epiphytes. Upper hill dipterocarp forest is found at elevations of 650 to about 1,500 m and contains dominant *Shorea polysperma* and oaks, chestnuts, and elaeocarps (Heaney and Regalado 1998).

Biodiversity Features

Mindanao and its neighbor, Basilan, situated adjacent to the Sulu Archipelago, have been influenced by immigration from Borneo, although in recent millennia movement has been primarily in the other direction (Dickinson et al. 1991). During the most recent ice ages, the Mindanao faunal region has developed its own unique fauna, with a large number of endemic vertebrates.

Tiny Camiguin Island (ca. 265 km^2) contains two strictly endemic and as yet undescribed mammal species: a small forest mouse (*Apomys* sp.) and a large moss-mouse (*Bullimus* sp.) (Heaney and Tabaranza 1995), in addition to an endemic frog. Several taxa found on Mindanao, only a short distance away, are absent from Camiguin, including squirrels, some murid rodents, flying lemurs, tarsiers, and deer. Camiguin is the smallest island in the Philippines known to have unique mammal species. Consisting of a series of active volcanic cones reaching a maximum elevation of 1,713 m, the island is surrounded by deep water. Fortunately, the island still has good forest cover (Heaney et al. 1998).

There is also variation within the island of Mindanao. Thirty-one bird species are polytypic on the island. Sixteen of these variations are based on differences between isolated mountain ranges, and seven species have races associated with the Zamboanga Peninsula and Basilan Island. There are three species that vary between the uplands and lowlands (Dickinson et al. 1991).

Approximately 80 percent of Greater Mindanao's nonvolant mammal species are found nowhere else in the world. Whereas flying lemurs, tree shrews, tree squirrels, and tarsiers are found on the islands of Greater Mindanao, they are not found on the other large Philippine island, Luzon, just 25 km from the northern tip of Samar (Heaney and Regalado 1998). More than 30 percent of nonvolant mammals in the ecoregion are endemic to Mindanao only, but the other islands in the ecoregion generally share their species with Mindanao. However, tiny Dinagat island, located just north of Mindanao, contains three of its own endemic mammals (Heaney 1986), including the endangered Dinagat Island cloud-rat (*Crateromys australis*). There are sixteen endemic or near-endemic mammal species in the ecoregion (table J101.1).

An endemic subspecies of Philippine deer (*Cervus mariannus nigricans*) is limited to Mindanao. Philippine deer are widespread (though patchily distributed) in the Philippines, being found on Luzon, Mindoro, Samar, Leyte, Mindanao, and the Basilan Islands. The subspecies is threatened by habitat loss and hunting (Wemmer 1998). It has been reported that the endangered Visayan or Philippine spotted deer (*Cervus alfredi*) was potentially found on Bohol Island, but it seems likely that these reports refer to *Cervus mariannus*. *Cervus alfredi* is not found on Bohol (Oliver et al. 1991; Wemmer 1998).

The *kagwang* (*Cyanocephalus volans*), or Philippine flying lemur, is also endemic to Greater Mindanao; the only other species of this unique order of mammals is found in Malaysia and Indonesia. These small nocturnal mammals glide between trees for distances up to 135 m. Fortunately, the kagwang actually prefers second-growth forests to old growth,

so they are more secure than other Philippine mammals (Heaney and Regalado 1998). However, the species is still considered vulnerable (Hilton-Taylor 2000).

The ecoregion also supports a population of the Philippine warty pig (*Sus philippensis*), which the IUCN considers rare and declining. The Philippine warty pig is widely but patchily distributed in the still-forested areas of Luzon, Mindoro, Samar, Leyte, Mindanao, and some of the smaller satellite islands. Many of these forested areas are found in existing national parks. The Philippine warty pig is closely related to *Sus barbatus* of the Greater Sundas and was once thought to be a subspecies, analogous to the Palawan bearded pig (*Sus barbatus ahoenobarbus*). This species is still threatened by hunting and habitat loss (Oliver 1993).

Greater Mindanao is also home to an endemic primate, the Philippine tarsier (*Tarsius sychrita*), which is found on Samar, Leyte, Dinagat, Siargao, Bohol, Mindanao, and Basilan. Although they are also found in primary forests and mangroves, these highly charismatic small mammals seem to prefer second-growth forests, and they are not considered threatened by the IUCN (Nowak 1999a).

The Philippine tree shrew (*Urogale everetti*), in the order Scandentia, which is found on Mindanao, Dinagat, and Siargao Islands, represents an endemic, monotypic genus. Worldwide there are sixteen species of tree shrew, a diurnal animal that resembles a squirrel but whose dentition, circulatory system, and large braincase are more like those of primates (Nowak 1999a). This species is considered vulnerable (Hilton-Taylor 2000).

Greater Mindanao also supports an endemic genus of Erinaceidae, *Podogymura*. There are two moonrat species in this genus, both of which are found in Greater Mindanao. One species is found in the adjacent montane ecoregion of Mindanao (*P. truei*), and the other (*P. aureospinula*) is found in the lowland forest of Dinagat, Siargao, and the Bucas Grande Islands (Heaney et al. 1998).

Lowland Greater Mindanao is home to endangered mammals also found in other parts of the Philippines, including the golden-capped fruit bat (*Acerodon jubatus*) and the mottle-winged flying-fox (*Pteropus leucopterus*) (found on Luzon and Dinagat) (Heaney et al. 1998; Hilton-Taylor 2000).

This ecoregion overlaps with the Mindanao and Eastern Visayas EBA, with the exception of the montane areas above 1,000 m, which have been given their own ecoregion. The EBA contains fifty-one restricted-range birds, twenty-four (or possibly twenty-five) of which are lowland and hill forest specialists and are thus resident in this ecoregion. All the restricted-range birds are forest species. The ecoregion contains thirty-six endemic or near-endemic bird species (Kennedy et al. 2000; table J101.2). Twelve of these species are threatened, including the endangered Mindanao bleeding-heart (*Gallicolumba criniger*). The remainder of the threatened species are considered vulnerable. This situation should be contrasted with the adjacent upland Mindanao montane rain forests ecoregion. Although the upland ecoregion contains more restricted-range species, only one of these is considered threatened (Stattersfield et al. 1998; Collar et al. 1999).

In addition to the restricted-range species, several widespread threatened species are found in the ecoregion,

TABLE J101.1. Endemic and Near-Endemic Mammal Species.

Family	Species
Erinaceidae	*Podogymnura aureospinula**
Soricidae	*Crocidura beatus*
Tupaiidae	*Urogale everetti*
Cynocephalidae	*Cynocephalus volans*
Pteropodidae	*Ptenochirus minor*
Rhinolophidae	*Hipposideros coronatus**
Sciuridae	*Sundasciurus philippinensis*
Muridae	*Bullimus bagobus*
Muridae	*Batomys salomonseni*
Muridae	*Batomys russatus** (Dinagat only)
Muridae	*Crateromys australis** (Dinagat only)
Muridae	*Crunomys melanius**
Muridae	*Apomys* sp. D* (Camiguin only)
Muridae	*Bullimus* sp. A* (Camiguin only)
Muridae	*Tarsomys echinatus**
Sciuridae	*Exilisciurus concinnus*

An asterisk signifies that the species' range is limited to this ecoregion.

TABLE J101.2. Endemic and Near-Endemic Bird Species.

Family	Common Name	Species
Rallidae	Brown-banded rail	Lewinia mirificus
Columbidae	Mindanao bleeding-heart	Gallicolumba criniger*
Columbidae	Mindanao brown-dove	Phapitreron brunneiceps
Columbidae	Grey imperial-pigeon	Ducula pickeringii
Cuculidae	Black-faced coucal	Centropus melanops*
Strigidae	Mindanao eagle-owl	Mimizuku gurneyi
Apodidae	Philippine needletail	Mearnsia picina
Alcedinidae	Silvery kingfisher	Alcedo argentata*
Alcedinidae	Blue-capped kingfisher	Actenoides hombroni
Bucconidae	Mindanao hornbill	Penelopides affinis
Bucconidae	Samar hornbill	Penelopides samarensis*
Bucconidae	Writhed hornbill	Aceros leucocephalus
Pittidae	Azure-breasted pitta	Pitta steerii*
Eurylaimidae	Wattled broadbill	Eurylaimus steerii*
Eurylaimidae	Visayan wattled broadbill	Eurylaimus samarensis*
Rhipiduridae	Blue fantail	Rhipidura superciliaris*
Monarchidae	Short-crested monarch	Hypothymis helenae
Monarchidae	Celestial monarch	Hypothymis coelestis
Muscicapidae	Little slaty flycatcher	Ficedula basilanica*
Muscicapidae	Cryptic flycatcher	Ficedula crypta
Pycnonotidae	Zamboanga bulbul	Ixos rufigularis*
Pycnonotidae	Yellowish bulbul	Ixos everetti
Sylviidae	Long-tailed bush-warbler	Bradypterus caudatus
Sylviidae	Rufous-headed tailorbird	Orthotomus heterolaemus
Sylviidae	Yellow-breasted tailorbird	Orthotomus samarensis*
Sylviidae	White-browed tailorbird	Orthotomus nigriceps*
Sylviidae	White-eared tailorbird	Orthotomus cinereiceps*
Timaliidae	Striated wren-babbler	Ptilocichla mindanensis*
Timaliidae	Pygmy babbler	Stachyris plateni
Timaliidae	Rusty-crowned babbler	Stachyris capitalis
Timaliidae	Brown tit-babbler	Macronous striaticeps
Timaliidae	Miniature tit-babbler	Micromacronus leytensis
Paridae	White-fronted tit	Parus semilarvatus
Dicaeidae	Whiskered flowerpecker	Dicaeum proprium
Dicaeidae	Olive-capped flowerpecker	Dicaeum nigrilore
Dicaeidae	Flame-crowned flowerpecker	Dicaeum anthonyi

An asterisk signifies that the species' range is limited to this ecoregion.

including the critically endangered Philippine eagle (Pithecophaga jeffreyi) and Philippine cockatoo (Cacatua haematuropygia). Four additional widespread but vulnerable species are also found in the ecoregion (Stattersfield et al. 1998; Collar et al. 1999).

The critically endangered Philippine crocodile (Crocodylus mindorensis) was historically found on Jolo, Luzon, Mindoro, Masbate, Samar, Negros, Busuanga, and Mindanao, but the only remaining populations are found on Mindoro, Negros, Mindanao, and Busuanga. The current wild population may be approximately 100 nonhatchlings (Ross 1998).

Mt. Apo, on Mindanao, is considered a Centre of Plant Diversity (Davis et al. 1995). This spectacular mountain in the southern portion of the Central Cordillera contains primary lowland forest and lower montane forests as well as montane forests found in the Mindanao Montane Rain Forests [103] ecoregion. Much of the lowland forest below 1,000 m has been cleared, but dipterocarp forest is found from 1,000 to 1,600 m.

Status and Threats

Current Status

All the islands in the ecoregion were once completely forested, but there is little forest left on most islands, and especially little lowland forest left. The dire situation in the lowlands of Mindanao and Eastern Visayas is highlighted by the contrast in conservation status between the lowland ecoregion and the adjacent upland Mindanao Montane Rain Forests [103] ecoregion. Although the upland ecoregion contains more restricted-range species, only one of these is considered threatened. In fact, the Mindanao and Eastern Visayas EBA contains more threatened birds than any other EBA in the southeast Asian island region, and all but one of these are found in the lowlands (Stattersfield et al. 1998; Collar et al. 1999).

Bohol is heavily deforested, and almost all of the island's natural forest is to be found in Rajah Sikatuna National Park (RSNP). The conditions in this 9,023-ha area are good, however, and the Philippine Department of Natural Resources is actively reforesting the edges of the park. Both of the Eastern Visayan endemic birds and all four of Bohol's endemic bird subspecies can be found in RSNP. Problems of firewood and rattan collection, hunting and trapping, and slash-and-burn agriculture are effectively limited to the eastern portions of the park (Brooks et al. 1995).

Samar and Leyte each have two areas of closed-canopy forest remaining. The largest blocks are found on Samar. Three of these patches are found in areas of suspended timber license agreements and the remaining forest block, on Leyte, is found in the Philippine National Oil Company Tungonan Forest Reserve (Development Alternatives 1992).

By 1988, approximately 29 percent of Mindanao's forest remained, including both primary and secondary forests (Stattersfield et al. 1998). There is much less today. The Zamboanga Peninsula on southwest Mindanao contains a number of isolated fragments, the largest of which is found in the watershed of Zamboanga City. The remaining patches are scattered in hill and montane areas around the peninsula.

These patches contained evidence of recent logging in 1992. In southern Mindanao, some large areas of forest remain in hill and montane areas. Political instability, lack of access, and poor commercial values have helped protect some of these areas. Ironically, some of the areas, which had been under now-suspended timber license agreements, are threatened by encroaching agriculture and fire. There are other large blocks of forest in the rest of Mindanao, but they are similarly limited to hill and montane areas; there is very little lowland dipterocarp forest remaining on Mindanao (Development Alternatives 1992). Southern Mindanao is faced with political instability that poses a challenge for active conservation.

Aerial surveys of Basilan in 1992 revealed less than 2 percent natural forest remaining. Unfortunately, Basilan is also subject to political insurgency that makes active conservation efforts quite difficult (Stattersfield et al. 1998).

Both the Philippine warty pig and Philippine deer suffer from intense hunting pressure and fragmentation of their remaining habitats. The pigs are in an especially poor situation because they tend to raid crops and are regarded as pests; consequently, no protections are in place for them (Oliver 1993). Table J101.3 details the existing protected areas on the island.

Types and Severity of Threats

Many of the factors that have contributed to the loss of habitat in the past still present threats to the future of these forests. They include firewood and rattan collection, hunting and trapping, slash-and-burn agriculture, and commercial forestry (Brooks et al. 1995).

Priority Conservation Actions

Short-Term Conservation Actions (1–5 Years)

- Protect all remaining old-growth forests.
- Expand existing protected areas to include more lowland rain forest and entire watersheds.

Oliver (1993) details a number of priority projects related to the Philippine warty pig. Pigs are known to exist on Bohol Island, but it is still unclear whether these individuals actually represent *Sus philippensis* or the critically endangered *Sus cebifrons* from the Western Visayas. Taxonomic material is needed to determine which species is present. Basilan is also in need of surveys on the status and distribution of *Sus philippensis*. Efforts to assist local conservation organizations in promoting awareness about the importance of the Philippines as a general center of suid endemicity are necessary. Finally, further support for field, taxonomic, and breeding research on Philippine pigs is considered a priority project.

Heaney and Tabaranza (1995) recommended several conservation measures for the Camiguin Island center of endemism, including the following:

- Maintain and expand protection efforts.
- Maintain and expand reforestation efforts geared toward sustainable timber products.
- Cancel existing timber salvage permits to maintain snags and reduce the temptation to kill trees.
- Declare remaining forest areas on the island, in addition to volcanic areas on Mt. Hibok-hibok, as a National Park.

The development of a national crocodile management program, for both the Philippine crocodile and the saltwater crocodile (*Crocodylous porosus*), is listed as a high priority in the Crocodile Status Survey and Conservation Action Plan (Ross 1998). Worldwide coordination of the Philippine crocodile captive breeding program is considered a medium priority (Ross 1998).

Collar (1996) calls for a new conservation strategy for the Philippine eagle, including the preservation and wise use of habitat; research into the ecology, population, and distribution of the species; and a major educational campaign on behalf of the Philippine eagle.

Longer-Term Conservation Actions (5–20 Years)

- Educate the public about watershed protection (L. Heaney, pers. comm., 2000).

Wemmer (1998) recommended several conservation actions related to Philippine deer, including a survey of populations on major islands to determine abundance and threats, reexamination of deer-farming initiatives to determine effects on wild populations, ecological studies of undisturbed populations, and the collection of biological materials for a taxonomic investigation of the species.

Collar et al. (1999) identified nineteen (incomplete and preliminary) key sites that are known to support several threatened bird species and retain sufficient areas of natural habitat for their survival. Some of these key sites are protected to some degree; others are not. A future effort will develop a comprehensive set of "Important Bird Areas" that would be sufficient to conserve the full range of bird species. For the Mindanao and the Eastern Visayas, the key sites are as follows:

- Samar: Mts. Cabalantian and Capoto-an
- Leyte: Mt. Lobi Range
- Dinagat: Mts. Kambinlio and Redondo
- Siargao Island: Siargo Island
- Bohol: Rajah Sikatuna National Park
- Mindanao: Mt. Hilong-hlong, Mt. Diwata, Mt. Dapiak, Mt. Malindang National Park, Agusan Marsh, Mt. Kitanglad National Park, Mt. Sugarloaf, Mt. Agtuuganon,

TABLE J101.3. WCMC (1997) Protected Areas That Overlap with the Ecoregion.

Protected Area	Area (km²)	IUCN Category
Mado Hot Spring	20	III
Sohoton Natural Bridge	40	III
Imelda Lake	40	II
Mahagnao Volcano	30	II
Lake Danao	5	IV
Rajah Sikatuna	110	II
Rizal	10	III
Initao	10	V
Mt. Malindang [106]	160	II
Mt. Apo [106]	130	II
Lake Butig	6	V
Liguasan March GRBS	410	IV
Lake Buluan	80	IV
Agusan Marsh	810	PRO
Basilan	90	II
Total	1,951	

Ecoregion numbers of protected areas that overlap with additional ecoregions are listed in brackets.

Mt. Piapayungan, Mt. Mayo, Mt. Apo National Park, Mt. Matutum, Mt. Three Kings
- Basilan: Central Basilan

Focal Species for Conservation Action

Endangered endemic species: Dinagat gymnure (*Podogymnura aureospinula*) (Heaney et al. 1998), Dinagat bushy-tailed cloud rat (*Crateromys australis*) (Heaney et al. 1998), Mindanao bleeding-heart (*Gallicolumba criniger*) (Collar et al. 1999)

Area-sensitive species: bearded pig (*Sus philippensis*) (Oliver 1993), fruit bats (*Acerodon jubatus, Dyacopterus spadiceus, Haplonycteris fischeri, Harpyionycteris whiteheadi, Megaerops wetmorei, Ptenochirus jagori, Ptenochirus minor, Pteropus hypomelanus, Pteropus leucopterus, Pteropus pumilus, Pteropus speciosus, Pteropus vampyrus, Eonycteris major, Eonycteris spelaea*) (Mickelburgh et al. 1992), writhed hornbill (*Aceros leucocephalus*) (Collar et al. 1999), Samar hornbill (*Penolipides samarensis*), Mindanao hornbill (*Penolipides affinis*) (Collar et al. 1999), Philippine eagle (*Pithecophaga jeffreyi*) (Collar 1996)

Top predator species: palm civet (*Paradoxurus hermaphroditus*) (Heaney et al. 1998), Malay civet (*Viverra tangalunga*) (Heaney et al. 1998), Philippine eagle (*Pithecophaga jeffreyi*) (Collar 1996), Philippine eagle-owl (*Bubo Philippenis*) (Collar et al. 1999), Philippine crocodile (*Crocodylus mindorensis*) (Ross 1998)

Selected Conservation Contacts

Green Mindanao

Haribon Foundation

Mindanao State University, Marawi

Philippine Eagle Foundation

Philippine Endangered Species Project

Protected Areas & Wildlife Bureau, Philippine DENR

Samar Island Biodiversity Project

WWF-Philippines

Justification for Ecoregion Delineation

MacKinnon (1997) identified seven subunits in the Philippines, and the Philippine Biodiversity Action Plan (Philippine BAP 1997) demarcated fifteen biogeographic units. Udvardy (1975) identified the Philippines as a single biogeographic province. We delineated nine ecoregions in the Philippine islands, including Palawan. We deviated from Udvardy (1975), MacKinnon (1997), Stattersfield (1998), and the Philippine BAP (1997) in varying degrees and based our delineation of the Philippine ecoregions on Heaney (1993), with the exception of Camiguin, which Heaney separated.

The islands of Leyte, Samar, Dinagat, and Bohol were combined with the lowland rain forests of Mindanao island to form the Mindanao–Eastern Visayas Rain Forests [101]. We also included the Basilan Islands off the southwest peninsula of Mindanao in this ecoregion, based on Heaney (1993). In Mindanao we used the 1,000-m contour from the DEM (USGS 1996) to delineate the montane forests from the lowland forests. The montane forests of Mindanao were placed into their own ecoregion, the Mindanao Montane Rain Forests [103]. In our delineation of the Mindanao–Eastern Visayas Rain Forests [101] and Mindanao Montane Rain Forests [103] ecoregions, we deviated from MacKinnon (1997). MacKinnon placed both of Mindanao's lowland and montane forests in a single subunit (26c). The Basilan Islands were part of subunit 26d, and the islands of Leyte and Samar made up subunit 26e.

Prepared by John Morrison

Ecoregion Number:	**102**
Ecoregion Name:	**Luzon Montane Rain Forests**
Bioregion:	**Sunda Shelf and Philippines**
Biome:	**Tropical and Subtropical Moist Broadleaf Forests**
Political Units:	**Philippines**
Ecoregion Size:	**8,300 km^2**
Biological Distinctiveness:	**Regionally Outstanding**
Conservation Status:	**Endangered**
Conservation Assessment:	**II**

The ecoregion has suffered from human exploitation but still contains some of the most extensive forests left in the Philippines. Therefore, the montane forests are extremely valuable for the wide range of endemic species they support and for their role in preventing soil erosion and protecting water quality. It is one of the biologically least known ecoregions in the Philippines.

Location and General Description

Luzon is located in the western Pacific Ocean. It is the largest island in the Philippines and lies at the northern end of the island group. The Luzon Montane Rain Forests [102] ecoregion comprises the high elevations of several mountain ranges including the Northern and Southern Sierra Madre, which parallels the northeastern coastline of Luzon. Also included in this ecoregion are Mt. Sapocoy, Mt. Magnas, and Mt. Agnamala in the northern Central Cordillera and the Zambales Mountains in the west.

The geologic history of the Philippines is very complex and has had tremendous influence on the biota found there. Luzon has developed many unique species of plants and animals as a result of its long-standing isolation from other landmasses. Parts of the Luzon highlands were established as a result of volcanic activity and the friction of the Australian and Asian plates at least 15 million years ago. The highlands began to take their current form over the next 10 million years. Luzon is therefore oceanic in character, having never been connected to mainland Asia. Even during the Pleistocene, as world sea levels fell 120 m, Luzon expanded to become a larger island including the modern islands of Polillo, Marinduque, and Catanduanes but never connected to other regions of the Philippines or to mainland Asia (Heaney and Regalado 1998).

Annual rainfall in the ecoregion can be as high as 10,000 mm in some areas, or about quadruple what the Luzon Rain Forests [98] receive. When the rain falls, varies with the mountain ranges. The Sierra Madres are only mildly seasonal, with a dry period occurring from December to April. The mountains of the northern Central Cordillera and the Zambales Mountains are more strongly seasonal, receiving a bit less rainfall and having a longer dry period. These forests are also affected by typhoons that sweep across the South China Sea, hitting the western side of Luzon every few years. These typhoons are a major element of disturbance.

The montane forests of this ecoregion begin at about 1,000 m and are characterized by the appearance of oak and laurel species. These trees gradually replace the dipterocarp trees that dominate at lower elevations. The oaks and laurels do not have the wide buttresses of lower-elevation trees. Montane forests in general are shorter in stature than lowland forests and have less undergrowth. Epiphytes and vines (particularly pandans of the genus *Freycinetia*) and moss-covered branches are very common in the montane forests. The decreased temperature that accompanies increasing elevation slows the decomposition of debris (Heaney and Regalado 1998). This makes the forest floor thick with humus.

The highest elevations of the montane forests sometimes are called upper montane forest or elfin forest. These forests are not treated as a separate ecoregion; rather, they are considered montane forests in extreme. Trees branches appear to be many times thicker than they actually are because the moss covering the branch is so thick. Tree height may be only a few meters, and plants that are not typically epiphytes become aerial because of the thick moisture and abundant organic material on and around trees here. Many of the endemic animal species of the Philippines are found as burrowers in the matty soil of this high-elevation forest (Heaney and Regalado 1998).

Biodiversity Features

The Luzon Montane Rain Forests [102] ecoregion has eight species of near-endemic mammals and one species that is strictly endemic (table J102.1). The strictly endemic Palanan shrew-mouse, *Archboldomys musseri*, is known only from two specimens taken at about 1,650 m from Mt. Cetaceo in the northern Sierra Madres (Danielsen et al. 1994; Heaney et al. 1998). Species listed as threatened (VU or above) include three near-endemic species: Luzon pygmy fruit bat (*Otopteropus cartilagonodus*), Luzon short-nosed rat (*Tryphomys adustus*), and long-nosed Luzon forest mouse (*Apomys sacobianus*).

Five relatively large mammals inhabit the ecoregion: long-tailed macaque (*Macaca fascicularis*), Philippine warty pig (*Sus philippensis*), Philippine brown deer (*Cervus mariannus*), Malay civet (*Viverra tangalunga*), and common palm civet (*Paradoxurus hermaphroditus*). All are fairly widespread, and none are listed as threatened by IUCN (2000). However, habitat destruction affects all these species, and hunting affects all but the civets (Heaney et al. 1998). Additionally,

TABLE J102.1. Endemic and Near-Endemic Mammal Species.

Family	Species
Pteropodidae	*Otopteropus cartilagonodus*
Muridae	*Apomys abrae*
Muridae	*Apomys datae*
Muridae	*Apomys microdon*
Muridae	*Apomys sacobianus*
Muridae	*Archboldomys musseri**
Muridae	*Bullimus luzonicus*
Muridae	*Phloeomys pallidus*
Muridae	*Tryphomys adustus*

An asterisk signifies that the species' range is limited to this ecoregion.

TABLE J102.2. Endemic and Near-Endemic Bird Species.

Family	Common Name	Species
Turnicidae	Spotted buttonquail	*Turnix ocellata*
Turnicidae	Luzon buttonquail	*Turnix worcesteri*
Rallidae	Brown-banded rail	*Lewina mirificus*
Columbidae	Luzon bleeding-heart	*Gallicolumba luzonica*
Columbidae	Flame-breasted fruit-dove	*Ptilinopus marchei*
Columbidae	Cream-breasted fruit-dove	*Ptilinopus merrilli*
Psittacidae	Luzon racquet-tail	*Prioniturus montanus*
Cuculidae	Scale-feathered malkoha	*Phaenicophaeus cumingi*
Cuculidae	Rufous coucal	*Centropus unirufus*
Strigidae	Luzon scops-owl	*Otus longicornis*
Bucconidae	Luzon hornbill	*Penelopides manilloe*
Pittidae	Whiskered pitta	*Pitta kochi*
Laniidae	Grey-capped shrike	*Lanius validirostris*
Turdidae	Ashy thrush	*Zoothera cinerea*
Muscicapidae	Luzon redstart	*Rhyacornis bicolor*
Timaliidae	Golden-crowned babbler	*Stachyris dennistouni*
Timaliidae	Chestnut-faced babbler	*Stachyris whiteheadi*
Sylviidae	Philippine bush-warbler	*Cettia seebohmi*
Sylviidae	Long-tailed bush-warbler	*Bradypterus caudatus*
Muscicapidae	Rusty-flanked jungle-flycatcher	*Rhinomyias insignis*
Muscicapidae	Ash-breasted flycatcher	*Muscicapa randi*
Muscicapidae	Blue-breasted flycatcher	*Cyornis herioti*
Pachycephalida	Green-backed whistler	*Pachycephala albiventris*
Rhabdornithidae	Long-billed rhabdornis	*Rhabdornis grandis*
Dicaeidae	Flame-crowned flowerpecker	*Dicaeum anthonyi*
Fringillidae	White-cheeked bullfinch	*Pyrrhula leucogenis*
Estrildidae	Green-faced parrotfinch	*Erythrura viridifacies*
Oriolidae	White-lored oriole	*Oriolus albiloris*

An asterisk signifies that the species' range is limited to this ecoregion.

the Philippine brown deer is said to be declining and is listed as data deficient by IUCN, although the species is not uncommon in appropriate habitat (Heaney et al. 1998; Wemmer 1998).

The ecoregion contains twenty-eight near-endemic bird species and no strict endemics (table J102.2). At least eleven threatened bird species (IUCN categories VU and above) occur in the ecoregion, and two others may be present, but their distributions are poorly known (brown-banded rail [*Lewinia mirificus*] and Luzon buttonquail [*Turnix worcesteri*]) (Collar et al. 1999). Whiskered pitta (*Pitta kochi*) is a species typical of mossy montane forests (Dickinson et al. 1991) and is listed as vulnerable (Collar et al. 1999). Whiskered pittas usually are found above 1,000 m in the Sierra Madre and Central Cordillera among oaks, 5–12 m high, with a fern and rhododendron understory. They hunt for invertebrate prey on the ground and are known to regularly forage where Philippine warty pigs have rooted over the soil and exposed prey (Poulsen 1995). The bird is sometimes described as uncommon to rare (Dickinson et al. 1991; Kennedy et al. 2000), although in appropriate habitat it may actually be quite common (Poulsen 1995). This discrepancy probably results from undersampling (and general lack of knowledge) of habitat throughout the pitta's range.

Another vulnerable species, typical of montane forests, is the flame-breasted fruit-dove (*Ptilinopus marchei*), a beautiful bird with a crimson orange breast and matching crown. It is the largest fruit-dove in the Philippines, and although it has probably always been uncommon and local, it appears to be particularly sensitive to the threats facing much of the ecoregion's biodiversity. The fruit-dove is not found in areas highly susceptible to habitat destruction, found in neither logged nor selectively logged areas (Poulsen 1995). The fruit-dove is also hunted for food and the pet trade (Collar et al. 1999).

Conservation Status

Current Status

There are no recent estimates of primary forest cover for the Philippines. If one adds the total forest cover for the provinces containing the bulk of the ecoregion, about 482,000 ha of forest remained in 1992 (Development Alternatives 1992). A small portion of the ecoregion's montane forest occurs outside these provinces, yet the overall figure is still too high because much of the forest is below 1,000 m. More importantly, the figure is too high because the amount of forest has declined a great deal since 1992. We know this because forest cover for the same provinces totaled more than a million hectares in the early to mid-1980s, for a decline of 55 percent since 1992, and deforestation has continued, with some slowdown since 1994 (figures based on 1981 and 1984 data, depending on the particular province) (Development Alternatives 1992).

The two largest remaining forested areas in the ecoregion are in the montane portions of the northern Sierra Madres, which have remained inaccessible, and the northern Central Cordillera. The Northern Sierra Madre Natural Park (also known as the Palanan complex or wilderness) is the main focus of conservation in the region, although the park covers only the lowland portions of the forest and should be extended to incorporate the higher elevations as well (Mallari and Jensen 1993; Poulsen 1995) (table J102.3). Other sites in the northern Sierra Madres that have been identified as important areas for biodiversity include Mt. Cetaceo and Mt. Los Dos Cuernos. Neither of these sites receives any formal protection, and both are being cleared (Collar et al. 1999). Recent field work in the northern Central Cordillera has documented extensive forest managed by the traditional cultural groups. Nearly unknown biologically until recently, it is now suspected of being one of the biologically richest and most important areas in the country (Heaney and Mallari in press).

Types and Severity of Threats

Habitat conversion is the primary threat to the ecoregion. Commercial logging (both legal and illegal) continues to have a devastating effect on biodiversity. Conversion of highland areas to "large-scale plantations is currently expanding, causing both displacement of subsistence farmers (who then move further upslope) and increased erosion, which is already a serious problem" (Heaney et al. 1999: 314).

New roads and mining projects directly threaten forests but also make forests more susceptible exploitation as a result of increased accessibility. Subsistence hunting and capture of species for a growing wildlife pet trade adversely affect many of the ecoregion's species.

Priority Conservation Actions

The conservation assessment and plan for the northern Sierra Madres (Danielsen et al. 1994) is excellent. Much of the work straddles two ecoregions but is equally important to both [98, 102]. Sponsored by DENR, BirdLife International, the Zoological Museum of Copenhagen University and the Danish Ornithological Society, the project documents the importance of the biodiversity of the region through careful field surveys. It also puts forward useful recommendations.

Conservation International–Philippines is also in the process of developing a large-scale plan for the entire Sierra Madres.

Short-Term Conservation Actions (1–5 Years)

- Establish biological surveys of the region.
- Protect remaining old-growth forests.
- Expand the existing protected area system to include additional lowland rain forest and to include entire watersheds.
- Increase public education about watershed protection (Heaney, pers. comm., 2000).
- Make montane forests more a part of the NIPA program.

TABLE J102.3. WCMC (1997) Protected Areas That Overlap with the Ecoregion.

Protected Area	Area (km²)	IUCN Category
PNOC 1636 [98]	200	?

Ecoregion numbers of protected areas that overlap with additional ecoregions are listed in brackets.

Longer-Term Conservation Actions (5–20 Years)

- Expand the existing protected area system to include additional lowland rain forest and to include entire watersheds.
- Increase public education about watershed protection (Heaney, pers. comm., 2000).
- Develop sound management plans for protected areas.

Focal Species for Conservation Action

Threatened endemic species: Luzon pygmy fruit bat (*Otopteropus cartilagonodus*) (Mickleburgh et al. 1992), Luzon short-nosed rat (*Tryphomys adustus*), long-nosed Luzon forest mouse (*Apomys sacobianus*), flame-breasted fruit-dove (*Ptilinopus marchei*), whiskered pitta (*Pitta kochi*) (Lambert and Woodcock 1996), ashy thrush (*Zoothera cinerea*), Luzon redstart (*Rhyacornis bicolor*), rusty-flanked jungle-flycatcher (*Rhinomyias insignis*), ash-breasted flycatcher (*Muscicapa randi*), green-faced parrotfinch (*Erythrura viridifacies*) (for all threatened birds see Collar et al. 1999 and Poulsen 1995)

Area-sensitive species: Pteropidae species (Mickleburgh et al. 1992), long-tailed macaque (*Macaca fascicularis*), Philippine warty pig (*Sus philippensis*) (Oliver 1993), Philippine brown deer (*Cervus mariannus*) (Wemmer 1998), Jerdon's baza (*Aviceda jerdoni*), oriental honey-buzzard (*Pernis ptilorhynchus*), barred honey-buzzard (*Pernis celebensis*), crested serpent-eagle (*Spilornis cheela*), Philippine eagle (*Pithecophaga jefferyi*) (Collar et al. 1999), rufous-bellied eagle (*Hieraaetus kienerii*), Philippine hawk-eagle (*Spizaetus philippensis*), rufous hornbill (*Buceros hydrocorax*) (Kemp 1995), Luzon hornbill (*Penelopides manilloe*) (Kemp 1995)

Top predator species: palm civet (*Paradoxurus hermaphroditus*) (Heaney et al. 1998), Malay civet (*Viverra tangalunga*) (Heaney et al. 1998), Jerdon's baza (*Aviceda jerdoni*), oriental honey-buzzard (*Pernis ptilorhynchus*), barred honey-buzzard (*Pernis celebensis*), crested serpent-eagle (*Spilornis cheela*), Philippine eagle (*Pithecophaga jefferyi*) (Collar et al. 1999), rufous-bellied eagle (*Hieraaetus kienerii*), Philippine hawk-eagle (*Spizaetus philippensis*)

Selected Conservation Contacts

CARE-Philippines

Conservation International

Haribon Foundation

Protected Areas & Wildlife Bureau, Philippine DENR

WWF-Philippines

Justification for Ecoregion Delineation

MacKinnon (1997) identified seven subunits in the Philippines, and the Philippine Biodiversity Action Plan (Philippine BAP 1997) demarcated fifteen biogeographic units. Udvardy (1975) identified the Philippines as a single biogeographic province. We delineated nine ecoregions in the Philippine islands, including Palawan. We deviated from Udvardy (1975), MacKinnon (1997), Stattersfield et al. (1998), and the Philippine BAP (1997) to varying degrees and based our delineation of the Philippine ecoregions primarily on Heaney (1993).

In Luzon we delineated three ecoregions, which correspond to MacKinnon's subunit 26a. First, we used the 1,000-m contour from the DEM (USGS 1996) to delineate the montane forests from the lowland forests. The Luzon Montane Rain Forests [102] are made up primarily of the montane moist evergreen forests along the Sierra Madre, northern Central Cordillera, and Zambales mountain ranges. MacKinnon (1997) showed an area of freshwater swamp forests as part of the original vegetation of Luzon Island, which we combined with the remaining lowland forest of Luzon to form the Luzon Rain Forests [98]. These freshwater swamps, in the valley to the east of the Zambales Mountain Range and in the Cagayan River plains, have been converted to rice fields (D. Madulid, pers. comm., 1999). Following Stattersfield et al. (1998) and Dickinson et al. (1991), we placed the Lubang Islands with the Luzon Rain Forests [98]. The Banguet pine *Pinus insularis* (also known as *P. kesiya*)–dominated conifer forests in the Central Cordillera were designated as the Luzon Tropical Pine Forests [106].

Prepared by John Lamoreux

Ecoregion Number:	**103**
Ecoregion Name:	**Mindanao Montane Rain Forests**
Bioregion:	**Sunda Shelf and Philippines**
Biome:	**Tropical and Subtropical Moist Broadleaf Forests**
Political Units:	**Philippines**
Ecoregion Size:	**17,800 km^2**
Biological Distinctiveness:	**Globally Outstanding**
Conservation Status:	**Endangered**
Conservation Assessment:	**I**

This ecoregion features the montane forests on the island of Mindanao. This disjunct ecoregion harbors Philippine warty pigs, Philippine deer, and some of the last strongholds of the charismatic Philippine eagle. Because most of the lowland forest on Mindanao has been cleared, the remaining montane forests are some of the last vestiges of wild Mindanao.

Location and General Description

The climate of the ecoregion is tropical wet (National Geographic Society 1999), with temperature and rainfall modified by the elevation, which reaches up to 2,700 m. There are extensive, disjunct areas of the island above 1,000 m. Mindanao generally is south of the typhoon track of the storms that usually hit the northern Philippines from July to November (Dickinson et al. 1991).

Mindanao and the Visayas have been transported across the western Pacific to their present location during the last 25 million years. Most of these islands have been uplifted above water only in the last 15 million years (Hall and Holloway 1998). During much of the Pleistocene, Mindanao and the Eastern Visayas (Samar, Leyte, and Bohol) were all one island—Greater Mindanao (Heaney 1986)—but the higher elevations of this larger island generally were limited to what is now Mindanao.

Vegetation types in the montane forests of Mindanao consist of hill dipterocarp forests, lower and upper montane forest, elfin woodland (mossy forest), and summit grasslands (Davis et al. 1995). The dominant forest type in Mindanao and the rest of the Philippines was dipterocarp forest. Whereas upper hill dipterocarp forest is found at elevations of 650–1,000, individual dipterocarps occur to 1,500 m, and on Mt. Apo, primary dipterocarp forest occurs from 1,000 to 1,600 m. Upper hill dipterocarp forest on Mt. Apo is dominated by the dipterocarps *Hopea plagata*, *Shorea guiso*, and *Dipterocarpus grandiflorus* and species of *Cinnamomum*, *Lithocarpus*, *Homalanthus*, and *Musa*. There are many epiphytes, mostly ferns and orchids. Tree ferns (*Cyathea*) and palms (*Areca*) are also found in the understory (Davis et al. 1995; Heaney and Regalado 1998).

The transition zone between dipterocarp forest and montane forest includes increasing numbers of tree ferns, pandans, rattans, and *Angiopteris*; the dipterocarps *Shorea almon*, *S. polysperma*, and *Lithocarpus* spp.; and *Agathis philippensis* (Davis et al. 1995).

On Mt. Apo, montane forest occurs above approximately 2,000 m. Dominant genera include *Lithocarpus*, *Cinnamomum*, *Melastoma*, *Caryota*, *Calamus*, *Ficus*, *Agathis*, and numerous Lauraceae (Davis et al. 1995). The mossy upper montane forest generally is found at elevations from 1,200 m to 1,500 m (Davis et al. 1995; Lewis 1988), where humidity is constantly high. This stunted, single-story, moss- and epiphyte-covered forest contains tree ferns up to 10 m high (Dickinson, Kennedy, and Parkes 1991). All surfaces are covered or draped with lichens, bryophytes, begonias, orchids, aroids, *Selaginella*, and *Nephrolepis* ferns (Davis et al. 1995).

Biodiversity Features

Mindanao and its neighbor, Basilan, situated adjacent to the Sulu Archipelago, have been influenced by animal dispersal from Borneo, although in recent millennia movement has been primarily in the other direction (Dickinson et al. 1991). Over the course of the most recent ice ages, the Mindanao faunal region has developed its own unique fauna, with a number of endemic vertebrates.

There is also variation within the island of Mindanao. Thirty-one bird species are polytypic on the island. Sixteen of these variations are based on differences between isolated mountain ranges, and seven species have races associated with the Zamboanga Peninsula and Basilan Island. There are three species that vary between the uplands and lowlands (Dickinson et al. 1991).

Approximately 80 percent of Greater Mindanao's nonvolant mammal species are found nowhere else in the world. Although flying lemurs, tree shrews, tree squirrels, and tarsiers are found on the islands of Greater Mindanao, they are not found on the other large Philippine island of Luzon, just 25 km away from the northern tip of Samar (Heaney and Regalado 1998). More than 30 percent of nonvolant mammals in the ecoregion are endemic to Mindanao only, but the other islands share their species with Mindanao. However, tiny Camiguin and Dinagat islands, located north of Mindanao, contain two and three, respectively, of their own endemic mammals (Heaney 1986; Heaney et al. 1995). Fourteen mammal species are endemic or near endemic to the ecoregion (table J103.1)

An endemic subspecies of Philippine deer (*Cervus mariannus nigricans*) is limited to Mindanao. Philippine deer are widespread (though distributed patchily) in the Philippines, being found on Luzon, Mindoro, Samar, Leyte, Mindanao, and the Basilan Islands. The subspecies (and species) is threatened by habitat loss and hunting (Wemmer 1998).

The Philippine tree shrew (*Urogale everetti*), which is found on Mindanao, Dinagat, and Siargao islands, represents an endemic, monotypic genus. There are sixteen species of tree shrews, a diurnal animal that resembles a squirrel but whose dentition, circulatory system, and large braincase are more like those of primates (Nowak 1999a). This species is considered vulnerable (Hilton-Taylor 2000).

The ecoregion also supports a population of the Philippine warty pig (*Sus philippensis*), which the IUCN considers rare and declining. The Philippine warty pig is widely distributed in the still-forested areas of Luzon, Mindoro, Samar, Leyte, Mindanao, and some of the smaller satellite islands. Many of these forested areas are found in existing national parks. The Philippine warty pig is closely related to *Sus barbatus* of the Greater Sundas and was once thought to be a subspecies, analogous to the Palawan bearded pig (*Sus barbatus ahoenobarbus*). The Philippine warty pig is still threatened by hunting and habitat loss (Oliver 1993).

Greater Mindanao also supports an endemic genus of Erinaceidae, *Podogymura*. There are two moonrat species in this genus, both of which are found in Greater Mindanao and one of which is found in the montane regions of Mindanao. The exclusively montane species *P. truei*, or Mindanao moonrat, is common in montane and mossy forest at elevations from 1,400 m to 2,800 m. Contrary to the IUCN listing, it is not threatened (Heaney et al. 1998; Heaney, pers. comm., 2000).

Montane Mindanao is also home to the endangered Greater Mindanao shrew (*Crocidura grandis*) and the widespread (within the Philippines) but endangered golden-crowned fruit bat (*Acerodon jubatus*) (Heaney et al. 1998).

TABLE J103.1. Endemic and Near-Endemic Mammal Species.

Family	Species
Erinaceidae	*Podogymnura truei**
Soricidae	*Crocidura beatus*
Soricidae	*Crocidura grandis**
Tupaiidae	*Urogale everetti*
Pteropodidae	*Alionycteris paucidentata**
Sciuridae	*Petinomys crinitus**
Sciuridae	*Sundasciurus philippinensis*
Sciuridae	*Exilisciurus concinnus*
Muridae	*Bullimus bagobus*
Muridae	*Limnomys sibuanus**
Muridae	*Tarsomys apoensis**
Muridae	*Batomys salomonseni*
Muridae	*Crunomys suncoides**
Muridae	*Limnomys* sp. B*

An asterisk signifies that the species' range is limited to this ecoregion.

TABLE J103.2. Endemic and Near-Endemic Bird Species.

Family	Common Name	Species
Columbidae	Mindanao brown-dove	Phapitreron brunneiceps
Psittacidae	Mindanao racquet-tail	Prioniturus waterstradti*
Loriidae	Mindanao lorikeet	Trichoglossus johnstoniae*
Strigidae	Mindanao scops-owl	Otus mirus*
Strigidae	Mindanao eagle-owl	Mimizuku gurneyi
Apodidae	Whitehead's swiftlet	Aerodramus whiteheadi
Apodidae	Philippine needletail	Mearnsia picina
Alcedinidae	Blue-capped kingfisher	Actenoides hombroni
Bucconidae	Mindanao hornbill	Penelopides affinis
Bucconidae	Writhed hornbill	Aceros leucocephalus
Rhipiduridae	Black-and-cinnamon fantail	Rhipidura nigrocinnamomea*
Campephagidae	McGregor's cuckoo-shrike	Coracina mcgregori*
Sturnidae	Apo myna	Basilornis miranda*
Muscicapidae	Mindanao jungle-flycatcher	Rhinomyias goodfellowi*
Muscicapidae	Cryptic flycatcher	Ficedula crypta
Laniidae	Mountain shrike	Lanius valdirostris
Zosteropidae	Mindanao white-eye	Lophozosterops goodfellowi*
Zosteropidae	Cinnamon white-eye	Hypocryptadius cinnamomeus*
Sylviidae	Long-tailed bush-warbler	Bradypterus caudatus
Sylviidae	Rufous-headed tailorbird	Orthotomus heterolaemus
Timaliidae	Bagobo babbler	Trichastoma woodi*
Timaliidae	Pygmy babbler	Stachyris plateni
Timaliidae	Rusty-crowned babbler	Stachyris capitalis
Timaliidae	Brown tit-babbler	Macronous striaticeps
Timaliidae	Miniature tit-babbler	Micromacronus leytensis
Estrildidae	Red-eared parrotfinch	Erythrura coloria*
Dicaeidae	Whiskered flowerpecker	Dicaeum proprium
Dicaeidae	Olive-capped flowerpecker	Dicaeum nigrilore
Dicaeidae	Flame-crowned flowerpecker	Dicaeum anthonyi
Nectariniidae	Grey-hooded sunbird	Aethopyga primigenius*
Nectariniidae	Mt. Apo sunbird	Aethopyga boltoni*
Nectariniidae	Linas sunbird	Aethopyga linaraborae*
Fringillidae	Mountain serin	Serinus estherae
Fringillidae	White-cheeked bullfinch	Pyrrhula leucogenis

An asterisk signifies that the species' range is limited to this ecoregion.

This ecoregion overlaps with the Mindanao and the Eastern Visayas EBA, but only the montane portions of it. The EBA contains fifty-one restricted-range birds, twenty-six (or possibly twenty-seven) of which are montane and mossy forest specialists and are thus resident in this ecoregion. All of the restricted-range birds are forest species. There are thirty-four endemic or near-endemic bird species in the ecoregion (Kennedy et al. 2000; table J103.2). Eight of these species are threatened, including the endangered Mindanao bleeding-heart (*Gallicolumba criniger*). This situation should be contrasted with the adjacent lowland Mindanao and Eastern Visayas rain forests. Although it supports fewer restricted range species, the lowland ecoregion contains twelve threatened species (Stattersfield et al. 1998; Collar et al. 1999).

In addition to the restricted-range species, the critically endangered Philippine eagle (*Pithecophaga jeffreyi*) and vulnerable spotted imperial-pigeon (*Ducula carola*) are found in the ecoregion (Stattersfield et al. 1998; Collar et al. 1999).

Mt. Apo on Mindanao is considered a Centre of Plant Diversity (Davis et al. 1995). This spectacular mountain in the southern portion of the Central Cordillera contains primary lowland forest and lower montane forests as well as montane forests found in the Mindanao Montane Rain Forests [103] ecoregion. Much of the lowland forest below 1,000 m has been cleared, but dipterocarp forest is found from 1,000 to 1,600 m.

Status and Threats

Current Status

Although the remaining forest is found in isolated patches, most forest remaining on the island of Mindanao is contained in this upland ecoregion. This is in contrast to the largely deforested lowlands of Mindanao and the Eastern Visayas. The conservation status of restricted-range birds is illustrative in this respect: although the upland ecoregion contains more restricted-range species than the lowlands, only one of the upland species is considered threatened, compared with nine threatened species in the lowlands (Stattersfield et al. 1998; Collar et al. 1999).

By 1988, approximately 29 percent of Mindanao's forest, including both primary and secondary forests, remained (Stattersfield et al. 1998). There is less today (Development Alternatives 1992). The Zamboanga Peninsula on southwest

TABLE J103.3. WCMC (1997) Protected Areas That Overlap with the Ecoregion.

Protected Area	Area (km²)	IUCN Category
Sacred Mountain	10	III
Rungkunan	10	V
Pantuwaraya Lake	8	V
Salikata	10	V
Lake Dapao	20	V
Mt. Kitanglad	250	II
Mt. Apo [104]	420	II
Mainit Hot Spring	30	PRO
Mt. Malindang [104]	300	II
Total	1,058	

Ecoregion numbers of protected areas that overlap with additional ecoregions are listed in brackets.

Mindanao contains a number of isolated fragments, the largest of which is found in the watershed of Zamboanga City. The remaining patches are scattered in hill and montane areas around the peninsula. These patches contained evidence of recent logging in 1992. In southern Mindanao, some large areas of forest remain in hill and montane areas. Political instability, lack of access, and poor commercial values have helped protect some of these areas. Ironically, some of the areas, which had been under now-suspended timber license agreements, are threatened by encroaching agriculture and fire. There are other large blocks of forest in the rest of Mindanao, but they are similarly limited to hill and montane areas (Development Alternatives 1992).

Varying levels of protection have been accorded to the protected areas in the ecoregion; by 1988, approximately 50 percent of Mt. Apo National Park had been deforested. In addition to the classic sequence of logging, invasion by *kaingineros*, and associated hunting and burning, the park is administered by more than one Bureau of Forest Development office and is sometimes occupied by rebel groups (Lewis 1988). Table J103.3 details the existing protected areas in the ecoregion.

Types and Severity of Threats

Habitat destruction is the main threat to biodiversity in the Philippines. Logging and shifting cultivation (*kaingin*) are cited as the primary forces of habitat conversion. Logging takes many forms, from industrial scale to smaller-scale operations that use water buffalo to haul logs out of the forest (Davis et al. 1995).

Both the Philippine warty pig and Philippine deer suffer from intense hunting pressure and fragmentation of their remaining habitats. The pigs are in an especially poor situation because they tend to raid crops and are regarded as pests; consequently, there are no effective protections in place for them (Oliver 1993).

Priority Conservation Actions

Short-Term Conservation Actions (1–5 Years)

- Protect all remaining old-growth forests.
- Expand existing protected areas to include more lowland rain forest and entire watersheds.

Oliver (1993) details a number of priority projects related to the Philippine warty pig. Pigs are known to exist on Bohol Island, but it is still unclear whether these individuals represent *Sus philippensis* or the critically endangered *Sus cebifrons* from the Western Visayas. Taxonomic material is needed to determine which species is present. Basilan is also in need of surveys on the status and distribution of *Sus philippensis*. Efforts to assist local conservation organizations in promoting awareness about the importance of the Philippines as a general center of suid endemicity are necessary. Finally, further support for field, taxonomic, and breeding research on Philippine pigs is considered a priority project.

Collar (1996) calls for a new conservation strategy for the Philippine eagle, including the preservation and wise use of habitat, research into the ecology, population and distribution of the species, and a major educational campaign on behalf of the Philippine eagle.

Provision of a management plan for the parrots of Mt. Kitanglad Range National Park is a priority conservation project recommended in the Parrot Action Plan (Snyder et al. 2000). Two species, the Mindanao lorikeet (*Trichoglossus johnstoniae*) and Mindanao racquet-tail (*Prioniturus waterstradti*), are known from the park, but current information on these species is patchy. Conservation action related to the Philippine cockatoo (*Cacatua haematuropygia*) is applicable to Dinagat and Siargao islands, including a program to raise awareness of the plight of the cockatoo and an export ban of wild-caught birds.

Longer-Term Conservation Actions (5–20 Years)

- Educate the public about watershed protection (L. Heaney, pers. comm., 2000).

Wemmer (1998) recommended several conservation actions related to Philippine deer, including a survey of populations on major islands to determine abundance and threats, reexamination of deer-farming initiatives to determine effects on wild populations, ecological studies of undisturbed populations, and the collection of biological materials for a taxonomic investigation of the species.

Collar et al. (1999) identified nineteen (incomplete and preliminary) key sites that are known to support several threatened bird species and retain sufficient areas of natural habitat for their survival. Some of these key sites are protected to some degree; others are not. A future effort will develop a comprehensive set of "Important Bird Areas" that would be sufficient to conserve the full range of bird species. There are twelve of these sites in the Mindanao Montane Rain Forests [103] ecoregion, most of which partially overlap with hill forests of the Mindanao and Eastern Visayas ecoregion. The key sites are Mt. Hilong-hlong, Mt. Diwata, Mt. Dapiak, Mt. Malindang National Park, Mt. Kitanglad National Park, Mt. Sugarloaf, Mt. Agtuuganon, Mt. Piapayungan, Mt. Mayo, Mt. Apo National Park, Mt. Matutum, and Mt. Three Kings.

Focal Species for Conservation Action

Endangered endemic species: Greater Mindanao shrew (*Crocidura grandis*) (Heaney et al. 1998), Mindanao bleeding-heart (*Gallicolumba criniger*) (Collar et al. 1998)

Area-sensitive species: Philippine warty pig (*Sus philippensis*) (Oliver 1993), Philippine deer (*Cervus mariannus*) (Wemmer 1998), fruit bats (*Acerodon jubatus, Alionycteris paucidentata, Dyacopterus spadiceus, Haplonycteris fischeri, Harpyionycteris whiteheadi, Ptenochirus jagori, Eonycteris major*) (Mickleburgh et al. 1992), Philippine eagle (*Pithecophaga jefferyi*) (Collar 1996), crested serpent-eagle (*Spilornis cheela*), rufous-bellied eagle (*Hieraaetus kienerii*), changeable hawk-eagle (*Spizaetus cirrhatus*), Philippine hawk-eagle (*Spizaetus philippensis*), peregrine falcon (*Falco peregrinus*)

Top predator species: palm civet (*Paradoxurus hermaphroditus*) (Heaney et al. 1998), Malay civet (*Viverra tangalunga*) (Heaney et al. 1998), Philippine eagle (*Pithecophaga jefferyi*) (Collar 1996), crested serpent-eagle (*Spilornis cheela*), rufous-bellied eagle (*Hieraaetus kienerii*), changeable hawk-eagle (*Spizaetus cirrhatus*),

Philippine hawk-eagle (*Spizaetus philippensis*), peregrine falcon (*Falco peregrinus*)

Selected Conservation Contacts

CARE-Philippines
Central Mindanao University, Bukidnon
Green Mindanao
Haribon Foundation
Mindanao State University, Marawi
Philippine Eagle Foundation
Protected Areas & Wildlife Bureau, Philippine DENR
WWF-Philippines

Justification for Ecoregion Delineation

MacKinnon (1997) identified seven subunits in the Philippines, and the Philippine Biodiversity Action Plan (Philippine BAP 1997) demarcated fifteen biogeographic units. Udvardy (1975) identified the Philippines as a single biogeographic province. We delineated nine ecoregions in the Philippine islands, including Palawan. We deviated from Udvardy (1975), MacKinnon (1997), Stattersfield et al. (1998), and the Philippine BAP (1997) to varying degrees and based our delineation of the Philippine ecoregions on Heaney (1993).

The islands of Leyte, Samar, Dinagat, and Bohol were combined with the lowland rain forests of Mindanao Island to form the Mindanao–Eastern Visayas Rain Forests [101]. We also included the Basilan Islands off the southwest peninsula of Mindanao in this ecoregion, based on Heaney (1993). In Mindanao we used the 1,000-m contour from the DEM (USGS 1996) to delineate the montane forests from the lowland forests. The montane forests of Mindanao were placed into their own ecoregion, the Mindanao Montane Rain Forests [103]. In our delineation of the Mindanao–Eastern Visayas Rain Forests [101] and Mindanao Montane Rain Forests [103] ecoregions, we deviated from MacKinnon (1997). MacKinnon placed both of Mindanao's lowland and montane forests in a single subunit (26c). The Basilan Islands were part of subunit 26d, and the islands of Leyte and Samar made up subunit 26e.

Prepared by John Morrison

Ecoregion Number:	**104**
Ecoregion Name:	**Sulu Archipelago Rain Forests**
Bioregion:	**Sunda Shelf and Philippines**
Biome:	**Tropical and Subtropical Moist Broadleaf Forests**
Political Units:	**Philippines**
Ecoregion Size:	**2,000 km²**
Biological Distinctiveness:	**Globally Outstanding**
Conservation Status:	**Critical**
Conservation Assessment:	**IV**

Although these islands represent transitional stepping stones from the island of Borneo to Mindanao in the Philippines, they have evolved their own distinctive faunas. Almost no forest remains on Sulu, and only the eastern portion of Tawitawi is forested. The islands are extremely politically unstable, which exacerbates a difficult conservation situation.

Location and General Description

This ecoregion includes the main islands of Jolo (Sulu) and Tawitawi and the surrounding smaller islands from Sibutu up to but not including Basilan Island. The climate of the ecoregion is tropical wet (National Geographic Society 1999). There are apparently short (two-week) dry seasons in January and May on Tawitawi (Allen 1998). The Sulus are located south of the main typhoon track that so strongly influences the more northerly Philippine islands (Dickinson et al. 1991).

The Philippines are essentially accreted terrains, an accretion of previously isolated island archipelagos that were brought together during the collision and partial subduction of large oceanic tectonic plates.

The precursors of the Sulu Islands were an arc of submarine volcanoes that have existed for at least 25 million years. However, the Sulus were not clearly above-water islands until within the last 15 million years (Hall and Holloway 1998). The islands are low-lying and coralline (limestone). Bongao Peak, on Bongao, reaches 300 m, and Mt. Sibangkok, the highest point on the central ridge that divides Tawitawi, reaches 532 m (Allen 1998).

During the Pleistocene, the majority of the present Sulu Archipelago was one island, separated from Basilan-Mindanao to the north and greater Sibutu (and Borneo) to the south by deepwater channels of 205 m and 290 m depths, respectively. The distances between these ice age islands were not great, however (Heaney 1986).

Vegetation types in the Sulu Archipelago originally included beach forest, lowland rain forest, scrub forest, and mangroves (Stattersfield et al. 1998). Beach forest is composed of *Barringtonia*, *Caesalpinia*, and *Terminalia*. A small patch of this forest type may be found on Simunul, but this is generally an endangered habitat because of coastal development (human habitation and cultivation, coconut plantations). Formerly the most prevalent forest type on the islands (as with the rest of the Philippines), lowland rain forest, or dipterocarp forest, is now mostly cleared. Represented dipterocarp genera include *Anisoptera*, *Dipterocarpus*, *Hopea*, and *Shorea*. Little information is available about the native scrub forest, which has been extensively cleared as well.

Mangroves are found on the coasts throughout the Archipelago but are especially extensive on Tawitawi; mangroves around Bongao have been cleared. The principal mangrove genera include *Rhizophora, Ceriops, Brugueira, Sonneratia, Avicennia,* and *Nypa* (palms) (Allen 1998).

Biodiversity Features

Unlike that of Palawan, which is located between Borneo and the Philippines, the Sulu Archipelago's fauna is not Sundaic (Allen 1998) and, though rather small, is poorly known biologically (L. Heaney, pers. comm., 2000). Palawan was the main pathway for immigrants from Borneo to the Philippines, and the Sulus have many taxa that are identical to or derived from taxa in Mindanao. Even Sibutu, close to Borneo and separated from the rest of the Sulus by the Sibutu Passage, contains an avifauna more closely related to the Sulus than to Borneo (Dickinson et al. 1991). Although there are some Sulu birds with Sundaic distributions, the avifauna of the Archipelago is essentially Philippine (Dutson et al. 1992). The Sulu hornbill (*Anthracoceros montani*) is one example of an animal whose likely closest relative, the black hornbill (*Anthracoceros malayanus*), is from Borneo. There is a cline of relatedness to Borneo as one moves north among the islands. Sibutu contains birds of Bornean origin that are not found on Tawitawi (Allen 1998). The Sulus (Sangasanga, Bongao, Simunul, Tawitawi) also support a population of slow loris (*Nycticebus coucang*), a Sundaic primate that is not found in the remainder of the Philippines (Heaney 1986). There is one endemic mammal in the ecoregion (table J104.1). The Tawitawi Island rat (*Rattus taitawiensis*) is considered vulnerable (Hilton-Taylor 2000).

TABLE J104.1. Endemic and Near-Endemic Mammal Species.

Family	Species
Muridae	*Rattus tawitawiensis**

An asterisk signifies that the species' range is limited to this ecoregion.

TABLE J104.2. Endemic and Near-Endemic Bird Species.

Family	Common Name	Species
Columbidae	Sulu bleeding-heart	*Gallicolumba menagei**
Columbidae	Dark-eared dove	*Phapitreron cinereiceps**
Columbidae	Grey imperial-pigeon	*Ducula pickeringii*
Psittacidae	Blue-winged racquet-tail	*Prioniturus verticalis**
Strigidae	Mantanani scops-owl	*Otus mantananensis*
Apodidae	Philippine needletail	*Mearnsia picina*
Bucconidae	Sulu hornbill	*Anthracoceros montani**
Monarchidae	Celestial monarch	*Hypothymis coelestis*
Pycnonotidae	Yellowish bulbul	*Ixos everetti*
Timaliidae	Brown tit-babbler	*Macronous striaticeps*

An asterisk signifies that the species' range is limited to this ecoregion.

A new pig species, *Sus spp. nov.*, is being described from the Sulus on the basis of MtDNA and skull measurements from a dead specimen (Rose and Grubb, in prep.). The same subspecies of bearded pig found on Borneo (*Sus barbatus barbatus*) is also found in the southwestern Sulus (Sibutu and Tawitawi), and this species can still be observed crossing open water to reach these islands from Borneo. It is unknown whether these over-water migrations are related to periodic eruptions on the Bornean mainland (Oliver 1993).

This ecoregion overlaps exactly with the Sulu Archipelago EBA. The EBA contains nine restricted-range birds, four of which are limited to the Sulus. All the restricted-range birds are forest species. Ten bird species qualify as endemic or near endemic to this ecoregion (Kennedy et al. 2000; table J104.2). Included in the ecoregion are the critically endangered Sulu bleeding-heart (*Gallicolumba menagei*), Tawitawi brown-dove (*Phapitreron cinereiceps*), and Sulu hornbill (*Anthracoceros montani*) and the endangered blue-winged racquet-tail (*Prionoturus verticalis*). Several endemic bird subspecies may warrant elevation to species status upon detailed review (Stattersfield et al. 1998; Collar et al. 1999).

Several widespread but threatened species also occur on the islands, including the critically endangered Philippine cockatoo (*Cacatua haematuropygia*) and vulnerable rufous-lored kingfisher (*Todirhamphus winchelli*) (Collar et al. 1999; Stattersfield et al. 1998).

The critically endangered Philippine crocodile (*Crocodylus mindorensis*) was historically found on Jolo (as well as Luzon, Mindoro, Masbate, Samar, Negros, Busuanga, and Mindanao), but the only remaining populations are found on Mindoro, Negros, Mindanao, and Busuanga. The current wild population may be approximately 100 nonhatchlings (Ross 1998).

Status and Threats

Current Status

There is almost no forest remaining on Jolo (Sulu) Island, and only the eastern and north-central portions of Tawitawi are forested (Stattersfield et al. 1998; Allen 1998). The majority of Tawitawi was selectively logged in the 1960s and early 1970s (Allen 1998). Apparently, there were plans to replace the remaining forests of Tawitawi with oil palm plantations. The situation on the smaller islands is mixed. Sibutu and Simunul have been largely cleared (Stattersfield et al. 1998). Simunul has some patches of forest remaining that support populations of Philippine cockatoo (*Cacatua haematuropygia*), blue-naped parrot (*Tanygnathus lucionensis*), and blue-backed parrot (*T. sumatranus*) (Dutson et al. 1992). Sibutu has considerable secondary forest, and the island supports numerous Sulu subspecies. The last forests of Sangasanga were cleared in 1992–1993. The island of Bongao still supports forests; an unidentified jungle flycatcher collected in 1973 has not been observed since (Dutson et al. 1992). Small islands in the Tandubas group (which also includes Tawitawi and Sangasanga) still have small forest tracts that reportedly maintain populations of the endemic Sulu bleeding-heart and Sulu hornbill (Stattersfield et al. 1998).

The main population center is on Bongao, where a busy port exists. Tawitawi is not heavily populated, but future economic development on the island is a concern (Allen 1998; Stattersfield et al. 1998). Table J104.3 details the existing protected area on the islands.

Types and Severity of Threats

In general, habitat loss is the main threat to wildlife, but hunting is also a problem. Small-scale logging continues to destroy the remaining habitat (Stattersfield et al. 1998).

Priority Conservation Actions

Short-Term Conservation Actions (1–5 Years)

- Protect all remaining old-growth forests.
- Expand existing protected areas to include more lowland rain forest and entire watersheds.
- This ecoregion is extremely poorly understood biologically, and there is an urgent need for biological surveys when this becomes feasible (L. Heaney, pers. comm., 2000).
- The development of a national crocodile management program for both the Philippine crocodile and the saltwater crocodile (*Crocodylous porosus*) is listed as a high priority in the Crocodile Status Survey and Conservation Action Plan (Ross 1998). Worldwide coordination of the Philippine crocodile captive breeding program is considered a medium priority (Ross 1998).

Longer-Term Conservation Actions (5–20 Years)

- Educate the public about watershed protection (L. Heaney, pers. comm., 2000).

Collar et al. (1999) identified (incomplete and preliminary) key sites that are known to support several threatened bird species and retain sufficient areas of natural habitat for their survival. Some of these key sites are protected to some degree; others are not. A future effort will develop a comprehensive set of "Important Bird Areas" that would be sufficient to conserve the full range of bird species. For the Sulu Archipelago, the key sites are Tawitawi Island and Sibutu and Tumindao Islands. Tawitawi is part of a proposed protected area.

Focal Species for Conservation Action

Endangered endemic species: Sulu hornbill (*Anthracoceros montani*) (Collar et al. 1999), Sulu bleeding-heart (*Gallicolumba menagei*) (Collar et al. 1999), Tawitawi brown-dove (*Phapitreron cinereiceps*) (Collar et al. 1999), blue-winged racquet-tail (*Prionoturus verticalis*) (Collar et al. 1999)

Area-sensitive species: bearded pig (*Sus barbatus barbatus*) (Oliver 1993), slow loris (*Nycticebus coucang*), fruit bats (*Acerodon jubatus, Cynopterus brachyotis, Ptenochirus jagori, Pteropus speciosus, Pteropus vampyrus, Rousettus amplexicaudatus, Eonycteris spelaea, Macroglossus minimus*) (Mickleburgh et al. 1992), Sulu hornbill (*Anthracoceros montani*)

Top predator species: palm civet (*Paradoxurus hermaphroditus*) (Heaney et al. 1998), Oriental honey-buzzard (*Pernis ptilorhynchus*), barred honey-buzzard (*Pernis celebensis*), crested serpent-eagle (*Spilornis cheela*), Philippine crocodile (*Crocodylus mindorensis*)

Selected Conservation Contacts

Protected Areas & Wildlife Bureau, Philippine DENR
WWF-Philippines

Justification for Ecoregion Delineation

MacKinnon (1997) identified seven subunits in the Philippines, and the Philippine BAP (Philippine DENR & UNEP 1997) demarcated fifteen biogeographic units. Udvardy (1975) identified the Philippines as a single biogeographic province. We delineated nine ecoregions in the Philippine islands, including Palawan. We deviated from Udvardy (1975), MacKinnon (1997), Stattersfield et al. (1998), and the Philippine BAP (Philippine DENR & UNEP 1997) to varying degrees and based our delineation of the Philippine ecoregions on Heaney (1993).

The islands of the Sulu Archipelago were delineated as a separate ecoregion, the Sulu Archipelago Rain Forests [104]. This ecoregion includes the Tawitawi Group, Tapul Group, Jolo Group, and Samales Group of islands. These islands, with a lowland moist or semi-evergreen moist forest vegetation (Whitmore 1984), are also an EBA (Stattersfield et al. 1998) and have been identified as a distinct biounit by MacKinnon (1997) and a biogeographic zone by the Philippine BAP (Philippine DENR & UNEP 1997).

Prepared by John Morrison

Ecoregion Number:	**105**
Ecoregion Name:	**Sumatran Tropical Pine Forests**
Bioregion:	**Sunda Shelf and Philippines**
Biome:	**Tropical and Subtropical Coniferous Forests**
Political Units:	**Indonesia**
Ecoregion Size:	**2,700 km²**
Biological Distinctiveness:	**Regionally Outstanding**
Conservation Status:	**Vulnerable**
Conservation Assessment:	**II**

Pine forests are not a vegetation type one would expect to find in a tropical region, but in a small area of mainly northern Sumatra, this is the dominant vegetation. The Sumatran Tropical Pine Forests [105] are not as species-rich as the surround montane forests but do contain similar species as well as those adapted to the vegetation.

Location and General Description

This ecoregion represents the tropical pine forests in northern Sumatra near Lake Toba and along the Barisan Mountain

---

TABLE J104.3. WCMC (1997) Protected Areas That Overlap with the Ecoregion.

Protected Area	Area (km²)	IUCN Category
Mt. Dajo	40	IV

Ecoregion numbers of protected areas that overlap with additional ecoregions are listed in brackets.

Range. These forests occur within the montane zone of Sumatra. About 150 million years ago Borneo, Sumatra, and western Sulawesi split off from Gondwanaland and drifted north. Around 70 million years ago India slammed into the Asian landmass, forming the Himalayas, and an associated thrust formed Sumatra's Barisan Mountains, which run the length of Sumatra (Whitten et al. 1987).

Based on the Köppen climate zone system, Sumatra falls in the tropical wet climate zone (National Geographic Society 1999). The montane forests of the Barisan Range receive more rainfall on their western slopes than their eastern slopes, which are in a rainshadow. However, most of Sumatra experiences less than three consecutive months of dry weather (less than 100 mm rainfall/month), and rainfall in the montane rain forests averages more than 2,500 mm/year (Whitten et al. 1987). However, the pines exploit the drier areas in the mountain range, mostly on the eastern slopes.

It is in the drier areas that forests are dominated by the Sumatran pine (*Pinus merkusii*). This species originally was an early pioneer of disturbed land (such as landslides). However, repeated burning of the montane forests by natural and human-made disturbance has caused thick pine forests with a pauce ground layer to become established.

Biodiversity Features

The flora and fauna of the pine forests are not as diverse as those of the surrounding montane or lowland rain forests. There are no endemic or near-endemic mammals in this ecoregion. Only 3 to 4 percent of the bird species found in the surrounding rain forests were also found in pine forests. A large majority of the bird species found in these forests are common to disturbed or secondary forests. Twelve near-endemic bird species are attributed to this ecoregion (table J105.1).

TABLE J105.1. Endemic and Near-Endemic Bird Species.

Family	Common Name	Species
Phasianidae	Bronze-tailed peacock-pheasant	Polyplectron chalcurum
Campephagidae	Sunda minivet	Pericrocotus miniatus
Irenidae	Blue-masked leafbird	Chloropsis venusta
Turdidae	Shiny whistling-thrush	Myiophonus melanurus
Muscicapidae	Rufous-vented niltava	Niltava sumatrana
Muscicapidae	Sunda robin	Cinclidium diana
Pycnonotidae	Cream-striped bulbul	Pycnonotus leucogrammicus
Pycnonotidae	Spot-necked bulbul	Pycnonotus tympanistrigus
Pycnonotidae	Sunda bulbul	Hypsipetes virescens
Zosteropidae	Black-capped white-eye	Zosterops atricapillus
Timaliidae	Sunda laughingthrush	Garrulax palliatus
Timaliidae	Rusty-breasted wren-babbler	Napothera rufipectus

An asterisk signifies that the species' range is restricted to this ecoregion.

Conservation Status

Current Status

The pine forests are found in montane areas, and large portions of the ecoregion are within two national parks, Kerinci Seblat and Lingga Isaq (table J105.2). This ecoregion burns frequently from anthropogenic and natural causes.

Types and Severity of Threats

The pine forests are under much less threat than the lowlands and surrounding montane forests. More than a third of the ecoregion is in protected areas, and the ecoregion lacks high-value dipterocarp tree species. Therefore, these forests are under less threat from logging than the surrounding landscape.

Priority Conservation Actions

With the rapid pace of forest destruction under way in Indonesia, quick conservation action is urgently needed to stave off the complete collapse of these forests. However, steps should be taken to lay the groundwork for longer-term conservation as well.

Short-Term Conservation Actions (1–5 Years)

- Initiate a major public awareness campaign on forest values and forest loss to drive political will.
- Revoke forest concessions in areas of high biodiversity outside protected areas and institute a moratorium on new concessions for oil palm, timber, and other plantations using instead the vast areas of deforested and cleared land.
- Mobilize major assistance for the more outstanding protected areas.

Longer-Term Conservation Actions (5–20 Years)

- Devise ways to maintain integrity of buffer zones of protected areas.
- Implement rigorous conservation actions in the Level I and Level II TCUs in the ecoregion.

Focal Species for Conservation Action

Other: *Pinus merkusii*

Selected Conservation Contacts

Lembaga Alam Tropika Indonesia (LATIN)

WWF-Indonesia

Yayasan Keanekaragaman Hayati

Yayasan Konsorsium Pembaruan Agraria (KPA)

TABLE J105.2. WCMC (1997) Protected Areas That Overlap with the Ecoregion.

Protected Area	Area (km²)	IUCN Category
Lingga Isaq [83]	230	VI
Kerinci Seblat [83]	760	II
Total	990	

Ecoregion numbers of protected areas that overlap with additional ecoregions are listed in brackets.

Justification for Ecoregion Delineation

The *Pinus merkusii*–dominated conifer forests along the Gunung Leuser Range are shown and identified as a distinct ecoregion, the Sumatran Tropical Pine Forests [105]. These forests are the only stands of *Pinus* found south of the equator (Whitmore 1984). Another tropical pine forest, the Luzon Tropical Pine Forests [106] ecoregion, also occurs in the Philippines, and it developed under similar conditions. MacKinnon's biounit 21 largely corresponds to Udvardy's Sumatra biogeographic province. However, Udvardy did not include the Nicobar Islands. There are eight ecoregions that overlap Udvardy's Sumatra biogeographic province: Sumatran Lowland Rain Forests [82], Sumatran Montane Rain Forests [83], Mentawai Islands Rain Forests [84], Sumatran Peat Swamp Forests [85], Sumatran Freshwater Swamp Forests [88], Sundaland Heath Forests [90], Sumatran Tropical Pine Forests [105], and Sunda Shelf Mangroves [107].

MacKinnon's biounit 21 largely corresponds to Udvardy's Sumatra biogeographic province. However, Udvardy did not include the Nicobar Islands. Eight ecoregions overlap Udvardy's Sumatra biogeographic province: Sumatran Lowland Rain Forests [82], Sumatran Montane Rain Forests [83], Mentawai Islands Rain Forests [84], Sumatran Peat Swamp Forests [85], Sumatran Freshwater Swamp Forests [88], Sundaland Heath Forests [90], Sumatran Tropical Pine Forests [105], and Sunda Shelf Mangroves [107].

Prepared by Colby Loucks and Tony Whitten

Ecoregion Number:	**106**
Ecoregion Name:	**Luzon Tropical Pine Forests**
Bioregion:	**Sunda Shelf and Philippines**
Biome:	**Tropical and Subtropical Coniferous Forests**
Political Units:	**Philippines**
Ecoregion Size:	**7,100 km²**
Biological Distinctiveness:	**Globally Outstanding**
Conservation Status:	**Endangered**
Conservation Assessment:	**I**

The mountainous, often cloudy Luzon Pine Forests [106] are a unique habitat in the Philippines. Regular fires have led to a parkland structure of grass with widely dispersed trees and prevented broadleaf trees from establishing over large areas. However, more typical montane forest covers so much of the ecoregion that some experts would lump the pine forests with the Luzon Montane Rain Forest [102] ecoregion. The ecoregion has been exploited for its trees and mineral resources in the past; shifting agriculture and mining are the greatest current threats.

Location and General Description

Luzon is located in the western Pacific Ocean. It is the largest island in the Philippines and lies at the northern end of the island group. The Luzon Pine Forests [106] ecoregion is entirely within the Central Cordillera Mountain Range of northwestern Luzon and includes all regions above 1,000 m, except a large tract of montane forest at the northern end (see ecoregion 102). Included in this ecoregion are some of Luzon's highest mountain peaks: Mt. Puguis, Mt. Polis, Mt. Data, and Mt. Pulog.

The geologic history of the Philippines is very complex and has had tremendous influence on the biota currently found there. Luzon has developed many unique species of plants and animals as a result of its long-standing isolation from other landmasses. Parts of the Luzon highlands were established as a result of volcanic activity and the friction of the Australian and Asian plates at least 15 million years ago. The highlands began to take their current form over the next 10 million years. Luzon is therefore oceanic in character, having never been connected to mainland Asia. Even during the Pleistocene, as world sea levels fell 120 m, Luzon expanded to become a larger island including the modern islands of Polillo, Marinduque, and Catanduanes but never connecting to other regions of the Philippines or to mainland Asia (Heaney and Regalado 1998).

The ecoregion receives about 2,500 mm of rain a year (even more than 4,000 mm on Mt. Pulog in some years), but this rain is highly seasonal. Often the pine forests of Luzon are categorized as monsoon forests because of the pronounced dry period (November–April) between rain (May–August, with July and August being the wettest). The temperature of the pine forests averages about 20°C and rarely exceeds 26°C. The forests are affected by typhoons that sweep across the South China Sea, hitting the western side of Luzon every few years. These typhoons are a major element of disturbance.

Most of the ecoregion's area is covered by grassland that is interspersed with trees. The pronounced dry periods and periodic fires favor the Benguet pine or *saleng* (*Pinus insularis*, also known as *P. kesiya*). This species ranges in elevation between 1,000 and 2,500 m and is also present in mainland Asia. However, montane forest interdigitates with the pine forests and covers much of the ecoregion's surface area. This is particularly true in Balbalasang-Balbalan protected area, which is nearly all montane forest. Species lists for the ecoregion therefore include many animals that are limited to montane forests within the ecoregion and are not even found in pine forests.

Biodiversity Features

The ecoregion is inhabited by nine near-endemic mammal species and seven strictly endemic mammals (table J106.1). All but one of these species belongs to the mouse and rat family, Muridae, which has undergone extensive adaptive radiation on Luzon. Cloud rats (genera *Crateromys* and *Phloeomys*) occur only in the northern Philippines and are perhaps the most unique and characteristic of mammals here. These large, bushy-tailed creatures look more like squirrels than typical rats (Heaney and Regalado 1998). All are subject to heavy hunting pressures (Heaney et al. 1998). Three relatively large mammals occur in the ecoregion: long-tailed macaque (*Macaca fascicularis*), Philippine warty pig (*Sus philippensis*), and Malay civet *Viverra tangalunga*). All are fairly widespread, and none are listed as threatened by IUCN (Hilton-Taylor 2000). However, habitat destruction affects all three species, and hunting affects all but the Malay civet.

The Luzon Pine Forests [106] are home to twenty-three near-endemic birds, none of which are strictly limited to the ecoregion (table J106.2). Perhaps the bird species most characteristic of the ecoregion is also one of its most widespread worldwide. Red crossbill (*Loxia curvirostra*), known primarily from high-latitude coniferous forests, reaches its southernmost extent in the Old World in the mountains of the Central Cordillera (the populations in Central America, notably Nicaragua, are further south). Dickinson et al. (1991) listed eleven other bird species that typify the pine forests, including one tit (*Parus elegans*), a nuthatch (*Sitta frontalis*), and thrush (*Turdus poliocephalus*). These groups of birds are familiar to many people of northern latitudes and are indicative of the unique habitat this ecoregion represents within the Philippines.

Conservation Status

Current Status

The status of the ecoregion is hard to assess from published sources. Pines from the ecoregion have been exploited for a long time. Resin from the pines provided an important commercial source of turpentine during the Spanish colonial period (Heaney and Regalado 1998).

One of the protected areas in the ecoregion, Mt. Pulog, is the highest peak in Luzon (table J106.3). Mt. Pulog is important globally as a center of plant diversity with many local endemics (Davis et al. 1995) and as a key site for threatened birds, with six (Collar et al. 1999). The mountain, like much of the ecoregion, is valuable for water quality in Luzon. The size of the park (11,500 ha) may be adequate, although there was initial difficulty in demarcating the park boundary when it was established in 1987 (Davis et al. 1995). The park continues to have problems with agricultural encroachment and wildlife exploitation (Collar et al. 1999). Mt. Polis is also listed as a key site for threatened birds, with six species having been recorded. Unfortunately, little forest remains on Mt. Polis, and it receives no legal protection (Collar et al. 1999).

TABLE J106.1. Endemic and Near-Endemic Mammal Species.

Family	Species
Pteropodidae	Otopteropus cartilagonodus
Muridae	Abditomys latidens
Muridae	Apomys abrae
Muridae	Apomys datae
Muridae	Apomys sacobianus
Muridae	Batomys dentatus*
Muridae	Batomys granti
Muridae	Bullimus luzonicus
Muridae	Carpomys melanurus*
Muridae	Carpomys phaeurus*
Muridae	Celaenomys silaceus*
Muridae	Chrotomys whiteheadi*
Muridae	Crateromys schadenbergi*
Muridae	Phloeomys pallidus
Muridae	Rhynchomys soricoides*
Muridae	Tryphomys adustus

An asterisk signifies that the species' range is limited to this ecoregion.

TABLE J106.2. Endemic and Near-Endemic Bird Species.

Family	Common Name	Species
Turnicidae	Luzon buttonquail	Turnix worcesteri
Rallidae	Brown-banded rail	Lewina mirificus
Columbidae	Luzon bleeding-heart	Gallicolumba luzonica
Columbidae	Flame-breasted fruit-dove	Ptilinopus marchei
Psittacidae	Luzon racquet-tail	Prioniturus montanus
Cuculidae	Scale-feathered malkoha	Phaenicophaeus cumingi
Strigidae	Luzon scops-owl	Otus longicornis
Apodidae	Whitehead's swiftlet	Aerodramus whiteheadi
Pittidae	Whiskered pitta	Pitta kochi
Laniidae	Grey-capped shrike	Lanius validirostris
Turdidae	Ashy thrush	Zoothera cinerea
Muscicapidae	Luzon redstart	Rhyacornis bicolor
Timaliidae	Golden-crowned babbler	Stachyris dennistouni
Timaliidae	Chestnut-faced babbler	Stachyris whiteheadi
Sylviidae	Philippine bush-warbler	Cettia seebohmi
Sylviidae	Long-tailed bush-warbler	Bradypterus caudatus
Muscicapidae	Rusty-flanked jungle-flycatcher	Rhinomyias insignis
Muscicapidae	Ash-breasted flycatcher	Muscicapa randi
Pachycephalida	Green-backed whistler	Pachycephala albiventris
Rhabdornithidae	Long-billed rhabdornis	Rhabdornis grandis
Dicaeidae	Flame-crowned flowerpecker	Dicaeum anthonyi
Fringillidae	White-cheeked bullfinch	Pyrrhula leucogenis
Oriolidae	White-lored oriole	Oriolus albiloris

An asterisk signifies that the species' range is limited to this ecoregion.

Types and Severity of Threats

The population growth of the Philippines and the extreme poverty of many has forced people to cultivate land at increasingly high altitudes. Population growth and rural poverty are definitely threats to the ecoregion. More typical threats to the biodiversity of the ecoregion exist but are perhaps symptoms of these overarching problems.

Fire and habitat conversion are the greatest threats, and they are in some ways the same problem. Fire is a natural process in the ecoregion, but human-induced fire is being used to clear extensive areas for growing vegetables and cut flowers. The increased number of fires limits regeneration of some forest plants; it also makes the region susceptible to invasive grasses (Davis et al. 1995). Logging (both legal and illegal) is a threat everywhere in the Philippines and is another means of habitat destruction.

Mining is an ever-present threat. The Central Cordillera is famous for its mineral wealth, including copper and gold. In fact, most of the mountain range is included within mining

TABLE J106.3. WCMC (1997) Protected Areas That Overlap with the Ecoregion.

Protected Area	Area (km²)	IUCN Category
Balbalasang-Balbalan	160	VI
Mt. Data	300	V
Bessang Pass	10	III
Mts. Pulog, Nueva Vizcaya, Ifugao, Benguet	80	II
Total	550	

Ecoregion numbers of protected areas that overlap with additional ecoregions are listed in brackets.

application and exploration areas. Wildlife exploitation is also a serious problem in many parts with species being sought for trade and consumption.

Priority Conservation Actions

Short-Term Conservation Actions (1–5 Years)

- Establish biological surveys of the region.
- Protect remaining old-growth forests.
- Expand the existing protected area system to include additional lowland rain forest and entire watersheds.
- Increase public education about watershed protection (Heaney, pers. comm., 2000).

Longer-Term Conservation Actions (5–20 Years)

- Expand the existing protected area system to include additional lowland rain forest and entire watersheds.
- Increase public education about watershed protection (Heaney, pers. comm., 2000).
- Develop sound management plans for protected areas.

Focal Species for Conservation Action

Threatened endemic species: Luzon pygmy fruit bat (*Otopteropus cartilagonodus*), Luzon short-nosed rat (*Tryphomys adustus*), long-nosed Luzon forest mouse (*Apomys sacobianus*), Luzon bushy-tailed cloud rat (*Crateromys schadenbergi*), Luzon montane striped shrew-rat (*Chrotomys whiteheadi*), flame-breasted fruit-dove (*Ptilinopus marchei*), whiskered pitta (*Pitta kochi*) (Lambert and Woodcock 1996), ashy thrush (*Zoothera cinerea*), Luzon redstart (*Rhyacornis bicolor*), rusty-flanked jungle-flycatcher (*Rhinomyias insignis*), ash-breasted flycatcher (*Muscicapa randi*)

Area-sensitive species: Fourteen Pteropidae species (Mickleburgh et al. 1992), long-tailed macaque (*Macaca fascicularis*), Philippine warty pig (*Sus philippensis*) (Oliver 1993), oriental honey-buzzard (*Pernis ptilorhynchus*), barred honey-buzzard (*Pernis celebensis*), crested serpent-eagle (*Spilornis cheela*), rufous-bellied eagle (*Hieraaetus kienerii*), Philippine hawk-eagle (*Spizaetus philippensis*), rufous hornbill (*Buceros hydrocorax*) (Kemp 1995)

Top predator species: Malay civet (*Viverra tangalunga*) (Heaney et al. 1998), oriental honey-buzzard (*Pernis ptilorhynchus*), barred honey-buzzard (*Pernis celebensis*), crested serpent-eagle (*Spilornis cheela*), rufous-bellied eagle (*Hieraaetus kienerii*), Philippine hawk-eagle (*Spizaetus philippensis*)

Selected Conservation Contacts

Haribon Foundation
WWF-Philippines

Justification for Ecoregion Delineation

MacKinnon (1997) identified seven subunits in the Philippines, and the Philippine BAP (1997) demarcated fifteen biogeographic units. Udvardy (1975) identified the Philippines as a single biogeographic province. We delineated nine ecoregions in the Philippine islands, including Palawan. We deviated from Udvardy (1975), MacKinnon (1997), Stattersfield et al. (1998), and the Philippine BAP (1997) to varying degrees and based our delineation of the Philippine ecoregions primarily on Heaney (1993).

In Luzon we delineated three ecoregions, which correspond to MacKinnon's subunit 26a. First, we used the 1,000-m contour from the DEM (USGS 1996) to delineate the montane forests from the lowland forests. The Luzon Montane Rain Forests [102] are made up primarily of the montane moist evergreen forests along the Sierra Madre, northern Central Cordillera, and Zambales mountain ranges. MacKinnon (1997) shows an area of freshwater swamp forests as part of the original vegetation of Luzon Island, which we combined with the remaining lowland forest of Luzon to form the Luzon Rain Forests [98]. These freshwater swamps, in the valley to the east of the Zambales Mountain Range and in the Cagayan River plains, have been converted to rice fields (D. Madulid, pers. comm., 1999). Following Stattersfield et al. (1998) and Dickinson et al. (1991), we placed the Lubang Islands with the Luzon Rain Forests [98]. The conifer forests in the Central Cordillera dominated by Banguet pine (*Pinus insularis*, also known as *P. kesiya*) were designated as the Luzon Tropical Pine Forests [106].

Prepared by John Lamoreux

Ecoregion Number: **107**
Ecoregion Name: **Sunda Shelf Mangroves**
Bioregion: **Sunda Shelf and Philippines**
Biome: **Mangroves**
Political Units: **Indonesia, Malaysia, Brunei**
Ecoregion Size: **37,200 km²**
Biological Distinctiveness: **Globally Outstanding**
Conservation Status: **Critical**
Conservation Assessment: **I**

The Sunda Shelf Mangroves [107] are some of the most biologically diverse mangroves in the world. They are home to the unique proboscis monkey. Like other mangrove forests in the region, they are under intense threats from logging, shrimp farming, and agriculture conversion.

Location and General Description

The Sunda Shelf Mangroves [107] are found on the island of Borneo and the east coast of Sumatra. The climate and physical conditions vary widely in this region, giving rise to a high diversity of plant and animal species found in these forests.

However, the region generally has high humidity, seasonal wind and precipitation, high temperatures, and high annual rainfall. Tidal fluctuations have large variations over short distances (Spalding et al. 1997).

There are five major mangrove types, or consociations, recognized in this region, based on the dominant species of *Avicennia*, *Rhizophora*, *Sonneratia*, *Bruguiera*, and *Nypa*. The relative occurrence of each type is based on fluctuations in soils, salinity, and the tidal regime. Typically mangroves display a zonation or succession of forests, with each zone being dominated by one of the consociations. On the seaward sediments, *Avicennia-Sonneratia* forest dominates. Moving inland, there is softer and deeper mud sediment dominated by *Rhizophora-Bruguiera* forests. Further inland, the soils become firmer and the forests display a greater species diversity. In areas with a substantial freshwater influence, *Nypa* palms dominate. Mangrove forests reach 50 m in height in many areas (Spalding et al. 1997).

Biodiversity Features

Mangrove diversity in terms of endemics or richness is not great. More than 250 birds are listed for this ecoregion, but many of them are transitory, some migrants, and some year-round inhabitants. Determining an exact count for this diverse ecoregion is difficult because of the transitional nature of the habitat.

The mangroves of Borneo are home to the proboscis monkey (*Nasalis larvatus*), which is one of the few large mammals limited to mangrove and peat swamp forest habitats (Spalding et al. 1997). Proboscis monkeys eat primarily young leaves and the seeds of unripe fruit (Bennett and Sebastian 1988; Yeager 1989). To digest this diet, they have developed highly complex sacculated stomachs with specialized bacteria (Bauchop 1978; Bauchop and Martucci 1968).

Although mangroves lack outstanding species diversity, they provide vital ecological functions by being at the interface between the terrestrial and marine realms. Mangroves stabilize coastlines from erosion, accumulate sediment, and provide a nursery for numerous coastal fishes.

Status and Threats

Current Status

Traditionally, mangroves have been harvested for fuelwood, charcoal, and timber, and in some instances this has been done sustainably. However, in recent decades mangroves have been severely degraded by deforestation, agriculture, urban development, fishing, and shrimp farming despite the many protected areas that include mangrove forests (table J107.1).

Protected areas have not addressed many of the conversion threats facing mangrove systems. Many protected areas have been encroached upon for consumptive uses and have not afforded real protection in recent decades.

Types and Severity of Threats

The threats to the habitat remain the same as those that already have claimed vast areas of land: logging, aquaculture, agriculture conversion, and urbanization.

Many mangroves reside in logging concessions or are being cut down for commercial charcoal production. Production of woodchips and pulp is increasing, and more chip mills are being built. Shrimp farming continues to threaten vast mangrove forests. Other aquaculture practices include cockle culture and exploitation of the finfish, bivalve, and crab fisheries. Pollution, agriculture conversion, and oil extraction also threatened mangrove forests (Spalding et al. 1997).

Priority Conservation Actions

With the rapid pace of mangrove destruction, quick conservation action is urgently needed to protect the remaining fragments of habitat. However, steps should be taken to lay the groundwork for longer-term conservation as well.

Short-Term Conservation Actions (1–5 Years)

- Promote sustainable forestry management practices.
- Initiate reforestation efforts.
- Cease development of shrimp farms.

Longer-Term Conservation Actions (5–20 Years)

- Restrict development of shrimp farming industry in sensitive mangrove areas.
- Increase and implement protection.

Focal Species for Conservation Action

Habitat specialists: proboscis monkey (Salter and MacKenzie 1985; Bennet and Gombek 1993; Yeager, 1989, 1990, 1991; Yeager and Blondal 1992)

TABLE J107.1. WCMC (1997) Protected Areas That Overlap with the Ecoregion.

Protected Area	Area (km^2)	IUCN Category
Kuala Jambu Aye/Air	140	PRO
Kuala Langsa	140	PRO
Bakau Selat Dumai [85]	190	PRO
Pulau Burung [85]	190	I
Berbak [85]	200	II
Bakau Muara Kampar	350	PRO
Tanjung Datuk	130	PRO
Kelompok Hutan Bakau Pantai Timur	90	I
Kulamba	230	VI
Muara Kayan [86]	320	PRO
SAR (Sanctuary Reserve) [96]	50	II
SAR (Sanctuary Reserve)	340	UA
SAR (Sanctuary Reserve)	30	PRO
SAR (Sanctuary Reserve)	40	II
Hutan Sambas [96]	20	PRO
SAR (Sanctuary Reserve) [96]	50	IV
Muara Kendawangan [86, 89]	290	PRO
Unnamed [89]	60	?
Tanjung Penghujan NR [89]	170	PRO
Tanjung Puting [86, 89, 90]	190	II
Kelompok Hutan Kahayan [87]	560	PRO
Pamukan	320	PRO
Teluk Kelumpang Selat Laut/Sebuku	880	I
Pleihari Tanah Laut	410	IV
Apar Besar [96]	1,000	PRO
Pantai Samarinda	140	PRO
Total	6,530	

Ecoregion numbers of protected areas that overlap with additional ecoregions are listed in brackets.

Selected Conservation Contacts

Konsorsium SHK Kalimantan Timur

Lembaga Alam Tropika Indonesia (LATIN)

WWF-Indonesia

WWF-Malaysia

Yayasan Keanekaragaman Hayati

Yayasan Konsorsium Pembaruan Agraria (KPA)

Justification for Ecoregion Delineation

Whitmore (1984) and MacKinnon (1997) show large extents of peat swamp forests along the northern coast of Sumatra, especially in Riau Province. We delineated the Sumatran Peat Swamp Forests [85] to represent these forests but extracted the smaller patches of freshwater swamp forests into the Sumatran Freshwater Swamp Forests [88] and the mangroves in the Sunda Shelf Mangroves [107].

Prepared by Colby Loucks

Ecoregion Number:	**108**
Ecoregion Name:	**Kinabalu Montane Alpine Meadows**
Bioregion:	**Sunda Shelf and Philippines**
Biome:	**Montane Grasslands and Shrublands**
Political Units:	**Malaysia**
Ecoregion Size:	**4,300 km²**
Biological Distinctiveness:	**Globally Outstanding**
Conservation Status:	**Relatively Stable**
Conservation Assessment:	**III**

The Kinabalu Montane Alpine Meadows [108] are unique to the region in that they have been isolated from other mountain chains for millions of years. This is one of only two ecoregions in the Indo-Pacific region to be globally outstanding for both bird and mammal richness and endemism (the other ecoregion is the Eastern Himalayan Broadleaf Forests [26]). This montane refuge supports a disjunct distribution of Himalayan, Australasian, and Indomalayan species. Although the very top of Mt. Kinabalu is devoid of vegetation, the slopes and surrounding area have an exceedingly rich flora of approximately 4,500 species in more than 180 families with 950 genera (Davis 1995; WWF and IUCN 1995). This represents one of the richest concentrations of endemic plant species in the world (WWF and IUCN 1995) and is the only Asian example of tropical alpine shrublands with high levels of endemism. This ecoregion supports the greatest concentration of wild orchids on Earth, with more than 750 species in more than sixty genera. This number accounts for more than one quarter of all orchid species found in Malesia.

Location and General Description

This ecoregion represents the upper montane habitat on Mt. Kinabalu and the Crocker Range and the surrounding upland areas in the Malaysian state of Sabah (Borneo). Thirty-five million years ago marine sediments were transformed to rock in the area where Mt. Kinabalu now stands. Approximately 25 million years ago these layers of shale and sandstone were uplifted to form a mountain range. The eroded remains of this range are now known as the Crocker Range. Approximately 15 million years ago a large mass of magma intruded between the folds of the Crocker Range and solidified into adamellite (a type of granite). This intrusion was uplifted rapidly, at a rate of an inch every five years. The exposed granite body, which is still growing, is Mt. Kinabalu. At 13,455 feet it is the highest mountain between the Himalayas and New Guinea (Myers 1978; Jacobson 1978). Based on the Köppen climate zone system, this ecoregion falls in the tropical wet climate zone (National Geographic Society 1999).

Kinabalu is the meeting place for plant species from the Himalayas, China, Australia, New Zealand, and the Indo-Malayan realm (Corner 1978). The vegetation of Kinabalu can be divided into zones based on altitude, but many factors may alter their distribution locally to either higher or lower elevations. Above 1,000 m a montane zone exists until about 2,600 m. This zone includes a mixing of lowland and montane families, giving this elevation zone a great diversity of life. The common lowland families, such as Dipterocarpaceae, Euphorbiaceae, Leuminosae, Myristicaceae, and Sapotaceae, begin to diminish, and they begin to be replaced by a great diversity of species from plant families such as Ericaceae, Myrtaceae, Fagaceae, Lauraceae, Magnoliaceae, and the majority of Bornean gymnosperms, including *Podocarpus*, *Agathis*, and *Phyllocladus*. The dipterocarp *Shorea monticola* occurs in this elevation belt but not in the lowlands. This elevation zone also supports a number of endemic species from *Rhododendron*, *Lithocarpus*, *Magnolia*, and *Rhamnus*. A large variety of pitcher plants, many endemic Mt. Kinabalu, occur in this zone (Cockburn 1978). The endemic species include *Nepenthes edwardsiana*, *N. rajah*, *N. villosa*, and *N. burbidgeae*. The most common species found in these forests is the spectacular *Nepenthes lowii*. Fig trees are also common in these forests. Borneo has 135 species of wild figs, with more than 78 species occurring on Kinabalu, including 13 endemic species (Corner 1952). Kinabalu may have one of the richest fig floras in the world (Corner 1978). This floristically rich area also supports *Rafflesia tengku-adlinii* (which occurs only on Trus Madi and in the Maliau Basin) (WWF and IUCN 1995).

Between 2,600 and 3,200 m is a band of ultrabasic rocks that give rise to a different type of vegetation. This vegetation ranges from the 10-m high *Dacrydium gibbsiae* to dwarf shrubs and includes moss, lichen, liverwort, and ferns. Above 3,200 m the soil cover soon disappears, giving way to granite. In areas where soil can support shrubs, species such as *Leptospermum recurvum*, *Coprosma hookeri*, and *Rhododendron buxifolium* dominate. On thinner soil, herbs such as *Diplocosia kinabaluensis*, *Machaerarina falcata*, and *Ranunculus lowii* are common (Cockburn 1978).

This ecoregion contains more than 750 orchid species in more than sixty genera. This number accounts for more than one-fourth of all orchid species found in Malesia (Lamb and Chan 1978). Perhaps the most famous orchid species found on Kinabalu are the several species of slipper orchid, *Paphiopedilum*. However, because of intensive collecting, they are rarely seen. The montane region of Kinabalu

TABLE J108.1. Endemic and Near-Endemic Mammal Species.

Family	Species
Crocidurinae	Suncus ater*
Sorcidae	Crocidura baluensis
Tupaiidae	Tupaia montana
Cercopithecidae	Presbytis comata
Mustelidae	Melogale everetti*
Sciuridae	Callosciurus baluensis
Sciuridae	Callosciurus orestes
Sciuridae	Glyphotes simus
Sciuridae	Sundasciurus brookei
Sciuridae	Petaurillus hosei
Sciuridae	Aeromys thomasi
Muridae	Rattus baluensis*
Muridae	Chiropodomys muroides*
Muridae	Maxomys alticola
Muridae	Maxomys ochraceiventer
Muridae	Maxomys baeodon

An asterisk signifies that the species' range is limited to this ecoregion.

presents a variety of microhabitats that have produced many orchid species common to *Eria, Bulbophyllum, Dendrobium, Liparis, Dendrochilum, Pholidota,* and *Coelogyne* (Lamb and Chan 1978).

Biodiversity Features

Most of the 114 mammal species found in this ecoregion live in the forest canopy. Only one-third of the mammal species are terrestrial. Common terrestrial species include three deer species, the Malaysian weasel (*Mustela nudipes*), small-clawed otter (*Aonyx cinerea*), and leopard cat (*Felis bengalensis*). The majority of the species live in the canopy of the forests. These species include many of the twenty-eight of the thirty-four squirrels known from Borneo, numerous bat species, tree shrews, slow loris, tarsier (*Tarsius bancanus*), grey leaf monkey (*Presbytis aygula*), red leaf monkey (*P. rubicunda*), orangutan (*P. pygmaeus*), Borneo gibbon (*Hylobates moloch*), linsang (*Prionodon linsang*), and binturong (*Arctictis binturong*). There are twelve near-endemic and four endemic mammal species, which include the Bornean black shrew (*Suncus ater*) and Bornean ferret-badger (*Melogale everetti*) (table J108.1).

This ecoregion also supports more than 180 bird species. The bird fauna includes twenty-four near-endemic species and one endemic species (table J108.2). The ecoregion overlaps with a portion of the Bornean Mountains EBA (157) (Stattersfield et al. 1998).

Status and Threats

Current Status

About one-third of this sensitive, high-altitude ecoregion has been cleared or degraded, mostly by agriculture or other practices associated with forest clearance. There are two protected areas in the ecoregion that cover a total of 1,440 km² (33 percent) (table J108.3). Kinabalu Park was gazetted in 1964, and the Crocker Range National Park was established in this ecoregion in 1984.

TABLE J108.2. Endemic and Near-Endemic Bird Species.

Family	Common Name	Species
Accipitridae	Mountain serpent-eagle	Spilornis kinabaluensis
Phasianidae	Red-breasted partridge	Arborophila hyperythra
Phasianidae	Crimson-headed partridge	Haematortyx sanguiniceps
Podargidae	Dulit frogmouth	Batrachostomus harterti
Trogonidae	Whitehead's trogon	Harpactes whiteheadi
Capitonidae	Mountain barbet	Megalaima monticola
Capitonidae	Golden-naped barbet	Megalaima pulcherrima
Capitonidae	Bornean barbet	Megalaima eximia
Eurylaimidae	Whitehead's broadbill	Calyptomena whiteheadi
Pachycephalidae	Bornean whistler	Pachycephala hypoxantha
Turdidae	Everett's thrush	Zoothera everetti
Turdidae	Fruit-hunter	Chlamydochaera jefferyi
Muscicapidae	Eyebrowed jungle-flycatcher	Rhinomyias gularis
Zosteropidae	Black-capped white-eye	Zosterops atricapillus
Zosteropidae	Pygmy white-eye	Oculocincta squamifrons
Zosteropidae	Mountain black-eye	Chlorocharis emiliae
Sylviidae	Bornean stubtail	Urosphena whiteheadi
Sylviidae	Friendly bush-warbler	Bradypterus accentor
Timaliidae	Sunda laughingthrush	Garrulax palliatus
Timaliidae	Bare-headed laughingthrush	Garrulax calvus
Timaliidae	Mountain wren-babbler	Napothera crassa
Timaliidae	Chestnut-crested yuhina	Yuhina everetti
Dicaeidae	Black-sided flowerpecker	Dicaeum monticolum
Nectariniidae	Bornean spiderhunter	Arachnothera everetti*
Nectariniidae	Whitehead's spiderhunter	Arachnothera juliae

An asterisk signifies that the species' range is limited to this ecoregion.

TABLE J108.3. WCMC (1997) Protected Areas That Overlap with the Ecoregion.

Protected Area	Area (km²)	IUCN Category
Kinabalu [98]	590	II
Crocker Range [98]	850	II
Total	1,440	

Ecoregion numbers of protected areas that overlap with

Types and Severity of Threats

The unique flora found on the slopes of Mt. Kinabalu is protected to some extent by the steepness of the terrain and poor soil conditions, which discourage logging and farming. Nevertheless, some of the surrounding slopes outside the park boundary are being cleared for farming, mainly of vegetables. Road construction has facilitated tourist access to Kinabalu Park, which has led to the construction of more facilities. Some of these developments have been poorly planned and even detrimental, such as the degazettement of a large area of alluvial Pinosuk Plateau in 1984 for government development projects such as a golf course and the 1984 redesignation of Trus Madi from a watershed protection forest to that of a commercial forest reserve to allow logging to take place. Commercial logging may have encroached into the park, and a section of the park was excised in 1974 for the development of a copper mine. A number of species, especially the rare and endemic species in the ecoregion, are being overcollected for the commercial wildlife and plant trade (WWF and IUCN 1995). These problems are exacerbated by the absence of buffer zones around the park and insufficient staff to enforce regulations.

Priority Conservation Actions

Short-Term Conservation Actions (1–5 Years)

- Stop illegal logging in and around protected areas.
- Revoke forest concessions in areas of high biodiversity outside protected areas.
- Stop illegal collection of flora (orchids).

Longer-Term Conservation Actions (5–20 Years)

- Develop a long-term management plan that provides for connection with lower-elevation forests.
- Establish additional protected areas.

Focal Species for Conservation Action

Endemic species: Suncus ater, Bornean spiderhunter, Kinabalu rat, Kinabalu ferret-badger, grey-bellied pencil-tailed tree mouse

Other: squirrel species, primates, orchids

Selected Conservation Contacts

WWF-Malaysia

Justification for Ecoregion Delineation

The large island of Borneo was divided into nine ecoregions. MacKinnon and MacKinnon (1986) divided the island's lowland forests into six subunits with a central subunit representing the montane forests. MacKinnon (1997) revised the boundaries of these seven subunits but retained the same general configuration. These authors used the major rivers, the Kapuas and Barito, to represent zoogeographic barriers to a few species of mammals and based subunits largely on these barriers but also used climatic regimes for the drier eastern biounits (MacKinnon and MacKinnon 1986; MacKinnon 1997).

Because ecoregions are based on biomes, we first isolated the central montane ecoregion—the Borneo Montane Rain Forests [95]—above the 1,000-m elevation contour using the DEM (USGS 1996). We then assigned the large patches of peat forests, heath forests, freshwater swamp forests, and mangroves, in the lowlands and along the periphery of the island, into their own ecoregions: the Borneo Peat Swamp Forests [86], Sundaland Heath Forests [90] (which also includes Belitung Island and the heath forests on Bangka Island), and Southern Borneo Freshwater Swamp Forests [89], and Sunda Shelf Mangroves [107], respectively. The alpine habitats of the Kinabalu Mountain Range were represented by the Kinabalu Montane Alpine Meadows [108]. Udvardy (1975) combined all of Borneo into the Borneo biogeographic province.

Prepared by Colby Loucks

Ecoregion Number:	**109**
Ecoregion Name:	**Sulawesi Lowland Rain Forests**
Bioregion:	**Wallacea**
Biome:	**Tropical and Subtropical Moist Broadleaf Forests**
Political Units:	**Indonesia**
Ecoregion Size:	**114,100 km^2**
Biological Distinctiveness:	**Globally Outstanding**
Conservation Status:	**Critical**
Conservation Assessment:	**I**

The Sulawesi Lowland Rain Forests [109] harbor some of the most unique animals on Earth. The islands are located in the region known as Wallacea, which contains a distinctive fauna representing a mix of Asian and Australasian species. A fruit-eating pig with huge tusks, a dwarf buffalo, endemic macaques, and cuscuses exemplify a truly unique mammal community. Sulawesi, like the hub of a wheel, is surrounded by a variety of exotic ocean basins, including the Flores Sea, the Banda Sea, the Molucca Sea, the Java Sea, and the Straits of Makassar, as well as the diverse islands of Borneo, Java, Flores, Halmahera, and the Philippines. More than half of the original forest has been cleared, and most of the remaining forests have been reduced to fragments.

Location and General Description

This ecoregion represents the lowland forests (less than 1,000 m) on Sulawesi and the surrounding islands of Banggai and Sula to the east and Talaud and Sangihe to the north. Sulawesi is almost completely mountainous. There are no extensive lowlands on Sulawesi, with large areas above 1,000 m and the highest elevation at 3,455 m on Mt. Rantemario. Sangihe is mountainous, reaching an elevation of 1,784 m, whereas Talaud is low-lying. The physiography of the Sula Islands is hilly, with mountains over 800 m only on the island of Taliabu (Stattersfield et al. 1998).

The upland areas (more than 1,000 m) of Sulawesi form a separate ecoregion, the Sulawesi montane rain forests. Based on the Köppen climate zone system, this ecoregion falls in the tropical wet climate zone (National Geographic Society 1999). Sulawesi has a complex geologic history and is composed of three geologic provinces based on that history. West and East Sulawesi form two of the geologic provinces, separated by the Palu-Koro fault, which runs from the town of Palu to the Gulf of Bone. The third geologic province consists of the Tokala region on the northeast peninsula, the Banggai Islands, Butung Island, and the Sula Islands. East and West Sulawesi collided approximately 13–19 million years ago, and ultrabasic rocks were exposed as East Sulawesi overrode the western portion. The forces that caused the collision are still at work, and Sulawesi is being torn apart today. The surface geology of Sulawesi is a diverse patchwork of ophiolites, Mesozoic sedimentary rocks, Tertiary sedimentary and igneous rocks, and Quaternary volcanics and sediments. Active volcanoes are located on the northern arm of Sulawesi (Whitten et al. 1987).

The lowland forest is predominantly tropical lowland evergreen and semi-evergreen rain forest, with some monsoon forests at the tip of the southeast peninsula and small areas of freshwater and peat swamp forest (Whitten et al. 1987; Stattersfield et al. 1998). Distinctive forest types on limestone are distributed around southern Sulawesi and on ultrabasic soils in scattered locations all around the island. The lowland and hill forests contain the most tree species, although these forests are not dominated by any one tree family; only seven dipterocarp species are found in Sulawesi (compared with 267 and 106 in Borneo and Sumatra, respectively). The dipterocarps include *Anisoptera costata, Hopea celebica, H. gregaria, Shorea assamica, Vatica rassak*, and *V. flavovirens*. Distinctive ebonies (*Diospyros* spp.) were common in dense clumps in the lowland forests. Palms are common in the lowland forest, including *Oncosperma horridum, Liculala celebensis, Pinanga, Areca, Caryota*, and *Livistona rotundifolia* (Whitten et al. 1987).

The isolated Sula Islands, just off Sulawesi's east coast, receive rain from both the northwest and the southeast and have volcanic soils that create excellent growth conditions (Monk et al. 1997).

Aopa Swamp, 100 km west of Kendari, is a major area of peat swamp that varies seasonally in extent from about 150 to 314 km^2. The dominant tree species in this forest include *Casuarina* spp., *Eugenia* spp., *Geunsia paloensis, Premna foetida, Metroxylon sagu, Pholidocarpus* spp., *Licuala* spp., *Arenga* spp., *Oncosperma* spp., and *Corypha* spp. Sedges such as *Scleria* spp. also occur along with 5-m tall *Pandanus* spp., at least two species of climbing rattan, and epiphytic *Lecanopteris* ant-ferns (Whitten et al. 1987).

Freshwater swamp forest is characterized by grassy areas near open water, with palms and pandans on firmer ground and ubiquitous pitcher plants (*Nepenthes*). Riverine forest dominated by tall *Eucalyptus deglupta* is found in the Sopu Valley northeast of Lake Lindu and Mt. Nokilalaki (Whitten et al. 1987).

This ecoregion also includes karst (limestone) areas that have a relative paucity of trees and tree species because of their shallow soils and steep slopes, resulting from the high solubility of limestone rocks. High calcium levels in the soil give rise to distinctive tolerant plant communities but support certain snail species limited to limestone forest as well as the large swallowtail butterfly (*Graphium androcles*) (Whitten et al. 1987).

Infertile ultrabasic substrates, with serpentine and peridotite rocks, contain unique forests with a high degree of plant endemism. Common species include ironwood (*Metrosideros*), *Agathis, Calophyllum*, Burseraceae, Sapotaceae, and dipterocarps (*Vatica* and *Hopea celebica*). Myrtaceae (*Eugenia, Kjellbergiodendron*, and *Metrosideros*) are dominant in the low and regular canopy. There is little marketable timber in such forests (Whitten et al. 1987).

Biodiversity Features

Wallace's Line, running from between Bali and Lombok and between Sulawesia and Borneo, marks the location of a deep oceanic trench and the point over which land animals and plants could not cross easily. Similarly, Lydekker's Line, running from between Timor and the Australian shelf to between Halmahera, Seram, and New Guinea, marks the point where Australasian flora and fauna could not easily pass. Sulawesi lies between these two lines. Sulawesi's location, geologic history, and long geographic isolation have created Sulawesi's distinctive fauna. There is variability, different among various animal and plant groups, in the amount of interchange between other biogeographic areas in the region, which led to the evolution of a large number of species endemic to the island. Although not species-rich relative to Borneo or Java, Sulawesi is high in endemicity because of its long isolation from Asia and Australia in Wallacea. This ecoregion exhibits high plant endemism, and the several distinct forest types provide habitat for the highest number of endemic mammals in Asia and several endemic birds (Whitten et al. 1987).

Of the 104 mammal species in the ecoregion, 29 are endemic or near endemic (table J109.1). Whereas the two cuscuses have Australasian affinities (i.e., the Peleng cuscus

TABLE J109.1. Endemic and Near-Endemic Mammal Species.

Family	Species
Phalangeridae	Phalanger pelengensis*
Sorcidae	Crocidura elongata
Sorcidae	Crocidura lea
Sorcidae	Crocidura levicula
Pteropodidae	Acerodon humilis*
Pteropodidae	Neopteryx frosti*
Pteropodidae	Nyctimene minutus
Rhinolophidae	Hipposideros inexpectatu
Vespertilionidae	Pipistrellus minahassae
Vespertilionidae	Hesperoptenus gaskelli
Tarsiidae	Tarsius pelengensis
Cercopithecidae	Macaca maura*
Cercopithecidae	Macaca ochreata*
Cercopithecidae	Macaca nigra
Suidae	Babyrousa babyrussa
Bovidae	Bubalus depressicornis*
Muridae	Rattus koopmani*
Muridae	Rattus xanthurus
Muridae	Rattus bontanus*
Muridae	Rattus elaphinus*
Muridae	Maxomys hellwaldii*
Muridae	Haeromys minahassae*
Muridae	Margaretamys beccarii*
Muridae	Taeromys celebensis*
Muridae	Taeromys punicans*
Muridae	Taeromys taerae*
Muridae	Echiothrix leucura*
Muridae	Melomys fulgens
Muridae	Melomys caurinus*

An asterisk signifies that the species' range is limited to this ecoregion.

TABLE J109.2. Endemic and Near-Endemic Bird Species.

Family	Common Name	Species	Family	Common Name	Species
Accipitridae	Small sparrowhawk	*Accipiter nanus*	Coraciidae	Purple-winged roller	*Coracias temminckii*
Megapodiidae	Sula scrubfowl	*Megapodius bernsteinii**	Bucconidae	Sulawesi hornbill	*Penelopides exarhatus**
Megapodiidae	Maleo	*Macrocephalon maleo*	Bucconidae	Knobbed hornbill	*Aceros cassidix*
Rallidae	Platen's rail	*Aramidopsis plateni*	Acanthizidae	Rufous-sided gerygone	*Gerygone dorsalis*
Rallidae	Bare-faced rail	*Gymnocrex rosenbergii*	Pachycephalida	Sulphur-bellied whistler	*Pachycephala sulfuriventer*
Rallidae	Isabelline waterhen	*Amaurornis isabellinus*	Pachycephalida	Drab whistler	*Pachycephala griseonota*
Columbidae	Dusky cuckoo-dove	*Macropygia magna*	Rhipiduridae	Rusty-flanked fantail	*Rhipidura teysmanni*
Columbidae	Sulawesi ground-dove	*Gallicolumba tristigmata*	Monarchidae	Cerulean paradise-flycatcher	*Eutrichomyias rowleyi**
Columbidae	Maroon-chinned fruit-dove	*Ptilinopus subgularis**	Monarchidae	White tipped monarch	*Monarcha everetti**
Columbidae	White-bellied imperial-pigeon	*Ducula forsteni*	Dicruridae	Sulawesi drongo	*Dicrurus montanus*
Columbidae	Pink-headed imperial pigeon	*Ducula rosacae*	Corvidae	Banggai crow	*Corvus unicolor**
Columbidae	Grey-headed imperial-pigeon	*Ducula radiata*	Campephagidae	Cerulean cuckoo-shrike	*Coracina temminckii*
Columbidae	Grey imperial-pigeon	*Ducula pickeringii*	Campephagidae	Pied cuckoo-shrike	*Coracina bicolor**
Columbidae	Silver-tipped imperial-pigeon	*Ducula luctuosa**	Campephagidae	White-rumped cuckoo-shrike	*Coracina leucopygia*
Psittacidae	Yellowish-breasted racquet-tail	*Prioniturus flavicans**	Campephagidae	Sula cuckoo-shrike	*Coracina sula**
Psittacidae	Moluccan hanging-parrot	*Loriculus amabilis*	Campephagidae	Slaty cuckoo-shrike	*Coracina schistacea**
Psittacidae	Sangihe hanging-parrot	*Loriculus catamene**	Campephagidae	White-rumped triller	*Lalage leucopygialis**
Psittacidae	Pygmy hanging-parrot	*Loriculus exilis**	Turdidae	Rusty-backed thrush	*Zoothera erythronota**
Loriidae	Red-and-blue lory	*Eos histrio**	Sturnidae	Pale-bellied myna	*Acridotheres cinereus*
Loriidae	Yellow-and-green lorikeet	*Trichoglossus flavoviridis*	Sturnidae	Sulawesi myna	*Basilornis celebensis*
Cuculidae	Sulawesi hawk-cuckoo	*Cuculus crassirostris*	Sturnidae	Helmeted myna	*Basilornis galeatus**
Cuculidae	Bay coucal	*Centropus celebensis**	Sturnidae	White-necked myna	*Streptocitta albicollis*
Strigidae	Ochre-bellied hawk-owl	*Ninox ochracea*	Sturnidae	Bare-eyed myna	*Streptocitta albertinae**
Strigidae	Speckled hawk-owl	*Ninox punctulata*	Sturnidae	Fiery-browed myna	*Enodes erythrophris*
Tytonidae	Minahassa owl	*Tyto inexspectata*	Sturnidae	Finch-billed myna	*Scissirostrum dubium**
Tytonidae	Taliabu owl	*Tyto nigrobrunnea**	Muscicapidae	Henna-tailed jungle-flycatcher	*Rhinomyias colonus**
Tytonidae	Sulawesi owl	*Tyto rosenbergii*	Muscicapidae	Rufous-throated flycatcher	*Ficedula rufigula**
Caprimulgidae	Diabolical nightjar	*Eurostopodus diabolicus*	Zosteropidae	Sulawesi white-eye	*Zosterops consobrinorum**
Caprimulgidae	Sulawesi nightjar	*Caprimulgus celebensis**	Zosteropidae	Black-ringed white-eye	*Zosterops anomalus*
Alcedinidae	Sulawesi kingfisher	*Ceyx fallax**	Sylviidae	Sulawesi leaf-warbler	*Phylloscopus sarasinorum*
Alcedinidae	Lilac kingfisher	*Cittura cyanotis**			
Alcedinidae	Black-billed kingfisher	*Pelargopsis melanorhyncha**	Dicaeidae	Crimson-crowned flowerpecker	*Dicaeum nehrkorni*
Alcedinidae	Talaud kingfisher	*Todirhamphus enigma**	Dicaeidae	Red-chested flowerpecker	*Dicaeum maugei*
Alcedinidae	Green-backed kingfisher	*Actenoides monachus**	Dicaeidae	Grey-sided flowerpecker	*Dicaeum celebicum**
Alcedinidae	Scaly kingfisher	*Actenoides princeps*			
Meropidae	Purple-bearded bee-eater	*Meropogon forsteni*	Nectariniidae	Elegant sunbird	*Aethopyga duyvenbodei**

An asterisk signifies that the species' range is limited to this ecoregion.

[*Phalanger pelengensis*] and dwarf cuscus [*Strigocuscus celebensis*]), the remainder of Sulawesi's mammals have Asian origins, including the crested macaque (*Macaca nigra*), moor macaque (*M. maura*), booted macaque (*M. ochreata*), lowland anoa (*Bubalus depressicornis*), spectral tarsier (*Tarsius spectrum*), and babirusa (*Babyrousa babyrussa*) (Flannery 1995). The crested macaque, moor macacque, and lowland anoa are considered endangered (Hilton-Taylor 2000).

Sulawesi contains a depauperate bird fauna but with high levels of endemicity (Stattersfield et al. 1998). The origin of Sulawesi's birds is predominantly Asian (Whitten et al. 1987). The bird fauna consists of about 337 species, of which 70 are endemic or near-endemic species (table J109.2). The ecoregion also overlaps the lowland portions of the Sulawesi EBA and completely overlaps both the Sangihe and Talaud and Banggai and Sula Islands EBAs and the Salayar and Bonerate Islands Secondary Area (Stattersfield et al. 1998). Of the seventy restricted-range birds in these three EBAs, thirty-two bird species are found nowhere else in the world but this lowland ecoregion. Thirteen additional species are found only in this ecoregion and the adjacent montane ecoregion (and nineteen more species are found only in the uplands of Sulawesi) (Stattersfield et al. 1998). One species found on Sangihe, the cerulean paradise flycatcher (*Eutrichomyias rowleyi*), is critically endangered, while the red-and-blue lory (*Eos histro*), Sangihe hanging-parrot (*Loriculus catamene*), Taliabu masked-owl (*Tyto nigrobrunnea*), white-tipped monarch (*Monarcha everetti*), Banggai crow (*Corvus unicolor*), and elegant sunbird (*Aethopyga duyvenbodei*), are endangered (BirdLife International 2000).

Four of Sulawesi's amphibians have Sundaland affinities, and two have Australasian roots. Thirty-eight of Sulawesi's sixty-three snake species are found on both sides of Wallace's Line. There are large reptiles of conservation significance: the sailfin lizard (*Hydrosaurus amboinensis*), saltwater crocodile (*Crocodylus porosus*), and reticulated python (*Python reticulatus*) (Whitten et al. 1987).

Sulawesi's flora is most closely related to the floras of dry areas in the Philippines, Moluccas, Lesser Sundas, and Java. The lowland forests have affinities to New Guinea, whereas the upland areas are more related to Borneo (Whitten et al. 1987). Three Centres of Plant Diversity are located in lowland Sulawesi: Dumoga-Bone National Park, Limestone Flora of Sulawesi, and Ultramafic Flora of Sulawesi (Davis et al. 1995).

Status and Threats

Current Status

More than half of the original forest has been cleared, and the remaining forests have been reduced to fragments except for a few fairly large blocks that are still intact. There are thirty-eight protected areas that cover 9,460 km^2 (8 percent) of the ecoregion area (table J109.3). Seven of these reserves are more than 500 km^2, but none are more than 1,100 km^2.

Sulawesi still supports some lowland moist forests on steep slopes that are unsuitable for agriculture. However, large areas in the south and some parts of the center and north of the island have been cleared for permanent and shifting cultivation (IUCN 1991). The lowland peneplain dry forest is completely gone because of large-scale agricultural plantations, transmigration, logging, and local clearance. The riverine forest in the Dumoga Valley is now the site of a major irrigation scheme, and some of the limestone vegetation has been destroyed by quarrying to supply the Tonasa cement factories. During the dry season, cattle farmers set fires to encourage the growth of young grass, and repeated burnings have resulted in a persistent grassland vegetation in some areas and a savanna with fire-resistant trees in others. Uncontrolled exploitation for the oil in its heartwood has depleted stands of the sandalwood tree *Santalum album*, even in protected areas such as Paboya Reserve (Whitten et al. 1987; Whitten, pers. comm., 2000).

Sangihe and Talaud were largely deforested by 1920, and there is minimal natural forest remaining on these islands. A survey has been proposed to determine appropriate locations for additional protected areas around remaining forest (Stattersfield et al. 1998).

Most of Taliabu, the largest island in the Sula Islands, is still forested, but there has been large-scale logging in the

TABLE J109.3. WCMC (1997) Protected Areas That Overlap with the Ecoregion.

Protected Area	Area (km^2)	IUCN Category
Karakelang Utara	260	VI
Karakelang Selatan	80	VI
Gunung Sahendaruman	100	PRO
Dua Saudara	120	I
Pulau Taliabu	700	PRO
Pulau Seho	20	I
Pati-Pati	20	IV
Lombuyan I, II	1,070	IV
Gunung Lokon	20	VI
Gunung Manembo–Nembo	50	IV
Dumoga [110]	940	?
Mas Popaya Raja	40	I
Tanggale	10	PRO
Panua	410	I
Marisa	30	PRO
Pulau Una-una	70	PRO
Tanjung Api	50	I
Bangkiriang	390	PRO
Morowali [110]	870	I
Mamuja/Tapalang	150	PRO
Danau Towuti	560	V
Lamiko-miko	280	PRO
Lasolo-Sampara	320	PRO
Lamedae	30	I
Rawa Aopa Watumohai	1,030	II
Tanjung Batikolo	30	IV
Lampoko Mampie	30	IV
Danau Tempe	310	PRO
Bontobahari	50	IV
Tanjung Peropa	300	IV
Polewai	100	PRO
Tanjung Amelango	10	IV
Kaya Kuku	40	PRO
Buton Utara	700	IV
Napabalano	10	I
Kokinawe	30	PRO
Lambu Sango NR	200	PRO
Tirta Rimba	30	V
Total	9,460	

Ecoregion numbers of protected areas that overlap with additional ecoregions are listed in brackets.

lowlands. The other main Sula Islands, Sanana and Mangole, have been heavily degraded. Extensive lowland forest still remains on the Banggai Islands (Stattersfield et al. 1998).

Types and Severity of Threats

Uncontrolled and illegal logging will continue to be the biggest threat to the integrity of the remaining forests. This situation has been and will be exacerbated by lack of authority and implementation of existing environmental laws.

Priority Conservation Actions

With the rapid pace of forest destruction under way in Indonesia, quick conservation action is urgently needed to stave off the complete collapse of these forests. However, steps should be taken to lay the groundwork for longer-term conservation as well.

Short-Term Conservation Actions (1–5 Years)

Oliver (1993) outlines objectives intended to safeguard populations of babirusa, including determining its present distribution and relative population sizes; establishing a network of protected areas representing all subspecies; conducting further research into systematics, biology, and conservation status; and promoting interest and awareness among local people.

- Initiate a major public awareness campaign on forest values and forest loss to drive political will.
- Stop illegal logging in and around protected areas by challenging the authorities and judiciary to act on existing laws and regulations.
- Revoke forest concessions in areas of high biodiversity outside protected areas and institute a moratorium on new concessions for oil palm, timber, and other plantations using instead the vast areas of deforested or cleared land.
- Place under active protection all remaining lowland forests.
- Mobilize major assistance for the more outstanding protected areas.
- Stop wildlife poaching, especially in key parks and reserves, by tightening surveillance around ports or airports and making the purchase of endangered species socially unacceptable.
- Instigate major biodiversity surveys or collecting expeditions in doomed areas to document the biodiversity that will soon be lost.

Longer-Term Conservation Actions (5–20 Years)

- Establish a long-term fund for conservation of lowland forests in Indonesia that would benefit Sulawesi.
- Ensure that redesign of the protected area system includes linkages between lowland and montane habitats.
- Devise ways to maintain the integrity of buffer zones of protected areas.

Focal Species for Conservation Action

Endangered endemic species: crested macaque (*Macaca nigra*) (Eudey 1987), moor macaque (*M. maura*) (Eudey 1987), lowland anoa (*Bubalus depressicornis*) (Hedges, in press), cerulean paradise flycatcher (*Eutrichomyias rowleyi*) (BirdLife International 2000), red-and-blue lory (*Eos histro*) (BirdLife International 2000), Sangihe hanging-parrot (*Loriculus catamene*) (BirdLife International 2000), elegant sunbird (*Aethopyga duyvenbodei*) (BirdLife International 2000), Taliabu masked-owl (*Tyto nigrobrunnea*) (BirdLife International 2000), white-tipped monarch (*Monarcha everetti*) (BirdLife International 2000), Banggai crow (*Corvus unicolor*) (BirdLife International 2000)

Area-sensitive species: lowland anoa (*Bubalus depressicornis*) (Hedges, in press), babirusa (*Babyrousa babyrussa*) (Oliver 1993), Sulawesi hornbill (*Penelopides exarhatus*) (Kemp 1995), knobbed hornbill (*Aceros cassidix*) (Kemp 1995)

Top predator species: Sulawesi serpent-eagle (*Spilornis rufipectus*), Sulawesi goshawk (*Accipiter griseiceps*), spot-tailed goshawk (*Accipiter trinotatus*), brown goshawk (*Accipiter fasciatus*), black eagle (*Ictinaetus malayensis*), rufous-bellied eagle (*Hieraaetus kienerii*), Sulawesi hawk-eagle (*Spizaetus lanceolatus*)

Selected Conservation Contacts

Lembaga Alam Tropika Indonesia (LATIN)
The Nature Conservancy
WWF-Indonesia
Yayasan Keanekaragaman Hayati
Yayasan Konsorsium Pembaruan Agraria (KPA)

Justification for Ecoregion Delineation

There have been several attempts to divide the bioregion into biogeographic units (MacKinnon 1997; Stattersfield et al. 1998; van Balgooy 1971, cited in Monk et al. 1997; MacKinnon and Artha 1981; MacKinnon and MacKinnon 1986; MacKinnon et al. 1982; van Steenis 1950; Udvardy 1975). Because many of the islands have distinct natural faunal communities and a high degree of endemism (Monk et al. 1997), the more recent attempts have used faunal dissimilarities—especially birds—to identify distinct biogeographic units (MacKinnon 1997; Stattersfield et al. 1998; MacKinnon and Artha 1981; MacKinnon and MacKinnon 1986). Because detailed floral data are largely unavailable across most of the bioregion, we followed these authors in delineating ecoregions based on distribution of biomes and vertebrate communities.

On Sulawesi Island we delineated two ecoregions: the Sulawesi Lowland Rain Forests [109] and Sulawesi Montane Rain Forests [110]. These represent the tropical lowland and montane tropical moist forests, respectively. The small patch of monsoon forests on the southwest peninsula of Sulawesi and on Butung Island (Whitmore 1984) were included in the Sulawesi Lowland Rain Forests [109] but should be considered a distinct habitat type in an ecoregion-based conservation assessment to ensure representation.

Prepared by John Morrison

Ecoregion Number: **110**
Ecoregion Name: **Sulawesi Montane Rain Forests**
Bioregion: **Wallacea**
Biome: **Tropical and Subtropical Moist Broadleaf Forests**
Political Units: **Indonesia**
Ecoregion Size: **75,500 km²**
Biological Distinctiveness: **Globally Outstanding**
Conservation Status: **Critical**
Conservation Assessment: **I**

The Sulawesi Montane Rain Forests [110] harbor some of the most unique animals on Earth. The islands are located in the region known as Wallacea, which contains a distinctive fauna representing a mix of Asian and Australasian species. A fruit-eating pig with large curly tusks, a dwarf buffalo, four monkey species, and cuscuses exemplify a truly unique mammal community. Like the hub of a wheel, Sulawesi is surrounded by a variety of exotic ocean basins, including the Flores Sea, the Banda Sea, the Moluccas Sea, the Java Sea, and the Straits of Makassar, as well as the diverse islands of Borneo, Java, Flores, Halmahera, and the Philippines. Although more than half of the original forest has been cleared, Sulawesi still supports tracts of montane moist forests in areas of steep slopes that are unsuitable for agriculture.

Location and General Description

This ecoregion represents these montane forests above 1,000 m, whereas the lowlands constitute a separate ecoregion. Most of Sulawesi lies above 500 m, and about 20 percent of the total land area—mostly the central region—is above 1,000 m (Whitten et al. 1987). Based on the Köppen climate zone system, this ecoregion falls in the tropical wet climate zone (National Geographic Society 1999). As might be surmised from its shape, Sulawesi has a complex geologic history and is composed of three geologic provinces based on that history. West and East Sulawesi form two of the geologic provinces, separated by the Palu-Koro fault, which runs from the town of Palu to the Gulf of Bone. The third geologic province consists of the Tokala region on the northeast peninsula, the Banggai Islands, Butung Island, and the Sula Islands. East and West Sulawesi collided approximately 13–19 million years ago, and ultrabasic rocks were exposed as East Sulawesi overrode the western portion. The forces that caused the collision are still at work, and Sulawesi is being torn apart today. The surface geology of Sulawesi is a diverse patchwork of ophiolites, Mesozoic sedimentary rocks, Tertiary sedimentary and igneous rocks, and Quaternary volcanics and sediments. Active volcanoes are located on the northern arm of Sulawesi (Whitten et al. 1987).

Above 1,000 m, forest trees become shorter and less massive, and epiphytes such as orchids become more common. Whereas the forests of Sulawesi's lowlands are not dominated by any particular tree family, the forests in the lower montane region are dominated by oaks (four species of *Lithocarpus*) and chestnut (two species of *Castanopsis*). An example association includes *Phyllocladus*, *Agathis dammara*, and *Eugenia* dominated by *Castanopsis*. Upper montane forest contains conifers (pines and related Gymnosperms such as *Podocarpus* spp., *Dacrycarpus* spp., *Dacrydium* spp., *Phyllocladus* spp.) and the magnificent and commercially important *Agathis* spp. (Whitten et al. 1987). The highest peaks have sub-alpine forests with yet smaller trees whose branches bear epiphytic lichens and a ground cover of shrubs, colorful herbs, and grasses (Whitten et al. 1987).

Biodiversity Features

Wallace's Line, running from between Bali and Lombok and between Sulawesia and Borneo, marks the location of a deep oceanic trench and the point over which land animals and plants could not cross easily. Similarly, Lydekker's Line, running from between Timor and the Australian shelf to between Halmahera, Seram, and New Guinea, marks the point where Australasian flora and fauna could not easily pass. Sulawesi lies between these two lines. Sulawesi's location, geologic history, and long geographic isolation have created Sulawesi's distinctive fauna. There is variability, different among various animal and plant groups, in the amount of interchange between other biogeographic areas in the region, which led to the evolution of a large number of species endemic to the island. Although not species-rich relative to Borneo or Java, Sulawesi is high in endemicity because of its long isolation from Asia and Australia in Wallacea. This ecoregion exhibits

TABLE J110.1. Endemic and Near-Endemic Mammal Species.

Family	Species
Sorcidae	*Crocidura elongata*
Sorcidae	*Crocidura lea*
Sorcidae	*Crocidura levicula*
Rhinolophidae	*Hipposideros inexpectatu*
Vespertilionidae	*Pipistrellus minahassae*
Vespertilionidae	*Hesperoptenus gaskelli*
Tarsiidae	*Tarsius pumilus**
Tarsiidae	*Tarsius dianae**
Cercopithecidae	*Macaca nigra*
Suidae	*Babyrousa babyrussa*
Bovidae	*Bubalus quarlesi**
Sciuridae	*Hyosciurus heinrichi**
Sciuridae	*Prosciurillus weberi**
Sciuridae	*Prosciurillus abstrusus**
Muridae	*Rattus mollicomulus**
Muridae	*Rattus xanthurus*
Muridae	*Rattus marmosurus**
Muridae	*Maxomys dollmani**
Muridae	*Maxomys wattsi**
Muridae	*Crunomys celebensis**
Muridae	*Bunomys coelestis**
Muridae	*Bunomys prolatus**
Muridae	*Bunomys fratrorum**
Muridae	*Bunomys heinrichi**
Muridae	*Bunomys penitus**
Muridae	*Eropeplus canus**
Muridae	*Margaretamys elegans**
Muridae	*Margaretamys parvus**
Muridae	*Taeromys hamatus**
Muridae	*Taeromys arcuatus**
Muridae	*Melasmothrix naso**
Muridae	*Melasmothrix rhinogradoi**
Muridae	*Melasmothrix macrocercus**

An asterisk signifies that the species' range is limited to this ecoregion.

high plant endemism, and the several distinct forest types provide habitat for the highest number of endemic mammals in Asia and several endemic birds (Whitten et al. 1987).

The ecoregion harbors 102 mammal species, of which 33 species are endemic or near endemic (table J110.1). Together with the lowland forests, the montane forests of Sulawesi have the highest recorded number of endemic mammals among the Indo-Pacific ecoregions. These endemic species include the endangered mountain anoa (*Bubalus quarlesi*) and crested macaque (*Macaca nigra*) and the vulnerable babirusa (*Babyrousa babyrussa*) and Sulawesi montane long-nosed squirrel (*Hyosciurus heinrichi*) (Flannery 1995; Hilton-Taylor 2000).

There are approximately 168 bird species listed as resident in the ecoregion, of which 44 species are endemic or near endemic (table J110.2). The ecoregion also overlaps the montane portions of the Sulawesi EBA (Stattersfield et al. 1998). Of the fifty-four restricted-range bird species found in the EBA, fourteen species are found in both lowland and montane Sulawesi, and twenty-two species are only found in the uplands of Sulawesi. Nineteen of these montane species are found nowhere else on Earth. Two montane bird species are classified as threatened: the endangered Lompobattang flycatcher (*Ficedula bonthaina*) and Matinan flycatcher (*Cyornis sanfordi*) (BirdLife International 2000).

Two Centres of Plant Diversity are found in the uplands of Sulwesi: Dumoga-Bone National Park and Pegunungan Latimojong. The montane forests of Dumoga-Bone National Park contain a rich gene pool of timber trees and rattans and are dominated by *Eugenia*, *Shorea*, and *Agathis*, with an abundance of rattans in the understory. The lower montane forests of Pegunungan Latimojong and contain *Lithocarpus*, *Phyllocladus hypophyllus*, *Podocarpus steupi*, and *Taxus sumatrana*, whereas the upper montane areas contain *Vaccinium* and *Rhododendron vanvuurenii*, *Hypericum leschenaultii*, and *Drimys piperata*. The area extends to 3,455 m and contains extensive sub-alpine vegetation above 3,200 m (Davis et al. 1995).

TABLE J110.2. Endemic and Near-Endemic Bird Species.

Family	Common Name	Species	Family	Common Name	Species
Accipitridae	Small sparrowhawk	*Accipiter nanus*	Pachycephalida	Maroon-backed whistler	*Coracornis raveni**
Megapodiidae	Maleo	*Macrocephalon maleo*	Pachycephalida	Sulphur-bellied whistler	*Pachycephala sulfuriventer*
Rallidae	Platen's rail	*Aramidopsis plateni*	Rhipiduridae	Rusty-flanked fantail	*Rhipidura teysmanni*
Rallidae	Bare-faced rail	*Gymnocrex rosenbergii*	Dicruridae	Sulawesi drongo	*Dicrurus montanus*
Rallidae	Isabelline waterhen	*Amaurornis isabellinus*	Campephagidae	Cerulean cuckoo-shrike	*Coracina temminckii*
Scolopacidae	Sulawesi woodcock	*Scolopax celebensis**	Campephagidae	Pygmy cuckoo-shrike	*Coracina abbotti**
Columbidae	Sulawesi ground-dove	*Gallicolumba tristigmata*	Turdidae	Geomalia	*Geomalia heinrichi**
Columbidae	Red-eared fruit-dove	*Ptilinopus fischeri**	Turdidae	Sulawesi thrush	*Cataponera turdoides**
Columbidae	White-bellied imperial-pigeon	*Ducula forsteni*	Turdidae	Great shortwing	*Heinrichia calligyna**
Columbidae	Grey-headed imperial-pigeon	*Ducula radiata*	Sturnidae	Pale-bellied myna	*Acridotheres cinereus*
Columbidae	Sombre pigeon	*Cryptophaps poecilorrhoa**	Sturnidae	Sulawesi myna	*Basilornis celebensis*
Loriidae	Yellow-and-green lorikeet	*Trichoglossus flavoviridis*	Sturnidae	Fiery-browed myna	*Enodes erythrophris*
Cuculidae	Sulawesi hawk-cuckoo	*Cuculus crassirostris*	Muscicapidae	Lompobattang flycatcher	*Ficedula bonthaina**
Strigidae	Ochre-bellied hawk-owl	*Ninox ochracea*	Muscicapidae	Matinan flycatcher	*Cyornis sanfordi**
Tytonidae	Minahassa owl	*Tyto inexspectata*	Muscicapidae	Blue-fronted flycatcher	*Cyornis hoevelli**
Caprimulgidae	Diabolical nightjar	*Eurostopodus diabolicus*	Zosteropidae	Black-ringed white-eye	*Zosterops anomalus*
Alcedinidae	Scaly kingfisher	*Actenoides princeps*	Zosteropidae	Streak-headed white-eye	*Lophozosterops squamiceps**
Meropidae	Purple-bearded bee-eater	*Meropogon forsteni*	Sylviidae	Chestnut-backed bush-warbler	*Bradypterus castaneus*
Coraciidae	Purple-winged roller	*Coracias temminckii*	Sylviidae	Sulawesi leaf-warbler	*Phylloscopus sarasinorum*
Meliphagidae	Dark-eared honeyeater	*Myza celebensis**	Timaliidae	Malia	*Malia grata**
Meliphagidae	Greater streaked honeyeater	*Myza sarasinorum**	Dicaeidae	Crimson-crowned flowerpecker	*Dicaeum nehrkorni*
Pachycephalida	Olive-flanked whistler	*Hylocitrea bonensis**	Fringillidae	Mountain serin	*Serinus estherae*

An asterisk signifies that the species' range is limited to this ecoregion.

Status and Threats

Current Status

This ecoregion is still largely intact, with about three-quarters of the original habitat remaining. Most of the habitat destruction has occurred in the southwestern portion, and large blocks of forest remain in the northern and eastern montane areas of the island. The twenty-nine protected areas cover 23 percent of the ecoregion (table J110.3). The average size of a protected area in this ecoregion is 602 km^2, and there are five protected areas that exceed 1,000 km^2.

Types and Severity of Threats

The steep slopes and the relative lack of commercially valuable tree species help to discourage logging activity. However, the logging that has occurred has had devastating effects on the landscape and the ecosystems; for instance, extensive erosion on surrounding deforested slopes has clogged the irrigation systems of the once fertile rice fields of Palu Valley (Whitten et al. 1987). Hunting and anthropogenic fires are also serious threats to the wildlife assemblages and habitat. Hunters set fires to facilitate hunting of anoa, creating montane meadows. Upper montane and sub-alpine forests are subject to periods of drought, during which the oil-rich leaves of *Rhododendron*, *Vaccinium*, and *Gaultheria* easily catch fire. With repeated burning, alang-alang grass (*Imperata cylindrica*) may become dominant. Other threats include transmigration and local clearance (Whitten et al. 1987).

TABLE J110.3. WCMC (1997) Protected Areas That Overlap with the Ecoregion.

Protected Area	Area (km^2)	IUCN Category
Gunung Kelabat	30	DE
Gunung Soputan	120	VI
Gunung Simbalang	270	PRO
Dumoga [109]	1,890	?
Dolongan	1	IV
Pinjan/Tanjung Matop	6	IV
Kelompok Hutan Buol Toli-toli	3,920	PRO
Kelompok extension	630	PRO
Gunung Sojol	690	PRO
Palu Mountains	3,190	PRO
Wera	4	V
Lore Lindu	2,220	II
Palu Mountains	320	PRO
Palu Mountains	130	PRO
Morowali [109]	1,150	I
Rompi	170	PRO
Rangkong	310	PRO
Lamiko-miko	280	PRO
Mambuliling	110	PRO
Pegunungan Latimojong	510	VI
Lampoko Mampie	20	IV
Peg. Feruhumpenai	860	I
Danau Matano	290	V
Bulu Saraung	50	I
Sungai Camba	10	IV
Karaenta	4	I
Bantimurung	5	I
Gunung Lompobatang	180	PRO
Tirta Rimba	90	V
Total	17,460	

Ecoregion numbers of protected areas that overlap with additional ecoregions are listed in brackets.

Appendix J: Ecoregion Descriptions

Priority Conservation Actions

Short-Term Conservation Actions (1–5 Years)

Oliver (1993) outlines objectives intended to safeguard populations of babirusa, including determining its present distribution and relative population sizes; establishing a network of protected areas representing all subspecies; conducting further research into systematics, biology, and conservation status; and promoting interest and awareness among local people.

- Initiate a major public awareness campaign on forest values and forest loss to drive political will.
- Stop illegal logging in and around protected areas by challenging the authorities and judiciary to act on existing laws and regulations.
- Revoke forest concessions in areas of high biodiversity outside protected areas and institute a moratorium on new concessions for oil palm, timber, and other plantations using instead the vast areas of deforested or cleared land.
- Mobilize major assistance for the more outstanding protected areas.
- Stop wildlife poaching, especially in key parks and reserves.

Longer-Term Conservation Actions (5–20 Years)

The Nature Conservancy outlines several priority actions in its work in Sulwesi, including the following:

- Train a strong team of Indonesian nationals to become an effective, long-term force for conservation and protected areas management.
- Develop business plans based on socioeconomic and ecological surveys to help launch community-based small businesses that both protect the environment and provide income to people who live in and around the parks.
- Provide dynamic, field-tested environmental education materials and training to teachers, community leaders, and resource managers.
- Develop management plans for Lore Lindu National Park, at the government's request.
- Ensure that redesign of the protected area system includes linkages between lowland and montane habitats.
- Devise ways to maintain the integrity of buffer zones of protected areas.

Focal Species for Conservation Action

Endangered endemic species: Sulawesian shrew-rat (*Melasmothrix naso*) (Heaney et al. 1998), Watt's spiny rat (*Maxomys wattsi*) (Heaney et al. 1998), heavenly hill rat (*Bunomys coelestis*), long-headed hill rat (*Bunomys prolatus*) (Heaney et al. 1998), Sulawesi soft-furred rat (*Eropeplus canus*) (Heaney et al. 1998), Celebes shrew-rat (*Crunomys celebensis*) (Heaney et al. 1998), mountain anoa (*Bubalus quarlesi*) (Hedges, in press), crested macaque (*Macaca nigra*) (Eudey 1987), Lompobattang flycatcher (*Ficedula bonthaina*) (BirdLife International 2000), Matinan flycatcher (*Cyornis sanfordi*) (BirdLife International 2000)

Area-sensitive species: mountain anoa (*Bubalus quarlesi*) (Hedges, in press), babirusa (*Babyrousa babyrussa*) (Oliver 1993), knobbed hornbill (*Aceros cassidix*) (Kemp 1995)

Top predator species: Sulawesi serpent-eagle (*Spilornis rufipectus*), Sulawesi goshawk (*Accipiter griseiceps*), spot-tailed goshawk (*Accipiter trinotatus*), black eagle (*Ictinaetus malayensis*), rufous-bellied eagle (*Hieraaetus kienerii*), Sulawesi hawk-eagle (*Spizaetus lanceolatus*)

Selected Conservation Contacts

Lembaga Alam Tropika Indonesia (LATIN)

The Nature Conservancy

WWF-Indonesia

Yayasan Keanekaragaman Hayati

Yayasan Konsorsium Pembaruan Agraria (KPA)

Justification for Ecoregion Delineation

There have been several attempts to divide the bioregion into biogeographic units (MacKinnon 1997; Stattersfield et al. 1998; van Balgooy 1971, cited in Monk et al. 1997; MacKinnon and Artha 1981; MacKinnon and MacKinnon 1986; MacKinnon et al. 1982; van Steenis 1950; Udvardy 1975). Because many of the islands have distinct natural faunal communities and a high degree of endemism (Monk et al. 1997), the more recent attempts have used faunal dissimilarities—especially birds—to identify distinct biogeographic units (MacKinnon 1997; Stattersfield et al. 1998; MacKinnon and Artha 1981; MacKinnon and MacKinnon 1986). Because detailed floral data are largely unavailable across most of the bioregion, we followed these authors in delineating ecoregions based on distribution of biomes and vertebrate communities.

On Sulawesi island we delineated two ecoregions: the Sulawesi Lowland Rain Forests [109] and Sulawesi Montane Rain Forests [110]. These represent the tropical lowland and montane tropical moist forests, respectively. The small patches of monsoon forests on the southwest peninsula of Sulawesi and on Butung Island (Whitmore 1984) were included in the Sulawesi Lowland Rain Forests [109] but should be considered a distinct habitat type in an ecoregion-based conservation assessment to ensure representation.

Prepared by John Morrison

Ecoregion Number:	**111**
Ecoregion Name:	**Lesser Sundas Deciduous Forests**
Bioregion:	**Wallacea**
Biome:	**Tropical and Subtropical Dry Broadleaf Forests**
Political Units:	**Indonesia**
Ecoregion Size:	**39,300 km²**
Biological Distinctiveness:	**Globally Outstanding**
Conservation Status:	**Endangered**
Conservation Assessment:	**I**

The Lesser Sundas Deciduous Forests [111] are found on a string of volcanic islands. They stretch across the Java Sea between Australia and Borneo. It is part of a unique biogeographic region known as Wallacea, which contains a very distinctive fauna representing a mix of Asian and Australasian species. These distinctive seasonal dry forests harbor unique species, including the Komodo dragon, the largest lizard in the world, and seventeen bird species found nowhere else on Earth. A combination of shifting agriculture and human-caused fires has significantly reduced the amount of natural forest in this ecoregion.

Location and General Description

This ecoregion represents the semi-evergreen dry forests in the Lesser Sunda Islands. It extends east from the islands of Lombok and Sumbawa to Flores and Alor in the Indonesian Archipelago. Rinjani volcano on Lombok is the highest mountain in the ecoregion, at 3,726 m. The Lesser Sundas are an inner volcanic island arc, created by the subduction and partial melting of the Australian tectonic plate below the Eurasian plate. The islands represent tertiary and quaternary volcanoes that have coalesced with lava and sediment. There is actually a geologic discontinuity between Lombok and Sumbawa, on the Sunda Arc, and the rest of the islands, part of the Banda Arc. With the exception of Komodo, which is Mesozoic, most of the islands were built during two pulses in the Tertiary (Mio-Pliocene) and Quaternary (recent) (Monk et al. 1997). This ecoregion is separated from Bali and Java to the west by Wallace's Line, which marks the end of the Sunda Shelf. With an average annual rainfall of 1,349 mm, this region is the driest but also the most seasonal in Indonesia. Based on the Köppen climate system, this ecoregion has a tropical dry climate zone (National Geographic Society 1999). This distinctive climate has given rise to a vegetation that is strikingly different from that of the rest of the archipelago. Much of the natural habitat is composed of monsoon forests and savanna woodlands (Whitten and Whitten 1992).

The monsoon forests consist of several forest subtypes, notably moist deciduous forest, dry deciduous forest, dry thorn forest, and dry evergreen forest. Moist deciduous forests also occur as a band of lowland forest at the base of the hills and as gallery forests along streams, especially on Komodo Island. Dominant trees include *Tamarindus indica* and *Sterculia foetida* (Monk et al. 1997). The dry deciduous forest at altitudes below 200 m is dominated by *Protium javanicum*, *Schleichera oleosa*, and *Schoutenia ovata*, whereas at medium altitudes, from 200 to 800 m, the dominant tree is *Tabernaemontana floribunda*. At these altitudes, lianas and climbers become common, especially the white-flowered liana *Bauhinia*. Above 1,000 m, Euphorbiaceae tend to become common and well represented (Monk et al. 1997).

Dry thorn forest is another type of monsoon forest in this ecoregion, although little is left because it has been cleared by setting fires. This forest formation still exists along the southeast coast of Lombok and the southwest coast of Sumbawa but is being cleared in the latter region for road building, mine development, and a transmigration site (Monk et al. 1997).

Dry evergreen forest occurs above dry deciduous forest and below the true evergreen montane forest, at 1,000 m above sea level on Mt. Batulante in northwest Sumbawa. Below 1,200 m on the north slopes, *Albizia chinensis* is a characteristic species. Other common species include *Chionanthus*, *Prunus*, and two *Cryptocarya* species. On many islands, drier areas in steep-sided valleys contain gallery forest. On

Sumbawa, for instance, gallery forest is found from sea level to 2,000 m above sea level and is also present in lower montane forests (Monk et al. 1997). By contrast, the southern hill slopes along the southern coasts are kept moist during the dry season by the southeast trade winds, and dipterocarp rain forest occurs on the southwest hills of both Lombok and Sumbawa. Lombok also contains one of the few remaining patches of tropical semi-evergreen rain forest, at volcanic Mt. Rinjani, which acts as the major water catchment area for the whole island (Monk et al. 1997).

Twenty-meter-high mixed montane forests of *Podocarpus* and *Engelhardia* are found from about 1,200 to 2,100 m, with lianas, epiphytes, and orchids such as *Corybas*, *Corymborkis*, and *Malaxis* very much in evidence. At higher elevations of up to 2,700 m, *Casuarina junghuhniana* forests occur. Toward the summit, from 3,300 to 3,400 m, the rocky ridges were once covered with lichens, mosses, grasses, herbs, and some ferns but are now being eroded. On Sumbawa, the south slopes of Mt. Batulante above 1,000 m are covered with a *Cryptocarya*-Meliaceae montane forest, although species composition varies with moisture. This forest is dominated by two species of *Cryptocarya*, one in the drier and usually lower forest (from 1,000 to 1,500 m above sea level) and the other at higher or moister sites. Drier, stonier slopes in poorer forest are the only places where lianas are common. Further east, from the eastern part of Flores to Alor, the forests are dominated by *Pterocarpus indicus* (Monk et al. 1997).

There are also two types of savanna in this ecoregion: a *Borassus flabellifer* savanna that occurs from sea level to 400 m on Komodo, Rinca, and the north and south coasts of Flores; and the *Ziziphus mauritiana* savanna, which occurs on more sandy clay alluvial, and sometimes water-logged, soil. The dominant grasses are *Eulalia leschenaultiana*, spear grass (*Heteropogon contortus*, *Themeda frondosa*), and *Themeda triandra* (Monk et al. 1997).

Biodiversity Features

This area, part of the Wallacean sub-region, includes a mix of Asian and Australian fauna, and because of the long years of isolation from the mainland it harbors many endemic mammals and birds. Most of the endemic mammals occur on Komodo and Flores eastward, rather than Lombok and Sumbawa. One of the important and better-known endemic species in this ecoregion is the Komodo dragon (*Varanus komodoensis*), the largest lizard in the world.

The mammal faunal in this ecoregion consists of fifty species, including five ecoregional endemics, including the critically endangered Flores shrew (*Suncus mertensi*) and the vulnerable Komodo rat (*Komodomys rintjanus*) (table J111.1) (Hilton-Taylor 2000). With the exception of the New Caledonia dry forests [140], with six endemic mammals, the five endemic mammals in this ecoregion are more than are found in any other dry forest ecoregion in the Indo-Pacific.

This ecoregion also harbors about 273 bird species, of which 29 are endemic or near endemic (table J111.2). The ecoregion is consistent with the Northern Nusa Tenggara EBA (Stattersfield et al. 1998). Of the twenty-nine restricted-range species in the EBA, seventeen are found nowhere else in the world. Four are endemic and endangered: Flores scops

TABLE J111.1. Endemic and Near-Endemic Mammal Species.

Family	Species
Soricidae	*Suncus mertensi**
Pteropodidae	*Pteropus lombocensis**
Vespertilionidae	*Nyctophilus heran**
Muridae	*Bunomys naso**
Muridae	*Komodomys rintjanus**

An asterisk signifies that the species' range is limited to this ecoregion.

TABLE J111.2. Endemic and Near-Endemic Bird Species.

Family	Common Name	Species
Columbidae	Dusky cuckoo-dove	*Macropygia magna*
Columbidae	Flores green-pigeon	*Treron floris**
Columbidae	Pink-headed imperial-pigeon	*Ducula rosacea*
Columbidae	Dark-backed imperial-pigeon	*Ducula lacernulata*
Psittacidae	Wallace's hanging-parrot	*Loriculus flosculus**
Loriidae	Olive-headed lorikeet	*Trichoglossus euteles*
Strigidae	Flores scops-owl	*Otus alfredi**
Strigidae	Wallace's scops-owl	*Otus silvicola**
Alcedinidae	Cinnamon-backed kingfisher	*Todirhamphus australasia*
Alcedinidae	White-rumped kingfisher	*Caridonax fulgidus**
Meliphagidae	Sunda honeyeater	*Lichmera lombokia**
Pachycephalida	Bare-throated whistler	*Pachycephala nudigula**
Rhipiduridae	Brown-capped fantail	*Rhipidura diluta**
Monarchidae	Flores monarch	*Monarcha sacerdotum**
Corvidae	Flores crow	*Corvus florensis**
Campephagidae	Sumba cuckoo-shrike	*Coracina dohertyi*
Campephagidae	Flores minivet	*Pericrocotus lansbergei**
Turdidae	Chestnut-backed thrush	*Zoothera dohertyi*
Muscicapidae	Flores jungle-flycatcher	*Rhinomyias oscillans*
Zosteropidae	Yellow-spectacled white-eye	*Zosterops wallacei*
Zosteropidae	White-browed white-eye	*Lophozosterops superciliaris**
Zosteropidae	Dark-crowned white-eye	*Lophozosterops dohertyi**
Zosteropidae	Flores white-eye	*Heleia crassirostris**
Sylviidae	Russet-capped tesia	*Tesia everetti**
Sylviidae	Timor leaf-warbler	*Phylloscopus presbytes*
Dicaeidae	Golden-rumped flowerpecker	*Dicaeum annae**
Dicaeidae	Black-fronted flowerpecker	*Dicaeum igniferum**
Dicaeidae	Red-chested flowerpecker	*Dicaeum maugei*
Nectariniidae	Flame-breasted sunbird	*Nectarinia solaris*

An asterisk signifies that the species' range is limited to this ecoregion.

owl (*Otus alfredi*), Fores monarch (*Monarcha sacerdotum*), Wallace's hanging-parrot (*Loriculus flosculus*), and Flores crow (*Corvus florensis*). In addition, the white-rumped kingfisher (*Caridonax fulgidus*) is the sole representative of an endemic monotypic genus (BirdLife International 2000).

The Komodo dragon deserves special mention. It is the largest lizard species in the world. *Varanus komodoensis* occupies five islands: Komodo, Padar, Rinca, Gili Motang, and Flores. These animals range from sea level to approximately 450 m in elevation, mainly in tropical deciduous monsoon forest, tropical savanna, and grassland. They feed on a wide variety of animal food, including insects, lizards, snakes, birds, deer, wild boar, monkeys, and bird eggs; they also feed on carrion. Adults may have a foraging range of 500 ha. There are approximately 4,000 protected individuals in Komodo National Park (Monk et al. 1997).

Status and Threats

Current Status

During World War II, logging and cultivation destroyed much of the forest cover of Lombok, east Flores, and the small islands of Adonara, Solor, Lomblen, Pantar, and Alor (Monk et al. 1997). The Lombok dipterocarp forest is almost depleted by commercial logging, and the forest of Sumbawa is partially covered by a mining concession (Monk et al. 1997).

More than half of this ecoregion's natural habitat has been cleared, mainly for agriculture. Except for the island of Sumbawa, which still contains a large block of intact forest, most of the islands in this group have only fragments of natural habitat remaining. The twenty-eight protected areas include about 10 percent of the ecoregion area, but most of the protected areas are small, with the average size being only 144 km^2 (table J111.3). Komodo National Park, the most famous wild area in the Lesser Sundas, a World Heritage Site, and an important tourist destination, is only one of two reserves that is greater than 500 km^2, but most of this park is marine. The park harbors the Komodo dragon. Lowlands are underrepresented in the protected areas system even though this habitat supports most of the ecoregion's species; for instance, most of Sumbawa's endemic birds are associated with lowland monsoon forest (Monk et al. 1997).

Increasing population pressure has also resulted in high rates of deforestation (WWF-Indonesia n.d.). In the dry season, fires often are set to clear the understory and to encourage new growth as forage for domestic animals. This has been done since prehistoric times—not for domestic animals, but for attracting game (introduced) into areas of new grass growth. This practice has resulted in the proliferation of fire-resistant trees such as *Casuarina junghuhniana* and formation of grassland over an extensive area of these islands. Most of the remaining forest is confined to the steepest slopes and the tops of mountains (Whitten and Whitten 1992). Poorly managed tourism, especially in the Komodo National Park and on Lombok, has also caused environmental degradation (WWF-Indonesia n.d.).

Types and Severity of Threats

The future threats will continue to be deforestation, an increasing population and their demands on the environment, intentionally set fires, and agricultural land development.

Priority Conservation Actions

Short-Term Conservation Actions (1–5 Years)

- Initiate a major public awareness campaign on forest values and forest loss to drive political will.
- Stop illegal logging in and around protected areas by challenging the authorities and judiciary to act on existing laws and regulations.
- Place under active protection all remaining forests.
- Stop wildlife poaching, especially in key parks and reserves, by tightening surveillance around ports or airports and making the purchase of endangered species socially unacceptable.
- Conserve Tanjung Kerita forest, *Varanus* reserves on north coast of Flores, and Hadakewa island forests

Longer-Term Conservation Actions (5–20 Years)

The Nature Conservancy outlines several priority actions in its work in Komodo, including the following:

- Strengthen the park authority's capacity for fishery management, enforcement, monitoring, and research.
- Create economic alternatives to destructive fishing, including ecotourism and environmentally sound mariculture.

Focal Species for Conservation Action

Endangered endemic species: Flores shrew (*Suncus mertensi*), (*Nyctophilus heran*) (Vespertilionidae),

TABLE J111.3. WCMC (1997) Protected Areas That Overlap with the Ecoregion.

Protected Area	Area (km^2)	IUCN Category
Pulau Sangiang	150	PRO
Pulau Moyo	160–222	VI
Tambora Utara GR	480	PRO
Tambora Selatan	150	VI
Gunung Rinjani	830 + 760 ext.	II
Pulau Panjang	200	PRO
Hutan Dompu Complex	110	PRO
Gunung Olet Sangenges NR	280–350	PRO
Suranadi	5	V
Pulau Rakit	20	PRO
Batu Gendang Forest	150	PRO
Pantai Palolowaru	5	PRO
Selahu Legini Complex	320–500	VI
Kurung Baya/Varanus	40	PRO
East Timor	210	PRO
Tanjung Kerita Mese	300	PRO
Danau Rana Mese	2	PRO
Danau Kelimutu	20	DE
Danau Sano	8	PRO
Gunung Ambu Lombo	40	PRO
Tuti	60	V
Tanjung Watupayung	90	PRO
Hadekawa-Labelakang	250	PRO
Gunung Muna	100–150	PRO
Adonara NR	20	PRO
Pulau Rusa	10	PRO
Egon-Iliwuli	30	PRO
Lewotobi	8	VI
Total	4,048+	

Ecoregion numbers of protected areas that overlap with additional ecoregions are listed in brackets.

Flores monarch (*Monarcha sacerdotum*) (BirdLife International 2000), Flores scops owl (*Otus alfredi*) (BirdLife International 2000), Wallace's hanging-parrot (*Loriculus flosculus*) (BirdLife International 2000), Flores crow (*Corvus florensis*) (BirdLife International 2000)

Area-sensitive species: Komodo dragon (*Varanus komodoensis*), Eurasian wild pig (*Sus scrofa*) (introduced?) (Oliver 1993), Sulawesi warty pig (*Sus celebensis*) (introduced?) (Oliver 1993), Sunda sambar (*Cervus timorensis*) (Wemmer 1998), barking deer (*Muntiacus muntjak*) (Wemmer 1998)

Top predator species: leopard cat (*Prionailurus bengalensis*) (Nowell and Jackson 1996), palm civet (*Paradoxurus hermaphroditus*) (Schreiber et al. 1989), Komodo dragon (*Varanus komodoensis*)

Selected Conservation Contacts

Lembaga Alam Tropika Indonesia (LATIN)

The Nature Conservancy

WWF-Indonesia

Yayasan Keanekaragaman Hayati

Yayasan Konsorsium Pembaruan Agraria (KPA)

Justification for Ecoregion Delineation

The drier forests in Nusa Tenggara were placed in three ecoregions that corresponded to the biogeographic units identified in Monk et al (1997). These are Lesser Sundas Deciduous Forests [111], which includes the chain of islands extending from Lombok, Sumbawa, Komodo, Flores, and the smaller satellite islands corresponding to the Flores biogeographic unit; Timor and Wetar Deciduous Forests [112], corresponding to the Timor biogeographic unit; and the Sumba Deciduous Forests [113], corresponding to the Sumba biogeographic unit. All three ecoregions belong to the tropical dry forests biome.

Prepared by John Morrison

Ecoregion Number:	**112**
Ecoregion Name:	**Timor and Wetar Deciduous Forests**
Bioregion:	**Wallacea**
Biome:	**Tropical and Subtropical Dry Broadleaf Forests**
Political Units:	**Indonesia**
Ecoregion Size:	**35,200 km^2**
Biological Distinctiveness:	**Globally Outstanding**
Conservation Status:	**Endangered**
Conservation Assessment:	**I**

The Timor and Wetar Deciduous Forests [112] are found on both inner and outer island arcs at the collision point of the Eurasian and Australian tectonic plates. The seasonally dry forests found in this dynamic geologic setting are part of the region known as Wallacea, which contains a very distinctive fauna representing a mix of Asian and Australasian species. Nearly two-thirds of the original extent of forest has been cleared, and the ecoregion contains only fragments of natural habitat, which are themselves threatened.

Location and General Description

This ecoregion represents the semi-evergreen dry forests of Timor, Wetar, and some smaller islands in the provinces of Nusa Tenggara and Maluku in the eastern Indonesian Archipelago. This ecoregion has a dry climate, with the most xeric being the mountains of Timor. Moa, in the Leti Islands, receives an average of 1,329 mm rainfall spread over just sixty-six days of the year. Based on the Köppen climate zone system, this ecoregion falls in the tropical dry climate zone (National Geographic Society 1999). The geology of the islands is a combination of inner and outer volcanic island arcs. Wetar, Romang, Damar, and the Banda Islands are part of the inner arc, and Timor, the Leti Islands, Sermata, and Babar are part of the outer arc. The inner arc islands are a result of the subduction and partial melting of the Australian tectonic plate below the Eurasian plate. With the exception of Wetar, the inner arc islands represent young volcanoes that have coalesced with lava and sediment. The basement rock of the outer islands, on the other hand, is composed of actual continental margin from the Australian plate that has not been subducted. These outer islands are less than 4 million years old. The resulting surface geology consists of complex sedimentary and metamorphic rocks: uplifted coral reefs over complex basement rocks (Monk et al. 1997).

The forest types in the ecoregion are dry deciduous, dry evergreen, and thorn forests. Below 1,000 m the common tree species include *Sterculia foetida* and *Calophyllum teysmannii* (both of which produce oil-bearing seeds) and *Aleurites moluccana*. The lowland monsoon forests are dominated by *Pterocarpus indicus*, especially in the lowland monsoon forest remnants of West Timor and in the well-drained, dry soils north of Oebelo on the Bena coastal plain in south Timor (Monk et al. 1997). Semi-evergreen rain forest is found on southern hill slopes at Buraen, which are kept moist by southeast trade winds, and on the Damar Islands (Monk et al. 1997). East Timor's few remaining forest patches contain the last natural stands of *Eucalyptus urophylla* (now widely used in plantations) and *Santalum album*, the sandalwood tree (Whitten and Whitten 1992). The shrub layer in these forests includes Verbenaceae, Rubiaceae, and Euphorbiaceae, and the herbs include Acanthaceae, *Tacca palmata*, the root parasite *Balanophora fungosa*, and ground orchids such as *Corymborkis* (Monk et al. 1997). Four types of savanna are found here, each characterized by palm, *Eucalyptus*, *Acacia* spp., and *Casuarina* spp. On Timor's larger coastal plains, the vegetation ranges from grassland to open stands of deciduous trees, with increasing forest cover toward the moister southern mountains.

Biodiversity Features

This ecoregion has the greatest number of bird species of any tropical dry forest ecoregion in the Indo-Pacific region. Because of the long isolation with the mainland communities, there are several endemic species from several taxonomic groups.

The ecoregion has thirty-eight mammal species, five of which are endemic or near endemic (table J112.1). Both

Asian species and an Australasian cuscus (*Phalanger orientalis timorensis*) are found on the islands. *Crocidura tenuis* (Soricidae), possibly introduced by man, and the Flores giant rat (*Papagomys armandvillei*) are considered vulnerable (Hilton-Taylor 2000).

The bird fauna consists of about 229 species. The bird fauna also represents a mix of mostly Asian species with some Australasian birds. Endemism is extremely high for these islands, with thirty-five species that are endemic or near endemic (table J112.2). The ecoregion encompasses with the Timor and Wetar EBA (Stattersfield et al. 1998). Thirty-five restricted-range bird species are found in the Timor and Wetar EBA, twenty-three of which are found nowhere else on Earth. Three of these species are endangered: the Wetar ground-dove (*Gallicolumba hoedtii*), Timor green-pigeon (*Treron psittacea*), and Timor imperial-pigeon (*Ducula cineracea*) (BirdLife International 2000).

Timor also harbors the endemic and rare Timor python (*Python timoriensis*) (Whitten and Whitten 1992).

Status and Threats

Current Status

Other than one remaining large block of forest near the center of Timor Island, this ecoregion contains only fragments of natural habitat. Nearly two-thirds of the original extent of forest has been cleared, mostly for agriculture. Most of the original monsoon forest on these islands has been replaced by savanna and grassland. On East Timor, the south escarpment of the Fuiloro limestone plateau originally was covered by primary rain forest, but in the 1950s this area was degraded to secondary forest. Wetar is threatened by poorly managed gold mines that have been passed from company to company, causing major environmental damage. There are twenty-four protected areas that include roughly 10 percent (3,661 km^2) of the ecoregion area, but all are small, with the average size being only 152 km^2 (Monk et al. 1997) (table J112.3).

Types and Severity of Threats

Deforestation is occurring very rapidly as people burn the forests for hunting, shifting cultivation, and fodder production (Whitten and Whitten 1992; Monk et al. 1997; WWF-Indonesia n.d.). Logging has also grown in importance; for instance, Damar Island was densely forested until the late 1980s, when logging began on a large scale to supply timber to the outer arc islands, where the forests had already been more heavily exploited. As a result, fire-resistant *Casuarina junghuhniana* grows in pure stands in cleared areas, and Mt.

TABLE J112.1. Endemic and Near-Endemic Mammal Species.

Family	Species
Sorcidae	*Crocidura tenuis**
Pteropodidae	*Pteropus chrysoproctus*
Rhinolophidae	*Rhinolophus canuti*
Muridae	*Papagomys armandvillei**
Muridae	*Rattus timorensis**

An asterisk signifies that the species' range is limited to this ecoregion.

TABLE J112.2. Endemic and Near-Endemic Bird Species.

Family	Common Name	Species
Columbidae	Dusky cuckoo-dove	*Macropygia magna*
Columbidae	Black cuckoo-dove	*Turacoena modesta**
Columbidae	Wetar ground-dove	*Gallicolumba hoedtii**
Columbidae	Timor green-pigeon	*Treron psittacea**
Columbidae	Pink-headed imperial-pigeon	*Ducula rosacea*
Columbidae	Timor imperial-pigeon	*Ducula cineracea**
Psittacidae	Olive-shouldered parrot	*Aprosmictus jonquillaceus**
Loriidae	Olive-headed lorikeet	*Trichoglossus euteles*
Loriidae	Iris lorikeet	*Psitteuteles iris**
Alcedinidae	Cinnamon-backed kingfisher	*Todirhamphus australasia*
Acanthizidae	Plain gerygone	*Gerygone inornata**
Meliphagidae	White-tufted honeyeater	*Lichmera squamata*
Meliphagidae	Yellow-eared honeyeater	*Lichmera flavicans**
Meliphagidae	Black-chested honeyeater	*Lichmera notabilis**
Meliphagidae	Crimson-hooded myzomela	*Myzomela kuehni**
Meliphagidae	Black-breasted myzomela	*Myzomela vulnerata**
Meliphagidae	Streak-breasted honeyeater	*Meliphaga reticulata**
Meliphagidae	Timor friarbird	*Philemon inornatus**
Pachycephalida	Fawn-breasted whistler	*Pachycephala orpheus**
Oriolidae	Timor oriole	*Oriolus melanotis**
Oriolidae	Timor figbird	*Sphecotheres viridis**
Oriolidae	Wetar figbird	*Sphecotheres hypoleucus**
Turdidae	Chestnut-backed thrush	*Zoothera dohertyi*
Turdidae	Orange-banded thrush	*Zoothera peronii*
Muscicapidae	Black-banded flycatcher	*Ficedula timorensis**
Muscicapidae	Timor blue-flycatcher	*Cyornis hyacinthinus**
Muscicapidae	Timor bushchat	*Saxicola gutturalis**
Zosteropidae	Timor white-eye	*Heleia muelleri**
Sylviidae	Timor stubtail	*Urosphena subulata*
Sylviidae	Timor leaf-warbler	*Phylloscopus presbytes*
Sylviidae	Buff-banded bushbird	*Buettikoferella bivittata**
Estrildidae	Tricolored parrotfinch	*Erythrura tricolor*
Estrildidae	Timor sparrow	*Padda fuscata**
Dicaeidae	Red-chested flowerpecker	*Dicaeum maugei*
Nectariniidae	Flame-breasted sunbird	*Nectarinia solaris*

An asterisk signifies that the species' range is limited to this ecoregion.

Mutis, on West Timor, is covered almost exclusively by *Eucalyptus urophylla* (Monk et al. 1997). This problem is worsening as the human populations expand. Savanna areas are especially prone to erosion. This ecoregion is highly threatened. In previous centuries, many forest resources such as sandalwood were depleted through uncontrolled exploitation (Monk et al. 1997).

Priority Conservation Actions

With the rapid pace of forest destruction under way in Indonesia, quick conservation action is urgently needed to stave off the complete collapse of these forests. However, steps should be taken to lay the groundwork for longer-term conservation as well.

Short-Term Conservation Actions (1–5 Years)

- Initiate a major public awareness campaign on forest values and forest loss to drive political will.
- Stop illegal logging in and around protected areas by challenging the authorities and judiciary to act on existing laws and regulations.
- Place all remaining forests under active protection.
- Stop wildlife poaching in key parks and reserves by tightening surveillance around ports or airports and making the purchase of endangered species socially unacceptable.
- Instigate major biodiversity surveys and collecting expeditions in doomed areas to document the biodiversity that will soon be lost.
- Conserve Mutis and remaining habitat on Wetar

Longer-Term Conservation Actions (5–20 Years)

- Develop and implement effective management plans for protected areas.

Focal Species for Conservation Action

Endangered endemic species: Wetar ground-dove (*Gallicolumba hoedtii*) (BirdLife International 2000), Timor green-pigeon (*Treron psittacea*) (BirdLife International 2000), Timor imperial-pigeon (*Ducula cineracea*) (BirdLife International 2000)

Area-sensitive species: Timor deer (*Cervus timorensis*) (Wemmer 1998), fruit bats (Pteropodidae, *Pteropus* spp., *Acerodon macklotii*, *Rousettus amplexicaudatus*, *Dobsonia peronii*, *Cynopterus titthaecheile*, *Nyctimene cephalotes*) (Mickleburgh et al. 1992), Eurasian wild pig (*Sus scrofa*) (introduced?) (Oliver 1993), Sulawesi warty pig (*Sus celebensis*) (introduced?) (Oliver 1993)

Top predator species: palm civet (*Paradoxurus hermaaphroditus*) (Schreiber et al. 1989)

Selected Conservation Contacts

Lembaga Alam Tropika Indonesia (LATIN)

WWF-Indonesia

Yayasan Keanekaragaman Hayati

Yayasan Konsorsium Pembaruan Agraria (KPA)

Justification for Ecoregion Delineation

The drier forests in Nusa Tenggara were placed in three ecoregions that corresponded to the biogeographic units identified in Monk et al (1997). These are Lesser Sundas Deciduous Forests [111], which includes the chain of islands extending from Lombok, Sumbawa, Komodo, Flores, and the smaller satellite islands corresponding to the Flores biogeographic unit; Timor and Wetar Deciduous Forests [112], corresponding to the Timor biogeographic unit; and the Sumba Deciduous Forests [113], corresponding to the Sumba biogeographic unit. All three ecoregions belong to the tropical dry forests biome.

Prepared by John Morrison

TABLE J112.3. WCMC (1997) Protected Areas That Overlap with the Ecoregion.

Protected Area	Area (km²)	IUCN Category
Gunung Api	1	I
Pulau Damar	200	PRO
Pulau Babar	620	PRO
Gunung Arnau	420	PRO
Pulau Kambing	20	PRO
Danau Ira Lalora–Pulau Yaco	120	PRO
Lore	110	?
Gunung Futumasin	30	PRO
Gunung Diatuto	40	PRO
Gunung Talamailu	200	?
Sungai Clere GR	300	?
Tilomar	160	PRO
Gunung Mutis	330	PRO
Gunung Timau	340	PRO
Maubesi	80	I
Keluk Kupang	730	I
Baun Forest	80	PRO
Dataran Bena	100	VI
Manipo	50	V
Teluk Pelikan	30	PRO
Watu Panggota/Bondokapu	30	PRO
Bakau Perhatu	20	PRO
Tanjung Pukuwatu	60	PRO
Pulau Dana	10	PRO
Total	4,081	

Ecoregion numbers of protected areas that overlap with additional ecoregions are listed in brackets.

Ecoregion Number: **113**
Ecoregion Name: **Sumba Deciduous Forests**
Bioregion: **Wallacea**
Biome: **Tropical and Subtropical Dry Broadleaf Forests**
Political Units: **Indonesia**
Ecoregion Size: **10,700 km^2**
Biological Distinctiveness: **Bioregionally Outstanding**
Conservation Status: **Endangered**
Conservation Assessment: **IV**

The Sumba Deciduous Forests [113] are found on the single island of Sumba and are part of the region known as Wallacea, which contains a distinctive fauna representing a mix of Asian and Australasian species. Although vertebrate diversity is low, the ecoregion contains seven bird species found nowhere else in the world and several other birds with very limited ranges. As a result of forest clearance and repeated burning for grazing and agriculture, the forested area of Sumba has declined significantly over the last century.

Location and General Description

This ecoregion represents the semi-evergreen forests on the island of Sumba, in the eastern Indonesian Archipelago. The surface geology of Sumba is composed primarily of sandstone and mudstone, with some igneous intrusions overlain by recent limestone (Whitten and Whitten 1992). Sumba is believed to be a fragment of the Australian continental crust that was separated some 20 million years ago, well before the neighboring outer arc island of Timor (Monk et al. 1997). The island is quite rugged, consisting of deeply dissected plateaus. There is very little area above 1,000 m, and the highest point on the island is 1,225 m (Stattersfield et al. 1998). Precipitation in Sumba is seasonal, and based on the Köppen climate zone system, this ecoregion falls in the tropical dry climate zone (National Geographic Society 1999).

The naturally dominant vegetation of the island was deciduous monsoon forest (Stattersfield et al. 1998). However, the southern hill slopes along the southern coasts, which remain moist during the dry season, are covered with lowland evergreen rain forest. The most extensive and important of these rain forest areas is the Mt. Wanggameti–Laiwanga forest complex in East Sumba, a major water catchment. In East Sumba there are extensive gallery forests in ravines and along rivers that form riparian corridors across open grasslands or savannas. The savanna understory includes an endemic insectivorous sundew (*Drosera indica*) (Monk et al. 1997).

Biodiversity Features

The ecoregion harbors seventeen mammal species, but none are considered to be endemic or even near endemic.

The avifauna of this ecoregion is highly distinctive, with both Asian and Australian influences, although the total diversity is low. There are approximately 180 bird species on the island, and 12 of these species are endemic or near endemic (table J113.1). The ecoregion corresponds to the Sumba EBA. The Sumba EBA contains twelve restricted-range bird species, seven of which are found nowhere else on

TABLE J113.1. Endemic and Near-Endemic Bird Species.

Family	Common Name	Species
Turnicidae	Sumba buttonquail	*Turnix everetti**
Columbidae	Sumba green-pigeon	*Treron teysmannii**
Columbidae	Red-naped fruit-dove	*Ptilinopus dohertyi**
Strigidae	Sumba boobook	*Ninox rudolfi**
Alcedinidae	Cinnamon-backed kingfisher	*Todirhamphus australasia*
Bucconidae	Sumba hornbill	*Aceros everetti**
Campephagidae	Sumba cuckoo-shrike	*Coracina dohertyi*
Turdidae	Chestnut-backed thrush	*Zoothera dohertyi*
Muscicapidae	Flores jungle-flycatcher	*Rhinomyias oscillans*
Muscicapidae	Sumba flycatcher	*Ficedula harterti**
Zosteropidae	Yellow-spectacled white-eye	*Zosterops wallacei*
Nectariniidae	Apricot-breasted sunbird	*Nectarinia buettikoferi**

An asterisk signifies that the species' range is limited to this ecoregion.

Earth. One of these species is endangered: Sumba buttonquail (*Turnix everetti*), and two others are considered vulnerable: red-naped fruit-dove (*Ptilinopus dohertyi*) and Sumba hornbill (*Aceros everetti*). These threatened species have specific habitat needs that make them susceptible to forest clearance (Stattersfield et al. 1998).

Status and Threats

Current Status

Almost three quarters of the ecoregion area has been burnt for hunting or cleared, mostly for agriculture or firewood extraction. A few small, intact patches exist but are scattered in isolated fragments. Most of the original monsoon forests have been replaced by savanna and grassland (Monk et al. 1997). The four small (average size 83 km^2) protected areas include about 3 percent (330 km^2) of the ecoregion area (table J113.2).

Pressures from the rapidly increasing, poor population are intense in this ecoregion (WWF-Indonesia n.d.), and nearly three-quarters of this ecoregion has been deforested, with only isolated fragments of natural habitat remaining.

TABLE J113.2. WCMC (1997) Protected Areas That Overlap with the Ecoregion.

Protected Area	Area (km^2)	IUCN Category
Watu Manggota	20	VI
Manupeu	180	VI
Luku Meloto	60	PRO
Laiwangi-Wanggameti NP	?	?
Gunung Wanggameti	70	DE
Manupea-Tanadaru NP	?	?
Total	330	

Ecoregion numbers of protected areas that overlap with additional ecoregions are listed in brackets.

Types and Severity of Threats

Threats include deforestation, burning of grasslands to establish agricultural fields, livestock grazing, and poaching (WWF-Indonesia n.d.). Much of the forest has already been replaced by fire-resistant casuarinas or eucalypts and extensive deciduous scrub. For instance, the ecoregion's dry thorny forest, which is especially vulnerable to clearance by fire, has almost completely disappeared (Monk et al. 1997).

Priority Conservation Actions

With the rapid pace of forest destruction under way in Indonesia, quick conservation action is urgently needed to stave off the complete collapse of these forests. However, steps should be taken to lay the groundwork for longer-term conservation as well.

Short-Term Conservation Actions (1–5 Years)

- Support conservation activities in national parks and Luku, Melolo, Yawila, Poronumbu, Lulundilu, and Pangadukakar
- Initiate a major public awareness campaign on forest values and forest loss to drive political will.
- Stop illegal logging in and around protected areas by challenging the authorities and judiciary to act on existing laws and regulations.
- Place all remaining forests under active protection.
- Stop wildlife poaching, especially in key parks and reserves, by tightening surveillance around ports or airports and making the purchase of endangered species socially unacceptable.
- Instigate major biodiversity surveys and collecting expeditions.

Longer-Term Conservation Actions (5–20 Years)

- Develop and implement effective management plans for protected areas.

Focal Species for Conservation Action

Threatened endemic species: Sumba buttonquail (*Turnix everetti*) (BirdLife International 2000), red-naped fruit-dove (*Ptilinopus dohertyi*) (BirdLife International 2000), Sumba hornbill (*Aceros everetti*) (BirdLife International 2000)

Area-sensitive species: fruit bats (Pteropodidae) (*Pteropus alecto, Acerodon macklotii, Rousettus amplexicaudatus, Cynopterus brachyotis*) (Mickleburgh et al. 1992), Timor deer (*Cervus timorensis*) (Wemmer 1998), Sumba hornbill (*Aceros everetti*) (Kemp 1995)

Top predator species: short-toed eagle (*Circaetus gallicus*), brown goshawk (*Accipiter fasciatus*)

Selected Conservation Contacts

BirdLife International

Lembaga Alam Tropika Indonesia (LATIN)

WWF-Indonesia

Yayasan Keanekaragaman Hayati

Yayasan Konsorsium Pembaruan Agraria (KPA)

Justification for Ecoregion Delineation

The drier forests in Nusa Tenggara were placed in three ecoregions that corresponded to the biogeographic units identified in Monk et al (1997): Lesser Sundas Deciduous Forests [111], which includes the chain of islands extending from Lombok, Sumbawa, Komodo, Flores, and the smaller satellite islands corresponding to the Flores biogeographic unit; Timor and Wetar Deciduous Forests [112], corresponding to the Timor biogeographic unit; and the Sumba Deciduous Forests [113], corresponding to the Sumba biogeographic unit. All three ecoregions belong to the tropical dry forests biome.

Prepared by John Morrison

Ecoregion Number: **114**
Ecoregion Name: **Halmahera Rain Forests**
Bioregion: **Wallacea**
Biome: **Tropical and Subtropical Moist Broadleaf Forests**
Political Units: **Indonesia**
Ecoregion Size: **26,800 km²**
Biological Distinctiveness: **Globally Outstanding**
Conservation Status: **Relatively Stable**
Conservation Assessment: **III**

This ecoregion comprises the original "Spice Islands." The tropical islands that constitute the complex and mountainous terrain of the Halmahera Rain Forests [114] are an important part of the region known as Wallacea, which contains a very distinctive fauna representing a mix of Asian and Australasian species. This small ecoregion contains an astounding twenty-six bird species, including four monotypic genera, which are found nowhere else in the world. Although there is some exploitation by logging and mining companies, extensive blocks of habitat still cover all the islands, and nearly 80 percent of its original forest still intact.

Location and General Description

This ecoregion represents the moist forests on Halmahera, Morotai, Obi, Bacan, and the other nearby Maluku Islands in the northeastern Indonesian Archipelago. Based on the Köppen climate zone system, this ecoregion falls in the tropical wet climate zone (National Geographic Society 1999). The geologic history of these islands is a very complex mixture of inner volcanic island arcs, outer volcanic island arcs, raised coral reefs, and fragments of continental crust. Halmahera is a product of a collision between two islands approximately 1–2 million years ago. The eastern half of the island was part of an outer arc on the Philippines tectonic plate and consists of sedimentary and intrusive igneous rocks. The western half of Halmahera and Morotai was part of an inner arc consisting of volcanic materials. Bacan is a mixture of volcanic inner island arc and some crustal materials (Monk et al. 1997).

The natural vegetation of these islands was tropical lowland evergreen and semi-evergreen forest (Stattersfield et al. 1998). Most of the remaining habitat in this ecoregion is semi-evergreen rain forest and includes eight characteristic dipterocarp species: *Anisoptera thurifera, Hopea gregaria, H. iriana, H. novoguineensis, Shorea assamica, S. montigena,*

S. selanica, and *Vatica rassak*. Volcanic soils and good aspect combine to produce almost optimal growth conditions. Most of the trees reach 30 m or more and carry thick-stemmed lianas and woody and herbaceous epiphytes. Rattans that grow to 130 m and other epiphytes are common in old-growth forests. The most luxuriant rain forests occur in northwest Morotai and north Halmahera, as opposed to the south arm of Halmahera, which is in the rain shadow of north Halmahera and Bacan. Low, shrubby vegetation is found in poor soil conditions on patches of ultrabasic rocks (Monk et al. 1997).

Biodiversity Features

Overall diversity is low in this ecoregion, but overall endemism is moderate to high when compared with that of other ecoregions in Indo-Malaysia. This ecoregion falls within the Wallacean biogeographic zone, and thus exhibits a mixture of Asian and Australian fauna. Together with Seram, Buru, and the Banda Sea Islands, this island group forms part of a bioregion with perhaps the highest levels of bird endemism for its size anywhere in the world and the highest number of endemic birds of any area in Asia.

The mammal fauna is depauperate, containing only thirty-eight species with both Asian and Australasian affinities (cuscuses), but includes eight ecoregional endemics (table J114.1). The Obi cuscus (*Phalanger rothschildi*) is considered vulnerable (Hilton-Taylor 2000).

TABLE J114.1. Endemic and Near-Endemic Mammal Species.

Family	Species
Phalangeridae	*Phalanger ornatus**
Phalangeridae	*Phalanger rothschildi**
Phalangeridae	*Phalanger* sp.*
Pteropodidae	*Pteropus chrysoproctus*
Pteropodidae	*Pteropus personatus**
Pteropodidae	*Nyctimene minutus*
Muridae	*Melomys obiensis**
Muridae	*Rattus* sp.*

An asterisk signifies that the species' range is limited to this ecoregion.

TABLE J114.2. Endemic and Near-Endemic Bird Species.

Family	Common Name	Species	Family	Common Name	Species
Accipitridae	Moluccan goshawk	*Accipiter henicogrammus**	Aegothelidae	Moluccan owlet-nightjar	*Aegotheles crinifrons**
Accipitridae	Rufous-necked sparrowhawk	*Accipiter erythrauchen*	Alcedinidae	Blue-and-white kingfisher	*Todirhamphus diops**
Megapodiidae	Moluccan scrubfowl	*Megapodius wallacei*	Alcedinidae	Sombre kingfisher	*Todirhamphus funebris**
Megapodiidae	Dusky scrubfowl	*Megapodius freycinet*	Coraciidae	Purple roller	*Eurystomus azureus**
Rallidae	Invisible rail	*Habroptila wallacii**	Pittidae	Ivory-breasted pitta	*Pitta maxima**
Scolopacidae	Moluccan woodcock	*Scolopax rochussenii**	Meliphagidae	Olive honeyeater	*Lichmera argentauris*
Columbidae	Scarlet-breasted fruit-dove	*Ptilinopus bernsteinii**	Meliphagidae	White-streaked friarbird	*Melitograis gilolensis**
Columbidae	Blue-capped fruit-dove	*Ptilinopus monacha**	Meliphagidae	Dusky friarbird	*Philemon fuscicapillus**
Columbidae	Grey-headed fruit-dove	*Ptilinopus hyogastra**	Pachycephalida	Drab whistler	*Pachycephala griseonota*
Columbidae	Carunculated fruit-dove	*Ptilinopus granulifrons**	Monarchidae	White-naped monarch	*Monarcha pileatus*
Columbidae	White-eyed imperial-pigeon	*Ducula perspicillata*	Monarchidae	Moluccan flycatcher	*Myiagra galeata*
Columbidae	Spice imperial-pigeon	*Ducula myristicivora*	Corvidae	Long-billed crow	*Corvus validus**
Columbidae	Pink-headed imperial-pigeon	*Ducula rosacea*	Paradisaeidae	Paradise-crow	*Lycocorax pyrrhopterus**
Columbidae	Cinnamon-bellied imperial-pigeon	*Ducula basilica**	Paradisaeidae	Wallace's standardwing	*Semioptera wallacii**
Psittacidae	Moluccan hanging-parrot	*Loriculus amabilis*	Oriolidae	Halmahera oriole	*Oriolus phaeochromus**
Cacatuidae	White cockatoo	*Cacatua alba**	Campephagidae	Moluccan cuckoo-shrike	*Coracina atriceps*
Loriidae	Violet-necked lory	*Eos squamata*	Campephagidae	Halmahera cuckoo-shrike	*Coracina parvula**
Loriidae	Chattering lory	*Lorius garrulus**	Campephagidae	Pale-grey cuckoo-shrike	*Coracina ceramensis*
Cuculidae	Moluccan cuckoo	*Cacomantis heinrichi*	Campephagidae	Rufous-bellied triller	*Lalage aurea**
Cuculidae	Pied bronze-cuckoo	*Chrysococcyx crassirostris*	Zosteropidae	Cream-throated white-eye	*Zosterops atriceps**
Cuculidae	Goliath coucal	*Centropus goliath**	Dicaeidae	Flame-breasted flowerpecker	*Dicaeum erythrothorax*
Strigidae	Moluccan hawk-owl	*Ninox squamipila*			

An asterisk signifies that the species' range is limited to this ecoregion.

The ecoregion supports approximately 223 bird species, including 43 ecoregional endemic species (table J114.2). The ecoregion corresponds with the Northern Maluku EBA. There are four endemic monotypic genera: *Habroptila*, *Melitorgrais*, *Lycocorax*, and *Semioptera*. These species include the invisible rail (*Habroptila wallacii*), white-streaked friarbird (*Melitorgrais gilolensis*), paradise-crow (*Lycocorax pyrrhopterus*), and the standardwing (*Semioptera wallacii*). Of the forty-three restricted-range species found in this ecoregion (and EBA), an astounding twenty-six are found nowhere else in the world. Five vulnerable species, four of which are found nowhere else, are found in the ecoregion: invisible rail (*Habroptila wallacii*), caranculated fruit-dove (*Ptilinopus granulifrons*), chattering lory (*Lorius garrulus*), and white cockatoo (*Cacatua alba*) (Stattersfield et al. 1998).

The world's largest bee—the rare, 4-cm Wallace's giant bee *Chalocodoma pluto*—is also found on Bacan, Tidore, and Halmahera. Wallace discovered this species in 1858, and it was thought to be extinct until 1981, when it was recollected. This ecoregion also has conservation importance for butterflies and includes *Troides aesacus*, which may be the most primitive member of the *T. priamus* species group (Whitten and Whitten 1992; K. Monk, pers. comm., 2000).

Status and Threats

Current Status

The rich volcanic soils of Ternate, Tidore, and nearby islands have been aggressively cultivated for cloves and other spices for centuries (Stattersfield et al. 1998). From the 1920s through the 1970s, commercial logging and enforced cultivation depleted the forests of Halmahera and Morotai (Monk et al. 1997). On Morotai, large tracts of lowland rain forest were cultivated with papaya (*Carica papaya*) during World War II (Monk et al. 1997). Currently, the wet evergreen lowland forests in the northwest of Halmahera are exploited by logging companies, primarily for the valuable damar trees (*Agathis*) (Whitten and Whitten 1992). The eastern forests are threatened by pulp plantations, especially using local transmigrants.

Extensive habitat blocks still cover all the islands, with only small areas near the coast cleared for human settlements (Monk et al. 1997). The seven protected areas cover 4,880 km^2 (18 percent) of the ecoregion area (table J114.3). Three protected areas are greater than 1,000 km^2 in area, and the average size is 697 km^2.

TABLE J114.3. WCMC (1997) Protected Areas That Overlap with the Ecoregion.

Protected Area	Area (km^2)	IUCN Category
Waya Bula	830	PRO
Lolabata	1,210	PRO
Gunung Gamkonara	110	PRO
Ake Tajawi	1,200	PRO
Saketa	1,100	PRO
Gunung Sibela	300	PRO
Pulau Obi	130	PRO
Total	4,880	

Ecoregion numbers of protected areas that overlap with additional ecoregions are listed in brackets.

Types and Severity of Threats

With nearly 80 percent of its original forest still intact, the Halmahera Rain Forests [114] ecoregion is largely free of intense habitat conversion threats. However, as the forests are lost on other Indonesian islands, there is an increasing potential for commercial forestry operations to move to Halmahera. A mining company, PT Halmahera Mineral (NHM), has already obtained an exploration license for Bacan and "neighboring islands" to look for gold and other minerals. A Canadian mining company has a license to mine nickel near Ake Tajawi on Halmahera (K. Monk, pers. comm., 2000).

Priority Conservation Actions

Short-Term Conservation Actions (1–5 Years)

- Stop illegal logging in and around protected areas by challenging the authorities and judiciary to act on existing laws and regulations.
- Instigate major biodiversity surveys and collecting expeditions.
- Develop a proactive strategy to protect the most biodiverse forests.
- Stop wildlife poaching in key parks and reserves.

Longer-Term Conservation Actions (5–20 Years)

- Conserve Lalobata, Ake Tajawi, Gunung Gamkonara, Pulau Obi, Bacan, and Gunung Sibela
- Increase the protected area system and enforce their protection.
- Develop and implement effective management plans for protected areas.

Focal Species for Conservation Action

Threatened endemic species: Obi cuscus (*Phalanger rothschildi*) (Flannery 1995), invisible rail (*Habroptila wallacii*) (BirdLife International 2000), caranculated fruit-dove (*Ptilinopus granulifrons*) (BirdLife International 2000), chattering lory (*Lorius garrulus*) (BirdLife International 2000), and white cockatoo (*Cacatua alba*) (BirdLife International 2000)

Area-sensitive species: fruit bats (Pteropodidae) (*Pteropus* spp., *Rousettus amplexicaudatus*, *Syconycteris carolinae*, *Dobsonia crenulata*) (Mickleburgh et al. 1992; Flannery 1995), Blyth's hornbill (*Aceros plicatus*) (Kemp 1995)

Top predator species: variable goshawk (*Accipiter hiogaster*), Moluccan goshawk (*Accipiter henicogrammus*), Meyer's goshawk (*Accipiter meyerianus*), grey-faced buzzard (*Butastur indicus*), black eagle (*Ictinaetus malayensis*), Gurney's eagle (*Aquila gurneyi*), little eagle (*Hieraaetus morphnoides*), rufous-bellied eagle (*Hieraaetus kienerii*)

Selected Conservation Contacts

BirdLife International
Lembaga Alam Tropika Indonesia (LATIN)
World Bank–Maluku Conservation and Natural Resource Management Project
WWF-Indonesia
Yayasan Keanekaragaman Hayati
Yayasan Konsorsium Pembaruan Agraria (KPA)

Justification for Ecoregion Delineation

The Sula Islands were included within the Sulawesi Lowland Rain Forests [109], and the Aru islands in the Vogelkop-Aru Lowland Rain Forests [119]. Buru Island, identified as a distinct subunit (13c) by MacKinnon (1997) and as an EBA (Stattersfield et al. 1998), was delineated as a distinct ecoregion, the Buru Rain Forests [115]. Seram, the larger island to the east of Buru, was also delineated as an ecoregion: Seram Rain Forests [116]. The larger Halmahera Rain Forests [114] ecoregion includes Obi Island, which MacKinnon (1997) recognized as a separate subunit (13b) from Halmahera Island (subunit 13a). We created the Banda Sea Islands Moist Deciduous Forests [117] by combining the islands in the Kai and Tanimbar archipelagos, which were distinguished as a biogeographic unit by Monk et al. (1997). The primary vegetation on the islands in both these archipelagos is moist deciduous forests and semi-evergreen forests, whereas the vegetation in the other, nearby large islands (Seram and Aru) is evergreen rain forests (Monk et al. 1997).

Prepared by John Morrison

Ecoregion Number:	**115**
Ecoregion Name:	**Buru Rain Forests**
Bioregion:	**Wallacea**
Biome:	**Tropical and Subtropical Moist Broadleaf Forests**
Political Units:	**Indonesia**
Ecoregion Size:	**8,600 km²**
Biological Distinctiveness:	**Bioregionally Outstanding**
Conservation Status:	**Vulnerable**
Conservation Assessment:	**V**

The Buru Rain Forests [115] are located on the small mountainous tropical island of Buru in the Banda Sea, part of the region known as Wallacea, which contains a distinctive fauna representing a mix of Asian and Australasian species. There are ten bird species in this ecoregion that are found nowhere else on Earth, including a monotypic bird genus. Although the northern portions of the island have been degraded by repeated burning and the coastal lowlands have been cleared, the remaining forest forms two large, contiguous blocks, current threats appear to be low, and the conservation outlook is relatively stable.

Location and General Description

This ecoregion represents the moist forests in the island of Buru. Based on the Köppen climate zone system, this ecoregion falls in the tropical wet climate zone (National Geographic Society 1999). Buru is part remnant crustal fragment, probably from the Australian continent, and part of the volcanic Inner Banda Arc. Consequently, the surface geology of Buru is complex, consisting of older metamorphic schists and gneiss, younger volcanics, and recent alluvium (Monk et al. 1997).

The natural vegetation of the island was tropical lowland evergreen and semi-evergreen rain forests (Stattersfield et al. 1998). The dominant tree species in this moist forest are the dipterocarps, *Anisoptera thurifera*, *Hopea gregaria*, *H. iriana*, *H. novoguineensis*, *Shorea assamica*, *S. montigena*, *S. selanica*, and *Vatica rassak* (Monk et al. 1997). In old-growth forests, the larger trees grow to more than 30 m in height and tend to be covered with thick-stemmed lianas and other epiphytes. Open forest, woodland, and savanna are also found in this ecoregion, with some being natural but most originating from human activity (Flannery 1995). The fire-resistant paper bark tree (*Melaleuca cajuputi*) is common and grows in nearly monotypic stands in dry areas (Whitten and Whitten 1992). The steep limestone cliffs in the northwestern part of the ecoregion are covered by mixed forests that include *Shorea* spp. (Monk et al. 1997). Exposed ridges between 1,800 and 2,000 m above sea level are characterized by stunted *Dacrydium novo-guineense* (Monk et al. 1997).

Biodiversity Features

Overall richness and endemism in this ecoregion are low to moderate when compared with those of other ecoregions in Indo-Malaysia. Being in the Wallacean biogeographic zone, the ecoregion contains a mixture of Asian and Australian fauna. The mountainous areas of this island are largely unexplored and may contain many undiscovered species (Flannery 1995).

The known mammal fauna of Buru consists of at least twenty-five species, including four near endemics (table J115.1). Three of these species are considered vulnerable: the Seram flying-fox (*Pteropus ocularis*), lesser tube-nosed fruit bat (*Nyctimene minutus*), and babirusa (*Barirousa babirussa*) (Hilton-Taylor 2000).

The bird fauna consists of 178 species, including twenty-nine endemic or near-endemic species (table J115.2). The ecoregion corresponds with the Buru EBA and contains twenty-eight restricted-range bird species, ten of which are found nowhere else on Earth. Among these species is the critically endangered blue-fronted lorikeet (*Charmosyna toxopei*), endangered rufous-throated white-eye (*Madanga ruficollis*), and vulnerable Moluccan scrubfowl (*Megapodius wallacei*) and black-lored parrot (*Tanygnathus gramineus*) (BirdLife International 2000).

Buru's butterflies include a large number of endemics and are therefore accorded highest conservation priority. Pifridae has 25 percent of the local species unique to Buru, and Papilionidae 7 percent (Vane-Wright and Peggae, in press).

Status and Threats

Current Status

The coastal lowland forests have been cleared, and the northern and northeastern portions of the island now contain

TABLE J115.1. Endemic and Near-Endemic Mammal Species.

Family	Species
Pteropodidae	*Pteropus chrysoproctus*
Pteropodidae	*Pteropus ocularis*
Pteropodidae	*Nyctimene minutus*
Suidae	*Babyrousa babyrussa*

An asterisk signifies that the species' range is limited to this ecoregion.

monsoon forest, gallery forest, and savannas as a result of repeated burning (Stattersfield et al. 1998). However, the remaining upland forest forms two large, contiguous blocks. Most of this forest is a mosaic of primary and secondary forest as a result of shifting cultivation (Monk 1997; Stattersfield et al. 1998).

The two protected areas—of which one is greater than 1,000 km^2—cover 17 percent of the ecoregion (table J115.3).

TABLE J115.2. Endemic and Near-Endemic Bird Species.

Family	Common Name	Species
Accipitridae	Rufous-necked sparrowhawk	Accipiter erythrauchen
Megapodiidae	Forsten's scrubfowl	Megapodius forstenii
Megapodiidae	Moluccan scrubfowl	Megapodius wallacei
Columbidae	White-eyed imperial-pigeon	Ducula perspicillata
Columbidae	Long-tailed mountain-pigeon	Gymnophaps mada
Psittacidae	Buru racquet-tail	Prioniturus mada*
Psittacidae	Black-lored parrot	Tanygnathus gramineus*
Loriidae	Red lory	Eos bornea
Loriidae	Blue-fronted lorikeet	Charmosyna toxopei*
Strigidae	Moluccan hawk-owl	Ninox squamipila
Tytonidae	Lesser masked-owl	Tyto sororcula
Meliphagidae	Buru honeyeater	Lichmera deningeri*
Meliphagidae	Wakolo myzomela	Myzomela wakoloensis
Meliphagidae	Black-faced friarbird	Philemon moluccensis
Pachycephalida	Drab whistler	Pachycephala griseonota
Rhipiduridae	Cinnamon-backed fantail	Rhipidura superflua*
Monarchidae	White-naped monarch	Monarcha pileatus
Monarchidae	Black-tipped monarch	Monarcha l oricatus*
Monarchidae	Moluccan flycatcher	Myiagra galeata
Oriolidae	Buru oriole	Oriolus bouroensis
Campephagidae	Buru cuckoo-shrike	Coracina fortis*
Campephagidae	Pale-grey cuckoo-shrike	Coracina ceramensis
Turdidae	Moluccan thrush	Zoothera dumasi
Muscicapidae	Streaky-breasted jungle-flycatcher	Rhinomyias addita*
Muscicapidae	Cinnamon-chested flycatcher	Ficedula buruensis
Zosteropidae	Buru white-eye	Zosterops buruensis*
Zosteropidae	Rufous-throated white-eye	Madanga ruficollis*
Sylviidae	Chestnut-backed bush-warbler	Bradypterus castaneus
Dicaeidae	Flame-breasted flowerpecker	Dicaeum erythrothorax

An asterisk signifies that the species' range is limited to this ecoregion.

Appendix J: Ecoregion Descriptions

Commercial logging on Buru intensified during the 1970s, but much of the island is still under extensive forest cover.

Types and Severity of Threats

Current threats to this ecoregion are low, causing its conservation status to remain vulnerable. Commercial logging and shifting cultivation are the primary threats to the remaining habitat.

Priority Conservation Actions

Short-Term Conservation Actions (1–5 Years)

- Initiate a major public awareness campaign on forest values and forest loss to drive political will.
- Increase protection of remaining habitat.
- Stop wildlife poaching in key parks and reserves.
- Instigate major biodiversity surveys and collecting expeditions.

Longer-Term Conservation Actions (5–20 Years)

- Develop and implement effective management plans for protected areas.

Focal Species for Conservation Action

Threatened endemic species: Seram flying-fox (*Pteropus ocularis*) (Mickleburgh et al. 1992; Flannery 1995), lesser tube-nosed fruit bat (*Nyctimene minutus*) (Mickleburgh et al. 1992; Flannery 1995), Moluccan scrubfowl (*Megapodius wallacei*) (BirdLife International 2000), blue-fronted lorikeet (*Charmosyna toxopei*) (BirdLife International 2000), black-lored parrot (*Tanygnathus gramineus*) (BirdLife International 2000), rufous-throated white-eye (*Madanga ruficollis*) (BirdLife International 2000)

Area-sensitive species: fruit bats (Pteropodidae) (*Pteropus* spp., *Nyctimene minutus*, *Rousettus amplexicaudatus*, *Dobsonia viridis*, *Thoopterus nigrescens*, *Nyctimene minutus*, *Nyctimene cephalotes*, *Dobsonia moluccensis*) (Mickleburgh et al. 1992; Flannery 1995)

Top predator species: variable goshawk (*Accipiter hiogaster*), brown goshawk (*Accipiter fasciatus*), black eagle (*Ictinaetus malayensis*), rufous-bellied eagle (*Hieraaetus kienerii*)

TABLE J115.3. WCMC (1997) Protected Areas That Overlap with the Ecoregion.

Protected Area	Area (km^2)	IUCN Category
Gunung Kelpat Muda	1,380	PRO
Waeapo	50	PRO
Total	1,430	

Ecoregion numbers of protected areas that overlap with additional ecoregions are listed in brackets.

Selected Conservation Contacts

Lembaga Alam Tropika Indonesia (LATIN)
WWF-Indonesia
Yayasan Keanekaragaman Hayati
Yayasan Konsorsium Pembaruan Agraria (KPA)

Justification for Ecoregion Delineation

The Sula Islands were included within the Sulawesi Lowland Rain Forests [109] and the Aru Islands in the Vogelkop-Aru Lowland Rain Forests [119]. Buru Island, identified as a distinct subunit (13c) by MacKinnon (1997) and as an EBA (Stattersfield et al. 1998), was delineated as a distinct ecoregion, the Buru Rain Forests [115]. Seram, the larger island to the east of Buru, was also delineated as an ecoregion: Seram Rain Forests [116]. The larger Halmahera Rain Forests [114] includes Obi Island, which MacKinnon (1997) recognized as a separate subunit (13b) from Halmahera Island (subunit 13a). We created the Banda Sea Islands Moist Deciduous Forests [117] by combining the islands in the Kai and Tanimbar archipelagos, which Monk et al. (1997) distinguished as a biogeographic unit. The primary vegetation on the islands in both these archipelagos is moist deciduous forests and semi-evergreen forests, whereas the vegetation in the other, nearby large islands (Seram and Aru) is evergreen rain forests (Monk et al. 1997).

Prepared by John Morrison

Ecoregion Number:	**116**
Ecoregion Name:	**Seram Rain Forests**
Bioregion:	**Wallacea**
Biome:	**Tropical and Subtropical Moist Broadleaf Forests**
Political Units:	**Indonesia**
Ecoregion Size:	**19,300 km²**
Biological Distinctiveness:	**Globally Outstanding**
Conservation Status:	**Relatively Stable**
Conservation Assessment:	**III**

The Seram Rain Forests [116] are part of the region known as Wallacea, which contains a very distinctive fauna representing a mix of Asian and Australasian species. This small island ecoregion contains sixteen bird species, including a monotypic bird genus, that are found nowhere else on Earth. Nearly a fifth of the original forest cover has been cleared, mostly along the northern coast. However, large areas of contiguous, intact forest still exist, and the conservation status of this ecoregion is relatively stable.

Location and General Description

This ecoregion represents the semi-evergreen and moist forests of Seram and associated islands in the easternmost section of the Indonesian Archipelago. Based on the Köppen climate zone system, this ecoregion falls in the tropical wet climate zone (National Geographic Society 1999). Seram is part remnant crustal fragment, probably from the Australian continent, and part of the volcanic Inner Banda Arc. Consequently, the surface geology of Seram is complex, consisting of older metamorphic schists and gneiss, younger volcanics, and recent alluvium (Monk et al. 1997). The interior of the island is mountainous, with several ranges reaching more than 1,000 m. The highest point on the island is the 3,027-m Merkele ridge (Stattersfield et al. 1998).

The natural vegetation of Seram is tropical lowland evergreen, semi-evergreen, and montane rain forest (Stattersfield et al. 1998). Semi-evergreen rain forest with trees that reach 30 m or more is a predominant forest type in this ecoregion. Rattans that exceed 100 m can be found in mature forests. The middle and lower layers include representatives of the Amaryllidaceae, sedges, and large ferns *Angiopteris* and *Marattia*, as well as climbers such as *Freycinetia*, *Gnetum*, *Mucuna*, *Bauhinia*, *Piper*, and *Smilax*. Most of the remaining dipterocarp forests are dominated by the endemic *Shorea selanica*, which can represent about 30 percent of the individual trees and 76 percent of the basal area in the forest. Also common are *Anisoptera thurifera*, *Hopea gregaria*, *H. iriana*, *H. novoguineensis*, *Shorea assamica*, *S. montigena*, *S. selanica*, and *Vatica rassak* (Monk et al. 1997).

This ecoregion also contains patches of ultrabasic rocks. The forests on these soils generally are poor in species, low, and shrubby. Tertiary limestone outcrops occur in the lowlands and on many mountains such as the Murkele Ridge and the top ridge of the central Mt. Binaiya, Seram's highest mountain (Monk et al. 1997).

In Seram's montane forests, the Fagaceae are represented by only two species. *Castanopsis buruana* dominates between 400 and 1,400 m above sea level, where individuals tend to clump together, and *Lithocarpus celebicus* is found along ridges. Above 2,400 m on Mt. Binaiya, a low, open scrubby woodland contains *Dacrydium* spp., *Myrica* spp., *Rapanea* spp., *Rhamnus* spp., *Rhododendron* spp., and *Vaccinium* spp. Tree ferns are also important and include *Cyathea binayana* and *C. pukuana*, which form distinctive groves that support many epiphytic ferns. Pockets of this tree-fern savanna extend to the summit along with low *Vaccinium* woodland. At the highest points, from 2,700 to 3,000 m above sea level, grassland dominates and is characterized by several endemic herbs such as *Viola binayensis*, *Pterostylis papuanum*, and *Euphrasia ceramensis* (Monk et al. 1997).

Biodiversity Features

The overall richness and endemism of this ecoregion are low to moderate when compared with those of other ecoregions in Indo-Malaya. The islands are part of Wallacea, a unique region that supports a mixture of Asian and Australian fauna.

TABLE J116.1. Endemic and Near-Endemic Mammal Species.

Family	Species
Perorictidae	*Rhyncholemes prattorum**
Pteropodidae	*Pteropus chrysoproctus*
Pteropodidae	*Pteropus ocularis*
Pteropodidae	*Pteropus argenatatus**
Muridae	*Rattus feliceus**
Muridae	*Melomys fulgens*
Muridae	*Melomys aerosus**
Muridae	*Melomys fraterculus**
Muridae	*Stenomys ceramicus**

An asterisk signifies that the species' range is limited to this ecoregion.

The montane area of Seram supports the greatest number of endemic mammals of any island in the region (Flannery 1995). The ecoregion harbors thirty-eight mammal species and includes nine species that are endemic or near endemic (table J116.1), several of which are limited to montane habitats (Flannery 1995). The Seram flying-fox (*Pteropus ocularis*) and spiny Seram rat (*Melomys feliceus*) are considered vulnerable (Hilton-Taylor 2000). The mammals found on Seram include Asian species (Murid rodents) as well as Australasian marsupials.

The ecoregion harbors more than 213 bird species (Wikramanyake et al. 2001), of which 33 are endemic or near endemic (table J116.2). The ecoregion corresponds to the Seram EBA. The EBA contains thirty restricted-range species, including fourteen that are found nowhere else on Earth. Four species are threatened. The vulnerable Moluccan scrubfowl (*Megapodius wallacei*) is also found on Buru and Halmahera. The remaining three species are found nowhere else: the critically endangered black-chinned monarch (*Monarcha boanensis*) and vulnerable salmon-crested cockatoo (*Cacatua moluccensis*) and purple-naped lory (*Lorius domicella*). The bicoloured white-eye (*Tephrozosterops stalkeri*), the sole member of its genus, is also found only on Seram. The fourteen endemic restricted-range birds can be divided into three groups: five species found generally in lowland forests (below 1,000 m), three species found in montane forests above 1,000 m, and six species found in both lowland and montane habitats (Stattersfield et al. 1998). The ecoregion also harbors the largest bird in the Moluccas, the two-wattled cassowary (*Casuarius casuarius*) (Whitten and Whitten 1992).

Status and Threats

Current Status

Nearly a fifth of the original forest of this ecoregion has been cleared, mostly along the northern coast. However, large areas of contiguous, intact forest still exist. Therefore, the conservation status of this ecoregion is relatively stable. Seven protected areas cover 3,121 km² (16 percent) of the ecoregion area, and one—Manusela National Park—is more than 2,000 km² (table J116.3). This last reserve, with a wide range of forest types, conserves the cassowary (*Casuarius casuarius*). However, wildlife trade has been strong since historical times, and it may threaten some bird species such as the salmon-crested cockatoo (*Cacatua moluccensis*). The Trans-Seram Highway also threatens forest habitat by illegal logging, land clearance, and soil erosion.

Seram's moist lowland forests are being exploited by logging companies, primarily for their valuable damar trees

TABLE J116.2. Endemic and Near-Endemic Bird Species.

Family	Common Name	Species
Accipitridae	Rufous-necked sparrowhawk	*Accipiter erythrauchen*
Megapodiidae	Forsten's scrubfowl	*Megapodius forstenii*
Megapodiidae	Moluccan scrubfowl	*Megapodius wallacei*
Columbidae	White-eyed imperial-pigeon	*Ducula perspicillata*
Columbidae	Long-tailed mountain-pigeon	*Gymnophaps mada*
Cacatuidae	Salmon-crested cockatoo	*Cacatua moluccensis**
Loriidae	Red lory	*Eos bornea*
Loriidae	Blue-eared lory	*Eos semilarvata**
Loriidae	Purple-naped lory	*Lorius domicella**
Cuculidae	Pied bronze-cuckoo	*Chrysococcyx crassirostris*
Strigidae	Moluccan hawk-owl	*Ninox squamipila*
Tytonidae	Lesser masked-owl	*Tyto sororcula*
Alcedinidae	Lazuli kingfisher	*Todirhamphus lazuli**
Meliphagidae	Olive honeyeater	*Lichmera argentauris*
Meliphagidae	Seram honeyeater	*Lichmera monticola**
Meliphagidae	Seram myzomela	*Myzomela blasii**
Meliphagidae	Wakolo myzomela	*Myzomela wakoloensis*
Meliphagidae	Seram friarbird	*Philemon subcorniculatus**
Pachycephalida	Drab whistler	*Pachycephala griseonota*
Rhipiduridae	Streaky-breasted fantail	*Rhipidura dedemi**
Monarchidae	Black-chinned monarch	*Monarcha boanensis**
Monarchidae	Moluccan flycatcher	*Myiagra galeata*
Oriolidae	Seram oriole	*Oriolus forsteni**
Campephagidae	Moluccan cuckoo-shrike	*Coracina atriceps*
Campephagidae	Pale-grey cuckoo-shrike	*Coracina ceramensis*
Turdidae	Moluccan thrush	*Zoothera dumasi*
Sturnidae	Long-crested myna	*Basilornis corythaix**
Muscicapidae	Cinnamon-chested flycatcher	*Ficedula buruensis*
Zosteropidae	Ambon white-eye	*Zosterops kuehni**
Zosteropidae	Bicoloured white-eye	*Tephrozosterops stalkeri**
Zosteropidae	Grey-hooded white-eye	*Lophozosterops pinaiae**
Sylviidae	Chestnut-backed bush-warbler	*Bradypterus castaneus*
Dicaeidae	Ashy flowerpecker	*Dicaeum vulneratum*

An asterisk signifies that the species' range is limited to this ecoregion.

TABLE J116.3. WCMC (1997) Protected Areas That Overlap with the Ecoregion.

Protected Area	Area (km²)	IUCN Category
Sabuda Tataruga	10	IV
Manusela	2,340	II
Wae Bula	600	PRO
Gunung Sahuai	120	PRO
Pulau Kassa	1	IV
Pulau Pombo	20	I
Laut Banda	30	I
Pulau Manuk	1	?
Total	3,122	

Ecoregion numbers of protected areas that overlap with additional ecoregions are listed in brackets.

(*Agathis*) (Whitten and Whitten 1992). The best dipterocarp stands were depleted by commercial loggers before the 1950s, and many other species were overexploited by intensive logging in the 1970s (Monk et al. 1997).

With no airport and only rudimentary ground transport, Seram is remote. Although this promotes conservation in many ways, it also prevents conservation employees from guarding boundaries, enlisting the support of local people, and conducting biological surveys (Whitten and Whitten 1992).

Types and Severity of Threats

The north Seram dipterocarp forests are still dominated by the endemic *Shorea selanica* and therefore are especially vulnerable to logging (Monk et al. 1997). The commercial wildlife trade is another significant threat. Parrots are captured and exported for the pet trade, with many casualties (Whitten and Whitten 1992).

Priority Conservation Actions

Short-Term Conservation Actions (1–5 Years)

- Initiate a major public awareness campaign on forest values and forest loss to drive political will.
- Stop illegal logging in and around protected areas.
- Increase forest protection, especially in northern Seram's dipterocarp forests.
- Stop wildlife poaching in key parks and reserves.
- Instigate biodiversity surveys and collecting expeditions.

Longer-Term Conservation Actions (5–20 Years)

- Develop and implement effective management plans for protected areas such as Manu Sela, Gunung Sahuai, Wae Bula, and Laut Banda

Focal Species for Conservation Action

Threatened endemic species: Seram flying-fox (*Pteropus ocularis*) (Mickleburgh et al. 1992; Flannery 1995), spiny Seram rat (*Melomys feliceus*) (Flannery 1995), black-chinned monarch (*Monarcha boanensis*), salmon-crested cockatoo (*Cacatua moluccensis*) (BirdLife International 2000), purple-naped lory (*Lorius domicella*) (BirdLife International 2000), Moluccan scrubfowl (*Megapodius wallacei*) (BirdLife International 2000)

Area-sensitive species: fruit bats (Pteropodidae) (*Pteropus* spp., *Dobsonia* spp., *Rousettus amplexicaudatus*, *Nyctimene cephalotes*, *Macroglossus minimus*, *Syconycteris australis*) (Mickleburgh et al. 1992; Flannery 1995), Blyth's hornbill (*Aceros plicatus*) (Kemp 1995)

Top predator species: variable goshawk (*Accipiter hiogaster*), Meyer's goshawk (*Accipiter meyerianus*), black eagle (*Ictinaetus malayensis*), Gurney's eagle (*Aquila gurneyi*), little eagle (*Hieraaetus morphnoides*)

Selected Conservation Contacts

Lembaga Alam Tropika Indonesia (LATIN)

WWF-Indonesia

Yayasan Keanekaragaman Hayati

Yayasan Konsorsium Pembaruan Agraria (KPA)

Justification for Ecoregion Delineation

The Sula Islands were included within the Sulawesi Lowland Rain Forests [109] and the Aru Islands in the Vogelkop-Aru Lowland Rain Forests [119]. Buru Island, identified as a distinct subunit (13c) by MacKinnon (1997) and as an EBA (Stattersfield et al. 1998), was delineated as a distinct ecoregion, the Buru Rain Forests [115]. Seram, the larger island to the east of Buru, was also delineated as an ecoregion: Seram Rain Forests [116]. The larger Halmahera Rain Forests [114] ecoregion includes Obi Island, which MacKinnon (1997) recognized as a separate subunit (13b) from Halmahera Island (subunit 13a). We created the Banda Sea Islands Moist Deciduous Forests [117] by combining the islands in the Kai and Tanimbar archipelagos, which were distinguished as a biogeographic unit by Monk et al. (1997). The primary vegetation on the islands in both these archipelagos is moist deciduous forests and semi-evergreen forests, whereas the vegetation in the other, nearby large islands (Seram and Aru) is evergreen rain forests (Monk et al. 1997).

Prepared by John Morrison

Ecoregion Number:	**117**
Ecoregion Name:	**Banda Sea Islands Moist Deciduous Forests**
Bioregion:	**Wallacea**
Biome:	**Tropical and Subtropical Moist Broadleaf Forests**
Political Units:	**Indonesia**
Ecoregion Size:	**7,500 km²**
Biological Distinctiveness:	**Regionally Outstanding**
Conservation Status:	**Vulnerable**
Conservation Assessment:	**II**

The Banda Sea Islands Moist Deciduous Forests [117] are found on small islands scattered across the Banda Sea and are part of the region known as Wallacea, which contains a distinctive fauna representing a mix of Asian and Australasian species. Active volcanoes are found on the Banda Islands, whereas other parts of the ecoregion represent portions of the Australian continent that have been torn off. The islands contain a remarkable twenty-one bird species found nowhere else on Earth. The forests in this ecoregion are still largely intact, but although many of these islands are tiny and uninhabitable, the bird populations are seriously threatened by accidentally released rats and cats and by the removal of their eggs for sale by fishers traveling through this area.

Location and General Description

This ecoregion represents the moist deciduous and limestone forests of Tanimbar, Kai, Banda, and smaller island groups in the Banda Sea, part of the eastern Indonesian Archipelago. Based on the Köppen climate zone system, this ecoregion falls in the tropical wet climate zone (National Geographic Society 1999). Geologically, the islands have a mixed history. The Banda Islands are part of the inner arc, whereas the rest of the ecoregion is part of the outer arc. The inner arc islands are a result of the subduction and partial melting of the Indo-Australian tectonic plate below the Eurasian plate. The inner arc islands represent young volcanoes that have coalesced with lava and sediment. The volcanically active Mt. Api is found in the Banda Islands, which represent the ruins of a very large volcano that erupted in prehistory. The basement rock of the outer islands, on the other hand, is composed of actual continental margin from the Australian plate that has not been subducted. These outer islands are less than 4 million years old. The resulting surface geology consists of complex sedimentary and metamorphic rocks: uplifted coral reefs over complex basement rocks (Monk et al. 1997).

In the south the forest biogeography of the Moluccas differs from that associated with the classic dipterocarp forests of Borneo or Sumatra. Northern Maluku has relatively similar dipterocarp forests. Many of the dipterocarp species have been replaced by dominants more typical of the Australo-Melanesian area. The forests of this ecoregion are varied but include evergreen rain forest (Kepulauan Kai), semi-evergreen rain forest, moist deciduous forest, and dry deciduous forest (Monk et al. 1997).

Biodiversity Features

The mammal fauna consists of twenty-two species with both Asian and Australasian affinities, and three species are endemic (table J117.1) (Flannery 1995). The Moluccan mouse-eared bat (*Myotis stalkeri*) is endangered, whereas the dusky pademelon (*Thylogale bruinii*) and brown-bearded sheathtail bat (*Taphozous achates*) are considered vulnerable (Hilton-Taylor 2000). The dusky pademelon is the only macropodid (kangaroo) found in the Banda Sea islands (Kai), although it is also found in the Aru Islands and the Trans Fly of New Guinea (Flannery 1995).

Together with Halmahera, Buru, and Seram islands, this ecoregion lies within an area with perhaps the highest levels of bird endemism for its size anywhere in the world and the highest number of endemic birds of any area in Asia. Even the smallest, uninhabitable islands are significant as breeding sites for large numbers of seabirds such as frigatebirds, tropicbirds, boobies, terns, and smaller species (Whitten and Whitten 1992). Manuk Island and Mt. Api (north of Wetar), two nature reserves in the Banda Sea, are the breeding and roosting sites for millions of seabirds. Active volcanoes, they are probably the greatest bird islands left in all southeast Asia (Whitten and Whitten 1992; Monk et al. 1997). This ecoregion contains more than 225 species of terrestrial birds, of which forty-three are endemic or near endemic (table J117.2). This ecoregion corresponds with the Banda Sea Islands EBA, which contains forty-one restricted-range species, and includes eighteen species that are found nowhere else on Earth (Stattersfield et al. 1998). Only one of these species, the Damar flycatcher (*Ficedula henrici*), is considered threatened.

TABLE J117.1. Endemic and Near-Endemic Mammal Species.

Family	Species
Macropodidae	*Thylogale bruinii*
Vespertilionidae	*Myotis stalkeri**
Emballonuridae	*Taphozous achates**

An asterisk signifies that the species' range is limited to this ecoregion.

Status and Threats

Current Status

This ecoregion consists of a chain of small islands. The forests in this ecoregion are still largely intact, with only about 20 percent of the habitat being lost. The island of Yamdena in the Tanimbars represents a fairly large block of undisturbed habitat. There are five protected areas that cover 1,500 km^2 (27 percent) of the ecoregion (table J117.3). However, the largest of the protected areas is still in a proposed state.

Types and Severity of Threats

Although many of these islands are tiny and uninhabitable, the bird populations are seriously threatened by predatory rats and cats that have been accidentally released onto these islands and by the removal of their eggs for sale by fishers traveling through this area (Whitten and Whitten 1992). On inhabited islands, such as Manuk Island Nature Reserve, small-scale agriculture poses another threat as farmers oust both tree-nesting and ground-nesting birds (Whitten and Whitten 1992).

Priority Conservation Actions

Short-Term Conservation Actions (1–5 Years)

- Initiate a major public awareness campaign on forest values and forest loss to drive political will.
- Stop wildlife poaching in key parks and reserves by tightening surveillance around ports or airports and making the purchase of endangered species socially unacceptable.
- Take steps to control the introduced rat and cat species.

Longer-Term Conservation Actions (5–20 Years)

- Develop and implement effective management plans for protected areas and important locations such as Jamdena (across the island of Yamdena), Kai Besar, and Pulau Angwarmase.

Focal Species for Conservation Action

Threatened endemic species: Moluccan mouse-eared bat (*Myotis stalkeri*) (Flannery 1995), dusky pademelon (*Thylogale bruinii*) (Flannery 1995), brown-bearded sheathtail bat (*Taphozous achates*) (Flannery 1995), Damar flycatcher (*Ficedula henrici*) (BirdLife International 2000)

Area-sensitive species: dusky pademelon (*Thylogale bruinii*) (Flannery 1995), fruit bats (Pteropodidae) (*Pteropus melanopogon, Dobsonia viridis, Macroglossus minimus, Syconycteris australis*) (Mickleburgh et al. 1992; Flannery 1995)

Top predator species: variable goshawk (*Accipiter hiogaster*), brown goshawk (*Accipiter fasciatus*), Bonelli's eagle (*Hieraaetus fasciatus*)

Selected Conservation Contacts

Lembaga Alam Tropika Indonesia (LATIN)

WWF-Indonesia

Yayasan Keanekaragaman Hayati

Yayasan Konsorsium Pembaruan Agraria (KPA)

Justification for Ecoregion Delineation

The Sula Islands were included within the Sulawesi Lowland Rain Forests [109] and the Aru Islands in the Vogelkop-Aru Lowland Rain Forests [119]. Buru Island, identified as a distinct subunit (13c) by MacKinnon (1997) and as an EBA (Stattersfield et al. 1998), was delineated as a distinct ecoregion, the Buru Rain Forests [115]. Seram, the larger island to the east of Buru, was also delineated as an ecoregion: Seram Rain Forests [116]. The larger Halmahera Rain Forests [114] includes Obi Island, which MacKinnon (1997) recognized as a separate subunit (13b) from Halmahera Island (subunit 13a). We created the Banda Sea Islands Moist Deciduous Forests [117] by combining the islands in the Kai and Tanimbar archipelagos, which were distinguished as a biogeographic unit by Monk et al. (1997). The primary vegetation on the islands in both these archipelagos is moist deciduous forests and semi-evergreen forests, whereas the vegetation in the other, nearby large islands (Seram and Aru) is evergreen rain forests (Monk et al. 1997).

Prepared by John Morrison

TABLE J117.3. WCMC (1997) Protected Areas That Overlap with the Ecoregion.

Protected Area	Area (km²)	IUCN Category
Kai Besar	290	PRO
Pulau Nuswotar	40–75	I
Jamdena	1,130	PRO
Pulau Nustaram	30	I
Pulau Lucipara	20	?
Pulau Angwarmase	10	I
Gunung Api (north of Wetar)	1	?
Total	1,521	

Ecoregion numbers of protected areas that overlap with additional ecoregions are listed in brackets.

TABLE J117.2. Endemic and Near-Endemic Bird Species.

Family	Common Name	Species
Megapodiidae	Tenimbar megapode	*Megapodius tenimberensis**
Megapodiidae	Forsten's scrubfowl	*Megapodius forstenii*
Columbidae	Dusky cuckoo-dove	*Macropygia magna*
Columbidae	Wallace's fruit-dove	*Ptilinopus wallacii*
Columbidae	Pink-headed imperial-pigeon	*Ducula rosacea*
Cacatuidae	Tanimbar cockatoo	*Cacatua goffini**
Loriidae	Red lory	*Eos bornea*
Loriidae	Blue-streaked lory	*Eos reticulata**
Loriidae	Olive-headed lorikeet	*Trichoglossus euteles*
Cuculidae	Green-cheeked bronze-cuckoo	*Chrysococcyx rufomerus**
Cuculidae	Pied bronze-cuckoo	*Chrysococcyx crassirostris*
Cuculidae	Kai coucal	*Centropus spilopterus**
Strigidae	Moluccan hawk-owl	*Ninox squamipila*
Tytonidae	Lesser masked-owl	*Tyto sororcula*
Alcedinidae	Cinnamon-backed kingfisher	*Todirhamphus australasia*
Acanthizidae	Rufous-sided gerygone	*Gerygone dorsalis*
Meliphagidae	White-tufted honeyeater	*Lichmera squamata*
Meliphagidae	Banda myzomela	*Myzomela boiei**
Meliphagidae	Black-faced friarbird	*Philemon moluccensis*
Eopsaltriidae	Golden-bellied flyrobin	*Microeca hemixantha**
Pachycephalida	Drab whistler	*Pachycephala griseonota*
Pachycephalida	Wallacean whistler	*Pachycephala arctitorquis**
Rhipiduridae	Cinnamon-tailed fantail	*Rhipidura fuscorufa**
Rhipiduridae	Long-tailed fantail	*Rhipidura opistherythra**
Monarchidae	White-naped monarch	*Monarcha pileatus*
Monarchidae	Black-bibbed monarch	*Monarcha mundus**
Monarchidae	White-tailed monarch	*Monarcha leucurus**
Monarchidae	Moluccan flycatcher	*Myiagra galeata*
Oriolidae	Buru oriole	*Oriolus bouroensis*
Campephagidae	Kai cuckoo-shrike	*Coracina dispar**
Turdidae	Slaty-backed thrush	*Zoothera schistacea**
Turdidae	Orange-banded thrush	*Zoothera peronii*
Turdidae	Fawn-breasted thrush	*Zoothera machiki**
Sturnidae	Tanimbar starling	*Aplonis crassa**
Muscicapidae	Cinnamon-chested flycatcher	*Ficedula buruensis*
Muscicapidae	Damar flycatcher	*Ficedula henrici**
Zosteropidae	Great Kai white-eye	*Zosterops grayi**
Zosteropidae	Little Kai white-eye	*Zosterops uropygialis**
Sylviidae	Timor stubtail	*Urosphena subulata*
Sylviidae	Tanimbar bush-warbler	*Cettia carolinae**
Estrildidae	Tricolored parrotfinch	*Erythrura tricolor*
Dicaeidae	Ashy flowerpecker	*Dicaeum vulneratum*
Dicaeidae	Red-chested flowerpecker	*Dicaeum maugei*

An asterisk signifies that the species' range is limited to this ecoregion.

General New Guinea Description

Introduction

Darwin's contemporary and independent developer of the theory of natural selection, Sir Alfred Russell Wallace, called New Guinea "a country which contains more strange and new and beautiful natural objects than any other part of the globe." The island of New Guinea, the world's largest tropical island (or, in the eyes of some biologists, the world's smallest continent), is located at the collision point of two large tectonic plates and the biogeographic meeting point of Asia, Australasia, and the Pacific Ocean. It is unique, a tropical island with glaciers covering its 4,000-m peaks, and its surrounding reefs support unique organisms, with some of the most diverse fish and coral populations on the planet. New Guinea's tropical forests have no monkeys or squirrels; instead, kangaroos, cuscuses, and birds live in the trees and fill the niche left vacant by a lack of primates and squirrels. Prehistoric flightless birds, as large as a man, still roam the undergrowth. Fabulously adorned birds-of-paradise perform bizarre and delightful courtship rituals on specially prepared stages. New Guinea is still swathed in the largest tracts of unbroken tropical rain forests in southeast Asia. Its peoples, occupying the island for almost 30,000 years, still live traditional lifestyles adapted to the landscape and have developed approximately 1,090 of the world's 6,000 independent languages (Diamond and Bishop 1998). What has led to this extraordinary uniqueness?

Geologic History

Portions of what is now New Guinea have existed as isolated islands since at least the middle Jurassic Period (160 m.y.). These islands were in close proximity to the ancient continent of Gondwanaland, and their flora and fauna showed this influence. Since that time, various portions of the island have emerged and submerged and have also been periodically joined to Australia. Biochemical evidence indicates three pulses of immigration from Australia when tropical wet forest would have been continuous: early Miocene (20 m.y.), middle Miocene (10–12 m.y.), and early Pliocene (3–5 m.y.). Until approximately 10–15 million years ago, New Guinea was a low-lying region, but at that time intensive mountain building began, which built up the Central Cordillera and continues to this day. Periodic connections with Australia continued, and during the most recent low sea-level stand approximately 20,000 years ago, New Guinea and Australia were connected by a broad band of dry savanna woodland, which only some taxa were able to cross. When the most recent ice age ended, sea levels rose, separating New Guinea and Australia until they took on their present configuration (Flannery 1995).

Biogeography

New Guinea has been buffered from animal and plant immigrations from Asia by deep-water barriers. Wallace's Line, just east of Bali, marks the eastern edge of the Sunda Shelf, which is exposed during periods of low sea level; this is the farthest that animals could move on land from Asia. Numerous small islands to the east of Wallace's Line, including those of the Lesser Sundas, Sulawesi, and the Moluccas, have provided stepping stones for Asian animal migrants moving east, but those migrants have been heavily filtered by the time one reaches New Guinea. In contrast, the Sahul Shelf connects New Guinea and Australia, and the area known as the Torres Straits provided a land bridge in the past: most of the island's large birds and mammals, along with most reptiles and amphibians, have their origins in Australia (Petocz 1989). Weber's Line, drawn just west of Halmahera and Buru but east of Timor and Babar in the Banda Sea, represents the approximate faunal balancing zone between the Oriental and Australasian faunas.

Consequently, New Guinea's fauna represents a unique mix of Indo-Malayan and Australasian elements. New Guinea's Australian influence is evidenced by the richest assemblage of forest-dwelling marsupials in the world (Beehler 1993). Although New Guinea's endemic monotremes and numerous endemic marsupials have evolved in the absence of southeast Asia's predators, more than half of New Guinea's mammals—rodents and bats—have an Asian origin (Petocz 1989). Of the seventy bird families in New Guinea, only four are absent from northern Australia, and only three of Australia's bird families are missing from New Guinea. However, New Guinea and Australia share thirteen bird families that are absent from southeast Asia (Beehler et al. 1986). New Guinea supports globally important radiations of birds-of-paradise (Paradisaeidae), kingfishers (Alcedinidae), fruit-pigeons (*Ducula, Ptilonopus*), parrots (Psittacidae), and honeyeaters (Meliphagidae). New Guinea's reptiles also are similar to those of Australia: all five lizard families (Agamidae, Gekkonidae, Pygopodidae, Scincidae, and Varanidae) and all seven snake families (Acrochordidae, Colubridae, Elapidae, Hydrophoiidae, Boidae, Pythonidae, Typhlopidae) are shared with Australia (Allison 1993). Pygopodidae is endemic to Australasia (O'Shea 1996). However, two of New Guinea's frog families (Hylidae and Hyobatrachidae) are Australian, and two (Ranidae and Microhylidae) are derived from Asia (Allison 1993).

New Guinea's position at the crossroads of Australasia and Indo-Malaya is partially responsible for a high floristic diversity, and the island is a mixture of Gondwanan and Malesian origins (R. Johns, pers. comm., 2000). Although there are at least eighty-four endemic plant genera, there are no endemic plant families (Johns 1993). Although the flora of New Guinea is among the most diverse in the tropics, it is poorly collected.

Physiography

Two physiographic factors of New Guinea contribute to its outstanding biodiversity: habitat diversity and geographic barriers to dispersal. New Guinea's position near the equator and its extreme elevational gradients are responsible for diverse habitats that include mangroves, freshwater swamp forest, dry savanna woodlands, tropical wet forests, alpine meadows, and glaciers. Rugged topography caused by remote tectonic forces means that populations of organisms are effectively isolated from each other.

The Central Cordillera, or Highlands of New Guinea, consists of a series of interconnected mountain ranges that run the length of the island, with extensive areas above 3,000 m. These ranges are flanked by vast lowlands to the north and south and several isolated mountain ranges to the north and to the west, on the Vogelkop and Bomberai peninsulas. The Central Ranges' elevation effectively prevents animal movement between the southern and northern lowland forests, and their ruggedness also isolates animals in distinct mountain ranges within the Central Cordillera (Petocz 1989). These ranges are cloaked in montane forest of epiphyte-covered southern beech (*Nothofagus*) and tree ferns on the outside, whereas the interior valleys have mostly given way to cultivation. These high valleys, with their pleasant, cooler climate, have been inhabited and cultivated continuously for 10,000 years. Nevertheless, hundreds of thousands of people lived here, unknown to the outside world until the their discovery in the 1930s.

General Vegetation

Lowland swamp forests are extensive in the northern and southern lowland portions of the island, associated with low-gradient river systems such as the Sepik, Strickland, and Fly rivers. The water table is near or above the ground surface but may fluctuate widely. The structure of swamp forests ranges from very small-crowned to medium-crowned and dense to open, with a 20- to 30-m canopy of an even height. Sago palm (*Metroxylon sagu*) and *Pandanus* spp. generally are present in the subcanopy. Canopy trees of the swamp forest include *Campnosperma brevipetiolata*, *C. auriculata*, *Terminalia canaliculata*, *Nauclea coadunata*, and *Syzygium* spp., with *Myristica hoolrungii* in delta areas. However, whereas some swamp forests are almost pure stands of *Campnosperma* or *Melaleuca*, others are species-rich, and many other tree species occur (Paijmans 1975).

The most extensive habitat on New Guinea is lowland broadleaf evergreen forest, which can be divided into alluvial and hill forest. Lowland alluvial forest has a canopy that is "multitiered and irregular, with many emergents, while the forest understory contains a shrub and herb layer and supports a variety of climbers, epiphytes, and ferns" (Petocz 1989, p. 23). Palms may be common in the shrub layer (Paijmans 1975). The very mixed floristic composition of the canopy trees includes *Pometia pinnata*, *Octomeles sumatrana*, *Ficus* spp., *Alstonia scholaris*, and *Terminalia* spp. Additional important genera include *Pterocarpus*, *Artocarpus*, *Planchonella*, *Canarium*, *Elaeocarpus*, *Cryptocarya*, *Celtis*, *Dracontomelum*, *Dysoxylum*, *Syzygium*, *Vitex*, *Spondias*, and *Intsia* (Paijmans 1975). Although present, dipterocarps do not form extensive forests in New Guinea, as they do in Malaysia and Borneo (Nightingale 1992). The somewhat lower-canopy, more closed lowland hill forest contains a more open shrub layer but a denser herbaceous layer. Palms are fewer in number. The dominant canopy trees include species of *Pometia*, *Canarium*, *Anisoptera*, *Cryptocarya*, *Terminalia*, *Syzygium*, *Ficus*, *Celtis*, *Dysoxylum*, and *Buchanania*. *Koompassia*, *Dillenia*, *Eucalyptopsis*, *Vatica*, and *Hopea* are locally abundant. Dense stands of *Araucaria*, the tallest tropical trees in the world, are present in scattered locations (Paijmans 1975; Nightingale 1992).

Although they are subject to variable climates and topography, montane forests are smaller-crowned and have more even canopies than lowland hill forest. Tree densities can be high, and the shrub density is also high. Predominant canopy trees include *Nothofagus*, Lauraceae, Cunoniaceae, Elaeocarpaceae, *Lithocarpus*, *Castanopsis*, *Syzygium*, *Illex*, and southern conifers. *Nothofagus* and *Araucari* may grow in pure dense stands. The numbers of Myrtaceae, Elaeocarpaceae, and conifers increases with altitude. The conifers, generally found above 2,000 m, include *Dacrycarpus*, *Podocarpus*, *Phyllocladus*, and *Papuacedrus* in the canopy and emergent layer (Paijmans 1975).

A strongly seasonal climate in the southern portions of New Guinea has fostered grasslands and savanna woodlands. The southern savannas are reminiscent of Australia, and indeed they were connected to Australia as recently as 6,000 years ago. Australian bustards, wallabies, and pademelons roam the grasslands and open *Eucalyptus* and *Acacia* woodlands in the dry season (Nightingale 1992). Portions of the southern savannas, or Trans Fly region, are seasonally inundated, and these areas are indicated by the presence of *Melaleuca*, which can withstand months of standing water on its roots. Salvadori's monitor (*Varanus salvadorii*), longer than the heavier Komodo dragon, is at home here in the wet or dry season.

Ecological Processes

Tropical forests often are considered a stable vegetation type. However, detailed studies in New Guinea reveal extensive forest areas dominated by secondary species, including *Anisoptera*, *Hopea*, *Intsia*, *Pterocarpus*, *Pometia*, and *Albizia*, which fail to regenerate in the subcanopy, except in canopy gaps. Extensive historical disturbance is suggested. Other forces, including periodic cyclonic winds and El Niño–related droughts and subsequent fire (1880s, 1915, 1941–1942) are also important natural disturbance features across the island (Johns 1993). The northern portions of New Guinea are tectonically active, and earthquakes, volcanoes, tsunamis, and landslides have been and will continue to be important natural features of the region.

Intraspecific competition between plants for nutrients and light and the threat of seed predation are intense in a tropical forest. Seedling survival generally increases the farther from the parent tree the seed is deposited (mortality up to 75 percent is found in the immediate proximity to the parent tree), so seed dispersal is a critical ecological process to maintain tree diversity. Animals and flowering plants have coevolved in this context: animals get a fruit meal, and seeds pass through and are dispersed. Animals disperse up to 90 percent of plants in the neotropics. In comparison with other tropical regions, Papua New Guinea has twice the number of frugivores as a proportion of the bird fauna. Such a critical ecological process needs a sufficiently large area in which to operate (Beehler 1982, in Sekhran and Miller 1994).

Status and Threats

The island of New Guinea is split into two political units. The western half of the island, once part of the Netherlands colonial system (until 1963), is now controlled by Indonesia and constitutes the province of Irian Jaya. The eastern portions of the island have a somewhat complicated colonial history but are now the independent nation of Papua New Guinea (PNG).

A number of protected areas, some of which are quite large, are found in Irian Jaya. However, the largest mine in the country is found outside one of Indonesia's premier national parks. The establishment of a comprehensive national park system in PNG is problematic because the vast majority of the country is in private hands, controlled by people themselves. Therefore, although there are some large protected areas in PNG, they are not administered or enforced, like reserves in other parts of the tropics.

Poorly managed logging probably is the main threat to habitats on New Guinea, and large areas of the countryside are covered by logging concessions. Two very large mining operations, one each in Indonesia and PNG, provide foreign exchange for the two countries but are associated with problems of toxic runoff.

One threat that is particular to the Indonesian (Irian Jaya) side of the island is transmigration. Transmigration represents the government-organized movement of large groups of people from overcrowded areas in other parts of Indonesia, mostly from the island of Java. As part of the program, preselected areas in Irian Jaya with agricultural potential are developed with housing and infrastructure for the new arrivals. Today approximately 25 percent of the population of Irian Jaya is from western Indonesia or is descended from transmigrants (Flannery 1995). Between 1969 and 1994, 8 million people were resettled in Irian Jaya, resulting in the loss of 1.7 million ha of forest (Bryant et al. 1997). Existing and proposed resettlement areas are located in the lowlands. Poor soils, difficult climate and topography, and health problems plague these schemes, but they continue because of extreme population pressures in the more heavily populated provinces. Environmental damage to the newly settled areas generally ensues, along with conflicts with the existing indigenous peoples.

Additional threats include exotic species. The introduction and spread of introduced rusa deer (*Cervus timorensis*) and macaques (*Macaca* sp.) can cause habitat degradation and competition and predation of vulnerable species, respectively. Additionally, introduced aquatic weeds such as salvinia (*Salvinia molesta*) and water hyacinth (*Eichhornia crassipes*) and introduced fish threaten native freshwater ecosystems (Miller et al. 1994).

Thus, although New Guinea harbors a large percentage of Asia's remaining frontier forests, large proportions of these forests are under medium or high threat, mostly from logging (Bryant et al. 1997).

Prepared by John Morrison

Ecoregion Number:	**118**
Ecoregion Name:	**Vogelkop Montane Rain Forests**
Bioregion:	**New Guinea and Melanesia**
Biome:	**Tropical and Subtropical Moist Broadleaf Forests**
Political Units:	**Indonesia**
Ecoregion Size:	**21,900 km^2**
Biological Distinctiveness:	**Globally Outstanding**
Conservation Status:	**Vulnerable**
Conservation Assessment:	**I**

The northwestern portion of the island of New Guinea is called the *Vogelkop*, or Bird's Head, Peninsula. Although this is meant to refer to the shape of the peninsula, it is also appropriate considering the large number of birds endemic to the area. The Vogelkop Montane Rain Forests [118] represent isolated tropical montane areas surrounded by ocean or lowland forest. A number of globally unique species that are adapted to upland conditions have evolved in this isolation, and although some of the mountain ranges are not large, they are still relatively pristine.

Location and General Description

This ecoregion consists of montane forests (greater than 1,000 m) in the Tamrau (to 2,582 m), Arfak (to 2,444 m), Fakfak (to 1,203 m), Kumawa (to 1,490 m), and Wandamen-Wondiwoi (to 2,552 m) mountains in northwestern Irian Jaya, Indonesia, on the island of New Guinea. The ecoregion itself is distributed in four disjunct areas, with the largest area in the northern Vogelkop Peninsula. The climate of the ecoregion is tropical wet, which is characteristic of this part of Melanesia, located in the western Pacific Ocean north of Australia (National Geographic Society 1999). Northern New Guinea is a very active tectonic area with a complex geologic history (Bleeker 1983). The surface geology of this scattered ecoregion is varied. The Wandamen-Wondiwoi Mountains are metamorphic, the Fakfak and Kumawas are composed of limestone, and the Arfak and Tamrau (Vogelkop) mountains are a diverse mix of sandstone, limestone, and volcanics (Petocz 1989).

The ecoregion is composed predominantly of tropical montane evergreen forest and tropical wet evergreen forest, with lesser amounts of tropical montane forest on limestone, limestone forest, and tropical semi-evergreen forest (MacKinnon 1997). The montane forest in this ecoregion is dominated by *Castanopsis* in the lower elevations, but with altitude the vegetation changes to moss-draped, Antarctic beech (*Nothofagus*) forests, which sometimes occur as monotypic stands, and then into coniferous forests of *Podocarpus*, *Dacrycarpus*, *Dacridium*, and *Papuacedrus* (Petocz 1989).

Biodiversity Features

The overall richness and endemism of this ecoregion are low to moderate compared with those of other ecoregions in the Indo-Pacific.

The ecoregion harbors forty-two mammal species, seven of which are endemic or near endemic (table J118.1). The Arfak ringtail (*Pseudocheirus schlegeli*) is known only from its type specimen from the Arfak Mountains and has never been seen anywhere else in the world (Flannery 1995; Flannery and Groves 1998; Bonaccorso et al., in press). The mammalian fauna consists of a wide variety of tropical Australasian marsupials, including tree kangaroos (Flannery 1995). The Arfak long-beaked echidna (*Zaglossus bruijni*) was considered endangered before it was split from the Papuan echidna (*Zaglossus bartoni*) (Hilton-Taylor 2000), and presumably it would still be considered so because it is a focal prey item for humans (Flannery 1995; Bonaccorso et al., in press). Doria's tree-kangaroo (*Dendrolagus dorianus*), a Central Cordillera species found only in the Wandammen Mountains within this ecoregion, is considered vulnerable (Hilton-Taylor 2000).

The avifauna of the ecoregion has a clear Australasian flavor, including representatives of several Australasian families including Ptilonorhynchidae, Eopsaltridae, Meliphagidae, and Paradisaeidae. There are 304 bird species in the ecoregion, of which 20 are endemic or near endemic (Beehler et al. 1986; Coates 1985). The ecoregion is equivalent to the West Papuan highlands EBA (Stattersfield et al. 1998) and harbors twenty restricted-range bird species, nine of which are found nowhere else on Earth (table J118.2). Several are limited to one small mountain range. The grey-banded munia (*Lonchura vana*) is considered vulnerable (Hilton-Taylor 2000).

Within this ecoregion, the Arfak Range, with twenty-three endemic species, and the Wamdammen Range, with seven endemic species, are both centers of butterfly endemicity on the island of New Guinea (Parsons 1999).

There are several endemic plants in the Arfak and Fakfak mountains, but in general the flora is poorly known. The ecoregions does encompass several Centres of Plant Diversity, including the Arfak Mountains, the Northern and Southern Tamrau Mountains, the Kumawa Mountains, and the Wandammen-Wondiwoi Mountains (Davis et al. 1995).

Status and Threats

Current Status

Except for a small area in the eastern part of the ecoregion that has been cleared, most of the habitat is still intact. The ten protected areas include 11,373 km^2 (52 percent) of the ecoregion (table J118.3). Two of the protected areas are more than 2,000 km^2, and two other large reserves (Pegunungan Fakfak and Pegunungan Kumawa) also extend into this ecoregion (MacKinnon 1997).

Types and Severity of Threats

The Arfak Mountains, famous for birdwing butterfly diversity, are surrounded by heavily populated areas, and the reserve itself is in danger from encroachment by population expansion (see Petocz 1989). The larger habitat block in the Tamarau

TABLE J118.2. Endemic and Near-Endemic Bird Species.

Family	Common Name	Species
Rallidae	White-striped forest-rail	*Rallina leucospila**
Rallidae	Chestnut forest-rail	*Rallina rubra*
Caprimulgidae	Mountain eared-nightjar	*Eurostopodus archboldi*
Psittacidae	Modest tiger-parrot	*Psittacella modesta*
Pachycephalidae	Vogelkop whistler	*Pachycephala meyeri**
Acanthizidae	Vogelkop scrubwren	*Sericornis rufescens**
Eopsaltriidae	Green-backed robin	*Pachycephalopsis hattamensis*
Eopsaltriidae	Smoky robin	*Peneothello cryptoleucus*
Estrildidae	Grey-banded munia	*Lonchura vana**
Melanocharitidae	Obscure berrypecker	*Melanocharis arfakiana*
Ptilonorhynchidae	Vogelkop bowerbird	*Amblyornis inornatus**
Meliphagidae	Rufous-sided honeyeater	*Ptiloprora erythropleura*
Meliphagidae	Black-backed honeyeater	*Ptiloprora perstriata*
Meliphagidae	Cinnamon-browed honeyeater	*Melidectes ochromelas*
Meliphagidae	Vogelkop honeyeater	*Melidectes leucostephes**
Meliphagidae	Arfak honeyeater	*Melipotes gymnops**
Paradisaeidae	Arfak astrapia	*Astrapia nigra**
Paradisaeidae	Long-tailed paradigalla	*Paradigalla carunculata*
Paradisaeidae	Western parotia	*Parotia sefilata**
Paradisaeidae	Greater melampitta	*Melampitta gigantea*

An asterisk signifies that the species' range is limited to this ecoregion.

TABLE J118.3. WCMC (1997) Protected Areas That Overlap with the Ecoregion.

Protected Area	Area (km^2)	IUCN Category
Jamursba-Mandi	40	PRO
Pegunungan Tamrau Utara	3,440	PRO
Mubrani Kaironi	20	IV
Gunung Meja	3	V
Pegunungan Tamrau Selatan	2,350	PRO
Mingima	40	IV
Pegunungan Arfak	720	IV
Pegunungan Fakfak	1,850	PRO
Pegunungan Kumawa	1,940	PRO
Wondiwoi	970	IV
Total	11,373	

Ecoregion numbers of protected areas that overlap with additional ecoregions are listed in brackets.

TABLE J118.1. Endemic and Near-Endemic Mammal Species.

Family	Species
Tachyglossidae	*Zaglossus bruijni*
Dasyuridae	*Phascolosorex doriae*
Peroryctidae	*Microperoryctes murina*
Macropodidae	*Dendrolagus ursinus*
Pseudocheiridae	*Pseudocheirus schlegeli**
Muridae	*Leptomys elegans*
Muridae	*Hyomys dammermani*

An asterisk signifies that the species' range is limited to this ecoregion.

Mountains is more remote and less threatened for the moment, although there are plans for logging concessions that seem to conflict with reserve gazettement (Petocz 1989). The Arfak, Fakfak, and Wandammen-Wondiwoi Mountains are all subject to potential population pressure, agricultural development, and sawmilling. Mineral deposits are small and low grade (R. Johns, pers. comm., 2000).

Priority Conservation Actions

Short-Term Conservation Actions (1–5 Years)

In 1997, Conservation International facilitated a priority-setting exercise for Irian Jaya (Conservation International 1999). This exercise mapped specific areas deemed to be important to conserve a variety of taxa. The southern Tamrau Mountains were delineated as an area in need of a major multidisciplinary field survey.

Longer-Term Conservation Actions (5–20 Years)

- Continue to implement conservation recommendations of Conservation International or other subsequent conservation priority exercises.

Focal Species for Conservation Action

Threatened endemic species: long-footed water rat (*Leptomys elegans*) (Flannery 1995), Arfak long-beaked echidna (*Zaglossus bruijnii*) (Kennedy 1992; Flannery and Groves 1998), grey-banded munia (*Lonchura vana*) (BirdLife International 2000)

Area-sensitive species: Blyth's hornbill (*Aceros plicatus*), Gurney's eagle (*Aquila gurneyi*), New Guinea eagle (*Harpyopsis novaeguineae*), peregrine falcon (*Falco peregrinus*), Doria's hawk (*Megatriorchis doriae*), tree kangaroos (Doria's tree-kangaroo [*Dendrolagus dorianus*], Vogelkop tree-kangaroo [*Dendrolagus ursinus*], grizzled tree-kangaroo [*Dendrolagus inustus*]), fruit bats (*Pteropus conspicillatus, Rousettus amplexicaudatus, Dobsonia minor, Dobsonia moluccense, Nyctimene aello, Nyctimene cyclotis, Macroglossus minimus, Syconycteris australis*) (Mickleburgh et al. 1992)

Top predator species: New Guinea quoll (*Dasyurus albopunctatus*) (Kennedy 1992), Gurney's eagle (*Aquila gurneyi*), New Guinea eagle (*Harpyopsis novaeguineae*), peregrine falcon (*Falco peregrinus*), Doria's hawk (*Megatriorchis doriae*)

Selected Conservation Contacts

Kew Gardens
Lembaga Alam Tropika Indonesia (LATIN)
WWF-Indonesia
Yayasan Keanekaragaman Hayati
Yayasan Konsorsium Pembaruan Agraria (KPA)

Justification for Ecoregion Delineation

Using Whitmore's (1984) map of the vegetation of Malesia and MacKinnon's (1997) reconstruction of the original vegetation, we delineated the large areas of distinct habitat types as ecoregions. Thus, the Vogelkop-Aru Lowland Rain Forests [119] ecoregion represents the tropical lowland moist forests in the Vogelkop region of New Guinea. The ecoregion largely corresponds to subunits P3d and P3b identified by MacKinnon (1997); however, we placed the tropical montane moist forests (more than 1,000) in the Vogelkop Montane Rain Forests [118]. Udvardy (1975) placed these ecoregions in the Papuan biogeographic province of the Oceanian Realm.

Prepared by John Morrison

Ecoregion Number:	**119**
Ecoregion Name:	**Vogelkop-Aru Lowland Rain Forests**
Bioregion:	**New Guinea and Melanesia**
Biome:	**Tropical and Subtropical Moist Broadleaf Forests**
Political Units:	**Indonesia**
Ecoregion Size:	**77,400 km²**
Biological Distinctiveness:	**Globally Outstanding**
Conservation Status:	**Critical**
Conservation Assessment:	**I**

The Vogelkop-Aru Lowland Rain Forests [119] are diverse in terms of both geography and biodiversity, and they constitute the majority of western Irian Jaya, the rest of the region being either montane forest or freshwater swamp forest. These relatively intact lowland tropical rain forests are among the largest and richest forests in the Australasian Realm. Limestone and ultramafic rock formations support unique and restricted-range floras.

Location and General Description

This ecoregion is made up of the lowland and hill (less than 1,000 m) moist forests of the Vogelkop and Bomberai peninsulas and the surrounding islands, including Misool, Salawati, Waigeo, and Kepulauan Aru in western Irian Jaya. The climate of the ecoregion is tropical wet, which is characteristic of this part of Melanesia, located in the western Pacific Ocean north of Australia (National Geographic Society 1999). Northern New Guinea is a very active tectonic area with a complex geologic history (Bleeker 1983). The surface geology of this ecoregion is composed predominantly of sedimentary rock and recent alluvium, with some large areas of limestone or ultramafics near Sorong and on Waigeo and Misool islands (Petocz 1989).

This ecoregion of plains and alluvial forests is among the most floristically rich in all of New Guinea and includes many important timber species (Petocz 1989). Most of the ecoregion is composed of a combination of alluvial and hill type tropical wet evergreen forest, with smaller amounts of limestone forest (MacKinnon 1997). Lowland alluvial forest has a canopy that is multitiered and irregular, with many emergents. The forest understory contains a shrub and herb layer with a variety of climbers, epiphytes, and ferns (Petocz 1989). Palms may be common in the shrub layer (Paijmans 1975). The very mixed floristic composition of the canopy trees includes *Pometia pinnata, Octomeles sumatrana, Ficus* spp., *Alstonia scholaris*, and *Terminalia* spp. Additional important genera include *Pterocarpus, Artocarpus, Planchonella, Canarium, Elaeocarpus, Cryptocarya, Celtis,*

Dracontomelum, Dysoxylum, Syzygium, Vitex, Spondias, and *Intsia* (Paijmans 1975). The somewhat lower-canopy, more closed lowland hill forest contains more open shrub layer but a denser herbaceous layer. Palms are fewer in number. The dominant canopy trees include species of *Pometia, Canarium, Anisoptera, Cryptocarya, Terminalia, Syzygium, Ficus, Celtis, Dysoxylum,* and *Buchanania. Koompassia, Dillenia, Eucalyptopsis, Vatica,* and *Hopea* are locally abundant. Dense stands of *Araucaria*, the tallest tropical trees in the world, are present in scattered locations (Paijmans 1975; Nightingale 1992).

On Waigeo Island and the adjacent northwest coast of New Guinea, ultramafic rocks result in a serpentine flora, a belt of low shrubby vegetation composed of *Alphitonia* spp., *Dillenia alata, Myrtella beccari,* and *Styphelia abnormis* (Brooks 1987).

Aru Island is composed of rain forest, savanna, and mangroves (R. Johns, pers. comm., 2000).

Biodiversity Features

Generally, this ecoregion exhibits low to moderate richness and endemism compared with those of other ecoregions in Indo-Malaysia. Reptile and amphibian richness, though poorly studied, is thought to be high, however.

Forty-seven mammal species are found in the ecoregion, of which eight are endemic or near endemic (Flannery 1995; Flannery and Groves 1998) (table J119.1). The mammalian fauna consists of a wide variety of tropical Australasian marsupials, including tree kangaroos (Flannery 1995). The Arfak long-beaked echidna (*Zaglossus bruijni*) was considered endangered before it was split from the Papuan echidna (*Zaglossus bartoni*) (Hilton-Taylor 2000) and presumably would still be considered so because it is a focal prey item for humans (Flannery 1995; Bonaccorso et al., in press). The lowland brush mouse (*Pogonomelomys bruijni*) is critically endangered while the dusky pademelon (*Thylogal bruinji*) is considered vulnerable (Hilton-Taylor 2000).

The avifauna of the ecoregion has a clear Australasian flavor, with representatives of several Australasian families including Ptilonorhynchidae, Eopsaltridae, Meliphagidae, and Paradisaeidae There are 366 bird species inhabiting the ecoregion (Beehler et al. 1986; Coates 1985). This ecoregion corresponds almost exactly with the West Papuan lowlands EBA (the EBA also includes the Southern New Guinea Freshwater Swamp Forests [127] ecoregion but does not include Aru Island), which includes nineteen species of restricted-range birds, nine of which are found nowhere else on Earth (Stattersfield et al. 1998). Twenty-one bird species are endemic or near endemic (table J119.2). Mollucan scrubfowl (*Megapodius wallacei*), Bruijn's brush-turkey (*Aepypodius bruijnii*), and the western crowned pigeon (*Goura cristata*) are considered vulnerable (Hilton-Taylor 2000).

The North Salawati Island Nature Reserve Centre of Plant Diversity is included in this ecoregion (Davis et al. 1995). Near Sorong several endemic plants have been collected, but the flora is poorly known. Areas of limestone and ultramafic rocks support high concentrations of unique plants near Sorong and on Waigeo and Misool islands (R. Johns, pers. comm., 2000).

Status and Threats

Current Status

About 90 percent of the natural habitat in the ecoregion is still intact. The eight protected areas cover 5,410 km^2 (7 percent) of the ecoregion (table J119.3). Three of these are large (more than 1,000 km^2) and are still linked by natural habitat (MacKinnon 1997).

TABLE J119.1. Endemic and Near-Endemic Mammal Species.

Family	Species
Tachyglossidae	*Zaglossus bruijni*
Dasyuridae	*Phascolosorex doriae*
Macropodidae	*Dendrolagus ursinus*
Macropodidae	*Thylogale brunii*
Macropodidae	*Dorcopsis muelleri*
Muridae	*Melomys lorentzi*
Muridae	*Pogonomelomys mayeri*
Muridae	*Pogonomelomys bruijni*

An asterisk signifies that the species' range is limited to this ecoregion.

TABLE J119.2. Endemic and Near-Endemic Bird Species.

Family	Common Name	Species
Megapodiidae	Bruijn's brush-turkey	*Aepypodius bruijnii**
Megapodiidae	Moluccan scrubfowl	*Megapodius wallacei*
Megapodiidae	Red-billed brush-turkey	*Talegalla cuvieri*
Megapodiidae	Dusky scrubfowl	*Megapodius freycinet*
Columbidae	Western crowned pigeon	*Goura cristata*
Columbidae	Wallace's fruit-dove	*Ptilinopus wallacii*
Columbidae	Spice imperial-pigeon	*Ducula myristicivora*
Loriidae	Violet-necked lory	*Eos squamata*
Loriidae	Black lory	*Chalcopsitta atra*
Alcedinidae	Spangled kookaburra	*Dacelo tyro*
Alcedinidae	Kofiau paradise-kingfisher	*Tanysiptera ellioti**
Alcedinidae	Red-breasted paradise-kingfisher	*Tanysiptera nympha*
Alcedinidae	Little paradise-kingfisher	*Tanysiptera hydrocharis*
Corvidae	Brown-headed crow	*Corvus fuscicapillus*
Monarchidae	Black-backed monarch	*Monarcha julianae**
Dicaeidae	Olive-crowned flowerpecker	*Dicaeum pectorale*
Meliphagidae	Olive honeyeater	*Lichmera argentauris*
Meliphagidae	Silver-eared honeyeater	*Lichmera alboauricularis*
Paradisaeidae	Wilson's bird-of-paradise	*Cicinnurus respublica**
Paradisaeidae	Red bird-of-paradise	*Paradisaea rubra**
Paradisaeidae	Greater bird-of-paradise	*Paradisaea apoda*

An asterisk signifies that the species' range is limited to this ecoregion.

Types and Severity of Threats

Logging concessions that overlap with protected areas are a major source of threat. These incursions into the protected area system from logging, when combined with the developments and infrastructure planned as part of the transmigration program (see Petocz 1989), exacerbate the threats to biodiversity in Irian Jaya, especially in the lowland forests, which are more accessible.

Hunting is a problem for some species, especially the western crowned pigeon (*Goura cristata*), northern cassowary (*Casuarius unappendiculatus*), and Nicobar pigeon (*Caloena nicobarica*) (Stattersfield et al. 1998).

On Misool Island, population pressure is responsible for the development and destruction of forests near villages for traditional agriculture, logging, and fire. The Sorong region is the petroleum center of Irian Jaya, and several government-sponsored resettlement initiatives are located in the vicinity (R. Johns, pers. comm., 2000).

Priority Conservation Actions

Short-Term Conservation Actions (1–5 Years)

In 1997, Conservation International facilitated a priority-setting exercise for Irian Jaya (Conservation International 1999). This exercise mapped specific areas deemed to be important to conserve a variety of taxa. The Bird's Neck karst peaks were delineated as a poorly known and physiographically unique area that merits a multidisciplinary field survey.

- Stop illegal logging in and around protected areas by challenging the authorities and judiciary to act on existing laws and regulations.

Longer-Term Conservation Actions (5–20 Years)

- Continue to implement conservation recommendations of Conservation International or other subsequent conservation priority exercises.
- Develop and implement effective management plans for protected areas.

Focal Species for Conservation Action

Threatened endemic species: dusky pademelon (*Thylogale bruinji*) (vulnerable) (Kennedy 1992), Bruijn's brush-turkey (*Aepypodius bruijnii*) (BirdLife International 2000), western crowned pigeon (*Goura cristata*) (BirdLife International 2000), lowland brush mouse (*Pogonomelomys bruijni*) (Flannery 1995), Mollucan scrubfowl (*Megapodius wallecei*) (BirdLife International 2000)

Area-sensitive species: Blyth's hornbill (*Aceros plicatus*) (Kemp 1995), Gurney's eagle (*Aquila gurneyi*), little eagle (*Hieraaetus morphnoides*), New Guinea eagle (*Harpyopsis novaeguineae*), Peregrine falcon (*Falco peregrinus*), fruit bats (*Pteropus melanopogon, Pteropus conspicillatus, Pteropus neohibernicus, Pteropus macrotis, Dobsonia moluccense, Nyctimene aello, Macroglossus minimus, Syconycteris australis*) (Mickleburgh et al. 1992)

Top predator species: New Guinea quoll (*Dasyurus albopunctatus*) (Kennedy 1992), Gurney's eagle (*Aquila gurneyi*), little eagle (*Hieraaetus morphnoides*), New Guinea eagle (*Harpyopsis novaeguineae*), Peregrine falcon (*Falco peregrinus*), Doria's hawk (*Megatriorchis doriae*)

Selected Conservation Contacts

Kew Gardens
Lembaga Alam Tropika Indonesia (LATIN)
WWF-Indonesia
Yayasan Keanekaragaman Hayati
Yayasan Konsorsium Pembaruan Agraria (KPA)

Justification for Ecoregion Delineation

Using Whitmore's (1984) map of the vegetation of Malesia and MacKinnon's (1997) reconstruction of the original vegetation, we delineated the large areas of distinct habitat types as ecoregions. Thus, the Vogelkop-Aru Lowland Rain Forests [119] ecoregion represents the tropical lowland moist forests in the Vogelkop region of New Guinea. The ecoregion largely corresponds to subunits P3d and P3b identified by MacKinnon (1997); however, we placed the tropical montane moist forests (more than 1,000) in the Vogelkop Montane Rain Forests [118]. Udvardy (1975) placed these ecoregions in the Papuan biogeographic province of the Oceanian Realm.

Prepared by John Morrison

TABLE J119.3. WCMC (1997) Protected Areas That Overlap with the Ecoregion.

Protected Area	Area (km²)	IUCN Category
Batanta Barat	70	I
Salawati Utara	620	I
Sidei-Wibain	30	IV
Misool Selatan	1,160	I
Pulau Waigeo	1,310	I
Pulau Kobroor	1,160	PRO
Pulau Baun	100	IV
Pegunungan Weyland [125, 131]	960	PRO
Total	5,410	

Ecoregion numbers of protected areas that overlap with additional ecoregions are listed in brackets.

Ecoregion Number: **120**
Ecoregion Name: **Biak-Numfoor Rain Forests**
Bioregion: **New Guinea and Melanesia**
Biome: **Tropical and Subtropical Moist Broadleaf Forests**
Political Units: **Indonesia**
Ecoregion Size: **2,900 km^2**
Biological Distinctiveness: **Regionally Outstanding**
Conservation Status: **Critical**
Conservation Assessment: **II**

The Biak-Numfoor Rain Forests [120], moderate-sized limestone islands guarding the entrance to Cenderawasih Bay, contain the most highly endemic avifauna of any single area in New Guinea. The islands have been heavily logged (Stattersfield et al. 1998).

Location and General Description

This small ecoregion is made up of the islands of Biak, Supiori (to 850 m), and Numfoor (to 204 m) (and small outlying islands) in Cenderawasih (Geelvink) Bay, approximately 50 km off the northwestern coast of Irian Jaya, Indonesia, on the island of New Guinea. The climate of the ecoregion is tropical wet, which is characteristic of this part of Melanesia, located in the western Pacific Ocean north of Australia (National Geographic Society 1999). The surface geology of this ecoregion consists mostly of extremely rugged limestone mountains, with an additional section of argillaceous sedimentary rock on Biak (Stattersfield et al. 1998; Petocz 1989). These oceanic islands, in contrast with nearby Yapen Island, have never been connected to the mainland, contributing to their high level of endemism.

The original lowland tropical wet evergreen forest of these islands was similar in structure and composition to the mainland lowland forest, which can be divided coarsely into alluvial and hill forest. Lowland alluvial forest has a multi-tiered and irregular canopy with many emergents. The forest understory contains a shrub and herb layer with a variety of climbers, epiphytes, and ferns (Petocz 1989). Palms may be common in the shrub layer. The somewhat lower-canopy, more closed lowland hill forest contains more open shrub layer but a denser herbaceous layer. Palms are fewer in number (Paijmans 1975). The dominant emergent trees on Biak and Numfoor include *Pometia*, *Ficus*, *Alstonia*, and *Terminalia* spp., and the lower-story trees consist of *Garcinia*, *Diospyros*, *Myristica*, *Maniltoa*, and *Microcos* spp. (Petocz 1989). Impressive coastal stands of *Calophyllum* are found in northern Biak (Beehler, pers. comm.).

Biodiversity Features

Overall richness and endemism are low to moderate when compared with those of other ecoregions in Indo-Malaysia, although for its size the ecoregion contains the most exclusively endemic avifauna of any single area in New Guinea (Stattersfield et al. 1998).

Twenty-nine mammal species are found in this ecoregion, including five endemic or near-endemic species (Wikramanyake et al. 2001; Flannery 1995; Flannery and Groves 1998) (table J120.1). The Biak bare-backed fruit bat (*Dobsonia emersa*) is considered vulnerable (Hilton-Taylor 2000).

The ecoregion harbors 107 bird species and matches the Geelvink Islands EBA (Stattersfield et al. 1998; Beehler et al. 1986; Coates 1985). The EBA contains fourteen restricted-range birds. The ecoregion contains thirteen endemic or near-endemic bird species (table J120.2). The Biak gerygone (*Gerygone hypoxantha*), Biak monarch (*Monarcha brehmii*), and Biak scops owl (*Otus beccarii*) are considered endangered while the black-winged lory (*Eos cyanogenia*) and Biak megapod (*Megapodius gellvinkianus*) are considered vulnerable (BirdLife International 2000).

This ecoregion is a center of butterfly endemicity in the New Guinea region, with eighteen endemic species (Parsons 1999).

The islands make up the Numfoor Island Nature Reserve–North Biak Island Nature Reserve Centre of Plant Diversity (Davis et al. 1995). Several endemic plants have been collected on the islands, but the flora is very poorly known (R. Johns, pers. comm., 2000).

TABLE J120.1. Endemic and Near-Endemic Mammal Species.

Family	Species
Petauridae	*Petaurus biacensis**
Pteropodidae	*Dobsonia emersa**
Muridae	*Rattus jobiensis*
Muridae	*Uromys boeadii**
Muridae	*Uromys emmae**

An asterisk signifies that the species' range is limited to this ecoregion.

TABLE J120.2. Endemic and Near-Endemic Bird Species.

Family	Common Name	Species
Megapodiidae	Geelvink scrubfowl	*Megapodius geelvinkianus**
Psittacidae	Geelvink pygmy-parrot	*Micropsitta geelvinkiana**
Loriidae	Black-winged lory	*Eos cyanogenia**
Columbidae	Spice imperial-pigeon	*Ducula myristicivora*
Columbidae	Yellow-bibbed fruit-dove	*Ptilinopus solomonensis*
Cuculidae	Biak coucal	*Centropus chalybeus**
Alcedinidae	Biak paradise-kingfisher	*Tanysiptera riedelii**
Alcedinidae	Numfoor paradise-kingfisher	*Tanysiptera carolinae**
Acanthizidae	Biak gerygone	*Gerygone hypoxantha**
Monarchidae	Biak monarch	*Monarcha brehmii**
Monarchidae	Biak flycatcher	*Myiagra atra**
Sturnidae	Long-tailed starling	*Aplonis magna**
Zosteropidae	Biak white-eye	*Zosterops mysorensis**

An asterisk signifies that the species' range is limited to this ecoregion.

Status and Threats

Current Status

The human population in Biak Island is the highest among the offshore islands, and the island has already undergone a phase of logging operations; in fact, further logging is economically unfeasible. Logging and subsistence farming have damaged or destroyed much of the forest on Biak and Numfoor. Biak and Supiori are both transmigration sites as well. As a result of the damage and poor growing conditions of the raised limestone substrate, Biak's southern plains are now stunted woodland and arid scrub (Petocz 1989 and Bishop 1982 in Stattersfield et al. 1998).

The three small protected areas cover 344 km^2, representing about 12 percent of the ecoregion (MacKinnon 1997) (table J120.3). The rugged topography of Supiori provides some degree of protection.

Types and Severity of Threats

If logging operations in other parts of Indonesia are hindered (for example, in the case of forest fires in Kalimantan), operations could shift to these islands to meet timber shortfalls. Continued subsistence farming to feed a growing population will further degrade the island's remaining habitats.

The birds that are limited to these islands are vulnerable simply because of the limited area of the islands. Hunting and trapping for trade are threats to several species (Stattersfield et al. 1998).

Priority Conservation Actions

Short-Term Conservation Actions (1–5 Years)

- Stop illegal logging in and around protected areas.
- Increase amount of habitat protection.
- Limit the gazettement of future logging concessions.

Longer-Term Conservation Actions (5–20 Years)

- Develop and implement effective management plans for protected areas.

Focal Species for Conservation Action

Threatened endemic species: Biak bare-backed fruit bat (*Dobsonia emersa*) (Mickleburgh et al. 1992), Biak gerygone (*Gerygone hypoxantha*) (BirdLife International 2000), Biak monarch (*Monarcha brehmii*) (BirdLife International 2000), Biak scops owl (*Otus beccarii*) (BirdLife International 2000)

Area-sensitive species: Gurney's eagle (*Aquila gurneyi*), peregrine falcon (*Falco peregrinus*), fruit bats (*Pteropus conspicillatus, Pteropus pohlei, Nyctimene cyclotis, Macroglossus minimus, Dobsonia beauforti, Dobsonia emersa*) (Mickleburgh et al. 1992)

Top predator species: Biak bare-backed fruit-bat (*Dobsonia emersa*) (Mickleburgh et al. 1992), variable goshawk (*Accipiter hiogaster*), Gurney's eagle (*Aquila gurneyi*), peregrine falcon (*Falco peregrinus*)

Selected Conservation Contacts

Lembaga Alam Tropika Indonesia (LATIN)
WWF-Indonesia
Yayasan Keanekaragaman Hayati
Yayasan Konsorsium Pembaruan Agraria (KPA)

Justification for Ecoregion Delineation

Using Whitmore's (1984) map of the vegetation of Malesia and MacKinnon's (1997) reconstruction of the original vegetation, we delineated the large areas of distinct habitat types as ecoregions. Yapen and Biak islands, which MacKinnon combined within biounit P3c, were delineated as separate ecoregions; Yapen Rain Forests [121] and Biak-Numfoor Rain Forests [120], respectively, were based on recommendations by Bob Johns (vegetation) and Bruce Beehler (birds) and the patterns of mammal distribution. Udvardy (1975) placed these ecoregions in the Papuan biogeographic province of the Oceanian Realm.

Prepared by John Morrison

Ecoregion Number:	**121**
Ecoregion Name:	**Yapen Rain Forests**
Bioregion:	**New Guinea and Melanesia**
Biome:	**Tropical and Subtropical Moist Broadleaf Forests**
Political Units:	**Indonesia**
Ecoregion Size:	**2,300 km^2**
Biological Distinctiveness:	**Critical**
Conservation Status:	**Relatively Stable**
Conservation Assessment:	**IV**

The Yapen Rain Forests [121] are important for their two restricted-range bird species and unique limestone and ultramafic floras. Although almost one-third of the ecoregion is under some form of protection, the island is subject to population pressure.

Location and General Description

This small ecoregion represents the lowland and montane rain forests of Yapen Island, off the northwestern coast of Irian Jaya, Indonesia, on the island of New Guinea. The climate of the ecoregion is tropical wet, which is characteristic of this part of Melanesia, located in the western Pacific Ocean north of Australia (National Geographic Society 1999). The surface geology of this ecoregion consists of low mountains of plutonic rock and limestone. The island extends to an elevation 1,430 m and is a land bridge island that was part of the New Guinea mainland during recent glacial periods.

TABLE J120.3. WCMC (1997) Protected Areas That Overlap with the Ecoregion.

Protected Area	Area (km^2)	IUCN Category
Pulau Supriori	270	I
Biak Utara	70	I
Pulau Biak	4	PRO
Total	344	

Ecoregion numbers of protected areas that overlap with additional ecoregions are listed in brackets.

The vegetation of Yapen Island is tropical lowland (alluvial and hill type) and montane forest.

Biodiversity Features

The overall richness and endemism of this ecoregion are low to moderate when compared with those of other ecoregions in Indo-Malaysia.

The mammal fauna consists of thirty-seven species, including a near endemic that Yapen shares with Biak and Numfoor islands (Yapen rat, *Rattus jobiensis*) (Flannery 1995; Flannery and Groves 1998; Bonaccorso et al., in press) (table J121.1).

Yapen is home to approximately 147 bird species (Beehler et al. 1986; Coates 1985), including two restricted-range species that qualify it as a Secondary EBA. These two near-endemic species (table J121.2), the spice imperial-pigeon (*Ducula myristicivora*) and the green-backed robin (*Pachycephalopsis hattamensis*), are also found on the mainland (Stattersfield et al. 1998).

Yapen and nearby Biak Island (part of Biak-Numfoor Rain Forests [120] ecoregion) share one endemic butterfly species (Parsons 1999).

The island constitutes the Yapen Island Nature Reserve Centre of Plant Diversity (Davis et al. 1995). Several endemic plants have been collected, but the flora of the island is poorly known. The island contains significant limestone and ultramafic floras.

Status and Threats

Current Status

Two protected areas, covering 790 km^2, protect 32 percent of the island's ecosystems (MacKinnon 1997) (table J121.3).

TABLE J121.1. Endemic and Near-Endemic Mammal Species.

Family	Species
Muridae	Rattus jobiensis

An asterisk signifies that the species' range is limited to this ecoregion.

TABLE J121.2. Endemic and Near-Endemic Bird Species.

Family	Common Name	Species
Columbidae	Spice imperial-pigeon	Ducula myristicivora
Eopsaltriidae	Green-backed robin	Pachycephalopsis hattamensis

An asterisk signifies that the species' range is limited to this ecoregion.

TABLE J121.3. WCMC (1997) Protected Areas That Overlap with the Ecoregion.

Protected Area	Area (km^2)	IUCN Category
Yapen Tengah	780	I
Inggresau	10	PRO
Total	790	

Ecoregion numbers of protected areas that overlap with additional ecoregions are listed in brackets.

Types and Severity of Threats

The island is subject to population pressure, agricultural development, local sawmilling operations, and human-made fire. Only small, low-grade mineral deposits are present (R. Johns, pers. comm., 2000).

Priority Conservation Actions

Short-Term Conservation Actions (1–5 Years)

- Stop illegal logging in and around protected areas.
- Conduct biological surveys to assess the island's endemic fauna and flora, especially in the ultramafic and limestone areas.

Longer-Term Conservation Actions (5–20 Years)

- Develop and implement effective management plans for protected areas.

Focal Species for Conservation Action

Area-sensitive species: Blyth's hornbill (*Aceros plicatus*) (Kemp 1995), Gurney's eagle (*Aquila gurneyi*), peregrine falcon (*Falco peregrinus*), fruit bats (*Pteropus conspicillatus, Pteropus pohlei, Rousettus amplexicaudatus, Dobsonia minor, Dobsonia magna, Syconycteris australis*) (Mickleburgh et al. 1992)

Top predator species: Gurney's eagle (*Aquila gurneyi*), Meyer's goshawk (*Accipiter meyerianus*), grey-headed goshawk (*Accipiter poliocephalus*), collared sparrowhawk (*Accipiter cirrocephalus*), long-tailed honey-buzzard (*Henicopernis longicauda*), Oriental hobby (*Falco severus*), peregrine falcon (*Falco peregrinus*)

Selected Conservation Contacts

Lembaga Alam Tropika Indonesia (LATIN)
WWF-Indonesia
Yayasan Keanekaragaman Hayati
Yayasan Konsorsium Pembaruan Agraria (KPA)

Justification for Ecoregion Delineation

Using Whitmore's (1984) map of the vegetation of Malesia and MacKinnon's (1997) reconstruction of the original vegetation, we delineated the large areas of distinct habitat types as ecoregions. Yapen and Biak islands, which MacKinnon combined within biounit P3c, were delineated as separate ecoregions; Yapen Rain Forests [121] and Biak-Numfoor Rain Forests [120], respectively, were based on recommendations by Bob Johns (vegetation) and the patterns of mammal distribution. Udvardy (1975) placed these ecoregions in the Papuan biogeographic province of the Oceanian Realm.

Prepared by John Morrison

Ecoregion Number:	**122**
Ecoregion Name:	**Northern New Guinea Montane Rain Forests**
Bioregion:	**New Guinea and Melanesia**
Biome:	**Tropical and Subtropical Moist Broadleaf Forests**
Political Units:	**Indonesia, Papua New Guinea**
Ecoregion Size:	**23,200 km²**
Biological Distinctiveness:	**Endangered**
Conservation Status:	**Critical**
Conservation Assessment:	**II**

The inaccessible, isolated mountain ranges of the Northern New Guinea Montane Rain Forests [122], surrounded by tropical lowland forest, are home to unique vertebrate species found nowhere else on Earth. Some portions of this ecoregion are poorly known. In fact, the Foya Mountains, a significant piece of upland east of the Mamberamo River, have no record of visitation of any kind before 1979 (Stattersfield et al. 1998). Because of their isolation, they are relatively intact.

Location and General Description

This ecoregion is composed of the isolated montane forests (more than 1,000 m) of the Van Rees (to 1,430 m) and Gauttier (Foya) (to 2,193 m), Cyclops (to 2,158 m), Denake, Bewani (to 2,000 m), Torricelli (to 1,650 m), Prince Alexander (to 1,240 m), and Adelbert Ranges (to 1,718 m) in Irian Jaya, Indonesia and PNG. These isolated mountain ranges are all on the northern side of the Central Cordillera of the island of New Guinea. The climate of the ecoregion is tropical wet, which is characteristic of this part of Melanesia, located in the western Pacific Ocean north of Australia (National Geographic Society 1999). Northern New Guinea is a very active tectonic area with a complex geologic history (Bleeker 1983). The geology of this mountainous ecoregion is a mixture of metamorphic and Pliocene fine-grained terrestrial and marine sediments (Bleeker 1983; Petocz 1989).

The vegetation of this ecoregion is generally tropical montane rain forest. Although they are subject to variable climates and topography, montane forests are smaller-crowned and have even more canopies than lowland hill forest. Tree densities can be high, and the shrub density is also high. Predominant canopy trees include *Nothofagus*, Lauraceae, Cunoniaceae, Elaeocarpaceae, *Lithocarpus*, *Castanopsis*, *Syzygium*, *Illex*, and southern conifers. *Nothofagus* and *Araucaria* may grow in pure, dense stands. The levels of Myrtaceae, Elaeocarpaceae, and conifers increase with altitude. The conifers generally found above 2,000 m include *Dacrycarpus*, *Podocarpus*, *Phyllocladus*, and *Papuacedrus* in the canopy and emergent layer (Paijmans 1975).

The open forests of the Cyclops Mountains, perhaps the most well-studied of the ranges, are dominated by *Kania*, *Metrrosideros*, and *Xanthmyrtus*, with *Lithocarpus* and *Nothofagus* at higher altitudes. Above 1,400 m, conifers (*Phyllocladus*, *Papuacedrus*, *Dacrydium*) dominate, with *Podocarpus* and *Rapanea* (R. Johns, pers. comm., 2000). At an elevation of 1,200 m, the Foya Mountains to the west are dominated by *Araucaria cunninghamii*, *Podocarpus neriifolius*, *Agathis labillardieri*, *Calophyllum*, and *Palaquium*. The Torricelli, Bewani, and Prince Alexander ranges consist of limestone and montane forest (Davis et al. 1995).

Biodiversity Features

Overall richness and endemism are low to moderate when compared with those of other ecoregions in Indo-Malaysia. There are fifty-one mammal species in the ecoregion, with six species that are endemic or near endemic (Flannery 1995; Flannery and Groves 1998) (table J122.1). The mammalian fauna consists of a wide variety of tropical Australasian marsupials, including tree kangaroos, and a glider (Flannery 1995). The Cyclops long-beaked echidna (*Zaglossus attenboroughi*) was considered endangered before it was split from the Papuan echidna (*Zaglossus bruijnii*) (Hilton-Taylor 2000), and presumably would still be considered so because it is a focal prey item for humans (Flannery 1995; Flannery and Groves 1998; Bonaccorso et al., in press). The northern glider (*Petaurus abidi*) is found nowhere else on Earth (Ziegler 1981). The highlands of the north coastal ranges also harbor Scott's tree kangaroo (*Dendrolagus scottae*), reputed to be the largest and most threatened native forest mammal in PNG (Beehler 1994).

This area also provides habitat for a number of isolated and taxonomically distinct bird populations. The avifauna of the ecoregion has a clear Australasian flavor, including representatives of several Australasian families including Ptilonorhynchidae, Eopsaltridae, Meliphagidae, and Paradisaeidae. This ecoregion includes all of the North Papuan Mountains EBA and portions of the Adelbert and Huon ranges EBA (the Adelbert Mountains) and the North Papuan lowlands EBA (the Van Rees Mountains) (Stattersfield et al. 1998). The ecoregion contains twelve endemic or near-endemic birds (Beehler et al. 1986; Coates 1985) (table J122.2). The North Papuan mountains EBA contains five restricted-range birds, including three found nowhere else on Earth. The Adelbert Range contains three restricted-range bird species. It shares Wahnes's parotia (*Parotia wahnesi*) and the olive-streaked honeyeater (*Ptiloprora guisei*) with the mountains of the Huon Peninsula, but the fire-maned bowerbird (*Sericulus bakeri*) is found nowhere else on Earth but this ecoregion (Stattersfield et al. 1998). The rarest bird species in PNG, fire-maned bowerbird has the most circumscribed geographic range known for any species on mainland PNG (Miller et al. 1994). Both the fire-maned bowerbird and Wahnes's parotia (*Parotia wahnesi*) are considered vulnerable (Hilton-Taylor 2000).

TABLE J122.1. Endemic and Near-Endemic Mammal Species.

Family	Species
Tachyglossidae	*Zaglossus attenboroughi**
Perorictidae	*Echymipera clara*
Petauridae	*Petaurus abidi**
Macropodidae	*Dendrolagus scottae**
Muridae	*Paraleptomys rufilatus*
Muridae	*Xenuromys barbatus*

An asterisk signifies that the species' range is limited to this ecoregion.

TABLE J122.2. Endemic and Near-Endemic Bird Species.

Family	Common Name	Species
Rallidae	Mayr's forest-rail	Rallina mayri*
Ptilonorhynchidae	Golden-fronted bowerbird	Amblyornis flavifrons*
Ptilonorhynchidae	Fire-maned bowerbird	Sericulus bakeri*
Meliphagidae	Mayr's honeyeater	Ptiloprora mayri*
Meliphagidae	Rufous-backed honeyeater	Ptiloprora guisei
Meliphagidae	Cinnamon-browed honeyeater	Melidectes ochromelas
Eopsaltriidae	Smoky robin	Peneothello cryptoleucus
Eopsaltriidae	Green-backed robin	Pachycephalopsis hattamensis
Cinclosomatidae	Brown-capped jewel-babbler	Ptilorrhoa geislerorum
Paradisaeidae	Greater melampitta	Melampitta gigantea
Paradisaeidae	Carola's parotia	Parotia carolae
Paradisaeidae	Wahnes's parotia	Parotia wahnesi

An asterisk signifies that the species' range is limited to this ecoregion.

Within this ecoregion, the Torricelli Range has one endemic butterfly species, making it a center of butterfly endemicity on the island of New Guinea (Parsons 1999).

Several Centres of Plant Diversity correspond with the various ranges of this ecoregion, including the Mamberamo-Pegunungan Jayawijay (Van Rees and Gauttier) and Cagar Alam Pegunungan Cyclops in Irian Jaya, the Torricelli Mountains–Bewani Mountains–Prince Alexander Range in PNG, and the Adlebert Range in PNG. The Torricelli, Bewani, and Prince Alexander ranges have a flora that is estimated to exceed 2,000 species and includes the only endemic fern genus on New Guinea (*Rheopteris cheesmannii*) (Davis et al. 1995).

Several endemic plants have been collected in the Cyclops Mountains, but in general the flora of this ecoregion is very poorly known. Ultrabasic formations are present in the Makanoi Range forests (R. Johns, pers. comm., 2000).

Status and Threats

Current Status

Much of the topography of this ecoregion is too steep for traditional logging activities, and the majority of the ecoregion is safe because of its inaccessibility. Twenty-six percent of the ecoregion is covered by five protected areas, mostly in Irian Jaya, although almost half of the ecoregion is in PNG (MacKinnon 1997) (table J122.3).

Types and Severity of Threats

The threats to this ecoregion include the potential for commercial logging if it becomes economically viable. The Cyclops Mountains are quite close to the main population center of Irian Jaya, Jayapura, however, and these hill forests are at risk from the town and a transmigration settlement in the area (Stattersfield et al. 1998).

Priority Conservation Actions

Short-Term Conservation Actions (1–5 Years)

Several priority-setting efforts have been conducted that include this ecoregion, including Beehler (1994) and Nix et al. (2000) for PNG and Conservation International (1999) for Irian Jaya. These exercises have mapped specific areas that are deemed important to conserve a variety of taxa. The Conservation International exercise indicated that the Foya and Van Rees mountains were in need of major multidisciplinary field surveys.

Longer-Term Conservation Actions (5–20 Years)

- Continue to implement conservation recommendations of Conservation International (1999), Beehler (1994), and Nix et al. (2000) and other subsequent conservation priority exercises.
- Develop and implement effective management plans for protected areas.
- Increase protection of PNG's forests.

Focal Species for Conservation Action

Endangered endemic species: Cyclops long-beaked echidna (*Zaglossus attenboroughi*) (Kennedy 1992), fire-maned bowerbird (*Sericulus bakeri*) (Stattersfield et al. 1998), Wahnes's parotia (*Parotia wahnesi*) (Frith and Beehler 1998; Stattersfield et al. 1998), Scott's tree-kangaroo (*Dendrolagus scottae*) (Flannery 1995)

Area-sensitive species: Scott's tree-kangaroo (*Dendrolagus scottae*) (Kennedy 1992), grizzled tree-kangaroo (*Dendrolagus inustus*) (Kennedy 1992), Gurney's eagle (*Aquila gurneyi*), little eagle (*Hieraaetus morphnoides*), New Guinea eagle (*Harpyopsis novaeguineae*), peregrine falcon (*Falco peregrinus*), fruit bats (*Pteropus hypomelanus, Pteropus neohibernicus, Rousettus amplexicaudatus, Nyctimene albiventer, Macroglossus minimus, Syconycteris australis, Dobsonia moluccensis*) (Mickleburgh et al. 1992)

Top predator species: New Guinea quoll (*Dasyurus albopunctatus*) (Flannery 1995), Gurney's eagle (*Aquila gurneyi*), little eagle (*Hieraaetus morphnoides*), New Guinea eagle (*Harpyopsis novaeguineae*), peregrine falcon (*Falco peregrinus*), chestnut-shouldered goshawk (*Erythrotriorchis buergersi*)

TABLE J122.3. WCMC (1997) Protected Areas That Overlap with the Ecoregion.

Protected Area	Area (km^2)	IUCN Category
Unnamed [123]	1,510	?
Mamberamo-Pegunungan Foya [123]	2,110	IV
Peg. Cycloop	210	I
Teluk Yotefa	90	IV
Mt. Menawa	2,150	?
Total	6,070	

Ecoregion numbers of protected areas that overlap with additional ecoregions are listed in brackets.

Selected Conservation Contacts

Lembaga Alam Tropika Indonesia (LATIN)
Tenkile Conservation Project
WWF-Indonesia
Yayasan Keanekaragaman Hayati
Yayasan Konsorsium Pembaruan Agraria (KPA)

Justification for Ecoregion Delineation

Using Whitmore's (1984) map of the vegetation of Malesia and MacKinnon's (1997) reconstruction of the original vegetation, we delineated the large areas of distinct habitat types as ecoregions. The tropical lowland moist and freshwater swamp forests to the north of the Central Cordillera were placed in the Northern New Guinea Lowland Rain and Freshwater Swamp Forests [123], and the montane forests in the Northern New Guinea Montane Rain Forests [122] (based largely on recommendations by Bob Johns, pers. comm., 1999). This ecoregion corresponds to MacKinnon's (1997) biounits P3e and P3j. Udvardy (1975) placed these ecoregions in the Papuan biogeographic province of the Oceanian Realm.

Prepared by John Morrison

Ecoregion Number:	**123**
Ecoregion Name:	**Northern New Guinea Lowland Rain and Freshwater Swamp Forests**
Bioregion:	**New Guinea and Melanesia**
Biome:	**Tropical and Subtropical Moist Broadleaf Forests**
Political Units:	**Indonesia, Papua New Guinea**
Ecoregion Size:	**134,700 km²**
Biological Distinctiveness:	**Regionally Outstanding**
Conservation Status:	**Critical**
Conservation Assessment:	**II**

Associated with the foothills north of the Central Ranges and some of New Guinea's great lowland river systems, the Northern New Guinea Lowland and Freshwater Swamp Forests [123] are an extensive, continuous tropical lowland and swamp forest that is still largely unexplored (Stattersfield et al. 1998).

Location and General Description

This ecoregion is made up of the lowland, freshwater, and peat swamp forests of Irian Jaya and PNG, from the foothills of the northern side of the Central Cordillera to the north coast of the island of New Guinea. The climate of the ecoregion is tropical wet, which is characteristic of this part of Melanesia, located in the western Pacific Ocean north of Australia (National Geographic Society 1999). Northern New Guinea is a very active tectonic area with a complex geologic history (Bleeker 1983). The Lakes-Plains depression, which forms this ecoregion, is squeezed between the foothills of the Central Cordillera to the south and the Van Rees and Foya Mountains to the north (Petocz 1989). The surface geology of this ecoregion consists of clastic sedimentary rocks and recent alluvium (Petocz 1989). The ecoregion is centered on three large river basins: the Mamberamo, Taritatu, and Tariku river basin in Irian Jaya and the Sepik and Ramu river basins in PNG. The Sepik River, one of the two largest watersheds in PNG, supports a large human population that is heavily dependent on the river (Miller et al. 1994). Although there is extensive area in the upper basins, large portions of the mainstems of these rivers flow through broad alluvial valleys, which consist of extensive wetland areas. Twenty-one of PNG's twenty-four largest lakes are found at elevations of 40 m or less, and many of these are associated with the Sepik River (Osborne 1993).

The lowland forests and freshwater swamps of this ecoregion contain diverse habitats, including lowland and hill forest, grass swamps, swamp forests, savannas, and woodlands. The most extensive habitat in this ecoregion is lowland broadleaf evergreen forest, which can be divided coarsely into alluvial and hill forest. Lowland alluvial forest has a canopy that is multitiered and irregular, with many emergents. The forest understory contains a shrub and herb layer with a variety of climbers, epiphytes, and ferns (Petocz 1989). Palms may be common in the shrub layer (Paijmans 1975). The very mixed floristic composition of the canopy trees includes *Pometia pinnata*, *Octomeles sumatrana*, *Ficus* spp., *Alstonia scholaris*, and *Terminalia* spp. Additional important genera include *Pterocarpous*, *Artocarpus*, *Planchonella*, *Canarium*, *Elaeocarpus*, *Cryptocarya*, *Celtis*, *Dracontomelum*, *Sysoxylum*, *Syzygium*, *Vitex*, *Spondias*, and *Intsia* (Paijmans 1975). The somewhat lower-canopy, more closed lowland hill forest contains more open shrub layer but a denser herbaceous layer. Palms are fewer in number. The dominant canopy trees include species of *Pometia*, *Canarium*, *Anisoptera*, *Cryptocarya*, *Terminalia*, *Syzygium*, *Ficus*, *Celtis*, *Dysoxylum*, and *Buchanania*. *Koompassia*, *Dillenia*, *Eucalyptopsis*, *Vatica*, and *Hopea* are locally abundant. Dense stands of *Araucaria*, the tallest tropical trees in the world, are present in scattered locations (Paijmans 1975; Nightingale 1992).

Because of its large geographic extent, this ecoregion is subject to local variation. In the Central Range and Sepik foothills, extensive stands of *Agathis labillardieri* support a highly diverse epiphytic fauna. The Ramu Basin supports extensive areas of lowland rain forest and swamp forest, some of which are developed on ultrabasic parent rock (Miller et al. 1994).

Lowland swamp forests are extensive in the northern lowland portions of the island, associated with low-gradient river systems such as the Sepik River. The water table is near or above the ground surface but may fluctuate widely. Much of the forest along the rivers is subject to inundation (Paijmans 1975; Davis et al. 1975). A dynamic environment such as these freshwater wetlands contains a mosaic of habitats. One list of the subhabitats of lowland freshwater swamps includes herbaceous swamp vegetation, *Leersia* grass swamp, *Saccharum-Phragmites* grass swamp, *Pseudoraphis* grass swamp, mixed swamp savanna, *Melaleuca* swamp savanna, mixed swamp woodland, sago swamp woodland, pandan swamp woodland, mixed swamp forest, *Campnosperma* swamp forest, *Teminalia* swamp forest, and *Melaleuca* swamp forest (Osborne 1993).

The structure of swamp forests ranges from very small-crowned to medium-crowned, dense to open, with a 20- to 30-m canopy of an even height. Sago palm (*Metroxylon sagu*) and *Pandanus* spp. generally are present in the subcanopy. Canopy trees of the swamp forest include *Campnosperma brevipetiolata*, *C. auriculata*, *Terminalia canaliculata*, *Nauclea coadunata*, and *Syzygium* spp., with *Myristica hollrungii* in delta areas. However, whereas some swamp forests are al-most pure stands of *Campnosperma* or *Melaleuca*, others are species-rich, and many other tree species are possible (Paijmans 1975). The canopy along the Mamberamo River is about 45 m high and includes *Ficus* and *Pittosporum ramiflorum*. Other inundated forests include *Timonius*, *Dillenia*, and *Nauclea* (Davis et al. 1995).

Grass habitats are dominated by *Leersia*, *Phragmites*, and *Saccharum* (Petocz 1989).

Biodiversity Features

Generally, this ecoregion exhibits low to moderate richness and endemism when compared with those of other ecoregions in Indo-Malaysia. Reptile and amphibian richness is thought to be high but is poorly documented.

The mammalian fauna consists of a wide variety of tropical Australasian marsupials, including tree kangaroos. The seventy-six mammal species in this ecoregion include thirteen species that are endemic or near endemic (Flannery 1995; Flannery and Groves 1998; Bonaccorso et al., in press) (table J123.1). The lesser tube-nosed bat (*Nyctimene draconilla*), Fly River trumpet-eared bat (*Kerivoula muscina*), mantled mastiff bat (*Otomops secundus*), and greater sheath-tailed bat (*Emballonura furax*) are considered vulnerable (Hilton-Taylor 2000). The western part of the ecoregion is the only known site in PNG for the western ringtail possum (*Pseudocheirus albertisi*) (Beehler 1994).

The avifauna of the ecoregion has a clear Australasian flavor, including representatives of several Australasian families such as Ptilonorhynchidae, Eopsaltriidae, Meliphagidae, and Paradisaeidae. This ecoregion corresponds very well with the North Papuan lowlands EBA, which has nine restricted-range bird species, including five species found nowhere else on Earth. All told, the ecoregion contains sixteen endemic or near-endemic species (Stattersfield et al. 1998; Beehler et al. 1986; Coates 1985) (table J123.2). Salvadori's fig-parrot (*Psittaculirostris salvadorii*) and black sicklebill (*Epimachus fastuosus*) are considered vulnerable (BirdLife International 2000).

A small portion of the ecoregion, along the lower Mamberamo River, is part of the Mamberamo–Pegunungan Jayawijaya Centre of Plant Diversity in Irian Jaya (Davis et al. 1995).

Status and Threats

Current Status

This ecoregion is still largely undisturbed. There are some transmigration sites in the lowland forest on the Indonesian side near Nabire and Jayapura. Nineteen percent of the ecoregion is covered by protected areas, mostly in Indonesia (MacKinnon 1997) (table J123.3).

Types and Severity of Threats

A large dam has been proposed for the Mamberamo Gorge, and timber and agricultural activities are a potential threat. A planned highway between Jayapura and Wamena is a threat because of improved access (Stattersfield et al. 1998).

TABLE J123.1. Endemic and Near-Endemic Mammal Species.

Family	Species
Perorictidae	Echymipera clara
Perorictidae	Echymipera echinista
Macropodidae	Dorcopsis hageni*
Macropodidae	Dorcopsis muelleri
Pteropodidae	Nyctimene draconilla
Emballonuridae	Emballonura furax
Rhinolophidae	Hipposideros wollastoni
Rhinolophidae	Hipposideros edwardshill*
Vespertilionidae	Kerivoula muscina
Molossidae	Otomops secundus
Muridae	Paraleptomys rufilatus
Muridae	Hydromys hussoni
Muridae	Pogonomelomys mayeri

An asterisk signifies that the species' range is limited to this ecoregion.

TABLE J123.2. Endemic and Near-Endemic Bird Species.

Family	Common Name	Species
Psittacidae	Edwards's fig-parrot	Psittaculirostris edwardsii
Psittacidae	Salvadori's fig-parrot	Psittaculirostris salvadorii
Loriidae	Brown lory	Chalcopsitta duivenbodei
Apodidae	Papuan swiftlet	Aerodramus papuensis
Alcedinidae	Red-breasted paradise-kingfisher	Tanysiptera nympha
Corvidae	Brown-headed crow	Corvus fuscicapillus
Monarchidae	Rufous monarch	Monarcha rubiensis
Monarchidae	Rufous-collared monarch	Arses insularis
Pachycephalida	White-bellied whistler	Pachycephala leucogastra
Cinclosomatidae	Brown-capped jewel-babbler	Ptilorrhoa geislerorum
Eopsaltriidae	Green-backed robin	Pachycephalopsis hattamensis
Meliphagidae	Silver-eared honeyeater	Lichmera alboauricularis
Meliphagidae	Brass's friarbird	Philemon brassi*
Paradisaeidae	Jobi manucode	Manucodia jobiensis
Paradisaeidae	Pale-billed sicklebill	Epimachus bruijnii*
Paradisaeidae	Greater melampitta	Melampitta gigantea

An asterisk signifies that the species' range is limited to this ecoregion.

Priority Conservation Actions

Short-Term Conservation Actions (1–5 Years)

Several priority-setting efforts have been conducted that include this ecoregion, including Beehler (1994) and Nix et al. (2000) for PNG and Conservation International (1999) for Irian Jaya. These exercises have mapped specific areas that are deemed important to conserve a variety of taxa.
- Limit access to biologically important areas.

Longer-Term Conservation Actions (5–20 Years)

- Continue to implement conservation recommendations of Conservation International (1999), Nix et al. (2000), or other subsequent conservation priority exercises.
- Develop and implement effective management plans for protected areas.
- Increase protection in PNG.

Focal Species for Conservation Action

Threatened endemic species: lesser tube-nosed bat (*Nyctimene draconilla*) (Mickleburgh et al. 1992), greater sheath-tailed bat (*Emballonuridae furax*) (Mickleburgh et al. 1992), Salvadori's fig-parrot (*Psittaculirostris salvadorii*) (BirdLife International 2000), Fly River trumpet-eared bat (*Kerivoula muscina*) (Flannery 1995), mantled mastiff bat (*Otomops secundus*) (Flannery 1995), black sicklebill (*Epimachus fastuosus*) (BirdLife International 2000)

Area-sensitive species: grizzled tree-kangaroo (*Dendrolagus inustus*), white-striped dorcopsis (*Dorcopsis hageni*), brown dorcopsis (*Dorcopsis muelleri*), fruit bats (*Pteropus neohibernicus, Pteropus macrotis, Pteropus conspicillatus, Rousettus amplexicaudatus, Dobsonia minor, Dobsonia moluccense, Nyctimene albiventer, Nyctimene aello, Nyctimene draconilla, Paranyctimene raptor, Macroglossus minimus, Syconycteris australis*) (Mickleburgh et al. 1992), Blyth's hornbill (*Aceros plicatus*) (Kemp 1995)

Top predator species: New Guinea quoll (*Dasyurus albopunctatus*), Gurney's eagle (*Aquila gurneyi*), little eagle (*Hieraaetus morphnoides*), peregrine falcon (*Falco peregrinus*), brown falcon (*Falco berigora*)

Selected Conservation Contacts

East Sepik Council of Women

Lembaga Alam Tropika Indonesia (LATIN)

Sepik Community Landcare Project

WWF-Indonesia

Yayasan Keanekaragaman Hayati

Yayasan Konsorsium Pembaruan Agraria (KPA)

Justification for Ecoregion Delineation

Using Whitmore's (1984) map of the vegetation of Malesia and MacKinnon's (1997) reconstruction of the original vegetation, we delineated the large areas of distinct habitat types as ecoregions. The tropical lowland moist and freshwater swamp forests to the north of the Central Cordillera were placed in the Northern New Guinea Lowland Rain and Freshwater Swamp Forests [123], and the montane forests were placed in the Northern New Guinea Montane Rain Forests [122] (based largely on recommendations by Bob Johns, pers. comm., 1999). This ecoregion corresponds to MacKinnon's (1997) biounits P3e and P3j. Udvardy (1975) placed these ecoregions in the Papuan biogeographic province of the Oceanian Realm.

Prepared by John Morrison

Ecoregion Number:	**124**
Ecoregion Name:	Huon Peninsula Montane Rain Forests
Bioregion:	New Guinea and Melanesia
Biome:	Tropical and Subtropical Moist Broadleaf Forests
Political Units:	Papua New Guinea
Ecoregion Size:	16,400 km²
Biological Distinctiveness:	Globally Outstanding
Conservation Status:	Relatively Stable
Conservation Assessment:	III

The Huon Peninsula Montane Rain Forests [124] consist of tropical montane forest surrounded by ocean and lowland forest on a rugged peninsula. The Finisterre Range, representing a third of the ecoregion, supports more mainland endemic species of warm-blooded vertebrates than any similar-sized area in PNG. The ecoregion's isolation has led to a high degree of endemism, and the area is still relatively intact.

Location and General Description

The Huon Peninsula Montane Rain Forests [124] are made up of the tropical montane moist forests (from 1,000 m to 3,000 m) of the Huon Peninsula in PNG, on the island of New Guinea. There are three mountain ranges on the peninsula: the Finisterre (to 4,176 m), Saruwaged (to 4,122 m), and Cromwell and Rawlinson ranges. The climate of the ecoregion is tropical wet, which is characteristic of this part of Melanesia, located in the western Pacific Ocean north of Australia (National Geographic Society 1999). This portion of New Guinea is a very active tectonic area with a complex geologic history (Bleeker 1983). The surface geology of this ecoregion is a combination of Miocene siltstone, conglomerate, volcanics, and limestone (Bleeker 1983). The Finisterre Range in particular consists of one steep ridge of limestone (Davis et al. 1995).

TABLE J123.3. WCMC (1997) Protected Areas That Overlap with the Ecoregion.

Protected Area	Area (km²)	IUCN Category
Unnamed [122]	3,190	?
Mamberamo-Pegunungan Foja [122]	9,130	IV
Foja (extension)	6,760	PRO
Cape Wom International Memorial Park	50	UA
Sepik River	3,850	?
Yakopi Nalenk Mountains [125]	2,180	?
Total	25,160	

Ecoregion numbers of protected areas that overlap with additional ecoregions are listed in brackets.

The vegetation of this ecoregion is mostly tropical wet evergreen forest (hill type), with a large percentage of tropical montane evergreen forest and a small amount of limestone forest (MacKinnon 1997; Paijmans 1975). Some of the higher peaks contain ecologically fragile high alpine areas, which are part of the adjoining Central Ranges sub-alpine grassland ecoregion.

The somewhat low-canopy, closed lowland hill forest contains more open shrub layer but a denser herbaceous layer than lower-elevation alluvial forest. Palms are fewer in number. The dominant canopy trees include species of *Pometia*, *Canarium*, *Anisoptera*, *Cryptocarya*, *Terminalia*, *Syzygium*, *Ficus*, *Celtis*, *Dysoxylum*, and *Buchanania*. *Koompassia*, *Dillenia*, *Eucalyptopsis*, *Vatica*, and *Hopea* are locally abundant. Dense stands of *Araucaria*, the tallest tropical trees in the world, are present in scattered locations (Paijmans 1975; Nightingale 1992).

Although they are subject to variable climates and topography, montane forests are smaller crowned and have more even canopies than lowland hill forest. Tree densities can be high, and the shrub density is also high. Predominant canopy trees include *Nothofagus*, Lauraceae, Cunoniaceae, Elaeocarpaceae, *Lithocarpus*, *Castanopsis*, *Syzygium*, *Ilex*, and southern conifers. *Nothofagus* and *Araucaria* may grow in pure, dense stands. The levels of Myrtaceae, Elaeocarpaceae, and conifers increase with altitude. The conifers, which are generally found above 2,000 m, include *Dacrycarpus*, *Podocarpus*, *Phyllocladus*, and *Papuacedrus* in the canopy and emergent layer (Paijmans 1975).

The Cromwell Range contains extensive *Dacrydium* forests (Miller et al. 1994). In the lower montane forests of the Cromwell Range, up to elevations of 2,000 m, *Castanopsis* and *Lithocarpus* predominate. Above 2,000 m, *Xanthomyrtus-Vaccinium-Rhododendron* communities are found, and *Lithocarpus-Elmerrillia* forest is present at approximately 2,300 m. Above 2,400 m, *Elaeocarpus* and conifers (*Phyllocladus*, *Podocarpus*, and *Dacrydium*) dominate (Davis et al. 1995).

Biodiversity Features

Overall richness is moderate to high and overall endemism is low to moderate when compared with those of other ecoregions in Indo-Malaysia; however, the Finisterre Range, representing a third of the ecoregion, supports more mainland endemic species of warm-blooded vertebrates than any similar-sized area in PNG (Beehler 1993).

The mammalian fauna consists of a wide variety of tropical Australasian marsupials, including tree kangaroos. There are eighty-one mammal species in this ecoregion, including six species that are endemic or near endemic (table J124.1). The Huon tree-kangaroo (*Dendrolagus matschiei*) is found no-where else on Earth and is considered endangered. The ecoregion also contains the widespread but endangered Papuan long-beaked echidna (*Zaglossus bartoni*) (Flannery 1995; Flannery and Groves 1998; Bonaccorso et al., in press; Hilton-Taylor 2000).

The avifauna of the ecoregion has a clear Australasian flavor, with representatives of several Australasian families including Ptilonorhynchidae, Eopsaltridae, Meliphagidae, and Paradisaeidae. The ecoregion basically corresponds with the Adelbert and Huon ranges EBA, although the small Adelbert Mountain Range is not part of this ecoregion. The EBA contains eleven restricted-range species, ten of which are found in the ecoregion. The ecoregion contains a total of sixteen endemic and near-endemic species (table J124.2), including the vulnerable Wahnes's parotia (*Parotia wahnesi*) (Beehler et al. 1986; Coates 1985; Stattersfield et al. 1998). The ecoregion is virtually unstudied ornithologically (Stattersfield et al. 1998).

This ecoregion, with one endemic butterfly species, is a center of butterfly endemicity on the island of New Guinea (Parsons 1999).

The Finisterre Range and Huon Peninsula Centres of Plant Diversity are located in this ecoregion. The Cromwell Ranges are the only extensive unlogged *Dacrydium* forests in the Southern Hemisphere (Davis et al. 1994; Miller et al. 1994).

TABLE J124.1. Endemic and Near-Endemic Mammal Species.

Family	Species
Macropodidae	*Dendrolagus matschiei**
Muridae	*Pogonomelomys mayeri*
Muridae	*Abeomelomys sevia*
Muridae	*Rattus novaeguineae*
Muridae	*Leptomys ernstmayeri*
Muridae	*Melomys gracilis*

An asterisk signifies that the species' range is limited to this ecoregion.

TABLE J124.2. Endemic and Near-Endemic Bird Species.

Family	Common Name	Species
Psittacidae	Edwards's fig-parrot	*Psittaculirostris edwardsii*
Psittacidae	Madarasz's tiger-parrot	*Psittacella madaraszi*
Loriidae	Brown lory	*Chalcopsitta duivenbodei*
Caprimulgidae	Mountain eared-nightjar	*Eurostopodus archboldi*
Apodidae	Papuan swiftlet	*Aerodramus papuensis*
Meliphagidae	Olive-streaked honeyeater	*Ptiloprora meekiana*
Meliphagidae	Rufous-backed honeyeater	*Ptiloprora guisei*
Meliphagidae	Cinnamon-browed honeyeater	*Melidectes ochromelas*
Meliphagidae	Huon wattled honeyeater	*Melidectes foersteri*
Meliphagidae	Spangled honeyeater	*Melipotes ater*
Cinclosomatidae	Brown-capped jewel-babbler	*Ptilorrhoa geislerorum*
Cinclosomatidae	Blue-capped ifrita	*Ifrita kowaldi*
Paradisaeidae	Wahnes's parotia	*Parotia wahnesi*
Paradisaeidae	Huon astrapia	*Astrapia rothschildi*
Paradisaeidae	Emperor bird-of-paradise	*Paradisaea guilielmi**
Motacillidae	Alpine pipit	*Anthus gutturalis*

An asterisk signifies that the species' range is limited to this ecoregion.

Status and Threats

Current Status

Except for some forest loss along the southern part and the Buweng Timber Rights Purchase (using helicopters), most of the ecoregion's natural habitat is intact (Johns 1993). The Huon Highlands are a major wilderness area (Beehler 1994). The two large protected areas (Finisterre and Mt. Bangeta) cover about 18 percent of the ecoregion area (table J124.3).

Types and Severity of Threats

The threats to this ecoregion are minimal at present. Most of the forests remain unthreatened by further degradation. However, certain alpine highlands and hill tracts are threatened by development (Miller et al. 1994).

Priority Conservation Actions

Short-Term Conservation Actions (1–5 Years)

Several priority-setting efforts have been conducted that include this ecoregion, including Beehler (1994) and Nix et al. (2000) for PNG. These exercises have mapped specific areas that are deemed important to conserve a variety of taxa.
- Conduct further biodiversity field surveys.

Longer-Term Conservation Actions (5–20 Years)

- Continue to implement conservation recommendations of Beehler (1994) and Nix et al. (2000) or other subsequent conservation priority exercises.
- Develop and implement effective management plans for protected areas.
- Limit development in alpine regions.

Focal Species for Conservation Action

Threatened endemic species: Huon tree-kangaroo (*Dendrolagus matschiei*) (Kennedy 1992), Wahnes's parotia (*Parotia wahnesi*) (Frith and Beehler 1998; BirdLife International 2000), Papuan long-beaked echidna (*Zaglossus bartoni*) (Flannery 1995)

Area-sensitive species: Huon tree-kangaroo (*Dendrolagus matschiei*) (Flannery 1995), small dorcopsis (*Dorcopsulus vanheurni*) (Flannery 1995), fruit bats (*Pteropus conspicillatus, Pteropus neohibernicus, Pteropus macrotis, Dobsonia minor, Dobsonia moluccense, Nyctimene albiventer, Nyctimene aello, Nyctimene cyclotis, Paranyctimene raptor, Macroglossus minimus, Syconycteris australis*) (Mickleburgh et al. 1992), Gurney's eagle (*Aquila gurneyi*), little eagle (*Hieraaetus morphnoides*), New Guinea eagle (*Harpyopsis novaeguineae*), peregrine falcon (*Falco peregrinus*), brown falcon (*Falco berigora*)

TABLE J124.3. WCMC (1997) Protected Areas That Overlap with the Ecoregion.

Protected Area	Area (km²)	IUCN Category
Finisterre [131]	2,290	?
Mt. Bangeta [131]	550	?
Total	2,840	

Ecoregion numbers of protected areas that overlap with additional ecoregions are listed in brackets.

Top predator species: New Guinea quoll (*Dasyurus albopunctatus*) (Flannery 1995), Gurney's eagle (*Aquila gurneyi*), little eagle (*Hieraaetus morphnoides*), New Guinea eagle (*Harpyopsis novaeguineae*), peregrine falcon (*Falco peregrinus*), brown falcon (*Falco berigora*)

Selected Conservation Contacts

Tree Kangaroo Conservation Project
Wildlife Conservation Society

Justification for Ecoregion Delineation

Using Whitmore's (1984) map of the vegetation of Malesia and MacKinnon's (1997) reconstruction of the original vegetation, we delineated the large areas of distinct habitat types as ecoregions. The tropical montane evergreen forests in the Huon Peninsula were delineated as another distinct ecoregion, the Huon Peninsula Montane Rain Forests [124], and correspond to MacKinnon's (1997) biounit P3k. Udvardy (1975) placed these ecoregions in the Papuan biogeographic province of the Oceanian Realm.

Prepared by John Morrison

Ecoregion Number:	**125**
Ecoregion Name:	**Central Range Montane Rain Forests**
Bioregion:	**New Guinea and Melanesia**
Biome:	**Tropical and Subtropical Moist Broadleaf Forests**
Political Units:	**Indonesia, Papua New Guinea**
Ecoregion Size:	**171,200 km²**
Biological Distinctiveness:	**Globally Outstanding**
Conservation Status:	**Relatively Stable**
Conservation Assessment:	**III**

The Central Range Montane Rain Forests [125], which form the mountain spine of the island of New Guinea, contain more than 100 endemic vertebrates. The Central Ranges separate adjoining lowland on each side of the cordillera, and several constituent mountain ranges are isolated such that an extraordinary level of speciation has occurred within the Central Ranges. Some species are shared with outlying mountain ranges, but there are a significant number of locally endemic plants that are known only from a single mountain or mountain range (R. Johns, pers. comm., 2000).

Location and General Description

This ecoregion is made up of the montane forests between 1,000 m and 3,000 m in the Central Cordillera of the island of New Guinea, in Irian Jaya, Indonesia and PNG, in the western Pacific Ocean north of Australia. The Central Cordillera is composed of a series of mountain ranges that are broadly grouped into the Snow Mountains in Irian Jaya, the Star Mountains in Irian Jaya and PNG, and the Central and Eastern Highlands in PNG. The climate of the ecoregion is tropical highland wet because of its elevation (National Geographic Society 1999). The surface geology of the Central Cordillera is composed of metamorphic and intrusive igneous rocks (Bleeker 1983). The metamorphic rocks were Cretaceous (100 m.y.) and Eocene (40 m.y.) ocean sediments that were folded between the Eocene and early Miocene Periods (20 m.y.) (Petocz 1989). Pleistocene stratovolcanoes are also found in the Central Ranges (Bleeker 1983).

There are three broad vegetation zones in the Central Ranges: lower montane forest, upper montane forest, and high mountain forest (although these are sometimes lumped into one lower montane rain forest; see Davis et al. 1995). Lower montane forest continues up from the lowlands to approximately 2,500 m. This zone is dominated by oaks, such as *Castanopsis acuminatissima*, elaeocarps, and laurels. *Araucaria* may form thick stands in lower areas. *Nothofagus*, sometimes in monotypic stands, is conspicuous in the moss-covered upper montane forest, which begins at about 1,500 m. High mountain forest begins at approximately 2,500 m and continues past the upper limits of the ecoregion, to 3,900 m. The species-poor, high mountain forest includes conifers (*Podocarpus, Dacrycarpus, Dacridium, Papuacedrus, Araucaria,* and *Libocedrus*) and Myrtacae, with a thin canopy and prominent understory (Beehler et al. 1986; Davis et al. 1995; FAO 1981; Petocz 1989; Stattersfield et al. 1998).

Biodiversity Features

The overall richness of this ecoregion is remarkable and ranges from moderate to high. The ecoregion contains some of the highest richness of vascular plants and herpetofauna in Indo-Malaysia and some of the highest endemism for mammals, birds, and vascular plants.

The mammalian fauna consists of a wide variety of tropical Australasian marsupials, including tree kangaroos (Flannery 1995). Ninety mammal species inhabit this ecoregion, of which an incredible forty-four are endemic or near endemic (Flannery 1995; Flannery and Groves 1998; Bonaccorso et al., in press) (table J125.1). Three of these species are considered critically endangered: large leptomys (*Leptomys elegans*), eastern shrew-mouse (*Pseudohydromys murinus*), and lesser small-toothed rat (*Macruromys elegans*) while another 15 species are considered vulnerable (Hilton-Taylor 2000).

The avifauna of the ecoregion has a clear Australasian flavor, including representatives of several Australasian families such as Ptilonorhynchidae, Eopsaltridae, Meliphagidae, and Paradisaeidae. The ecoregion harbors 348 bird species, of which 55 are endemic or near endemic (table J125.2). This ecoregion forms the majority of the Central Papuan mountains EBA, which contains fifty-three restricted-range bird species, eight of which are found in the adjacent Central Range sub-alpine grasslands, thirteen of which are shared with the adjacent Central Range sub-alpine grasslands, and seventeen of which are found nowhere else on Earth. Four of these species represent endemic genera (Stattersfield et al. 1998; Beehler et al. 1986; Coates 1985). The ribbon-tailed astrapia (*Astrapia mayeri*), and the blue bird-of-paradise (*Paradisaea rudolphi*) are considered vulnerable (Hilton-Taylor 2000).

Within this ecoregion, the Weyland Range (with nine endemic species) and the Hagen–Sepik-Wahgi Divide (with five endemic species) are both centers of butterfly endemism on the island of New Guinea (Parsons 1999).

Seven Centres of Plant Diversity are shared between this ecoregion and the adjacent Central Ranges sub-alpine grassland ecoregion. The Star Mountains–Telefomin–Tifalmin–Strickland Gorge CPD in PNG contains very rich (more than 3,000 vascular plant species) montane and high-altitude vegetation. The Hunstein Range–Bürgers Mountain–Schatteburg, Mt. Giluwe–Tari Gap–Doma Peaks CPD in PNG contains more than 2,500 vascular plant species and extensive stands of *Agathis labillardieri* and associated epiphytic flora. More than 3,000 vascular plant species are found in the Mt. Giluwe–Tari Gap–Doma Peaks CPD in PNG, including a unique *Dacrydium* swamp forest. The poorly known Kubor Ranges in PNG are a fragile ecosystem that probably contains many endemics on limestone and volcanic ash. The Bismarck Falls–Mt. Wilhelm–Mt. Otto–Schrader Range–Mt. Hellwig–Gahavisuka CPD has a wide variety of vegetation

TABLE J125.1. Endemic and Near-Endemic Mammal Species.

Family	Species
Dasyuridae	Neophascogale lorentzi*
Dasyuridae	Phascolosorex doriae
Dasyuridae	Antechinus wilhelmina
Phalangeridae	Phalanger vestitus*
Phalangeridae	Phalanger matanim*
Macropodidae	Dorcopsulus macleayi
Pseudocheiridae	Pseudocheirus caroli*
Pseudocheiridae	Pseudocheirus mayeri*
Petauridae	Dactylopsila megalura*
Peroryctidae	Microperoryctes murina
Peroryctidae	Echymipera clara
Pteropodidae	Aproteles bulmerae
Pteropodidae	Syconycteris hobbit
Emballonuridae	Emballonura furax
Rhinolophidae	Hipposideros corynophyll*
Rhinolophidae	Hipposideros muscinus
Vespertilionidae	Pipistrellus collinus
Vespertilionidae	Nyctophilus microdon*
Vespertilionidae	Kerivoula muscina
Molossidae	Otomops secundus
Muridae	Leptomys elegans
Muridae	Paraleptomys wilhelmina*
Muridae	Neohydromys fuscus
Muridae	Pseudohydromys murinus
Muridae	Mayermys ellermani
Muridae	Hydromys hussoni
Muridae	Hydromys habbema*
Muridae	Crossomys moncktoni
Muridae	Macruromys elegans*
Muridae	Xenuromys barbatus
Muridae	Melomys fellowsi*
Muridae	Melomys lorentzi
Muridae	Melomys lanosus*
Muridae	Pogonomelomys mayeri
Muridae	Pogonomelomys bruijni
Muridae	Abeomelomys sevia
Muridae	Coccymys albidens*
Muridae	Rattus giluwensis*
Muridae	Rattus novaeguineae
Muridae	Hyomys dammermani
Muridae	Hydromys shawmayeri
Muridae	Leptomys ernstmayeri
Muridae	Melomys gracilis
Muridae	Pogonomys championi*

An asterisk signifies that the species' range is limited to this ecoregion.

types and contains more than 5,000 vascular plant species. Important *Araucaria cunninghamii*, *A. hunsteinii*, and *Castanopsis* forests are found in the Mt. Michael–Okapa–Crater Mountain CPDs in PNG (Davis et al. 1995).

Status and Threats

Current Status

The montane rain forests are generally undisturbed because of low population densities and traditional lifestyles. Some highland valleys are heavily populated, and this has resulted in local deforestation. A large and well-known hardrock mine is found in this ecoregion: the Freeport copper mine, located within the Lorentz Strict Nature Reserve in Irian Jaya. This is a large facility, and its location in a pristine area has caused concern about sedimentation and toxic runoff into adjacent stream and river systems. Petroleum extraction also occurs in Southern Highlands Province in PNG, but the environmental effects are minimal (Diamond and Bishop 1998). More than half of the Lorentz Nature Reserve is under petroleum concessions.

TABLE J125.2. Endemic and Near-Endemic Bird Species.

Family	Common Name	Species	Family	Common Name	Species
Rallidae	Chestnut forest-rail	*Rallina rubra*	Meliphagidae	Spot-breasted meliphaga	*Meliphaga mimikae*
Loriidae	Orange-billed lorikeet	*Neopsittacus pullicauda*	Meliphagidae	Olive-streaked honeyeater	*Ptiloprora meekiana*
Loriidae	Striated lorikeet	*Charmosyna multistriata*	Meliphagidae	Yellow-browed honeyeater	*Melidectes rufocrissalis*
Psittacidae	Painted tiger-parrot	*Psittacella picta*	Meliphagidae	Sooty honeyeater	*Melidectes fuscus*
Psittacidae	Madarasz's tiger-parrot	*Psittacella madaraszi*	Meliphagidae	Belford's honeyeater	*Melidectes belfordi*
Psittacidae	Modest tiger-parrot	*Psittacella modesta*	Meliphagidae	Rufous-backed honeyeater	*Ptiloprora guisei*
Aegothelidae	Archbold's owlet-nightjar	*Aegotheles archboldi*	Meliphagidae	Cinnamon-browed honeyeater	*Melidectes ochromelas*
Caprimulgidae	Mountain eared-nightjar	*Eurostopodus archboldi*	Estrildidae	Snow Mountain munia	*Lonchura montana**
Apodidae	Bare-legged swiftlet	*Aerodramus nuditarsus*	Estrildidae	Black-breasted munia	*Lonchura teerinki**
Apodidae	Papuan swiftlet	*Aerodramus papuensis*	Estrildidae	Mountain firetail	*Oreostruthus fuliginosus*
Campephagidae	Hooded cuckoo-shrike	*Coracina longicauda*	Ptilonorhynchidae	Archbold's bowerbird	*Archboldia papuensis**
Acanthizidae	Papuan thornbill	*Acanthiza murina*	Paradisaeidae	Short-tailed paradigalla	*Paradigalla brevicauda**
Cinclosomatidae	Blue-capped ifrita	*Ifrita kowaldi*			
Cinclosomatidae	Papuan whipbird	*Androphobus viridis**	Paradisaeidae	King-of-Saxony bird-of-paradise	*Pteridophora alberti**
Melanocharitidae	Streaked berrypecker	*Melanocharis striativentris*	Paradisaeidae	Yellow-breasted bird-of-paradise	*Loboparadisea sericea*
Eopsaltriidae	White-winged robin	*Peneothello sigillatus*	Paradisaeidae	Loria's bird-of-paradise	*Cnemophilus loriae*
Eopsaltriidae	Green-backed robin	*Pachycephalopsis hattamensis*	Paradisaeidae	Long-tailed paradigalla	*Paradigalla carunculata*
Eopsaltriidae	Greater ground-robin	*Amalocichla sclateriana*	Paradisaeidae	Carola's parotia	*Parotia carolae*
Eopsaltriidae	Alpine robin	*Petroica bivittata*	Paradisaeidae	Lawes's parotia	*Parotia lawesii*
Eopsaltriidae	Smoky robin	*Peneothello cryptoleucus*	Paradisaeidae	Splendid astrapia	*Astrapia splendidissima*
Pachycephalidae	Sooty shrike-thrush	*Colluricincla umbrina**	Paradisaeidae	Ribbon-tailed astrapia	*Astrapia mayeri*
Pachycephalidae	Lorentz's whistler	*Pachycephala lorentzi*	Paradisaeidae	Blue bird-of-paradise	*Paradisaea rudolphi*
Pachycephalidae	Wattled ploughbill	*Eulacestoma nigropectus*	Paradisaeidae	Crested bird-of-paradise	*Cnemophilus macgregorii*
Pachycephalidae	Black sittella	*Daphoenositta miranda*	Paradisaeidae	MacGregor's bird-of-paradise	*Macgregoria pulchra*
Climacteridae	Papuan treecreeper	*Cormobates placens*	Paradisaeidae	Brown sicklebill	*Epimachus meyeri*
Meliphagidae	Orange-cheeked honeyeater	*Oreornis chrysogenys*	Paradisaeidae	Princess Stephanie's astrapia	*Astrapia stephaniae*
Meliphagidae	Leaden honeyeater	*Ptiloprora plumbea*	Paradisaeidae	Greater melampitta	*Melampitta gigantea*
Meliphagidae	Rufous-sided honeyeater	*Ptiloprora erythropleura*			
Meliphagidae	Black-backed honeyeater	*Ptiloprora perstriata*			

An asterisk signifies that the species' range is limited to this ecoregion.

Almost 20 percent of the ecoregion is covered by eleven protected areas (table J125.3). The bulk of the protected area is in Irian Jaya, however (MacKinnon 1997). The largest protected area in the Central Ranges is the 21,500 km² Gunung Lorentz Nature Reserve in the Snow Mountains of Irian Jaya, although only 7,350 km² of the area is in this ecoregion (Stattersfield et al. 1998).

Types and Severity of Threats

Logging concessions have been granted for large areas of the ecoregion. The threat of increased access (and subsequent hunting and illegal logging) via new roads is a significant concern (Stattersfield et al. 1998). Mining poses threats in restricted locations.

Priority Conservation Actions

Short-Term Conservation Actions (1–5 Years)

Several priority-setting efforts have been conducted that include this ecoregion, including Beehler (1994) and Nix et al. (2000) for PNG and Conservation International (1999) for Irian Jaya. These exercises have mapped specific areas that are deemed important to conserve a variety of taxa.
- Stop illegal logging in and around protected areas.

Longer-Term Conservation Actions (5–20 Years)

- Continue to implement conservation recommendations of Beehler (1994), Conservation International (1999), and Nix et al. (2000) or other subsequent conservation priority exercises.
- Develop and implement effective management plans for protected areas.
- Stop implementation of logging concessions.

Focal Species for Conservation Action

Endangered endemic species: large leptomys (*Leptomys elegans*), eastern shrew-mouse (*Pseudohydromys murinus*), lesser small-toothed rat (*Macruromys elegans*)

Area-sensitive species: Doria's tree-kangaroo (*Dendrolagus dorianus*) (Flannery 1995), Macleay's dorcopsis (*Dorcopsulus macleayi*) (Flannery 1995), small dorcopsis (*Dorcopsulus vanheurni*) (Flannery 1995), fruit bats (*Aproteles bulmerae, Macroglossus minimus, Syconycteris australis, Syconycteris hobbit*) (Mickleburgh et al. 1992), Gurney's eagle (*Aquila gurneyi*), little eagle (*Hieraaetus morphnoides*), New Guinea eagle (*Harpyopsis novaeguineae*), peregrine falcon (*Falco peregrinus*)

Top predator species: New Guinea quoll (*Dasyurus albopunctatus*) (Flannery 1995), Gurney's eagle (*Aquila gurneyi*), little eagle (*Hieraaetus morphnoides*), New Guinea eagle (*Harpyopsis novaeguineae*), peregrine falcon (*Falco peregrinus*)

Selected Conservation Contacts

Lembaga Alam Tropika Indonesia (LATIN)

Research and Conservation Foundation of PNG

Wildlife Conservation Society

WWF Kikori Integrated Conservation and Development Area

Yayasan Keanekaragaman Hayati

Yayasan Konsorsium Pembaruan Agraria (KPA)

Justification for Ecoregion Delineation

Using Whitmore's (1984) map of the vegetation of Malesia and MacKinnon's (1997) reconstruction of the original vegetation, we delineated the large areas of distinct habitat types as ecoregions. The montane evergreen moist forests along the Central Cordillera, including the Snow Mountains, Star Mountains, Central Highlands, and Eastern Highlands, were placed in the Central Range Montane Rain Forests [125]. This ecoregion roughly corresponds to MacKinnon's subunits P3g, P3h, and P3i. The moist forests in the southeastern peninsula were distinguished as the Southeastern Papuan Rain Forests [126]. This ecoregion consists mostly of montane forests but also includes some lowland forests along the coasts and is roughly equivalent to MacKinnon's (1997) biounit P3n. We used the 1,000-m contour from a DEM (USGS 1996) to define the montane-lowland transition. All along the Central Cordillera and in the Huon Peninsula, we separated the alpine habitat into a distinct (Central Range Sub-Alpine Grasslands [131]) —ecoregion. Udvardy (1975) placed these ecoregions in the Papuan biogeographic province of the Oceanian Realm.

Prepared by John Morrison

TABLE J125.3. WCMC (1997) Protected Areas That Overlap with the Ecoregion.

Protected Area	Area (km²)	IUCN Category
Pegunungan Weyland [119, 131]	1,830	PRO
Enarotali	3,540	IV
Gunung Lorentz [127, 128, 129, 131]	7,350	I
Jayawijaya	7,710	IV
Jayawijaya extension [131]	4,260	PRO
Mt. Capella [131]	1,500	?
Yakopi Nalenk Mts. [123]	4,050	?
Mt. Wilhelm [131]	380	?
Mt. Onuare [131]	620	?
Mt. Michael	1,160	?
Mt. Bosavi [128]	1,260	?
Total	33,660	

Ecoregion numbers of protected areas that overlap with additional ecoregions are listed in brackets.

Ecoregion Number: **126**
Ecoregion Name: **Southeastern Papuan Rain Forests**
Bioregion: **New Guinea and Melanesia**
Biome: **Tropical and Subtropical Moist Broadleaf Forests**
Political Units: **Papua New Guinea**
Ecoregion Size: **77,000 km^2**
Biological Distinctiveness: **Globally Outstanding**
Conservation Status: **Relatively Stable**
Conservation Assessment: **III**

The Southeastern Papuan Rain Forests [126]—dominated by the Owen Stanley highlands, the major mountain range in the ecoregion—contain vast tracts of pristine montane forests (Miller et al. 1994). Because of the dissected landscape and edaphic variations, the ecoregion is rich in endemic species with very local distributions. Although some of PNG's major population centers, including the capital, Port Moresby, are located in this ecoregion, major wilderness areas are still present. This ecoregion is extremely rich because of the diversity of its habitats: it includes coastal, lowland, and montane habitats.

Location and General Description

This lowland and montane ecoregion is made up of the Owen Stanley Range and surrounding lowland and coastal areas in southeastern PNG. The climate of the ecoregion is tropical wet, which is characteristic of this part of Melanesia, located in the western Pacific Ocean north of Australia (National Geographic Society 1999). The surface geology of the Central Cordillera, of which the Owen Stanleys are an extension, is generally composed of metamorphic and intrusive igneous rocks. More specifically, this ecoregion is composed of metamorphosed Mesozoic greywacke sandstone, siltstone, and marine volcanics overlain by Miocene intrusives, Pliocene marine and terrestrial fine-grained sediments, and Quaternary lavas and pyroclastics (Bleeker 1983).

Most of this ecoregion is composed of tropical wet evergreen forest, with a significant (25 percent) percentage of tropical montane evergreen forest. Smaller percentages of upper montane and freshwater swamp forest are also found (MacKinnon 1997). Coastal vegetation contains *Casuarina*, whereas mixed coastal vegetation contains *Calophyllum*, *Terminalia*, and *Anisoptera* (MacKinnon 1997).

Lowland forest up to 1,400 m on the north side of the Owen Stanleys is made up of *Pometia*, *Terminalia*, *Myristica*, *Horsfieldia*, *Celtis*, and *Ficus* (MacKinnon 1997). Lowland forest is made up of both alluvial and hill types (Paijmans 1975). Lowland alluvial forest has a canopy that is multitiered and irregular, with many emergents. The forest understory contains a shrub-and-herb layer with a variety of climbers, epiphytes, and ferns (Petocz 1989). Palms may be common in the shrub layer (Paijmans 1975). The somewhat lower-canopy, more closed lowland hill forest contains more open shrub layer but a denser herbaceous layer. Palms are fewer in number. The dominant canopy trees include species of *Pometia*, *Canarium*, *Anisoptera*, *Cryptocarya*, *Terminalia*, *Syzygium*, *Ficus*, *Celtis*, *Dysoxylum*, and *Buchanania*.

Koompassia, *Dillenia*, *Eucalyptopsis*, *Vatica*, and *Hopea* are locally abundant. Dense stands of *Araucaria*, the tallest tropical trees in the world, are present in scattered locations (Paijmans 1975; Nightingale 1992).

Although they are subject to variable climates and topography, montane forests are smaller crowned and have more even canopies than lowland hill forest. Tree densities can be high, and the shrub density is also high (Paijmans 1975). Lower montane forest transitioning in from the lowlands is dominated by oaks such as *Castanopsis acuminatissima*, *Lithocarpus*, elaeocarps, and laurels. Seventy-meter *Araucaria* may form thick stands in lower areas. *Nothofagus*, sometimes in monotypic stands, is conspicuous in the moss-covered mid and upper zones of the ecoregion (Davis et al. 1995).

Biodiversity Features

Overall richness is generally high and endemism is generally moderate to high when compared with those of other ecoregions in Indo-Malaysia. Bird richness and reptile and amphibian richness and endemism are particularly high in this ecoregion.

The mammalian fauna consists of a wide variety of tropical Australasian marsupials, including tree kangaroos (Flannery 1995). There are 138 mammal species in the ecoregion, of which 28 are endemic or near endemic (Flannery 1995; Flannery and Groves 1998; Bonaccorso et al., in press) (table J126.1). Included are the critically endangered large-eared nyctophilus (*Pharotis imogene*), Eastern shrew-mouse

TABLE J126.1. Endemic and Near-Endemic Mammal Species.

Family	Species
Dasyuridae	*Murexia rothschildi* *
Dasyuridae	*Planigale novaeguineae*
Peroryctidae	*Peroryctes broadbenti* *
Peroryctidae	*Microperoryctes papuensi* *
Macropodidae	*Dorcopsulus macleayi*
Macropodidae	*Dorcopsis luctuosa*
Macropodidae	*Thylogale brunii*
Pteropodidae	*Syconycteris hobbit*
Vespertilionidae	*Pipistrellus collinus*
Vespertilionidae	*Pharotis imogene* *
Vespertilionidae	*Kerivoula muscina*
Molossidae	*Otomops papuensis*
Molossidae	*Otomops secundus*
Rhinolophidae	*Hipposideros muscinus*
Muridae	*Leptomys elegans*
Muridae	*Neohydromys fuscus*
Muridae	*Pseudohydromys murinus*
Muridae	*Mayermys ellermani*
Muridae	*Crossomys moncktoni*
Muridae	*Chiruromys forbesi*
Muridae	*Chiruromys lamia* *
Muridae	*Xenuromys barbatus*
Muridae	*Melomys levipes* *
Muridae	*Rattus novaeguineae*
Muridae	*Hydromys shawmayeri*
Muridae	*Leptomys ernstmayeri*
Muridae	*Melomys gracilis*
Muridae	*Stenomys vandeuseni* *

An asterisk signifies that the species' range is limited to this ecoregion.

(*Pseudohydromys murinus*), and long-footed hydromine (*Leptomys elegans*) and endangered long-beaked echidna (*Zaglossus bartoni*) and Van Deusen's rat (*Stenomys vandeuseni*) (Hilton-Taylor 2000; Flannery 1995).

The avifauna of the ecoregion has a clear Australasian flavor, including representatives of several Australasian families such as Ptilonorhynchidae, Eopsaltriidae, Meliphagidae, and Paradisaeidae. Because the ecoregion includes coastal, lowland, and montane areas, the number of birds found in the ecoregion is quite large (510 bird species). Forty of these are endemic or near endemic (table J126.2). This ecoregion constitutes the eastern end of the Central Papuan mountains EBA. Whereas the EBA contains a total of fifty-three restricted-range birds, this ecoregion contains only some of them, twenty-seven of which are shared with the Central Ranges montane rain forests ecoregion, one of which is also found in the Central Ranges sub-alpine grasslands, and two of which are found nowhere else on Earth: the streaked bowerbird (*Amblyornis subalaris*) and eastern parotia (*Parotia helenae*) (Stattersfield et al. 1998; Beehler et al. 1986; Coates 1985).

Within this ecoregion, the Kodama Range, with eight endemic species; the Western Owen Stanley Range, with seven endemic species; the Central Owen Stanley Range, with five endemic species; and the Southeastern Coastal area, with one endemic species, are all centers of butterfly endemicity on the island of New Guinea (Parsons 1999).

There are seven recognized Centres of Plant Diversity located in this ecoregion, some of which are shared with the higher Central Ranges sub-alpine grasslands ecoregion. The Galley Reach CPD contains mangrove, lowland swamp, and *Nypa* communities. Galley Reach contains most of the mangrove species in Papuasia, which in turn has the most mangrove diversity. The Menyamya–Aseki–Amungwiwa–Bowutu Mountains–Lasanga Island CPD ranges from sea level to 3,278 m and represents the diversity of the altitudinal gradient on the north side of the Owen Stanley Range. Lowland rain forests (with extensive dipterocarp forest), lowland swamp forests, lower to upper montane forests, and sub-alpine forests are all represented here. Ultramafic vegetation is found in the Bowutu Mountains. Important lowland forest and ultramafic substrate endemics are found in the Milne

TABLE J126.2. Endemic and Near-Endemic Bird Species.

Family	Common Name	Species	Family	Common Name	Species
Psittacidae	Painted tiger-parrot	*Psittacella picta*	Meliphagidae	Spot-breasted meliphaga	*Meliphaga mimikae*
Psittacidae	Madarasz's tiger-parrot	*Psittacella madaraszi*	Meliphagidae	Rufous-backed honeyeater	*Ptiloprora guisei*
Caprimulgidae	Mountain eared-nightjar	*Eurostopodus archboldi*	Meliphagidae	Black-backed honeyeater	*Ptiloprora perstriata*
Apodidae	Bare-legged swiftlet	*Aerodramus nuditarsus*	Meliphagidae	Olive-streaked honeyeater	*Ptiloprora meekiana*
Alcedinidae	Brown-headed paradise-kingfisher	*Tanysiptera danae**	Meliphagidae	Yellow-browed honeyeater	*Melidectes rufocrissalis*
Motacillidae	Alpine pipit	*Anthus gutturalis*	Meliphagidae	Cinnamon-browed honeyeater	*Melidectes ochromelas*
Acanthizidae	Papuan thornbill	*Acanthiza murina*	Meliphagidae	Sooty honeyeater	*Melidectes fuscus*
Pachycephalidae	Wattled ploughbill	*Eulacestoma nigropectus*	Meliphagidae	Belford's honeyeater	*Melidectes belfordi*
Eopsaltriidae	Greater ground-robin	*Amalocichla sclateriana*	Ptilonorhynchidae	Streaked bowerbird	*Amblyornis subalaris**
Eopsaltriidae	Alpine robin	*Petroica bivittata*	Paradisaeidae	Crested bird-of-paradise	*Cnemophilus macgregorii*
Pachycephalidae	White-bellied whistler	*Pachycephala leucogastra*	Paradisaeidae	MacGregor's bird-of-paradise	*Macgregoria pulchra*
Pachycephalidae	Black sittella	*Daphoenositta miranda*	Paradisaeidae	Brown sicklebill	*Epimachus meyeri*
Climacteridae	Papuan treecreeper	*Cormobates placens*	Paradisaeidae	Princess Stephanie's astrapia	*Astrapia stephaniae*
Cinclosomatidae	Brown-capped jewel-babbler	*Ptilorrhoa geislerorum*	Paradisaeidae	Yellow-breasted bird-of-paradise	*Loboparadisea sericea*
Cinclosomatidae	Blue-capped ifrita	*Ifrita kowaldi*	Paradisaeidae	Loria's bird-of-paradise	*Cnemophilus loriae*
Melanocharitidae	Obscure berrypecker	*Melanocharis arfakiana*	Paradisaeidae	Eastern parotia	*Parotia helenae**
Estrildidae	Alpine munia	*Lonchura monticola**	Paradisaeidae	Lawes's parotia	*Parotia lawesii*
Estrildidae	Grey-headed munia	*Lonchura caniceps**	Paradisaeidae	Blue bird-of-paradise	*Paradisaea rudolphi*
Estrildidae	Mountain firetail	*Oreostruthus fuliginosus*	Paradisaeidae	Greater melampitta	*Melampitta gigantea*
Meliphagidae	Silver-eared honeyeater	*Lichmera albo auricularis*			
Meliphagidae	Leaden honeyeater	*Ptiloprora plumbea*			

An asterisk signifies that the species' range is limited to this ecoregion.

Bay–Collinwood Bay to southern coast CPD. The Owen Stanley Mountains CPD contains many local endemics and is the center of diversity for *Agapetes* (Ericacae). The Varirata and Astrolabe ranges, Safia Savanna, and Topographer's Range CPDs are all little-known areas that merit further study (Davis et al. 1995).

The lowland forests are home to the world's largest butterfly, *Ornithoptera alexandrae*, a globally threatened species (Miller et al. 1994).

Status and Threats

Current Status

The East Peninsula Highlands and the North Peninsula Highlands constitute major wilderness areas (Beehler 1994). Five protected areas make up only about 6 percent of the ecoregion area (table J126.3). Two of the protected areas, Mts. Albert Edward/Victoria and Morobe, are large, more than 1,000 km^2 in size (MacKinnon 1997).

Types and Severity of Threats

Although the threats to this ecoregion are low, potential threats include logging, nickel exploitation, and traditional agriculture (Johns 1993). The extension of the highway from the capital, Port Moresby, through to Milne Bay will increase accessibility of the coastal plain south of the Owen Stanley Ranges, opening this extensive forest area to exploitation (Beehler 1994).

Priority Conservation Actions

Short-Term Conservation Actions (1–5 Years)

Several priority-setting efforts have been conducted that include this ecoregion, including Beehler (1994) and Nix et al. (2000) for PNG. These exercises have mapped specific areas that are deemed important to conserve a variety of taxa.
- Limit accessibility into ecoregion.

Longer-Term Conservation Actions (5–20 Years)

- Continue to implement conservation recommendations of Beehler (1994) and Nix et al. (2000) or other subsequent conservation priority exercises.
- Increase the protected area system.
- Develop and implement effective management plans for protected areas.
- Stop gazettement of logging concessions.

TABLE J126.3. WCMC (1997) Protected Areas That Overlap with the Ecoregion.

Protected Area	Area (km^2)	IUCN Category
Morobe	1,430	?
Popondetta	750	?
Mt. Suckling	600	?
Abau	550	?
Mts. Albert Edward/ Victoria [131]	1,600	?
Total	4,930	

Ecoregion numbers of protected areas that overlap with additional ecoregions are listed in brackets.

Focal Species for Conservation Action

Endangered endemic species: large-eared nyctophilus (*Pharotis imogene*) (Flannery 1995), long-footed hydromine (*Leptomys elegans*) (Flannery 1995), Papuan long-beaked echidna (*Zaglossus bartoni*) (Flannery 1995), Van Deusen's rat (*Stenomys vandeuseni*), Eastern shrew-mouse (*Pseudohydromys murinus*) (Flannery 1995)

Area-sensitive species: Doria's tree-kangaroo (*Dendrolagus dorianus*) (Flannery 1995), Goodfellow's tree-kangaroo (*Dendrolagus goodfellowi*) (Flannery 1995), Macleay's dorcopsis (*Dorcopsulus macleayi*) (Flannery 1995), small dorcopis (*Dorcopsulus vanheurni*) (Flannery 1995), white-striped dorcopsis (*Dorcopsis hageni*) (Flannery 1995), grey dorcopsis (*Dorcopsis luctuosa*) (Flannery 1995), New Guinea pademelon (*Thylogale brunii*) (Flannery 1995), agile wallaby (*Macropus agilis*) (Flannery 1995), Blyth's hornbill (*Aceros plicatus*) (Kemp 1995), Gurney's eagle (*Aquila gurneyi*), little eagle (*Hieraaetus morphnoides*), New Guinea eagle (*Harpyopsis novaeguineae*), peregrine falcon (*Falco peregrinus*)

Top predator species: New Guinea quoll (*Dasyurus albopunctatus*), Gurney's eagle (*Aquila gurneyi*), little eagle (*Hieraaetus morphnoides*), New Guinea eagle (*Harpyopsis novaeguineae*), peregrine falcon (*Falco peregrinus*)

Selected Conservation Contacts

Conservation International
Conservation Melanesia
Foundation for People and Community Development
Village Development Trust

Justification for Ecoregion Delineation

Using Whitmore's (1984) map of the vegetation of Malesia and MacKinnon's (1997) reconstruction of the original vegetation, we delineated the large areas of distinct habitat types as ecoregions. The montane evergreen moist forests along the Central Cordillera, including the Snow Mountains, Star Mountains, Central Highlands, and Eastern Highlands, were placed in the Central Range Montane Rain Forests [125]. This ecoregion roughly corresponds to MacKinnon's subunits P3g, P3h, and P3i. The moist forests in the southeastern peninsula were distinguished as the Southeastern Papuan Rain Forests [126]. This ecoregion consists mostly of montane forests but also includes some lowland forests along the coasts and is roughly equivalent to MacKinnon's (1997) biounit P3n. We used the 1,000-m contour from a DEM (USGS 1996) to define the montane-lowland transition. All along the Central Cordillera and in the Huon Peninsula, we separated the alpine habitat into a distinct—Central Range Sub-Alpine Grasslands [131]—ecoregion. Udvardy (1975) placed these ecoregions in the Papuan biogeographic province of the Oceanian Realm.

Prepared by John Morrison

Ecoregion Number: **127**
Ecoregion Name: **Southern New Guinea Freshwater Swamp Forests**
Bioregion: **New Guinea and Melanesia**
Biome: **Tropical and Subtropical Moist Broadleaf Forests**
Political Units: **Indonesia, Papua New Guinea**
Ecoregion Size: **99,400 km²**
Biological Distinctiveness: **Regionally Outstanding**
Conservation Status: **Relatively Intact**
Conservation Assessment: **III**

The Southern New Guinea Freshwater Swamp Forests [127], a unique mosaic of swamp grassland, savanna, gallery forest, and lakes (including PNG's largest lake, Lake Murray), represent wetlands for resident waterfowl and staging areas for those migrating from Australia (Miller et al. 1994). With the exception of the largest mine in PNG, much of this lightly populated ecoregion is still relatively intact. However, this ecoregion faces challenges from exotic species.

Location and General Description

This ecoregion is made up of the freshwater swamp forests associated with low-gradient river systems (Purari and Fly rivers) along the southern coast of New Guinea. The ecoregion consists of a large extent in south-central Irian Jaya and PNG and a smaller, disjunct patch along the southern coast of the Vogelkop Peninsula in Irian Jaya. The climate of the ecoregion is tropical wet, which is characteristic of this part of Melanesia, located in the western Pacific Ocean north of Australia (National Geographic Society 1999). Although the rainfall here is lower than in most parts of New Guinea, the area's low relief and large rivers draining the nearby Central Ranges result in this ecoregion's inundation during the wet season. The wetland system associated with the Fly River is the largest watershed in PNG (Miller et al. 1994). Although New Guinea is a very active tectonic area with a complex geologic history, the geology of the West Papuan shelf, where this ecoregion is located, shows little folding, an indication of relative stability. The surface geology of the ecoregion consists of alluvium on active and relict alluvial plains and fans (Bleeker 1983).

The freshwater swamps along the southern lowlands are made up of several habitat types, ranging from purely aquatic to herbaceous vegetation, grass swamps, savannas, and woodlands to swamp forests (Petocz 1989). The mosaic of habitats in lowland freshwater swamps includes herbaceous swamp vegetation, *Leersia* grass swamp, *Saccharum-Phragmites* grass swamp, *Pseudoraphis* grass swamp, mixed swamp savanna, *Melaleuca* swamp savanna, mixed swamp woodland, sago swamp woodland, pandan swamp woodland, mixed swamp forest, *Campnosperma* swamp forest, *Teminalia* swamp forest, and *Melaleuca* swamp forest (Osborne 1993). The ecoregion is composed of approximately 32 percent freshwater swamp forest, 28 percent grasslands, and 13 percent peat swamp forest, with the remainder being wet forest, dry forest, or cleared land (MacKinnon 1997).

The water table is near or above the ground surface but may fluctuate widely. The structure of swamp forests range from very small-crowned to medium-crowned, dense to open, with a 20- to 30-m canopy of an even height. Sago palm (*Metroxylon sagu*) and *Pandanus* spp. generally are present in the subcanopy. Canopy trees of the swamp forest include *Campnosperma brevipetiolata*, *C. auriculata*, *Terminalia canaliculata*, *Nauclea coadunata*, and *Syzygium* spp., with *Myristica hollrungii* in delta areas. However, although some swamp forests are almost pure stands of *Campnosperma* or *Melaleuca*, others are species-rich, and many other tree species are possible (Paijmans 1975).

Two sets of a total of five small, 1-ha plots have been inventoried in Lakekamu Basin in PNG (Oatham and Beehler 1997; Mack 1998). One of these forest plots was in hill forest (175–260 m elevation), which was surveyed on the margins of the ecoregion, and this plot may or may not be more representative of the Southern New Guinea lowland rain forests. The hill forest plot contained less diversity than the alluvial plots. The hill forest plot was dominated by members of the Myrtaceae and Lauraceae families (25 percent of the species and 40 percent of the stems), whereas the corresponding lowland plot did not contain such dominants. Both plots contained many rare species and a few fairly common ones. Although all five of the plots are found in continuous, closed forest and none of the plots are more than 20 km apart, they all exhibit considerable differences in species composition and structure (Mack 1998).

The taller grasses usually are *Leersia*, *Saccharum*, and *Phragmites*. Sago palms (*Metroxylon sagu*) form woodlands, especially in areas that have freshwater influxes, with *Pandanus* being more common in areas that are more saline. Seasonally inundated areas, especially in the fluctuating backswamps of the Middle Fly and Strickland rivers, are characterized by *Melaleuca* woodlands (Miller et al. 1994; Petocz 1989). The dominant tree species is *Melaleuca cajuputi*, a highly fire-resistant species that forms an even canopy (Petocz 1989). Other common trees in the open woodlands include *Nauclea*, *Campnosperma*, *Syzygium*, *Melaleuca* (Miller et al. 1994), and *Carallia* (Petocz 1989), mixed with *Terminalia*, *Alstonia*, *Barringtonia*, *Diospyros*, *Pandanus*, and *Myristica* in deltaic regions (Petocz 1989).

Biodiversity Features

Overall richness and endemism generally are low to moderate when compared with those of other ecoregions in Indo-Malaysia. Reptile and amphibian richness, though poorly studied, is thought to be high.

The mammalian fauna consists of a wide variety of tropical Australasian marsupials, including a tree kangaroo (Flannery 1995). Fifty mammal species are found in the ecoregion, including six endemics and near endemics (Flannery 1995; Flannery and Groves 1998; Bonaccorso et al., in press) (table J127.1). The Fly River leptomys (*Leptomys signatus*) is considered critically endangered, whereas the dusky pademelon (*Thylogale bruinii*) and Fly River trumpet-eared bat (*Kerivoula muscina*) are considered vulnerable (Hilton-Taylor 2000).

The avifauna of the ecoregion has a clear Australasian flavor, including representatives of several Australasian families such as Ptilonorhynchidae, Eopsaltriidae, Meliphagidae, and Paradisaeidae. There are 339 bird species in the ecoregion,

TABLE J127.1. Endemic and Near-Endemic Mammal Species.

Family	Species
Macropodidae	Dendrolagus spadix
Macropodidae	Dorcopsis luctuosa
Macropodidae	Thylogale brunii
Peroryctidae	Echymipera echinista
Vespertilionidae	Kerivoula muscina
Muridae	Leptomys signatus*

An asterisk signifies that the species' range is limited to this ecoregion.

TABLE J127.2. Endemic and Near-Endemic Bird Species.

Family	Common Name	Species
Megapodiidae	Red-billed brush-turkey	Talegalla cuvieri
Columbidae	Wallace's fruit-dove	Ptilinopus wallacii
Columbidae	Western crowned-pigeon	Goura cristata
Loriidae	Black lory	Chalcopsitta atra
Alcedinidae	Little paradise-kingfisher	Tanysiptera hydrocharis
Pachycephalidae	White-bellied pitohui	Pitohui incertus
Sylviidae	Fly River grassbird	Megalurus albolimbatus
Estrildidae	Grey-crowned munia	Lonchura nevermanni
Estrildidae	Black munia	Lonchura stygia
Dicaeidae	Olive-crowned flowerpecker	Dicaeum pectorale
Paradisaeidae	Greater bird-of-paradise	Paradisaea apoda

An asterisk signifies that the species' range is limited to this ecoregion.

including eleven endemics and near endemics (Beehler et al. 1986; Coates 1985) (table J127.2). This ecoregion intersects two separate EBAs: the South Papuan lowlands and the Trans Fly. All six of the restricted-range birds from the South Papuan lowlands EBA are found in this ecoregion (Beehler et al. 1986; Diamond and Bishop 1998). One species, the Fly River grassbird (*Megalurus albolimbatus*), and possibly two munias (*Lonchura* spp.) from the Trans Fly EBA are also found in this ecoregion (Beehler et al. 1986; Stattersfield et al. 1998). The vulnerable Fly River grassbird and southern crowned pigeon (*Goura cristata*) are the only threatened bird species in the ecoregion (Hilton-Taylor 2000).

The ecoregion overlaps with the lowland swamps of the Southern Fly Platform Centre of Plant Diversity (Davis et al. 1995).

Status and Threats

Current Status

About 80 percent of the natural habitat in this ecoregion is still intact. The nine protected areas cover 14,450 km^2, or 15 percent of the ecoregion (table J127.3). However, most of this coverage is in Indonesia. Several of the protected areas are large (more than 1,000 km^2), and these core conservation areas are distributed across the length of the ecoregion to cover most of the major watersheds.

Like most of PNG, much of this ecoregion is still in good condition (Miller et al. 1994). Human population density is low, and use is mainly for subsistence.

Types and Severity of Threats

A number of areas within this ecoregion have been altered. A large and well-known mine is found in this ecoregion: the OK Tedi copper and gold mine in PNG. This is a large facility, and its location in a pristine area has caused concern about sedimentation and toxic runoff into adjacent stream and river systems. The wetlands of Western Province are under severe threat from mine waste discharges that have greatly reduced fish catches from the Ok Tedi and Fly River systems (Miller et al. 1994). The Ok Tedi mine apparently was able to gain a special exemption from constructing a tailings dam that could have drastically reduced waste discharge (Allison 1994). After the mine is closed in approximately fifteen years, recovery may take another twenty years (Miller et al. 1994). The Laiagap-Strickland River system is also threatened by mining; high levels of mercury have been detected in fish here (Miller et al. 1994). In Irian Jaya, many of the huge swamp forest areas are slated for conversion to wetland rice fields, and transmigration from overpopulated Java and Madura threatens to double the population (Whitmore 1989).

Exotic species are a concern in this ecoregion. The introduction of rusa deer (*Cervus timorensis*) from other areas of Indonesia has had serious impacts in this ecoregion (Allison 1994), and other exotics, including fish, have been introduced (Beehler, pers. comm., 2000).

Priority Conservation Actions

Short-Term Conservation Actions (1–5 Years)

Several priority-setting efforts have been conducted that include this ecoregion, including Beehler (1994) and Nix et al. (2000) for PNG and Conservation International (1999) for Irian Jaya. These exercises have mapped specific areas that are deemed important to conserve a variety of taxa.

- There is a need to identify, map, and control the introduction of exotic species in this ecoregion (Beehler, pers. comm., 2000).
- Work to make the OK Tedi mining operation more environmentally friendly.

TABLE J127.3. WCMC (1997) Protected Areas That Overlap with the Ecoregion.

Protected Area	Area (km^2)	IUCN Category
Gunung Lorentz [125, 128, 129, 131]	1,780	I
Pulau Dolok [129]	4,960	IV
Pulau Dolok extension	490	PRO
Pulau Pombo	70	I
Danau Bian	1,010	IV
Kumbe-Merauke	1,040	PRO
Strickland River	3,520	?
Bamu River	1,440	?
Kikori River [128, 129]	140	?
Total	14,450	

Ecoregion numbers of protected areas that overlap with additional ecoregions are listed in brackets.

Longer-Term Conservation Actions (5–20 Years)

- Continue to implement conservation recommendations of Conservation International (1999), Nix et al. (2000), and other subsequent conservation priority exercises.
- Develop and implement effective management plans for protected areas.
- Increase protection in PNG.
- Limit mining operations and mercury use.

Focal Species for Conservation Action

Threatened endemic species: Fly River leptomys (*Leptomys signatus*) (Flannery 1995), dusky pademelon (*Thylogale brunii*) (Flannery 1995), Fly River trumpet-eared bat (*Kerivoula muscina*) (Flannery 1995), Fly River grassbird (*Megalurus albolimbatus*) (BirdLife International 2000), southern crowned pigeon (*Goura cristata*) (BirdLife International 2000)

Area-sensitive species: grey dorcopsis (*Dorcopsis luctuosa*) (Flannery 1995), lowlands tree kangaroo (*Dendrolagus spadix*) (Flannery 1995), dusky pademelon (*Thylogale brunii*) (Flannery 1995), agile wallaby (*Macropus agilis*) (Flannery 1995), fruit bats (*Pteropus alecto, Pteropus conspicillatus, Pteropus neohibernicus, Nyctimene cephalotes, Nyctimene albiventer, Nyctimene aello, Nyctimene cyclotis, Paranyctimene raptor, Macroglossus minimus, Syconycteris australis*) (Mickleburgh et al. 1992), Blyth's hornbill (*Aceros plicatus*) (Kemp 1995), Gurney's eagle (*Aquila gurneyi*), New Guinea eagle (*Harpyopsis novaeguineae*), wedge-tailed eagle (*Aquila audax*), peregrine falcon (*Falco peregrinus*)

Top predator species: Gurney's eagle (*Aquila gurneyi*), New Guinea eagle (*Harpyopsis novaeguineae*), wedge-tailed eagle (*Aquila audax*), peregrine falcon (*Falco peregrinus*)

Selected Conservation Contacts

Indo-Pacific Conservation Alliance (Asmat area, Irian Jaya)

Lembaga Alam Tropika Indonesia (LATIN)

WWF Kikori Integrated Conservation and Development Area

Yayasan Keanekaragaman Hayati

Yayasan Konsorsium Pembaruan Agraria (KPA)

Justification for Ecoregion Delineation

Using Whitmore's (1984) map of the vegetation of Malesia and MacKinnon's (1997) reconstruction of the original vegetation, we delineated the large areas of distinct habitat types as ecoregions. The tropical lowland moist evergreen forests to the south of the Central Cordillera were placed in the Southern New Guinea Lowland Rain Forests [128], and the extensive freshwater swamp forests were placed in the Southern New Guinea Freshwater Swamp Forests [127]. The freshwater swamp forests in the southern Vogelkop were also included in this ecoregion. Udvardy (1975) placed these ecoregions in the Papuan biogeographic province of the Oceanian Realm.

Prepared by John Morrison

Ecoregion Number:	**128**
Ecoregion Name:	Southern New Guinea Lowland Rain Forests
Bioregion:	New Guinea and Melanesia
Biome:	Tropical and Subtropical Moist Broadleaf Forests
Political Units:	Indonesia, Papua New Guinea
Ecoregion Size:	122,300 km^2
Biological Distinctiveness:	Globally Outstanding
Conservation Status:	Critical
Conservation Assessment:	I

The Southern New Guinea Lowland Rain Forests [128] are some of the largest tropical wilderness areas in Australasia, with huge swaths of wet evergreen forest and important mosaics of other forest types. It is one of the richest areas for vascular plants in Indo-Malaysia.

Location and General Description

The Southern New Guinea Lowland Rain Forests [128] ecoregion stretches across the lowland forests south of the Central Cordillera of Indonesian Irian Jaya and PNG. The climate of the ecoregion is tropical wet, which is characteristic of this part of Melanesia, located in the western Pacific Ocean north of Australia (National Geographic Society 1999). The wettest portions of New Guinea are found in the Central Cordillera and from 100 to 1,000 m on the southern slopes of the Cordillera in this ecoregion. Although the remainder of the ecoregion is lower than most parts of New Guinea, the area's low relief and large rivers draining the nearby Central Ranges result in this ecoregion's inundation during the wet season. The wetland system associated with the Fly River is the largest watershed in PNG (Miller et al. 1994). Although New Guinea is a very active tectonic area with a complex geologic history, the geology of the West Papuan shelf, where this ecoregion is located, shows little folding, an indication of stability. The surface geology of the ecoregion consists of alluvium on active and relict alluvial plains and fans (Bleeker 1983).

The vegetation is primarily tropical lowland broadleaf forest, of both alluvial and hill type, in addition to a (17 percent) portion of savanna. Small (less than 10 percent each) percentages of semi-evergreen rain forest, tropical montane evergreen forest, limestone forest, peat swamp forest, and freshwater swamp forest are also present in the ecoregion (MacKinnon 1997).

The most extensive habitat in the ecoregion is lowland broadleaf evergreen forest, which can be divided coarsely into alluvial and hill forest. Lowland alluvial forest has a canopy that is multitiered and irregular, with many emergents. The forest understory contains a shrub and herb layer with a variety of climbers, epiphytes, and ferns (Petocz 1989). Palms may be common in the shrub layer (Paijmans 1975). The mixed floristic composition of the canopy trees includes *Pometia pinnata, Octomeles sumatrana, Ficus* spp., *Alstonia scholaris*, and *Terminalia* spp. Additional important genera include *Pterocarpous, Artocarpus, Planchonella, Canarium, Elaeocarpus, Cryptocarya, Celtis*,

Dracontomelum, Dysoxylum, Syzygium, Vitex, Spondias, and *Intsia* (Paijmans 1975). Although present, dipterocarps do not form extensive forests in New Guinea, as they do in Malaysia and Borneo (Nightingale 1992). The somewhat lower-canopy, more closed lowland hill forest contains more open shrub layer but a denser herbaceous layer. Palms are fewer in number. The dominant canopy trees include species of *Pometia, Canarium, Anisoptera, Cryptocarya, Terminalia, Syzygium, Ficus, Celtis, Dysoxylum,* and *Buchanania. Koompassia, Dillenia, Eucalyptopsis, Vatica,* and *Hopea* are locally abundant. Dense stands of *Araucaria,* the tallest tropical trees in the world, are present in scattered locations (Paijmans 1975; Nightingale 1992).

The lowland rain forests of this ecoregion are ecologically critical to the great Fly River watershed that is born in this high rainfall zone (Beehler 1993). The lowland hill forests are dominated by *Hopea celtidifolia* and *Vatica russak,* whereas the lower montane elements include *Podocarpus* and *Lithocarpus* (Johns 1993). The Upper Fly lowlands, except for extensive settlements around the Ok Tedi mine in the west, represent a large expanse of sparsely populated, old-growth wet rain forest that is characteristic of the extraordinarily rich biota of the Upper Fly platform (Miller et al. 1994). This ecoregion also includes an enormous extent of botanically unknown tower limestone as well as *Araucaria* forest, *Castanopsis, Quercus,* traditional medicinal and food plants, nutmeg, and traditional spirit trees (Johns 1993).

Lake Kutubu, PNG's largest lake, is also found in this ecoregion. This lake supports a diverse aquatic plant flora, and eleven of its fourteen fish species are endemic to it (Miller et al. 1994).

Biodiversity Features

Overall richness and endemism for this ecoregion are low to moderate when compared with those of other ecoregions in Indo-Malaysia, although plant richness is high.

The mammalian fauna consists of a wide variety of tropical Australasian marsupials, including a tree kangaroo (Flannery 1995). The ecoregion contains sixty-nine mammal species, thirteen of which are endemic or near endemic (Flannery 1995; Flannery and Groves 1998; Bonaccorso et al., in press) (table J128.1). Bulmer's fruit bat (*Aproteles bulmerae*) and large pogonomelomys (*Pogonomelomys bruijni*) are critically endangered, whereas the lesser tube-nosed bat (*Nyctimene draconilla*), Fly River trumpet-eared bat (*Kerivoula muscina*), New Guinea sheathtail-bat (*Emballonura furax*), and Papuan mastiff bat (*Otomops papuensis*) are considered vulnerable (Hilton-Taylor 2000).

The avifauna of the ecoregion has a clear Australasian flavor, including representatives of several Australasian families such as Ptilonorhynchidae, Eopsaltridae, Meliphagidae, and Paradisaeidae. There are 344 bird species in the ecoregion, including 5 endemic and near-endemic species (table J128.2). The ecoregion intersects two separate EBAs, the South Papuan lowlands and the Trans Fly (Beehler et al. 1986; Coates 1985; Diamond and Bishop 1998; Stattersfield et al. 1998). No bird species in this ecoregion are currently threatened.

Status and Threats

Current Status

The Papuan Plateau and the Purari River Basin are major wilderness areas in this ecoregion (Beehler 1993). Most of the habitat remains in contiguous blocks, and the three protected areas, which cover 9,330 km², or 8 percent of the ecoregion area, are well distributed between Indonesia and PNG (table J128.3). All the protected areas are large (more than 1,000 km²). An additional large protected area (Kumbe-Merauke, 1,268 km²) also extends into this ecoregion.

Types and Severity of Threats

The ecoregion suffers from the potential major threat of logging, and several large timber concessions are allotted throughout the ecoregion. Additional threats include traditional agriculture and expansion of subsistence coffee areas (Johns 1993). Traditional hunting for the Raggiana bird-of-paradise for sale and trade also occur (Johns 1993).

TABLE J128.1. Endemic and Near-Endemic Mammal Species.

Family	Species
Macropodidae	Dendrolagus spadix
Macropodidae	Dorcopsis luctuosa
Macropodidae	Dorcopsis muelleri
Pteropodidae	Aproteles bulmerae
Pteropodidae	Nyctimene draconilla
Emballonuridae	Emballonura furax
Rhinolophidae	Hipposideros wollastoni
Rhinolophidae	Hipposideros muscinus
Vespertilionidae	Kerivoula muscina
Molossidae	Otomops papuensis
Muridae	Melomys lorentzi
Muridae	Melomys gracilis
Muridae	Pogonomelomys bruijni

An asterisk signifies that the species' range is limited to this ecoregion.

TABLE J128.2. Endemic and Near-Endemic Bird Species.

Family	Common Name	Species
Megapodiidae	Red-billed brush-turkey	Talegalla cuvieri
Loriidae	Striated lorikeet	Charmosyna multistriata
Alcedinidae	Little paradise-kingfisher	Tanysiptera hydrocharis
Meliphagidae	Spot-breasted meliphaga	Meliphaga mimikae
Paradisaeidae	Greater bird-of-paradise	Paradisaea apoda

An asterisk signifies that the species' range is limited to this ecoregion.

TABLE J128.3. WCMC (1997) Protected Areas That Overlap with the Ecoregion.

Protected Area	Area (km²)	IUCN Category
Gunung Lorentz [125, 127, 129, 131]	5,360	I
Mt. Bosavi [125]	2,280	?
Kikori River [127, 129]	1,690	?
Total	9,330	

Ecoregion numbers of protected areas that overlap with additional ecoregions are listed in brackets.

Priority Conservation Actions

Short-Term Conservation Actions (1–5 Years)

Several priority-setting efforts have been conducted that include this ecoregion, including Beehler (1994) and Nix et al. (2000) for PNG and Conservation International (1999) for Irian Jaya. These exercises have mapped specific areas that are deemed important to conserve a variety of taxa.

Longer-Term Conservation Actions (5–20 Years)

- Continue to implement conservation recommendations of Beehler (1994), Conservation International (1999), and Nix et al. (2000) or other subsequent conservation priority exercises.
- Develop and implement effective management plans for protected areas.
- Halt allocation of new timber concessions in conservation priority areas previously identified.

Focal Species for Conservation Action

Threatened endemic species: Fly river trumpet-eared bat (*Kerivoula muscina*) (Flannery 1995), Bulmer's fruit bat (*Aproteles bulmerae*) (Flannery 1995; Mickleburgh et al. 1992), large pogonomelomys (*Pogonomelomys bruijni*) (Flannery 1995), lesser tube-nosed bat (*Nyctimene draconilla*) (Flannery 1995; Mickleburgh et al. 1992), New Guinea sheathtail-bat (*Emballonura furax*) (Flannery 1995), Papuan mastiff bat (*Otomops papuensis*) (Flannery 1995)

Area-sensitive species: lowlands tree kangaroo (*Dendrolagus spadix*) (Flannery 1995), grey dorcopsis (*Dorcopsis luctuosa*) (Flannery 1995), brown dorcopsis (*Dorcopsis muelleri*) (Flannery 1995), fruit bats (*Pteropus neohibernicus, Pteropus macrotis, Rousettus amplexicaudatus, Dobsonia minor, Dobsonia moluccense, Aproteles bulmerae, Nyctimene albiventer, Nyctimene aello, Nyctimene cyclotis, Nyctimene draconilla, Paranyctimene raptor, Macroglossus minimus, Syconycteris australis*) (Mickelburgh 1992), Blyth's hornbill (*Aceros plicatus*) (Kemp 1995), Gurney's eagle (*Aquila gurneyi*), little eagle (*Hieraaetus morphnoides*), New Guinea eagle (*Harpyopsis novaeguineae*), peregrine falcon (*Falco peregrinus*)

Top predator species: New Guinea quoll (*Dasyurus albopunctatus*), Gurney's eagle (*Aquila gurneyi*), little eagle (*Hieraaetus morphnoides*), New Guinea eagle (*Harpyopsis novaeguineae*), peregrine falcon (*Falco peregrinus*)

Selected Conservation Contacts

Conservation International
Foundation for People and Community Development
Lembaga Alam Tropika Indonesia (LATIN)
WWF Kikori Integrated Conservation and Development Area
Yayasan Keanekaragaman Hayati
Yayasan Konsorsium Pembaruan Agraria (KPA)

Justification for Ecoregion Delineation

Using Whitmore's (1984) map of the vegetation of Malesia and MacKinnon's (1997) reconstruction of the original vegetation, we delineated the large areas of distinct habitat types as ecoregions. The tropical lowland moist evergreen forests to the south of the Central Cordillera were placed in the Southern New Guinea Lowland Rain Forests [128], and the extensive freshwater swamp forests were placed in the Southern New Guinea Freshwater Swamp Forests [127]. The freshwater swamp forests in the southern Vogelkop were also included in this ecoregion. Udvardy (1975) placed these ecoregions in the Papuan biogeographic province of the Oceanian Realm.

Prepared by John Morrison

Ecoregion Number:	**129**
Ecoregion Name:	**New Guinea Mangroves**
Bioregion:	**New Guinea and Melanesia**
Biome:	**Mangroves**
Political Units:	**Indonesia, Papua New Guinea**
Ecoregion Size:	**26,700 km^2**
Biological Distinctiveness:	**Globally Outstanding**
Conservation Status:	**Critical**
Conservation Assessment:	**I**

A dynamic and essential vegetation type at the interface of terrestrial and marine environments, New Guinea's mangroves are among the most diverse in the world. In some areas, monodominant stands of 1-m-diameter *Xylocarpus granatum* resemble North American bald-cypress swamps.

Location and General Description

New Guinea mangroves are found along extensive lengths of its coastline. There are several disjunct sections along the north coast, including the eastern side of Cenderwasih Bay, adjacent to the mouths of the Sepik and Ramu rivers, and Dyke Ackland Bay and Ward Hunt Strait. The longest and deepest stretches of mangroves are found on the south side of the island, especially at the mouths of the Purari, Kikori, Fly, Northwest, and Otakwa rivers, Bintuni Bay, and the southern portions of the Vogelkop Peninsula. With the exception of the coast along the Trans Fly region of southern New Guinea, the climate of the ecoregion is tropical wet, which is characteristic of this part of Melanesia, located in the western Pacific Ocean north of Australia (National Geographic Society 1999). The surface geology of this widespread, dynamic ecoregion consists of alluvium on active alluvial plains and fans (Bleeker 1983).

Located in the Indo-Malayan center of diversity, southern PNG has the highest diversity of mangroves in the world (Duke 1992, in Ellison 1997). In the dynamic tidal environment of the delta, a general succession process occurs through time and space. The pioneering species that establish on sheltered coastal shores are *Avicennia alba* or *Avicennia marina*. On tidal creeks, *Sonneratia* species are the first to establish. The complex root network of these species encourages further sedimentation in the immediate area. As the land surface continues to rise relative to water levels, *Rhizophora mucronata* invades, overshadows the shade-intolerant *Avicennia* and *Sonneratia* species, and takes over. *Rhizophora apiculata* and *Bruguiera parviflora* are the next colonists, with occasional *Bruguiera gymnorrhiza*. *B. gymnorrhiza* develops into dominant stands away from the water. *Xylocarpus*, *Lunmitzera*, and *Heritiera* also establish late in succession (Percival and Womersley 1975).

Within this simplified successional picture there are more subtle patterns. Mangrove areas transitional between upland or coastal environments contain the greatest variety of species, growth forms, and forest structures (Cragg 1987). Where there is substantial freshwater influence within the mangrove, stands of *Nypa fruticans* develop (Cragg 1987). Mangrove areas adjacent to swamp forest contain *Bruguiera sexangula*, *Camptostemon schultzii*, *Dolichandrone spathacea*, *Diospyros* spp., *Excoecaria agallocha*, *Heritiera littoralis*, *Rhizophora apiculata*, and *Xylocarpus granatum*. However, typical swamp forest trees such as *Calophyllum*, *Instia bijuga*, *Myristica hollrungii*, and *Amoora cucullata* are also part of the mix in transitional areas (Floyd 1977, in Cragg 1987). The upper limit of *Sonneratia lanceolata* is taken to be the boundary between brackish and freshwater reaches of the Purari delta (Petr 1983).

Mangrove vegetation is sensitive to soil and water salinity, drainage, and inundation period. Because of a larger tidal influence (resulting from smaller freshwater flows), some areas, such as the Kikori delta, have a more extensive mangrove coverage than others (Cragg 1987).

A simpler description of zonation of the nearby Fly River mangroves is given by Robertson et al. (1992, in Ellison 1997). In this zonation there are three zones as one moves inland: a *Sonneratia-Avicennia* zone on actively accreting shores, a *Rhizophora-Bruguiera* zone where salinities generally are greater than 10 percent (subdominant species include *B. gymnorrhiza*, *Avicennia marina*, *Ceriops decandro*, and *Heritiera littoralis*), and a *Nypa* zone. *Nypa fruticans* may form vast monotypic stands, but the *Nypa* zone also includes *Xylocarpus granatum* and *Heritiera littoralis*. Salinities in this zone generally are between 1 and 10 percent.

Biodiversity Features

Mangroves provide a keystone habitat that provides for vital ecological functions in the terrestrial–marine interface. They accumulate sediment, stabilizing and protecting the coastline from erosion, and provide a buffering exchange of nutrients between the terrestrial and marine environments. Mangroves also provide a nursery for many coastal fishes.

Generally, this ecoregion exhibits low richness and endemism when compared with other ecoregions in Indo-Malaysia. Exact species lists are difficult to derive for mangroves because of the transitional nature of the habitat. Some species are clear inhabitants of mangroves.

The ecoregion harbors one near-endemic mammal species: New Guinea sheathtail bat (*Emballonura furax*), which is considered vulnerable (Flannery 1995; Hilton-Taylor 2000) (table J129.1).

TABLE J129.1. Endemic and Near-Endemic Mammal Species.

Family	Species
Emballonuridae	Emballonura furax

An asterisk signifies that the species' range is limited to this ecoregion.

Mangroves are only a component of the several EBAs identified by Birdlife International. New Guinea mangroves contain ten endemic or near-endemic bird species (Beehler et al. 1986; Coates 1985; Stattersfield et al. 1998) (table J129.2). Two of these species, Salvadori's fig-parrot (*Psittaculirostris salvadorii*) and the southern crowned pigeon (*Goura cristata*) are considered vulnerable (Hilton-Taylor 2000).

The key ecological process in mangrove areas is disturbance related to the dynamism of the geology and geomorphology of the coastal and delta systems on which the mangroves are built. The regeneration of various tree species is closely tied to subtle changes in elevation, drainage, and inundation periods, which are in turn the result of larger-scale geomorphologic processes. Delta development is controlled by the geologic environment, sea-level change, tectonics, fluvial contributions, climate, vegetation, and the receiving basin (Thom and Wright 1982).

Although New Guinea's tropical evergreen forests may appear unbroken and stable, closer inspection reveals widespread past episodes of periodic large-scale natural disturbance, including extreme wind events and El Niño–mediated drought-associated fires. River meandering is a major force in coastal mangrove areas. Lightning strikes are common in New Guinea, and analysis of lightning strikes in the mangroves of PNG's Gulf Province indicates that gaps from lightning strikes affect the entire forest for about a 200- to 300-year period. In a plot established in a 50-m-diameter lightning strike patch in mangroves near Labu (Central Province), all canopy trees were killed. Lightning is thought to kill trees by traveling through the ground via root grafts, disrupting membrane structure and salt uptake abilities. A dense ground flora subsequently developed. The ground flora was dominated by 5- to 6-m-high *Xylocarpus*, with small patches of *Acrostichum aureum* and many *Rhizophora apiculata* seedlings (Johns 1993).

TABLE J129.2. Endemic and Near-Endemic Bird Species.

Family	Common Name	Species
Megapodiidae	Red-billed brush-turkey	Talegalla cuvieri
Columbidae	Wallace's fruit-dove	Ptilinopus wallacii
Columbidae	Western crowned-pigeon	Goura cristata
Psittacidae	Salvadori's fig-parrot	Psittaculirostris salvadorii
Loriidae	Black lory	Chalcopsitta atra
Loriidae	Brown lory	Chalcopsitta duivenbodei
Apodidae	Papuan swiftlet	Aerodramus papuensis
Alcedinidae	Red-breasted paradise-kingfisher	Tanysiptera nympha
Pachycephalida	White-bellied pitohui	Pitohui incertus
Dicaeidae	Olive-crowned flowerpecker	Dicaeum pectorale

An asterisk signifies that the species' range is limited to this ecoregion.

Status and Threats

Current Status

Four protected areas in Indonesia and PNG protect approximately 33 percent of the ecoregion (MacKinnon 1997) (table J129.3).

Harvestable quantities of timber species are present in some of the larger mangrove areas, including Kwila (*Instia bijuga*, Fabaceae) and Papuan mahogany (*Xylocarpus granatum*, Meliaceae). Some woody species have been exploited to make charcoal for cooking in Indochina, including *Rhizophora apiculata*, *R. mucronata*, *Avicennia marina*, and *Xylocarpus* (Spalding et al. 1997; FAO 1981).

Types and Severity of Threats

The mangroves are a difficult environment in which to work, and so far mostly small walkabout mills have been used in the mangroves to cut timber. Large-scale timber operations exist in Bintuni Bay, Irian Jaya (Beehler, pers. comm., 2000). Estuarine crocodiles were heavily exploited, to the point of population collapse, in the mangroves of PNG's Gulf Province.

Priority Conservation Actions

Short-Term Conservation Actions (1–5 Years)

Several priority-setting efforts have been conducted that include this ecoregion, including Beehler (1994) and Nix et al. (2000) for PNG and Conservation International (1999) for Irian Jaya. These exercises have mapped specific areas that are deemed important to conserve a variety of taxa.

- Limit exploitation of estuarine crocodiles and establish protection in critical habitat areas.

Longer-Term Conservation Actions (5–20 Years)

- Continue to implement conservation recommendations of Beehler (1994), Conservation International (1999), and Nix et al. (2000) or other subsequent conservation priority exercises.
- Develop and implement effective management plans for protected areas.
- Stop implementation of logging concessions.

Focal Species for Conservation Action

Threatened endemic species: New Guinea sheathtail bat (*Emballonura furax*) (Flannery 1995), Salvadori's fig-parrot (*Psittaculirostris salvadorii*) (BirdLife International 2000), southern crowned pigeon (*Goura cristata*) (BirdLife International 2000)

Habitat specialists: hornbills (mature forests)

Area-sensitive species: Gurney's eagle (*Aquila gurneyi*), little eagle (*Hieraaetus morphnoides*), New Guinea eagle (*Harpyopsis novaeguineae*), peregrine falcon (*Falco peregrinus*), fruit bats

Top predator species: Gurney's eagle (*Aquila gurneyi*), little eagle (*Hieraaetus morphnoides*), New Guinea eagle (*Harpyopsis novaeguineae*), peregrine falcon (*Falco peregrinus*), saltwater crocodile (*Crocodylus porosus*)

Selected Conservation Contacts

Lembaga Alam Tropika Indonesia (LATIN)
WWF Kikori Integrated Conservation and Development Area
Yayasan Keanekaragaman Hayati
Yayasan Konsorsium Pembaruan Agraria (KPA)

Justification for Ecoregion Delineation

The mangroves along the coast of New Guinea were delineated as the New Guinea Mangroves [129]. We used the World Mangrove Atlas (Spalding et al. 1997) to determine the distribution and boundaries of mangroves. Udvardy (1975) placed these ecoregions in the Papuan biogeographic province of the Oceanian Realm.

Prepared by John Morrison

Ecoregion Number:	**130**
Ecoregion Name:	**Trans Fly Savanna and Grasslands**
Bioregion:	**New Guinea and Melanesia**
Biome:	**Tropical and Subtropical Grasslands, Savannas, and Shrublands**
Political Units:	**Indonesia, Papua New Guinea**
Ecoregion Size:	**26,600 km2**
Biological Distinctiveness:	**Globally Outstanding**
Conservation Status:	**Intact**
Conservation Assessment:	**III**

The Trans Fly Savanna and Grasslands [130] ecoregion is one of the most extensive lowlands on the island of New Guinea. Its seasonally dry climate is unusual for the island of New Guinea and more like that of northern Australia. The habitats in this ecoregion are still relatively intact.

Location and General Description

This ecoregion is made up of the monsoonal savanna and grassland habitat along the southern coast of New Guinea, in both Indonesia and PNG. The climate of the ecoregion is strongly seasonal tropical dry, a climate that only this portion of New Guinea shares with much of Australia (National Geographic Society 1999; Stattersfield et al. 1998). Although the island of New Guinea is an active tectonic area with a complex geologic history, the geology of the West Papuan shelf, where this ecoregion is located, shows little folding, an indication of relative stability. The surface geology of the

TABLE J129.3. WCMC (1997) Protected Areas That Overlap with the Ecoregion.

Protected Area	Area (km2)	IUCN Category
Teluk Bintuni	2,670	PRO
Gunung Lorentz [125, 127, 128, 131]	3,130	I
Pulau Dolok [127]	1,780	IV
Kikori River [127, 128]	1,190	?
Total	8,770	

Ecoregion numbers of protected areas that overlap with additional ecoregions are listed in brackets.

ecoregion consists of alluvium on active and relict alluvial plains and fans (Bleeker 1983).

This ecoregion is composed primarily of grasslands, although almost a third of the region is savanna, and there are areas of dry evergreen forest (Paijmans 1975; MacKinnon 1997). The pronounced seasonal rainfall, local relief, drainage, and the frequency of burning contribute to the variation in floristic structure (Miller et al. 1994). The savannas have strong structural and floristic affinities with those of northern Australia (Gillison 1983, in Miller et al. 1994). The dominant trees in the savannas include *Eucalyptus*, *Albizia*, and *Melaleuca* (Miller et al. 1994). The *Melaleuca* forest dominates areas that are submerged in up to 1 m of water during the wet season (R. Johns, pers. comm., 2000). An extensive belt of bamboo dominated by *Schizistachyum* occurs along the transitional stage between the adjoining forests and the savanna vegetation of the monsoonal area (Petocz 1989). The dune and beach communities in this ecoregion contain the uncommon *Barringtonia asiatica* (R. Johns, pers. comm., 2000).

Biodiversity Features

The overall richness and endemism of this ecoregion are low to moderate when compared with those of other ecoregions in Indo-Malaysia. There are forty-four mammal species in the ecoregion, including five species that are endemic or near endemic (Flannery 1995; Flannery and Groves 1998; Bonaccorso et al., in press) (table J130.1). Three of these species, the Papuan planigale (*Planigale novaeguineae*), the bronze quoll (*Dasyurus spartacus*), and the dusky pademelon (*Thylogale brunii*) are considered vulnerable (Hilton-Taylor 2000).

The ecoregion represents a portion of the Trans Fly EBA, which contains six restricted-range bird species. Five bird species are endemic or near endemic, including the vulnerable Fly River grassbird (*Megalurus albolimbatus*) (Stattersfield et al. 1998; Beehler et al. 1986; Coates 1985; Hilton-Taylor 2000) (table J130.2). Tonda Wildlife Management Area is a globally significant wintering ground for migratory waders and waterfowl from Australia and the Palearctic (Beehler 1994).

This ecoregion forms the heart of the Southern Fly Platform Centre of Plant Diversity. Its flora is closely related to that of Australia (Davis et al. 1995).

The Trans Fly region is also critical habitat for several species of endemic amphibians and reptiles and is the only location of the pitted turtle (*Carettochelys insculpta*), a unique species in its own family (Allison 1994).

The key ecological process in savannas is fire. Although fires can occur during any rainless period, most savannas burn at the end of the dry season, when conditions are most favorable (Archibold 1995).

Status and Threats

Current Status

Although the region is inhabited by a large number of sparsely distributed tribal groups, population pressure is low. Access from the outside of the ecoregion is poor, and there is generally little disturbance (Stattersfield et al. 1998; Davis et al. 1995). More than 90 percent of the original habitat is still intact in this ecoregion. The five protected areas, well-distributed between Indonesia and PNG, cover 9,530 km^2, representing about 36 percent of the ecoregion area (table J130.3). Tonda, in PNG, and Wasur, in Indonesia, form a transboundary protected area complex that covers most of the coastal habitat (MacKinnon 1997).

Types and Severity of Threats

There is some threat on the Indonesian side from transmigration settlements, which result in increased hunting, wildlife trade, agricultural conversion, and unsustainable forestry practices. This would only be exacerbated by new roads, such as the planned Trans-Irian Highway (Stattersfield et al. 1998).

The introduction of rusa deer (*Cervus timorensis*) from other areas of Indonesia has had serious impacts on the grasslands of this ecoregion (Allison 1994).

Priority Conservation Actions

Short-Term Conservation Actions (1–5 Years)

Several priority-setting efforts have been conducted that include this ecoregion, including Beehler (1994) and Nix et al. (2000) for PNG and Conservation International (1999) for Irian Jaya. These exercises have mapped specific areas that are deemed important to conserve a variety of taxa.

TABLE J130.1. Endemic and Near-Endemic Mammal Species.

Family	Species
Dasyuridae	*Planigale novaeguineae*
Dasyuridae	*Dasyurus spartacus**
Dasyuridae	*Sminthopsis archeri**
Macropodidae	*Dorcopsis luctuosa*
Macropodidae	*Thylogale brunii*

An asterisk signifies that the species' range is limited to this ecoregion.

TABLE J130.2. Endemic and Near-Endemic Bird Species.

Family	Common Name	Species
Alcedinidae	Spangled kookaburra	*Dacelo tyro*
Alcedinidae	Little paradise-kingfisher	*Tanysiptera hydrocharis*
Sylviidae	Fly River grassbird	*Megalurus albolimbatus*
Estrildidae	Grey-crowned munia	*Lonchura nevermanni*
Estrildidae	Black munia	*Lonchura stygia*

An asterisk signifies that the species' range is limited to this ecoregion.

TABLE J130.3. WCMC (1997) Protected Areas That Overlap with the Ecoregion.

Protected Area	Area (km^2)	IUCN Category
Wasur	2,430	II
Wasur (extension)	1,020	PRO
Wasur (extension)	690	PRO
Rawa Biru	100	DE
Tonda	5,290	VIII
Total	9,530	

Ecoregion numbers of protected areas that overlap with additional ecoregions are listed in brackets.

- Stop transmigration settlements in Indonesia.
- Halt development of the Trans-Irian Highway in Indonesia.

Longer-Term Conservation Actions (5–20 Years)

- Continue to implement conservation recommendations of Beehler (1994), Conservation International (1999), and Nix et al. (2000) or other subsequent conservation priority exercises.
- Develop and implement effective management plans for protected areas.
- Limit or halt introduction of exotic species into the habitats.

Focal Species for Conservation Action

Threatened endemic species: Papuan planigale (*Planigale novaeguineae*) (Flannery 1995), bronze quoll (*Dasyurus spartacus*) (Flannery 1995), dusky pademelon (*Thylogale brunii*) (Flannery 1995), Fly River grassbird (*Megalurus albolimbatus*) (Stattersfield et al. 1998)

Area-sensitive species: grey dorcopsis (*Dorcopsis luctuosa*) (Flannery 1995), dusky pademelon (*Thylogale brunii*) (Flannery 1995), red-legged pademelon (*Thylogale stigmatica*) (Flannery 1995), agile wallaby (*Macropus agilis*) (Flannery 1995), fruit bats (*Pteropus alecto, Pteropus scapulatus, Pteropus neohibernicus, Pteropus macrotis, Rousettus amplexicaudatus, Dobsonia minor, Dobsonia moluccense, Nyctimene cephalotes, Nyctimene albiventer, Nyctimene aello, Macroglossus minimus, Syconycteris australis*) (Mickleburgh et al. 1992), Blyth's hornbill (*Aceros plicatus*) (Kemp 1995), New Guinea eagle (*Harpyopsis novaeguineae*), wedge-tailed eagle (*Aquila audax*), peregrine falcon (*Falco peregrinus*), brown falcon (*Falco berigora*)

Top predator species: bronze quoll (*Dasyurus spartacus*) (Flannery 1995), Gurney's eagle (*Aquila gurneyi*), New Guinea eagle (*Harpyopsis novaeguineae*), wedge-tailed eagle (*Aquila audax*), peregrine falcon (*Falco peregrinus*), brown falcon (*Falco berigora*)

Selected Conservation Contacts

Lembaga Alam Tropika Indonesia (LATIN)

WWF-Tonda Transfly Conservation Area

Yayasan Keanekaragaman Hayati

Yayasan Konsorsium Pembaruan Agraria (KPA)

Justification for Ecoregion Delineation

Using Whitmore's (1984) map of the vegetation of Malesia and MacKinnon's (1997) reconstruction of the original vegetation, we delineated the large areas of distinct habitat types as ecoregions. The savanna and grasslands in the Trans Fly region were placed in the Trans Fly Savanna and Grasslands [130] under the Grasslands, Savannas, and Shrublands biome. This ecoregion also extends across the Arafura Sea to Australia, which was outside the region of analysis. Udvardy (1975) placed these ecoregions in the Papuan biogeographic province of the Oceanian Realm.

Prepared by John Morrison

Ecoregion Number:	**131**
Ecoregion Name:	Central Range Sub-Alpine Grasslands
Bioregion:	New Guinea and Melanesia
Biome:	Montane Grasslands and Shrublands
Political Units:	Indonesia, Papua New Guinea
Ecoregion Size:	15,500 km²
Biological Distinctiveness:	Globally Outstanding
Conservation Status:	Relatively Stable
Conservation Assessment:	III

The Central Range Sub-Alpine Grasslands [131] are a unique example of alpine shrublands surrounding the highest peaks (up to 4,884 m) in Australasia. These fragile environments are fairly well protected by a combination of formal protection and remoteness.

Location and General Description

This ecoregion is made up of scattered alpine meadow habitats above 3,000 m along the Central Cordillera in Irian Jaya, Indonesia and PNG. The Central Cordillera is composed of a series of mountain ranges, which are broadly grouped into the Snow Mountains in Irian Jaya, the Star Mountains in Irian Jaya and PNG, and the Central and Eastern Highlands PNG. Although most of New Guinea has a tropical wet climate, in the case of this ecoregion it is modified by extreme altitude (National Geographic Society 1999). Another classification system puts this ecoregion in a humid upper montane climate (Bleeker 1983). The surface geology of the Central Cordillera is composed of metamorphic and intrusive igneous rocks (Bleeker 1983), along with some sedimentary formations.

The vegetation of the ecoregion consists of alpine meadow, montane, and upper montane tropical evergreen forest (MacKinnon 1997). Vegetation in the Central Ranges varies with elevation, local climate, aspect, and substrate. Upper montane forest consists of conifers (*Podocarpus, Dacrycarpus, Dacridium, Papuacedrus, Araucaria,* and *Libocedrus*) and Myrtacae, with a thin canopy and prominent understory (Davis et al. 1995). Above the montane forest, at approximately 3,000 m, the vegetation abruptly changes, and the high plateau areas of the cordillera are interspersed with tree fern (*Cyathea*) savannas, bogs, and grasslands. Immediately below the alpine zone, the vegetation typically is low shrubs and *Deschampsia* tussock grasslands (Petocz 1989). Heaths of *Rhododendron, Vaccinium, Coprosma, Rapanea,* and *Saurauia* form the limit of sub-alpine forest (Petocz 1989). The alpine habitat above 4,000 m consists of compact herbs such as *Ranuculus, Potentilla, Gentiana,* and *Epilobium,* the grasses *Poa* and *Deschampsia,* and bryophytes and lichens (Petocz 1989). Rosette and cushion herbs, mosses, lichens, and low ferns become progressively more abundant with altitude and replace grasses above 4,300 m (Miller et al. 1994). The highest areas are capped by snow and ice fields (Petocz 1989).

TABLE J131.1. Endemic and Near-Endemic Mammal Species.

Family	Species
Dasyuridae	Antechinus wilhelmina
Muridae	Pseudohydromys occidentalis*
Muridae	Stenomys richardsoni*
Muridae	Mallomys gunung*

An asterisk signifies that the species' range is limited to this ecoregion.

Biodiversity Features

With the exception of plant endemism, overall richness and endemism of this ecoregion are low when compared with those of other ecoregions in Indo-Malaysia.

This extreme environment is inhabited by only nine mammals, consisting of four murid rodents, two microchiropteran bats, and three marsupials: a tree kangaroo, a cuscus, and an antechinus. Of these mammals, four are endemic or near endemic (Flannery 1995) (table J131.1). The alpine wooly rat (Mallomys gunung) is critically endangered (Hilton-Taylor 2000).

Eighty-four bird species inhabit this ecoregion, of which an amazing twenty-eight species are endemic or near endemic (Beehler et al. 1986; Coates 1985) (table J131.2). The Central Range Sub-Alpine Grasslands [131] constitute the upper elevations of the Central Papuan Mountains EBA, which it shares with the Central Range montane rain forest ecoregion. Although the entire EBA contains fifty-four restricted-range species, only twenty-four of these range into the sub-alpine grasslands. The long-bearded honeyeater (Melidectes princeps) and MacGregor's bird-of-paradise (Macgregoria pulchra) are considered vulnerable (BirdLife International 2000).

Five Centres of Plant Diversity are shared between this ecoregion and the adjacent Central Range montane rain forests ecoregion. The Star Mountains–Telefomin–Tifalmin–Strickland Gorge CPD in PNG contains very rich (more than 3,000 vascular plant species) montane and high-altitude vegetation. Alpine communities are found on the summit of Mt. Giluwe, in the Mt. Giluwe–Tari Gap–Doma Peaks CPD. The poorly known Kubor Ranges in PNG contain extensive areas of high-altitude vegetation, and many endemics are likely to exist on the limestone and volcanic ash. The Bismarck Falls–Mt. Wilhelm–Mt. Otto–Schrader Range–Mt. Hellwig–Gahavisuka CPD has a wide variety of vegetation types and contains more than 5,000 vascular plant species. There are numerous endemics already known from the sub-alpine and alpine areas of Mt. Wilhelm, the highest point in PNG (Davis et al. 1995).

Status and Threats

Current Status

The eleven protected areas that extend into this small ecoregion cover 7,290 km^2, representing more than 47 percent of the ecoregion area (table J131.3). A gap analysis, based on detailed vegetation and habitat type mapping, has never been performed to determine whether the existing protected area network adequately covers all habitats with protected areas that are large enough to maintain all critical ecological processes.

Types and Severity of Threats

Mining exploration and access afforded by mining operations have become significant sources of threats to this sensitive alpine habitat. Visitation has also increased the level of pollution and litter in the area (Petocz 1989).

TABLE J131.2. Endemic and Near-Endemic Bird Species.

Family	Common Name	Species
Phasianidae	Snow Mountain quail	Anurophasis monorthonyx*
Psittacidae	Painted tiger-parrot	Psittacella picta
Aegothelidae	Archbold's owlet-nightjar	Aegotheles archboldi
Motacillidae	Alpine pipit	Anthus gutturalis
Acanthizidae	Papuan thornbill	Acanthiza murina
Eopsaltriidae	Greater ground-robin	Amalocichla sclateriana
Eopsaltriidae	Snow Mountain robin	Petroica archboldi*
Eopsaltriidae	Alpine robin	Petroica bivittata
Cinclosomatidae	Blue-capped ifrita	Ifrita kowaldi
Pachycephalida	Black sittella	Daphoenositta miranda
Pachycephalida	Lorentz's whistler	Pachycephala lorentzi
Meliphagidae	Orange-cheeked honeyeater	Oreornis chrysogenys
Meliphagidae	Short-bearded honeyeater	Melidectes nouhuysi*
Meliphagidae	Long-bearded honeyeater	Melidectes princeps*
Meliphagidae	Huon wattled honeyeater	Melidectes foersteri
Meliphagidae	Spangled honeyeater	Melipotes ater
Meliphagidae	Sooty honeyeater	Melidectes fuscus
Meliphagidae	Belford's honeyeater	Melidectes belfordi
Meliphagidae	Rufous-backed honeyeater	Ptiloprora guisei
Meliphagidae	Black-backed honeyeater	Ptiloprora perstriata
Estrildidae	Mountain firetail	Oreostruthus fuliginosus
Paradisaeidae	Crested bird-of-paradise	Cnemophilus macgregorii
Paradisaeidae	MacGregor's bird-of-paradise	Macgregoria pulchra
Paradisaeidae	Brown sicklebill	Epimachus meyeri
Paradisaeidae	Princess Stephanie's astrapia	Astrapia stephaniae
Paradisaeidae	Splendid astrapia	Astrapia splendidissima
Paradisaeidae	Ribbon-tailed astrapia	Astrapia mayeri
Paradisaeidae	Huon astrapia	Astrapia rothschildi

An asterisk signifies that the species' range is limited to this ecoregion.

Priority Conservation Actions

Short-Term Conservation Actions (1–5 Years)

Several priority-setting efforts have been conducted that include this ecoregion, including Beehler (1994) and Nix et al. (2000) for PNG and Conservation International (1999) for Irian Jaya. These exercises have mapped specific areas that are deemed important to conserve a variety of taxa.
- Conduct gap analyses to determine whether protected area system adequately protects the ecoregion's biodiversity.

Longer-Term Conservation Actions (5–20 Years)
- Continue to implement conservation recommendations of Beehler (1994), Conservation International (1999), and Nix et al. (2000) or other subsequent conservation priority exercises.
- Develop and implement effective management plans for protected areas.

Focal Species for Conservation Action

Threatened endemic species: alpine woolly rat (*Mallomys gunung*) (Flannery 1995), longbearded honeyeater (*Melidectes princeps*) (BirdLife International 2000), MacGregor's bird-of-paradise (*Macgregoria pulchra*) (Frith and Beehler 1998; BirdLife International 2000)

Area-sensitive species: black-mantled goshawk (*Accipiter melanochlamys*), New Guinea eagle (*Harpyopsis novaeguineae*), Australian kestrel (*Falco cenchroides*), peregrine falcon (*Falco peregrinus*), greater sooty-owl (*Tyto tenebricosa*)

Top predator species: black-mantled goshawk (*Accipiter melanochlamys*), New Guinea eagle (*Harpyopsis novaeguineae*), Australian kestrel (*Falco cenchroides*), peregrine falcon (*Falco peregrinus*), greater sooty-owl (*Tyto tenebricosa*)

TABLE J131.3. WCMC (1997) Protected Areas That Overlap with the Ecoregion.

Protected Area	Area (km^2)	IUCN Category
Pegunungan Weyland [119, 125]	300	PRO
Gunung Lorentz [125, 127, 128, 129]	3,190	I
Gunung Lorentz Addition	250	PRO
Gunung Lorentz Addition	130	PRO
Mt. Capella [125]	600	?
Jayawijaya extension [125]	1,130	PRO
Mt. Wilhelm [125]	260	?
Mt. Onuare [125]	350	?
Finisterre [124]	310	?
Mt. Bangeta [124]	410	?
Mts. Albert Edward/Victoria [126]	360	?
Total	7,290	

Ecoregion numbers of protected areas that overlap with additional ecoregions are listed in brackets.

Selected Conservation Contacts

Lembaga Alam Tropika Indonesia (LATIN)
Office of Environment and Conservation
WWF-Indonesia
Yayasan Keanekaragaman Hayati
Yayasan Konsorsium Pembaruan Agraria (KPA)

Justification for Ecoregion Delineation

Using Whitmore's (1984) map of the vegetation of Malesia and MacKinnon's (1997) reconstruction of the original vegetation, we delineated the large areas of distinct habitat types as ecoregions. The montane evergreen moist forests along the Central Cordillera, including the Snow Mountains, Star Mountains, Central Highlands, and Eastern Highlands, were placed in the Central Range Montane Rain Forests [125]. This ecoregion roughly corresponds to MacKinnon's subunits P3g, P3h, and P3i. The moist forests in the southeastern peninsula were distinguished as the Southeastern Papuan Rain Forests [126]. This ecoregion consists mostly of montane forests but also includes some lowland forests along the coasts and is roughly equivalent to MacKinnon's (1997) biounit P3n. We used the 1,000-m contour from a DEM (USGS 1996) to define the montane-lowland transition. All along the Central Cordillera and in the Huon Peninsula, we separated the alpine habitat into a distinct (Central Range Sub-Alpine Grasslands [131]) ecoregion. Udvardy (1975) placed these ecoregions in the Papuan biogeographic province of the Oceanian Realm.

Prepared by John Morrison

Ecoregion Number:	**132**
Ecoregion Name:	**Admiralty Islands Lowland Rain Forests**
Bioregion:	**New Guinea and Melanesia**
Biome:	**Tropical and Subtropical Moist Broadleaf Forests**
Political Units:	**Papua New Guinea**
Ecoregion Size:	**2,100 km^2**
Biological Distinctiveness:	**Critical**
Conservation Status:	**Regionally Outstanding**
Conservation Assessment:	**II**

The Admiralty Islands Lowland Rain Forests [132] contain several endemic species, yet the biodiversity of these islands is still poorly known. Commercial logging and conversion of forests to agriculture are the greatest threats to the ecoregion.

Location and General Description

The Admiralty Islands are located just north of PNG in the southwest Pacific Ocean and are often grouped together with New Britain and New Ireland to make up the Bismarck Archipelago. The Admiralty Islands form the political unit of Manus Province, PNG. Manus Province is the smallest province of PNG in both land area and population (32,713). The temperature of the Admiralty Islands varies little throughout the year, reaching daily highs of 30–32°C and

20–24°C at night. Average annual rainfall is 3,382 mm and is somewhat seasonal, with June–August being the wettest months (McAlpine et al. 1983).

Manus is the main island that reaches an elevation of 700 m, although there appears to be no discernible change in biota with altitude. Manus is volcanic in origin and probably broke through the ocean's surface in the late Miocene, 8–10 million years ago (Allison 1996). The substrate of the island is either directly volcanic or from uplifted coral limestone (Bleeker 1983). The vegetation of the Admiralty Islands is broadly described as lowland tropical rain forest. Johns (1993) highlighted the need to study the forests of central Manus, from Mt. Dremsel to the northern coast, and to protect the *Calophyllum* forests as an area of high biological importance.

Biodiversity Features

The Admiralty Islands are distinctive and contain endemic plant species because of their isolation from other landmasses. Characteristic species include tree species of the *Calophyllum* and *Sararanga* forests.

The Admiralty Islands contain several species with limited distribution, including five mammals species, two of which are found only in this ecoregion (table J132.1). Six bird species are endemic to the ecoregion, and seven more are near endemics (table J132.2). There are two birds listed by IUCN (Hilton-Taylor 2000) as vulnerable (*Pitta superba*, and *Rhipidura semirubra*). *P. superba* is quite beautiful and certainly a candidate for becoming a flagship species. *R. semirubra* is a forest-dwelling species that may be overlooked on Manus. Small islands off Manus are thought to be important to *R. semirubra* (Stattersfield et al. 1998).

The Admiralty Islands Lowland Rain Forests [132] have an endemic *Platymantis* frog and four endemic lizards (Allison 1993). Additionally, the first land snail to be listed by IUCN, the green tree snail (*Papustyla pucherrima*), is endemic to Manus (Parkinson et al. 1987; now considered data deficient, Hilton-Taylor 2000).

Status and Threats

Current Status

There is little information about the status of the Admiralty Islands Lowland Rain Forests [132] ecoregion. Rannells (1995) stated that four-fifths of Manus is forested, although this figure includes both primary and secondary forest cover. The interior forests around Mt. Dremsel are still intact and have been listed as an important area of terrestrial biodiversity in PNG by Beehler (1993). The Mt. Dremsel forests are listed as a protected area by WCMC (1997), but the category of protected area has not been determined (table J132.3). Meanwhile, many of the smaller islands have been converted into coconut plantations (Rannells 1995).

Types and Severity of Threats

Threats to the Admiralty Island moist forests include commercial timber extraction and destruction of habitat caused by shifting agriculture (Stattersfield et al. 1998).

Priority Conservation Actions

Short-Term Conservation Actions (1–5 Years)

- Establish biological surveys of the region.
- Protect remaining old-growth forests.

Longer-Term Conservation Actions (5–20 Years)

- Continue short-term actions.

Focal Species for Conservation Action

Threatened endemic species: superb pitta (*Pitta superba*) (Lambert and Woodcock 1996; Stattersfield et al. 1998), Manus fantail (*Rhipidura semirubra*) (Stattersfield et al. 1998), green tree snail (*Papustyla pulcherrima*), *Calophyllum* and *Sararanga* forests (Johns 1993)

TABLE J132.1. Endemic and Near-Endemic Mammal Species.

Family	Species
Phalangeridae	*Spilocuscus kraemeri**
Pteropodidae	*Dobsonia anderseni*
Pteropodidae	*Pteropus admiralitatum*
Emballonuridae	*Emballonura serii*
Muridae	*Melomys matambuai**

An asterisk signifies that the species' range is limited to this ecoregion.

TABLE J132.2. Endemic and Near-Endemic Bird Species.

Family	Common Name	Species
Megapodiidae	Melanesian scrubfowl	*Megapodius eremita*
Columbidae	Yellow-bibbed fruit-dove	*Ptilinopus solomonensis*
Columbidae	Yellow-tinted imperial-pigeon	*Ducula subflavescens*
Columbidae	Pied cuckoo-dove	*Reinwardtoena browni*
Psittacidae	Meek's pygmy-parrot	*Micropsitta meeki*
Tytonidae	Manus owl	*Tyto manusi**
Strigidae	Manus hawk-owl	*Ninox meeki**
Pittidae	Black-headed pitta	*Pitta superba**
Monarchidae	Manus monarch	*Monarcha infelix**
Rhipiduridae	Manus fantail	*Rhipidura semirubra**
Zosteropidae	Black-headed white-eye	*Zosterops hypoxanthus*
Meliphagidae	Ebony myzomela	*Myzomela pammelaena*
Meliphagidae	White-naped friarbird	*Philemon albitorques**

An asterisk signifies that the species' range is limited to this ecoregion.

TABLE J132.3. WCMC (1997) Protected Areas That Overlap with the Ecoregion.

Protected Area	Area (km²)	IUCN Category
Ndrolowa	60	VIII
Mt. Dremsel	240	?
Total	300	

Ecoregion numbers of protected areas that overlap with additional ecoregions are listed in brackets.

Area-sensitive species: Admiralty flying-fox (*Pteropus admiralitatum*) (Mickleburgh et al. 1992), variable flying-fox (*Pteropus hypomelanus*) (Mickleburgh et al. 1992), great flying-fox (*Pteropus neohibernicus*) (Mickleburgh et al. 1992), rousette bat (*Rousettus amplexicaudatus*) (Mickleburgh et al. 1992), common tube-nosed bat (*Nyctimene albiventer*) (Mickleburgh et al. 1992), northern blossom-bat (*Macroglossus minimus*) (Mickleburgh et al. 1992), common blossom-bat (*Syconycteris australis*) (Mickleburgh et al. 1992), Andersen's bare-backed fruit-bat (*Dobsonia anderseni*) (Mickleburgh et al. 1992), Pacific baza (*Aviceda subcristata*), white-bellied sea-eagle (*Haliaeetus leucogaster*), variable goshawk (*Accipiter hiogaster*)

Top predator species: Pacific baza (*Aviceda subcristata*), white-bellied sea-eagle (*Haliaeetus leucogaster*), variable goshawk (*Accipiter hiogaster*)

Selected Conservation Contacts

BirdLife International

Justification for Ecoregion Delineation

We delineated two ecoregions to represent the montane and lowland evergreen moist forests in the New Britain and New Ireland island complex; the New Britain–New Ireland Lowland Rain Forests [133] and the New Britain–New Ireland Montane Rain Forests [134]. The 1,000-m contour of the DEM (USGS 1996) was used as the transition between lowland and montane ecoregions. We placed the Admiralty Islands Lowland Rain Forests [132] into a distinct ecoregion, following Stattersfield et al. (1998). MacKinnon (1997) combined these three ecoregions into a single subunit (P3p). Udvardy (1975) placed these ecoregions in the Papuan biogeographic province of the Oceanian Realm.

Prepared by John Lamoreux

Ecoregion Number:	**133**
Ecoregion Name:	**New Britain–New Ireland Lowland Rain Forests**
Bioregion:	**New Guinea and Melanesia**
Biome:	**Tropical and Subtropical Moist Broadleaf Forests**
Political Units:	**Papua New Guinea**
Ecoregion Size:	**34,700 km2**
Biological Distinctiveness:	**Globally Outstanding**
Conservation Status:	**Critical**
Conservation Assessment:	**I**

Past volcanic eruptions have been tremendous in the lowlands of New Britain and New Ireland. The New Britain city of Rabaul is surrounded by six volcanoes, and in September 1994 one of these forced the abandonment of the city. The numbers of animal endemics of the New Britain–New Ireland Lowland Rain Forests [133] are as remarkable as the volcanoes that mark the landscape. Commercial logging and conversion of forests to agriculture have altered much of the ecoregion.

Location and General Description

The narrow Vitiaz Strait separates the Huon Peninsula of northeastern New Guinea from the island chain known as the Bismarck Archipelago, which is dominated by two islands: New Britain and New Ireland (both exceed 400 km in length). St. Matthias Islands, New Hanover, and many satellite islands are also part of the archipelago. The lowland rain forests ecoregion includes all of the Bismarck Archipelago below 1,000 m. New Britain and New Ireland are both long and narrow and contain several mountain ranges that trap rainfall. The climate of the ecoregion is tropical wet but varies dramatically in amount of average annual rainfall from about 1,500 to more than 6,000 mm depending on the location.

Despite the proximity of the Bismarck Archipelago to New Guinea and the existence of small islands that appear to be the remnants of a land bridge, the island arc was never connected to the mainland. The islands breached the ocean surface in the late Miocene (8–10 million years ago) as the result of volcanic uplift, and many active volcanoes still exist (particularly on New Britain). Most of the islands are made up of volcanic (acidic) soils and limestone. Limestone makes up 30 percent of New Britain and nearly 40 percent of New Ireland, or the entire northern half.

Soils with limestone substrates are different from volcanic soils because the former lack nutrients and drain quickly. The vegetation of the Bismarcks therefore is unusual in that there appears to be no noticeable difference between the two main substrates in species composition (Mueller-Dombois and Fosberg 1998). Overall diversity of tree species is not impressive when compared with that of mainland New Guinea. Major lowland rain forest tree genera include *Pometia* (Sapindaceae), *Octomeles* (Datiscaceae), *Alstonia* (Apocynaceae), *Campnosperma* (Anacardiaceae), *Canarium* (Burseraceae), *Dracontomelon* (Anacardiaceae), *Pterocymbium* (Sterculiaceae), *Crytocarya* (Lauraceae), *Intsia* (Leguminosae), *Ficus* (Moraceae), and *Terminalia* (Combretaceae) (Mueller-Dombois and Fosberg 1998). The vegetation of the Bismarck Archipelago is interesting for species that are not dominant. *Araucaria hunsteinii* and *A. cunninghamii* are two conifers that tower well above the lowland broadleaf forests in New Guinea but are not present in this ecoregion. Also, the dipterocarps that dominate much of Indonesia have only three species in New Guinea, and although one of these is reported from the Bismarcks, it has never been sufficiently documented. Other forest types in the lowlands include freshwater swamp and mangrove forests. The species composition of mangrove forests is specialized and occurs in zones beginning with *Avicennia* and *Sonneratia* spp. and moving inland to *Rhizophora* and *Bruguiera* spp., adding taller legumes and other species. Freshwater swamp forests are less specialized but include some notable species: *Campnosperma brevipetiolata*, *Terminalia brassii*, sago palm (*Metroxylon sagu*), and species of the genus *Pandus* (Mueller-Dombois and Fosberg 1998). Limestone forests near the coast of southern New Ireland and along the coast and interior of New Britain are dominated by *Vitex cofassus* (Verbenaceae) (Foster 2001).

Biodiversity Features

No comprehensive, modern botanical datasets exist for the Bismarcks, and much of the area is unknown in terms of biodiversity for any taxa (Keast 1996; Sekhram and Miller 1994). However, southern New Ireland has been surveyed botanically for a lowland site at about 300 m elevation in the Weitin Valley and two montane sites at 1,200 and 1,800 m (Takeuchi and Wiakabu 2001). Findings from these surveys show that tree species diversity decreases steadily with increasing elevation, although the mid-elevational forests between 250 and 800 m that were not sampled by Takeuchi and Wiakabu are thought to be the most diverse (Foster 2001). Foster (2001) found that even the dominant species of midelevational forests of southern New Ireland differ greatly with each ridge, although *Pometia pinnata* was overall the most abundant large species at lower elevations. Johns (1993) listed several regions within New Britain and New Ireland as areas of high biological importance based on their flora. Most of these regions are montane, but they include lowland portions: Lelet Plateau, southern Namatanai, Hans Meyer Range of New Ireland, and Willaumez Peninsula, Whiteman Range, Nakanai Mountains of New Britain.

There are forty-seven mammal species in the ecoregion. Most of these species are bats (thirty-six) in four families (Pteropodidae, Emballonuridae, Rhinolophidae, and Vespertilionidae), followed by rodent species (Muridae). Nine mammal species are near endemic to the ecoregion; none are strictly endemic (table J133.1). Several species are listed as threatened (VU or higher) by IUCN (1996): New Guinea pademelon (*Thylogale brownii*), Gilliard's flying-fox (*Pteropus gilliardorum*), large-eared sheathtail-bat (*Emballonura dianae*), Bismarck trumpet-eared bat (*Kerivoula myrella*), and New Britain water-rat (*Hydromys neobrittanicus*).

The ecoregion includes the St. Matthias Islands EBA and the lowland portions of the New Britain and New Ireland EBA (Stattersfield et al. 1998). The lowland portions of New Britain and New Ireland contain nineteen endemic and thirty-six near-endemic species (table J133.2). Two of these species are endemic to the St. Matthias Islands, one to New Hanover, one to Feni, ten to New Britain, and five to New Ireland. The rest are found on a combination of islands. Eight of the restricted-range species are listed as threatened (VU or higher) by IUCN (Hilton-Taylor 2000): black honey-buzzard (*Henicopernis infuscatus*), slaty-mantled sparrowhawk (*Accipiter luteoschistaceus*), Bismarck sparrowhawk (*Accipiter brachyurus*), yellow-legged pigeon (*Columba pallidiceps*), New Britain bronzewing (*Henicophaps foersteri*), Bismarck owl (*Tyto aurantia*), Bismarck kingfisher (*Alcedo websteri*), and Atoll starling (*Aplonis feadensis*). There is little doubt that the Bismarck Archipelago contains undescribed birds, but most of the areas for which information is lacking for birds lie outside the ecoregion at higher elevations.

Umboi Island is part of the ecoregion and lies between New Britain and mainland New Guinea. It is noteworthy for containing an amazing number of fruit bats (eight) and in having one of the most important waterbird sites in the Bismarcks (Beehler 1993).

Status and Threats

Current Status

The ecoregion has been largely logged and replaced by forest plantation, copra, or oil palm production. Both provinces of New Britain (East and West) are among the leading producers of oil palm, copra, and timber. New Ireland is also a major producer of copra and timber (Rannells 1995). Much of the population of the Bismarcks consists of migrant workers. The province of West New Britain has the highest growth rate of any province, at 4.0 percent (Rannells 1995). As early as 1993, the few remaining natural portions of lowland forest on New Britain's north coast were thought to be in danger (Stattersfield et al. 1998). Likewise, surveys conducted in 1994 predicted that without conservation action all of the lowland forest of New Ireland would be selectively logged within a few years (Beehler and Alonso 2001). There is almost no primary forest left on the St. Matthias Islands (Stattersfield et al. 1998).

The current haphazard method of logging forests for any large, straight tree has many adverse effects. Beehler and Alonso (in press) highlighted three main considerations specifically for southern New Ireland, although they apply to the ecoregion as a whole:

- Species dependent on large trees, such as Blyth's hornbills, are unable to nest in logged areas.
- Non-native species (e.g., cats, feral dogs, Polynesian rats, and cane toads) are introduced into forests through logging operations and are often detrimental to native species.
- The long-term effects of removing large tree species may be detrimental to the regeneration of forests.

Feral pigs have been singled out as particularly harmful introduced species in New Ireland (Foster 2001). Pigs eat certain plants selectively to the point that they have been nearly eliminated (e.g., a large *Marattia* fern) (Foster 2001).

Two protected areas on New Britain contain nearly all of the lowland birds, including nesting sites of the Melanesian scrubfowl (*Megapodius eremita*): Pokili (98 km^2) and Garu Wildlife Management Areas (87 km^2) (Stattersfield et al. 1998). Two other protected areas overlap with the ecoregion (table J133.3). Several key areas of lowland forest remain located at the base of important montane regions. Protected areas connecting the lowland and montane forests are needed for conservation.

TABLE J133.1. Endemic and Near-Endemic Mammal Species.

Family	Species
Pteropodidae	*Dobsonia anderseni*
Pteropodidae	*Dobsonia praedatrix*
Pteropodidae	*Melonycteris melanops*
Pteropodidae	*Nyctimene major*
Pteropodidae	*Pteropus admiralitatum*
Pteropodidae	*Pteropus gilliardorum*
Emballonuridae	*Emballonura serii*
Muridae	*Hydromys neobritannicus*
Muridae	*Uromys neobritannicus*

An asterisk signifies that the species' range is limited to this ecoregion.

Priority Conservation Actions

In 1994 Conservation International sponsored a rapid assessment of southern New Ireland (Beehler and Alonso 2001). This study provides the best in depth look at southern New Ireland and provides numerous conservation recommendations for New Ireland. Some of the recommendations that are relevant to the ecoregion as a whole follow.

Short-Term Conservation Actions (1–5 Years)

- Establish protected areas that connect lowland and montane habitats. Currently, lowland habitats face the most severe threats of biodiversity loss. Beehler and Alonso (2001) recommended preserving a representative assemblage of lowland forest in New Ireland before it has been cut over, with an eye toward preserving any large tracts that remain. This recommendation entails immediate action and applies to the ecoregion as a whole.
- Land-use planning should be coordinated in an organized fashion so that timber extraction and biodiversity conservation are balanced (Beehler and Alonso 2001).

Longer-Term Conservation Actions (5–20 Years)

- Continue to promote a representative protected area system and work toward building linkages between key sites for conservation.

TABLE J133.2. Endemic and Near-Endemic Bird Species.

Family	Common Name	Species	Family	Common Name	Species
Accipitridae	Black honey-buzzard	Henicopernis infuscatus*	Strigidae	Russet hawk-owl	Ninox odiosa
Accipitridae	Pied goshawk	Accipiter albogularis	Apodidae	Mayr's swiftlet	Aerodramus orientalis
Accipitridae	Slaty-mantled sparrowhawk	Accipiter luteoschistaceus*	Alcedinidae	Bismarck kingfisher	Alcedo websteri*
Accipitridae	Bismarck sparrowhawk	Accipiter brachyurus*	Alcedinidae	New Britain kingfisher	Todirhamphus albonotatus*
Megapodiidae	Melanesian scrubfowl	Megapodius eremita	Turdidae	New Britain thrush	Zoothera talaseae
Rallidae	New Britain rail	Gallirallus insignis	Sylviidae	Rusty thicketbird	Megalurulus rubiginosus*
Columbidae	Yellow-legged pigeon	Columba pallidiceps	Monarchidae	Black-tailed monarch	Monarcha verticalis
Columbidae	Pied cuckoo-dove	Reinwardtoena browni	Monarchidae	White-breasted monarch	Monarcha menckei*
Columbidae	New Britain bronzewing	Henicophaps foersteri*	Monarchidae	Dull flycatcher	Myiagra hebetior
Columbidae	Yellow-bibbed fruit-dove	Ptilinopus solomonensis	Rhipiduridae	Bismarck fantail	Rhipidura dahli
Columbidae	Knob-billed fruit-dove	Ptilinopus insolitus	Rhipiduridae	Matthias fantail	Rhipidura matthiae*
Columbidae	Red-knobbed imperial-pigeon	Ducula rubricera	Dicaeidae	Red-banded flowerpecker	Dicaeum eximium
Columbidae	Finsch's imperial-pigeon	Ducula finschii	Zosteropidae	Black-headed white-eye	Zosterops hypoxanthus
Columbidae	Bismarck imperial-pigeon	Ducula melanochroa	Zosteropidae	Louisiade white-eye	Zosterops griseotinctus
Columbidae	Yellow-tinted imperial-pigeon	Ducula subflavescens	Meliphagidae	New Ireland myzomela	Myzomela pulchella
Loriidae	Cardinal lory	Chalcopsitta cardinalis	Meliphagidae	Ebony myzomela	Myzomela pammelaena
Loriidae	White-naped lory	Lorius albidinuchus	Meliphagidae	Black-bellied myzomela	Myzomela erythromelas*
Loriidae	Red-chinned lorikeet	Charmosyna rubrigularis	Meliphagidae	Ashy myzomela	Myzomela cineracea*
Cacatuidae	Blue-eyed cockatoo	Cacatua ophthalmica*	Meliphagidae	Scarlet-bibbed myzomela	Myzomela sclateri*
Psittacidae	Finsch's pygmy-parrot	Micropsitta finschii	Meliphagidae	New Britain friarbird	Philemon cockerelli
Psittacidae	Meek's pygmy-parrot	Micropsitta meeki	Meliphagidae	New Ireland friarbird	Philemon eichhorni
Psittacidae	Singing parrot	Geoffroyus heteroclitus	Estrildidae	Mottled munia	Lonchura hunsteini*
Psittacidae	Green-fronted hanging-parrot	Loriculus tener*	Estrildidae	New Ireland munia	Lonchura forbesi*
Cuculidae	Violaceous coucal	Centropus violaceus	Estrildidae	New Hanover munia	Lonchura nigerrima*
Cuculidae	Pied coucal	Centropus ateralbus	Estrildidae	Bismarck munia	Lonchura melaena
Tytonidae	Bismarck owl	Tyto aurantia	Sturnidae	Atoll starling	Aplonis feadensis
Strigidae	Bismarck hawk-owl	Ninox variegata*	Dicruridae	Ribbon-tailed drongo	Dicrurus megarhynchus
			Artamidae	Bismarck woodswallow	Artamus insignis*

An asterisk signifies that the species' range is limited to this ecoregion.

- Work with local communities to ensure that resource extraction is compatible with both the long-term needs of people and the preservation of the full set of habitat and native species within the ecoregion.

Focal Species for Conservation Action

Threatened endemic species: black honey-buzzard (*Henicopernis infuscatus*), slaty-mantled sparrowhawk (*Accipiter luteoschistaceus*), Bismarck sparrowhawk (*Accipiter brachyurus*), yellow-legged pigeon (*Columba pallidiceps*), New Britain bronzewing (*Henicophaps foersteri*), Bismarck owl (*Tyto aurantia*), Bismarck kingfisher (*Alcedo websteri*), and Atoll starling (*Aplonis feadensis*) (Stattersfield et al. 1998; Feare and Craig 1999).

Area-sensitive species: Andersen's bare-backed fruit-bat (*Dobsonia anderseni*) (Mickleburgh et al. 1992), Bismarck bare-backed fruit-bat (*Dobsonia praedatrix*) (Mickleburgh et al. 1992), common tube-nosed bat (*Nyctimene albiventer*) (Mickleburgh et al. 1992), island tube-nosed bat (*Nyctimene major*) (Mickleburgh et al. 1992), northern blossom-bat (*Macroglossus minimus*) (Mickleburgh et al. 1992), Admiralty flying-fox (*Pteropus admiralitatum*) (Mickleburgh et al. 1992), New Britain flying-fox (*Pteropus gilliardorum*) (Mickleburgh et al. 1992), variable flying-fox (*Pteropus hypomelanus*) (Mickleburgh et al. 1992), great flying-fox (*Pteropus neohibernicus*) (Mickleburgh et al. 1992), Temminck's flying-fox (*Pteropus temmincki*) (Mickleburgh et al. 1992), rousette bat (*Rousettus amplexicaudatus*) (Mickleburgh et al. 1992), Bismarck blossom-bat (*Melonycteris melanops*) (Mickleburgh et al. 1992), common blossom-bat (*Syconycteris australis*) (Mickleburgh et al. 1992), Pacific baza (*Aviceda subcristata*), white-bellied sea-eagle (*Haliaeetus leucogaster*), variable goshawk (*Accipiter hiogaster*), Meyer's goshawk (*Accipiter meyerianus*), pied goshawk (*Accipiter albogularis*), black honey-buzzard (*Henicopernis infuscatus*) (Stattersfield et al. 1998), peregrine falcon (*Falco peregrinus*), Bismarck imperial-pigeon (*Ducula melanochroa*) (Stattersfield et al. 1998), Blyth's hornbill (*Aceros plicatus*) (Kemp 1995)

Top predator species: Pacific baza (*Aviceda subcristata*), white-bellied sea-eagle (*Haliaeetus leucogaster*), variable goshawk (*Accipiter hiogaster*), Meyer's goshawk (*Accipiter meyerianus*), pied goshawk (*Accipiter albogularis*) (Stattersfield et al. 1998), black honey-buzzard (*Henicopernis infuscatus*) (Stattersfield et al. 1998), peregrine falcon (*Falco peregrinus*)

Table J133.3. WCMC (1997) Protected Areas That Overlap with the Ecoregion.

Protected Area	Area (km²)	IUCN Category
Long Island	410	IV
Whiteman Mts. [134]	1,690	?
Total	2,100	

Ecoregion numbers of protected areas that overlap with additional ecoregions are listed in brackets.

Selected Conservation Contacts

Conservation International

East New Britain Sosal Eksen Komite

Mahonia na Dari, Research and Conservation Center, PNG

The Nature Conservancy (in New Britain)

Pacific Heritage Foundation

Justification for Ecoregion Delineation

We delineated two ecoregions to represent the montane and lowland evergreen moist forests in the New Britain and New Ireland island complex: the New Britain–New Ireland Lowland Rain Forests [133] and the New Britain–New Ireland Montane Rain Forests [134]. We used the 1,000-m contour of the DEM (USGS 1996) as the transition between lowland and montane ecoregions. We defined the Admiralty Islands Lowland Rain Forests [132] as a distinct ecoregion, following Stattersfield et al. (1998). MacKinnon (1997) combined these three ecoregions into a single subunit (P3p). Udvardy (1975) placed these ecoregions in the Papuan biogeographic province of the Oceanian Realm.

Prepared by John Lamoreux

Ecoregion Number:	**134**
Ecoregion Name:	**New Britain–New Ireland Montane Rain Forests**
Bioregion:	**New Guinea and Melanesia**
Biome:	**Tropical and Subtropical Moist Broadleaf Forests**
Political Units:	**Papua New Guinea**
Ecoregion Size:	**12,100 km²**
Biological Distinctiveness:	**Globally Outstanding**
Conservation Status:	**Critical**
Conservation Assessment:	**I**

Like the lowland rain forests, the montane forests of New Britain and New Ireland are rich in endemic species. However, unlike the lowlands, the karst topography of the montane forests is too steep for plantations. The montane forests therefore are relatively intact yet under increasing threat of being logged or degraded as a result of increasing populations.

Location and General Description

The narrow Vitiaz Strait separates the Huon Peninsula of northeastern New Guinea from the island chain known as the Bismarck Archipelago, which is dominated by two islands: New Britain and New Ireland (both exceed 400 km in length). The montane rain forests ecoregion includes the mountainous regions above 1,000 m of New Britain and New Ireland. New Britain and New Ireland are both long and narrow and contain several mountain ranges that trap rainfall. The climate of the ecoregion is tropical wet but varies in amount of average annual rainfall from about 3,000 to more than 6,000 mm.

Despite the close proximity of the Bismarck Archipelago to New Guinea and the existence of small islands that appear to be the remnants of a land bridge, the island archipelago was never connected to the mainland (Tyler 1999; Allison 1996). The islands breached the ocean surface in the late Miocene (8–10 million years ago) as the result of volcanic uplift, and many active volcanoes still exist (particularly on New Britain). Most of the islands are made up of both volcanic (acidic) soils and limestone. Limestone makes up 30 percent of New Britain and nearly 40 percent of New Ireland, or the entire northern half (Mueller-Dombois and Fosberg 1998). The mountain ranges on New Britain and New Ireland often are isolated from each other by lowlands. The key mountain ranges of New Britain are the Whiteman, Nakanai, Baining, and Willaumez ranges. Key ranges in New Ireland are the Hans Meyer, Verron, and Lelet ranges.

The boundary between lowland and montane vegetation is gradual and placed at different altitudes by various authors. The discrepancies arise because the authors look at combinations of changes in forest structure, species composition, and degree of cloud cover. The boundary for the ecoregion is 1,000 m, marking a point of transition where the height of the forest diminishes, tree leaves become smaller and thicker, and tree crowns become smaller (Mueller-Dombois and Fosberg 1998). Temperatures in the montane forest decrease with altitude, and humidity generally increases. Paijamans (1975) listed the occurrence of tree genera characteristic of lower montane forests, including *Araucaria*, *Lithocarpus*, *Castanopsis*, *Syzygium*, and *Ilex*. Beech (*Nothofagus*) and oak trees are found on mainland New Guinea and New Britain but are absent from New Ireland (Beehler and Alonso, in press). The high-elevation forests of New Ireland are dominated by *Metrosideros salomonensis* (Myrtaceae) 10–20 tall (Foster 2001).

Biodiversity Features

No comprehensive, modern botanical datasets exist for the Bismarcks, and much of the area is unknown in terms of biodiversity for any taxa (Keast 1996; Sekhran and Miller 1994). However, southern New Ireland has been surveyed botanically at two montane sites at 1,200 and 1,800 m (Takeuchi and Wiakabu 2001). Findings from these two sites suggest that the trees of New Ireland montane forests are less diverse than expected, but the diversity and abundance of epiphytes from 1,000 to 1,600 m are impressive (Takeuchi and Wiakabu 2001; Foster 2001). Johns (1993) listed several montane regions within New Britain and New Ireland as areas of high biological importance based on their flora: Lelet Plateau, southern Namatanai, Hans Meyer Range of New Ireland and Willaumez Peninsula, Whiteman Range, Nakanai Mountain, and Mts. Sinewit and Burringa of New Britain.

TABLE J134.2. Endemic and Near-Endemic Bird Species.

Family	Common Name	Species
Accipitridae	New Britain goshawk	*Accipiter princeps**
Megapodiidae	Melanesian scrubfowl	*Megapodius eremita*
Rallidae	New Britain rail	*Gallirallus insignis*
Columbidae	Yellow-legged pigeon	*Columba pallidiceps*
Columbidae	Yellow-bibbed fruit-dove	*Ptilinopus solomonensis*
Columbidae	Knob-billed fruit-dove	*Ptilinopus insolitus*
Columbidae	Red-knobbed imperial-pigeon	*Ducula rubricera*
Columbidae	Finsch's imperial-pigeon	*Ducula finschii*
Columbidae	Bismarck imperial-pigeon	*Ducula melanochroa*
Loriidae	White-naped lory	*Lorius albidinuchus*
Loriidae	Red-chinned lorikeet	*Charmosyna rubrigularis*
Psittacidae	Singing parrot	*Geoffroyus heteroclitus*
Cuculidae	Violaceous coucal	*Centropus violaceus*
Cuculidae	Pied coucal	*Centropus ateralbus*
Tytonidae	Bismarck owl	*Tyto aurantia*
Strigidae	Russet hawk-owl	*Ninox odiosa*
Apodidae	Mayr's swiftlet	*Aerodramus orientalis*
Turdidae	New Britain thrush	*Zoothera talaseae*
Sylviidae	Bismarck thicketbird	*Megalurulus grosvenori**
Monarchidae	Black-tailed monarch	*Monarcha verticalis*
Monarchidae	Dull flycatcher	*Myiagra hebetior*
Rhipiduridae	Bismarck fantail	*Rhipidura dahli*
Dicaeidae	Red-banded flowerpecker	*Dicaeum eximium*
Zosteropidae	Black-headed white-eye	*Zosterops hypoxanthus*
Meliphagidae	New Ireland myzomela	*Myzomela pulchella*
Meliphagidae	New Britain friarbird	*Philemon cockerelli*
Meliphagidae	New Ireland friarbird	*Philemon eichhorni*
Meliphagidae	Bismarck honeyeater	*Melidectes whitemanensis**
Estrildidae	Bismarck munia	*Lonchura melaena*
Dicruridae	Ribbon-tailed drongo	*Dicrurus megarhynchus*

TABLE J134.1. Endemic and Near-Endemic Mammal Species.

Family	Species
Pteropodidae	*Dobsonia anderseni*
Pteropodidae	*Dobsonia praedatrix*
Pteropodidae	*Melonycteris melanops*
Pteropodidae	*Nyctimene major*
Pteropodidae	*Pteropus admiralitatum*
Pteropodidae	*Pteropus gilliardorum*
Muridae	*Hydromys neobritannicus*
Muridae	*Uromys neobritannicus*

An asterisk signifies that the species' range is limited to this ecoregion.

An asterisk signifies that the species' range is limited to this ecoregion.

There are forty-five mammal species in the ecoregion. Most of these species are bats (thirty-six) in four families (Pteropodidae, Emballonuridae, Rhinolophidae, Vespertilionidae), followed by rodent species (Muridae). No mammal species is strictly endemic to the ecoregion, but eight are near endemics (table J134.1). Several species are listed as threatened (VU or higher) by IUCN (Hilton-Taylor 2000): New Guinea pademelon (*Thylogale brownii*), *Hipposideros demissus*, large-eared sheathtail-bat (*Emballonura dianae*), Bismarck trumpet-eared bat (*Kerivoula myrella*), New Britain water-rat (*Hydromys neobrittanicus*).

The ecoregion includes the highland portions of the New Britain and New Ireland EBA (Stattersfield et al. 1998). The New Britain–New Ireland Montane Rain Forests [134] contain thirty endemic and near-endemic bird species (table J134.2). One of these species is listed as endangered by IUCN (Hilton-Taylor 2000), yellow-legged pigeon (*Columba pallidiceps*). There is little doubt that the Bismarck Archipelago contains undescribed birds, especially at higher elevations in the Hans Meyer, Nakanai, Baining, and Whiteman ranges. All four ranges are listed by Beehler (1993) as biologically important areas for PNG.

Status and Threats

Current Status

The ecoregion is largely intact. Steep slopes and smaller trees probably are the reason these forests are in good shape. The montane forests of the Bismarcks are crucial for capturing fresh water from the clouds and supplying it to streams and thus for maintaining water resources for communities year-round (Foster 2001). The thick humus layer in montane forests acts as a sponge that collects moisture and releases it in a more uniform fashion than a bare hillside would so that even in dry periods water reaches the streams. Two protected areas overlap with the ecoregion (table J134.3).

Types and Severity of Threats

There is the threat of logging operations extending into the montane forests from the lowlands. Logging also leads to the proliferation of nonnative species, which often outcompete native species.

Priority Conservation Actions

In 1994 Conservation International sponsored a rapid assessment of southern New Ireland (Beehler and Alonso 2001). This study provides the best in-depth look at southern New Ireland and provides numerous conservation recommendations for New Ireland. Some of the recommendations that are relevant to the ecoregion as a whole are as follows:

Short-Term Conservation Actions (1–5 Years)

- Establish protected areas that include both montane and lowland forest.

Beehler and Alonso (2001) recommended establishing a core protected area in southern New Ireland. Several similar large tracts of forest remain in the mountain ranges throughout the ecoregion, and all should receive similar consideration. The value of the montane forests for protecting freshwater resources is immense.

Additional field studies are needed. Beehler and Alonso (in 2001) noted that the Verron Range and the summit of the Hans Meyer Range in New Ireland are still unsurveyed. However, no mountain range within the ecoregion is well known.

Longer-Term Conservation Actions (5–20 Years)

- Continue to promote a representative protected area system and work toward building linkages between key sites for conservation.
- Inform local communities of the value of montane forests for protecting biodiversity and regulating water flow.

Focal Species for Conservation Action

Threatened endemic species: New Guinea pademelon (*Thylogale brownii*), *Hipposideros demissus* (Mickleburgh et al. 1992), large-eared sheathtail-bat (*Emballonura dianae*), Bismarck trumpet-eared bat (*Kerivoula myrella*), New Britain water-rat (*Hydromys neobrittanicus*), yellow-legged pigeon (*Columba pallidiceps*) (Stattersfield et al. 1998), Bismarck owl (*Tyto aurantia*) (Stattersfield et al. 1998)

Area-sensitive species: Andersen's bare-backed fruit-bat (*Dobsonia anderseni*) (Mickleburgh et al. 1992), Bismarck bare-backed fruit-bat (*Dobsonia praedatrix*) (Mickleburgh et al. 1992), common tube-nosed bat (*Nyctimene albiventer*) (Mickleburgh et al. 1992), round-eared tube-nosed bat (*Nyctimene cyclotis*) (Mickleburgh et al. 1992), island tube-nosed bat (*Nyctimene major*) (Mickleburgh et al. 1992), northern blossom-bat (*Macroglossus minimus*) (Mickleburgh et al. 1992), Admiralty flying-fox (*Pteropus admiralitatum*) (Mickleburgh et al. 1992), New Britain flying-fox (*Pteropus gilliardorum*) (Mickleburgh et al. 1992), variable flying-fox (*Pteropus hypomelanus*) (Mickleburgh et al. 1992), great flying-fox (*Pteropus neohibernicus*) (Mickleburgh et al. 1992), Temminck's flying-fox (*Pteropus temmincki*) (Mickleburgh et al. 1992), rousette bat (*Rousettus amplexicaudatus*) (Mickleburgh et al. 1992), Bismarck blossom-bat (*Melonycteris melanops*) (Mickleburgh et al. 1992), common blossom-bat (*Syconycteris australis*) (Mickleburgh et al. 1992), Pacific baza (*Aviceda subcristata*), variable goshawk (*Accipiter hiogaster*), Meyer's goshawk (*Accipiter meyerianus*), New Britain goshawk (*Accipiter princeps*) (Stattersfield et al. 1998), peregrine falcon (*Falco peregrinus*), Bismarck imperial-pigeon (*Ducula melanochroa*) (Stattersfield et al. 1998), Blyth's hornbill (*Aceros plicatus*) (Kemp 1995)

TABLE J134.3. WCMC (1997) Protected Areas That Overlap with the Ecoregion.

Protected Area	Area (km²)	IUCN Category
Mt. Bamus	1,090	?
Whiteman Mts. [133]	440	?
Total	1,530	

Ecoregion numbers of protected areas that overlap with additional ecoregions are listed in brackets.

Top predator species: Pacific baza (*Aviceda subcristata*), variable goshawk (*Accipiter hiogaster*), Meyer's goshawk (*Accipiter meyerianus*), New Britain goshawk (*Accipiter princeps*) (Stattersfield et al. 1998), peregrine falcon (*Falco peregrinus*)

Selected Conservation Contacts

Conservation International

East New Britain Sosal Eksen Komite

Mahonia na Dari, Research and Conservation Center, PNG

The Nature Conservancy (in New Britain)

Pacific Heritage Foundation

Justification for Ecoregion Delineation

We delineated two ecoregions to represent the montane and lowland evergreen moist forests in the New Britain and New Ireland island complex: the New Britain–New Ireland Lowland Rain Forests [133] and the New Britain–New Ireland Montane Rain Forests [134]. The 1,000-m contour of the DEM (USGS 1996) was used as the transition between lowland and montane ecoregions. We placed the Admiralty Islands Lowland Rain Forests [132] into a distinct ecoregion, following Stattersfield et al. (1998). MacKinnon (1997) combined these three ecoregions into a single subunit (P3p). Udvardy (1975) placed these ecoregions in the Papuan biogeographic province of the Oceanian Realm.

Prepared by John Lamoreux

Ecoregion Number:	**135**
Ecoregion Name:	**Trobriand Islands Rain Forests**
Bioregion:	**New Guinea and Melanesia**
Biome:	**Tropical and Subtropical Moist Broadleaf Forests**
Political Units:	**Papua New Guinea**
Ecoregion Size:	**4,200 km²**
Biological Distinctiveness:	**Relatively Stable**
Conservation Status:	**Regionally Outstanding**
Conservation Assessment:	**III**

The islands of this ecoregion have been separated from mainland New Guinea since the late Pleistocene, and much of the biota is unique, including four mammal species and two birds-of-paradise. The main threats to the ecoregion include logging by foreign companies and conversion of habitat into agricultural lands.

Location and General Description

The ecoregion lies just off the southeastern tip of PNG in the southwest Pacific and includes Woodlark Island and two island groups: the D'Entrecasteaux and the Trobriand. The largest portion of the ecoregion and the nearest to the New Guinea mainland is made up of three principal islands of the D'Entrecasteaux group: Goodenough, Fergusson, and Normanby. Goodenough is the highest island in the ecoregion, reaching an altitude of 2,750 m, and Fergusson is the largest. Kaileuna and Kiriwina islands in the Trobriand group lie further out in the Pacific, and Woodlark Island is even further to the southwest. The ecoregion is warm and moist tropical, typical of most of New Guinea.

According to Whitmore (1984), Woodlark and the Trobriand Islands consist primarily of lowland rain forest on limestone substrates. Goodenough, Fergusson, and Normanby islands consist mainly of lowland rain forest on acid soil; however, Normanby has one area of ultrabasic soils. The major rain forest tree genera include *Pometia, Octomeles, Alstonia, Campnosperma, Canarium* (Burseraceae), *Dracontomelon* (Anacardiaceae), *Pterocymbium* (Sterculiaceae), *Crytocarya* (Lauraceae), *Intsia, Ficus,* and *Terminalia* (Mueller-Dombois and Fosberg 1998). The emergent the hoop pine (*A. cunninghami*) is present in the ecoregion on Fergusson (Mueller-Dombois and Fosberg 1998). The northwest portions of Goodenough and Fergusson islands are anthropogenic grasslands and agricultural lands.

Biodiversity Features

The ecoregion has many endemic species, but more certainly await discovery. The D'Entrecasteaux Islands in particular are thought to contain numerous endemic plant species (Johns 1993). The ultrabasic soils of Normanby have turned up exciting new finds, including two endemic ant plant species (Johns 1993). The forested mountains of Fergusson Island have never been surveyed biologically and are considered one of PNG's great biological unknowns (Swartzendruber 1993). Beehler (1993) emphasized this point, saying that the massif on Fergusson "promises to be a treasure trove to the first vertebrate biologists who climb its heights," and "one can only imagine what undescribed populations of vertebrates inhabit the wet montane forests above 1500 meters there." Goodenough Island contains an endemic bat (*Dobsonia pan-*

TABLE J135.1. Endemic and Near-Endemic Mammal Species.

Family	Species
Perorictidae	*Echymipera davidi**
Phalangeridae	*Phalanger lullulae**
Macropodidae	*Dorcopsis atrata**
Petauridae	*Dactylopsila tatei**
Pteropodidae	*Nyctimene major*
Pteropodidae	*Pteropus pannietensis*
Vespertilionidae	*Kerivoula agnella*
Vespertilionidae	*Pipistrellus collinus*
Muridae	*Chiruromys forbesi*

An asterisk signifies that the species' range is limited to this ecoregion.

TABLE J135.2. Endemic and Near-Endemic Bird Species.

Family	Common Name	Species
Paradisaeidae	Curl-crested manucode	*Manucodia comrii**
Paradisaeidae	Goldie's bird-of-paradise	*Paradisaea decora**

An asterisk signifies that the species' range is limited to this ecoregion.

nietensis) and an endemic forest wallaby, the black dorcopsis (*Dorcopsis atrata*). Woodlark Island is also considered to be very unusual botanically and contains an endemic cuscus (*Phalanger lullulae*).

There are thirty-eight mammal species in the Trobriand Islands Rain Forests [135]. Most of these species are bats (twenty-four) in four families (Pteropodidae, Molossidae, Rhinolophidae, and Vespertilionidae), followed by rodent species (Muridae). Four species are listed as threatened (VU or higher) by IUCN (Hilton-Taylor 2000): black dorcopsis (*Dorcopsis atrata*), St. Aignan's trumpet-eared bat (*Kerivoula agnella*), Tate's triok (*Dactylopsila tatei*), and *Hipposideros demissus*. Four mammal species are limited to the ecoregion, and five more are near endemics (table J135.1).

Most of the ecoregion, with the notable exception of Woodlark Island, is part of the D'Entrecasteaux and Trobriand Islands EBA (Stattersfield et al. 1998). Two bird species are endemic to the ecoregion (table J135.2). One of these birds, the curl-crested manucode, is found on all three major islands of the D'Entrecasteaux group and in the Trobriand Islands. The other, Goldie's bird-of-paradise, is found only on Fergusson and Normandy.

Status and Threats

Current Status

As mentioned earlier, the northwest portions of Goodenough and Fergusson islands are anthropogenic grasslands and agricultural lands. A scheme for a possible large-scale agricultural project on Normanby Island is mentioned by Stattersfield et al. (1998), but its status is unknown. The only protected area found in these islands is Lake Lavu (table J135.3).

Types and Severity of Threats

The main threats to the biodiversity of the ecoregion are logging and conversion of forest to agriculture. However, it is hard to obtain any detailed information on the current or future threats to the islands.

Priority Conservation Actions

Short-Term Conservation Actions (1–5 Years)

- Establish additional protected areas.
- Conduct biological surveys.

Longer-Term Conservation Actions (5–20 Years)

- Increase the protected area system and develop effective management.

TABLE J135.3. WCMC (1997) Protected Areas That Overlap with the Ecoregion.

Protected Area	Area (km²)	IUCN Category
Lake Lavu	40	VIII

Ecoregion numbers of protected areas that overlap with additional ecoregions are listed in brackets.

Focal Species for Conservation Action

Threatened endemic species: black dorcopsis (*Dorcopsis atrata*), St. Aignan's trumpet-eared bat (*Kerivoula agnella*), *Dactylopsila tatei*

Area-sensitive species: variable flying-fox (*Pteropus hypomelanus*) (Mickleburgh et al. 1992), *Pteropus pannietensis* (Mickleburgh et al. 1992), rousette bat (*Rousettus amplexicaudatus*) (Mickleburgh et al. 1992), northern blossom-bat (*Macroglossus minimus*) (Mickleburgh et al. 1992), common blossom-bat (*Syconycteris australis*) (Mickleburgh et al. 1992), island tube-nosed bat (*Nyctimene major*) (Mickleburgh et al. 1992), De Vis's bare-backed fruit-bat (*Dobsonia pannietensis*) (Mickleburgh et al. 1992), Pacific baza (*Aviceda subcristata*), white-bellied sea-eagle (*Haliaeetus leucogaster*), Gurney's eagle (*Aquila gurneyi*), variable goshawk (*Accipiter hiogaster*), grey-headed goshawk (*Accipiter poliocephalus*), peregrine falcon (*Falco peregrinus*)

Top predator species: Pacific baza (*Aviceda subcristata*), white-bellied sea-eagle (*Haliaeetus leucogaster*), Gurney's eagle (*Aquila gurneyi*), variable goshawk (*Accipiter hiogaster*), grey-headed goshawk (*Accipiter poliocephalus*), peregrine falcon (*Falco peregrinus*)

Selected Conservation Contacts

Conservation International

Justification for Ecoregion Delineation

The Trobriand Islands Rain Forests [135] and Louisiade Archipelago Rain Forests [136] were made distinct ecoregions based on Stattersfield et al. (1998). MacKinnon (1997) placed these two ecoregions together into subunit P3o. Udvardy (1975) placed these ecoregions in the Papuan biogeographic province of the Oceanian Realm.

Prepared by John Lamoreux

Ecoregion Number:	**136**
Ecoregion Name:	**Louisiade Archipelago Rain Forests**
Bioregion:	**New Guinea and Melanesia**
Biome:	**Tropical and Subtropical Moist Broadleaf Forests**
Political Units:	**Papua New Guinea**
Ecoregion Size:	**1,600 km²**
Biological Distinctiveness:	**Bioregionally Outstanding**
Conservation Status:	**Vulnerable**
Conservation Assessment:	**V**

The Louisiade Archipelago Rain Forests [136] contain many endemic species of plants and animals, particularly birds, that help define it as a distinct unit. The main threats to the islands are logging, conversion of habitat into agricultural lands, and gold mining.

Location and General Description

The ecoregion includes a group island chain that lies off the southeastern tip of PNG. The first islands moving eastward in the chain are close to the mainland and include Sideia and Basilaki islands. Further east are the major islands in the archipelago: Misima, Sudest (or Tagula), and Rossel. Sudest Island is the largest (800 km^2). All of the major islands of the Louisiades are volcanic, although there are numerous smaller islands that are coral formations.

The climate of the Louisiade Archipelago is moist tropical, and the vegetation consists of rain forest, although some of the low-lying smaller islands receive less rainfall. Paijmans (1975) categorized most of the ecoregion as small crowned lowland hill forest. This type of forest is shorter (20–30 m in height) than that found in other areas in New Guinea because of either poor soil conditions or less rainfall. The former is probably the case, as Johns (1993) remarked on the very poor soils of Rossel and Misima. Tree genera in pure forest stands include *Casuarina*, *Castanopsis*, and *Hopea*. Paijames (1975) listed mixed forest genera for low rainfall areas as *Pometia*, *Canarium*, *Anisoptera*, *Cryptocarya*, *Terminalia*, *Syzygium*, *Ficus*, and *Celtis*.

Biodiversity Features

The ecoregion is distinctive botanically. Johns (1993) noted that the Louisiade Archipelago has long been known as an area with many endemic species. Misima stands out for having several ant plant species (Rubiaceae) and an endemic *Pandanus*. Rossel contains an undescribed *Diospyros* species and several *Hopea* species. Perhaps most impressive is that Rossel also contains an undescribed genus in the family Burseraceae. The archipelago is also noteworthy for its endemic reptiles and amphibians. Allison (1993) listed eight total endemics, including five frogs and two lizards.

There are twenty-four mammal species in the Louisiade Archipelago rain forests. Most of these species are bats (eighteen) in three families (Pteropodidae, Rhinolophidae, and Vespertilionidae), followed by rodent species (Muridae). No mammal species are endemic to the Louisiade Archipelago, but three species are near endemics (table J136.1). St. Aignan's trumpet-eared bat (*Kerivoula agnella*) and *Hipposideros demissus* are the only threatened mammals in the ecoregion (Hilton-Taylor 2000).

For the most part, the ecoregion matches the Louisiade Archipelago EBA (Stattersfield et al. 1998). There are five endemic bird species in the ecoregion and two near endemics (table J136.2). Stattersfield et al. (1998) stated that very little is known about any of these birds, and four of the seven species are listed as data deficient (DD) by IUCN (1996), meaning that a category of threat cannot be determined.

TABLE J136.1. Endemic and Near-Endemic Mammal Species.

Family	Species
Pteropodidae	*Nyctimene major*
Pteropodidae	*Pteropus pannietensis*
Vespertilionidae	*Kerivoula agnella*

An asterisk signifies that the species' range is limited to this ecoregion.

TABLE J136.2. Endemic and Near-Endemic Bird Species.

Family	Common Name	Species
Meliphagidae	White-chinned myzomela	*Myzomela albigula**
Meliphagidae	Tagula honeyeater	*Meliphaga vicina**
Pachycephalida	White-bellied whistler	*Pachycephala leucogastra*
Cracticidae	Tagula butcherbird	*Cracticus louisiadensis**
Zosteropidae	White-throated white-eye	*Zosterops meeki**
Zosteropidae	Louisiade white-eye	*Zosterops griseotinctus*
Dicaeidae	Louisiade flowerpecker	*Dicaeum nitidum**

An asterisk signifies that the species' range is limited to this ecoregion.

Status and Threats

Current Status

Vegetation maps show that as early as a quarter century ago the habitats of Deboyne, northwest Sudest Island, and nearby Pana Tinani were all degraded as either anthropogenic grasslands or agricultural lands (Paijames 1975). Allison (1993) noted that much of the forest is gone from Sudest and that the forests around Mt. Riu must be protected if many of the endemic species of the Louisiades are to remain. Beehler (1993) also stated that the forests from Mt. Riu eastward are very important to the survival of the Tagula honeyeater and the Tagula butcherbird. Gold mining has been very destructive on Misima (Johns 1993). There are no protected areas in the ecoregion.

Types and Severity of Threats

The main threats to the ecoregion are logging, conversion of habitat into agricultural lands, and gold mining.

Priority Conservation Actions

Short-Term Conservation Actions (1–5 Years)

- Establish biological surveys of the region.
- Establish protected areas in the ecoregion, especially around the Mt. Riu region.

Longer-Term Conservation Actions (5–20 Years)

- Develop a protected area system on the islands with effective management plans.

Focal Species for Conservation Action

Endangered endemic species: St. Aignan's trumpet-eared bat (*Kerivoula agnella*)

Area-sensitive species: variable flying-fox (*Pteropus hypomelanus*) (Mickleburgh et al. 1992), *Pteropus pannietensis* (Mickleburgh et al. 1992), rousette bat (*Rousettus amplexicaudatus*) (Mickleburgh et al. 1992), northern blossom-bat (*Macroglossus minimus*) (Mickleburgh et al. 1992), common blossom-bat (*Syconycteris australis*) (Mickleburgh et al. 1992), island tube-nosed bat (*Nyctimene major*) (Mickleburgh et al. 1992), De Vis's bare-backed fruit-bat (*Dobsonia pannietensis*) (Mickleburgh et al. 1992), white-bellied sea-eagle

(*Haliaeetus leucogaster*), variable goshawk (*Accipiter hiogaster*), grey-headed goshawk (*Accipiter poliocephalus*)

Top predator species: white-bellied sea-eagle (*Haliaeetus leucogaster*), variable goshawk (*Accipiter hiogaster*), grey-headed goshawk (*Accipiter poliocephalus*)

Selected Conservation Contacts

Conservation International

Justification for Ecoregion Delineation

The Trobriand Islands Rain Forests [135] and Louisiade Archipelago Rain Forests [136] were made distinct ecoregions based on Stattersfield et al. (1998). MacKinnon (1997) placed these two ecoregions together into subunit P3o. Udvardy (1975) placed these ecoregions in the Papuan biogeographic province of the Oceanian Realm.

Prepared by John Lamoreux

Ecoregion Number:	**137**
Ecoregion Name:	**Solomon Islands Rain Forests**
Bioregion:	**New Guinea and Melanesia**
Biome:	**Tropical and Subtropical Moist Broadleaf Forests**
Political Units:	**Papua New Guinea, Solomon Islands**
Ecoregion Size:	**35,700 km²**
Biological Distinctiveness:	**Globally Outstanding**
Conservation Status:	**Vulnerable**
Conservation Assessment:	**I**

The Solomon Islands Rain Forests [137] are true oceanic islands with high vertebrate endemism, including single-island endemics, restricted-range mammals, and an astounding sixty-nine bird species found nowhere else in the world. Large areas of naturally restricted lowlands below 400 m either have been or are under threat of logging or clearance for subsistence agriculture. Introduced cats have eliminated most native mammals on Guadalcanal.

Location and General Description

This ecoregion consists of tropical lowland and montane forests on Bougainville and Buka Islands in PNG and most of the island nation of the Solomon Islands (not including the Santa Cruz Group). The climate of the Solomon Island is tropical wet (National Geographic Society 1999). The islands are predominantly hill forest, although only small portions of a few of the islands extend beyond 1,000 m in elevation. The mountains on Guadalcanal reach past 2,000 m. The Solomons are the result of the subduction of the Australian tectonic plate beneath the Pacific tectonic plate, and the islands are a very active tectonic area. The surface geology of the islands consists predominantly of volcanic rocks, with some metamorphic rocks, uplifted coral islands (Rennell, Bellona, and Ontong Java), and recent (Pliocene to recent) alluvium in the lowlands. The islands increase in age from northwest to southeast (Mueller-Dombois and Fosberg 1998).

Mueller-Dombois and Fosberg (1998) outlined seven broad natural vegetation types in the ecoregion, including coastal strand vegetation, mangrove forests, freshwater swamp forests, two types of lowland rain forests, seasonally dry forest and grassland (only on Guadalcanal), and montane rain forest. Bougainville also contains floodplain forest, a transitional submontane rain forest, forest on ancient limestone, and vegetation on recent volcanic surfaces.

Coastal strand vegetation consists of mixed *Spinifex-Canavalia* containing *Ipmoea*, *Spinifex*, *Canavalia*, *Thuarea*, *Cyperus*, *Scaevola*, *Hibiscus*, *Pandanus*, *Tournefortia*, *Cerbera*, *Calophyllum*, *Barringtonia*, *Terminalia*, and *Casuarina*. Two types of mangrove vegetation are identified for the Solomons: a low forest dominated by *Rhizophora apiculata* and a tall forest dominated by *Rhizophora* spp. and *Brugiera* spp. *Brugiera sexangula*, *B. parviflora*, and *Ceriops tagal* reach their eastern limits in the Solomons. Most of the islands have large areas of freshwater swamp forest. Easily recognized subunits of freshwater swamp forest include *Campnosperma brevipetiolata* forests, closed-canopy *Terminalia brassi* forests, sago swamp (*Metroxlyon solomonense*), low-canopy *Pandanus* spp., and mixed swamp forest (Mueller-Dombois and Fosberg 1998).

The most widespread vegetation type is lowland rain forest. The canopy is uneven as a result of frequent natural disturbance (tropical storms, landslips, treefalls). The twelve most common tree species are *Calophylum kajewskii*, *C. vitiense*, *Dillenia salomonensis*, *Elaeocarpus sphaericus*, *Endospermum medullosum*, *Parinari salomonensis*, *Maranthes corymbosa*, *Pometia pinnata*, *Gmelina mollucana*, *Schizomeria serrata*, *Terminalia calamansanai*, and *Campnosperma brevipetiolata*. Whitmore (1974) recognized six lowland rain forest types, distinguished by whether the forest was on the northern or western side of the islands, elevation, and level of disturbance (Whitmore 1974; Mueller-Dombois and Fosberg 1998). Two groups of low-diversity lowland rain forest are recognized. The first group consists of monodominant forests of *Campnosperma brevipetiolata* (Santa Isabel, New Georgia, Choisel) or those with co-dominant *C. brevipetiolata* and *Dillenia* or *C. brevipetiolata*, *Pometia pinnata*, and *Teysmanniodendron* (Verbenaceae). It is thought that these forest types are related to disturbance. The second group of low-diversity forests is associated unusual soils, including limestone (*Vitex cofassus* and *Pometia pinnata*), flooding (*Pterocarpus indicus* and *Terminalia brassi*), or ultramafic soils (*Casuarina papuana*, *Dillenia crenata*, *Syzygium*, or *Dacrydium*) (Mueller-Dombois and Fosberg 1998).

Seasonally dry forest is found only on the leeward (north) side of Guadalcanal. These forests consist of mixed deciduous forest and *Themeda australis* grassland. The canopy is composed of *Pometia pinnata*, *Vitex cofassus*, and *Kleinhovia hospita*. The deciduous species are *Pterocarpus indicus*, *Antiais toxicaria* (Moraceae), *Ficus* spp., and *Sterculia* spp. The grassland probably is related to periodic burning by humans (Mueller-Dombois and Fosberg 1998).

The Fagaceae species that generally mark montane rain forest in the region (*Castanopsis*, *Nothofagus*, and *Lithocarpus*) are absent in the Solomons. Instead, a reduction in stature (from 25 to 35 m in the lowlands to 15 to 20 m in the uplands) is apparent. *Syzygium*, *Metrosideros*, *Ardisia*,

TABLE J137.1. Endemic and Near-Endemic Mammal Species.

Family	Species	Family	Species
Pteropodidae	Melonycteris fardoulisi*	Pteropodidae	Pteropus woodfordi*
Pteropodidae	Melonycteris woodfordi*	Rhinolophidae	Anthops ornatus*
Pteropodidae	Dobsonia inermis*	Molossidae	Chaerephon solomonis*
Pteropodidae	Nyctimene vizcaccia*	Muridae	Melomys bougainville*
Pteropodidae	Nyctimene major	Muridae	Melomys spechti*
Pteropodidae	Pteralopex anceps*	Muridae	Solomys ponceleti*
Pteropodidae	Pteralopex atrata*	Muridae	Solomys salamonis*
Pteropodidae	Pteralopex pulchra*	Muridae	Solomys salebrosus*
Pteropodidae	Pteralopex sp.*	Muridae	Solomys sapientis*
Pteropodidae	Pteropus admiralitatum	Muridae	Uromys imperator*
Pteropodidae	Pteropus howensis*	Muridae	Uromys porculus*
Pteropodidae	Pteropus mahaganus*	Muridae	Uromys rex*
Pteropodidae	Pteropus rayneri*		
Pteropodidae	Pteropus rennelli*		

An asterisk signifies that the species' range is limited to this ecoregion.

TABLE J137.2. Endemic and Near-Endemic Bird Species.

Family	Common Name	Species	Family	Common Name	Species
Accipitridae	Imitator sparrowhawk	Accipiter imitator*	Psittacidae	Singing parrot	Geoffroyus heteroclitus
Accipitridae	Solomon sea-eagle	Haliaeetus sanfordi*	Psittacidae	Finsch's pygmy-parrot	Micropsitta finschii
Accipitridae	Pied goshawk	Accipiter albogularis	Cuculidae	Buff-headed coucal	Centropus milo*
Megapodiidae	Melanesian scrubfowl	Megapodius eremita	Strigidae	Solomon hawk-owl	Ninox jacquinoti*
Rallidae	Woodford's rail	Nesoclopeus woodfordi*	Strigidae	Fearful owl	Nesasio solomonensis*
Rallidae	Roviana rail	Gallirallus rovianae*	Apodidae	Mayr's swiftlet	Aerodramus orientalis
Rallidae	San Cristobal moorhen	Gallinula silvestris*	Alcedinidae	Ultramarine kingfisher	Todirhamphus leucopygius*
Columbidae	Yellow-bibbed fruit-dove	Ptilinopus solomonensis	Alcedinidae	Moustached kingfisher	Actenoides bougainvillei*
Columbidae	Yellow-legged pigeon	Columba pallidiceps	Pittidae	Black-faced pitta	Pitta anerythra*
Columbidae	Red-knobbed imperial-pigeon	Ducula rubricera	Meliphagidae	Cardinal myzomela	Myzomela cardinalis
Columbidae	Crested cuckoo-dove	Reinwardtoena crassirostris*	Meliphagidae	Bougainville honeyeater	Stresemannia bougainvillei*
Columbidae	Thick-billed ground-dove	Gallicolumba salamonis*	Meliphagidae	Scarlet-naped myzomela	Myzomela lafargei*
Columbidae	Choiseul pigeon	Microgoura meeki*	Meliphagidae	Yellow-vented myzomela	Myzomela eichhorni*
Columbidae	Silver-capped fruit-dove	Ptilinopus richardsii*	Meliphagidae	Red-bellied myzomela	Myzomela malaitae*
Columbidae	White-headed fruit-dove	Ptilinopus eugeniae*	Meliphagidae	Black-headed myzomela	Myzomela melanocephala*
Columbidae	Chestnut-bellied imperial-pigeon	Ducula brenchleyi*	Meliphagidae	Sooty myzomela	Myzomela tristrami*
Columbidae	Pale mountain-pigeon	Gymnophaps solomonensis*	Meliphagidae	Guadalcanal honeyeater	Guadalcanaria inexpectata*
Cacatuidae	Ducorps's cockatoo	Cacatua ducorpsii*	Meliphagidae	San Cristobal honeyeater	Melidectes sclateri*
Loriidae	Cardinal lory	Chalcopsitta cardinalis	Pachycephalida	Mountain whistler	Pachycephala implicata*
Loriidae	Yellow-bibbed lory	Lorius chlorocercus*	Rhipiduridae	White-winged fantail	Rhipidura cockerelli*
Loriidae	Meek's lorikeet	Charmosyna meeki*	Rhipiduridae	Brown fantail	Rhipidura drownei*
Loriidae	Duchess lorikeet	Charmosyna margarethae*	Rhipiduridae	Dusky fantail	Rhipidura tenebrosa*
			Rhipiduridae	Rennell fantail	Rhipidura rennelliana*

Psychotria, Schefflera, Ficus, Rhododendron, Dacrydium, and *Podcarpus pilgeri* have been collected in the mountains of the Solomons (Mueller-Dombois and Fosberg 1998).

The outlying coral atolls support depleted lowland rain forest, remnant coastal and swamp vegetation, and *Pandanus* thickets. The vegetation is a product of generally poor soils combined with human alteration (Mueller-Dombois and Fosberg 1998).

Biodiversity Features

Overall richness and endemism in the Solomon Islands range from low to high when compared with those of other ecoregions in Indo-Malaysia. Bird and mammal endemism are high.

There is a clear difference between the mammalian faunas of the Solomon Islands and the Bismarck Archipelago and richer New Guinea to the west. Except for pteropodid bats, the Solomons and Bismarcks have many fewer mammals than New Guinea, and the Solomons, unlike New Britain, contain no marsupials. East beyond the Solomons there are even fewer mammal species. Almost all the mammal species have their origins in or via New Guinea (Flannery 1990).

Although the Solomon Islands contain only forty-seven mammal species, a remarkable twenty-six of those species are endemic or near endemic, including nine murid rodents (*Melomys, Solomys, Uromys*), fifteen pteropodid bats (*Dobsonia, Melonycteris, Nyctimene, Pteralopex, Pteropus*), a horseshoe bat (*Anthops*), and one molossid bat (*Chaerephon*) (Flannery 1995) (table J137.1). Three of the fruit bats—Bougainville monkey-faced bat (*Pteralopex ancep*), Guadalcanal monkey-faced bat (*Pteralopex atrata*), and montane monkey-faced bat (*Pteralopex pulchra*)—are critically endangered, and three of the rodents—Specht's mosaic-tailed rat (*Melomys spechti*), Poncelet's giant rat (*Solomys ponceleti*), and emperor rat (*Uromys imperator*)—are endangered (Hilton-Taylor 2000; Wilson and Cole 2000;

Family	Common Name	Species	Family	Common Name	Species
Rhipiduridae	Malaita fantail	*Rhipidura malaitae**	Sturnidae	White-eyed starling	*Aplonis brunneicapilla**
Monarchidae	Rennell shrikebill	*Clytorhynchus hamlini**	Zosteropidae	Louisiade white-eye	*Zosterops griseotinctus*
Monarchidae	Bougainville monarch	*Monarcha erythrostictus**	Zosteropidae	Rennell white-eye	*Zosterops rennellianus**
Monarchidae	Chestnut-bellied monarch	*Monarcha castaneiventris**	Zosteropidae	Banded white-eye	*Zosterops vellalavella**
Monarchidae	White-capped monarch	*Monarcha richardsii**	Zosteropidae	Ganongga white-eye	*Zosterops splendidus**
Monarchidae	Black-and-white monarch	*Monarcha barbatus**	Zosteropidae	Splendid white-eye	*Zosterops luteirostris**
Monarchidae	Kulambangra monarch	*Monarcha browni**	Zosteropidae	Solomon Islands white-eye	*Zosterops kulambangrae**
Monarchidae	White-collared monarch	*Monarcha viduus**	Zosteropidae	Kulambangra white-eye	*Zosterops murphyi**
Monarchidae	New Caledonian flycatcher	*Myiagra caledonica*	Zosteropidae	Yellow-throated white-eye	*Zosterops metcalfii**
Monarchidae	Steel-blue flycatcher	*Myiagra ferrocyanea**	Zosteropidae	Grey-throated white-eye	*Zosterops rendovae**
Monarchidae	Ochre-headed flycatcher	*Myiagra cervinicauda**	Zosteropidae	Malaita white-eye	*Zosterops stresemanni**
Dicruridae	Solomon Islands drongo	*Dicrurus solomenensis**	Zosteropidae	Bare-eyed white-eye	*Woodfordia superciliosa**
Corvidae	White-billed crow	*Corvus woodfordi**	Sylviidae	Shade warbler	*Cettia parens**
Corvidae	Bougainville crow	*Corvus meeki**	Sylviidae	San Cristobal leaf-warbler	*Phylloscopus makirensis**
Campephagidae	Melanesian cuckoo-shrike	*Coracina caledonica*	Sylviidae	Kulambangra leaf-warbler	*Megalurulus X amoenus**
Campephagidae	Long-tailed triller	*Lalage leucopyga**	Sylviidae	Guadalcanal thicketbird	*Megalurulus whitneyi*
Campephagidae	Solomon cuckoo-shrike	*Coracina holopolia*	Sylviidae	Bougainville thicketbird	*Megalurulus llaneae**
Turdidae	New Britain thrush	*Zoothera talaseae*	Dicaeidae	Midget flowerpecker	*Dicaeum aeneum**
Turdidae	Olive-tailed thrush	*Zoothera lunulata**	Dicaeidae	Mottled flowerpecker	*Dicaeum tristrami**
Turdidae	San Cristobal thrush	*Zoothera margaretae**	Estrildidae	Bismarck munia	*Lonchura melaena*
Sturnidae	Rennell starling	*Aplonis insularis**	Acanthizidae	Fan-tailed gerygone	*Gerygone flavolateralis*
Sturnidae	Atoll starling	*Aplonis feadensis*			
Sturnidae	Brown-winged starling	*Aplonis grandis**			
Sturnidae	San Cristobal starling	*Aplonis dichroa**			

An asterisk signifies that the species' range is limited to this ecoregion.

Flannery 1995).

Bird diversity drops off sharply from New Guinea as one moves east across the Pacific to the Solomons. Whereas New Guinea has seventy-one families and subfamilies of birds, the Solomons have forty-four. The Solomons are considered a center of bird endemism, with at least seven endemic genera. The dropoff in diversity seen in other animal groups as one moves east from New Guinea is also consistent with that seen in birds. Whereas New Guinea has seventy-one families and subfamilies of birds, and the Solomons have forty-four, Vanuatu has thirty-one (Keast 1996).

A total of 199 bird species inhabit the Solomons (Doughty et al. 1999). The Solomon Islands ecoregion has an almost exact correspondence with the Solomon group EBA (Stattersfield et al. 1998). This EBA contains more restricted-range bird species (seventy-eight) than any other EBA. Several of the islands, especially Makira (San Cristobal) and the New Georgia group, have their own endemic species and would qualify as important EBAs by themselves. The unique islands of Rennell and Bellona, separated from the rest of the Solomons by a submarine trench, are also an EBA, containing a total of twelve endemic species, and seven additional species. The atolls of Ontong Java, which are also part of this ecoregion, qualify as a Secondary Area because they provide habitat for an additional species, the atoll starling (*Aplonis feadensis*), which is also found off of small islands in the Bismarck Archipelago. Of these ninety-one restricted-range bird species, an incredible sixty-nine species are found nowhere else in the world; thus the Solomons are a global priority for bird conservation. Ninety species are endemic or near endemic (table J137.2). Three bird species are critically endangered: Makira moorhen (*Gallinula silvestris*), yellow-legged pigeon (*Columba pallidiceps*), and thick-billed ground-dove (*Gallicolumba salamonis*). Four additional bird species are endangered: imitator sparrowhawk (*Accipiter imitator*), Woodford's rail (*Nesoclopeus woodfordi*), chestnut-bellied imperial pigeon (*Ducula rubricera*), and white-eyed starling (*Aplonis brunneicapilla*) (Stattersfield et al. 1998; Hilton-Taylor 2000). The Choisel pigeon (*Microgoura meeki*) was last reliably seen in 1904 and is presumed extinct (Stattersfield et al. 1998).

Buka, Bougainville, and the rest of the Solomon Islands (excluding the Santa Cruz Group) form a distinct and rather uniform phytogeographic unit. About a third of the Solomon Islands' flora is of Malesian (southeast Asian) origin, a third has Paleotropical origins, and a third is cosmopolitan, with a small Pacific contribution. There is a distinct break in floristic compositions with the nearby Bismarck Archipelago to the west, corresponding with the New Britain Trench that separates the submarine platforms. Two important Indo-Malayan tree families, Fagacae and Dipterocarpaceae, are not present in the Solomons. There are only about a dozen common tree species in the Solomons (Mueller-Dombois and Fosberg 1998).

Between November and April of each year the Solomon Islands are subject to tropical cyclones, which are an important source of natural disturbance to the islands' forests. Extreme droughts are also a natural event and occur irregularly at intervals of six to twenty years (Mueller-Dombois and Fosberg 1998).

Davis et al. (1995) identified two Centres of Plant Diversity on Bougainville Island: Mt. Balbi to southern coast, containing the largest stands of bamboo forest in Papuasia and remnant stands of *Terminalia brassii*, and Mt. Takuan–Tonolei Harbour, containing natural stands of *Terminalia brassii* and more than 1,000 vascular plant species.

Status and Threats

Current Status

A large Australian-run copper mine was located in Bougainville, but it was shut down because of civil unrest several years ago. Introduced species are a special concern here, and most native mammals have been eliminated from Guadalcanal by cats. Hunting native species is common (Stattersfield et al. 1998,). Many bird species in the Solomons are vulnerable simply because of their small natural ranges (Stattersfield et al. 1998).

Only one protected area, 930 km^2 surrounding Mt. Balbi on Bougainville, exists in the ecoregion (table J137.3). A gap analysis, based on detailed vegetation and habitat type mapping, has never been performed to determine whether the existing protected area network adequately covers all habitats with protected areas that are large enough to maintain all critical ecological processes.

Types and Severity of Threats

Large areas of the naturally limited natural forest below 400 m have been logged or are planned to be logged. An adequate survey of timber resources has not been conducted (Stattersfield et al. 1998; Thistlethwait and Votaw 1992).

Forest clearing for subsistence agriculture is an ongoing threat. Most households are self-sufficient (seven out of eight), and because population growth is high there is pressure to clear land. This is especially true around urban areas because the population is mobile and many people move to the outskirts of overcrowded urban centers. Satellite imagery indicates that the area under cultivation doubled between 1972 and 1992 (Thistlethwait and Votaw 1992).

Priority Conservation Actions

Short-Term Conservation Actions (1–5 Years)

- Identify hotspot regions of endemicity or other biological priority sites.
- Increase the terrestrial protected area system based on priority-setting exercises.
- Stop logging of lowland forests.
- Limit the spread of subsistence agriculture land uses.
- Educate the public about the ecoregion's tremendous endemic flora and fauna.

TABLE J137.3. WCMC (1997) Protected Areas That Overlap with the Ecoregion.

Protected Area	Area (km^2)	IUCN Category
Mt. Balbi	930	?

Ecoregion numbers of protected areas that overlap with additional ecoregions are listed in brackets.

Longer-Term Conservation Actions (5–20 Years)

- Develop and implement effective management plans for protected areas.

Focal Species for Conservation Action

Endangered endemic species: Bougainville monkey-faced bat (*Pteralopex ancep*) (Mickleburgh et al. 1992; Flannery 1995), Guadalcanal monkey-faced bat (*Pteralopex atrata*) (Mickleburgh et al. 1992; Flannery 1995), montane monkey-faced bat (*Pteralopex pulchra*) (Mickleburgh et al. 1992; Flannery 1995), Specht's mosaic-tailed rat (*Melomys spechti*) (Flannery 1995), Poncelet's giant rat (*Solomys poncelet*) (Flannery 1995), emperor rat (*Uromys imperator*) (Flannery 1995), Makira moorhen (*Gallinula silvestris*) (Stattersfield et al. 1998), yellow-legged pigeon (*Columba pallidiceps*) (Stattersfield et al. 1998), thick-billed ground-dove (*Gallicolumba salamonis*) (Stattersfield et al. 1998), imitator sparrowhawk (*Accipiter imitator*) (Stattersfield et al. 1998), Woodford's rail (*Nesoclopeus woodfordi*) (Stattersfield et al. 1998), chestnut-bellied imperial pigeon (*Ducula rubricera*) (Stattersfield et al. 1998), white-eyed starling (*Aplonis brunneicapilla*) (Stattersfield et al. 1998)

Area-sensitive species: fruit bats (*Rousettus amplexicaudatus, Macroglossus minimus, Melonycteris fardoulisi, Melonycteris woodfordi, Pteralopex anceps, Pteralopex atrata, Pteralopex pulchra, Pteralopex sp., Pteropus hypomelanus, Pteropus admiralitatum, Pteropus howensis, Pteropus mahaganus, Pteropus rayneri, Pteropus rennelli, Pteropus tonganus, Pteropus tuberculatus, Pteropus woodfordi, Dobsonia inermis, Nyctimene vizcaccia, Nyctimene major*) (Mickleburgh et al. 1992; Flannery 1995), Pacific baza (*Aviceda subcristata*), variable goshawk (*Accipiter hiogaster*), brown goshawk (*Accipiter fasciatus*), Meyer's goshawk (*Accipiter meyerianus*), pied goshawk (*Accipiter albogularis*), imitator sparrowhawk (*Accipiter imitator*)

Top predator species: Pacific baza (*Aviceda subcristata*), variable goshawk (*Accipiter hiogaster*), brown goshawk (*Accipiter fasciatus*), Meyer's goshawk (*Accipiter meyerianus*), pied goshawk (*Accipiter albogularis*), imitator sparrowhawk (*Accipiter imitator*)

Selected Conservation Contacts

BirdLife International

WWF-South Pacific

Justification for Ecoregion Delineation

Distinct island groups were placed in their own ecoregions: the Solomon Islands Rain Forests [137] and the Vanuatu Rain Forests [138]. We followed Stattersfield et al. (1998) in delineating these ecoregions. MacKinnon (1997) did not extend his assessment beyond the island of Bougainville in the Solomon Islands. However, we followed Bouchet et al. (1995) and separated the distinctive dry forests in New Caledonia from the moist forests to delineate the New Caledonia Rain Forests [139] and the New Caledonia Dry Forests [140]. Stattersfield et al. (1998) did not show this distinction.

Udvardy (1975) placed all the ecoregions in the New Guinea and Melanesia bioregion, with the exception of New Caledonia, into the Papuan biogeographic province of the Oceanian Realm. New Caledonia was placed in the New Caledonian biogeographic province.

Prepared by John Morrison

Ecoregion Number:	**138**
Ecoregion Name:	**Vanuatu Rain Forests**
Bioregion:	**New Guinea and Melanesia**
Biome:	**Tropical and Subtropical Moist Broadleaf Forests**
Political Units:	**Solomon Islands, Vanuatu**
Ecoregion Size:	**13,200 km²**
Biological Distinctiveness:	**Globally Outstanding**
Conservation Status:	**Endangered**
Conservation Assessment:	**I**

The Vanuatu Rain Forests [138] consist of more than eighty true oceanic islands, in two groups, at the edge of both the Australasian realm and the Pacific Basin. They contain fifteen bird species and several mammal species found nowhere else in the world. Although it is faced with population pressures and regular visits by destructive cyclones, with few exceptions Vanuatu's natural heritage is nearly intact.

Location and General Description

This ecoregion consists of tropical lowland and montane forests in the island nation of Vanuatu (formerly New Hebrides), with the addition of the Santa Cruz Group, or Temotu Province of the Solomon Islands. The climate of Vanuatu is tropical wet (National Geographic Society 1999), although leeward slopes of islands experience a distinct dry season that may last from April to October (Mueller-Dombois and Fosberg 1998). Destructive tropical cyclones are a regular event. The highest point in Vanuatu is 1,879 m (Espiritu Santo Island), but most of the islands are low-lying. Vanuatu is the result of the subduction of the northward-moving Australian tectonic plate beneath the Pacific plate, and the islands are a very active tectonic area. Most of the surface geology of the islands consists of Pliocene-Pleistocene volcanic rocks and uplifted coral. The Santa Cruz Islands contain areas of uplifted limestone or volcanic ash over limestone. Very few rocks are older than 38 m.y., and the older material is found in the northern islands, with the exception of the Santa Cruz Islands, which are less than 5 million years old (Mueller-Dombois and Fosberg 1998). There are active volcanoes (1988) in Vanuatu (Stattersfield et al. 1998).

Geology and vegetation can be differentiated between the northern Santa Cruz Islands group and the rest of Vanuatu.

When not converted to agricultural use, the predominant vegetation type in the Santa Cruz Islands is lowland rain forest. The Santa Cruz Islands have two of the twelve common tree species found in the Solomons (*Campnosperma brevipetiolata* and *Calophyllum vitiense*). The islands differ phytogeographically from the rest of Vanuatu. Because the highest point on Vanikoro (in the Santa Cruz Islands) is only

924 m, there is no well-developed montane rain forest. However, some montane species, such as *Metrosideros ornata*, are found in the lowlands. Other prominent species include *Gmelina solomoensis* (Verbenaceae), *Parinari corymbosa* (Rosaceae), *Paraserianthes (Albizia) falcataria* and *Pterocarpus indicus* (Fabaceae), and *Endospermum medullosum* (Euphorbiaceae). *Agathis* (kauri pine) is found in the Santa Cruz Islands, as are *Dacrydium elatum* and several *Syzygium* (Myrtaceae) species, which are generally montane species elsewhere (Mueller-Dombois and Fosberg 1998).

Natural vegetation in the rest of Vanuatu changes with altitude, substrate, and aspect and can be broadly classified into lowland forest, montane forest, seasonal forest and scrub, vegetation on new volcanic surfaces, coastal vegetation, and secondary vegetation (Mueller-Dombois and Fosberg 1998).

Lowland rain forest is the natural vegetation on all southeastern, or windward, sides of Vanuatu's islands, and it can be further subdivided into high- and medium-stature forests, complex forest scrub densely covered with lianas, alluvial and floodplain forests, *Agathis-Calophyllum* forest, and mixed-species forests without gymnosperms and *Calophyllum*. The complex forest scrub densely covered with lianas is the most widespread forest type on the larger northern islands; it is related to cyclone disturbance and is structurally heterogeneous. *Agathis-Calophyllum* forest is found only in the southern islands (Erromango and Anatom), although scattered *Agathis* spp. are reported in western Espiritu Santo (Mueller-Dombois and Fosberg 1998).

Montane forest types consisting of *Agathis* and *Podocarpus*—growing in a matrix of *Metrosideros, Syzygium, Weinmannia, Geissois, Quintania,* and *Ascarina*—begin at altitudes of approximately 500 m and grade into stunted, patchy cloud forest up to the summits of Vanuatu's mountain peaks (Mueller-Dombois and Fosberg 1998).

Seasonal forest, scrub, and grassland are associated with the leeward sides of the islands and can be further subdivided based on a moisture gradient. Semideciduous *Kleinhovia-Castanospermum* forest contains some of the rain forest species and represents a transition from dry to rain forest. *Acacia spirorbis* (or *gaiac*, as it is locally known) forest is found in somewhat drier habitats. Finally, in well-sheltered locations on the west and northwest sides of the islands, *Leucaena* thickets, savannas, and grasslands are found. These areas are subject to burning by humans (Mueller-Dombois and Fosberg 1998).

The coastal vegetation consists of both strand and mangroves. Tree species in the littoral forest include *Casuarina equiseifolia, Pandanus, Barringtonia asiatica, Terminalia catappa, Henandia* spp., and *Thespesia populnea*. The mangroves contain *Rhizophora, Avicennia, Sonneratia, Xylocarpus,* and *Ceriops* and are not found on all of the islands (Mueller-Dombois and Fosberg 1998).

Such vegetation is typical of the main islands, but the outlying islands, including Torres Islands, Banks Islands, and Aoba, Ambrym, Epi, and the Shepards, have their own unique vegetations (Mueller-Dombois and Fosberg 1998).

Biodiversity Features

Overall richness and endemism in Vanuatu range from low to moderate when compared with those of other ecoregions in Indo-Malaysia.

There is a clear gradient of the mammalian faunas from New Guinea, to the Bismarck Archipelago, through the Solomon Islands, to Vanuatu; areas of open ocean have acted as an effective filter. Except for pteropodid bats, the Solomons and Bismarcks (New Britain, New Ireland) have many fewer mammals than New Guinea. Unlike New Britain, the Solomons, contain no marsupials, and east beyond the Solomons there are even fewer mammal species. However, almost all mammal species in Vanuatu have their origins in or via New Guinea. The only mammals in Vanuatu are four pteropodid bats and eight microchiroptera. (Flannery 1995). Six of these species are endemic or near endemic (table J138.1). Of these endemic and near-endemic species, the Fijian blossom-bat (*Notopteris macdonaldi*) and Banks flying-fox (*Pteropus fundatus*) are considered vulnerable, and the Nendö tube-nosed bat (*Nyctimene sanctacrucis*) is presumed to be extinct (Hilton-Taylor 2000).

There are seventy-nine bird species in Vanuatu, of which an amazing thirty species are endemic or near endemic (table J138.2). The dropoff in diversity seen in other animal groups as one moves east from New Guinea is consistent with that seen in birds in Vanuatu. Whereas New Guinea has seventy-one families and subfamilies of birds, and the Solomons have forty-four, Vanuatu has thirty-one (Keast 1996).

Vanuatu has an almost exact correspondence with the Vanuatu and Temotu EBA. This EBA contains thirty restricted-range bird species. The Santa Cruz Islands (Temotu Province) have enough endemic species to qualify as an important EBAs by themselves. Fifteen bird species in the EBA are found nowhere else in the world. The Vanuatu scrubfowl (*Megapodius layardi*), Santa Cruz ground-dove (*Gallicolumba sanctaecrucis*), Vanuatu imperial-pigeon (*Ducula bakeri*), chestnut-bellied kingfisher (*Todirhampus farquhari*), royal parrotfinch (*Erythrura regia*), and Santo Mountain starling (*Aplonis santovestris*) are all considered vulnerable (Stattersfield et al. 1998). The Tanna ground-dove (*Gallicolumba ferruginea*) is presumed to be extinct; no specimens have ever been collected (Doughty et al. 1999)

Floristically, Vanuatu follows the pattern of the other outlying Melanesian islands: a decrease in diversity as one moves east from New Guinea. The Santa Cruz islands are a different phytogeographic unit from the rest of Vanuatu. Whereas the Santa Cruz Islands have two of the twelve common tree species found in the Solomons (*Campnosperma brevipetiolata* and *Calophyllum vitiense*), Vanuatu has only one: *Pometia pinnata*. Vanuatu is also more closely related to Fiji than to

TABLE J138.1. Endemic and Near-Endemic Mammal Species.

Family	Species
Pteropodidae	*Notopteris macdonaldi*
Pteropodidae	*Nyctimene sanctacrucis**
Pteropodidae	*Pteropus anetianus**
Pteropodidae	*Pteropus nitendiensis**
Pteropodidae	*Pteropus tuberculatus**
Pteropodidae	*Pteropus fundatus**

An asterisk signifies that the species' range is limited to this ecoregion.

New Caledonia despite the current physical positions of the islands (Mueller-Dombois and Fosberg 1998).

Between November and April of each year Vanuatu is subject to tropical cyclones, which are an important source of natural disturbance to the islands' forests. Cyclones and associated flooding occur at intervals of approximately every thirty years, and damage is enormous: up to 30 percent of forests are affected on some islands (Stattersfield et al. 1998).

Vanuatu is the easternmost limit of the range of the saltwater crocodile, Crocodylus porosus (Bregulla 1992), and is also home to the endangered Fiji banded iguana, Brachylophus fasciatus (Hilton-Taylor 2000).

Status and Threats

Current Status

Ninety percent of the population is engaged in subsistence farming in rural areas, and 41 percent of the nation of Vanuatu is considered to have average or better agricultural suitability. This fertile land is not spread evenly among the islands, however (Brugulla 1992). There is a growing emphasis on cash crops (cocoa, copra), and clearing for plantations and cattle has occurred (Bregulla 1992; Stattersfield et al. 1998). Commercial logging has intensively focused on two species: kauri (Agathis macrophylla) and sandalwood (Santalum austrocaledonicum). Sandalwood was almost eliminated by the end of the nineteenth century (Bregulla 1992).

Severe cyclones, volcanic eruptions, and earthquakes are regular events in Vanuatu, and although these are natural events, their effects can be exacerbated by human disturbance of watersheds and remaining forest areas.

There are no protected areas in Vanuatu (MacKinnon 1997). Establishing protected areas is problematic because of customary land rights: 99 percent of the land is owned by village people (Bregulla 1992; Stattersfield et al. 1998).

Types and Severity of Threats

Although the potential area available for commercial forestry is limited by topography, available commercial species, cultivation, and cyclone damage (Bregulla 1992), remaining lowland forests are under great pressure from logging companies (Stattersfield et al. 1998).

Several bird species in Vanuatu are vulnerable simply because of their small natural ranges (Stattersfield et al. 1998). Though subject to reduction by cyclones and consumed by almost all rural Vanuatuans, flying-fox populations seem healthy. The saltwater crocodile population has been reduced by hunting and cyclones (Bregulla 1992).

Priority Conservation Actions

Short-Term Conservation Actions (1–5 Years)

- Identify hotspot regions of endemicity or other biological priority sites.
- Increase the terrestrial protected area system, based on priority-setting exercises.
- Stop logging of lowland forests.
- Limit the spread of subsistence agriculture land uses.
- Educate the public about the ecoregion's tremendous endemic flora and fauna.

Longer-Term Conservation Actions (5–20 Years)

- Develop and implement effective management plans for protected areas.

TABLE J138.2. Endemic and Near-Endemic Bird Species.

Family	Common Name	Species
Megapodiidae	New Hebrides scrubfowl	Megapodius layardi*
Accipitridae	Pied goshawk	Accipiter albogularis
Columbidae	Santa Cruz ground-dove	Gallicolumba sanctaecrucis*
Columbidae	Tanna ground-dove	Gallicolumba ferruginea*
Columbidae	Tanna fruit-dove	Ptilinopus tannensis*
Columbidae	Baker's imperial-pigeon	Ducula bakeri*
Columbidae	Red-bellied fruit-dove	Ptilinopus greyii
Loriidae	Palm lorikeet	Charmosyna palmarum*
Alcedinidae	Chestnut-bellied kingfisher	Todirhamphus farquhari*
Campephagidae	Melanesian cuckoo-shrike	Coracina caledonica
Campephagidae	Long-tailed triller	Lalage leucopyga
Sylviidae	Guadalcanal thicketbird	Megalurulus whitneyi
Acanthizidae	Fan-tailed gerygone	Gerygone flavolateralis
Monarchidae	Vanikoro monarch	Mayrornis schistaceus*
Monarchidae	Buff-bellied monarch	Neolalage banksiana*
Monarchidae	Black-throated shrikebill	Clytorhynchus nigrogularis*
Monarchidae	Southern shrikebill	Clytorhynchus pachycephaloides
Monarchidae	Vanikoro flycatcher	Myiagra vanikorensis*
Monarchidae	New Caledonian flycatcher	Myiagra caledonica
Rhipiduridae	Streaked fantail	Rhipidura spilodera
Zosteropidae	Santa Cruz white-eye	Zosterops santaecrucis*
Zosteropidae	Yellow-fronted white-eye	Zosterops flavifrons*
Zosteropidae	Sanford's white-eye	Woodfordia lacertosa*
Meliphagidae	New Hebrides honeyeater	Phylidonyris notabilis*
Meliphagidae	Dark-brown honeyeater	Lichmera incana
Meliphagidae	Cardinal myzomela	Myzomela cardinalis
Estrildidae	Royal parrotfinch	Erythrura regia*
Sturnidae	Polynesian starling	Aplonis tabuensis*
Sturnidae	Rusty-winged starling	Aplonis zelandica*
Sturnidae	Mountain starling	Aplonis santovestris*

An asterisk signifies that the species' range is limited to this ecoregion.

Focal Species for Conservation Action

Endangered endemic species: Fijian blossom-bat (*Notopteris macdonaldi*) (Stattersfield et al. 1998), Banks flying-fox (*Pteropus fundatus*) (Stattersfield et al. 1998), Vanuatu scrubfowl (*Megapodius layardi*) (Stattersfield et al. 1998), Santa Cruz ground-dove (*Gallicolumba sanctaecrucis*) (Stattersfield et al. 1998), Vanuatu imperial-pigeon (*Ducula bakeri*) (Stattersfield et al. 1998), chestnut-bellied kingfisher (*Todirhampus farquhari*) (Stattersfield et al. 1998), royal parrotfinch (*Erythrura regia*) (Stattersfield et al. 1998), Santo Mountain starling (*Aplonis santovestris*) (Stattersfield et al. 1998), Tanna ground-dove (*Gallicolumba ferruginea*) (Stattersfield et al. 1998)

Area-sensitive species: fruit bats (*Notopteris macdonaldi, Nyctimene sanctacrucis, Pteropus anetianus, Pteropus nitendiensis, Pteropus tonganus, Pteropus fundatus*) (Mickleburgh et al. 1992; Flannery 1995), Pacific baza (*Aviceda subcristata*), variable goshawk (*Accipiter hiogaster*), brown goshawk (*Accipiter fasciatus*), Meyer's goshawk (*Accipiter meyerianus*), pied goshawk (*Accipiter albogularis*), imitator sparrowhawk (*Accipiter imitator*), Oriental hobby (*Falco severus*)

Top predator species: Pacific baza (*Aviceda subcristata*), variable goshawk (*Accipiter hiogaster*), brown goshawk (*Accipiter fasciatus*), Meyer's goshawk (*Accipiter meyerianus*), pied goshawk (*Accipiter albogularis*), imitator sparrowhawk (*Accipiter imitator*), Oriental hobby (*Falco severus*)

Selected Conservation Contacts

BirdLife International
WWF-South Pacific

Justification for Ecoregion Delineation

Distinct island groups were placed in their own ecoregions: the Solomon Islands Rain Forests [137], and the Vanuatu Rain Forests [138]. We followed Stattersfield et al. (1998) in delineating these ecoregions. MacKinnon (1997) did not extend his assessment beyond the island of Bougainville in the Solomon Islands. However, we followed Bouchet et al. (1995) and separated the distinctive dry forests in New Caledonia from the moist forests to delineate the New Caledonia Rain Forests [139] and the New Caledonia Dry Forests [140]. Stattersfield et al. (1998) did not show this distinction.

Udvardy (1975) placed all the ecoregions in the New Guinea and Melanesia bioregion, with the exception of New Caledonia, into the Papuan biogeographic province of the Oceanian Realm. New Caledonia was placed in the New Caledonian biogeographic province.

Prepared by John Morrison

Ecoregion Number:	**139**
Ecoregion Name:	**New Caledonia Rain Forests**
Bioregion:	**New Guinea and Melanesia**
Biome:	**Tropical and Subtropical Moist Broadleaf Forests**
Political Units:	**New Caledonia (France)**
Ecoregion Size:	**14,500 km²**
Biological Distinctiveness:	**Globally Outstanding**
Conservation Status:	**Endangered**
Conservation Assessment:	**I**

The islands of New Caledonia contain some of the most distinctive plants in the world, with a large number of species, endemics, and an ancient character to much of the flora. The New Caledonia Rain Forests [139] are the richest part of the French territory, but they have suffered large losses of native habitat.

Location and General Description

New Caledonia is located in the southwest Pacific Ocean about 1,200 km east of Australia and 1,500 km northeast of New Zealand. The main island of Grande Terre runs in a north-south orientation and is 16,372 km². Unlike the much smaller neighboring islands, which are volcanic and relatively recent in origin, Grand Terre is an original piece of Gondwanaland. It separated from Australia 85 million years ago and has maintained its current isolation from other landmasses for more than 55 million years (Kroenke 1996). Isolation and an ancient source of plant life are major factors leading to its diverse flora, but they are not the only factors. Grand Terre has an extremely diverse soil substrate, with ultramafics forming about one-third of the island. It is also diverse topographically and climatically. Grand Terre is the only high island of New Caledonia, with a mountain chain running down the center of the island and five peaks exceeding 1,500 m. Many smaller ranges and valleys run counter to the island's north-south orientation. The soils of the Loyalty Islands to the east and Iles des Pines to the south of Grand Terre are largely from limestone substrates that resulted from the volcanic uplifting of corals when the islands were formed (Mueller-Dombois 1998).

Rainfall in New Caledonia is highly seasonal. Trade winds bring the rains, which usually come from the east. The average annual rainfall is about 1,500 mm for the Loyalty Islands, 2,000 mm for the low elevation eastern Grand Terre, and 2,000–4,000 mm at high elevations (Mueller-Dombois 1998). The western side of Grand Terre receives much less rainfall because of the orographic nature of the island's weather (see description for New Caledonia Dry Forests [140]).

The tropical moist forest of this ecoregion is generally subdivided into lowland rain forests of the Loyalty Islands and Grand Terre, the montane forests of Grand Terre, and Grand Terre's wet maquis forest. The lowland tropical rain forests are of a mixed-species composition, with the prevalent gymnosperms being *Agathis lanceolata, A. ovata, Araucaria columnaris, A. bernieri, Dacrydium araucarioides, Dacrycarpus vieillardii,* and *Falcatifolium taxoides*. The main angiosperm trees include *Montrouziera cauliflora,*

Calophyllum neocaledonicum, Dysoxylum spp., *Neogullauminia cleopatra, Hernandia cordigera*, and species of the genera *Kermadecia, Macadamia*, and *Sleumerodendron*. In some places single-species dominant stands of *Araucaria, Callistemon*, and *Nothofagus* occur. Montane rain forest species include *Araucaria, Agathis, Podocarpus, Dacrydium, Libocedrus, Acmopyle, Metrosideros, Weinmannia, Quintinia*, and *Nothofagus* (Mueller-Dombois 1998). The unique maquis forests on Grande Terre are dominated by *Araucaria* species and are unique. Their scrublike structure resembles that of Mediterranean climate woodlands. However, the maquis of New Caledonia are a response to the ultramafic substrate rather than the climate (Mueller-Dombois 1998).

Biodiversity Features

New Caledonia has five endemic plant families (Amborellaceae, Oncothecaceae, Papracrypyiaceae, Phellinaceae, and Strasburgiaceae) out of a total of 196 families found on the islands. Nearly 14 percent of the plant genera and 79.5 percent of the species are endemic. The percentage of endemic species is greater than that of all other Pacific island groups, with the exception of Hawaii (89 percent) and New Zealand (81.9 percent), and is comparable to continental levels of endemism. However, Hawaii contains only 956 species, compared with New Caledonia's 2,973 (Sohmer 1990, in Jaffré 1993). New Zealand shares New Caledonia's Gondwanaland history, but even New Zealand has fewer total species and fewer total endemics, although the North Island alone is seven times larger than New Caledonia (Mueller-Dombois 1998). The lower numbers for New Zealand probably result from a less diverse substrate and its location outside the tropics.

Plant numbers obscure an underlying theme to New Caledonia's unique biodiversity. Part of what makes New Caledonia so unique is the large number of ancient lineages and absence of widespread genera and families. The ancient nature of plants in particular is exemplified by *Amborella trichopoda*, the only species in the family Amborellaceae, thought to be one of the closest living relatives to the first angiosperms (flowering plants) and the high number of woody species that lack vessels, a feature typical of primitive families. New Caledonia has a remarkable diversity of gymnosperms (primitive nonflowering plants that include conifers), with forty-four species (forty-three of which are endemic) out of fifteen genera (at least three of which are endemic) (Keast 1996; Morat 1993). Eric Dinerstein, chief scientist for WWF, commented in a recent *National Geographic* article that when in New Caledonia, "I feel like I'm walking in a forest the dinosaurs knew" (O'Neil 2000, p. 65).

TABLE J139.1. Endemic and Near-Endemic Mammal Species.

Family	Species
Pteropodidae	*Notopteris macdonaldi*
Pteropodidae	*Pteropus ornatus*
Pteropodidae	*Pteropus vetulus*
Vespertilionidae	*Chalinolobus neocaledonicus*
Vespertilionidae	*Miniopterus robustior**
Vespertilionidae	*Nyctophilus* sp.

An asterisk signifies that the species' range is limited to this ecoregion.

TABLE J139.2. Endemic and Near-Endemic Bird Species.

Family	Common Name	Species
Accipitridae	White-bellied goshawk	*Accipiter haplochrous*
Rallidae	New Caledonian rail	*Gallirallus lafresnayanus**
Rhynochetidae	Kagu	*Rhynochetos jubatus**
Columbidae	Red-bellied fruit-dove	*Ptilinopus greyii*
Columbidae	Cloven-feathered dove	*Drepanoptila holosericea**
Columbidae	New Caledonian imperial-pigeon	*Ducula goliath*
Loriidae	New Caledonian lorikeet	*Charmosyna diadema**
Psittacidae	Horned parakeet	*Eunymphicus cornutus*
Aegothelidae	New Caledonian owlet-nightjar	*Aegotheles savesi*
Campephagidae	Melanesian cuckoo-shrike	*Coracina caledonica*
Campephagidae	New Caledonian cuckoo-shrike	*Coracina analis*
Campephagidae	Long-tailed triller	*Lalage leucopyga*
Sylviidae	New Caledonian grassbird	*Megalurulus mariei*
Acanthizidae	Fan-tailed gerygone	*Gerygone flavolateralis*
Eopsaltriidae	Yellow-bellied robin	*Eopsaltria flaviventris*
Monarchidae	Southern shrikebill	*Clytorhynchus pachycephaloides*
Monarchidae	New Caledonian flycatcher	*Myiagra caledonica*
Rhipiduridae	Streaked fantail	*Rhipidura spilodera*
Pachycephalida	New Caledonian whistler	*Pachycephala caledonica*
Zosteropidae	Large Lifou white-eye	*Zosterops inornatus**
Zosteropidae	Green-backed white-eye	*Zosterops xanthochrous*
Zosteropidae	Small Lifou white-eye	*Zosterops minutus**
Meliphagidae	New Caledonian myzomela	*Myzomela caledonica*
Meliphagidae	Cardinal myzomela	*Myzomela cardinalis*
Meliphagidae	Dark-brown honeyeater	*Lichmera incana*
Meliphagidae	New Caledonian friarbird	*Philemon diemenensis*
Meliphagidae	Crow honeyeater	*Gymnomyza aubryana**
Meliphagidae	Barred honeyeater	*Phylidonyris undulata*
Estrildidae	Red-throated parrotfinch	*Erythrura psittacea*
Sturnidae	Striated starling	*Aplonis striata*
Corvidae	New Caledonian crow	*Corvus moneduloides*

An asterisk signifies that the species' range is limited to this ecoregion.

Like New Caledonia's plant life, terrestrial animals are represented often by unique species and ancient lineages, but many types of widespread species are missing (Keast 1996). There are no native amphibians, three snakes (none of which is on Grand Terre), and only nine mammals species, all of which are bats (six of which are either endemic or near endemic; table J139.1). All of New Caledonia's sixty-eight lizards (sixty of which are endemic) are from just three families: geckos (Gekkonidae and Diplodactylidae) and skinks (Scincidae) (Bauer 1999). Both families of reptiles contain recent arrivals as well as ancient Gondwanaland groups. The birds of New Caledonia consist mainly of modern forms (Mayr 1940). However, one ancient family, Rynochetidae, is endemic to New Caledonia and is currently represented by one species, the kagu (*Rhynochetos jubatus*). The kagu is the national bird of New Caledonia and is listed by IUCN (1996) as endangered (EN), along with the Australasian bittern (*Botaurus poiciloptilus*), New Caledonian lorikeet (*Charmosyna diadema*), and New Caledonian owlet-nightjar (*Aegotheles savesi*). The most precarious existence, however, may belong to the New Caledonian rail (*Gallirallus lafresnayanus*), which IUCN (1996) listed as critical (CR). IUCN (1996) listed five other bird species as vulnerable to extinction. A total of seven birds are endemic to the ecoregion, but there are twenty-four near endemics (table J139.2). Addition-ally, two bats are considered endangered (*Chalinolobus neocaledonicus* and *Miniopterus robustior*), and two more are vulnerable (long-tailed fruit bat, *Notopteris macdonaldi*, and ornate flying-fox, *Pteropus ornatus*).

Some of the vertebrate species of New Caledonia stand out for being unique in size: the New Caledonian imperial-pigeon (*Ducula goliath*) is the largest arboreal pigeon in the world, *Rhacodactylus leachianus* is the world's largest gecko, and the giant skink (*Phoboscincus bocourti*) is the largest skink, although it has not been seen since the 1870s and may be extinct (Mittermeier et al. 1999).

Status and Threats

Current Status

The New Caledonia rain forests have suffered large losses of native habitat. Rain forests in New Caledonia used to occupy 70 percent of the land area and now occupy 21.5 percent (Mittermeier et al. 1999). Logging and mining are decreasing as logging operations are becoming localized, and the degree of mining has been scaled back from the boom years of the 1960s and 1970s (Lowry 1996). Still, New Caledonia produces about half of the world's nickel and contains 40 percent of the world's known nickel deposits. The past impacts of these land uses are severe. Deforestation and large-scale open mines have given New Caledonia some of the worst soil erosion in the world (Stattersfield et al. 1998). The accessibility of forests to hunting is increased by both logging and prospecting and threatens some species such as the New Caledonian imperial-pigeon.

Maquis makes up the remainder of the ecoregion and naturally covers much of the ultramafic substrates that contain high concentrations of nickel, iron, magnesium, olivine, and chromium. Mining therefore has been the main land use destroying maquis, but this has been localized. The soil of maquis will not support agriculture, and the vegetation is too scrublike for timber (Lowry 1996). Maquis vegetation is expanding primarily into disturbed areas at mid- to low elevation (Mittermeier et al. 1999).

The protected area network of New Caledonia is poor both in size (covering 2.8 percent of the land area) and in resources, making the protected areas little more than paper parks (table J139.3). The Rivière Bleue Park is an exception, being well managed and having some of the only resident park personnel (Jaffré et al. 1998). The park also has the only kagu populations that are on the rise as a result of controlling introduced predators (Stattersfield et al. 1998). One glaring gap in New Caledonia's biodiversity protection is the lack of protected areas in the Loyalty Islands (Stattersfield et al. 1998).

Types and Severity of Threats

The main threats to rain forests in New Caledonia have been and will continue to be widespread logging, mining, and wildlife hunting. However, introduced species are a growing problem in New Caledonia. Pigs, goats, cats, dogs, and rats present problems for native species here, as they do on many islands throughout the world. New Caledonia also has Java deer (*Cervus timorensis*) that are widely hunted. In addition to deer trampling and grazing understory plants, people often start fires to attract deer to the new growth that follows (Bouchet et al. 1995; Lowry 1996). In addition to setting fire for deer, Bouchet et al. (1995, p. 420) explained, "lighting fire has also become an expression of protest from young rural unemployed males. It is not exaggerated to write that fires plague New Caledonia, west and east coast alike, from July to December." Many of the native species are not adapted to be fire resistant, and as a result some introduced species and native species that are fire resistant are taking over. The Neotropical ant (*Wassmannia auropunctata*) that was brought in with Caribbean pine cultivation is diminishing native lizard and invertebrate abundance and diversity (Mittermeier et al. 1999; WWF-France 1997). The severe impacts of this ant may determine the long-term persistence of native communities in this ecoregion.

New Caledonia is a prosperous territory of France, and this prosperity affects the future of its biodiversity. The per capita income of New Caledonia is similar to that of New Zealand and Australia. The prosperity means that many of the problems of rapid population increases found in other tropical forested areas are not prevalent. However, because New Caledonia, as part of France, is considered a developed country, it does not qualify for funds to protect biodiversity through traditional international sources. Meanwhile, the French government has paid little attention to the conservation of New Caledonia's wealth of biodiversity (Mittermeier et al. 1999).

TABLE J139.3. WCMC (1997) Protected Areas That Overlap with the Ecoregion.

Protected Area	Area (km²)	IUCN Category
Twenty nature reserves (no names in database)	370	?

Ecoregion numbers of protected areas that overlap with additional ecoregions are listed in brackets.

Priority Conservation Actions

Short-Term Conservation Actions (1–5 Years)

- Protect remaining forests.
- Increase public education about the ecoregion's native biodiversity.
- Work to limit mining in remaining maquis vegetation.
- Stop or limit expansion of exotic species.

Longer-Term Conservation Actions (5–20 Years)

- Work with France to increase biodiversity protection (increase protected areas, stop mining and logging in key biodiversity areas, and draw international attention to ecoregion's biodiversity).

Focal Species for Conservation Action

Endangered endemic species: New Caledonia wattled bat (*Chalinolobus neocaledonicus*) (Mickleburgh et al. 1992), Loyalty bentwing-bat (*Miniopterus robustior*) (Mickleburgh et al. 1992), long-tailed fruit bat (*Notopteris macdonaldi*) (Mickleburgh et al. 1992), ornate flying-fox (*Pteropus ornatus*) (Mickleburgh et al. 1992), New Caledonian rail (*Gallirallus lafresnayanus*), kagu (*Rhynochetos jubatus*) (Hunt 1996; Hunt et al. 1996; Stattersfield et al. 1998), cloven-feathered dove (*Drepanoptila holosericea*) (Stattersfield et al. 1998), New Caledonian imperial-pigeon (*Ducula goliath*) (Stattersfield et al. 1998), New Caledonian lorikeet (*Charmosyna diadema*) (Juniper and Parr 1998; Snyder et al. 2000; Stattersfield et al. 1998), horned parakeet (*Eunymphicus cornutus*) (Juniper and Parr 1998; Snyder et al. 2000; Stattersfield et al. 1998), New Caledonian owlet-nightjar (*Aegotheles savesi*) (Stattersfield et al. 1998), crow honeyeater (*Gymnomyza aubryana*) (Stattersfield et al. 1998)

Area-sensitive species: New Caledonia flying-fox (*Pteropus vetulus*) (Mickleburgh et al. 1992), ornate flying-fox (*P. ornatus*) (Mickleburgh et al. 1992), Pacific flying-fox (*P. tonganus*) (Mickleburgh et al. 1992), long-tailed fruit bat (*Notopteris macdonaldi*) (Mickleburgh et al. 1992), brown goshawk (*Accipiter fasciatus*), white-bellied goshawk (*Accipiter haplochrous*), peregrine falcon (*Falco peregrinus*)

Top predator species: brown goshawk (*Accipiter fasciatus*), white-bellied goshawk (*Accipiter haplochrous*), peregrine falcon (*Falco peregrinus*)

Selected Conservation Contacts

Centre d'Initiation à l'Environnement

CIRAD

Direction de l'Agriculture, de la Forêt et de l'Environnement (Haut Commissariat)

Governments of Province Nord and Province Sud

Kanak (Ti-Va-Ouere)

The Maruia Society

Overseas Scientific Research Organization (supported by the French Government)

WWF-France

Justification for Ecoregion Delineation

The New Caledonia Rain Forests [139] ecoregion is based on the original extent of humid forests appearing in Jaffré and Veillon (1994). All other islands of New Caledonia (including the Loyalty Islands and Iles des Pines) have been included as part of the ecoregion based on assumptions of rainfall vegetation descriptions appearing in Mueller-Dombois (1998). Habitats within the ecoregion include subregions of maquis forests, montane and lowland forests, and savannas.

Prepared by John Lamoreux

Ecoregion Number: **140**
Ecoregion Name: **New Caledonia Dry Forests**
Bioregion: **New Guinea and Melanesia**
Biome: **Tropical and Subtropical Dry Broadleaf Forests**
Political Units: **New Caledonia (France)**
Ecoregion Size: **4,400 km^2**
Biological Distinctiveness: **Globally Outstanding**
Conservation Status: **Critical**
Conservation Assessment: **I**

The dry forests of New Caledonia contain a unique assemblage of plants with fifty-nine endemic species, including the famous *Captaincookia margaretae*. The ecoregion has been severely degraded by human activity, and immediate conservation action is needed to ensure the long-term survival of many species.

Location and General Description

The islands of New Caledonia are remarkable for their number of plant species, plant endemics, and the ancient character of much of the flora. New Caledonia is located in the southwest Pacific Ocean about 1,200 km east of Australia and 1,500 km northeast of New Zealand. The main island of Grande Terre runs in a north-south orientation and is 16,372 km^2. Unlike the much smaller neighboring islands, which are volcanic and recent in origin, Grand Terre is an original piece of Gondwanaland. It separated from Australia 85 million years ago and has maintained its current isolation from other landmasses for more than 55 million years (Kroenke 1996). Isolation and an ancient source of plant life are major factors leading to the island's diverse plant life, but they are not the only factors. Grand Terre has an extremely diverse soil substrate, with ultramafics forming about one-third of the island. It is also very diverse topographically and climatically. Grand Terre is the only high island of New Caledonia, with a mountain chain running down the center of the island and five peaks exceeding 1,500 m. Many smaller ranges and valleys run counter to the island's north-south orientation.

The New Caledonia Dry Forests [140] ecoregion is found only on the western side of Grand Terre, where the average annual rainfall is much lower than on the rest of the island. Rainfall on Grand Terre is highly seasonal. Trade winds bring the rains, which usually come from the east. The western side of the island receives about 1,200 mm of rainfall annually, although in some years it may be as low as 250 mm. However, because of the orographic nature of the island's rainfall, the average annual rainfall is about 2,000 mm for the low-elevation eastern Grand Terre and 2,000–4,000 mm at high elevations (Mueller-Dombois 1998).

Dry forest vegetation used to make up most of the lowland portions (below 300 m) of the western side of Grand Terre, although this ecoregion has been greatly reduced and exists mainly in disjunct forest patches. In the natural state, these dry forests are dense, contain many vines, and reach heights of 5–15 m. The vegetation is often dominated by *Acacia spirorbis* and *Leucaena leucocephala* (an introduced species). Other dominants include Rubiaceae of the genera *Canthium* and *Gardenia*, *Pittosporum*, *Dodonaea*, and *Premna*, and Verbenaceae genera (Mueller-Dombois 1998). The dry forests also have a thick, unstratified understory of shrubs and grasses. Near the coast the typical dry forests give way to stands of *Cycas circinalis*, the only gymnosperm typical of the ecoregion.

Biodiversity Features

The New Caledonia Dry Forests [140] ecoregion contains 379 native plant species (phanerogam), 59 of which are found only in the dry forests. The total number of species and number of endemics are substantial but still less than the numbers for the New Caledonia Rain Forests [139]. The dry forests also differ from the moist forests in their composition. The latter are famous for having the greatest diversity of gymnosperms anywhere, a high number of plants in the primitive flowering plants of the genus *Pandanus* and family Winteraceae, and a great number of endemic palms. The dry forests have only one gymnosperm and none of these other taxa. Furthermore, the dry forests lack all five of the plant families endemic to New Caledonia. Still, the dry forests contain many extremely interesting plants, such as *Captaincookia margaretae* (Rubiaceae), the only member of its genus.

Few if any of the vertebrates of New Caledonia are endemic to the dry forests. However, there are five near-endemic mammals (table J140.1) and twenty-three near-endemic bird species (table J140.2), most of which are shared with the New Caledonia moist forests. Invertebrates are less numerous in dry forests than in the rain forests, but many that live there are likely to be endemic.

Status and Threats

Current Status

Tropical dry forests are the most threatened tropical forest type worldwide (Janzen 1988), and the dry forests of New Caledonia are no exception. With less than 2 percent remaining in isolated patches, the dry forests are by far the most endangered vegetation type in the territory. Comparisons between the biological features of the dry forests and the rain forests of New Caledonia are almost irrelevant because the remaining dry forests are so tiny: 10,000 ha, an area slightly larger than New York's Manhattan Island. The high rates of endemism also occur with a high turnover, meaning that almost all remaining dry forest patches are likely to contain complementary assemblages of species found nowhere else and should be preserved. The rest of the dry forest ecoregion has been replaced by ranchland or introduced species.

Out of 117 dry forest plant species evaluated for IUCN classification, 59 (or 50 percent) are threatened (VU or higher). The first recorded plant extinction in New Caledonia occurred within the dry forests when a fire wiped out the only known population of *Pittosporum tanianum*, a very distinct species discovered in the mid-1980s. Given the reduced size of the dry forests, it is almost certain that many plants have gone

TABLE J140.1. Endemic and Near-Endemic Mammal Species.

Family	Species
Pteropodidae	*Pteropus vetulus*
Pteropodidae	*Notopteris macdonaldi*
Pteropodidae	*Pteropus ornatus*
Vespertilionidae	*Chalinolobus neocaledonicus*
Vespertilionidae	*Nyctophilus* sp.

An asterisk signifies that the species' range is limited to this ecoregion.

TABLE J140.2. Endemic and Near-Endemic Bird Species.

Family	Common Name	Species
Accipitridae	White-bellied goshawk	*Accipiter haplochrous*
Columbidae	Red-bellied fruit-dove	*Ptilinopus greyii*
Columbidae	New Caledonian imperial-pigeon	*Ducula goliath*
Psittacidae	Horned parakeet	*Eunymphicus cornutus*
Aegothelidae	New Caledonian owlet-nightjar	*Aegotheles savesi*
Campephagidae	Melanesian cuckoo-shrike	*Coracina caledonica*
Campephagidae	New Caledonian cuckoo-shrike	*Coracina analis*
Campephagidae	Long-tailed triller	*Lalage leucopyga*
Sylviidae	New Caledonian grassbird	*Megalurulus mariei*
Acanthizidae	Fan-tailed gerygone	*Gerygone flavolateralis*
Eopsaltriidae	Yellow-bellied robin	*Eopsaltria flaviventris*
Monarchidae	Southern shrikebill	*Clytorhynchus pachycephaloides*
Monarchidae	New Caledonian flycatcher	*Myiagra caledonica*
Rhipiduridae	Streaked fantail	*Rhipidura spilodera*
Pachycephalida	New Caledonian whistler	*Pachycephala caledonica*
Zosteropidae	Green-backed white-eye	*Zosterops xanthochrous*
Meliphagidae	New Caledonian myzomela	*Myzomela caledonica*
Meliphagidae	Dark-brown honeyeater	*Lichmera incana*
Meliphagidae	New Caledonian friarbird	*Philemon diemenensis*
Meliphagidae	Barred honeyeater	*Phylidonyris undulata*
Estrildidae	Red-throated parrotfinch	*Erythrura psittacea*
Sturnidae	Striated starling	*Aplonis striata*
Corvidae	New Caledonian crow	*Corvus moneduloides*

An asterisk signifies that the species' range is limited to this ecoregion.

extinct that were never described. This is likely given that all the current dry forests of New Caledonia are on top of sedimentary substrates and that large portions of dry forest on basaltic substrates have been completely destroyed.

The protected area network of New Caledonia is poor in both scope (covering 2.8 percent of the land area) and in the resources necessary to make the protected areas effective (table J140.3). Almost all of the dry forests are in private hands, and of the six provincial parks in the dry forests, none are improving the conservation of the species living there. Jaffré et al. (1998) consider the dry forests completely unprotected. Even the traditional act of putting up fences would not be sufficient to protect species from fires such as the one that wiped out the only known populations of *Pittosporum tanianum*.

Types and Severity of Threats

With so little habitat remaining, these forests are vulnerable to conversion threats (agriculture, rangelands), fire, introduced species, and natural disturbances. Their ability to survive is highly threatened even with immediate intervention.

Introduced species pose a grave threat to the biodiversity of New Caledonia. Pigs, goats, cats, dogs, and rats present problems for native species here, as they do on many islands throughout the world. The introduced rusa deer (*Cervus timorensis*) is widely hunted. In addition to deer trampling and grazing understory plants, people often start fires to attract deer to the new growth that follows (Bouchet et al. 1995; Lowry 1996). In addition to setting fire for deer, Bouchet et al. (1995, p. 420) explained, "lighting fire has also become an expression of protest from young rural unemployed males. It is not exaggerated to write that fires plague New Caledonia, west and east coast alike, from July to December." Many of the native species are not adapted to be fire resistant, and as a result some introduced species as well as native species that are fire resistant are taking over. The Neotropical ant (*Wassmannia auropunctata*) that was brought in with Caribbean pine cultivation is greatly diminishing native lizard and invertebrate abundance and diversity (Mittermeier et al. 1999; WWF-France 1997). The severe impacts of this ant may determine the long-term persistence of native communities in this ecoregion.

New Caledonia is a prosperous territory of France, and this prosperity affects the future of its biodiversity. The per capita income of New Caledonia is similar to that of New Zealand and Australia. The prosperity means that many of the problems of rapid population increases found in other tropical forested areas are not prevalent. However, because New Caledonia, as part of France, is considered a developed country, it does not qualify for funds to protect biodiversity through traditional international sources. Meanwhile, the French government has paid little attention to the conservation of New Caledonia's wealth of biodiversity (Mittermeier et al. 1999).

Priority Conservation Actions

Short-Term Conservation Actions (1–5 Years)

- Establish biological surveys of the remaining forests habitat.
- Protect all remaining forests; work with landowners to protect forests on private lands.
- Increase public education about plight of dry forests and its biodiversity.
- Remove all introduced species or stop them from adversely affecting remaining forest patches (fencing off remaining forests).

Longer-Term Conservation Actions (5–20 Years)

- Work with France and local communities to conserve and protect remaining forests.

Focal Species for Conservation Action

Endangered endemic species: New Caledonia wattled bat (*Chalinolobus neocaledonicus*) (Mickleburgh et al. 1992), long-tailed fruit bat (*Notopteris macdonaldi*) (Mickleburgh et al. 1992), ornate flying-fox (*Pteropus ornatus*) (Mickleburgh et al. 1992), New Caledonian imperial-pigeon (*Ducula goliath*) (Stattersfield et al. 1998), horned parakeet (*Eunymphicus cornutus*) (Juniper and Parr 1998; Snyder et al. 2000; Stattersfield et al. 1998), New Caledonian owlet-nightjar (*Aegotheles savesi*) (Stattersfield et al. 1998), *Captincookia margaretae*

Area-sensitive species: New Caledonia flying-fox (*Pteropus vetulus*) (Mickleburgh et al. 1992), ornate flying-fox (*P. ornatus*) (Mickleburgh et al. 1992), Pacific flying-fox (*P. tonganus*) (Mickleburgh et al. 1992), long-tailed fruit bat (*Notopteris macdonaldi*) (Mickleburgh et al. 1992), brown goshawk (*Accipiter fasciatus*), white-bellied goshawk (*Accipiter haplochrous*), peregrine falcon (*Falco peregrinus*)

Top predator species: brown goshawk (*Accipiter fasciatus*), white-bellied goshawk (*Accipiter haplochrous*), peregrine falcon (*Falco peregrinus*)

Selected Conservation Contacts

Centre d'Initiation à l'Environnement
CIRAD
Direction de l'Agriculture, de la Forêt et de l'Environnement (Haut Commissariat)
Governments of Province Nord and Province Sud
Kanak (Ti-Va-Ouere)
The Maruia Society
Overseas Scientific Research Organization (supported by the French Government)
WWF-France

Justification for Ecoregion Delineation

The New Caledonia Rain Forests [140] ecoregion is based on the original extent of sclerophyllous forests appearing in Jaffré and Veillon (1994).

Prepared by John Lamoreux

TABLE J140.3. WCMC (1997) Protected Areas That Overlap with the Ecoregion.

Protected Area	Area (km²)	IUCN Category
2 nature reserves (unnamed in database)	20	?

Ecoregion numbers of protected areas that overlap with additional ecoregions are listed in brackets.

Glossary

adaptive radiation The evolution of a single species into many species that occupy diverse ways of life within the same geographic range.

alpha diversity Species diversity within a single site.

amphibian A member of the vertebrate class Amphibia (frogs and toads, salamanders, and caecilians).

anadromous Describes species that spawn in fresh water and migrate to marine habitats to mature (e.g., salmon).

anthropogenic Human induced.

aquatic Growing in, living in, or frequenting water.

aquifer A formation, group of formations, or part of a formation that contains sufficient saturated permeable material to yield significant quantities of water to wells and springs.

assemblage In conservation biology, a predictable and particular collection of species within a biogeographic unit (e.g., ecoregion or habitat).

barrens A colloquial name given to habitats with sparse vegetation or low agricultural productivity.

basin See catchment.

beta-diversity Species diversity between habitats (thus reflecting changes in species assemblages along environmental gradients).

biodiversity (Also called biotic or biological diversity.) The variety of organisms considered at all levels, from genetic variants belonging to the same species through arrays of species to arrays of genera, families, and still higher taxonomic levels; includes the variety of ecosystems, which comprise both communities of organisms within particular habitats and the physical conditions under which they live.

biodiversity conservation Five classes of biodiversity conservation priorities were determined in this priority study by integrating biological distinctiveness with conservation status (matrix in Chapter 6). The five classes roughly reflect the concern with which we should view the erosion of biodiversity in different ecoregions and the timing and sequence of response by governments and donors to the loss of biodiversity.

biogeographic unit A delineated area based on biogeographic parameters.

biogeography The study of the geographic distribution of organisms, both past and present.

biological distinctiveness Scale-dependent assessment of the biological importance of an ecoregion based on species richness, endemism, relative scarcity of ecoregion, and rarity of ecological phenomena. Biological distinctiveness classes are globally outstanding, regionally outstanding, bioregionally outstanding, and nationally important.

biome A global classification of natural communities in a particular region based on dominant or major vegetation types and climate.

bioregion A geographically related assemblage of ecoregions that share a similar biogeographic history and thus have strong affinities at higher taxonomic levels (e.g., genera, families). There are six biogeographic divisions of North America, consisting of contiguous ecoregions, designed to better address the biogeographic distinctiveness of ecoregions.

bioregionally outstanding Biological distinctiveness category.

biota The combined flora, fauna, and microorganisms of a given region.

biotic Biological, especially referring to the characteristics of faunas, floras, and ecosystems.

bog A poorly drained area rich in plant residues, usually surrounded by an area of open water and having characteristic flora.

canebrake A thicket of cane.

catadromous Describes diadromous species that spawn in marine habitats and migrate to fresh water to mature (e.g., eels).

catchment All lands enclosed by a continuous hydrologic-surface drainage divide and lying upslope from a specified point on a stream or, in the case of closed-basin systems, all lands draining to a lake.

centinelan extinction The phenomenon of species going extinct before they have been discovered or described by the scientific community.

clearcut A logged area where all or almost all the forest canopy trees have been eliminated.

community Collection of organisms of different species that co-occur in the same habitat or region and interact through trophic and spatial relationships.

conifer A tree or shrub in the phylum Gymnospermae whose seeds are borne in woody cones. There are 500–600 species of living conifers.

conservation biology Discipline that treats the content of biodiversity, the natural processes that produce it, and the techniques used to sustain it in the face of human-caused environmental disturbance.

conservation status Assessment of the status of ecological processes and the viability of species populations in an ecoregion. The different status categories used are extinct, critical, endangered, vulnerable, relatively stable, and relatively intact. The snapshot conservation status is based on an index derived from values of four landscape-level variables. The final conservation status is the snapshot assessment modified by an analysis of threats to the ecoregion over the next twenty years.

conversion Habitat alteration by human activities to such an extent that it no longer supports most characteristic native species and ecological processes.

critical Conservation status category characterized by low probability of persistence of remaining intact habitat.

deciduous forest Habitat type dominated by trees whose leaves last a year or less; they drop and replace their leaves over periods sufficiently distinct that they are leafless for some portion of the year.

degradation The loss of native species and processes caused by human activities such that only certain components of the original biodiversity still persist, often including significantly altered natural communities.

diadromous Describes species that migrate between freshwater and marine habitats while spawning in one habitat and maturing in another.

disturbance Any discrete event that disrupts an ecosystem, community, or population structure and changes resources, substrate availability, or the physical environment.

drainage basin See catchment.

ecological processes Complex mix of interactions between animals, plants, and their environment that ensure that an ecosystem's full range of biodiversity is adequately maintained. Examples include population and predator-prey dynamics, pollination and seed dispersal, nutrient cycling, migration, and dispersal.

ecoregion A large area of land or water that contains a geographically distinct assemblage of natural communities that share a large majority of their species and ecological dynamics, share similar environmental conditions, and interact ecologically in ways that are critical for their long-term persistence.

ecoregion conservation Conservation strategies and activities whose efficacy are enhanced through close attention to larger (landscape- or aquascape-level) spatial and temporal scale patterns of biodiversity, ecological dynamics, threats, and strong linkages of these issues to fundamental goals and targets of biodiversity conservation.

ecosystem A system resulting from the integration of all living and nonliving factors of the environment.

ecosystem service Service provided free by an ecosystem or the environment, such as clean air, clean water, and flood amelioration.

endangered Conservation status category characterized by medium to low probability of persistence of remaining intact habitat.

endemic Native to a particular place and found only there.

endemism Degree to which a geographically circumscribed area, such as an ecoregion or a country, contains species not naturally occurring elsewhere.

enduring feature A landform complex or geographic unit within a natural region characterized by uniform origin and texture of surface material and topography-relief patterns.

environmental gradients Changes in biophysical parameters such as rainfall, elevation, or soil type over distance.

estuarine Associated with an estuary.

estuary A deepwater tidal habitat and its adjacent tidal wetlands, which are usually semi-enclosed by land but have open, partly obstructed, or sporadic access to the ocean and in which ocean water is at least occasionally diluted from freshwater runoff from the land.

evolutionary phenomenon Within the context of WWF regional conservation assessments, an evolutionary phenomenon is a pattern of community structure and taxonomic composition that is the result of extraordinary examples of evolutionary processes, such as pronounced adaptive radiations.

evolutionary radiation See radiation.

exotic species A species that is not native to an area and has been introduced intentionally or unintentionally by humans; not all exotics become successfully established.

extinct Describes a species or population (or any lineage) with no surviving individuals.

extinction The termination of any lineage of organisms, from subspecies to species and higher taxonomic categories from genera to phyla. Extinction can be local, in which one or more populations of a species or other unit vanish but others survive elsewhere, or total (global), in which all the populations vanish.

extirpated Status of a species or population that has vanished from a given area but continues to exist in some other location.

extirpation Process by which an individual, population, or species is destroyed.

family In the hierarchical classification of organisms, a group of species of common descent higher than the genus and lower than the order; a related group of genera.

fauna   All the animals found in a particular place.

fire regime   The characteristic frequency, intensity, and spatial distribution of natural fire events in a given ecoregion or habitat.

flooded grassland   A grassland habitat that experiences regular inundation by water.

flora   All the plants found in a particular place.

fragmentation   Landscape-level variable measuring the degree to which remaining habitat is separated into smaller discrete blocks; process by which habitats are increasingly subdivided into smaller discrete blocks.

fresh water   In the strictest sense, water that has less than 0.5 percent of salt concentration; in this study, refers to rivers, streams, creeks, springs, and lakes.

genera   The plural of *genus*.

genus   A group of similar species with common descent, ranked below the family.

Global 200   A set of approximately 200 terrestrial, freshwater, and marine ecoregions around the world that support globally outstanding or representative biodiversity as identified through analyses by World Wildlife Fund–United States. One component of the Living Planet Campaign.

globally outstanding   Biological distinctiveness category for units of biodiversity whose biodiversity features are equaled or surpassed in only a few other areas around the world.

grassland   A habitat type with landscapes dominated by grasses and with biodiversity characterized by species with wide distributions; communities are resilient to short-term disturbances but not to prolonged, intensive burning or grazing. In such systems larger vertebrates, birds, and invertebrates display extensive movement to track seasonal or patchy resources.

groundwater   Water in the ground that is in the zone of saturation, from which wells, springs, and groundwater runoff are supplied.

guild   Group of organisms, not necessarily taxonomically related, that are ecologically similar in characteristics such as diet, behavior, or microhabitat preference or with respect to their ecological role in general.

gymnosperm   Any of a class or subdivision of woody vascular seed plants that produce naked seeds, not enclosed in an ovary. Conifers and cycads are examples of gymnosperms.

habitat   An environment of a particular kind, often used to describe the environmental needs of a certain species or community.

habitat blocks   Landscape-level variable that assesses the number and extent of blocks of contiguous habitat, taking into account size needs of populations and ecosystems to function naturally. It is measured here by a habitat-dependent and ecoregion size–dependent system.

habitat loss   Landscape-level variable that refers to the percentage of the original land area of the ecoregion that has been lost (converted). It underscores the rapid loss of species and disruption of ecological processes predicted to occur in ecosystems when the total area of remaining habitat declines.

habitat type   In this study, a habitat type is defined by the structure and processes associated with one or more natural communities. An ecoregion is classified under one major habitat type but may encompass multiple habitat types.

headwater   The source of a stream or river.

herbivore   A plant-eating animal, especially ungulates.

herpetofauna   All the species of amphibians and reptiles inhabiting a specified region.

hibernacula   Microhabitats where organisms hibernate during the winter.

indigenous   Native to an area.

intact habitat   Undisturbed areas characterized by the maintenance of most original ecological processes and by communities with most of their original native species still present.

introduced species   See exotic species.

invasive species   Exotic species (i.e., alien or introduced) that rapidly establish themselves and spread through the natural communities into which they are introduced.

invertebrate   Any animal lacking a backbone or bony segment that encloses the central nerve cord.

karst   Areas underlain by gypsum, anhydrite, rock salt, dolomite, quartzite (in tropical moist areas), and limestone, often highly eroded, complex landscapes with high levels of plant endemism.

keystone species   Species that are critically important for maintaining ecological processes or the diversity of their ecosystems.

landform   The physical shape of the land reflecting geologic structure and processes of geomorphology that have sculptured the structure.

landscape   An aggregate of landforms, together with its biological communities.

landscape ecology   Branch of ecology concerned with the relationship between landscape-level features, patterns, and processes and the conservation and maintenance of ecological processes and biodiversity in entire ecosystems.

late-successional   Species, assemblages, structures, and processes associated with mature natural communities that have not experienced significant disturbance for a long time.

life cycle   The entire lifespan of an organism from the moment it is conceived to the time it reproduces.

macroinvertebrates Invertebrates large enough to be seen with the naked eye (e.g., most aquatic insects, snails, and amphipods).

marine Living in salt water.

mesic Mesic habitats are moist, wet areas.

mesophytic Describes plants that grow under conditions of abundant moisture.

meso-predators Intermediate-sized predators, typically about 5–25 kg in body size.

nationally important Biological distinctiveness category.

natural disturbance event Any natural event that significantly alters the structure, composition, or dynamics of a natural community. Floods, fire, and storms are examples.

natural range of variation A characteristic range of levels, intensities, and periodicities associated with disturbances, population levels, or frequency in undisturbed habitats or communities.

non-native species See exotic species.

obligate species A species that must have access to particular habitat type to persist.

old-growth forest A late-successional or climax stage in forest development exhibiting characteristic structural features, species assemblages, and ecological processes.

phylum Primary classification of animals that share similar body plans and development patterns.

population In biology, any group of organisms belonging to the same species at the same time and place.

population sink An area where a species displays negative population growth, often because of insufficient resources and habitat or high mortality.

predator-prey system An assemblage of predators and prey species and the ecological interactions and conditions that permit their long-term coexistence.

protection Landscape-level variable that assesses how well humans have conserved large blocks of intact habitat and the biodiversity they contain. It is measured here by the number of protected blocks and their sizes in a habitat-dependent and ecoregion size–dependent system.

pyrogenic Describes communities or habitats that develop after fire events.

radiation The diversification of a group of organisms into multiple species, caused by intense isolating mechanisms or opportunities to exploit diverse resources.

rarity Seldom occurring either in absolute number of individuals or in space.

refugia Habitats that have allowed the persistence of species or communities because of the stability of favorable environmental conditions over time.

regionally outstanding Biological distinctiveness category.

relatively intact Conservation status category indicating the least possible disruption of ecosystem processes. Natural communities are largely intact, with species and ecosystem processes occurring within their natural ranges of variation.

relatively stable Conservation status category between vulnerable and relatively intact in which extensive areas of intact habitat remain but in which local species declines and disruptions of ecological processes have occurred.

relictual taxa A species or group of organisms largely characteristic of a past environment or ancient biota.

representation The protection of the full range of biodiversity of a given biogeographic unit within a system of protected areas.

restoration Management of a disturbed or degraded habitat that results in recovery of its original state.

riparian Describes the interface between freshwater streams and lakes and the terrestrial landscape.

savanna A habitat dominated by grasslands but with woodland and gallery forest elements.

sclerophyll Type of vegetation characterized by hard, leathery, evergreen foliage that is specially adapted to prevent moisture loss; generally characteristic of regions with Mediterranean climates.

sclerophyllous Relating to sclerophyll.

seral The stages a natural community experiences after a disturbance event.

shrublands Habitats dominated by various species of shrubs, often with many grass and forb elements.

silviculture The management of forest trees, usually to enhance timber production.

sinkholes Depressions or cavities created by dissolution of limestone bedrock or collapse of caves. Typically found in karst landscapes.

source pool A habitat that provides individuals or propagules that disperse to and colonize adjacent or neighboring habitats.

species The basic unit of biological classification, consisting of a population or series of populations of closely related and similar organisms.

species richness A simple measure of species diversity calculated as the total number of species in a habitat or community.

spring A natural discharge of water as leakage or overflow from an aquifer through a natural opening in the soil or rock onto the land surface or into a body of water.

steppe Arid land with xerophilous vegetation, usually found in regions with extreme temperature range and less soils.

subspecies Subdivision of a species. Usually defined as a population or series of populations occupying a discrete range and differing genetically from other geographic races of the same species.

subtropical An area in which the mean annual temperature ranges from 13° C to 20° C.

taxon (pl. taxa) A general term for any taxonomic category, such as a species, genus, family, or order.

temperate An area in which the mean annual temperature ranges from 10° C to 13° C.

terrestrial Living on land.

umbrella species A species whose effective conservation will benefit many other species and habitats, often because of their large area needs or sensitivity to disturbance.

ungulate A member of the group of mammals with hoofs, of which most are herbivorous.

vagile Able to be transported or to move actively from one place to another.

vascular plant A plant with a specialized vascular system for supplying its tissues with water and nutrients from the roots and food from the leaves.

vulnerable Conservation status category characterized by good probability of persistence of remaining intact habitat (assuming adequate protection) but also by loss of some sensitive or exploited species.

watershed See catchment.

wetlands Lands transitional between terrestrial and aquatic systems, where the water table usually is at or near the surface or the land is covered by shallow water; areas inundated or saturated by surface or ground water at a frequency and duration sufficient to support vegetation typically adapted for life in saturated soil conditions.

xeric Describes dryland or desert areas.

xerophilous Thriving in or tolerant of xeric climates.

zoogeography The study of the distributions of animals.

Literature Cited and Consulted

Abdulali, H. 1949. Some peculiarities of avi-faunal distribution in peninsular India. *Proceedings of the National Institute of Science, India* 15: 387–393.

Abeywickrama, B. A. 1980. The flora of Sri Lanka. *Spolia Zeylanica* 35: 1–8.

Abrams, P. A. 1994. Should prey overestimate the risk of predation? *American Naturalist* 144: 317–328.

Adler, G. H., C. C. Austin, and R. Dudley. 1995. Dispersal and speciation of skinks among archipelagos in the tropical Pacific Ocean. *Evolutionary Ecology* 9: 529–541.

Alcala, A. C., and W. C. Brown. 1998. *Philippines amphibians: An illustrated fieldguide*. Makati City: Bookmark, Inc.

Alcala, A. C., W. C. Brown, and A. C. Diesmos. 1998. Two new species of the genus *Platymantis* (Amphibia: Ranidae) from Luzon Island, Philippines. *Proceedings of the California Academy of Science* 50: 381–388.

Alcover, J. A., A. Sans, and M. Palmer. 1998. The extent of extinctions of mammals on islands. *Journal of Biogeography* 25: 913–918.

Ali, S. I. 1978. The flora of Pakistan: Some general and analytical remarks. *Notes from the Royal Botanic Garden, Edinburgh* 36: 427–439.

Allan, T., and A. Warren. 1993. *Deserts. The encroaching wilderness. A world conservation atlas*. New York: Oxford University Press.

Allen, D. A. 1998. On the birds of Tawitawi province in the Philippines. *Bulletin of Tsurumi University* 35: 73–154.

Allen, D. L. 1979. *Wolves of Minong: Their vital role in a wild community*. Boston: Houghton-Mifflin.

Allison, A. 1993. Biodiversity and conservation of the fishes, amphibians, and reptiles of Papua New Guinea. Pages 157–224 in B. M. Beehler (editor), *Papua New Guinea conservation needs assessment*, Vol. 2. Washington, DC: Biodiversity Support Program.

———. 1994. Chapter 16: Biodiversity and conservation of the fishes, amphibians, and reptiles of Papua New Guinea. Pages 157–225 in *Papua New Guinea conservation needs assessment*, Vol. 2. Government of Papua New Guinea Department of Environment and Conservation (DEC). Boroko: The Biodiversity Support Program.

———. 1996. Zoogeography of amphibians and reptiles of New Guinea and the Pacific region. Pages 407–436 in J. A. Keast and S. E. Miller (editors), *The origin and evolution of Pacific island biotas, New Guinea to eastern Polynesia: Patterns and processes*. Amsterdam: SPB Academic Publishing.

Alonso, L. E., and A. L. Mack (editors). 2000. *A biological assessment of the Wapoga River area of northwestern Irian Jaya, Indonesia*. RAP Bulletin of Biological Assessment 14. Washington, DC: Conservation International.

Alvard, M. 2000. The impact of traditional subsistence hunting and trapping on prey populations: Data from Wana horticulturalists of upland central Sulawesi, Indonesia. Pages 214–230 in J. G. Robinson and E. L. Bennett (editors), *Hunting for sustainability in tropical forests*. New York: Columbia University Press.

Amato, G., M. G. Egan, and A. Rabinowitz. 1999. A new species of muntjac, *Muntiacus putaoensis* (Artiodactyla: Cervidae) from northern Myanmar. *Animal Conservation* 2: 1–7.

Anderson, J. A. R. 1980. *A check list of the trees of Sarawak*. Kuching, Sarawak: Vanguard Press Sdn. Bdn.

———. 1983. Chapter 6: The tropical peat swamps of western Malesia. Pages 181–199 in A. J. P. Gore (editor), *Ecosystems of the world: Mires: Swamp, bog, fen and moor*. New York: Elsevier.

Angermeier, P. L. 2000. The natural imperative for biological conservation. *Conservation Biology* 14: 373–381.

Annandale, N. 1918. Fish and fisheries of the Inle Lake. *Rec. Ind. Mus.* 14: 33–64, pls. 1–7.

Anonymous. 1994. *Impact assessment studies of Narmada Sagar and Omkareshwar projects on flora and fauna with attendant human aspects*. WII-EIA Technical Report #9. Dehra Dun, India: WII.

Archer, S., and F. E. Smeins. 1991. Ecosystem-level processes. Pages 109–139 in R. K. Heidtschmidt and J. W. Stuth (editors), *Grazing management: An ecological perspective*. Portland, OR: Timber Press.

Archibold, O. W. 1995. *Ecology of world vegetation*. New York: Chapman & Hall.

Arulchelvam, K. 1969. Mangroves. *The Ceylon Forester* 8: 1–34.

Ashton, M., S. Gunatilleke, N. De Zoysa, M. D. Dassanayake, N. Gunatilleke, and S. Wijesundara. 1997. *A field guide to the common trees and shrubs of Sri Lanka*. Sri Lanka: Wildlife Heritage Trust Publications.

Ashton, P. M. S., C. V. S. Gunatilleke, and I. A. U. N. Gunatilleke. 1992. A shelterwood method of regeneration for sustained timber production in Mesua-Shorea forest of southwest Sri Lanka. In W. Erdelen, C. Preu, N. Ishwaran, and C. M. Madduma Bandara (editors), *Proceedings of the International and Interdisciplinary Symposium, Ecology and Landscape Management in Sri Lanka*. Margraf Scientific Books.

Ashton, P. S. 1980. Dipterocarpaceae. Pages 364–423 in M. D. Dassanayake and F. R. Fosberg (editors), *Flora of Ceylon*, Vol. 1. New Dehli: Oxford and IBH Publishing.

———.1982. Dipterocarpaceae. Pages 237–552 in C. G. G. J. Van Steeni, (editor), *Flora Malesiana*, Vol. 9. *Spermatophyta*. 1. The Hague: Martinus Nijoff.

———. 1989. Sundaland. Pages 91–99 in D. G. Campbell and H. D. Hammond (editors), *Floristic inventory of tropical countries*. New York: The New York Botanical Garden.

———. 1995. Towards a regional forest classification for the humid tropical of Asia. Pages 435–464 in E. O. Box et al. (editors), *Vegetation science in forestry*. Dordrecht, The Netherlands: Kluwer Academic Publishers.

Ashton, P. S., and C. V. S. Gunatilleke. 1987. New light on the plant geography of Ceylon I. Historical plant geography. *Journal of Biogeography* 14: 249–285.

Audley-Charles, M. G., A. M. Hurley, and A. G. Smith. 1981. Continental movements in the Mesozoic and Cenozoic. Pages 9–23 in T. C. Whitmore (editor), *Wallace's line and plate tectonics*. Oxford, England: Clarendon.

Austin, C. C. 1995. Molecular and morphological evolution in South Pacific scincid lizards: Morphological conservatism and phylogenetic relations of Papuan *Lipinia* (Scincidae). *Herpetologica* 51: 291–300.

Avise, J. C. 1998. The history and purview of phylogeography: A personal reflection. *Molecular Ecology* 7: 371–379.

———. 2000. *Phylogeography: The history and formation of species*. Cambridge, MA: Harvard University Press.

Bagarinao, T. 1998. Nature parks, museums, gardens, and zoos for biodiversity conservation and environment education: The Philippines. *Ambio* 27(3): 230–237.

Bahir, M. M. 1998. Three new species of crabs of the genus *Perbrinckia* (Crustacea: Paathelphusidae) from the central mountains of Sri Lanka. *Journal of South Asian Natural History* 3: 197–212.

Baig, K. J. 1998. Amphibian fauna of Azad Jammu and Kashmir with new record of *Paa liebigii*. *Proc. Pakistan Acad. Sci.* 35: 117–121.

Baig, K. J., and L. Gvozdik. 1998. *Uperodon systoma*: Record of a new microhylid frog from Pakistan. *Pakistan J. Zool.* 30: 155–156.

Bailey, R. G. 1998. *Ecoregions: The ecosystem geography of the oceans and continents*. New York: Springer Verlag.

Balakrishnan, N. P. 1989. Andaman Islands: Vegetation and floristics. Pages 55–68 in C. J. Saldanha (editor), *Andaman, Nicobar and Lakshadweep: An environmental impact assessment*. New Delhi: Oxford & IBH Publishing.

Balgooy, M. M. J. van. 1971. Plant geography of the Pacific, based on a census of Phanerogram genera. Blumea, Supplement 6. Cited in K. A. Monk, Y. de Fretes, and G. Reksodiharjo-Lilley et al. 1997. *The ecology of Nusa Tenggara and Maluku*. Hong Kong: Periplus Editions.

Baltzer, M. C., T. D. Nguyen, and R. G. Shore. 2001. *Towards a vision for biodiversity conservation in the forests of the lower Mekong ecoregion complex*. Washington, DC: World Wildlife Fund.

BAPPENAS. 1993. *Biodiversity action plan for Indonesia*. Jakarta.

———. 1999. *Final report, annex I: Causes, extent, impact and costs of 1997/98 fires and drought*. Asian Development Bank technical assistance grant TA 2999-INO, Planning for fire prevention and drought management project. Jakarta.

Barber, C. V., and J. Schweithelm. 2000. *Trial by fire*. Washington, DC: World Resources Institute.

Barbier, E. B., J. C. Burgess, and C. Folke. 1994. *Paradise lost? The ecological economics of biodiversity*. London: Earthscan.

Bartlett, H. H. 1955. *Fire in relation to primitive agriculture in the tropics*. Annotated bibliography, Vol. 1. Ann Arbor: University of Michigan Botanic Garden.

Bauchop, T. 1978. Digestion of leaves in vertebrate arboreal folivores. Pages 193–204 in G. G. Montgomery (editor), *The ecology of arboreal folivores*. Washington, DC: Smithsonian Institution.

Bauchop, T., and R. W. Martucci. 1968. Ruminant-like digestion of the langur monkey. *Science* 161: 698–700.

Bauer, A. M. 1988. Hypothesis: A geological basis for some herpetofaunal disjunctions in the southwest Pacific, with special reference to Vanuatu. *Herp. J.* 1: 259–263.

———. 1999. The terrestrial reptiles of New Caledonia: The origin and evolution of a highly endemic herpetofauna. Pages 3–25 in H. Ota (editor), *Tropical island herpetofauna: Origin, current diversity, and conservation*. Amsterdam: Elsevier.

Bauer, A. M., and R. A. Sadlier. 1993. Systematics, biogeography and conservation of the lizards of New Caledonia. *Biodiversity Letters* 1: 107–122.

———. 1994. The terrestrial herpetofauna of the Ile des Pins, New Caledonia. *Pacific Science* 48: 353–366.

Bauer, A. M., and J. V. Vindum. 1990. A checklist and key to the herpetofauna of New Caledonia, with remarks on biogeography. *Proc. Calif. Acad. Sci.* 47: 17–45.

Bauer, J. J. 1990. The analysis of plant-herbivore interactions between ungulates and vegetation on alpine grasslands in the Himalayan region of Nepal. *Vegetatio* 90: 15–34.

Beebee, T. J. C. 1990. Identification of closely related anuran early life-stages by electrophoretic fingerprinting. *Herpetol. J.* 1: 454–457.

Beehler, B. M. 1993a. Biodiversity and conservation of the warm-blooded vertebrates of Papua New Guinea. Pages 77–155 in B. M. Beehler (editor), *Papua New Guinea conservation needs assessment*, Vol. 2. Washington, DC: Biodiversity Support Program.

———. 1993b. Mapping PNG's biodiversity. Pages 193–209 in J. B. Alcorn (editor), *Papua New Guinea conservation needs assessment*, Vol. 1. Washington, DC: Biodiversity Support Program.

——— (editor). 1994. *Papua New Guinea conservation needs assessment*. Government of Papua New Guinea Department of Environment and Conservation (DEC). Boroko: The Biodiversity Support Program.

Beehler, B. M., and L. E. Alonso (editors). 2001. *Southern New Ireland, Papua New Guinea: A biodiversity assessment*. RAP Bulletin of Biological Assessment 22. Washington, DC: Conservation International.

Beehler, B. M., T. K. Pratt, and D. A. Zimmerman. 1986. *Birds of New Guinea*. Princeton, NJ: Princeton University Press.

Behnke, R. H., and I. Scoones. 1993. Rethinking range ecology: Implications for rangeland management in Africa. Pages 1–30 in R. H. Behnke, I. Scoones, and C. Kerven (editors), *Range ecology at disequilibrium: New models of natural variability and pastoral adaptation in African savannas*. London: ODI.

Belsky, A. J. 1986. Does herbivory benefit plants? A review of the evidence. *American Naturalist* 127: 870–892.

Bennett, E. L. 1998. *The natural history of Orang-Utan*. Kota Kinabalu, Sabah, Malaysia: Natural History Publications (Borneo) Sdn. Bhn.

Bennett, E. L., and F. Gombek. 1993. *Proboscis monkeys of Borneo*. Borneo: Natural History Publications.

Bennett, E. L., A. J. Nyaoi, and J. Sompud. 2000. Saving Borneo's bacon: The sustainability of hunting in Sarawak and Sabah. Pages 305–324 in J. G. Robinson and E. L. Bennett (editors), *Hunting for sustainability in tropical forests*. New York: Columbia University Press.

Bennett, E. L., and J. G. Robinson. In press. *Hunting of wildlife in tropical forests: Implications for biodiversity and forest peoples*. Washington, DC: Global Environment Division, The World Bank.

Bennett, E. L., and A. C. Sebastian. 1988. Social organization and ecology of proboscis monkeys (*Nasalis larvatus*) in mixed coastal forest in Sarawak. *International Journal of Primatology* 9: 233–255.

Berger, J. 1990. Persistence of different-sized populations: An empirical assessment of rapid extinctions in bighorn sheep. *Conservation Biology* 4: 91–98.

Bermingham, E., and C. Moritz. 1998. Comparative phylogeography: Concepts and applications. *Molecular Ecology* 7: 367–369.

Biodiversity Action Plan. 1995. *Biodiversity action plan for Vietnam*. Government of Vietnam and GEF.

Bishop, K. D. 1982. *Endemic birds of Biak Island*. International Council for Bird Preservation survey. Unpublished report.

Bjonness, I. M. 1990. Animal husbandry and grazing: A conservation and management problem in Sagarmatha (Mt. Everest) National Park, Nepal. *Norsk Geogr. Tidsskr* 34: 59–76.

Blair, W. F. 1958. Mating call in the speciation of anuran amphibians. *American Naturalist* 92: 27–51.

Blanford, W. T. 1901. The distribution of vertebrate animals of India, Ceylon, and Burma. *Philosophical Transactions of the Royal Society of London (B)* 194: 335–436. Cited in J. MacKinnon and K. MacKinnon. 1986. *Review of the protected areas system in the Indo-Malayan Realm*. IUCN/UNEP publication.

Blasco, F. 1975. *Mangroves of India*. Pondicherry, India: French Institute.

Blasco, F., M. F. Bellan, and M. Aizpuru. 1996. A vegetation map of tropical continental Asia at scale 1:5 million. *Journal of Vegetation Science* 7: 623–634.

Blaustein, A. R. 1994. Chicken Little or Nero's fiddle: A perspective on declining amphibian populations. *Herpetologica* 50: 85–97.

Blaustein, A. R., and D. B. Wake. 1990. Declining amphibian populations: A global phenomenon? *Trends in Ecology and Evolution* 5: 203–204.

———. 1995. The puzzle of declining amphibian populations. *Scientific American* 1995: 56–61.

Bleeker, P. 1983. *Soils of Papua New Guinea*. Canberra: Australian National University Press.

Blower, J. n.d. *Species conservation priorities in Burma*. Burma: UNDP/FAO Nature Conservation and National Parks Project.

Blume, K. L. 1825. *Dipterocarpaceae in Bijdragen tot de Flora van Nederlanisch Indie*. Batavia.

Böhme, W., and T. Ziegler. 1997. *Varanus melinus* sp. n., ein neuer Waran aus der *V. indicus*-Gruppe von den Molukken, Indonesien. *Herpetofauna* 19: 26–34.

Bonaccorso, F. 1998. *Bats of Papua New Guinea*. Washington, DC: Conservation International.

Bonaccorso, F., L. Salas, and S. Stephens. In prep. *Marsupials and monotremes of New Guinea and satellite islands*.

Boonratana, R. 1997. *Field training in wildlife conservation research techniques and large mammal survey at Nam Poui National Biodiversity Conservation Area*. Vientiane, Laos: IUCN/LSFP.

———. 1998a. *Wildlife survey training at Dong Hua Sao and Phou Xiang Thong National Biodiversity Conservation Areas, Lao PDR*. Vientiane, Laos: IUCN.

———. 1998b. *Field management of Nam Poui and Phou Xang He National Biodiversity Conservation Areas*. Vientiane, Laos: IUCN/LSFP.

———. 1998c. *Nakai–Nam Theun Conservation Project [Phase 2]: Wildlife monitoring techniques and participatory conservation at Nakai–Nam Theun NBCA*. Vientiane, Laos: IUCN/WCS.

———. 1999a. *Establishment and development of Tan Phu Elephant Sanctuary*. Hanoi: FFI.

———. 1999b. *Fauna & Flora International–Indochina Programme: Na Hang Rainforest Conservation Project*. Hanoi: FFI.

———. 1999c. *A preliminary wildlife survey in the Kravanh Range of southwestern Cambodia*. Hanoi: FFI.

———. 2000a. *The Nam Theun Social and Environmental Project (year 2000 activities): Wildlife movements/seasonal migration study and training*. Vientiane, Laos: IUCN.

———. 2000b. *A rapid participatory assessment of wildlife diversity in Dong Sithouane Production Forest*. Vientiane, Laos: FOMACOP.

Boonratana, R., and X. C. Le. 1994. *A report on the ecology, status and conservation of the Tonkin snub-nosed monkey (Rhinopithecus avunculus) in northern Vietnam*. New York/Hanoi: WCS/IEBR.

———. 1998a. Preliminary observations on the ecology and behaviour of the Tonkin snub-nosed monkey (*Rhinopithecus [Presbyticus] avunculus*) in northern Vietnam. Pages 207–215 in N. G. Jablonski (editor), *The natural history of the doucs and snub-nosed monkeys*. Singapore: World Scientific Publishing.

———. 1998b. Conservation of Tonkin snub-nosed monkeys (*Rhinopithecus [Presbyticus] avunculus*) in Vietnam. Pages 315–322 in N. G. Jablonski (editor), *The natural history of the doucs and snub-nosed monkeys*. Singapore: World Scientific Publishing.

Bork, E. W., N. E. West, and J. W. Walker. 1998. Cover components on long-term seasonal sheep grazing treatments in three-tip sagebrush steppe. *Journal of Range Management* 51(3): 293–300.

Bortone, S. A., and W. P. Davis. 1994. Fish intersexuality as indicator of environmental stress: Monitoring fish reproductive systems can serve to alert humans to potential harm. *BioScience* 44: 165–172.

Bosheng, L. 1995. *Biodiversity of the Qinghai-Tibetan plateau and its conservation*. Discussion Paper, Series No. MNR 95/3. Katmandu, Nepal: International Centre for Integrated Mountain Development.

Bouchet, P., T. Jaffré, and J.-M. Veillon. 1995. Plant extinction in New Caledonia: Protection of sclerophyll forests urgently needed. *Biodiversity and Conservation* 4: 415–428.

Boulenger, G. A. 1892. Concluding report on the reptiles and batrachians obtained in Burma by Signor L. Fea dealing with the collection made in Pegu and the Karin Hills in 1887–1888. *Ann. Mus. Civ. Genoa ser. 2* 13: 1–14, pls. I–VI.

Box, E. O., et al. (editors). 1995. *Vegetation science in forestry*. Dordrecht, The Netherlands: Kluwer Academic Publishers.

Bradley, J. 1876. *A narrative of travel and sport*. London: Samual Tinsley.

Bregulla, H. 1992. *Birds of Vanuatu*. Shropshire, England: Anthony Nelson.

Brinkman, W. J., and V. T. Xuan. 1986. *Melaleuca leucadendron, a useful and versatile tree for acid sulfate soils and some other poor environments*. M.S. thesis, Wageningen Agricultural University.

Briske, D. D., and J. H. Richards. 1995. Plant responses to defoliation: A physiological, morphological and demographic evaluation. Pages 635–710 in D. J. Bedunah and R. E. Sosebee (editors), *Wildland plants: Physiological ecology and developmental morphology*. Denver: Society for Range Management.

Brooks, R. R. 1987. *Serpentine and its vegetation*. Portland: Dioscorides Press.

Brooks, T., G. Dutson, B. King, and P. M. Magsalay. 1995. An annotated check-list of the forest birds of Rajah Sikatuna National Park, Bohol, Philippines. *Forktail* 11(1995): 121–134.

Brooks, T. M., T. D. Evans, G. C. L. Dutson, G. Q. A. Anderson, D. C. Asane, R. J. Timmins, and A. G. Toledo. 1992. The conservation status of the birds of Negros, Philippines. *Bird Conservation International* 2: 273–302.

Brooks, T. M., P. Magsalay, G. Dutson, and R. Allen. 1995. Forest loss, extinctions and last hope for birds on Cebu. *Oriental Bird Club Bulletin* 21: 24–26.

Brooks, T. M., S. L. Pimm, and N. J. Collar. 1997. Deforestation predicts the number of threatened birds in insular south-east Asia. *Conservation Biology* 11: 382–394.

Brooks, T. M., S. L. Pimm, V. Kapos, and C. Ravilious. 1999. Threat from deforestation to montane lowland birds and mammals in insular south-east Asia. *Journal of Animal Ecology* 68: 1061–1078.

Brower, B. 1990. Range conservation and Sherpa livestock management in Khumbu, Nepal. *Mountain Research and Development* 10(1): 34–42.

Brown, J. H., and A. C. Gibson. 1983. *Biogeography*. St. Louis, MO: C. V. Mosby.

Brown, J. H., and W. McDonald. 1995. Livestock grazing and conservation on southwestern rangelands. *Conservation Biology* 9: 1644–1647.

Brown, J. S., J. W. Laundre, and M. Gurung. 1999. The ecology of fear: Optimal foraging, game theory, and trophic interactions. *Journal of Mammalogy* 80: 385–399.

Brown, M., and B. Wyckoff-Baird. 1992. *Designing integrated conservation and development projects*. Washington, DC: Biodiversity Support Program.

Brown, R. W. 1995. The water relations of range plants: Adaptations to water deficits. Pages 291–413 in D. J. Bedunah and R. Sosebee (editors), *Wildland plants: Physiological ecology and developmental morphology*. Denver, CO: Society for Range Management.

Brown, W. C. 1991. Lizards of the genus *Emoia* (Scincidae) with observations on their evolution and biogeography. *Mem. Calif. Acad. Sci.* 15: 1–92.

Brown, W. C., and A. C. Alcala. 1970. The zoogeography of the herpetofauna of the Philippine Islands, a fringing archipelago. *Proc. Calif. Acad. Sci.* 38(6) (4.s.): 105–130.

Brown, W. C., A. C. Alcala, and A. C. Diesmos. 1997. A new species of the genus *Platymantis* (Amphibia: Ranidae) from Luzon Island, Philippines. *Proc. Biol. Soc. Washington* 110: 18–23.

Brown, W. C., A. C. Alcala, A. C. Diesmos, and E. Alcala. 1997. Species of the *guentheri* group of *Platymantis* (Amphibia: Ranidae) from the Philippines, with descriptions of four new species. *Proc. California Acad. Sci.* 50: 1–20.

Brown, W. C., R. M. Brown, and A. C. Alcala. 1997. Species of the *hazelae* group of *Platymantis* (Amphibia: Ranidae) from the Philippines, with descriptions of two new species. *Proc. California Acad. Sci.* 49: 405–421.

Brown, W. C., R. M. Brown, V. Kapos, and C. Ravilious. 1999. Threat from deforestation to montane lowland birds and mammals in insular south-east Asia. *Journal of Animal Ecology* 68: 1061–1078.

Brown, W. C., R. M. Brown, P. S. Ong, and A. C. Diesmos. 1999. A new species of *Platymantis* (Amphibia: Ranidae) from the Sierra Madre Mountains, Luzon Island, Philippines. *Proc. Biol. Soc. Washington* 112: 510–514.

Brown, W. C., J. A. McGuire, and A. C. Diesmos. 2000. Status of some Philippine frogs referred to *Rana everetti* (Anura: Ranidae), descriptions of a new species, and resurrection of *Rana igorota* Taylor 1922. *Herpetologica* 56: 81–104.

Bruna, E. M., R. N. Fisher, and T. J. Case. 1996. Morphological and genetic evolution appear decoupled in Pacific skinks (Squamata: Scincidae: *Emoia*). *Proceedings of the Royal Society of London B* 263: 681–688.

Brünig, E. F. 1973. Species richness and stand diversity in relation to site and succession of forests in Sarawak and Brunei (Borneo). *Amazoniana* 4: 293–320.

———. 1974. *Ecological studies in the kerangas forests of Sarawak and Brunei*. Kuching: Borneo Literature Bureau.

Brussard, P. F., D. D. Murphy, and D. R. Tracy. 1994. Cattle and conservation biology: Another view. *Conservation Biology* 8(4): 919–921.

Bryant, D., D. Nielsen, and L. Tangley. 1997. *The last frontier forests: Ecosystems and economies on the edge*. Washington, DC: World Resources Institute.

Burgess, N., et al. In press. *Terrestrial ecoregions of Africa: A conservation assessment*. WWF-USA.

Burley, F. W. 1988. Monitoring biological diversity for setting priorities in conservation. Pages 227–230 in E. O. Wilson (editor), *Biodiversity*. Washington, DC: National Academy Press.

Burley, J. 1991. *A floral inventory of West Kalimantan*. Project proposal.

Burnham, C. P. 1984. Soils. Pages 137–152 in T. C. Whitmore (editor), *Tropical rain forests of the Far East*. New York: Oxford University Press.

Bury, R. B., H. W. Campbell, and N. J. Scott. 1980. Role and importance of nongame wildlife. *Trans. North American Wildl. & Nat. Res. Conf.* 45: 197–207.

Caldecott, J. O. 1980. Habitat quality and populations of two sympathetic gibbons (Hylobatidae) on a mountain in Malaya. *Folia Primatol.* 33: 291–309.

Caldwell, G. S. 1986. Predation as a selective force on foraging herons: Effects of plumage color and flocking. *Auk* 103: 494–505.

Calvo, A., and G. Navarrete. 1999. *El desarrollo del sistema de pago de servicios ambientales en Costa Rica*. San Jose, Costa Rica: FONAFIFO and UNDP.

Campbell, E. J. F. 1994. *A walk through the lowland rain forest of Sabah*. K. M. Wong (editor). Kota Kinabalu, Borneo: Natural History Publications, Sdn. Bhd.

Cannatella, D. 1985. *A phylogeny of primitive frogs (Archaeobatrachians)*. Unpublished doctoral dissertation, University of Kansas, Lawrence.

Cao, M., and J. Zhang. 1997. Tree species diversity of tropical forest vegetation in Xishuangbanna, SW China. *Biodiversity and Conservation* 6: 995–1006.

Carbone, C., S. Christie, K. Conforti, T. Coulson, N. Franklin, J. Ginsberg, M. Griffiths, J. Holden, K. Kawanishi, M. Kinnaird, R. Laidlaw, A. Lynam, D. Martyr, C. McDougal, L. Nath, T. O'Brien, J. Seidensticker, L. J. D. Smith, M. Sunquist, R. Tilson, and W. N. Wan Shahruddin. (2001). The use of photographic rates to estimate densities of tigers and other cryptic mammals. *Animal Conservation* 4: 75–79.

Carbone, C., G. M. Mace, S. C. Roberts, and D. W. MacDonald. 1999. Energetic constraints on the diet of terrestrial carnivores. *Nature* 402: 286–288.

Case, T. J., and M. R. Gilpin. 1974. Inference competition and niche theory. *Proc. National Acad. Sci. U.S.A.* 71: 3073–3077.

Castro, R., L. Gamez, N. Olson, and F. Tattenbach. 1998. *The Costa Rican experience with market instruments to mitigate climate change and conserve biodiversity*. San Jose, Costa Rica: FUNDECOR, MINAE, and the World Bank.

Center for Borneo Studies. 1992. *Forest biology and conservation in Borneo*. Kota Kinabalu, Malaysia: Yayasan Sabah.

Center for Conservation Biology. 1992. *Rapid assessment of forest/wildlife/river ecology in area affected by Kaeng Sua Ten Dam*. Report to World Bank.

Central Statistical Organization, Ministry of National Planning and Economic Development. 1998. *Statistical yearbook*. Rangoon, Union of Myanmar.

Champion, H. G., and S. K. Seth. 1968. *A revised survey of the forest types of India*. Government of India Press.

Chan-Ard, T., W. Grossmann, A. Gumprecht, and K.-D. Schulz. 1999. *Amphibians and reptiles of peninsular Malaysia and Thailand: An illustrated checklist/Amphibien und Reptilien der Halbinsel Malaysia und Thailands: Eine illustrierte Checkliste*. Würselen: Bushmaster.

Chang, D. H. S. 1981. The vegetation zonation of the Tibetan Plateau. *Mountain Research and Development* 1(1): 29–48.

Chaudhary, R. P. 1998. *Biodiversity in Nepal. Status and conservation*. Know Nepal Series No 17. Thailand: Craftsman Press.

Chaudhry, A. A., A. Hussain, M. Hameed, and R. Ahmad. 1997. Biodiversity in Cholistan desert, Punjab, Pakistan. Pages 81–100 in S. A. Mufti, C. A. (editors), *Biodiversity of Pakistan*.

Chazeau, J. 1993. Research on New Caledonian terrestrial fauna: achievements and prospects. *Letters* 1: 123–129.

Chazée, L. 1990. *La province d'Attopeu (Monographie provinciale et étude de districts et villages)*. Vientiane, Laos: UNDP.

Chundawat, R. S., and G. S. Rawat. 1994. *Indian cold desert: A status report on biodiversity*. Dehra Dun, India: WII.

Cincotta, R. P., and R. Engelman. 2000. *Nature's place: Human population and the future of biological diversity*. Washington, DC: Population Action International.

Clark, T. W., A. P. Curlee, S. C. Minta, and P. M. Kareiva (editors). 1999. *Carnivores in ecosystems: The Yellowstone experience.* New Haven: Yale University Press.

Clawson, M. E., T. S. Baskett, and M. J. Armbruster. 1984. An approach to habitat modeling for herpetofauna. *Wildl. Soc. Bull.* 12: 61–69.

Clayton, L., and E. J. Milner-Gulland. 2000. The trade in wildlife in North Sulawesi, Indonesia. Pages 473–495 in J. G. Robinson and E. L. Bennett (editors), *Hunting for sustainability in tropical forests.* New York: Columbia University Press.

Clements, F. E., and V. E. Shelford. 1939. *Bio-Ecology.* New York: John Wiley & Sons.

Coates, B. J. 1985. *The birds of Papua New Guinea: Including the Bismarck Archipelago and Bougainville.* Alderley, Australia: Dove Publications.

Coates, B. J., and K. D. Bishop. 1997. *A guide to the birds of Wallacea.* Alderley, Australia: Dove Publications.

Cockburn, P. F. 1978. The flora. Pages 179–198 in D. M. Luping, C. Wen, and E. R. Dingley (editors), *Kinabalu: Summit of Borneo.* Sabah, Malaysia: The Sabah Society.

Cogger, H., E. Cameron, R. Sadlier, and P. Eggler. 1993. *The action plan for Australian reptiles.* Canberra: Australian Nature Conservation Agency.

Collar, N. J. 1996. The Philippine eagle on its hundredth birthday. *The Raptor* 1996/97.

Collar, N. J., N. Aldrin, D. Mallari, and B. R. Tabaranza, Jr. 1999. *Threatened birds of the Philippines: The Haribon Foundation/BirdLife International red data book.* Makati City, Philippines: Bookmark.

Collar, N. J., M. J. Crosby, and A. J. Stattersfield. 1994. *Birds to watch 2: The world list of threatened birds.* Cambridge, England: BirdLife International.

Collins, N. M., J. A. Sayer, and T. C. Whitmore (editors). 1991. *The conservation atlas of tropical forests: Asia and the Pacific.* London: Macmillan.

Compton, J. 1998. *Borderline: A report on wildlife trade in Vietnam.* Washington, DC: World Wildlife Fund—US.

Conservation International. 1999. *The Irian Jaya Biodiversity Conservation Priority-Setting Workshop final report.* Washington, DC: Conservation International.

Coppedge, B. R., and J. H. Shaw. 1998. Bison grazing practices on seasonally burned tallgrass prairie. *Journal of Range Management* 51(3): 258–264.

Corbet, G. B., and J. E. Hill. 1986. *A world list of mammalian species.* London: British Natural History Museum.

———. 1992. *The Mammals of the Indomalayan Region.* Oxford: Oxford University Press.

Corner, E. J. H. 1952. *Wayside trees of Malaya.* 2 vols. Singapore: n.p.

———. 1978. The plant life. Pages 112–178 in D. M. Luping, C. Wen, and E. R. Dingley (editors), *Kinabalu: Summit of Borneo.* Sabah, Malaysia: The Sabah Society.

Coughenour, M. B. 1991. Spatial components of plant-herbivore interactions in pastoral, ranching and native ungulate ecosystems. *Journal of Range Management* 44: 530–542.

Cox, R., A. Laurie, and M. Woodford. 1992. *The results of four field surveys for Kouprey Bos sauveli in Viet Nam and Lao PDR.* Kouprey Conservation Trust.

Coyne, P. I., M. J. Trlica, and C. E. Owensby. 1995. Carbon and nitrogen dynamics in range plants. Pages 59–167 in D. J. Bedunah and R. E. Sosebee (editors), *Wildland plants: Physiological ecology and developmental morphology.* Denver: Society for Range Management.

Cragg, S. 1987. Papua New Guinea. In R. M. Umali et al. (editors), *Mangroves of Asia and the Pacific: Status and management.* Manila: Natural Resources Management Center and National Mangrove Committee, Ministry of Natural Resources.

Cronin, E. W., Jr. 1979. *The Arun: A natural history of the world's deepest valley.* Boston: Houghton Mifflin.

Cronk, Q. C. B. 1997. Islands: Stability, diversity, conservation. *Biodiversity and Conservation* 6: 477–493.

Cubitt and Stewart-Cox. 1995. *Wild Thailand.* London: New Holland.

Dahl, B. E., and D. N. Hyder. 1977. Developmental morphology and management implications. Pages 257–290 in R. E. Sosebee (editor), *Rangeland plant physiology.* Range Science Series No. 4. Denver: Society for Range Management.

Daily, G. C. 1997. Introduction: What are ecosystem services? Pages 1–10 in G. Daily (editor), *Nature's services: Societal dependence on natural ecosystems.* Washington, DC: Island Press.

Daniels, R. J. R. 1992. Geographical distribution patterns of amphibians in the Western Ghats, India. *Journal of Biogeography* 19: 521–529.

———. 1996. The vanishing aborigines of the Andaman and Nicobar Islands. *Current Science* 70: 775–776.

Danielsen, F., D. S. Balete, T. D. Christensen, M. Heegaard, O. F. Jakobsen, A. Jensen, T. Lund, and M. K. Poulsen. 1994. *Conservation of biological diversity in the Sierra Madre Mountains of Isabela and southern Cagayan Province, the Philippines.* Manila and Copenhagen: BirdLife International.

Darevsky, I. S. 1992. Two new species of the worm-like lizard *Dibamus* (Sauria: Dibamidae) with remarks on distribution and ecology of *Dibamus* in Vietnam. *Asiat. Herpetol. Res.* 4: 1–12.

Darevsky, I. S., and L. A. Kupriyanova. 1993. Two new all-female lizard species of the genus *Leiolepis* Cuvier, 1829 from Thailand and Vietnam (Squamata: Sauria: Uromastycinae). *Herpetozoa* 6: 3–20.

Darevsky, I. S., and N. L. Orlov. 1994. *Vietnascincus rugosus*, a new genus and species of *Dasia*-like arboreal skinks (Sauria, Scincidae) from Vietnam. *Russ. J. Herpetol.* 1: 37.

———. 1997. A new genus and species of scincid lizard from Vietnam: First Asiatic skink with double rows of basal subdigital pads. *Journal of Herpetology* 31: 323–326.

Das, I. 1996a. *Biogeography of the reptiles of south Asia.* Malabar, FL: Krieger.

———. 1996b. Folivory and seasonal change in diet in *Rana hexadactyla* (Anura: Ranidae). *Journal of Zoology, London* 238: 785–794.

———. 1996c. Spatio-temporal resource utilization by a Bornean rainforest herpetofauna: Preliminary results. Pages 315–323 in D. Edwards, W. Booth, and S. Choy (editors), *Tropical rainforest research: Current issues.* Dordrecht, The Netherlands: Kluwer Academic Publishers.

———. 1997. Identifying areas of high herpetological diversity in the Western Ghats, southwestern India. Pages 414–416 in J. van Abbema (editor), *Proceedings: Conservation, restoration and management of tortoises and turtles: An International Conference.* New York: American Museum of Natural History.

———. 1998. A remarkable new species of ranid (Anura: Ranidae) with phytotelmonous larvae from Mount Harriet, Andaman Islands. *Hamadryad* 23: 41–49.

———. 1999. Biogeography of the amphibians and reptiles of the Andaman and Nicobar Islands, India. In H. Ota (editor), *Tropical island herpetofauna: Origin, current diversity, and conservation.* Oxford: Elsevier Science.

———. 2000. A noteworthy collection of mammals from Mount Harriet, Andaman Islands, India. *Journal of South Asian Natural History* 4: 181–185.

———. In press. *The South Asian herpetofauna: A conservation action plan*. Compiled by Indraneil Das and Members of the IUCN/SSC South Asian Reptile and Amphibian Specialist Group.

Das, I., and A. M. Bauer. 1998. Systematics and biogeography of Bornean geckos of the genus *Cnemaspis* Strauch, 1887 (Sauria: Gekkonidae), with the description of a new species. *Raffles Bull. Zool.* 46: 11–28.

Das, I., and S. K. Chanda. 1997. *Philautus sanctisilvaticus* (Anura: Rhacophoridae), a new frog from the sacred groves of Amarkantak, central India. *Hamadryad* 22: 21–27.

———. 1998. A new species of *Philautus* (Anura: Rhacophoridae) from the Eastern Ghats, south-eastern India. *J. South Asian Nat. Hist.* 3: 103–112.

———. 2000. A new species of *Scutiger* (Anura: Megophryidae) from Nagaland, north-eastern India. *Herpetol. J.* 10: in press.

Das, I., and J. Palden. In press. A herpetological collection from Bhutan, with new country records. *Herpetological Review*

Das, I., and M. S. Ravichandran. 1998. A new species of *Polypedates* (Anura: Rhacophoridae) from the Western Ghats, India, allied to the Sri Lankan *P. cruciger* Blyth, 1852. *Hamadryad* 22: 88–94.

Dasmann, R. F. 1972. Towards a system for classifying natural regions of the world and their representation by national parks and reserves. *Biological Conservation* 4: 247–255.

———. 1973. *Classification and use of protected natural and cultural areas*. Morges, UK: IUCN.

———. 1974. *Biotic provinces of the world. Further development of a system for defining and classifying natural regions for purposes of conservation*. IUCN Occasional Paper No 9.

Dassanayake, M. D., and F. R. Fosberg. 1980. *Flora of Ceylon*, Vol. 1. Oxford and IBH Publishing Co.

Daugherty, C. H. A., A. Cree, J. M. Hay, and M. B. Thompson. 1990. Neglected taxonomy and continuing extinctions of tuatara (*Sphenodon*). *Nature, London* 347: 177–179.

Davidson, P. 1998. *A wildlife and habitat survey of Nam Et and Phou Loeuy NBCAs, Houaphanh Province, Lao PDR*. Vientiane, Laos: CPAWM/WCS.

Davies, G., and J. Payne 1982. *A faunal survey of Sabah*. Kuala Lumpur: WWF, Malaysia.

Davis et al. 1995. *Centres of Plant Diversity: A guide and strategy for their conservation*. Vol. 2, Asia, Australasia and the Pacific. Cambridge, England: IUCN.

Davis, M. B. 1981. Quaternary history and the stability of forest communities. Pages 132–153 in D. C. West and H. H. Shugart (editors), *Forest succession*. New York: Springer-Verlag.

———. 1986. Climatic instability, time lags, and community disequilibrium.Pages 269–284 in J. Diamond and T. J. Case (editors), *Community ecology*. New York: Harper and Row.

Davis, P. H. 1971. Distribution patterns in Anatolia with particular reference to endemism. Pages 15–27 in P. H. Davis, P. C. Harper, and I. C. Hedge (editors), *Plant life of south-west Asia*. Edinburgh: The Botanical Society of Edinburgh.

Daws, G., and M. Fujita. 1999. *Archipelago: The islands of Indonesia*. Berkeley: The Nature Conservancy, University of California Press.

Deignan, H. G. 1939. Three new birds of the genus *Stachyris*. *Zool. Series Field Mus. Nat. Hist.* 34: 109–114.

———. 1945. The birds of northern Thailand. *U.S. Nat. Mus. Bull* 186. Washington, DC: Smithsonian Institution.

Denburgh, J. van. 1917. Notes on the herpetology of Guam, Mariana Islands. *Proc. Calif. Acad. Sci.* VII: 37–39.

DENR and UNEP. 1997. *Philippine biodiversity: An assessment and plan of action*. Makati City, Philippines: Bookmark, Inc.

Deraniyagala, P. E. P. 1958. *The Pleistocene of Ceylon*. Ceylon Museum Publication.

Deraniyagala, S. U. 1992. *The prehistory of Sri Lanka. An ecological perspective*. Part I. Department of Archeological Survey. Government of Sri Lanka.

de Rosayro, R. A. 1942. The soils and ecology of the wet evergreen forests of Ceylon. *The Tropical Agriculturist* 98: 4–14.

———. 1950. Ecological conceptions and vegetational types with special reference to Ceylon. *The Tropical Agriculturist* 56: 108–131.

de Silva, A. 1998. Biology and conservation of the amphibians, reptiles and their habitats in South Asia. *Proceedings of the International Conference on the Biology and Conservation of the Amphibians and Reptiles of South Asia*. N.p.: Institute of Fundamental Studies, Kandy, and University of Peradeniya, Sri Lanka.

Dessauer, H. C. 1966. Taxonomic significance of electrophoretic patterns of animal sera. *Rutgers Univ., Serol. Mus. Bull.* 34: 4–8.

Detling, J. K. 1988. Grasslands and savannas: Regulation of energy flow and nutrient cycling by herbivores. Pages 131–154 in L. R. Pomeroy and J. J. Alberts (editors), *Concepts of ecosystem ecology, a comparative view*. New York: Springer-Verlag.

Development Alternatives. 1992. *An aerial reconnaissance of closed canopy forests*. USAID/Philippines Program Document PI 7–92.

Diamond, J., and K. D. Bishop. 1998. *Rivers in the sky: The avifauna of the Kikori pipeline area, Papua New Guinea*. World Wildlife Fund Kikori Integrated Conservation and Development Project, unpublished.

Diamond, J. M., and M. E. Gilpin. 1983. Biogeographic umbilici and the origin of the Philippine avifauna. *Oikos* 41: 307–321.

Dickinson, E. C., R. S. Kennedy, and K. C. Parkes. 1991. *The birds of the Philippines: An annotated check-list*. Tring, UK: British Ornithologists' Union (Checklist no. 12).

Dijuwantoko. 1991. *Orangutan habitat conservation*. The Great Apes Conference, December 15–22, 1991. Indonesia.

Dillon, T. C., and E. D. Wikramanayake. 1997. Parks, peace and progress: A forum for transboundary conservation in Indo-china. *PARKS* 7(3): 36–51.

Dinerstein, E. 1980. An ecological survey of the Royal Karnali-Bardia Wildlife Reserve, Nepal. Part III: Ungulate populations. *Biological Conservation* 18: 5–38.

———. 2002. The return of the unicorns: A success story in the conservation of Asian rhinoceros. New York: Columbia University Press.

Dinerstein, E., and J. N. Mehta. 1989. The clouded leopard in Nepal. *Oryx* 23: 199–201.

Dinerstein, E., D. M. Olson, D. Graham, A. Webster, S. Primm, M. Bookbinder, and G. Ledec. 1995. *A conservation assessment of the terrestrial ecoregions of Latin America and the Caribbean*. Washington, DC: World Bank.

Dinerstein, E., and C. L. Price. 1991. Demography and habitat use by greater one-horned rhinoceros in Nepal. *Journal of Wildlife Management* 55: 401–411.

Dinerstein, E., A. Rijal, M. Bookbinder, and B. Kattel. 1999. Tigers as neighbours: Efforts to promote local guardianship of endangered species in lowland Nepal. Page 393 in J. Seidensticker, S. Christie, and P. Jackson (editors), *Riding the tiger: Tiger conservation in human-dominated landscapes*. Cambridge, England: Cambridge University Press.

Dinerstein, E., and E. D. Wikramanayake. 1993. Beyond "hotspots": How to prioritize investments in biodiversity in the Indo-Pacific region. *Conservation Biology* 7: 53–65.

Dinerstein, E., E. Wikramanayake, J. Robinson, U. Karanth, A. Rabinowitz, D. Olson, T. Mathew, P. Hedao, M. Connor, G. Hemley, and D. Bolze. 1997. *A framework for identifying high priority areas and actions for conservation of tigers in the wild*. Washington, DC: WWF-US and Wildlife Conservation Society.

Dodd, C. K., Jr. 1997. Imperiled amphibians: A historical perspective. Pages 165–200 in G. W. Benz and D. E. Collins (editors), *Aquatic fauna in peril: The southeastern perspective*. Special Publication 1. Decatur, GA: Southeast Aquatic Research Institute.

Dormaar, J. F., B. W. Adams, and W. D. Willms. 1997. Impacts of rotational grazing on mixed prairie soils and vegetation. *Journal of Range Management* 50(6): 647–651.

Doughty, C., N. Day, A. Plant. 1999. *Birds of the Solomons, Vanuatu & New Caledonia*. London: Christopher Helm.

Driessen, P. M. 1977. Peat soils. Pages 763–779 in *Soils and rice*. Los Banos, Philippines: IRRI.

Dubois, A. 1975. Un nouveau complexe d'especies jumelles distinguées par le chant: Les grenouilles du Nepal voisines de *Rana limnocharis* Boie (Amphibiens, Anoures). *C. R. Acad. Sci. Paris* 281: 1717–1720.

———. 1987. Miscellanea taxinomica batrachologica (I). *Alytes* 5(1–2): 7–95.

———. 1992. Notes sur la classification des Ranidae (Amphibiens, Anoures). *Bull. Mens. Soc. Linn. Lyon* 61: 305–352.

———. 1999. South Asian Amphibia: A new frontier for taxonomists. *J. South Asian Nat. Hist.* 4: 1–11.

Dubois, A., and A. Ohler. 1998. A new species of *Leptobrachium* (*Vibrissaphora*) from northern Vietnam, with a review of the taxonomy of the genus *Leptobrachium* (Pelobatidae, Megophryinae). *Dumerilia* 4: 1–32.

Duckworth, J. W. 1996. Bird and mammal records from the Sangthong district, Vientiane municipality. *Nat. Hist. Bull. Siam Soc.* 44: 217–242.

Duckworth, J. W., R. E. Salter, and K. Khounboline. 1999. *Wildlife in Lao PDR: 1999 status report*. Vientiane, Laos: IUCN, Wildlife Conservation Society, and Centre for Protected Areas and Watershed Management.

Duckworth, J. W., R. J. Timmins, R. M. Thewlis, T. D. Evans, and G. Q. A. Anderson. 1994. Field observations of mammals in Lao, 1992–1993. *Nat. Hist. Bull. Siam Soc.* 42: 177205.

Duke, N. C., M. C. Ball, and J. C. Ellison. 1998. Factors influencing biodiversity and distributional gradients in mangroves. *Global Ecology and Biogeography Letters* 7: 27–47.

Dumbacher, J. P., B. M. Beehler, T. F. Spande, H. M. Garraffo, and J. W. Daly. 1992. Homobatrachotoxin in the genus *Pitohui*: Chemical defense in birds? *Science* 258: 799–801.

Dung, V. V., P. M. Gaio, N. N. Chinh, D. Tuoc, P. Arctander, and J. MacKinnon. 1993. A new species of living bovid from Vietnam. *Nature* 363: 443–445.

Dutson, C. L., T. D. Evans, T. M. Brooks, D. C. Asane, R. J. Timmins, and A. Toledo. 1992. Conservation status of birds on Mindoro, Philippines. *Bird Conservation International* 2: 303–325.

Dutta, S. K. 1997. *Amphibians of India and Sri Lanka (checklist and bibliography)*. Bhubaneswar, India: Odyssey Publishing House.

Dutta, S. K., M. F. Ahmed, and I. Das. In press. *Kalophrynus* (Anura: Microhylidae), a new genus for India, with the description of a new species from Assam State. *Hamadryad* 25.

Dutta, S., and K. Manamendra-Arachchi. 1996. *The amphibian fauna of Sri Lanka*. Sri Lanka: The Wildlife Heritage Trust of Sri Lanka.

Eisenberg, J. F. 1989. An introduction to the Carnivora. Pages 1–9 in J. Gittleman (editor), *Carnivore behavior, ecology, and evolution*. Ithaca, NY: Cornell University Press.

Eisenberg, J. F., and J. Seidensticker. 1976. Ungulates in southern Asia: A consideration of biomass estimates for selected habitats. *Biological Conservation* 10: 293–308.

Ellis, J. L., S. N. Yoganarasimhan, M. R. Gurudeva, and P. Ramanujam. 2000. Prioritization of biodiversity rich sites of conservation significance in the Andaman and Nicobar Islands. Pages 75–81 in S. Singh, A. R. K. Sastry, R. Mehta, and V. Uppal (editors), *Setting biodiversity conservation priorities for India*. New Delhi: WWF-India.

Ellison, J. C. 1997. Mangrove ecosystems of the Western and Gulf Provinces, Papua New Guinea: A review. *Science in New Guinea* 23: 1–15.

Ellison, L. 1960. Influence of grazing on plant succession of rangelands. *Botanical Review* 26: 1–78.

Emerson, S. B. 1992. Courtship and nest-building behavior of a Bornean frog, *Rana blythi*. *Copeia* 1992: 1123–1127.

Emmons, L. 1991. Body size and feeding tactics. Page 62 in J. Seidensticker and S. Lumpkin (editors), *Great cats*. Emmaus, PA: Rodale Press.

Environmental Systems Research Institute. 1993. *Digital chart of the world*. CD-ROM. Redlands, CA: ESRI.

Erdelen, W. R. 1988. Forest ecosystems and nature conservation in Sri Lanka. *Biological Conservation* 43: 115–135.

———. 1993. *Biogeographical patterns and processes: The tropical continental island Sri Lanka*. [In German]. Saarland University, unpublished.

———. 1996. Tropical rain forest in Sri Lanka: Characteristics, history of human impact, and the protected area system. Pages 503–511 in D. S. Edwards, W. E. Booth, and S. C. Choy (editors), *Tropical rainforest research: Current issues*. Dordrecht, The Netherlands: Kluwer Academic Publishers.

Eudey, A. A. 1987. *Action plan for Asian primate conservation*. Gland, Switzerland: IUCN.

Evans, T., A. J. Stones, and R. C. M. Thewlis. 1996. *A wildlife and habitat survey of the Dong Hua Sao National Biodiversity Conservation Area, Pakxe*. Vientiane, Laos: IUCN.

Evans, T. D., J. W. Duckworth, and R. J. Timmins. In prep. *Field observations of mammals in Laos during 1994–1995*.

Fairlie, S., M. Hagler, and B. O'Riordan. 1995. The politics of overfishing. *The Ecologist* 25(2/3): 46–73.

Fall, J. J. 1999. Transboundary biosphere reserves: Applying landscape ecological arguments to protected area planning. *Landskabsøkologiske Skrifter*, special issue, University of Roskilde, Denmark.

FAO. 1974. *A land capability appraisal, Indonesia. Interim report*. AGL/INS/72/011. Rome: UNDP, FAO. Cited in T. Whitmore. 1984. A vegetation map of Malesia at scale 1:5 million. *Journal of Biogeography* 11: 461–471.

———. 1981. *Tropical Forest Resources Assessment Project: Forest resources of tropical Asia*. Rome: FAO and UNEP.

FAO/UNDP. 1983. *Thamihla Kyun and southern Arakan. Preliminary survey*. FO/BUR/80/006. Nature Conservation and National Parks Project.

Fauna & Flora International. 2000. Asia's best kept secret. *Fauna & Flora News* 13: 1–2.

Federal Interagency Stream Restoration Working Group. 1998. *Stream corridor restoration*.

Ferguson, S. H., A. T. Bergerud, and R. Ferguson. 1988. Predation risk and habitat selection in the persistence of a remnant caribou population. *Oecologia* 76: 236–245.

Fernando, P., E. Wikramanayake, and D. Weerakoon. In prep. *Developing an elephant conservation strategy in Sri Lanka: Field research, community-based problem solving, and training for local conservation scientists*.

Ferraro, P. J. 2000. *Global habitat protection: Limitations of development interventions and a role for conservation performance payments*. Department of Agricultural, Resource, and Managerial Economics Working Paper no. 2000-03. Ithaca, NY: Cornell University. (forthcoming in *Conservation Biology*).

Fiedler, P. L., and P. M. Kareiva (editors). 1998. *Conservation biology: For the coming decade*, 2d ed. New York: Chapman & Hall.

Flannery, T. 1990. *Mammal of New Guinea*. Carina: Robert Brown and Associates.

———. 1995. *Mammals of the southwest Pacific & Moluccan Islands*. Ithaca, NY: Cornell University Press.

Flannery, T. F., and C. P. Groves. 1998. A revision of the genus *Zaglossus* with description of new species and subspecies. *Mammalia* 62: 367–396.

Fleischner, T. L. 1994. Ecological costs of livestock grazing in western North America. *Conservation Biology* 8: 629–644.

Foose, T. J., and N. J. van Strien (editors). 1997. *Asian rhinos: Status survey and conservation action plan*. Gland, Switzerland: IUCN.

———. 2000. *Report on the regional meeting for India and Nepal of the IUCN/SSC Asian Rhino Specialist Group (AsRSG): 21–27 February 1999*. Kaziranga, Assam, India: The World Conservation Union.

Foster, R. B. 2001. The forest vegetation of southern New Ireland. In B. M. Beehler and L. E. Alonso (editors), *Southern New Ireland, Papua New Guinea: A biodiversity assessment*. Washington, DC: Conservation International.

Fox, J. L. 1989. *A review of the status and ecology of the snow leopard (Panthera uncia)*. Seattle: International Snow Leopard Trust.

———. 1994. Snow leopard conservation in the wild—a comprehensive perspective on a low density and highly fragmented population. In *Proceedings of the Seventh International Snow Leopard Symposium*. Seattle: International Snow Leopard Trust.

———. 1997. Rangeland management and wildlife conservation in the Hindu Kush–Himalayas. Pages 53–57 in D. J. Miller and S. R. Craig (editors), *Rangelands and pastoral development in the Hindu Kush–Himalayas*. Proceedings of a regional expert's meeting, November 5–7, 1996, Kathmandu, Nepal. Kathmandu: International Centre for Integrated Mountain Development.

Fox, J. L., C. Norbu, S. Bhatt, and A. Chandola. 1994. Wildlife conservation and land-use changes in the Transhimalayan region of Ladakh, India. *Mountain Research and Development* 14(1): 39–60.

Fox, J. L., S. P. Sinha, R. S. Chundawat, and P. K. Das. 1991. Status of the snow leopard *Panthera uncia* in Northwest India. *Biological Conservation* 55: 283–298.

Franklin, N., Bastoni, Sriyanto, S. Dwiatmo, J. Manansang, and R. Tilson. 1999. Last of the Indonesian tigers: A cause for optimism. In J. Seidensticker, S. Christie, and P. Jackson (editors), *Riding the tiger: Tiger conservation in human-dominated landscapes*. Cambridge, England: Cambridge University Press.

Freitag, H., 1971. Die naturliche vegetation Afghanistan. *Vegetation* 22: 285–344.

Frith, C. B., and B. M. Beehler. 1998. *The birds of paradise: Paradisaeidae*. New York: Oxford University Press.

Frost, D. R. 1999. *Amphibian species of the world: An online reference*. Version 2.1 (15 November 1999). New York: American Museum of Natural History. Available at http://research.amnh.org/herpetology/amphibia/index.html.

Galdikas, B. M. F. 1971–1977. *Wild orangutan studies at Tanjung Puting Reserve, Central Indonesian Borneo*. Research Reports. National Geographic Society.

Galdikas, B. M. F., and N. Briggs. 1999. *Orangutan odyssey*. New York: Orangutan Foundation International and Harry N. Abrams Inc.

Gamauf, A., and S. Tebbich. 1995. Re-discovery of the Isabela oriole *Oriolus isabellae*. *Forktail* 11: 170–171.

Ganesh, T., R. Ganeshan, S. M. Devy, P. Davidar, and K. S. Bawa. 1996. Assessment of plant biodiversity at a mid-elevation evergreen forests of Kalakad–Mundanthurai Tiger Reserve, Western Ghats, India. *Current Science* 17: 389–392.

Gans, C. 1993. Fossorial amphibians and reptiles: Their distributions as environmental indicators. Pages 189–199 in W. Erdelen, C. Preu, N. Ishwaran, and C. M. Madduma Bandara (editors), *Ecology and landscape management in Sri Lanka*. Weikersheim: Margraf.

Gascon, C., B. Williamson, and G. da Fonseca. 2000. Receding forest edges and vanishing reserves. *Science* 288: 1356–1358.

Gaston, K. J. 2000. Global patterns in biodiversity. *Nature* 405: 220–227.

Gaussen H., V. M. Meher-Homji, F. Blasco, A. Delacourt, J. Fontanel, P. Legris, and J. P. Troy. 1969. *International map of the vegetation and of environmental conditions*. Wainganga sheet. Pondicherry, India: French Institute.

———. 1973. *International map of the vegetation and of environmental conditions*. Orissa sheet. Pondicherry, India: French Institute.

Georgiadis, N. J., and S. J. McNaughton. 1990. Elemental and fibre contents of savanna grasses: Variation with grazing, soil type, season, and species. *Journal of Applied Ecology* 27: 623–634.

Giao, P. M., V. V. Dung, D. Touc, E. D. Wikramanayake, G. Amato, P. Arctander, and J. M. MacKinnon. 1998. Description of *Muntiacus truongsonensis*, a new species of muntjac (Artiodactyla: Muntiacidae) from central Vietnam, and implications for conservation. *Animal Conservation* 1: 61–68.

Gibbons, J. R. H. 1981. The biogeography of *Brachylophus* including the description of a new species, *B. vitiensis*, from Fiji. *Journal of Herpetology* 15: 255–273.

———. 1985. The biogeography and evolution of Pacific island reptiles and amphibians. Pages 125–142 in G. Grigg, R. Shine, and H. Ehmann (editors), *Biology of Australasian frogs and reptiles*. Sydney: Royal Society of New South Wales.

Giesen, W. 1987. *Danau Sentarum Wildlife Reserve. Inventory, ecology and management guidelines*. Bogor, Indonesia: WWF/PHPA.

Gilg, E. 1925. Dipterocarpaceae in Engler. *Naturlichen Pflanzen-familien* 121: 237–269.

Ginsberg, J. R., and D. W. MacDonald. 1990. Foxes, wolves, jackals, and dogs: An action plan for the conservation of canids. Gland, Switzerland: IUCN/SSC Canid and Wolf Specialist Groups.

Global Witness. 1999. *The untouchables: Forest crimes and concessionaires—can Cambodia afford to keep them?* London: Global Witness.

GOI. 1997. *Status report on biodiversity in India*. New Delhi: Ministry of Environment and Forests, Government of India.

Goldstein, M. C., C. M. Beall, and R. P. Cincotta. 1990. Traditional nomadic pastoralism and ecological conservation on Tibet's "Northern Plateau." *National Geographic Research* 6: 139–156.

Gonzalez, P. C., and C. P. Rees. 1988. *Birds of the Philippines*. Manila: Haribon Foundation for the Conservation of Natural Resources.

Goodman, S. M., and N. R. Ingle. 1993. Sibuyan Island in the Philippines: Threatened and in need of conservation. *Oryx* 27: 174–180.

Graham, R. W. 1986. Response of mammalian communities to environmental changes during the late Quaternary. Pages 330–313 in J. Diamond and T. J. Case (editors), *Community ecology*. New York: Harper & Row.

Greene, M., and J. Paine. 1997. State of the world's protected areas. Presented at the IUCN World Commission on Protected Areas Symposium on Protected Areas in the 21st Century: From Islands to Networks. Available online at http://www.wcmc.org.uk/protected-areas/albany.htm.

Grewal, B. (editor) 1992. *Insight guides: Indian wildlife*. Singapore: APA Publications.

Grimmett, R., C. Inskipp, and T. Inskipp. 1998. *Birds of the Indian subcontinent*. London: Christopher Helm.

———. 1999. *A guide to the birds of India, Pakistan, Nepal, Bangladesh, Bhutan, Sri Lanka, and the Maldives*. Princeton, NJ: Princeton University Press.

Grip, H. 1986. *A short description of the experimental watershed study at Sipitang, Sabah*. Paper presented at the Workshop on Hydrological Studies in Sabah, Kota Kinabalu, April 28.

Groombridge, B. (editor). 1992. *Global biodiversity: State of the earth's living resources*. New York: Chapman & Hall.

Groves, C. P., G. B. Schaller, G. Amato, and K. Khounboline. 1997. Rediscovery of the wild pig *Sus bucculentus*. *Nature* 386: 335.

Groves, C. P., G. Schaller, K. Khounboline, and G. Amato. 1998. Phylogenetic and conservation significance of the rediscovery of *Sus bucculentus* (Mammalia, Suidae). *Nature*.

Gunatilleke, I. A. U. N., and C. V. S. Gunatilleke. 1981. The floristic composition of Sinharaja. A rain forest in Sri Lanka with special reference to endemics and dipterocarps. *Malaysian Forester* 44: 386–396.

———. 1990. Distribution of floristic richness and its conservation in Sri Lanka. *Conservation Biology* 4: 21–31.

Günther, R. 1994. A new species of the genus *Gekko* (Reptilia, Squamata, Gekkonidae) from southern Vietnam [in German]. *Zool. Anzeiger* 233: 57–67.

Gupta, R. K. 1994. Arcto-alpine and boreal elements in the high altitude flora of northwest Himalaya. Pages 11–32 in Y. Pangtey and R. Rawal (editors), *High altitudes of the Himalaya: Biogeography, ecology and conservation*. Gyanodaya, Nainital.

Gyamtsho, P. 1996. *Assessment of the condition and potential for improvement of high altitude rangelands of Bhutan*. Unpublished Ph.D. dissertation. Zurich: Swiss Federal Institute of Technology.

Hagmeier, E. M. 1966. A numerical analysis of the distributional patterns of North American mammals. II. Re-evaluation of the provinces. *Systematic Zoology* 15: 279–299.

Hagmeier, E. M., and C. D. Stults. 1964. A numerical analysis of the distributional patterns of North American mammals. *Systematic Zoology* 13: 125–155.

Halim, H., G. Hatta, K. MacKinnon, and A. Mangalik. 1996. *The ecology of Kalimantan: Indonesia Borneo*. The Ecology of Indonesia Series, Vol. 3. Hong Kong: Periplus Editions.

Hall, R., and J. D. Holloway (editors). 1998. *Biogeography and geological evolution of SE Asia*. Leiden, The Netherlands: Backhuys Publishers.

Hannecart, F., and Y. Letocart. 1980. *Oiseaux de Nouvelle Caledonie et des Loyautes*, [Birds of New Caledonia and the Loyalty Islands], Vol. 1. Auckland, New Zealand: Clark and Matheson Ltd.

———. 1983. *Oiseaux de Nouvelle Caledonie et des Loyautes*, [Birds of New Caledonia and the Loyalty Islands], Vol. 2. Hong Kong: Dai Nippon.

Haridasan, K., and R. R. Rao. 1985. *Forest flora of Meghalaya*. Dehra Dun, India: Bishen SinghMahendra Pal Singh.

Harper, P., and L. Fullerton. 1993. *Philippine handbook*, 2d ed. Chico, CA: Moon Publications.

Harrison, B. 1987. *Orang-utan*. New York: Oxford University Press.

Harvey, M. B., and D. G. Barker 1998. A new species of blue-tailed monitor lizard (genus *Varanus*) from Halmahera Island, Indonesia. *Herpetologica* 54: 34–44.

Hassinger, J. D. 1968. *Introduction to the mammal survey of the 1965 Street expedition to Afghanistan*. Chicago: Field Museum of Natural History.

Hawkins, R. E. (editor). 1986. *Encyclopedia of Indian natural history*. Bombay Natural History Society. Delhi: Oxford University Press.

Heaney, L. R. 1985. Zoogeographic evidence for Middle and Late Pleistocene land bridges to the Philippine Islands. *Modern Quaternary Research in Southeast Asia* 9: 127–143.

———. 1986. Biogeography of mammals in southeast Asia: Estimates of rates of colonization, extinction, and speciation. *Biological Journal of the Linnean Society* 28: 127–165.

———. 1991a. A synopsis of climatic and vegetational change in southeast Asia. *Climatic Change* 19: 53–61.

———. 1991b. An analysis of patterns of distribution and species richness among Philippine fruit bats (Pteropodidae). *Bulletin of the American Museum of Natural History* 206: 145–167.

———. 1993. Biodiversity patterns and the conservation of mammals in the Philippines. *The Asian International Journal of Life Sciences* 2(2): 261–274.

Heaney, L. R., D. S. Balete, L. Dolar, A. C. Alcala, A. Dans, P. C. Gonzales, N. Ingle, M. Lepiten, W. Oliver, E. A. Rickart, B. R. Tabaranza, Jr., and R. C. B. Utzurrum. 1998. A synopsis of the mammalian fauna of the Philippine Islands. *Fieldiana Zoology* new series, 88: 1–61.

Heaney, L. R., and N. A. D. Mallari. 2001. A preliminary analysis of current gaps in the protection of threatened Philippine terrestrial mammals. *Sylvatrop* 9: in press.

Heaney, L. R., and R. A. Mittermeier. 1997. The Philippines. Pages 236–255 in R. A. Mittermeier, P. Robles Gil, and C. G. Mittermeier (editors), *Megadiversity. Earth's biologically wealthiest nations*. Monterrey, Mexico: CEMEX.

Heaney, L. R., P. S. Ong, R. A. Mittermeier, and C. G. Mittermeier. 1999. Philippines. Pages 309–315 in R. A. Mittermeier, N. Meyers, P. R. Gil, and C. G. Mittermeier (editors), *Hotspots: Earth's biologically richest and most endangered terrestrial ecoregions*. Mexico City: CEMEX, S.A.

Heaney, L. R., and J. C. Regalado, Jr. 1998. *Vanishing treasures of the Philippine rain forest*. Chicago: The Field Museum.

Heaney, L. R., and E. A. Rickart. 1990. Correlations of clades and clines: Geographic, elevational, and phylogenetic distribution patterns among Philippine mammals. Pages 321–332 in G. Peters and R. Hutterer (editors), *Vertebrates in the Tropics*. Bonn, Germany: Mus. Alexander Koenig.

Heaney, L. R., and B. R. Tabaranza, Jr. 1997. A preliminary report on mammalian diversity and conservation status of Camiguin Island, Philippines. *Sylvatrop* 5(1995): 57–64.

Heaney, L. et al. 1998. A synopsis of the mammalian fauna of the Philippine Islands. *Fieldiana: Zoology*, n.s., no. 88, Field Museum of Natural History, Chicago.

Hedges, S. In press. *Asian wild cattle and buffaloes status survey and conservation action plan*. Gland, Switzerland: IUCN.

Heinen, J. T., and P. B. Yonzon. 1994. A review of conservation issues and programs in Nepal: From a single species focus toward biodiversity protection. *Mountain Research and Development* 14: 61–76.

Hemmer, H. 1972. Uncia uncia. *Mammalian Species* 20: 1–5.

Henry, G. M. 1998. *A guide to the birds of Sri Lanka*. Calcutta: Oxford University Press.

Heude, P. M. 1892. Etude sur les Suilliens. *Memoires de l'Histoire Naturelle de l'Empire Chinois* 3: 85–111, 212–222. In G. B. Schaller and E. S. Vrba. 1996. Description of the giant muntjac (*Megamuntiacus vuquangensis*) in Laos. *Journal of Mammalogy* 77(3): 675–683.

Hillis, D. M., C. Moritz, and B. K. Mable (editors). 1996. *Molecular systematics*. Sunderland, MA: Sinauer.

Hilton-Taylor, C. (compiler). 2000. *2000 IUCN red list of threatened species*. Viewed November 2000. The IUCN Species Survival Commission and the International Union for Conservation of Nature and Natural Resources. Available at http://www.redlist.org.

His Majesty's Government of Nepal/Asian Development Bank/ANZDEC. 1992. *Nepal livestock master plan*. Kathmandu.

Hogarth, P. J. 1999. *The biology of mangroves*. Oxford, England: Oxford University Press.

Holechek, J. L., R. D. Pieper, and C. H. Herbel. 1995. *Range management, principles and practices*. Englewood Cliffs, NJ: Prentice Hall.

Holloway, J. D. 1987. Lepidoptera patterns involving Sulawesi: What do they indicate of past geography? Pages 103–118 in T. C. Whitmore (editor), *Biogeographical evolution of the Malay Archipelago*. Oxford, England: Clarendon.

———. 1998. Geological signal and dispersal noise in two contrasting insect groups in the Indo-Australian tropics: R-mode analysis of pattern in Lepidoptera and cicadas. Pages 291–314 in R. Hall and J. D. Holloway (editors), *Bio-geography and geological evolution of SE Asia*. Leiden, The Netherlands: Backhuys Publishers.

Hoogerwerf, A. 1970. *Udjung Kulon: The land of the last Javan rhinoceros*. Leiden, The Netherlands: E. J. Brill.

Hora, S. L. 1949. Satpura hypothesis of the distribution of the Malayan fauna and flora to Peninsular India. *Proceedings of the National Institute of Science, India* 15: 309–314.

How, R. A., and D. J. Kitchener. 1997. Biogeography of Indonesian snakes. *Journal of Biogeography* 24: 725–735.

How, R. A., and L. H. Schmitt, Maharadatunkamsi 1996a. Geographical variation in the genus *Dendrelaphis* (Serpentes: Colubridae) within the islands of south-eastern Indonesia. *J. Zool. Lond.* 238: 351–363.

How, R. A., L. H. Schmitt, and A. Suyanto. 1996b. Geographical variation in the morphology of four snake species from the Lesser Sunda Islands, eastern Indonesia. *Biol. J. Linn. Soc.* 59: 439–456.

Huen-pu, W., C. Sing-chi, and W. Si-yu. 1989. China. Pages in D. G. Campbell and H. D. Hammond (editors), *Floristic inventory of tropical countries*. New York: The New York Botanical Garden.

Hundley, H. G. 1964. *Checklist of Burmese reptiles*. Burma: KYI.

Hunt, G. R. 1996. Environmental variables associated with population patterns of the kagu *Rhynochetos jubatus*. *Ibis* 138: 778–785.

Hunt, G. R., R. Hay, and C. J. Veltman. 1996. Multiple kagu *Rhynochetos jubatus* deaths caused by dog attacks at a high-altitude study site on Pic Ningua, New Caledonia. *Bird Conservation International* 6: 295–306.

Hutchinson, C. S. 1989. *Geological evolution of southeast Asia*. Oxford, England: Clarendon.

IFFM/GTZ. 1998. IFFM/GTZ interim report about the fire situation in Kalimantan Timur and the on-going activities. Available online at http://www.iffm.or.id/firesit.html.

Inger, R. F. 1999. Distribution of amphibians in southern Asia and adjacent islands. Pages 445–482 in W. E. Duellman (editor), *Patterns of distribution of amphibians*. Baltimore: The John Hopkins University Press.

Inger, R. F., Boeadi, and A. Taufik. 1996. New species of ranid frogs (Amphibia: Anura) from central Kalimantan, Borneo. *Raffles Bull. Zool.* 44: 363–369.

Inger, R. F., N. Orlov, and I. Darevsky. 1999. Frogs of Vietnam: A report on new collections. *Fieldiana Zool.* n.s. (92): i–iv + 1–46.

Inger, R. F., H. B. Shaffer, M. Koshy, and R. Bakde. 1987. Ecological structure of a herpetological assemblage in south India. *Amphibia-Reptilia* 8: 189–202.

Inger, R. F., and R. Stuebing. 1997. *A field guide to the frogs of Borneo*. Kota Kinabalu, Malaysia: Natural History Publications.

Inger, R. F., H. K. Voris, and P. Walker. 1986. Larval transport in Bornean Indonesia ranid frog. *Copeia* 1986(2): 523–525.

Inskipp, C. 1989. *Nepal's forest birds: Their status and conservation*. Cambridge, England: International Council for Bird Preservation.

Iremonger, S., C. Ravilious, and T. Quinton (editors). 1997. *A global review of forest conservation*. CD-ROM. Cambridge, England: WCMC/CIFOR.

Iskandar, D. T. 1996. The biodiversity of the amphibians and reptiles of the Indo-Australian Archipelago: Assessment for future studies and conservation. Pages 353–365 in I. M. Turner, C. H. Diong, S. S. L. Lim, and P. K. L. Ng (editors), *Biodiversity and the dynamics of ecosystems*. DIWPA Series Vol. 1.

———. 1998. *Amfibi Jawa dan Bali*. Puslitbang Biologi-LIPI and GEF-Biodiversity Collections Project, Bogor. English edition, 1998. *The amphibians of Java and Bali*. Bogor, Indonesia: Research and Development Centre for Biology-LIPI and GEF-Biodiversity Collections Project.

Iskandar, D., T. Boeadi, and M. Sancoyo. 1996. *Limnonectes kadarsani* (Amphibia: Anura: Ranidae), a new frog from the Nusa Tenggara Islands. *Raffles Bull. Zool.* 44: 21–28.

Iskandar, D. T., and W. R. Erdelen. *Conservation of amphibians and reptiles in Indonesia: Issues and problems*. Unpublished manuscript.

Iskandar, D. T., and K. N. Tjan. 1996. The amphibians and reptiles of Sulawesi, with notes on the distribution and chromosomal numbers of frogs. Pages 39–46 in D. J. Kitchener and A. Suyanto (editors), *Proceedings of the First International Conference on Eastern Indonesian–Australian Vertebrate Fauna*. Perth: Western Australian Museum.

IUCN. 1991. *The conservation atlas of tropical forests: Asia and the Pacific*. London: Macmillan.

———. 1992. *National conservation review*. Interim Progress Report: January–June 1992. IUCN/EMD Report No 13. IUCN-UNDP-FAO.

———. 1993. *Nature reserves of the Himalaya and mountains of central Asia*. Prepared by the World Conservation Monitoring Centre. Gland, Switzerland: IUCN.

———. 1994. *1994 IUCN red list of threatened animals*. Gland, Switzerland: IUCN.

———. 1994. *Partnership in conservation*. IUCN Sri Lanka's Programme. 1988–1994. Colombo, Sri Lanka: IUCN.

———. 1996. *1996 IUCN red list of threatened animals*. Gland, Switzerland: IUCN.

Jackson, P., and E. Kemf. 1994. *Wanted alive! Tigers in the wild*. Gland, Switzerland: WWF.

Jackson, R. 1979. Snow leopards in Nepal. *Oryx* 15: 191–195.

———. 1992. Snow leopard. Unpublished data sheet, IUCN/SSC/Cat Specialist Group, Bougy-Villars, Switzerland.

———. 1998. *Snow-leopard habitat maps*. Unpublished digital maps. Seattle, WA: International Snow-Leopard Trust.

Jacobs, M. 1988. *The tropical rainforest. A first encounter*. Berlin: Springer-Verlag.

Jacobson, G. 1978. The geology. Pages 101–111 in D. M. Luping, C. Wen, and E. R. Dingley (editors), *Kinabalu: Summit of Borneo*. Sabah, Malaysia: The Sabah Society.

Jaffré, T. 1993. The relationship between ecological diversity and floristic diversity in New Caledonia. *Biodiversity Letters* 1: 82–87.

Jaffré, T., P. Bouchet, and J.-M. Veillon. 1998. Threatened plants of New Caledonia: Is the system of protected areas adequate? *Biodiversity and Conservation* 7: 107–135.

Jaffré, T., and J.-M. Veillon. 1994. Les principales formations végétales autochtones en Nouvelle-Calédonie: Caractéristiques, vulnérabilité, mesures de sauvegarde. *Rapports de synthèses, Sciences de la Vie Biodiversité*, no. 2, Nouméa, Nouvelle-Calédonie: ORSTOM.

James, A. N., M. J. B. Green, and J. R. Paine, 1999. *A global review of protected area budgets and staff*. WCMC Biodiversity Series no. 10. Cambridge: WCMC Conservation Press.

Jameson, S. C., J. W. McManus, and M. D. Spalding. 1995. *State of the reefs: Regional and global perspectives*. International Coral Reef Initiative Executive Secretariat. Background Paper. Washington, DC: US Department of State.

Janzen, D. H. 1986. *Guanacaste National Park: Tropical, ecological and cultural restoration*. San José, Costa Rica: Editorial Universidad Estatal a Distancia.

———. 1988. Tropical dry forests: The most endangered major tropical ecosystem. Pages 130–137 in E. O. Wilson (editor), *Biodiversity*. Washington, DC: National Academy Press.

Jayasuriya, A. H. M. 1984. Flora of Ritigala Strict Natural Reserve. *The Sri Lanka Forester* XVI: 61–155.

Johns, R. J. 1993. Biodiversity and conservation of native flora of Papua New Guinea. Pages 15–75 in B. M. Beehler (editor), *Papua New Guinea conservation needs assessment*, Vol. 2. Washington, DC: Biodiversity Support Program.

Johnsingh, A. J. T., H. S. Panwar, and W. A. Rodgers. 1991. Ecology and conservation of large felids in India. Pages 160–166 in N. Marayuma (editor), *Wildlife conservation: Present trends and perspectives for the 21st century*. Yushima, Bankyo-Ku, Tokyo: Japan Wildlife Research Centre.

Johnsingh, A. J. T., S. N. Prasad, and S. P. Goyal. 1990. Conservation status of the Chila-Motichur corridor for elephant movement in Rajaji-Corbett National Parks area, India. *Biological Conservation* 51: 125–138.

Johnsingh, A. J. T., and A. C. Williams. 1998. Elephant corridors in India: Lessons for other elephant range countries. *Oryx* 33: 210–214.

———. 1999. Elephant corridors in India: Lessons for other elephant range countries. *Oryx* 33: 210–214.

Joshi, A. R., J. L. D. Smith, and F. J. Cuthbert. 1995. Influence of food distribution and predation pressure on spacing behavior in palm civets. *Journal of Mammalogy* 76: 1205–1212.

Joshi, A. R., J. L. D. Smith, and D. L. Garshelis. 1999. Sociobiology of the myrmecophagous sloth bear in Nepal. *Canadian Journal of Zoology* 77: 1690–1704.

Juniper, T., and M. Parr. 1998. *Parrots: A guide to parrots of the world*. New Haven: Yale University Press.

Kachroo, P. 1993. *Plant diversity in northwest Himalaya: A preliminary survey*. Pages 111–132. in U. Dhar (editor), *Himalayan biodiversity: Conservation strategies*. Almora, India: G. P. Pant Institute.

Karanth, K. U. 1995. Estimating tiger *panthera tigris* populations from camera-trap data using capture-recapture models. *Biological Conservation* 71: 333–338.

Karanth, K. U., and B. M. Stith. 1999. Prey depletion as a critical determinant of tiger population viability. Pages 100–113 in J. Seidensticker, S. Christie, and P. Jackson (editors), *Riding the tiger: Tiger conservation in human-dominated landscapes*. Cambridge, England: Cambridge University Press.

Karanth, K. U., and M. Sunquist. 1995. Prey selection by tiger, leopard and dhole in tropical forests. *Journal of Animal Ecology* 64: 439–450.

———. 2000. Behavioural correlates of predation by tiger (*Panthera tigris*), leopard (*Panthera pardus*) and dhole (*Cuon alpinus*) in Nagarahole, India. *Journal of Zoology, London* 250: 255–265.

Karanth, K. U., M. Sunquist, and K. M. Chinnappa. 1999. Long-term monitoring of tigers: Lessons from Nagarahole. Pages 114–122 in J. Seidensticker, S. Christie, and P. Jackson (editors), *Riding the tiger: Tiger conservation in human-dominated landscapes*. Cambridge, England: Cambridge University Press.

Kareiva, P., and U. Wennergren. 1995. Connecting landscape patterns to ecosystem and population processes. *Nature* 373(6512): 299–302.

Kartawinata, K., R. Abdulhadi, and T. Partomiharjo. 1981. Composition and structure of a lowland dipterocarp forest at Wanariset, East Kalimantan. *Malaysian Forester* 44(2): 397–406.

Karunakaran, P. V., G. S. Rawat, and V. K. Uniyal. 1998. *Ecology and conservation of the grasslands of Eravikulam National Park, Western Ghats*. RR-98/001. August 1998. Dehra Dun, India: WII.

Keast, A. 1996. Pacific biogeography: Patterns and processes. Pages 477–512 in A. Keast and S. E. Miller (editors), *The origin and evolution of Pacific island biotas, New Guinea to eastern Polynesia: Patterns and processes*. Amsterdam: SPB Academic Publishing.

Kemp, A. 1995. *The hornbills*. Oxford, England: Oxford University Press.

Kenderick, K. 1989. Sri Lanka. In D. G. Campbell and H. D. Hammond (editors), *Floristic inventory of tropical countries*. New York: The New York Botanical Garden.

Kennedy, M. 1992. *Australasian marsupials and monotremes. An action plan for their conservation*. Gland, Switzerland: IUCN/SSC Australasian Marsupial and Monotreme Specialist Group.

Kennedy, R. S., P. C. Gonzales, E. C. Dickinson, H. C. Miranda, Jr., and T. H. Fisher. 2000. *A guide to the birds of the Philippines*. Oxford, England: Oxford University Press.

Kenney, J. S., J. L. D. Smith, A. M. Starfield, and C. W. McDougal. 1995. The long-term effects of tiger poaching on population viability. *Conservation Biology* 9: 1127–1133.

Kerley, L. L., J. M. Goodrich, D. G. Miquelle, E. N. Smirnov, H. B. Quigley, and M. G. Hornocker. *Effects of roads and human disturbance on Amur tigers*. In review.

Khan, M. S. 1994. A revised checklist and key to the amphibians of Pakistan. *Hamadryad* 19: 11–14.

———. 1997. A new toad of genus *Bufo* from the foot of Siachin Glacier, Baltistan, north-eastern Pakistan. *Pakistan Journal of Zoology* 29: 43–48.

King, B., and S. Kanwanich. 1978. First wild sighting of the white-eyed river martin. *Biological Conservation* 13: 183–186.

King, B., M. Woodcock, and E. C. Dickinson. 1975. *Birds of South-East Asia*. London: Collins.

Kingdon-Ward, F. 1921. *In farthest Burma: The record of an arduous journey of exploration and research through the unknown frontier territory of Burma and Tibet*. Philadelphia: Lippincott.

———. 1930. *Plant hunting on the edge of the world*. London: V. Gollancz Ltd.

———. 1952. *Plant hunter in Manipur*. London: Cape.

Kinghorn, J. R. 1928. Herpetology of the Solomon Islands. *Rec. Austral. Mus.* 16: 123–178.

Kirtisinghe P. 1957. *The Amphibia of Ceylon*.

Kitchener, D. J., Boedi, L. Charlton, and Maharadatunkamsi. 1990. Wild mammals of Lombok Island. *Records of the Western Australian Museum Supplement* 33: 1–129.

Kiyasetuo, and M. K. Khare. 1986. A new genus of frog (Anura: Ranidae) from Nagaland at the north-eastern hills of India. *Asian J. Expl. Sci.* 1: 12–17.

Klepper, O., G. Hatta, G. Chairuddin, Sunardi, and Iriansyah. 1990. *Acid sulphate soils in the humid tropics. Ecology component.* First interim report. RIN/KPSL-UNLAM.

Knight, W. J., and J. D. Holloway (editors). 1990. *Insects and the rain forests of south east Asia (Wallacea).* London: The Royal Entomological Society.

Knott, C. 1998. Changes in orangutan diet, caloric intake and ketones in response to fluctuating fruit availability. *International Journal of Primatology* 19: 1061–1079.

Kostermans, A. J. G. H. 1992. *A handbook of the Dipterocarpaceae of Sri Lanka.* Wildlife Heritage Trust of Sri Lanka.

Kottelat, M. 1989. Zoogeography of the fishes from Indochinese inland waters with an annotated check-list. *Bull. Zool. Mus. Univ. Amst.* 12: 1–54.

———. 1990a. The ricefishes (Oryziidae) of the Malili Lakes, Sulawesi, Indonesia, with description of a new species. *Ichthyol. Explor. Freshwaters* 1: 151–166.

———. 1990b. Sailfin silversides (Pisces: Telmatherinidae) of Lakes Towuti, Mahalona and Wawontoa (Sulawesi, Indonesia) with descriptions of two new genera and two new species. *Ichthyol. Explor. Freshwaters* 1: 227–246.

———. 1991. Sailfin silversides (Pisces: Telmatherinidae) of Lake Matano, Sulawesi, Indonesia, with descriptions of six new species. *Ichthyol. Explor. Freshwaters* 1: 321–344.

———. 1995. The fishes of the Mahakam River, East Borneo: An example of the limitations of zoogeographic analyses and the need for extensive fish surveys in Indonesia. *Trop. Biodiv.* 2: 401–426.

———. 1998. Fishes of the Nam Theun and Xe Bangfai basins, Laos, with diagnoses of twenty-two new species (Teleostei: Cyprinidae, Balitoridae, Cobitidae, Coiidae and Odontobutidae). *Ichthyol. Explor. Freshwaters* 9: 1–128.

———. 1999. *Ecoregions and fish diversity in the Indochinese area.* Washington, DC: World Wildlife Fund.

———. 2000. *Fishes of Laos.* Colombo, Sri Lanka: Wildlife Heritage Trust.

Kottelat, M., A. J. Whitten, S. N. Kartikasari, and S. Wirjoatmodjo. 1993. *Freshwater fishes of western Indonesia and Sulawesi.* Hong Kong: Periplus Editions.

Kottelat, M., and T. Whitten. 1996. Freshwater biodiversity in Asia with special reference to fish. *World Bank Technical Paper* 343: i–ix + 1–59.

Kroenke, L. W. 1996. Place tectonic development of the western and south western Pacific: Mesozoic to the present. Pages 19–34 in A. Keast and S. E. Miller (editors), *The origin and evolution of Pacific Island biotas, New Guinea to eastern Polynesia: Patterns and processes.* Amsterdam: SPB Academic Publishing.

Lamb, A., and C. L. Chan. 1978. The orchids. Pages 219–254 in D. M. Luping, C. Wen, and E. R. Dingley (editors), *Kinabalu: Summit of Borneo.* Sabah, Malaysia: The Sabah Society.

Lambert, F., and M. Woodcock. 1996. *Pittas, broadbills and asities.* Mountfield, England: Pica Press.

Langholz, J., and J. Lassoie. In press. Perils and promise of privately owned protected areas. *BioScience.*

Lathrop, A. 1997. Taxonomic review of the megophryid frogs (Anura: Pelobatoidea). *Asiatic Herpetol. Res.* 7: 68–79.

Lauenroth, W. K., D. G. Micchunas, J. L. Dodd, R. H. Hart, R. K. Heitschmidt, and L. R. Rittenhous. 1994. Effects of grazing on ecosystems of the Great Plains. Pages 69–100 in M. Vavra, W. Laycock, and R. Pieper (editors), *Ecological implications of livestock herbivory in the West.* Denver, CO: Society for Range Management.

Launbenfels, D. J. de. 1975. *Mapping the world's vegetation. Regionalization of formations and flora.* Syracuse, NY: Syracuse University Press.

Laurance, W. F. 1991. Ecological correlates of extinction proneness in Australian tropical rain forest mammals. *Conservation Biology* 5: 79–89.

———. 1995. Is a virus decimating frog populations? *Aliens* (newsletter of the Invasive Species Specialist Group of the IUCN/SSC), March: 7.

———. 2000. Do edge effects occur over large spatial scales? *Trends in Ecology and Evolution* 15: 134–135.

Laurance, W. F., and R. O. Bierregaard (editors). 1997. *Tropical forest remnants. Ecology, management, and conservation of fragmented communities.* Chicago: University of Chicago Press.

Lawlor, T. E. 1986. Comparative biogeography of mammals on islands. *Biological Journal of the Linnean Society* 28: 99–125.

Lawrance, W. F. 1995. Is a virus decimating frog populations? *Aliens* March (1995): 7.

Laycock, W. A. 1994. Implications of grazing vs. no grazing on today's rangelands. Pages 250–280 in M. Vavra, W. A. Laycock, and R. D. Pieper (editors), *Ecological implications of livestock herbivory in the West.* Denver, CO: Society for Range Management.

Lazell, J. 1992. New flying lizards and predictive biogeography of two Asian archipelagos. *Bull. Mus. Comp. Zool.* 152: 475–505.

Lekagul, B., and J. A. McNeely. 1977. *Mammals of Thailand.* Bangkok: Sahakarnbhat.

Lekagul, B., and P. D. Round. 1991. *A guide to the birds of Thailand.* Bangkok: Saha Karn Bhaet.

Lepiten, M. V. 1997. The mammals of Siquijor Island, central Philippines. *Sylvitrop* 5(1995): 1–17.

Leslie, A. J. 1999. "For whom the bell tolls." *Tropical Forest Update* 9: 4. Yokohama, Japan: International Tropical Timber Organization.

Lewis, R. E. 1988. Mt. Apo and other national parks in the Philippines. *Oryx* 22(2): 100–109.

Liao, Y. F., and Tan, B. J. 1988. A preliminary study on the geographical distribution of snow leopards in China. In H. Freeman (editor), *Proceedings of the Fifth International Snow Leopard Symposium.* Seattle: International Snow Leopard Trust.

Lima, S. L., and L. M. Dill. 1990. Behavioral decisions made under the risk of predation: A review and prospectus. *Canadian Journal of Zoology* 69: 619–640.

Liu, D. S., L. R. Iverson, and S. Brown. 1993. Rates and patterns of deforestation in the Philippines: Application of geographic information system analysis. *Forest Ecology and Management* 57: 1–16.

Lloyd, M., R. F. Inger, and F. W. King. 1968. On the diversity of reptile and amphibian species in a Bornean rainforest. *American Naturalist* 102: 497–516.

Lomolino, M. V. 1986. Mammalian community structure on islands: The importance of immigration, extinction and interactive effects. *Biological Journal of the Linnean Society* 28: 1–21.

Londoño, A. C., E. Alvarez, E. Forero, and C. M. Morton. 1995. A new genus and species of Dipterocarpaceae from the Neotropics. I. Introduction, taxonomy, ecology and distribution. *Brittonia* 47: 225–236.

Longino, J. T. 1993. Scientific naming. *Natl. Geogr. Res. & Explor.* 9: 80–85.

Lovejoy, T. E. 1980. A projection of species extinctions. In Council on Environmental Quality, *The Global 2000 Report to the President,* vol. 2. Washington, DC: Council on Environmental Quality.

Lovich, J., and J. W. Gibbons. 1997. Conservation of covert species: Protecting species we don't even know. Pages 426–429 in J. van Abbema (editor), *Proceedings: Conservation, restoration and management of turtles and tortoises—An international conference.* Purchase: State University of New York.

Low, J. 1991. *The smuggling of endangered wildlife across the Taiwan Strait*. Cambridge: TRAFFIC International.

Lowry, P. P. 1996. Diversity, endemism, and extinction in the flora of New Caledonia. Viewed April 25, 2000. Available at http://www.mobot.org/MOBOT/research/newcaledonia/caledonia.html#toc; INTERNET.

Luping, D. M., C. Wen, and E. R. Dingley (editors). 1978. *Kinabalu: Summit of Borneo*. Kota Kinabalu, Sabah, Malaysia: The Sabah Society.

Lwin, T. U. 1995. Biodiversity conservation and management in Myanmar. Pages 259–326 in *Banking on biodiversity. Report on the regional consultation on biodiversity assessment in the Hindu Kush–Himalayas*. Nepal: ICIMOD.

Lynam, A. J. 1997. Rapid decline of small mammal diversity in monsoon evergreen forest fragments in Thailand. Pages 222–239 in W. F. Laurance and R. O. Bierregaard (editors), *Tropical forest remnants. Ecology, management, and conservation of fragmented communities*. Chicago: University of Chicago Press.

Ma, Y., L. Baining, and L. Qingyun. 1997. Improve yak productivity through resuming "black soil type" deteriorated grassland. Pages 291–294 in Yang Rongzhen, H. Xingtai, and L. Xiaolin (editors), *Yak production in central Asian highlands, Proceedings of the Second International Congress on Yak*. September 1–6, 1997, Xining, Qinghai Province, China. Xining: Qinghai People's Publishing House.

Mabberley, D. J. 1987. *The plant book: A portable dictionary of the higher plants*. Cambridge: Cambridge University Press.

MacArthur, R. H., and E. O. Wilson. 1967. The theory of island biogeography. Princeton, NJ: Princeton University Press.

Mace, G. M., A. Balmford, L. Boitani, G. Cowlishaw, A. P. Dobson, D. P. Faith, K. J. Gaston, C. J. Humphries, R. I. Vane-Wright, P. H. Williams, J. H. Lawton, C. R. Margules, R. M. May, A. O. Nicholls, H. P. Possingham, C. Rahbek, and A. S. van Jaarsveld. 2000. It's time to work together and stop duplicating conservation efforts. *Nature* 405(6785): 393.

Mace, G. M., N. Collar, J. Cooke, K. Gaston, J. Ginsberg, N. Leader-Williams, N. Maunder, and E.J. Milner-Gulland. 1992. The development of new criteria for listing species on the IUCN Red List. *Species* 19: 16–22.

Mace, G. M., and S. Stuart. 1994. Draft IUCN red list categories. *Species* 22–23: 13–34.

Mack, A. (editor). 1998. *A biological assessment of the Lakekamu Basin, Papua New Guinea*. RAP Working Paper No. 9. Washington, DC: Conservation International.

MacKinnon, J. 1987. *Ecological guidelines for the development of Xishuangbanna Prefecture. Yunnan Province, China*. Gland, Switzerland: WWF.

———. 1997. *Protected areas systems review of the Indo-Malayan realm*. Canterbury, England: ABC/WCMC/ World Bank.

MacKinnon, J. M., and M. B. Artha. 1981. *National conservation plan for Indonesia*. Vol. 7, *Maluku and Irian Jaya*. UNDP/FAO National Parks Development Project. Bogor, Indonesia: FAO.

MacKinnon, J. M., R. Beudels, A. Robinson, and M. B. Artha. 1982. *National conservation plan for Indonesia*. Vol. 4, *Nusa Tenggara*. UNDP/FAO National Parks Development Project. Bogor, Indonesia: FAO.

MacKinnon, J., and K. MacKinnon. 1986. *Review of the protected areas system in the Indo-Malayan Realm*. Gland, Switzerland: IUCN/UNEP.

MacKinnon, J., K. MacKinnon, G. Child, and J. Thorsell. 1986. *Managing protected areas in the tropics*. Gland, Switzerland: IUCN.

MacKinnon, J., and K. Phillipps. 1993. *A field guide to the birds of Borneo, Sumatra, Java, and Bali*. Oxford: Oxford University Press.

MacKinnon, K. 1983. Report of a WHO consultancy to Indonesia to determine population estimates of the cynomolgus of long-tailed macaque *Macaca fascicularis* (and other primates) and the feasibility of semi-wild breeding projects of this species. *WHO Primate Resources Programme feasibility study: Phase II*.

———. 1986. The conservation status of nonhuman primates in Indonesia. Pages 99–126 in K. Benirschke (editor), *Primates, the road to self-sustaining populations*. New York: Springer-Verlag.

MacKinnon, K., G. Hatta, H. Halim, and A. Mangalik. 1996. *The ecology of Kalimantan. Indonesian Borneo*. Singapore: Periplus Editions.

Macnae, W. 1968. A general account of the fauna and flora of mangrove swamps and forests in the Indo–West Pacific region. *Adv. Mar. Biol.* 6: 73–270.

Madhusudan, M. D., and K. U. Karanth. 2000. Hunting for an answer: Is local hunting compatible with large mammal conservation in India? Pages 339–355 in J. G. Robinson and E. L. Bennett (editors), *Hunting for sustainability in tropical forests*. New York: Columbia University Press.

Madoc, G. 1950. Field notes on some Siamese birds. *Bull. Raffles Mus.* 23: 129–190.

Maehr, D. S. 1997. *The Florida panther*. Washington, DC: Island Press.

Maguire, B. P. C., and P. S. Ashton. 1977. Pakaramoideae, Dipterocarpaceae of the Western Hemisphere. II. Systematic, geographic, and phyletic considerations. *Taxon* 26: 359–368.

Mallari, N. A. D., and A. Jensen. 1993. Biological diversity in northern Sierra Madre, Philippines: Its implication for conservation and management. *Asia Life Sciences* 2(2): 101–112.

Mallon, D. 1983. The status of Ladakh urial *Ovis orientalis vignei* in Ladakh, India. *Biological Conservation* 27: 373–381.

Maloney, B. K. 1985. Man's impact on the rainforests of West Malesia: The palynological record. *Journal of Biogeography* 12: 537–558.

Maluku and Irian Jaya. UNDP/FAO National Parks Development Project. Bogor, Indonesia: FAO.

Manamendra-Arachchi, K., and S. Liyanage. 1994. Conservation of the agamid lizards of Sri Lanka with illustration of the extant species. *Journal of South Asian Natural History* 1: 77–96.

Mani, M. S. 1974. *Ecology and biogeography of India*. The Hague: Junk.

———. 1978. *Ecology and phytogeography of high altitude plants of north west Himalaya*. New Delhi: Oxford and IBH Publishing.

———. 1994. The Himalaya, its ecology and biogeography: A review. Pages 1–10 in Y. P. S. Pangtey and R. S. Rawal (editors), *High altitudes of the Himalaya (biogeography, ecology and conservation)*. Gyanodaya Prakashan, Nainital.

Manthey, U., and C. Steiof. 1998. *Rhacophorus cyanopunctatus* sp. n. (Anura: Rhacophoridae), ein neuer Flugfrosch von der Malaiischen Halbinsel, Sumatra und Borneo. *Sauria, Berlin* 20: 37–42.

Mares, M. A. 1999. *Encylopedia of deserts*. Norman: University of Oklahoma Press.

Margules, C. R., and R. L. Pressey. 2000. Systematic conservation planning. *Nature* 405(6783): 243–253.

Marsh, C. W., and W. L. Wilson. 1981. *A survey of primates in peninsular Malaysian forests*. Universiti Kebangsaan Malaysia and University of Cambridge, U.K.

Maskey, T. M. 1979. *Royal Chitwan National Park: Report on Gharial*. Kathmandu, Nepal: Department of National Parks and Wildlife Conservation.

Matsui, M., T. Chan-Ard, and J. Nabjitabhata. 1996. Distinct specific status of *Kalophrynus pleurostigma interlineatus* (Anura, Microhylidae). *Copeia* 1996: 440–445.

Matsui, M., J. Nabhitabhata, and S. Panha. 1999. On *Leptobrachium* from Thailand with a description of a new species (Anura: Pelobatidae). *Japanese Journal of Herpetology* 18: 19–29.

Mayr, E. 1940. The origin and history of the bird fauna of Polynesia. *Proc. 6 Pacific Science Congress* 4: 197–216.

McAlpine, J. R., G. Keig, and R. Falls. 1983. *Climate of Papua New Guinea*. Canberra: Australian National University Press.

McClung, R. M. 1997. *Last of the wild: Vanished and vanishing giants of the animal world*. North Haven, CT: Linnet Books.

McClure, H. E., and Lekagul. 1961. Some birds of the Mae Ping River in northern Thailand. *Nat. Hist. Bull. Siam Soc.* 20(1): 1–8.

McCoy, M. 1980. *Reptiles of the Solomon Islands*. Wau Ecology Institute Handbook No 7. Papua New Guinea: Wau.

———. 2000. *Reptiles of the Solomon Islands*. CD-ROM.

McGregor, R. C. 1910. Birds collected in the island of Polillo, Philippines. *Phil. J. Sci.* 5: 103–114. Cited in N. J. Collar, N. Aldrin, D. Mallari, and B. R. Tabaranza, Jr. III. 1999. *Threatened birds of the Philippines: The Haribon Foundation/BirdLife International red data book*. Makati City, Philippines: Bookmark, Inc.

McKay, G. M. 1973. *Behavior and ecology of the Asiatic elephant in southeastern Ceylon*. Smithsonian Contributions to Zoology. 125. Washington, DC: Smithsonian Institution Press.

McPhee, R. D. E., and P. A. Marx. 1997. The 40,000-year plague. Humans, hyperdisease, and first-contact extinctions. Pages 169–217 in S. M. Goodman and B. D. Patterson (editors), *Natural change and human impact in Madagascar*. Washington, DC: Smithsonian Institution Press.

Md. Nor, S. In press. Elevational diversity patterns of small mammals on Mount Kinabalu, Sabah, Malaysia. *Global Ecology and Biogeography*.

Mech, D. L. 1966. *The wolves of Isle Royale*. U.S. National Park Service Faunal Series No 7: 210.

Meffe, G. K., and C. R. Carroll. 1994. *Principles of conservation biology*. Sunderland, MA: Sinauer Associates.

Meher-Homji, V. M. 1989. History of vegetation of peninsular India. *Man and Environment* 13: 1–10.

Memmott, K. L., V. J. Anderson, and S. B. Monsen. 1998. Seasonal grazing impact on cryptogamic crusts in a cold desert ecosystem. *Journal of Range Management* 51(5): 547–550.

Menasveta, P. 1985. Current fish disease epidemic in Thailand. *J. Sci. Soc. Thailand* 11: 147–160.

Menon, V. 1996. *Under siege: Poaching and protection of greater one-horned rhinoceroses in India*. Washington, DC: WWF-IUCN.

Meteorological Department. 1987. *Climatological data of Thailand, 30-year period (1956–1985)*. Bangkok: Ministry of Communications.

Mickleburgh, S., A. M. Hutson, and P. A. Racey. 1992. *Old world fruit bats. An action plan for their conservation*. Gland, Switzerland: IUCN/SSC Chiroptera Specialist Group.

Miehe, G. 1988. Geoecological reconnaissance in the alpine belt of southern Tibet. *GeoJournal* 17(4): 635–648.

———. 1989. Vegetation patterns on Mt. Everest as influenced by monsoon and föhn. *Vegetatio* 79: 21–32.

Milchaunas, D. G., O. E. Sala, and W. K. Lauenroth. 1988. A generalized model of the effects of grazing by large herbivores on grassland community structure. *American Naturalist* 132: 87–106.

Miller, D. J. 1997. Conserving biological diversity in the Hindu Kush–Himalayas–Tibetan Plateau rangelands. Pages 47–52 in D. J. Miller and S. R. Craig (editors), *Rangelands and pastoral development in the Hindu Kush–Himalayas, proceedings of a regional expert's meeting*. November 5–7, 1996, Kathmandu, Nepal. Kathmandu: International Centre for Integrated Mountain Development.

———. 1998a. *Fields of grass: Portraits of the pastoral landscape and nomads of the Himalaya and Tibetan Plateau*. Kathmandu: ICIMOD.

———. 1998b. Conserving biodiversity in Himalayan and Tibetan Plateau rangelands. Pages 291–320 in *Ecoregional co-operation for biodiversity conservation in the Himalaya, Report on the International Meeting on Himalaya Ecoregional Co-operation*. February 16–18, 1998, Kathmandu, Nepal. New York: UNDP.

Miller, D. J., and R. Jackson. 1994. Livestock and snow leopards: Making room for competing users on the Tibetan Plateau. Pages 315–328 in J. L. Fox and D. Jizeng (editors), *Proceedings of the Seventh International Snow Leopard Symposium*, July 25–30, 1992, Xining, China. Seattle: International Snow Leopard Trust.

Miller, R. F., T. J. Svejcar, and N. E. West. 1994. Implications of livestock grazing in the intermountain sagebrush region: Plant composition. Pages 101–146 in M. Vavra, W. A. Laycock, and R. D. Pieper (editors), *Ecological implications of livestock herbivory in the West*. Denver: Society for Range Management.

Miller, S., P. Osborne, W. Asigau, and A. J. Mungkage. 1994. Chapter 7: Environments in Papua New Guinea. Pages 97–124 in *Papua New Guinea country study on biological diversity*. A report to the United Nations Environment Program. Waigani, Papua New Guinea: Department of Environment and Conservation, Conservation Resource Centre; and Nairobi, Kenya: Africa Centre for Resources and Environment (ACRE).

Mills, J. A., and P. Jackson. 1994. *Killed for a cure: A review of the world-wide trade in tiger bone*. Cambridge, England: TRAFFIC International.

Milton, O., and R. D. Estels. 1963. *Burma wildlife survey*. Special publication No. 15. New York: American Committee for International Wild Life Protection.

Miquelle, D. G., E. N. Smirnov, T. W. Merrill, A. E. Myselenkov, H. B. Quigley, M. G. Hornocker, and B. Schleyer. 1999. Hierarchical spatial analysis of Amur tiger relationships to habitat and prey. Pages 71–99 in J. Seidensticker, S. Christie, and P. Jackson (editors), *Riding the tiger: Tiger conservation in human-dominated landscapes*. Cambridge, England: Cambridge University Press.

Mishra, C., and A. J. T. Johnsingh. 1998. Population and conservation status of the Nilgiri tahr *Hemitragus hylocrius* in Anamalai Hills, South India. *Biological Conservation* 86: 199–206.

Mishra, H. R., and M. Jeffries. 1991. *Royal Chitwan National Park: Wildlife heritage of Nepal*. Seattle: The Mountaineers Books.

Mitchell, B. A. 1963. Forestry and tanah beris. *Malaysian Forester* 26: 160–170.

Mittermeier, R. A., P. Bouchet, A. M. Bauer, T. Werner, and A. Lees. 1999. New Caledonia. Pages 367–376 in R. A. Mittermeier, N. Meyers, P. R. Gil, and C. G. Mittermeier (editors), *Hotspots: Earth's biologically richest and most endangered terrestrial ecoregions*. Mexico City: CEMEX, S.A.

Mittermeier, R. A., J. L. Carr, I. R. Swingland, T. B. Werner, and R. B. Mast. 1992. Conservation of amphibians and reptiles. Pages 59–80 in K. Adler (editor), *Herpetology: Current research on the biology of amphibians and reptiles*. Oxford, OH: Society for the Study of Amphibians and Reptiles.

Mittermeier, R. A., and C. G. Mittermeier. 1997. *Megadiversity: Earth's biologically wealthiest nations*. Mexico City: CEMEX, S.A.

Mittermeier, R. A., N. Myers, P. Robles, and C. G. Mittermeier (editors). 1999. *Hotspots, Earth's biologically richest and most endangered terrestrial ecosystems*. Mexico City: CEMEX.

Monk, K. A., Y. de Fretes, and G. Reksodiharjo-Lilley. 1997. *The ecology of Indonesia*. Vol. 5, *The ecology of Nusa Tenggara and Maluku*. Hong Kong: Periplus Editions.

Morat, P. 1993. Our knowledge of the flora of New Caledonia: Endemism and diversity in relation to vegetation types and substrates. *Biodiversity Letters* 1: 69–71.

Morley, R. J. 1981. Development and vegetation dynamics of a lowland ombrogenous peat swamp in Kalimantan Tengah, Indonesia. *J. Biog.* 8: 383–404.

Moudud, H. J. 1999. *Tiger conservation in the Sunderbans forest of Bangladesh*. Coastal Area Resource Development and Management Association Report.

Mueller-Dombois, D. 1998. Chapter 3: Eastern Melanesia. Pages 84–161 in D. Mueller-Dombois and F. R. Fosberg (editors), *Vegetation of the tropical Pacific islands*. New York: Springer-Verlag.

Mueller-Dombois, D., and F. R. Fosberg (editors). 1998. *Vegetation of the tropical Pacific islands*. New York: Springer-Verlag.

Murray, A. 1995. An annotated list of birds recorded in the Bhumipol Dam Area, Tak province, Thailand. *Nat. Hist. Bull. Siam Soc.* 43: 85–107.

Musser, G. G. 1987. The mammals of Sulawesi. Pages 73–93 in T. C. Whitmore (editor), *Biogeographical evolution of the Malay Archipelago*. Oxford, England: Clarendon.

Myers, C. W., and A. S. Rand. 1969. Checklist of amphibians and reptiles of Barro Colorado Island, Panama, with comments on faunal change and sampling. *Smithsonian Contrib. Zool.* (10): 1–11.

Myers, L. C. 1978. The geomorphology. Pages 91–100 in D. M. Luping, C. Wen, and E. R. Dingley (editors), *Kinabalu: Summit of Borneo*. Sabah, Malaysia: The Sabah Society.

Myers, N. 1988. Threatened biotas: "Hotspots" in tropical forests. *Environmentalist* 8: 187–208.

———. 1990. The biodiversity challenge: Expanded hot-spot analysis. *Environmentalist* 10: 243–256.

———. 1994. Tropical deforestation: Rates and patterns. Pages 27–40 in K. Brown and D. W. Pierce (editors), *The causes of tropical deforestation*. London: University College Press.

Myers, N., R. A. Mittermeier, C. G. Mittermeier, G. A. B. da Fonseca, and J. Kent. 2000. Biodiversity hotspots for conservation priorities. *Nature* 403: 853–858.

Narins, P. M., A. S. Feng, H.-S. Yong, and J. Christensen-Dalsgaard. 1998. Morphological, behavioral, and genetic divergence of sympatric morphotypes of the treefrog, *Polypedates leucomystax* in peninsular Malaysia. *Herpetologica* 54: 129–142.

Nash, S. V. 1993. *Sold for a song: The trade in southeast Asian non-CITES birds*. Cambridge, England: TRAFFIC-International.

National Geographic Society. 1999. *National Geographic atlas of the world*, 7th ed. Washington, DC: National Geographic Society.

National Research Council. 1992. *Grasslands and grassland sciences in northern China*. Washington, DC: National Academy Press.

Newbery, D. M., E. J. F. Campbell, Y. F. Lee, C. E. Ridsdale, and M. J. Still. 1992. Primary lowland dipterocarp forest at Danum Valley, Sabah, Malaysia: Structure, relative abundance and family composition. Pages 341–356 in A. G. Marshall and M. D. Swaine (editors), *Tropical rainforest: Disturbance and recovery*. London: The Royal Society.

Newman, M. F., P. F. Burgess, and T. C. Whitmore. 1996. *Sumatra light hardwoods: Anisoptera, Parashorea, Shorea (red, white and yellow meranti)*. Manuals of Dipterocarps for Foresters. Edinburgh: Royal Botanic Garden, and Jakarta, Indonesia: Center for International Forestry Research.

———. 1998a. *Borneo Island medium and heavy hardwoods: Dipterocarpus, Dryobalanops, Hopea, Shorea (balau/selangan batu), Upuna*. Manuals of Dipterocarps for Foresters. Edinburgh: Royal Botanic Garden, and Jakarta, Indonesia: Center for International Forestry Research.

———. 1998b. *Java to New Guinea*. 1998. Manuals of Dipterocarps for Foresters. Edinburgh: Royal Botanic Garden, and Jakarta, Indonesia: Center for International Forestry Research.

———. 1998c. *Sumatra medium and heavy hardwoods: Dipterocarpus, Dryobalanops, Hopea, Shorea (balau)*. Manuals of Dipterocarps for Foresters. Edinburgh: Royal Botanic Garden, and Jakarta, Indonesia: Center for International Forestry Research.

Newmark, W. D. 1991. Tropical forest fragmentation and the local extinction of understory birds in the Eastern Usambara Mountains, Tanzania. *Conservation Biology* 5: 67–78.

Ng, P. K. L. 1994. Peat swamp fishes of southeast Asia: Diversity under threat. *Wallaceana* 73: 1–5.

———. 1995. A revision of the Sri Lankan montane crabs of the genus Perbrinckia Bott, 1969 (Crustacea: Decapoda: Brachyura: Parathelphusidae). *Journal of South Asian Natural History* 1(2): 175–184.

Ng, P. K. L., and K. K. P. Lim. 1994. Freshwater fishes of Batam Island, Indonesia. *Malay. Nat.* 48 (2/3): 6–8.

Nightingale, N. 1992. *New Guinea, an island apart*. London: BBC Books.

Nix, H. A., D. P. Faith, M. F. Hutchinson, C. R. Margules, J. West, A. Allison, J. L. Kesteven, G. Natera, W. Slater, J. L. Stein, and P. Walker. 2000. *The BioRap Toolbox, a national study of biodiversity assessment and planning for Papua New Guinea*. Canberra: Centre for Resource and Environmental Studies, Australian National University.

Noss, R. F. 1991. Landscape connectivity: Different functions at different scales. Pages 27–39 in W. E. Hudson (editor), *Landscape linkages and biodiversity*. Washington, DC: Island Press.

———. 1992. The Wildlands Project land conservation strategy. *Wild Earth* Special Issue: 10–25.

Noss, R. F., and A. Y. Cooperrider. 1994. *Saving nature's legacy, protecting and restoring biodiversity*. Washington, DC: Island Press.

Nowak, R. M. 1991. *Walker's mammals of the world*, Vols. 1 and 2. Baltimore: Johns Hopkins University Press.

———. 1999a. *Walker's Mammals of the World*, vol. 1. Baltimore: Johns Hopkins University Press.

———. 1999b. *Walker's Mammals of the World*, vol. 2. Baltimore: Johns Hopkins University Press.

Nowell, K., and P. Jackson. 1996. *Wild cats: Status report and conservation action plan*. Gland, Switzerland: IUCN/SSC Cat Specialist Group.

Oates, J. 1999. *Myth and reality in the rain forest: How conservation strategies are failing west Africa*. Berkeley: University of California Press.

Oates, J., M. Abedi-Lartey, W. S. McGraw, T. T. Struhsaker, and G. H. Whitesides. 2000. Extinction of the west African red colobus monkey. *Conservation Biology* 14(5): 1526–1532.

Oatham, M., and B. M. Beehler. 1997. Richness, taxonomic composition, and species patchiness in three lowland forest tree plots in Papua New Guinea. In *Proceedings of the International Symposium for Measuring and Monitoring Forests and Biological Diversity; The International Networks of Biodiversity Plots*, Smithsonian Institute/Man and Biosphere Biodiversity Program (SI/MAB).

OECD. 1997. *The environmental effects of agricultural land diversion schemes*. Paris: OECD.

Oli, M. K., I. R. Taylor, and M. E. Rogers. 1994. Snow leopard *Panthera unica* predation of livestock: An assessment of local perceptions in the Annapurna conservation area, Nepal. *Biological Conservation* 68: 63–68.

Oliver, W. L. R. 1980. The pygmy hog: The biology and conservation of the pigmy hog (*Susporcula salvanius*), and the Hispid hare (*Capralagus hispidus*). *Special Scientific Reports of the Jersey Wildife Preservation Trust* 1: 1–80.

———. 1993. *Status survey and conservation action plan: Pigs, peccaries, and hippos*. Gland, Switzerland: IUCN/SSC Pigs, Peccaries, and Hippo Specialist Groups.

Oliver, W. L. R. (editor). 1993. *Pigs, peccaries, and hippos status survey and conservation action plan*. Gland, Switzerland: IUCN.

Oliver, W. L. R., C. R. Cox, and L. L. Dolar. 1991. The Philippine Spotted Deer Conservation Project. *Oryx* 25(4): 199–205.

Olson, D. M., and E. Dinerstein. 1998. The Global 200: A representation approach to conserving the earth's distinctive ecoregions. *Conservation Biology* 12: 502–515.

Olson, D. M., E. Dinerstein, R. Abell, T. Allnutt, C. Carpenter, L. McClenachan, J. D'Amico, P. Hurley, K. Kassem, H. Strand, M. Taye, and M. Thieme. 2000. *The global 200: A representation approach to conserving the earth's distinctive ecoregions*. Washington, DC: World Wildlife Fund.

Olson, D. M., E. Dinerstein, E. D. Wikramanayake, N. D. Burgess, G. V. N. Powell, E. C. Underwood, J. A. D'Amico, I. Itoua, H. E. Strand, J. C. Morrison, C. J. Loucks, T. F. Allnutt, T. H. Ricketts, Y. Kura, J. F. Lamoreux, W. W. Wettengel, P. Hedao, and K. R. Kassem. 2001. Terrestrial ecoregions of the world: A new map of life on Earth. *BioScience* 51: 933–938.

O'Neil, T. 2000. New Caledonia. *National Geographic* 197(5): 54–75.

Orians, G. H. 1993. Endangered at what level? *Ecological Applications* 3: 206–208.

Orlov, N. L., and I. S. Darevsky. 1999. Description of a new mainland species of *Goniurosaurus* genus, from north-eastern Vietnam. *Russ. J. Herpetol.* 6: 72–78.

Orlov, N., I. S. Darevsky, and R. W. Murphy. 1998. A new species of mountain stream snake, genus *Opisthotropis* Günther, 1872 (Serpentes: Colubridae: Natricinae), from the tropical rain forest of southern Vietnam. *Russ. J. Herpetol.* 5: 61–64.

Osborne, P. L. 1993. Chapter 18: Biodiversity and conservation of freshwater wetlands in Papua New Guinea. Pages 327–380 in *Papua New Guinea conservation needs assessment*, Vol. 2. Government of Papua New Guinea Department of Environment and Conservation (DEC). Boroko: The Biodiversity Support Program.

O'Shea, M. 1996. *A guide to the snakes of Papua New Guinea*. Port Moresby: Independent Publishing.

Ota, H., T. Hikida, M. Matsui, M. Hasegawa, D. Labang, and J. Nabhitabhata. 1996. Chromosomal variation in the scincid genus *Mabuya* and its arboreal relatives (Reptilia: Squamata). *Genetica* 98: 87–94.

Ota, H., and T. Weidenhöfer. 1992. A first male specimen of the poorly known agamid lizard *Japalura chapensis* Bourret, 1937 (Reptilia: Sauria) from the northern Vietnam, with notes on its taxonomic status. *Raffles Bull. Zool.* 40: 193–199.

Paijmans, K. 1975. Explanatory notes to the vegetation map of Papua New Guinea. *CSIRO Land Research Series* 35: 1–25, pls. 1–20, 4 sheets of maps.

Palomares, F., and T. M. Caro. 1999. Interspecific killing among mammalian carnivores. *American Naturalist* 153: 492–508.

Pandav, B., and B. C. Choudhury. 1996. Diurnal and seasonal activity patterns of water monitor (*Varanus salvator*) in the Bhitarkanika mangroves, Orissa, India. *Hamadryad* 21: 4–12.

Panwar, H. S. 1990. *Status of management of protected areas in India: Problems and prospects (revised)*. Regional Expert Consultation on Management of Protected Areas in the Asia-Pacific Region, 10–14 December 1990. Bangkok: FAO Regional Office for Asia and Pacific.

Panzer, R., and M. W. Schwartz. 1998. Effectiveness of a vegetation-based approach to insect conservation. *Conservation Biology* 12: 693–702.

Parent, G. H. 1992. L'utilisation des batrachians et reptiles comme bio-indicateurs [The use of batrachians and reptiles as bioindicators]. *Les Naturalistes Belges* 73: 33–63.

Parsons, M. 1999. *The butterflies of Papua New Guinea, their systematics and biology*. New York: Academic Press.

Pascal, J. P. 1988. *Wet evergreen forests of the Western Ghats of India: Ecology, structure, floristic composition and succession*. Pondicherry, India: French Institute.

Pascal, J. P., S. S. Sunder, and V. M. Meher-Homji. 1982. *Forest map of south India*. Shimoga sheet. Pondicherry, India: French Institute.

Payne, J., P. Bernazzani, and W. Duckworth. 1995. *A preliminary wildlife and habitat survey of Phou Khao Khouay National Biodiversity Conservation Area, Vientiane Prefecture, Vientiane Province and Bolikhamsai Province, Lao PDR*. Vientiane, Laos: CPAWM/WCS.

Payne, J., and G. Cubitt. 1990. *Wild Malaysia: The wildlife and scenery of peninsular Malaysia, Sarawak and Sabah*. Cambridge, MA: The MIT Press.

Payne J., G. Cubitt, D. Lau, and G. Cubitt. 1994. *This is Borneo*. London: New Holland Publishers.

Peeris, C. V. S. 1975. *The ecology of the endemic tree species of Sri Lanka in relation to their conservation*. Ph.D. thesis. Aberdeen, UK: University of Aberdeen.

Percival, M., and J. S. Womersley. 1975. *Floristics and ecology of the mangrove vegetation of Papua New Guinea*. Botany Bulletin no. 8. Lae, Papua New Guinea: Department of Forests, Division of Botany.

Peres, C. A., and J. W. Terborgh. 1995. Amazonian nature reserves: An analysis of the defensibility status of existing conservation units and design criteria for the future. *Conservation Biology* 9: 34–46.

Peter, W. P., and A. Feiler. 1994. Eine neue Bovidenart aus Vietnam und Cambodia (Mammalia: Ruminantia). *Zoologische Abhandlungen, Museum fur Tierkunde Dresden* 48: 169–177.

Peterson, A. T., and L. R. Heaney. 1993. Genetic differentiation in Philippine bats of the genera *Cynopterus* and *Haplonycteris*. *Biological Journal of the Linnean Society* 49: 203–218.

Peterson, R. O. 1977. Wolf ecology and prey relationships on Isle Royale. *U.S. National Park Service. Sci. Monogr. Series* 7: 210.

Pethiyagoda, R. 1994. Threats to the indigenous freshwater fishes of Sri Lanka and remarks on their conservation. *Hydrobiologia* 285: 189–201.

Pethiyagoda, R., and K. Manamendra-Arachchi. 1998a. Evaluating Sri Lanka's amphibian diversity. *Occasional Papers of the Wildlife Heritage Trust* 2: 1–12.

———. 1998b. A revision of the endemic Sri Lankan agamid genus *Ceratophora* Gray, 1835, with description of two new species. *Journal of South Asian Natural History* 3: 1–50.

Petocz, R. G. 1989. *Conservation and development in Irian Jaya: A strategy for rational resource utilization*. Leiden, The Netherlands: E.J. Brill.

Petr, T. 1983. Limnology of the Purari Basin, Part 2. The delta. In T. Petr, (editor), *The Purari: Tropical environment of a high rainfall river basin*. The Hague: Dr. W. Junk Publishers.

Philipp, K. M., W. Böhme, and T. Ziegler. 1999. The identity of *Varanus indicus*: Redefinition and description of a sibling species coexisting at the type locality (Sauria, Varanidae, *Varanus indicus* group). *Spixiana* 22: 273–287.

Philippine BAP. 1997. *Philippine biodiversity. An assessment and action plan*. Department of Environment and Natural Resources and UNEP. Bookmark Inc.

Phillips, W. W. A. 1958. Some observations on the fauna of the Maldive Islands. Part IV: Amphibians and reptiles. *J. Bombay Nat. Hist. Soc.* 55: 217–220.

———. 1980. *Manual of the mammals of Sri Lanka*, 2d ed. Wildlife and Nature Protection Society of Sri Lanka.

Pieper, R. D. 1994. Ecological implications of livestock grazing. Pages 177–211 in M. Vavra, W. A. Laycock, and R. D. Pieper (editors), *Ecological implications of livestock herbivory in the West*. Denver: Society for Range Management.

Pillai, R. S., and M. S. Ravichandran. 1999. Gymnophiona (Amphibia) of India. A taxonomic study. *Occ. Pap. Zool. Surv. India* (172): i–vi, 1–117.

Plant form, diversity, communities and succession. New Delhi: Oxford and IBH Publishing Co.

Polis, G. A., C. A. Meyers, and R. D. Holt. 1989. The ecology and evolution of intraguild predation: Potential competitors that eat each other. *Annu. Rev. Ecol. Sust.* 20: 297–330.

Polunin, O., and A. Stainton. 1997. *Flowers of the Himalaya*. Delhi: Oxford India Paperbacks.

Poulsen, M. K. 1995. The threatened and near-threatened birds of Luzon, Philippines, and the role of the Sierra Madre mountains in their conservation. *Bird Conservation International* 5: 79–115.

Power, M. 1992. Top-down and bottom-up forces in food webs: Do plants have primacy? *Ecology* 73: 733–746.

Prakash, U., Du Nai-zheng, and P. P. Tripathi. 1992. Fossil woods from Tipam sandstones of northeast India with remarks on palaeoenvironment of the region during the middle Miocene. *Biological Memoirs* 18: 1–26.

Prance, G. T., H. Beentje, J. Dransfield, and R. Johns. 2000. The tropical flora remains undercollected. *Ann. Mo. Bot. Garden* 87(1): 67–71.

Prasad, S. N. 1989. *An ecological reconnaissance of mangals in Krishna estuary, peninsular India: A plea for conservation*. Unpublished manuscript, WII.

Prater, S. H. 1971. *The book of Indian mammals*, 3d ed. Bombay: Bombay Natural History Society.

Primack, R. B. 1993. *Essentials of conservation biology*. Sunderland, MA: Sinauer Associates.

Puri, G. S., R. K. Gupta, and V. M. P. S. Meher-Homji. 1989. *Forest ecology*, Vol. 2. New Delhi: Oxford & IBH Publishing Company.

Quinnell, R., and A. Balmford. 1988. A future for Palawan's forests? *Oryx* 22: 32–35.

Rabinowitz, A. 1998. Killed for a cure. *Natural History* April.

———. 1999. The status of the Indochinese tiger: Separating fact from fiction. Pages 148–165 in J. Seidensticker, S. Christie, and P. Jackson (editors), *Riding the tiger: Tiger conservation in human-dominated landscapes*. Cambridge, England: Cambridge University Press.

Ram, J., and S. P. Singh. 1994. Ecology and conservation of alpine meadows in Central Himalaya, India. Pages 33–55 in Y. Pangtey and R. Rawal (editors), *High altitudes of the Himalaya: Biogeography, ecology and conservation*. Gyanodaya, Nainital.

Ramesh, B. R. 1999. Physiognomy and classification of vegetation using remote sensing techniques. In WII, *Ecological and socio-economic studies on the Kalakad-Mundanthurai tiger reserve: An eco-development approach*. WII.

Ramesh, B. R., and J. P. Pascal. 1999. *Atlas of endemics of the Western Ghats (India)*. Pondicherry, India: French Institute.

Randhawa, M. S. 1945. Progressive desiccation of northern India. *J. Bombay Nat. Hist. Soc.* 45: 558–565.

Rannells, J. 1995. *PNG: A fact book on modern Papua New Guinea*, 2d ed. Oxford, England: Oxford University Press.

Rao, A., and M. J. Casimir. 1990. Perspectives on pastoral economy and ecology in the Western Himalaya. Pages 386–402 in S. D. Bhatt and R. K. Pande (editors), *Himalaya: Environment, resources and development*. Almora, India: Shree Almora Book Depot.

Rao, A. N. J. 1991. *Atlas to India's wildlife*. Madras: TT Maps and Publications.

Rao, P. S. N. 1996. Phytogeography of the Andaman and Nicobar Islands, India. *Malayan Nature Journal* 50: 57–79.

Rao, R. R. 1994. *Biodiversity in India. Floristic aspects*. Dehra Dun, India: Bishen Sing Mahendra Pal Singh.

Rao, R. R., and K. Haridasan. 1982. Notes on the distribution of certain rare, endangered, and endemic plants of Meghalaya with a brief remark on the flora. *Journal of Bombay Natural History Society* 79: 93–99.

Rao, R. S., P. Venkanna, and T. A. Reddy. 1986. *Flora of West Godavari District, Andhra Pradesh*. I.B.S. Meerut.

Rao, S. 1995. *Treasured islands! An environmental handbook for teachers in the Andaman and Nicobar Islands*. Madras: Kalpvriksh and Andaman and Nicobar Islands Environmental Team.

Ravichandran, M. S. 1997. A new frog of the genus *Nyctibatrachus* (Anura: Ranidae) from southern India. *Hamadryad* 22: 9–12.

Rawat, G. S. 1997. Conservation status of forests and wildlife in the Eastern Ghats, India. *Environmental Conservation* 24(4): 307–315.

———. 1998. Temperate and alpine grasslands of the Himalayas: Ecology and conservation. *PARKS* 8: 27–36.

Rawat, G. S., and M. M. Babu. 1995. *Ecological status of forests in and around protected areas of Andhra Pradesh. A report on Andhra Pradesh Forestry Project*. Dehra Dun, India: WII.

Rawat, G. S., and S. K. Mukherjee. 1999. *Foot-hill forests of the Himalayas: Patterns of biodiversity and conservation status*. Unpublished manuscript.

Rawat, G. S., and W. A. Rodgers. 1988. The alpine meadows of Uttar Pradesh: An ecological review. Pages 119–137 in P. Singh and P. S. Pathak (editors), *Rangelands, resource and management, proceedings of the National Rangeland Symposium*. November 9–12, 1987. Jhansi: Range Management Society of India.

Redford, K. H. 1992. The empty forest. *BioScience* 42: 412–422.

RePPProT. 1990. *The land resources of Indonesia*. Jakarta: ODA/Ministry of Transmigration.

Ricciardi, A., and J. B. Rasmussen. 1999. Extinction rates of North American freshwater fauna. *Conservation Biology* 13: 1220–1222.

Richards, P. W. 1936. Ecological observations on the rainforest of Mount Dulit, Sarawak. Parts I and II. *Journal of Ecology* 24: 1–37, 340–363.

———. 1952. *The tropical rain forest*. Cambridge: Cambridge University Press.

Rickart, E. A., L. R. Heaney, D. S. Balete, and B. R. Tabaranza, Jr. 1998. A review of the genera *Crunomys* and *Archboldomys* (Rodentia, Muridae, Murinae) with descriptions of two new species from the Philippines. *Fieldiana Zoology* (new series) 89: 1–24.

Rickart, E. A., L. R. Heaney, P. D. Heideman, and R. C. B. Utzurrum. 1993. The distribution and ecology of mammals on Leyte, Biliran, and Maripipi islands, Philippines. *Fieldiana Zoology* (new series) 72: 1–62.

Ricketts, T. H., E. Dinerstein, D. M. Olson, C. J. Loucks, W. Eichbaum, D. DellaSala, K. Kavanagh, P. Hedao, P. T. Hurley, K. M. Carney, R. Abell, and S. Walters. 1999. *Terrestrial ecoregions of North America: A conservation assessment*. Washington, DC: Island Press.

Ricklefs, R. E., and R. E. Latham. 1993. Global patterns of diversity in mangrove floras. Pages 215–229 in R. E. Ricklefs and D. Schluter (editors), *Species diversity in ecological communities*. Chicago: University of Chicago.

Rijksen, H. D. D., and E. Meijaard. 1999. *Our vanishing relative: The status of wild orang-utans at the close of the twentieth century*. Dordrecht, The Netherlands: Kluwer Academic Publishers.

Rikhari, H. C., G. C. S. Negi, and S. P. Singh. 1993. Species and community diversity patterns in an alpine meadow of Central Himalaya. Pages 205–218 in U. Dhar (editor), *Himalayan biodiversity: Conservation strategies*. Almora, India: G.P. Pant Institute.

Risser, P. G. 1988. Diversity in and among grasslands. Pages 176–180 in E. O. Wilson (editor), *Biodiversity*. Washington, DC: National Academy Press.

Roberts, T. J. 1997. *The mammals of Pakistan*. London: Ernest Benn.

Roberts, T. R. 1993. Just another damned river? Negative impacts of Pak Mun dam on fishes of the Mekong basin. *Nat. Hist. Bull. Siam Soc*. 41: 105–133.

Robichaud, W. 1999. *Saola conservation action plan for Lao PDR: 1999 revision*. Vientiane, Laos: WCS/IUCN.

Robinson, J. G. 1994. Carving up tomorrow's planet. *Int. Wildlife* 24: 30–37.

Robinson, J., and E. L. Bennett (editors). 2000. *Hunting for sustainability in tropical forests*. New York: Columbia University Press.

Robson, C. 2000. *A guide to the birds of southeast Asia*. Princeton, NJ: Princeton University Press.

Roca, R., L. Adkins, C. M. Wurschy, and K. Skerl. 1996. Transboundary conservation: An ecoregional approach to protect neotropical migratory birds in South America. *Environmental Management* 20: 849–863.

Rodgers, W. A. 1985. Biogeography and protected area planning in India. In J. Thorsell, *Conserving Asia's natural heritage*. IUCN.

Rodgers, W. A., and H. S. Panwar. 1988. *Planning a wildlife protected areas network in India*, Vols. 1 and 2. Department of Environment, Forests, and Wildlife/Wildlife Institute of India report. Wildlife Institute of India.

Rodgers, W. A., H. S. Panwar, and V. B. Mathur. 2000. *Wildlife protected area network in India: A review (executive summary)*. Dehra Dun: Wildlife Institute of India.

Rollet, B. 1962. *Inventaire forestier de l'Est Mekong* [Forest inventory of the eastern Mekong]. Rome: UN/FAO.

Ross, J. P. 1998. *Status survey and conservation action plan: Revised action plan for crocodiles*. Gland, Switzerland: IUCN/SSC.

Round, P. D. 1988. *Wildlife, habitats and priorities for conservation in Dong Khanthung proposed National Biodiversity Conservation Area, Champask Province, Lao PDR*. Vientiane, Laos: CPAWM and WCS.

———. 1990. Bangkok Bird Club survey of the bird and mammal trade in the Bangkok Weekend Market. *Nat. Hist. Bull. Siam Soc*. 3: 81–43.

———. 2000. *Field check-list of Thai birds*. Bangkok: Bird Conservation Society of Thailand.

Rundel, P. W. 1999. *Forest habitats and flora in Lao PDR, Cambodia, and Vietnam*. Los Angeles: University of California.

Rundel, P. W., and K. Boonpragob. 1995. Dry forests of Thailand. Pages 93–123 in S. H. Bullock, H. A. Mooney, and E. Medina (editors), *Seasonally dry tropical forests*. Cambridge, England: Cambridge University Press.

Saberwal, V. 1996. The politicisation of Gaddi access to grazing resources in Kangra, Himachal Pradesh, 1960–1994. *Himalayan Research Bulletin* 16: 7–11.

Sadlier, R. A., and A. M. Bauer. 1997. The terrestrial herpetofauna of the Loyalty Islands. *Pacific Science* 51: 76–90.

Saldanha, C. J. 1989. *Andaman, Nicobar, and Lakshadweep: An environmental impact assessment*. New Delhi: Oxford University Press.

Salleh, K. M. 1991. *Rafflesia: Magnificent flower of Sabah*. Kota Kinabalu, Sabah, Malaysia: Borneo Publishing Company.

Salter, R. E. 1982. *Summary of currently available information on internationally threatened wildlife species in Burma*. Field Document 7/83. FO:BUR/80/006. Nature Conservation and National Parks Project.

Salter, R. E., and N. A. MacKenzie. 1985. Conservation status of the proboscis monkey in Sarawak. *Biological Conservation* 33: 119–132.

Sankaran, R. 1995. The decline of the edible-nest swiftlet in the Nicobar Islands. *Oriental Bird Club Bulletin* 22: 51–54.

———. 1997. Developing a protected area network in the Nicobar Islands: The perspective of endemic avifauna. *Biodiversity and Conservation* 6: 797–815.

———. 1998. An annotated list of the endemic avifauna of the Nicobar Islands. *Forktail* 13: 17–22.

Santiapillai, C. 1997. *The Asian elephant conservation: A global strategy*. Gajah. July.

———. 1999. *Tiger conservation in the Sundarbans through improved transborder cooperation by Bangladesh and India*. WWF Project BD0007.01 Report.

Santiapillai, C., P. M. Giao, and V. V. Dung. 1993. Conservation and management of Javan rhino (*Rhinoceros sondaicus annamiticus*) in Vietnam. *Tiger Paper* XX(4).

Santiapillai, C., and P. Jackson. 1990. *The Asian elephant: An action plan for its conservation*. IUCN SSC, Asian Elephant Specialist Group.

Sargeant, A. B., and S. H. Allen. 1989. Observed interactions between coyotes and red foxes. *Journal of Mammalogy* 70: 631–633.

Saunders, D. A., R. J. Hobbs, and C. R. Margules. 1991. Biological consequences of ecosystem fragmentation. A review. *Conservation Biology* 5: 18.

Saw, H. 1982. *Better than machine: Elephants in Burma*. Paper presented at the World Congress on National Parks, Bali, 11–22 October 1982.

———. 1992. *Comparative study on Ayeyarwaddy and Rakhine mangroves*. UNDP/FAO: MYA/90/003.

———. 1995. *Wildlife conservation and the trade in wildlife*. A report prepared for the Ministry of Forestry in October 1995.

———. 1997. *Proposed Pegu Yomas National Park* FO: BUR/80/006. Field Report, February 1982. Foresters' Compendium. JICA/F.D.

———. 2000. *Myanmar National Forest management plan*. (2000–2001 to 2030–2031).

Saxena, H. O., and P. K. Dutta. 1975. Studies on the ethnobotany of Orissa. *Bull. Bot. Surv. India* 17: 124–131.

Sayer, J., N. Ishwaran, J. Thorsell, and T. Sigaty. 2000. Tropical forest biodiversity and the World Heritage Convention. *Ambio* 29: 302–309.

Schaller, G. B. 1967. *The deer and the tiger*. Chicago: University of Chicago Press.

———. 1977. *Mountain monarchs: Wild sheep and goats of the Himalaya*. Chicago: University of Chicago Press.

———. 1995. *A wildlife survey of the Annamite Mountains of Laos, December 1994–January 1995*. Vientiane, Laos: WCS.

———. 1998a. On the trail of new species. *International Wildlife* July/August.

———. 1998b. *Wildlife of the Tibetan steppe*. Chicago: University of Chicago Press.

Schaller, G. B., H. Li, H. Lu, J. R. Ren, M. J. Qiu, and H. B. Wang. 1987. Status of large mammals in the Taxkorgan Reserve, Xinjiang, China. *Biological Conservation* 42: 53–71.

Schaller, G. B., and A. R. Rabinowitz. 1995. The saola or spindlehorn bovid *Pseudooryx nghetinhensis* in Laos. *Oryx* 29: 107–114.

Schaller, G. B., and E. S. Vrba. 1996. Description of the giant muntjac (*Megamuntiacus vuquangensis*) in Laos. *Journal of Mammalogy* 77: 675–683.

Schneider, H., U. Sinsch, and E. Nevo. 1992. The lake frogs of Israel represent a new species. *Zool. Anz.* 228: 97–106.

Schreiber, A., R. Wirth, M. Riffel, and H. van Rompaey. 1989. *Weasels, civets, mongooses and their relatives. An action plan for the conservation of mustelids and viverrids*. Gland, Switzerland: IUCN/SSC Mustelid and Viverrid Specialist Group.

Schultz, J. 1995. *The ecozones of the world. The ecological divisions of the geosphere*. Berlin: Springer Verlag.

Schwartz, J. H. (editor). 1988. *Orang-utan biology*. New York: Oxford University Press.

Scott, D. A. 1989. *A field guide to the waterbirds of Asia. A directory of Asian wetlands*. Wildbird Society of Japan.

——— (editor). 1989. *A directory of Asian wetlands*. Gland, Switzerland: IUCN.

Scott, J. M., E. Davis, B. Csuti, R. F. Noss, B. Butterfield, C. Groves, H. Anderson, S. Caicco, F. D'Erchia, T. C. Edwards Jr., J. Ulliman, and G. Wright. 1993. Gap analysis: A geographic approach to protection of biological diversity. *Wildlife Monographs* No. 123.

Scott, J. M., F. W. Davis, G. McGhie, and R. G. Wright. *Natural reserves: Do they capture the full range of America's biological diversity?* In preparation.

Scott, N. J., Jr., and R. A. Seigel. 1991. The management of amphibian and reptile populations: Species priorities and methodological and theoretical constraints. Pages 343–368 in D. R. McCullough and R. H. Barrett (editors), *Wildlife 2001: Populations*. New York: Elsevier Applied Science.

Seidensticker, J. 1976. On the ecological separation between tigers and leopards. *Biotropica* 8: 225–234.

———. 1986. Large carnivores and the consequences of habitat insularization: Ecology and conservation of tigers in Indonesia and Bangladesh. Pages 1–80 in S. D. Miller and D. D. Everett (editors), *Cats of the world: Biology, conservation, and management*. Washington, DC: National Wildlife Federation.

———. 1991. Leopards. Pages 106–114 in J. Seidensticker and S. Lumpkin (editors), *Great cats*. Emmaus, PA: Rodale Press.

Seidensticker, J., S. Christie, and P. Jackson. 1999. *Riding the Tiger. Tiger conservation in human-dominated landscapes*. Cambridge, England: Cambridge University Press.

Seidensticker, J., and A. Hai. 1983. *The Sundarbans wildlife management plan: Conservation in the Bangladesh coastal zone*. IUCN and WWF.

Seidensticker, J., and C. McDougal. 1993. Tiger predatory behaviour, ecology and conservation. *Symp. Zool. Soc. Lond.* 65: 105–125.

Sekhran, N., and S. Miller. 1994. *Papua New Guinea country study on biological diversity*. Waigani, Papua New Guinea: Department of Environment and Conservation, Conservation Resource Centre, and the African Centre for Resources and Environment (ACRE).

Senanayake, F. R. 1993. The evolution of the major landscape categories in Sri Lanka and distribution patterns of some selected taxa: Ecological implications. Pages 201–220 in W. Erdelen, C. Preu, N. Ishwaran, and C. M. Madduma Bandara (editors), *Proceedings of the International and Interdisciplinary Symposium, Ecology and Landscape Management in Sri Lanka*. Germany: Margraf Scientific Books.

Senanayake, F. R., M. Soulé, and J. W. Senner. 1977. Habitat values and endemicity in the vanishing rain forests of Sri Lanka. *Nature* 265: 351–354.

Shafer, C. L. 1995. Values and shortcomings of small reserves. *Bioscience* 45: 80–88.

Shah, K. B. 1998. Checklist of the herpetofauna of Nepal with English and vernacular names. *Nat. Hist. Soc. Nepal Bull.* 8: 1–4.

Sharma, D., and A. J. T. Johnsingh. 1996. *Impacts of management practices on lion and ungulate habitats in Gir Protected Area*. RR-96/001. Dehra Dun: Wildlife Institute of India.

Shcherbak, N. N., and O. D. Nekrasova. 1994. A contribution to the knowledge of the lizards of southern Vietnam with description of a new species (Reptilia, Gekkonidae). [In Russian]. *Vestnik Zoologii* 1: 48–52.

Shengji, P. (editor). 1996. *Banking on biodiversity: Proceedings of a workshop on biodiversity conservation, monitoring, and management in the Hindu Kush-Himalayas*. Kathmandu, Nepal: International Centre for Integrated Mountain Development.

Sherpa, M. N., and U. P. Norbu. 1999. Linking protected areas for ecosystem conservation: A case study from Bhutan. *PARKS* 9: 35–45.

Showler, D. A., P. Davidson, K. Khounboline, and K. Salivong. 1998. *A wildlife and habitat survey of Nam Xam NBCA, Houaphanh Province, Lao PDR*. Vientiane, Laos: CPAWM/WCS.

Shrestha, T. B., and R. B. Joshi. 1996. *Rare, endemic and endangered plants of Nepal*. WWF Nepal.

———. 1997. *Biodiversity gap analysis: Terrestrial ecoregions of the Himalaya (Nepal)*. Draft report submitted to WWF-Nepal Program, Lal Durbar, Kathmandu, Nepal, November 1997.

Shrestha, T. K. 1998. Country report for Nepal. Herpetofauna of Nepal: Present status, distribution and conservation. Pages 26–46 in A. de Silva (editor), *Biology and conservation of the amphibians, reptiles and their habitats in south Asia*. Peradeniya, Sri Lanka: Amphibia and Reptile Research Organization of Sri Lanka (ARROS).

Shuttleworth, C. 1981. *Malaysia's green and timeless world*. Singapore: Heinemann Educational Books.

Sih, A., B. G. Jonsson, and G. Luikart. 2000. Habitat loss: Ecological, evolutionary and genetic consequences. *Trends in Ecology and Evolution* 15: 132–134.

Silva, A. de (editor). 1998. *Biology and conservation of the amphibians, reptiles and their habitats in south Asia*. Peradeniya, Sri Lanka: Amphibia and Reptile Research Organization of Sri Lanka (ARROS).

Singh, S. P. 1991. *Structure and function of low and high altitude grazingland ecosystems and the impact of the livestock component in the central Himalaya*. Final Technical Report. New Delhi: D.O En., Government of India.

Singh, S., A. R. K. Sastry, R. Mehta, and V. Uppal (editors). 2000. *Setting biodiversity conservation priorities for India*. New Delhi: WWF-India.

Singhakumara, B. M. P. 1995. *Floristic survey of Adam's Peak Wilderness*. Sri Lanka: Sri Lanka Forest Department Publication.

Sinha, A. R. P. 1992. Impacts of growing population and tourism on the endemic flora of Andaman and Nicobar Islands. *Environmental Conservation* 19 (2): 173–174, 182.

Sinsakul, S. 1997. *Country report: Late quaternary geology of the Lower Central Plain, Thailand*. Paper presented at the International Symposium on Quaternary Environmental Change in the Asia and Western Pacific Region, October 14–17, 1997, University of Tokyo, Japan.

Sites, J. W., Jr., and K. Crandall. 1997. Testing species boundaries in biodiversity studies. *Conservation Biology* 11: 1289–1297.

Skole, D., and C. Tucker. 1993. Tropical deforestation and habitat fragmentation in the Amazon: Satellite data from 1978 to 1988. *Science* 260: 1905–1910.

Smith, J. L. D. 1993. The role of dispersal in structuring the Chitwan tiger population. *Behaviour* 124: 165–195.

Smith, J. L. D., S. C. Ahearn, and C. McDougal. 1998. Landscape analysis of tiger distribution and habitat quality in Nepal. *Conservation Biology* 12: 1338–1346.

Smith, J. L. D., C. McDougal, S. C. Ahearn, A. R. Joshi, and K. Conforti. 1999. Metapopulation structure of tigers in Nepal. Pages 176–191 in J. Seidensticker, S. Christie, and P. Jackson (editors), *Riding the tiger: Tiger conservation in human-dominated landscapes*. Cambridge, England: Cambridge University Press.

Smith, J. L. D., C. McDougal, and M. E. Sunquist. 1987. Land tenure system in female tigers. Pages 464–474 in R. L. Tilson and U. S. Seal (editors), *Tigers of the world: The biology, biopolitics, management and conservation of an endangered species*. Park Ridge, NJ: Noyes Publications.

Smith, M. A. 1931. *The fauna of British India, including Ceylon and Burma. Reptilia and Amphibia*. Vol. I, *Loricata, Testudines*. London: Taylor & Francis.

———. 1935. *The fauna of British India, including Ceylon and Burma. Reptilia and Amphibia*. Vol. II, *Sauria*. London: Taylor & Francis.

———. 1943. *The fauna of British India, including Ceylon and Burma. Reptilia and Amphibia*. Vol. III, *Serpentes*. London: Taylor & Francis.

Smith, T. B., M. W. Bruford, and R. K. Wayne. 1993. The preservation of process: The missing element of conservation programs. *Biodiversity Letters* 1: 164–167.

Smitinand, T. 1989. Thailand. In D. G. Campbell and H. D. Hammond (editors), *Floristic inventory of tropical countries*. New York: The New York Botanical Garden.

Smitinand, T., T. Santisuk, and C. Phengklai. 1980. *The manual of Dipterocarpaceae of mainland south-east Asia*. Bangkok: The Forest Herbarium, Royal Forest Department.

Smitinand, T., J. E. Vidal, and P. H. Hô. 1990. Diptérocarpacées [Dipterocarpaceae]. Pages 1–123 in P. Morat (editor), *Flore du Cambodge, du Laos, et du Vietnam* [Flora of Cambodia, Laos, and Vietnam], Vol. 25. Paris: Laboratoire de Phanerogamie.

Smythies, B. E. 1965. *Common Sarawak trees*. Borneo Literature Bureau.

———. 1986. *The birds of Burma*, 3d ed. Liss, Hampshire, England: Nimrod.

Snyder, N., P. McGowan, J. Gilardi, and A. Grajal. 2000. *Parrots status survey and conservation action plan 2000–2004*. Gland, Switzerland: IUCN.

Soepadmo, E. 1995. Plant diversity of the Malesian tropical rainforest and its phytogeographical and economic significance. Pages 19–40 in R. B. Primack and T. E. Lovejoy (editors), *Ecology, conservation, and management of southeast Asian rainforests*. New Haven: Yale University Press.

Sohmer, S. H. 1990. Element of Pacific phytodiversity. In P. Baas, K. Kalkman, and R. Geesink (editors), *The plant diversity of Malesia*. Dordrecht, The Netherlands: Kluwer Academic Publishers.

Soulé, M. E., D. T. Bolger, A. C. Alberts, J. Wright, M. Sorice, and S. Hill. 1988. Reconstruction dynamics of rapid extinction of chaparral-requiring birds in urban habitat islands. *Conservation Biology* 2: 75–92.

Soulé, M. E., and M. A. Sanjayan. 1998. Conservation targets: do they help? *Science* 279: 2060–2061.

Soulé, M. E., and J. Terborgh. 1999. *Continental conservation. Scientific foundations of regional reserve networks*. Washington, DC: Island Press.

Spalding, M. D., F. Blasco, and C. D. Field (editors). 1997. *World mangrove atlas*. Okinawa, Japan: The International Society for Mangrove Ecosystems.

Spellerberg, I. F., and J. W. D. Sawyer. 1999. *An introduction to applied biogeography*. Cambridge, England: Cambridge University Press.

Sprackland, R. G. 1999. New species of monitor (Squamata: Varanidae) from Indonesia. *Reptile Hobbyist* Feb. 1999: 20–27.

Srikosamatara, S., B. Siripholdej, and V. Suteethorn. 1992. Wildlife trade in Lao P.D.R. and between Lao P.D.R. and Thailand. *Nat. Hist. Bull. Siam Soc.* 40: 1–47.

Srikosamatara, S., and V. Suteethorn. 1995. Populations of gaur and banteng and their management in Thailand. *Nat. Hist. Bull. Siam Soc.* 43: 55–83.

Sri Lanka National Report. 1991. *Sri Lanka national report to the United National Conference on Environment and Development*. Prepared under the auspices of the Ministry of Environment and Parliamentary Affairs, Government of Sri Lanka.

SSC. 1988. *Mapping the natural conditions of the Philippines*. Final report. Solna: Swedish Space Corporation.

Statistical Yearbook 1998. Central Statistical Organization, Ministry of National Planning and Economic Development, Myanmar.

Stattersfield, A. J., M. J. Corsby, A. J. Long, and D. C. Wege. 1998. *Global directory of endemic bird areas*. Cambridge, England: BirdLife International.

Steinmetz, R. 1998. *A survey of habitats and mammals around Khammouan limestone National Biodiversity Conservation Area, Lao PDR*. Bangkok: WWF.

Steinmetz, R., T. Stones, and T. Chan-ard. 1999. *An ecological survey of habitats, wildlife and people in Xe Sap National Biodiversity Conservation Area, Saravan Province, Lao PDR*. Bangkok: World Wide Fund for Nature.

Stevens, S. F. 1996. *Claiming the High Ground: Sherpas, subsistence, and environmental change in the highest Himalaya*. Delhi: Motilal Banarsidass Publishers.

Stewart-Cox, B., and G. Cubitt. 1995. *Wild Thailand*. Cambridge, MA: The MIT Press.

Stone, D. 1994. *Biodiversity of Indonesia: Tanah Air*. Singapore: Archipelago Press.

Stone, D., and A. Compost. 1997. *Biodiversity of Indonesia: Tanah Air*. Singapore: Tien Wah Press.

Stott, P. 1986. The spatial pattern of dry season fires in the savanna forests of Thailand. *Journal of Biogeography* 13(4): 345–358.

———. 1988. The forest as phoenix: Towards a biogeography of fire in mainland southeast Asia. *Geographical Journal* 154(3): 337–350.

Stuart, B. L. 1999. Amphibians and reptiles. Pages 43–67 in J. W. Duckworth, R. E. Salter, and K. Khounboline (compilers), *Wildlife in Lao PDR. 1999 Status report*. Vientiane, Laos: IUCN–The World Conservation Union/Wildlife Conservation Society/Centre for Protected Areas and Watershed Management.

Sukumar, R. 1989. *The Asian elephant: Ecology and management*. Cambridge Studies in Applied Ecology and Resource Management. New York: Cambridge University Press.

Sundriyal, R. C., and J. P. Joshi. 1990. Effect of grazing on standing crop, productivity, and efficiency of energy capture in an alpine grassland ecosystem at Tungnath (Garhwal Himalaya), India. *Applied Tropical Ecology* 31: 84–97.

Sunquist, M. E. 1981. *The social organization of tigers (Panthera tigris) in Royal Chitawan National Park, Nepal*. Washington, DC: Smithsonian Institution Press.

Sunquist, M., K. U. Karanth, and F. Sunquist. 1999. Ecology, behaviour and resilience of the tiger and its conservation needs. Pages 5–18 in J. Seidensticker, S. Christie, and P. Jackson (editors), *Riding the tiger: Tiger conservation in human-dominated landscapes*. Cambridge, England: Cambridge University Press.

Surridge, A. K., R. J. Timmins, G. M. Hewitt, and D. J. Bell. 1999. Striped rabbits in southeast Asia. *Nature* 400(6746): 726.

Survey Department. 1988. *The national atlas of Sri Lanka*. Survey Department of Sri Lanka.

Swartzendruber, J. F. 1993. *Executive summary of Papua conservation needs assessment*. Washington, DC: Biodiversity Support Program.

Symington, C. F. 1974. *Malayan forest records no. 16: Foresters' manual of dipterocarps*. Kuala Lumpur, Malaysia: Penerbit Universiti Malaya.

Synder, N., P. McGowan, J. Gilardi, and A. Grajal (editors). 2000. *Parrots. Status survey and conservation action plan 2000–2004*. Gland, Switzerland: IUCN.

Takeuchi, W., and J. Wiakabu. 2001. A transect-based floristic reconnaissance of Southern New Ireland. In B. M. Beehler and L. E. Alonso (editors), *Southern New Ireland, Papua New Guinea: A biodiversity assessment*. Washington, DC: Conservation International.

Taki, Y. 1978. An analytical study of the fish fauna of the Mekong basin as a biological production system in nature. *Res. Inst. Evol. Biol. Spec. Publ.* 1: 1–74, 3 pls.

Tan, S. H., and H. H. Tan. 1994. The freshwater fishes of Pulau Bintan, Riau Archipelago, Sumatera, Indonesia. *Trop. Biodiv.* 2: 351–367.

Taylor, E. H. 1963. The lizards of Thailand. *Univ. Kansas Science Bull.* 44: 687–1077.

———. 1965. The serpents of Thailand and adjacent waters. *Univ. Kansas Science Bull.* 45: 609–1096.

———. 1970. The turtles and crocodiles of Thailand and adjacent waters. *Univ. Kansas Science Bull.* 49: 87–179.

Taylor, R. J. 1984. *Predation*. New York: Chapman & Hall.

Terborgh, J. 1991. *Diversity and the tropical rain forest*. New York: Freeman.

———. 1999. *Requiem for nature*. Washington, DC: Island Press.

Terborgh, J., and M. Soulé. 1999. *Continental conservation: Scientific foundations of regional reserve networks*. Washington, DC: Island Press.

Thackaway, R., and I. D. Creswell (editors). 1995. *An interim biogeographic regionalisation for Australia: A framework for establishing the national system of reserves*, Version 4.0. Canberra: Australian National Conservation Agency.

Thewlis, R. M., R. J. Timmins, T. D. Evans, and J. W. Duckworth. 1998. The conservation status of birds in Laos: A review of key species. *Bird Conservation International* 8 (Suppl.): 1–159.

Thistlethwait, R., and G. Votaw. 1992. *Environment and development: A Pacific island perspective*. Manilla: Asian Development Bank.

Thom, B. G., and L. D. Wright. 1982. *Geomorphology of the Purari Delta, Papua New Guinea*. Purari River (Wabo) Hydroelectric Scheme, Environmental Studies, Vol. 17. Waigani, Papua New Guinea: Office of Environment and Conservation.

Tilson, R. L., K. Soemarna, W. Ramono, S. Lusli, K. Traylor-Holzer, and U. S. Seal (editors). 1994. *Sumatran tiger population and habitat viability analysis report*. Apple Valley, MN: Indonesian Forest Protection and Nature Conservation and IUCN/SSC Captive Breeding Specialist Group.

Tilson, R., K. Traylor-Holzer, and U. Seal. 1993. *Briefing book of the Orangutan Population and Habitat Viability Analysis Workshop*. Medan, Indonesia.

Timmins, R. J., and W. V. Bleisch. 1995. *A wildlife and habitat survey of Xe Bang Nouan National Biodiversity Conservation Area, Savannakhet and Salavan Provinces, Lao PDR*. Vientiane, Laos: CPAWM/WCS.

Timmins, R. J., and K. Khounboline. 1996. *A preliminary wildlife and habitat survey of Hin Namno National Biodiversity Conservation Area, Khammouane Province, Lao PDR*. Vientiane, Laos: WCS.

Tobias, J. 1997. *Environmental and social action plan for the Nakai–Nam Theun catchment and corridor areas: Report of the Wildlife Survey*. Vientiane, Laos: WCS.

Tomlinson, P. B. 1986. *The botany of mangroves*. New York: Cambridge University Press.

Tucker, R. P. 1986. The evolution of transhuman grazing in the Punjab Himalaya. *Mountain Research and Development* 6: 17–28.

Tuoc, D., V. V. Dung, S. Dawson, P. Arctander, and J. MacKinnon. 1994. Introduction of a new large mammal species in Vietnam. [In Vietnamese]. *Science and Technology News*. Forest Inventory and Planning Institute (Hanoi), Vietnam.

Turner, I. M. 1996. Species loss in fragments of tropical rain forest: A review of the evidence. *Journal of Applied Ecology* 33: 200–209.

Tyler, M. J. 1999. Distribution patterns of amphibians in the Australo-Papuan region. Pages 541–563 in W. E. Duellman (editor). *Patterns of distribution of amphibians: A global perspective*. Baltimore, MD: The Johns Hopkins University Press.

Udvardy, M. D. F. 1975. *A classification of the biogeographical provinces of the world*. IUCN Occasional Paper No. 18.

U Ga. 2000. *Conservation and use of wild Asian elephants*. N.p.: Myanmar Forest Department.

UNDP/FAO. 1984. *Alaungdaw Kathapa National Park. preliminary master plan (revised)*. FO: BUR/80/006. Field Document No. 8. Nature Conservation and National Parks Project.

UNDP/FAO. 1986. *Burma. Forest fire management*. FO:DP/BUR/81/001. Field Document No. 5. Rome.

United Nations Environment Programme, Department of Environment and Natural Resources. 1997. *Philippine biodiversity: An assessment and action plan*. Makati City, Philippines: Bookmark, Inc.

U.S. Geological Survey. 1997. *Global 30 arc second elevation data (Gtopo30)*. EROS Data Center.

Vallentine, J. F. 1979. Grazing systems as a management tool. Pages 214–219 in *The sagebrush ecosystem: A symposium*. Logan: Utah State University.

van der Linde, M. 1995. A further record of the Isabela oriole *Oriolus isabellae* from Baggao, Cagayan Province, northern Philippines. *Fortail* 11: 171.

Vane-Wright, R. I., and D. Peggae. In press. The butterflies of northern and central Maluku: diversity, endemism, biogeography, and conservation priorities. *Tropical Biodiversity*.

Van Steenis, C. G. G. J. 1950. *Flora Malesiana, being an illustrated systematic account of the Malaysian flora*. Djakarta: Noordhoff-Kolff.

———. 1987. *Checklist of generic names in Malesian botany (Spermatophytes)*. Leiden, The Netherlands: Flora Malesiana Foundation.

Vasudevan, K. 1996. Effect of rainforest fragmentation on Western Ghats amphibians. *Frogleg* 2: 1.

Vasudevan, K., and S. K. Dutta. 2000. A new species of *Rhacophorus* (Anura: Rhacophoridae) from the Western Ghats, India. *Hamadryad* 25: 21–28.

Veillon, J.-M. 1993. Protection of floristic diversity in New Caledonia. *Biodiversity Letters* 1: 88–91.

Vences, M., F. Glaw, M. Veith, and I. Das. 2000. Polyphyly of *Tomopterna* (Amphibia: Ranidae) and ecological biogeography of Malagasy relict amphibian groups. In W. Lourenco (editor), *Symposium on the biogeography of Madagascar. Memoires de biogeographie*. Paris: University of Paris.

Vidal, J. 1960. La vegetation du Laos [The vegetation of Laos]. 4 Vol. Toulouse, France: Souladoure. Cited in J. MacKinnon. 1997. *Protected areas systems review of the Indo-Malayan Realm*. Asian Bureau for Conservation/World Conservation Monitoring Center/World Bank.

Vogt, R. C. 1987. You *can* set drift fences in the canopy! *Herpetological Review* 18: 13–14.

Wadia, D. N. 1966. *The geology of India*, 3d ed. London: Macmillan.

Wake, D. B. 1998. Action on amphibians. *Trends in Ecology and Evolution* 13: 379–380.

Walker, B. H. 1993. Rangeland ecology: Understanding and managing change. *Ambio* 22(2–3): 80–87.

Wallace, A. R. 1860. On the zoological geography of the Malay Archipelago. *Journal of the Linnean Society of London* 4: 172–184.

———. 1876. The geographical distribution of animals. 2 vols. New York: Harper. Cited in M. D. F. Udvardy. 1969. *Dynamic zoogeography*. New York: Van Nostrand Reinhold.

———. 1880. *Island life, or the phenomena and causes of insular faunas and floras*. London: Macmillan.

Walter, K. S., and H. J. Gillett (editors). 1998. *1997 IUCN red list of threatened plants. Compiled by the World Conservation Monitoring Centre*. Gland, Switzerland: IUCN.

Wang, J.-T. 1988. The steppes and deserts of Xizang Plateau (Tibet). *Vegetatio* 75: 135–142.

WCMC. 1992. *Global biodiversity: Status of the earth's living resources*. London: Chapman & Hall.

———. 1997. *Protected areas of the world*.

WCMC and Asian Bureau for Conservation. 1997. *Protected areas system review of the Indo-Malayan realm*. (edited by John MacKinnon). Canterbury, England: Asian Bureau for Conservation Limited.

WCMC and IUCN. 1993. *1993 United Nations list of national parks and protected areas*. Gland, Switzerland: IUCN.

Wegge, P. 1991. Khunjerab National Park: Ecological status management recommendations. In B. G. Bell (editor), *Proceedings of the International Workshop on the Management Planning of Khunjerab National Park, June 7–16, 1989*. Washington, DC: U.S. National Park Service, Office of International Affairs.

Weins, J. A. 1993. Fat times, lean times, and competition among predators. *Trends in Ecology and Evolution* 8: 348–349.

Wells, M., S. Guggenheim, A. Khan, W. Wardojo, and P. Jepson, World Bank. 1999. *Investing in biodiversity: A review of Indonesia's integrated conservation and development projects*. Directions in Development Series. Washington, DC: World Bank (Indonesia and Pacific Islands Country Department).

Welzen, P. C. van. 1992. *Species richness and speciation in Melesia. Second Flora Malesiana Symposium: Programme and summary of papers and posters*. Yogyakarta, Indonesia.

Wemmer, C. (editor). 1998. Deer. Status survey and conservation action plan. IUCN/SSC Deer Specialist Group. Gland, Switzerland: IUCN.

West, N. E., K. H. Rea, and R. O. Harniss. 1979. Plant demographic studies in sagebrush-grass communities of southeastern Idaho. *Ecology* 60: 376–388.

Wharton, C. 1957. Man, fire and cattle in north Cambodia. An ecological study of the Kouprey *Novibos sauveli*. *Proceedings of the Annual Tall Timbers Forest Ecology Conference* 6: 23–65.

Wheeler, O. D. 1995. Systematics, the scientific basis for inventories of biodiversity. *Biodiv. & Conserv.* 4: 476–489.

Whitmore, T. C. 1974. *Change with time and the role of cyclones in tropical rain forest on Kolombangara, Solomon Islands*. Commonwealth For. Inst. paper no. 46. Oxford, England: Holywell.

——— (editor). 1984a. *Tropical rain forests of the Far East*, 2d ed. Oxford, England: Oxford University Press.

———. 1984b. A vegetation map of Malesia at scale 1:5 million. *Journal of Biogeography* 11: 461–471.

——— (editor). 1987. *Biogeographical evolution of the Malay Archipelago*. Oxford, England: Clarendon.

———. 1989. Chapter 10: Southeast Asian tropical forests. Pages 195–218 in H. Lieth and M. J. A. Werger (editors), *Ecosystems of the world 14B: Tropical rain forest ecosystems*. Amsterdam: Elsevier.

———. 1995. Comparing southeast Asian and other tropical rainforests. Pages 5–15 in R. B. Primack and T. E. Lovejoy (editors), *Ecology, conservation, and management of southeast Asian rainforests*. New Haven, CT: Yale University Press.

———. 1997. Tropical forest disturbance, disappearance, and species loss. Pages 3–12 in W. F. Lawrance and R. O. Bierregaard, Jr. (editors), *Tropical forest remnants*. Chicago: University Press.

Whitmore, T. C., and I. G. M. Tantra (editors). 1987. *The flora of Indonesia. Draft check list for Borneo*. Bogor, Indonesia: Forest Research and Development Centre.

Whitten, A. J., S. J. Damanik, J. Anwar, and N. Hisyam. 1987. *The ecology of Sumatra*. Yogyakarta, Indonesia: Gajah Mada University Press.

Whitten, A. J., M. Mustafa, and G. S. Henderson. 1987. *The ecology of Sulawesi*. Yogyakarta, Indonesia: Gadjah Mada University Press.

Whitten, T. 1980. Arenga fruit as a food for gibbons. *Principes* 24: 143–146.

———. 1981. Notes on the ecology of *Myrmecodia tuberosa* Jack on Siberut Island, Indonesia. *Annals of Botany* 47: 525–526.

———. 1982a. Diet and feeding behaviour of Kloss gibbons on Siberut Island, Indonesia. *Folia Primatologica* 37: 177–209.

———. 1982b. The ecology of singing in Kloss gibbons on Siberut Island, Indonesia. *International Journal of Primatology* 3: 33–51.

———. 1982c. *The gibbons of Siberut*. London: Dent.

———. 1982d. Home range use by Kloss gibbons (*Hylobates klossii*) on Siberut Island, Indonesia. *Animal Behaviour* 30: 182–198.

———. 1982e. A numerical analysis of tropical rain forest using floristic and structural data and its application to an analysis of gibbon ranging behaviour. *Journal of Ecology* 70: 249–271.

———. 1982f. Possible niche expansion in the spangled drongo on Siberut Island, Indonesia. *Ibis* 124: 192–193.

———. 1982g. The role of ants in selection of night trees by gibbons. *Biotropica* 14: 237–238.

———. 1984a. Defence by singing. Pages 420–421 in D. W. MacDonald (editor), *The encyclopaedia of mammals*. London: Allen & Unwin.

———. 1984b. Ecological and behavioural comparisons between *Hylobates klossii* and some other gibbon species. Pages 219–227 in H. Preuschoft, D. J. Chivers, W. Y. Brokelman, and N. Creel (editors), *The lesser apes: Evolutionary and behavioural biology*. Edinburgh: Edinburgh University Press.

———. 1984c. Gibbon management: Techniques and alternatives. Pages 32–43 in H. Preuschoft, D. J. Chivers, W. Y. Brokelman, and N. Creel (editors), *The lesser apes: Evolutionary and behavioural biology*. Edinburgh: Edinburgh University Press.

———. 1984d. The trilling handicap in gibbons. Pages 416–419 in H. Preuschoft, D. J. Chivers, W. Y. Brokelman, and N. Creel (editors), *The lesser apes: Evolutionary and behavioural biology*. Edinburgh: Edinburgh University Press.

———. 1987. One or more extinctions from Sulawesi? *Conservation Biology* 1: 42–48.

———. 1988. The Presbytis of Sumatra. *Newsletter of the IUCN Primate Specialist Group* 8: 46–47.

———. 1991. *Biokrisis: Kehilangan Keanekaragaman Biologi*. Jakarta: Obar.

———. 1996a. Conservation and ecotourism. In J. Rigg (editor), *Indonesian heritage: Human environment*. Singapore: Archipelago Press.

———. 1996b. *Field guides: Useful tools in environmental planning and management*. Environment Department Dissemination Note No. 15. Washington, DC: The World Bank.

Whitten, T., J. Anwar, S. J. Damanik, and N. Hisyam. 1984. *The ecology of Sumatra*. Yogyakarta, Indonesia: Gadjah Mada University Press.

Whitten, T., and S. J. Damanik. 1986. Mass defoliation of mangrove in Sumatra, Indonesia. *Biotropica* 18: 176.

Whitten, T., J. Dring, and C. McCarthy. 1990. The herpetofauna of the Mentawai Islands, Indonesia. *Indo-Australian Zoology* 1.

Whitten, T., E. H. Haimoff, S. P. Gittins, and D. J. Chivers. 1982. A phylogeny of gibbons (*Hylobates* spp.) based on morphology and behavioural characters. *Folia Primatologica* 39: 213–237.

———. 1984. A phylogeny and classifications of gibbons based on morphology and ethology. Pages 416–419 in H. Preuschoft, D. J. Chivers, W. Y. Brokelman, and N. Creel (editors), *The lesser apes: Evolutionary and behavioural biology*. Edinburgh: Edinburgh University Press.

Whitten, T., and K. MacKinnon. 1987. Indonesia ablaze again. *Wallaceana* 48: 18–19.

Whitten, T., and C. McCarthy. 1993. List of the amphibians and reptiles of Jawa and Bali. *Tropical Biodiversity* 1: 169–177.

Whitten, T., M. Mustafa, and G. S. Henderson. 1987. *The ecology of Sulawesi*. Yogyakarta, Indonesia: Gadjah Mada University Press.

Whitten, T., and J. Ranger. 1986. Logging at Bohorok. *Oryx* 20: 246–248.

Whitten, T., and Z. Sardar. 1981. Masterplan for a tropical paradise. *New Scientist* 91: 230–235.

Whitten, T., R. E. Soeriaatmadja, and S. A. Afiff. 1996. *The ecology of Java and Bali*. Singapore: Periplus Editions.

Whitten, T., and J. Whitten. 1980. Solution for Siberut? *Oryx* 15: 166–169.

———. 1982. Preliminary observation of the Mentawai macaque on Siberut Island, Indonesia. *International Journal of Primatology* 3: 445–459.

———. 1992. *Wild Indonesia*. London: New Holland.

——— (editors). 1996. *Indonesian heritage: Plants*. Singapore: Archipelago Press.

Whitten, T., J. Whitten, C. Goettsch, J. Supria-tna, and R. Mittermeier. 1997. Pages 74–77 in R. Mittermeier, P. Gil, and C. Goettsch-Mittermeier (editors), *Megadiversity: Earth's biologically wealthiest nations*. Prado Norte: CEMEX.

Whitten, T., J. Whitten, A. P. N. House, and J. McNeely. 1980. *Saving Siberut: A conservation masterplan*. Bogor, Indonesia: World Wildlife Fund.

Whitten, T., J. Whitten, J. Supriatna, and R. Mittermeier. 1998a. Sundaland. Pages 279–290 in R. Mittermeier, P. Gil, and C. Goettsch-Mittermeier (editors), *Biodiversity hotspots of the world*. Prado Norte: CEMEX.

———. 1998b. Wallacea. Pages 297–307 in R. Mittermeier, P. Gil, and C. Goettsch-Mittermeier (editors), *Biodiversity hotspots of the world*. Prado Norte: CEMEX.

WHT. 2000. World Heritage Trust of Sri Lanka: http://www.wht.org.

WII. 1993. *Ecology of wild ass (Equus hemionus khur) in Little Rann of Kutch*. Wildlife Institute of India Report.

———. 1994. EIA technical report 9. *Impact assessment studies of Narmada Sagar and Omkareshwar projects on flora and fauna with attendant human aspects*. Dehra Dun, India: WWI.

———. 1997. *Study on the management of rhinoceros in northwest Bengal*. WII Report.

———. 1999. *Management of elephant populations in West Bengal for mitigating man-elephant conflicts*. WII Report.

Wikramanayake, E. D. 1990. Ecomorphology and biogeography of a tropical stream fish assemblage: Evolution of assemblage structure. *Ecology* 71(5): 1756–1764.

———. 2000. *A biological vision for the eastern Himalayas*. In preparation.

Wikramanayake, E. D., E. Dinerstein, T. Allnutt, C. Loucks, and W. Wettengel. 1998. *A biodiversity assessment and gap analysis of the Himalayas*. Conservation Science Program, World Wildlife Fund–US Report.

———. 1999. *A biodiversity assessment and gap analysis of the Himalayas*. Washington, DC: World Wildlife Fund.

Wikramanayake, E. D., E. Dinerstein, J. G. Robinson, K. U. Karanth, A. Rabinowitz, D. Olson, T. Mathew, P. Hedao, M. Connor, G. Hemley, and D. Bolze. 1999. Where can tigers live in the future? A framework for identifying high-priority areas for the conservation of tigers in the wild. Pages 255–272 in J. Seidensticker, S. Christie, and P. Jackson (editors), *Riding the tiger: Tiger conservation in human-dominated landscapes*. Cambridge, UK: Cambridge University Press.

Wilcove, D., C. McClellan, and A. Dobson. 1986. Habitat fragmentation in the temperate zone. Pages 237–256 in M. Soulé (editor), *Conservation biology, the science of scarcity and diversity*. Sunderland, MA: Sinauer.

Wildlife Conservation Society and Sarawak Forest Department. 1996. *A master plan for wildlife in Sarawak*. Kuching, Sarawak: Wildlife Conservation Society and Sarawak Forest Department.

Wilson, C. C., and W. L. Wilson. 1975. The influence of selective logging on primates and some other animals in East Kalimantan. *Fol. Primatol.* 23: 245–274.

Wilson, D., and F. R. Cole. 2000. *Common names of mammals of the world*. Washington, DC: Smithsonian Institution Press.

Wilson, D. E., and D. M. Reeder. 1993. *Mammal species of the world*. Washington, DC: Smithsonian Institution Press.

Wilson, E. O. 1988. *Biodiversity*. Washington, DC: National Academy Press.

———. 1993. *The diversity of life*. New York: W. W. Norton.

Wollaston, A. L. R. 1921. *Mount Everest, the reconnaissance*.

Wood, A., P. Stedman-Edwards, and J. Mang. 2000. *The root causes of biodiversity loss*. Sterling, VA: Earthscan Publications.

Wooding, J. B. 1984. *Coyote food habits and the spatial relationship of coyotes and foxes in Mississippi and Alabama*. M.S. thesis, Mississippi State University, State College.

Woodroffe, R., and J. R. Ginsberg. 1998. Edge effects and the extinction of populations inside protected areas. *Science* 260: 2126–2128.

Woods, A. A. H. Pakistan Museum of Natural History, Islamabad and Florida Museum of Natural History, Gainesville.

World Bank. 1996a. *The Kerinci-Seblat National Park integrated conservation and development project*. Washington, DC: World Bank.

———. 1996b. *India ecodevelopment project*. Washington, DC: World Bank.

World Bank and WWF. 1999. *World Bank WWF alliance for forest conservation and sustainable use*. Annual report.

Wu Ning. 1997. *Ecological situation of high-frigid rangeland and its sustainability: A case study on the constraints and approaches in pastoral Western Sichuan, China*. Berlin: Dietrich Remier Verlag.

WWF and ICIMOD. 2001. *Ecoregion-based conservation in the eastern Himalayas: Identifying important areas for biodiversity conservation*. Kathmandu, Nepal: WWF Nepal Program.

WWF and ICIMOD. In prep. *A biological vision for the eastern Himalayas*.

WWF and IUCN. 1995. *Centres of plant diversity: A guide and strategy for their conservation*. Vol. 2, Asia, Australasia and the Pacific. Cambridge, UK: IUCN.

WWF and WCS. 1997. *A framework for identifying high-priority areas and action of the conservation of tigers in the wild*. Washington, DC: National Fish and Wildlife Foundation.

WWF-France. 1997. *Emergency conservation measures for a critically endangered Global 200 ecoregion: Tropical dry forest of New Caledonia*. Project concept (unpublished manuscript).

WWF-Indochina. 1998. *Borderline: An assessment of wildlife trade in Vietnam*. Hanoi, Vietnam: WWF-Indochina.

WWF-Indonesia. n.d. *Gunung Leuser National Park*. Unpublished document.

Yahner, R. H. 1988. Changes in wildlife communities near edges. *Conservation Biology* 2: 333–339.

Yeager, C. P. 1989. Feeding behavior and ecology of the proboscis monkey (*Nasalis larvatus*). *International Journal of Primatology* 10: 497–530.

———. 1990. Proboscis monkey (*Nasalis larvatus*) social organization: Group structure. *American Journal of Primatology* 20: 95–106.

———. 1991. Possible antipredator behavior associated with river crossings by proboscis monkeys (*Nasalis larvatus*). *American Journal of Primatology* 24: 61–66.

———. 1991. Proboscis monkey (*Nasalis larvatus*) social organization: Intergroup patterns of association. *American Journal of Primatology* 23: 73–86.

———. 1996. Feeding ecology of the long-tailed macaque (*Macaca fascicularis*) in Kalimantan Tengah, Indonesia. *International Journal of Primatology* 17: 51–62.

———. 1998. *Interim report on fire impacts on Tanjung Puting National Park*. Unpublished report, WWF-Indonesia.

———. 1999a. *Fire impacts on vegetational diversity and abundance in Kalimantan, Indonesia during 1997/1998*. Jakarta: WWF-Indonesia.

———. (chief editor). 1999b. *Orangutan action plan for conservation*. Jakarta: WWF-Indonesia.

Yeager, C. P., and T. K. Blondal. 1992. Conservation status of the proboscis monkey (*Nasalis larvatus*) at Tanjung Puting National Park, Kalimantan Tengah, Indonesia. In G. Ismail, M. Mohamed, and S. Omar (editors), *Forest biology and conservation in Borneo*. Yayasan Sabah Centre for Borneo Studies publication no. 2, 220–228.

Yeager, C. P., and G. Fredriksson. 1999a. *Draft paper: Fire impacts on primate and other wildlife in Kalimantan, Indonesia during 1997–98*. Jakarta: WWF-Indonesia.

———. 1999b. *Fire impacts on primates and other wildlife in Kalimantan, Indonesia during 1997/1998*. Jakarta: WWF-Indonesia.

Yeager, C. P., and A. Russon. 1998. Trial by fire. *International Primatological Society Newsletter* 25: 8–9.

Yongzu, Z., J. Sangke, Q. Guoqiang, L. Shihau, Y. Zhongyao, W. Fenggui, and Z. Manli. 1997. *Distribution of mammalian species in China*. CITES Management Authority of China. China Forestry Publishing House.

Yonzon P. B., and M. L. Hunter. 1991. Conservation of the red panda *Ailurus fulgens*. *Biological Conservation* 57: 1–11.

Zhou, L. 1990. Economic development in China's pastoral regions: Problems and solutions. Pages 43–56 in J. W. Longworth (editor), *The wool industry in China: Some Chinese perspectives*. Mt. Waverly, Australia: Inkata Press.

Ziegler, A. C. 1981. *Petaurus abidi*, a new species of glider (Marsupialia: Petauridae) from Papua New Guinea. *Australian Mammalogy* 4(2): 81–88.

Ziegler, T., W. Böhme, and K. M. Philipp. 1999. *Varanus caerulivirens* sp. n., a new monitor lizard of the V. indicus group from Halmahera, Moluccas, Indonesia (Squamata: Sauria: Varanidae). *Herpetozoa* 12: 45–56.

Zoysa, N., and R. Raheem. 1990. *Sinharaja. A rain forest in Sri Lanka*. Sri Lanka: March for Conservation.

Zug, G. R., and Mitchell, J. C. 1995. Amphibians and reptiles of Royal Chitwan National Park, Nepal. *Asiatic Herpetol. Res.* 6: 172–180.

Zug, G. R., H. Win, T. Thin, T. Z. Min, W. Z. Lhon, and K. Kyaw. 1998. Herpetofauna of the Chhathin Wildlife Sanctuary, north-central Myanmar, with preliminary observations on their natural history. *Hamadryad* 23: 111–120.

Authors

Eric D. Wikramanayake, Ph.D.
Senior Conservation Scientist
Conservation Science Program
World Wildlife Fund–United States

Eric Dinerstein, Ph.D.
Chief Scientist
Conservation Science Program
World Wildlife Fund–United States

Colby Loucks, M.E.M.
Senior Conservation Specialist
Conservation Science Program
World Wildlife Fund–United States

David Olson, Ph.D.
Director Conservation Science Program
Conservation Science Program
World Wildlife Fund–United States

John Morrison, M.Sc.
Senior Conservation Specialist
Conservation Science Program
World Wildlife Fund–United States

John Lamoreux
Conservation Specialist
Conservation Science Program
World Wildlife Fund–United States

Meghan McKnight
GIS/Conservation Specialist
Conservation Science Program
World Wildlife Fund–United States

Preshant Hedao, M.L.A.
Environmental Systems Research Institute, Inc.

CONTRIBUTORS

Nantiya Aggimarangsee
PO Box 54
Chiang Mai University
Chiang Mai 50202, Thailand

Tom Allnutt
Conservation Science Program
World Wildlife Fund
1250 24th Street NW
Washington, DC 20037, USA

Peter Ashton
Harvard University
1 Eliot Street Building
Cambridge, MA 02138, USA

Bruce M. Beehler
Counterpart International
1200 18th Street NW, Suite 1100
Washington, DC 20036, USA

Elizabeth Bennett
Wildlife Conservation Society
2300 Southern Boulevard
Bronx, NY 10460, USA

Frank Bonaccorso
National Museum
Port Moreby, Papua New Guinea

Ramesh Boonratana
PO Box 54
Chiang Mai University
Chiang Mai 50202, Thailand

Thomas M. Brooks
Center for Applied Biodiversity Science
Conservation International
2501 M Street NW, Suite 200
Washington, DC 20037, USA

Chris Carpenter
Wildlands Studies Program
San Francisco State University
College of Extended Learning
San Francisco, CA 94132, USA

Jason Clay
Research and Development Program
World Wildlife Fund
1250 24th Street NW
Washington, DC 20037, USA

Arnaud Collin
Worldwide Fund for Nature–France
188, Rue de la Roquette
Paris 75011, France

Indraneil Das
Institute of Biodiversity and Environmental Conservation
Universiti Malaysia Sarawak
94300 Kota Samarahan
Sarawak, Malaysia

Geoffrey Davison
WWF-Malaysia
Suite 1-6-W11
6th Floor, CPS Tower
Centrepoint
88000 Kota Kinabalu
Sabah, Malaysia

Ajay Desai
BC 65, Camp
Belgaum 590001, India

Walter Erdelen
Department of Biology
Institute of Technology
Jl. Ganesha 10
Bandung 40132
West Java, Indonesia

Paul J. Ferraro
Andrew Young School of Policy Studies
Georgia State University
University Plaza
Atlanta, GA 30303, USA

Arnaud Greth
Worldwide Fund for Nature–France
188, Rue de la Roquette
Paris 75011, France

Savithri Gunatilleke
Department of Botany
Faculty of Science
University of Peradeniya
Peradeniya, Sri Lanka

U Saw Han
Nature and Wildlife Conservation Division
Forest Department
No. 35 D, Kyaikwaing Pagoda Road
Mayangone Township
Yangon, Mynamar

Lawrence R. Heaney
Division of Mammals
The Field Museum
Roosevelt Road at Lake Shore Drive
Chicago, IL 60605, USA

Derek Holmes (*deceased*)

Don Hunter
United States Geological Survey
Midcontinent Ecological Science Center
4512 McMurry Avenue
Fort Collins, CO 80525, USA

Rodney Jackson
International Snow Leopard Trust
4649 Sunnyside Avenue North, Suite 325
Seattle, WA 98103, USA

Tim Jessup
WWF-Indonesia
Jl. Hayam Wuruk 179
Denpasar, Bali 80235, Indonesia

Robert Johns
Herbarium
Royal Botanic Gardens, Kew
Richmond, Surrey
TW9 3AE, United Kingdom

A. J. T. Johnsingh
Wildlife Institute of India
PB 18, Chandrabani
Dehra Dun 248 001
Uttar Pradesh, India

Anup Joshi
Department of Fisheries and Wildlife
University of Minnesota
St. Paul, MN 55108, USA

Saw Tun Khaing
WCS Mynamar Program
Building C-1, Ayeyeikmon 1st Street
Ward (3), Hlaing Township
Yangon, Mynamar

Maurice Kottelat
Route de la Baroche 12
Case Postale 57
CH-2952, Switzerland

Gary A. Krupnick
Department of Botany MRC-166
National Museum of Natural History
Smithsonian Institution
Washington, DC 20560-0166, USA

Jim LaFrankie
Center for Tropical Forest Science
University of Singapore
469 Bukit Timar Road
Singapore 1025

Alf J. Leslie
School of Forestry
University of Melbourne
Victoria, Australia

Peter Lowry
Missouri Botanical Garden
PO Box 299
St. Louis, MO 63166-0299, USA

Kathy MacKinnon
Global Environment Division
The World Bank
1818 H Street NW
Washington, DC 20433, USA

Domingo Madulid
National Museum
Botany Division
P. Burgos Drive (Near Manila City Hall)
P.O. Box 2659, CPO
Manila, Philippines

Andy Maxwell
Virachey Conservation Project
WWF-Indochina Program, Cambodia Office
Phnom Penh, Cambodia

John Maxwell
Chiangmai University Herbarium
Chiang Mai University
Chiang Mai 50200, Thailand

Daniel J. Miller
5235 Western Avenue NW
Washington, DC 20015, USA

Judy Mills
Species Program
World Wildlife Fund
1250 24th Street NW
Washington, DC 20037, USA

Kathryn Monk
Research, Monitoring, and Information Division
Leuser Management Unit
Jl. Dr. Mansyur 68
Medan 20154
Sumatera Utara, Indonesia

Rohan Pethiyagoda
Wildlife Heritage Trust of Sri Lanka
95 Cotta Road, Colombo 8, Sri Lanka

Ajay Rastogi
Eastern Himalayan Programme
Ashoka Trust for Research in Ecology and Environment
Bungalow 2, Bhujiapani
Bagdogra 734 422
Darjeeling, West Bengal, India

Gopal S. Rawat
Wildlife Institute of India
PO Box 18, Chadrabani
Dehra Dun 248001
Uttar Pradesh, India

Tom Roberts
Cae Gors
Rhoscefnhir
Nr. Pentraeth
Anglesey LL 75 8YU, Wales

John G. Robinson
Wildlife Conservation Society
2300 Southern Boulevard
Bronx, NY 10460, USA

Phillip D. Round
Centre for Conservation Biology
Faculty of Science
Mahidol University
Rama 6 Road
Bangkok 10400, Thailand

June Rubis
302 Batu Kawa Road
Sarawak, Malaysia

Phillip Rundel
Department of Organismic Biology, Ecology and Evolution
University of California
Los Angeles, CA 90095, USA

John Seidensticker
Smithsonian National Zoological Park
3001 Connecticut Avenue NW
Washington, DC 20008-2598, USA

David Smith
Department of Fisheries and Wildlife
University of Minnesota
St. Paul, MN 55108, USA

Joyotee Smith
CIFOR
PO Box 6596 JKPWB
Jakarta 10065, Indonesia

Hema Somanathan
c/o Bombay Natural History Society
S.B. Singh Road
Mumbai 400 023, India

Sompoad Srikosamatara
Centre for Conservation Biology
Faculty of Science
Mahidol University
Rama 6 Road
Bangkok 10400, Thailand

Meseret Taye
Conservation Science Program
World Wildlife Fund
1250 24th Street NW
Washington, DC 20037, USA

Seng Teak
Ministry of Environment
48 Samdech Preah Sihanouk Street
Phnom Penh, Cambodia

U Tin Than
WWF-Thailand Program
Asian Institute of Technology
PO Box 4 Klong Luang
Pathumatani 12120, Thailand

Robert Tizard
1901 Nueces Drive
College Station, TX 77840, USA

Max van Balgooy
Rijksherbarium
Van Steenisgebouw Einsteinweg 2
PO Box 9514 2300
RA Leiden, Netherlands

Peter Paul van Dijk
Biology Department Science Faculty
Chulalongkorn University
Phaya Thai Road
Bangkok 10330, Thailand

Meeta Vyas
WWF-India
172-B Lodi Road
Max Mueller Marg
New Delhi 110 003, India

T. C. Whitmore
Geography Department
Cambridge University
Downing Place
Cambridge CB2 3EN, England

Tony Whitten
Environment and Social Development Sector
East Asia and Pacific Region
The World Bank
1818 H St NW
Washington, DC 20433, USA

Pralad Yonzon
Resources Nepal
GPO Box 2448
Kumaripati
Lalipur, Nepal

INDEX

Admiralty Islands Lowland Rain Forests ecoregion, 576–78
Afghanistan. *See* Indian Subcontinent bioregion
agricultural land diversion, 186
agriculture, biodiversity and, 1–2
amphibians, 47–49
Andaman Islands Rain Forests ecoregion, 400–402
Annamite Mountains, 62, 90–91, 154–55
anthropogenic impacts. *See* biodiversity, threats to
aquatic birds, 34–35
aquatic systems, 31–36
Asian elephants (*Elaphas maximus*), 168, 170–72
Asian Rhinoceros and Elephant Strategies (AREAS), 171
Atlantic–Caribbean–East Pacific (ACEP) region, 55
Australasian zoogeographic realm, 64

babirusa (*Babyrousa babyrussa*), 64, 125, 149
Baghmara Community Forest Project (Nepal), 180, 181. *See also* Terai Arc Wildlife Corridor
Baluchistan Xeric Woodlands ecoregion, 369–71
bamboo rat (*Rhynchomys soricoides*), 148
Banda Sea Islands Moist Deciduous Forests ecoregion, 540–42
Bangladesh. *See* Indian Subcontinent bioregion
Bardia National Park (Nepal), 178–80
Basauti Forest (Nepal), 179
bats, 37
beta-diversity, 24
Bhutan, 135, 178. *See also* Himalayas; Indian Subcontinent bioregion
Biak-Numfoor Rain Forests ecoregion, 550–51
Big Conservation, 7, 165
biodiversity, proxies for, 139
biodiversity, threats to. *See also* conservation status
    agriculture, 1–2
    in aquatic systems, 33–34
    degradation vs. conversion, 109
    exotic species, 33, 34
    fire, 1–2, 8–9, 37, 106
    high-impact, 116
    logging, 109–10, 116–18, 119
    oil palm plantations, 122–24
    overfishing, 34
    overgrazing, 110–11, 173
    population density, human, 111–15, 172, 179
    types of, 109
    wildlife exploitation, 110
    wildlife trade, 110, 124–27
Biodiversity Action Plan (Indonesia), 12–13
biodiversity hotspots, 85, 87, 88, 89
biogeographic regions, 17, 25, 47, 50–51, 62–64. *See also* bioregions
Biological Distinctiveness Index (BDI). *See also* Conservation Status Index (CSI); endemism; globally outstanding ecoregions; species richness
    biodiversity hotspots and, 85, 88
    by biome and bioregion, 66, 85
    categories, 40–41
    conservation status and, 45–46
    criteria, 39
    database construction, 39–41
    description of, 6
    endemic bird areas (EBAs) and, 88–89
    general patterns, 63, 66
biomes, 5, 21–25, 26–28. *See also individual biomes*
bioregions, 4–5, 6, 18–20, 21–24
biounits, 17, 25

BirdLife International, 40, 88–89
birds
    aquatic, 34–35
    endemic bird areas (EBAs), 88–89
    endemism, patterns of, 72–73, 82
    species richness, 68, 70, 71, 72
    swiftlets, 37
Borneo. *See also* Sunda Shelf and Philippines bioregion
    deforestation, 8–13
    fires, 9
    Gunung Subis coral reef, 36
    logging threats, 116, 117
    species richness, 65
    timber trade, 14
    wildlife trade, 125, 126
Borneo Lowland Rain Forests ecoregion, 145, 176, 475–79
Borneo Montane Rain Forests ecoregion, 156, 472–74
Borneo Peat Swamp Forests ecoregion, 452–54
bottom-up control processes, 56–59
Brahmaputra Valley Semi-Evergreen Rain Forests ecoregion, 298–301
Burma. *See* Indochina bioregion
Buru Rain Forests ecoregion, 536–38
Buxa reserve (India), 189–90

Cambodia, 116, 117. *See also* Indochina bioregion
canals, aquatic biodiversity and, 33
Captaincookia margaretae, 152, 162
carbon credits, 173–74, 182
Cardamom Mountains Rain Forests ecoregion, 84, 155, 389–91
carnivores, 56–59. *See also* tigers (*Panthera tigris*)
cave ecosystems, 33, 36, 37
Central Deccan Plateau Dry Deciduous Forests ecoregion, 324–26
Central Indochina Dry Forests ecoregion, 159, 176, 421–24
Central Range Montane Rain Forests ecoregion, 157, 559–62
Central Range Sub-Alpine Grasslands ecoregion, 84, 158, 574–76
Centres of Plant Diversity, 89–90
Chao Phraya Freshwater Swamp Forests ecoregion, 407–10
Chao Phraya Lowland Moist Deciduous Forests ecoregion, 418–19
Chhota-Nagpur Dry Deciduous Forests ecoregion, 159, 319–20
China, 184, 185. *See also* Himalayas; Indian Subcontinent bioregion; Indochina bioregion
Chin Hills–Arakan Yoma Montane Rain Forests ecoregion, 374–75
Chitwan National Park (Nepal), 129–30, 172, 178–81
classes, priority. *See* priority portfolio
classification systems, biogeographical, 17, 25. *See also* biomes; bioregions; ecoregion approach
Clean Development Mechanism (CDM), 182–85
community forestry projects, 180, 181
competition, interspecies, 58–59
connectivity, 172, 178–81
conservation incentives, 172–76
conservation initiatives. *See also* integrated conservation and development projects (ICDPs)
    community forestry projects, 180, 181
    direct vs. indirect, 186–87
    ecological service compensation, 174, 181, 184–87
    ecoregion-scale, 165–68
    landscape-scale, 168, 170–73
    Lower Mekong forests workshop, 166–68, 169
conservation performance payments, 174, 181, 184–87
conservation planning, ecoregion-scale, 165–66
conservation status
    fifty-percent target, 108–9

637

conservation status (*continued*)
    fragmentation, 104
    habitat block size, 104
    habitat loss, 101, 103
    habitat protection, 104, 107, 108
    habitat remaining, by elevation, 105
    human population threats and, 114–15
    logging threats and, 116–18, 119
    lowland vs. montane ecoregions, 120
    patterns by category, 100–101, 102
    threat-modified, 118, 120, 121
Conservation Status Index (CSI), 6, 41–46, 99. *See also* Biological Distinctiveness Index (BDI)
continents, islands vs., 68, 71–73, 79, 80
contracting approaches. *See* conservation performance payments
Convention on Biodiversity (CBD), 183
conversion threats, 109. *See also* logging; oil palm plantations
Corbett National Park (India), 178–80
Corbett-to-Chitwan Corridor, 171–73
corridor restoration, 171–73, 178–81
Costa Rica, 186
cryptic species, 48, 50

dam construction, 32
Deccan Thorn Scrub Forests ecoregion, 328–30
deforestation. *See also* conservation status; habitat loss; logging
    aquatic habitats and, 32
    by elevation, 103, 105
    Indonesia, 7–13
    Sumatra, 104, 106
degradation threats, 109. *See also* livestock grazing; population density, human
deserts and xeric shrublands biome, 28
development programs. *See* integrated conservation and development projects (ICDPs)
Dipterocarpaceae, 92–98
dispersal
    barriers to, 61–65
    corridor restoration, 171–73, 178–81
    fragmentation and, 57–58
    species richness and, 71
Doria's tree kangaroo (*Dendrolagus dorianus*), 157
Dudhwa National Park (India), 178–80

East Afghan Montane Coniferous Forests ecoregion, 345–47
East Deccan Dry Evergreen Forests ecoregion, 295–96
Eastern Highlands Moist Deciduous Forests ecoregion, 84, 143, 176, 306–8
Eastern Himalayan Alpine Shrub and Meadows ecoregion, 153, 154, 362–65
Eastern Himalayan Broadleaf Forests ecoregion, 152, 153, 335–38
Eastern Himalayan Sub-Alpine Conifer Forests ecoregion, 159, 176, 340–43
Eastern Java–Bali Montane Rain Forests ecoregion, 470–72
Eastern Java–Bali Rain Forests ecoregion, 465–67
ecodevelopment projects. *See* integrated conservation and development projects (ICDPs)
ecological processes
    Biological Distinctiveness Index (BDI) and, 41
    globally outstanding ecoregions, 66, 80, 83, 84
    hunting as threat to, 125
    mangrove systems, 40, 55–56
    predation, 56–59
ecological services compensation, 174, 181, 184–87
economic incentives, 173–75
ecoregion approach. *See also* globally outstanding ecoregions
    biodiversity hotspots, overlap with, 85, 88
    biomes and bioregions, 18–25
    conservation planning, 165–66
    conservation units, 18, 166
    definitions, 6, 17–18
    delineation criteria, 25–26, 28
    endemic bird areas (EBAs), overlap with, 88–89
    Global 200 analysis, 84–85, 86
ecoregions, Indo-Pacific, 4–5, 18–20, 25–26
ecosystem approach, 183
ecosystem services compensation, 174, 181, 184–87
ecotones, 62–64
ecotourism, 172, 175, 176–77, 180, 181
elephants, Asian (*Elaphas maximus*), 168, 170–72
El Niño, 8
empty forest syndrome, 110, 125, 178
endemic bird areas (EBAs), 88–89, 89–90
endemism. *See also* Biological Distinctiveness Index (BDI)
    amphibians, 47–48
    bats, 29–30
    Biological Distinctiveness Index (BDI) and, 39–40
    birds, 29–30, 72–73, 75, 82
    continental vs. island, 71–73, 79, 80
    criteria for, 40
    Dipterocarpaceae, 93, 95, 96, 97
    dispersal barriers and, 61–64
    globally outstanding status and, 64–65
    habitat loss and, 30
    islands and, 64
    limestone habitats, 36–38
    mammals, 72–73, 74, 78–80, 81
    New Caledonia, 61–62, 162
    new species, discovery of, 62, 90–92
    patterns of, 29–31, 65–71
    peat swamps, 35
    plant species, 73, 77, 78
    primates, 78, 79
    reptiles, 50, 51
    in Sulawesi (Indonesia), 64
    terrestrial islands and, 65
environmental services compensation, 174, 181, 184–87
evolutionary processes, 41, 66, 80, 83, 84
exotic species, 33, 34

fire, 1–2, 8–9, 37, 106
fish. *See* aquatic systems
flagship species
    megavertebrates, 168–71
    as proxies, 46
    reptiles, 53–54
    top carnivores, 56–59
flooded grasslands and savannas biome, 28
focal species. *See* flagship species
forestry
    community projects, 180, 181
    Kyoto Protocol Clean Development Mechanism (CDM) and, 182–85
    sustainable, 15
    timber market, 14–16
forests. *See also* logging
    frontier, 109–10
    lowland, 7–13, 10, 25–26, 103, 104, 105
    montane, 10–11, 25–26, 103, 105, 173
    submontane, 103, 105
    use and exploitation, 183
Forests of the Lower Mekong workshop, 166–68, 169
fragmentation, habitat
    Conservation Status Index (CSI) and, 42
    hunted species and, 124
    logging and, 116
    megavertebrates, 168, 170–72
    patterns, 105
    Terai Arc, 180
    tigers, 57–58
freshwater systems, 31–36
frogs, 47–49
frontier forests, 109–10

Gandhi, Indira, 177
globally outstanding ecoregions
    by biome, 80, 84
        distribution patterns, 66–67
        ecological and evolutionary distinctiveness, 83, 84
        ecotones and, 62–64
        endemism and, 64–65, 71
        species richness, 71
globally rare habitats, 83, 84
Global 200 ecoregions, 84–85, 86
Godavari-Krishna Mangroves ecoregion, 176, 352–53
Gondwanaland, 162
Greater Negros–Panay Rain Forests ecoregion, 146–47, 176, 493–97
Gua Salukkan Kallang–Gua Tanette cave system (Sulawesi, Indonesia), 36
Gunung Lorentz (Indonesia), 128
Gunung Subis coral reef, 36

habitat blocks, 42, 104, 105. *See also* fragmentation, habitat
habitat loss. *See also* conservation status; deforestation
    aquatic, 34
    by biome, 103
    chronic and catastrophic threats, 44
    Conservation Status Index (CSI) and, 41–42
    endemism and, 30
    forests, by elevation, 103, 105
    human population density and, 113–14, 172
    logging and, 116–18
    patterns of, 101
habitats, aquatic, 31–33
habitats, globally rare, 83, 84
habitats, limestone, 36–37
Halmahera Rain Forests ecoregion, 156, 533–36
hardwoods, tropical, 14–16
Hasuliya Forest (Nepal), 179
herpetofauna
    amphibians, 47–49
    reptiles, 49–54
Himalayan Subtropical Broadleaf Forests ecoregion, 332–35
Himalayan Subtropical Pine Forests ecoregion, 159, 347–49
Himalayas
    alpine protected areas, 131–34
    endemism, 62
    livestock grazing, 110–11, 134–37, 173
    Siwalik Hills megafauna, 172
    Terai Arc Wildlife Corridor, 171–73, 175, 178–81
Holarctic flora, 64
hotspots, biodiversity, 85, 87, 88, 89
human population density, 111–15, 172, 179
hunting. *See* wildlife trade
Huon Peninsula Montane Rain Forests ecoregion, 156–57, 557–59
hydroelectricity micro-generators, 33

ICDPs. *See* integrated conservation and development projects
India. *See also* Himalayas; Indian Subcontinent bioregion
    forest management programs, 184, 185
    Rumbak-Rumchang catchment, 135
    Terai Arc Wildlife Corridor, 171–73, 178–81
India Ecodevelopment Project, 189–91
Indian Subcontinent bioregion. *See also* Himalayas
    amphibians, Western Ghats, 47–48
    conservation status, 118, 120
    demarcation of, 21
    ecoregional map, 18
    endemic bird areas, ecoregion overlap with, 89
    globally outstanding ecoregions, 65–66
    human population threats, 111–15, 172
    priority ecoregions, 142–44, 152–53, 159, 161
    protected areas, 127, 128
    reptiles, 50–51, 54
    World Heritage Sites, 130

Indochina bioregion
    conservation status, 118
    demarcation of, 21
    ecoregional map, 19
    Forests of the Lower Mekong workshop, 166–68, 169
    globally outstanding ecoregions, 65–66
    human population threats, 111–15
    mammal and bird richness, 68, 71
    priority ecoregions, 144, 153–55, 159, 161
    reptiles, 51–52, 54
    wildlife exploitation, 110
    World Heritage Sites, 130
Indochina Mangroves ecoregion, 159, 176, 435–36
Indo-Malayan zoogeographic realm, 64
Indonesia. *See also* Borneo; New Guinea and Melanesia bioregion; Sunda Shelf and Philippines bioregion; Wallacea bioregion
    Biodiversity Action Plan, 12–13
    forest cover, 10–11
    lowland forests, 7–13
    oil palm plantations, 122–23, 145
    priority analyses, 89–90
    protected areas, 129
Indo-Pacific region
    biodiversity hotspots, 87
    biological distinctiveness status, 63
    biomes, 22–23, 24–25
    bioregions, 21–24
    bird endemism, 75, 82
    bird richness, 70
    conservation status, 102, 121
    definition, 6, 18–19
    Dipterocarpaceae distributions, 94–95, 97–98
    ecological and evolutionary distinctiveness, 83
    ecoregions, 4–5, 25–26
    globally outstanding ecoregions, 66–67
    habitat protection, 107
    habitats remaining, by elevation, 105
    human population densities, 112
    mammal endemism, 74, 81
    mammal richness, 69
    plant endemism, 77
    plant richness, 76
    priority classes, 141
    species richness patterns, 67
Indus River Delta–Arabian Sea Mangroves ecoregion, 176, 349–50
Indus Valley Desert ecoregion, 373–74
insect diversity, 143
integrated conservation and development projects (ICDPs)
    contracting approaches vs., 174, 185–87
    criticism of, 129–30, 175–76
    India Ecodevelopment Project, 189–91
    Kerinci-Seblat National Park (Sumatra), 188–89
integration matrix, 45–46, 140, 141
International Tropical Timber Organization, 14
investment screens, 123
Iran. *See* Indian Subcontinent bioregion
Irian Jaya. *See* Indonesia; New Guinea and Melanesia bioregion
Irrawaddy Dry Forests ecoregion, 420–21
Irrawaddy Freshwater Swamp Forests ecoregion, 406–7
Irrawaddy Moist Deciduous Forests ecoregion, 412–14
islands
    colonization, 52
    continents vs., 68, 71–73, 79, 80
    endemism, 64, 71, 79, 80
    species richness, 65, 71–72

Java. *See* Indonesia; Sunda Shelf and Philippines bioregion
Joint Forest Management programs (India), 184, 185

Kalimantan, Indonesia, 8–13. *See also* Borneo; Indonesia; Sunda Shelf and Philippines bioregion
Karakoram–West Tibetan Plateau Alpine Steppe ecoregion, 365–67

Katarniaghat Wildlife Sanctuary (India), 178–80
Kathiarbar-Gir Dry Deciduous Forests ecoregion, 315–17
Kayah-Karen Montane Rain Forests ecoregion, 153–54, 379–81
Kayan Mentarang reserve (Indonesia), 128
Kerinci-Seblat National Park (Sumatra), 188–89
keystone habitats, 56
Kinabalu Montane Alpine Meadows ecoregion, 84, 156, 516–18
Kishanpur Wildlife Sanctuary (India), 178–80
Komodo dragon (*Varanus komodoensis*), 149
Koshi Tappu Wildlife Reserve (Nepal), 178–80
Kyoto Protocol, 182–85

landscape-level approach, 41, 188. *See also* corridor restoration
Laos. *See* Indochina bioregion
leopards (*Panthera pardus*), 58–59
Lesser Sundas Deciduous Forests ecoregion, 149, 176, 526–29
limestone biodiversity, 36–38
livestock grazing, 134–37, 173
lizards, 51, 52. *See also* reptiles
local guardianship, 176–77, 181
logging
    forest management and, 183
    large-scale, 44, 109–10, 116–18, 119, 146
    oil palm plantations and, 122
    reduced-impact, 184
    threats from, 119, 146
    timber market prospects, 14–16
Lorentz National Park (Indonesia), 128
Louisiade Archipelago Rain Forests ecoregion, 585–87
Lower Gangetic Plains Moist Deciduous Forests ecoregion, 302–4
lowland forests, 7–13, 10, 25–26, 103, 104, 105
Luang Prabang Montane Rain Forests ecoregion, 382–83
Luzon Montane Rain Forests ecoregion, 501–4
Luzon Rain Forests ecoregion, 146, 176, 486–89
Luzon Tropical Pine Forests ecoregion, 148, 176, 512–14

Malabar Coast Moist Deciduous Forests ecoregion, 309–11
Malaysia. *See also* Indochina bioregion; Sunda Shelf and Philippines bioregion
    Gunung Subis coral reef, 36
    oil palm plantations, 122–23
    timber market and, 14
    wildlife trade, 125, 126
Malesia botanical region, 61
Malili lakes (Sulawesi, Indonesia), 35
Maluku, Indonesia, 8, 12
mammal endemism
    bird endemism, concordance with, 72–73
    Indo-Pacific distribution, 74
    strict endemics, 78–80
    Wallace's Line, 29, 30
mammal species richness, 68, 69, 71, 72
mangroves, biological distinctiveness of, 40
mangroves, ecology and distribution of, 55–56
mangroves biome, 28, 101
market economy, forest-dwelling people and, 124
marshes, 32–33
mass-specific energy needs, 56–57
matrix, priority-setting, 45–46, 140, 141
medicine, traditional, 125, 126–27
megavertebrates, 46, 168, 170–72. *See also* tigers (*Panthera tigris*)
Meghalaya Subtropical Forests ecoregion, 159, 176, 304–6
Mekong, Lower, 166–68, 169
Mentawai Islands Rain Forests ecoregion, 448–50
meso-predator release, 58–59
metapopulation management, 171–72, 178–81
Mindanao–Eastern Visayas Rain Forests ecoregion, 147, 176, 497–501
Mindanao Montane Rain Forests ecoregion, 147–48, 176, 504–8
Mindoro Rain Forests ecoregion, 146, 176, 490–93
Ministry of Forestry and Estate Crops (Indonesia), 7, 8
Mizoram-Manipur-Kachin Rain Forests ecoregion, 159, 176, 377–79
Moluccas (Indonesia), 8, 12

montane forest, 10–11, 25–26, 103, 105, 173
montane grasslands and shrublands biome, 28
Myanmar. *See* Indochina bioregion
Myanmar Coastal Mangroves ecoregion, 176, 433–35
Myanmar Coastal Rain Forests ecoregion, 375–77

Nagarhole reserve (India), 189–90
Narmada Valley Dry Deciduous Forests ecoregion, 322–24
National Forestry Financial Fund (FONAFIFO), 186
national parks. *See* protected areas
Nepal. *See also* Himalayas; Indian Subcontinent bioregion
    conservation incentives, 172–73, 177
    rangelands, 135, 137
    reforestation programs, 3
    Terai Arc Wildlife Corridor, 171–73, 178–81
New Britain–New Ireland Lowland Rain Forests ecoregion, 150, 176, 578–81
New Britain–New Ireland Montane Rain Forests ecoregion, 150, 176, 581–84
New Caledonia, 52, 61–62, 64–65, 162–63. *See also* New Guinea and Melanesia bioregion
New Caledonia Dry Forests ecoregion, 84, 151–52, 162–63, 176, 597–99
New Caledonia Rain Forests ecoregion, 84, 151, 176, 594–97
New Guinea, 65, 116, 117, 157
New Guinea and Melanesia bioregion. *See also* Indonesia; New Caledonia
    conservation status, 101, 120
    demarcation of, 24
    ecoregional map, 20
    endemism, 73
    globally outstanding ecoregions, 65–66
    habitat protection, 108
    human population threats, 111–15
    mammal and bird richness, 71
    priority ecoregions, 150–52, 156–58, 159, 161
    reptiles, 52, 54
    World Heritage Sites, 130–31
New Guinea Mangroves ecoregion, 150–51, 176, 570–72
Nicobar Islands Rain Forests ecoregion, 402–4
Niligiri Biosphere Reserve (India), 189
Northeast India–Myanmar Pine Forests ecoregion, 159, 432–33
Northern Annamites Rain Forests ecoregion, 154–55, 386–89. *See also* Annamite Mountains
Northern Dry Deciduous Forests ecoregion, 321–22
Northern Indochina Subtropical Forests ecoregion, 159, 176, 426–28
Northern Khorat Plateau Moist Deciduous Forests ecoregion, 416–18
Northern New Guinea Lowland Rain and Freshwater Swamp Forests ecoregion, 555–57
Northern New Guinea Montane Rain Forests ecoregion, 553–55
Northern Thailand–Laos Moist Deciduous Forests ecoregion, 414–16
Northern Triangle Subtropical Forests ecoregion, 155, 424–26
Northern Triangle Temperate Forests ecoregion, 155, 430–32
Northern Vietnam Lowland Rain Forests ecoregion, 391–93
North Western Ghats Moist Deciduous Forests ecoregion, 311–13
North Western Ghats Montane Rain Forests ecoregion, 281–84
Northwestern Himalayan Alpine Shrub and Meadows ecoregion, 358–59
Northwestern Thorn Scrub Forests ecoregion, 159, 330–32

oil palm plantations, 122–24, 145
Oriental biogeographic region, 47, 50–51, 64
Orissa Semi-Evergreen Rain Forests ecoregion, 293–94
Oryza neocalidonea, 162
overfishing, 34

Pakistan. *See* Indian Subcontinent bioregion
Palamau reserve (India), 189–90
Palawan Rain Forests ecoregion, 145–46, 176, 482–85
Palearctic biogeographic region, 50–51, 64
palm oil, 122–24
Papua New Guinea. *See* New Guinea and Melanesia bioregion
parks. *See* protected areas

Parsa Wildlife Reserve (Nepal), 178–80
peat swamps, 34, 35, 147, 176
Pench reserve (India), 189
Peninsular Malaysian Montane Rain Forests ecoregion, 439–41
Peninsular Malaysian Peat Swamp Forests ecoregion, 147, 176, 455–56
Peninsular Malaysian Rain Forests ecoregion, 159, 176, 436–38
Periyar reserve (India), 189
philanthropists, 175
Philippines, 29–30. *See also* Sunda Shelf and Philippines bioregion
plants
    data sources, 92, 96
    Dipterocarpaceae, 92–98
    endemism patterns, 73, 77, 78
    Holarctic/Tropical ecotones, 64
    reptile richness, link with, 53
    species richness, 73, 76, 78, 151
Pleistocene land patterns, 28–31
poaching. *See* wildlife trade
pollution, aquatic, 33–34
population density, human, 111–15, 172, 179
predator-prey ecology, 56–59
primates, 58, 78, 79
prioritization. *See also* Biological Distinctiveness Index (BDI); Conservation Status Index (CSI)
    assessment methods, 6
    Biodiversity Action Plan (Indonesia), 12–13
    Centres of Plant Diversity, 89–90
    ecoregions by biome and priority class, 142
    endemic bird areas (EBAs), 88–90
    Forests of the Lower Mekong workshop, 166–68, 169
    Global 200 ecoregions, 84–85, 86
    guiding principles, 39
    hotspots, biodiversity, 85, 87, 88, 89
    integration matrix, 45–46, 140, 141
    reptiles, 50
priority portfolio
    classes, 140
    distribution patterns, 158–61
    ecoregions by biome, 142, 161
    high-priority ecoregions, 152–58
    priority ecoregions, 158, 159
    protected areas in, 176
    urgent-priority ecoregions, 142–52
proboscis monkey (*Nasalis larvatus*), 148
protected areas
    boundary threat exposure, 57
    conservation issues, 127–31
    corridor linkages, 172
    ecotourism, 172, 176–77
    fifty-percent target, 108–9
    Himalayan alpine, 131–34
    ICDPs and, 129–30
    network expansion, 175
    New Caledonia, 163
    oil palm plantations and, 122
    patterns of, 104, 107, 108
    in priority ecoregions, 176
    private, 174–75
    subsidies for, 174
    Terai Arc Wildlife Corridor, 171–73, 178–81
    World Heritage Sites, 130–31
proxies, 46, 139

quarrying, limestone, 37, 38

rangelands, Himalayan, 134–37
Rann of Kutch Seasonal Salt Marsh ecoregion, 153, 356–57
Ranthambore reserve (India), 189–90
rapids habitats, 32. *See also* river and stream systems
red-bellied pitta (*Pitta erythrogaster*), 147
Red River Freshwater Swamp Forests ecoregion, 412
reforestation programs, 3

regional-scale analysis, 6
representation, ecoregional, 21
reptiles, 49–54, 149
reserves. *See* protected areas
reservoirs as habitat, 33
resource limitations, competition and, 58–59
revenue sharing, 172–73, 177
rhinoceros (*Rhinoceros sp.*), 168, 170–72
rhododendron forests, 153
river and stream systems, 31–34
roads, 58, 109

Sarawak, Malaysia. *See also* Borneo; Sunda Shelf and Philippines bioregion
    Gunung Subis coral reef, 36
    timber market and, 14
    wildlife trade, 125, 126
sediment load, 34
Seram Rain Forests ecoregion, 156, 538–40
shifting cultivation, 1–2
Siwalik Hills, Nepal, 172, 180
slipper orchid (*Paphiopedilum sanderianum*), 37
snails, 37–38
snakes, 51, 52. *See also* reptiles
snow leopard (*Uncia uncia*), 154
Sohelwa Wildlife Sanctuary (India), 178–80
Solomon Islands Rain Forests ecoregion, 176, 587–91
Soulé, Michael, 2
source sink conditions, 57–58
South Asia. *See* Indian Subcontinent bioregion
South China–Vietnam Subtropical Evergreen Forests ecoregion, 176, 429–30
South Deccan Plateau Dry Deciduous Forests ecoregion, 326–28
Southeast Asia, mainland. *See* Indochina bioregion
Southeastern Indochina Dry Evergreen Forests ecoregion, 144, 176, 397–400
Southeastern Papuan Rain Forests ecoregion, 157, 563–65
Southern Annamites Montane Rain Forests ecoregion, 159, 176, 393–95
Southern Borneo Freshwater Swamp Forests ecoregion, 458–60
Southern New Guinea Freshwater Swamp Forests ecoregion, 158, 566–68
Southern New Guinea Lowland Rain Forests ecoregion, 84, 150, 176, 568–70
Southern Vietnam Lowland Dry Forests ecoregion, 396–97
South Western Ghats Moist Deciduous Forests ecoregion, 143, 313–15
South Western Ghats Montane Rain Forests ecoregion, 65, 84, 142, 284–87
species-area effect, 65, 71–72
species flocks, 35
species radiations, extraordinary, 83, 84
species richness
    amphibian, 48
    aquatic organisms, 35
    Biological Distinctiveness Index (BDI) and, 39–40
    birds, 68–71, 72
    continental vs. island, 68, 71–73
    Dipterocarpaceae, 93, 94, 98
    distribution patterns, 65–71
    island size and, 65
    Kerinci-Seblat National Park (Sumatra), 188
    mammals, 68–71, 72
    mangrove systems, 55–56
    New Caledonia, 162
    plants, 73, 76, 78
    reptiles and plants, links between, 53
Sri Lanka, 61. *See also* Indian Subcontinent bioregion
Sri Lanka Dry-Zone Dry Evergreen Forests ecoregion, 296–98
Sri Lanka Lowland Rain Forests ecoregion, 84, 143, 176, 287–90
Sri Lanka Montane Rain Forests ecoregion, 84, 143, 176, 290–93
streams. *See* river and stream systems
submontane forests, 103, 105

subsidies, 174
Sukla Phanta Wildlife Reserve (Nepal), 178–80
Sulaiman Range Alpine Meadows ecoregion, 367–69
Sulawesi, Indonesia. *See also* Indonesia; Wallacea bioregion
    Biodiversity Action Plan, 12–13
    conservation status, 120
    deforestation, 8, 10–13, 116
    endemism, high level of, 64
    forest cover, 10–11
    freshwater biodiversity, 35
    limestone caves, 36
    wildlife trade, 125
Sulawesi Lowland Rain Forests ecoregion, 148, 518–22
Sulawesi Montane Rain Forests ecoregion, 148–49, 523–26
Sulu Archipelago Rain Forests ecoregion, 508–10
Sumatra, Indonesia. *See also* Indonesia; Sunda Shelf and Philippines bioregion
    Biodiversity Action Plan, 12–13
    deforestation, 2, 8–10, 104, 106, 116, 117
    fires, 9
    forest cover, 10
    Kerinci-Seblat National Park, 188–89
    species richness, 65
Sumatran Freshwater Swamp Forests ecoregion, 456–58
Sumatran Lowland Rain Forests ecoregion, 144–45, 176, 441–44
Sumatran Montane Rain Forests ecoregion, 155, 444–48
Sumatran Peat Swamp Forests ecoregion, 450–52
Sumatran Tropical Pine Forests ecoregion, 510–12
Sumba Deciduous Forests ecoregion, 532–33
Sundaland Heath Forests ecoregion, 460–62
Sundarbans Freshwater Swamp Forests ecoregion, 301–2
Sundarbans Mangroves ecoregion, 144, 176, 350–52
Sunda Shelf and Philippines bioregion. *See also* Borneo; Indonesia; Malaysia
    conservation status, 118, 120
    demarcation of, 24
    ecoregional map, 19
    endemism, 29–31, 72
    globally outstanding ecoregions, 65–66
    habitat protection, 108
    human population threats, 111–15
    mammal and bird richness, 71
    Pleistocene land patterns, 29–30
    priority ecoregions, 144–48, 155–56, 159, 161
    reptile flagship species, 54
    species richness, 72
    Wallace's Line, 29
    World Heritage Sites, 130
Sunda Shelf Mangroves ecoregion, 147, 148, 176, 514–16
swamp forest, 10–11
swamps, low biodiversity value of, 32–33
swiftlets, 37
systematics, 48, 49–50

taxonomic uniqueness, higher, 83, 84
temperate broadleaf and mixed forests biome, 27
temperate coniferous forests biome, 28
Tenasserim–South Thailand Semi-Evergreen Rain Forests ecoregion, 154, 383–86
Terai Arc Wildlife Corridor, 171–73, 175, 178–81
Terai-Duar Savanna and Grasslands ecoregion, 84, 143–44, 176, 353–56
terrestrial ecoregions. *See* ecoregion approach; ecoregions, Indo-Pacific
terrestrial islands, 65
Thailand. *See* Indochina bioregion
Thar Desert ecoregion, 371–73
threat analysis, 44, 118–20
Tiger Conservation Units (TCUs), 171, 189
tigers (*Panthera tigris*)
    India Ecodevelopment Project, 189–90
    Kerinci-Seblat National Park (Sumatra), 188
    poaching, 125, 126–27
    ranges, 168, 170–71
    Terai Arc Wildlife Corridor, 171–73, 178–81
    as top carnivores, 56–59
timber plantations, 14–16. *See also* logging
Timor and Wetar Deciduous Forests ecoregion, 149, 176, 529–31
Tonle Sap Freshwater Swamp Forests ecoregion, 410–12
Tonle Sap–Mekong Peat Swamp Forests ecoregion, 404–6
top-down control processes, 56–57, 58–59
trade in wildlife. *See* wildlife trade
Trans Fly Savanna and Grasslands ecoregion, 158, 572–74
Trobriand Islands Rain Forests ecoregion, 159, 584–85
tropical and subtropical coniferous forests biome, 27
tropical and subtropical dry broadleaf forests biome
    conservation status, 101
    definition, 25
    ecoregions represented in, 27
    globally outstanding ecoregions, 84
    habitat fragmentation, 105
    habitat protection, 108
    human population threats, 115
    priority classes, 140
    World Heritage Sites, 130
tropical and subtropical grasslands biome, 28
tropical and subtropical moist broadleaf forests biome
    conservation status, 100
    definition, 25
    ecoregions represented in, 26–27
    globally outstanding ecoregions, 80, 84
    habitat blocks, size of, 104
    habitat fragmentation, 105
    habitat loss, 101
    habitat protection, 108
    human population threats, 115
    priority classes, 140
    World Heritage Sites, 130
Tropical flora, 64
turtles, 51, 53. *See also* reptiles

ungulates
    corridor restoration and, 181
    endemic species, 78–79
    Himalayan diversity, 62
    in prey assemblage, 57, 58–59
Upper Gangetic Plains Moist Deciduous Forests ecoregion, 317–19

Valmiki Wildlife Sanctuary (India), 178–80
Vanuatu, 52. *See also* New Guinea and Melanesia bioregion
Vanuatu Rain Forests ecoregion, 151, 176, 591–94
vertebrates
    as classification criteria, 25
    intact assemblages of, 83, 84
Vietnam. *See* Indochina bioregion
Vogelkop-Aru Lowland Rain Forests ecoregion, 150, 176, 547–49
Vogelkop Montane Rain Forests ecoregion, 176, 545–47

Wallace, Alfred Russell, 28–29, 61
Wallacea bioregion. *See also* Indonesia
    conservation status, 120
    demarcation of, 24
    ecoregional map, 20
    endemic bird areas, ecoregion overlap with, 89
    endemism, 29–31, 73
    globally outstanding ecoregions, 65–66
    human population threats, 111–15
    Indo-Malayan/Australasian ecotone, 64
    mammal and bird richness, 71
    Pleistocene land patterns, 29–30
    priority ecoregions, 148–49, 156, 161
    reptile flagship species, 54
    species richness, 73
    World Heritage Sites, 130
Wallace's Line, 28–29, 61
watershed resources, 173

Western Ghats range, 47–48
Western Himalayan Alpine Shrub and Meadows ecoregion, 360–62
Western Himalayan Broadleaf Forests ecoregion, 338–40
Western Himalayan Sub-Alpine Conifer Forests ecoregion, 343–45
Western Java Montane Rain Forests ecoregion, 145, 176, 467–70
Western Java Rain Forests ecoregion, 462–65
wetlands, 34–35

Wildlife Conservation Society, 171
wildlife sanctuaries. *See* protected areas
wildlife trade, 110, 124–26, 126–27
World Heritage Sites (WHSs), 130–31
World Wildlife Fund (WWF), 165, 171, 175

Yapen Rain Forests ecoregion, 551–52

ISLAND PRESS BOARD OF DIRECTORS

CHAIR Henry Reath
President, Collector's Reprints, Inc.

VICE-CHAIR Victor M. Sher
Environmental Lawyer

SECRETARY Dane Nichols
Chairman, The Natural Step, U.S.

TREASURER Drummond Pike
President, The Tides Foundation

Robert E. Baensch
Professor of Publishing, New York University

Mabel H. Cabot

David C. Cole
Sunnyside Farms, LLC

Catherine M. Conover

Gene E. Likens
Director, The Institute of Ecosystem Studies

Carolyn Peachey
Campbell Peachey & Associates

Will Rogers
Trust for Public Lands

Charles C. Savitt
President, Center for Resource Economics/Island Press

Susan E. Sechler
Director of Global Programs, The Rockefeller Foundation

Peter R. Stein
Managing Partner, The Lyme Timber Company

Richard Trudell
Executive Director, American Indian Resources Institute

Wren Wirth
President, The Winslow Foundation